George R. Newkome, Charles N. Moorefield, Fritz Vögtle

Dendrimers and Dendrons

Concepts, Syntheses, Applications

⊛WILEY-VCH

Related Titles from WILEY-VCH:

J. Otera (Ed.)
Modern Carbonyl Chemistry
2000. ISBN 3-527-29871-1

A. Ricci (Ed.)
Modern Amination Methods
2000. ISBN 3-527-29976-9

F. Diederich / P. J. Stang (Eds.)
Templated Organic Synthesis
2000. ISBN 3-527-29666-2

F. Diederich / P. J. Stang (Eds.)
Metal-catalyzed Cross-coupling Reactions
1997. ISBN 3-527-29421-X

H. Yamamoto (Ed.)
Lewis Acids in Organic Synthesis
2 Vols., 2000, ISBN 3-527-29579-8

George R. Newkome, Charles N. Moorefield, Fritz Vögtle

Dendrimers and Dendrons

Concepts, Syntheses, Applications

 **WILEY-VCH**

Weinheim · New York · Chichester · Brisbane · Singapore · Toronto

Prof. Dr. G.R. Newkome
The University of Akron
Goodyear Polymer Center
Akron, OH, 44325-4717
USA

Prof. Dr. C. N. Moorefield
University of South Florida
Center of Molecular Design
and Recognition
Tampa, FL
USA

Prof. Dr. F. Vögtle
University of Bonn
Dept. of Organic Chemistry
and Biochemistry
Gerhard-Domagk-Str. 1
53121 Bonn
Germany

Library of Congress Card No. applied for.

A catalogue record for this book is available from the British Library.

Deutsche Bibliothek Cataloguing-in-Publication Data:
A catalogue record for this book is available from Die Deutsche Bibliothek
ISBN 3-527-29997-1

© WILEY-VCH Verlag GmbH, D-69469 Weinheim (Federal Republic of Germany), 2001

Printed on acid-free paper.

Composition: Mitterweger & Partner GmbH, D-68723 Plankstadt
Printing: betz-druck GmbH, D-64291 Darmstadt
Bookbinding: J. Schäffer GmbH & Co. KG, D-67269 Grünstadt
Printed in the Federal Republic of Germany

Preface

Since the publication of our 1st monograph in 1996, which admittedly now is out-dated, dendrimer chemistry has transcended aesthetically-driven, preliminary synthetic protocols to become integrated with nearly all contemporary scientific disciplines. From the material arts to medicinal chemistry to the supramolecular and nanoscopic regimes, branched architectures have permeated our lives. Thus, our discussion of the subject at a meeting in Japan at the Seahawk Hotel in Fukuoka, Japan in September, 1998 focused on a new and comprehensive manuscript on the topic. The addition of almost 1800 new references to the previous 750 make this essentially a new book and not just a simple update. The format of this book is essentially the same with greatly expanded coverage of each chapter. We have moved the previous chapter dealing with nomenclature to the appendices and divided the divergent section into two parts stressing 1→2 and 1→3 branching architectures each individually.

One goal while writing this book was to act as a repository for the now interdisciplinary field so that researchers interested in the topic could have an easily accessible reference for all related works and publications. It is not an easy task to stay current in burgeoning areas where manuscripts appear in a myriad of magazines and books. It is hoped that this was realized and at some point it had to "go to press"; even in the galley proofs, over 100 new references were added with minimal textual changes in order not to delay production. This book contained so much additional new information – including the book authors own research that we have given it the new title of "Dendrimers and Dendrons" in order to stress the rapid development and all encompassing nature of this now entrenched field.

As with all efforts requiring copious amounts of energy, there are numerous individuals that with out their help the 'effort' would not have been possible. First and foremost – we thank our patient and understanding wives, who not only helped with some of the editing but also allowed this endeavor to come to fruition. Of course, it is a pleasure to thank Dr. Peter Gölitz at Wiley-VCH for his persistent enthusiasm and support of the project as well as Dr. Gudrun Walter and Maike Petersen, also of Wiley-VCH, for their continued professional support and efforts. As well, our coworkers responsible for the monumental task of newly preparing many of the drawings: Dr. Ansgar Affeld, Kaweh Beizai, Marius Gorka, Annette Hünten, Qian-Yi Li, Oliver Mermagen, Dr. André Mohry, Dr. Guido Nachtsheim, Dr. Amir Parham, Gregor Pawlitzki, Janosch Recker, Dr. Carin Reuter, Frauke Schelhase, Dr. Holger Schwierz, Nicole Werner, and Dr. Björna Windisch all in Bonn and Mr. Jason Hair and Dr. Enfei He at the University of South Florida, are all gratefully acknowledged and commended for their Herculean efforts. Dr. Gregory R. Baker of the University of South Florida is acknowledged and thanked for his numerous contributions and the initial establishment of a comprehensive database in this field. We would also like to thank the numerous research associates, graduate students and collaborators – worldwide – who have been a constant source of interdisciplinary thoughts and efforts so necessary for entry into the nanoregime. Lastly, we would like to acknowledge the efforts of all the researchers herein cited. Their efforts and work continue to make this topic a truly exciting discipline to explore.

Also, we have tried to be as comprehensive, complete, and up-to-date as possible but inevitably there will be the 'missed' reference. To these authors, we apologize and ask that they bring the matter to our attention for inclusion in future monographs.

Finally, we hope that you will enjoy perusal of this book and find some useful information for your ongoing research, in which fractal designs can afford a logical framework to construct large materials possessing specific composition.

George R. Newkome   June, 2001
Charles N. Moorefield
Fritz Vögtle

Contents

1 Masses, Sizes, and Shapes of Macromolecules from Multifunctional Monomers*

1.1 Introduction

Macromolecules have large molecular masses, but their bulk densities are comparable with those of small molecules of similar composition. Therefore, macromolecules must have large molar volumes and large extensions in space. Furthermore, the presence of many atoms within each macromolecule provides potential access to an enormous variety of conformations with different shapes, as well as sizes. For an unbranched macromolecule, most of the conformations are asymmetric and relatively open, so that segments from other molecules (either solvent or other polymers) have ready access to the interior of the macromolecule. More compact conformations, with a closer approach to spherical symmetry, may become more likely if the macromolecule has a high density of branches that appear in a regular pattern. Each of these topics is discussed in this chapter, first in the context of macromolecules in general, and then specifically for the class in which the macromolecule is highly branched, as in a dendrimer.

The discussion of molecular weights in this chapter is focused on the polydispersity index, defined as the ratio of the weight- (Equation 1.2) and number-average (Equation 1.1) molecular weights, $\overline{M}_w / \overline{M}_n$.

$$\overline{M}_n \equiv \frac{\sum_i W_i}{\sum_i N_i} = \frac{\sum_i N_i M_i}{\sum_i N_i} \tag{1.1}$$

$$\overline{M}_w \equiv \frac{\sum_i W_i M_i}{\sum_i W_i} = \frac{\sum_i N_i M_i^2}{\sum_i N_i M_i} \tag{1.2}$$

The summations run over all species present, and N_i and W_i are the number and mass, respectively, of molecules with molecular weight M_i. The polydispersity index has a value of one if all molecules are of the same molecular weight; the polydispersity index of most polymers is larger than one. Section 1.2.1 defines conditions under which the polydispersity is small. The polydispersity index is also pertinent to the identification of the requirements for the avoidance of gelation when many multifunctional monomers are incorporated into the macromolecule.

The discussion of molecular sizes throughout the chapter is focused on even moments (averaged even powers) of the radius of gyration. Shapes are evaluated using the principal moments of the radius of gyration tensor, with emphasis on their use for the detection of asphericity in a population of flexible macromolecules. The solvent-induced collapse of unbranched chains to a dense globular state is then described.

The chapter closes with a brief review of several recent computer simulations of the extensions, shapes, and internal structures of dendrimers. Accurate simulations of these molecules are not easily performed. Dense, amorphous, linear (unbranched) polymers present a considerable challenge to simulation when the models are expressed with atomistic detail, due to the broad time scales for relaxation of these systems.[1] The problem is confounded when the macromolecule has a high degree of branching, as in the case of a dendrimer. Nevertheless, the recent literature provides evidence of useful progress in the simulation of these interesting molecules.

* The introductory chapter was written by Professor Wayne L. Mattice, the Alex Schulman Professor, and Professor Coleen Pugh at the Maurice Morton Institute of Polymer Science, University of Akron, Akron, Ohio. They acknowledge the financial support by the National Science Foundation (DMR 95–23278).

1.2   Molecular Mass and Chain Connectivity in Macromolecules

As implied by the name, a fundamental distinction between a macromolecule and a small molecule lies in their mass, or molecular weight, M. The lower limit for the molecular weight of a macromolecule is ill-defined, but it is often considered to lie roughly in the range $M = 10^3–10^4$. As a consequence of the strong tendency for chemical reactions to proceed in a random manner, the simplest and most common procedures used for the polymerization of ordinary monomers do not produce a set of molecules of exactly the same mass. Instead, they produce a population of similarly constituted macromolecules that have a distribution of molecular weights. The physical properties of the sample depend not only on the average molecular weight, but also on the shape of the distribution of molecular weights. Therefore, descriptions of the molecular weight of a macromolecule require a characterization of the distribution of the masses present. Usually, the width of the distribution is characterized by the dimensionless ratio of two different average molecular weights that are accessible from experiment. The most commonly used molecular weights for this purpose are the number-average molecular weight, $\overline{M}_n$, which is measured with colligative properties, and the weight-average molecular weight, $\overline{M}_w$, which can be measured by light scattering.

1.2.1   Distributions of Molecular Weight in Unbranched Chains

The "most probable" distribution of molecular weight is expected in condensation polymerizations of bifunctional monomers when all functional groups have equal reactivity.[2] The reactants can be A–B or an equimolar mixture of A–A and B–B, where A and B represent functional groups that are capable of reaction to form an AB covalent bond. The reactivity of each terminal functional group is assumed to be independent of the length of the unspecified internal part of each molecule, which is denoted here merely by "–". When only linear products are formed, the mole fraction and weight fraction, respectively, of x-mer are given by Equations 1.3 and 1.4.

$$X_x = (1 - p)p^{x-1} \tag{1.3}$$

$$W_x = x(1 - p)^2 p^{x-1} \tag{1.4}$$

The symbol p denotes the fraction of the functional groups that have reacted. These distributions of x-mers yield number- and weight-average degrees of polymerization given by Equations 1.5 and 1.6.

$$\overline{x}_n = \frac{1}{1 - p} \tag{1.5}$$

$$\overline{x}_w = \frac{1 + p}{1 - p} \tag{1.6}$$

Ignoring any differences in the masses of the terminal units and those in the interior of the chain, the ratio of the weight- and number-average molecular weights, or polydispersity index, for the "most probable" distribution is defined by Equation 1.7.

$$\frac{\overline{M}_w}{\overline{M}_n} = 1 + p \tag{1.7}$$

This ratio becomes experimentally indistinguishable from 2 well before the average degree of polymerization becomes 10^2 (before p rises above 0.99). Polymer scientists reserve the word "monodisperse" for samples with much narrower distributions. Some might use this word for samples with a polydispersity index below 1.1; others would require that it be lower than 1.02. But all would agree that "monodisperse" does not characterize the "most probable" distribution.

Other polymerization mechanisms can produce distributions of molecular weights that are either broader or narrower than those obtained with the "most probable" distribution. Much narrower distributions are obtained if a specified number of chains is initiated at the same time, no other chains are initiated, and the chains grow exclusively by the addition of monomer to the reactive end. If ξ denotes the ratio of the number of monomers that have reacted to the number of growing chains, we obtain Equation 1.8.[3]

$$\frac{\overline{M}_w}{\overline{M}_n} = 1 + \frac{\xi}{(\xi + 1)^2} \tag{1.8}$$

For large chains (where $\xi > 10^2$) prepared by this mechanism, $\overline{M}_w/\overline{M}_n$ may become experimentally indistinguishable from 1.

When the polymerization utilizes monomers that contain more than two reactive groups, branched macromolecules are formed. Often, the introduction of branches is accompanied by profound effects on the polydispersity. An example is the condensation of an excess of A–B with a small amount of an f-functional monomer, denoted R–A$_f$. The f-functional monomer contains f equivalent functional groups of type A, but no functional groups of type B.[4] Linear chains are obtained when f is 1 or 2, but multichain condensation polymers are produced when $f > 2$. At high conversion, the polydispersity index depends only on f.

$$\frac{\overline{M}_w}{\overline{M}_n} \simeq 1 + \frac{1}{f} \tag{1.9}$$

The distribution of molecular weights becomes narrower as the functionality of the multifunctional monomer increases, and approaches 1 in the limit as $f \rightarrow \infty$. The formation of an infinite network is prevented because all of the molecules derived from each R–A$_f$ monomer contain only the functional group A at the reactive ends, which reacts only with B, and not with another A. Another result (and a very different polydispersity) is obtained if the system contains B–B, or if the multifunctional monomer contains both A and B at its reactive ends. This topic is developed in the next section.

1.2.2 Branched Structures and Gelation with Multifunctional Monomers

The last paragraph in Section 1.2.1 described the formation of multichain condensation polymers of low polydispersity, using an excess of A–B with a small amount of the multifunctional monomer R–A$_f$, under conditions where A cannot react with another A. Polyfunctional condensation polymerization using multifunctional monomers in other contexts can produce infinite networks.[5, 6] If the multifunctional R–A$_f$ is polymerized with a mixture of both A–A and B–B, the system is different from the simpler polymerization of R–A$_f$ with A–B in two important ways. At various stages in the polymerization, the molecules derived from R–A$_f$ can now have some arms that terminate with functional group B, and other arms that terminate with functional group A. When both functional groups are simultaneously present, intramolecular reaction, leading to the formation of rings of various sizes, becomes possible. This intramolecular cyclization occurs, but another process has a more dramatic consequence. The second process is the formation of macromolecules that contain more than one f-functional site derived from R–A$_f$, as the consequence of intermolecular reaction. Eventually, this second process can produce a single gigantic molecule, with many branch points, that extends throughout the macroscopic volume available to the polymerizing system. The system forms a gel.

The point in the reaction at which gelation occurs has been deduced by Flory.[5] The gel point is developed in terms of the branching coefficient, α, which is the probability that a given functional group on the multifunctional monomer leads, via a chain that can contain any number of bifunctional units, to another multifunctional monomer. Gelation occurs at a critical value of the branching coefficient, denoted by α_c.

$$\alpha_c = \frac{1}{f - 1} \tag{1.10}$$

This critical value of α is obtained when the expected number of chains that will succeed any of the A in the multifunctional R–A$_f$, due to branching of one or more of them, exceeds one. The ratio $\overline{M}_w/\overline{M}_n$ diverges when the system gels. The divergence arises because some of the material is not incorporated into the enormously large, highly branched network. The gel contains extractable monomer, oligomers, and polymers. The simultaneous presence of these small extractable molecules and the huge gel specifies an enormous value for the polydispersity index.

Gelation will also occur in the polycondensation of R–A$_f$ and A–A if the chemistry of the functional groups permits the formation of a covalent bond between two A groups. An interesting variant of this process is the gelation that occurs in a thermoreversible manner when the groups denoted by A can aggregate into clusters stabilized by noncovalent interactions.[7, 8] Thermoreversible gelation has been simulated in systems containing symmetric triblock copolymers, A$_{NA}$–B$_{NB}$–A$_{NA}$, in a medium that is a good solvent for block B, but a poor solvent for block A.[9] Aggregation of $x + f$ molecules of the triblock copolymer into a micelle, with x molecules placing both of their A blocks in the micellar core, and the remaining f molecules placing only one of their A blocks in the core, produces a micelle that has the characteristics of the f-functional monomer described by Flory, in the sense that there are f dangling chains with "sticky" ends comprised of their A blocks. The critical gel point determined by simulation[9] of the triblock copolymers is in good agreement with the result predicted by the application of Flory's theory to this reversibly associating system.

1.2.3 Branching without Gelation

Another type of behavior occurs in the polymerization of a multifunctional monomer that has one functional group of type A and $f - 1$ functional groups of type B, with the only reaction being between A and B. This multifunctional monomer can be written as A–R–B$_{f-1}$. The reaction produces highly branched structures, but without gelation.[10] For this system, the branching probability is given by Equation 1.11.

$$\alpha = \frac{p}{f - 1} \tag{1.11}$$

Here, p denotes the fraction of the A groups that have reacted. Since p cannot be larger than 1, α cannot exceed the critical value for gelation, specified by Equation 1.10.

If the polyfunctional condensation of A–R–B$_{f-1}$ proceeds under the condition of equal reactivity, polydispersity increases as the reaction proceeds (as p increases), and the polydispersity index approaches infinity as $p \rightarrow 1$. The uncontrolled polycondensation of A–R–B$_{f-1}$ does not produce a unique molecular species, but instead produces highly branched molecules with an extremely broad range of molecular weights.

A remarkable narrowing of the distribution of molecular weights is possible in a polymerization composed of a repeating sequence of different reactions. Let R–B$_{f-1}$ initially react with $f - 1$ molecules of A–R–C$_{f-1}$ under conditions where the only reaction is of A with B. Then the protecting group C is converted to B, and the cycle is repeated again and again. If all the reactions are quantitative, the branched molecules obtained after each cycle all have exactly the same covalent structure, and $\overline{M}_w/\overline{M}_n = 1$. Incomplete reaction at any stage will introduce polydispersity. The broadening of the molecular weight distribution that is expected if the reaction is less than quantitative during some or all cycles has been considered by Mansfield.[11] The influence of a less than quantitative reaction on the polydispersity is least if it occurs in the final cycle and greatest if it occurs in the first cycle. Therefore, the production of molecules with $\overline{M}_w/\overline{M}_n$ very close to 1 requires that the synthesis must be kept perfect for as many cycles as possible. With any real system, a point will be reached at which future generations cannot remain perfect, because the mass of the molecule grows faster than the available volume as the number of cycles increases, implying the eventual attainment of a stage beyond which subsequent reactions can no longer be quantitative.

1.3 Overall Extension in Space: Molecular Sizes

The overall molecular shape, or conformation, is one of the most important properties of a macromolecule. Indeed, their large extension in space is an important way in which macromolecules differ from small molecules. With large unbranched flexible chains, the overall extension is frequently discussed in terms of the mean square end-to-end distance, $\langle r^2 \rangle$, or the mean square radius of gyration, $\langle s^2 \rangle$.[12–21] For very long flexible homopolymers, these two properties are approximately related as $\langle r^2 \rangle = 6 \langle s^2 \rangle$. This approximation becomes exact when a very long flexible homopolymer is unperturbed by long-range interactions, as it is in the θ state defined by Flory.[6, 16, 22] The θ state is obtained in pure bulk amorphous polymers, a condition of enormous practical importance. It is also obtained in dilute solution at the temperature at which excluded volume effects disappear. The disappearance of excluded volume can be identified experimentally with the point where the second virial coefficient becomes equal to zero.

When $\langle r^2 \rangle$ and $\langle s^2 \rangle$ are equally pertinent to a discussion of the size of an unbranched chain, there is often a preference for the mean square end-to-end distance because theoretical expressions involving $\langle r^2 \rangle$ usually have a simpler appearance than their counterparts that involve $\langle s^2 \rangle$. For example, in the very simple model of the freely jointed chain of n identical bonds of length l connecting $n + 1$ atoms of identical mass, the expressions for the mean square unperturbed dimensions are given in Equations 1.12 and 1.13.

$$\langle r^2 \rangle_0 = nl^2 \tag{1.12}$$

$$\langle s^2 \rangle_0 = \left(\frac{n+2}{n+1}\right) \frac{nl^2}{6} \tag{1.13}$$

Both equations are valid at any value of n. The subscript zero denotes the absence of any influence of excluded volume, as found in Flory's θ state. Equation 1.12 is clearly of more compact form than Equation 1.13. It also shows a simpler dependence of the mean square dimensions on the degree of polymerization, n.

Although the first expression is of simpler form than the second, it has important disadvantages that limit its usefulness in the discussion of macromolecules with architectures other than the unbranched chain, because these molecules contain a number of ends different from two. The mean square end-to-end distance becomes ambiguous when the macromolecule is branched, because each molecule contains more than two ends. An end-to-end distance cannot be defined when the macromolecule is cyclic, because there are no ends at all. In both of these cases, the mean square radius of gyration remains unambiguous and well defined. This attribute of the mean square radius of gyration arises because its definition does not demand any specification of the connectivity of the collection of particles that constitute the molecule. Since this book is devoted to highly branched molecules, we will henceforth discuss $\langle s^2 \rangle$, in preference to $\langle r^2 \rangle$.

1.3.1 Radius of Gyration

The most fundamental characterization of the extension of a macromolecule in space is its radius of gyration, s. The squared radius of gyration of an immobile collection of $n + 1$ particles indexed by i running from 0 to n, and with particle i weighted as m_i, is given by the expression labeled Equation 1.14.

$$s^2 \equiv \frac{\sum_{i=0}^{n} m_i (\mathbf{r}_i - \mathbf{g}) \cdot (\mathbf{r}_i - \mathbf{g})}{\sum_{i=0}^{n} m_i} \tag{1.14}$$

Here, $\mathbf{g}$ denotes the vector to the center of mass, expressed in the same (arbitrarily chosen!) coordinate system used for expression of the vector to atom i, $\mathbf{r}_i$ (Equation 1.15).

$$\mathbf{g} \equiv \frac{\sum\limits_{i=0}^{n} m_i \mathbf{r}_i}{\sum\limits_{i=0}^{n} m_i} \tag{1.15}$$

The weighting factor m_i may be useful in the study of copolymers, and especially so when an apparent mean square radius of gyration of a copolymer is determined by light-scattering measurements in which the effective scattering power differs strongly for the various components of the macromolecule. Here, however, we shall assume that these differences are not important. We will henceforth adopt the simpler definition of s^2 that arises when all of the m_i are identical.

$$s^2 = \frac{1}{n+1} \sum_{i=0}^{n} (\mathbf{r}_i - \mathbf{g}) \cdot (\mathbf{r}_i - \mathbf{g}) \tag{1.16}$$

$$\mathbf{g} = \frac{1}{n+1} \sum_{i=0}^{n} \mathbf{r}_i \tag{1.17}$$

1.3.1.1   Unbranched Chains

One of the simplest models for a flexible macromolecule is the freely jointed chain comprised of n bonds of length l between $n + 1$ atoms of identical mass. The expression for $\langle s^2 \rangle_0$ for this model is given by Equation 1.13. This same model has $\langle r^2 \rangle_0 = nl^2$ (Equation 1.12), which, with Equation 1.13, yields the limiting behavior $\langle r^2 \rangle_0 \rightarrow 6 \langle s^2 \rangle_0$ as $n \rightarrow \infty$.

This very simple model correctly leads to the conclusion that long, flexible, unbranched chains do not prefer the compact globular state. The globular state of uniform density is characterized by $\nu = 1/3$ in Equation 1.18.

$$\langle s^2 \rangle \propto n^{2\nu} \tag{1.18}$$

The freely jointed unperturbed chain has a larger value of ν, $\nu = 1/2$, implying that it is more expanded than the globule. Asymptotic ($n \rightarrow \infty$) values of ν associated with four simple models for macromolecules are presented as the first four entries in Table 1.1. The values of ν for these four classic simple models range from a low of $\nu = 1/3$ for the dense uniform globule to a high of $\nu = 1$ for the rigid rod of constant diameter and length proportional to n.

Table 1.1. Values of ν for various conformations ($\langle s^2 \rangle \propto n^{2\nu}$)

Conformation	ν
Dense Uniform Globule	1/3
Random Coil (θ solvent)	1/2
Random Coil (Good Solvent)	~3/5[a]
Rigid Rod of Constant Diameter	1
Coarse-Grained[b] Dendrimers (MC, Good Solvent)	0.15–0.21[c]
Coarse-Grained[b] Dendrimers (MD, Various Solvents)	~0.30[d]

[a] The best estimate for an unbranched random coil in a good solvent is $\nu = 0.588 \pm 0.001$ (Le Guillou and Zinn-Justin[23] and Oono.[24])

[b] Each fundamental particle in the model represents a collection of several atoms.

[c] Estimated from the radii of gyration obtained in Monte Carlo (MC) simulations of well-equilibrated coarse-grained dendrimers.[25] The value of ν is not constant, but instead tends to decrease as the generation number increases. The largest dendrimer considered is generation nine.

[d] From radii of gyration obtained in Molecular Dynamics (MD) simulations of well-equilibrated coarse-grained dendrimers of generations 5–8 in solvents of widely varying quality.[26]

Extremely detailed descriptions of the conformations of unperturbed flexible homopolymers can be constructed using the rotational isomeric state model.[17–19] These models incorporate differences in the lengths of the bonds, realistic values of the bond angles, and restrictions on the torsion angles that are imposed by short-range intramolecular interactions. The literature contains detailed rotational isomeric state models for an enormous number of polymers, varying in complexity from polymeric sulfur to polynucleotides. Many of these rotational isomer state models have been compiled in a standard format in a recent review.[27] In the limit as $n \to \infty$, all of these detailed models yield Equations 1.19 and 1.20.

$$\lim_{n \to \infty} \frac{\langle r^2 \rangle_0}{\langle s^2 \rangle_0} = 6 \tag{1.19}$$

$$\lim_{n \to \infty} \frac{\langle s^2 \rangle_0}{nl^2} = \frac{C}{6} \tag{1.20}$$

Equation 1.19 specifies the same limiting behavior for the dimensionless ratio $\langle r^2 \rangle_0 / \langle s^2 \rangle_0$ as was obtained with the much simpler freely jointed chain. Equation 1.20 specifies the same limiting $\langle s^2 \rangle_0$ as the freely jointed chain only in the special case where the characteristic ratio, C, is equal to one (Nature may not provide any real chain that meets the requirements of this special case!). In general, the restrictions on bond angles and torsion angles in real molecules produce $C > 1$, meaning that real flexible unperturbed polymers have larger mean square dimensions than those suggested by the random flight chain with bonds of the same number and length. For example, measurements on long polyethylene chains lead to $C = 6.7$ at $140\,°C$.[28] The restriction on the bond angle alone raises the value of C from 1 for the freely jointed chain to 2.20 in the freely rotating chain with a bond angle of 112°. The additional restrictions on the torsion angles due to first- and second-order interactions bring the result to $C = 6.7$ for polyethylene.[29] The first-order interaction causes *gauche* states to be of higher energy than *trans* states. The most important second-order interaction strongly penalizes two successive *gauche* states of opposite sign, due to the repulsive interaction known as the "pentane effect".

1.3.1.2 Macrocycles

The average extension of the $n + 1$ chain atoms in space can be reduced by changing their connectivity. A conceptually simple means by which this reduction can be obtained is by the introduction of a covalent bond that links the two ends of the chain. This macrocyclization leads to a reduction in $\langle s^2 \rangle$, which, in the case of large unperturbed chains, is a factor of 1/2.[30]

$$\lim_{n \to \infty} \frac{\langle s^2 \rangle_{0,\text{cyclic}}}{\langle s^2 \rangle_{0,\text{linear}}} = \frac{1}{2} \tag{1.21}$$

The exponent ν (Equation 1.18) remains 1/2. The value of $\langle s^2 \rangle_0$ for the large unperturbed macrocycle is far in excess of the result expected for the same collection of $n + 1$ atoms in the dense globular state.

1.3.1.3 Branched Macromolecules

A larger reduction in the mean square unperturbed dimensions can be achieved by rearrangement of the $n + 1$ atoms into an *f*-functional star-branched polymer. In this architecture, the macromolecule contains f branches ($f > 2$), each with n/f bonds that emanate from a common atom. The same terminology is frequently used for branched macromolecules with very large f, if the branches emanate from a collection of atoms that are constrained to remain close together, so that the origin of all of the branches is clustered into a volume much smaller than $\langle s^2 \rangle^{3/2}$. The influence of the star-branched architecture on the mean square dimensions is traditionally designated by a factor g that is defined according to Equation 1.22.[31]

$$g \equiv \frac{\langle s^2 \rangle_{0,\text{branched}}}{\langle s^2 \rangle_{0,\text{linear}}} \tag{1.22}$$

(The g in Equation 1.22 should not be confused with the **g** in Equations 1.14–1.17. By coincidence, the prior literature[31,32] has selected the same letter of the alphabet for these two unrelated properties.) In Equation 1.22, it is understood that the branched and linear molecules contain the same number of bonds. If n_j denotes the number of bonds in branch j, the application of random flight statistics leads to a very simple expression for g.[31]

$$g = \frac{1}{n^3} \sum_j (3nn_j^2 - 2n_j^3) \tag{1.23}$$

For the special case where all of the branches contain the same number of bonds, with that number being n/f, g is given by Equation 1.24.

$$g = \frac{3f - 2}{f^2} \tag{1.24}$$

This equation suggests that the mean square dimensions of the macrocycle are approximated by the star-branched polymer with five arms, for which $g = 13/25$. It also suggests that a macromolecule with a closer approach to the collapsed globular state could be obtained by using $f > 5$.

As f increases without limit, Equation 1.24 predicts a value of g that approaches 0, implying the attainment of a globular state of arbitrarily high density. This limiting behavior is not achieved with real systems due to a breakdown in the validity of the assumptions behind Equation 1.23. Deviations between the prediction from Equation 1.24 and experimental measurements are significant at $f \sim 6$ and become more important as f increases. As the density of segments near the branch point increases, one must abandon the assumption of equivalent random flight statistics for the unbranched chain and for the star-branched polymer. The real ratio of the two mean square radii of gyration in Equation 1.22 becomes larger than the prediction in Equation 1.24 as f increases.

$$\left(\frac{\langle s^2 \rangle_{0,\text{branched}}}{\langle s^2 \rangle_{0,\text{linear}}} \right)_{\text{experiment}} > \frac{3f - 2}{f^2} \quad \text{at large } f \tag{1.25}$$

Star-branched polymers with f larger than 10^2 have been prepared.[33,34] Their dimensions are much larger than the prediction from Equation 1.24. These highly branched structures have been called "fuzzy spheres",[34] because the comparison of the thermodynamic radii deduced from equilibrium measurements with the hydrodynamic radii from transport measurements suggests the macromolecules behave as spheres with a hydrodynamically penetrable surface layer.

1.3.2   Breadth of the Distribution for the Squared Radius of Gyration

Information about the shape (width, skewness, etc.) of the distribution function for the squared radius of gyration, $P(s^2)$, can be deduced from ratios that involve the even moments of the radius of gyration, $\langle s^{2p} \rangle_0$, $p = 1, 2, \ldots$.

$$\langle s^{2p} \rangle_0 = \frac{\int_0^\infty s^{2p} P(s^2) ds^2}{\int_0^\infty P(s^2) ds^2} \tag{1.26}$$

The value of $\langle s^2 \rangle_0$ is the average over this distribution function. By itself, $\langle s^2 \rangle_0$ provides no information about the shape of this distribution function. In particular, it does not reveal whether the distribution is broad, implying that the macromolecule can populate conformations with very different extensions, or whether it is narrow, as would be the case if a single conformation were to be populated in preference to all others. Information about the shape of the distribution function for the squared dimensions can be assessed by evaluation of the higher even moments of s. The breadth of the distribution function is deduced from $\langle s^4 \rangle_0 / \langle s^2 \rangle_0^2$. If only a single conformation is accessible, this dimensionless ratio has a value of 1. The freely jointed chain provides a useful benchmark for interpreting larger values of $\langle s^4 \rangle_0 / \langle s^2 \rangle_0^2$. At any value of n, this simple model for a flexible chain obeys Equation 1.27.[34]

$$\frac{\langle s^4 \rangle_0}{\langle s^2 \rangle_0} = \frac{19n^3 + 45n^2 + 32n - 6}{15n(n+1)(n+2)} \tag{1.27}$$

The dimensionless ratio in Equation 1.27 is equal to 1 for $n = 1$, and it is larger than 1 for $n \geq 2$. It increases with n to a limiting value of 19/15, which is 1.267 ... when expressed as a real number.[36] This limit is associated with any flexible chain that obeys Gaussian statistics.

The freely jointed chain has a distribution function for s^2 that is narrower than the distribution function for r^2, as shown in Equation 1.28.

$$\frac{\langle r^4 \rangle_0}{\langle r^2 \rangle_0^2} = \frac{5n - 2}{3n} \tag{1.28}$$

The limit as $n \to \infty$ is $\langle r^4 \rangle_0 / \langle r^2 \rangle_0^2 \to 5/3$, which is larger than the limit $\langle s^4 \rangle_0 / \langle s^2 \rangle_0^2 \to 19/15$.

The dimensionless ratios of the form $\langle s^{2p} \rangle_0 / \langle s^2 \rangle_0^p$ are easily evaluated from an integratable theoretical model for the distribution function, $P(s^2)$, using Equation 1.26. For small p, they can also be calculated for unperturbed rotational isomeric state chains by efficient generator matrix methods.[32]

1.4 Shape Analysis from the Radius of Gyration Tensor

Analysis based on the averaged principal moments of the radius of gyration tensor provides information about the shapes of the accessible conformations. This information is different from that contained in the $\langle s^{2p} \rangle_0 / \langle s^2 \rangle_0^p$ ratio. Consider, for example, a rather strange molecule that can adopt any of three shapes (a sphere, a disk, and a cylinder), all three of which have exactly the same radius of gyration. The value of $\langle s^{2p} \rangle_0 / \langle s^2 \rangle_0^p$ provides no insight into which of these three conformations might be preferred, because $\langle s^{2p} \rangle_0 / \langle s^2 \rangle_0^p = 1$ for any combination of preferences due to the postulate that the sphere, disk, and cylinder have the same radius of gyration. The distinction between the three conformations can be drawn by exploiting the fact that spheres, disks, and rods have distinguishable symmetries. Information about the symmetry is lost in s^2, which is a scalar quantity, but it is retained in the tensorial representation of the squared radius of gyration.

For a rigid (static) array of $n + 1$ particles, the radius of gyration tensor, **S**, can be expressed as a symmetric 3×3 matrix.

$$\mathbf{S} = \begin{bmatrix} X^2 & XY & XZ \\ XY & Y^2 & YZ \\ XZ & YZ & Z^2 \end{bmatrix} \tag{1.29}$$

Alternatively, it can be written as a column in which the nine elements of Equation 1.29 are listed in "reading order".[32]

$$\mathbf{S}^{\text{col}} = \begin{bmatrix} X^2 \\ XY \\ XZ \\ XY \\ Y^2 \\ YZ \\ XZ \\ YZ \\ Z^2 \end{bmatrix} \tag{1.30}$$

When the $n + 1$ particles are all of the same mass, the latter representation is generated from the vectors $\mathbf{s}_i$ (each from the center of mass to atom i, $\mathbf{s}_i \equiv \mathbf{r}_{0i} - \mathbf{g}$) according to Equation 1.31.

$$\mathbf{S}^{\text{col}} = \frac{1}{n+1} \sum_{i=0}^{n} \mathbf{s}_i^{\times 2} = \frac{1}{n+1} \sum_{i=0}^{n} \mathbf{r}_{0i}^{\times 2} - \mathbf{g}^{\times 2} \tag{1.31}$$

Here, $\times 2$ as a superscript denotes the self-direct (Kronecker) product. For the specific case of two vectors in three-dimensional space, the direct product is shown in Equation 1.32.

$$\begin{bmatrix} a_1 \\ a_2 \\ a_3 \end{bmatrix} \otimes \begin{bmatrix} b_1 \\ b_2 \\ b_3 \end{bmatrix} \equiv \begin{bmatrix} a_1 b_1 \\ a_1 b_2 \\ a_1 b_3 \\ a_2 b_1 \\ a_2 b_2 \\ a_2 b_3 \\ a_3 b_1 \\ a_3 b_2 \\ a_3 b_3 \end{bmatrix} \tag{1.32}$$

The squared radius of gyration that was the subject of Equation 1.16 is the trace of $\mathbf{S}$, $s^2 = X^2 + Y^2 + Z^2$. This trace is, of course, independent of the orientation of the coordinate system used for the expression of $\mathbf{S}$. However, in the absence of spherical symmetry, the sizes of the off-diagonal elements, as well as each of X^2, Y^2, and Z^2 individually (but not their sum) depend on the orientation of the coordinate system used for the expression of $\mathbf{S}$. This dependence on the coordinate system must be dealt with before drawing conclusions about the asymmetry from the relationship between the elements of $\mathbf{S}$.

1.4.1 Principal Moments

Preparatory to the shape analysis, the coordinate system in which $\mathbf{S}$ is expressed is rotated by a similarity transformation so that the 3×3 representation for each individual conformation is rendered in diagonal form.

$$\mathbf{S}_{\text{diag}} = \mathbf{T}\mathbf{S}\mathbf{T}^{\text{T}} = \begin{bmatrix} L_1^2 & 0 & 0 \\ 0 & L_2^2 & 0 \\ 0 & 0 & L_3^2 \end{bmatrix} = \text{diag}(L_1^2, L_2^2, L_3^2) \tag{1.33}$$

$\mathbf{T}$ is the transformation matrix, $\mathbf{T}^{\text{T}}$ is its transpose, and $L_1^2 + L_2^2 + L_3^2 = X^2 + Y^2 + Z^2 = s^2$ because the trace is an invariant. Let us further stipulate that the subscripts on the principal moments, L_i^2, should be assigned so that $L_1^2 \geq L_2^2 \geq L_3^2$.

The asymmetry of any one of the conformations is characterized by the dimensionless ratios $1 \geq L_2^2/L_1^2 \geq L_3^2/L_1^2 \geq 0$.[37, 38] Spherical symmetry requires $L_2^2/L_1^2 = L_3^2/L_1^2 = 1$.

Averaging of the corresponding principal moments over all conformations permits discussion of the asymmetry of the population of conformations in terms of $\langle L_2^2 \rangle / \langle L_1^2 \rangle$ and $\langle L_3^2 \rangle / \langle L_1^2 \rangle$. Examples of these dimensionless ratios of averaged principle moments

Table 1.2 $\langle L_2^2 \rangle_0 / \langle L_1^2 \rangle_0$ and $\langle L_3^2 \rangle_0 / \langle L_1^2 \rangle_0$ for large unperturbed macromolecules

Architecture	$\langle L_2^2 \rangle_0 / \langle L_1^2 \rangle_0$	$\langle L_3^2 \rangle_0 / \langle L_1^2 \rangle_0$	References
Linear Chain	0.23	0.08	37, 38
Trifunctional Star[a]	0.33	0.12	39, 40
Macrocycle	0.36–0.37	0.15–0.16	39, 40
Tetrafunctional Star[a]	0.39–0.41	0.15–0.16	39, 40

[a] All arms contain the same number of bonds.

are presented in Table 1.2 for four types of macromolecules, in the state where they are unperturbed by long-range interactions. In none of the four cases do the conformations have spherical symmetry. Deviations from spherical symmetry are largest for the linear chain,[37, 38] and smallest (but still appreciable, with values of the dimensionless ratios that are much smaller than 1) for the tetrafunctional star.[39,40] The macrocycle occupies a position intermediate between the tri- and tetrafunctional stars.[39, 40]

1.4.2 Asphericity, Acylindricity, and Relative Shape Anisotropy

Three additional measurements that are useful in detecting other types of symmetry can be derived from the traceless form of the diagonal tensor defined in Equation 1.34.

$$\mathbf{S}_{\text{diag,traceless}} \equiv \text{diag}\left(L_1^2, L_2^2, L_3^2\right) - \frac{s^2}{3} \, \text{diag}(1, 1, 1) \tag{1.34}$$

The use of this traceless diagonal tensor for the analysis of the asymmetry of macromolecular conformations stems from comparison with an analogous traceless tensor with a long history of use in the treatment of the polarizability.[40a]

$$\mathbf{S}_{\text{diag,traceless}} = b \, \text{diag}(2/3, -1/3, -1/3) + c \, \text{diag}(0, 1/2, -1/2) \tag{1.35}$$

Comparison of Equations 1.34 and 1.35 defines the asphericity b and the acylindricity c in terms of the principal moments.[41]

$$b = L_1^2 - \frac{1}{2}\left(L_2^2 + L_3^2\right) \tag{1.36}$$

$$c = L_2^2 - L_3^2 \tag{1.37}$$

The value of b is zero if the collection of $n + 1$ points has tetrahedral or higher symmetry; otherwise $b > 0$. For long linear unperturbed chains, $\langle b \rangle_0 / \langle s^2 \rangle_0 = 0.66$.[41] If the shape is cylindrically symmetric, $c = 0$, otherwise $c > 0$. For long linear unperturbed chains, $\langle c \rangle_0 / \langle s^2 \rangle_0 = 0.11$.

Another parameter, called the relative shape anisotropy, is denoted by κ^2 and defined in Equation 1.38.[41]

$$\kappa^2 \equiv \frac{b^2 + (3/4)c^2}{s^4} \tag{1.38}$$

It has the property that $\kappa^2 = 0$ when the structure has tetrahedral or higher symmetry, $\kappa^2 = 1$ when the points describe a linear array, and $0 < \kappa^2 < 1$ otherwise. For long linear unperturbed chains, $\langle \kappa^2 \rangle_0 = 0.41$.

For a perfectly spherical globule of any density, we anticipate $\langle L_2^2 \rangle_0 / \langle L_1^2 \rangle_0 = \langle L_3^2 \rangle_0 / \langle L_1^2 \rangle_0 = 1$, which will automatically produce $\langle b \rangle_0 = \langle c \rangle_0 = \langle \kappa^2 \rangle_0 = 0$.

1.5   Approach to the Globular State by Unbranched Homopolymers

For a relatively small linear polyethylene chain of molecular weight 10,000, Equation 1.20 provides an estimate for $\langle s^2 \rangle_0^{1/2}$ of 4.3 nm. This value is several times larger than the value (~1 nm) expected for the same collection of atoms if they were arranged into the most compact globule consistent with the density of liquid n-alkanes. This globule has the shape and size of a sphere with the volume given by the product of partial specific volume and mass, $\bar{v}M/L$.

$$s^2_{\min} = \frac{3}{5} \left(\frac{3\bar{v}M}{4\pi L} \right)^{2/3} \tag{1.39}$$

Here, $\bar{v}$ is the partial specific volume and L is Avogadro's number. A similar conclusion is obtained for virtually all unperturbed linear chains; $\langle s^2 \rangle_0$ is much larger than the result expected for a compact globule.

The average extension of the flexible chain in space can be modified by its interaction with the environment. This modification will occur in dilute solution, except in the special case where the chain is dissolved in a θ solvent. The typical dilute solution is prepared using a "good" solvent, in which case the polymer-solvent interaction produces a positive excluded volume that causes expansion of a flexible chain. The introduction of the consequences of long-range interactions into the conformational description of a linear chain molecule greatly complicates the theoretical description.[15, 21, 42] The focus is on an expansion factor defined in Equation 1.40.

$$\alpha_s^2 \equiv \frac{\langle s^2 \rangle}{\langle s^2 \rangle_0} \tag{1.40}$$

Here, $\langle s^2 \rangle$ without the zero as a subscript denotes the mean square radius of gyration in the presence of the long-range interactions. In good solvents, both experiment and theory show that $\alpha_s^2 > 1$, with α_s^2 increasing (meaning the chain has a greater expansion) as n or the solvating ability of the solvent increases. Therefore, typical flexible polymer chains in the usual solvents adopt a swollen conformation that is very much different from a globule.

Collapse of the chain toward dimensions approaching the small-size characteristic of a compact globule of uniform density requires the use of a solvent that is poorer than a θ solvent. These solutions are handled with difficulty in experiments. The reduction in the dimensions of the chain requires that the quality of the solvent must become so poor that suppression of aggregation due to intermolecular attractions and the maintenance of stable solutions with measurable concentrations of the homopolymer becomes a formidable challenge. The transition of a linear homopolymer toward the globular state is studied with least difficulty when the chain is of very high molecular weight. Under these circumstances, the chains in the sample are, on average, separated by large distances when the concentration (expressed as mass/volume) is in the lowest range where measurements of the mean dimensions are possible. Studies of these systems (notably high molecular weight polystyrene in poor solvents) show that the collapse to the globular state is a sharp transition.[41,44,45] The behavior typical of a globule (ν ~ 1/3) can be achieved in terms of the scaling of $\langle s^2 \rangle$ with n, although $\langle s^2 \rangle$ itself remains substantially larger than $s^2_{\min}$, the value expected for completion of the coil → globule transition. The smallest α_s obtained experimentally by this procedure is ~0.7. It is very difficult to obtain stable solutions of the globules formed by the intramolecular collapse of linear homopolymers.

The collapse of a linear chain to a structure closely approximating a globule can more easily be achieved in simulations of a single chain,[46,47] because intermolecular aggregation is more easily avoided in a computer simulation than in a laboratory experiment. A simulation can be performed with a system that contains only a single chain, and intermolecular aggregation is (by definition) impossible if a system contains only a single molecule. In contrast, detection of a measurable signal from a laboratory experiment (such as light scattering) with a sample in dilute solution requires a system that contains many chains, opening up the possibility of intermolecular aggregation. Values of α_s as small as

0.33 have been achieved in the simulation of the collapse of a poly(vinyl chloride) chain with a degree of polymerization of 300. The complete collapse, to a structure with radius of gyration given by Equation 1.39, would require $\alpha_s = 0.28$ for this chain. The radius of gyration achieved in the simulation was within 20 % of the value expected for the idealized space-filling sphere. The values of $\langle L_2^2 \rangle / \langle L_1^2 \rangle$ and $\langle L_3^2 \rangle / \langle L_1^2 \rangle$ for this collapsed poly(vinyl chloride) chain were 0.78 and 0.56, respectively. The collapsed structure does not have spherical symmetry, but it is closer to being spherically symmetrical than any of the structures considered in Table 1.2. Qualitatively similar behavior is seen in the molecular dynamics simulation of the collapse of an isolated chain of poly(1,4-*trans*-butadiene).[48]

Simulations of coarse-grained chains, in which a collection of atoms is represented by a single particle, can produce a more complete equilibration of collapsed chains with large numbers of segments than is possible when the macromolecule is represented with fully atomistic detail. The complete equilibration is achieved at the expense of an identification of the macromolecule with any particular real chain. Simulations of coarse-grained chains with as many as 1000 segments find a very close approximation to the idealized globular state. Not only is ν equal to 1/3, but $\langle s^2 \rangle$ itself is in excellent agreement with the expectation from Equation 1.39.[47]

The collapse produced in these simulations arises from the intramolecular short-range attractive two-body interactions experienced by a polymer in a poor solvent. Another interesting type of collapse was recently proposed[49] and then identified in a simulation of chains with conformations that could be mapped onto a diamond lattice.[50] This collapse mechanism operates in grafted layers of chains when the short-range binary interactions are repulsive, but higher-order interactions in "*N*-clusters" are attractive. The value $N = 3$ was assumed in the simulation.[50]

1.6   Simulations of the Conformations of Dendrimers

As with all areas of scientific endeavor, the discipline of the simulation of dense systems composed of macromolecules has its own set of jargon. Some of the terms may be unfamiliar to scientists whose principal expertise lies in other areas of science, or these terms may assume special nuances. A few terms that are peculiar to the study of simulations, or which have special meaning in the field of simulations, are defined in Section 1.6.7. They are also highlighted in italic font when first used in Section 1.6. Readers who seek definitions of these terms can consult Section 1.6.7 as needed; readers who are comfortable with the meanings of these terms should simply read onward.

Simulations of dense amorphous macromolecular materials are conveniently separated into two types, based on the level of structural detail incorporated in the model. Sometimes the molecules are expressed with fully atomistic detail (or nearly atomistic detail, as when a CH, CH_2, or CH_3 group is represented by a single "united atom"). Unfortunately, the *equilibration* of atomistic models of dendrimers is achieved only when they are of lower molecular weight than the molecules of greatest interest to the experimentalist. This limitation arises because the time scale of the simulation of a system expressed at fully atomistic detail is of the order of nanoseconds, and the longest relaxation time of large dendrimers is orders of magnitude longer.[25] Equilibration is more easily obtained with *coarse-grained models*, which may be expressed either in continuous space or in the *discrete space* of a lattice. Unfortunately, the coarse-grained models (in which each fundamental particle in the simulation represents a collection of many atoms) are not easily related to a system composed of a specific type of monomer unit, organized into macromolecules with a specific degree of polymerization. Methods are currently being developed for *bridging* between fully atomistic and coarse-grained models of macromolecules,[1] which will permit complete equilibration using the coarse-grained models, followed by restoration of fully atomistic detail to the equilibrated coarse-grained structure. These new methods have not yet been applied to dendrimers.

Several computer simulations of dendrimers are listed in Table 1.3. The simulations reported thus far seem to reach a consensus on some properties of dendrimers, but there

Table 1.3 Several computer simulations of dendrimers

Model	Generations	Evidence for Equilibration	Reference
Atomistic			
	1–7	None	Naylor et al.[51]
	1–5	Incomplete	Scherrenberg et al.[53]
	1–5	Incomplete	Naidoo et al.[52]
Coarse-grained, MC, lattice			
	1–9	Excellent	Mansfield and Klushin[25] and Mansfield[60]
Coarse-grained, MC, off-lattice			
	1–11	None	Lescanec and Muthukumar[57]
	1–8	Good	Carl[54]
	1–9	Good	Chen and Cui[56]
	1–5	Good	Lue and Prausnitz[59]
	1–5	Good	Welch and Muthukumar[58]
	1–6	Good	Cai and Chen[55]
Coarse-grained, MD, off-lattice			
	1–8	Excellent	Murat and Grest[26]

is disagreement on other properties. At least part of the disagreement may arise from incomplete equilibration in some of the simulations, or the unfortunate adoption of simulation methods that are known to be unlikely to produce equilibrated samples. Comparison of unequilibrated samples is unlikely to lead to useful conclusions in simulations, as well as in experiments. Evidence for equilibration in several simulations of dendrimers expressed with atomistic detail is either absent[51] or unconvincing.[52, 53] Coarse-grained simulations of isolated dendrimers on a lattice[25, 54] or in continuous space[26, 54–58] have a greater likelihood of being fully equilibrated. Impressive evidence for equilibration of coarse-grained models on a lattice has been provided by the *Monte Carlo* (MC) simulation of Mansfield and Klushin. Here, the system was subjected to random displacements, which were accepted or rejected by a rule that incorporated the change in energy produced by the random displacement.[25] Murat and Grest[26] presented impressive evidence for equilibration of coarse-grained models in continuous space using *Molecular Dynamics* (MD), where the simulation was performed by integration of Newton's equations of motion using a *classical description* of the interactions in the system. Their simulations were performed for times well in excess of the longest relaxation time, and they demonstrated reversible behavior when the system was subjected to a perturbation. Nearly all of the simulations concern single dendrimers, but the recent work of Lue and Prausnitz examines the interaction of pairs of dendrimer molecules in solution,[59] with the objective of understanding the thermodynamic properties of dilute solutions.

Analytical models are, by their nature, fully equilibrated. The issue with the analytical models is whether the equilibration was in response to unrealistic assumptions, as was unfortunately the case in the early work of de Gennes and Hervet.[61] Virtually none of the simulations support their hypothesis that segments in successive generations are confined in spherical shells located at increasing distances from the center of mass. The more recent analytical work of Boris and Rubinstein[62] is in better agreement with the results from simulations. However, there are still important differences in the shape of the *radial distribution function*, with regard to whether it exhibits a monotonic decay (as found by Boris and Rubinstein[62]) or has a more complex shape (as is usually found in simulations).

1.6.1 Mean Square Radius of Gyration

In the simulations of Mansfield and Klushin,[25] the exponent ν in Equation 1.18 is not constant, but instead decreases as the number of generations increases. It is always smaller than the values of ν for the more common architectures of macromolecules, as shown by comparison of the entries in Table 1.1. Extremely small values of ν (smaller than the ν = 1/3 for a series of dense uniform spheres, all of the same density) can occur if the smaller molecules adopt a more open structure than the larger molecules, so that the densities in the core increase with the number of generations.

The results of Murat and Grest,[26] presented in the same Table, were evaluated for model dendrimers of generations 5, 6, 7, and 8, each equilibrated under conditions that mimic immersion of a single dendrimer in solvents of widely varying quality. Their value of ν is close to the expectation of 1/3 for molecules with a compact spherical conformation.

1.6.2 Shapes Deduced from the Radius of Gyration Tensor

Some conclusions about the shapes of highly branched molecules, based on successive perfect generations of the controlled condensation of A–R–B$_2$, are accessible from simulations on a *diamond lattice*.[25] The shape deduced from the analysis of the radius of gyration tensor changes as the generation number increases, as shown in Table 1.4. The entries in this Table were calculated from the results for the moments of inertia that were reported by Mansfield and Klushin.[25] Even as soon as the second generation, these model dendrimers are closer to being spherically symmetric than are any of the conformations listed in Table 1.2. However, perfect adherence to spherical symmetry is not attained until a much higher generation number (beyond the range covered in the simulations), if indeed it is attained at all.

Table 1.4 Ratios of the averaged principal moments of the radius of gyration tensor, asphericity, acylindricity, and anisotropic shape factor for model dendrimers[a]

Generation	$\langle L_2^2 \rangle / \langle L_1^2 \rangle$	$\langle L_3^2 \rangle / \langle L_1^2 \rangle$	$\langle b \rangle / \langle s^2 \rangle$	$\langle c \rangle / \langle s^2 \rangle$	$\langle \varkappa^2 \rangle / \langle s^2 \rangle^2$
1	0.42	0.14	0.46	0.18	0.24
2	0.49	0.20	0.39	0.17	0.17
3	0.58	0.28	0.31	0.16	0.12
4	0.64	0.38	0.24	0.13	0.07
5	0.68	0.44	0.21	0.11	0.05
6	0.77	0.55	0.15	0.09	0.03
Ideal Sphere[b]	1	1	0	0	0

[a] Numerical results for generations 1–6 were calculated from the moments of inertia obtained in the simulations reported by Mansfield and Klushin.[25]
[b] Limiting results of idealized spheres, either of uniform density, or with a density that is a function of *r*.

The behavior of the asphericity with generation number is depicted in Figure 1.1. The best linear extrapolation of the six points suggests that the asphericity would fall to zero by generation 9. However, there is no reason to expect the linear extrapolation to be valid. Instead, one anticipates that the asphericity should exhibit positive curvature as the generation number increases, as is evident from close inspection of Figure 1.1. Therefore, the proper interpretation is that generations 1–9 are all aspherical, although less so than more common flexible polymers. The asphericity decreases as the generation number increases. Perhaps the asphericity would fall to zero at some generation above the ninth.

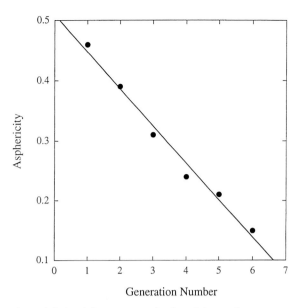

Figure 1.1 Asphericity of model dendrimers, as calculated from the moments of inertia described by Mansfield and Klushin.[25]

1.6.3   Radial Distribution Functions: Hollow Core?

The possibility of a hollow core in well-equilibrated models of dendrimers is important for their potential use as binding agents (host molecules). This important issue is more easily addressed by replacing the words "hollow core" by two related concepts that are more amenable to a reliable quantitative evaluation from radial distribution functions produced in the simulations.

Firstly, does the *radial distribution function* exhibit a local minimum, followed by a local maximum at some larger displacement from the center of mass, as sketched in Figure 1.2? The answer must be "no" if the radial distribution function is a monotonically decreasing function, as found in the analytical work of Boris and Rubinstein.[62] This result denies the existence of a hollow core in the dendrimer. However, well-equilibrated computer simulations frequently produce radial distribution functions that exhibit a local minimum followed by a local maximum, if the model dendrimer contains a sufficiently large number of generations. This type of radial distribution function opens the possibility of a "hollow core".

Secondly, if a local minimum is observed in the radial distribution function, how strong is it? One measure of the strength is the ratio of the density at the local minimum, ϱ_{min}, to the density at the local maximum observed somewhat further from the center of mass, ϱ_{max}. These two densities are depicted in the sketch in Figure 1.2. By definition, $0 \leq \varrho_{min}/\varrho_{max} < 1$. But just how small is this ratio in the well-equilibrated models in which the density exhibits a local minimum and a local maximum?

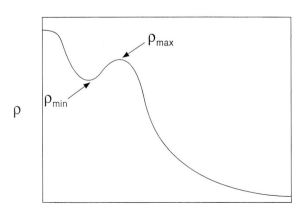

Figure 1.2 A schematic radial distribution function, showing ϱ_{min} and ϱ_{max} as they are used in the text.

Simulations do not find very small values of $\varrho_{min}/\varrho_{max}$. Instead, these ratios are only slightly smaller than one. The smallest value of $\varrho_{min}/\varrho_{max}$ in the dendrimers examined by Mansfield and Klushin[25] is 0.84. This number is comparable with the smallest value (about 0.83) found in the well-equilibrated dendrimers studied by Murat and Grest.[26] In both types of simulations, the local minima in the radial distribution functions are quite narrow. Since the local minima are shallow and narrow, it may be misleading to use the words "hollow core" in the description of the internal structure of these well-equilibrated, coarse-grained dendrimers.

Molecular dynamics simulations[26] find a plateau region in the density profile, beyond which the density drops slowly. The length of this plateau region (but not the density within it) increases as the generation number of the dendrimer increases. The density in the plateau region decreases as the quality of the solvent increases.

Two issues remain outstanding: Would the value of $\varrho_{min}/\varrho_{max}$ become significantly smaller in dendrimers with a higher number of generations than those studied thus far in simulations? Moreover, at a specific number of generations, how does the value of $\varrho_{min}/\varrho_{max}$ respond to variations in the atomistic structures (width, length, flexibility) of the spacers, the bulk of the branch points, and the introduction of strong non-covalent interactions within the dendrimer? Newer simulations, of the type that include bridging between atomistic and coarse-grained representations of the same dense structure,[1,63] may eventually provide some answers to these questions.

1.6.4 Locations of the Terminal Groups of the Last Generation

Mansfield and Klushin,[25] and also Murat and Grest,[26] found that for the early generations (roughly the first four generations in their simulations), the segments of successive generations are localized in the radial direction, being displaced further from the core as the generation number increases within this range. Backfolding becomes pronounced in the fifth generation. By the eighth generation, segments from the last generation added to the molecule are nearly uniformly distributed throughout the dendrimer. Segments belonging to the last generation are dispersed completely throughout the molecules in the large dendrimers. Functional groups placed at the free ends of the monomers in the last generation of a large dendrimer in the simulations are not confined to the surface, but instead occur throughout the molecule. This conclusion might be modified if the terminal functional groups were designed so that they had markedly different intermolecular interactions than the remainder of the dendrimer. One circumstance that might produce this situation is the placement of identical ionized terminal functional groups on an otherwise non-ionic dendrimer.

1.6.5 Stretching of the Spacers from Early Generations

Mansfield and Klushin[25] found that the flexible spacers of the inner generations become increasingly stretched (have bonds with an increasingly strong population of *trans* states) as one proceeds to larger and larger dendrimers. The spacers from the inner generation are almost completely stretched in the seventh generation dendrimer. Meanwhile, the outer tier of spacers is almost completely relaxed, even in the largest dendrimer examined.

1.6.6 Angular Segregation of Dendrons

Radial distribution functions (Section 1.6.3) indicate the distribution of segments along the radial distance, r, from the center of the dendrimer. A spherical coordinate system contains two other variables, the angles θ and ϕ. The issue of angular segregation asks whether the segments from a particular dendron are randomly distributed over θ and ϕ, or whether they are restricted in their distribution. Both MC simulation[60] and

MD simulation[26] find angular segregation of dendrons when the dendrimer contains a sufficiently large number of generations.

1.6.7   Comments on Several Terms Used in the Description of Simulations

This section presents a few comments on terms that are highlighted in italic font when first used in Section 1.6.

Analytical model: A model represented by a closed-form equation that can be solved for its exact answer in a finite amount of time. This is in contrast with another model, not in closed form, for which approximate solutions can be obtained only by use of a computer, with the accuracy of the approximate answer increasing continuously as the computer runs longer and longer.

Bridging: A procedure that allows mapping of a model expressed in fully atomistic detail onto a coarse-grained model, and also allows mapping of this coarse-grained model (after equilibration) back to a model with fully atomistic detail.

Classical description: Employed in a model that does not explicitly include quantum effects, but instead contains particles that interact with energies from a classical description. An example is a chain of beads connected by Hooke's Law springs, with interactions between non-bonded pairs of beads obeying a Lennard–Jones potential energy function, $E = -a/r^6 + b/r^{12}$.

Coarse-grained model: A model in which the individual particles do not represent atoms, but represent instead a collection (often one or more monomer units) of atoms.

Diamond lattice: A lattice where all steps are of the same length, all angles between two successive steps have the tetrahedral value, with $\cos \theta = -1/3$, and three consecutive bonds adopt one of three torsion angles, with a value of exactly $\pm 60°$ or $180°$ (in the convention where a *cis* placement has a torsion angle of $0°$).

Discrete space: Certain regions of space, described by a simple repeating pattern, are allowed, and all other regions of space are absolutely prohibited.

Equilibration: Several methods are used for examination of the extent of equilibration. If the longest relaxation time for the system is known (or can be estimated), the simulation can be run for a time that is several multiples of this relaxation time. In melts of linear chains, this procedure often involves running the simulation long enough so that the center of mass of a chain has translated a distance several multiples of $\langle s^2 \rangle^{1/2}$, and/or the correlation function for **r**, $\langle \mathbf{r}(t + t_0) \cdot \mathbf{r}(t_0) \rangle / \langle r^2 \rangle$, has decayed to a value close to zero. For dendrimers, one might look for the decay in the correlation function defined by the radius of gyration, $\langle s(t + t_0)s(t_0) \rangle / \langle s^2 \rangle$, to a constant value smaller than one. One can also perturb the system (as by raising its temperature), wait for a new steady state, and then remove the perturbation and assess whether the initial structure is recovered. Finally, one should avoid simulation methods that have clearly been demonstrated to produce poorly equilibrated samples.

Molecular Dynamics (MD): A method whereby Newton's equations of motion are integrated in order to compute the trajectory of the system through time and space. Each particle is characterized by six degrees of freedom (three coordinates define its instantaneous position, and three momenta define its instantaneous velocity). The integration involves an approximation that is acceptable only when the time step is very small, typically on the order of 10^{-15} s. Therefore, a trajectory of 1 ns requires 10^6 computationally intensive successive integration steps. The upper limit on the length of the trajectory is imposed by the patience of the simulator (how long should we let the computer run?), and by the accumulation of errors that arise because the integration step is not infinitesimally small; although it is a small Δt, it is not the mathematician's dt.

Monte Carlo (MC): A method that uses random displacements of the system in order to generate an ensemble that mimics the static properties of the system. Each particle is characterized by three degrees of freedom, which are the three coordinates that define its instantaneous position. MD also incorporates the momenta, but these are not used in MC. This attribute of MC often can be exploited to advantage in the equilibration of large dense systems, because the random moves employed in the MC simulation need not be restricted to those moves that will reproduce "sensible" dynamics. A subset of MC is Metropolis MC[64], in which the criteria for acceptance of a move are: (a) accept uncon-ditionally if the proposed move decreases the energy of the system, but (b) accept with probability $\exp(-\Delta E/kT)$ if the energy of the system increases. If the application of (b) causes us to reject the proposed move, the original configuration must be counted again as the new configuration if one wants to generate an equilibrated system at constant *NVT*. One must resist the temptation to keep trying until a move is finally accepted.

Radial Distribution Function: A function that gives the probability of finding a particle at a distance $r + dr$ from a reference particle, divided by the value of this probability if the density is uniform everywhere. If the system is an ideal gas (no excluded volume and no intermolecular interactions), the radial distribution function has a value of 1 at all val-ues of r. By its definition, the radial distribution function for any amorphous systems must approach 1 as $r \to \infty$. If the particles cannot overlap (and therefore have some degree of excluded volume), then the radial distribution function is 0 when $r = 0$ and has at least one maximum at finite r, before converging to the asymptotic limit of 1.

1.7   References

[1] J. Baschnagel, K. Binder, P. Doruker, A. A. Gusev, O. Hahn, W. L. Mattice, F. Müller-Plathe, M. Murat, W. Paul, S. Santos, U. W. Suter, V. Tries, "Bridging the Gap between Atomistic and Coarse-Grained Models of Polymers: Status and Perspectives", *Adv. Polym. Sci.* **2000**, *152*, 41–156.

[2] P. J. Flory, "Molecular Size Distribution in Linear Condensation Polymers", *J. Am. Chem. Soc.* **1936**, *58*, 1877–1885.

[3] P. J. Flory, "Molecular Size Distribution in Ethylene Oxide Polymers", *J. Am. Chem. Soc.* **1940**, *62*, 1561–1565.

[4] J. R. Schaefgen, P. J. Flory, "Synthesis of Multichain Polymers and Investigation of Their Vis-cosities", *J. Am. Chem. Soc.* **1948**, *70*, 2709–2718.

[5] P. J. Flory, "Molecular Size Distribution in Three-Dimensional Polymers; I. Gelation", *J. Am. Chem. Soc.* **1941**, *63*, 3083–3090.

[6] P. J. Flory, "Fundamental Principles of Condensation Polymerization", *Chem. Rev.* **1946**, *39*, 137–197.

[7] W. H. Stockmayer, "Thermoreversible Gelation via Multichain Junctions", *Macromolecules* **1991**, *24*, 6358–6368.

[8] F. Tanaka, W. H. Stockmayer, "Thermoreversible Gelation with Junctions of Variable Multi-plicity", *Macromolecules* **1994**, *27*, 3943–3954.

[9] M. Nguyen-Misra, W. L. Mattice, "Micellization and Gelation of Symmetric Triblock Copoly-mers with Insoluble End Blocks", *Macromolecules* **1995**, *28*, 1444–1457.

[10] P. J. Flory, "Molecular Size Distribution in Three-Dimensional Polymers; IV. Branched Poly-mers Containing A–R–B$_{f-1}$ Type Units", *J. Am. Chem. Soc.* **1952**, *74*, 2718–2723.

[11] M. L. Mansfield, "Molecular Weight Distributions of Imperfect Dendrimers", *Macromole-cules* **1993**, *26*, 3811–3814.

[12] T. M. Birshtein, O. B. Ptitsyn, *Conformations of Macromolecules*, Wiley-Interscience, New York, **1966**.

[13] P. G. de Gennes, *Concepts in Polymer Physics*, Cornell University Press, Ithaca, New York, **1979**.

[14] J. des Cloizeaux, G. Jannink, *Polymers in Solution – Their Modelling and Structure*, Claren-don Press, Oxford, **1990**.

[15] M. Doi, S. F. Edwards, *The Theory of Polymer Dynamics*, Clarendon Press, Oxford, **1986** .

[16] P. J. Flory, *Principles of Polymer Chemistry*, Cornell University Press, Ithaca, New York, **1953**.

[17] P. J. Flory, *Statistical Mechanics of Chain Molecules*, Wiley, New York, **1969**.

[18] P. J. Flory, *Statistical Mechanics of Chain Molecules*, Hanser, München, **1989**.

[19] W. L. Mattice, U. W. Suter, *Conformational Theory of Large Molecules – The Rotational Iso-meric State Model in Macromolecular Systems*, Wiley, New York, **1994**.

[20] M. V. Volkenstein, *Configurational Statistics of Polymer Chains*, Wiley-Interscience, New York, **1963**.

[21] H. Yamakawa, *Modern Theory of Polymer Solutions*, Harper & Row, New York, **1999**.

[22] P. J. Flory, "The Configuration of Real Polymer Chains", *J. Chem. Phys.* **1949**, *17*, 303–310.

[23] J. C. Le Guillou, J. Zinn-Justin, "Critical Exponents from Field Theory", *Phys. Rev. B: Con-dens. Matter* **1980**, *21*, 3976–3998.

[24]   Y. Oono, "Statistical Physics of Polymer Solutions – Conformational-Space Renormalization-Group Approach", *Adv. Chem. Phys.* **1985**, *61*, 301–437.

[25]   M. L. Mansfield, L. I. Klushin, "Monte Carlo Studies of Dendrimer Macromolecules", *Macromolecules* **1993**, *26*, 4262–4268.

[26]   M. Murat, G. S. Grest, "Molecular Dynamics Study of Dendrimer Molecules in Solvents of Varying Quality", *Macromolecules* **1996**, *29*, 1278–1285.

[27]   M. Rehahn, W. L. Mattice, U. W. Suter, "Rotational Isomeric State Models in Macromolecular Systems", *Adv. Polym. Sci.* **1997**, *131 / 132*, 1–460.

[28]   R. Chiang, "Intrinsic Viscosity–Molecular Weight Relation for Fractions of Linear Polyethylene", *J. Phys. Chem.* **1965**, *69*, 1645–1653.

[29]   A. Abe, R. L. Jernigan, P. J. Flory, "Conformational Energies of *n*-Alkanes and the Random Configuration of Higher Homologs Including Polymethylene", *J. Am. Chem. Soc.* **1966**, *88*, 631–639.

[30]   H. A. Kramers, "The Behavior of Macromolecules in Inhomogeneous Flow", *J. Chem. Phys.* **1946**, *14*, 415–424.

[31]   B. H. Zimm, W. H. Stockmayer, "The Dimensions of Chain Molecules Containing Branches and Rings", *J. Chem. Phys.* **1949**, *17*, 1302–1314.

[32]   P. J. Flory, "Foundations of Rotational Isomeric State Theory and General Methods for Generating Configurational Averages", *Macromolecules* **1974**, *7*, 381–392.

[33]   L. Willner, O. Jucknischke, D. Richter, J. Roovers, L.-L. Zhou, P. M. Toporowski, L. J. Fetters, J. S. Huang, M. Y. Lin, N. Hadjichristidis, "Structural Investigation of Star Polymers in Solution by Small-Angle Neutron Scattering", *Macromolecules* **1994**, *27*, 3821–3829.

[34]   J. Roovers, L.-L. Zhou, P. M. Toporowski, M. van der Zwan, H. Iatrou, N. Hadjichristidis, "Regular Star Polymers with 64 and 128 Arms – Models for Polymeric Micelles", *Macromolecules* **1993**, *26*, 4324–4331.

[35]   W. L. Mattice, K. Sienicki, "Extent of the Correlation Between the Squared Radius of Gyration and Squared End-to-End Distance in Random Flight Chains", *J. Chem. Phys.* **1989**, *90*, 1956–1959.

[36]   M. Fixman, "Radius of Gyration of Polymer Chains", *J. Chem. Phys.* **1962**, *36*, 306–310.

[37]   K. Solc, W. H. Stockmayer, "Shape of a Random-Flight Chain", *J. Chem. Phys.* **1971**, *54*, 2756–2757.

[38]   K. Solc, "Shape of a Random-Flight Chain", *J. Chem. Phys.* **1971**, *55*, 335–344.

[39]   W. L. Mattice, "Asymmetry of Flexible Chains, Macrocycles, and Stars", *Macromolecules* **1980**, *13*, 506–511.

[40]   K. Solc, "Statistical Mechanics of Random-Flight Chains; IV. Size and Shape Parameters of Cyclic, Star-like, and Comb-like Chains", *Macromolecules* **1973**, *6*, 378–385.

[40a] R. P. Smith, E. M. Mortensen, "Calculation of Depolarization Ratios, Anisotropies, and Average Dimensions of *n*-Alkanes", *J. Chem. Phys.* **1961**, *35*, 714–721.

[41]   D. N. Theodorou, U. W. Suter, "Shape of Unperturbed Linear Polymers: Polypropylene", *Macromolecules* **1985**, *18*, 1206–1214.

[42]   P. G. de Gennes, *Scaling Concepts in Polymer Physics*, Cornell University Press, Ithaca and London, **1979**.

[43]   I. H. Park, Q.-W. Wang, B. Chu, "Transition of Linear Polymer Dimensions from the θ to Collapsed Regime; 1. Polystyrene/Cyclohexane System", *Macromolecules* **1987**, *20*, 1965–1975.

[44]   I. H. Park, Q.-W. Wang, B. Chu, "Transition of Linear Polymer Dimensions from the θ to the Collapsed Regime; 2. Polystyrene/Methyl Methacrylate System", *Macromolecules* **1987**, *20*, 2883–2840.

[45]   B. Chu, R. Xu, J. Zuo, "Transition of Polystyrene in Cyclohexane from the θ to the Collapsed State", *Macromolecules* **1988**, *21*, 273–274.

[46]   G. Tanaka, W. L. Mattice, "Chain Collapse by Atomistic Simulation", *Macromolecules* **1995**, *28*, 1049–1059.

[47]   G. Tanaka, W. L. Mattice, "Chain Collapse by Lattice Simulation", *Macromol. Theory Simul.* **1996**, *6*, 499–523.

[48]   Y. Zhan, W. L. Mattice, "Molecular Dynamics Simulation of the Collapse of Poly(1,4-*trans*-butadiene) to Globule and to a Thin Film", *Macromolecules* **1994**, *27*, 7056–7062.

[49]   P. G. de Gennes, "A Second Type of Phase Separation in Polymer Solutions", *C. R. Acad. Sci. Paris II* **1991**, *313*, 1117–1122.

[50]   W. L. Mattice, S. Misra, D. H. Napper, "Collapse of Tethered Chains Due to N-Clusters when Binary Interactions are Weakly Repulsive but Ternary Interactions are Weakly Attractive", *Europhys. Lett.* **1994**, *28*, 603–608.

[51]   A. M. Naylor, W. A. Goddard, III, G. E. Kiefer, D. A. Tomalia, "Starburst Dendrimers; 5. Molecular Shape Control", *J. Am. Chem. Soc.* **1989**, *111*, 2339–2341.

[52] K. J. Naidoo, S. J. Hughes, J. R. Moss, "Computational Investigations into the Potential Use of Poly(benzyl phenyl ether) Dendrimers as Supports for Organometallic Catalysts", *Macromolecules* **1999**, *32*, 331–341.

[53] R. Scherrenberg, B. Coussens, P. van Vliet, G. Edouard, J. Brackman, E. de Brabander, K. Mortensen, "The Molecular Characteristics of Poly(propyleneimine) Dendrimers as Studied with Small-Angle Neutron Scattering, Viscosimetry, and Molecular Dynamics", *Macromolecules* **1998**, *31*, 456–461.

[54] W. Carl, "A Monte Carlo Study of Model Dendrimers", *J. Chem. Soc., Faraday Trans.* **1996**, *92*, 4151–4154.

[55] C. Cai, Z. Y. Chen, "Intrinsic Viscosity of Starburst Dendrimers", *Macromolecules* **1998**, *31*, 6393–6396.

[56] Z. Y. Chen, S.-M. Cui, "Monte Carlo Simulations of Starburst Dendrimers", *Macromolecules* **1996**, *29*, 7943–7952.

[57] R. L. Lescanec, M. Muthukumar, "Configurational Characteristics and Scaling Behavior of Starburst Molecules: A Computational Study", *Macromolecules* **1990**, *23*, 2280–2288.

[58] P. Welch, M. Muthukumar, "Tuning the Density Profiles of Dendritic Polyelectrolytes", *Macromolecules* **1998**, *31*, 5892–5897.

[59] L. Lue, J. M. Prausnitz, "Structure and Thermodynamics of Homogeneous Dendritic Polymer Solutions: Computer Simulation, Integral-Equation, and Lattice-Cluster Theory", *Macromolecules* **1997**, *30*, 6650–6657.

[60] M. L. Mansfield, "Dendron Segregation in Model Dendrimers", *Polymer* **1994**, *35*, 1827–1830.

[61] P. G. de Gennes, H. Hervet, "Statistics of "Starburst" Polymers", *J. Phys. Lett.* **1983**, *44*, L351–L360.

[62] D. Boris, M. Rubinstein, "A Self-Consistent Mean Field Model of a Starburst Dendrimer: Dense Core *vs.* Dense Shell", *Macromolecules* **1996**, *29*, 7251–7260.

[63] P. Doruker, W. L. Mattice, "A Second Generation of Mapping/Reverse Mapping of Coarse-Grained and Fully Atomistic Models of Polymer Melts", *Macromol. Theory Simul.* **1999**, *8*, 463–478.

[64] N. Metropolis, A. N. Rosenbluth, M. N. Rosenbluth, A. H. Teller, E. Teller, "Equation of State Calculations by Fast Computing Machines", *J. Chem. Phys.* **1953**, *21*, 1087–1092.

2 From Theory to Practice: Historical Perspectives, Self-Similarity, and Preparative Considerations

2.1 Introduction

The foundations for a general understanding of macroassemblies by considering molecular masses, sizes, and shapes were presented in Chapter 1. Mathematical analyses of these physical properties were extended from an examination of classically prepared, long chain, linear polymers to highly branched polymers and finally to iteratively constructed macromolecules commonly known today as dendrimers or cascade macromolecules. With respect to their macromolecular assembly and utility, a notable concept was established by investigation of the averaged principal moments of the radius of gyration tensors, which allowed analysis of the asphericity, acylindricity, and shape anisotropy for the particular macromolecular models examined.

This ability to predict, as elucidated in the discussion, suggests precise and deliberate control over macromolecular geometry. This far-reaching idea can be gleaned from the general trend toward pseudospherical or globular symmetry that was observed for dendrimers synthesized by repetitive condensations with $1 \rightarrow 2$ branching AB_2 monomers. Extension of this type of analysis to cascades constructed with symmetrical, four-directional cores and $1 \rightarrow 3$ branching AB_3 monomers would presumably lead to even more spherical assemblies due to an enhancement of space-filling considerations. Control over macromolecular configuration is therefore a key goal in dendritic chemistry. Ramifications extend far beyond command of the overall molecular shape to include choice over such parameters as internal and external rigidity, lipophilicity and hydrophilicity, degrees of void and excluded volumes, density gradients, complementary functionalities, and environmental cooperativity. The iterative synthetic method with its extreme flexibility is, therefore, the quintessential macromolecular assembly process.

Development of the dendritic approaches to macroassembly did not simply passively follow the first examples of deliberately prepared branched molecules, but rather, it has been a logical progression of synthetic approaches building upon the efforts of countless researchers to realize new materials with novel properties and utilities.

2.2 Branched Architectures

From a historical perspective, progress towards the deliberate construction of macromolecules possessing branched architecture can be considered to have occurred during three general eras. The first such period was roughly from the late 1860's to the early 1940's, when macrostructures were considered as responsible for insoluble and intractable materials formed in polymerization processes. Synthetic control, mechanical separations, and physical characterization were primitive, at best, as judged by current standards; isolation and proof of structure were simply not feasible. The need for new synthetic resources to meet the increased demand for materials driven by World War II was an underlying driving force to replace traditional natural materials, e.g. latex, with synthetic or "unnatural" substitutes.

The early 1940's to the late 1970's defines the second period, in which branched structures were considered primarily from a theoretical viewpoint with initial attempts at preparation via classical, or single-pot, polymerization of functionally differentiated monomers. As noted by Flory[1] in 1952: "The breadth of the distribution coupled with the impossibility of selectivity fractionating 'branching' and 'molecular weight' separately make this approach impractical. Attempts to investigate 'branching' by such means consequently have been notably fruitless."

The late 1970's and early 1980's recorded initial successful progress toward macromolecular assembly, based on the iterative methodology that has become the cornerstone of dendritic chemistry and thus defined the start of the third period of development. At this point in time, the concept of control over macroassembly was proposed and demonstrated. Advances in physical isolation and purification, as well as the introduction of diverse spectroscopic procedures, had reached the level of sophistication necessary to support this budding field. In essence, the expansion of synthetic limits of single, stepwise prepared, chemical structures beyond the approximate 2000 amu historical threshold was effected. This discovery lifted the glass ceiling for molecular construction of single, monomolecular macromolecules possessing a predetermined composition. Applying the synthetic precision of the traditional, natural product chemist to the polymeric as well as biochemical areas has opened the door to chemical nanoscience.

2.2.1 Early Observations

Initial reports leading to speculation about nonlinear or branched polymeric connectivity were provided by researchers such as Zincke,[2] who, in 1869, isolated an insoluble hydrocarbon-based material when benzyl chloride was treated with copper. Later, in 1885, reaction of benzyl chloride with $AlCl_3$ led Friedel and Crafts[3] to report similar observations. When benzyl chloride was subjected to the action of a zinc–copper couple, analogous materials were obtained.[4]

A significant aspect of chemical research at this time relating to the basic understanding of materials was the idea that virtually any substance could exist in a "colloidal state",[5, 6] analogous to the gaseous, liquid, and solid states. The term "colloid", coined by Graham[7] in 1861 and synonymous with "glue-like", was introduced to describe polymers possessing a negligible diffusion ratio in solution and which did not pass through semi-impermeable membranes. Materials that could be obtained in crystalline form were described as "crystalloids". Extended use of "colloid" terminology[5, 6] was unfortunate from the perspective of covalent molecular assembly because a simple physical change of state will not, in general, disassemble a polymer into its starting components. For many years, the notion that polymers obtained their properties through non-covalent small molecule intermolecular interactions supplanted the view of a polymeric molecule consisting of covalently connected units.

Regardless of the misleading colloid concept, the study of polymers continued unabated. In 1871, Hlasiwetz and Habermann[8] concluded that proteins and carbohydrates were comprised of polymeric units with differing degrees of condensation. Certain examples of these substances were noted as "soluble and unorganized", while others were described as "insoluble and organized". Much later, Flory[9] likened these categories to crystalline and noncrystalline polymers.

Thirty-five years after Gladstone and Tribe's observation[4] concerning the reaction of benzyl chloride with the zinc–copper couple, reports concerning the preparation of unclassifiable substances continued. Hunter and Woollet[10] described the synthesis of an amorphous material by polymerization of triiodophenolic salts and salicylic acid. Interestingly, in 1922, Ingold and Nickolls[11] reported the preparation of the branched, small molecule "methanetetraacetic acid"; this is perhaps the earliest example of a deliberately constructed, symmetrically substituted, polyfunctional, dendritic paradigm, or archetype. In a prescient forecast of potential utility of branched architecture with respect to supramolecular host–guest chemistry, the authors noted that crystals of this tetraacid "contained 2–3 percent of nitrogen, which was only slowly removed by boiling with 50% sulfuric acid or with 40% potassium hydroxide."

Staudinger[12] (Nobel Prize in Chemistry, 1953), at about the same time, suggested that materials such as natural rubber were composed of long-chain, high molecular weight molecules and should not be considered as small molecule aggregates in the colloidal state. Carothers' studies[13] on condensation polymerizations promoted the formulation of Staudinger's theory of a macromolecular composition. The intractability of substances such as Bakelite-C was attributed to a three-dimensional network-like structure,[13] while Jacobson[14] suggested the presence of the three-dimensional structure to account for insoluble materials that were present after polymerization processes.

2.2.2 Prelude to Practice

The second distinct era in the development of branched macromolecular architectures encompassed the 1940 to 1978 time frame, or approximately the next four decades. Before 1940, Kuhn[15] first reported the use of statistical methods for analysis of a polymer problem. Equations were derived for degraded cellulose molecular weight distributions. Thereafter, mathematical analyses of polymer properties and interactions flourished. Perhaps no single person has affected linear and nonlinear polymer chemistry as profoundly as Professor P. J. Flory (Nobel Prize in Chemistry, 1974).

With respect to understanding the historical development of polymer chemistry in general, it is interesting to note a prevailing attitude[16] that material preparation was acceptable mainly from a perspective of isolation of discrete molecules with definite structure. As Flory[1] noted in his seminal treatise on polymers: "To be eligible for acceptance in the chemical kingdom, a newly created substance or a material of natural origin had to be separated in such a state that it could be characterized by a molecular formula." This was largely the result of the progress in synthetic organic chemistry.

Evolution of dendritic chemistry, and of material science in general, might well have occurred more rapidly had the field of polydisperse polymer chemistry been more widely recognized and accepted earlier. For example, during the 1860's, Lourenco[17, 18] and Kraut[19] observed and reported the synthesis of materials consisting of difficult to separate molecular mixtures upon condensation of ethylene glycol with an ethylene dihalide and thermal condensation of acetylsalicylic acid, respectively. Their papers noted the basic nature of condensation polymer connectivity and formulation. It was not until many years later (ca. 1910) that the general scientific community began to significantly recognize and accept polydisperse macromolecular materials. Flory even suggests[20] that "evidence of retrogression could be cited."

During 1941 and 1942, Flory[21–24] disseminated theoretical and experimental evidence for the appearance of branched-chain, three-dimensional[25] macromolecules. These papers discussed a feature of polymerization reactions called "gelation". Descriptive terminology used by Flory to categorize differing polymeric fractions included the terms "gel" and "sol", referring to polymers that were insoluble or soluble, respectively. Flory showed statistically that branched polymeric products began to appear after polymerization had progressed to a definite extent. Molecular size distributions, the number average degree of polymerization (DP), as well as derivations relating to tri- and tetrafunctional branching units (monomers) were also addressed. Flory's concept of a branched macromolecule is illustrated in Figure 2.1 by a two-dimensional drawing of a 2,3,2,3,1,0 macromolecule (**1**). The prefix numbering denotes the number of branch points occurring at each successive generation.

Stockmayer[26] subsequently developed equations related to branched-chain polymer size distributions and "gel" formation, whereby branch connectors were of unspecified length and branch functionality was undefined. An equation was derived for the determination of the extent of reaction where a three-dimensional network ("gel") forms; this relation was similar to Flory's, although it was derived by an alternative approach. Stockmayer likened gel formation to a phase transition and noted the need to consider: (a) intramolecular reactions, and (b) unequal reactivity of differing functional groups. This work substantially corroborated Flory's earlier studies.

In 1949, Flory[27] examined branched polymer "scaling" properties, e.g., the number of chain monomers relative to the mean squared end-to-end chain distance. Pursuing synthetic experiments with branched polymers, he reported[1] the preparation of a highly branched polymer without "insoluble gel formation". Flory employed a $1 \rightarrow 2$ branching AB$_2$ monomer, which generated an envisaged branched structure **2** (Figure 2.2). With respect to separating monodisperse (or nearly so) fractions of these polymers, he noted the difficulties inherent in the fractionation of high molecular weight polydisperse materials. This observation was the predecessor to present day hyperbranched polymers.

In 1952, Goldberg[28] presented a theory for the reaction of a multivalent antigen with a bivalent or univalent antibody. Like Stockmayer[26] and Flory,[1, 27] he considered the extent of reaction to be critical and suggested that it was "the point at which the system changes from one chiefly composed of small aggregates into one composed chiefly of rel-

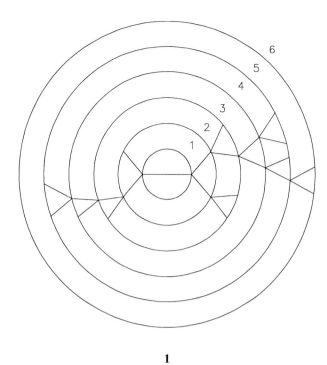

1

Figure 2.1 Flory's depiction of a branched 2,3,2,3,1,0 macromolecule.[1]

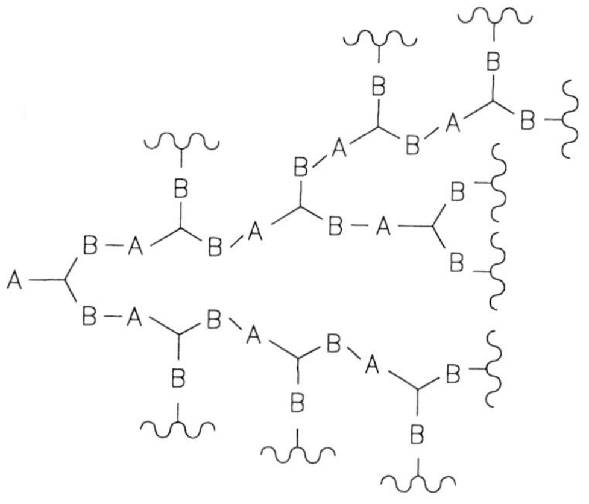

2

Figure 2.2 Branched polymer architecture[1] by the assembly of AB_2-type monomers.

atively few exceedingly large aggregates" (visually observed as the precipitation point in some systems). Accordingly, Goldberg[29] extended the theory (1953) to encompass multivalent antigen and multivalent antibody reactions. Many other researchers contributed to the theoretical analysis of macromolecules. For example, Rouse[30] based a theory of the linear viscoelastic properties of dilute solutions of coiling polymers on the coordinated motions of different polymer units. In contrast, Zimm[31] considered the "problem of motions of a chain molecule diffusing in a viscous fluid under the influence of external forces or currents".

2.2.3 Incipient Macromolecular Progression

2.2.3.1 Initial Concept and Practice

"For the construction of large molecular cavities and pseudocavities that are capable of binding ionic guests and molecules (as [in] a complex or inclusion compounds) in a host–guest interaction, synthetic pathways allowing a frequent repetition of similar steps would be advantageous." The third and current era of cascade or dendrimer chemistry came to life when Vögtle published this introductory sentence in his 1978 paper entitled "Cascade and Nonskid-Chain-like Syntheses of Molecular Cavity Topologies".[32] Ramifications of this statement extended far beyond the use of iterative methodology for the construction of host–guest assemblies. Repetition of similar and complementary synthetic steps has since been used for the preparation of many new and exciting materials.

The essence of the "cascade" synthesis[33] is depicted in Scheme 2.1. Two procedures, alkylation and reduction, comprised the [a → b → a → b →...] sequence; thus, treatment of a diamine with acrylonitrile afforded tetranitrile **3**, which was followed by Co-mediated reduction to give tetraamine **4**. Further amine alkylation provided the 2nd generation octanitrile **5**. The "nonskid-chain-like" synthesis[32] is shown in Scheme 2.2. Construction of polycycle **6** was accomplished through a repetitive (a → b → c → d → a → b → ...) sequence. Again, repetitive and multiple reaction sequences were employed for the generation of new molecular assemblies. Most notable about these syntheses is that, for the first time, 'generational' molecules were prepared and characterized at each stage of the construction process; a synthetic organic chemists approach to repetitive growth.

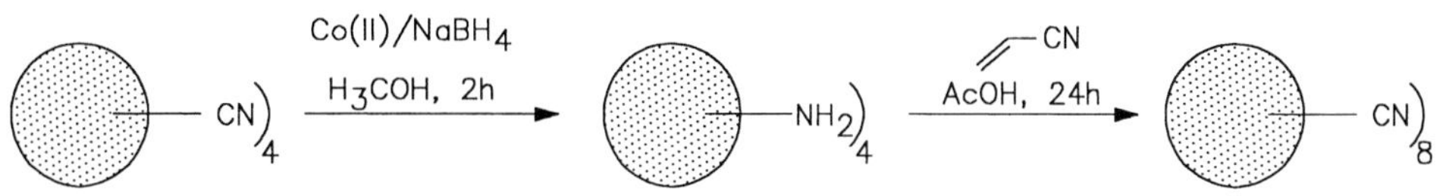

Scheme 2.1 The concept of "cascade" or "repeating" synthesis.[32] For details see Scheme 3.2.

2.2.3.2 Implied Dendritic Construction

As is perhaps the case with many scientific and technological advances, earlier accounts of synthetic efforts towards novel (macro)molecular assemblies logically implied the utility of the iterative method. Consider Figure 2.3, whereby Lehn[34] (Nobel Prize in Chemistry, 1987) pictorially delineates stepwise synthetic strategies for the construction of macrocyclic organic complexing agents. Multiple component attachment (in this case, two) is used for ring and cage preparation of the "molecular cavities." Divergent as well as convergent methods are envisaged. Similar strategies for ligand construction can be found, as described in the work of Cram et al.[35] (Nobel Prize in Chemistry, 1987) entitled "Chiral Recognition in Complexation of Guests by Designed Host Molecules". These two accounts were pivotal in promoting the seminal cascade methodology as proposed by Vögtle.[32]

Finally, to underscore and stress the impact of seemingly unrelated works on this field, an examination of critical reviews by pioneering scientists, such as Lehn,[34, 36, 37] Ringsdorf,[38] and Lindsey,[39] reveals unparalleled insight into the potential utility of cascade-related macroassemblies, including, of course, dendrimers.

2.2.3.3 Alternative Architectures

After the initial disclosure of "a viable" iterative synthetic method for the construction of polyfunctional macromolecules, a small number of articles explored the use of repetitive chemistry for the preparation of dendritic materials. Figure 2.4 illustrates the early branched architectures that were constructed, in most cases, by employing protection-

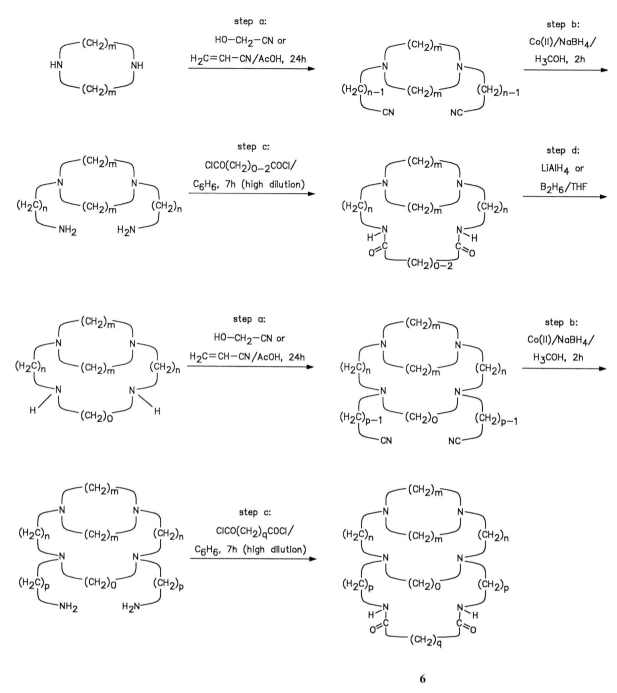

6

Scheme 2.2 Repetitive a → b → c → d... sequences formed the basis of the so-called "nonskid-chain-like" synthesis.[32]

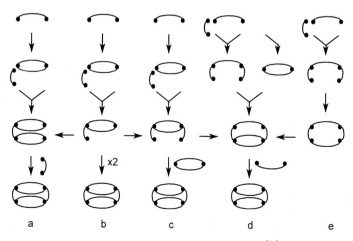

a   b   c   d   e

Figure 2.3 Construction pathways for "molecular cavities".[34] Redrawn with permission of Professor J.-M. Lehn.

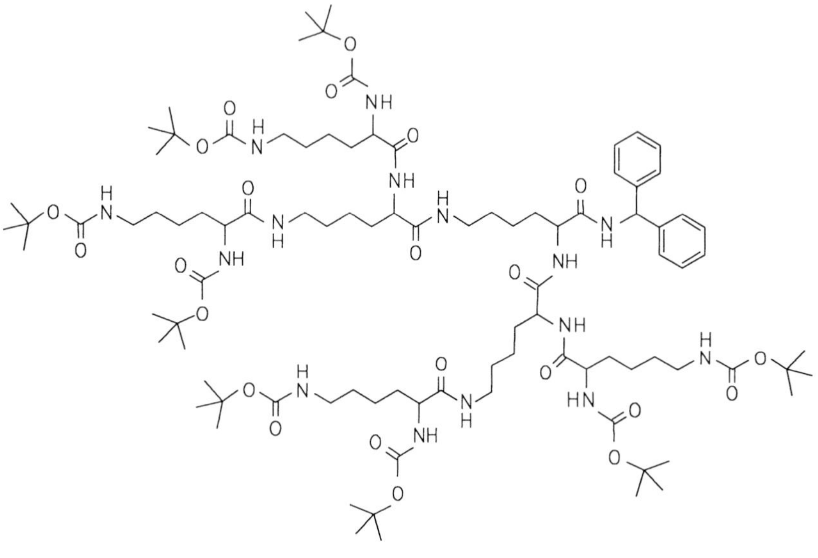

7

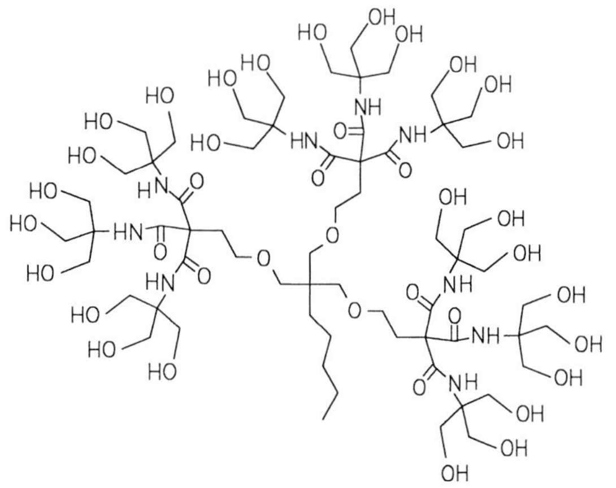

8

Figure 2.4 Illustrations of different branched architectures.

deprotection schemes. Detailed descriptions of the synthetic procedures can be found in Chapters 3, 4 (structures **7–9**) and 5 (structures **10** and **11**).

Denkewalter, Kolc, and Lukasavage[40] patented a method for the synthesis of polylysine-based dendrimers (**7**). Interesting features of these dendritic polymers include a $1 \rightarrow 2$ asymmetric branching pattern and the incorporation of multiple chiral centers at each tier. Aharoni, Crosby, and Walsh[41] studied these lysine dendrimers and reported that each member of the series was monodisperse; higher generations behaved as non-draining spheres (i.e., they trapped solvent within the void regions). Shortly thereafter, Kricheldorf, Zang, and Schwarz[42] demonstrated a renewed interest in branched topologies prepared by single-pot procedures; the products are currently referred to as 'hyper-branched' polymers (see Chapter 6).

In 1985, two new architectures were published in which the iterative synthetic protocol was used in the preparation of large tree-like macromolecules ("arborols" and "dendrimers"). Newkome et al.[43] relied on triester amidation with 'tris' for the construction of polyols (**8**) possessing maximal $1 \rightarrow 3$ C-based branching. Repetition of the sequence was precluded by surface steric inhibition towards nucleophilic substitution; however, the concept that such branched materials can be "unimolecular micelles" was therein suggested and proven several years later.[44, 45] Tomalia et al.[46] reported the first preparation of an entire series of dendrimers [**9** (up to the 7th tier)] possessing trigonal, $1 \rightarrow 2$ N-based, branching centers. Their early commercial availability has made this series one of the most widely investigated and has undoubtedly served to promote the field of dendrimer chemistry by ubiquity. These procedures are classified as divergent in nature, in that

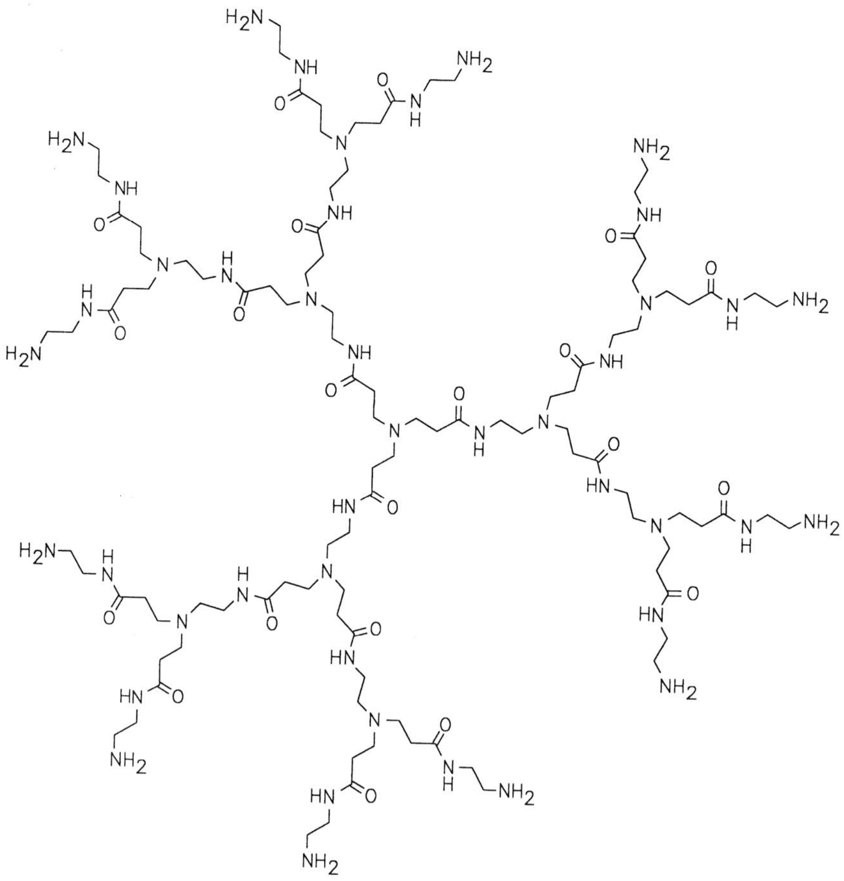

9

Figure 2.4 Illustrations of different branched architectures (continued).

they are constructed from the inside out. It is interesting to note that, also in 1985, Bidd and Whiting[47] described the iterative-based synthesis of a series of pure, *linear* alkyl hydrocarbons possessing 102, 150, 198, 246, and 390 carbon atoms; their preparative methodology involved the use of repetitive coupling and cyclic acetal hydrolysis reactions.

In 1990, the first convergent, i.e. built from the outside in, preparation of dendrimers resulted in a poly(aryl ether) architecture (**10**), as reported by Fréchet and Hawker.[48] Their innovative use of a pivotal phenoxide-based, benzylic bromide displacement sequence has since led to many creative and novel macromolecular assemblies. During that same year (1990), Miller and Neenan[49] published their efforts with respect to the convergent preparation of the first series of aromatic-based, all-hydrocarbon dendrimers (**11**). High rigidity was inherent in this series; other prominent pioneers in the area of convergently constructed rigid branched topologies included Moore and Xu.[50]

2.2.4   Accompanying Theoretical Framework

After Flory[51] and Stockmayer[52] published books describing the 'Statistical Mechanics of Chain Molecules' and 'Molecular Fluids', respectively, which further laid the foundation of modern polymer analyses, theoretical examination of branched macromolecules continued to advance. Notable theoretical investigations of branched polymers at this time included the treatise by Maciejewski[53], which examined "trapping topologically by shell molecules." This author alludes to numerous dendritic properties such as the concepts of *dense packing*, whereby monomer connectivity can be precluded due to steric hindrance, and *void volume entrapment* of solvent (or guest) molecules as noted for the polylysine series of non-draining spheres.[40] Speculation about the preparation of

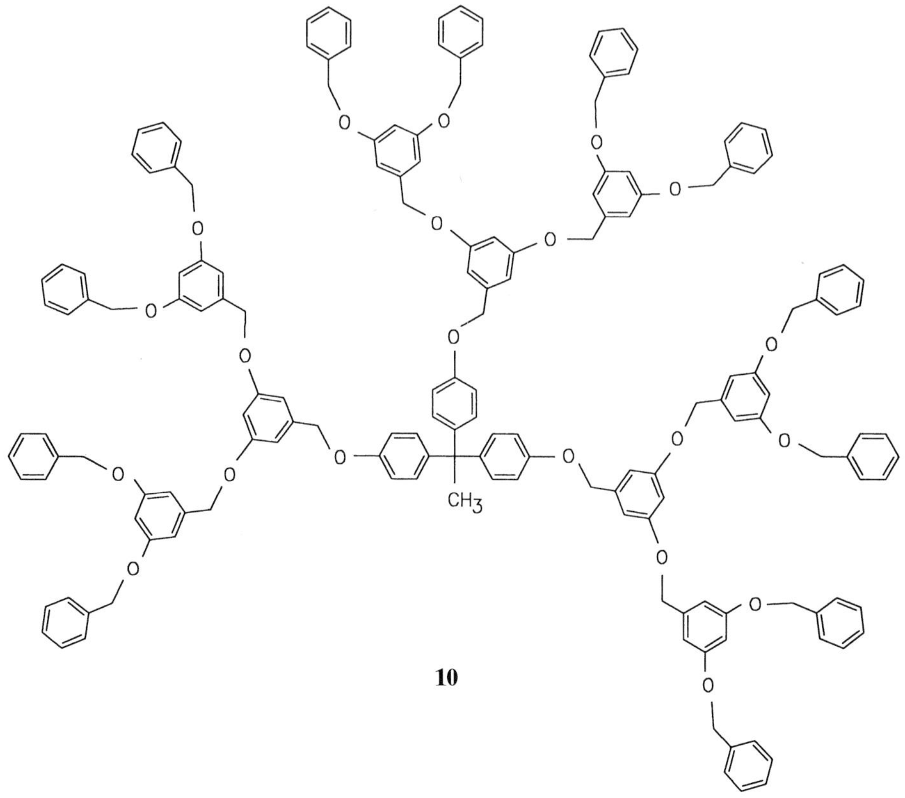

10

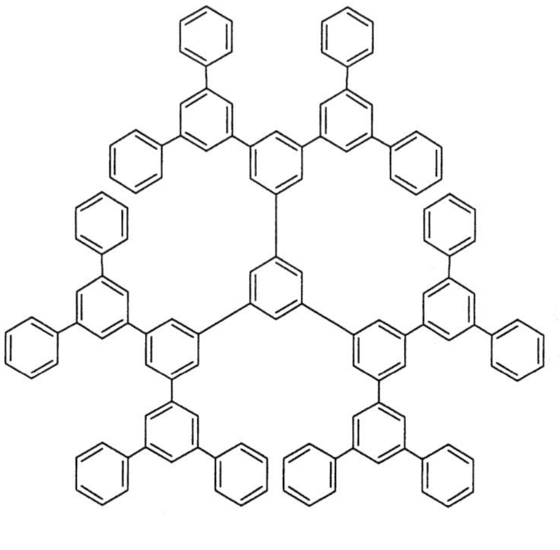

11

Figure 2.4 Illustrations of different branched architectures (continued).

macromolecules with infrastructures consisting of all quaternary carbons (e.g., a 'diamond' polymer) as well as assemblies possessing cylindrical topologies was further included. Moreover, earlier papers by Seitz and Klein[54, 55] demonstrated through Monte Carlo methods[55] and the Cayley–Polya[56] isomer model the importance of considering the excluded volume of branched polymers in dimensional analysis. Buchard, Kajiwara, and Nerger[57] examined dynamic and static light scattering of "regularly branched chain molecules" and formulated a soft-sphere microgel model, while Mikeš and Dušek[58] employed the Monte Carlo method to simulate "tree-like" polymer structures. Results thus obtained demonstrated the similarity of this statistical treatment to that of the kinetic method. Spouge[59] investigated particle aggregation whereby pairwise bonding formed "tree structures." Consideration was given to multibranching processes: finite aggregation distribution (a *sol*), infinite aggregate (a *gel*) existence criteria, and relationships for the determination of mole- and weight-average molecular masses of the finite aggregates were presented.

De Gennes (Nobel Prize in Physics, 1991) and Hervet[60] published a statistical treatment of starburst dendrimers, concluding from a mathematical growth model that steric hindrance (limiting continued tier addition) was dictated by the length of the spacer units that connect the branching centers. This reinforced the dense packing concept.

Dhar and Ramaswamy[61] reported on the "classical diffusion on 'Eden trees'". Essentially, they studied an aggregation process resulting in cluster formation that was applicable to branched tree-like graphs. Monte Carlo simulations were then used for spectral dimension determinations. Kim and Ree[62] developed a new method for determination of the ratio of intrinsic viscosities of branched and linear polymers possessing identical molecular weights. Their theory accounted for excluded volume effects. Mattice and Sienicki[63] later considered the correlation between the squared radius of gyration and the squared end-to-end distance for freely jointed random flight chains comprised of n bonds and length l. Equilibrium and dynamic properties of dendrimers under excluded-volume conditions have recently been calculated, in particular the radius of gyration and the viscometric radius as based on the intrinsic viscosity.[64] Biswas and Cherayil[65] examined the radial properties of starburst polymers based on renormalization group techniques starting with the Edwards chain model. The asphericity of star polymers was examined[66] using renormalized perturbation theory.

Diudea[67] has reported on orbital and wedgeal subgraph enumeration in irregular dendrimers. Molecular weight and volume were also evaluated. A "congesting criterion" (as in dense packing) for families of isodiametric dendrimers was developed. This author later published monographs delineating formulas for the determination of the Wiener index[68] and the Randiæ Extension – the so-called hyper-Wiener index[69] for dendrimers. These quantities have been shown to correlate molecular topology with physicochemical and biological aspects of organic molecules. Diudea[70] further described the related Wiener-type number D_Δ (delta number) for regular dendrimers; an excellent review[71] of this theoretical work is available. Considering a family of trees with n vertices and a fixed maximum vertex degree, methods were derived[72] that strictly reduce the Wiener number by shifting leaves; thus, a dendrimer with n vertices is a unique graph attaining the minimum Wiener number.

A unified lattice theory of homopolymers with general architectures including dendrimers, stars, and branched polymers has appeared,[73] in which expressions were developed for entropy, free energy, and phase diagrams. Using Monte Carlo methods, Carl[74] has analyzed the structural properties of "freely jointed, bead–rod" model dendrimers. The dendritic internal structure was studied by considering an internally located test particle, with position-dependent, Helmholtz free energy. Interior void regions were suggested, close to the center, for generations 3 and 4, while the dendrimers remained ellipsoidal with increasing generation. A "bead–spring" model possessing more interior degrees of freedom was suggested to simulate higher generation dendrimers. Monte Carlo simulations for dilute to concentrated solutions of athermal, homogeneous dendrimers have been applied in investigating their structure and thermodynamics.[75]

Cai and Chen[76] employed the Rouse approximation to evaluate the dynamic properties of dendrimers (such as the internal rotational and elastic motions). In this instance, intrinsic viscosity was found to vary linearly with generation. They[77] also performed a "more rigorous" analysis of the Mansfield–Klushin[78] theory and concluded that it does not yield the anticipated maxima. Excluded volume interactions used for equilibrium averages were represented by a smaller diameter than the "hard-core" diameter employed in hydrodynamic interaction tensors.

Yeh[79] has considered and delineated a theoretical treatment of the isomerism of asymmetric dendrimers. A new category of graph-theoretical tree was modeled and analyzed using the Cayley scheme.

Stutz[80] has described a theoretical study of the glass transition temperature (T_g) of dendritic polymers examining parameters such as end group, core, branch unit, and functionality. T_gs were found to be a function of "dendron" generation and its molecular weight, and not of the molecular weight of the entire molecule; thus, they are governed by the "backbone" glass temperature. Branching was found to affect the T_gs only minimally as it is compensated by end group influences.

Using time-dependent quantum mechanical methods, Elicker and Evans[81] have examined dendritic electron-transfer dynamics; a split operator method exploiting Cayley tree topology was employed. Electron transfer was found to be asymmetric depending on the initial excitation site and electron-transfer efficiency to the central node was shown to involve dendrimer structure as well as solvent fluctuation characteristics.

More recently, Gorman and Smith[82, 83] used atomistic molecular dynamics simulations to study the effect of repeat unit flexibility on dendrimer conformation. For all architectures, the different generations were found to be radially distributed throughout the interior; this was attributed to arm back-folding or the effect of branching angle in the rigid case; however, this does not preclude a preponderance of the termini being located at or near the molecular surface.

The first theoretical[84] analysis of the dynamics of branched polymers in 1959, subsequent theoretical investigations, and early alternative synthetic architectures serve as an exordium for the expanding and synergistic discourse on dendritic material science. Further discussion pertaining to synthetic and theoretical reports can be found later in this book.

2.3 The Fractal Geometry of Macromolecules

2.3.1 Introduction

Understanding (molecular) geometry leads to a better understanding of (molecular) physical properties, and vice versa. The intrinsic geometrical beauty of the hexagonal symmetry of benzene is inextricably related to its physical properties. For example, the chemical and magnetic equivalence of the constituent carbon and hydrogen nuclei (for unsubstituted benzene) are directly related to the atomic juxtaposition. Hence, just as classical Euclidean geometrical shapes such as the regular hexagon, circle, or cone are indispensable to a discussion of small molecules, 'fractal' geometry allows primal insight into the structures and properties of macromolecules, such as dendrimers.

This section attempts to examine macromolecular geometry, and in particular dendritic surface characteristics, from the perspectives of self-similarity and surface irregularity, or complexity, which are fundamental properties of basic fractal objects. It is further suggested that analyses of dendritic surface fractality can lead to a greater understanding of molecule/solvent/dendrimer interactions, based on analogous examinations of other materials (e.g., porous silica and chemically reactive surfaces such as those found in heterogeneous catalysts).[85]

2.3.2 Fractal Geometry

'Fractal' geometry was introduced and pioneered in the mid-1970s by the brilliant mathematician Professor Benoit B. Mandelbrot.[86–88] His development of this new mathematical language has led to a greater understanding of seemingly highly disordered objects and shapes. In short, fractal geometry provides a realistic description of complicated structures. Intuitively, objects such as lines, squares, and cubes possess dimensionalities of 1.0, 2.0, and 3.0, respectively. It is also rational to expect that many natural, as well as man-made, objects possess non-integral, or fractional, dimensionalities due to complicated patterns. Classical examples of common fractal patterns and forms include naturally occurring objects, such as coastlines, clouds, mountains, and snowflakes.[86–88] A rigorous treatment of fractal mathematics is well beyond the scope of this treatise; therefore, only essential concepts relating to a descriptive understanding of fractal shapes and analyses will be discussed. More complete and comprehensive discussions and reviews of this growing area are available.[85,88–95]

2.3.2.1 Self-similarity or Scale-invariance

For a limited discussion of fractal geometry, some simple descriptive definitions should suffice. *Self-similarity* is a characteristic of basic fractal objects. As described by Mandelbrot,[88] "When each piece of a shape is geometrically similar to the whole, both the shape and the cascade that generate it are called *self-similar*." Another term that is synonymous with self-similarity is *scale-invariance*, which also describes shapes that remain constant regardless of the scale of observation. Thus, a *self-similar* or *scale-invariant* macromolecular assembly possesses the same topology, or pattern of atomic connectivity,[96] in small as well as large segments. Self-similar objects are thus said to be invariant under dilation.

2.3.2.2 Fractal Dimension (*D*)

Fractal dimension, D, is another crucial property that is used to describe fractal objects and shapes. It is a measure of the amount of irregularity, or complexity, possessed by an object: for lines, $1 \leq D < 2$, while for surfaces, $2 \leq D < 3$. The greater the value of D, the more complex the object. As described by Avnir,[97] "*D* is obtained from a resolution analysis: the rate of appearance of new features of the irregularity as a function of the size of the probing yardstick (or degree of magnification) is measured. An object is fractal if the rate is given by the power law, $n \propto r^D$, where n is the number of yardsticks of the size r needed to measure the total length of the wiggly line."

"More generally, a power-law scaling relation characterizes one or more of the properties of an object or of a process carried out near the object:

$$property \propto scale^\beta \tag{2.1}$$

Examples of <<property>> are the surface area, the rate of a heterogeneous reaction, or the shape of an adsorption isotherm. The scales, or yardsticks, would be pore diameter, cross-sectional area of an adsorbate, particle size, or layer thickness. The exponent β is an empirical parameter, which indicates how sensitive the property is to changes in scale, and, depending on the case, can be either negative (e.g., in length measurement) or positive (e.g., in measurements of mass distribution)."[98]

Wegner and Tyler[99] define "the fractal dimension of an object as a measure of its degree of irregularity considered at all scales, and it can be a fractional amount greater than the classical geometrical dimension of the object. The fractal dimension is related to how fast the estimated measurement of the object increases as the measurement device becomes smaller. A higher fractal dimension means the fractal is more irregular, and the estimated measurement increases more rapidly. For objects of classical geometry, the dimension of the object and its fractal dimension are the same. A fractal is an object that has a fractal dimension that is strictly greater than its classical dimensions." Thus, "if the estimated property of an object becomes arbitrarily large as the measuring stick, or scale, becomes smaller and smaller, then the object is called a fractal object."

2.3.3 Applied Fractal Geometry

Fractal analysis provides an indication of complexity and a convenient method of categorization. In chemistry, it has been applied to the interpretation and understanding of macromolecular surface phenomena. Fractal geometry thus offers new insights and perspectives relating to nanoscale chemistry. It has further been applied to modeling of growth in plants and biological objects such as arterial and bronchial organs.

2.3.3.1 Fractals in Chemistry

During 1979, de Gennes[100] emphasized the importance of fractal geometry for the study of macromolecules with the publication of "Scaling Concepts in Polymer Physics." His work delineated the potential to obtain information relating to polymeric properties through examination of scaling relationships. Later, in 1983, de Gennes and Harvet[60] reported that the <<property>> of dendritic radius (R) was proportional to the corresponding molecular weight (M) raised to a fractional exponent (i.e., $R \approx M^{0.2}$ for low M; $R \approx M^{0.33}$ for M at or near the growth limit). No specific reference to dendritic fractality was mentioned; however, a fractional exponential, or scaling, relationship between dendritic properties was established. Dendritic surfaces were thus described as "showing some interesting cusps, which may become a natural locus for stereochemically active sites."

Pfeifer, Wely, and Wippermann[101] examined surface portions of a lysozyme protein molecule. A fractal dimension of $D = 2.17$ was determined and related to the rate of substrate trapping at the enzyme active site. The root-mean-square substrate displacement by diffusion at the surface increases with time as $(t)^{1/D}$, whereas surface capture of substrates is greater as compared to a smooth ($D = 2$) surface. From their results, the observed value of $D = 2.17$ was suggested to correspond to an optimum overall capture rate.

Lewis and Rees[102] determined values of $D = 2.44$, 2.44, and 2.43 for the proteins lysozyme, ribonuclease A, and superoxide dismutase, respectively. Protein regions associated with tight complexes (i.e., interfaces and antibody-binding regions) were shown to be more irregular than transient complex areas, such as active sites.

Muthukumar[103] described a theory of a fractal polymer possessing solution viscosity. Solutions containing dilute, semidilute, and high concentrations of fractal polymers were examined; intra- and interfractal hydrodynamic interactions as well as excluded volume effects were included in the analysis. Klein, Cravey, and Hite[104] delineated the fractality of benzenoid hydrocarbons constructed from regularly repeating hexagonal patterns. They suggested that these fractal hydrocarbons might serve as "zero-order" models for carbonaceous materials such as coal, lignite, char, and soot. Generalized protection–deprotection schemes were described for the construction of various fractal ring systems without specifically proposing the complementary functional groups to be employed.

Avnir et al.[97] employed an improved computerized image analysis of boundary lines of objects possessing irregular surfaces. Standard fractal line analyses were demonstrated to be insensitive, possibly leading to erroneous conclusions. Objects analyzed included the protein α-cobratoxin ($D = 1.11$), Pt-black catalyst ($D = 1.20$), the Koch curve ($D = 1.34$), and an arbitrary object – a *rabbit* ($D = 1.06$). The authors point out that caution should be exercised in the interpretation of experimental values of $D < 1.2$, as demonstrated by obtaining low D values for any "low irregularity line, even for lines that are neither fractals nor self-similar" (e.g., as evidenced by the implication of the fractal nature of a *rabbit* outline). As Avnir suggests, prudence is necessary due to a general "smoothing effect" that log–log plots impart on data, as well as due to the fact that the plotted points are sometimes indicative of limited or localized regions only.

Aharoni et al.[105] prepared hyperbranched polymers possessing rigid $1 \rightarrow 2$ branching centers and stiff "rod-like" spacers, or connectors. Scanning electron microscopy of dried material revealed fractal morphology. Porosimetry experiments supported the fractal supposition. These authors proposed that single-step polymerizations form fractal polymers and that as the polymerization proceeds, a contiguous network is formed, described as an "infinite cluster of polymeric fractals"; gelation occurs at this point.

Abad-Zapatero and Lin[106] examined globular protein surfaces and suggested that the exponent of the Box–Cox transformation[107] is a function of the fractal dimension D and the shape parameter S. D was approximated as 2.2 for two lysozymes and 2.4 for superoxide dismutase, which agrees well with previously reported values of 2.19[108] and 2.43,[102] respectively.

Avnir and Farin[98,109–111] published numerous articles examining the fractal dimensions of reactive and adsorptive surfaces. The authors reported[109] that, in many situations, the reaction rate, v, with respect to substrates diffusing from the surrounding solution and a surface, is scaled with the reactant radius, R, where $R^{(D_R-3)} - 3 \propto v$, and D_R is the "reaction dimension". D_R, it was suggested, could be thought of as the fractal dimension of the surface reaction sites. It was demonstrated that reaction efficiency could be enhanced by controlling the geometrical parameters of the reacting material, i.e. decreasing particle size for $D_R < D, m > 1$ and increasing particle size for $D_R > D, m < 1$ reactions, where $m = (D_R - 3)/(D - 3)$. Also, D_R is not necessarily equal to D, the surface fractal dimension; this could perhaps be explained in terms of a multifractal concept.[112]

Catalytic activity of a variety of dispersed metals, such as Pt, Pd, Ir, Ag, Rh, Fe, and Ni, as well as bimetallic catalysts dispersed on supports such as SiO_2 and Al_2O_3, was examined.[110] A wide range of D_R values were found, ranging from $D_R = 0.2$ for ethylene oxidation with Ag on SiO_2, suggesting little activity dependence on particle size and a low concentration of reactive surface sites, to $D_R = 5.8$ for NH_3 synthesis on Fe/MgO, suggesting severe sensitivity to structure. It should be noted that D_R is derived through consideration of an 'effective' surface site geometry that has implications for the reaction under investigation but does not necessarily equate with a 'true' geometry. Therefore, D_R can vary over a greater range than that observed for D.

Fractal geometry was employed to study surface geometry effects on adsorption conformations of polymers.[111] Using polystyrene for the case study, it was shown for highly porous objects that solution conformation changed very little after solvent adsorption. Farin and Avnir[98] extended and reviewed the fractal geometry studies of molecule–surface interactions by evaluating surface accessibility to physisorption of small molecules, surface accessibility of proteins, surface geometry effects on the adsorption of polymers, and surface accessibility with respect to adsorbate energy transfer.

Mansfield[113] performed Monte Carlo calculations on model dendrimers and determined that, as a result of the unique architecture of the branches (even when similar chemically), they are well segregated. Further, he concluded that dendrimers are fractal (D ranges from 2.4 to 2.8) and self-similar only over a rather narrow scale of lengths.

2.3.3.2 Fractals in Biology

Fractal geometry is not only ubiquitous in dendritic topologies, but is abundantly obvious in naturally occurring objects; this is acutely evident in physiological objects. West and Goldberger[114] assert "the mathematical concept of fractal scaling brings an elegant new logic to the irregular structure, growth, and function of complex biological forms". Their review includes an excellent discussion of the fractal properties of motifs created by such objects as the human 'bronchial tree' that is essentially the lung superstructure, and the human heart, which can be considered as being comprised of a 'fractal hierarchy' (i.e., it possesses layered fractal objects).

Rabouille, Cortassa, and Aon[115] dried protein-, glycoprotein-, and polysaccharide-containing brine solutions, which resulted in dendritic-like fractal patterns. The fractal dimension, $D = 1.79$, was determined for the pattern afforded by an ovomucin-ovalbumin mixture (0.1 M NaCl). Similar D values were obtained for dried solutions of fetuin, ovalbumin, albumin, and starch; the authors subsequently suggest that 'fractal patterning' is characteristic of biological polymers.

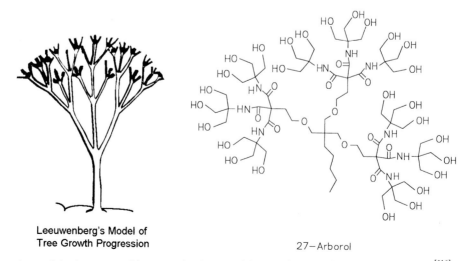

Leeuwenberg's Model of
Tree Growth Progression

27–Arborol

Figure 2.5 Leeuwenberg's model of a tree architecture (redrawn with permission of *American Scientist*[116]) as compared to a 1 → 3 branched dendrimer.[43]

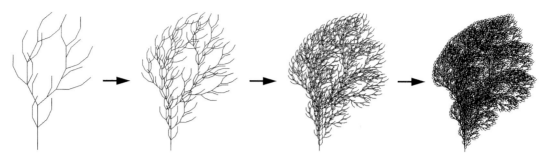

Figure 2.6 A bush-like geometry generated by an "L-system" algorithm.

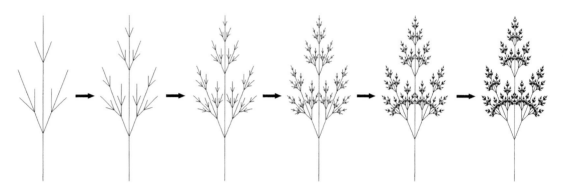

Figure 2.7 An "L-system" algorithm was employed for the generation of this tree-like architecture with $1 \rightarrow 3$ branching.

Plants also exhibit fractal patterns. Figure 2.5 illustrates the Leeuwenberg model of tree architecture as depicted by Professor P. B. Tomlinson[116] in a treatise on tree motifs and a primitive 27-arborol, prepared by Newkome et al.[43] Features of the Leeuwenberg model include $1 \rightarrow 3$ branching centers and a crowded 'canopy', or periphery, much like the morphology of a dendrimer constructed from building blocks that incorporate $1 \rightarrow 3$ C-based branching. It is interesting to note that this model is but one of the architectures delineated in Tomlinson's fascinating article.

During 1990, Lindenmayer systems, or L-systems, were developed as a means of graphically modeling the fractal geometry of plants. Iteration of 'compressed' mathematical descriptions of plant architecture generates the models that are delineated in the text entitled "The Algorithmic Beauty of Plants".[117] Figures 2.6 and 2.7 illustrate the use of L-system algorithms for the creation of "tree-like" patterns.[118] Algorithmic iteration resembles divergent, generational, dendritic growth resulting in dense packed surfaces and internal void regions.

2.3.3.3 Solvent Accessible Surface Area (A_{SAS})

By 1990, it was proposed[119] that the repetitive, generational, branching topology inherent in dendritic structures could be characterized as being fractal. Support for this conjecture was provided by a computational[120] examination of the _s_olvent _a_ccessible _s_urface area (A_{SAS}) of dendrimers at different generations. Figure 2.8 illustrates the SAS concept as well as the method used for its determination. SASs are essentially computed by generating a three-dimensional, graphical representation of the dendrimer and computationally 'rolling' probes (p) of various radii (r) over the surface. Intuitively, as well as physically, the larger the probe radius the less chance for contact the probe has within the internal void region of the dendrimer. For probes with a small radius, and in particular at the theoretical limit of $r = 0$ Å, the total internal surface area can be determined (Figure 2.9). Typically, ($\sqrt{A_{SAS}}$) _is plotted versus_ probe radius or diameter; thus, a measure of dendritic porosity can be derived.

In Figure 2.9, the porosity of a series of polyamido, acid-terminated dendrimers[121, 122] (generation 1 through to 4) as determined by a plot of $(A_{SAS})^{1/2}$ vs. probe radius is illustrated. For an ideal, completely space-filled sphere, $A_{SAS} = 4\pi(R + P)^2$, where $R =$

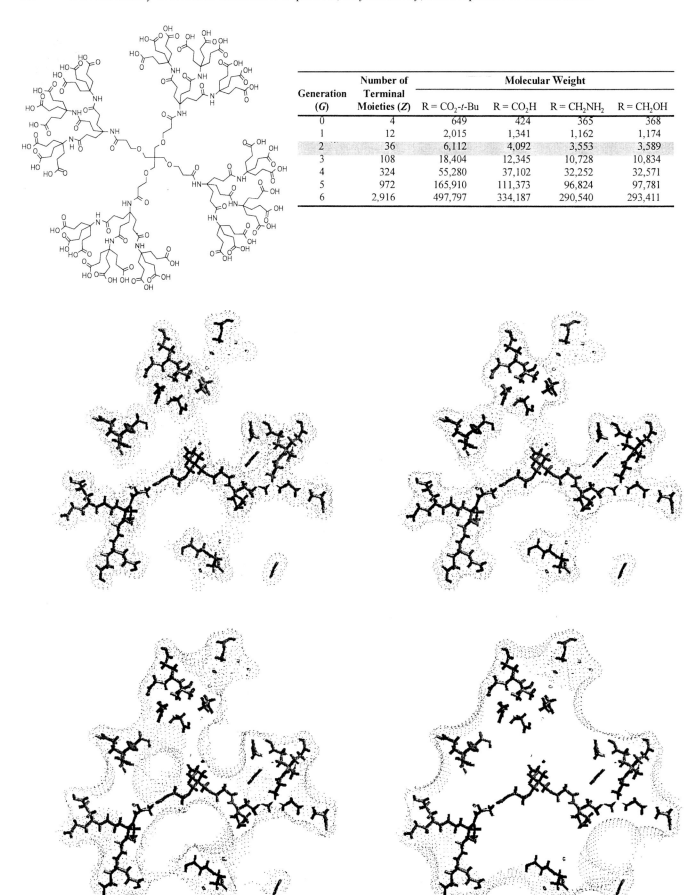

Generation (*G*)	Number of Terminal Moieties (*Z*)	Molecular Weight			
		R = CO$_2$-*t*-Bu	R = CO$_2$H	R = CH$_2$NH$_2$	R = CH$_2$OH
0	4	649	424	365	368
1	12	2,015	1,341	1,162	1,174
2	36	6,112	4,092	3,553	3,589
3	108	18,404	12,345	10,728	10,834
4	324	55,280	37,102	32,252	32,571
5	972	165,910	111,373	96,824	97,781
6	2,916	497,797	334,187	290,540	293,411

Figure 2.8 Solvent accessible surface calculated for a 2nd generation amide-based dendrimer using probes with increasing radii.

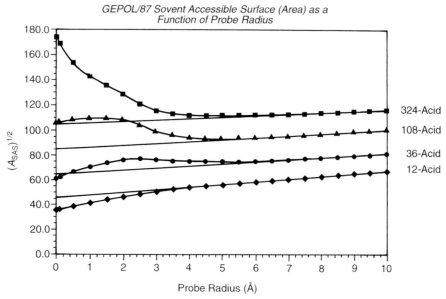

Porosity of Acid Terminated Polyamido Cascades

Figure 2.9 Graphical depiction of porosity for generations 1–4 of polyamido, carboxylic acid terminated dendrimers.[121]

radius of the sphere and P = radius of the probe. Thus, a perfect sphere gives a linear plot of $(A_{SAS})^{1/2}$ *vs.* P with a slope of $2\pi^{1/2}$ and an ordinate intercept proportional to R at $P = 0$. Extrapolation of the linear portion of the curves to $P = 0$ (linear regression analysis) gives lines that would be obtained upon A_{SAS} determination for a smooth surfaced sphere (fractal dimension $D = 2$). Plotted solid curves, at generations 3 and 4 (108 and 324 carboxylic acid moieties, respectively), reveal a greater *internal* surface area than *external* surface area (with radius R at $P = 0$).

These curves thus suggest an open and vacuous structure characterized by dynamic channels and pockets. The A_{SAS}, as measured for the 1st generation 12-acid, indicates a non-porous structure possessing little or no void regions, as evidenced by experimental values falling below the ideal surface area. Data for the 2nd generation 36-acid reveal a structure intermediate between "dendritic" and simply "branched".

These data are corroborated[119] by S_{SAS} measurements on polyamidoamine (PAMAM) dendrimers as well as a series of compact polyethereal dendrimers. Comparison of the PAMAM series, which is structurally characterized by a $1 \rightarrow 2$ *N*-branching pattern, with the $1 \rightarrow 3$ *C*-branched polyamide acid series reveals that a $1 \rightarrow 3$ branching pattern reaches fractal dendritic status more rapidly [i.e., generation 3 for the $1 \rightarrow 3$ growth series *vs.* generation 5 for the $1 \rightarrow 2$ series]. It should be further noted that these measurements are based on static, computer generated models; as usual, experimental data are needed to fully validate, or substantiate, the findings.

2.3.3.4 Dendritic Fractality

The self-similarity (see Section 2.3.2.1) of a dendrimer is readily apparent when each generation is viewed consecutively. Figure 2.10 shows three-dimensional, computer-generated space-filling models of the 1st through to the 4th tier polyamido, acid-terminated dendrimers.[121, 122] The appearance of each generation is strikingly similar to the next with respect to such features as $1 \rightarrow 3$ branching, distance between branching centers, and nuclei connectivity. The requirement of self-similarity does not necessitate the strict use of similar monomers at each generation. Each tier is gray coded in order to visually demonstrate the openness and accessibility.

Generational self-similarity is also evident in Figures 2.11 and 2.12. Figure 2.11 shows a computer-generated quadrant of a 3rd tier, hydrocarbon-based, carboxylic acid terminated dendrimer[44, 123] with the van der Waals surface added. Repetition of the branching pattern at different generations results in large, superstructure-bounded void volumes,

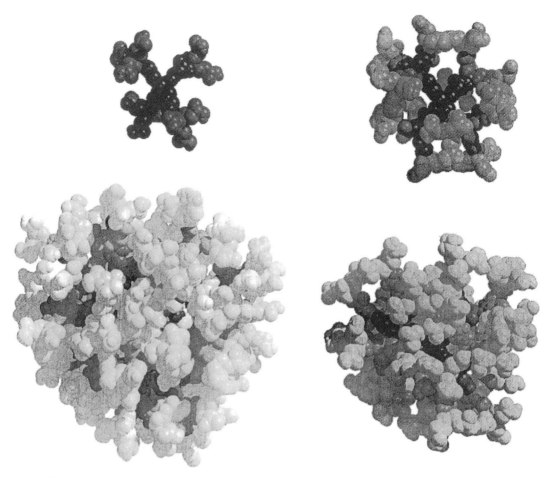

Figure 2.10 Gray coded, space-filling models of dendrimers at generations 1–4. Internal accessibility via a dynamic porosity is readily apparent.

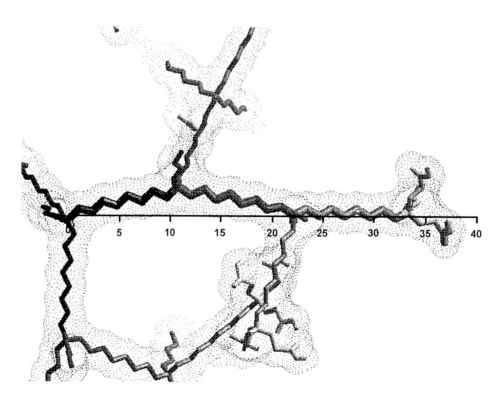

Figure 2.11 Computer-generated extended quadrant of a hydrocarbon-based 36-Micellanoate™ (4.82) illustrating segmented self-similarity.

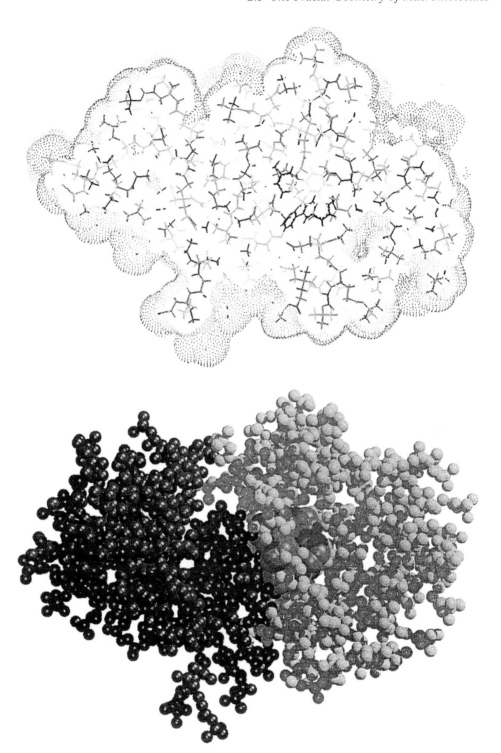

Figure 2.12 Wire-frame and space-filling representations of two metal-connected dendrons.

as indicated by the atomic ruler juxtaposed along an extended arm. Figure 2.12 depicts a cross-sectioned and space-filling view of two dendrimers specifically connected[124] through a terpyridine–ruthenium–terpyridine complex centrally located in each diagram. Scale-invariance at differing levels of observation is apparent within the limiting physical boundaries of the macroassembly. The diagrams (Figures 2.10–2.12) support the implications derived from SAS calculations in the previous section.

Determination of the fractal nature of a dendritic surface was first reported[125] employing data from A_{SAS} analysis (see Section 2.3.3.3) for the PAMAMs (a surface fractal dimension, D, was derived). Two different methods were used, the first of which applied the relationship:

$$A \approx \sigma^{2-D/2} \tag{2.2}$$

where A and σ correspond to the accessible dendritic surface area and the probe radius, respectively. For a 6th generation dendrimer (192 termini) in the A_{SAS} region of probe radius 1.5–7.0 Å, $D = 2.41$ (as determined from a log–log plot of A vs. σ). This result compared favorably to that obtained by the application of another proportional relationship:

$$A \approx d^D \qquad (2.3)$$

where d represents the size of an object with surface area A. A log–log plot of A vs. d gives $D = 2.42$.

Similar analyses of A_{SAS} data for a much more compacted pentaerythritol-based polyether dendrimer[119] gave $D = 1.96$. Comparison of the fractal dimensions for both series, i.e. the PAMAMs and polyethers, suggests structural differences. For $D = 1.96$, surface complexity and irregularity are minimized and it is relatively smooth. This is corroborated by the difficulty in forcing all of the desired surface reactions to go to completion when attempting to prepare higher generations; functional group surface congestion is most likely to be responsible, due to short spacers, or connectors, between tetrahedrally-substituted, C-branching centers. For the 6th generation PAMAM, the fractal dimension ($D = 2.42$) suggests an open and complex surface comprised of repeating peaks and valleys.

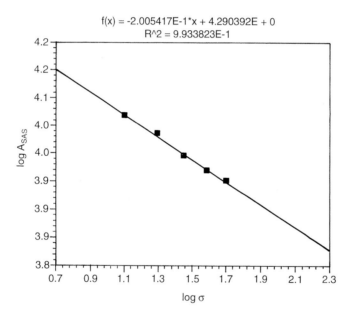

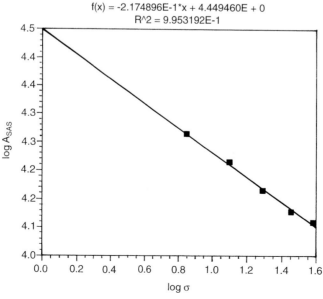

Figure 2.13 Determination of the "fractal dimension" ($D = 2.40$ and $D = 2.43$) for the 3rd (top) and 4th (bottom) generation amide-based dendrimers [see the Table in Figure 2.8]

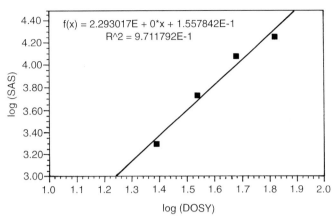

$f(x) = 2.293017E + 0*x + 1.557842E-1$
$R^2 = 9.711792E-1$

Figure 2.14 Employing the log of experimentally determined dendritic diameters, the fractal dimension ($D = 2.29$) was determined for a series of acid-terminated dendrimers (see the Table presented in Figure 2.8).

A similar analysis of the surface fractal dimension for the polyamido, acid-terminated dendrimers has been conducted.[122] Analysis of A_{SAS} for the generations 1 through to 4 with probes of various sizes produced the GEPOL/87 curves depicted in Figure 2.9 (see also Figure 2.8). Application of Equation 2.2 resulted in the construction of a best-fit straight-line relation for a log–log plot of A vs. σ for the 3rd tier 108-acid and the 4th tier 324-acid (Figure 2.13). A first-order fit of the data afforded slopes corresponding to –2.00 and –2.17 (solving for fractal dimensions [i.e., slope = $(2 - D/2)$]: $D = 2.40$ and $D = 2.43$), respectively. These numbers compare well with those determined by Farin and Avnir[125] for the higher level PAMAMs. Employing relationship (Eq. 2.3), $A = d^D$, where d is represented by the log of the experimentally determined DOSY NMR[121] dendritic diameter, $D = 2.29$ (Figure 2.14).

2.3.4 Fractal Summation

According to Farin and Avnir,[110] an important "contribution of fractal geometry has been a clarification of the physical meaning of *non-integer* dimensions and the creation of a continuous scale of dimensions." With respect to dendritic chemistry, fractal geometry affords a novel method of comparison of branched architecture as well as, and perhaps more importantly, a new way to envisage these unique macromolecules and their properties. Dendritic surface complexity is derived from the features of the supporting superstructure, which, in turn, derive from the inherent nature of the monomers used for its construction. Thus, the fractal geometry of dendritic macroassemblies implies translation, or transcription, of molecular level information to much larger superassemblies. Conveyance of information is based on topological 'scaffolding' connectivity of dendritic structures and this affords an enhanced perspective of this emerging discipline.

2.4 Dendritic Construction and Properties

Prior to a survey of branched macromolecular architectures in the literature, a brief discussion of the general methods available for dendritic construction will be examined. Relative to these methods, consideration will be given to the physicochemical ramifications and pertinent questions arising from design criteria of the differing methods. A more comprehensive discussion of the structural and physical aspects of dendrimers can be found in a review by Meijer et al.[126]

2.4.1 Construction Methods

In general, synthetic methods for the preparation of branched architectures rely on two similar procedures described as *divergent* and *convergent* (Scheme 2.3). Both procedures usually rely on mutually compatible and complementary protection and deprotection sequences. Since each method is discussed in detail in their respective chapters, a diagrammatic representation of the methods in question is given as an introduction to not only synthetic protocols, but also to generally accepted terminology and concepts, such as: *core, monomer, generation, termini, focal site, spacer or connector, periphery,* and *void region.* Furthermore, in contrast to the notable differences inherent in the divergent and convergent techniques, there is some similarity. The convergent protocol can be considered a "higher order" divergent method whereby "complex" monomers are attached to a central unit whether it is formally a final core or simply another self-similar monomer. However, this is where the similarity ends, as there are major differences in the outcome of using each technique.

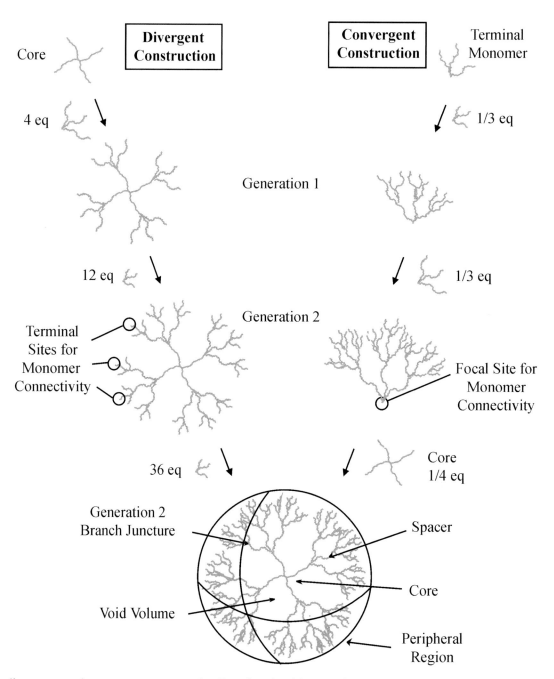

Scheme 2.3 Idealized divergent and convergent protocols. Reprinted with permission of the American Chemical Society.[126a]

2.4.2 Monodispersity *versus* Unimolecular

Questions that invariably arise during a discussion of dendrimers concern the concept of molecular (or structural) purity: "How structurally perfect are the final products?" "Are there any branch defects, and if so, what is the extent of the "imperfection(s)"? "If there are imperfections, can they be eliminated?" "Which method, *convergent* or *divergent*, offers a better avenue to a "unimolecular" or "monomolecular" product as opposed to a "monodisperse" mixture?"

These types of questions are extremely pertinent with respect to the construction of dendrimers that are to be structurally perfect, i.e. where the product consists of a single, "monomolecular" species. This is in contrast to a product that possesses a certain degree of defect, caused by incomplete reactions (i.e., monomer attachment), and is thus a "mixture of molecular structures" or a monodisperse mixture.

A search of the literature for answers to these questions reveals that early in the development of dendritic chemistry it was generally assumed that almost all dendrimers were structurally perfect or nearly so. This assumption was predicated on the methods of characterization that were available to chemists at that time – primarily NMR techniques. It is now well known and generally accepted that this characterization technique (NMR) needs to be employed with caution and in conjunction with additional methods such as MS (i.e., MALDI-TOF MS and ESI MS) to ascertain dendritic purity. NMR, in many cases, has been found to be of little use in determining structural defects, especially at higher generations, due to the "self-similarity" inherent in the dendritic architecture. Also, elemental analysis is of little use for defect analysis, due to the similar elemental composition of the higher generations.

With the advent of improved MS techniques, major differences in the divergent and convergent procedures of dendrimer construction are evident. Essentially, the divergent process is verified as a "statistical" method, whereby increasing numbers of reactions are required for successive generation construction. Consider, for example, as pointed out by Meijer et al.,[126] a dendrimer with a branch multiplicity of *three* (3) and an average reaction selectivity of 99 %; thus, with an inherent requirement of a total of 100 reactions to reach the product, at only generation 2 the percent dendritic purity will be 0.99^{100} or 0.60 (i.e., 60 % of defect-free dendrimer). Clearly, based solely on statistical grounds, the chance of obtaining only structurally pure (monomolecular) material by the divergent protocol at higher generations essentially approaches zero. Through the use of large excesses of reagents and/or forcing (temperature and pressure) reaction conditions, this problem can be minimized but not circumvented. Meijer et al.[126] have described the divergent protocol as a macromolecular procedure referring to the resulting monodisperse mixture.

In sharp contrast to the divergent method is the convergent protocol. Again, the use of MS has verified[127, 128] the potential to obtain perfect, defect-free (assuming the dendrons or dendritic wedges are structurally monomolecular in composition) dendritic structures. This results from the necessity of only a limited number of reactions for generational construction (i.e., 2–4 usually). Further, the convergent method allows for product purification due to the size difference of the side or intermediary products. Meijer et al.[126] referred to the convergent protocol as an "organic chemistry approach" for dendrimer preparation, alluding to a more controlled avenue, although the requirement for reaction at a sterically hindered focal site must also be considered. In general, each method has its advantages and disadvantages and must be considered on the basis of the chemistry employed as well as the desired target and attendant properties. Individual examples are explored in the following chapters.

2.5   References

[1] P. J. Flory, "Molecular Size Distribution in Three-Dimensional Polymers; IV. Branched Polymers Containing A–R–B$_{f-1}$ Type Units", *J. Am. Chem. Soc.* **1952**, *74*, 2718–2723.

[2] T. Zincke, "Zur Synthese aromatischer Säuren", *Chem. Ber.* **1869**, *2*, 737–740.

[3] C. Friedel, J. M. Crafts, *Bull. Soc. Chim. Fr.* **1885**, *43*, 53.

[4] J. H. Gladstone, J. Tribe, *J. Chem. Soc.* **1885**, *47*, 448.

[5] W. Ostwald, "The Classification of Colloids", *Z. Chem. Ind. Kolloide* **1907**, *1*, 331–341.

[6] W. Ostwald, "The Classification of Colloids", *Z. Chem. Ind. Kolloide* **1909**, *3*, 28–30.

[7] T. Graham, *Trans. Roy. Soc. (London)* **1861**, *151*, 183.

[8] H. Hlasiwetz, J. Habermann, *Ann. Chem. Pharm.* **1871**, *159*, 304.

[9] P. J. Flory, *Principles of Polymer Chemistry*, Cornell University Press, Ithaca, New York, **1953**.

[10] W. H. Hunter, G. H. Woollett, "A Catalytic Decomposition of Certain Phenol Silver Salts; IV. The Constitution of the Amorphous Oxides", *J. Am. Chem. Soc.* **1921**, *43*, 135–151.

[11] C. K. Ingold, L. C. Nickolls, "Experiments on the Synthesis of the Polyacetic Acids of Methane; Part VII: iso-Butylene-α,γ,γ'-tricarboxylic Acid and Methanetetraacetic Acid", *J. Chem. Soc.* **1922**, *121*, 1638–1648.

[12] R. Ferdinand, *Principles of Polymer Systems*, McGraw Hill, New York, **1970**, p. 8.

[13] W. H. Carothers, "Polymerization", *Chem. Rev.* **1931**, *8*, 353–426.

[14] R. A. Jacobson, "Polymers from Benzyl Chloride and Related Compounds", *J. Am. Chem. Soc.* **1932**, *54*, 1513–1518.

[15] W. Kuhn, "Über die Kinetik des Abbaues hochmolekularer Ketten", *Chem. Ber.* **1930**, *63*, 1503–1509.

[16] P. J. Flory, *Principles of Polymer Chemistry*, Cornell University Press, Ithaca, New York, **1953**, pp. 19.

[17] A.-V. Lourenco, "Sur les Alcohols Polyéthyléniques", *C. R. l'Academie. Sci., Ser. III* **1860**, *51*, 365–367.

[18] A.-V. Lourenco, *Ann. Chim. Phys.* **1863**, *67*, 273.

[19] K. Kraut, *Ann. Chim. (Paris)* **1869**, *150*, 1.

[20] P. J. Flory, *Principles of Polymer Chemistry*, Cornell University Press, Ithaca, New York, **1953**, pp. 14.

[21] P. J. Flory, "Molecular Size Distribution in Three-Dimensional Polymers; III. Tetrafunctional Branching Units", *J. Am. Chem. Soc.* **1941**, *63*, 3096–3100.

[22] P. J. Flory, "Molecular Size Distribution in Three-Dimensional Polymers; I. Gelation", *J. Am. Chem. Soc.* **1941**, *63*, 3083–3090.

[23] P. J. Flory, "Molecular Size Distribution in Three-Dimensional Polymers; II. Trifunctional Branching Units", *J. Am. Chem. Soc.* **1941**, *63*, 3091–3096.

[24] P. J. Flory, "Random Reorganization of Molecular Weight Distribution in Linear Condensation Polymers", *J. Am. Chem. Soc.* **1942**, *64*, 2205–2212.

[25] The term "three-dimensional", as used in this context, loosely refers to substantial molecular dimension in the direction of each of the *x, y,* and *z* coordinates that define the geometry of a particular macroassembly.

[26] W. H. Stockmayer, "Theory of Molecular Size Distribution and Gel Formation in Branched-Chain Polymers", *J. Chem. Phys.* **1943**, *11*, 45–55.

[27] P. J. Flory, "The Configuration of Real Polymer Chains", *J. Chem. Phys.* **1949**, *17*, 303–310.

[28] R. J. Goldberg, "A Theory of Antibody–Antigen Reactions; I. Theory for Reactions of Multivalent Antigen with Bivalent and Univalent Antibody", *J. Am. Chem. Soc.* **1952**, *74*, 5715–5725.

[29] R. J. Goldberg, "A Theory of Antibody–Antigen Reactions; II. Theory for Reactions of Multivalent Antigen with Multivalent Antibody", *J. Am. Chem. Soc.* **1953**, *75*, 3127–3131.

[30] P. E. Rouse, Jr., "A Theory of the Linear Viscoelastic Properties of Dilute Solutions of Coiling Polymers", *J. Chem. Phys.* **1953**, *21*, 1272–1280.

[31] B. H. Zimm, "Dynamics of Polymer Molecules in Dilute Solution: Viscoelasticity, Flow Birefringence, and Dielectric Loss", *J. Chem. Phys.* **1956**, *24*, 269–278.

[32] E. Buhleier, W. Wehner, F. Vögtle, "'Cascade' and 'Nonskid-Chain-like'" Syntheses of Molecular Cavity Topologies", *Synthesis* **1978**, 155–158.

[33] "'Cascade syntheses' meant reaction sequences that could be carried out repeatedly, whereby a functional group is made to react in such a way as to appear twice in the subsequent molecule. However, 'nonskid synthesis' meant stepwise construction of polycyclic ring compounds by repeatedly occurring reaction sequences; thus, a ring system is connected by a new bridge, which possesses functional groups for the annexation of further bridges." For a review on repetitive (or iterative) syntheses, see: N. Feuerbacher, F. Vögtle, "Iterative Synthesis in Organic Chemistry," *Top. Curr. Chem.* **1998**, *197*, 2–18.

[34] J.-M. Lehn, "Design of Organic Complexing Agents – Strategies Towards Properties" in *Structure and Bonding* (Eds.: J. D. Dunitz, P. Hemmerich, C. K. Jørgensen, J. B. Neilands, D. Reinen, R. J. P. Williams), Springer, New York, **1973**.

[35] D. J. Cram, R. C. Helgeson, L. R. Sousa, J. M. Timko, M. Newcomb, P. Moreau, F. de Jong, G. W. Gokel, D. H. Hoffman, L. A. Domeier, S. C. Peacock, K. Madan, L. Kaplan, "Chiral Recognition in Complexation of Guests by Designed Host Molecules", *Pure Appl. Chem.* **1975**, *43*, 327–349.

[36] J.-M. Lehn, "Supramolecular Chemistry – Scope and Perspectives. Molecules, Supermolecules, and Molecular Devices (Nobel lecture)", *Angew. Chem.* **1988**, *100*, 91–114; *Int. Ed. Engl.* **1988**, *27*, 89–112.

[37] J.-M. Lehn, "Perspectives in Supramolecular Chemistry – From Molecular Recognition Toward Molecular Information Processing and Self-Organization", *Angew. Chem.* **1990**, *102*, 1347–1362; *Int. Ed. Engl.* **1990**, *29*, 1304–1319.

[38] H. Ringsdorf, B. Schlarb, J. Venzmer, "Molecular Architecture and Function of Polymeric Oriented Systems: Models for the Study of Organization, Surface Recognition, and Dynamics of Biomembranes", *Angew. Chem.* **1988**, *100*, 117–162; *Int. Ed. Engl.* **1988**, *27*, 113–158.

[39] J. S. Lindsey, "Self-Assembly in Synthetic Routes to Molecular Devices. Biological Principles and Chemical Perspectives: A Review", *New J. Chem.* **1991**, *15*, 153–180.

[40] R. G. Denkewalter, J. F. Kolc, W. J. Lukasavage, "Macromolecular Highly Branched Homogeneous Compounds", **1983**, U. S. Pat., *4, 410, 688*.

[41] S. M. Aharoni, C. R. Crosby, III, E. K. Walsh, "Size and Solution Properties of Globular *tert*-Butyloxycarbonyl-poly(α,ε-L-lysine)", *Macromolecules* **1982**, *15*, 1093–1098.

[42] H. R. Kricheldorf, Q.-Z. Zang, G. Schwarz, "New Polymer Syntheses: 6. Linear and Branched Poly(3-hydroxybenzoates)", *Polymer* **1982**, *23*, 1821–1829.

[43] G. R. Newkome, Z. Yao, G. R. Baker, V. K. Gupta, "Cascade Molecules: A New Approach to Micelles. A [27]-Arborol", *J. Org. Chem.* **1985**, *50*, 2003–2004.

[44] G. R. Newkome, C. N. Moorefield, G. R. Baker, M. J. Saunders, S. H. Grossman, "Unimolecular Micelles", *Angew. Chem., Int. Ed. Engl.* **1991**, *30*, 1178–1180.

[45] G. R. Newkome, C. N. Moorefield, "Unimolecular Micelles and Method of Making the Same", **1992**, U. S. Pat., *5, 154, 853*.

[46] D. A. Tomalia, H. Baker, J. Dewald, M. Hall, G. Kallos, S. Martin, J. Roeck, J. Ryder, P. Smith, "A New Class of Polymers: Starburst-Dendritic Macromolecules", *Polym. J. (Tokyo)* **1985**, *17*, 117–132.

[47] I. Bidd, M. C. Whiting, "The Synthesis of Pure *n*-Paraffins with Chain Lengths Between One and Four-hundred", *J. Chem. Soc., Chem. Commun.* **1985**, 543–544.

[48] C. Hawker, J. M. J. Fréchet, "A New Convergent Approach to Monodisperse Dendritic Macromolecules", *J. Chem. Soc., Chem. Commun.* **1990**, 1010–1013.

[49] T. M. Miller, T. X. Neenan, "Convergent Synthesis of Monodisperse Dendrimers Based upon 1,3,5-Trisubstituted Benzenes", *Chem. Mater.* **1990**, *2*, 346–349.

[50] J. S. Moore, Z. Xu, "Synthesis of Rigid Dendritic Macromolecules: Enlarging the Repeat Unit Size as a Function of Generation Permits Growth to Continue", *Macromolecules* **1991**, *24*, 5893–5894.

[51] P. J. Flory, *Statistical Mechanics of Chain Molecules*, Wiley, New York, **1969**.

[52] W. H. Stockmayer, "Dynamics of Chain Molecules" in *Molecular Fluids* (Eds.: R. Balian, G. Weill), Gordon and Branch, London, **1976**, pp. 107–149.

[53] M. Maciejewski, "Concepts of Trapping Topologically by Shell Molecules", *J. Macromol. Sci., Chem.* **1982**, *A17*, 689–703.

[54] D. J. Klein, "Rigorous Results for Branched Polymer Models with Excluded Volume", *J. Chem. Phys.* **1981**, *75*, 5186–5193.

[55] W. A. Seitz, D. J. Klein, "Excluded Volume Effects for Branched Polymers: Monte Carlo Results", *J. Chem. Phys.* **1981**, *75*, 5190–5193.

[56] Refs. 1–3 in D. J. Klein, *J. Chem. Phys.* **1981**, *75* (10), 5186–5189.

[57] W. Burchard, K. Kajiwara, D. Nerger, "Static and Dynamic Scattering Behavior of Regularly Branched Chains: A Model of Soft-Sphere Microgels", *J. Polym. Sci., Polym. Phys. Ed.* **1982**, *20*, 157–171.

[58] J. Mikes, K. Dusek, "Simulation of Polymer Network Formation by the Monte Carlo Method", *Macromolecules* **1982**, *15*, 93–99.

[59] J. L. Spouge, "Tree Models of Aggregation: Multiple Particle and Bond-Types", *Proc. Roy. Soc. (London), A: Math. Phys. Sci.* **1983**, *387*, 351–365.

[60] P. G. de Gennes, H. Hervet, "Statistics of <<Starburst>> Polymers", *J. Phys. Lett.* **1983**, *44*, L351–L360.

[61] D. Dhar, R. Ramaswamy, "Classical Diffusion on Eden Trees", *Phys. Rev. Lett.* **1985**, *54*, 1346–1349.

[62] J. R. Kim, T. Ree, "Dilute Solution Properties of Branched Polymers", *J. Polym. Sci., Polym. Chem. Ed.* **1985**, *23*, 1119–1124.

[63] W. L. Mattice, K. Sienicki, "Extent of the Correlation Between the Squared Radius of Gyration and Squared End-to-End Distance in Random Flight Chains", *J. Chem. Phys.* **1989**, *90*, 1956–1959.

[64] F. Ganazzoli, R. La Ferla, G. Terragni, "Conformational Properties and Intrinsic Viscosity of Dendrimers Under Excluded-Volume Conditions", *Macromolecules* **2000**, *33*, 6611–6620.

[65] P. Biswas, B. J. Cherayil, "Radial Dimensions of Starburst Polymers", *J. Chem. Phys.* **1994**, *100*, 3201–3209.

[66] O. Jagodzinski, "The Asphericity of Star Polymers: A Renormalization Group Study", *J. Phys. A: Math. Gen.* **1994**, *27*, 1471–1494.

[67] M. V. Diudea, "Molecular Topology; 19. Orbital and Wedgeal Subgraph Enumeration in Dendrimers", *MATCH* **1994**, *30*, 79–91.

[68] M. V. Diudea, "Molecular Topology; 21. Wiener Index of Dendrimers", *MATCH* **1995**, *32*, 71–83.

[69] M. V. Diudea, B. Parv, "Molecular Topology; 25. Hyper-Wiener Index of Dendrimers", *J. Chem. Inf. Comput. Sci.* **1995**, *35*, 1015–1018.

[70] M. V. Diudea, G. Katona, B. Pârv, "Delta Number, D_Δ, of Dendrimers", *Croat. Chem. Acta* **1997**, *70*, 509–519.

[71] M. V. Diudea, G. Katona, "Molecular Topology of Dendrimers" in *Advances in Dendritic Macromolecules* (Ed.: G. R. Newkome), JAI Press, Inc., Stamford, CN, **1999**, pp. 135–201.

[72] S.-C. Liu, L.-D. Tong, Y.-N. Yeh, "Trees with the Minimum Wiener Number", *Int. J. Quantum Chem.* **2000**, *78*, 331–340.

[73] P. D. Gujrati, "Unified Lattice Theory of Homopolymers of General Architecture: Dendrimers, Stars, and Branched Polymers", *Phys. Rev. Lett.* **1995**, *74*, 1367–1370.

[74] W. Carl, "A Monte Carlo Study of Model Dendrimers", *J. Chem. Soc., Faraday Trans.* **1996**, *92*, 4151–4154.

[75] L. Lue, "Volumetric Behavior of Athermal Dendritic Polymers: Monte Carlo Simulation", *Macromolecules* **2000**, *33*, 2266–2272.

[76] C. Cai, Z. Y. Chen, "Rouse Dynamics of a Dendrimer Model in the θ Condition", *Macromolecules* **1997**, *30*, 5104–5117.

[77] C. Cai, Z. Y. Chen, "Intrinsic Viscosity of Starburst Dendrimers", *Macromolecules* **1998**, *31*, 6393–6396.

[78] M. L. Mansfield, L. I. Klushin, "Intrinsic Viscosity of Model Starburst Dendrimers", *J. Phys. Chem.* **1992**, *96*, 3994–3998.

[79] C. Yeh, "Isomerism of Asymmetric Dendrimers and Stereoisomerism of Alkanes", *J. Mol. Struct.* **1998**, *432*, 153–159.

[80] H. Stutz, "The Glass Temperature of Dendritic Polymers", *J. Polym. Sci., Part B: Polym. Phys.* **1995**, *33*, 333–340.

[81] T. S. Elicker, D. G. Evans, "Electron Dynamics in Dendrimers", *J. Phys. Chem. A.* **1999**, *103*, 9423–9431.

[82] C. B. Gorman, J. C. Smith, "Structure-Property Relationships in Dendritic Encapsulation", *Acc. Chem. Res.* **2001**, *34*, 60–71.

[83] C. B. Gorman, J. C. Smith, "Effect of Repeat Unit Flexibility on Dendrimer Conformation as Studied by Atomistic Molecular Dynamics Simulations", *Polymer* **2000**, *41*, 675–683.

[84] B. H. Zimm, R. W. Kilb, "Dynamics of Branched Polymer Molecules in Dilute Solution", *J. Polym. Sci.* **1959**, *37*, 19–42.

[85] *The Fractal Approach to Heterogenous Chemistry: Surface, Colloids, Polymers* (Ed.: D. Avnir), Wiley, New York, **1989**.

[86] B. B. Mandelbrot, *Les Objets Fractals: Forme, Hasard et Dimension*, Flammarion, Paris, **1975**.

[87] B. B. Mandelbrot, *Fractals: Form, Chance and Dimension*, Freeman, San Francisco, **1977**.

[88] B. B. Mandelbrot, *The Fractal Geometry of Nature*, Freeman, San Francisco, **1982**.

[89] *Fractals, Quasicrystals, Chaos, Knots, and Algebraic Quantum Mechanics* (Eds.: A. Amann, L. Cederbaum, W. Gans), Kluwer, Dordrecht, The Netherlands, **1988**.

[90] A. Blumen, H. Schnörer, "Fractals and Related Hierarchical Models in Polymer Science", *Angew. Chem.* **1990**, *102*, 158–169; *Int. Ed. Engl.* **1990**, *29*, 113–222.

[91] A. Harrison, *Fractals in Chemistry*, Oxford University Press, New York, **1995**.

[92] D. H. Rouvray, "Similarity in Chemistry: Past, Present, and Future", *Top. Curr. Chem.* **1995**, Vol. *173*, 1.

[93] H. Takayasu, *Fractals in the Physical Sciences*, Manchester University Press, Manchester, **1990**.

[94] G. Zumofen, A. Blumen, J. Klafter, "The Role of Fractals in Chemistry", *New J. Chem.* **2000**, *14*, 189–196.

[95] P. Biswas, R. Kant, A. Blumen, "Polymer Dynamics and Topology: Extension of Stars and Dendrimers in External Fields", *Macromol. Theory Simul.* **2000**, *9*, 56–67.

[96] M. Zander, "Molecular Topology and Chemical Reactivity of Polynuclear Benzenoid Hydrocarbons", *Top. Curr. Chem.* **1990**, *153*, 101–122.

[97] D. Farin, S. Peleg, D. Yavin, D. Avnir, "Applications and Limitations of Boundary-Line Fractal Analysis of Irregular Surfaces: Proteins, Aggregates, and Porous Materials", *Langmuir* **1985**, *1*, 399–407.

[98] D. Farin, D. Avnir, "Fractal Scaling Laws in Heterogeneous Chemistry: Adsorptions, Chemisorptions, and Interactions Between Adsorbates", *New J. Chem.* **1990**, *14*, 197–205.

[99] T. Wegner, B. Tyler, *Fractal Creations*, 2nd ed., Waite Group Press, Corte Madera, California, **1993**, p. 16.

[100] P. G. de Gennes, *Scaling Concepts in Polymer Physics*, Cornell University Press, Ithaca and London, **1979**.

[101] P. Pfeifer, U. Wely, H. Wippermann, "Fractal Surface Dimension of Proteins: Lysozyme", *Chem. Phys. Lett.* **1985**, *113*, 535–540.

[102] M. Lewis, D. C. Rees, "Fractal Surfaces of Proteins", *Science* **1985**, *230*, 1163–1165.

[103] M. Muthukumar, "Dynamics Of Polymeric Fractals", *J. Chem. Phys.* **1985**, *83*, 3161–3168.

[104] D. J. Klein, M. J. Cravey, G. E. Hite, "Fractal Benzenoids" in *Polycyclic Aromatic Compounds*, Gordon and Breach, New York, **1991**, pp. 163–182.

[105] S. M. Aharoni, N. S. Murthy, K. Zero, S. F. Edwards, "Fractal Nature of One-Step Highly Branched Rigid Rod-like Macromolecules and Their Gelled-Network Progenies", *Macromolecules* **1990**, *23*, 2533–2549.

[106] C. Abad-Zapatero, C. T. Lin, "Statistical Descriptors for the Size and Shape of Globular Proteins", *Biopolym.* **1990**, *29*, 1745–1754.

[107] G. E. P. Box, D. R. Cox, "An Analysis of Transformation", *J. Royal Stat. Soc., Series B* **1964**, *26*, 211–252.

[108] J. Aqvist, O. Tapia, "Surface Fractality as a Guide for Studying Protein–Protein Interactions", *J. Mol. Graphics* **1987**, *5*, 30–34.

[109] D. Farin, D. Avnir, "Reactive Fractal Surfaces", *J. Phys. Chem.* **1987**, *91*, 5517–5521.

[110] D. Farin, D. Avnir, "The Reaction Dimension in Catalysis on Dispersed Metals", *J. Am. Chem. Soc.* **1988**, *110*, 2039–2045.

[111] D. Farin, D. Avnir, "The Effects of the Fractal Geometry of Surfaces on the Absorption Conformation of Polymers at Monolayer Coverage; Part 1. The Case of Polystyrene", *Colloids Surf.* **1989**, *37*, 155–170.

[112] J. Nittmann, H. E. Stanley, E. Touboul, G. Daccord, "Experimental Evidence for Multifractality", *Phys. Rev. Lett.* **1987**, *58*, 619.

[113] M. L. Mansfield, "Dendron Segregation in Model Dendrimers", *Polymer* **1994**, *35*, 1827–1830.

[114] B. J. West, A. L. Goldberger, "Physiology in Fractal Dimensions", *Am. Sci.* **1987**, *75*, 354–365.

[115] C. Rabouille, S. Cortassa, M. A. Aon, "Fractal Organization in Biological Macromolecular Lattices", *J. Biomol. Struct. Dyn.* **1992**, *9*, 1013–1024.

[116] P. B. Tomlinson, "Tree Architecture", *Am. Sci.* **1983**, *71*, 141–149.

[117] P. Prusinkiewicz, A. Lindenmayer, *The Algorithmic Beauty of Plants*, Springer, New York, **1990**.

[118] The computer-generated "trees" were constructed using the Fractint program available from the Waite Group in Fractal Creations: see T. Wegner, B. Tyler, *Fractal Creations*, 2nd ed., Waite Group Press, Corte Madera, California, **1993**.

[119] D. A. Tomalia, A. M. Naylor, W. A. Goddard, III, "Starburst Dendrimers: Molecular-Level Control of Size, Shape, Surface Chemistry, Topology and Flexibility in the Conversion of Atoms to Macroscopic Materials", *Angew. Chem.* **1990**, *102*, 119–156; *Int. Ed. Engl.* **1990**, *29*, 113–150.

[120] J. L. Pascal-Ahuir, E. Silla, J. Tomasi, R. Bonaccors, "*GEPOL, QCPE Program No. 554*", Quantum Chemistry Program Exchange Center, Bloomington, IN, **1987**.

[121] G. R. Newkome, J. K. Young, G. R. Baker, R. L. Potter, L. Audoly, D. Cooper, C. D. Weis, K. F. Morris, C. S. Johnson, Jr., "Cascade Polymers: pH Dependence of Hydrodynamic Radii of Acid-Terminated Dendrimers", *Macromolecules* **1993**, *26*, 2394–2396.

[122] J. K. Young, "Nomenclature, Synthesis, and "Smart" Behavior of Cascade Polymers", Ph.D. Dissertation, University of South Florida, **1993**.

[123] G. R. Newkome, C. N. Moorefield, G. R. Baker, A. L. Johnson, R. K. Behera, "Alkane Cascade Polymers Possessing Micellar Topology: Micellanoic Acid Derivatives", *Angew. Chem.* **1991**, *103*, 1205–1207; *Int. Ed. Engl.* **1991**, *30*, 1176–1178.

[124] G. R. Newkome, R. Güther, C. N. Moorefield, F. Cardullo, L. Echegoyen, E. Pérez-Cordero, H. Luftmann, "Routes to Dendritic Networks: Bis-Dendrimers by Coupling of Cascade Macromolecules through Metal Centers", *Angew. Chem.* **1995**, *107*, 2159–2162; *Int. Ed. Engl.* **1995**, *34*, 2023–2026.

[125] D. Farin, D. Avnir, "Surface Fractality of Dendrimers", *Angew. Chem. Int. Ed. Engl.* **1991**, *30*, 1379–1380.

[126] A. W. Bosman, H. M. Janssen, E. W. Meijer, "About Dendrimers: Structure, Physical Properties, and Applications", *Chem. Rev.* **1999**, *99*, 1665–1688.

[126a] G. R. Newkome, E. He, C. N. Moorefield "Suprasupermolecules with Novel Properties: Metallodendrimers" *Chem. Rev.* **1999**, *99*, 1689–1746.

[127] J. W. Leon, J. M. J. Fréchet, "Analysis of Aromatic Polyether Dendrimers and Dendrimer-Linear Block Copolymers by Matrix-Assisted Laser Desorption Ionization Mass Spectrometry", *Polym. Bull.* **1995**, *35*, 449–455.

[128] K. L. Walker, M. S. Kahr, C. L. Wilkins, Z. Xu, J. S. Moore, "Analysis of Hydrocarbon Dendrimers by Laser Desorption Time-of-Flight and Fourier Transform Mass Spectrometry", *J. Am. Soc. Mass Spectrom.* **1994**, *5*, 731–739.

3 Dendrimers Derived by Divergent Procedures Using 1 → 2 Branching Motifs

3.1 General Concepts

Divergent dendritic construction results from sequential monomer addition beginning from a core and proceeding outward toward the macromolecular surface. This methodology is illustrated in Scheme 3.1. To a respective core representing the 0th generation and possessing one or more reactive site(s), a generation or layer of monomeric building blocks is covalently connected. The number of building blocks that can be added is dependent upon the available reactive sites on the particular core, assuming that parameters such as monomer functional group steric hindrance and core reactive site accessibility are not a concern. Repetitive addition of similar, or for that matter dissimilar, building blocks (usually but not always effected by a protection–deprotection protocol) affords successive generations. A key feature of the divergent method is the exponentially increasing number of reactions that are required for the attachment of each subsequent tier (layer or generation).

Faultless (structure perfect) growth is realized by the complete reaction of all the available reactive groups and thus results in the attachment of the maximum number of monomers possible. Defective (structure imperfect) growth or incomplete reactions results in branch errors, which, if they occur in the early stages of growth, are generally more problematic than those occurring at higher generations from a dendritic property viewpoint.

Branching is dependent on monomer valency (this includes the cores since they are a special class of building block). Thus, a core possessing one reactive moiety, such as a primary amine, is divalent and will accommodate two monomers assuming a neutral trisubstituted amine product; branching therefore proceeds in a 1 → 2 manner. With three monomers, the resultant product is an ammonium salt, in which branching proceeds in a 1 → 3 fashion. For neutral amine products, conversion of the new terminal groups (e.g., cyano moieties) to primary amines and repetition of monomer addition procedure (e.g., amine alkylation) results in a general 1 → 2 → 4 → 8 → 16 → 32 → ... branching pattern. A tetravalent, four-directional core that is reacted with 4 equivalents of a 1 → 2 branching monomer will result in a progression 4 → 8 → 16 → 32 → 64 → 128 → ... (Scheme 3.1(a)), whereas employing the same core with a 1 → 3 branching monomer will give a dendritic series with an increasing peripheral multiplicity of 4 → 12 → 36 → 108 → 324 → 972 → ... (Scheme 3.1(b)).

3.2 Early Procedures

In 1978, Vögtle and coworkers in Bonn reported[1] the first preparation, separation, and mass spectrometric characterization of simple fractal-like structures by an iterative methodology (Scheme 3.2). This was defined as a *cascade synthesis*, being described as "reaction sequences, which can be carried out repeatedly." The key features of polyamine **4** include the trigonal *N*-centers of branching and the critical distance imposed by the $-(CH_2)_3-$ linkage between the branching centers. Treatment of primary amine **1** with acrylonitrile in a Michael-type addition afforded the desired dinitrile **2**, which was reduced to terminal diamine **3**. After purification, **3** was subjected to the same reaction sequence to generate heptaamine **4**, the structure of which was fully characterized. The original reduction conditions have proven difficult in certain laboratories, possibly on account of incomplete extraction of the water-soluble polyamine into lipophilic solvents. Repetition[2] of this sequence, using diisobutylaluminum hydride for nitrile reduction rather

(a) 1 -> 2 Branching

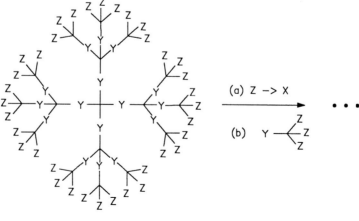

Scheme 3.1 Examples of divergent synthetic procedures for macromolecular construction.

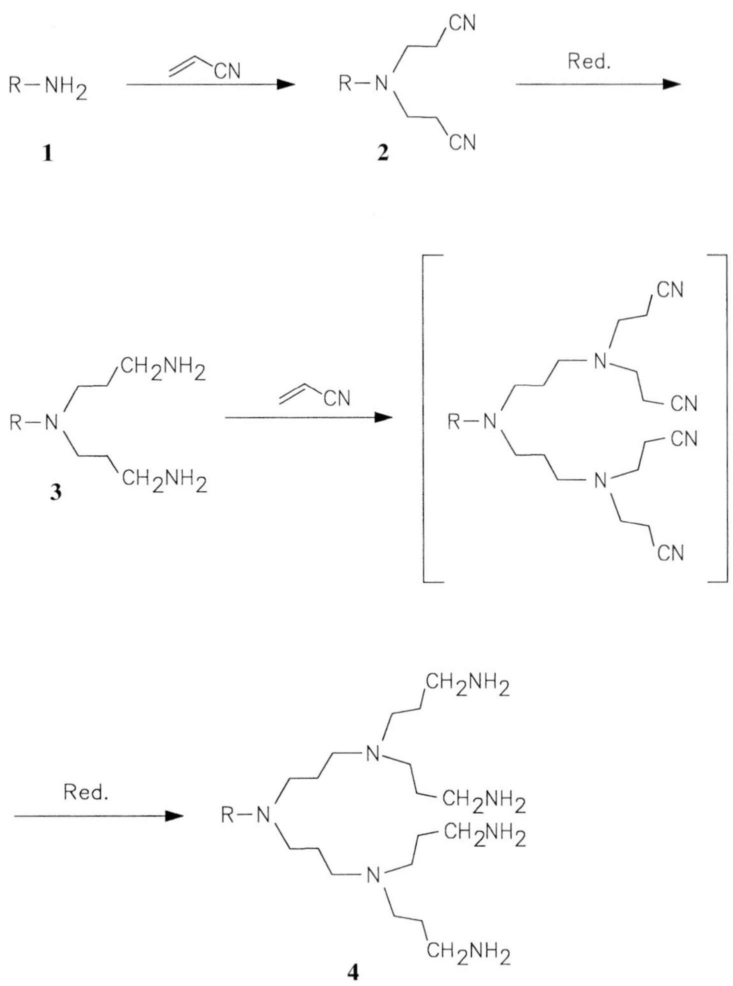

Scheme 3.2 Vögtle et al.'s[1] original repetitive synthesis.

than a Co(II)-borohydride complex, supported the original[1] structural conclusions. Technically, heptaamine **4** is not a macromolecule, but rather an example of a multi-armed,[3] meso-[4] or medio-molecule. The most important aspect of this synthetic sequence was that Vögtle presented the first application of an iterative reaction sequence and thus introduced the cascade approach to the synthetic chemist's repertoire.

Between 1979 and 1981, Denkewalter et al.[5] reported in three patents the first divergent preparation of dendritic polypeptides utilizing the protected amino acid *N,N'-bis(tert*-butyloxycarbonyl)-L-lysine as the monomeric building block. The two-directional, asymmetric core **5** was constructed from L-lysine and benzhydrylamine. Coupling was accomplished by the use of an activated *p*-nitrophenyl ester **6**, with subsequent removal of the *tert*-butyloxycarbonyl (*t*-Boc) protecting groups; see Scheme 3.3. The free polyamine moieties of **7** were then available for the construction of the next generation. Repetition of the simple coupling and deprotection sequences, utilizing ester **6**, led to lysine polymers containing 1023 terminal (*t*-Boc) protected lysine groups, and the creation of a series of generations of increasing size and molecular weight. These biopolymers were inherently asymmetrical, used peptide coupling technology, and contained a 1 → 2 *C*-branching pattern that resulted in an approximate generational doubling of the molecular weight through to the 10th generation (mol. wt. at the 10th generation: 233,600 amu). Their original[5] and related patents,[6, 7] however, offered little insight into the purification, characterization, or physical/chiral properties of these macromolecules.

In 1985, Newkome et al.[8] and Tomalia et al.[9] published different divergent routes to 1 → 3 *C*-branching arborols (*Arbor*: Latin: tree) and 1 → 2 *N*-branching dendrimers (*Dendro*-: Greek: tree-like), respectively (Scheme 3.4). These works described the construction of polyfunctional molecules that possessed multiple branching centers *and* offered spectral characterization supporting the structural assignments. There is another subtle but important difference between these two divergent processes: in the initial

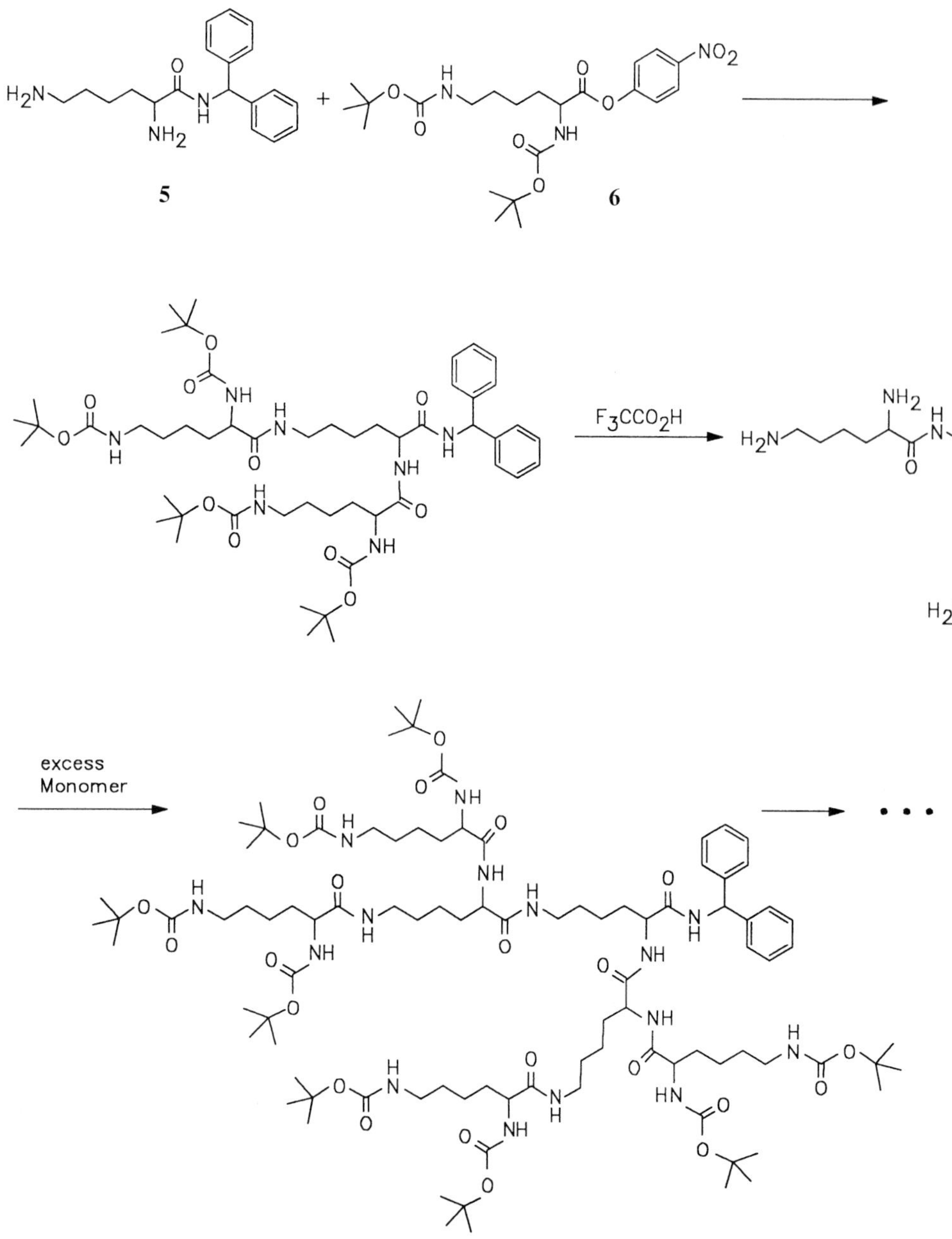

Scheme 3.3 Denkewalter et al.'s[5] preparation of dendritic polypeptides.

Vögtle–Tomalia routes branching occurs at the surface substituent(s), whereas the Denkewalter–Newkome approaches involve a *single* surface attachment and the monomer possesses the 2- or 3-branching center, respectively. These latter examples can be considered as both divergent (with simple branched monomers) or convergent (with complex monomer addition) processes. These surface branching protocols and all subsequent divergent routes using the 1 → 2 branching motifs will be described in this Chapter under the appropriate branching pattern and mode of monomer connectivity. The 1 → 3 branching methodology will be considered in Chapter 4.

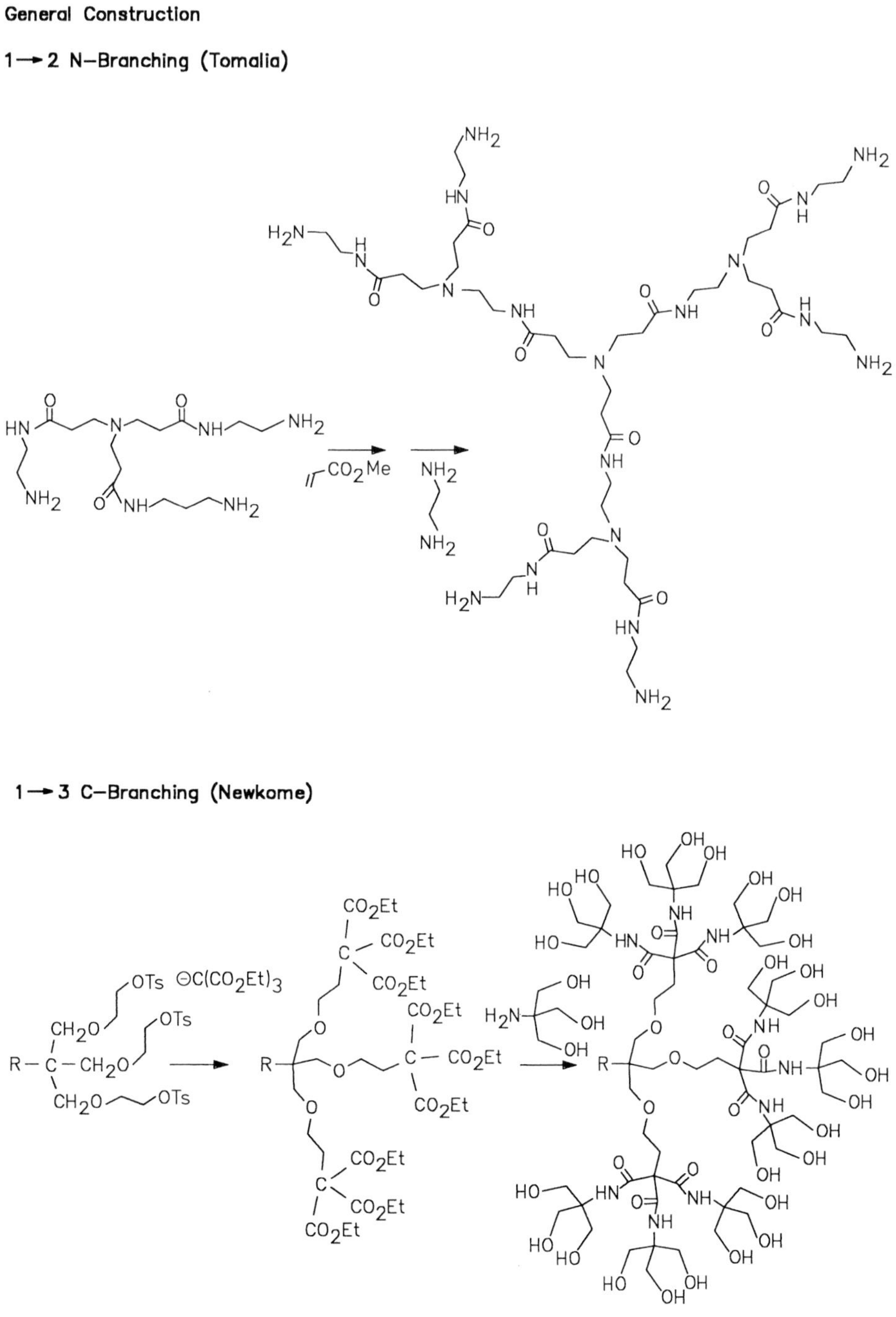

General Construction

1 → 2 N–Branching (Tomalia)

1 → 3 C–Branching (Newkome)

Scheme 3.4 Tomalia et al.'s[9] (1 → 2 *N*-branching) and Newkome et al.'s[8] (1 → 3 *C*-branching) original motifs.

3.3   1 → 2 Branched

3.3.1   1 → 2 *N*-Branched and Connectivity

In their quest for large molecular cavities capable of binding molecules or ionic guests, Vögtle et al.[1] synthesized noncyclic polyaza compounds by a repetitive stepwise process (one-directional case: Scheme 3.2). The use of two-directional cores (Scheme 3.5), e.g., ethylenediamine, 2,6-bis(aminomethyl)pyridine (**8**), or 1,3-bis(aminomethyl)benzene, in conjunction with acrylonitrile afforded tetranitriles **9**, which were reduced with borohydride under Co(II)-catalysis to produce hexaamines **10**. These terminal tetraamines were subsequently reacted with excess acrylonitrile to generate the octanitriles **11**; although the process was terminated at this stage, a foundation for dendrimer construction was established.

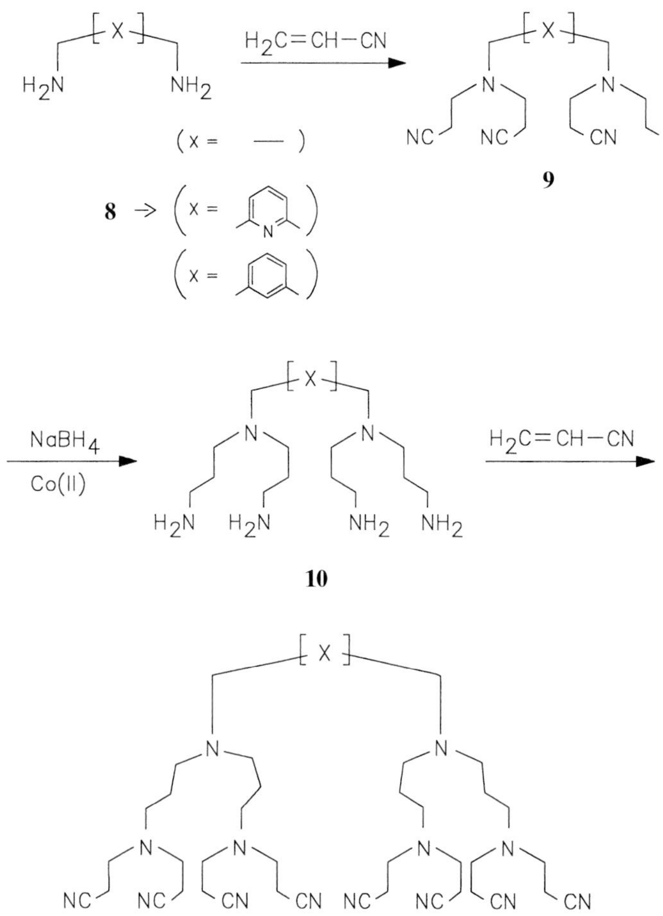

$H_2C=CH-CN$

$8 \rightarrow$ (x = —) (x = pyridine) (x = benzene)

9

$\xrightarrow{\text{NaBH}_4 \quad \text{Co(II)}}$

10

$\xrightarrow{H_2C=CH-CN}$

11

Scheme 3.5 Vögtle's repetitive *N*-alkylation and reduction for the construction of polyamines and polynitriles.[1]

Wörner and Mülhaupt[10] improved Vögtle's procedure by modifications that do not require the use of excess reagents and complicated purifications. They first improved the nitrile reduction step by the use of Raney nickel[11] at ambient temperature (8 bar H_2) in the presence of a trace of NaOH in EtOH. Then, they found that when the cyanoethylation was conducted in MeOH, it occurred without the generation of monosubstituted side products. By application of these synthetic modifications,[10] the original route could be extended to the 5th generation nitrile, with excellent overall yields for most steps.

Meijer and de Brabander-van den Berg[12, 13] also reported (in back-to-back manuscripts with Wörner and Mülhaupt[10]) procedures for the large-scale preparation of poly(propylene imine) dendrimers (herein denoted as PPIs also known as POPAM) using a sequence of reactions analogous to those originally employed.[1] Thus, repetitive addition of a primary amine to two equivalents of acrylonitrile, followed by catalytic (Raney cobalt) hydrogenation, afforded the desired polyamine-terminated macromolecules (Scheme 3.6). The ^{15}N NMR spectra of these dendrimers confirmed their highly branched and well-defined structures.[14, 15] Recently, the use of double-resonance 3D HMQC-TOCSY NMR on the 3rd generation PPI (16-terminal amino groups) has appeared.[16] Starting with 1,4-diaminobutane (**12**), the 5th tier polynitrile (**13**) was produced in kilogram quantities[17] and is now commercially available.[18–20] Ammonium addition during hydrogenation facilitated an 8-fold increase in substrate concentration and allowed a 10-fold decrease in the catalyst concentration. Conversions of > 99.5 % were still observed. Further, the 5th generation, amine-terminated PPI was formed after 248 consecutive steps, and possessed an approximate dendritic purity of 20 %; this corresponds to a polydispersity of ca. 1.002.[21] Based on SAXS and SANS data, the hydrodynamic radii of generations 1–5 were determined as 5.2, 7.3, 9.4, 11.7, and 14.0 Å, respectively;[18, 22] the 5th generation of a coated PPI was also analyzed by SANS in solution.[23] The data compare well with molecular modeling values of 4.8, 6.2, 8.1, 9.8, and 11.9 Å, respectively. Related SANS studies have been conducted[24] to determine the size changes of the 4th and 5th generation PPIs in concentrated solution; preliminary small-angle X-ray scattering data support the hypothesis that these dendrimers can collapse. Scherrenberg et al.[25] studied these ubiquitous materials by SANS, viscometry, and molecular dynamics, from which it was concluded that dendrimer dimensions increase linearly with generation as well as with ca. (molar mass)$^{1/3}$. Fractal dimensions were found to be ca. 3. The amino termini were found to exhibit a substantial degree of back-folding resulting in a monotonic density decrease toward the periphery and a homogeneous radial density distribution. The box counting method has also been applied[26] to DAB(CN)$_{64}$ (where DAB represents the *dia*mino*bu*tane core). Interdendrimer interactions have been examined[27] by means of

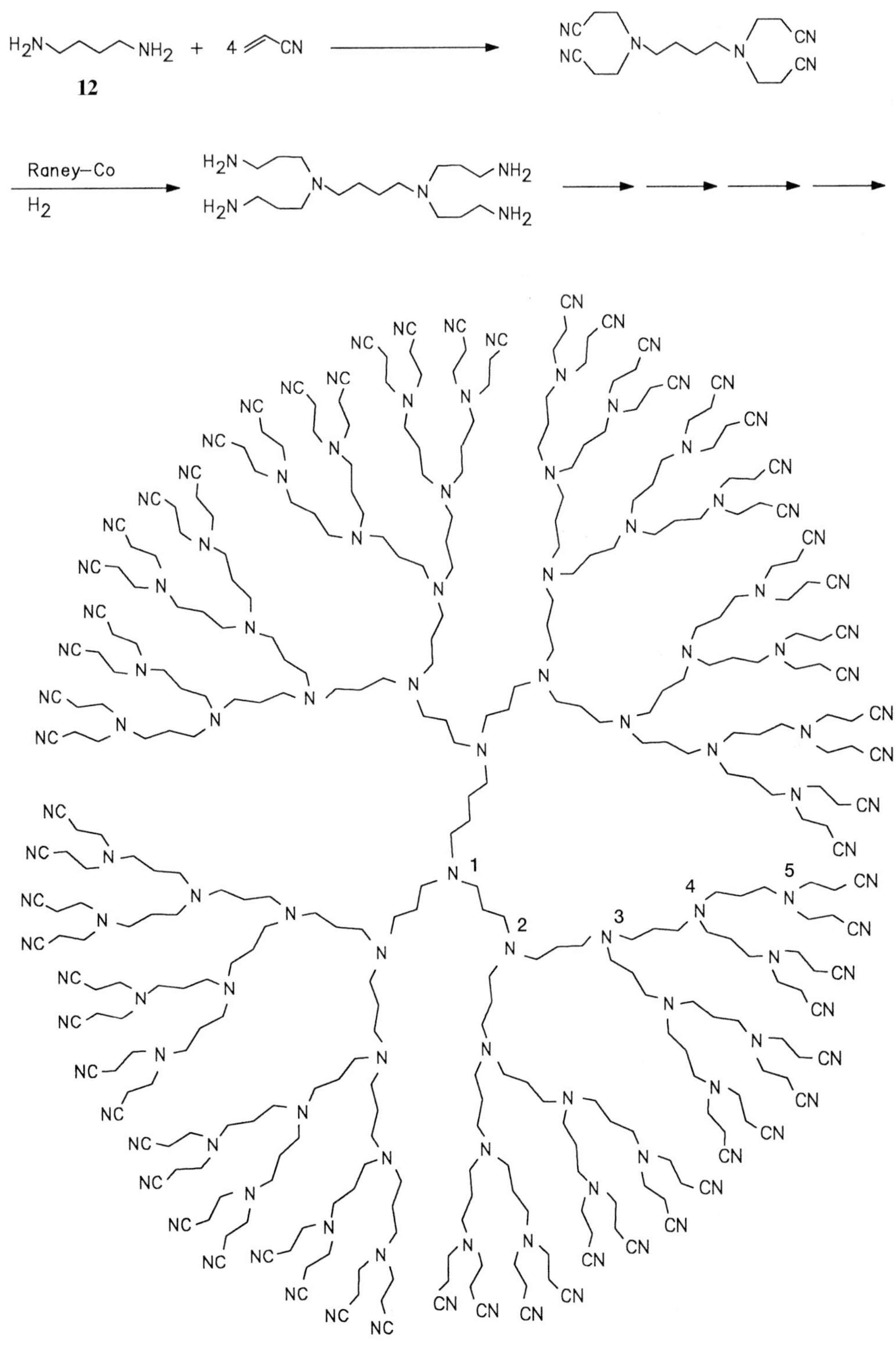

13 [DAB(CN)$_{64}$]

Scheme 3.6 Procedure for the large-scale synthesis of poly(propylene imine) dendrimers (PPI).[12]

SANS experiments; at high concentration, dendrimers have been shown to behave as soft molecules capable of interpenetration; however, when protonated, hard particle behavior was observed.[27a] MALDI-TOF-MS, SEC, and dilute solution viscosity were used to analyze three generations of these nitrile-terminated PPI materials; the data indicated the degree of imperfection and confirmed their compact spherical composition.[28] These nitrile-coated dendrimers were subsequently converted to the corresponding acid series by treatment with hydrochloric acid. Several excellent overviews of the PPI series have appeared.[29,30] Preliminary data on the optical properties of PPI colloidal films using scanning angle reflectometry have recently appeared, also see.[31a]

Astruc and coworkers[32] employed a slightly modified procedure for the construction of related higher generation dendrimers. Nitrile reduction was effected using $BH_3 \cdot THF$ or $BH_3 \cdot Me_2S$ instead of the Co(II)-promoted transformation. Starting with an aryl nona-alcohol, which was treated with acrylonitrile and then reduced to give the corresponding nonaamine, the 1 → 2 branching procedure was applied to give amine and nitrile dendrimers through to the 4th generation (144 CN moieties). Interestingly, the higher generation polyamines showed poor water solubility. Generation of the corresponding metallocenes afforded efficient sensors for molecular recognition. Similarly, Sakai and Matile[33] introduced branching onto cholestamine to investigate the potential of polyamines to enhance transmembrane ion transport; the results suggested that the protonated polyamines increasingly promoted ion transport relative to systems with more free amine moieties.

In a related scheme for the preparation of an even more compact series possessing the shorter ethano spacer, the polyethylene imines (SPEI) were constructed[34] by nucleophilic addition of an amine to *N*-mesylaziridine (i.e., 14). Scheme 3.7 shows the addition of excess 14 to a simple branched tetraamine 15 to generate protected polyamine 16. Deprotection of 16 afforded terminal primary amines, which were further treated with excess 14 to create successive generations (e.g., 17). Due to the diminished distance between branching centers in these SPEI dendrimers when compared with the PAMAM series (see Section 3.4), the resultant cascades reached dense packing limits at lower generations,[34] where "divergence from branching ideality becomes significant as one

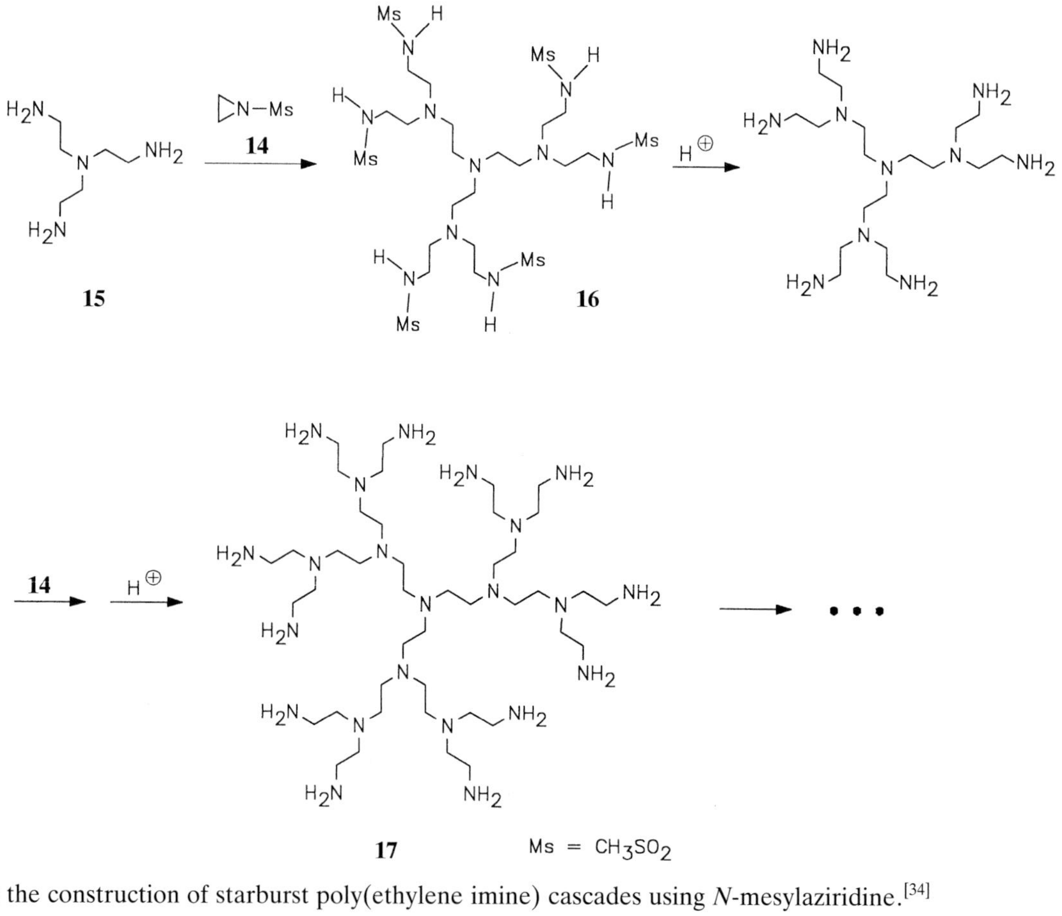

Scheme 3.7 Method for the construction of starburst poly(ethylene imine) cascades using *N*-mesylaziridine.[34]

approaches generation 3 or 4 and especially at generation 5." The 3rd generation SPEIs have been treated with stoichiometric amounts of octanoic acid to form lamellar liquid crystals;[35] their presence was evidenced by polarized optical microscopy and supported by IR data, which showed the absorbances of carboxylate moieties (formed by amine–carboxylic acid interaction). Supramolecular ordering of the linear alkyl chains above and below the plane of the flattened dendrimer was postulated to give rise to the liquid-crystalline phase. For PAMAM examples see ref. 35a.

Suh, Hah, and Lee[36] connected single β-cyclodextrin (CD) units to the surfaces of 3rd and 4th generation poly(ethylene imine)s, prepared form the N-tosylaziridine, by terminal amine reactions with mono-6-(p-toluenesulfonyl)-β-cyclodextrin. ^{1}H NMR data indicated the attachment of 0.87 and 1.05 CD moieties, respectively. p-Nitrophenyl esters possessing CD cavity affinity were observed to undergo rapid aminolysis when complexed to the CD-dendrimer complement; kinetic data suggested that diacylation occurs rapidly and that two amino moieties are involved.

Fructose modification for heptacyte apopotosis suppression has been examined.[36a]

3.3.1.1 Surface Modifications

Meijer et al.[37] utilized the 5th generation polyamine (denoted as **18**), generated by reduction of the corresponding nitrile, for the construction (Scheme 3.8) of host dendritic 'boxes' that allowed the sterically induced entrapment of guest molecules. The concept of "trapping topologically by shell molecules" was considered theoretically by Maciejewski[38] and was previously reported by Denkewalter et al.[5–7] for their lysine-based dendrimers, which were classified as "non-draining" spheres. Meijer's dendritic boxes (**20**; a sense of the large interior void volume and congested surface can be gleaned from the full 2D drawing) with molecular weights up to 24,000 amu were prepared[37] by surface modification of **18** with activated chiral esters, for example the N-hydroxysuccinimide ester of a t-Boc-protected amino acid (phenylalanine shown; **19**); further details concerning chirality are presented in Chapter 7. The concept of a dendritic box has been expanded[39, 40] by the introduction of other surface amino acid groups, e.g. L-alanine, L-t-Bu-serine, L-Tyr-cysteine, and L-t-Bu-aspartic ester. Additional studies concerning induced chirality of encapsulated guests,[41] the encapsulation of Rose Bengal,[42, 43] and triplet radical pairs[44] have appeared; an excellent review of the host–guest ramifications of dendrimers has also appeared.[45] Goddard et al.[46] described a molecular dynamics analysis of the encapsulation of Rose Bengal within the dendritic box.[37, 39, 40] A concentration-dependent equilibrium between surface and interior solvent regions was found for dendrimers not possessing the "capping" periphery. Theory supported the experimental results with respect to the number of guests residing within the interior regions. Moreover, a "box" terminated with bulky capping groups was found to be completely impermeable to the molecular guests. Cavallo and Fraternali[47] also performed a molecular dynamics study detailing such properties as H-bonding, solvent-accessible surface areas, and excluded volumes for the first five generations of these N-t-Boc-L-phenylalanine-terminated frameworks.

Verlhac et al.[48] prepared branched perfluorinated alkanes (Scheme 3.9; **23** and **25**) by reaction of the tosylated, linear monomer **22** with either polyamine **21** or **24**, respectively, under basic conditions (K_2CO_3, CH_3CN). These small dendritic molecules were subsequently employed as perfluorinated Cu(I) ligands for the catalytic, intramolecular cyclization of alkenyl trichloromethyl esters under "fluoro biphasic" conditions,[49] which allow for ease of work-up and catalyst recovery. PPIs have also been coated with freon 113 (a perfluorinated PEG-like material, the heptameric acid fluoride of hexafluoropropylene oxide) to yield a "CO_2-philic" shell;[50] these materials were shown to facilitate the transfer of the CO_2-insoluble, ionic, methyl orange dye from aqueous media to CO_2.

Moszner et al.[51] reported the surface modification of PPIs with 2-(acryloyloxy)ethyl methacrylate. The methacrylic-modified dendrimers (generations 2, 4, and 5) were polymerized both in solution and in bulk to yield cross-linked polymers exhibiting T_gs near, or below, room temperature. Dendrimer and polymer properties were further examined by treating the polyamine surfaces with mixtures of methacrylate, resulting in "dendritic" melting points slightly higher than 25°C. The introduction of hexafluorobutyl or 2-(tri-

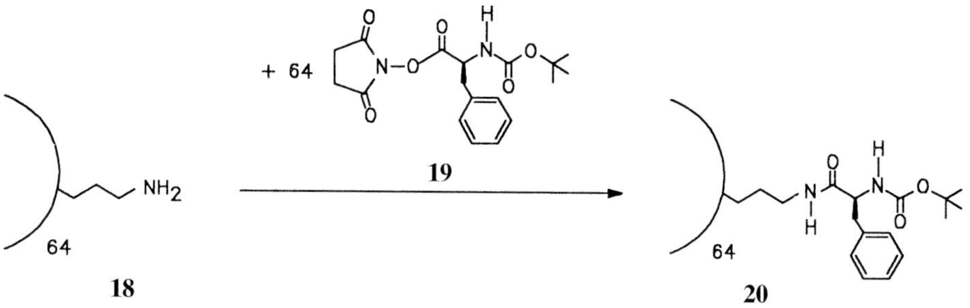

18 **19** **20**

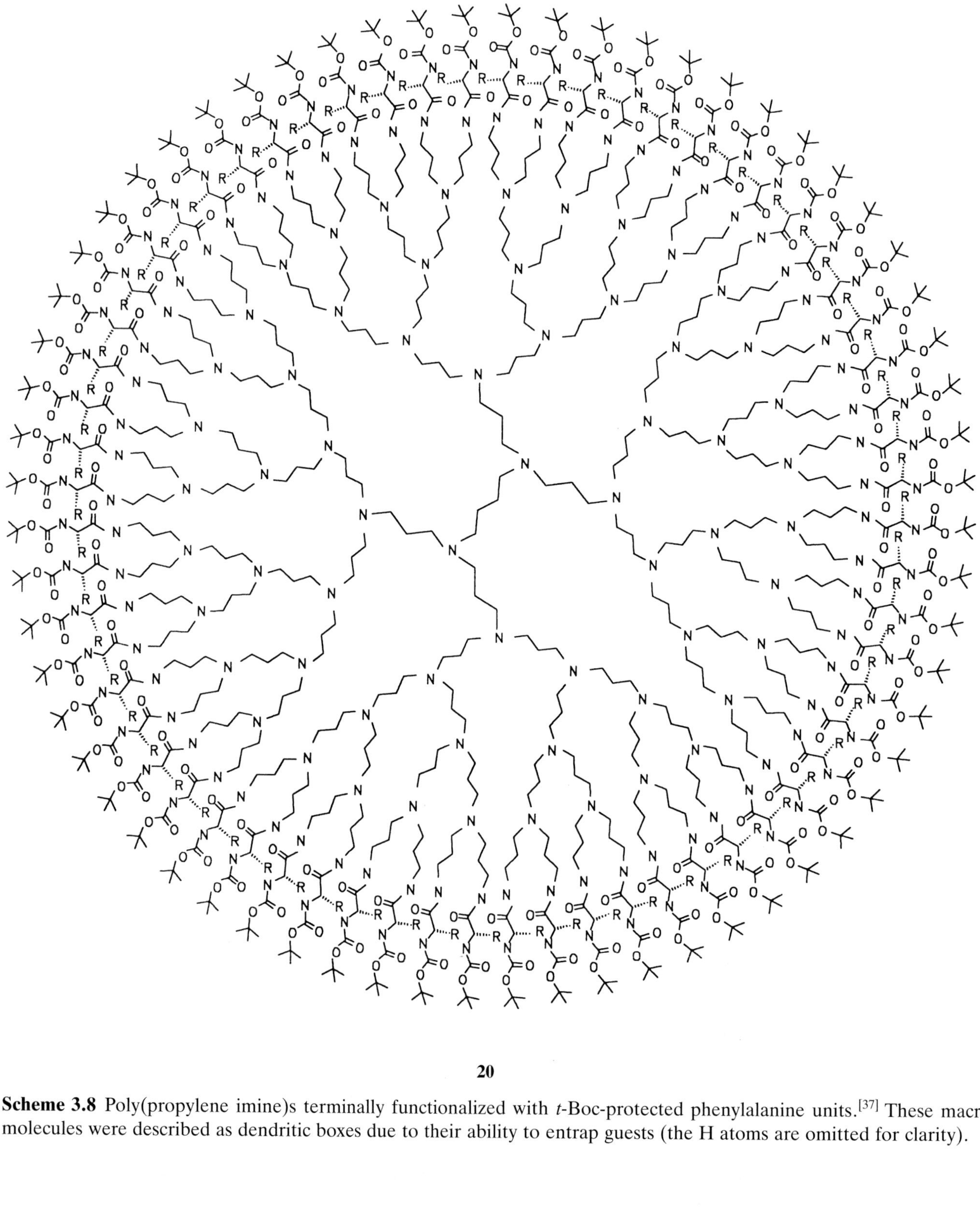

20

Scheme 3.8 Poly(propylene imine)s terminally functionalized with *t*-Boc-protected phenylalanine units.[37] These macromolecules were described as dendritic boxes due to their ability to entrap guests (the H atoms are omitted for clarity).

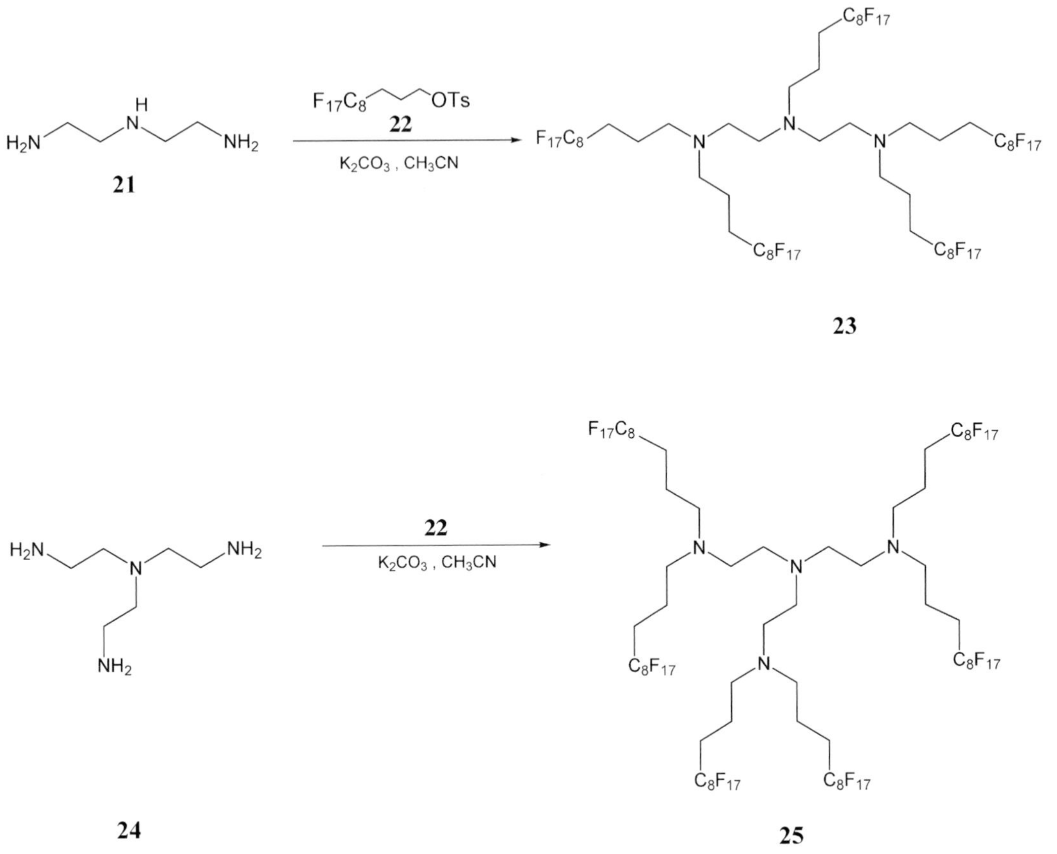

Scheme 3.9 Synthetic route to perfluorinated *N*-dendrimers.[48]

methylsilyl)ethyl acrylate in surface mixtures was also reported. Treatment of the terminal amines of these PPIs with methyl acrylate followed by hydrolysis has facilitated the examination of their protonation behavior;[52] a characteristic "onion-like" shell protonation behavior was observed.

Put, Meijer, and coworkers[53] functionalized the PPI family at the termini with second-order nonlinear optical chromophores, e.g., 4-(*N*,*N*-dimethyl)phenylcarboxamide end groups prepared by reaction with the corresponding acyl chloride. Employing the hyper-Rayleigh scattering technique to study hyperpolarizabilities, dendritic solution structure and symmetry were found to be flexible at lower generations, and, on average, sphere-like. At higher generations, however, the dendrimers were found to be even more spherical and became more rigid. This is in agreement with results reported[34] for the higher generations of the PAMAM series, which became "globular" after generation 3.5. A related series of coated PPIs was also prepared, by Michael reactions with o-(4'-cyanobiphenyloxy)alkyl acrylates, to afford liquid-crystalline dendrimers.[54]

Wagner et al.[55] modified the surfaces of the 4th and 5th generation PPIs with acetyl and deuterated acetyl chloride to study their rheology; Newtonian properties were observed at all concentrations, whereas non-acetylated materials proved to be less viscous. Data from SANS measurements suggested significant solvent penetration and chain back-folding, as well as dendrimer clustering and interdigitation. Similar experiments were performed[56] with acetylated and hydroxy-terminated PPIs in aqueous poly(ethylene oxide) solutions.

A series of star-shaped polymers based on the 1st to 5th generation PPI cores has been prepared[57] for investigating miscibility properties with linear poly(styrene). After coating each member with tyrosine moieties, possessing phenolic units, a "redistribution reaction" with poly(2,6-dimethyl-1,4-phenylene ether) (PPE) was carried out. The star-branched PPE length was regulated by control of the phenol/PPE ratio; the average arm length was determined to be 90 repeat units for each member. For star-polymers constructed from generations 3–5 (16, 32, and 64 theoretical termini, respectively) inhomogeneous blends with poly(styrene) were obtained, while miscible blends were obtained using 1st and 2nd generation-based stars. T_g values for all the members of this series were ca. 210 °C.

Meijer et al.[58] described the terminal modification of PPIs using simple, long-chain (C_n, where $n = 5$, 9–15) acid chlorides to afford inverted unimolecular micelles. Dynamic light-scattering experiments evidenced single-particle behavior (i.e., no aggregation) and suggested a 2–3 nm hydrodynamic diameter using CH_2Cl_2 as the solvent. Micellar characteristics were verified by entrapment, or host–guest complexation using the 5, 9, and 15 C-chain, modified 2nd and 5th generation PPIs. Upon addition of Rose Bengal and the dendrimer to EtOH, followed by precipitation and copious washing, it was determined that the 2nd generation inverted micelles entrapped an average of 1 dye molecule, whereas the 5th generation micelles trapped ca. 7 to 8. Vapor-liquid equilibria for PPIs coated with dodecyl or octadecyl amides, or with polyisobutylene, were measured by a classical gravimetric–sorption procedure; solvent absorption was found to depend strongly on dendrimer composition and generation number.[59]

Inverted micelles, formed from these PPIs with palmitoyl chloride, were found[60, 61a] to be very effective extractants of anionic dyes from water into organic solvents. At low pH, a 5th generation modified PPI extracted fluorescein, 4,5,6,7-tetrachlorofluorescein, erythrosine B, and Rose Bengal, whereas at higher pH (slightly above pH 6.0) extraction yields fell dramatically. The extraction behavior was attributed to the internal tertiary amine moieties. Similar host–guest characteristics were reported[62] using pyrene; the PPI–pyrene binding constants were ascertained by fluorescence spectroscopy. Ramzi et al.[63] coated these PPIs with fatty acids and conducted SANS experiments to elucidate single-chain factors of both the fatty acid and dendrimer components. Blends have also been made by solution and melt-mixing fatty-acid-modified PPIs with diverse olefins;[64] SANS was used to determine the degree of miscibility. The coating of PPI surfaces with oligo(*p*-phenylene vinylene)s (OPVs) has been successful[65] and the resultant amphiphilic globular species have been found to self-assemble at the air–water interface forming stable monolayers. These OPV-coated PPIs can be loaded with dyes and their ratios easily tuned; the dendrimer–dye system can be mixed with poly(*p*-phenylene vinylene)s affording quality thin films, which suggests the possibility of tuning the emission wavelength. Ballauff and Vögtle et al.[66] coated a 4th generation polyamine with diphenyl ether moieties and evaluated the structure by SANS, which led to the conclusion that the density distribution maximum occurs at the center of the molecule. Partial end group modification (stearic acid, 150 °C) of the PPI family generated two types (compositional and positional) of heterogeneity; lower generations showed a random distribution of dyads (end group substitution patterns), whereas at higher generations a marked preference for a single substitution of dyads was realized.[67]

Meijer and coworkers[68] generated *N-t*-Boc-protected glycine-coated PPIs to study the question of termini and branch back-folding. An X-ray structure of the 1st generation tetraglycine revealed a high degree of termini H-bonding; thus, it was concluded that end group localization was dependent on the dendritic structure and that secondary interactions need to be considered when using models to address this question. Meijer, Stoddart, et al.[69] reported the preparation of a related series of glycodendrimers based on PPIs of generations 1 through to 5 (see Chapter 7).

Bosman, Janssen, and Meijer[70] modified the surfaces of PPIs of generations 1 through to 5 with nitroxyl radicals by reaction of the terminal amines with 3-carboxy-2,2,5,5-tetramethyl-1-pyrrolidinyloxy (3-carboxy-proxyl) radicals. The polyradicals were shown to exhibit strong exchange interactions with lower generations, directly showing the number of hyperfine transitions, and thus the number of termini, by EPR experiments. FT-IR confirmed the presence of an amide-based, H-bonded network at the periphery.

Vögtle, Balzani, et al.[71] described the terminal modification of these dendrimers with diazobenzene moieties. This was the first reported use of these materials as holographic materials. Thin films of the poly(azobenzene) give holographic gratings with diffraction efficiencies up to ca. 20 %. These authors[72] further examined the use of these poly(azobenzene)-coated materials as molecular hosts for eosin Y (2',4',5',7'-tetrabromofluorescein dianion). Quenching of the hosted eosin fluorescence, as well as the hosting potential, was found to be more efficient with the *Z* form of the construct.

Cyanobiphenyl mesogens have been attached to the surfaces of these polyamines by Meijer et al.[73] in order to study the resultant liquid-crystalline properties. Mesogen attachment was effected using either a 5- or 10-carbon activated ester spacer chain. All of the resulting mesogenic dendrimers were observed to form a smectic A mesophase, while the thermal properties, dependent on spacer length, exhibited $g \rightarrow S_n \rightarrow I$ transitions (C_{10}). In terms of transition enthalpies and kinetics, mesophase formation was found to be more favorable for the C_{10} series.

One of the more intriguing chemical transformations in this field was that reported by Peerlings and Meijer,[74] where all of the amino groups of the surface of a 5th generation PPI were converted to the corresponding isocyanates. This was accomplished under very mild conditions using di-*tert*-butyltricarbonate, which can be prepared[75] (84%) by the insertion of CO_2 into potassium *tert*-butoxide and subsequent reaction with phosgene.

Termini modifications with long (C_{15}) alkyl chains, C_{10} chains possessing diazobenzenoid moieties, and adamantane[75a, b] groups for five generations of PPIs have been reported.[76] Monolayer formation at the air–water interface, as studied by X-ray diffraction of cast films, as well as electron microscopy and dynamic light scattering of acidic aqueous solutions, revealed distortion of the dendritic core, thereby illustrating the high flexibility of these materials. Randomly substituted PPIs, with palmitoyl and azobenzene-containing alkyl chains in a 1:1 ratio, designed to exhibit reversible, photo-induced switching have recently been reported by Weener and Meijer.[77]

Chen et al.[78] functionalized the 3rd generation PPI with dimethyldodecylammonium chloride units and demonstrated strong antibacterial properties of the resulting products. These novel dendrimer biocides were studied using Gram-negative *E.coli* employing a bioluminescence protocol. Quaternized 32 or 64 PPIs, obtained by reaction with glycidyltrimethylammonium chloride, have been prepared as pH-sensitive controlled-release systems by Paleos et al.[79]

Vögtle, Balzani, De Cola et al.[80, 80a] customized the surface of these PPIs, through to the 5th generation, with dansyl groups. The products were studied with regard to their protonation, absorption, and photophysical properties, as well as intradendrimer quenching and sensitizing processes. The 4th generation of these coated PPIs was shown[81] to exhibit a strong fluorescence, which was quenched when a Co^{2+} ion was N-coordinated within its interior; a Co^{2+} concentration of 4.6×10^{-7} M resulted in a 5% decrease in fluorescence intensity revealing the potential to fine-tune this effect. Azobenzene-coated PPIs have been shown to self-assemble in aqueous solution below pH 8 to give large vesicles with an onion-like structure.[82] Vögtle et al.[83] also coated the surfaces of the PPIs with related methyl orange moieties; molecular inclusion and subsequent release of guests was controlled by modifying the pH.

Aoi et al.[84] coated the surface of 64-PPI by the ring-opening polymerization of sarcosine N-carboxyanhydride; the resultant PPI–*block*–(polysarcosine)$_{64}$ possessed a narrow $\overline{M}_w / \overline{M}_n$ of 1.0_1–1.0_3 (by SEC) and a controlled polysarcosine chain length was realized by varying the molar feed ratio of the reactants.

Using various alkyl isocyanates (i.e., hexyl, octyl, dodecyl, and phenyl), Vögtle et al.[85] terminally modified these dendrimers to produce efficient organic media hosts for oxyanions, specifically pertechnetate, perrhenate, AMP, ADP, and ATP. Extractability rates were determined and controlled release of the guests was demonstrated by pH-dependent behavior (e.g., greater host–guest binding occurred at a lower pH).

Pan and Ford[86, 87] reported the creation of an amphiphilic architecture (Figure 3.1; **26**) possessing hydrophobic and hydrophilic terminal chains. These materials, prepared as water-soluble catalysts, were accessed in five steps from the 32-PPI. Terminal modification with octanoyl chloride, reduction (LAH) of the resulting amide to give the terminal secondary amine, subsequent reaction of the latter with the triethylene glycol acid chloride, and further reduction, gave the polyamine precursor of the methyl iodide treated polyammonium catalyst.[87a] These cationic chloride dendrimers[87] act as hosts for guests, such as Reichardt's dye and pyrene, in an aqueous environment and enhance (500 × over water alone) the rate of decarboxylation of 6-nitrobenzisoxazole-3-carboxylic acid.

Water-soluble, oligo(ethyleneoxy)-terminated PPIs, based on a 3,4,5-(PEG)benzoate moiety, have been shown to act as unimolecular micelles.[88] Coating of the PPIs with porphyrin groups has been accomplished[88, 89] and the products were employed in a time-resolved fluorescence study to examine the dynamics of electronic energy transfer.

Noble and McCarley[90] treated these PPIs with 2,5-dimethoxytetrahydrofuran (glacial AcOH, MeCN) to convert the amine termini to pyrrole moieties. Exposure of gold to

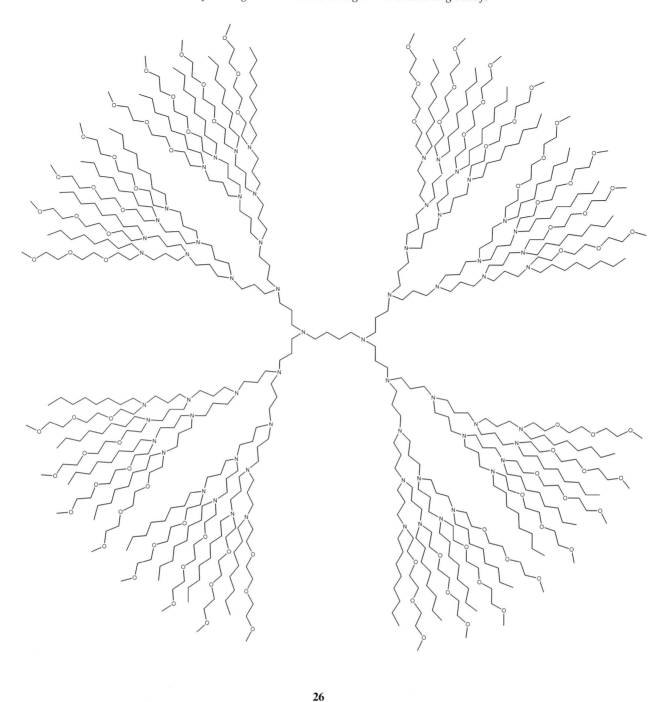

26

Figure 3.1 Pan and Ford's amphiphilic dendrimers for use as catalysts in aqueous media.[86]

solutions of these functionalized dendrimers resulted in surface adsorption, as evidenced by reflection-adsorption infrared (RAIR) spectra. Electrochemical oxidation of the polypyrrole films facilitated *intra*molecular pyrrole oligomerization. Films formed on DAB-py$_{16}$/Au were found to be extremely scratch resistant, but could be partially removed by repeated application and removal of Scotch tape.[91] For other recent examples see.[91a–c]

3.3.1.2 Dendron Attachment

Poly(propylene imine)-like wedges through to the 5th generation have been constructed on one end of a polystyrene, thereby forming a macromolecular surfactant or an envisaged macroamphiphile. For polystyrene–PPI hybrids possessing multiple CO$_2$H termini, a pH-dependent[92] aggregation was observed [e.g., PS-dendri-(CO$_2$H)$_8$ formed "worm-like" micelles at high pH; PS = polystyrene, dendri = branched head group]. Amine-coated head groups [i.e., PS-dendri-(NH$_2$)$_n$] led to generation-dependent aggre-

27

Figure 3.2 PPI-capped polystyrene (**27**).[94, 95]

gation;[93] aqueous solutions were found to contain micellar spheres, rods, and vesicles for –(NH$_2$)$_8$, –(NH$_2$)$_{16}$, and –(NH$_2$)$_{32}$, respectively. Meijer et al.[94, 95] attached a PPI head to one end of a linear polystyrene (M_n = 3.2 × 10^3, M_w/M_n = 1.04) to produce "macroamphiphiles", as exemplified by **27** in Figure 3.2. The carboxylic acid terminus of the polystyrene was transformed to an amine by reduction (LAH), cyanoethylation (CH$_2$=CHCN), and nitrile reduction (Raney Co, H$_2$). Standard iterative divergent PPI synthesis was exploited, giving access to generations 1–5 of the PPI dendritic head group. Perhaps one of the more interesting aspects of these copolymers was their aggregation behavior. Aqueous aggregate formation for the 3rd, 4th, and 5th tiers led, as revealed by TEM imaging, to flexible bilayers, rod-like micelles, and spherical micelles, respectively. Further study of these materials by SAXS and TEM supported the aggregation phenomena.[96]

Tokuhisa and Crooks[97] covalently linked PPIs to self-assembled monolayers to prepare chemically sensitive interfaces with the potential to detect volatile organic compounds. Dendrimer prefunctionalization, prior to monolayer attachment, was found to generate more effective surfaces due to greater dendritic surface densities of the desired "active" termini.

These PPIs have also been incorporated as (1) polycations into linear poly(phenylene vinylene)-based self-assembled polymer superlattices for the investigation of controlled unidirectional energy transfer[98] as well as (2) carring trivalent galactosides as cell targeting ligands.[98a]

3.3.1.3   Physical Properties

Stöckigt, Lohmer, and Belder[99] investigated the composition and purity of a 2nd generation, diaminobutane-based PPI through the use of capillary electrophoresis in conjunction with a sector mass spectrometer. Numerous by-products were observed, including materials possessing branch defects, cyclized arms, incomplete reduction products from the previous reactions, and dimers. It is worth noting that procedural modifications in PPI synthesis have, in many cases, facilitated the development to purer materials. For example, analysis of a commercial sample of the corresponding poly(amine) [DAB-dendri-(NH$_2$)$_8$] by MALDI-TOF MS[100] showed it to be defect-free. High-performance liquid chromatography in conjunction with electrospray mass spectrometry has been used for compositional analysis of the nitrile counterparts of the PPI dendrimers.[101] A detailed description of dendrimer generational defects was presented. An electrospray mass spectral study of the PPI family has been completed.[102] Their gas-phase chemistry generates a unique fingerprint of small fragmentation adducts, which corresponds well with observations made upon their prolonged heating in aqueous solution. The findings support the generally accepted globular shape of the dendrimers based on the observed linear relationship between the extent of protonation and molecular mass as well as the unique fragmentation patterns. The fragmentation of various charge states of protonated PPIs with a butane core has been investigated[103] by surface-induced dissociation; the charged species have been proposed to fragment according to an $S_{N}i$ mechanism. The ions derived from electrospray ionization of these PPIs of generations 1–5 based on a butane core were also subjected to ion-trap tandem MS;[104] all products derived from the dendrimer parent ions were rationalized in terms of dissociation processes stemming from intramolecular nucleophilic attack by the nearest neighboring atom on a carbon α to the site of protonation.

In 1996, van der Wal and coworkers[105] reported on size-exclusion chromatography of PPI (as well as POPAM) dendrimers. Nitrile-terminated intermediate generations were chromatographed using a polystyrene-divinylbenzene stationary phase and THF as the mobile phase. For the amine-terminated "full" generations, the optimal systems were based on reversed-phase silica packing (deactivated with tetrazacyclotetradecane) with 0.25 M formic acid as the mobile phase at 60 °C.

Koper et al.[106] examined the protonation mechanisms of the PPIs (1st to 3rd generations, based on the 1,4-diaminobutane core) using ^{15}N NMR spectroscopy. "Shells" of N nuclei, corresponding to each nitrogen tier, in which the core nitrogens were the α-shell, the 1st tier of nitrogens the β-shell, and so forth, were analyzed for their degree of protonation based on the number or availability of protons. Further, the distribution of Z protons on a structure possessing six protonation sites constitutes a microstate (denoted as $n = 2$) comprised of the possible positions of protonation for that structure. For generation 1, possessing six amine moieties, it was found that the first shell to be fully protonated is the β-shell; at $n = 4$, 90 % of the microstate is comprised of complete terminal amine protonation. The interior is then left to protonate. For the $n = 1$ state, 80 % of the possible structures are terminally protonated, while for $n = 2$ and 3 proton distribution is more widespread. Potentiometric titration experiments with the PPI family were also conducted at different ionic strengths;[107a,b] the Ising model permitted a quantitative analysis of the titration curve around pH 6 and 10, which supported the stability of an onion-like structure, where alternate shells of the PPIs are protonated.

The 2nd generation PPI, composed of three shells, is initially protonated at the γ-shell (outer) and the innermost α-shell; the β-shell then follows. The 3rd generation PPI, at low proton concentration, exhibits some protonation at all levels other than the β-shell. The δ-shell is first to be fully protonated; at pH 8 the δ- and β-shells are nearly completely protonated; at still higher proton concentrations the γ-shell begins to fill.

Polyelectrolyte behavior[107c] of these PPIs has been examined[108] and, in general, potentiometric titrations reveal the presence of two amine types possessing differing basicities and electrostatic interactions (protonated forms). Separate pH regions were noted and, for higher generation dendrimers, a polyelectrolyte effect dominates the protonated form such that a strong decrease in $pK_{(\alpha)}$ was observed during protonation.

Welch and Muthukumar,[109] using Monte Carlo methods, demonstrated that by varying the ionic strength of the medium, the intramolecular density profile for dendritic polyelectrolytes (in this case PPIs) can be cycled between the expanded, "hollow-core dense-shell" model **28** (low ionic strength) and the contracted, "dense-core" model (**29**) (high ionic strength) (Figure 3.3). This is analogous to the expanded *versus* contracted forms observed for ionic-terminated *versus* neutral terminated forms (e.g., carboxylate *versus* carboxylic acid).[110] Employing simulation procedures, a molecular-level picture of these host–guest interactions and the conditions necessary for their manifestation has recently appeared.[111]

Employing inverse gas chromatography, Polese, Mio, and Bertucco[112] reported infinite-dilution activity coefficients of polar and nonpolar solvents in solutions of these polyamines (generations 2 through to 5). Solvent activity coefficients were observed to change with dendrimer generation. For example, a maximum in solvent solubility was seen for the 4th generation; protic solvents, such as EtOH and MeOH, were found to be better solvents than their non-polar counterparts such as THF, toluene, or EtOAc. Flory's interaction parameters were also determined, as was solvent activity for combburst polymers.

Joosten et al.[113] assessed the interaction of these materials with linear polyanions such as poly(sodium acrylate), poly(acrylic acid), poly(sodium styrenesulfonate), and native DNA by means of potentiometric, argentometric, and turbidimetric titrations. For flexible polyanions, interpenetration of the dendrimers was observed.

Solid electrolytes have been prepared with Li[CF_3So_2]_2N and analyzed for ion transport potential.[113a]

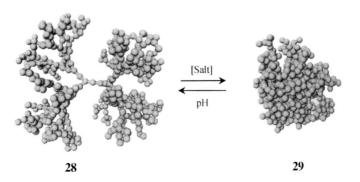

[Salt]

pH

28   **29**

Figure 3.3 Expanded (**28**) and contracted (**29**) forms under low and high ionic strength conditions, respectively. Reprinted with permission from the American Chemical Society.[109]

Wyn-Jones et al.[114] reported a study of the interaction of these poly(amine)s, as well as their corresponding pyrrolidone-modified analogs, with sodium dodecylsulfate (SDS), tetradecyltrimethylammonium bromide (TTAB), and hexaethylene glycol monodecyl ether ($C_{12}EO_6$) surfactants, based on electromotive force (EMF) measurements and isothermal titration microcalorimetry (ITM). SDS was found to bind to all the dendrimers at concentrations of 10^{-5} mol dm^{-3}, while $C_{12}EO_6$ and TTAB showed only very limited binding.

Colligative properties and viscosity characteristics of the PPIs in MeOH have been studied by Rietveld and Smit[115a], along with osmotic compressibility,[115b] using low-angle laser light scattering. Effective radii and partial specific volumes were also determined. Solvent was shown to reside within the interior of the dendrimers. Comparisons of the rheological behavior of PPIs with those of hyperbranched polyesters and poly(ether imide)s have also been reported.[116]

PPIs have also been examined as pseudostationary phases for electrokinetic chromatography.[117] Comparison of the technique using SDS micelles to that using the PPIs (G5; 64 CO_2H) as applied to mixtures of hydroquinone, resorcinol, phenol, benzyl alcohol, *o*-cresol, and 2,6-xylenol generally revealed reversed retention times for the analytes.

3.3.2   1 → 2 *N*-Branched, *Amide* Connectivity

In 1985 and 1986, Tomalia et al. described[9, 118] the preparation of polyamidoamines (denoted herein simply as PAMAMs), which were generated from a three-directional core (e.g., ammonia) and possessed three-directional *N*-branching centers as well as amide connectivity (Scheme 3.10). Each generation was synthesized by exhaustive Michael-type addition of methyl acrylate to amines (e.g., for an ammonia core, **30**) to generate a β-amino acid ester (e.g. **31**), followed by amidation with excess ethylenediamine to produce a new-branched polyamine **32**. The general procedure was repeated to create the higher generations (e.g., **34**). Similar dendrimers were prepared by employing related cores, such as ethylenediamine, as well as amino alcohols and other functionalizable groups, such as amino and thiol moieties.[119] This procedure is applicable to most primary amines, resulting in the 1 → 2 branching pattern.[120] It was noted that aryl esters could be utilized as the initiator core.[121, 122]

In order to achieve a high degree of product monodispersity, the potential synthetic problems associated with amine additions to esters [e.g., *intra*molecular cyclization (lactam formation), *retro*-Michael reactions,[123] incomplete addition, and *inter*molecular coupling] were minimized by the use of excess diamine, maintaining moderate (< 80 °C) reaction temperatures, and avoiding aqueous solvents.[9] Under optimized conditions, defects produced by these undesired reactions can, for the most part, be suppressed. Interestingly, a recent ESI-MS study on PAMAM at the 4th generation indicated that the sample under analysis possessed a purity of no more than 8 %;[30] this may bear out the statement "...the excess EDA required to make 95 % or greater purity at generations higher than 4.0 becomes prohibitive experimentally."[124] This synthetic procedure was especially noteworthy because, for the first time, an iterative synthesis was reported that allowed the preparation of high molecular weight cascade polymers. Recently, PAMAMs were conveniently and efficiently synthesized on a solid support and the products were found to possess good homogeneity.[125]

Structural, computer simulated,[126–128] molecular modeling,[129, 129a] comparative and electron microscopy,[130] Brownian dynamics simulation,[131] and physical characterizations[124, 128] of these macromolecules have included standard spectroscopic methods, e.g. ^1H, ^2H,[132, 132a] and ^{13}C[128, 133] NMR,[134, 135] IR, as well as mass spectrometry (electrospray),[136–140] HPLC, GPC, DSC, TGH, PAGE, and CE electrophoretic methods,[141] intrinsic viscosity measurements,[142] vapor-liquid equilibrium data by classical isothermal gravimetric-sorption,[143] small-angle X-ray scattering,[144] and SANS analysis of the 5th [145] or 7th generations.[146] The rates of hindered diffusion of PAMAMs and linear polystyrenes in porous glasses have been evaluated;[147, 148] the dendrimers conform to the hydrodynamic theory for a hard sphere in a cylindrical pore. Detailed evaluations of the molecular, solution, and bulk properties of 5th generation PAMAM have appeared.[149, 150] Further, the rheological behavior of the first eight generations of these PAMAMs has

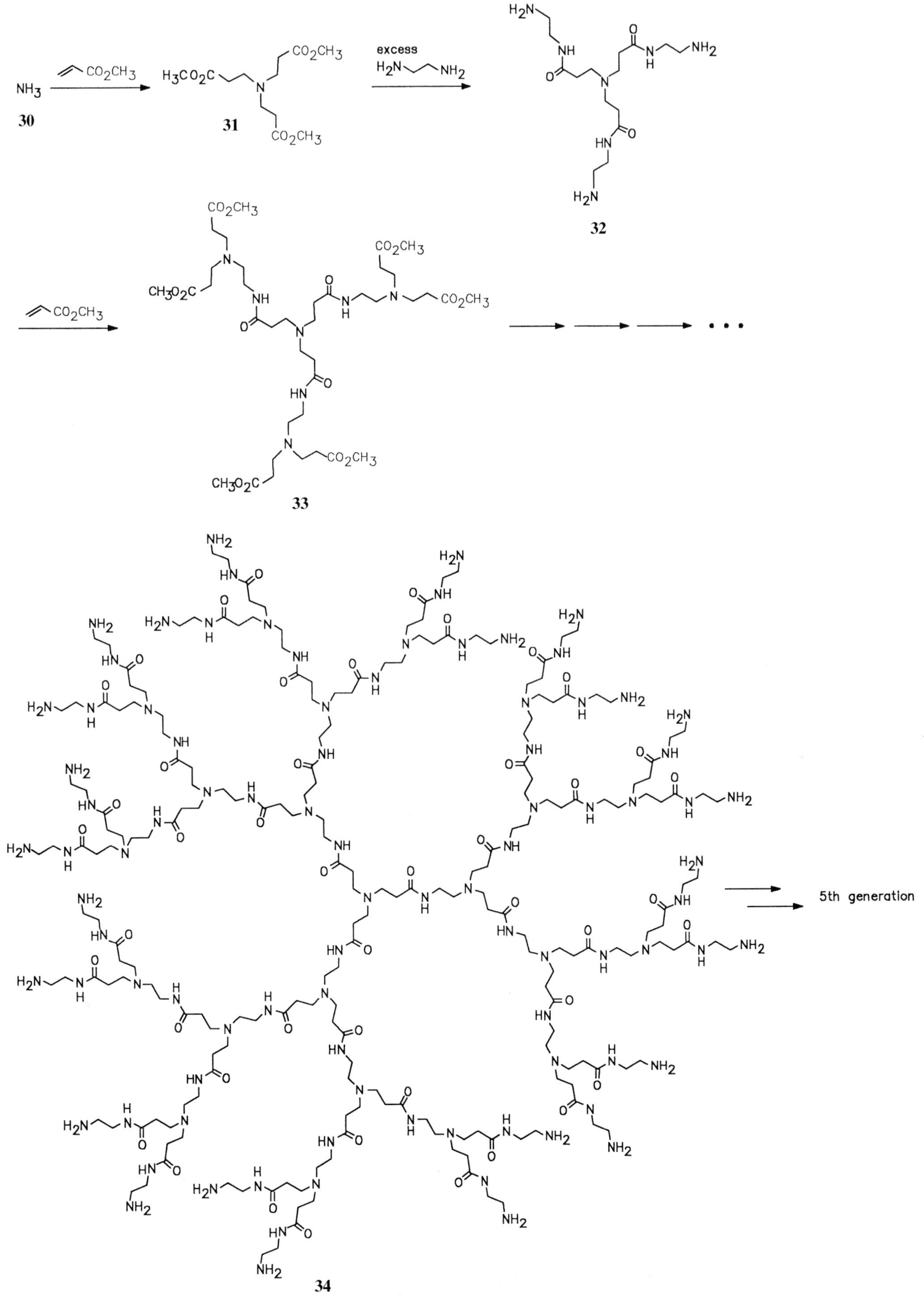

Scheme 3.10 Procedure for the preparation of PAMAM polymers.[118]

been investigated[151] under steady shear, creep, and dynamic oscillatory shear at various temperatures; a model was proposed that "involves dynamics of structural elements that are smaller than the overall dendrimer molecules."

Dvornic and Tomalia[137] elaborated on the "genealogically directed" synthetic nature of the PAMAM preparative protocol. Essentially, this protocol is comprised of an "excess monomer method" that facilitates the isolation of (high purity and high yield) dendritic intermediates (i.e., generations) without loss due to potential side reactions that may occur with reagents that were not intended to be incorporated into the dendritic structure. Thus, the true "genealogy" of the series can be examined from generation to generation; electrospray mass spectrometry was used for the purpose. These authors[152, 152a, b] further published a treatise describing the use of PAMAMs as well as other dendritic systems in a conceptual approach to nanoscopic chemistry and architectures.

Morgan, Stejskal, and Andrady[153] investigated the free volume of cross-linked polymers [poly(propylene glycol) cross-linked with tris(*p*-isocyanatophenyl)thiophosphate] and PAMAMs by the absorption of ^{129}Xe. A measure of the fractional free volume ("void volume") in macromolecular architectures can be obtained from xenon chemical shift values, which are proportional to collision rates within free-volume regions. Cross-linked poly(propylene glycol) exhibited typical polymer network-based Xe chemical shifts (δ = 217.2 to 222.2 for prepolymer average molecular weights increasing from 670 to 2470 amu), while the PAMAMs showed an essentially linear xenon chemical shift increase from δ = 214.8 to 229.2 for generations 3 through to 8. Thus, the fractional free-volume in the PAMAM series was shown to decrease with increasing generation. However, a need for caution was noted because these data are not indicative of the location of the free-volume within the dendritic structure [i.e., core *vs.* outer region(s)].

Related dendrimers possessing primary amine functions were synthesized[154] from hexaacrylonitrile, a by-product in acrylonitrile polymerization, which was first converted to the corresponding hexaester. The PAMAM-like process involved aminolysis with ethylenediamine and Michael addition with ethyl acrylate; the sequence was repeated to create the higher generations. Unsymmetrical PAMAMs have been crafted *via* a divergent/divergent strategy[154a] and carbamate analogs are known.[154b]

The synthesis of a series of PAMAM wedges on a 1,4,7,10-tetraazacyclododecane core has been accomplished by the standard two-step approach; the acetylated dendrimers were treated with 1-bromoacetyl-5-uracil to generate the dendrimer-5FU conjugates, subsequent hydrolysis (pH 7.4; 37 °C) of which released the free 5-fluorouracil.[155]

3.3.2.1 Surface Modifications

Surface functionalization has most commonly been performed on PAMAMs, since these were among the first commercially available dendrimers. The attachment procedures generally follow traditional routes.

Shinkai and coworkers[156] coated PAMAMs with anthracene and phenylboronic acid units (Scheme 3.11); these dendrimers (**37**) act as "saccharide sponges". Pertinent transformations include imine formation by reaction of anthracenecarboxyaldehyde with the terminal amine units, followed by reduction (NaBH₄) to afford saturated C–N bonds (**35**), and benzyl halide displacement with the attachment of the boronic acid moiety (**36**). The resulting dendritic boronic acid was shown to form stable complexes with D-fructose, D-galactose, and D-glucose. In contrast to other flexible diboronic acid binding sites that show weaker binding compared to more rigid, preorganized hosts, the dendritic boronic acids form remarkably stable complexes. This was rationalized by proposing that the increased number of binding sites allows for "assisted" stabilization from any one of the seven remaining sites.

Miller et al.[157,158] described the surface-coating of the PAMAM series with aryl diimides capable of forming anion radicals[159, 160] The vis/IR spectra of the generated anion radicals in this series suggested the formation of diimide "π-stacks," and cyclic voltammograms further supported a diimide aggregation phenomena. Subsequent investigation[161] indicated that the dendrimer-based π-stacked diimide network established a path for electrical conductivity, as opposed to the phenomenon being simply ionic. Films cast

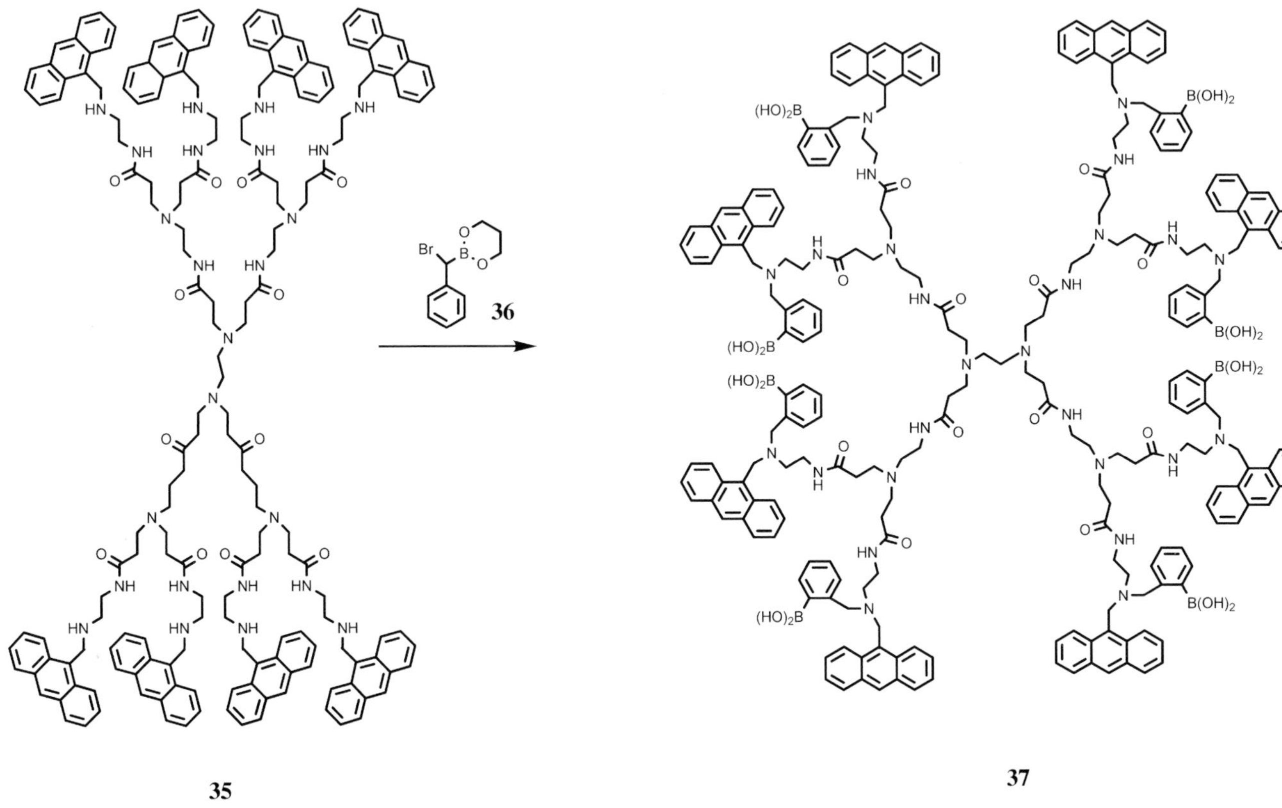

Scheme 3.11 Synthesis of a dendritic "saccharide sponge" (**37**).[156]

at 60 °C possessed conductivity values ($\sigma = 2 \times 10^{-3}$ S cm^{-1}) ten times greater than those cast at 120 °C, suggesting that stacking is improved at lower temperatures. It was postulated that the "3-D" features of dendrimers render the "isotropic nature of these films of particular interest."

PAMAMs have been peripherally modified[162, 163] using radical anions that form π-dimers and π-stacks. The anion radical units were prepared by reaction of naphthalene dianhydride with one equivalent of 4-aminomethylpyridine (125 °C, DMA, 18 h) followed by reaction with SOCl$_2$ (DMF) to afford the pyridinium hydrochloride monoamide, which was subsequently treated with MeI to give the methylpyridinium iodide salt. This monoanhydride pyridinium salt was then attached, by a similar anhydride to imide conversion, to the PAMAM's terminal amines at generations 1–6. Terminal conversions, or "loadings", greater than 70 % were realized. Following electrochemical or sodium dithionite reduction of the modified dendrimers, near-infrared (NIR) spectroscopy indicated π-dimer and π-stack anion radical associations in D$_2$O and formamide solutions. Cyclic voltammograms of these materials further suggested the occurrence of aggregation; however, the size (generation) of the dendrimer was found to have little effect on the extent of this.

Miller et al.[164] later reported additional results from investigations of the electrically conducting PAMAMs. Fully reduced (1.1e/diimide) films of these cationically substituted, naphthalene diimide modified, 3rd generation PAMAMs exhibited conductivities of 10^{-3} S cm^{-1} (σ), while half-reduced films (0.55 e/diimide) showed $\sigma = 10^{-2}$ S cm^{-1}. Conductivity increased as a function of humidity approaching 18 S cm^{-1} at 90 % humidity. NIR spectroscopy, X-ray powder diffraction, and quartz crystal microbalance assessment suggested that water absorption plasticizes the film and facilitates more rapid stack-to-stack electron transfer thus increasing conductivity. Similar studies were later reported for films prepared using the 1st through to the 5th generation PAMAMs.[165]

Turro et al.[166] studied the aggregation behavior of methylene blue adsorbed at the surface of half-generation, anionic PAMAMs. The dye was proposed to aggregate in a perpendicular fashion relative to the dendritic surface.

Antibodies have been coupled[167] to PAMAMs; these reagents offer an attractive approach to the development of immunoassays, since they are stable and retain full immunological activity, both in solution as well as when bound to a solid substrate; their analytical sensitivity is equal to, or better than, that in established methods. The dendritic approach to radial partition immunoassays has been shown to possess the best features of both homogeneous and heterogeneous immunoassay formats. Dendrimer thin films biosensors for live bacteria detection have been described.[167a]

Haensler and Szoka[168] reported the use of PAMAMs for the "transfection" of, or the introduction of DNA into, cultured cells. PAMAM-mediated transfection thus has ramifications in the area of hereditary disease treatment by gene therapy. Transfection was found to depend on the dendrimer to DNA ratio and dendritic diameter; a 6:1 terminal amine to phosphate charge ratio and a cascade diameter of 68 Å was found to maximize transfection of firefly luciferase, used as an expression vector. When membrane disrupting amphipathic peptides (GALA[169]) were attached to the periphery of the cascade (e.g., an average of 13 GALA residues/5th tier, with the 96 amine dendrimer), DNA transfections increased significantly using a 1:1 dendrimer/DNA complex.

Later, Szoka and coworkers[168, 170] reported that "degraded" or "fractured" cationic PAMAMs mediate high levels of gene transfection in a variety of cells. Fractured PAMAMs were prepared by heating them in a solvolytic solvent, such as H_2O, *n*-BuOH, *sec*-BuOH, or 2-ethoxyethanol. This process introduces branch defects by cleavage of the amide connectivity or by retro-Michael reactions to afford a material possessing a wide range of molecular weights (<1500 to >10,000 amu). It was determined that the higher molecular weight components were responsible for facilitating the observed transfections, suggesting that the fracturing process increased dendrimer flexibility allowing the contraction and expansion necessary for DNA capture and release, respectively. DeLong et al.[171] obtained similar results pertaining to intracellular delivery, except that the 3rd generation PAMAMs were found to be efficient oligonucleotide coupling agents. These specifically altered PAMAM mixtures have further been demonstrated to efficiently transfer genes into mammalian cells,[172, 173] murine cardiac grafts,[174, 174a] *Plasmodium falciparum*,[175] lung epithelial cells,[176] hairless mouse skin,[177] as well as selectable and nonselectable marker genes into RPE host cell DNA.[178] DNA dosage and charge ratios to the dendrimer, as well as dendrimer generation, were found to be critical parameters that need to be optimized for each model system under consideration. In a related report,[179] the efficiency of PAMAMs as masking membrane modifiers for DNA-based gene transfer due to their membrane-destabilizing action, was found to be only slightly enhanced by the use of endosomolytic agents. Fractured 6th generation PAMAMs have also been used in the study of gene delivery systems (i.e., transfection and related physicochemical investigations) with extracellular glycosaminoglycans.[180] Although transfection efficiencies of cationic polymers, e.g., the PAMAMs[168] and poly(lysines),[181] depend on the degree of polymerization, the sizes (10,000–60,000 amu) of the related poly(allylamine)s do not interfere with their ability to mediate gene transfer into cells *in vitro*.[182]

Baker and coworkers[183] reported studies concerning complex formation between PAMAMs and DNA based on electrostatic interactions between negative phosphate groups and positive protonated amine moieties. Attraction and repulsion appeared to be a function of generation. Transfection was found to be effected by low-density soluble complexes that constituted only 10–20 % of the total DNA aggregate. Hélin et al.[184] also investigated the use of PAMAMs as vectors for the intracellular distribution of oligodeoxynucleotides; a cell-cycle phase dependence was found. DNA–PAMAM complexes have been studied[185] using ethidium bromide as a probe for the binding interactions. DNA was observed to "wrap" around higher generation materials, i.e., generation 7, but not lower generations, i.e., 2 and 4.

Baker et al.[186] studied gene expression regulation by PAMAM-mediated *in vitro* transfection of antisense oligonucleotides and antisense expression plasmids. Using this technique, cell lines that permanently expressed the luciferase gene were obtained. On the other hand, antisense oligonucleotide or *c*DNA plasmid transfection led to dose-dependent luciferase expression inhibition. These PAMAMs were also evaluated[174] as to whether they could augment plasmid-mediated gene-transfer efficiency in a murine cardiac transplantation model; they were shown to increase the efficiency of transfer and expression.

Tumor-specific targeting monoclonal antibodies (MoAbs) have been coupled to boronated starburst dendrimers and evaluated for their effectiveness against the murine B16 melanoma using BNCT techniques.[187, 188] The synthesis was effected employing amine-terminated PAMAMs and an isocyanate-based boron reagent, namely $NaMe_3NB_{10}H_8NCO$. However, it was concluded that "starburst dendrimers do not seem to be well-suited for the linkage of boron to MoAbs for immunotargeting if they have to be administered systemically". Nevertheless, it was noted that they might be more useful as "linkers" in immuno-electron microscopy.

The preparation of an amphipathic peptide-modified cascade (**40**, Scheme 3.12) was facilitated[167] by reaction of the polyamine dendrimer with N-succinimidyl 3-(2-pyridinyldithio)propionate (SPDP; **38**) to give the polypyridinyldisulfide (**39**). Subsequent treatment of the pyridinyldisulfide moieties with the GALA peptide possessing a cysteine amino acid afforded the modified cascade (**40**). Wu, Gansow, and coworkers[189] reported metal-chelate–dendrimer–antibody hybrids for use in radioimmunotherapy and imaging. PAMAMs were peripherally modified with a slightly less than stoichiometric amount of metal chelator (i.e., DOTA or DTPA). The tumor-targeting antibody 2E4 was then connected to the remaining amine moieties.

Barth et al.[188] reported the preparation of boronated dendrimer–monoclonal antibody immunoconjugates as a potential delivery system for BNCT.[187, 188, 190–193] Dendrimers have also been employed as "linker molecules" for the covalent connection of synthetic porphyrins to antibodies.[194, 195] The commercially available PAMAMs have been utilized as a dendrocore for surface modification, e.g. for glucodendrimers (see Chap-

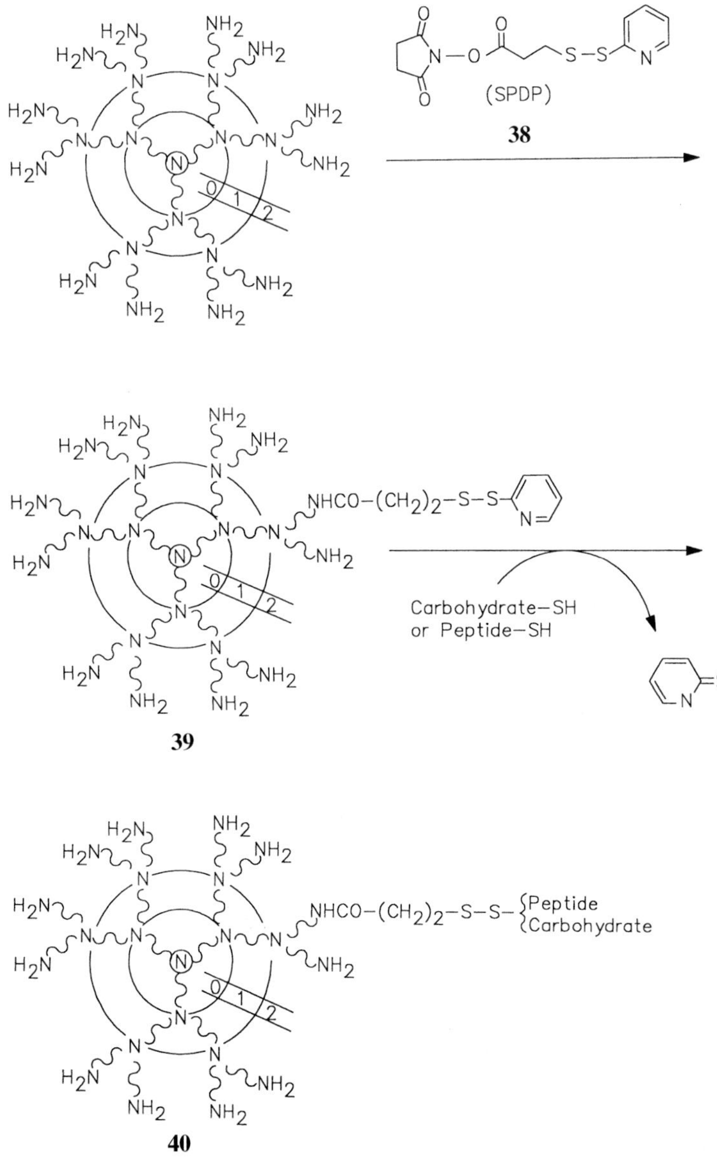

Scheme 3.12 Use of a disulfide linker for the non-site-specific connection of amphipathic peptides and carbohydrates to amine-terminated dendrimers.[167]

ter 7) from *p*-isothiocyanatophenyl sialoside;[196] these constructs were successfully employed for probing multivalent carbohydrate–lectin binding properties. In order to eliminate the retro-Michael reaction associated with the PAMAM family, Zanini and Roy[197, 198] devised and introduced a cleaver modification into the α-thiosialo dendrimers.

Kim et al.[199] used PAMAMs as scaffolding for combinatorial library construction (Scheme 3.13); the potential of this technique, termed "dendrimer-supported combinatorial chemistry," was demonstrated through the production of a single species and a small library of modifications on the dendrimer periphery. To facilitate simple ester cleavage of the surface constructs after their preparation, 4-hydroxymethylbenzoic acid was coupled (EDC) to the eight dendritic terminal amines (generation 1); this provided the base-labile terminal hydroxy attachment starting points. Single molecule construction was exemplified by a sequential three-component synthesis of a biologically active indole. Thus, *N*-Fmoc-protected-L-phenylalanine was coupled to the hydroxy-terminated dendrimers (EDC/DMAP) and then deprotected (piperidine/DMF) to yield the "supported" amino acid **41**. Following reaction with 4-benzoylbutyric acid (pyBOP; benzotriazole-tris-pyrrolidinophosphonium hexafluorophosphate/HOBt; 1-hydroxy-benzotriazole/DIPEA), phenylhydrazine hydrochloride was added (Fischer indole conditions; AcOH/ZnCl$_2$) to complete the multi-component synthesis (i.e., of **42**). After isolation, indole cleavage (MeOH/NEt$_3$) afforded the dendrimer, which was removed by filtration, and the desired phenylalanine-based indole **43**.

In an analogous manner, a split synthesis protocol[200] was employed to generate a 3 × 3 × 3 (27 component) combinatorial library. Pertinent features of these methods include solution-phase chemistry, homogeneous purification, intermediate characterization, and high support loadings.

Small hydrazide-terminated PAMAMs were prepared[201] for use as cross-linkers in hydrogels created from hyaluronic acid.

Margerum et al.[202] treated the surface of PAMAMs (generations 2–5) with gadolinium chelating moieties and polyethylene glycol units for an investigation of the effect of molecular weight on the biological and physical properties of MRI contrast agents.[203]

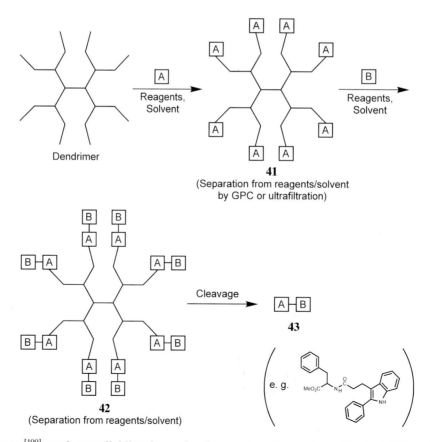

Scheme 3.13 Dendrimers[199] used as scaffolding for molecular construction and combinatorial libraries.

A microreview concerning MRI contrast agents has recently appeared.[204] Isothiocyanate-based coupling of the chelates [1-(4-isothiocyanatobenzyl)amido-4,7,10-tetraacetic acid–tetraazacyclododecane; DO3A-bz-NCS] to the dendrimers followed by Gd^{3+} complexation yielded water-soluble materials possessing, for example, 11 Gd^{3+} complexes/3rd generation PAMAM and 57 Gd^{3+} complexes/5th generation PAMAM, based on wt. % Gd as determined by inductively coupled plasma (ICP) absorption spectroscopy. For generations 2 and 3, unreacted terminal amines were treated with poly(ethylene glycol) (both the PEG$_{2000}$ and PEG$_{5000}$ analogs were prepared for each generation) in an effort to enhance biocompatibility. Using NMR dispersion, non-PEG substituted dendrimers exhibited peak relaxivities as high as 18.8 $mM^{-1}s^{-1}$ (25 MHz); the magnitude showed a linear increase with molecular weight. Half-life blood elimination in rats also increased with molecular weight, ranging from 11 (±5) min. for the 3rd generation to 115 (±8) min. for the 5th generation. Liver retention over a seven day period also showed an increase with molecular weight (i.e., 1–40 %). PEG-modified polychelated relaxivities ranged from 11 to 14.9 $mM^{-1}s^{-1}$ (2nd and 3rd generation only), while blood elimination half-lives dramatically increased to as long as 1219 min; seven day liver retention decreased to 1–8 % per dose. In this case, PEG modifiers enhanced the biodistribution and pharmacokinetics of the MRI contrast agents. PAMAMs peripherally modified with macrocyclic gadolinium(III) complexes for use as MRI contrast agents have been examined by Merbach et al.[205] Rotational correlation times were found to be 4 to 8 times greater for these dendritic agents when compared to smaller mono- or bis-complexes; however, the observed low H_2O exchange rates prevented a corresponding increase in proton relaxivities. The synthesis, and evaluation as MSI agents, of a related series of polymeric gadolinium complexes, including the surface-coated, 3rd generation PAMAM, have been reported;[206] the blood half-life for the globular polymers was found to be longer than that of Magnevist, but considerably shorter than that of the linear DOTA polymer. An alternative reagent [2-(4-isothiocyanatobenzyl)amido-1,4,7,10-tetraacetic acid-tetraazacyclododecane; p-NCS-bz-DOTA][207] was also used to functionalize the higher level PAMAMs.[189, 208]

Layer-by-layer film formation based on biomolecule interactions and dendrimer architecture has been reported.[209] Films were composed of the branched architectures of either biotin-labeled PAMAM (4th generation, 14,215 amu) or biotin-labeled PEI ($\overline{M}_w$ of 45,000 amu) and avidin. PAMAM-based films were shown to facilitate *mono*layer deposition of avidin, whereas randomly branched PEI facilitated only *multi*layer avidin deposition. Biotinylated PAMAMs were prepared and evaluated[210] for use in cancer "pretargeting" protocols aimed at increasing the amount of bound radionuclei on cancer cells. *In vivo* studies showed that the modified dendrimers had cleared the bloodstream within 4 hours of administration. Additional animal studies were deemed useful.

Cross et al.[211] terminally modified PAMAMs using tetracyanoquinodimethane (TCNQ), 2-naphthoyl chloride, and 4-nitrobenzyl chloride to prepare materials with potentially novel electronic or nonlinear optical properties.

Gray and Hsu[212] synthesized sulfonic acid-terminated PAMAMs (generation 2) by reacting the amine termini with 1,4-butane sulfone. Their modified materials were then used as a pseudostationary phase in aqueous electrokinetic capillary electrophoresis to separate a series of neutral isomeric phenols. Better separation was achieved compared to that using sodium dodecyl sulfate as the pseudostationary phase.

The modification of PAMAM carboxylate ester- and amino-terminal moieties has been described[213] in the context of using these modified peripheries for attachment to alkaline phosphatase.

Balogh et al.[214] polymerized N-methyl-2-ethynylpyridinium triflate on the surface of generation 3 and 4 amine-terminated PAMAMs resulting in macromolecular "sea urchins". These core–shell copolymers showed conductivities in the range of 10^{-4} to 10^{-5} S cm^{-1} when doped, as well as enhanced dendrimer thermal stability.

The 2nd and 3rd generations of PAMAMs have been stiochiometrically methylated (MeI) to afford the internal and external quaternized materials. These were found to act as conductors, exhibiting conductivities of 10^{-5} to 10^{-6} S cm^{-1} at ambient temperature.[215] Plots of conductivity *vs.* temperature were presented; their thermal stability was not addressed.

Mitchell et al.[216, 217] converted (tris, DMSO, K_2CO_3) the ester termini of half-generation PAMAMs to their polyhydroxy derivatives to study their solubilizing potential for use in drug delivery, in essence transforming the PAMAM dendrimers to arborols.[8] Inclusion complexes with a variety of hydrophobic molecular guests were reported. For example, a 1:1 complex with benzoic acid was found to increase the aqueous solubility of the guest from 2.9 mg/mL to 305 mg/mL at neutral pH. Treatment of the simple ester-terminated PAMAMs with diethanolamine generated a water-soluble dendrimer, which, upon treatment with benzyl bromide, underwent random quaternization at only four of the six internal nitrogens.[218] As well, amine-terminated analogs have been used for aminolysis accelaration.[218a]

Leblanc et al.[219] reported the surface functionalization of the 3rd generation PAMAM with 10,12-pentacosadiynoic acid; the resulting materials formed colloidal particles in $CHCl_3$ and were polymerized under UV irradiation.[219a, b]

Gong et al.[220] described ionic conductive PAMAMs of generations 2.5 and 3.5, prepared by terminal alkali metal (Li^+, Na^+, or K^+) carboxylation. These exhibited a conductivities of 10^{-5} to 10^{-6} S cm^{-1} at 30 °C; the temperature dependence conformed neither to the WLF mechanism nor to the Arrhenius equation.

A temperature-sensitive dendritic host has been formed by Kimura et al.[221] Essentially, a 32-amine dendrimer was coated with 11-(thioacetyl)undecanoic acid, the thiol was deprotected, and then subjected to N-isopropylacrylamide polymerization to impart temperature-dependent solubility. Encapsulation of water-soluble Co(II) phthalocyanine complexes was studied with respect to catalytic activity.

An interesting comparative study[222] of PAMAMs, PPIs (with two different cores), and poly(ethylene oxide) grafted carbosilane dendrimers was systematically carried out in order to probe the relationship between generation size and surface functionality and biological properties *in vitro*. This evaluation of the individual structural components of the various dendrimers provided a first insight into the parameters that are critical with regard to the rational design and tailoring of macromolecules for drug delivery. Lipid carriers with a membrane-bound active component and small complex size proved necessary for the efficient cellular delivery of phosphorothioate oligonucleotides.[223]

The use of PAMAMs as a spherical core that is coated with organosilicon groups has generated a family of materials possessing a hydrophilic core with a hydrophobic exterior;[224] this novel family of dendrimers has been abbreviated as PAMAMOS. Their synthesis has been accomplished by either (a) Michael addition of organosilicon acrylates or methacrylates to the PAMAM polyamine surface, or (b) haloalkylation with chloroalkyl- or iodoalkylsilanes.

Crooks et al.[225] also demonstrated the fascinating self-assembly of fatty acids about the surface of PAMAMs to afford an ordered ionic array of terminally arranged aliphatic chains. Ramifications included solubility in nonpolar solvents as well as dye and metal encapsulation.

3.3.2.2   Physical Properties

Turro et al.[226–228] characterized the PAMAMs utilizing fluorescence spectroscopy, whereby pyrene was used as a photoluminescent probe to evaluate the internal hydrophobic sites. Their results were consistent with the theoretically predicted morphology of this family of dendrimers; thus, at generations 0.5–3.5 (carboxylate surface) the structures are open, but at generations 4.5–9.5, they are closed, and the surface becomes congested with increasing generation. Photoinduced electron transfer between species associated with the surface carboxylate groups also supports the structural change at the 3.5th generation.[229] The dynamics of electron-transfer quenching of photoexcited Ru(phen)$_3^{2+}$ were evaluated using methyl viologen in solution with the anionic PAMAMs.[230] These results further supported the structural change at 3.5th generation, and despite the structural differences with micellar aggregates, striking similarities were demonstrated. Reviews[231, 232] of these studies have appeared and should be consulted for details.

Binding constant data of Ru(phen)$_3^{2+}$ to carboxylate-terminated PAMAMs have been collected.[233] Increased excited-state lifetimes were attributed to a decreased O_2 concentration at the dendritic surface relative to that in the aqueous media. Quenching rates of excited Ru(phen)$_3$$^{2+}$ by dendrimer-bound Co(phen)$_3$$^{3+}$ were found to be independent of the quencher concentration and the process to be *intra*dendrimer in nature.

A statistical analysis of luminescence quenching of Ru(II)-tris(phenanthroline) by Co(III)-tris(phenanthroline) at the PAMAM surface has been carried out by Bossmann et al.[234] Their results suggest that the donor and acceptor complexes bind in a non-random manner preferring adjacent juxtaposition. The necessary "energy of attraction" for this phenomenon results from hydrophobic interactions. The emission intensity and emission lifetimes of Ru[4,7-(SO$_3$C$_6$H$_4$)$_2$-phen]$_3$$^{4-}$ in solutions containing cationic PAMAMs in the presence and absence of potential quenchers have been investigated;[235] quenching constants between this complex with the dendrimer and methyl viologen, K$_4$Fe(CN)$_6$, or K$_3$Fe(CN)$_6$ were determined by laser flash-photolysis.

Dubin et al.[236] studied complex formation between carboxylated PAMAMs and poly(dimethyldiallylammonium chloride) to gain insight into colloid diameter, or surface curvature, upon polyelectrolyte binding. Observations based on turbidity experiments were consistent with those of Turro et al.;[230–232] higher generation dendrimers behave as "hindered Stern layer structures", whereas early generations exhibit "simple electrolyte characteristics". Dubin and coworkers[237] investigated the binding interactions of carboxylic acid-terminated PAMAMs (generations 0.5, 3.5, 5.5, and 7.5) with the same ammonium salt, whereby complex formation was observed to occur readily with generation 7.5. This was attributed to its high charge density compared to that in the lower generations. Analysis of pH titration curves yielded estimates of the free energy of complexation. The use of half-generation PAMAMs as calibration standards for aqueous size-exclusion chromatography has also been examined.[238, 239] A correlation between the chromatographic partition coefficient and the generation number was observed, due, in part, to the relationship between generation number and molecular volume. Ionic liquid crystalline properties have been described.[239a]

Ottaviani et al.[240] conducted extensive EPR studies on half-generation PAMAMs possessing carboxylate surface functionalities. Positively charged nitroxide radicals, attached to carbon chains of variable length, were used to evaluate the hydrophobic and hydrophilic binding loci. Mobility (τ_c) and polarity (A_n) parameters as a function of pH demonstrated electrostatic interactions in the binding at the dendritic surface–water interface. The radical chain was shown to intercalate in the inner regions of the dendrimer and to interact at hydrophobic sites. Activation energies for the rotational motion of the probe were also determined. The aggregation behavior of a positively charged, TEMPO-functionalized, C$_{16}$ surfactant (CAT16) in the presence of half-generation carboxylic acid terminated PAMAMs has been evaluated by EPR.[241] For low aqueous concentrations of early generation dendrimers (G < 3.5) smaller than or of comparable size to the CAT16 micelles, dendrimer "guests" were postulated to bind to micellar "hosts", whereas for low concentrations of later generation dendrimers (G > 3.5), micelles can act as "guests" of the "host" dendrimers. Bilayer surfactant-based aggregates at the dendritic periphery were proposed to extend along bridged dendrimer–bilayer–dendrimer complexes. Interactions of anionic (half-generation, carboxylic acid-terminated) PAMAMs and cationic surfactants have been investigated by an EPR spin probe method.[242] The probes used were alkylated (C$_{12}$, C$_9$, or C$_{16}$) TEMPO derivatives, while surfactants included dodecyltrimethylammonium bromide (DTAB) and cetyltrimethylammonium bromide (CTAB). The data suggested dendrimer–substrate models of two types, that being preferred depending on substrate concentrations, PAMAM size, and temperature. Surface adsorption of molecular probes by the PAMAM constitutes a "primary structure", while monomer adsorption in surfactant aggregates adhered to the PAMAM periphery gives a "secondary structure."

Turro and coworkers[243] employed their alkyl core-based PAMAMs in host–guest complexation of the dye Nile red. A hydrophobic core created by a C$_{12}$ alkyl chain was observed to significantly enhance Nile red fluorescence emission in aqueous media, while PAMAMs created with shorter chains or ammonia cores showed no effect. The addition of anionic surfactants greatly increased the accommodation of Nile red due to the formation of a dendrimer–surfactant supramolecular assembly. Santo and Fox[244] analyzed the interactions between smaller PAMAMs and several biologically relevant guests; association constants suggested that two interaction sites were possible, in the interior and at the periphery.

Ottaviani et al.[245] examined the interaction of PAMAMs with dimyristoylphosphatidylcholine vesicles. Structural modification of the vesicles was monitored by continuous wave (CW)- and pulsed-electron paramagnetic resonance (EPR), using doxyl-modified stearic acids as probes. Vesicle bilayer integrity was found, supporting their potential use as drug and gene carriers. Ottaviani et al.[246] characterized these vesicles, as well as mixtures with the PAMAMs, again by CW- and pulsed-EPR, and also by TEM and dynamic light-scattering with the goal of obtaining information on the potential use of PAMAMs as gene carriers. Nitroxyl radical probes were attached to the surfaces of the 2nd and 6th generation dendrimers to facilitate EPR measurements. Results showed that dendrimer–vesicle interactions (1) modify the fast rotation axis of the radical, (2) are stronger for higher generation dendrimers, (3) are such that the mean vesicle size seemingly remains unchanged, and (4) are stronger with protonated dendrimer forms.

Ottaviani et al.[247] further reported an ESR investigation of the interactions of polynucleotides with nitroxide-labeled PAMAMs. Double-stranded polynucleotides used included Calf thymus DNA, Poly(AT), Poly(GC), and 12mer-DNA. Results showed (1) a better "wrap" of DNA with large, high generation structures, (2) that increasing protonation of large dendrimers decreased DNA interaction, and (3) that increasing protonation of small, more flexible structures led to enhanced DNA interactions; also see.[247a]

De Gennes and Harvet[248] statistically found that these "cauliflower polymers"[9] (dendrimers) exhibit restricted idealized growth, also known as 'dense packing', when the number of generations $m = m_1$, where $m_1 \cong 2.88(\ln P + 1.5)$. This relates in space to the limiting radius R_1, which increases linearly with P monomers. Below this limit, the radius $R(M)$ of the dendrimer, when plotted as a function of molecular weight (M), should increase $(m^{0.2})$, whereas above this limit $(R < R_1)$ compact structures $(R \approx M^{0.33})$ should result.

Murat and Grest[249] performed molecular dynamics studies of dendrimers using a course-grained model in solvents of varying quality. The dendrimers were found to possess a space-filling, or compact structure under all solvent conditions and the radius of gyration was found to be proportional to the number of monomers (i.e., $R_G \propto N^{1/3}$). High generation dendrimer density profiles, under all solvent conditions, showed core regions of high density, probably due to back-folding of the outer segments. At maximum size, density was nearly uniform throughout, while for low generations $(0 \leq g \leq 4)$ higher density was found to be localized near the surface, or termini; this is in agreement with earlier work.[250, 251]

Using Gaussian monomer–monomer interactions, Hammouda determined structure factors for dendrimers,[252] polymer gels, and networks.[253] Considerations included intrabranch self-correlation, intrabranch cross-correlation, and interbranch correlations.

La Ferla[254] examined dendrimers and cascade macromolecules using an extended Rouse–Zimm discrete hydrodynamic model. Pertinent dynamic parameters were studied on the basis of criteria such as local stiffness, topology, and generation. Good agreement with the results of Mansfield and Klushin[142] was also found with respect to moderate local increases in stiffness, accounting for dendritic molecular dimensions and intrinsic viscosity.

Mansfield[255] also investigated, by means of Monte Carlo calculations, the segregation of individual branches at equilibrium. It was postulated that in solutions of the highest attainable generation in poor solvents, or in neat dendrimer fluids, branch segregation disappears. Fractal dimensions for dendrimers were determined to be in the 2.4 to 2.8 range. Chen et al.[256, 257] reported the results of Monte Carlo simulations on starburst dendrimers. The size of these branched macromolecules, in the scaling regime, was determined to be proportional to $(P_g)^{1-\nu}N^{2\nu-1}$, where P is the number of spacer bonds, g the generation, N the total number of monomers, and ν the scaling exponent. The scaling exponent for the starburst dendrimers was found to have a similar value as that for linear polymers. Physical properties in the high-concentration regime were discussed, i.e. in the regime where the scaling law is not valid.

Mansfield[258] examined molecular weight distributions of imperfect dendrimers and thus their relationship to hyperbranched materials. Fourier analysis of these distributions revealed that essentially monodisperse molecular weight ranges could be obtained (for divergent growth) if, at the early stages of growth, perfection was maintained, or nearly so, while later generations inevitably possess arbitrary amounts of defects. This hypothesis has been borne out by the recent report of a one-step preparation of PAMAM-like dendrimers by Hobson, Kenwright, and Feast.[259]

These PAMAMs have been subjected[260] to two different fractal analyses:[261, 262] (a) $A \approx \sigma^{(2-D)/2}$, where A is the surface area accessible to probe spheres of cross-sectional area, σ, and D is the surface fractal dimension, which quantifies the degree of surface irregularity, and (b) $A \approx d^D$, *where d is the object size.* Both methods give similar results with $D = 2.41 \pm 0.04$ (correlation coefficient = 0.988) and 2.42 ± 0.07 (0.998), respectively. Essentially, the higher generation dendrimers are porous structures with a rough surface. For additional information on dendritic fractality, see Section 2.3.

Yu and Russo[263, 264] reported the fluorescence photobleaching recovery and dynamic light-scattering characterization of these PAMAMs.[9] Agreement between the two techniques suggests that the attachment of a fluorescent dye does not significantly change the diffusion coefficient of the 5th generation dendrimer. At high salt concentrations, the measured hydrodynamic diameters obtained by application of the Stokes–Einstein equation are close to the reported diameters as determined by size-exclusion chromatography (SEC).

Optimal spline cut-offs for coulombic and van der Waals interactions have been addressed[265] using computational methods, including those implemented in CHARMM and AMBER.

Boris and Rubinstein,[266] employing a self-consistent mean field model for starburst dendrimers, examined the dense core *vs.* dense shell models. Their data show that for flexible dendrimers the cores are dense, not hollow, and that the density decreases on progressing outward toward the surface. This is in agreement with other pertinent simulations.[142, 249–251] The authors further noted that flexible spacers distribute the density and facilitate the construction of larger species (i.e., push dense-packing limitations to higher generations – a phenomenon that has been addressed and demonstrated by Moore[267] by means of his SYNDROME construction approach).[268] Also, a Flory theory was invoked to account for the correspondence between dendrimer size and generation.

Mansfield[269, 269a] further studied flexible model dendrimers and their surface adsorption characteristics. Computer simulations of lattice model dendrimers were analyzed for their interaction with an adsorbing planar surface. It was observed that for an increase in interaction strength (A), dendrimer flattening and spreading occurred. For low dendrimer generation (G) and high A, most of the dendrimer components can access and contact the surface due to the ease of molecular deformation. For high G, however, a smaller percentage of the overall dendrimer is able to access the surface due to decreased deformability. Different adsorption states were observed; thus, a tridendron-based dendrimer was either completely adhered to the surface (S_3 state), or two of the dendron components were adsorbed while the third was arranged more perpendicular to the surface plane (S_2 state).

Amis et al.[146, 270] reported on the spatial distribution and location of PAMAM termini employing SANS. Results obtained using partially deuterated 7th generation dendrimers showed a larger radius of gyration of the terminal groups than that of the entire dendrimer (i.e., 39.3 ± 1.0 Å *vs.* 34.4 ± 0.2 Å, respectively), thus indicating that the termini are located near the outer surface. These findings are at variance with many computer simulations,[250, 251, 255] which suggest that a significant number of termini are back-folded into the interior framework. The effect of solvent quality on the molecular dimensions of these PAMAMs was also investigated;[271] for example, using the solvents $D(CH_2)_nOD$, where $n = 0, 1, 2,$ and 4, the radius of gyration for an 8th generation PAMAM was found to be decreased by 10 % on going from $n = 0$ to 4. Funayama and Imae[272] reported SANS data for a 5th generation PAMAM possessing hydroxyl end groups, which indicated that the greatest density occurs at the 4th generation and that penetrated water reaches a maximum at the 5th generation.

A preliminary biological evaluation of PAMAMs has been conducted.[195] These authors investigated *in vitro* and *in vivo* toxicity along with immunogenicity and biodistribution. Studies were conducted using either V79 cells or Swiss–Webster mice. Of the generations studied (G3, 24 terminal amines; G5, 96 amines; G7, 384 amines), the 7th generation dendrimer was the only construct that exhibited potential biological compli-cations. Biodistribution properties were unexpected, with G3 preferentially accumulat-ing in the kidneys and G5 and G7 showing highest localizations in the pancreas. It was concluded that the use of PAMAMs in biological applications was warranted with close attention to the generation employed, dose, etc., along with further study of the biodis-tribution.

Duncan[273] and Malik[274] described preliminary experiments focused on the biocom-patibility of dendrimers. The 3rd and 4th generation, amine-terminated PAMAMs were found to be cytotoxic towards CCRF and B16F10 cell lines, whereas the corresponding carboxylate analogs were non-toxic. Complexes of generation 3.5 with doxorubicin and cisplatin also displayed *in vitro* cytotoxicity.

Spindler, Tomalia, and coworkers[275] reported the functionalization of the terminal amines of PAMAMs, prepared using diaminoalkyl [with $(CH_2)_n$, $n = 2, 4, 8$, or 12] cores, with various epoxyalkanes, thus rendering them hydrocarbon-soluble, and their behavior as inverse micelles. Micellar behavior was evident from the transport of $Cu(SO_4)_2$ from the aqueous phase into an organic phase (toluene). In the absence of the two-directional dendrimer, no copper ions were transported, as evidenced by a clear organic phase instead of a characteristic blue coloration. Also, Langmuir isotherm data for these materi-als were examined to gain an understanding of the properties at the air–water interface. Typical isotherm data reveal increasing surface pressure with decreasing available area until monolayer collapse resulting in multilayer formation. With additional compression, surface pressure remained constant. This is in contrast to observations by Fréchet,[276, 277] where a nucleation phenomenon was seen, albeit with different dendron architectures.

PAMAMs have been demonstrated[167] to enhance immunoassay sensitivity and, in some cases, to reduce instrumental analysis time.

Solvent-dependent swelling of PAMAMs has been studied[278] by means of holographic relaxation spectroscopy, whereby the influence of solvent quality on molecular dimen-sions was characterized. Low generation dendrimers were observed to possess similar hydrodynamic radii in the various solvents studied, whereas higher generations (G > 4) showed significant swelling in "good" solvents. This was suggested as a tool for the con-trolled trapping and release of guest molecules.

Low generation PAMAMs have been examined[279] by fluorescence experiments with pyrene. Excimer fluorescence was observed at [pyrene]/[dendrimer] concentration ratios as low as 10^{-3}. It was further concluded that the size of the dendrimer varies with the amount of pyrene dissolved in aqueous media.

Esumi and Goino[280] examined the adsorption of PAMAMs at alumina/water and sil-ica/water interfaces. The weight of dendrimer adsorbed was found to increase with increasing generation for both systems. For both alumina and silica dispersions, lower generations behaved as electrolytes, while higher generations exhibited ionic surfactant or polyelectrolyte behavior. The swelling response of the 8th generation PAMAM to poly-electrolytes in D_2O has been evaluated by means of SANS; it was concluded that the PAMAM's size was independent of the charge density or ionic strength of the solvent.[281] Goino and Esumi[282] studied the interaction of these PAMAMs (1/2 generation, CO_2H-terminated) with positively charged alumina particles. Their results correspond well with other reports,[237] indicating that early generation dendrimers behave as electrolytes, while polyelectrolyte or surfactant-like behavior is observed for higher generations. Interactions in aqueous solutions between PAMAMs possessing surface carboxyl moie-ties were investigated using cationic surfactants.[283]

Jackson et al.[284] imaged these dendrimers by conventional TEM; by staining with sodium phosphotungstate, *single* PAMAMs of generations 5 through to 10 were observed. A cryo-TEM technique was also used for generation 10. Circular appearances were noted, with diameters following a Gaussian distribution with increasing generation, although some broadening was observed at higher generations. Cryo-TEM generally supported the standard staining-based TEM, but also suggested dendrimer shape vari-ability (e.g., polyhedral motifs frequently occur).

PAMAMs have also been employed[285] as buffer additives in electropherograms of chicken sarcoplasmic proteins. Full- and half-generation materials were used through to tier 3.5 (anionic). The resolution of the protein electropherograms was shown to improve with concentration and generation of the anionic form; the generation zero cationic form also improved resolution.

Hammond et al.[286, 287] attached PAMAM architecture to one end of poly(ethylene oxide) (2,000 amu) through to the 4th generation and examined the Langmuir thin-film behavior. Dendron peripheries were modified with stearic acid and an aryl vinyl acid. Transfer of monolayers onto hydrophobically functionalized surfaces afforded smooth, continuous, defect-free (holes in the film) films. "Z-type" multilayer films were also examined.

Naka et al.[288, 288a] used carboxylate-terminated PAMAMs as additives in the crystallization of $CaCO_3$ and observed the formation of spherical valerite crystals in contrast to the rhombohedral calcite crystals formed in the absence of the additive. A linear poly(-carboxylic acid) was found to inhibit crystallization.

Wade et al.[289, 289a] have conducted spectrochemical investigations with carboxylate-terminated PAMAMs to evaluate nitromethane as a selective fluorescence-quenching agent. Nitromethane selective quenching of alternate polyaromatic hydrocarbons (PAHs), as opposed to that of non-alternating PAHs, was observed. This was rationalized in terms of differing PAH locations in the dendrimer. Ramifications regarding HPLC separations were postulated.

Photochemical and spectroscopic probes have been utilized for the comparison of nitrogen (trivalent) *vs.* ethylenediamine (tetravalent) core-based PAMAMs.[290] Similar surface characteristics were observed, such as a switch from an "open" to a "closed" architecture at generation 3 to 3.5, thereby suggesting ready extrapolation of earlier findings[166, 226, 229, 233, 241, 291] predicated on the *N*-based dendrimer to the ethylenediamine-based constructs.

Striegel et al.[292] examined and compared the dilute solution characteristics of PAMAMs and PPIs to those of polysaccharides by means of SEC, ESI-MS, and computer modeling. Intrinsic viscosities were observed to decrease in the order $dextran_{[\eta]}$ > $dextrin_{[\eta]}$ > $dendrimer_{[\eta]}$. Solution radii and molecular weights were found to correlate well with commercial literature values.

Solid-state deuterium NMR studies on PAMAM salts aimed at elucidating their structures and H-bonding have been conducted by Malyarenko, Vold, and Hoatson.[293] Variable-temperature dependent studies of generations 1–3, 5, 7, and 9 gave spectra characteristic of amorphous substances showing T_gs in the range 25–65 °C. Interior H-bond lengths were estimated as 2.2 ± 0.15 Å and were independent of generation.

Dvornic et al.[294] reported on the rheological properties of PAMAMs. Medium to high dendrimer concentrations in ethylenediamine solution exhibited typical Newtonian flow behavior. Interpenetration entanglements and surface sticking interactions were not observed.

A simple method to produce dendrimer "nanodots" via solvent evaporation has been considered.[294a]

3.3.2.3 Dendrimers as Attachments

Watanabe and Regen reported[295] the use of these PAMAMs in the preparation of Iler-like arrays.[147, 148, 296] These arrays were constructed on a (3-aminopropyl)triethoxysilane activated silicon wafer by a sequence of exposure to K_2PtCl_4, rinsing, treatment with a solution of the dendrimer, further exposure to K_2PtCl_4, and rinsing. Multilayers were constructed by repetition of the last three steps. Examination of a multilayer coating after five cycles by atomic force microscopy showed the surface to be smooth at the molecular level, with an average roughness of 7.1 Å.

Wells and Crooks[297, 298] attached PAMAMs to *self-assembled monolayers* (SAMs) and subsequently demonstrated their usefulness in the construction of *surface acoustic wave* (SAW) devices. Generation 4 was determined to be the most useful as a mass balance detector due to its globular architecture and readily accessible interior endoreceptors. Crooks et al.[299] subsequently demonstrated that PAMAMs could form high-density monolayers on gold platforms based on metal-terminated amine interactions without the requirement of an interfacial supporting monolayer. Mixed dendrimer–alkanethiol monolayers have also been reported.[300] Single-component dendrimer monolayers were described as "unlike the spherical form", while the presence of hexade-cylthiol appeared to compress the monolayer such that the dendrimer conformation changed to an "end-on oblate spheroid". At pH 11.0, the redox probe $[Ru(NH_3)_6]^{3+}$ is reduced to the Ru^{2+} species on gaining access to the Au surface through the dendritic interior; at lower pH values (e.g., 6.3), the protonated dendritic terminal amines repel the Ru complex, thereby denying it access to the Au surface. Intradendrimer probe transfer was examined by cyclic voltammetry following deactivation of the amine mono-layer surface towards pH effects. These self-assembled dendrimers were described as "molecular gates". Self-assembled films based on small PAMAM carboxylates (G 1.5) and a nitro-containing diazoresin have been fabricated;[301] UV irradiation was shown to cause linkages between layers to switch from ionic to covalent.

Crooks, Bergbreiter, et al.[302] also reported the preparation of highly cross-linked dendrimer–polyanhydride composite thin films. Preparation involved the combination of either amine- or hydroxyl-terminated PAMAMs or PPIs with poly(maleic anhydride)-*c*-poly(methyl vinyl ether) (also known as "Gantrez"). Essentially, the dendrimers were used as building blocks to cross-link the Gantrez copolymer and then as *in situ* thermo-setting agents. Prior to heating, the film permeability was found to be pH-dependent, whereas after heat treatment the films became "highly blocking". Surfaces used for film preparation included Au, Si, and Al. Crooks et al.[303] subsequently reported the use of PAMAMs as adhesion promoters between vapor-deposited gold films and silicon-based materials. AFM has been used to examine PAMAMs of generations 4 and 8 absorbed on gold surfaces.[304] Dendrimer-coated surfaces, upon exposure to the more strongly bind-ing hexadecanethiol, showed a dendrimer morphology change from oblate to prolate; monolayer surfaces showed gradual agglomeration ultimately producing dendrimer pil-lars up to 30 nm in height upon alkyl thiol exposure.[304a]

Bar et al.[305] described a protocol using dendrimer-coated silicon oxide surfaces for the for-mation of Au and Ag colloid monolayers. The 4th generation PAMAMs were adsorbed onto glass, silicon, or indium tin oxide glass surfaces and then treated with the colloidal metals, obtained by reduction (trisodium citrate dihydrate) of $HAuCl_4 \cdot 3H_2O$ or $AgNO_3$ in aqueous solution. AFM, SEM, XPS, SERS, and UV/vis spectroscopy were used to characterize these materials. Dendrimer layer thickness was determined to range from 14 to 25 Å, while intercolloidal spacing could be controlled over a wide range (74–829 nm) by variation of the particle size, concentration, and substrate immersion time. Adsorp-tion on clean gold has also been reported.[305a]

Barth et al.[306] utilized isocyanatododecaborane to modify the surface of a 4th genera-tion PAMAM in a sub-stoichiometric manner, and then subjected the product to maleimide–sulfide coupling of *e*pidermal *g*rowth *f*actor (EGF). The *b*oronated *s*tarburst *d*endrimer (BSD)–EGF conjugate was prepared as a reagent for use in BNCT, which is used to destroy cancer cells by the emission of tissue-destroying low-energy alpha (α) particles at the tumor site. Since increased numbers of EGF receptors are accumulated at the cancer cell surface, it was reasoned that high local concentrations of boron could be delivered by attachment to EFG. Binding of the BSD–EFG conjugate was shown to be EFG receptor specific, although the binding constant (K_A) was found to decrease slightly, presumably due to dendrimer-based steric hindrance.

Singh et al.[307] coupled multiple antibodies to simple PAMAMs for use as multifunc-tional reagents in immunoassays. The lower generation PAMAMs were found to be the most useful, while the 5th generation afforded a "product of unacceptable performance"; this highlights the critical size dependence when dealing with bioproducts.

PAMAMs have been prepared by solid-phase synthesis using a polystyrene–PEG resin by Bradley and coworkers.[308] PAMAM–resin conjugate synthesis involved initial reac-tion of methyl acrylate (2.50 equiv) with a diamine linker followed by removal of the excess reactant and treatment with a 1,*n*-diaminoalkane (250 equiv.; where *n* = 2 or 3). Dendrimers up to the 5th generation were realized. The 3rd generation hybrid was termi-nated with the super acid sensitive linker 4-[4-(hydroxymethyl)-3-methoxyphenoxy] butyric acid via Fmoc chemistry and sequentially treated with lysine and glycine in 1 % TFA/CH_2Cl_2 to produce a dendrimer-bound dipeptide. This demonstrates the utility of these materials in applications such as combinatorial chemistry and chromatography based on high bead loading. Cleavage of the dendrimer from the resin was achieved by treatment with 50 % TFA in CH_2Cl_2. Bradley and coworkers[309] subsequently used these TentaGel(Polystyrene–PEG) bound dendrimers as high-loading solid-phase scaffolds for the synthesis of a library of aryl ethers; ramifications include the potential to significantly increase bead loading.

Tsukruk et al.[310] described the self-assembly of multilayer films built-up of alternating layers of amine- (4th, 6th, or 10th generations) and carboxylic acid- (generations 3.5, 5.5, or 9.5) terminated PAMAMs. Full-generation PAMAMs were shown to form stable homogeneous monolayers on silicon surfaces. By alternately immersing a clean silicon substrate in 1 % dendrimer solutions, adjusted to an appropriate pH, films up to twenty layers thick were prepared. Scanning probe microscopy and X-ray reflectivity were used for film characterization. Monolayers were observed to possess smooth surfaces at the molecular level. Dendrimer conformation in the monolayer was seen to be "collapsed" or "highly compressed", which compares well with the "lateral compression" detected for other flexible dendrimers by means of neutron reflectometry.[277] Molecular dynamics simulations corroborated the compressed model, which compared well with the measured film thickness. Film thickness varied linearly with increasing layer-by-layer deposition, evidencing "organized multilayer films". The structural states of PAMAMs at the air–water and air–solid interfaces have been evaluated;[311] for deformation-prone constructs, high interaction strength between "sticky" surface moieties and substrates was deemed responsible for compact monolayers and macromolecular compression.

Tanaka et al.[312] reported and Palmer reviewed[313, 314] the use of dendrimers as carriers, i.e., a pseudostationary phase, in micellar electrokinetic chromatography. Uncharged aromatic analytes were separated using water and water/MeOH mixtures as mobile phases. Later, Tanaka et al.[315] modified the surfaces of half-generation PAMAMs with *n*-octylamine and examined the electrokinetic chromatography potential of the products. These carriers facilitated efficient separations of aryl alkyl ketones and aldehydes and exhibited reversed-phase liquid chromatography characteristics in that they allowed separation optimization by organic solvent content manipulation. Analytes were further examined[316] using PAMAMs constructed from *p*-xylylenediamine cores. A clear propensity for separation of rigid aromatics from aliphatics was observed.

Chujo et al.[317] employed methyl ester-terminated PAMAMs to control pore size in porous silica. Organic–inorganic hybrid materials were obtained by the acid-promoted "sol-gel" reaction of $Si(OMe)_4$ with added dendrimer. Full-generation dendrimers (amine-terminated) led to phase-separated hybrids, whereas the use of ester-terminated substrates produced transparent, homogeneous blends. Subjecting the polymer hybrids to pyrolysis at 600 °C for 24 h led to complete elimination of the PAMAM frameworks, as evidenced by elemental analysis. Pore size distribution correlated well with the size of the dendrimer employed. SiO_2–PAMAM dendrimer hybrids have been prepared[318] by a multi-step procedure and have been shown to possess metal ion complexing ability.

Okada and coworkers[319] prepared poly(2-methyl-2-oxazoline) by "living" ring-opening polymerization, transformed the reactive terminus to a terminal amine moiety, and constructed PAMAMs through to the 5th generation using the polymer as the starting core (Figure 3.4). Aggregation behavior was studied by means of surface-tension measurements and small-angle neutron scattering analysis. Fujiki et al.,[320] and others,[321] similarly constructed PAMAMs on amine-modified silica in order to modify surface characteristics.

The mechanical properties of blends of PAMAMs with poly(vinyl chloride)s and poly(vinyl acetate) have been assessed by Xe NMR, dynamic mechanical analysis, and tensile property measurements.[322] Phase-separated dendrimer–PVC matrices facilitated mechanical relaxation, while the opposite was observed for the dendrimer–PVAc hybrid suggesting better compatibility.

Bliznyuk et al.[323] electrostatically prepared self-assembled PAMAM monolayers on silver wafers using generations 3.5–10. Monolayer thickness, morphology, and stability were studied by scanning probe microscopy, while a proposed model assumed compressed ellipsoidal dendrimer architectures. Imae et al.[324] used SANS and surface force studies to investigate mica-adsorbed PAMAMs terminated with hydroxy groups. The coatings were found to behave as surface-improvement agents by promoting fine particle dispersion stability. Yoon and Kim[325, 325a] prepared a thickness-controlled biosensing inter-

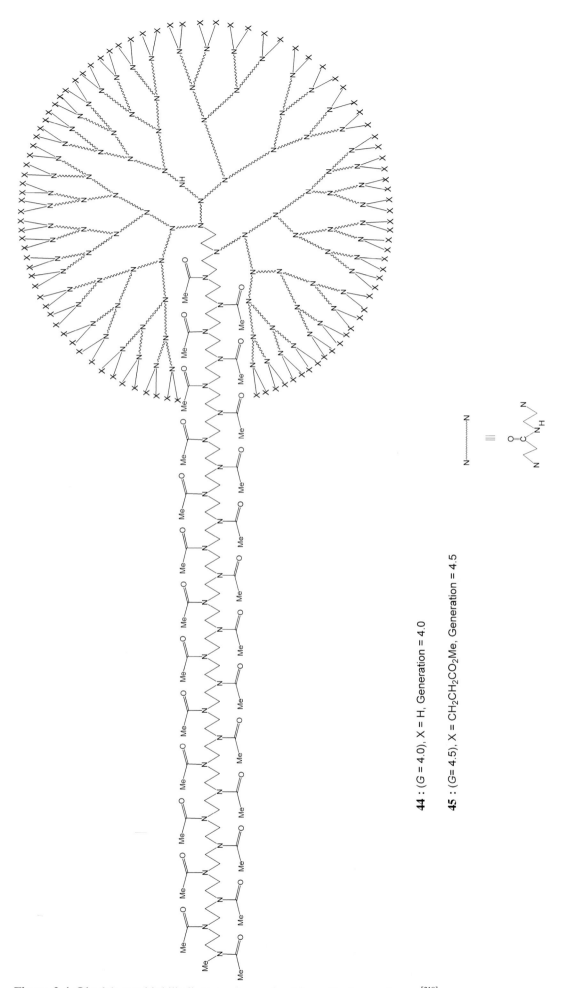

44 : (G = 4.0), X = H, Generation = 4.0

45 : (G= 4.5), X = CH₂CH₂CO₂Me, Generation = 4.5

Figure 3.4 Okada's amphiphilic linear polymer-dendrimer block copolymer.[319]

face by multilayer assembly of PAMAMs (G4) and periodate-oxidase glucose oxidase on an Au electrode surface. PAMAM–PEG–peptide conjugates have been reported[326] for the specific control of cell aggregation.

Linear PAMAM diblock copolymers incorporating PEGs with molecular weights of 2,000 and 5,000 amu have been prepared.[327] For the shorter chain copolymer, intrinsic viscosity data resembled that for linear polymers, whereas data collected for the longer chain derivative suggested the formation of "unimolecular micelles".

The attachment of PAMAM architecture to the reactive amines found in low molecular weight linear poly(ethylene imine) polymers (accessed by "living" cationic polymerization of 2-ethyl-2-oxazoline) to produce rod-shaped cylindrical structures has been reported.[328] These hybridized materials were termed "architectural copolymers" and their synthesis described as a "divergent, *in situ* branched cell" strategy. The authors noted the requirement for large excesses of reagents and long reaction times (e.g., for ethylenediamine addition, 1,260–10,000 equiv./ester group and 5–8 days, respectively) to achieve complete reactions. Cross-linked products were also observed following long-term storage of the amine-terminated hybrids.

The biodistribution of indium (^{111}In)- and yttrium (^{88}Y)-labeled 2nd generation PAMAM modified with 2-(*p*-isothiocyanatobenzyl)-6-methyl-diethylenetriaminepenta-acetic acid (1B4M) was reported,[329, 329a, b] where humanized anti-Tac IgC (HuTac) was also conjugated to the PAMAM–1B4M complex and analyzed for its *in vitro* and *in vivo* properties. The dendrimer conjugates were shown to give greater liver and spleen accumulation than that seen with native materials, suggesting detrimental effects on biodistribution.

3.3.3 1 → 2 *N*-Branched, *Aryl* Connectivity

A simple series of 1 → 2 *N*-branching cascade molecules, based on 4,4',4''-tris(*N*,*N*-diphenylamino)triphenylamine, has been prepared and characterized.[330] The ESR spectrum of the cationic triradical of the related 1,3,5-tris(diphenylamino)benzene has been studied in detail.[331] 1,3,5-Tris[*N*-(4'-methylbiphenyl-4-yl)-*N*-(diphenylaminophenyl)amino]- and 1,3,5-tris{*N*-[4-bis(4-methylphenyl)aminophenyl]-*N*-(4-diphenylaminophenyl)amino}benzene have found to be thermally and morphologically stable amorphous materials that exhibit unique multi-redox properties.[332]

3.3.4 1 → 2 *N*-Branched, *Si*-Connectivity

Hu and Son[333, 334] employed the now standard hydrosilylation–chlorosilane substitution protocol in concert with silazane monomers (Scheme 3.14) for the preparation of *N*-branched, silane-based structures. This iterative synthesis is exemplified by the Pt-mediated reaction of core **46** with HMe$_2$SiCl, followed by treatment with lithium (dimethylvinylsilyl)amide (**47**); repetition then afforded the 1st and 2nd tier constructs **48** and **49**. Notably, the hydrosilylations required three days to proceed to completion, although quantitative yields were realized. Complete reaction for access to a defect-free 3rd generation was not realized. Interesting features of this architecture include the known planarity of the N(Si)$_x$ units ($x \geq 2$)[335] and the facile degradation of the Si–N moieties.

3.3.5 1 → 2 *N*-Branched, *N*- & *Amide*-Connectivity

Beer and Gao[336] created poly-1,4,7-triazacyclononane-based architectures of the 1st and 2nd generations, which were demonstrated to form Cu(II) and Ni(II) complexes. The triazacyclononane heteromacrocycles formed the branching units, which were connected divergently using traditional acid–amine coupling procedures (EDC, HOBt, DMF).

299

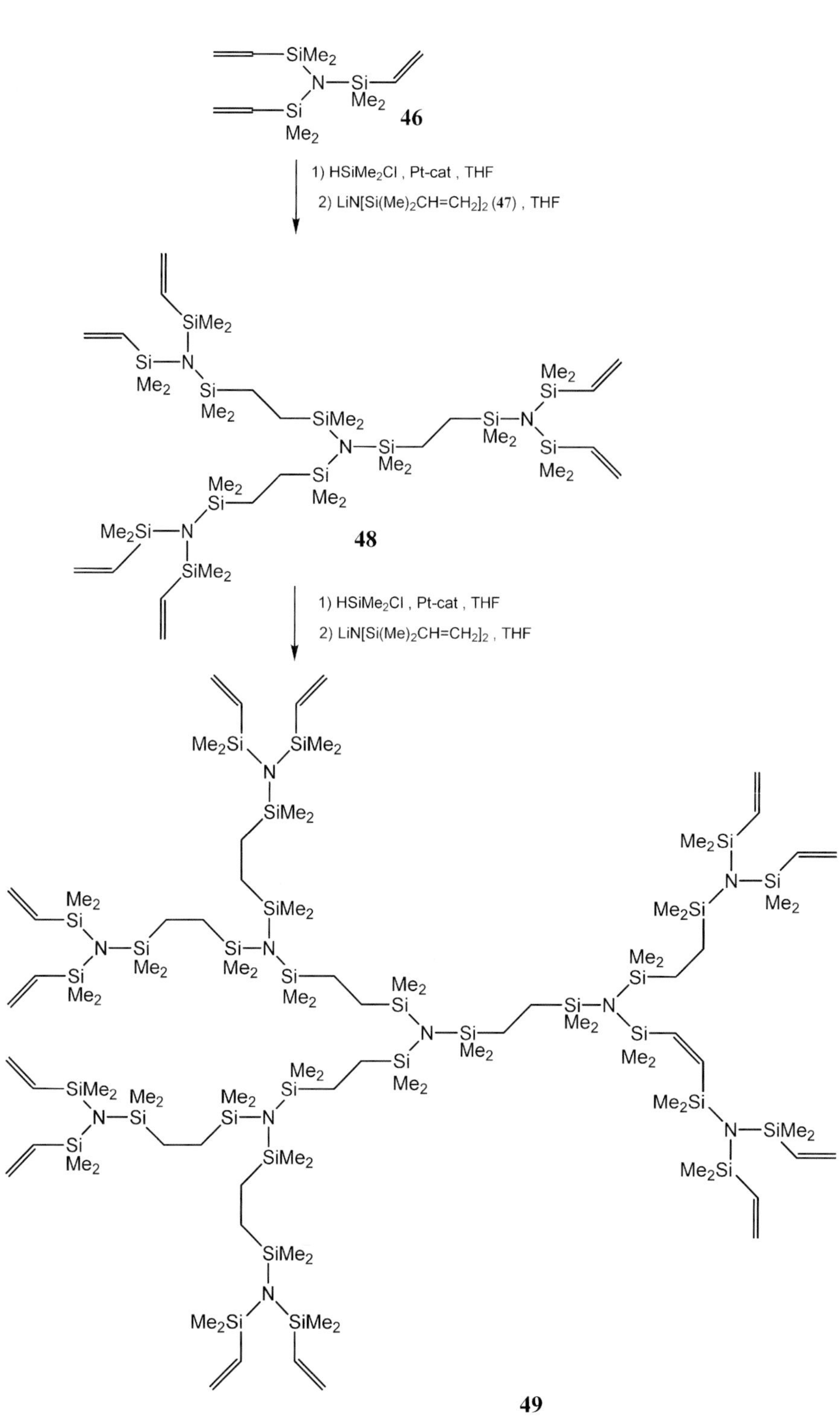

49

Scheme 3.14 Creation of carbosilazane dendrimers (**49**).[333, 334]

3.4 1 → 2 *Aryl*-Branched

3.4.1 1 → 2 *Aryl*-Branched, *Amide* Connectivity

Vögtle et al.[337–340] devised a simple route to very bulky cascade molecules by employing the *N*-tosylate of dimethyl 5-aminoisophthalate (**51**; Scheme 3.15). Utilizing 1,3,5-tris(bromomethyl)benzene[341] (**50**) as the core and tosylate **51** as the building block, they were able to generate hexaester **52** (X = CO_2CH_3), which was reduced and transformed to the hexakis(bromomethyl) derivative, which, in turn, was treated with 6 equivalents of **51**, generating the corresponding dodecaester **53**. This three-step divergent reaction sequence was repeated, ultimately achieving three generations, e.g., **54**. An X-ray structure (Figure 3.5) of the intermediate hexaester **52** was obtained, from which an appreciation of the internal rigidity can be gained. The structural homogeneity associated with further tier construction beyond the 3rd tier (**54**) can be expected to be problematic due to the steric demands of the bulky monomer (**51**) and the diminished spatial region available for reaction of each peripheral bromomethyl moiety.

Wang et al.[342] prepared branched aryl imides comprised of tetrahydro[5]helicene units (Scheme 3.16). Essentially, monomer **55**, obtained by nitration of anhydride **56** followed by treatment with 2-aminoethanol and subsequent NO_2 reduction, was reacted (*m*-cresol, 200 °C) with anhydride **56** to give dendron **57**. Further elaboration of monomer **55** with the corresponding dinitro anhydride, followed by reduction (Fe, HCl) afforded tetraamine **58**. Capping with **56** gave dendron **59**. Linear analogs (up to 10 units) were also reported. All of these materials exhibited reversible redox behavior as well as generation-dependent fluorescent emission.

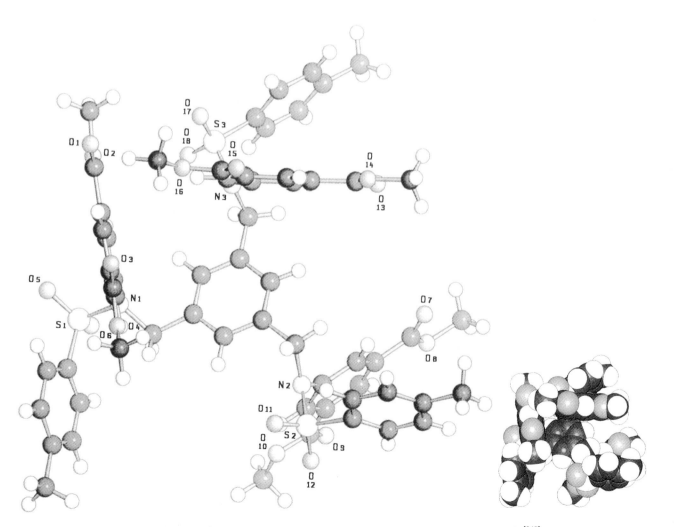

Figure 3.5 X-ray structure of the 1st generation tris-*N*-tosylated dendrimer (**52**; Scheme 3.15).[340]

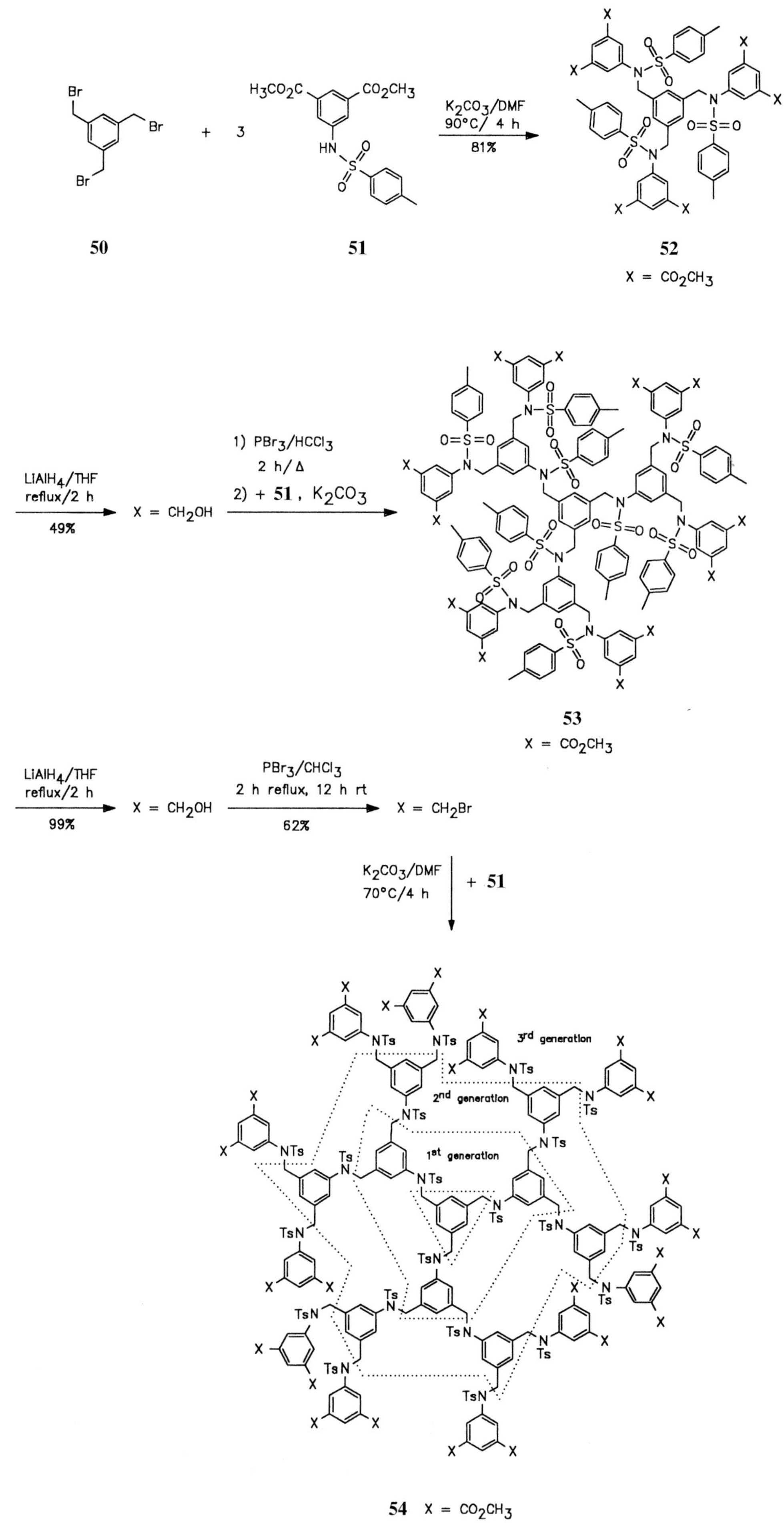

Scheme 3.15 Synthesis of "bulky" dendrimers (e.g., **54**) by the sequential addition of *N*-tosylated aminoisophthalate diester monomer (**51**).[337–340]

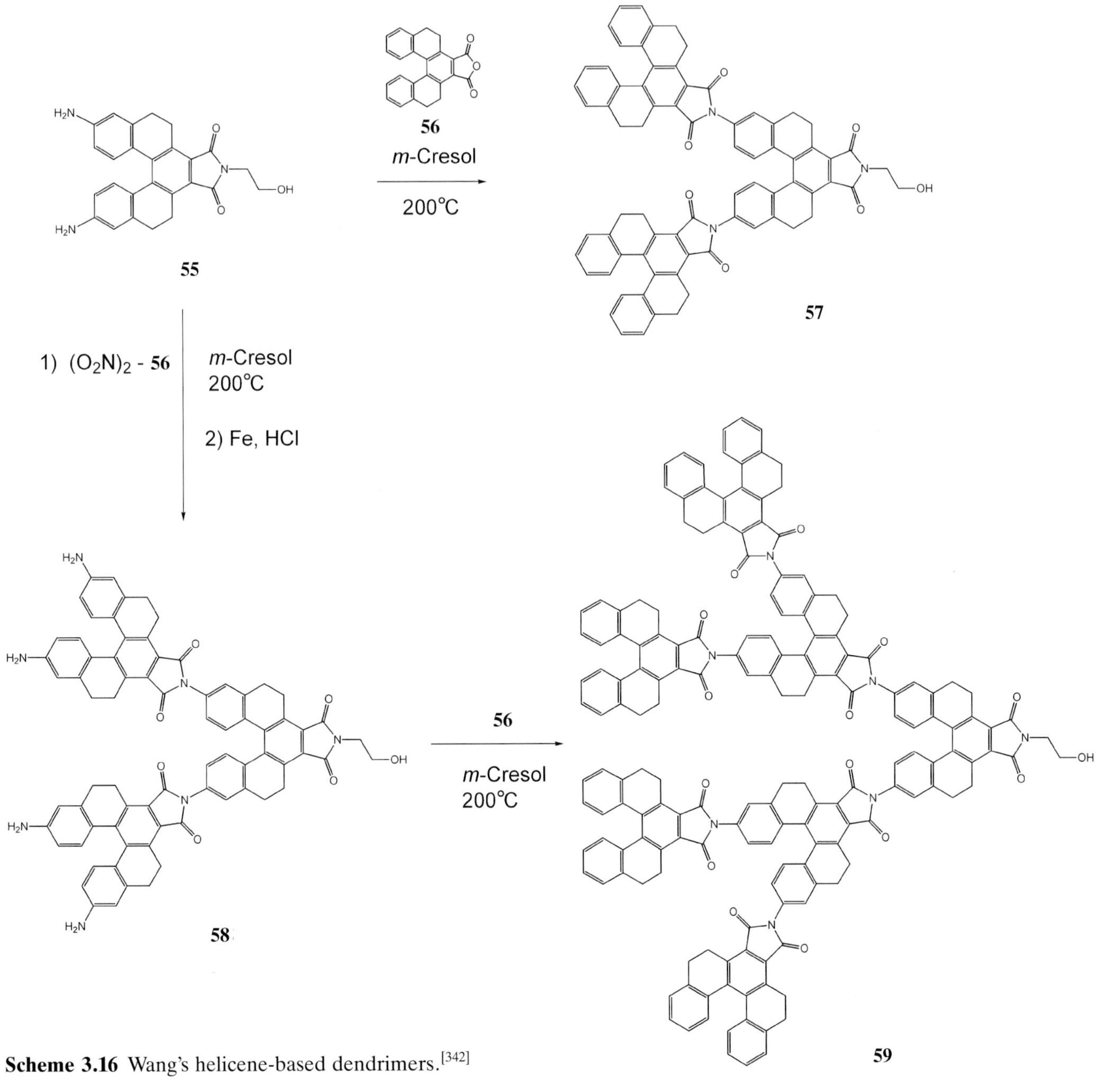

Scheme 3.16 Wang's helicene-based dendrimers.[342]

3.4.2   1 → 2 *Aryl*-Branched, *Ester* Connectivity

Haddleton and coworkers[343] reported a divergent mode of construction utilizing either phloroglucinol (1,3,5-trihydroxybenzene) or hydroquinone (**60**) as the core, and the benzyl protected 3,5-dihydroxybenzoic acid (**61**) as the building block (Scheme 3.17). Treatment (DCC) of diol **60** with two equivalents of acid **61** gave the bis(ester) **62**, which was deprotected (catalytic hydrogenolysis) to liberate tetrahydroxy diester **63**. Repetition of the esterification process gave rise to hexaester **64**; three tiers were constructed and characterized by MALDI MS studies, which showed no evidence for dimer or trimer formation either during the synthesis or within the mass spectrometer.

Taylor et al.[344] later extended their synthetic protocol to include dendrimers constructed with naphthalene-2,6-diol as the core (i.e., **65**; Figure 3.6). For the phloroglucinol-based dendrimer **66**, where branching initially grows from a 3-directional core, the dense-packed, de Gennes limit[248] appeared to occur between the 3rd and 4th generations, while for the 2-directional cored dendrimers (including the hydroquinone-based dendrimer in Scheme 3.17) the limit was reached between the 4th and 5th generations, presumably due to diminished congestion. Notably, these authors observed that

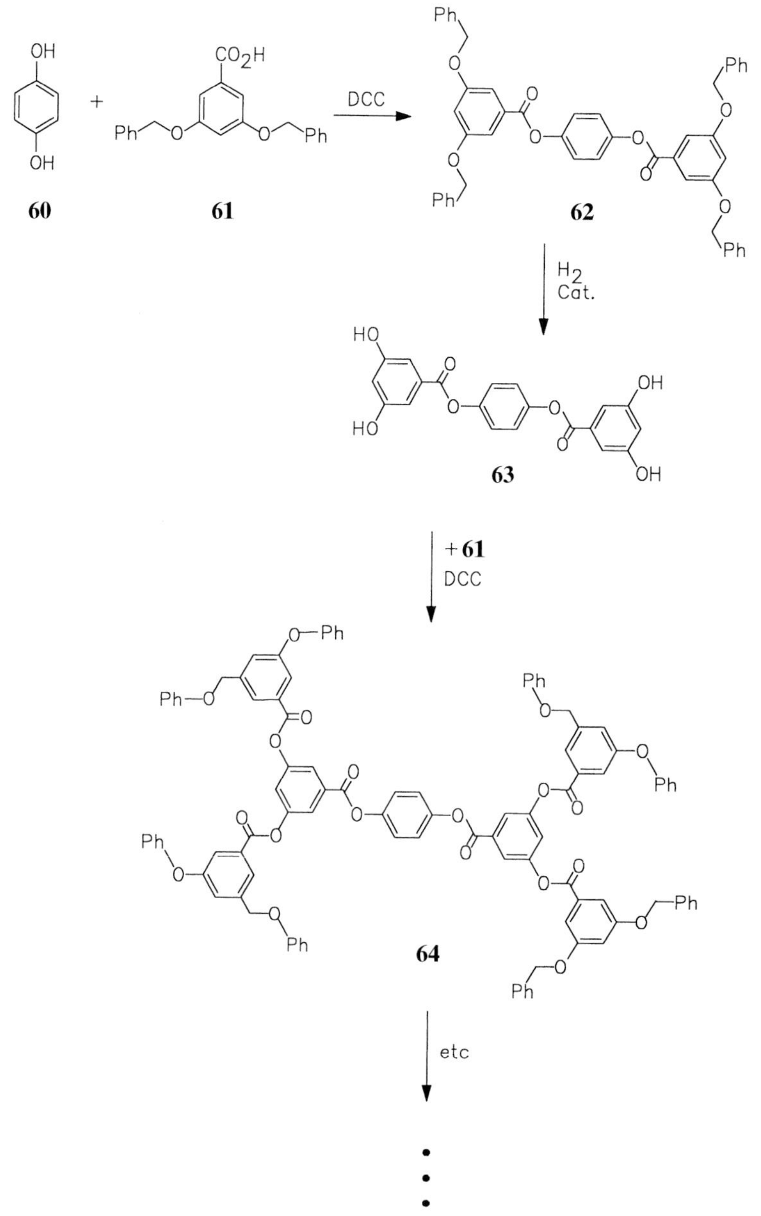

Scheme 3.17 Construction of 1 → 2 aryl-branched polyesters via phenol acylation.[343]

different coupling procedures were more useful during different stages of dendrimer construction; acid halide coupling of building blocks afforded better yields during higher generation steps, while DCC-based coupling gave superior results at the lower generations.

Shi and Rånby[345–347] published a series of papers concerning the photopolymerization of methacrylate-terminated, polyester dendrimers. Preparation of these materials began with the esterification (SnCl$_2$ promotion) of pentaerythritol with 1,2,4-benzenetricarboxylic anhydride followed by reaction with glycidyl methacrylate in the presence of benzyldimethylamine and hydroquinone, as a radical inhibitor, to give octakis(hydroxymethacrylate)-terminated dendrimers. Reaction of the hydroxyl moieties with methacrylic anhydride afforded the corresponding approximately 12 (incomplete reaction) and 16 alkene-terminated dendrimers. The T_gs of UV-cured films of these materials were found to increase with increasing terminal functionality.[345] In the presence of 3 % benzyldimethyl ketal as a photofragmenting initiator, UV polymerization was found to occur rapidly (75 % conversion after 0.15 s). On the other hand, 10–40 wt. % addition of the multifunctional monomer trimethylolpropane triacrylate produced little effect with regard to curing efficiency, which is in contrast to the effect seen with conventional linear acrylate-based oligomers.[346] Mixtures of the polymethacrylate dendrimers, TMPTA, and BDF were poured over Mylar-coated, glass fiber mats and subjected to UV radiation

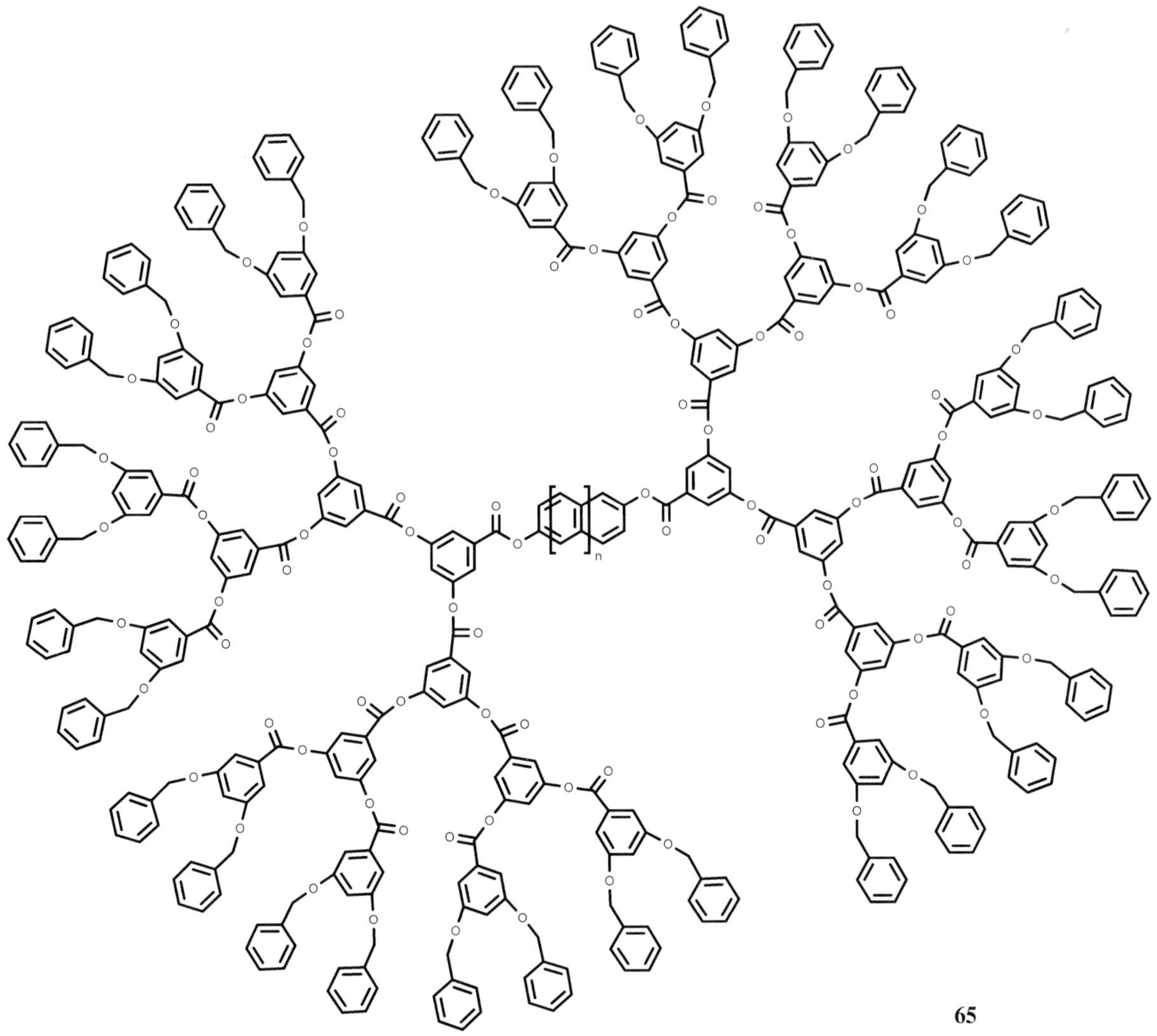

65

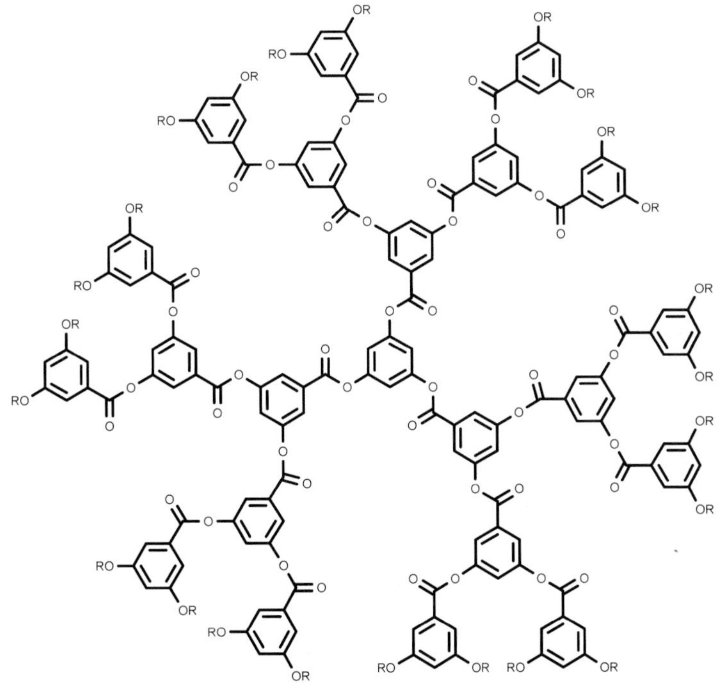

66

Figure 3.6 Poly(ester) dendrimers possessing naphthalenediol (**65**) and phloroglucinol (**66**) core units.[344]

to produce polyester laminates for investigation as composite materials that could potentially replace metals in the vehicle, furniture, and construction industries.[347] Liquid
crystalline fulleropyrrolidines based on poly(aryl ester)s are also known.[347a]

3.4.3   1 → 2 *Aryl*-Branched, *N*-Connectivity

Hall and Polis[348] prepared a series of polyarylamines by an aromatic nucleophilic
substitution–reduction sequence (Scheme 3.18). Thus, 2,4-dinitrofluorobenzene **68** was
reacted with *p*-diaminobenzene (**67**) to afford tetranitrodiamine **69**, which was reduced
to give the corresponding 1st generation amine **70**. Repetition of this sequence afforded
the 2nd generation octanitro and tetradecaamine constructs **71** and **72**. These polyarylamines were complexed with iodine to form semiconducting materials and were the first
dendrimers to be examined by cyclic voltammetry.

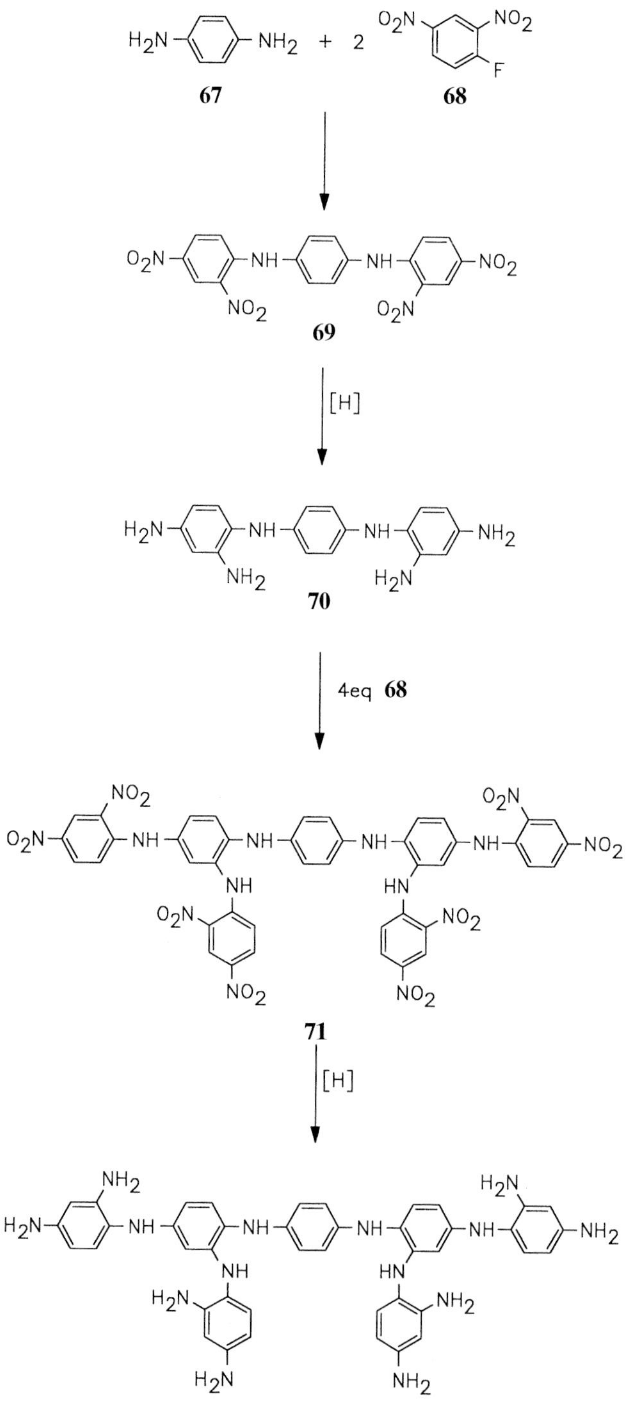

Scheme 3.18 Synthesis of poly(aryl amines)[348] through aromatic nucleophilic displacement of fluoride ion.

Blackstock et al.[349] described the properties of stable, isolable polyradical cations based on a small poly(phenylenediamine). This poly(aryl amine) was constructed (overall yield 21%) by phloroglucinol condensation with *N*-phenyl-*p*-phenylenediamine, followed by an Ullmann reaction with iodobenzene or *p*-iodomethoxybenzene affording the methoxy-terminated analogue. A larger redox-gradient poly(aryl amine) dendrimer was prepared[350] by means of similar chemistry. The gradient arises from a shell of relatively difficult to oxidize aryl amines surrounding the more readily oxidized interior phenylenediamino moieties, as evidenced by cyclic voltammetry. Other small, branched triarylamines have been reported.[351, 352]

Lambert and Nöll[353] reported one- and two-electron transfer processes in triarylamines with multiple redox centers. Co-mediated benzene formation from a bis(triarylamine)-substituted alkyne led to the formation of a hexakis(triarylamine)benzene. Thelakkat and coworkers[354] have also reported the construction of a variety of triarylamines possessing photoconductive and nonlinear optical properties.

Heinen and Walder[355] prepared a dendritic "electron sponge" that exhibits generation-dependent intramolecular charge-transfer complexation; the architecture was comprised of viologen (4,4'-bipyridinium) spacers with aryl branching centers. Up to 3 generations were reported. Ramifications for molecular recognition, signal transduction, and charge trapping[355a] were considered.

3.4.4 1 → 2 *Aryl*-Branched, *O*-Connectivity

Poly(aryl ether) dendrons of the 1st and 2nd generations have been divergently incorporated onto a calix[4]resorcinarene (Scheme 3.19; **73**) as a core unit. Thus, the bis(allyl) benzyl bromide monomer **74** was attached (K$_2$CO$_3$, 18-crown-6) to give the 1st generation polyether **75**.[356] Following deallylation [(Ph$_3$P)$_2$PdCl$_2$, HCO$_2$NH$_4$] to give the poly(phenoxyl) intermediate **76** and reaction with bromide **74**, the 2nd generation construct **77** was created. Due to the large number of hydroxyl groups on the core, M_ws of 7171 and 9345 amu were realized for the smaller and larger materials, respectively. Employing a calix[4]resorcinarene core, Udea et al.[357] constructed a 1st generation poly(aryl ether) motif designed to function as negative-working, alkaline-developable photoresist material, which exhibited an unmistakable negative pattern following a sequence of post-baking (110 °C), UV irradiation, and subsequent treatment with aqueous Me$_4$N$^+$OH$^-$ (0.3%, 25 °C).

3.4.5 1 → 2 *Aryl* & *C*-Branching and Connectivity

Veciana et al.[358–362a, b] reported the synthesis of perchlorinated polyradicals through to the 2nd generation (Scheme 3.20). The 1st tier triradical **82** was obtained by subjecting 1,3,5-trichlorobenzene **78** to dihalomethylation (AlCl$_3$, CHCl$_3$) to give the Reimer–Tiemann intermediate, i.e. tris(α,α-dichloromethylbenzene) (**79**), which, upon treatment with pentachlorobenzene in the presence of AlCl$_3$, afforded the polychlorinated heptaaryl radical precursor **80**. Deprotonation at the triphenylmethane sites (*n*-Bu$_4$N$^+$OH$^-$, 35 days) gave trianion **81**, which was converted (excess *p*-chloranil) to the corresponding triradical **82**.

Triradical **82** was isolated in two isomeric forms possessing D_3 and C_2 symmetries. Due to steric shielding provided by the chloro groups, the polyradicals showed exceptional stability in the solid state up to 250 °C. The 2nd generation (**83**; Figure 3.7) of these perchlorinated polyradicals was prepared, although the authors reported, "several structural defects disrupted some of the desired ferromagnetic couplings."

Rovira et al.[363] were later successful in their quest to prepare a 2nd generation, perchlorinated tetraradical. Using modified Friedel–Crafts reaction conditions (Scheme 3.21) that included high temperatures (reflux) and a large excess of benzene and AlCl$_3$ (molar ratios > 60:1 and 3:1, respectively) yields of nonaarylmethane **86** approached 90% starting from the tris(dichloromethyl) precursor **84**. Exhaustive chlorination

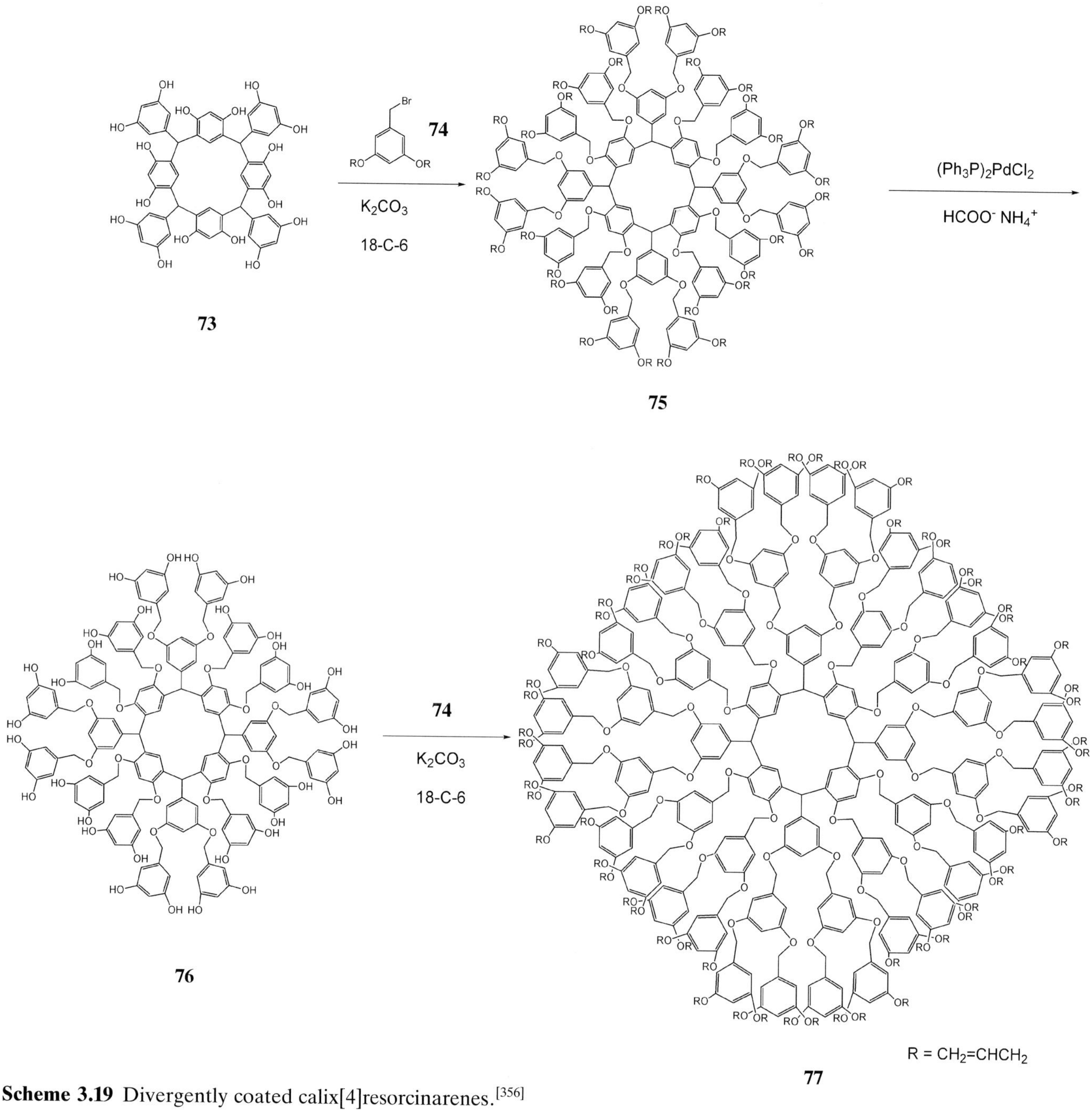

Scheme 3.19 Divergently coated calix[4]resorcinarenes.[356]

(SO$_2$Cl$_2$, S$_2$Cl$_2$, AlCl$_3$) of the Friedel–Crafts product **85** gave the desired, highly congested dendrimer **86**. Polyradical formation (i.e., **86a–d**) was achieved by carbanion generation (*i*-Bu$_4$NOH) followed by oxidation (*p*-chloranil). These radicals were studied by X-band ESR. A crystal structure has been determined[364] for a small, high-spin triradical, which revealed the presence of diastereoisomers, one with C_2 symmetry and the other with D_3 symmetry. The molecular surface characteristics of the quartet 2,4,6-trichloro-α,α,α',α',α'',α''-hexa(pentachlorophenyl)mesitylene, which exists in two atropisomeric forms, and its interaction with adjacent solvent molecules was studied[365] employing linear free-energy relationships for solvation; the shape and fractality were shown to be the most important factors, whereas the cavitational effects were found to be unimportant.

Iwamura et al.[366] reported the preparation of a "branched-chain" nonacarbene possessing a nonadecet ground state with the goal of constructing superparamagnetic polycarbenes. Bock et al.[367] reported the preparation of three-directional diradicals, while a tetrahedral tetraradical has been synthesized by Kirste et al.[368] The synthesis, characterization, and physical properties of the perchloro-2,6-bis(diphenylmethyl)pyridine-α,α'-ylene biradical have also been reported.[369]

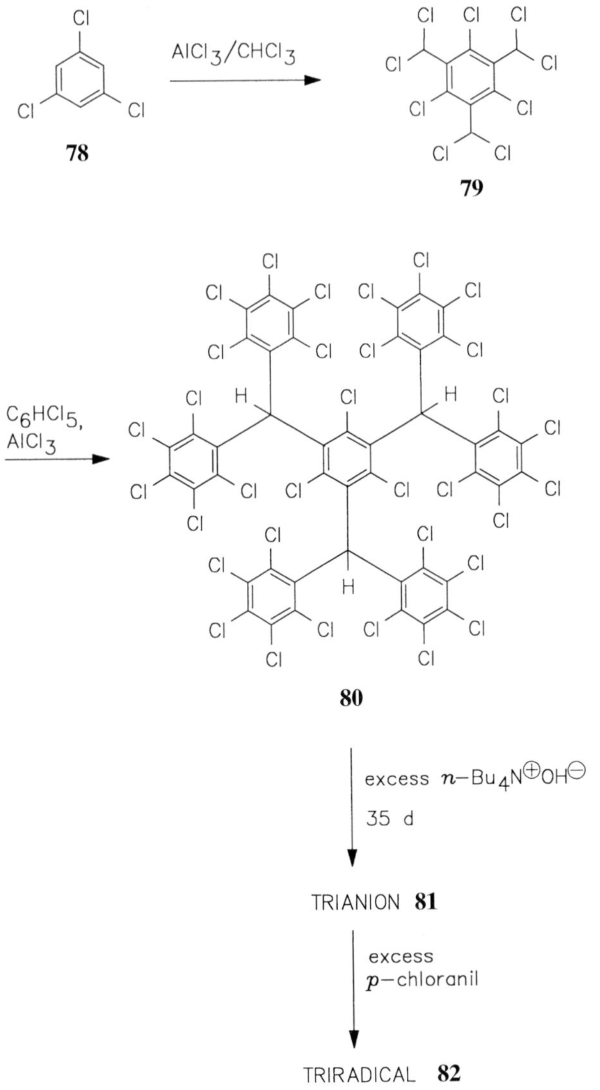

Scheme 3.20 Preparation of a perchlorinated triradical[358–362] by *AlCl₃*-activated arylation; the radical was isolated in two isomeric forms having D_3 and C_2 symmetries.

Müllen et al.[370–372] described the synthesis of polyphenylene dendrimers predicated on the Diels–Alder cycloaddition reaction of an alkyne dienophile to an activated diene. Key monomers included dienone **87**, accessed by coupling [Pd(PPh₃)₂Cl₂, CuI, Et₃N, toluene][373] of triisopropylsilylacetylene (TiPSA) to 4,4'-dibromobenzil followed by condensation (KOH, EtOH) with 1,3-diphenylacetone, and diyne **88**, prepared by [4+2]-cycloaddition of **87b** with diphenylacetylene and subsequent alkyne deprotection [Bu₄NF]. Preparative iteration is illustrated in Scheme 3.22, whereby 2 equivalents of diene **88** are reacted with bis(alkyne) **87b** to give, after silyl deprotection, tetraalkyne **89**; repetition of the sequence yields octaalkyne **90**. Analogously, rigid dendrimer **92** (Scheme 3.23) was constructed starting from the tetraalkyne core **91**. All the silyl-protected intermediates, as well as the free polyalkynes, were found to be freely soluble in common organic solvents. Notably, the dienophile was added portionwise to an excess of the diene to obtain the large polyalkynes.

Morgenroth, Kübel, and Müllen[374] later described improved procedures for the synthesis of their polyphenylene dendrimers using a [2 + 4]cycloaddition–deprotection protocol; reviews are also available.[375, 376, 376a] The 1st–3rd generations were prepared consisting of 22, 62, and 142 benzene rings, respectively; dense packing for the 2nd and 3rd generations was shown by molecular mechanics to limit the conformational degrees of freedom, thus providing support for structural "shape-persistence". The diameters of these structures were determined to be 7, 21, 38, and 55 Å for the 0–3rd generations, respectively.

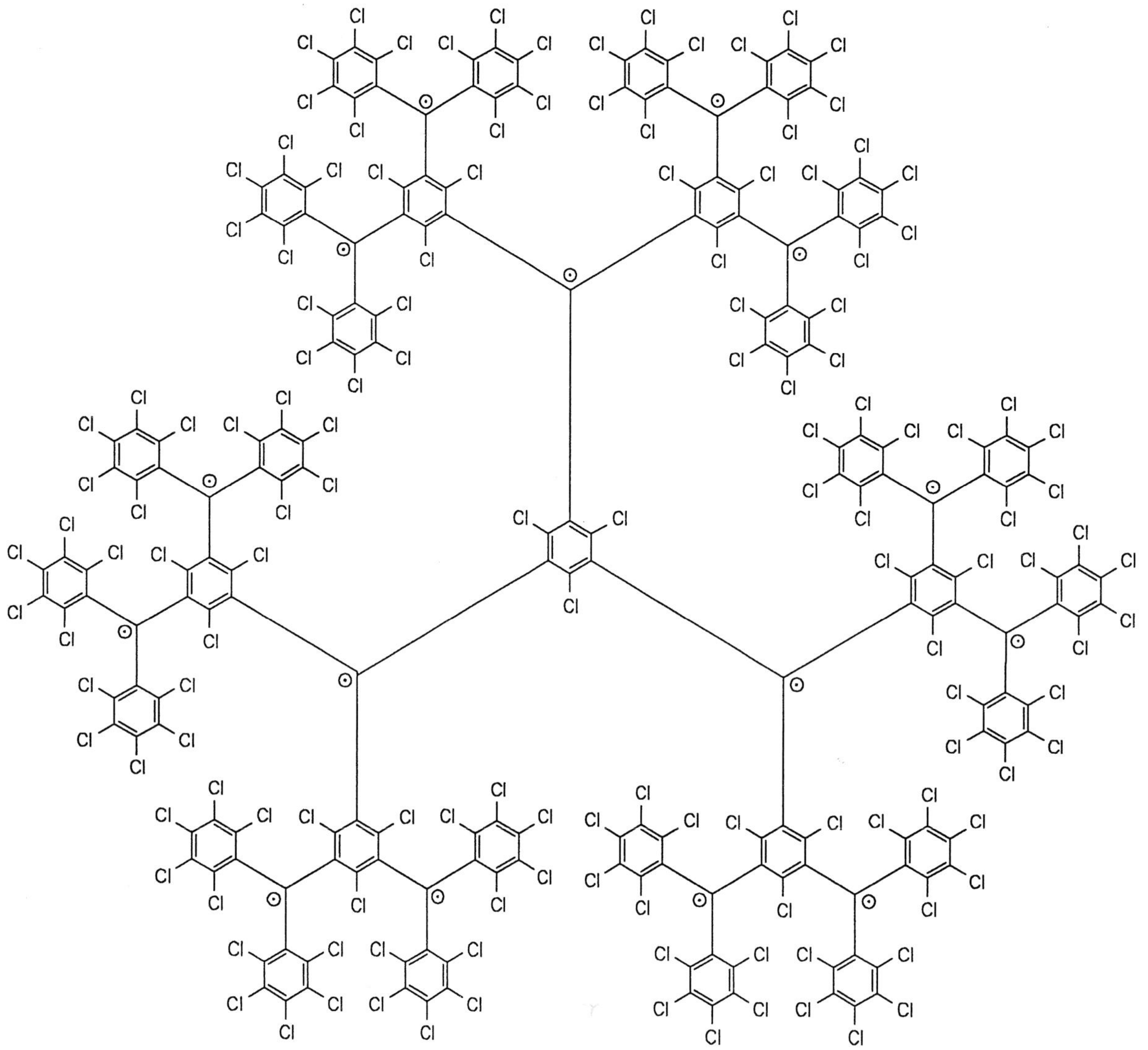

83

Figure 3.7 A 2nd generation, perchlorinated polyradical constructed to investigate solid-state radical stability.

Scheme 3.24 depicts the synthesis of the PAHs **93** and **94** by reaction of dienone **87a**, possessing only terminal H-moieties, with core units **88** and **91**, respectively, followed by cyclodehydrogenation under Kovacic[377] conditions. Spectroscopic analysis of these materials was hindered by poor solubility; however, the M^+ signal in the mass spectrum of **94** was 56 amu less than that of the precursor Diels–Alder adduct (i.e., 2×28 H atoms are lost during the formation of the 28 new C–C bonds). Müllen et al.[378] later extended their work in this area to include construction using the AB_4 monomer 2,3,4,5-tetrakis(4-triisopropylsilylethynylphenyl)cyclopenta-2,4-dienone, which facilitated a more rapid synthesis of these polyphenylenes. Müllen et al.[379] have also used their synthetic protocol to prepare organic-soluble C_{60} graphite segments bearing dodecaalkyl chains, and light-emitting polymer.[379a]. Highly ordered monolayers of these segments were prepared and subsequently analyzed by STM. The generation of large polycyclic aromatic hydrocarbons,[380] e.g., "superbiphenyl",[381] has recently been reported.[381a]

Suzuki et al.[382] reported the synthesis of perfluorinated poly(aryl) dendrimers using ArCu-promoted cross-coupling with aryl bromides. A 3rd generation dendrimer was created, along with smaller, less branched materials; these constructs, due to their electron-transport properties, were prepared in order to examine their potential as field-effect transistors and light-emitting diodes.

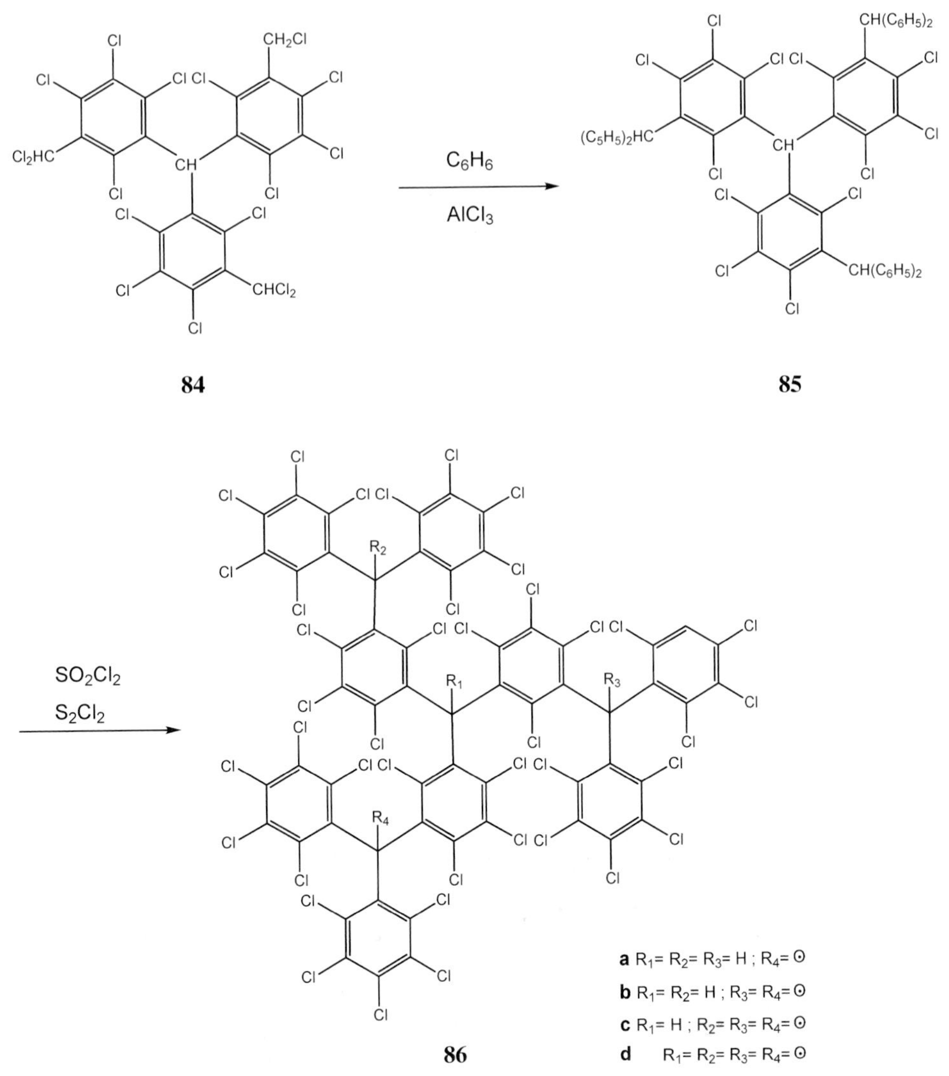

Scheme 3.21 Synthesis of a highly congested perchlorinated poly(aryl methane).[363]

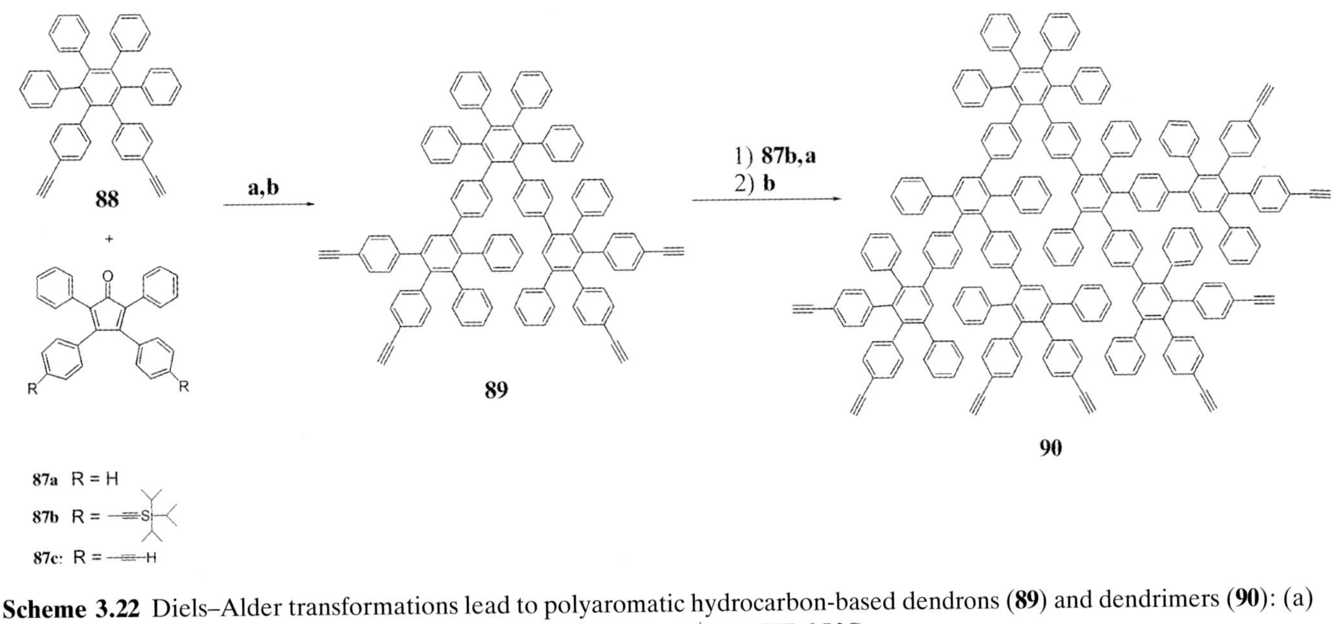

Scheme 3.22 Diels–Alder transformations lead to polyaromatic hydrocarbon-based dendrons (**89**) and dendrimers (**90**): (a) diphenyl ether/α-methylnaphthylene, 180–200 °C, 4 h; (b) Bu₄N⁺F⁻, THF, 25 °C.

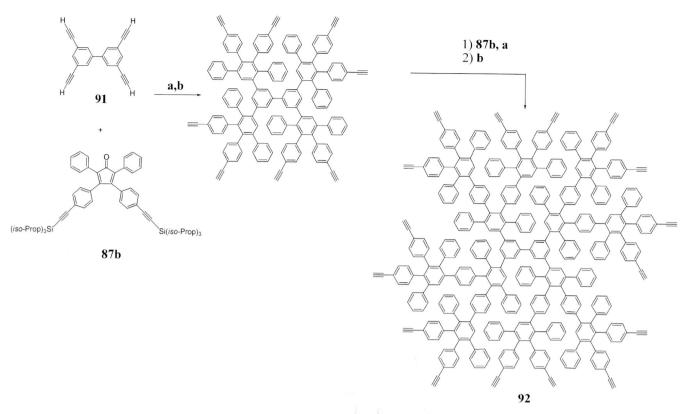

Scheme 3.23 Polyaromatic hydrocarbon synthesis based on a tetravalent core.

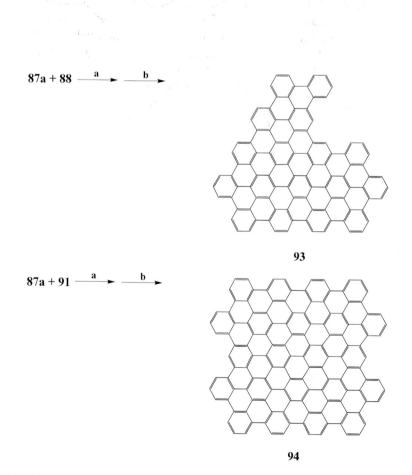

Scheme 3.24 Cyclodehydrogenation of Diels–Alder adducts affords unique 2-D architectures: a) diphenyl ether/α-methylnaphthalene, 180–200 °C; b) CuCl$_2$, AlCl$_3$, CS$_2$, 25 °C, 100 h.

3.5   1 → 2 *C*-Branched

3.5.1   1 → 2 *C*-Branched, *Amide* Connectivity

Denkewalter et al.[6] synthesized a series of *tert*-butyloxycarbonyl-protected poly(α,ε-L-lysine) macromolecules (see Scheme 3.3), the molecular models of which suggested that they were globular, dense spheres, and the molecular weight distributions for which were determined to be very narrow ($M_w/M_n \cong 1.0$). Since each generation in this series was synthesized in a stepwise manner, each member in the series was predicted to have a monodisperse molecular weight.

Aharoni and coworkers[383] characterized Denkewalter's cascade macromolecules[6] by employing classical polymer techniques: viscosity determinations, photo correlation spectroscopy, and size-exclusion chromatography. It was concluded that at each tier (2 through to 10) these globular polymers were, in fact, monodisperse and behaved as nondraining spheres. The purity of these molecules was not ascertained and the dense packing limits were either not realized or simply not noted, nor was the issue of chirality addressed. [For details concerning the applications of the lysine dendrimers, see Chapter 7.]

3.5.2   1 → 2 *C*-Branched and Connectivity

Hart et al.[384–389] reported the construction of numerous branched polyaromatic, all-hydrocarbon cascades termed "iptycenes", which are extended triptycenes based on bicyclo[2.2.2]octane moieties as the branching groups. Connection of the bicyclic moieties with benzene units resulted in a three-directional pattern. Although their construction was not strictly iterative, multiple Diels–Alder transformations with bis(9,10-anthradiyl)-substituted butadiene monomer (**98**) were employed (Scheme 3.25). Treatment of the highly substituted chlorobutadiene **95** with quinone **96** afforded the bis(triptycene) intermediate **97**, reaction of which with diene **98** generated dione **99**. Transformation of the core dione moiety of **99** to the anthracene nucleus **100** was achieved in three steps (NBS; LAH; DDQ); treatment of the tetrakis(triptycene) **100** with 1,2-dichloroethene, then with Li, and subsequent Diels–Alder reaction with diene **98** afforded an intermediate (not shown), which was finally aromatized to generate "super-iptycene" **101**. These highly rigid superstructures possess unique molecular cavities, as evidenced by the fact that crystals of iptycene **101** "include" solvent molecules within these cavities; solvent inclusion was suggested as a factor affecting X-ray structure determination.

Hart et al.[390, 391] also reported the preparation of symmetric iptycenes possessing benzene cores, for example, "nonadecaiptycene" (**103**) by trimerization of the bicyclic vinyl halide triptycene dimer **102**, as well as the synthesis of the related asymmetric iptycenes[390] (Scheme 3.26).

Although the initial crystallographic study of the simple heptiptycene[392] did not afford a structural model,[393] the structure of the crystalline 1:1 heptiptycene–chlorobenzene clathrate, in which the solvent molecule was found to be packed in the channels between ribbons of the heptiptycene, was later ascertained.[394] The molecular geometry calculated by Hartree–Fock [6–31G(D)] and local density methods compared well with the X-ray data.

Webster[395] described the preparation of a water-soluble tritriptycene and examined the ^{1}H NMR chemical shift changes of various substrates due to interactions with the aromatic ring currents. For example, a D_2O solution of the tritriptycene and *p*-toluidine exhibited an upfield shift ($\Delta v = 55$ Hz) of the substrate methyl group absorption.

Zefirov et al.[396] described a general synthetic strategy for the preparation of *branched triangulanes* or spiro-condensed polycyclopropanes (Scheme 3.27); an overall review has appeared.[397] Key features of this strategy include chloromethylcarbene addition to methylenecyclopropanes[398] and subsequent dehydrohalogenation. Thus, treatment of bicyclopropylidene **104** with the chloromethylcarbene generated from dichloride **105** gave the tricyclopropane **106**. Dehydrochlorination followed by cyclopropanation

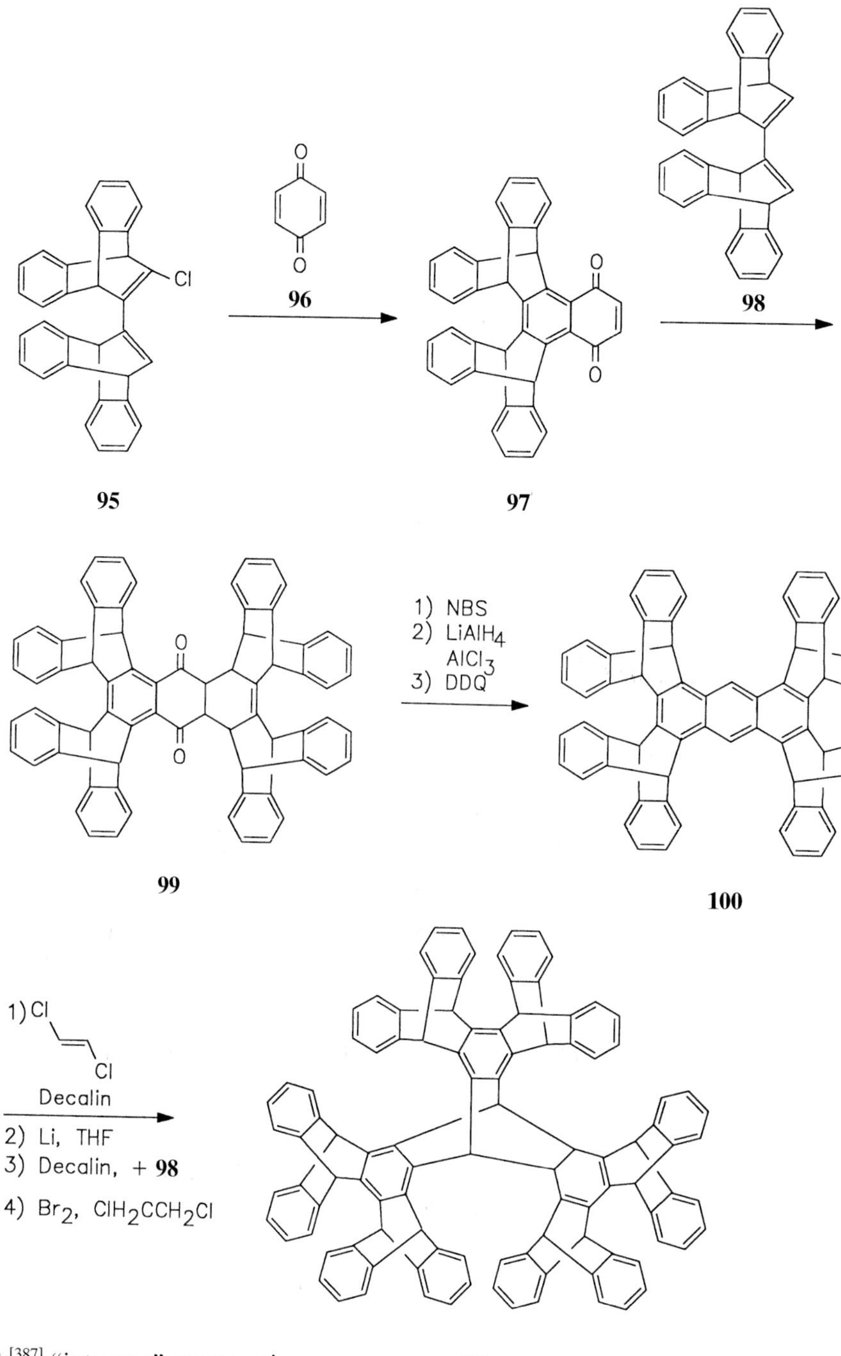

Scheme 3.25 Hart et al.'s[387] "iptycene" construction.

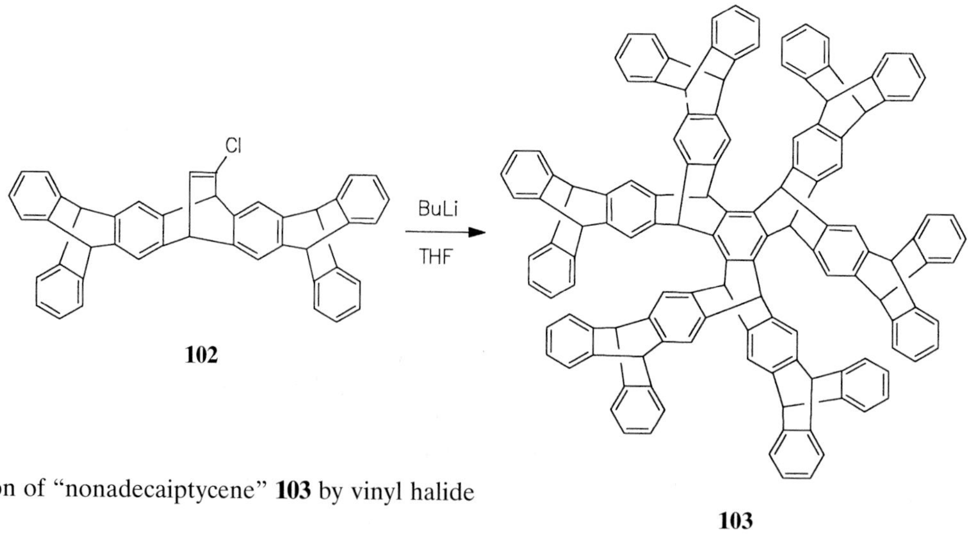

Scheme 3.26 Preparation of "nonadecaiptycene" **103** by vinyl halide trimerization.[390, 391]

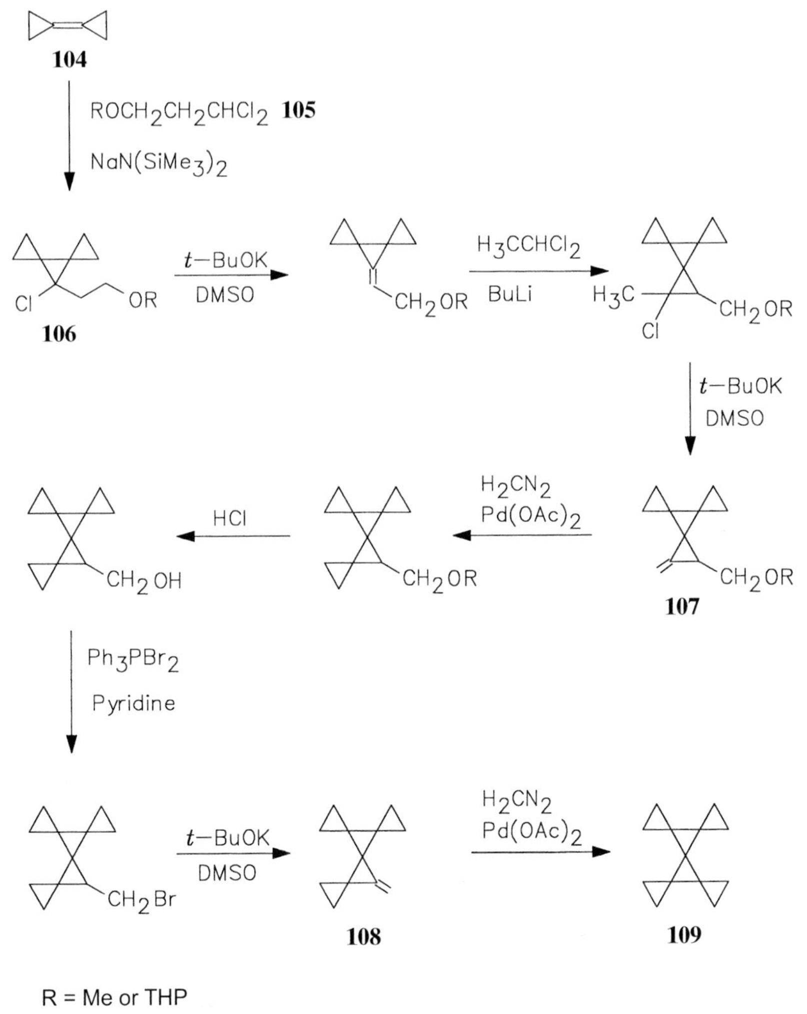

104

ROCH$_2$CH$_2$CHCl$_2$ **105**

NaN(SiMe$_3$)$_2$

106

t–BuOK
DMSO

H$_3$CCHCl$_2$
BuLi

t–BuOK
DMSO

107

H$_2$CN$_2$
Pd(OAc)$_2$

HCl

Ph$_3$PBr$_2$

Pyridine

t–BuOK
DMSO

108

H$_2$CN$_2$
Pd(OAc)$_2$

109

R = Me or THP

Scheme 3.27 Synthesis of "branched triangulanes" possessing adjacent quaternary carbon moieties[396] (R = Me or THP).

(CH$_3$CHCl$_2$, *n*-BuLi) and alkene formation (*t*-BuOK, DMSO) gave unsaturated ether **107**. After Pd(OAc)$_2$-mediated methyl carbene addition (CH$_2$N$_2$), acidic alcohol deprotection (HCl), bromination (Ph$_3$PBr$_2$, pyr), and β-elimination (*t*-BuOK, DMSO), alkene **108** was converted to hexakis(spirocyclopropane) (**109**).

The preparation of other branched triangulanes with varying symmetries has also been demonstrated. A notable feature of this series of small hydrocarbon cascades is that the framework is composed entirely of quaternary, tetraalkyl-substituted carbons. This architecture closely resembles, or is at least reminiscent of, Maciejewski's[38] proposed cascade molecule comprised of an all 1 → 3 C-branched interior framework (i.e., without spacers between branching centers).

De Meijere et al. reported the preparation of symmetrical branched triangulanes constructed from 10[399] (Scheme 3.28) and 14[400] fused cyclopropane moieties. Although the synthesis was not repeating or iterative, the structurally rigid spirocyclopropane **112** was obtained (14 %) by the reaction of nitrosourea **111** with bicyclopropylidene and NaOMe, which, in turn, was accessed from alkene **110**. An unequivocal structure determination of **112** was provided by X-ray crystallography, which demonstrated its D_{3h} molecular symmetry. [10]Triangulane **112** showed high thermal stability, even though its thermal strain energy ($\approx$ 1130 kJ mol^{-1}) indicates that it is more strained than cubane.[401, 402] The authors speculated on the potential for a carbon network based on spiro-linked, three-membered rings. Strain energies in [*n*]triangulenes and spirocyclopropanated cyclobutanes have been determined.[403]

Díez-Barra et al.[404] reported the benzylation of aryl acetyl groups for the preparation of small dendritic polyketones. A crystal structure obtained for a 1st generation tetraphenyl diketone showed aryl π-stacking in the solid state.

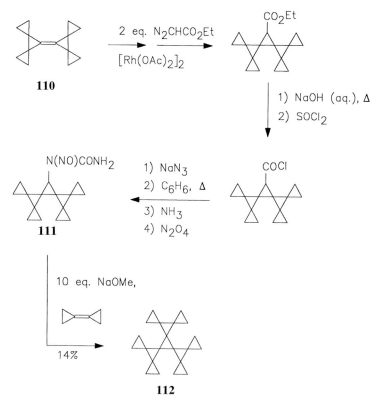

Scheme 3.28 Non-iterative preparation of [10]triangulane[399] possessing D_{3h} symmetry.

3.5.3   1 → 2 *C*-Branched, *Ether & Sulfone* Connectivity

Using activated aryl fluoride sites for monomer attachment, Martínez and Hay[405] constructed a series of poly(aryl ether) dendrimers. Key fluoro-sulfide-containing phenolic monomers (Scheme 3.29; **113**), prepared in three steps by the reaction of 4,4'-dichlorobenzophenone with 2 equiv. of 4-fluorothiophenol, bromobenzene Grignard addition, and acid-catalyzed carbonium ion addition to the phenol moiety, were dimerized (triphosgene) to give an aryl carbonate **114**. The carbonate moiety was then cleaved *in situ* (155–165 °C) using a metal carbonate, which essentially eliminates water formation (e.g., as occurs on generation of a phenoxide using hydroxide as base) as well as the need to remove the initial water from the starting material. Thus, reaction of carbonate **114** with bis(4-fluorophenyl)sulfone (**116**) in the presence of a CaCO$_3$/CsCO$_3$ mixture yielded the 1st generation dendrimer **117** through phenoxide (i.e., **115**) displacement of the sulfoxide-activated aryl fluoride. Oxidation of the sulfide moieties to sulfone groups (wet alumina/ozone) facilitated repetition of the strategy to give the 3rd generation dendrimer (**118**). Formula weights of these materials ranged from 1,392 to 19,239 amu for generations 1–4, while the measured T_gs increased accordingly from 92 to 231 °C.

3.5.4   1 → 2 *C*-Branched, *Ether* Connectivity

Haag et al.[406a] developed an approach (Scheme 3.30) for the synthesis of glycerol-type dendrimers. Starting with tris(hydroxymethyl)ethane (**119**), a three-fold iteration of allylation (allyl chloride, TBAB, NaOH, H$_2$O) followed by dihydroxylation [OsO$_4$ (cat.), NMO, H$_2$O, acetone, *t*-BuOH] generated polyol **120**. After one iteration of allylation and oxidation, reaction of the core with epihydroxyhydrin afforded hyperbranched analog **121**, which was further subjected to allylation and oxidation to yield the "pseudo-dendrimer" **122**. Notably, these water-soluble materials were accessed using aqueous reaction media for each step.

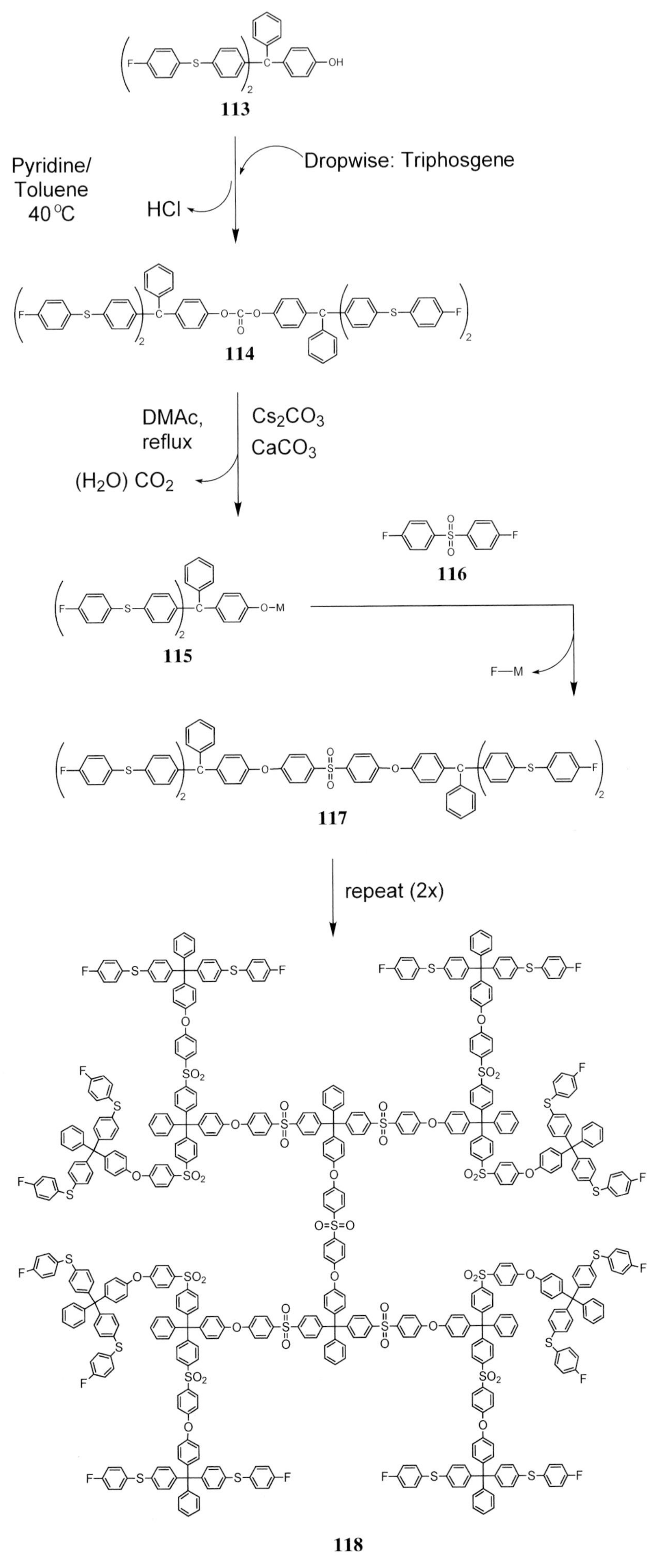

Scheme 3.29 Construction of fluoro-sulfide-based dendrimers[405] via carbonate masked monomers.

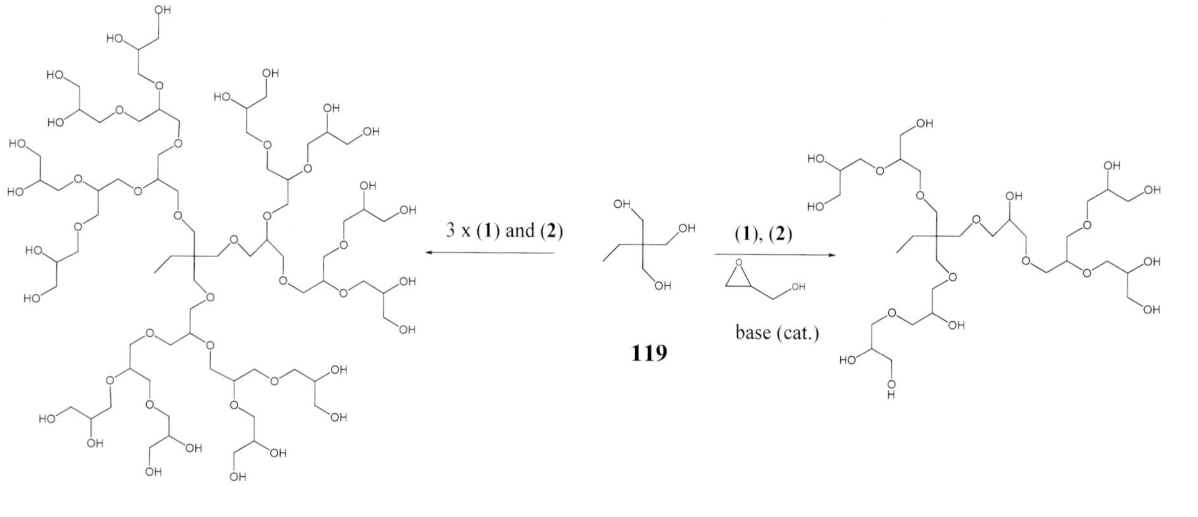

120:[G-3]Dendrimer

121: Hyperbranched polymer

(**1**) = [structure], NaOH, TBAB, H$_2$O

(**2**) = NMO, OsO$_4$ (cat), H$_2$O, Acetone, *t*-BuOH

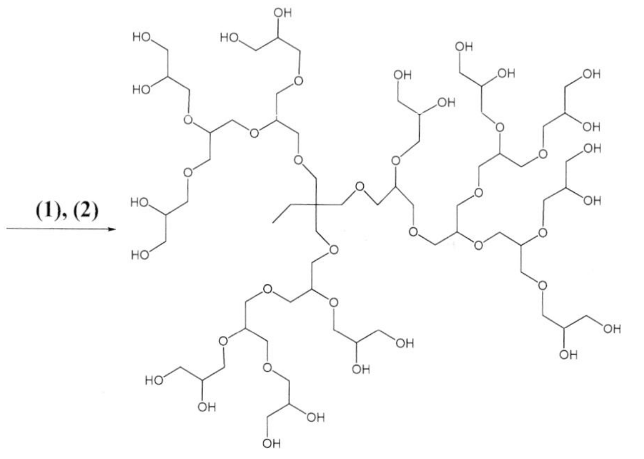

122: Pseudo-dendrimer

Scheme 3.30 "Glycerol" architectures prepared in aqueous media.[406]

3.6   1 → 2 *Si*-Branched and Connectivity

Masamune et al.[407] reported the preparation of the first series of high molecular weight, silicon-branching macromolecules by means of the procedure shown in Scheme 3.31, although low molecular weight poly(siloxanes) were known.[408] Their iterative procedure utilized two differently branched synthetic equivalents: a trifunctional, hydrido-terminated core **123** and a trigonal monomer **125**. Syntheses of the polysiloxane core **123** and building block **125** were each accomplished by the treatment of trichloromethylsilane with three or two equivalents of the siloxane oligomers, HO[Si(Me)$_2$O]$_5$Si(Me)$_2$H and HO[Si(Me)$_2$O]$_3$Si(Me)$_2$H, respectively.

Repetitive Si-based transformations were then employed for dendritic construction. Palladium-catalyzed silane hydroxylation of the core **123** afforded triol **124**, which was then treated with three equivalents of monochloropolysiloxane **125** to generate the hexa-hydrido, 1st generation dendrimer **126**. Further application of the Pd-mediated hydroxyl-ation, followed by attachment of monochloro monomer **125**, led to the 2nd (**127**) and 3rd (**128**) generation polysiloxane cascades.

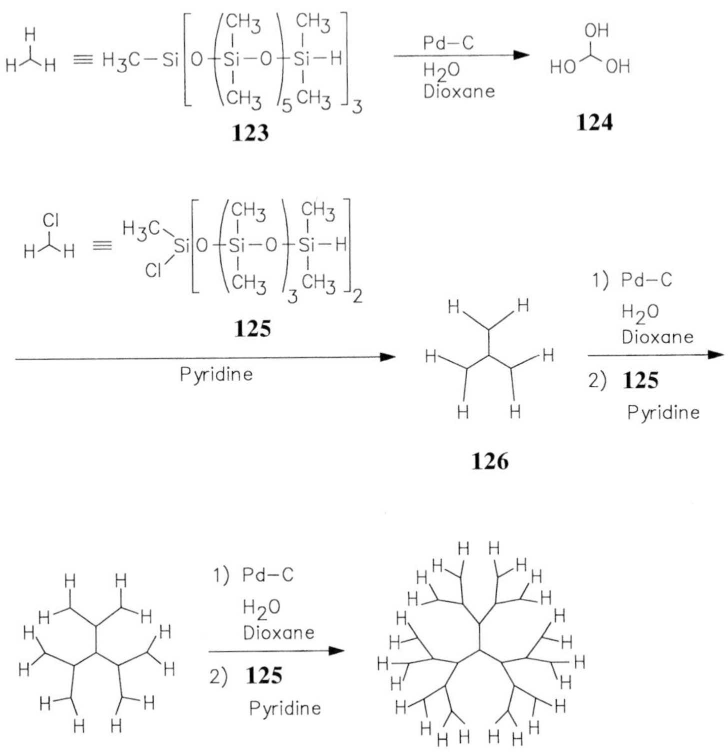

Scheme 3.31 Construction of polysiloxane dendrimers[407] prepared by an iterative silane hydroxylation and chloride displacement.

Morikawa, Kakimoto, and Imai[409, 410] employed divergent methodology for the construction of a series of siloxane-based dendrimers possessing dimethylamino, phenyl, benzyl, or hydroxy peripheral moieties. Sequential tier addition involved transformation of phenylsilane termini to the corresponding silyl bromide (Br$_2$), treatment with HNEt$_2$ to generate the silylamine moiety, and hydroxysilyl monomer displacement of the amino group. Characterization by gel permeation chromatography and ^{1}H NMR was discussed. Polydispersity indices were found to range from 1.30 for the 2nd generation, phenyl-terminated polysiloxane to 1.71 for the 3rd generation, hydroxyl-terminated dendrimer. The water-soluble poly(hydrochloride) salt of the dimethylsilylamine-terminated dendrimer (3rd tier) was compared to a unimolecular micelle as a result of structural similarities.

Roovers et al.[411] synthesized a series of carbosilane dendrimers using Pt-catalyzed addition of methyldichlorosilane (**130**) to an alkene followed by nucleophilic substitution with vinylmagnesium bromide at the terminal dichlorosilane moieties (Scheme 3.32) as the iterative method. Thus, using tetravinylsilane[412–415] (**129**) as the initial tetrafunctional core, the 1st generation tetrakis(methyldichlorosilane) **131** was generated after addition of four equivalents of monomer **130**. Reaction of eight equivalents of vinylmagnesium bromide with pentasilane **131** generated octaolefin **132**. Continued iteration gave rise to the polyalkene **133**, possessing a molecular weight of 6016 amu at the 4th tier and 64 terminal vinyl groups. These dendritic carbosilanes,[413, 416] with 64 and 128 surface Si–Cl bonds, were used as coupling reagents for monodisperse poly(butadienyl)lithium. Two series of regular star polymers with molecular weights between 6,400 and 72,000 amu were prepared, which constituted good models for polymeric micelles. SANS measurements on this series of carbosilanes supported the tendency for enhanced spherical-like behavior with increasing growth.[417]

Comanita and Roovers[418] reported the modification of their protocol to include the terminal attachment of a methyl bis(alkyl THP ether) silane monomer, which facilitated the construction of carbosilanes with hydrophilic termini. These materials were proposed as useful multifunctional anionic initiators for the synthesis of dendrimer–polymer hybrids. Generation zero, one, and two hydroxy-terminated carbosilanes have been employed as cores for the anionic polymerization of ethylene oxide.[419] The "star" character of these polymers was confirmed by analysis of their molecular weights, intrinsic

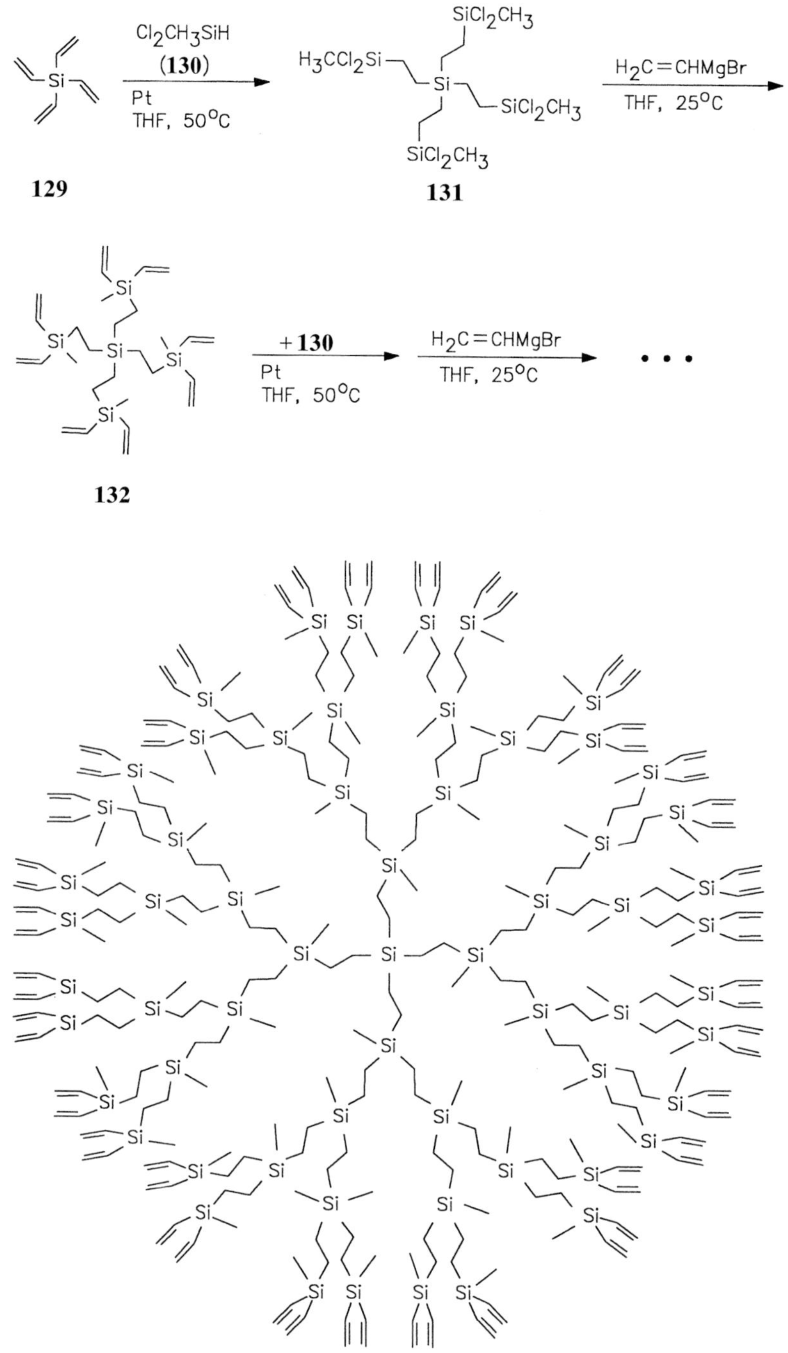

Scheme 3.32 Alkylsilane dendrimer construction by Pt-mediated silane alkenylation and vinylation.[411]

viscosities, and translational diffusion. Coupar, Jaffrès, and Morris[420] have also converted numerous carbosilanes (1 → 2 branched as well as 1 → 3 branched motifs) to their SiOH-terminated derivatives by hydrolysis of terminal SiCl groups. The crystal structure of a 1st generation carbosilane based on a tetravinylsilane core was reported.

Morán et al.[421] described the utilization of a similar[411] procedure (Scheme 3.33), except that allyl spacers were incorporated. Thus, tetra(allyl)silane[421] (**134**), as the core, and a simple allyl Grignard reagent were used. When silane **134** was hydrosilylated with MeCl$_2$SiH (**130**) under Pt-catalyzed conditions, the pentasilane **135** was generated. Subsequent branching was accomplished by reaction of the octachlorosilane (**135**) with allyl-magnesium bromide to afford octaene **136**, which was hydrosilylated (Me$_2$ClSiH) to give the capped chlorosilane **137**. Treatment of this polychlorosilane with either lithio- or aminoethyl-ferrocene gave the corresponding Si-dendrimers coated with ferrocenyl moieties (**138** and **139**, respectively), which were described as non-interacting redox centers (see also Chapter 8, Section 8.5); for Ru(II)tris(bipyridine) peripheral units, see.[421a].

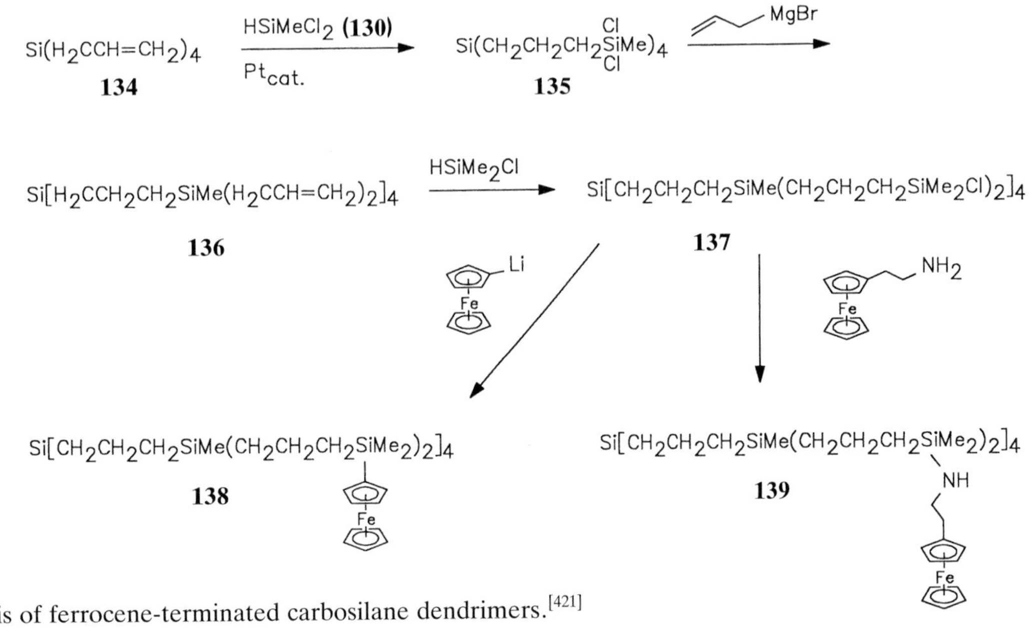

Scheme 3.33 Synthesis of ferrocene-terminated carbosilane dendrimers.[421]

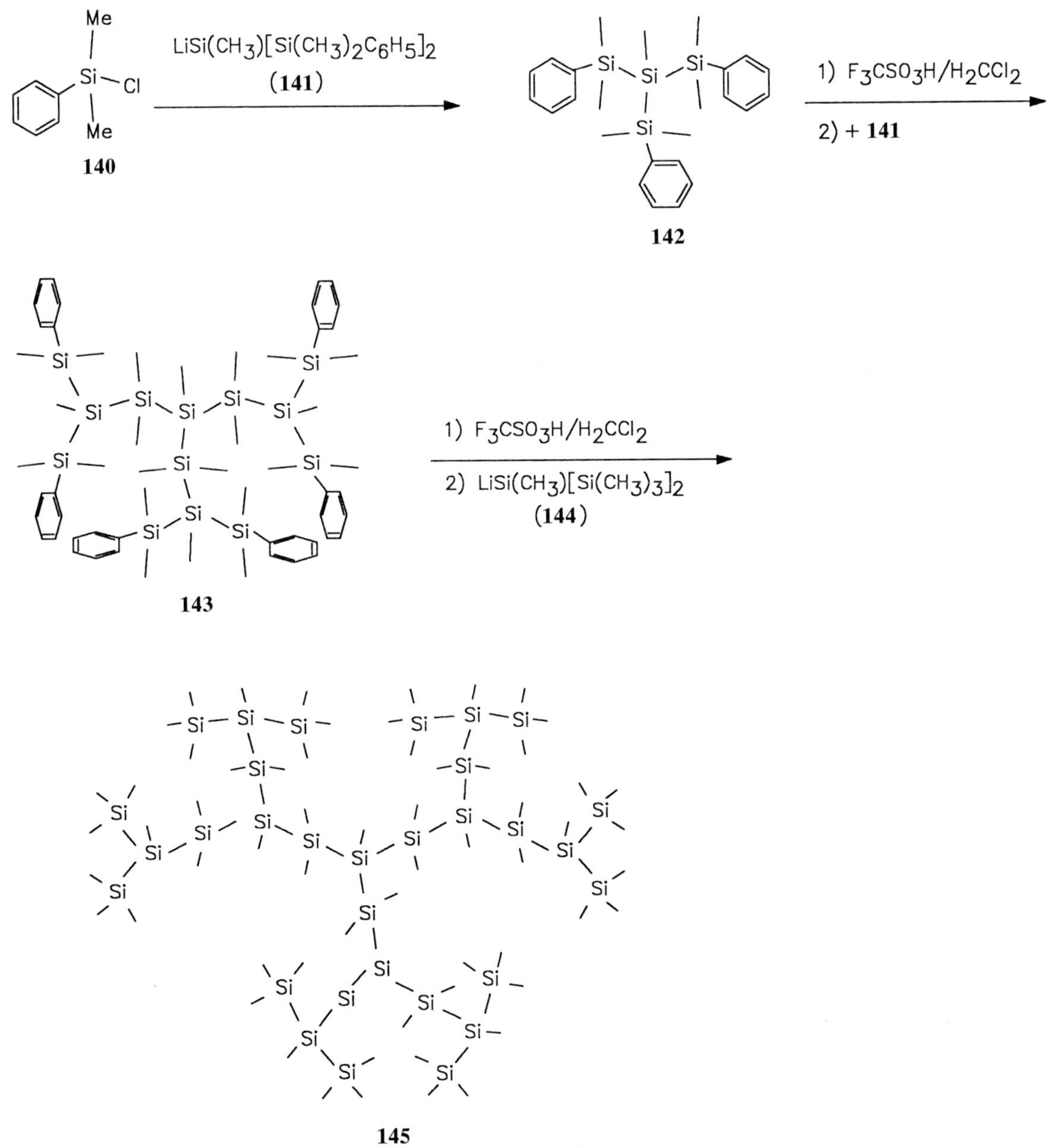

Scheme 3.34 Preparation of Si-based dendrimers[422] with contiguous Si–Si connectivity.

Construction of a new series of polysilane dendrimers, in which the structure of the 2nd generation product was unambiguously confirmed by X-ray diffraction analysis, has been reported.[422] The divergent procedure for the synthesis of polysilane dendrimer **145** is shown in Scheme 3.34. Treatment of core **140** with highly inflammable silyllithium **141**, prepared (80%) by the reaction of bis(1,3-diphenylpentamethyltrisilanyl)mercury with lithium, afforded tetrasilane **142** as colorless crystals. Treatment of the latter with trifluoromethanesulfonic acid, followed by reaction with monomer **141**, generated the next higher level dendrimer **143**. The permethylated polysilane **145** was then prepared (29%) from tridecasilane **143** by a similar two-step sequence, utilizing the capping reagent 2-lithioheptamethyltrisilane [**144**, (Me₃Si)₂Si(Me)Li].

Kim, Park, and Kang[423] reported the construction of chlorosilane dendrimers through to the 3rd generation. Scheme 3.35 shows the general procedure of their repetitive method, whereby the 1st generation dodecaallyl dendrimer **146** was subjected to Pt-mediated hydrosilylation (MeCl₂SiH) to give dodeca(dichlorosilane) **147**, which was then reacted with allylmagnesium bromide to afford the 2nd generation polyalkene **148**. These authors noted difficulties in attempting to obtain the 4th generation from the preceding generation (48 termini) by this hydrosilylation procedure. Kim et al. subsequently reported the preparation of higher generation silane-based dendrimers possessing 64[424] and 96[425] allylic termini using similar technology. This methodology has been expanded to include the preparation of cylindrical dendrimers possessing a polycarbosilane backbone[426] (see also Chapter 9).

Kim et al.[427–429] prepared poly(unsaturated) carbosilane dendrimers employing a modification of the standard Pt-mediated hydrosilylation–chlorosilane allylation protocol (Scheme 3.36); here, lithium phenylacetylide was used as the alkylating agent. Thus, hydrosilylation of the rigid core **149** gave tetraalkene **150**, which was subsequently treated with lithium alkynide **153** to yield the enyne construct **151**. Repetition of the sequence afforded the corresponding hexadecachloride **152** and higher generation poly(enyne)s **154** and **155**. Defects were noted on proceeding to the 3rd generation poly(chlorosilane). Nevertheless, use of the monochlorosilane reagent (Me₂ClSiH) proved successful and

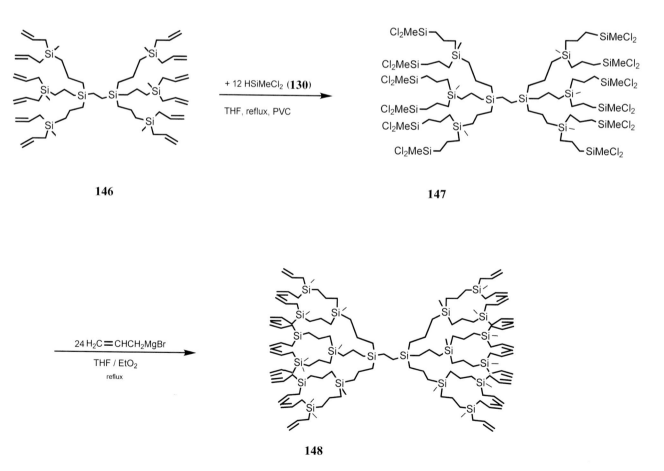

Scheme 3.35 Kim's method for the preparation of carbosilane dendrimers.[423]

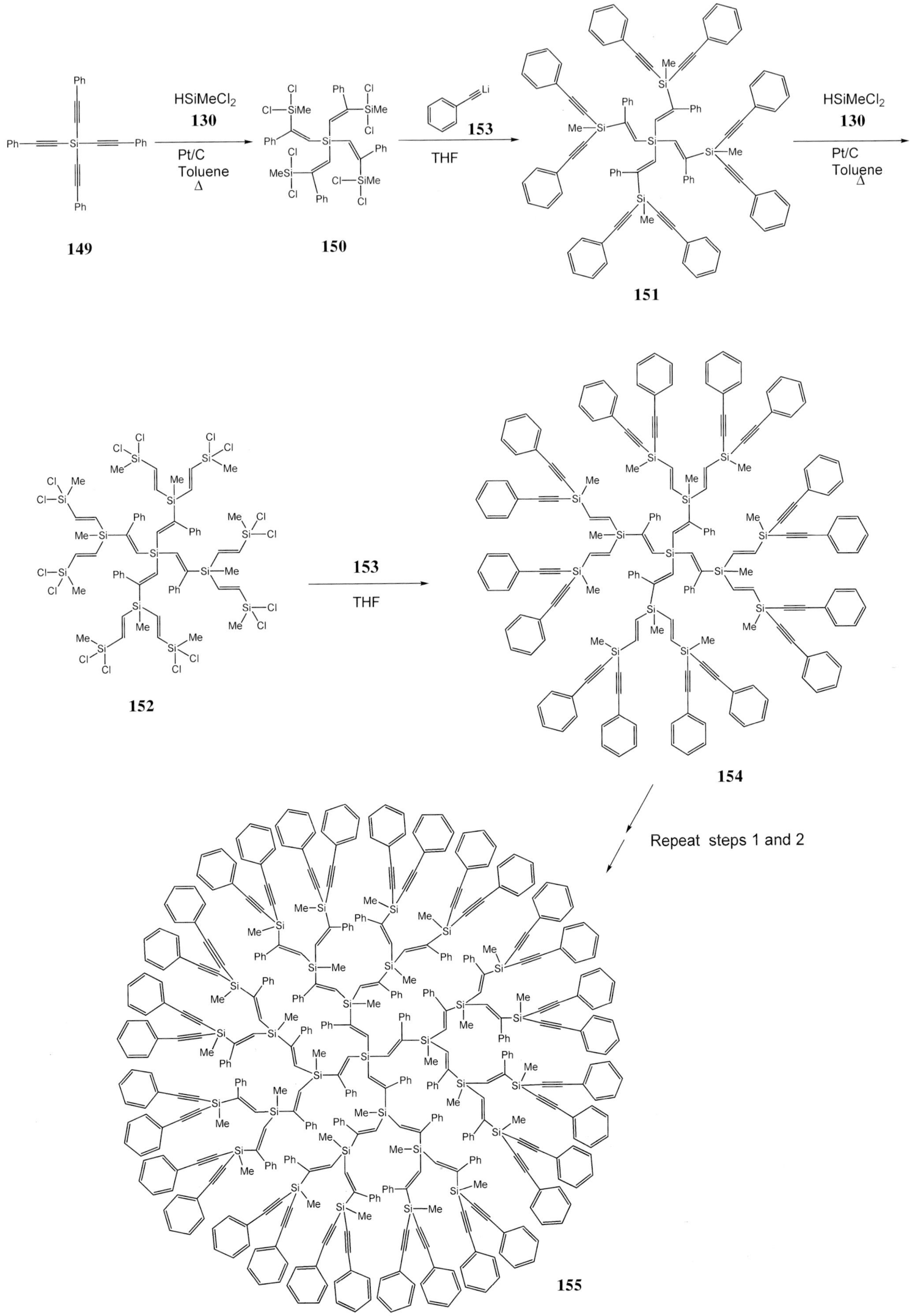

Scheme 3.36 Kim's novel poly(unsaturated)carbosilanes.[427]

allowed final capping with the acetylide. Similar lithium acetylide-based chemistry was performed on a siloxane polymer core to produce rod-shaped architectures.[430] Later, Kim and Ryu[431] combined the lithium acetylide protocol with that of the allyloxy-based method to yield "double-layered" (i.e., diblock architectured) dendrimers. A dendritic macromolecule possessing 144 phenylethynyl moieties has been prepared;[432] this construct was derived from 1,3,5-tris(dimethylvinylsilyl)benzene and bis(phenylethynyl)methylsilyl groups.

Kim and Park[433] further modified this protocol to include propargyl alcohol as the alkylating agent, thereby producing carbosilane architectures with internal as well as external unsaturated sites (Figure 3.8; **156**); dendrimers up to the 4th generation were reported. The description of double-layered carbosilane dendrimers possessing 96 phenylethynyl groups at the periphery, constructed by hydrosilylation and alkenylation, as well as by alkynylation, has recently appeared.[434]

Hydroxyl-terminated carbosilanes possessing excellent amphiphilic properties have been prepared and reported by Getmanova et al.[435] The spreading of these carbosilane constructs, possessing trimethylsilyl or hydroxymethyl termini, at the air/water interface has been examined.[436, 437] Kim et al.[438] reported the conversion of the allylic termini of this scaffolding to the corresponding hydroxyl moieties by hydroboration (9-BBN).

Shibaev and coworkers[439] described the synthesis of a 1st generation, liquid-crystalline dendrimer[440] employing carbosilane-based scaffolding. Scheme 3.37 depicts the hydroxyl-based, mesogen modification,[440a] i.e., cyanobiphenyl,[441] methoxyphenyl benzoate, cholesteryl, undertaken to facilitate dendrimer attachment. 10-Undecylenic acid chloride was esterified by reaction with the appropriate alcohol, the ester was hydrosilylated [{(C₈H₁₇)₃PhCH₂N}₂Pt(NO₂)₄] using Me₂ClSiH, and then dual hydrolysis of the mesogenic chlorosilane and more Me₂ClSiH afforded the desired tetramethyldisiloxane mesogen **157**. Modification of the carbosilane **136** (Scheme 3.38) with the mesogenic siloxane **157** was affected by silylation (Pt catalysis) of the unsaturated termini to afford octamesogen **158**. On the basis of DSC, optical polarizing microscopy, and X-ray diffraction data,

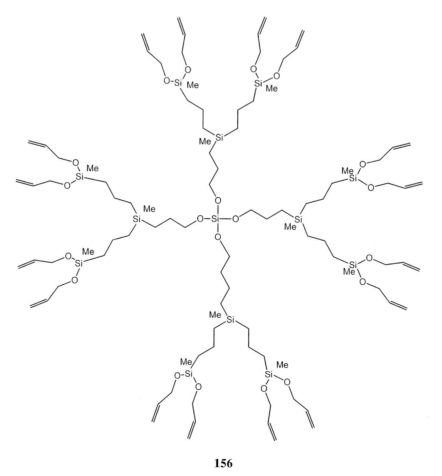

156

Figure 3.8 Carbosilane architectures prepared by propargyl alcohol displacement of Cl–Si moieties.[433]

it was surmised that different smectic mesophases were formed by the different substituted dendrimers. This type of carbosilane architecture has also been prepared starting with the cyclic siloxane tetramer [Me(CH$_2$=CH)SiO]$_4$.[442] Terminal allyl functionalization has been partially extended by hydrosilylation–addition of didecylmethylsilane followed by partial lithiation with butyllithium; the internal carbanionic sites were shielded from intermolecular interactions and did not aggregate.[443]

Ignat'eva et al.[444] employed a repetitive Grignard–hydrosilylation sequence for the construction of a series of poly(allyl-terminated) carbosilanes through to the 7th generation. Thermomechanical properties were examined and the T_g was found to attain a constant value from generation four onwards. The excitation dynamics of these poly(allylcarbosilane)s labeled with pyrene moieties has appeared.[445]

Carbosilanes of the 1st and 2nd generation terminated with chloro, amino, ammonium, and sulfonic acid sodium salt moieties have been analyzed by MALDI-TOF mass spectrometry by Wu and Biemann;[446] while others have introduced lactose and sialyllactose moieties.[446a]

Brüning and Lang[447] employed alternating allylmagnesium bromide and allyl alcohol additions to dichlorosilane termini to afford alternating silane–siloxane generational branching. The authors later reported[448] the repetitive use of a hydrosilylation–alcoholysis protocol for the preparation of linear as well as branched carbosiloxane architectures.

Boysen and Lindhorst[449] selectively protected a single saccharide alcohol moiety and elaborated the remaining hydroxyl group to construct a 1st tier carbosilane motif.

The 1 → 2 branched, chloromethyl-terminated dendrimers have been converted to their alcohol, dimethylamino, and sodium sulfonate derivatives.[450] The water-soluble

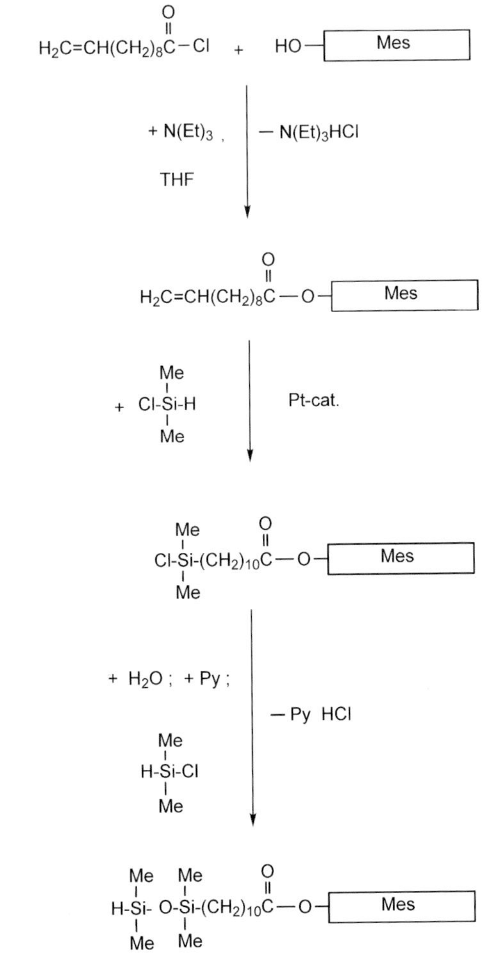

Scheme 3.37 Mesogen modification for attachment to alkene-terminated dendrimers.[439]

sulfonates and poly(ammonium) salts were demonstrated to enhance the solubility of lipophilic alkyl-substituted benzene derivatives, thereby illustrating their micellar potential.

These types of carbosilanes have also been characterized by employing ^{1}H/^{13}C/^{29}Si triple-resonance 3D NMR techniques.[451]

This silane-based architecture has been terminally modified with nonpolar fluorinated alkyl chains and characterized by ApcI mass spectrometry and small-angle X-ray scattering.[452]

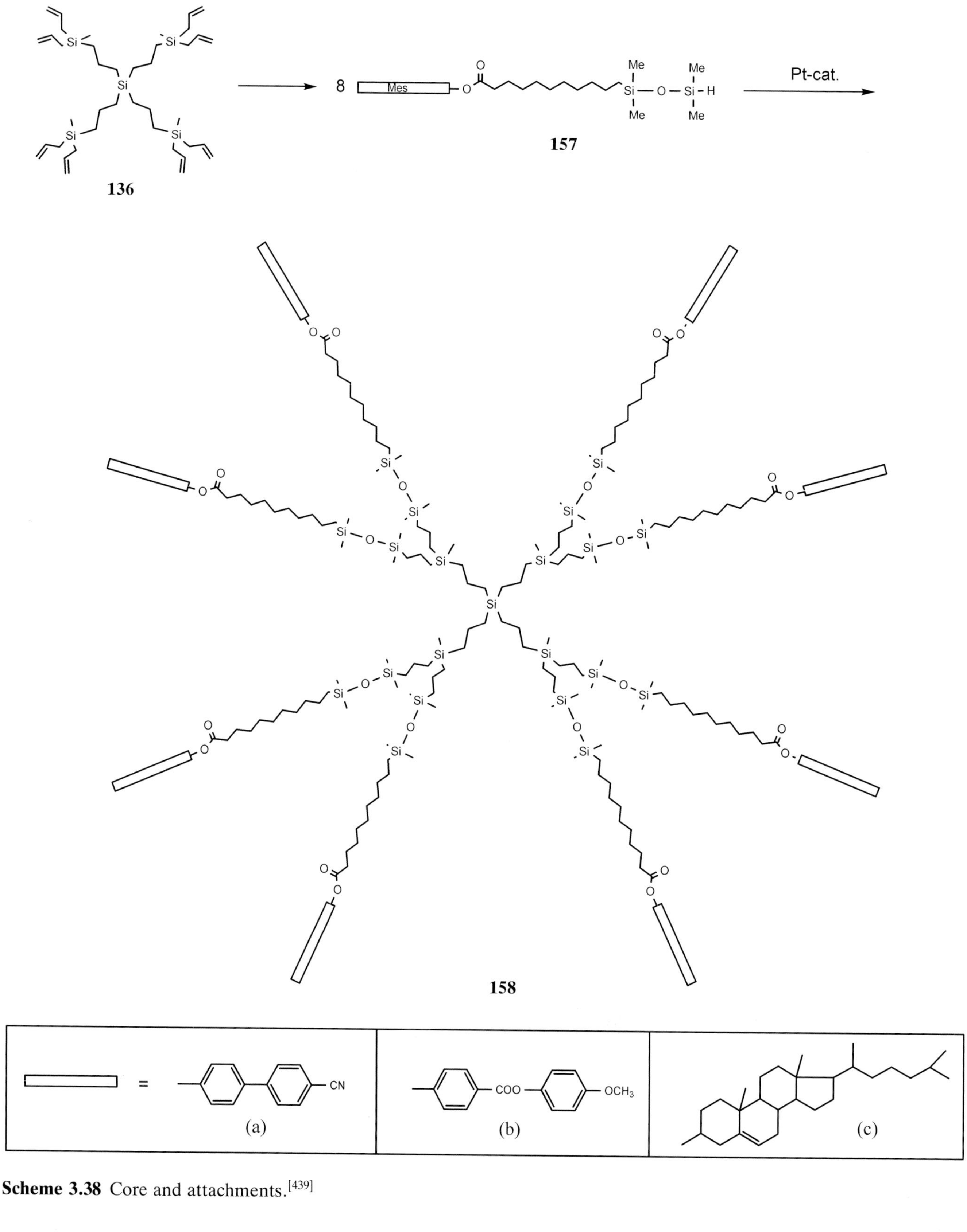

Scheme 3.38 Core and attachments.[439]

Casado and Stobart[453] created monomers adapted for sequential divergent and convergent steps for the construction of carbosilanes.

Recently, van Koten et al.[454] described a general method for the terminal modification of Si–Cl coated carbosilanes using organolithium or organomagnesium reagents and the development of polycationic core-shell derivatives as phase-transfer catalysts; host-guest chemistry was also described.[454a]

3.7   1 → 2 *P*-Branched and Connectivity

DuBois and coworkers[455] reported the simple construction of *P*-dendrimers by the free radical addition of primary phosphines to diethyl vinylphosphonate, followed by reduction (LAH) of the phosphonate moieties to give the corresponding polyphosphines (Scheme 3.39). Thus, tris(phosphine) (**160**), prepared by known procedures from diphosphonate **159**,[456, 457] was subjected to this method to generate heptaphosphine **161** or was treated with vinyldiphenylphosphine to afford tetrakis(diphenylphosphine) **162**. This termination procedure was also used on phosphine **163** to give the phenyl-capped **164**, whereas treatment of the corresponding ethyl-terminated dendrimer **165** with five equivalents of [Pd(MeCN)$_4$](BF$_4$)$_2$ resulted in pentametallation. The corresponding site-specific metal complex is described in more detail in Chapter 8 (Section 8.5).

The application of this approach to *P*-dendrimers could readily be applied[455] to the creation of a tetrahedral series through the use of a four-directional silane core (Scheme 3.40). Treatment of tetravinylsilane (**129**) with phosphine **166** quantitatively afforded the

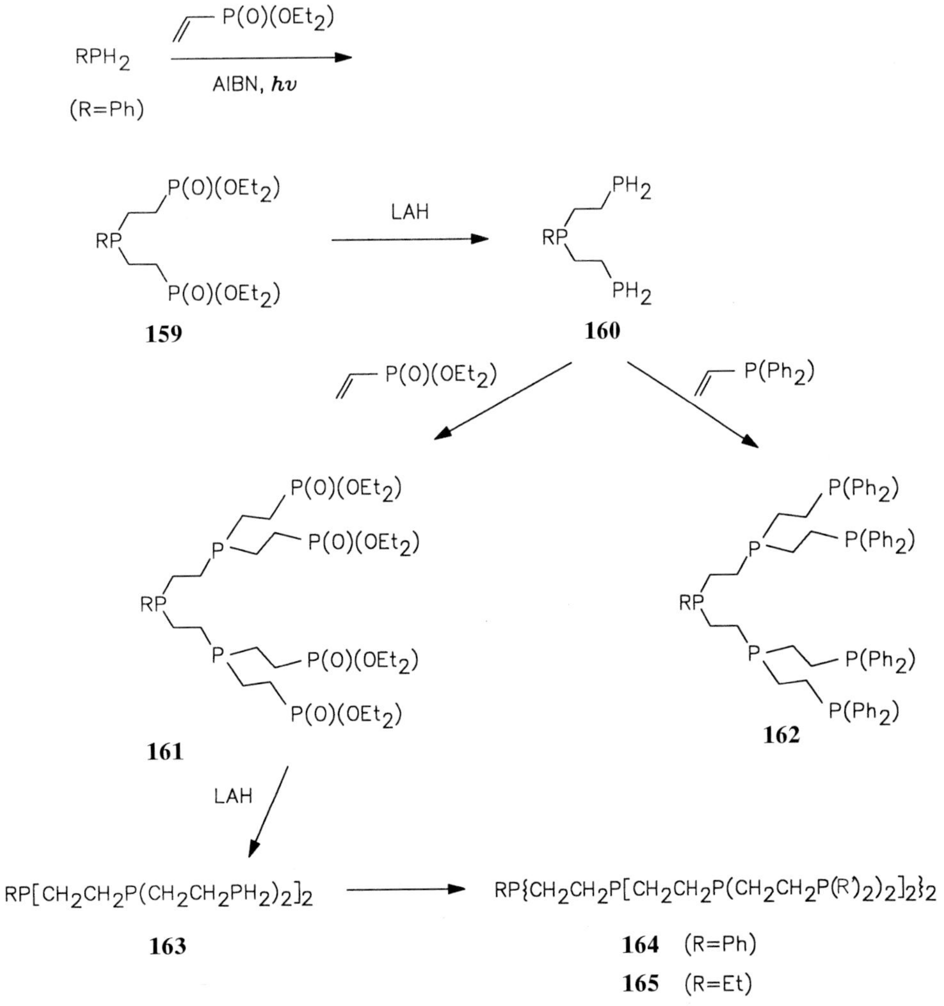

Scheme 3.39 DuBois et al.[455] prepared P-based cascades to examine their potential in metal ion complexation.

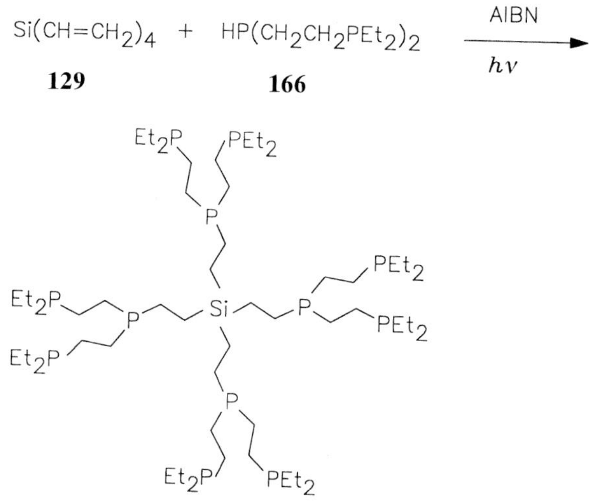

$$\text{Si(CH=CH}_2)_4 \quad + \quad \text{HP(CH}_2\text{CH}_2\text{PEt}_2)_2 \quad \xrightarrow[hv]{\text{AIBN}}$$

129   **166**

167

Scheme 3.40 Free radical mediated construction[455] of mixed Si/P-based dendrimers.

desired "small" dendrimer **167**, which could be transformed to the tetrakis(square planar palladium) complex; see Chapter 8.

Majoral et al.[458] reported the facile divergent synthesis of a novel *P*-dendrimer series (Scheme 3.41). Treatment of the sodium salt of 4-hydroxybenzaldehyde (**168**) with trichlorothiophosphorus(V) gave the trialdehyde **169**, reaction of which with three equivalents of the hydrazine derivative **170** quantitatively afforded the 1st generation dendrimer **171** possessing six P–Cl bonds suitably arranged for repetition of the sequence. Construction of the 2nd, 3rd, and 4th (e.g., **172**) generations was achieved following this iterative sequence. Key features of this sequence were that no protection–deprotection procedures were required and that the only by-products were NaCl and water, assuming quantitative transformations. In a subsequent paper, the expansion of these P-dendrimers to the 5th, 6th, and 7th generations possessing up to 384 functional groups was reported.[459] Facile functional group manipulation at the periphery allowed the attachment of α,β-unsaturated ketones, crown ethers, and alcohols. Treatment of the PCl$_2$ moieties with bis(allyl)amine afforded the monosubstituted termini [i.e., P(Cl)N(CH$_2$CH=CH$_2$)$_2$]. The surface-incorporated crown ethers interestingly acted as "shields" with respect to attempted imine hydrolysis [THF, H$_2$O (4:1), 25 °C, 48 h]. Spectral characterization of the structure included ^{31}P NMR data; no overlapping ^{31}P resonances were observed until the 4th tier.

The synthetic and spectral details of *P*-dendrimers up to the 3rd generation based on a cyclotriphosphazene core [N$_3$P$_3$(OC$_6$H$_4$CHO)$_6$][460, 461] possessing six formyl moieties have also been reported.[462] The simple procedures led to a spherical surface bearing electrophilic or nucleophilic reactive moieties, such as aldehydes, hydrazones, and aminophosphines. Small *P*-based dendrimers starting from (S)P[N(Me)NH$_2$]$_3$ have also been reported.[463] Additional notable chemistry of interest associated with these novel dendrimers includes:[464–469] phosphate-, phosphite-, ylide-, and phosphonate-surface modification;[470] amino acid termination by means of the Horner–Wadsworth–Emmons reaction,[471] dipole moment measurements,[472] synthesis based on phosphoryl group chemistry,[473] chemoselective polyalkylations,[474] chiroptical properties of stereogenically terminated materials,[475] coating of the surface with different types of tetraazamacrocycles,[476] regiospecific functionalization following construction,[477] chemoselective internal functionalization,[478] and layer-block construction with regular alternation of repeat units [RP(S)/RP(O)].[479] More recent reports include hybrid materials,[479a–e] redox-active,[479d–f] and multidentritic systems.[479g] These *P*-based dendrimers have further been terminated with conjugated poly(thiophene) chains affording electroactive materials, as demonstrated by cyclic voltammetry, and electronic absorption spectra have been recorded.[480] *N*-Thiophosphorylated and *N*-phosphorylated iminophosphoranes have also been reported[481] as models for these dendrimers. Schmid et al.[482] used thiol-terminated, Majoral-type dendrimers as matrices for perfect crystal growth. Gold clus-

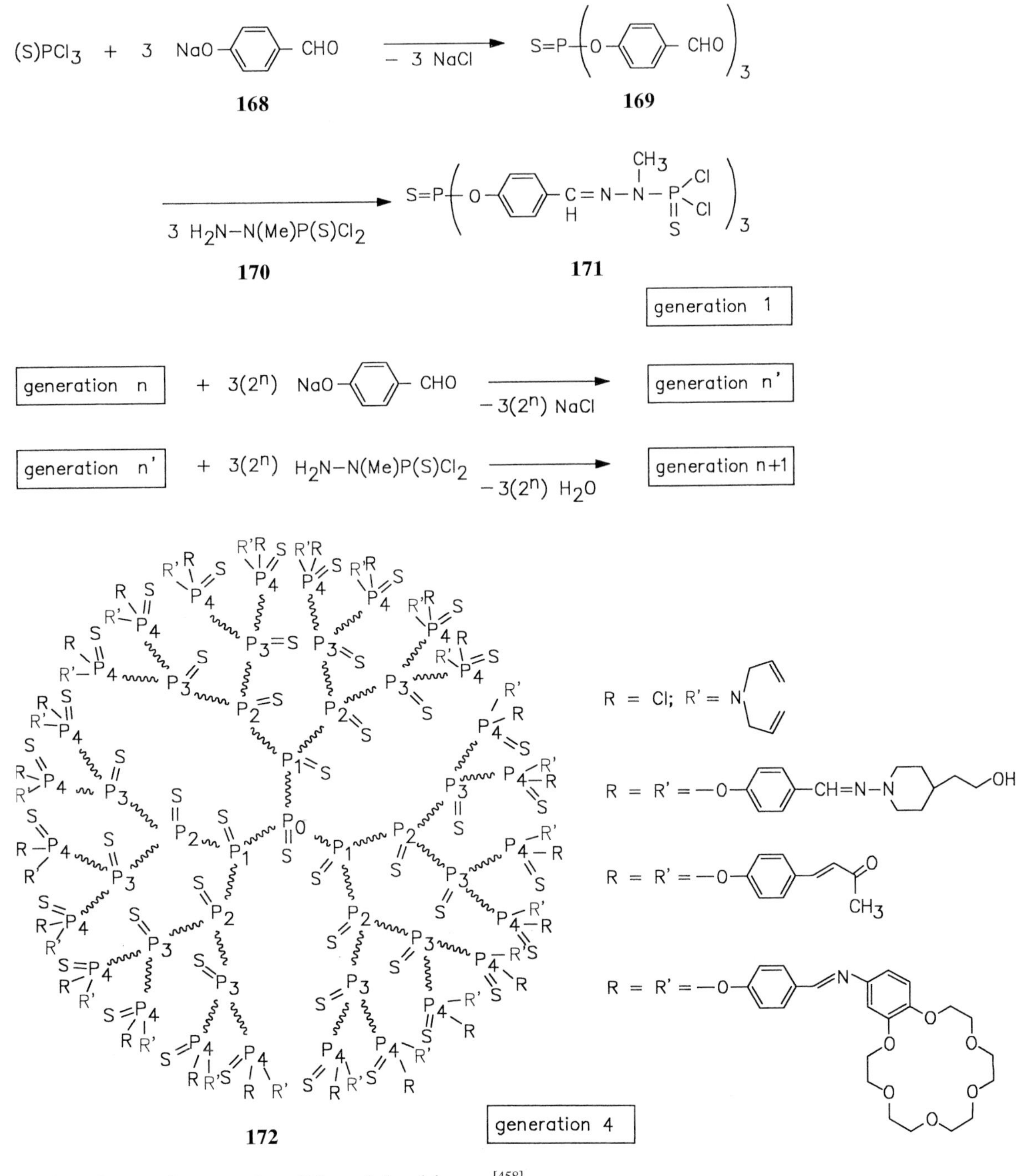

$(S)PCl_3$ + 3 NaO—⟨ ⟩—CHO → (− 3 NaCl) → $S=P(-O-⟨ ⟩-CHO)_3$

168   **169**

→ (3 $H_2N-N(Me)P(S)Cl_2$) → $S=P\left(-O-⟨ ⟩-\underset{H}{C}=N-\underset{CH_3}{N}-\underset{S}{\overset{Cl}{\underset{\|}{P}}}Cl\right)_3$

170   **171**

generation 1

| generation n | + 3(2^n) NaO—⟨ ⟩—CHO | → (−3(2^n) NaCl) → | generation n' |

| generation n' | + 3(2^n) $H_2N-N(Me)P(S)Cl_2$ | → (−3(2^n) H_2O) → | generation n+1 |

generation 4

172

R = Cl; R' = N(...)

R = R' = —O—⟨ ⟩—CH=N—N(piperidine)—CH₂CH₂OH

R = R' = —O—⟨ ⟩—CH=CH—C(=O)—CH₃

R = R' = —O—⟨ ⟩—=N—(benzo-crown ether)

Scheme 3.41 A series of neutral pentavalent P-based dendrimers.[458]

ters were formed, which coalesced into "well-formed" microcrystals. Examples of surface-block, layer-block, and segment-block architectures have recently been described.[483]

Majoral et al.[484, 485] demonstrated the versatility of another cyclotriphosphazene (**173**; Scheme 3.42) as a starting core for the synthesis of dendrimers. This core was obtained in high yields by treatment of hexachlorocyclotriphosphazene with methylhydrazine. Hexaamine **173** was condensed with *p*-hydroxybenzaldehyde hexaphenol (**174**), and then the product was treated with PPh₂Cl to give the corresponding hexadiphenylphosphene **175**. Reaction of **175** with azidophosphodihydrazide[486] **176** yielded the 12-amino-terminated dendrimer **177**, which was "not isolated in a pure state", but nevertheless fully characterized by ³¹P NMR, the data being consistent with the assigned structure. Similar chemistry has been employed for the synthesis through 8 generations of "bowl-shaped" dendrimers,[487] the largest of which possessing a purported 1536 peripheral aldehyde moieties.

A family of poly(organophosphazenes) possessing ethyleneoxy side-chains has been reported for investigation of their solid-electrolyte properties.[488]

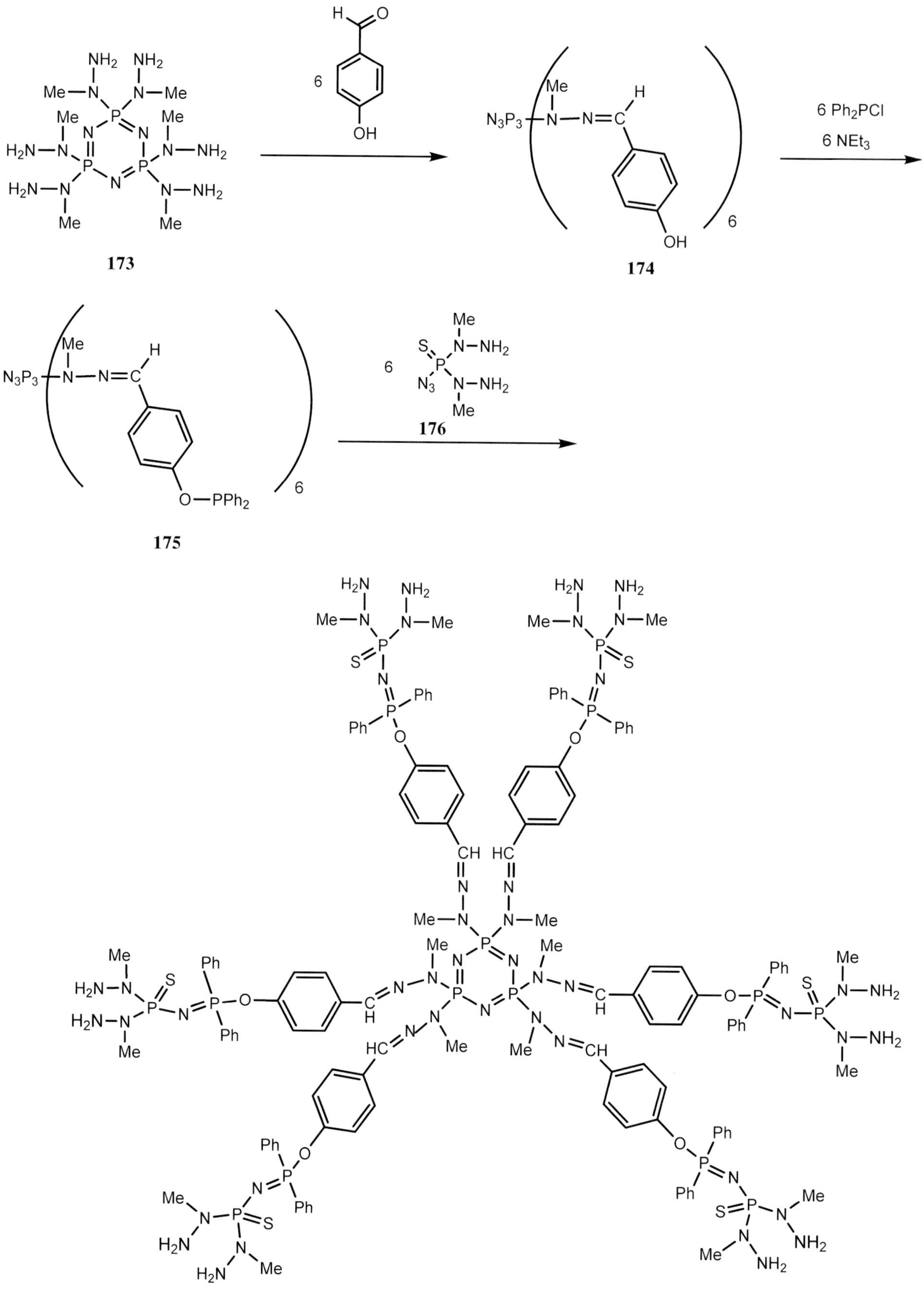

Scheme 3.42 Use of a cyclotriphosphazene core for dendrimer construction.[484]

Majoral, Caminade et al.[489] also reported tri- and tetrafunctionalization at the terminal units of their *P*-based dendrimers (Scheme 3.43). The introduction of multiple substituents at each chain end [P(S)Cl$_2$ or P(O)Cl$_2$] was predicated on quantitative and selective monosubstitution at each terminus prior to the observation of disubstitution. Thus, reaction of the dichlorothiophosphoryl moieties (**178**) with one equivalent each of allylamine afforded the trifunctional chlorophosphoryl amido alkene **179**. Reaction of dichlorooxaphosphoryl termini (**180**) proceeded with similar selectivity to give the desired monoalkene **181**, while similar reactivities were also observed when propargylamine was used instead of allylamine (e.g., **182**). Treatment with one equivalent of allylamine followed by one equivalent of propargylamine yielded the desired ene-yne moiety **183**. Attempts to react excess bis(allyl)amine so as to obtain the tetraalkene gave only the monosubstituted product **184**, which was further treated with propargylamine to smoothly afford the diene-yne **185**. Nitrile surface groups were also introduced by the Wittig reaction of phosphate aldehyde **186** with nitrile **187** to give the ene-nitrile **188**. In all, 40 new dendrimers were reported.

Majoral et al.[490] further reported the terminal modification of these *P*-based dendrimers, such as that shown in Scheme 3.41; benzaldehyde-coated dendrimers were reacted with a variety of amine- and hydrazine-based reagents. These included hydrazine, methylhydrazine, 1-amino-4-(2-hydroxyethyl)piperazine, fluorenone hydrazone, and 4-aminobenzo-15-crown-5. Wittig transformations affording α,β-unsubstituted ketone and nitrile termini were effected by reaction with (acetylmethylene)- or (cyanomethylene)triphenylphosphorane, respectively. Finally, exhaustive substitution of the P(S)Cl$_2$ termini was achieved by reaction with allylamine, propargylamine, and *N*-(trimethylsilyl)imida-

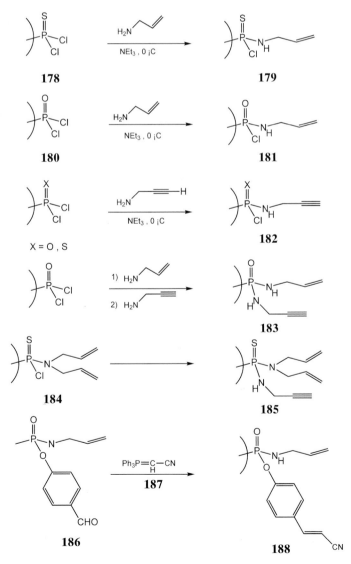

Scheme 3.43 Selective surface functionalization of dichlorophosphoryl termini.[489]

zole. Selective alkylation of an internal sulfur moiety [P=N–P(S)] by treatment with methyl trifluoromethylsulfonate, followed by reaction with tris(dimethylamino)phosphane, produced highly reactive P=N–P units within these *P*-dendrimers.[491] Addition of the azide $N_3P(S)(OC_6H_4CHO)$ generated a starting point for branched construction on the internal framework. The preparation of thiophosphate-based dendrimers up to the 5th generation, predicated on phosphitylation of propanediol derivatives and reaction with elemental sulfur, has been reported.[492]

3.8 References

[1]   E. Buhleier, W. Wehner, F. Vögtle, "'Cascade' and 'Nonskid-Chain-like' Syntheses of Molecular Cavity Topologies", *Synthesis* **1978**, 155–158.

[2]   R. Moors, F. Vögtle, "Dendrimere Polyamine", *Chem. Ber.* **1993**, *126*, 2133–2135.

[3]   F. M. Menger, "Chemistry of Multi-Armed Organic Compounds", *Top. Curr. Chem.* **1986**, *136*, 1–15.

[4]   G. R. Newkome, C. N. Moorefield, "Cascade Molecules", in *Mesomolecules From Molecules to Materials, Struct. Energ. React. Chem. Ser.* (Eds.: G. D. Mendenhall, A. Greenberg, J. F. Liebman), Chapman & Hall, New York, **1995**, pp. 27–68.

[5]   R. G. Denkewalter, J. F. Kolc, W. J. Lukasavage, "Preparation of Lysine-Based Macromolecular Highly Branched Homogeneous Compound", **1979**, *U. S. Pat.*, 4, 360, 646.

[6]   R. G. Denkewalter, J. F. Kolc, W. J. Lukasavage, "Macromolecular Highly Branched Homogeneous Compound Based on Lysine Units", **1981**, *U. S. Pat.*, 4, 289, 872.

[7]   R. G. Denkewalter, J. F. Kolc, W. J. Lukasavage, "Macromolecular Highly Branched Homogeneous Compound", **1983**, *U. S. Pat.*, 4, 410, 688.

[8]   G. R. Newkome, Z. Yao, G. R. Baker, V. K. Gupta, "Cascade Molecules: A New Approach to Micelles. A [27]-Arborol", *J. Org. Chem.* **1985**, *50*, 2003–2004.

[9]   D. A. Tomalia, H. Baker, J. Dewald, M. Hall, G. Kallos, S. Martin, J. Roeck, J. Ryder, P. Smith, "A New Class of Polymers: Starburst-Dendritic Macromolecules", *Polym. J. (Tokyo)* **1985**, *17*, 117–132.

[10]   C. Wörner, R. Mülhaupt, "Polynitrile- and Polyamine-Functional Poly(trimethylene imine) Dendrimers", *Angew. Chem.* **1993**, *105*, 1367–1369; *Int. Ed. Engl.* **1993**, *32*, 1306–1308.

[11]   R. J. Bergeron, J. R. Garlich, "Amines and Polyamines from Nitriles", *Synthesis* **1984**, 782–784.

[12]   E. M. M. de Brabander-van den Berg, E. W. Meijer, "Poly(propylene imine) Dendrimers: Large-Scale Synthesis via Heterogeneously Catalyzed Hydrogenation", *Angew. Chem.* **1993**, *105*, 1370–1373; *Int. Ed. Engl.* **1993**, *32*, 1308–1311.

[13]   E. W. Meijer, H. J. M. Bosman, H. A. M. J. Vandenbooren, E. M. M. de Brabander-van den Berg, A. M. C. F. Castelijns, H. C. J. De Man, R. W. E. C. Reintjens, C. J. C. Stoelwinder, A. J. Nijenhuis, "Dendritic Macromolecule and the Preparation Thereof", **1996**, *U. S. Pat.*, 5, 610, 268.

[14]   M. H. P. van Genderen, M. W. P. L. Baars, J. C. M. van Hest, E. M. M. de Brabander-van den Berg, E. W. Meijer, "Observing Individual Generations in Poly(propylene imine) Dendrimers with Natural Abundance ^{15}N-NMR Spectroscopy", *Recl. Trav. Chim. Pays-Bas* **1994**, *113*, 573–574.

[15]   M. H. P. van Genderen, M. W. P. L. Baars, C. Elissen-Román, E. M. M. de Brabander-van den Berg, E. W. Meijer, "Natural Abundance ^{15}N-NMR Spectroscopic Investigations of Poly(propylene imine) Dendrimers", *Polym. Mater. Sci. Eng.* **1995**, *73*, 336–337.

[16]   M. Chai, Y. Niu, W. J. Youngs, P. L. Rinaldi, "3D NMR Studies of DAB-16 Dendrimer", *Macromolecules* **2000**, *33*, 5395–5398.

[17]   E. M. M. de Brabander-van den Berg, A. Nijenhuis, M. Mure, J. Keulen, R. Reintjens, F. Vandenbooren, B. Bosman, R. de Raat, T. Frijns, S. v. d. Wal, M. Castelijns, J. Put, E. W. Meijer, "Large-Scale Production of Polypropylenimine Dendrimers", *Makromol. Chem., Macromol. Symp.* **1994**, *77*, 51–62.

[18]   E. M. M. de Brabander, J. Brackman, M. Mure-Mak, H. de Man, M. Hogeweg, J. Keulen, R. Scherrenberg, B. Coussens, Y. Mengerink, Sj. van der Wal, "Polypropylenimine Dendrimers: Improved Synthesis and Characterization", *Macromol. Symp.* **1996**, *102*, 9–17.

[19]   E. W. Meijer, H. J. M. Bosman, F. H. A. M. J. Vandenbooren, E. M. M. de Brabander-van den Berg, A. M. C. F. Castelijns, H. C. J. De Man, R. W. E. C. Reintjens, C. J. C. Stoelwinder, A. J. Nijenhuis, "Dendritic Macromolecules and the Preparation Thereof", **1996**, *U. S. Pat.*, 5, 530, 092.

[20] C. J. C. Stoelwinder, E. M. M. de Brabander-van den Berg, A. J. Nijenhuis, "Dendritic Macromolecule and Process for the Preparation Thereof", **1997**, *U. S. Pat.*, 5, 698, 662.

[21] J. C. Hummelen, J. L. J. van Dongen, E. W. Meijer, "Electrospray Mass Spectrometry of Poly(propylene imine) Dendrimers – The Issue of Dendritic Purity or Polydispersity", *Chem. Eur. J.* **1997**, *3*, 1489–1493.

[22] E. M. M. de Brabander, J. A. Put, "Polypropylenimine Dendrimers: Synthesis, Characterization and Applications" *Polym. Mater. Sci. Eng.* **1995**, *73*, 79.

[23] D. Pötschke, M. Ballauff, P. Lindner, M. Fischer, F. Vögtle, "Analysis of the Structure of Dendrimers in Solution by Small-Angle Neutron Scattering Including Contrast Variation", *Macromolecules* **1999**, *32*, 4079–4087.

[24] A. Topp, B. J. Bauer, T. J. Prosa, R. Scherrenberg, E. J. Amis, "Size Change of Dendrimers in Concentrated Solution", *Macromolecules* **1999**, *32*, 8923–8931.

[25] R. Scherrenberg, B. Coussens, P. van Vliet, G. Edouard, J. Brackman, E. de Brabander, K. Mortensen, "The Molecular Characteristics of Poly(propyleneimine) Dendrimers as Studied with Small-Angle Neutron Scattering, Viscosimetry, and Molecular Dynamics", *Macromolecules* **1998**, *31*, 456–461.

[26] W. G. Rothschild, "Fractality and its Measurements" in *Fractals in Chemistry*, John Wiley & Sons, Inc., New York, **1998**, pp. 170–179.

[27] A. Ramzi, R. Scherrenberg, J. Brackman, J. Joosten, K. Mortensen, "Intermolecular Interactions between Dendrimer Molecules in Solution Studied by Small-Angle Neutron Scattering", *Macromolecules* **1998**, *31*, 1621–1626.

[27a] C. N. Likos, M. Schmidt, H. Löwen, M. Ballauff, D. Pötschke, P. Lindner, "Soft Interaction between Dissolved Flexible Dendrimers: Theory and Experiment", *Macromolecules* **2001**, *34*, 2914–2920.

[28] L. Bu, W. K. Nonidez, J. W. Mays, N. B. Tan, "MALDI/TOF/MS and SEC Study of Astromol Dendrimers Having Cyano End Groups", *Macromolecules* **2000**, *33*, 4445–4452.

[29] M. H. P. van Genderen, E. M. M. de Brabander-van den Berg, E. W. Meijer, "Poly(propylene imine) Dendrimers" in *Advances in Dendritic Macromolecules* (Ed.: G. R. Newkome), JAI Press, Inc., Stamford, CN, **1999**, pp. 61–105.

[30] A. W. Bosman, H. M. Janssen, E. W. Meijer, "About Dendrimers: Structure, Physical Properties, and Applications", *Chem. Rev.* **1999**, *99*, 1665–1688.

[31] G. J. M. Koper, "Optical Properties of Colloidal Films", *Colloids Surf. A.: Physicochem. Eng. Aspects* **2000**, *165*, 39–57.

[31a] Y. Li, C. A. McMillan, D. M. Bloor, J. Penfold, J. Warr, J. F. Holzwarth, E. Wyn-Jones, "Small-Angle Neutron Scattering and Fluorescence Quenching Studies of Aggregated Ionic and Nonionic Surfactants in the Presence of Poly(1,4-diaminobutane) Dendrimers", *Langmuir* **2000**, *16*, 7999–8004.

[32] C. Valério, J. Ruiz, E. Alonso, P. Boussaguet, J. Guittard, J.-C. Blais, D. Astruc, "Syntheses of Polyamine and Polynitrile Dendrimers from a Nona-Arm Core up to 144-Nitrile using Vögtle's Iteration", *Bull. Soc. Chim. Fr.* **1997**, *134*, 907–914.

[33] N. Sakai, S. Matile, "Transmembrane Ion Transport Mediated by Amphiphilic Polyamine Dendrimers", *Tetrahedron Lett.* **1997**, *38*, 2613–2616.

[34] D. A. Tomalia, A. M. Naylor, W. A. Goddard, III, "Starburst Dendrimers: Molecular-Level Control of Size, Shape, Surface Chemistry, Topology, and Flexibility in the Conversion of Atoms to Macroscopic Materials", *Angew. Chem.* **1990**, *102*, 119–156; *Int. Ed. Engl.* **1990**, *29*, 113–150.

[35] S. E. Friberg, M. Podzimek, D. A. Tomalia, D. M. Hedstrand, "A Non-Aqueous Lyotropic Liquid Crystal with a Starburst Dendrimer as a Solvent", *Mol. Cryst. Liq. Cryst.* **1988**, *164*, 157–165.

[35a] M. Marcos, R. Giménez, J. L. Serrano, B. Donnio, B. Heinrich, D. Guillon, "Dendromesogens: Liquid Crystal Organizations of Poly(amidoamine) Dendrimers versus Starburst Structures", *Chem. Eur. J.* **2001**, *7*, 1006–1013.

[36] J. Suh, S. S. Hah, S. H. Lee, "Dendrimer Poly(ethylenimine)s Linked to β-Cyclodextrin", *Bioorg. Chem.* **1997**, *25*, 63–75.

[36a] M. Kawase, T. Shiomi, H. Matsui, Y. Ouji, S. Higashiyama, T. Tsutsui, K. Yagi, "Suppression of apoptosis in heptocytes by fructose-modified dendrimers", *J. Biomed. Mater. Res.* **2001**, *54*, 519–524.

[37] J. F. G. A. Jansen, E. M. M. de Brabander-van den Berg, E. W. Meijer, "Encapsulation of Guest Molecules into a Dendritic Box", *Science* **1994**, *266*, 1226–1229.

[38] M. Maciejewski, "Concepts of Trapping Topologically by Shell Molecules", *J. Macromol. Sci., Chem.* **1982**, *A17*, 689–703.

[39] J. F. G. A. Jansen, E. W. Meijer, E. M. M. de Brabander-van den Berg, "The Dendritic Box: Shape-Selective Liberation of Encapsulated Guests", *J. Am. Chem. Soc.* **1995**, *117*, 4417–4418.

[40] J. F. G. A. Jansen, E. M. M. de Brabander-van den Berg, E. W. Meijer, "The Dendritic Box: Synthesis, Properties, and Applications" in *New Molecular Architectures and Functions, Proceedings of the OUMS 1995* (Eds.: M. Kamachi, A. Nakamura), Springer-Verlag, Berlin, Heidelberg, **1996**, pp. 99–108.

[41] J. F. G. A. Jansen, E. M. M. de Brabander-van den Berg, E. W. Meijer, "Induced Chirality of Guest Molecules Encapsulated into a Dendritic Box", *Recl. Trav. Chim. Pays-Bas* **1995**, *114*, 225–230.

[42] J. F. G. A. Jansen, E. M. M. de Brabander-van den Berg, E. W. Meijer, "The Dendritic Box and Bengal Rose" *Polym. Mater. Sci. Eng.* **1995**, *73*, 123–124.

[43] J. F. G. A. Jansen, E. W. Meijer, E. M. M. de Brabander-van den Berg, "Bengal Rose@Dendritic Box", *Macromol. Symp.* **1996**, *102*, 27–33.

[44] J. F. G. A. Jansen, R. A. J. Janssen, E. M. M. de Brabander-van den Berg, E. W. Meijer, "Triplet Radical Pairs of 3-Carboxyproxyl Encapsulated in a Dendritic Box", *Adv. Mater. (Weinheim, Fed. Repub. Ger.)* **1995**, *7*, 561–564.

[45] M. W. P. L. Baars, E. W. Meijer, "Host–Guest Chemistry of Dendritic Molecules", *Top. Curr. Chem.* **2000**, *210*, 131–227.

[46] P. Miklis, T. Cagin, W. A. Goddard, III, "Dynamics of Bengal Rose Encapsulated in the Meijer Dendrimer Box", *J. Am. Chem. Soc.* **1997**, *119*, 7458–7462.

[47] L. Cavallo, F. Fraternali, "A Molecular Dynamics Study of the First Five Generations of Poly(propylene imine) Dendrimers Modified with *N-t*Boc-L-Phenylalanines", *Chem. Eur. J.* **1998**, *4*, 927–934.

[48] F. De Campo, D. Lastécouères, J.-M. Vincent, J.-B. Verlhac, "Copper(I) Complexes Mediated Cyclization Reaction of Unsaturated Ester under Fluoro Biphasic Procedure", *J. Org. Chem.* **1999**, *64*, 4969–4971.

[49] I. T. Horváth, J. Rabái, "Fractal Catalyst Separation Without Water: Fluorous Biphase Hydroformylation of Olefins", *Science* **1994**, *266*, 72–75.

[50] A. I. Cooper, J. D. Londono, G. Wignall, J. B. McClain, E. T. Samulski, J. S. Lin, A. Dobrynin, M. Rubinstein, A. L. C. Burke, J. M. J. Fréchet, J. M. DeSimone, "Extraction of a Hydrophilic Compound from Water into Liquid CO_2 using Dendritic Surfactants", *Nature* **1997**, *389*, 368–371.

[51] N. Moszner, T. Völkel, V. Rheinberger, "Synthesis, Characterization and Polymerization of Dendrimers with Methacrylic End Groups", *Macromol. Chem. Phys.* **1996**, *197*, 621–631.

[52] R. C. van Duijvenbode, A. Rajanayagam, G. J. M. Koper, M. W. P. L. Baars, B. F. M. de Waal, E. W. Meijer, "Synthesis and Protonation Behavior of Carboxylate-Functionalized Poly(propyleneimine) Dendrimers", *Macromolecules* **2000**, *33*, 46–52.

[53] E. J. H. Put, K. Clays, A. Persoons, H. A. M. Biemans, C. P. Luijkx, E. W. Meijer, "The Symmetry of Functionalized Poly(propylene imine) Dendrimers Probed with Hyper-Rayleigh Scattering", *Chem. Phys. Lett.* **1996**, *260*, 136–141.

[54] K. Yonetake, T. Masuko, T. Morishita, K. Suzuki, M. Ueda, R. Nagahata, "Poly(propyleneimine) Dendrimers Peripherally Modified with Mesogens", *Macromolecules* **1999**, *32*, 6578–6586.

[55] I. Bodnár, A. S. Silva, R. W. Deitcher, N. E. Weisman, Y. H. Kim, N. J. Wagner, "Structure and Rheology of Hyperbranched and Dendritic Polymers; I. Modification and Characterization of Poly(propyleneimine) Dendrimers with Acetyl Groups", *J. Polym. Sci., Part B: Polym. Phys.* **2000**, *38*, 857–873.

[56] I. Bodnár, A. S. Silva, Y. H. Kim, N. J. Wagner, "Structure and Rheology of Hyperbranched and Dendritic Polymers; II. Effects of Blending Acetylated and Hydroxy-Terminated Poly(propyleneimine) Dendrimers with Aqueous Poly(ethylene oxide) Solutions", *J. Polym. Sci., Part B: Polym. Phys.* **2000**, *38*, 874–882.

[57] H. A. M. van Aert, M. H. P. van Genderen, E. W. Meijer, "Star-Shaped Poly(2,6-dimethyl-1,4-phenylene ether)", *Polym. Bull.* **1996**, *37(3)*, 273–280.

[58] S. Stevelmans, J. C. M. van Hest, J. F. G. A. Jansen, D. A. F. J. van Boxtel, E. M. M. de Brabander-van den Berg, E. W. Meijer, "Synthesis, Characterization, and Guest–Host Properties of Inverted Unimolecular Dendritic Micelles", *J. Am. Chem. Soc.* **1996**, *118*, 7398–7399.

[59] J. G. Lieu, M. Liu, J. M. J. Fréchet, J. M. Prausnitz, "Vapor-Liquid Equilibria for Dendritic-Polymer Solutions", *J. Chem. Eng. Data* **1999**, *44*, 613–620.

[60] M. W. P. L. Baars, P. E. Froehling, E. W. Meijer, "Liquid–Liquid Extractions using Poly(propylene imine) Dendrimers with an Apolar Periphery", *Chem. Commun.* **1997**, 1959–1960.

[61] M. W. P. L. Baars, D. A. F. J. van Boxtel, C. W. M. Bastiaansen, D. J. Broer, S. H. M. Söntjens, E. W. Meijer, "A Scattering Electro-Optical Switch Based on Dendrimers Dispersed in Liquid Crystals", *Adv. Mater.* **2000**, *12*, 715–719.

[61a] J. J. J. M. Donners, B. R. Heywood, E. W. Meijer, R. J. M. Nolte, C. Roman, A. P. H. J. Schenning, N. A. J. M. Sommerdijk, "Amorphous calcium carbonate stabilised by poly(propylene imine) dendrimers", *Chem. Commun.* **2000**, 1937–1938.

[62] G. Pistolis, A. Malliaris, D. Tsiourvas, C. M. Paleos, "Poly(propyleneimine) Dendrimers as pH-Sensitive Controlled-Release Systems", *Chem. Eur. J.* **1999**, *5*, 1440–1444.

[63] A. Ramzi, B. J. Bauer, R. Scherrenberg, P. Froehling, J. Joosten, E. J. Amis, "Fatty Acid Modified Dendrimers in Bulk and Solution: Single-Chain Neutron Scattering from Dendrimer Core and Fatty Acid Shell", *Macromolecules* **1999**, *32*, 4983–4988.

[64] B. J. Bauer, A. Ramzi, D.-W. Liu, R. L. Scherrenberg, P. Froehling, J. Joosten, "Blends of Fatty Acid Modified Dendrimers with Polyolefins", *J. Polym. Sci., Part B: Polym. Phys.* **2000**, *38*, 95–100.

[65] A. P. H. J. Schenning, E. Peeters, E. W. Meijer, "Energy Transfer in Supramolecular Assemblies of Oligo(*p*-phenylene vinylene)s Terminated Poly(propylene imine) Dendrimers", *J. Am. Chem. Soc.* **2000**, *122*, 4489–4495.

[66] D. Pötschke, M. Ballauff, P. Lindner, M. Fischer, F. Vögtle, "The Structure of Dendritic Molecules in Solution as Investigated by Small-Angle Neutron Scattering", *Macromol. Chem. Phys.* **2000**, *201*, 330–339.

[67] P. E. Froehling, H. A. J. Linssen, "Positional and Compositional Heterogeneity of Partially Modified Poly(propyleneimine) Dendrimers", *Macromol. Chem. Phys.* **1998**, *199*, 1691–1695.

[68] A. W. Bosman, M. J. Bruining, H. Kooijman, A. L. Spek, R. A. J. Janssen, E. W. Meijer, "Concerning the Localization of End Groups in Dendrimers", *J. Am. Chem. Soc.* **1998**, *120*, 8547–8548.

[69] P. R. Ashton, S. E. Boyd, C. L. Brown, S. A. Nepogodiev, E. W. Meijer, H. W. I. Peerlings, J. F. Stoddart, "Synthesis of Glycodendrimers by Modification of Poly(propylene imine) Dendrimers", *Chem. Eur. J.* **1997**, *3*, 974–984.

[70] A. W. Bosman, R. A. J. Janssen, E. W. Meijer, "Five Generations of Nitroxyl-Functionalized Dendrimers", *Macromolecules* **1997**, *30*, 3606–3611.

[71] A. Archut, F. Vögtle, L. De Cola, G. C. Azzellini, V. Balzani, P. S. Ramanujam, R. H. Berg, "Azobenzene-Functionalized Cascade Molecules: Photoswitchable Supramolecular Systems", *Chem. Eur. J.* **1998**, *4*, 699–706.

[72] A. Archut, G. C. Azzellini, V. Balzani, L. De Cola, F. Vögtle, "Toward Photoswitchable Dendritic Hosts. Interaction between Azobenzene-Functionalized Dendrimers and Eosin", *J. Am. Chem. Soc.* **1998**, *120*, 12187–12191.

[73] M. W. P. L. Baars, S. H. M. Söntjens, H. M. Fischer, H. W. I. Peerlings, E. W. Meijer, "Liquid-Crystalline Properties of Poly(propylene imine) Dendrimers Functionalized with Cyanobiphenyl Mesogens at the Periphery", *Chem. Eur. J.* **1998**, *4*, 2456–2466.

[74] H. W. I. Peerlings, E. W. Meijer, "A Mild and Convenient Method for the Preparation of Multi-Isocyanates starting from Primary Amines", *Tetrahedron Lett.* **1999**, *40*, 1021–1024.

[75] B. M. Pope, Y. Yamamoto, D. S. Tarbell, "Di-*tert*-Butyl Dicarbonate", *Org. Synth.* **1977**, *57*, 45–50.

[75a] M. W. P. L. Baars, A. J. Karlsson, V. Sorokin, B. F. M. de Waal, E. W. Meijer, "Supramolecular Modification of the Periphery of Dendrimers Resulting in Rigidity and Functionality", *Angew. Chem. Int. Ed.* **2000**, *39*, 4262–4265.

[75b] U. Boas, A. J. Karlsson, B. F. M. de Waal, E. W. Meijer, "Synthesis and Properties of New Thiourea-Functionalized Poly(propylene imine) Dendrimers and Their Role as Hosts for Urea Functionalized Guests", *J. Org. Chem.* **2001**, *66*, 2136–2145.

[76] A. P. H. J. Schenning, C. Elissen-Román, J.-W. Weener, M. W. P. L. Baars, S. J. van der Gaast, E. W. Meijer, "Amphiphilic Dendrimers as Building Blocks in Supramolecular Assemblies", *J. Am. Chem. Soc.* **1998**, *120*, 8199–8208.

[77] J.-W. Weener, E. W. Meijer, "Photoresponsive Dendritic Monolayers", *Adv. Mater.* **2000**, *12*, 741–746.

[78] C. Z. Chen, N. C. B. Tan, S. L. Cooper, "Incorporation of Dimethyldodecylammonium Chloride Functionalities onto Poly(propylene imine) Dendrimers Significantly Enhances their Antibacterial Properties", *Chem. Commun.* **1999**, 1585–1586.

[79] Z. Sideratou, D. Tsiourvas, C. M. Paleos, "Quaternized Poly(propylene imine) Dendrimers as Novel pH-Sensitive Controlled-Release Systems", *Langmuir* **2000**, *16*, 1766–1769.

[80] F. Vögtle, S. Gestermann, C. Kauffmann, P. Ceroni, V. Vicinelli, L. De Cola, V. Balzani, "Poly(propylene amine) Dendrimers with Peripheral Dansyl Units: Protonation, Absorption Spectra, Photophysical Properties, Intradendrimer Quenching, and Sensitization Processes", *J. Am. Chem. Soc.* **1999**, *121*, 12161–12166.

[80a] F. Vögtle, S. Gestermann, C. Kauffmann, P. Ceroni, V. Vicinelli, V. Balzani, "Coordination of Co^{2+} Ions in the Interior of Poly(propylene amine) Dendrimers Containing Fleorescent Dansyl Units in the Periphery", *J. Am. Chem. Soc.* **2000**, *122*, 10398–10404.

[81] V. Balzani, P. Ceroni, S. Gestermann, C. Kauffmann, M. Gorka, F. Vögtle, "Dendrimers as Fluorescent Sensors with Signal Amplification", *Chem. Commun.* **2000**, 853–854.

[82] K. Tsuda, G. C. Dol, T. Gensch, J. Hofkens, L. Latterini, J.-W. Weener, E. W. Meijer, F. C. De Schryver, "Fluorescence from Azobenzene-Functionalized Poly(propylene imine) Dendrimers in Self-Assembled Supramolecular Structures", *J. Am. Chem. Soc.* **2000**, *122*, 3445–3452.

[83] H. Stephan, H. Spies, B. Johannsen, C. Kauffmann, F. Vögtle, "pH-Controlled Inclusion and Release of Oxyanions by Dendrimers Bearing Methyl Orange Moieties", *Org. Lett.* **2000**, *2*, 2343–2346.

[84] K. Aoi, T. Hatanaka, K. Tsutsumiuchi, M. Okada, T. Imae, "Synthesis of a Novel Star-Shaped Dendrimer by Radial-Growth Polymerization of Sarcosine *N*-Carboxyanhydride Initiated with Poly(trimethyleneimine) Dendrimer", *Macromol. Rapid Commun.* **1999**, *20*, 378–382.

[85] H. Stephan, H. Spies, B. Johannsen, L. Klein, F. Vögtle, "Lipophilic Urea-Functionalized Dendrimers as Efficient Carriers for Oxyanions", *Chem. Commun.* **1999**, 1875–1876.

[86] Y. Pan, W. T. Ford, "Dendrimers with Both Hydrophilic and Hydrophobic Chains at Every End", *Macromolecules* **1999**, *32*, 5468–5470.

[87] Y. Pan, W. T. Ford, "Amphiphilic Dendrimers with Both Octyl and Triethylenoxy Methyl Ether Chain Ends", *Macromolecules* **2000**, *33*, 3731–3738.

[87a] J. L. Kreider, W. T. Ford, "Quaternary ammonium ion dendrimers from methylation of poly(propylene imine)s", *J. Polym. Sci., Part A: Polym. Chem.* **2001**, *39*, 821–832.

[88] M. W. P. L. Baars, R. Kleppinger, M. H. J. Koch, S.-L. Yeu, E. W. Meijer, "The Localization of Guests in Water-Soluble Oligoethyleneoxy-Modified Poly(propylene imine) Dendrimers", *Angew. Chem.* **2000**, *112*, 1341–1344; *Int. Ed.* **2000**, *39*, 1285–1288.

[89] E. K. L. Yeow, K. P. Ghiggino, J. N. H. Reek, M. J. Crossley, A. W. Bosman, A. P. H. J. Schenning, E. W. Meijer, "The Dynamics of Electronic Energy Transfer in Novel Multiporphyrin Functionalized Dendrimers: A Time-Resolved Fluorescence Anisotropy Study", *J. Phys. Chem. B* **2000**, *104*, 2596–2606.

[90] C. O. Noble, IV, R. L. McCarley, "Surface-Confined Monomers on Electrode Surfaces; 7. Synthesis of Pyrrole-Terminated Poly(propylene imine) Dendrimers", *Org. Lett.* **1999**, *1*, 1021–1023.

[91] C. O. Noble, IV, R. L. McCarley, "Pyrrole-Terminated Diaminobutane (DAB) Dendrimer Monolayers on Gold: Oligomerization of Peripheral Group and Adhesion Promotion of Poly-(pyrrole) Films", *J. Am. Chem. Soc.* **2000**, *122*, 6518–6519.

[91a] J. W. Lee, Y. H. Ko, S.-H. Park, K. Yamaguchi, K. Kim, "Novel Pseudorotaxane-Terminated Dendrimers: Supramolecular Modification of Dendrimer Periphery", *Angew. Chem. Int. Ed.* **2001**, *40*, 746–749.

[91b] S. M. Cohen, S. Petoud, K. N. Raymond, "Synthesis and Metal Binding Properties of Salicylate-, Catecholate-, and Hydroxypyridinonate-Functionalized Dendrimers", *Chem. Eur. J.* **2001**, *7*, 272–279.

[91c] L. A. Baker, R. M. Crooks, "Photophysical Properties of Pyrene-Functionalized Poly(propylene imine) Dendrimers", *Macromolecules* **2000**, *33*, 9034–9039.

[92] J. C. M. van Hest, M. W. P. L. Baars, C. Elissen-Román, M. H. P. van Genderen, E. W. Meijer, "Acid-Functionalized Amphiphiles Derived from Polystyrene–Poly(propylene imine) Dendrimers, with a pH-Dependent Aggregation", *Macromolecules* **1995**, *28*, 6689–6691.

[93] J. C. M. van Hest, D. A. P. Delnoye, M. W. P. L. Baars, M. H. P. van Genderen, E. W. Meijer, "Polystyrene-Dendrimer Amphiphilic Block Copolymers with a Generation-Dependent Aggregation", *Science* **1995**, *268*, 1592–1595.

[94] J. C. M. van Hest, C. Elissen-Román, M. W. P. L. Baars, D. A. P. Delnoye, M. H. P. van Genderen, E. W. Meijer, "Polystyrene–Poly(propylene imine) Dendrimer Block Copolymers: A New Class of Amphiphiles", *Polym. Mater. Sci. Eng.* **1995**, *73*, 281–282.

[95] J. C. M. van Hest, D. A. P. Delnoye, M. W. P. L. Baars, C. Elissen-Román, M. H. P. van Genderen, E. W. Meijer, "Polystyrene–Poly(propylene imine) Dendrimers: Synthesis, Characterization, and Association Behavior of a New Class of Amphiphiles", *Chem. Eur. J.* **1996**, *2*, 1616–1626.

[96] C. Román, H. R. Fischer, E. W. Meijer, "Microphase Separation of Diblock Copolymers Consisting of Polystyrene and Acid-Functionalized Poly(propylene imine) Dendrimers", *Macromolecules* **1999**, *32*, 5525–5531.

[97] H. Tokuhisa, R. M. Crooks, "Interactions between Organized, Surface-Confined Monolayers and Vapor-Phase Probe Molecules; 12. Two New Methods for Surface-Immobilization and Functionalization of Chemically-Sensitive Dendrimer Surfaces", *Langmuir* **1997**, *13*, 5608–5612.

[98] H.-L. Wang, D. W. McBranch, V. I. Klimov, R. Helgeson, F. Wudl, "Controlled Unidirectional Energy Transfer in Luminescent Self-Assembled Conjugated Polymer Superlattices", *Chem. Phys. Lett.* **1999**, *315*, 173–180.

[98a] T. Ren, G. Zhang, D. Liu, "Synthesis of bifunctional cationic compounds for gene delivery", *Tetrahedron Lett.* **2001**, *42*, 1007–1010.

[99] D. Stöckigt, G. Lohmer, D. Belder, "Separation and Identification of Basic Dendrimers Using Capillary Electrophoresis On-line Coupled to a Sector Mass Spectrometer", *Rapid Commun. Mass Spectrom.* **1996**, *10*, 521–526.

[100] G. R. Baker, personal communication, University of South Florida, 1999.

[101] Sj. van der Wal, Y. Mengerink, J. C. Brackman, E. M. M. de Brabander, C. M. Jeronimus-Stratingh, A. P. Bruins, "Compositional Analysis of Nitrile-Terminated Poly(propylene imine) Dendrimers by High-Performance Liquid Chromatography Combined with Electrospray Mass Spectrometry", *J. Chromatogr. A* **1998**, *825*, 135–147.

[102] J.-W. Weener, J. L. J. van Dongen, E. W. Meijer, "Electrospray Mass Spectrometry Studies of Poly(propylene imine) Dendrimers: Probing Reactivity in the Gas Phase", *J. Am. Chem. Soc.* **1999**, *121*, 10355.

[103] J. de Maaijer-Gielbert, C. Gu, A. Somogyi, V. H. Wysocki, P. G. Kistemaker, T. L. Weeding, "Surface-Induced Dissociation of Singly and Multiply Protonated Polypropylenamine Dendrimers", *J. Am. Soc. Mass Spectrom.* **1999**, *10*, 414–422.

[104] S. A. McLuckey, K. G. Asano, T. G. Schaaff, J. L. Stephenson, Jr., "Ion Trap Collisional Activation of Protonated Poly(propylene imine) Dendrimers: Generations 1–5", *Int. J. Mass Spectrom.* **2000**, *195/196*, 419–437.

[105] Y. Mengerink, M. Mure, E. M. M. de Brabander, Sj. van der Wal, "Exclusion Chromatography of Polypropylenamine Dendrimers", *J. Chromatogr. A* **1996**, *730*, 75–81.

[106] G. J. M. Koper, M. H. P. van Genderen, C. Elissen-Román, M. W. P. L. Baars, E. W. Meijer, M. Borkovec, "Protonation Mechanism of Poly(propylene imine) Dendrimers and Some Associated Oligo Amines", *J. Am. Chem. Soc.* **1997**, *119*, 6512–6521.

[107] R. C. van Duijvenbode, M. Borkovec, G. J. M. Koper, "Acid-Base Properties of Poly(propylene imine) Dendrimers", *Polymer* **1998**, *39*, 2657–2664.

[107a] R. C. van Duijvenbode, G. J. M. Koper, M. R. Böhmer, "Adsorption of Poly(propylene imine) Dendrimers on Glass. An Interplay between Surface and Particle Properties.", *Langmuir* **2000**, *16*, 7713–7719.

[107b] R. C. van Duijvenbode, I. B. Rietveld, G. J. M. Koper, "Light Reflectivity Study on Adsorption Kinetics of Poly(propylene imine) Dendrimers on Glass", *Langmuir* **2000**, *16*, 7720–7725.

[107c] J. L. Casson, D. W. McBranch, J. M. Robinson, H.-L. Wang, J. B. Roberts, P. A. Chiarelli, M. S. Johal, "Reversal of interfacial dipole orientation in polyelectrolyte superlattices due to polycationic layers", *J. Polym. Sci. , Part B: Polym. Phys.* **2000**, *104*, 11996–12001.

[108] V. A. Kabanov, A. B. Zezin, V. B. Rogacheva, Zh. G. Gulyaeva, M. F. Zansochova, J. G. H. Joosten, J. Brackman, "Polyelectrolyte Behavior of Astramol Poly(propyleneimine) Dendrimers", *Macromolecules* **1998**, *31*, 5142–5144.

[109] P. Welch, M. Muthukumar, "Tuning the Density Profiles of Dendritic Polyelectrolytes", *Macromolecules* **1998**, *31*, 5892–5897.

[110] J. K. Young, G. R. Baker, G. R. Newkome, K. F. Morris, C. S. Johnson, Jr., "'Smart' Cascade Polymers. Modular Syntheses of Four-Directional Dendritic Macromolecules with Acidic, Neutral, or Basic Terminal Groups and the Effect of pH Changes on their Hydrodynamic Radii", *Macromolecules* **1994**, *27*, 3464–3471.

[111] P. Welch, M. Muthukumar, "Dendrimer–Polyelectrolyte Complexation: A Model Guest–Host System", *Macromolecules* **2000**, *33*, 6159–6167.

[112] A. Polese, C. Mio, A. Bertucco, "Infinite-Dilution Activity Coefficients of Polar and Nonpolar Solvents in Solutions of Hyperbranched Polymers", *J. Chem. Eng. Data* **1999**, *44*, 839–845.

[113] V. A. Kabanov, A. B. Zezin, V. B. Rogacheva, Zh. G. Gulyaeva, M. F. Zansochova, J. G. H. Joosten, J. Brackman, "Interaction of Astramol Poly(propyleneimine) Dendrimers with Linear Polyanions", *Macromolecules* **1999**, *32*, 1904–1909.

[113a] R. E. A. Dillon, D. F. Shriver, "Ion Transfer and Vibrational Spectra of Branched Polymer and Dendrimer Electrolytes", *Chem. Mater.* **2001**, *13*, 1369–1373.

[114] S. M. Ghoreishi, Y. Li, J. F. Holzwarth, E. Khoshdel, J. Warr, D. M. Bloor, E. Wyn-Jones, "The Interaction between Nonionic Dendrimers and Surfactants – Electromotive Force and Microcalorimetry Studies", *Langmuir* **1999**, *15*, 1938–1944.

[115] I. B. Rietveld, J. A. M. Smit, "Colligative and Viscosity Properties of Poly(propylene imine) Dendrimers in Methanol", *Macromolecules* **1999**, *32*, 4608–4614.

[115a] I. B. Rietveld, D. Bedeaux, "Self-Diffusion of Poly(propylene imine) Dendrimers in Methanol", *Macromolecules* **2000**, *33*, 7912–7917.

[115b] I. B. Rietveld, D. Bedeaux, J. A. M. Smit, "Osmotic Compressibility of Poly(propylene imine) Dendrimers in Deuterated Methanol", *J. Colloid Interface Sci.* **2000**, *232*, 317–325.

[116] I. Sendijarevic, A. J. McHugh, "Effects of Molecular Variables and Architecture on the Rheological Behavior of Dendritic Polymers", *Macromolecules* **2000**, *33*, 590–596.

[117] P. G. H. M. Muijselaar, H. A. Claessens, C. A. Cramers, J. F. G. A. Jansen, E. W. Meijer, E. M. M. de Brabander-van den Berg, S. van der Wal, "Dendrimers as Pseudo-Stationary Phases in Electrokinetic Chromatography", *J. High Resolut. Chromatogr.* **1995**, *18*, 121–123.

[118] D. A. Tomalia, H. Baker, J. Dewald, M. Hall, G. Kallos, S. Martin, J. Roeck, J. Ryder, P. Smith, "Dendritic Macromolecules: Synthesis of Starburst Dendrimers", *Macromolecules* **1986**, *19*, 2466–2468.

[119] D. A. Tomalia, D. R. Swanson, J. W. Klimash, H. M. Brothers, III, "Cascade (Starburst™) Dendrimer Synthesis by the Divergent Dendron/Divergent Core Anchoring Methods" *Polym. Prepr.* **1993**, *34*, 52–53.

[120] D. A. Tomalia, M. Hall, D. M. Hedstrand, "Starburst Dendrimers; 3. The Importance of Branch Junction Symmetry in the Development of Topological Shell Molecules", *J. Am. Chem. Soc.* **1987**, *109*, 1601–1603.

[121] D. A. Tomalia, J. R. Dewald, "Dense Star Polymer Having Core, Core Branches, Terminal Groups", **1985**, *U. S. Pat.*, 4, 507, 466.

[122] D. A. Tomalia, J. R. Dewald, "Dense Star Polymer", **1985**, *U. S. Pat.*, 4, 558, 120.

[123] M. B. Smith, J. March, *March's Advanced Organic Chemistry*, 5th ed., Wiley-Interstate, New York, **2001**, pp 975 & 1022.

[124] P. B. Smith, S. J. Martin, M. J. Hall, D. A. Tomalia, "A Characterization of the Structure and Synthetic Reactions of Polyamidoamine 'Starburst' Polymers" (Ed.: J. Mitchell, Jr.), Hanser, Munich, **1987**, pp. 357–385.

[125] N. J. Wells, A. Basso, M. Bradley, "Solid-Phase Support", *Biopolym.* **1998**, *47*, 381–396.

[126] A. M. Naylor, W. A. Goddard, III, "Simulations of Starburst Dendrimer Polymers" *Polym. Prepr.* **1988**, *29*, 215–216.

[127] A. M. Naylor, W. A. Goddard, III, "Application of Simulation and Theory to Biocatalysis and Biomimetics" *ACS Symposium Series* (Eds.: J. D. Burrington, D. S. Clark), American Chemical Society, Washington DC, **1989**, pp. 65–87.

[128] A. M. Naylor, W. A. Goddard, III, G. E. Kiefer, D. A. Tomalia, "Starburst Dendrimers; 5. Molecular Shape Control", *J. Am. Chem. Soc.* **1989**, *111*, 2339–2341.

[129] M. K. Bhalgat, J. C. Roberts, "Molecular Modeling of Polyamidoamine (PAMAM) Starburst™ Dendrimers", *Eur. Polym. J.* **2000**, *36*, 647–651.

[129a] N. Hurduc, D. Scutaru, V. Toader, S. Alazaroaie, L. Petraru, "Molecular Modelling and properties simulations", *Materiale Plastice* **2001**, *37*, 205–209.

[130] D. A. Tomalia, V. Berry, M. Hall, D. M. Hedstrand, "Starburst Dendrimers; 4. Covalently Fixed Unimolecular Assemblages Reminiscent of Spheroidal Micelles", *Macromolecules* **1987**, *20*, 1164–1167.

[131] A. V. Lyulin, G. R. Davies, D. B. Adolf, "Location of Terminal Groups of Dendrimers: Brownian Dynamics Simulation", *Macromolecules* **2000**, *33*, 6899–6900.

[132] A. D. Meltzer, D. A. Tirrell, A. A. Jones, P. T. Inglefield, "Chain Dynamics in Poly(amidoamine) Dendrimers. A Study of ^{2}H NMR Relaxation Parameters", *Macromolecules* **1992**, *25*, 4549–4552.

[132a] D. I. Malyarenko, R. L. Vold, G. L. Hoatson, "Solid State Deuteron NMR Studies of Polyamidoamine Dendrimer Salts. 2. Relaxation and Molecular Motion", *Macromolecules* **2000**, *33*, 7508–7520.

[133] A. D. Meltzer, D. A. Tirrell, A. A. Jones, P. T. Inglefield, D. M. Hedstrand, D. A. Tomalia, "Chain Dynamics in Poly(amido amine) Dendrimers. A Study of ^{13}C NMR Relaxation Parameters", *Macromolecules* **1992**, *25*, 4541–4548.

[134] A. D. Meltzer, D. A. Tirrell, A. A. Jones, P. T. Inglefield, D. M. Downing, D. A. Tomalia, "^{13}C NMR Relaxation in Poly(amido amine) Starburst Dendrimers" *Polym. Prepr.* **1989**, *30*, 121–122.

[135] A. D. Meltzer, "Mobility of Poly(amidoamine) Dendrimers – A Study of NMR Relaxation Times" , **1990**.

[136] G. J. Kallos, D. A. Tomalia, D. M. Hedstrand, S. Lewis, J. Zhou, "Molecular Weight Determination of a Polyamidoamine Starburst Polymer by Electrospray Ionization Mass Spectrometry", *Rapid Commun. Mass Spectrom.* **1991**, *5*, 383–386.

[137] P. R. Dvornic, D. A. Tomalia, "Geneologically Directed Syntheses (Polymerizations): Direct Evidence by Electrospray Mass Spectroscopy", *Macromol. Symp.* **1995**, *98*, 403–428.

[138] B. L. Schwartz, A. L. Rockwood, R. D. Smith, D. A. Tomalia, R. Spindler, "Detection of High Molecular Weight Starburst Dendrimers by Electrospray Ionization Mass Spectrometry", *Rapid Commun. Mass Spectrom.* **1995**, *9*, 1552–1555.

[139] P. T. Tolic, G. A. Anderson, R. D. Smith, H. M. Brothers, II, R. Spindler, D. A. Tomalia, "Electrospray Ionization Fourier-Transform Ion Cyclotron Resonance Mass Spectrometric Characterization of High Molecular Mass Starburst™ Dendrimers", *Int. J. Mass Spectrom. Ion Processes* **1997**, *165/166*, 405–418.

[140] W. Henderson, B. K. Nicholson, L. J. McCaffrey, "Applications of Electrospray Mass Spectrometry in Organometallic Chemistry", *Polyhedron* **1999**, *17*, 4291–4313.

[141] H. M. Brothers, II, L. T. Piehler, D. A. Tomalia, "Slab-Gel and Capillary Electrophoretic Characterization of Polyamidoamine Dendrimers", *J. Chromatogr. A* **1998**, *814*, 233–246.

[142] M. L. Mansfield, L. I. Klushin, "Intrinsic Viscosity of Model Starburst Dendrimers", *J. Phys. Chem.* **1992**, *96*, 3994–3998.

[143] C. Mio, S. Kiritsov, Y. Thio, R. Brafman, J. Prausnitz, C. J. Hawker, E. Malmström, "Vapor-Liquid Equilibrium for Solutions of Dendritic Polymers", *J. Chem. Eng. Data* **1998**, *43*, 541–550.

[144] T. J. Prosa, B. J. Bauer, E. J. Amis, D. A. Tomalia, R. Scherrenberg, "A SAXS Study of the Internal Structure of Dendritic Polymer Systems", *J. Polym. Sci., Part B: Polym. Phys.* **1998**, *35*, 2913–2924.

[145] G. Nisato, R. Ivkov, E. J. Amis, "Structure of Charged Dendrimer Solutions as seen by Small-Angle Neutron Scattering", *Macromolecules* **1999**, *32*, 5895–5900.

[146] A. Topp, B. J. Bauer, J. W. Klimash, R. Spindler, D. A. Tomalia, E. J. Amis, "Probing the Location of the Terminal Groups of Dendrimers in Dilute Solution", *Macromolecules* **1999**, *32*, 7226–7231.

[147] K. S. Åkerfeldt, J. D. Lear, Z. R. Wasserman, L. A. Chung, W. F. DeGrado, "Synthetic Peptides as Models for Ion Channel Proteins", *Acc. Chem. Res.* **1993**, *26*, 191–197.

[148] Y. Guo, K. H. Langley, F. E. Karasz, "Restricted Diffusion of Highly Dense Starburst-Dendritic Poly(amido amine) in Porous Glasses", *Macromolecules* **1992**, *25*, 4902–4904.

[149] S. Uppuluri, P. R. Dvornic, J. W. Klimash, P. I. Carver, N. C. B. Tan, "The Properties of Dendritic Polymers I: Generation 5 Poly(amidoamine) Dendrimers" *ARL-TR-1606*, 5-1-1998, p. 1–21.

[150] A. Nourse, D. B. Millar, A. P. Minton, "Physicochemical Characterization of Generation 5 Polyamidoamine Dendrimers", *Biopolym.* **2000**, *53*, 316–328.

[151] S. Uppuluri, F. A. Morrison, P. R. Dvornic, "Rheology of Dendrimers; 2. Bulk Polyamidoamine Dendrimers under Steady Shear, Creep, and Dynamic Oscillatory Shear", *Macromolecules* **2000**, *33*, 2551–2560.

[152] P. R. Dvornic, D. A. Tomalia, "Starburst™ Dendrimers: A Conceptual Approach to Nanoscopic Chemistry and Architecture", *Makromol. Chem., Macromol. Symp.* **1994**, *88*, 123–148.

[152a] D. A. Tomalia, S. Uppuluri, D. R. Swanson, J. Li, "Dendrimers as reactive modules for the synthesis of new structure-controlled, higher-complexity megamers", *Pure Appl. Chem.* **2000**, *72*, 2342–2358.

[152b] S. Uppuluri, D. R. Swanson, L. T. Piehler, J. Li, G. L. Hagnauer, D. A. Tomalia, "Core-Shell Tecto(dendrimers): I. Synthesis and Characterization of Saturated Shell Models", *Adv. Mater.* **2000**, *12*, 796–800.

[153] D. R. Morgan, E. O. Stejskal, A. L. Andrady, "^{129}Xe NMR Investigation of the Free Volume in Dendritic and Cross-Linked Polymers", *Macromolecules* **1999**, *32*, 1897–1903.

[154] D. J. Evans, A. Kanagasooriam, A. Williams, R. J. Pryce, "Aminolysis of Phenyl Esters by Microgel and Dendrimer Molecules Possessing Primary Amines", *J. Mol. Catal.* **1993**, *85*, 21–32.

[154a] I. K. Martin, L. J. Twyman, "The synthesis of unsymmetrical PAMAM dendrimers using a divergent/divergent approach", *Tetrahedron Lett.* **2001**, *42*, 1119–1121.

[154b] D. S. Jones, M. E. Tedder, C. A. Gamino, J. R. Hammaker, H.-T. Ton-Nu, "Synthesis of multivalent carbonate esters by divergent growth of branched carbamates", *Tetrahedron Lett.* **2001**, *42*, 2069–2072.

[155] R. X. Zhuo, B. Du, Z. R. Lu, "In vitro Release of 5-Fluorouracil with Cyclic Core Dendritic Polymer", *J. Controlled Release* **1999**, *57*, 249–257.

[156] T. D. James, H. Shinmori, M. Takeuchi, S. Shinkai, "A Saccharide 'Sponge'. Synthesis and Properties of a Dendritic Boronic Acid", *Chem. Commun.* **1996**, 705–706.

[157] L. L. Miller, T. Hashimoto, I. Tabakovic, D. R. Swanson, D. A. Tomalia, "Delocalized π-Stacks Formed on Dendrimers", *Chem. Mater.* **1995**, *7*, 9–11.

[158] L. L. Miller, K. R. Mann, "π-Dimers and π-Stacks in Solution and in Conducting Polymers", *Acc. Chem. Res.* **1996**, *29*, 417–423.

[159] J.-F. Penneau, B. J. Stallman, P. H. Kasai, L. L. Miller, "An Imide Anion Radical That Dimerizes and Assembles into π-Stacks in Solution", *Chem. Mater.* **1991**, *3*, 791–706.

[160] C. J. Zhong, W. S. V. Kwan, L. L. Miller, "Self-Assembly of Delocalized π-Stacks in Solution. Assessment of Structural Effects", *Chem. Mater.* **1992**, *4*, 1423–1428.

[161] R. G. Duan, L. L. Miller, D. A. Tomalia, "An Electrically Conducting Dendrimer", *J. Am. Chem. Soc.* **1995**, *117*, 10783–10784.

[162] S. Bhattacharya, S. N. G. Acharya, A. R. Raju, "Exceptional Adhesive and Gelling Properties of Fibrous Nanoscopic Tapes of Self-Assembled Bipolar Urethane Amides of L-Phenylalanine", *Chem. Commun.* **1996**, 2101–2102.

[163] I. Tabakovic, L. L. Miller, R. G. Duan, D. C. Tully, D. A. Tomalia, "Dendrimers Peripherally Modified with Anion Radicals that form π-Dimers and π-Stacks", *Chem. Mater.* **1997**, *9*, 736–745.

[164] L. L. Miller, D. A. Tomalia, R. G. Duan, "Electrically Conducting Dendrimers and their Preparation" *PCT Int. Appl.* **1997**.

[165] L. L. Miller, R. G. Duan, D. C. Tully, D. A. Tomalia, "Electrically Conducting Dendrimers", *J. Am. Chem. Soc.* **1997**, *119*, 1005–1010.

[166] S. Jockusch, N. J. Turro, D. A. Tomalia, "Aggregation of Methylene Blue Adsorbed on Starburst Dendrimers", *Macromolecules* **1995**, *28*, 7416–7418.

[167] P. Singh, F. Moll, III, S. H. Lin, C. Ferzli, K. S. Yu, R. K. Koski, R. G. Saul, P. Cronin, "Starburst™ Dendrimers: Enhanced Performance and Flexibility for Immunoassays", *Clin. Chem.* **1994**, *40*, 1845–1849.

[167a] A. C. Chang, J. B. Gillespie, M. B. Tabacco, "Enhanced detection of live bacteria using a dendrimer thin film in an optical biosensor", *Anal. Chem.* **2001**, *73*, 467–470.

[168] J. Haensler, F. C. Szoka, Jr., "Polyamidoamine Cascade Polymers Mediate Efficient Transfection of Cells in Culture", *Bioconj. Chem.* **1993**, *4*, 372–379.

[169] N. K. Subbarao, R. A. Parente, F. C. Szoka, Jr., L. Nadasdi, K. Pongracz, "The pH-Dependent Bilayer Destabilization by an Amphipathic Peptide", *Biochem.* **1987**, *26*, 2964.

[170] M. X. Tang, C. T. Redemann, F. C. Szoka, Jr., "*In vitro* Gene Delivery by Degraded Polyamidoamine Dendrimers", *Bioconj. Chem.* **1996**, *7*, 703–714.

[171] R. DeLong, K. Stephenson, T. Loftus, M. Fisher, S. Alahari, A. Nolting, R. L. Juliano, "Characterization of Complexes of Oligonucleotides with Polyamidoamine Starburst Dendrimers and Effects on Intracellular Delivery", *J. Pharm. Sci.* **1997**, *86*, 762–764.

[172] J. F. Kukowska-Latallo, A. U. Bielinska, J. Johnson, R. Spindler, D. A. Tomalia, J. R. Baker, Jr., "Efficient Transfer of Genetic Material into Mammalian Cells using Starburst Polyamidoamine Dendrimers", *Proc. Natl. Acad. Sci. U. S. A.* **1996**, *93*, 4897–4902.

[173] A. U. Bielinska, J. F. Kukowska-Latallo, J. R. Baker, Jr., "The Interaction of Plasmid DNA with Polyamidoamine Dendrimers: Mechanism of Complex Formation and Analysis of Alterations Induced in Nuclease Sensitivity and Transcription Activity of the Complexed DNA", *Biochim. Biophys. Acta* **1997**, *1352*, 180–190.

[174] L. Qin, D. R. Pahud, Y. Ding, A. U. Bielinska, J. F. Kukowska-Latallo, J. R. Baker, Jr., J. S. Bromberg, "Efficient Transfer of Genes into Murine Cardiac Grafts by Starburst Polyamidoamine Dendrimers", *Human Gene Therapy* **1998**, *9*, 553–560.

[174a] Y. O. Wang, P. Boros, J. H. Liu, L. H. Qin, Y. L. Bai, A. U. Bielinska, J. F. Kukowska-Latallo, J. R. Baker, Jr., J. S. Bromberg, "DNA/dendrimer complexes mediate gene transfer into murine cardiac transplants ex vivo", *Molecular Therapy* **2000**, *2*, 602–608.

[175] C. B. Mamoun, R. Truong, I. Gluzman, N. S. Akopyants, A. Oksman, D. E. Goldberg, "Transfer of Genes into *Plasmodium falciparum* by Polyamidoamine Dendrimers", *Mol. Biochem. Parasitol.* **1999**, *103*, 117–121.

[176] J. F. Kukowska-Latallo, C. Chen, J. Eichman, A. U. Bielinska, J. R. Baker, Jr., "Enhancement of Dendrimer-Mediated Transfection Using Synthetic Lung Surfactant Exosurf Neonatal *in Vitro*", *Biochem. Biophys. Res. Commun.* **1999**, *264*, 253–261.

[177] A. U. Bielinska, A. Yen, H. L. Wu, K. M. Zahos, R. Sun, N. D. Weiner, J. R. Baker, Jr., B. J. Roessler, "Application of Membrane-Based Dendrimer/DNA Complexes for Solid-Phase Transfection *in Vitro* and *in Vivo*", *Biomaterials* **2000**, *21*, 877–887.

[178] E. Chaum, M. P. Hatton, G. Stein, "Polyplex-Mediated Gene Transfer into Human Retinal Pigment Epithelial Cells *in Vitro*", *J. Cell. Biochem.* **1999**, *76*, 153–160.

[179] E. Wagner, "Effects of Membrane-Active Agents in Gene Delivery", *J. Controlled Release* **1998**, *53*, 155–158.

[180] M. Ruponen, S. Ylä-Herttuala, A. Urtti, "Interactions of Polymeric and Liposomal Gene Delivery Systems with Extracellular Glycosaminoglycans: Physicochemical and Transfection Studies", *Biochim. Biophys. Acta* **1999**, *1415*, 331–341.

[181] P. Erbacher, A. C. Roche, M. Monsigny, P. Midoux, "Glycosylated Polylysine/DNA Complexes: Gene Transfer Efficiency in Relation with the Size and the Sugar Substitution Level of Glycosylated Polylysines and with the Plasmid Size", *Bioconj. Chem.* **1995**, *6*, 401–410.

[182] O. Boussif, T. Delair, C. Brua, L. Veron, A. Pavirani, H. V. J. Kolbe, "Synthesis of Polyallylamine Derivatives and Their Use as Gene Transfer Vectors *in Vitro*", *Bioconj. Chem.* **1999**, *10*, 877–883.

[183] A. U. Bielinska, C. Chen, J. Johnson, J. R. Baker, Jr., "DNA Complexing with Polyamidoamine Dendrimers: Implications for Transfection", *Bioconj. Chem.* **1999**, *9*, 843–850.

[184] V. Hélin, M. Gottikh, Z. Mishal, F. Subra, C. Malvy, M. Lavignon, "Cell Cycle-Dependent Distribution and Specific Inhibitory Effect of Vectorized Antisense Oligonucleotides in Cell Culture", *Biochem. Pharmacol.* **1999**, *58*, 95–107.

[185] W. Chen, N. J. Turro, D. A. Tomalia, "Using Ethidium Bromide to Probe the Interactions between DNA and Dendrimers", *Langmuir* **2000**, *16*, 15–19.

[186] A. Bielinska, J. F. Kukowska-Latallo, J. Johnson, D. A. Tomalia, J. R. Baker, Jr., "Regulation of *in vitro* Gene Expression using Antisense Oligonucleotides or Antisense Expression Plasmids Transfected using Starburst PAMAM Dendrimers", *Nucleic Acids Res.* **1996**, *24*, 2176–2182.

[187] R. F. Barth, D. M. Adams, A. H. Soloway, M. V. Darby, "*In Vivo* Distribution of Boronated Monoclonal Antibodies and Starburst Dendrimers" in *Adv. Neutron Capture Ther. [Proc. Int. Symp.]*, 5th ed. (Eds.: A. H. Soloway, R. F. Barth, D. E. Carpenter), Plenum, New York, N.Y., **1993**, pp. 351–355.

[188] R. F. Barth, D. M. Adams, A. H. Soloway, F. Alam, M. V. Darby, "Boronated Starburst Dendrimer–Monoclonal Antibody Immunoconjugates: Evaluation as a Potential Delivery System for Neutron Capture Therapy", *Bioconj. Chem.* **1994**, *5*, 58–66.

[189] C. Wu, M. W. Brechbiel, R. W. Kozak, O. A. Gansow, "Metal-Chelate–Dendrimer–Antibody Constructs for use in Radioimmunotherapy and Imaging", *Bioorg. Med. Chem. Lett.* **1994**, *4*, 449–454.

[190] R. F. Barth, A. H. Soloway, R. G. Fairchild, "Boron Neutron Capture Therapy of Cancer", *Cancer Res.* **1990**, *50*, 1061–1070.

[191] R. F. Barth, A. H. Soloway, D. M. Adams, F. Alam, "Delivery of Boron-10 for Neutron Capture Therapy by means of Monoclonal Antibody–Starburst Dendrimer Immunoconjugates", in *SO Prog. Neutron Capture Ther. Cancer [Proc. Int. Symp.]*, 4th ed. (Eds.: B. J. Allen, D. E. Moore, B. V. Harrington), Plenum, New York, N.Y., **1992**, pp. 265–268.

[192] R. F. Barth, A. H. Soloway, R. G. Fairchild, R. M. Brugger, "Boron Neutron Capture Therapy for Cancer", *Cancer* **1992**, *70*, 2955–3007.

[193] R. F. Barth, A. H. Soloway, "Boron Neutron Capture Therapy of Primary and Metastatic Brain Tumors", *Mol. Chem. Neuropath.* **1994**, *21*, 139–154.

[194] J. C. Roberts, Y. E. Adams, D. Tomalia, J. A. Mercer-Smith, D. K. Lavallee, "Using Starburst Dendrimers as Linker Molecules to Radiolabel Antibodies", *Bioconj. Chem.* **1990**, *1*, 305–308.

[195] J. C. Roberts, M. K. Bhalgat, R. T. Zera, "Preliminary Biological Evaluation of Polyamidoamine (PAMAM) Starburst™ Dendrimers", *J. Biomed. Mater. Res.* **1996**, *30*, 53–65.

[196] D. Zanini, R. Roy, "Practical Synthesis of Starburst PAMAM α-Thiosialodendrimers for Probing Multivalent Carbohydrate-Lectin Binding Properties", *J. Org. Chem.* **1998**, *63*, 3486–3491.

[197] D. Zanini, R. Roy, "Synthesis of New α-Thiosialodendrimers and Their Binding Properties to the Sialic Acid Specific Lectin from *Limax flavus*", *J. Am. Chem. Soc.* **1997**, *119*, 2088–2095.

[198] D. Zanini, R. Roy, "Novel Dendritic α-Sialosides: Synthesis of Glycodendrimers Based on a 3,3'-Iminobis(propylamine) Core", *J. Org. Chem.* **1996**, *61*, 7348–7354.

[199] R. M. Kim, M. Manna, S. M. Hutchins, P. R. Griffin, N. A. Yates, A. M. Bernick, K. T. Chapman, "Dendrimer-Supported Combinatorial Chemistry", *Proc. Natl. Acad. Sci. U. S. A.* **1996**, *93*, 10012–10017.

[200] E. Erb, K. D. Janda, S. Brenner, "Recursive Deconvolution of Combinatorial Chemical Libraries", *Proc. Natl. Acad. Sci. U. S. A.* **1994**, *91*, 11422–11426.

[201] K. P. Vercruysse, D. M. Marecak, J. F. Marecek, G. D. Prestwich, "Synthesis and *in vitro* Degradation of New Polyvalent Hydrazide Cross-Linked Hydrogels of Hyaluronic Acid", *Bioconj. Chem.* **1997**, *8*, 686–694.

[202] L. D. Margerum, B. K. Campion, M. Koo, N. Shargill, J.-J. Lai, A. Marumoto, P. C. Sontum, "Gadolinium(III) DO3A Macrocycles and Polyethylene Glycol Coupled to Dendrimers. Effect of Molecular Weight on Physical and Biological Properties of Macromolecular Magnetic Resonance Imaging Contrast Agents", *J. Alloys Compd.* **1997**, *249*, 185–190.

[203] P. Caravan, J. J. Ellison, T. J. McMurry, R. B. Lauffer, "Gadolinium(III) Chelates as MRI Contrast Agents: Structure, Dynamics, and Applications", *Chem. Rev.* **1999**, *99*, 2293–2352.

[204] M. Botta, "Second Coordination Sphere Water Molecules and Relaxivity of Gadolinium(III) Complexes: Implications for MRI Contrast Agents", *Eur. J. Inorg. Chem* **2000**, 399–407.

[205] É. Tóth, D. Pubanz, S. Vauthey, L. Helm, A. E. Merbach, "The Role of Water Exchange in Attaining Maximum Relaxivities for Dendrimeric MRI Contrast Agents", *Chem. Eur. J.* **1996**, *2*, 1607–1615.

[206] D. L. Ladd, R. Hollister, X. Peng, D. Wei, G. Wu, D. Delecki, R. A. Snow, J. L. Toner, K. Kellar, J. Eck, V. C. Desai, G. Raymond, L. B. Kinter, T. S. Desser, D. L. Rubin, "Polymeric Gadolinium Chelate Magnetic Resonance Imaging Contrast Agents: Design, Synthesis, and Properties", *Bioconj. Chem.* **1999**, *10*, 361–370.

[207] T. J. McMurry, M. Brechbiel, K. Kumar, O. A. Gansow, "Convenient Synthesis of Bifunctional Tetraaza Macrocycles", *Bioconj. Chem.* **1992**, *3*, 108–117.

[208] L. H. Bryant, Jr., M. W. Brechbiel, C. Wu, J. W. M. Bulte, V. Herynek, J. A. Frank, "Synthesis and Relaxometry of High-Generation (G = 5, 7, 9, and 10) PAMAM Dendrimer–DOTA--Gadolinium Chelates", *J. Magn. Reson. Imag.* **1999**, *9*, 348–352.

[209] J. Anzai, M. Nishimura, "Layer-by-Layer Deposition of Avidin and Polymers on a Solid Surface to Prepare Thin Films: Significant Effects of Molecular Geometry of the Polymers on the Deposition Behaviour", *J. Chem. Soc., Perkin Trans. 2* **1997**, 1887–1889.

[210] D. S. Wilbur, P. M. Pathare, D. K. Hamlin, K. R. Buhler, R. L. Vessella, "Biotin Reagents for Antibody Pretargeting; 3. Synthesis, Radioiodination, and Evaluation of Biotinylated Starburst Dendrimers", *Bioconj. Chem.* **1998**, *9*, 813–825.

[211] A. Thornton, D. Bloor, G. H. Cross, M. Szablewski, "Novel Functionalized Poly(amido amine) (PAMAM) Dendrimers: Synthesis and Physical Properties", *Macromolecules* **1997**, *30*, 7600–7603.

[212] A. L. Gray, J. T. Hsu, "Novel Sulfonic Acid-Modified Starburst Dendrimer used as a Pseudostationary Phase in Electrokinetic Chromatography", *J. Chromatogr. A* **1998**, *824*, 119–124.

[213] P. Singh, "Terminal Groups in Starburst Dendrimers: Activation and Reactions with Proteins", *Bioconj. Chem.* **1999**, *9*, 54–63.

[214] L. Balogh, A. de Leuze-Jallouli, P. Dvornic, Y. Kunugi, A. Blumstein, D. A. Tomalia, "Architectural Copolymers of PAMAM Dendrimers and Ionic Polyacetylenes", *Macromolecules* **1999**, *32*, 1036–1042.

[215] A. Gong, C. Liu, Y. Chen, X. Zhang, C. Chen, F. Xi, "A Novel Dendritic Anion Conductor: Quaternary Ammonium Salt of Poly(amidoamine) (PAMAM)", *Macromol. Rapid Commun.* **1999**, *20*, 492–496.

[216] L. J. Twyman, A. E. Beezer, R. Esfand, M. J. Hardy, J. C. Mitchell, "The Synthesis of Water-Soluble Dendrimers, and their Application as Possible Drug Delivery Systems", *Tetrahedron Lett.* **1999**, *40*, 1743–1746.

[217] R. Esfand, D. A. Tomalia, A. E. Beezer, J. C. Mitchell, M. Hardy, C. Orford, "Dendripore and Dendrilock Concepts, New Controlled Delivery Strategies", *Polym. Prepr.* **2000**, *41*, 1324–1325.

[218] L. J. Twyman, "Post-Synthetic Modification of the Hydrophobic Interior of a Water-Soluble Dendrimer", *Tetrahedron Lett.* **2000**, *41*, 6875–6878.

[218a] I. K. Martin, L. J. Twyman, "Acceleration of an aminolysis reaction using a PAMAM dendrimer with 64 terminal amine groups", *Tetrahedron Lett.* **2001**, *42*, 1123–1126.

[219] G. Sui, M. Micic, Q. Huo, R. M. Leblanc, "Studies of a Novel Polymerizable Amphiphilic Dendrimer", *Colloids Surf. A.: Physicochem. Eng. Aspects* **2000**, *171*, 185–197.

[219a] G. Sui, M. Micic, Q. Huo, R. M. Leblanc, "Synthesis and Surface Chemistry Study of a New Amphiphilic PAMAM Dendrimer", *Langmuir* **2000**, *16*, 7847–7851.

[219b] J. Wang, X. Jia, H. Zhong, H. Wu, Y. Li, X. Xu, M. Li, Y. Wei, "Cinnamoyl Shell-Modified Poly(amidoamine) Dendrimers", *J. Polym. Sci., Part A: Polym. Chem.* **2000**, *38*, 4147–4153.

[220] A. Gong, C. Liu, Y. Chen, C. Chen, F. Xi, "Ionic Conductivity of Alkyl-Metal Carboxylated Dendritic Poly(amidoamine) Electrolytes and their Lithium Perchlorate Salt Complex", *Polymer* **2000**, *41*, 6103–6111.

[221] M. Kimura, M. Kato, T. Muto, K. Hanabusa, H. Shirai, "Temperature-Sensitive Dendritic Hosts: Synthesis, Characterization, and Control of Catalytic Activity", *Macromolecules* **2000**, *33*, 1117–1119.

[222] N. Malik, R. Wiwattanapatapee, R. Klopsch, K. Lorenz, H. Frey, J.-W. Weener, E. W. Meijer, W. Paulus, R. Duncan, "Dendrimers: Relationship Between Structure and Biocompatibility *in Vitro*, and Preliminary Studies on the Biodistribution of ^{125}I-Labelled Polyamidoamine Dendrimers *in Vivo*", *J. Controlled Release* **2000**, *65*, 113–148.

[223] I. Jääskeläinen, S. Peltola, P. Honkakoski, J. Mönkkönen, A. Urtti, "A Lipid Carrier with a Membrane Active Component and a Small Complex Size are Required for Efficient Cellular Delivery of Anti-Sense Phosphorothioate Oligonucleotides", *Eur. J. Pharm. Sci.* **2000**, *10*, 187–193.

[224] P. R. Dvornic, A. M. de Leuze-Jallouli, M. J. Owen, S. V. Perz, "Radially Layered Poly(amidoamine-organosilicon) Dendrimers", *Macromolecules* **2000**, *33*, 5366–5378.

[225] V. Chechik, M. Zhao, R. M. Crooks, "Self-Assembled Inverted Micelles Prepared from a Dendrimer Template: Phase Transfer of Encapsulated Guests", *J. Am. Chem. Soc.* **1999**, *121*, 4910–4911.

[226] G. Caminati, N. J. Turro, D. A. Tomalia, "Photophysical Investigation of Starburst Dendrimers and Their Interactions with Anionic and Cationic Surfactants", *J. Am. Chem. Soc.* **1990**, *112*, 8515–8522.

[227] G. Caminati, D. A. Tomalia, N. J. Turro, "Photo-induced Electron Transfer at Polyelectrolyte–Water Interface", *Prog. Colloid Polym. Sci.* **1991**, *84*, 219–222.

[228] G. Caminati, K. Gopidas, A. R. Leheny, N. J. Turro, D. A. Tomalia, "Photochemical Probes of Starburst Dendrimers and Their Utilization as Restricted Reaction Spaces for Electron Transfer Processes" *Polym. Prepr.* **1991**, *32*, 602–603.

[229] M. C. Moreno-Bondi, G. Orellana, N. J. Turro, D. A. Tomalia, "Photoinduced Electron-Transfer Reactions to Probe the Structure of Starburst Dendrimers", *Macromolecules* **1990**, *23*, 910–912.

[230] K. R. Gopidas, A. R. Leheny, G. Caminati, N. J. Turro, D. A. Tomalia, "Photophysical Investigation of Similarities between Starburst Dendrimers and Anionic Micelles", *J. Am. Chem. Soc.* **1991**, *113*, 7335–7342.

[231] N. J. Turro, J. K. Barton, D. A. Tomalia, "Molecular Recognition and Chemistry in Restricted Reaction Spaces. Photophysics and Photoinduced Electron Transfer on the Surfaces of Micelles, Dendrimers, and DNA", *Acc. Chem. Res.* **1991**, *24*, 332–340.

[232] N. J. Turro, J. K. Barton, D. Tomalia, Photoelectron Transfer Between Molecules Adsorbed in Restricted Spaces. Photochem. Convers. Storage Sol. Energy, Proc. Int. Conf., 8th (1991), E. Pelizzetti & M. Schiavello, eds., Kluwer, Dordrecht, The Netherlands

[233] C. Turro, S. Niu, S. H. Bossmann, D. A. Tomalia, N. J. Turro, "Binding of *Ru(phen)$_3^{2+}$ to Starburst Dendrimers and Its Quenching by Co(phen)$_3^{3+}$: Generation Dependence of the Quenching Rate Constant", *J. Phys. Chem.* **1995**, *99*, 5512–5517.

[234] D. Ben-Avraham, L. S. Schulman, S. H. Bossmann, C. Turro, N. J. Turro, "Luminescence Quenching of Ruthenium(II)-tris(phenanthroline) by Cobalt(III)-tris(phenanthroline) Bound to the Surface of Starburst Dendrimers", *J. Phys. Chem. B.* **1998**, *102*, 5088–5093.

[235] P. F. Schwarz, N. J. Turro, D. A. Tomalia, "Interactions Between Positively Charged Starburst Dendrimers and Ru[4,7-(SO$_3$C$_6$H$_5$)-phen]$_3^{4-}$", *J. Photochem. Photobiol., A* **1998**, *112*, 47–52.

[236] Y. Li, P. L. Dubin, R. Spindler, D. A. Tomalia, "Complex Formation between Poly(dimethyl-diallylammonium chloride) and Carboxylated Starburst Dendrimers", *Macromolecules* **1995**, *28*, 8426–8428.

[237] H. Zhang, P. L. Dubin, R. Spindler, D. Tomalia, "Binding of Carboxylated Starburst Dendrimers to Poly(diallyldimethylammonium Chloride)", *Ber. Bunsen-Ges. Phys. Chem.* **1996**, *100*, 923–928.

[238] P. L. Dubin, S. L. Edwards, J. I. Kaplan, M. S. Mehta, D. Tomalia, J. Xia, "Carboxylated Starburst Dendrimers as Calibration Standards for Aqueous Size-Exclusion Chromatography", *Anal. Chem.* **1992**, *64*, 2344–2347.

[239] P. L. Dubin, S. L. Edwards, M. S. Mehta, D. Tomalia, "Quantitation of Non-Ideal Behavior in Protein Size-Exclusion Chromatography", *J. Chromatogr.* **1993**, *635*, 51–60.

[239a] S. Ujiie, M. Osaka, Y. Yano, K. Iimura, "Ionic liquid crystalline systems with branched or hyperbranched polymers", *Kobunshi Ronbunshu* **2000**, *57*, 797–802.

[240] M. F. Ottaviani, E. Cossu, N. J. Turro, D. A. Tomalia, "Characterization of Starburst Dendrimers by Electron Paramagnetic Resonance; 2. Positively Charged Nitroxide Radicals of Variable Chain Length Used as Spin Probes", *J. Am. Chem. Soc.* **1995**, *117*, 4387–4398.

[241] M. F. Ottaviani, N. J. Turro, S. Jockusch, D. A. Tomalia, "Characterization of Starburst Dendrimers by EPR; 3. Aggregational Processes of a Positively Charged Nitroxide Surfactant", *J. Phys. Chem.* **1996**, *100*, 13675–13686.

[242] M. F. Ottaviani, P. Andechaga, N. J. Turro, D. A. Tomalia, "Model for the Interactions between Anionic Dendrimers and Cationic Surfactants by Means of the Spin Probe Method", *J. Phys. Chem. B* **1997**, *101*, 6057–6065.

[243] D. M. Watkins, Y. Sayed-Sweet, J. W. Klimash, N. J. Turro, D. A. Tomalia, "Dendrimers with Hydrophobic Cores and the Formation of Supramolecular Dendrimer–Surfactant Assemblies", *Langmuir* **1997**, *13*, 3136–3141.

[244] M. Santo, M. A. Fox, "Hydrogen Bonding Interactions Between Starburst Dendrimers and Several Molecules of Biological Interest", *J. Phys. Org. Chem.* **1999**, *12*, 293–307.

[245] M. F. Ottaviani, R. Daddi, M. Brustolon, N. J. Turro, D. A. Tomalia, "Structural Modifications of DMPC Vesicles upon Interaction with Poly(amidoamine) Dendrimers Studied by CW-Electron Paramagnetic Resonance and Electron Spin-Echo Techniques", *Langmuir* **1999**, *15*, 1973–1980.

[246] M. F. Ottaviani, P. Matteini, M. Brustolon, N. J. Turro, S. Jockusch, D. A. Tomalia, "Characterization of Starburst Dendrimers and Vesicle Solutions and Their Interactions by CW- and Pulsed-EPR, TEM, and Dynamic Light Scattering", *J. Phys. Chem. B.* **1998**, *102*, 6029–6039.

[247] M. F. Ottaviani, B. Sacchi, N. J. Turro, W. Chen, S. Jockusch, D. A. Tomalia, "An EPR Study of the Interactions between Starburst Dendrimers and Polynucleotides", *Macromolecules* **1999**, *32*, 2275–2282.

[247a] M. F. Ottaviani, F. Furini, A. Casini, N. J. Turro, S. Jockusch, D. A. Tomalia, L. Messori, "Formation of Supramolecular Structures between DNA and Starburst Dendrimers Studied by EPR, CD, UV, and Melting Profiles", *Macromolecules* **2000**, *33*, 7842–7851.

[248] P. G. de Gennes, H. Hervet, "Statistics of <<Starburst>> Polymers", *J. Phys. Lett.* **1983**, *44*, L351–L360.

[249] M. Murat, G. S. Grest, "Molecular Dynamics Study of Dendrimer Molecules in Solvents of Varying Quality", *Macromolecules* **1996**, *29*, 1278–1285.

[250] R. L. Lescanec, M. Muthukumar, "Configurational Characteristics and Scaling Behavior of Starburst Molecules: A Computational Study", *Macromolecules* **1990**, *23*, 2280–2288.

[251] M. L. Mansfield, L. I. Klushin, "Monte Carlo Studies of Dendrimer Macromolecules", *Macromolecules* **1993**, *26*, 4262–4268.

[252] B. Hammouda, "Structure Factor for Starburst Dendrimers", *J. Polym. Sci., Part B: Polym. Phys.* **1992**, *30*, 1387–1390.

[253] B. Hammouda, "Structure Factors for Regular Polymer Gels and Networks", *J. Chem. Phys.* **1993**, *99*, 9182–9187.

[254] R. La Ferla, "Conformations and Dynamics of Dendrimers and Cascade Macromolecules", *J. Chem. Phys.* **1997**, *106*, 688–700.

[255] M. L. Mansfield, "Dendron Segregation in Model Dendrimers", *Polymer* **1994**, *35*, 1827–1830.

[256] Z. Y. Chen, S.-M. Cui, "Monte Carlo Simulations of Starburst Dendrimers", *Macromolecules* **1996**, *29*, 7943–7952.

[257] Z. Y. Chen, C. Cai, "Dynamics of Starburst Dendrimers", *Macromolecules* **1999**, *32*, 5423–5434.

[258] M. L. Mansfield, "Molecular Weight Distributions of Imperfect Dendrimers", *Macromolecules* **1993**, *26*, 3811–3814.

[259] L. J. Hobson, A. M. Kenwright, W. J. Feast, "A Simple 'One Pot' Route to the Hyperbranched Analogues of Tomalia's Poly(amidoamine) Dendrimers", *Chem. Commun.* **1997**, 1877–1878.

[260] D. Farin, D. Avnir, "Surface Fractality of Dendrimers", *Angew. Chem.* **1991**, *103*, 1409–1410; *Int. Ed. Engl.* **1991**, *30*, 1379–1380.

[261] P. Pfeifer, D. Avnir, "Chemistry in Noninteger Dimensions Between Two and Three; I. Fractal Theory of Heterogeneous Surfaces", *J. Chem. Phys.* **1983**, *79*, 3558–3565.

[262] P. Pfeifer, D. Avnir, "Erratum: Chemistry in Noninteger Dimensions Between Two and Three; I. Fractal Theory of Heterogeneous Surfaces [*J. Chem. Phys. 79*, 3558 (1983)]", *J. Chem. Phys.* **1984**, *80*, 4573.

[263] K. Yu, P. S. Russo, "The Characterization of a Polyamidoamine Cascade Polymer by Fluorescence Photobleaching Recovery and Dynamic Light Scattering" *Polym. Prepr.* **1994**, *35*, 773–774.

[264] K. Yu, P. S. Russo, "Light Scattering and Fluorescence Photobleaching Recovery Study of Poly(amidoamine) Cascade Polymers in Aqueous Solution", *J. Polym. Sci., Part B: Polym. Phys.* **1996**, *34*, 1467–1475.

[265] H.-Q. Ding, N. Karasawa, W. A. Goddard, III, "Optimal Spline Cutoffs for Coulomb and van der Waals Interactions", *Chem. Phys. Lett.* **1992**, *193*, 197–201.

[266] D. Boris, M. Rubinstein, "A Self-Consistent Mean Field Model of a Starburst Dendrimer: Dense Core *vs.* Dense Shell", *Macromolecules* **1996**, *29*, 7251–7260.

[267] Z. Xu, J. S. Moore, "Rapid Construction of Large-size Phenylacetylene Dendrimers up to 12.5 Nanometers in Molecular Diameter", *Angew. Chem.* **1993**, *105*, 1394–1397; *Int. Ed. Engl.* **1993**, *32*, 1354–1357.

[268] Z. Xu, "Syntheses, Characterizations and Physical Properties of Stiff Dendritic Macromolecules", Diss. Abstr. Int. **1993**, B, 53(II) pp 234.

[269] M. L. Mansfield, "Surface Adsorption of Model Dendrimers", *Polymer* **1996**, *37*, 3835–3841.

[269a] M. L. Mansfield, "Monte Carlo Studies of Dendrimers. Additional Results for the Diamond Lattice Model", *Macromolecules* **2000**, *33*, 8043–8049.

[270] E. J. Amis, A. Topp, B. J. Bauer, D. A. Tomalia, "SANS Study of Labeled PAMAM Dendrimer" *Polym. Mater. Sci. Eng.* **1997**, *77*, 183–184.

[271] A. Topp, B. J. Bauer, D. A. Tomalia, E. J. Amis, "Effect of Solvent Quality on the Molecular Dimensions of PAMAM Dendrimers", *Macromolecules* **1999**, *32*, 7232–7237.

[272] K. Funayama, T. Imae, "Structural Analysis of Spherical Water-Soluble Dendrimer by SANS", *J. Phys. Chem. Solids* **1999**, *60*, 1355–1357.

[273] R. Duncan, "Polymer Therapeutics into the 21st Century" *Polym. Prepr.* **1999**, *40*, 285.

[274] R. Duncan, N. Malik, "Dendrimers: Biocompatibility and Potential for Delivery of Anticancer Agents" *Proc. Int. Symp. Controlled Relat. Bioact. Mater.* **1996**, *23*, 105–106.

[275] Y. Sayed-Sweet, D. M. Hedstrand, R. Spindler, D. A. Tomalia, "Hydrophobically Modified Poly(amidoamine) (PAMAM) Dendrimers: Their Properties at the Air–Water Interface and Use as Nanoscopic Container Molecules", *J. Mater. Chem.* **1997**, *7*, 1199–1205.

[276] P. M. Saville, J. W. White, C. J. Hawker, K. L. Wooley, J. M. J. Fréchet, "Dendrimer and Polystyrene Surfactant Structure at the Air–Water Interface", *J. Phys. Chem.* **1993**, *97*, 293–294.

[277] P. M. Saville, P. A. Reynolds, J. W. White, C. J. Hawker, J. M. J. Fréchet, K. L. Wooley, J. Penfold, J. R. P. Webster, "Neutron Reflectivity and Structure of Polyether Dendrimers as Langmuir Films", *J. Phys. Chem.* **1995**, *99*, 8283–8289.

[278] S. Stechemesser, W. Eimer, "Solvent-Dependent Swelling of Poly(amido amine) Starburst Dendrimers", *Macromolecules* **1997**, *30*, 2204–2206.

[279] G. Pistolis, A. Malliaris, C. M. Paleos, D. Tsiourvas, "Study of Poly(amidoamine) Starburst Dendrimers by Fluorescence Probing", *Langmuir* **1997**, *13*, 5870–5875.

[280] K. Esumi, M. Goino, "Adsorption of Poly(amidoamine) Dendrimers on Alumina/Water and Silica/Water Interfaces", *Langmuir* **1998**, *14*, 4466–4470.

[281] G. Nisato, R. Ivkov, E. J. Amis, "Size Invariance of Polyelectrolyte Dendrimers", *Macromolecules* **2000**, *33*, 4172–4176.

[282] M. Goino, K. Esumi, "Interactions of Poly(amidoamine) Dendrimers with Alumina Particles", *J. Colloid Interface Sci.* **1998**, *203*, 214–217.

[283] K. Esumi, R. Saika, M. Miyazaki, K. Torigoe, Y. Koide, "Interactions of Poly(amidoamine) Dendrimers Having Surface Carboxyl Groups with Cationic Surfactants", *Colloids Surf. A.: Physicochem. Eng. Aspects* **2000**, *166*, 115–121.

[284] C. L. Jackson, H. D. Chanzy, F. P. Booy, B. J. Drake, D. A. Tomalia, B. J. Bauer, E. J. Amis, "Visualization of Dendrimer Molecules by Transmission Electron Microscopy (TEM): Staining Methods and Cryo-TEM of Vitrified Solutions", *Macromolecules* **1998**, *31*, 6259–6265.

[285] C. Stathakis, E. A. Arriaga, N. J. Dovichi, "Protein Profiling Employing Capillary Electrophoresis with Dendrimers as Pseudostationary Phase Media", *J. Chromatogr. A* **1998**, *817*, 233–238.

[286] J. Iyer, P. T. Hammond, "Langmuir Behavior and Ultrathin Films of New Linear-Dendritic Diblock Copolymers", *Langmuir* **1999**, *15*, 1299–1306.

[287] M. Johnson, C. Santini, P. T. Hammond, "Morphological Behavior and Self-Assembly of Semicrystalline Linear-Dendritic Block Copolymers", *Polym. Prepr.* **2000**, *41*, 1519–1520.

[288] K. Naka, Y. Tanaka, Y. Chujo, Y. Ito, "The Effect of an Anionic Starburst Dendrimer on the Crystallization of $CaCO_3$ in Aqueous Solution", *Chem. Commun.* **1999**, 1931–1932.

[288a] Y. Tanaka, T. Nemoto, K. Naka, Y. Chujo, "Preparation of $CaCO_3$/polymer composite films via interaction of anionic starburst dendrimer with poly(ethylenimine)", *Polym. Bull.* **2000**, *45*, 447–450.

[289] D. A. Wade, P. A. Torres, S. A. Tucker, "Spectrochemical Investigations in Dendritic Media: Evaluation of Nitromethane as a Selective Fluorescence Quenching Agent in Aqueous Carboxylate-Terminated Polyamido Amine (PAMAM) Dendrimers", *Anal. Chim. Acta* **1999**, *397*, 17–31.

[289a] D. A. Wade, C. Mao, A. C. Hollenbeck, S. A. Tucker, "Spectrochemical investigations in molecularly organized solvent media: Evaluation of pyridinium chloride as a selective fluorescence quenching agent of polycyclic aromatic hydrocarbons in aqueous carboxylate-terminated poly(amido)amine dendrimers and anionic micelles", *Fresenius. J. Anal. Chem.* **2001**, *369*, 378–384.

[290] S. Jockusch, J. Ramirez, K. Sanghvi, R. Nociti, N. J. Turro, D. A. Tomalia, "Comparison of Nitrogen Core and Ethylenediamine Core Starburst Dendrimers through Photochemical and Spectroscopic Probes", *Macromolecules* **1999**, *32*, 4419–4423.

[291] M. F. Ottaviani, S. Bossmann, N. J. Turro, D. A. Tomalia, "Characterization of Starburst Dendrimers by the EPR Technique; 1. Copper Complexes in Water Solution", *J. Am. Chem. Soc.* **1994**, *116*, 661–671.

[292] A. M. Striegel, R. D. Plattner, J. L. Willett, "Dilute Solution Behavior of Dendrimers and Polysaccharides: SEC, ESI-MS, and Computer Modeling", *Anal. Chem.* **1999**, *71*, 978–986.

[293] D. I. Malyarenko, R. L. Vold, G. L. Hoatson, "Solid-State Deuteron NMR Studies of Polyamidoamine Dendrimer Salts; 1. Structure and Hydrogen Bonding", *Macromolecules* **2000**, *33*, 1268–1279.

[294] S. Uppuluri, S. E. Keinath, D. A. Tomalia, P. R. Dvornic, "Rheology of Dendrimers; I. Newtonian Flow Behavior of Medium and Highly Concentrated Solutions of Polyamidoamine (PAMAM) Dendrimers in Ethylenediamine (EDA) Solvent", *Macromolecules* **1998**, *31*, 4498–4510.

[294a] M. Sano, J. Okamura, A. Ikeda, S. Shinkai, "A Simple Method To Produce Dendrimer Nanodots over Centimeter Scales by Rapid Evaporation of Solvents", *Langmuir* **2001**, *17*, 1807–1810.

[295] S. Watanabe, S. L. Regen, "Dendrimers as Building Blocks for Multilayer Construction", *J. Am. Chem. Soc.* **1994**, *116*, 8855–8856.

[296] R. K. Iler, "Multilayers of Colloidal Particles", *Colloid Interface Sci.* **1966**, *21*, 569–594.

[297] M. Wells, R. M. Crooks, "Interactions between Organized, Surface-Confined Monolayers and Vapor-Phase Probe Molecules; 10. Preparation and Properties of Chemically-Sensitive Dendrimer Surfaces", *J. Am. Chem. Soc.* **1996**, *118*, 3988–3989.

[298] L. Sun, R. M. Crooks, A. J. Ricco, "Molecular Interactions between Organized, Surface-Confined Monolayers and Vapor-Phase Probe Molecules; 5. Acid-Base Interactions", *Langmuir* **1999**, *9*, 1775–1780.

[299] M. Zhao, H. Tokuhisa, R. M. Crooks, "Molecule-Sized Gates Based on Surface-Confined Dendrimers", *Angew. Chem.* **1997**, *109*, 2708–2710; *Int. Ed. Engl.* **1997**, *36*, 2596–2598.

[300] H. Tokuhisa, M. Zhao, L. A. Baker, V. T. Phan, D. L. Dermody, M. E. Garcia, R. F. Peez, R. M. Crooks, T. M. Mayer, "Preparation and Characterization of Dendrimer Monolayers and Dendrimer–Alkanethiol Mixed Monolayers Adsorbed to Gold", *J. Am. Chem. Soc.* **1998**, *120*, 4492–4501.

[301] J. Wang, J. Chen, X. Jia, W. Cao, M. Li, "Self-Assembly Ultrafilms Based on Dendrimers", *Chem. Commun.* **2000**, 511–512.

[302] M. Zhao, Y. Liu, R. M. Crooks, D. E. Bergbreiter, "Preparation of Highly Impermeable Hyperbranched Polymer Thin-Film Coatings Using Dendrimers First as Building Blocks and then as *in situ* Thermosetting Agents", *J. Am. Chem. Soc.* **1999**, *121*, 923–930.

[303] L. A. Baker, F. P. Zamborini, L. Sun, R. M. Crooks, "Dendrimer-Mediated Adhesion between Vapor-Deposited Au and Glass or Si wafers", *Anal. Chem.* **1999**, *71*, 4403–4406.

[304] A. Hierlemann, J. K. Campbell, L. A. Baker, R. M. Crooks, A. J. Ricco, "Structural Distortion of Dendrimers on Gold Surfaces: A Tapping-Mode AFM Investigation", *J. Am. Chem. Soc.* **1998**, *120*, 5323–5324.

[304a] W. M. Lackowski, J. K. Campbell, G. Edwards, V. Chechik, R. M. Crooks, "Time-Dependent Phase Segregation of Dendrimer/n-Alkylthiol Mixed-Monolayers on Au(111): An Atomic Force Microscopy Study", *Langmuir* **1999**, *15*, 7632–7638.

[305] G. Bar, S. Rubin, R. W. Cutts, T. N. Taylor, T. A. Zawodzinski, Jr., "Dendrimer-Modified Silicon Oxide Surfaces as Platforms for the Deposition of Gold and Silver Colloid Monolayers: Preparation Method, Characterization, and Correlation between Microstructure and Optical Properties", *Langmuir* **1996**, *12*, 1172–1179.

[305a] K. M. A. Rahman, C. J. Durning, N. J. Turro, D. A. Tomalia, "Adsorption of poly(amidoamine) dendrimers on gold", *Langmuir* **2000**, *16*, 10154–10160.

[306] J. Capala, R. F. Barth, M. Bendayan, M. Lauzon, D. M. Adams, A. H. Soloway, R. A. Fenstermaker, J. Carlsson, "Boronated Epidermal Growth Factor as a Potential Targeting Agent for Boron Neutron Capture Therapy of Brain Tumors", *Bioconj. Chem.* **1996**, *7*, 7–15.

[307] P. Singh, F. Moll, III, S. H. Lin, C. Ferzli, "Starburst™ Dendrimers: A Novel Matrix for Multifunctional Reagents in Immunoassays", *Clin. Chem.* **1996**, *42*, 1567–1569.

[308] V. Swali, N. J. Wells, G. J. Langley, M. Bradley, "Solid-Phase Dendrimer Synthesis and the Generation of Super-High-Loading Resin Beads for Combinatorial Chemistry", *J. Org. Chem.* **1997**, *62*, 4902–4903.

[309] A. Basso, B. Evans, N. Pegg, M. Bradley, "Solid-Phase Synthesis of Aryl Ethers on High Loading Dendrimer Resin", *Tetrahedron Lett.* **2000**, *41*, 3763–3767.

[310] V. V. Tsukruk, F. Rinderspacher, V. N. Bliznyuk, "Self-Assembled Multilayer Films from Dendrimers", *Langmuir* **1997**, *13*, 2171–2176.

[311] V. V. Tsukruk, "Dendritic Macromolecules at Interfaces", *Adv. Mater.* **1998**, *10*, 253–257.

[312] N. Tanaka, T. Tanigawa, K. Hosoya, K. Kimata, S. Terabe, "Starburst Dendrimers as Carriers in Electrokinetic Chromatography", *Chem. Lett.* **1992**, 959–962.

[313] C. P. Palmer, N. Tanaka, "Selectivity of Polymeric and Polymer-Supported Pseudostationary Phases in Micellar Electrokinetic Chromatography", *J. Chromatogr. A* **1997**, *792*, 105–124.

[314] C. P. Palmer, "Micelle Polymers, Polymer Surfactants and Dendrimers as Pseudostationary Phases in Micellar Electrokinetic Chromatography", *J. Chromatogr. A* **1997**, *780*, 75–92.

[315] N. Tanaka, T. Fukutome, K. Hosoya, K. Kimata, T. Araki, "Polymer-Supported Pseudo-stationary Phase for Electrokinetic Chromatography. Electrokinetic Chromatography in a Full Range of Methanol–Water Mixtures with Alkylated Starburst Dendrimers", *J. Chromatogr. A* **1995**, *716*, 57–67.

[316] N. Tanaka, T. Fukutome, T. Tanigawa, K. Hosoya, K. Kimata, T. Araki, K. K. Unger, "Structural Selectivity Provided by Starburst Dendrimers as Pseudostationary Phase in Electrokinetic Chromatography", *J. Chromatogr. A* **1995**, *699*, 331–341.

[317] Y. Chujo, H. Matsuki, S. Kure, T. Saegusa, T. Yazawa, "Control of Pore Size of Porous Silica by Means of Pyrolysis of an Organic–Inorganic Polymer Hybrid", *J. Chem. Soc., Chem. Commun.* **1994**, 635–636.

[318] E. Ruckenstein, W. Yin, "SiO2–Poly(amidoamine) Dendrimer Inorganic/Organic Hybrids", *J. Polym. Sci., Part A: Polym. Chem.* **2000**, *38*, 1443–1449.

[319] K. Aoi, A. Motoda, M. Okada, T. Imae, "Novel Amphiphilic Linear Polymer/Dendrimer Block Copolymer: Synthesis of Poly(2-methyl-2-oxazoline)-*block*-poly(amido amine) Dendrimer", *Macromol. Rapid Commun.* **1997**, *16*, 943–952.

[320] K. Fujiki, M. Sakamoto, T. Sato, N. Tsubokawa, "Postgrafting of Hyperbranched Dendritic Polymer from Terminal Amino Groups of Polymer Chains Grafted onto Silica Surface", *J. Macromol. Sci. – Pure Appld. Chem.* **2000**, *A37*, 357–377.

[321] S. Yoshikawa, T. Satoh, N. Tsubokawa, "Post-Grafting of Polymer with Controlled Molecular Weight onto Silica Surface by Termination of Living Polymer Cation with Terminal Amino Groups of Dendrimer-Grafted Ultrafine Silica", *Colloids Surf. A.: Physicochem. Eng. Aspects* **1999**, *153*, 395–399.

[322] C. M. Nunez, A. L. Andrady, R. K. Guo, J. N. Baskir, D. R. Morgan, "Mechanical Properties of Blends of PAMAM Dendrimers with Poly(vinyl chloride) and Poly(vinyl acetate)", *J. Polym. Sci., Part A: Polym. Chem.* **1998**, *36*, 2111–2117.

[323] V. N. Bliznyuk, F. Rinderspacher, V. V. Tsukruk, "On the Structure of Polyamidoamine Dendrimer Monolayers", *Polymer* **1998**, *39*, 5249–5252.

[324] T. Imae, K. Funayama, K. Aoi, K. Tsutsumiuchi, M. Okada, M. Furusaka, "Small-Angle Neutron Scattering and Surface Force Investigations of Poly(amido amine) Dendrimer with Hydroxyl End Groups", *Langmuir* **1999**, *15*, 4076–4084.

[325] H. C. Yoon, H.-S. Kim, "Multilayer Assembly of Dendrimers with Enzymes on Gold: Thickness-Controlled Biosensing Interface", *Anal. Chem.* **2000**, *72*, 922–926.

[325a] H. C. Yoon, M. Y. Hong, H. S. Kim, "Reversible association/dissociation reaction of avidin on the dendrimer monolayer functionalized with a biotin analogue for a regenerable affinity-sensing surface", *Langmuir* **2001**, *17*, 1234–1239.

[326] N. Belcheva, S. P. Baldwin, W. M. Saltzman, "Synthesis and Characterization of Polymer-(multi)-peptide Conjugates for Control of Specific Cell Aggregation", *Biomater. Sci. Polym. Ed.* **1998**, *9*, 207–226.

[327] J. Iyer, K. Fleming, P. T. Hammond, "Synthesis and Solution Properties of New Linear-Dendritic Diblock Copolymers", *Macromolecules* **1998**, *31*, 8757–8765.

[328] R. Yin, Y. Zhu, D. A. Tomalia, H. Ibuki, "Architectural Copolymers: Rod-Shaped, Cylindrical Dendrimers", *J. Am. Chem. Soc.* **1998**, *120*, 2678–2679.

[329] H. Kobayashi, C. Wu, M.-K. Kim, C. H. Paik, J. A. Carrasquillo, M. W. Brechbiel, "Evaluation of the *in Vivo* Biodistribution of Indium-111 and Yttrium-88 Labeled Dendrimer-1B4M-DTPA and Its Conjugation with Anti-Tac Monoclinic Antibody", *Bioconj. Chem.* **1999**, *9*, 103–111.

[329a] H. Kobayashi, N. Sato, S. Kawamoto, T. Saga, A. Hiraga, T. L. Haque, T. Ishimori, J. Konishi, K. Togashi, M. W. Brechbiel, "Comparison of the macromolecular MR contrast agents with ethylenediamine-core versus ammonia-core generation-6 polyamidoamine dendrimer", *Bioconj. Chem.* **2001**, *12*, 100–107.

[329b] H. Kobayashi, N. Sato, A. Hiraga, T. Saga, Y. Nakamoto, H. Ueda, J. Konishi, K. Togashi, M. W. Brechbiel, "3D-Micro-MR angiography of mice using macromolecular MR contrast agents with polyamidoamine dendrimer core with reference to their pharmakinetic properties", *Magn. Reson. Med.* **2001**, *45*, 454–460.

[330] Y. Shirota, T. Kobata, N. Noma, "Starburst Molecules for Amorphous Molecular Materials. 4,4',4''-Tris(*N,N*-diphenylamino)triphenylamine and 4,4',4''-Tris[*N*-(3-methylphenyl)-*N*-phenylamino]triphenylamine", *Chem. Lett.* **1989**, 1145–1148.

[331] K. Yoshizawa, A. Chano, A. Ito, K. Tanaka, T. Yamabe, H. Fujita, J. Yamauchi, M. Shiro, "ESR of the Cationic Triradical of 1,3,5-Tris(diphenylamino)benzene", *J. Am. Chem. Soc.* **1992**, *114*, 5994–5998.

[332] K. Katsuma, Y. Shirota, "A Novel Class of π-Electron Dendrimers for Thermally and Morphologically Stable Amorphous Molecular Materials", *Adv. Mater.* **1998**, *10*, 223–226.

[333] J. Hu, D. Y. Son, "Carbosilazane Dendrimers – Synthesis and Preliminary Characterization Studies", *Macromolecules* **1998**, *31*, 8644–8646.

[334] J. Hu, D. Y. Son, "Synthesis of Novel Carbosilazane Dendrimers", *Polym. Prepr.* **1998**, *39*, 410–411.

[335] E. A. V. Ebsworth, "Shapes of Simple Silyl Compounds in Different Phases", *Acc. Chem. Res.* **1987**, *20*, 295–301.

[336] P. D. Beer, D. Gao, "Dendrimers Based on Multiple 1,4,7-Triazacyclononane Derivatives", *Chem. Commun.* **2000**, 443–444.

[337] K. Kadei, R. Moors, F. Vögtle, "Dendrimere und Dendrimer-Bausteine mit trisubstituiertem Benzol und 'Hexacyclen'" als Kern", *Chem. Ber.* **1994**, *127*, 897–903.

[338] H.-B. Mekelburger, K. Rissanen, F. Vögtle, "Repetitive Synthesis of Bulky Dendrimers – A Reversibly Photoactive Dendrimer with Six Azobenzene Side Chains", *Chem. Ber.* **1993**, *126*, 1161–1169.

[339] H.-B. Mekelburger, F. Vögtle, "Dendrimers with Bulky Repeat Units using a New Repetitive Synthetic Strategy", *Supramol. Chem.* **1993**, *1*, 187–189.

[340] R. Moors, F. Vögtle, "Cascade Molecules: Building Blocks, Multiple Functionalization, Complexing Units, Photoswitches" in *Advances in Dendritic Macromolecules* (Ed.: G. R. Newkome), JAI, Greenwich, Conn., **1995**, pp. 41–71.

[341] F. Vögtle, M. Zuber, R. G. Lichtenthaler, "Notiz über ein vereinfachtes Verfahren zur Darstellung von 1,3,5-Tris(brommethyl)benzol", *Chem. Ber.* **1973**, *106*, 717–718.

[342] T. P. Bender, Y. Qi, P. Desjardins, Z. Y. Wang, "Dendritic Aryl Imides Containing the Tetrahydro[5]helicene Unit: Synthesis, Characterization, Electrochemical Behavior, and Comparison with a Linear Oligoimide Analogue", *Can. J. Chem.* **1999**, *77*, 1444–1452.

[343] H. S. Sahota, P. M. Lloyd, S. G. Yeates, P. J. Derrick, P. C. Taylor, D. M. Haddleton, "Characterisation of Aromatic Polyester Dendrimers by Matrix-Assisted Laser Desorption Ionisation Mass Spectroscopy", *J. Chem. Soc., Chem. Commun.* **1994**, 2445–2446.

[344] D. M. Haddleton, H. S. Sahota, P. C. Taylor, S. G. Yeates, "Synthesis of Polyester Dendrimers", *J. Chem. Soc., Perkin Trans. 1* **1996**, 649–656.

[345] W. Shi, B. Rånby, "Photopolymerization of Dendritic Methacrylated Polyesters, I. Synthesis and Properties", *J. Appl. Polym. Sci.* **1996**, *59*, 1937–1944.

[346] W. Shi, B. Rånby, "Photopolymerization of Dendritic Methacrylated Polyesters, II. Characteristics and Kinetics", *J. Appl. Polym. Sci.* **1996**, *59*, 1945–1950.

[347] W. Shi, B. Rånby, "Photopolymerization of Dendritic Methacrylated Polyesters, III. FRP Composites", *J. Appl. Polym. Sci.* **1997**, *59*, 1951–1956.

[347a] S. Campidelli, R. Deschenaux, "Liquid-Crystalline Fulleropyrrolidines", *Helv. Chim. Acta* **2001**, *84*, 589–593.

[348] H. K. Hall, Jr., D. W. Polis, "'Starburst' Polyarylamines and their Semiconducting Complexes as Potentially Electroactive Materials", *Polym. Bull.* **1987**, *17*, 409–416.

[349] K. R. Stickley, T. D. Selby, S. C. Blackstock, "Isolable Polyradical Cations of Polyphenylenediamines with Populated High-Spin States", *J. Org. Chem.* **1997**, *62*, 448–449.

[350] T. D. Selby, S. C. Blackstock, "Preparation of a Redox-Gradient Dendrimer. Polyamines Designed for One-Way Electron Transfer and Charge Capture", *J. Am. Chem. Soc.* **1998**, *120*, 12155–12156.

[351] T. Yamamoto, M. Nishiyama, Y. Koie, "Palladium-Catalyzed Synthesis of Triarylamines from Aryl Halides and Diarylamines", *Tetrahedron Lett.* **1998**, *39*, 2367–2370.

[352] M. J. Plater, M. McKay, T. Jackson, "Synthesis of 1,3,5-Tris[4-(diarylamino)phenyl]benzene and 1,3,5-Tris(diarylamino)benzene Derivatives", *J. Chem. Soc., Perkin Trans. 1* **2000**, 2695–2701.

[353] C. Lambert, G. Nöll, "One- and Two-Dimensional Electron Transfer Processes in Triarylamines with Multiple Redox Centers", *Angew. Chem.* **1998**, *110*, 2239–2242; *Int. Ed.* **1998**, *37*, 2107–2110.

[354] M. Thelakkat, C. Schmitz, C. Hohle, P. Strihriegl, H.-W. Schmidt, U. Hofmann, S. Schloter, D. Haarer, "Novel Functional Materials Based on Triarylamines – Synthesis and Application in Electroluminescent Devices and Photorefractive Systems", *Phys. Chem. Chem. Phys.* **1999**, *1*, 1693–1698.

[355] S. Heinen, L. Walder, "Generation-Dependent Intramolecular CT Complexation in a Dendrimer Electron Sponge Consisting of a Viologen Skeleton", *Angew. Chem.* **2000**, *112*, 811–814; *Int. Ed.* **2000**, *39*, 806–809.

[355a] S. Heinen, W. Meyer, L. Walder, "Charge trapping in dendrimers with a viologen skeleton and a radial redox gradient", *J. Electroanal. Chem.* **2001**, *498*, 34–43.

[356] Y. Yamakawa, M. Ueda, R. Nagahata, K. Takeuchi, M. Asai, "Rapid Synthesis of Dendrimers Based on Calix[4]resorcinarenes", *J. Chem. Soc., Perkin Trans. 1* **1998**, 4135–4139.

[357] O. Haba, K. Haga, M. Ueda, O. Morikawa, H. Konishi, "A New Photoresist Based on Calix[4]resorcinarene Dendrimer", *Chem. Mater.* **1999**, *11*, 427–432.

[358] J. Veciana, C. Rovira, M. I. Crespo, O. Armet, V. M. Domingo, F. Palacio, "Stable Polyradicals with High-Spin Ground States. 1. Synthesis, Separation, and Magnetic Characterization of the Stereoisomers of 2,4,5,6-Tetrachloro-α,α,α',α'- tetrakis(pentachlorophenyl)-*m*-xylylene Biradical", *J. Am. Chem. Soc.* **1991**, *113*, 2552–2561.

[359] J. Veciana, C. Rovira, "Stable Polyradicals with High-Spin Ground States" in *Magnetic Molecular Materials* (Ed.: D. Gatteschi), Kluwer Academic Publishers, Dordrecht, The Netherlands, **1991**, pp. 121–132.

[360] J. Veciana, C. Rovira, E. Hernandez, N. Ventosa, "Towards the Preparation of Purely Organic Ferromagnets. Super High-Spin Polymeric Materials versus Open-Shell Molecular Solids", *Anales Quimica* **1993**, *89*, 73–78.

[361] J. Veciana, C. Rovira, N. Ventosa, M. I. Crespo, F. Palacio, "Stable Polyradicals with High-Spin Ground States; 2. Synthesis and Characterization of a Complete Series of Polyradicals Derived from 2,4,6-Trichloro-α,α,α',α',α'',α''- hexakis(pentachlorophenyl)mesitylene with $S = 1/2$, 1, and 3/2 Ground States", *J. Am. Chem. Soc.* **1993**, *115*, 57–64.

[362] N. Ventosa, D. Ruiz, C. Rovira, J. Veciana, "Dendrimeric Hyperbranched Alkylaromatic Polyradicals with Mesoscopic Dimensions and High-Spin Ground States", *Mol. Cryst. Liq. Cryst. Sci. Technol., Sect. A* **1993**, *232*, 333–342.

[362a] J. Sedó, N. Ventosa, M. A. Molins, M. Pons, C. Rovira, J. Veciana, "Stereoisomerism of Molecular Multipropellers. 1. Static Stereochemistry of Bis- and Tris-Triaryl Systems", *J. Org. Chem.* **2001**, *66*, 1567–1578.

[362b] J. Sedó, N. Ventosa, M. A. Molins, M. Pons, C. Rovira, J. Veciana, "Stereoisomerism of Molecular Multipropellers. 2. Dynamic Stereochemistry of Bis- and Tris-Triaryl Systems", *J. Org. Chem.* **2001**, *66*, 1579–1589.

[363] D. Ruiz-Molina, J. Veciana, F. Palacio, C. Rovira, "Drawbacks Arising from the High Steric Congestion in the Synthesis of New Dendritic Polyalkylaromatic Polyradicals", *J. Org. Chem.* **1997**, *62*, 9009–9017.

[364] J. Sedó, N. Ventosa, D. Ruiz-Molina, M. Mas, E. Molins, C. Rovira, J. Veciana, "Crystal Structures of Chiral Diastereoisomers of a Carbon-Based High-Spin Molecule", *Angew. Chem.* **1998**, *110*, 344–347; *Int. Ed.* **1998**, *37*, 320–333.

[365] N. Ventosa, D. Ruiz-Molina, J. Sedó, C. Rovira, X. Tomas, J.-J. André, A. Bieber, J. Veciana, "Influence of the Molecular Surface Characteristics of the Diastereoisomers of a Quartet Molecule on their Physicochemical Properties: A Linear Solvation Free-Energy Study", *Chem. Eur. J.* **1999**, *5*, 3533–3548.

[366] N. Nakamura, K. Inoue, H. Iwamura, "A Branched-Chain Nonacarbene with a Nonadecet Ground State: A Step Nearer to Superparamagnetic Polycarbenes", *Angew. Chem.* **1993**, *105*, 900–902; *Int. Ed. Engl.* **1993**, *32*, 872–874.

[367] H. Bock, A. John, Z. Havlas, J. W. Bats, "The Triplet Biradical Tris(3,5-di-*tert*-butyl-4-oxophenylene)methane: Crystal Structure, and Spin and Charge Distribution", *Angew. Chem.* **1993**, *105*, 416–418; *Int. Ed. Engl.* **1993**, *32*, 416–418.

[368] B. Kirste, M. Grimm, H. Kurreck, "EPR, ^{1}H, and ^{13}C ENDOR Studies of a Quintet-State ^{13}C-Labeled Galvinoxyl-Type Tetraradical", *J. Am. Chem. Soc.* **1989**, *111*, 108–114.

[369] R. Chaler, J. Carilla, E. Brillas, A. Labarta, Ll. Fajarí, J. Riera, L. Juliá, "Trichloro-2,6-pyridylene, A Good Ferromagnetic Coupling Unit between Two Persistent Carbon Radical Centers", *J. Org. Chem.* **1994**, *59*, 4107–4113.

[370] F. Morgenroth, E. Reuther, K. Müllen, "Polyphenylene Dendrimers: From Three-Dimensional to Two-Dimensional Structures", *Angew. Chem.* **1997**, *109*, 647–649; *Int. Ed. Engl.* **1997**, *36*, 631–634.

[371] F. Morgenroth, K. Müllen, "Dendritic and Hyperbranched Polyphenylenes via a Simple Diels–Alder Route", *Tetrahedron* **1997**, *53*, 15349–15366.

[372] Z. B. Shifrina, M. S. Averina, A. L. Rusanov, M. Wagner, K. Müllen, "Branched Polyphenylenes by Repetitive Diels–Alder Cycloaddition", *Macromolecules* **2000**, *33*, 3525–3529.

[373] S. Takahashi, Y. Kuroyama, K. Sonogashira, N. Hagihara, "A Convenient Synthesis of Ethynylarenes and Diethynylarenes", *Synthesis* **1980**, 627–630.

[374] F. Morgenroth, C. Kübel, K. Müllen, "Nanosized Polyphenylene Dendrimers Based upon Pentaphenylbenzene Units", *J. Mater. Chem.* **1997**, *7*, 1207–1211.

[375] A. J. Berresheim, M. Müller, K. Müllen, "Polyphenylene Nanostructures", *Chem. Rev.* **1999**, *99*, 1747–1785.

[376] M. Müller, C. Kübel, K. Müllen, "Giant Polycyclic Aromatic Hydrocarbon", *Chem. Eur. J.* **1998**, *4*, 2099–2109.

[376a] U.-M. Wiesler, T. Weil, K. Müllen, "Nanosized Polyphenylene Dendrimers", *Top. Curr. Chem.* **2001**, 1–40.

[377] P. Kovacic, F. W. Koch, "Coupling of Naphthalene Nuclei by Lewis Acid Catalyst-Oxidant", *J. Org. Chem.* **1965**, *30*, 3176–3181.

[378] F. Morgenroth, A. J. Berresheim, M. Wagner, K. Müllen, "Spherical Polyphenylene Dendrimers *via* Diels–Alder Reactions: The First Example of an A$_4$B Building Block in Dendrimer Chemistry", *Chem. Commun.* **1998**, 1139–1140.

[379] V. S. Iyer, K. Yoshimura, V. Enkelmann, R. Epsch, J. P. Rabe, K. Müllen, "A Soluble C$_{60}$ Graphite Segment", *Angew. Chem.* **1998**, *110*, 2843–2846; *Int. Ed.* **1998**, *37*, 2696–2699.

[379a] S. Setayesh, A. C. Grimsdale, T. Weil, V. Enkelmann, K. Müllen, F. Meghdadi, E. J. W. List, G. Leising, "Polyfluorenes with Polyphenylene Dendron Side Chains: Toward Non-Aggregating, Light-Emitting Polymers", *J. Am. Chem. Soc.* **2001**, *123*, 946–953.

[380] F. Dötz, J. D. Brand, S. Ito, L. Gherghel, K. Müllen, "Synthesis of Large Polycyclic Aromatic Hydrocarbons: Variation of Size and Periphery", *J. Am. Chem. Soc.* **2000**, *122*, 7707–7717.

[381] S. Ito, P. T. Herwig, T. Böhme, J. P. Rabe, W. Rettig, K. Müllen, "Bishexa-*peri*-hexabenzocoronenyl: A 'Superbiphenyl'", *J. Am. Chem. Soc.* **2000**, *122*, 7698–7706.

[381a] U.-M. Wiesler, A. J. Berresheim, F. Morgenroth, G. Lieser, K. Müllen, "Divergent Synthesis of Polyphenylene Dendrimers: The Role of Core and Branching Reagents upon Size and Shape", *Macromolecules* **2001**, *34*, 187–199.

[382] Y. Sakamoto, T. Suzuki, A. Miura, H. Fujikawa, S. Tokito, Y. Taga, "Synthesis, Characterization, and Electron-Transport Property of Perfluorinated Phenylene Dendrimers", *J. Am. Chem. Soc.* **2000**, *122*, 1832–1833.

[383] S. M. Aharoni, C. R. Crosby, III, E. K. Walsh, "Size and Solution Properties of Globular *tert*-Butyloxycarbonyl-poly(α,ε-L-lysine)", *Macromolecules* **1982**, *15*, 1093–1098.

[384] H. Hart, A. Bashir-Hashemi, J. Luo, M. A. Meador, "Iptycenes: Extended Triptycenes", *Tetrahedron* **1986**, *42*, 1641–1654.

[385]   H. Hart, N. Raju, M. A. Meador, D. L. Ward, "Synthesis of Heptiptycenes with Face-to-Face Arene Rings via a 2,3:6,7-Anthradiyne Equivalent", *J. Org. Chem.* **1983**, *48*, 4357–4360.

[386]   A. Bashir-Hashemi, H. Hart, D. L. Ward, "Tritriptycene: A D_{3h} C_{62} Hydrocarbon with Three U-Shaped Cavities", *J. Am. Chem. Soc.* **1986**, *108*, 6675–6679.

[387]   H. Hart, "Iptycenes, Cuppedophanes and Cappedophanes", *Pure Appl. Chem.* **1993**, *65*, 27–34.

[388]   K. Shahlai, H. Hart, "Supertriptycene, $C_{104}H_{62}$", *J. Am. Chem. Soc.* **1990**, *112*, 3687–3688.

[389]   K. Shahlai, H. Hart, "Synthesis of Supertriptycene and Two Related Iptycenes", *J. Org. Chem.* **1991**, *56*, 6905–6912.

[390]   K. Shahlai, H. Hart, A. Bashir-Hashemi, "Synthesis of Three Helically Chiral Iptycenes", *J. Org. Chem.* **1991**, *56*, 6912–6916.

[391]   S. B. Singh, H. Hart, "Extensions of Bicycloalkyne Trimerizations", *J. Org. Chem.* **1990**, *55*, 3412–3415.

[392]   H. Hart, S. Shamouilian, Y. Takehira, "Generalization of the Triptycene Concept. Use of Diaryne Equivalents in the Synthesis of Iptycenes", *J. Org. Chem.* **1981**, *46*, 4427–4432.

[393]   C. F. Huebner, R. T. Puckett, M. Brzechffa, S. L. Schwartz, "A Trimeric $C_{48}H_{30}$ Hydrocarbon of Unusual Structural Interest Derived from 9,10-Dihydro-9,10-ethenoanthracene", *Tetrahedron Lett.* **1970**, 359–362.

[394]   P. Venugopalan, H.-B. Bürgi, N. L. Frank, K. K. Baldridge, J. S. Seigel, "The Crystal Structure of a Heptiptycene–Chlorobenzene Clathrate", *Tetrahedron Lett.* **1995**, *36*, 2419–2422.

[395]   O. W. Webster, "Synthesis and Hydrophobic Binding Studies on a Water Soluble Tritriptycene", *Polym. Prepr.* **1993**, *34*, 98–99.

[396]   N. S. Zefirov, S. I. Kozhushkov, B. I. Ugrak, K. A. Lukin, O. V. Kokoreva, D. S. Yufit, Y. T. Struchkov, S. Zoellner, R. Boese, A. de Meijere, "Branched Triangulanes: General Strategy of Synthesis", *J. Org. Chem.* **1992**, *57*, 701–708.

[397]   A. de Meijere, S. I. Kozhushkov, "The Chemistry of Highly Strained Oligospirocyclopropane Systems", *Chem. Rev.* **2000**, *100*, 93–142.

[398]   A. de Meijere, S. I. Kozhushkov, T. Spaeth, N. S. Zefirov, "A New General Approach to Bicyclopropylidenes", *J. Org. Chem.* **1993**, *58*, 502–505.

[399]   S. I. Kozhushkov, T. Haumann, R. Boese, A. de Meijere, "Perspirocyclopropanated [3]Rotane – A Section of a Carbon Network Containing Spirocyclopropane Units?", *Angew. Chem.* **1993**, *105*, 426–428; *Int. Ed. Engl.* **1993**, *32*, 401–403.

[400]   M. von Seebach, S. I. Kozhushkov, R. Boese, J. Benet-Buchholz, D. S. Yufit, J. A. K. Howard, A. de Meijere, "A Third-Generation Bicyclopropylidene: Straightforward Preparation of 15,15'-Bis(hexaspiro[2.0.2.0.0.0.2.0.2.01.0]pentadecylidene) and a C_2-Symmetric Branched [15]Triangulene", *Angew. Chem.* **2000**, *112*, 2617–2620; *Int. Ed.* **2000**, *39*, 2495–2498.

[401]   G. W. Griffin, A. P. Marchand, "Synthesis and Chemistry of Cubanes", *Chem. Rev.* **1989**, *89*, 997–1010.

[402]   P. E. Eaton, "Cubanes: Starting Materials for the Chemistry of the 1990s and the New Century", *Angew. Chem.* **1992**, *104*, 1447–1462; *Int. Ed. Engl.* **1992**, *31*, 1421–1436.

[403]   H.-D. Beckhaus, C. Rüchardt, S. I. Kozhushkov, V. N. Belov, S. P. Verevkin, A. de Meijere, "Strain Energies in [*n*]Triangulanes and Spirocyclopropanated Cyclobutanes: An Experimental Study", *J. Am. Chem. Soc.* **1995**, *117*, 11854–11860.

[404]   E. Díez-Barra, R. González, A. de la Hoz, A. Rodríguez, P. Sánchez-Verdú, "Acetyl Substituted Benzenes. Useful Cores for the Synthesis of Dendrimeric Polyketones", *Tetrahedron Lett.* **1997**, *38*, 8557–8560.

[405]   C. A. Martínez, A. S. Hay, "Synthesis of Poly(aryl ether) Dendrimers Using an Aryl Carbonate and Mixtures of Metal Carbonates and Metal Hydroxides", *J. Polym. Sci., Part A: Polym. Chem.* **1997**, *35*, 1781–1798.

[406]   R. Haag, A. Sunder, J.-F. Stumbé, "An Approach to Glycerol Dendrimer and Pseudo-Dendritic Polyglycerols", *J. Am. Chem. Soc.* **2000**, *122*, 2954–2955.

[406a] M. A. Carnahan, M. W. Grinstaff, "Synthesis and Characterization of Polyether – Ester Dendrimers from Glycerol and Lactic Acid", *J. Am. Chem. Soc.* **2001**, *123*, 2905–2906.

[407]   H. Uchida, Y. Kabe, K. Yoshino, A. Kawamata, T. Tsumuraya, S. Masamune, "General Strategy for the Systematic Synthesis of Oligosiloxanes. Silicone Dendrimers", *J. Am. Chem. Soc.* **1990**, *112*, 7077–7079.

[408]   E. A. Rebrov, A. M. Muzafarov, V. S. Papkov, A. A. Zhdanov, "Three-Dimensionally Propagating Polyorganosiloxanes", *Doklady Akademii Nauk SSSR* **1990**, *309*, 376–380.

[409]   A. Morikawa, M. Kakimoto, Y. Imai, "Synthesis and Characterization of New Polysiloxane Starburst Polymers", *Macromolecules* **1991**, *24*, 3469–3474.

[410]   A. Morikawa, M. Kakimoto, Y. Imai, "Introduction of Functional Groups into Divergent Starburst Polysiloxanes", *Polym. J. (Tokyo)* **1992**, *24*, 573–581.

[411] J. Roovers, P. M. Toporowski, L.-L. Zhou, "Synthesis of Carbosilane Dendrimers and its Application on the Preparation of 32 Arms and 64 Arms Star Polymers", *Polym. Prepr.* **1992**, *33*, 182–183.

[412] P.-S. Chang, T. S. Hughes, Y. Zhang, G. R. Webster, Jr., D. Poczynok, M. A. Buese, "Synthesis and Characterization of Oligocyclosiloxanes via the Hydrosilation of Vinylsilanes and Vinylsiloxanes with Heptamethylcyclotetrasiloxane", *J. Polym. Sci., Part A: Polym. Chem.* **1993**, *31*, 891–900.

[413] J. Roovers, L.-L. Zhou, P. M. Toporowski, M. van der Zwan, H. Iatrou, N. Hadjichristidis, "Regular Star Polymers with 64 and 128 Arms. Models for Polymeric Micelles", *Macromolecules* **1993**, *26*, 4324–4331.

[414] D. Seyferth, D. Y. Son, A. L. Rheingold, R. L. Ostrander, "Synthesis of an Organosilicon Dendrimer Containing 324 Si–H Bonds", *Organometallics* **1994**, *13*, 2682–2690.

[415] L.-L. Zhou, J. Roovers, "Synthesis of Novel Carbosilane Dendritic Macromolecules", *Macromolecules* **1993**, *26*, 963–968.

[416] L.-L. Zhou, N. Hadjichristidis, P. M. Toporowski, J. Roovers, "Synthesis and Properties of Regular Star Polybutadienes with 32 Arms", *Rubber Chem. Technol.* **1992**, *65*, 303–314.

[417] L. Willner, O. Jucknischke, D. Richter, J. Roovers, L.-L. Zhou, P. M. Toporowski, L. J. Fetters, J. S. Huang, M. Y. Lin, N. Hadjichristidis, "Structural Investigation of Star Polymers in Solution by Small-Angle Neutron Scattering", *Macromolecules* **1994**, *27*, 3821–3829.

[418] B. Comanita, J. Roovers, "Synthesis of New Carbosilane Dendrimers with Hydrophilic End-Groups. Polyols", *Designed Monomers and Polymers* **1999**, *2*, 111–124.

[419] B. Comanita, B. Noren, J. Roovers, "Star Poly(ethylene oxide)s from Carbosilane Dendrimers", *Macromolecules* **1999**, *32*, 1069–1072.

[420] P. I. Coupar, P.-A. Jaffrès, R. E. Morris, "Synthesis and Characterisation of Silanol-Functionalised Dendrimers", *J. Chem. Soc., Dalton Trans.* **1999**, 2183–2187.

[421] B. Alonso, I. Cuadrado, M. Morán, J. Losada, "Organometallic Silicon Dendrimers", *J. Chem. Soc., Chem. Commun.* **1994**, 2575–2576.

[421a] M. Zhou, J. Roovers, "Dendritic Supramolecular Assemblies with Multiple Ru(II) Tris(bipyridine) Units at the Periphery: Synthesis, Spectroscopic, and Electrochemical Study", *Macromolecules* **2001**, *34*, 244–252.

[422] A. Sekiguchi, M. Nanjo, C. Kabuto, H. Sakurai, "Polysilane Dendrimers", *J. Am. Chem. Soc.* **1995**, *117*, 4195–4196.

[423] C. Kim, E. Park, E. Kang, "Synthesis and Characterization of a Carbosilane Dendrimer Containing Allylic End Groups", *Bull. Korean Chem. Soc.* **1996**, *17*, 592–595.

[424] C. Kim, D.-D. Sung, D. Chung, E. Park, E. Kang, "Preparation of Silane Dendrimer (I)", *J. Korean Chem. Soc.* **1995**, *39*, 789–798.

[425] C. Kim, E. Park, E. Kang, "Preparation of Silane Dendrimer (II)", *J. Korean Chem. Soc.* **1995**, *39*, 799–805.

[426] C. Kim, E. Park, I. Jung, "Silane Arborols (V). The Formation of Dendrimeric Silane on Poly(carbosilane): Silane Arborols", *J. Korean Chem. Soc.* **1996**, *40*, 347–356.

[427] C. Kim, M. Kim, "Synthesis of carbosilane dendrimers based on tetrakis(phenylethynyl)silane", *J. Organomet. Chem.* **1998**, *563*, 43–51.

[428] C. Kim, Y. Jeong, I. Jung, "Preparation and Identification of Dendritic Carbosilanes Containing Allyloxy Groups Derived from 2,4,6,8-Tetramethyl-2,4,6,8-tetravinylcyclotetrasiloxane [Me(CH$_2$=CH)SiO]$_4$ and 1,2-Bis(triallyloxysilyl)ethane [(CH$_2$=CHCH$_2$O)$_3$SiCH$_2$]$_2$", *J. Organomet. Chem.* **1998**, *570*, 9–22.

[429] C. Kim, K. Jeong, I. Jung, "Progress Toward Limiting Generation of Dendritic Ethynylsilanes (PhC≡C)$_{4-n}$MenSi$_{(n = 0-2)}$", *J. Polym. Sci., Part A: Polym. Chem.* **2000**, *38*, 2749–2759.

[430] C. Kim, S. Kang, "Carbosilane Dendrimers Based on Siloxane Polymer", *J. Polym. Sci., Part A: Polym. Chem.* **2000**, *38*, 724–729.

[431] C. Kim, M. Ryu, "Synthesis of Double-Layered Dendritic Carbosilanes Based on Allyloxy and Phenylethynyl Groups", *J. Polym. Sci., Part A: Polym. Chem.* **2000**, *38*, 764–774.

[432] C. Kim, I. Jung, "Preparation of Ethynylsilane Dendrimers", *J. Organomet. Chem.* **2000**, *599*, 208–215.

[433] C. Kim, J. Park, "Preparation of Dendritic Carbosilanes Containing Propargyloxy Groups", *Synthesis* **1999**, 1804–1808.

[434] C. Kim, S. Son, "Preparation of Double-Layered Dendritic Carbosilanes", *J. Organomet. Chem.* **2000**, *599*, 123–127.

[435] E. V. Getmanova, E. A. Rebrov, N. G. Vasilenko, A. M. Muzafarov, "Synthesis of Organosilicon Dendrimer with Specific Hydrophobic-Hydrophilic Properties", *Polym. Prepr.* **1998**, *39*, 581–582.

[436] S. S. Sheiko, A. I. Buzin, A. M. Muzafarov, E. A. Rebrov, E. V. Getmanova, "Spreading of Carbosilane Dendrimers at the Air/Water Interface", *Polym. Prepr.* **1998**, *39*, 481–482.

[437] S. S. Sheiko, A. I. Buzin, A. M. Muzafarov, E. A. Rebrov, E. V. Getmanova, "Spreading of Carbosilane Dendrimers at the Air/Water Interface", *Langmuir* **1998**, *14*, 7468–7474.

[438] C. Kim, S. Son, B. Kim, "Dendritic Carbosilanes Containing Hydroxy Groups on the Periphery", *J. Organomet. Chem.* **1999**, *588*, 1–8.

[439] S. A. Ponomarenko, E. A. Rebrov, A. Yu. Bobrovsky, N. I. Boiko, A. M. Muzafarov, V. P. Shibaev, "Liquid-Crystalline Carbosilane Dendrimers: First Generation", *Liq. Cryst.* **1996**, *21*, 1–12.

[440] S. A. Ponomarenko, E. A. Rebrov, N. I. Boiko, N. G. Vasilenko, A. M. Muzafarov, Y. S. Freidzon, V. P. Shibaev, "Synthesis of Cholesterol-Containing Polyorganosiloxane Dendrimers", *Vysokomol. Soedin., Ser. A* **1994**, *36*, 1086–1092.

[440a] X. M. Zhu, R. A. Vinokur, S. A. Ponomarenko, E. A. Rebrov, A. M. Muzafarov, N. I. Boiko, V. P. Shibaev, "Synthesis of new carbosilane ferroelectric liquid-crystalline dendrimers", *Polym. Sci. A* **2000**, *42*, 1263–1271.

[441] S. A. Ponomarenko, N. I. Boiko, V. P. Shibaev, S. M. Richardson, I. J. Whitehouse, E. A. Rebrov, A. M. Muzafarov, "Carbosilane Liquid-Crystalline Dendrimers: From Molecular Architecture to Supramolecular Nanostructures", *Macromolecules* **2000**, *33*, 5549–5558.

[442] C. Kim, K. An, "Preparation and Termination of Carbosilane Dendrimers Based on a Siloxane Tetramer as a Core Molecule: Silane Arborols, Part VIII", *J. Organomet. Chem.* **1997**, *547*, 55–63.

[443] N. G. Vasilenko, E. A. Rebrov, A. M. Muzafarov, B. Eβwein, B. Striegel, M. Möller, "Preparation of Multi-Arm Star Polymers with Polylithiated Carbosilane Dendrimers", *Macromol. Chem. Phys.* **1998**, *199*, 889–895.

[444] G. Ignat'eva, E. A. Rebrov, V. D. Myakushev, A. M. Muzafarov, M. N. Il'ina, I. I. Dubovik, V. S. Papkov, "Polyallylcarbosilane Dendrimers: Synthesis and Glass Transition", *Polym. Sci. A* **1997**, *39*, 874–831.

[445] M. I. Sluch, I. G. Scheblykin, O. P. Varnavsky, A. G. Vitukhnovsky, V. G. Krasovskii, O. B. Gorbatsevich, A. M. Muzafarov, "Excitation Dynamics of Pyrenyl Labeled Polyallylcarbosilane Dendrimers", *J. Lumin.* **1998**, *76&77*, 246–251.

[446] Z. Wu, K. Biemann, "The MALDI Mass Spectra of Carbosilane-Based Dendrimers Containing up to Eight Fixed Positive or 16 Negative Charges", *Int. J. Mass Spectrom. Ion Processes* **1997**, *165/166*, 349–361.

[446a] K. Matsuoka, H. Oka, T. Koyama, Y. Esumi, D. Terunuma, "An alternative route for the construction of carbosilane dendrimers uniformly functionalized with lactose or sialyllactose moieties", *Tetrahedron Lett.* **2001**, *42*, 3327–3330.

[447] K. Brüning, H. Lang, "Silicumorganische Dendrimere mit verschiedenen Dendronen", *J. Organomet. Chem.* **1998**, *571*, 145–148.

[448] K. Brüning, H. Lang, "Linear and Branched Carbosiloxane Dendrimers by Repetitive Hydrosilylation–Alcoholysis Cycles", *Synthesis* **1999**, 1931–1936.

[449] M. M. K. Boysen, T. K. Lindhorst, "Synthesis of Selectively Functionalized Carbosilane Dendrimers with a Carbohydrate Core", *Org. Lett.* **1999**, *1*, 1925–1927.

[450] S. W. Krska, D. Seyferth, "Synthesis of Water-Soluble Carbosilane Dendrimers", *J. Am. Chem. Soc.* **1998**, *120*, 3604–3612.

[451] M. Chai, Z. Pi, C. Tessier, P. L. Rinaldi, "Preparation of Carbosilane Dendrimers and Their Characterization Using ^{1}H/^{13}C/^{29}Si Triple Resonance 3D NMR Methods", *J. Am. Chem. Soc.* **1999**, *121*, 273–279.

[452] B. A. Omotowa, K. D. Keefer, R. L. Kirchmeier, J. M. Shreeve, "Preparation and Characterization of Nonpolar Fluorinated Carbosilane Dendrimers by APcI Mass Spectrometry and Small-Angle X-ray Scattering", *J. Am. Chem. Soc.* **1999**, *121*, 11130–11138.

[453] M. A. Casado, S. R. Stobart, "Modular Construction of Dendritic Carbosilanes. Organization of Dendrimer Connectivity around Bifunctional Precursors that are Adapted for Sequential Convergent and Divergent Progative Steps", *Org. Lett.* **2000**, *2*, 1549–1552.

[454] P. Wijkens, J. T. B. H. Jastrzebski, P. A. van der Schaaf, R. Kolly, A. Hafner, G. van Koten, "Synthesis of Periphery-Functionalized Dendritic Molecules Using Polylithiated Dendrimers as Starting Material", *Org. Lett.* **2000**, *2*, 1612–1624.

[454a] A. W. Kleij, R. van de Coevering, R. J. M. K. Gebbink, A.-M. Noordman, A. L. Spek, G. van Koten, "Polycationic (Mixed) Core – Shell Dendrimers for Binding and Delivery of Inorganic/Organic Substrates", *Chem. Eur. J.* **2001**, *7*, 181–192.

[455] A. Miedaner, C. J. Curtis, R. M. Barkley, D. L. DuBois, "Electrochemical Reduction of CO_2 Catalyzed by Small Organophosphine Dendrimers Containing Palladium", *Inorg. Chem.* **1994**, *33*, 5482–5490.

[456] R. B. King, J. C. Cloyd, Jr., "Some 1:1 Base-Catalyzed Addition Reactions of Compounds Containing Two or More Phosphorus–Hydrogen Bonds to Various Vinylphosphorus Derivatives", *J. Am. Chem. Soc.* **1975**, *97*, 46–52.

[457] R. B. King, J. C. Cloyd, Jr., P. N. Kapoor, "Syntheses and Properties of Novel Polyphosphines Containing Various Combinations of Primary, Secondary, and Tertiary Phosphorus Atoms", *J. Chem. Soc., Perkin Trans. 1* **1973**, 2226–2229.

[458] N. Launay, A.-M. Caminade, R. Lahana, J.-P. Majoral, "A General Synthetic Strategy for Neutral Phosphorus-Containing Dendrimers", *Angew. Chem.* **1994**, *106*, 1682–1685; *Int. Ed. Engl.* **1994**, *33*, 1589–1592.

[459] N. Launay, A.-M. Caminade, J.-P. Majoral, "Synthesis and Reactivity of Unusual Phosphorus Dendrimers. A Useful Divergent Growth Approach up to the Seventh Generation", *J. Am. Chem. Soc.* **1995**, *117*, 3282–3283.

[460] J. Y. Chang, H. J. Ji, M. J. Han, S. B. Rhee, S. Cheong, M. Yoon, "Preparation of Star-Branched Polymers with Cyclotriphosphazene Cores", *Macromolecules* **1994**, *27*, 1376–1380.

[461] J. Mitjaville, A.-M. Caminade, J.-P. Majoral, "Facile Syntheses of Phosphorus-Containing Multisite Receptors", *Tetrahedron Lett.* **1994**, *35*, 6865–6866.

[462] C. Galliot, D. Prévoté, A.-M. Caminade, J.-P. Majoral, "Polyaminophosphine-Containing Dendrimers. Syntheses and Characterizations", *J. Am. Chem. Soc.* **1995**, *117*, 5470–5476.

[463] N. Launay, C. Galliot, A.-M. Caminade, J.-P. Majoral, "Synthesis of Small Phosphorus Dendrimers from (S)P[N(Me)-NH$_2$]$_3$", *Bull. Soc. Chim. Fr.* **1995**, *132*, 1149–1155.

[464] M. Slany, M. Bardají, A.-M. Caminade, B. Chaudret, J.-P. Majoral, "Versatile Complexation Ability of Very Large Phosphino-Terminated Dendrimers", *Inorg. Chem.* **1997**, *36*, 1939–1945.

[465] A.-M. Caminade, R. Laurent, B. Chaudret, J.-P. Majoral, "Phosphine-Terminated Dendrimers: Synthesis and Complexation Properties", *Coord. Chem. Rev.* **1998**, *178–180*, 793–821.

[466] J.-P. Majoral, A.-M. Caminade, "Divergent Approaches to Phosphorus-Containing Dendrimers and their Functionalization", *Top. Curr. Chem.* **1998**, *197*, 79–124.

[467] J.-P. Majoral, A.-M. Caminade, "Dendrimers Containing Heteroatoms (Si, P, B, Ge, or Bi)", *Chem. Rev.* **1999**, *99*, 845–880.

[468] J.-P. Majoral, C. Larré, R. Laurent, A.-M. Caminade, "Chemistry in the Internal Voids of Dendrimers", *Coord. Chem. Rev.* **1999**, *190–192*, 3–18.

[469] V. Maraval, D. Prévoté-Pinet, L. Régis, A.-M. Caminade, J.-P. Majoral, "Choice of Strategies for the Divergent Synthesis of Phosphorus-Containing Dendrons, Depending on the Function Located at the Core", *New J. Chem.* **2000**, *24*, 561–566.

[470] D. Prévoté, A.-M. Caminade, J.-P. Majoral, "Phosphate-, Phosphite-, Ylide-, and Phosphonate-Terminated Dendrimers", *J. Org. Chem.* **1997**, *62*, 4834–4841.

[471] D. Prévoté, S. Le Roy-Gourvennec, A.-M. Caminade, S. Masson, J.-P. Majoral, "Application of the Horner–Wadsworth–Emmons Reaction to the Functionalization of Dendrimers: Synthesis of Amino Acid Terminated Dendrimers", *Synthesis* **1997**, 1199–1207.

[472] M.-L. Lartigue, B. Donnadieu, C. Galliot, A.-M. Caminade, J.-P. Majoral, J.-P. Fayet, "Large Dipole Moments of Phosphorus-Containing Dendrimers", *Macromolecules* **1997**, *30*, 7335–7337.

[473] M.-L. Lartigue, A.-M. Caminade, J.-P. Majoral, "Synthesis and Reactivity of Dendrimers Based on Phosphoryl (P=O) Groups", *Phosphorus, Sulfur Silicon Relat. Elem.* **1997**, *123*, 21–34.

[474] C. Larré, A.-M. Caminade, J.-P. Majoral, "Chemoselective Polyalkylations of Phosphorus-Containing Dendrimers", *Angew. Chem.* **1997**, *109*, 614–617; *Int. Ed. Engl.* **1997**, *36*, 596–598.

[475] M.-L. Lartigue, A.-M. Caminade, J.-P. Majoral, "Chiroptical Properties of Dendrimers with Stereogenic End Groups", *Tetrahedron: Asymmetry* **1997**, *8*, 2697–2708.

[476] D. Prévoté, B. Donnadieu, M. Moreno-Mañas, A.-M. Caminade, J.-P. Majoral, "Grafting of Tetraazamacrocycles on the Surface of Phosphorus-Containing Dendrimers", *Eur. J. Org. Chem.* **1999**, 1701–1708.

[477] C. Larré, D. Bressolles, C. Turrin, B. Donnadieu, A.-M. Caminade, J.-P. Majoral, "Chemistry within Megamolecules: Regiospecific Functionalization after Construction of Phosphorus Dendrimers", *J. Am. Chem. Soc.* **1998**, *120*, 13070–13082.

[478] C. Larré, B. Donnadieu, A.-M. Caminade, J.-P. Majoral, "Phosphorus-Containing Dendrimers: Chemoselective Functionalization of Internal Layers", *J. Am. Chem. Soc.* **1998**, *120*, 4029–4030.

[479] M.-L. Lartigue, N. Launay, B. Donnadieu, A.-M. Caminade, J.-P. Majoral, "First 'Layer-Block' Dendrimer Built with a Regular Alternation of Two Types of Repeat Units up to the Fourth Generation", *Bull. Soc. Chim. Fr.* **1997**, 981–988.

[479a] G. J. d. A. A. Soler-Illia, L. Rozes, M. K. Boggiano, C. Sanchez, C.-O. Turrin, A.-M. Caminade, J.-P. Majoral, "New Mesotextured Hybrid Materials Made from Assemblies of Dendrimers and Titanium(IV)-Oxo-Organo Clusters", *Angew. Chem. Int. Ed.* **2000**, *39*, 4250–4254.

[479b] C.-O. Turrin, V. Maraval, A.-M. Caminade, J.-P. Majoral, A. Mehdi, C. Reyé, "Organic-Inorganic Hybrid Materials Incorporating Phosphorus-Containing Dendrimers", *Chem. Mater.* **2000**, *12*, 3848–3856.

[479c] M. Benito, O. Rossell, M. Seco, G. Segalés, V. Maraval, R. Laurent, A.-M. Caminade, J.-P. Majoral, "Very large neutral and polyanionic Fe/Au cluster-containing dendrimers", *J. Organomet. Chem.* **2001**, *622*, 33–37.

[479d] C.-O. Turrin, J. Chiffre, D. de Montauzon, J.-C. Daran, A.-M. Caminade, E. Manoury, G. Balavoine, J.-P. Majoral, "Phosphorus-Containing Dendrimers with Ferrocenyl Units at the Core, within the Branches, and on the Periphery", *Macromolecules* **2000**, *33*, 7328–7336.

[479e] F. Le Derf, E. Levillain, G. Trippé, A. Gorgues, M. Sallé, R.-M. Sebastían, A.-M. Caminade, J.-P. Majoral, "Immobilization of Redox-Active Ligands on an Electrode: The Dendrimer Route", *Angew. Chem. Int. Ed.* **2001**, *40*, 224–227.

[479f] C.-O. Turrin, J. Chiffre, J.-C. Daran, D. de Montauzon, A.-M. Caminade, E. Manoury, G. Balavoine, J.-P. Majoral, "New chiral phosphorus-containing dendrimers with ferrocenes on the periphery", *Tetrahedron* **2001**, *57*, 2521–2536.

[479g] V. Maraval, R. Laurent, S. Merino, A.-M. Caminade, J.-P. Majoral, "Michael-Type Addition of Amines to the Vinyl Core of Dendrons – Application to the Synthesis of Multidendritic Systems", *Eur. J. Org. Chem.* **2000**, 3555–3568.

[480] R.-M. Sebastian, A.-M. Caminade, J.-P. Majoral, E. Levillain, L. Huchet, J. Roncali, "Electrogenerated Poly(dendrimers) Containing Conjugated Poly(thiophene) Chains", *Chem. Commun.* **2000**, 507–508.

[481] C. Larré, B. Donnadieu, A.-M. Caminade, J.-P. Majoral, "*N*-Thiophosphorylated and *N*-Phosphorylated Iminophosphoranes [R$_3$P=N–P(X)R'$_2$; X = O, S] as Models for Dendrimers: Synthesis, Reactivity and Crystal Structures", *Eur. J. Inorg. Chem.* **1999**, 601–611.

[482] G. Schmid, W. Meyer-Zaika, R. Pugin, T. Sawitowski, J.-P. Majoral, A.-M. Caminade, C.-O. Turrin, "Naked Au$_{55}$ Clusters: Dramatic Effect of a Thiol-Terminated Dendrimer", *Chem. Eur. J.* **2000**, *6*, 1693–1697.

[483] V. Maraval, R. Laurent, B. Donnadieu, M. Mauzac, A.-M. Caminade, J.-P. Majoral, "Rapid Synthesis of Phosphorus-Containing Dendrimers with Controlled Molecular Architectures: First Example of Surface-Block, Layer-Block, and Segment-Block Dendrimers Issued from the Same Dendron", *J. Am. Chem. Soc.* **2000**, *122*, 2499–2511.

[484] R. Kraemer, C. Galliot, J. Mitjaville, A.-M. Caminade, J.-P. Majoral, "Hexamethylhydrazinocyclotriphosphazene N$_3$P$_3$(NMeNH$_2$)$_6$: Starting Reagent for the Synthesis of Multifunctionalized Species, Macrocycles, and Small Dendrimers", *Heteroat. Chem.* **1996**, *7*, 149–154.

[485] M. Slany, A.-M. Caminade, J.-P. Majoral, "Specific Functionalization on the Surface of Dendrimers", *Tetrahedron Lett.* **1996**, *37*, 9053–9056.

[486] J. Mitjaville, A.-M. Caminade, J.-C. Daran, B. Donnadieu, J.-P. Majoral, "Phosphorylated Hydrazines and Aldehydes as Precursors of Phosphorus-Containing Multimacrocycles", *J. Am. Chem. Soc.* **1995**, *117*, 1712–1721.

[487] N. Launay, A.-M. Caminade, J.-P. Majoral, "Synthesis of Bowl-Shaped Dendrimers from Generation 1 to Generation 8", *J. Organomet. Chem.* **1997**, *529*, 51–58.

[488] H. R. Allcock, S. E. Kuharcik, C. S. Reed, M. E. Napierala, "Synthesis of Polyphosphazenes with Ethyleneoxy-Containing Side Groups: New Solid Electrolyte Materials", *Macromolecules* **1996**, *29*, 3384–3389.

[489] M.-L. Lartigue, M. Slany, A.-M. Caminade, J.-P. Majoral, "Phosphorus-Containing Dendrimers: Synthesis of Macromolecules with Multiple Tri- and Tetrafunctionalization", *Chem. Eur. J.* **1996**, *2*, 1417–1426.

[490] N. Launay, M. Slany, A.-M. Caminade, J.-P. Majoral, "Phosphorus-Containing Dendrimers. Easy Access to New Multi-Difunctionalized Macromolecules", *J. Org. Chem.* **1996**, *61*, 3799–3805.

[491] C. Galliot, C. Larré, A.-M. Caminade, J.-P. Majoral, "Regioselective Stepwise Growth of Dendrimer Units in the Internal Voids of a Main Dendrimer", *Science* **1997**, *277*, 1981–1984.

[492] G. M. Salmonczyk, M. Kuznikowski, A. Skowronska, "A Divergent Synthesis of Thiophosphate-Based Dendrimers", *Tetrahedron Lett.* **2000**, *41*, 1643–1645.

4 Dendrimers Derived by Divergent Procedures Using 1 → 3 Branching Motifs

4.1 1 → 3 *C*-Branched, *Amide* Connectivity

4.1.1 1 → 3 *C*-Branched, *Amide* ('Tris') Connectivity

In 1985, Newkome et al. reported[1] the first example of divergently constructed cascade spherical macromolecules utilizing sp^3-carbon atoms as 1 → 3 branching centers for the monomeric dendrons. Although these initial syntheses were not strictly iterative, notable dendritic preparative features were explored and exploited. The incorporated building blocks possessed tetrahedral, tetrasubstituted *C*-branching centers that maximized branching for a *C*-based system, and differential monomer layering (analogous to block copolymer construction). The molecular architecture resembled the Leeuwenberg model[2, 3] for trees, as described by Tomlinson,[4] and since this original series was terminated by hydroxyl moieties, the simple descriptive term "arbor*ol*s" was coined. The initial core[1] consisted of a one-directional 1,1,1-tris(hydroxymethyl)alkane (**1**) and two readily available building blocks were used, namely trialkyl methanetricarboxylates[5–7] or their sodium salts (the "triester", **4**) and tris(hydroxymethyl)aminomethane ("tris", **6**; Scheme 4.1). The use of an appropriate spacer between branching centers was found to be necessary due to steric hindrance associated with the quaternary carbon center, as subsequent studies have shown.[8] To circumvent retardation of these S_N2-type chemical transformations, a three-atom distance was demonstrated to be needed between the branch point and the reactive chemical center. Thus, triol **1** was treated with chloroacetic acid, esterified (MeOH, H^+) to produce polyester **2**, reduced (LAH), and transformed (TsCl) to the tris(tosylate) **3**, treatment of which with the sodio anion of triester **4** generated the nonaester **5**. Subsequent amidation of this ester **5** with "tris" (**6**) afforded the desired 27-arborol (**7**), which was fully characterized and shown to be water soluble, thus allowing access to "unimolecular" micelles.[1]

Application of this two-step procedure (Scheme 4.2) of nucleophilic substitution of a substrate possessing an appropriate leaving group with the anion of a trialkyl methanetricarboxylate (**4**) to generate a polyester, followed by amidation with "tris", was extended to the preparation of dumbbell-shaped, two-directional arborols (**9**).[9] Treatment of 1,ω-dibromo- or dimesyloxy-alkanes (**8**) with anion **4**, followed by reaction with tris (**6**) afforded the bis(nonaol)s (**9**), which, when $n = 8–12$, possess unique structural features, as a result of which they stack in an orthogonal array (Figure 4.1(a)) to form spaghetti-like aggregates. These aggregates form aqueous, thermally reversible gels,[10] based on the maximization of lipophilic–lipophilic and hydrophilic–hydrophilic interactions. Fluorescence and electron microscopies, as well as light-scattering experiments, provided evidence for supramolecular stacking and a rod-like micellar topology of these aggregates at low concentrations.

Since these arborols (**9**) are comprised of two hydrophilic groups connected by a hydrophobic linkage, they fit the simple definition of a bolaamphiphile, a term derived from "bolaform amphiphile" originally introduced in 1951 by Fuoss and Edelson.[11] In 1984, when Fuhrhop and Mathieu[12] reported the synthesis and self-assembly of several bolaamphiphiles, these two-directional surfactant-like macromolecules offered a simple entrance to the bolaamphiphile field; this subject has been highlighted[13, 14] and reviewed.[15]

Since these two-directional arborols stack in an organized manner such that the lipophilic alkyl chain moieties are orthogonally juxtaposed, a functionality incorporated on this linkage would by necessity be preorganized for subsequent interactions. The introduction of a central alkyne bond (e.g., **10**) was accomplished by the transformations

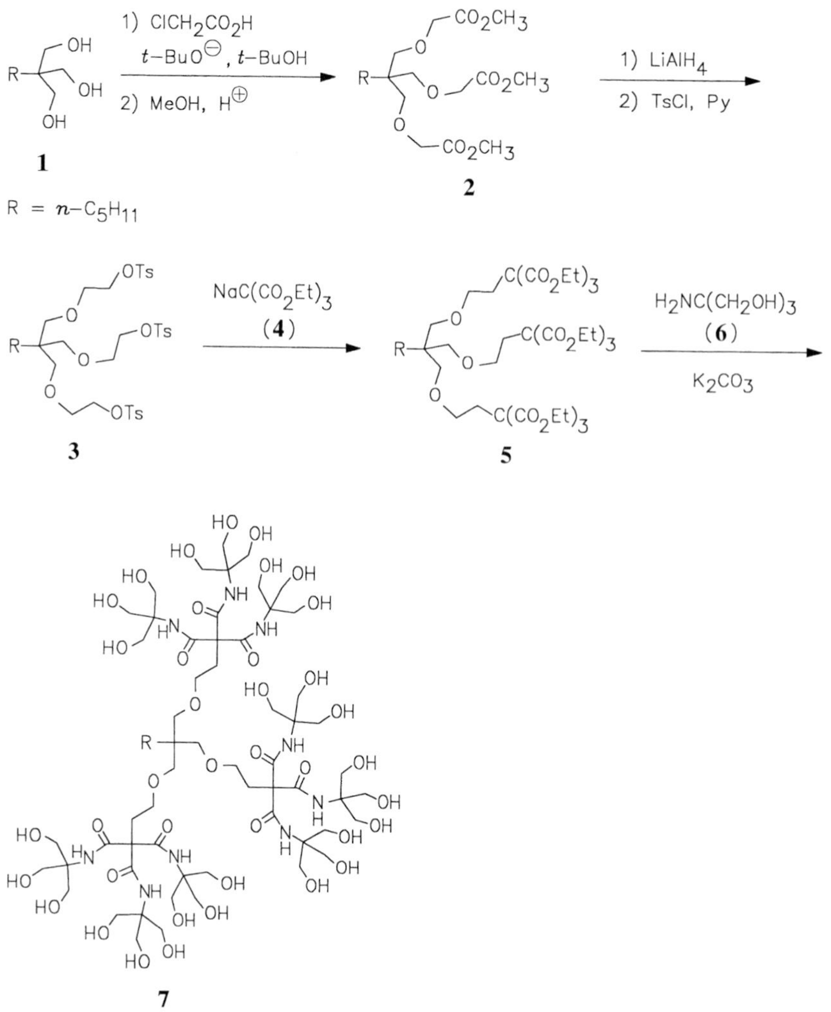

Scheme 4.1 Utilization of methanetriester and aminotriol monomers to construct a 27-arborol.[1]

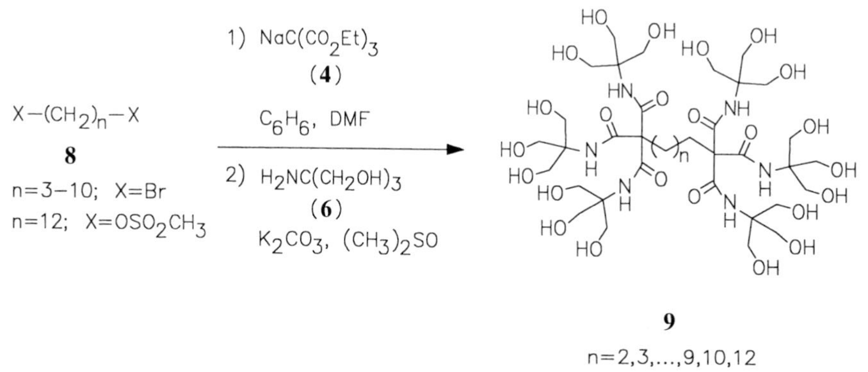

Scheme 4.2 Cascade construction of dumbbell-shaped molecules (**9**)[1] that form stacked aggregates in aqueous environments.

shown in Scheme 4.3. Following transformation of diol **10** to the corresponding mesylate **11**, application of the simple two-step procedure gave rise to polyols **12** or **13** depending on the ester reagents used (i.e., malonic ester or triester). Upon dissolution in water, the resultant alkyne **12** formed a gel in a manner analogous to that seen with alkane-bridged bolaamphiphile **9**. Figure 4.1(b) shows an electron micrograph of **12**, which supports the stacking phenomenon, but the presence of the rigid, linear, central alkyne moiety induces a less than orthogonal chain alignment giving rise to the formation of a helical morphology.[16] The large diameters of the twisted aggregates (Figure 4.1(b)) probably result from the packing of individual rods into the groves of adjacent helical rods, or aggregates, thus producing a "super-coil" or "molecular rope". Pre-determined self-assembly has been termed "automorphogenesis" by Lehn.[17]

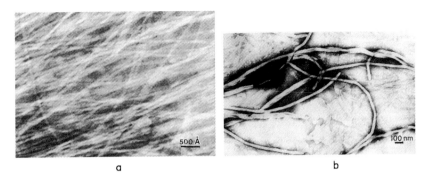

Figure 4.1 Electron micrographs of two-dimensional cascades.[9] Linear aggregates are formed with flexible alkyl bridges (4.1a),[10] whereas curved, rope-like structures result from the incorporation of bridge structural rigidity as provided by an alkyne moiety (4.1b).

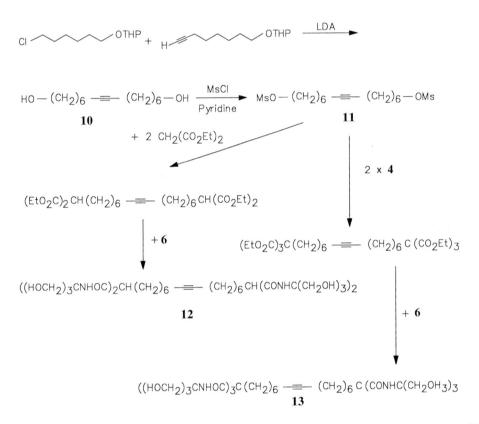

Scheme 4.3 Synthetic methodology for the incorporation of an alkyne unit into a two-dimensional cascade.[16]

Newkome et al.[18] probed the inner lipophilic region of the two-directional arborols by the incorporation of spirane and biphenyl cores. This introduced various intermolecular interactions, which disrupted the aggregation process. Based on computationally generated pictures, a better understanding of the molecular interactions during the initial stages of preorganization prior to gelation has been proposed. Encapsulation or guest inclusion during the gelation process has also been examined, modeled, and demonstrated.

Russo et al.[19] examined the two-directional arborols **9** ($n = 10$) in MeOH/water mixtures by light and small-angle X-ray scattering, differential scanning calorimetry, and freeze-fracture transmission electron microscopy. The self-assembled arborols were seen to be laterally aligned so as to form bundles. The involvement of individual self-assembled "fibers" in multiple bundles leads to the formation of an extended three-dimensional gel network.

Related bolaamphiphiles have been described: Bhattacharya et al.[20] reported the preparation of L-phenylalanine-derivatized alkanes that self-assemble to form fibrous gels. Numerous other related examples have appeared,[21, 22] based on butadiyne capped with 1-glucosamide,[23] *O*-(alkane-α,ω-diyl)bis(glycerol),[24] ω-hydroxy quaternary

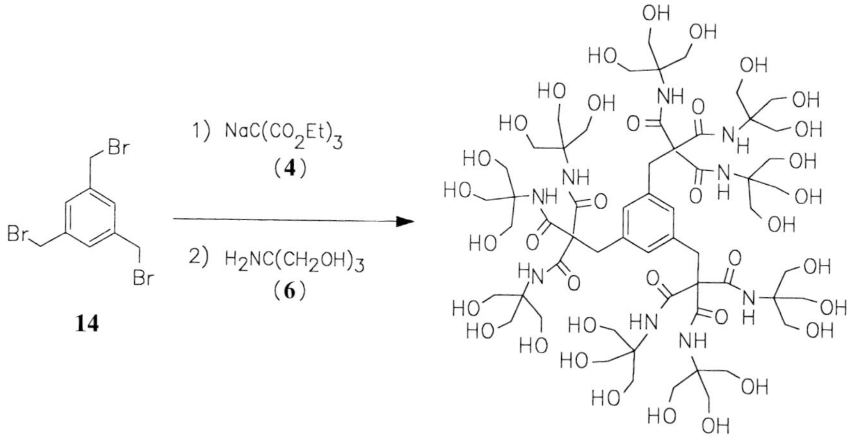

Scheme 4.4 Construction of a cascade triad (**15**).[37]   **15**

ammonium bolaform surfactants,[25] multi-armed "geminis",[26] *N*,*N*'-bis(2-deoxy-D-glucopyranosid-2-yl)-[27] or bis(*O*-galactopyranosyl)- and bis(*O*-lactosyl)-[28] alkane-1,ω-dicarboxamides, bis(glucuroamido)- (as well as 1- or 2-glucosamido-) 1,11-undecane,[29] diacetylenes with terminal *s*-triazines,[30] $Me_3N^+(CH_2)_nO$–Aryl–$O(CH_2)_nN^+Me_3$ 2Br⁻,[31] 1,1'-di-*O*-decyl (or hexadecamethylene) -2,2'-di-*O*-decyl (or octyl) bis(glycerol-3-phosphocholine),[32] peptide-based cationic "gemini" surfactants,[33] azobenzene-4,4'-dicarboxylic acid bis(pyridinohexyl [or undecyl] ester) dibromide,[34] $[Me_3N^+(CH_2)_2$-$OCH_2]_3CCH_2OCH_2$–Aryl–$CH_2OCH_2C[CH_2O(CH_2)_2N^+Me_3]_3$,[35] and boronic acid-appended bolaamphiphiles.[29] Menger and Keiper[36] presented an interesting review of related "gemini surfactants" or bis-surfactants, which self-assemble at low concentrations. Other related examples have appeared.[36a–c]

A related three-directional member of this series[37] was the benzene [9]³-arborol [**15**; 27-Cascade:benzene[3–1,3,5]:(ethylidene):(3-oxo-2-azapropylidene):methanol], which was prepared by application of the two-step (alkylation–amidation or triester–tris) reaction sequence to 1,3,5-tris(bromomethyl)benzene (**14**; Scheme 4.4). Electron microscopy and subsequent light-scattering data suggested that **15** aggregated through stacking of its hydrophilic triad of three small spheres into a spherical array of ca. 20 nm (diameter), very reminiscent of globular micelles.

The results of dynamic light-scattering experiments on aqueous solutions of **15** have been reported.[38] This benzene[9]³-arborol forms aggregates in water, which exhibit dynamic properties very similar to those of single polymer chains in solvents in the crossover region (qr_h) ≈ 1, where q is the absolute value of the scattering vector, and r_h is the hydrodynamic radius of the scattering particles. The size of these particles appears to be concentration independent in the concentration range studied ($3.5 \times 10^{-3}\, c^+ \le c \le 13.37 \times 10^{-3}\, c^+$; where c^+ = 1 mol dm⁻³). From the ratio of the scattered light intensity to the square of the absolute value of the scattering vector ($\Gamma_{max}/q^2)_o$ at the limit $q \Rightarrow$ o, a hydrodynamic radius of r_h = O (100 nm) was calculated.

Expansion of this technology to other more complex systems was demonstrated[39] by the synthesis of water-soluble calixarenes (**18**; Scheme 4.5), termed "silvanols", which possess dendritic polyhydroxy spheres on the upper rims of [1$_n$]metacyclophanes. The initial polytrimethylammonium calixarene (i.e., **16a**)[40, 41] was converted by established procedures to the crystalline dodecaester [1$_4$]metacyclophane **17a** (*n* = 1), the X-ray structure of which confirmed the assigned structure. Treatment of polyester **17a** with tris generated the [36]-silvanol **18a**. Electron micrographs of **18a** showed it to possess a diameter of 57 Å, corresponding to six aggregating macromolecules, as predicted by molecular modeling. The use of [1$_8$]metacyclophane **16b** as the starting material gave rise to the corresponding [72]-silvanol **18b** via the same series of steps.

In 1985, application of this simple dendritic construction to a polymer core, specifically chloromethyl-functionalized polystyrene was reported.[42] A further example of this procedure was reported,[43] which differed in that another polymeric core backbone, derived from a functionalized vinyl ether monomer, was utilized. These systems represented very early examples of dendritic "comb" macromolecules.

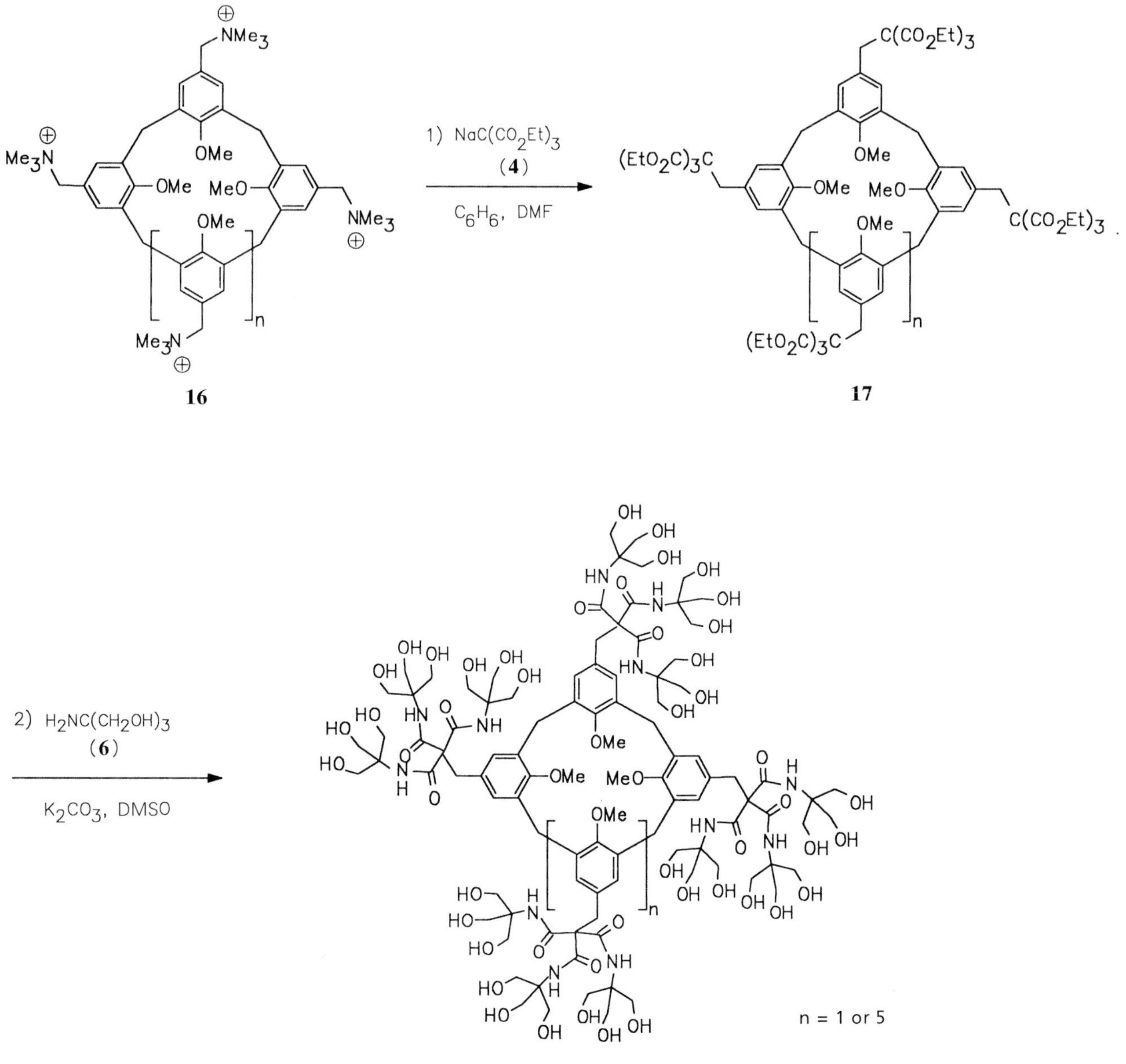

Scheme 4.5 Silvanol construction[39] of molecular forests atop a calixarene plateau.

Lhotak and Shinkai[44] reported the synthesis of oligocalix[4]arenes based on mono-functionalization of calixarene units with bis(alkyl bromides). For example, as depicted in Scheme 4.6, monobromide **19** (4 equiv.) was reacted (NaH, DMF, THF) with tetraol **20** to afford pentakiscalixarene **21**. All calixarenes used in this series were in the cone conformation. Bis- and tris-linear analogues were also prepared and the metal (Li$^+$ and Na$^+$) complexation properties of these materials were investigated by ^{1}H NMR spectroscopy.

Sugawara and Matsuda[45] developed "tris"-based cascades that were modified to include photoactive azide groups for attachment to polyethylene (PE) surfaces, thereby allowing the creation of new hydrophilic surfaces. Synthesis of these focally reactive polyols relied on the above protocol for polyalcohol formation. Prior to cascade amide formation, azide introduction was effected by reaction of *p*-azidobenzoyl chloride with the appropriate monoalcohol di- or triester extended ω-hydroxyalkyl derivatives of **6**. Subsequent treatment with tris generated the corresponding amide with a hydroxy surface. Reaction with the PE surfaces was effected under UV radiation. PE surfaces modified with these molecules were found to be wettable with water. Biomolecular surface assemblies based on branched as well as non-branched coating alcohol moieties were shown to have "well-structured" molecular organization and high packing densities. This simple conversion of an ester to a triol has been utilized[46] to transform polyesters, e.g.,

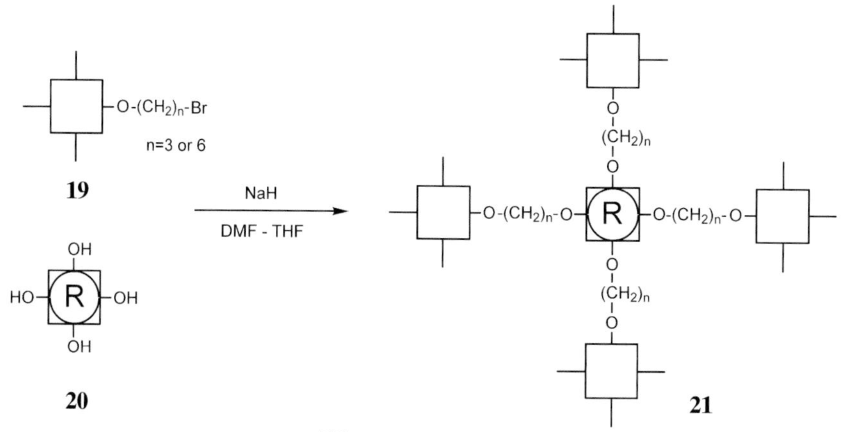

Scheme 4.6 Shinkai's branched oligocalixarenes (**21**).[44]

PAMAMs, to the arborol surface; such materials possess high water solubility and uni-molecular micellar properties.[1]

Alvarez and Strumia[47] studied two protocols for obtaining hydrophilic acrylic acid–ethylene glycol dimethylacrylate matrices, namely carbodiimide-based coupling with tris and copolymerization of hydrophilic monomers.

4.1.2   1 → 3 (1 → 2) *C*-Branched, *Amide* ('Tris') Connectivity

Although Newkome et al. reported[10] a series of two-directional arborols, the original process was a 1 → 3 *C*-branching scheme; under more drastic amidation conditions, the triester, especially the methyl ester, readily decomposed to the 1 → 2 *C*-branched products – the same as those derived from the monoalkylation of malonates. Subsequent treatment of the malonate esters with tris gave rise to the [6]–(X)$_n$–[6] arborol series (e.g., **12**; see Scheme 4.3).

Jørgensen et al.[48] utilized this molecular organization process (dumbbell-like stacking) to incorporate tetrathiafulvalene (TTF; a substrate currently of interest[49] in such areas as molecular electronics and organomagnetism) within the central lipophilic region of the self-assembled, supramolecular structures (Scheme 4.7). A multi-step synthesis was undertaken to assemble the derivatized TTF core **22**. Treatment of tetraester **22** with tris generated the desired TTF-bis-arborol **23**. Calculations based on an orthogonal stacking with the TTF cores adopting the *trans* conformation indicated that the diameter of the stack of molecules should be ca. 3.5 nm. Aggregates derived from dodecaol **23** clearly reveal (microscopy) thin string-like assemblages with lengths in the order of tens of microns and diameters in the order of ca. 100 nm. These structures are therefore super-structures of the single strands, an observation analogous to that previously reported.[9, 10]

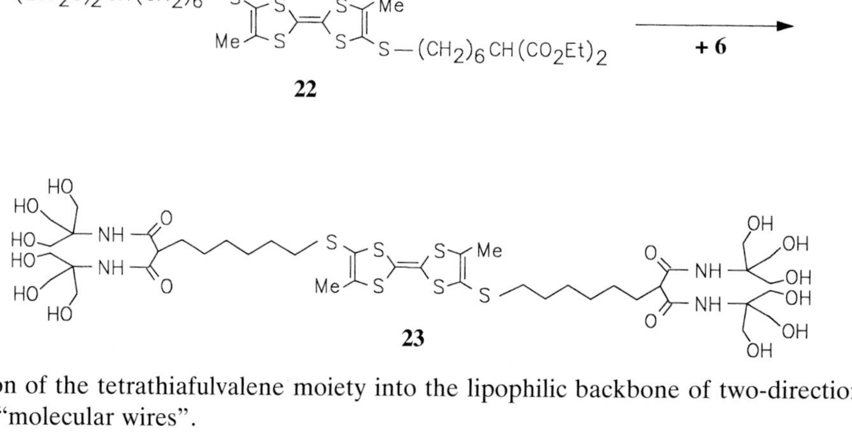

Scheme 4.7 Introduction of the tetrathiafulvalene moiety into the lipophilic backbone of two-directional arborols (**23**)[48] for the construction of "molecular wires".

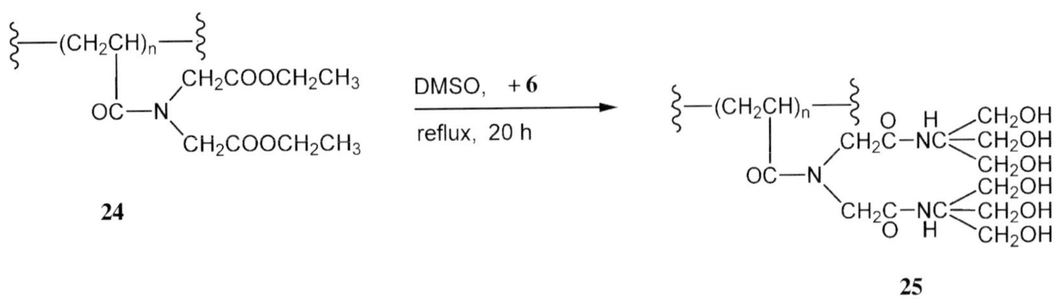

Scheme 4.8 Polyhydroxylation of polyacrylamide.[50]

Relying on this procedure[9, 10] to create arborols, Saito et al.[50] reported the preparation of hydroxylated poly(acrylamide)s. Iminodiacetic diethyl ester was reacted with acryloyl chloride and polymerized (AIBN) to yield polyester **24**. Subsequent treatment with tris gave polyol **25** (Scheme 4.8). Based on surface tension data, the hydrophilicity of these materials unexpectedly decreased as the number of hydroxyl groups per repeat unit was increased. This was attributed to increased intramolecular *H*-bonding effectively reducing the number of OH moieties available for interaction with the aqueous interface. Matsuda and Sugawara[51] later applied a similar strategy (i.e., reaction of tris with either a di- or triester to afford the corresponding hexa- or nona-alcohol units) for the modification of a nonionic poly(vinyl ether).

The use of the monomer "tris" in the above simple amidation process to convert an ester to the corresponding amide has been demonstrated to instill hydrophilic properties into numerous nondendritic, initially hydrophobic materials (e.g., a neutral cyclophane-dodecaalcohol).[52] Other non-ionic, "tris"-based polyols useful as sugar macronutrient substrates in low-calorie food formulations have been reported by Yalpani et al.[53]

Zeroth generation synthetic triglycerides have been investigated[54] as vehicles for peptide delivery across the "blood–brain" barrier.

4.1.3   1 → 3 *C*-Branched, *Amide* ('Bis*homo*tris') Connectivity

In order to circumvent unfavored S_N2-type substitution at neopentyl positions preventing continued iteration with 'triester–tris' methodology,[55] a 'bis*homo*tris' monomer **29** was prepared (Scheme 4.9).[56–59] The monomer was obtained by first reacting nitromethane with acrylonitrile to give trinitrile **26**, which was then hydrolyzed to the triacid **27**, reduction of which gave the corresponding triol **28**. Heterogeneous reduction then afforded the desired amine **29**. The use of 'bis*homo*tris' **29** in place of tris **6** in the original alkylation–amidation sequence gave rise to transesterification products. This suggested that the amidation procedure using tris proceeded via a five-membered intermediate ester **31** to give amide **32** through an intramolecular rearrangement (Scheme 4.10). It was therefore postulated[10] that an unfavorable seven-membered transition state (**30**) precluded amide formation. Treatment of this intermediate ester **30** with base (KOH) in DMSO forced the amidation to completion, albeit in extremely poor (< 10 %) yields.

Whitesell and Chang[60] reported the preparation of the monomer H₂NC(CH₂CH₂CH₂SH)₃ from bis*homo*tris in four steps. This aminotrithiol was used to directionally align helical peptide polymerization on gold and indium/tin oxide (ITO) glass surfaces. Unidirectional alignment of macromolecules and their polarizability are of interest in the area of supramolecular chemistry and molecular electronic devices.[61] Hence, dendritic branching combined with "anchoring" units (e.g., sulfur affinity for gold) are logical choices to assist in non-covalent as well as covalent molecular organization.

Recently, von Kiedrowski et al.[62] described the synthesis of trigonal "trisoligonucleotidyls" based on a solid-phase phosphoramidite protocol where the requisite branching was incorporated through *the use of the "bishomotris" monomer*. Bimolecular complexes of these tripodal DNA constructs were described as forming nano-scale acetylene and cyclobutadiene architectures by self-assembly.

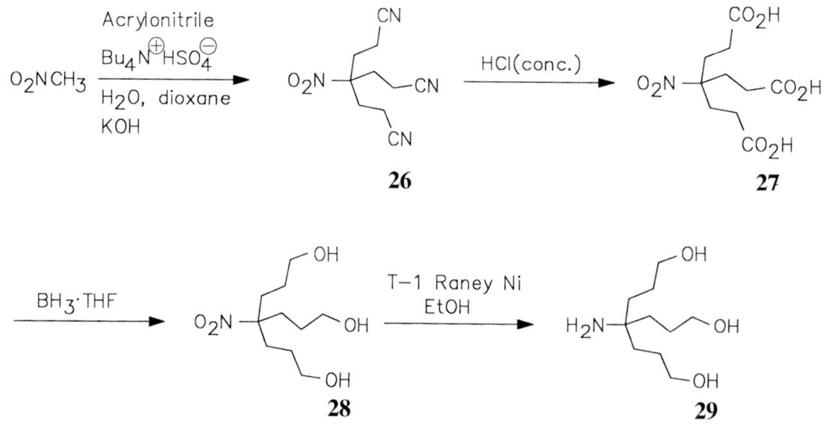

Scheme 4.9 Synthesis of the bis-homologated analog of tris(hydroxymethyl)aminomethane (bis*homo*tris).[10]

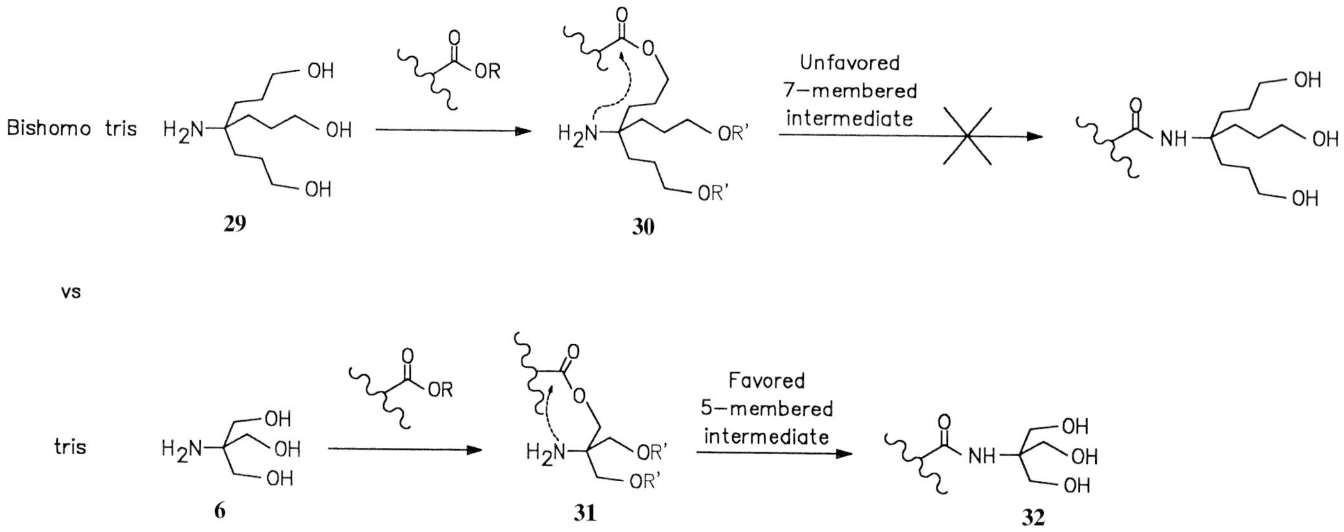

Scheme 4.10 Rationale for the difficulties encountered in the amino acylation of bis*homo*tris.[10]

4.1.4   1 → 3 *C*-Branched, *Amide* ("Behera's Amine") Connectivity

Addition of the anion of nitromethane to α,β-unsaturated carbonyls and nitriles followed by reduction of the nitro group to an amine (Scheme 4.11) provided the basis for the preparation[63] of "Behera's amine"(**35**); this name was derived from Dr. Rajani K. Behera, who was the first to prepare the aminotriester. Treatment of nitromethane with *tert*-butyl acrylate (**33**) in the presence of base in a modified literature procedure[64] gave di-*tert*-butyl 4-[(2-*tert*-butyloxycarbonyl)ethyl]-4-nitroheptanedioate (**34**). Catalytic reduction[65] of nitrotriester **34** quantitatively afforded Behera's amine **35**. Uniquely, Behera's amine does not undergo facile intramolecular lactam formation during the hydrogenation process, as is predominant in other related but less branched esters.[66] The X-ray structure of Behera's amine confirms the extended conformation, with 15 of the 16 torsion angles in the *anti* orientation (mean value of 176.6°).[67]

The use of amine **35** with diverse cores[68–70] has demonstrated its utility in divergent dendritic construction (Scheme 4.12). For example, when 1,3,5,7-tetrakis(chlorocarbonyl)adamantane[71] {prepared from the corresponding tetraacid **36**, in turn synthesized in one step [(COCl)$_2$, hv; 20–30 % yield][72] from 1-adamantanecarboxylic acid} was treated with aminotriester **35**, the dodecaester **37** was isolated. Quantitative hydrolysis (HCO$_2$H) of the ester groups yielded the corresponding acid **38**. Amidation of dodeca-acid **38** under peptide coupling conditions[73] using a slight excess of Behera's amine **35** afforded the 2nd tier 36-ester **39**, treatment of which with formic acid gave 36-Cascade:tri-

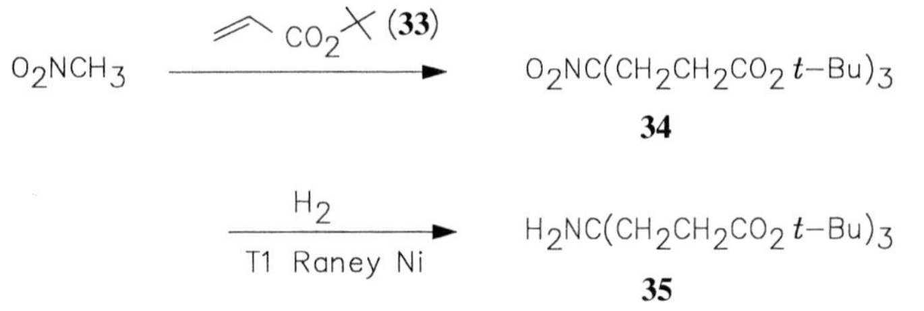

$$O_2NCH_3 \xrightarrow{\displaystyle \diagup\!\!\diagdown CO_2X \ (33)} O_2NC(CH_2CH_2CO_2\,t\text{–Bu})_3$$

34

$$\xrightarrow[\text{T1 Raney Ni}]{H_2} H_2NC(CH_2CH_2CO_2\,t\text{–Bu})_3$$

35

Scheme 4.11 Two-step construction[63] of an aminotriester monomer (**35**).

cyclo[3.3.1.1^{3,7}]decane[4–1,3,5,7] : (3-oxo-2-azapropylidyne) : (3-oxo-2-azapentylidyne) : propanoic acid (**40**).

Newkome et al.[74] reported the use of Behera's amine **35** in the synthesis of the [12]- (**42**), [36]- (**43**), [108]- (**44**), [324]- (**45**), and [972]- (**46**) polyamido cascade series by an iterative, divergent procedure (Scheme 4.13) based on the ethereal core **41**.[75] This was constructed by Bruson's method[76] of exhaustive 1,4-addition of acrylonitrile to penta-erythritol[77] followed by hydrolysis. Repetition of the amidation[73]–deprotection sequence (DCC coupling; HCO$_2$H hydrolysis) allowed the construction of the 5th generation dendrimers with purported molecular weights of 165,909 ([972]-ester) and 111,373 amu for

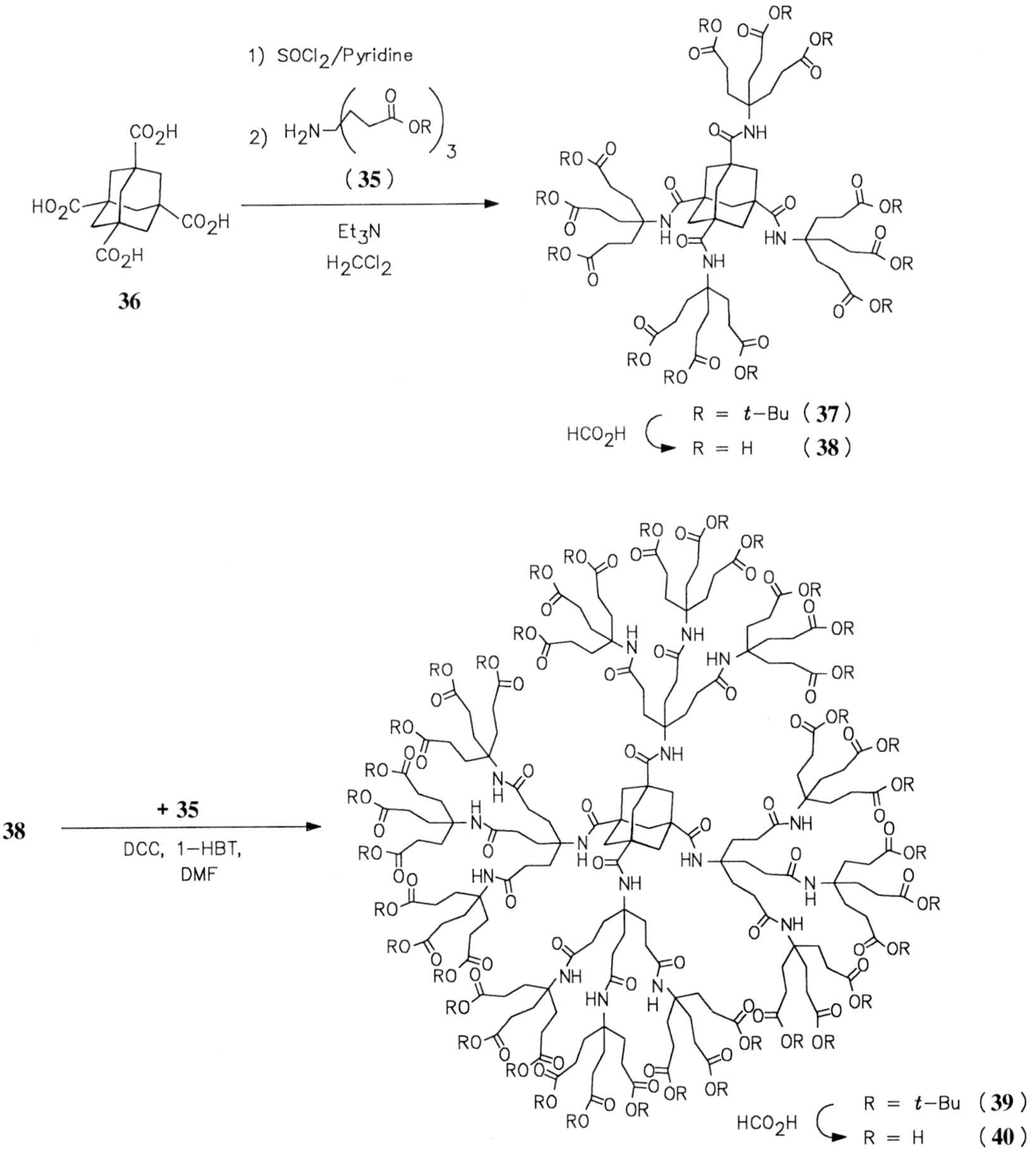

Scheme 4.12 Construction of dendrimers[68, 69] using adamantane tetracarboxylic acid as a tetravalent core.

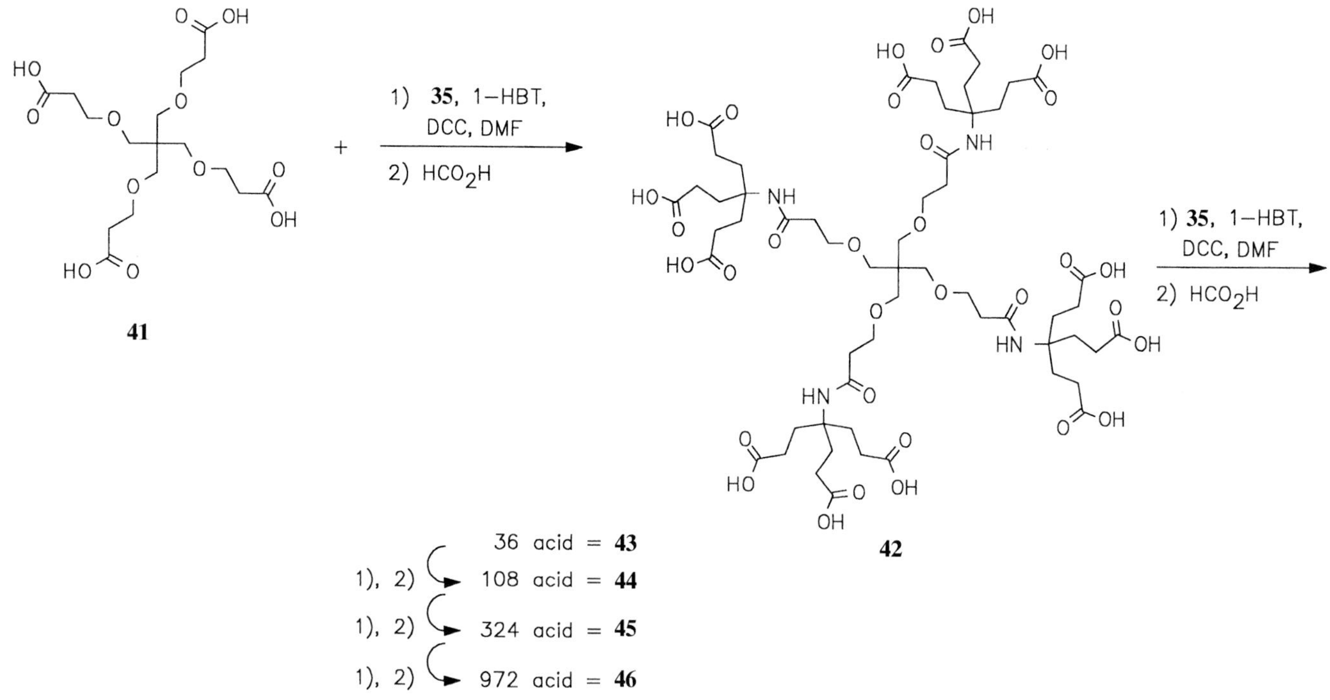

1) **35**, 1–HBT, DCC, DMF
2) HCO₂H

+

41

1) **35**, 1–HBT, DCC, DMF
2) HCO₂H

42

```
                  36 acid = 43
1), 2)  ⤷  108 acid = 44
1), 2)  ⤷  324 acid = 45
1), 2)  ⤷  972 acid = 46
```

Scheme 4.13 Iterative procedure for the preparation of dendrimers with flexible pentaerythritol-based cores.[74]

972-Cascade:methane[4]:(3-oxo-6-oxa-2-azaheptylidene):(3-oxo-2-azapentylidyne)[4]: propanoic acid (**46**). Structural support for these amide-based dendrimers was provided by standard spectroscopic procedures as well as by two-dimensional diffusion-ordered spectroscopy (DOSY) NMR,[78] whereby diffusion coefficients were ascertained by pulsed field gradient NMR for each generation of the water-soluble, carboxylic acid dendrimers. Both the [36]- (**43**) and [108]- (**44**) polyesters have been prepared by a convergent approach using the 9- and 27-counterparts of **35**; the larger dendrimers in this series are not monomolecular in character, as has been noted for other divergently generated dendrimers (see Section 2.4.2). Application of the Stokes–Einstein equation gave dendritic hydrodynamic radii (tabulated at acid, neutral, and basic pH) that correlate well with those obtained by SEC measurements and computer-generated molecular modeling.[74] Hydrodynamic radius changes in the acid-terminated series at different solution pH values are notable; for example, the size of the [108]-acid **44** increases by 35 % on going from pH 3.64 to pH 7.04, corresponding to a 264 % increase in overall dendritic volume. Similar results have been obtained by others.[78a] Monnig and Kuzdzal[79] extended these experiments to estimate analyte–dendrimer/solvent distribution coefficients (K) and capacity factors. Thermodynamic parameters (i.e., H, S, G) pertaining to analyte solubilization within dendritic structures were also obtained by examining the variation in K in relation to temperature. Capacity factors were seen to increase linearly as a function of increasing dendrimer concentration. It was also determined that as the polyacid dendrimer size increases, entropy becomes the dominant force for analyte solubilization. This dendritic macromolecule series (**42–46**) was examined[80] as micellar substitutes in electrokinetic capillary chromatography[81] employing aqueous mobile phase conditions; a series of alkyl parabens was separated with significantly enhanced efficiency and resolution using these dendrimers when compared to more traditional methods using surfactants such as SDS. Furthermore, molecular relaxation studies have demonstrated that the 1st generation *tert*-butyl ester terminated dendrimer possesses rheological properties similar to those of high polymers.[82] Strumia et al.[82a] have used Behera's amine for the surface modification of activated polymeric matrices with 2,6-di(acylamino)pyridine units that are capable of molecular recognition. The use of these modified beads in affinity chromatography was described.

Dubin et al.[83] studied the dissociation of Newkome's[74] carboxylic acid terminated dendrimers by means of potentiometric titrations. Theoretical surface potentials, obtained by applying the nonlinearized Poisson–Boltzmann equation, were found to be larger than those determined by experiment for the 2nd to 4th generations in NaCl. This was rationalized in terms of a counterion binding effect. The observation of even larger surface potentials on changing the counterion to Me$_4$N$^+$ supports this explanation. The effect of pH and ionic strength on the mobility of these dendrimers has also been examined.[83a]

Miller et al.[84] terminally modified these polyacids (2nd generation; 36 acid groups) with oligothiophenes and examined their "vapoconductivity". Unlike their PAMAM counterparts,[85, 86] which bear tertiary amines capable of undergoing facile oxidation, these materials are stable to oxidation conditions. Cast films of the polyoligothiophenes were oxidized with I$_2$ vapor to give electron conductivities of σ = 10^{-3} s cm^{-1}. Exposure to organic vapors dramatically enhanced the conductivity of the films. For example, acetone vapor led to an 800-fold increase in conductivity compared to the unsaturated materials.

Three generations of the related alcohol- and amine-terminated polyamide cascades were prepared by coupling the appropriate polyacid with either the aminotris(*tert*-butyl carbamate) 48 or aminotris(acetate) 52 monomers (Scheme 4.14). Trinitrile 26 was reduced with borane to give the triamine 47, treatment of which with di-*tert*-butyl dicarbonate[87, 88] followed by catalytic reduction with T-1 Raney nickel[89] gave the desired amine 48 in excellent overall yield. The precursor to bis*homo*tris (Scheme 4.9), i.e. triol 50, prepared by hydrolysis of nitrile 26 followed by reduction of triacid 49, was acylated (Ac$_2$O) to afford triester 51, catalytic reduction of which with T-1 Raney nickel gave the corresponding amine 52 in high overall yield.

The 'modular' syntheses[90] of three related cascade families from the corresponding polyacid and appropriately designed monomers are shown in Scheme 4.15. In each case, building block coupling utilized standard DCC[91]/DMF/1-HBT[92] peptide coupling conditions. Removal (HCO$_2$H) of the protecting *tert*-butyl groups afforded the corresponding polyacids or polyamines; transesterification (K$_2$CO$_3$, EtOH) of the acetate-coated cascades liberated the hydroxy-terminated dendrimers.

Newkome et al.[93] employed their modular monomers to incorporate sites capable of molecular recognition within the internal molecular framework. Taking advantage of a three-component, high-dilution reaction, the monomer (Scheme 4.16) was prepared by reaction of Behera's amine 35 (1 equiv.) with glutaryl dichloride (54), followed by reaction with excess 2,6-diaminopyridine (53) in a single-pot process to afford the extended aminotriester 55. Treatment of this pyridylamine with the tetraacyl halide 56 gave the 1st tier dendrimer 57, which was divergently expanded by the above two-step process. This series of dendrimers was shown to exhibit *H*-bonding-based molecular recognition of guests possessing imide functionalities (e.g., 58).[94] ^{1}H NMR titration experiments revealed consistent, albeit small (0.5 to 0.7 ppm), downfield shifts of the pertinent

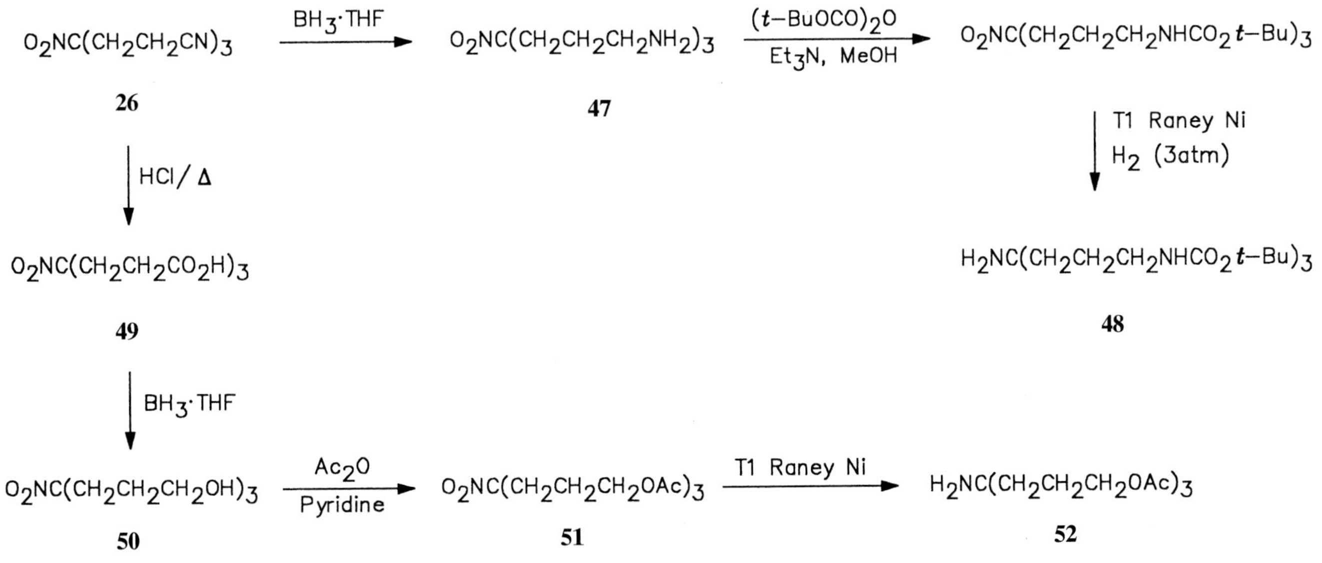

Scheme 4.14 Preparation of monomers for the modular introduction of terminal amines and alcohols.

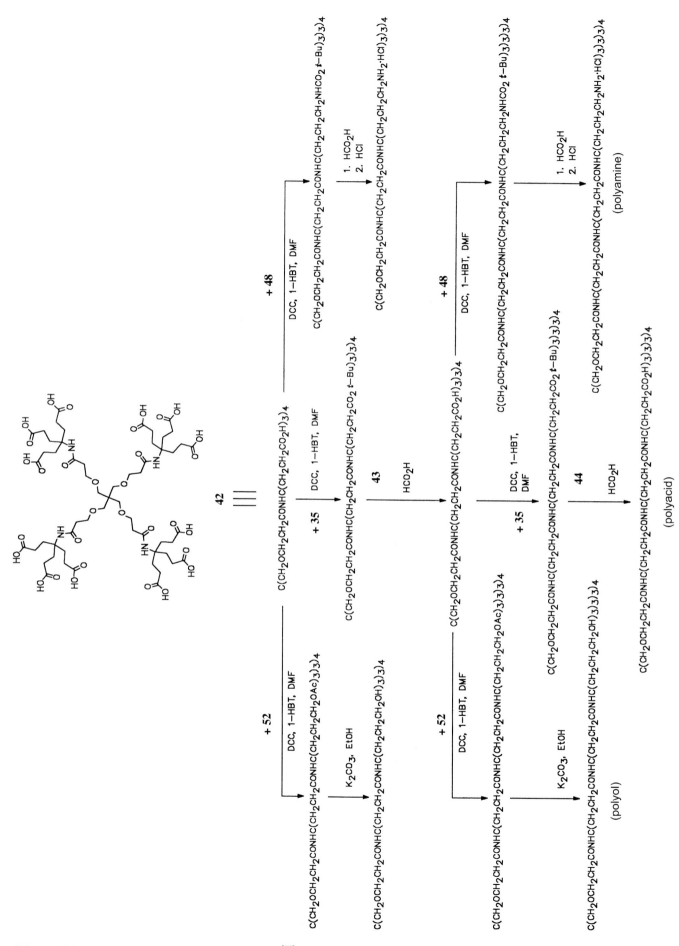

Scheme 4.15 Synthesis of a complementary series[90] of hydroxyl-, amine-, and carboxy-terminated dendrimers.

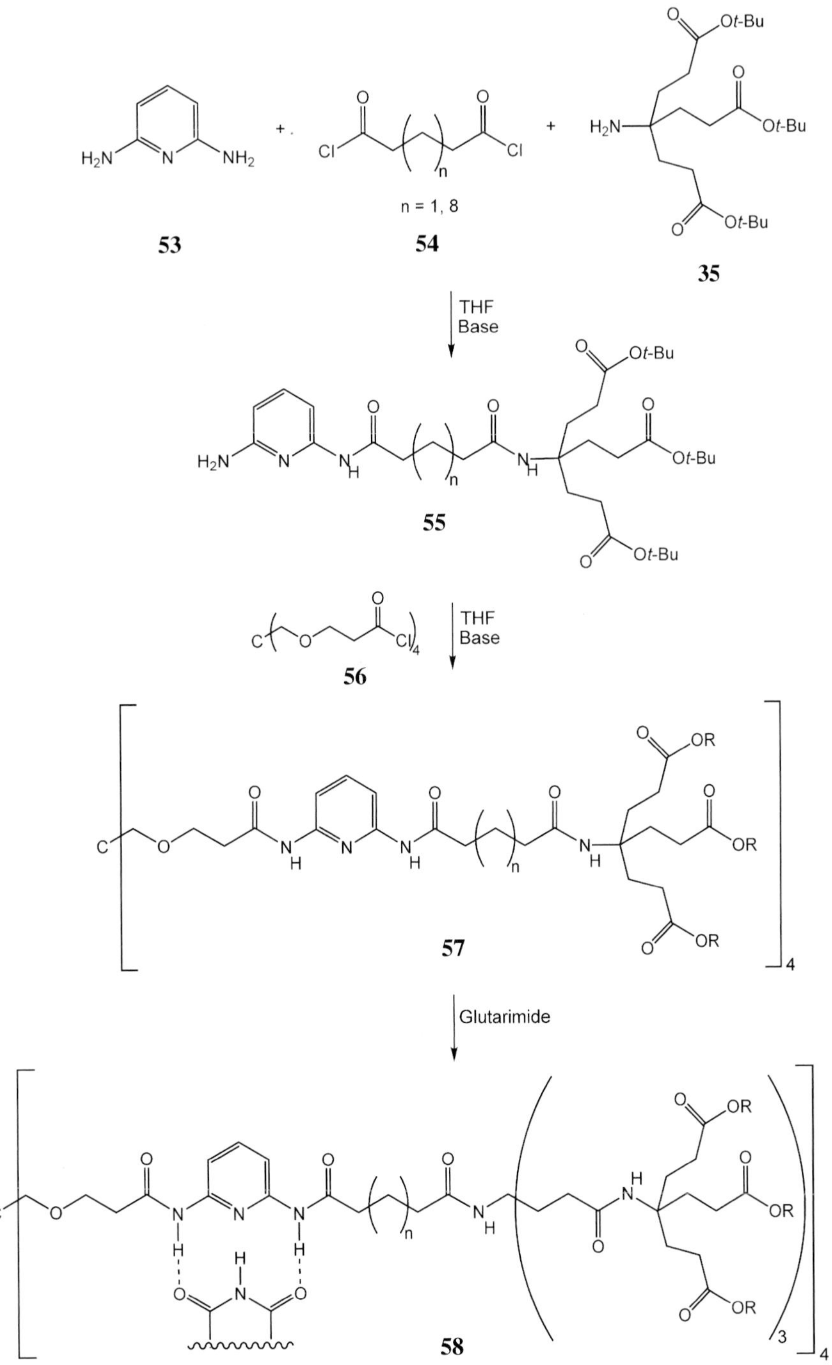

Scheme 4.16 Site-specific molecular recognition within the dendritic superstructure.[93]

diamide proton signals using glutarimide as the molecular guest. The cancer therapeutic drug AZT (3'-azido-3'-deoxythimidine) was also used as a molecular guest to demonstrate the utility of site-specific incorporation of *H*-bonding receptors. The construction of related linear and convergent wedges possessing 2,6-di(acylamino)pyridine subunits capable of molecular recognition has also been reported;[95] these moieties were attached to an activated agarose matrix by surface modification, then evaluated with regard to the formation of *H*-bonded complexes. Strumia and Halabi[95a] also reported a novel dendritic acrylic monomer using this core.

Employing application-oriented monomers, Newkome et al.[96, 97, 97a] created anthraquinone-based, redox-active dendrimers (Scheme 4.17). Synthesis of the requisite monomers, e.g. of the homologated aminotriester **60**, was achieved by a three-component, single-pot reaction of glutaryl dichloride (**54**), aminotriester **35**, and 1,4-diaminoanthraquinone (**59**). Connection of amine **60** to the flexible core **56** generated the 1st generation ester **61**, which was then transformed to the acid (HCO₂H) and reacted with excess amine **35** to afford the 2nd generation [36]-ester **62**; iteration generated the higher-tiered mem-

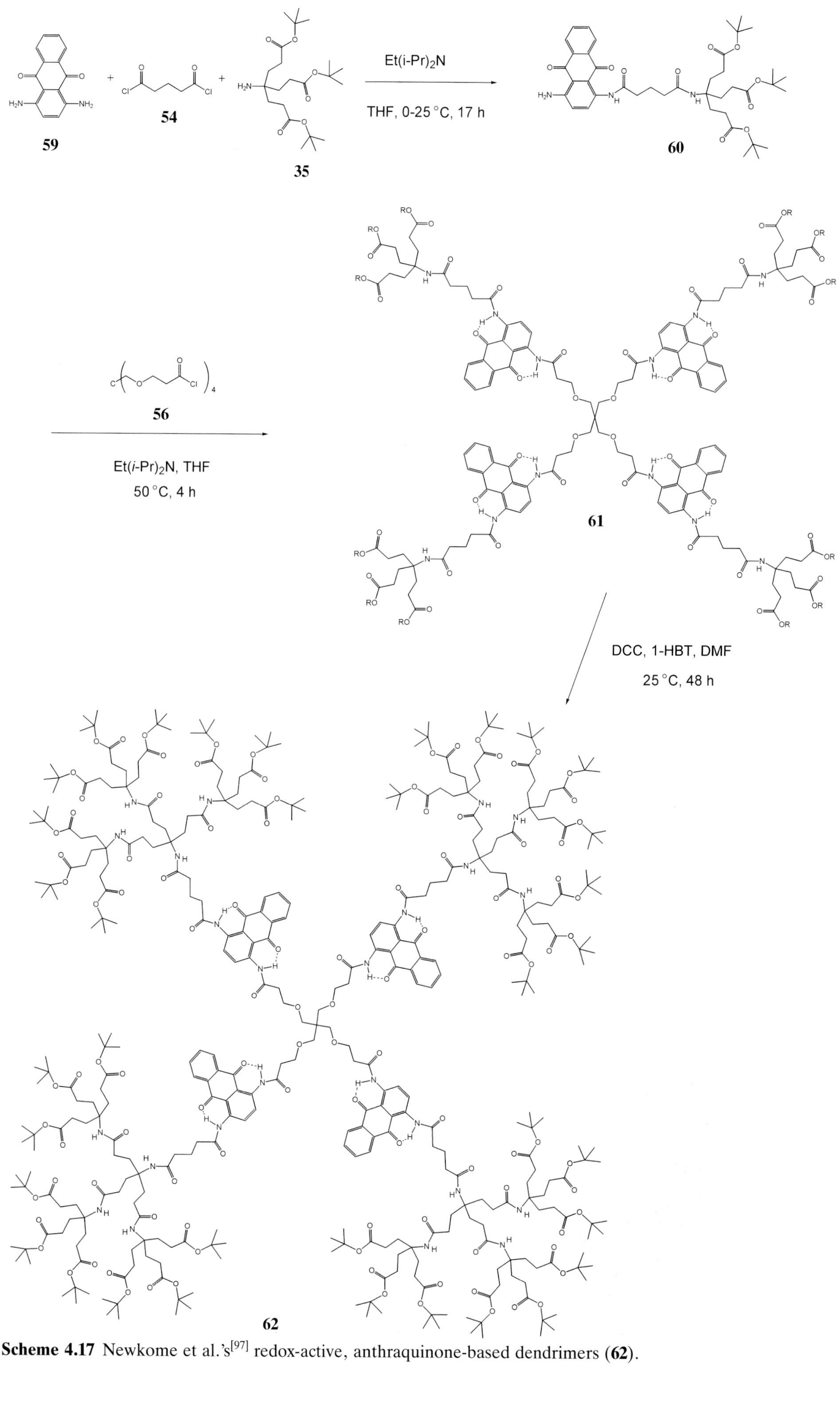

Scheme 4.17 Newkome et al.'s[97] redox-active, anthraquinone-based dendrimers (**62**).

bers. Cyclic voltammetry showed that progressive steric congestion of the redox centers retards electron-transfer kinetics and eventually leads to irreversible electrochemistry. This work was later extended to the study of similar dianthraquinone-based constructs (e.g., 1,4- and 1,5-isomers) as well as the incorporation of a more rigid adamantane core intended to effect better separation of the redox centers.[98] Incorporation of more rigid spacer units between the branching centers and the redox sites through the use of aryl diacid dichlorides has also been examined, along with the attendant electrochemistry.[99] Electrochemical comparisons of these anthraquinoid architectures have been reported.[100]

A series of complementary isocyanate-based 1 → 3 branched monomers – predicated on Behera's amine triester construction – has been reported.[101, 102] These stable isocyanates can react with many isolated functional groups. For example, they can be grafted (Scheme 4.18) onto the narrower ring side of β-cyclodextrin to create a new macromolecular building block for use in the convergent self-assembly of dendrimer-based networks.[103] Construction began by selective conversion of the upper rim primary hydroxyl moieties to amine groups (**63**),[104] followed by treatment with the isocyanate triester (**64**) to yield the 21-acid **65** after treatment with formic acid. Subsequent coupling with the

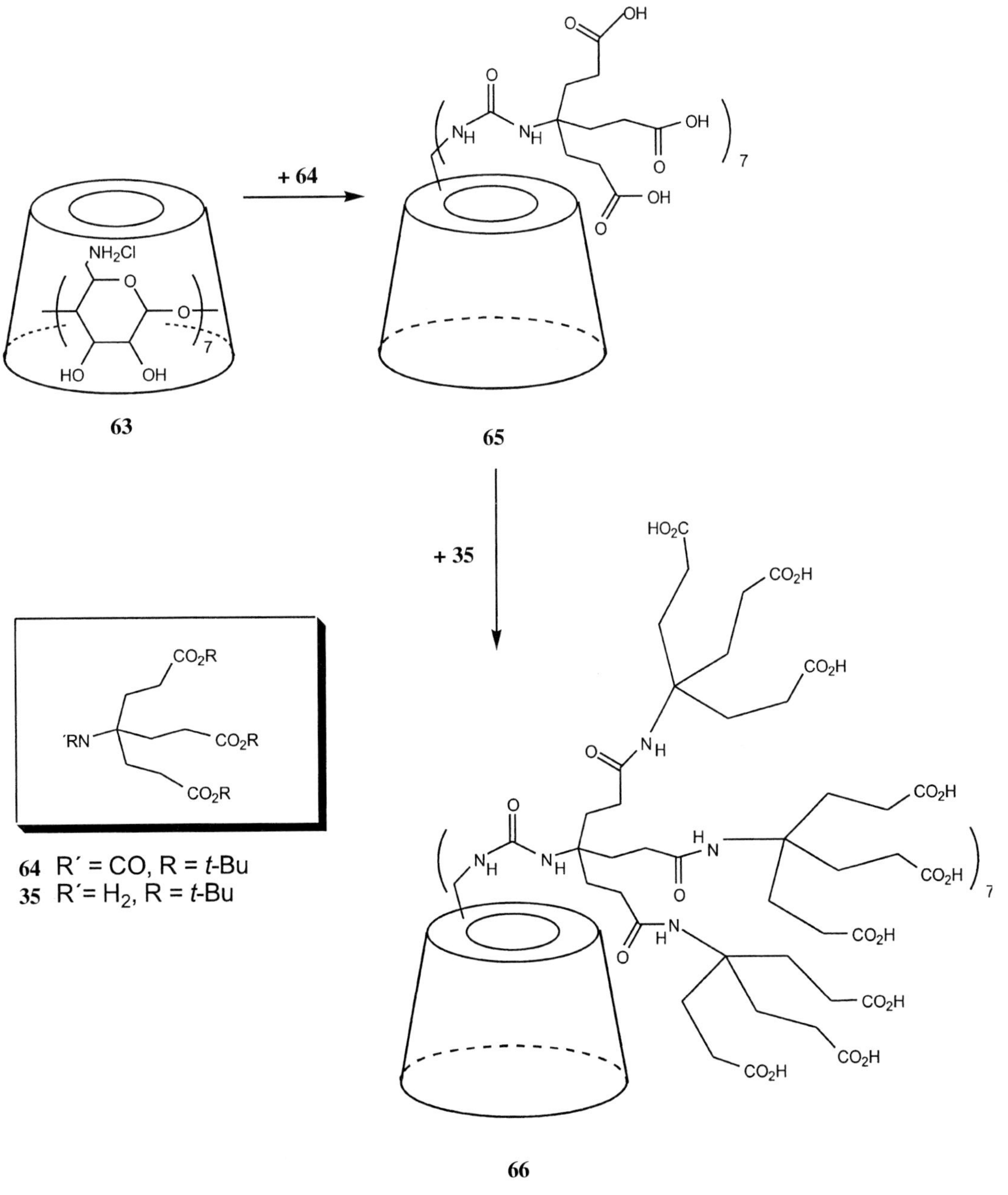

64 R′ = CO, R = *t*-Bu
35 R′= H₂, R = *t*-Bu

Scheme 4.18 "Dendronized" cyclodextrin and molecular assembly.[103]

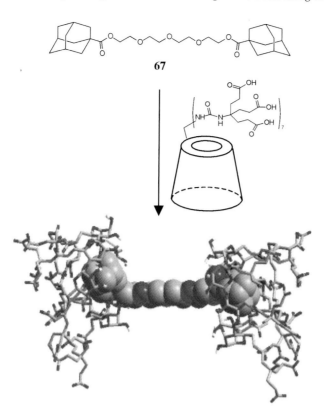

67

Figure 4.2 Adamantane-cyclodextrin host-guest encapsulation facilitatas dendrimer self-assembly. Reprinted with permission of the Royal Society of Chemistry.[103]

corresponding aminotriester (**35**) followed by deesterification produced the 2nd generation polyacid **66**. Molecular recognition properties of the cyclodextrin moiety were conserved, as demonstrated by the inclusion and subsequent forced displacement of phenolphthalein by adamantane. The potential for self-assembly was demonstrated by the coordination of two dendritic cyclodextrins to the adamantane-terminated ends of a tetraethylene glycol chain (e.g., **67**; Figure 4.2); dendritic self-assembly has recently appeared.[104a]

This family of monomers has also been used to functionalize the termini of the PPI dendrimers delivered by DSM company with protected alcohols, amines, esters, and nitriles.[105] As an extension to this work, these isocyanates, all of which possess similar reactivity, have been used combinatorially for the construction of multifunctional dendrimers.[105–107] Essentially, stoichiometric mixtures of monomers are reacted at the same time to produce a multifunctional material such as **68**, which may be further elaborated by the deprotection of a specific set of functional group(s), as in polyamine **69**, and subsequent reaction with another set of logically chosen monomers to give unsymmetrical materials (i.e., **70**; Scheme 4.19). These architectures can be considered as being intermediate between hyperbranched polymers and dendrimers. Ramifications of this protocol include rapid macromolecular property modification and the construction of dynamic heterogeneous surfaces.[108, 109]

Brütting et al.[110] modified C_{60} by incorporating 2nd generation Behera's amine dendrons (ester-terminated), thereby affording good solubility in linear poly(*p*-phenylenevinylene), with a view to enhancing the photovoltaic properties of this conjugated polymer. Essentially, the C_{60} dendrimer was added to suppress recombination of photogenerated charge carriers. Investigation of these blends as films layered between ITO and Al was reported. Hirsch et al.[111] convergently attached two 2nd generation Behera's amine dendrons to a lipophilic C_{60} unit; aggregation was studied by cryo-TEM. Narayanan and Wiener[112] divergently constructed a two-directional dendron based on a protected ethylenediamine core using established procedures; after deprotection of the core, self-assembly around a Co(III) center occurred in a convergent manner. Other cores have been dendrimerized with these Behera's amine dendrons.[112a]

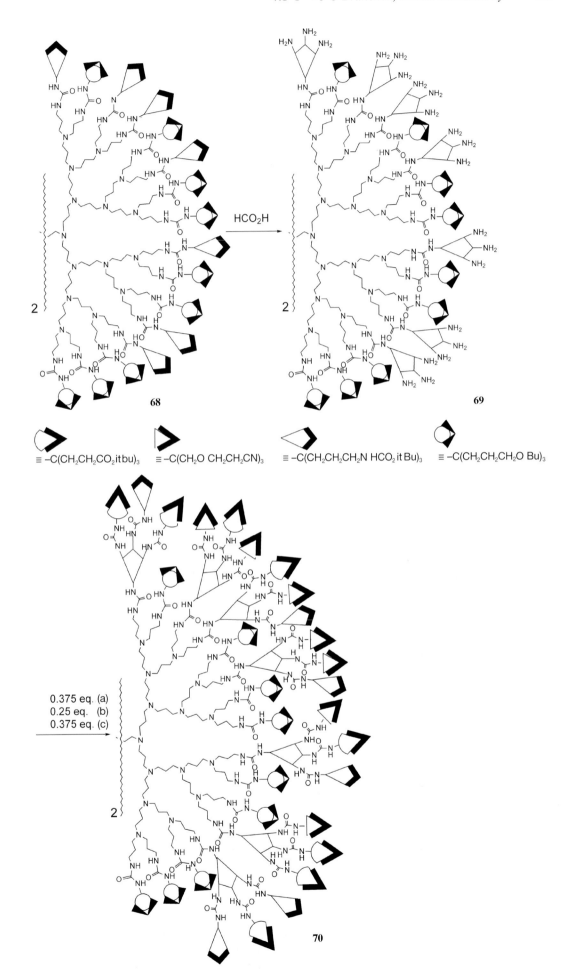

HCO₂H

68

69

≡ –C(CH₂CH₂CO₂itbu)₃   ≡ –C(CH₂O CH₂CH₂CN)₃   ≡ –C(CH₂CH₂CH₂N HCO₂ it Bu)₃   ≡ –C(CH₂CH₂CH₂O Bu)₃

0.375 eq. (a)
0.25 eq. (b)
0.375 eq. (c)

70

Scheme 4.19 Combinatorial layering (**68 → 69 → 70**) of monomers resulting in asymmetric dendrimer construction:[105, 106] (a) OCNC(CH₂CH₂CH₂NHCO₂ *t* Bu)₃; (b) OCNC(CH₂OCH₂CH₂CN)₃; (c) OCNC(CH₂CH₂CO₂ *t* Bu)₃

Chemical attachment of a 9-anthryldiazomethane moiety has been employed to gain insight into the number of unreacted carboxylic acid termini left following tier elaboration.[113] Qualitative analysis of structural integrity could be realized and the potential to internally functionalize the slightly imperfect dendrimer can be envisaged.

4.1.5   Physical Properties

Shah, Dubin et al.[114] examined the size-exclusion chromatography of the 1st to the 5th generations of Behera's amine-based, carboxylic acid-terminated dendrimers as a permeation model for charged particles into like-charged cavities. Chromatography was performed using a porous glass stationary phase and partition coefficients were determined for solute diameters ranging from 2 to 8 nm in ionic strengths ranging from 0.01 to 0.09 M at neutral pH. Observed degrees of particle permeation were generally 20–100 % greater than the values calculated from the theory of Smith and Deen[115] for charged spheres permeating cylindrical pores of like charge, which was found to overestimate repulsive forces. It was postulated that this arises from employment of the linearized form of the Poisson–Boltzmann equation.

Dubin et al.[116] also examined complex formation between these carboxyl-terminated dendrimers and charged poly(dimethyldiallylammonium chloride) using turbidimetry, dynamic light scattering, viscometry, and potentiometric titration. A discontinuity observed with all these methods at a well-defined pH corresponded to incipient complex formation. A model was presented for polyelectrolyte backbone distortion. Similar studies using copolymers of [(methacrylamido)propyl]trimethylammonium chloride and acrylamide have also been reported.[117]

Harmon et al.[118] examined molecular relaxations, as well as viscoelastic properties,[118a] of the *tert*-butyl and methyl ester terminated derivatives of these polyamides. The study included TGA, DSC, X-ray diffraction, and electric analysis. Glass transition temperatures and apparent activation energies were both observed to increase with generation as well as with the size of the termini. Secondary transitions also increased in temperature with generation. Ionic conductivity was found to dominate electrical properties at high temperatures.

Dubin et al.[119] used capillary electrophoresis to examine counterion-binding effects on these carboxylate-terminated polyamides. Titration studies showed that the effective surface charge density of the generation 5 dendrimer was lower than the geometric surface charge density, which was attributed to counterion binding.

4.2   1 → 3 *C*-Branched and Connectivity

Although initial efforts utilizing bis*homo*tris as a dendritic building block were disappointing due to the inability to effect complete amine acylation, the nitro intermediates proved to be excellent precursors to a diverse series of *C*-based, alkyl building blocks. The construction of a cascade polymer possessing an all-saturated, symmetrical, tetrahedrally-branched hydrocarbon interior framework was designed on the basis of the bis*homo*tris nitro precursor **28**. The preparation of Micellanol™ cascades **79** and **81** (see Scheme 4.21) was subsequently accomplished by Newkome et al.[120–122] Scheme 4.20 shows the synthesis of the core **76** and the key building block **77** required for their preparation. The nitrotriol **28** was protected with benzyl chloride to give triether **71**, which underwent denitration–cyanoethylation[123] upon treatment with *n*-Bu₃SnH and AIBN in the presence of acrylonitrile to give the key nitrile triether intermediate **72**. Nitrile **72** played a critical role in the synthesis since it could be uniquely converted to both the core **76** [previously prepared from tetrakis(2-bromoethyl)methane and citric acid[124] in 17 steps (ca. 1 % overall yield), from tetrakis(β-carbethoxyethyl)methane[125] (ca. 70 %), or from γ-pyrone[126] in 8 steps (24 % overall yield)], as well as the alkyne building block **77**. Hydrolysis of nitrile **72** gave the corresponding acid **73**, which was quantitatively reduced (BH₃·THF) to give alcohol **74**. This was transformed by concomitant deprotection (HBr)

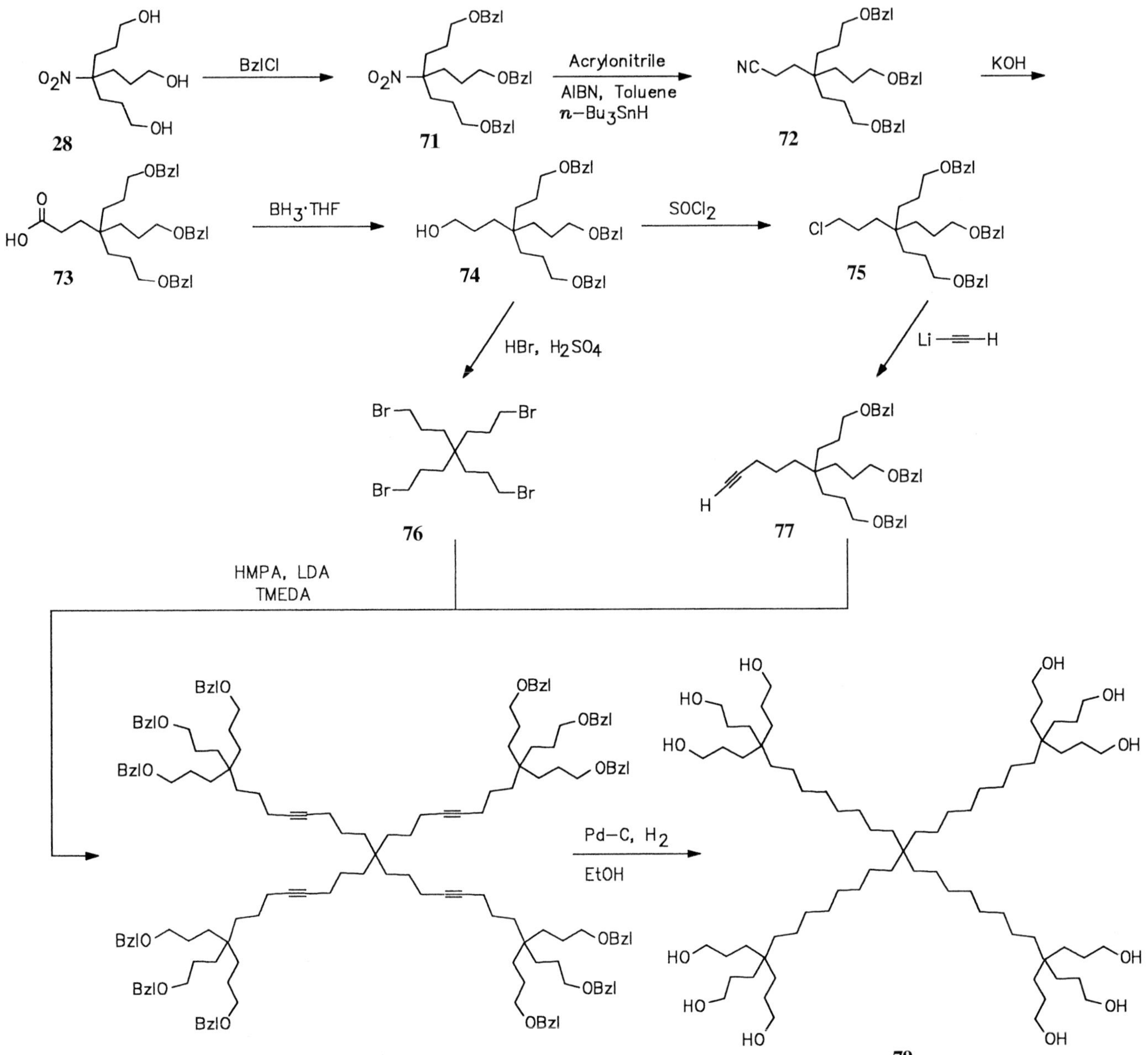

Scheme 4.20 Sequence for the preparation of alkyl monomers (**76** and **77**) used in the construction of unimolecular micelles.[120–122]

and dehydroxylation–bromination to give core **76** in excellent overall yield.[127] Treatment of further alcohol **74** with SOCl$_2$ gave the corresponding monochloride **75**, reaction of which with lithium acetylide gave the desired functionally differentiated alkyne building block **77**.

Although tetrakis(2-bromoethyl)methane reacted sluggishly with hindered nucleophiles, reaction with azide or cyanide afforded the respective azide and nitrile, both of which could be reduced to corresponding amines.[128] On the other hand, homologue **76** appeared to be unaffected by steric problems. Thus, treatment of tetrabromide **76** with four equivalents of the lithium salt of alkyne **77** afforded the desired tetraalkyne **78**, which was concomitantly reduced and deprotected to give the saturated dodecaalcohol **79**. Polyol **79** was then converted to the corresponding polychloride (Scheme 4.21), which was treated with slightly in excess of 12 equivalents of alkyne building block **77** to yield the 36-benzyl ether **80**. Reduction and hydrogenolysis of tetraalkyne **80** gave rise to the 36-Micellanol™ [**81**; 36-Cascade:methane[4]:(nonylidyne)^2:propanol].[57–59] Its water solubility was further enhanced by oxidation (RuO$_4$) and conversion (Me$_4$NOH) to the corresponding polytetramethylammonium 36-Micellanoate™ **82**.

The "unimolecular" micellar character[1] of this poly(ammonium carboxylate) **82**[57, 122, 129] was established by UV analysis of guest molecules such as pinacyanol chloride, phe-

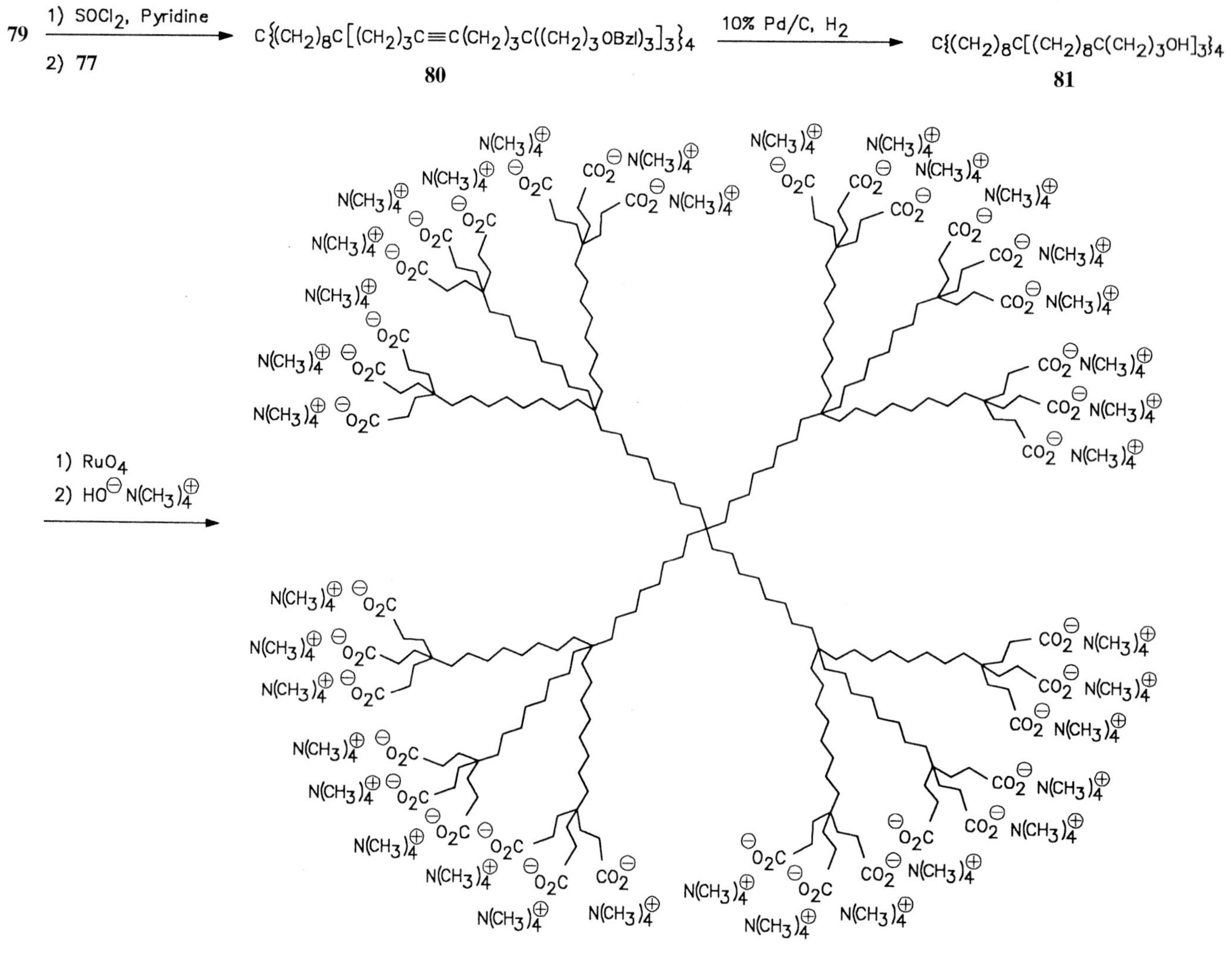

79 $\xrightarrow[\text{2) 77}]{\text{1) SOCl}_2, \text{ Pyridine}}$ $C\{(CH_2)_8C[(CH_2)_3C\equiv C(CH_2)_3C((CH_2)_3OBzl)_3]_3\}_4$ $\xrightarrow{\text{10\% Pd/C, H}_2}$ $C\{(CH_2)_8C[(CH_2)_8C(CH_2)_3OH]_3\}_4$

80   **81**

$\xrightarrow[\text{2) HO}^{\ominus}\text{N(CH}_3)_4^{\oplus}]{\text{1) RuO}_4}$

82

Scheme 4.21 Preparation of the poly(tetramethylammonium salt) of Micellanoic™ Acid (**82**). [120, 129]

nol blue, and naphthalene, in conjunction with fluorescence lifetime decay experiments employing diphenylhexatriene as a molecular probe. The monodispersity, or absence of intermolecular aggregation, and molecular size were determined by electron microscopy.

The polyalkyne precursors of the hydrocarbon-based unimolecular micelles[129] allowed examination of the effects of chemical modification at specific sites within the interior of a cascade infrastructure.[130] Thus, treatment of the alkynes **78** or **80** with deca-borane afforded excellent yields of the 1,2-dicarba-*closo*-dodecaboranes[131] (*o*-carbo-ranes), while reaction with Co$_2$(CO)$_8$ afforded the desired poly(dicobalt carbonyl) clus-ters;[132] details are given in Chapter 8.

4.3   1 → 3 *C*-Branched, *Alkene & Ester* Connectivity

Sengupta and Sadhukhan[133] reported initial studies aimed at the construction of tetraphenylmethane-based architectures (Scheme 4.22). Employing a fourfold Heck reaction [Jeffery's conditions[134] gave the best results; Pd(OAc)$_2$, Bu$_4$NCl, NaHCO$_3$, DMF, 80 °C], tetrakis(aryliodo)methane (**83**) was coupled to the activated alkene **84** to afford the small dendrimer **85**. The reaction was also performed using the corresponding tetradiazonium salt in place of the iodo monomer. Other functionally differentiated monomers were also reported (e.g., **86–88**).

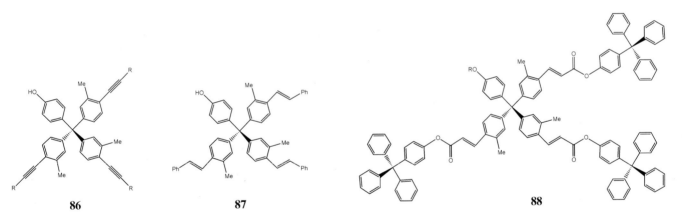

Scheme 4.22 Heck-type reactions have afforded access to tetraphenylmethane-based structures.[133]

4.4 1 → 3 *C*-Branched, *Ether* Connectivity

Hall et al.[135, 136] reported the synthesis of starburst polyethereal and polythioethereal dendrimers, as shown in Scheme 4.23. These cascades possess the shortest distance between branching centers yet reported. The use of pentaerythrityl tetrabromide[137] (**89**) as the core, with the anion of the potassium salt of the corresponding orthoester of pentaerythritol **90** as the building block, afforded the protected dodecaol **91**. Deprotection with acid and subsequent two-step conversion of the hydroxy groups to the dodecabromide via the dodecatosylate provided the precursor for the construction of the next tier. The 36-polyol **92** was subsequently transformed by this simple procedure to the 108-polyol **93**, which is the most densely packed cascade polymer yet reported, as evidenced by the branching defects encountered after the formation of the 2ⁿᵈ and 3ʳᵈ tiers,[138, 139] these presumably resulting from an increased number of neopentyl displacements required for tier construction.

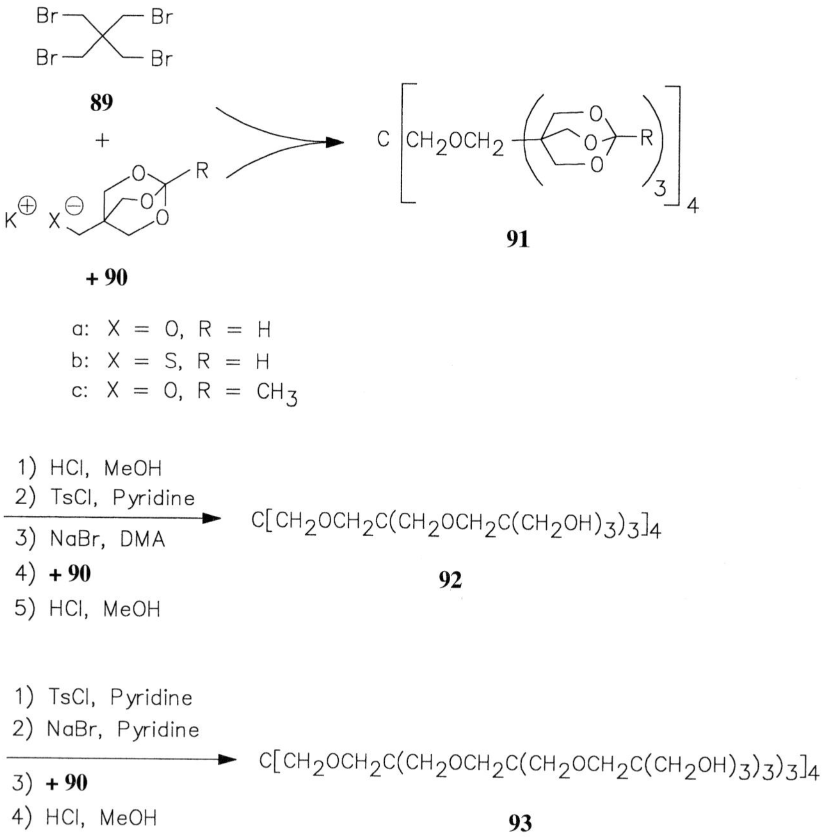

89

+

+ 90

a: X = O, R = H
b: X = S, R = H
c: X = O, R = CH$_3$

C[CH$_2$OCH$_2$ ... R]$_4$ **91**

1) HCl, MeOH
2) TsCl, Pyridine
3) NaBr, DMA
4) **+ 90**
5) HCl, MeOH

C[CH$_2$OCH$_2$C(CH$_2$OCH$_2$C(CH$_2$OH)$_3$)$_3$]$_4$ **92**

1) TsCl, Pyridine
2) NaBr, Pyridine
3) **+ 90**
4) HCl, MeOH

C[CH$_2$OCH$_2$C(CH$_2$OCH$_2$C(CH$_2$OCH$_2$C(CH$_2$OH)$_3$)$_3$)$_3$]$_4$ **93**

Scheme 4.23 Construction of highly compact pentaerythrityl-based dendrimers (**91–93**).[135, 136]

Ford et al.[140] reported an improved synthesis of this series of ethereal dendrimers. This polyol series was converted to the corresponding homologated polyamines, which were then alkylated (excess CH$_3$I) to generate the polyammonium salts, for example, the 36-tetraalkylammonium salt **94** [PE-TMAI(36); Scheme 4.24]. Moore also noted[141] that the use of the orthoacetate of pentaerythritol **90c** in place of the orthoformate **90a** improves the stability of this building block, thus making it more versatile. Use of these 'tied-back' building blocks aids facile nucleophilic substitution, even at hindered neopentyl centers. The polyionic constructs have been employed in "ion-exchange displacement chromatography".[142]

Early on, a series of related "cascadols", prepared from pentaerythritol, was reported,[143] but little supportive data were provided.

Kim et al.[144] also employed branched architecture for a process that they have termed Combinatorial Synthesis on Multivalent Oligomeric Supports (COSMOS), whereby multiple compound copies are constructed on soluble scaffolds by solution-phase synthesis, and are then separated according to size. Although not strictly a dendrimer, tetra-amine (**95**; Figure 4.3) was prepared from the commercially available tetravalent PEG oligomer possessing an average molecular weight of 2kDa. The hydroxyl termini were transformed to amines and coupled (diisopropyl carbodiimide) to an acid-functionalized aminobenzylidine unit. These newly introduced amines were then used to prepare a library of di- and tri-substituted guanidines.

Perylene diimides possessing pentaerythritol-branched allyl moieties have been reported by Gregg and Cormier[145] to form liquid-crystalline phases.[146]

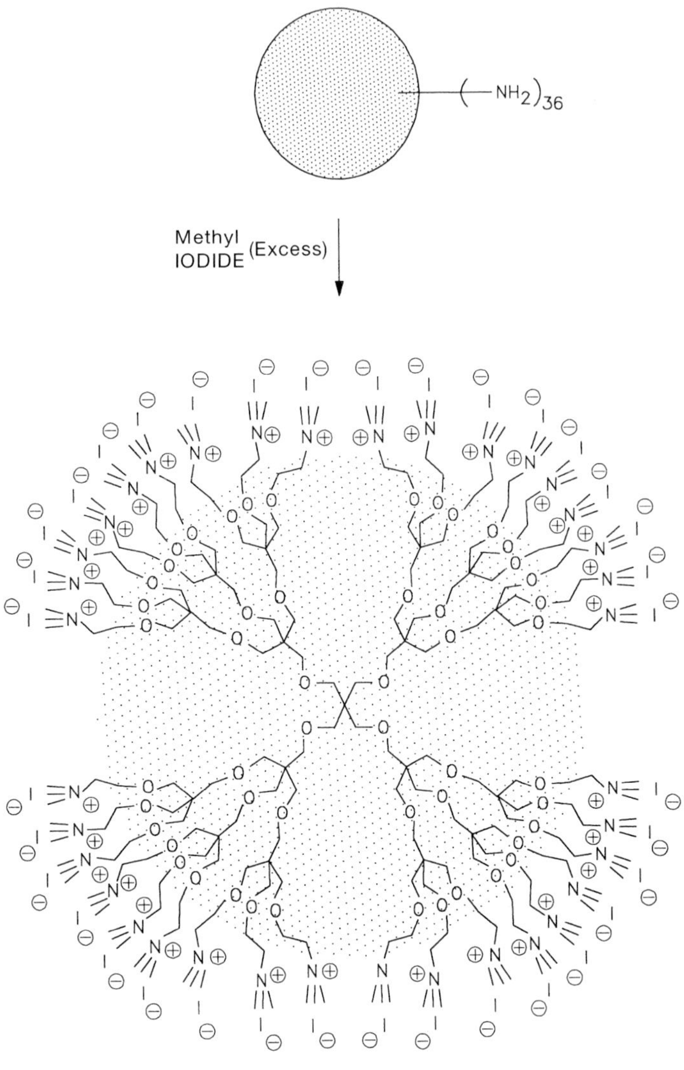

94

Scheme 4.24 Poly(ammonium iodide) dendrimers[140] prepared for catalytic ester hydrolysis.

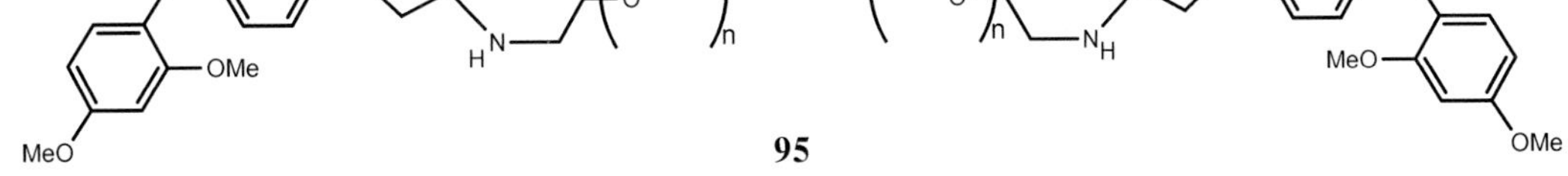

95

Figure 4.3 Kim's multivalent soluble supports.[144]

4.5   1 → 3 *C*-Branched, *Ether & Amide* Connectivity

Using a series of simple monomers,[8] a diverse array of dendritic macromolecules can be devised and easily prepared. The ethereal amine building block **96** (i.e., Lin's amine, after the student that first synthesized it) is but one example, which is readily prepared[147] in two steps (Scheme 4.25) from "tris" (**6**) by Michael-type addition of acrylonitrile followed by methanolysis. A small amount of *N*-addition occurs, but the major product of the reaction stems from exclusive *O*-addition. Treatment of the tetraacyl chloride **56** with the ethereal amine **96** gives the dodecaester in high yields; subsequent saponification affords the corresponding dodecaacid **97** (Scheme 4.26). Higher-tiered systems, such as the 36-Cascade:methane[4]:(3-oxo-6-oxa-2-azaheptylidyne)²:4-oxapentanoic acid (**98**), were prepared under standard peptide coupling conditions (DCC/1-HBT/DMF) giving rise to the poly(ethereal-amido) cascade series.[147]

Diederich et al.[148] reported the divergent synthesis of dendrimers possessing porphyrin cores with the aim of modeling redox potentials of electroactive chromophores through environmental polarity modification. The dendrimers can thus be considered as electron-transfer protein mimics for proteins such as cytochrome *c*; oxidation potentials for cytochrome *c* in aqueous solution are known to be 300–400 mV more positive than those reported for similarly ligated heme mimics lacking hydrophobic peptide encapsulation.[148]

The iterative route to these porphyrin-core dendrimers employed the readily available ethereal building block **96**.[147] Thus, amine acylation (DCC, 1-HBT, THF) of monomer **96** (Scheme 4.27) with tetraacid **99** afforded the 1st tier dodecaester **100a**; transformation [LiOH, MeOH/H₂O (1:1)] of the termini of this polyester then gave the polyacid **100b**. Repetition of this sequence allowed the construction of two additional tiers, e.g. ester **101**. The dendrimers were characterized by ^{13}C and ^{1}H NMR, FT-IR spectroscopy, as well as mass spectrometry using the FAB and MALDI-TOF techniques. Molecular-ion base-peaks were observed in the MALDI-TOF MS of polyester **101** [m/z = 18,900 (calcd. 19,044)] along with minor peaks at m/z ≈ 37,000 and ≈ 54,000 corresponding to ionic gas-phase dimer and trimer complexes.

Small C_{60} adducts, incorporating this monomer, have been reported by Diederich et al.[149]

Examination of the cyclic voltammetry (CV) of the Zn-porphyrin dendrimers in THF and CH_2Cl_2 with 0.1 M $Bu_4N^+PF_6^-$ as the electrolyte revealed first oxidation potentials up to 300 mV (THF) less positive than the corresponding values obtained for the "unshielded" tetraester Zn-porphyrin core. These preliminary electrochemical experiments suggested that dendritic encapsulation of redox-active chromophores can effectively influence the electrophoric environment; controlled and well-conceived cascade architecture may thus lead to new avenues of selective redox catalyst design.

Similar dendritic building blocks have been used to prepare "dendrophanes"[150] (*dendr*imer + cycl*ophane*), which possess internal cyclophane cores. Key reactions for core construction[151, 152] included a Cs_2CO_3-promoted dimerization of hydroxynaphthalene benzyl ether to form the cyclic cavity. A fourfold Suzuki cross-coupling introduced four phenolic moieties, which were subsequently treated with $BrCH_2CO_2Me$ and saponified to yield the desired tetravalent core (**102**; Scheme 4.28). Elaboration of the tetraacid with monomer **96** gave rise to generations 1–3 (i.e., **103**). Inclusion complexes (1:1) with steroids were examined and it was found that the dendritic shell did not affect the binding constants.

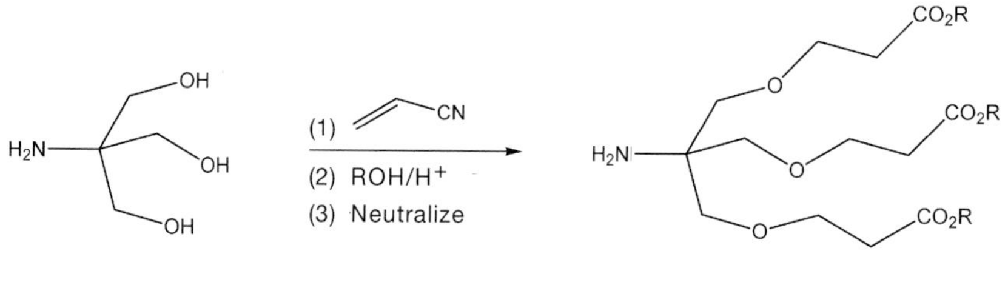

Scheme 4.25 Preparation of Lin's amine (**96**).[147]

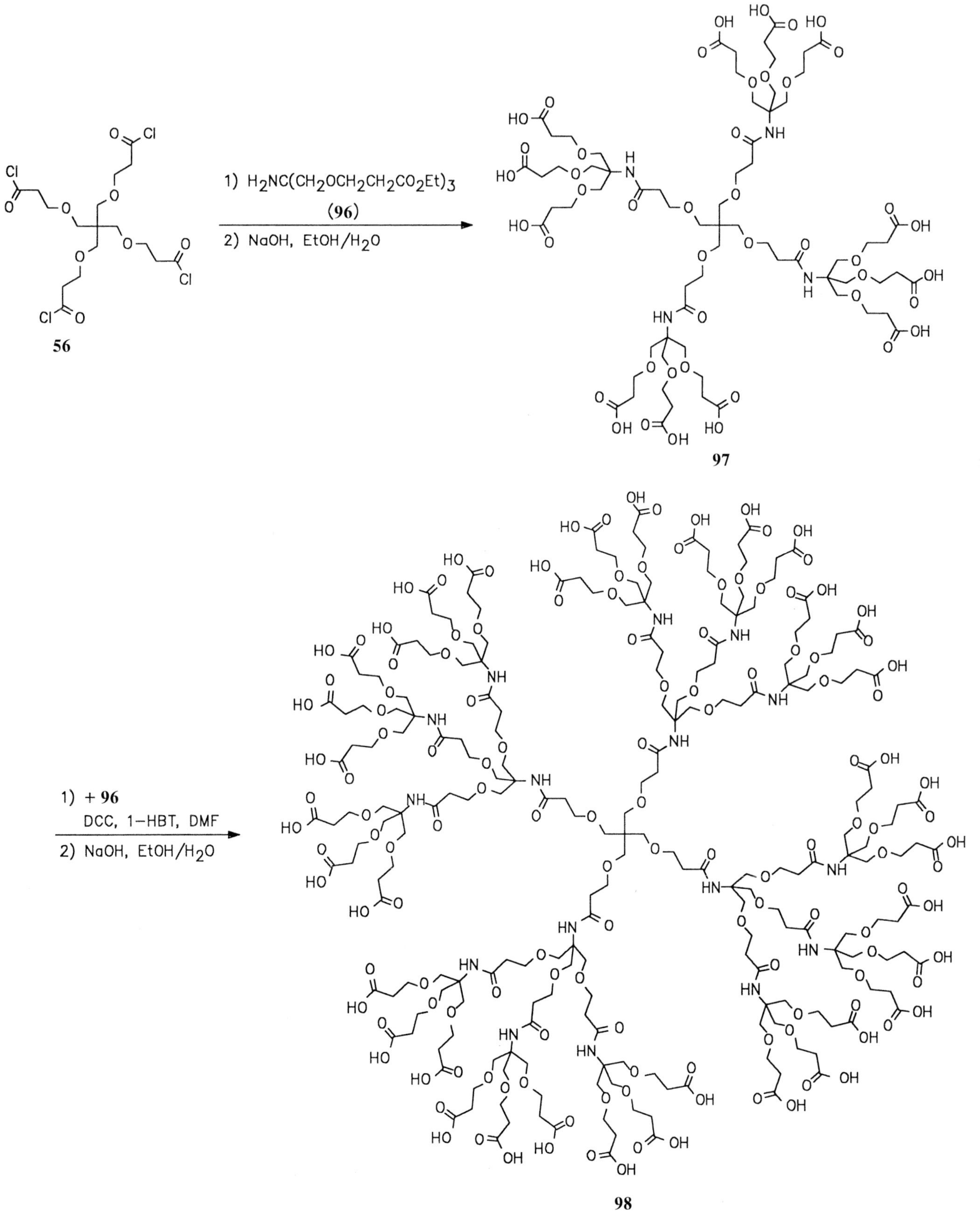

Scheme 4.26 Preparation of poly(ether-amido) cascades[147] employing a "tris"-based aminotriester monomer.

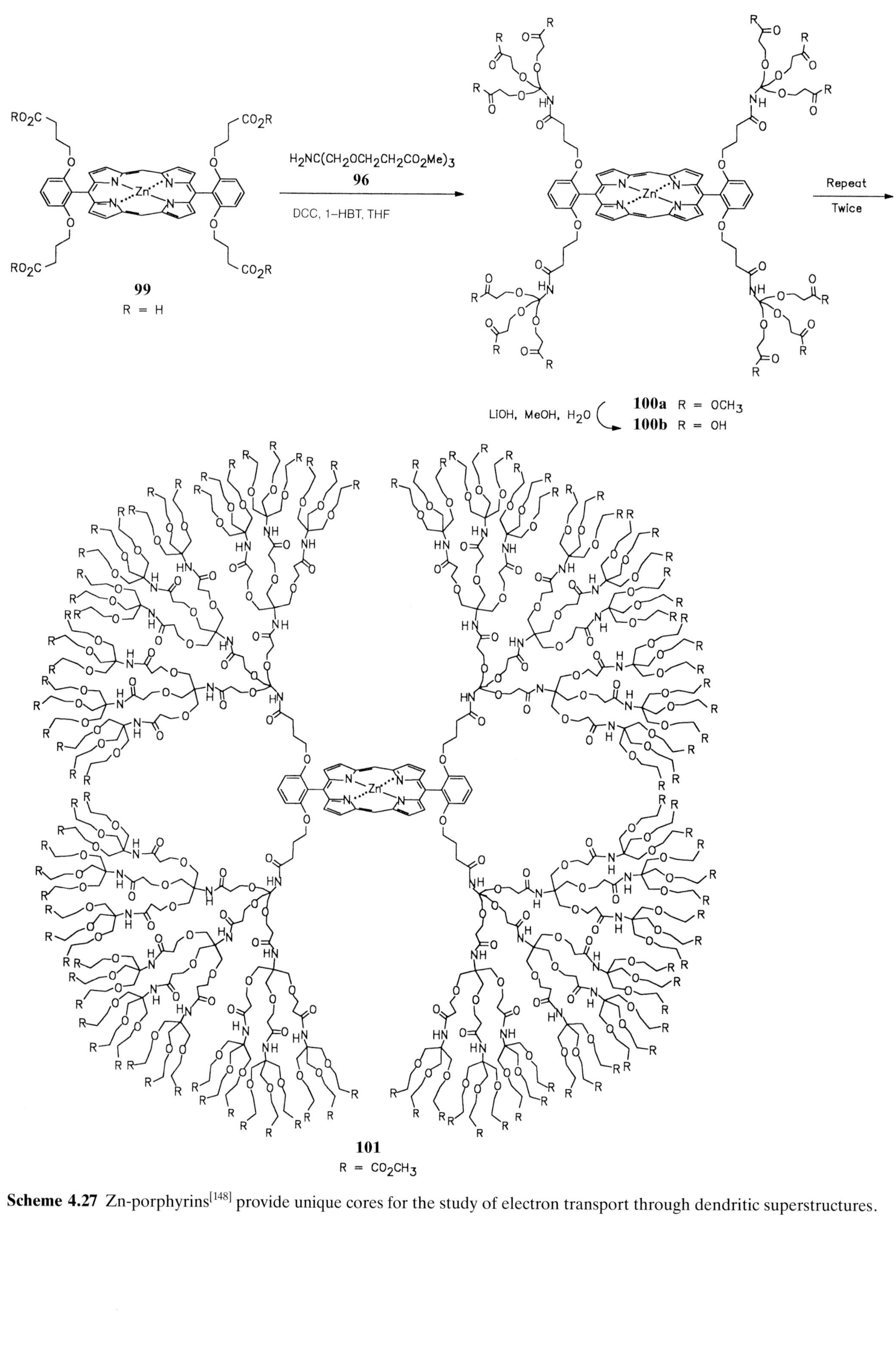

Scheme 4.27 Zn-porphyrins[148] provide unique cores for the study of electron transport through dendritic superstructures.

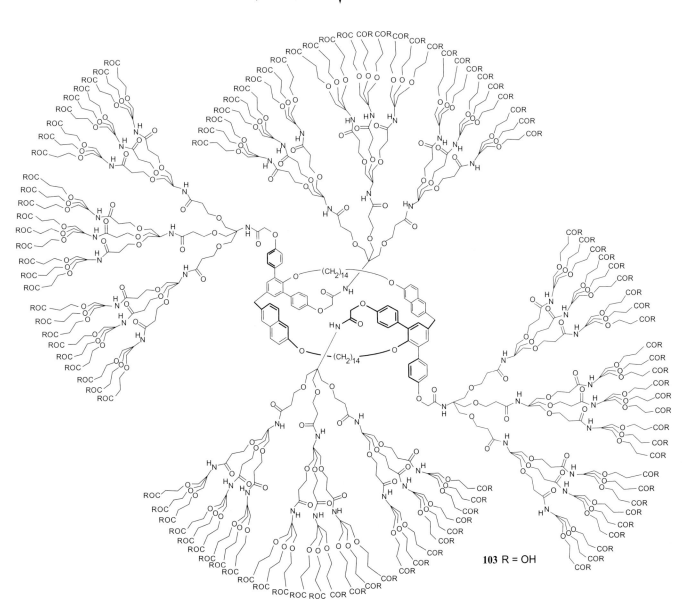

102

+ **96**   1. DCC, 1-HBT, THF
2. LiOH, H₂O, THF, MeOH

G1

Repeat Steps 1. + 2.

G2

Repeat Steps 1. + 2.

103 R = OH

Scheme 4.28 Dendrophane construction.[150, 151]

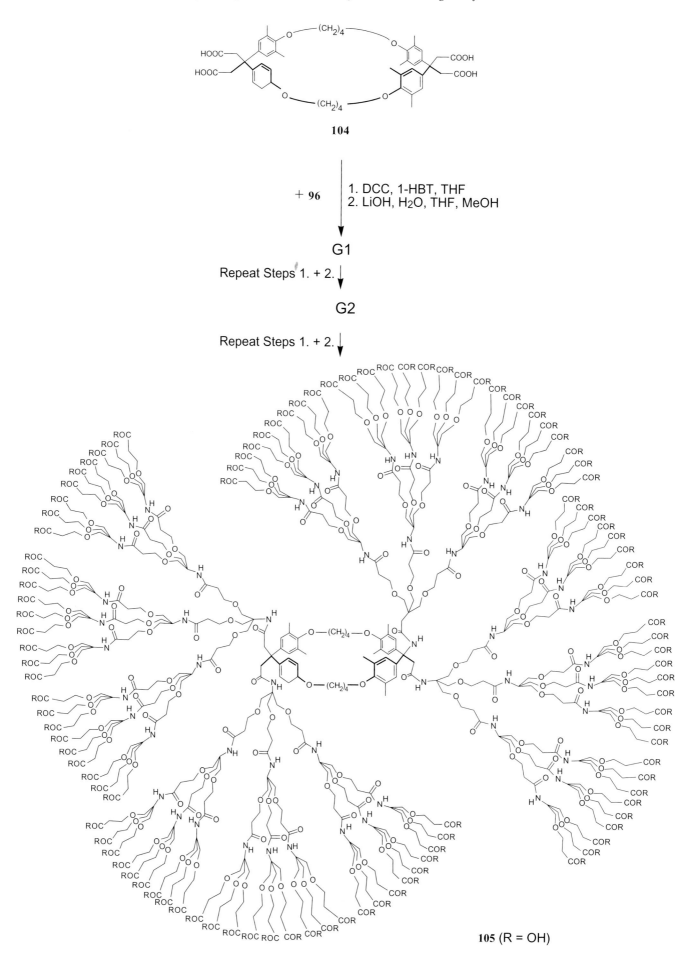

Scheme 4.29 Dendrophanes capable of complexing flat aromatic-type guests.[151]

Synthesis, binding properties, and crystallographic data of the core (Scheme 4.29), as well as those of other cyclophanes such as tetraacid **104**, were subsequently reported.[153] Their use in the synthesis of 3rd generation dendrimers (e.g., **105**) capable of acting as water-soluble receptor models for globular protein recognition sites was later described.[151] Dendrons incorporating tris-based monomer **96**[147] were prepared[154] by protection (BzOCOCl) of the amine to give the protected triester, hydrolysis (NaOH, H₂O, MeOH) to give the corresponding triacid, and coupling (DCC, 1-HBT, THF) with three equivalents of amine **96** to afford the Boc-protected nonaester. This could either be hydrogenated (HCO₂NH₄, 10 % Pd/C, EtOH) to give the 2nd generation dendron **107**, or transformed to the nonaacid and then treated with more amine **96** to give the 3rd generation wedge, which could be converted to the free 27-amino dendron. The 3rd generation dendrimer **105** was also prepared by a "semi-convergent" route (Scheme 4.30), where the 2nd tier dendron[154] **107** was reacted with the 1st generation dendrophane **106**; the spectra of this material were identical to those of that produced by the divergent route. Attempted divergent preparation of the 4th generation dendrophanes was unsuccessful.

Using 6-(p-toluidino)naphthalene-2-sulfonate as a fluorescent probe for the series of dendrophanes incorporating cyclophane core **104**, it was found that core micropolarity decreased as dendrimer size increased (i.e., on going from H₂O to MeOH to EtOH for the 1st to the 3rd generations). A 1st generation water-soluble, non-ionic dendrimer based on core **104**[151] and possessing terminal triethylene glycol monomethyl ether moieties was constructed to compare its complexation abilities with those of its carboxylic acid analogue; the stabilities of naphthalenediol complexes in buffered D₂O were found to be markedly reduced, presumably due to cavity occupation by the less polar PEG units. Dendrophanes prepared using core **102** were examined for their capacity to complex testosterone. Binding constants were determined in all cases.[151]

Kayser, Altman, and Beck[155] constructed a hexaalkynyl α-amino acid by Pd-mediated coupling of p-ethynylphenylalanine to hexabromobenzene, and modified the exterior with ethereal aminotriester **96**,[147] as well as with lysine monomers analogous to those used by Denkewalter[156] and Shao and Tam.[157]

A pyrene-tethered tripodal triether-acid chelator incorporating Lin's amine has been reported;[158] complexation with Eu³⁺ led to an antenna effect between the pyrene chromophore and the Eu³⁺ ion. Other materials have been dendrimerized with these ethereal dendrons[158a] and the photopolymerization kinetics of alkene-terminated poly(ether-amide)s.[159]

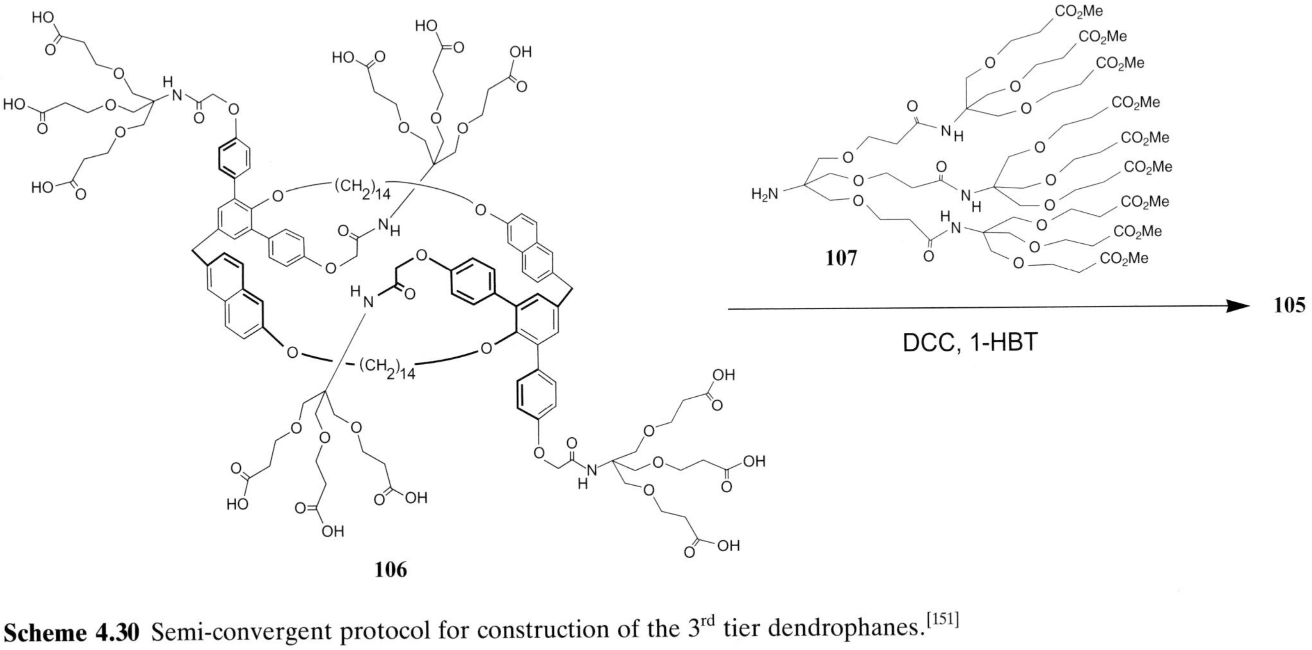

Scheme 4.30 Semi-convergent protocol for construction of the 3rd tier dendrophanes.[151]

4.6   1 → 3 *C*-Branched, *Ether, Amide, & Urea* Connectivity

Bradley and Fromont[160] created an isocyanate triester, and used it to generate high-loading resin beads for solid-phase synthesis.

4.7   1 → 3 *C* & 1 → 2 *N*-Branched and *Amide* Connectivity

Sommerdijk and Wright[161] prepared a tris-terminated PAMAM to study matrix effects on chemical sensing, e.g., of ibuprofen, by sol-gel entrapped complexing agents.

4.8   1 → 3 *N*-Branched and Connectivity

Rengan and Engel[162] reported and reviewed[163] the preparation of polyammonium cascade polymers (Scheme 4.31), which commenced with the quaternization of triethanolamine (**108**) with an alkyl halide or 2-chloroethanol to give a three- (**109**) or four- (**110**) directional core. Tosylation (TsCl, pyr) of the terminal alcohol groups of ammonium chloride **110** and treatment with excess **108** afforded the 1st generation pentaammonium dendrimer. Following two iterations, the 2nd tier, four-directional dendrimer **111** possessing 17-ammonium branching centers and 36 terminal hydroxy groups was prepared.[164] Attachment of these polyammonium polyols to a polymeric backbone generated a new, high capacity ion-exchange substrate.[165] Similar polyammonium architectures have been grafted onto chloromethyl styrene–methyl methacrylate supported on montmorrillonite.[166] Examination of these materials as phase-transfer catalysts for nucleophilic substitution reactions (e.g., SCN$^-$ reacting with BuBr) revealed them to be highly activating.

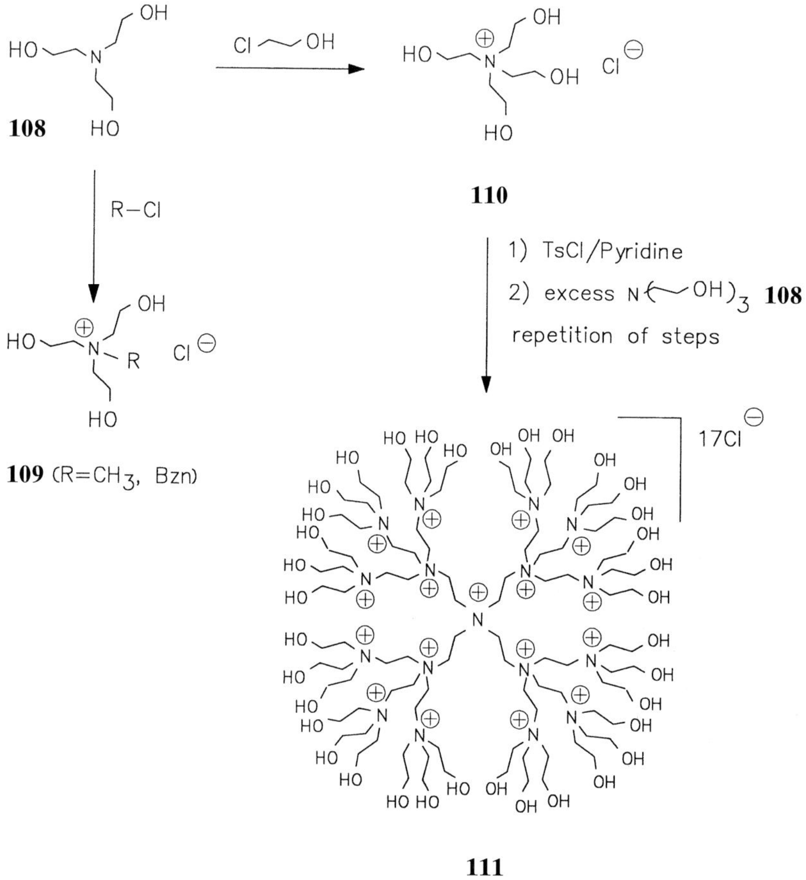

Scheme 4.31 Iterative synthesis of dendrimers[162] using tetrahedral alkylammonium moieties as branching centers.

4.9 1 → 3 *P*-Branched and Connectivity

Engel and Rengan[162, 167–169] also reported and reviewed[163] the preparation of poly-phosphonium cascade polymers (Scheme 4.32). A tetradirectional phosphonium core **114** was synthesized by treatment of phosphine **112** with 4-methoxymethylbromobenzene (**113**) in the presence of NiBr$_2$. The tetramethoxy core **114** was then transformed (Me$_3$SiI) to the tetraiodide and subsequently reacted with phosphine monomer **112** to generate the 1st tier pentaphosphonium dendrimer **115**. The 2nd tier cascade **116**, possessing 17 phosphonium moieties, was generated by repetition. Good solubility of dendrimer **116** in common organic solvents (e.g., MeCN, CHCl$_3$) was reported. Related three-directional cascades were similarly prepared by reaction of phosphine **112** with methyl, benzyl, or C$_{18}$H$_{37}$ halides to yield the requisite starting core.

A further series of phosphonium dendrimers was constructed by Engel and Rengan[170] starting from tris(*p*-methoxymethyl)phenylphosphine (**112**). Their goal was to construct dendrimers possessing (trivalent) phosphine and (pentavalent) phosphorane core moieties. These *P*-based cascades were prepared by oxidation (H$_2$O$_2$, AcOH) of building block **112** to the corresponding *P*-oxide **117** (Scheme 4.33). Divergent elaboration was then accomplished by repetitive benzyl ether transformation (Me$_3$SiI/MeCN) to the benzyl iodide, followed by addition of phosphine monomer **112**. After two iterations, the central phosphine oxide of dendrimer **118** was reduced (Cl$_3$SiH) to afford the trivalent phosphine core dendrimer **119**. Treatment of phosphine **119** with NaAuCl$_4$ gave (97 % yield) the 1:1 gold chloride–phosphorus dendrimer complex.

A neutral phosphorane core **121** was generated by treating tetrakis(*p*-methoxymethyl)phenylphosphonium bromide (**114**)[170] with 4-lithiobenzylmethyl ether (**120**). This pentavalent core was subjected to the previously described procedures to generate the five-directional dendritic molecule **122** (Scheme 4.34).

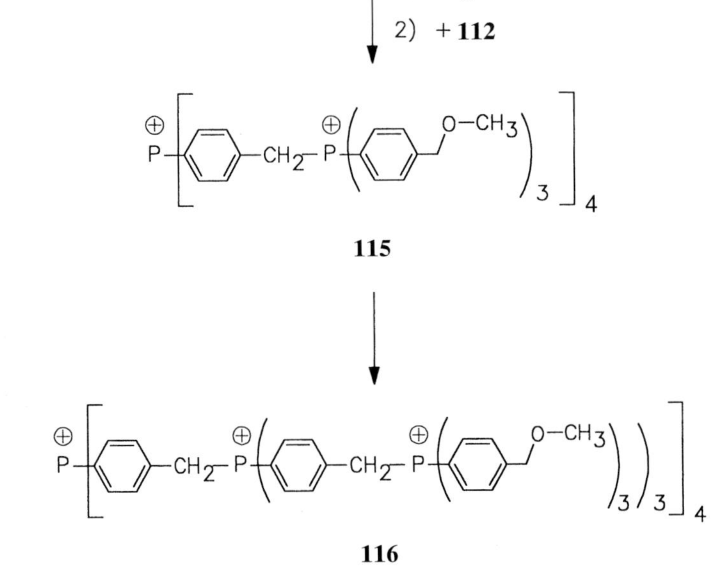

Scheme 4.32 Construction of novel charged *P*-based dendrimers.[162, 167, 168]

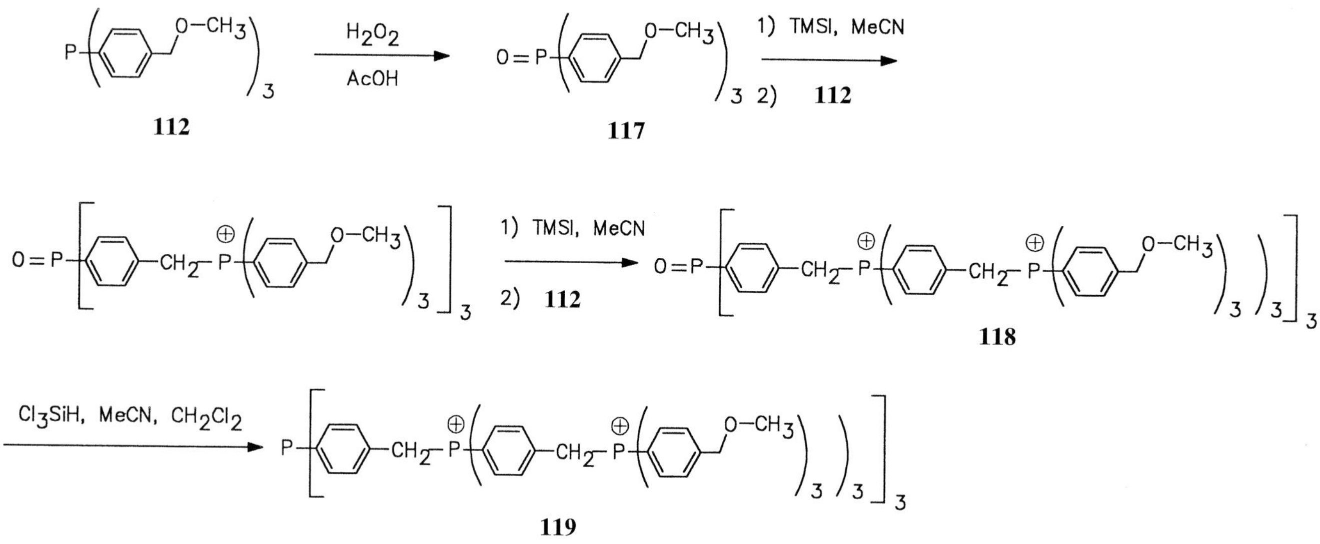

Scheme 4.33 Construction of P-based dendrimers[170] with neutral trivalent *P*-cores.

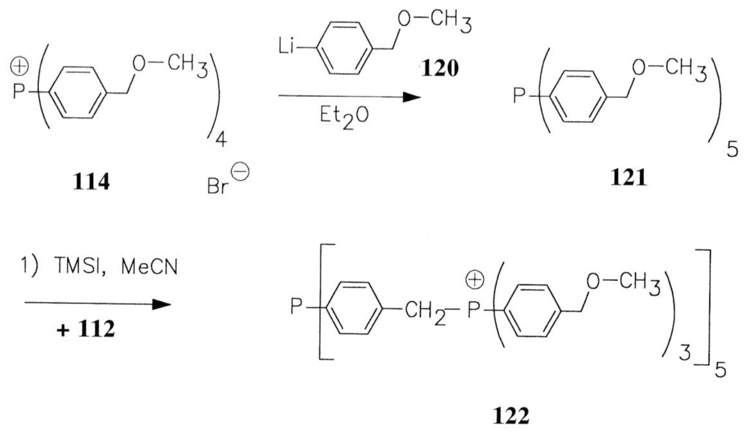

Scheme 4.34 Cascade synthesis employing a pentavalent, neutral *P*-core (**121**).[170]

4.10   1 → 3 *Si*-Branched and Connectivity

In the early 1990s, van der Made and van Leeuwen reported[171, 172] the synthesis of Si-based dendrimers by a repetitive hydrosilylation and alkenylation sequence (Scheme 4.35). Thus, the 0th tier tetraalkene **123**, prepared from tetrachlorosilane and allylmagnesium bromide, was treated with Cl₃SiH in the presence of a Pt catalyst to give the 1st generation tetrakis(trichlorosilane) **124**. Subsequent exhaustive alkylation with a 10% excess of allylmagnesium bromide afforded dodecaalkene **125**. Up to the 5th generation dendrimer was prepared by this procedure. Branching (1 → 3) employing tetravalent Si gave the 5th tier cascade a molecular weight of 73,912 amu with 972 peripheral groups.

Employing similar chemistry for carbosilane construction, van Leeuwen et al.[173] prepared dendritic wedges possessing an alkyl bromide focal group up to the 3rd generation. Reaction with excess ammonia afforded the corresponding alkylamine moieties, which were reacted with 1,3,5-tris(chlorocarbonyl)benzene to give the desired tris(amide) core dendrimer. Binding studies of these materials using Fmoc-glycine, *Z*-glutamic acid 1-methyl ester, and propionic acid as guests were performed; 1:1 complexes were observed based on *H*-bonding.

Frey et al.[174, 175] employed the same method,[171, 172] as did others,[176, 177] for the preparation of the 1st through to the 3rd generations of the 1 → 3 Si-branched dendrimers possessing 12, 36, and 108 termini, respectively. An excellent review by Frey and Schlenk[178] that deals exclusively with Si-dendrimers has recently appeared. Each dendrimer was

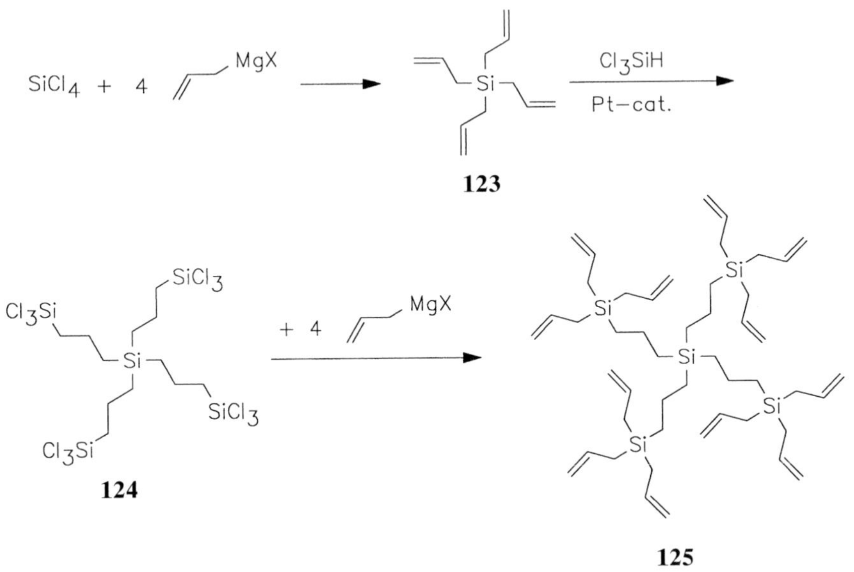

Scheme 4.35 Dendrimers prepared using tetraalkyl-substituted silicon as a branching center.[171, 172]

peripherally modified with mesogenic cholesteryl groups[174, 179] (surface hydroboration to give terminal OH groups, followed by reaction with cholesteryl chloroformate, resulting in carbonate-based attachments); G1 was completely substituted, while G2 and G3 were determined to possess 32–36 and 92–108 moieties, respectively. Ultrathin films (5–15 nm) of these carbosilane dendrimers were then examined by atomic force microscopy. Films formed from concentrated solutions possessed a thickness of 2–4 dendrimers, whereas those obtained from dilute solutions possessed monolayer thickness and an irregular pattern of holes. For films formed from both G1 and G2 (in the liquid-crystalline phase), a molecular "reorientation" upon annealing was observed, as manifested in a gradual coalescence of the holes. It was postulated that the mesogenic groups undergo reorientation from perpendicular to parallel juxtaposition relative to the surface due to more favorable carbon–mica interaction(s).

Seyferth and coworkers[176, 180] prepared a series of Si-based dendrimers up to the 4th generation. By virtue of the tetravalent nature of silicon, these macromolecules possessed a tetrahedral, four-directional core and 1 → 3 branching centers. The divergent strategy (Scheme 4.36) utilized two repetitive transformations: displacement of halide at silicon with CH_2=CHMgBr and Pt-catalyzed alkyl trichlorosilane formation. Reaction of tetravinylsilane **126** with $H_2PtCl_6 \cdot 6H_2O$ and Cl_3SiH afforded tetrakis(trichlorosilane) **127**, which was treated with CH_2=CHMgBr to produce the corresponding dodecavinylsilane **128**. Hydrosilylation of silane **128** under similar catalytic conditions was unsuccessful, resulting in the formation of impure products. Acceptable yields of the 1st generation dodeca(trichlorosilane) **130** were obtained, however, by using the Karstedt catalyst[181] {i.e., the product of $[(CH_2$=CH$)(CH_3)_2Si]_2$ and $H_2PtCl_6 \cdot 6H_2O$}; further vinylation gave good yields of the 2nd generation dodeca(trivinylsilane) **133**.

Attempted transformation of vinylsilane **132** to the corresponding 36-trichlorosilane **133** using the Karstedt catalyzed hydrosilylation procedure led to "grossly impure" products, as indicated by ¹H NMR spectroscopy. However, a simple solvent change from THF to diethyl ether gave the desired 2nd tier dendrimer **133** and suppressed impurity formation. Conversion of the 2nd tier trichlorosilane to the corresponding trivinylsilane proceeded smoothly, while the construction of the 3rd tier, 108-(trichlorosilane) **134** (Scheme 4.37) required forcing reaction conditions (excess $HSiCl_3$, Karstedt catalyst, Et_2O, 140 °C, 45 h, sealed Pyrex glass vessel).

Reduction ($LiAlH_4$) of the 3rd tier poly(trichlorosilane) **134** gave the terminal poly(trihydridosilane) **135** as a clear, hard solid. Similar reduction of the 1st through to the 3rd tier trichlorosilanes afforded the related hydrido-terminated silanes (e.g., **129** and **131**; see Scheme 4.36). Interest in the utility of these materials for ceramic applications provided an impetus for cross-linking experiments with the hydrido-terminated series; in particular, the 1st generation dendrimer **129** was examined due, in part, to its ready availability.

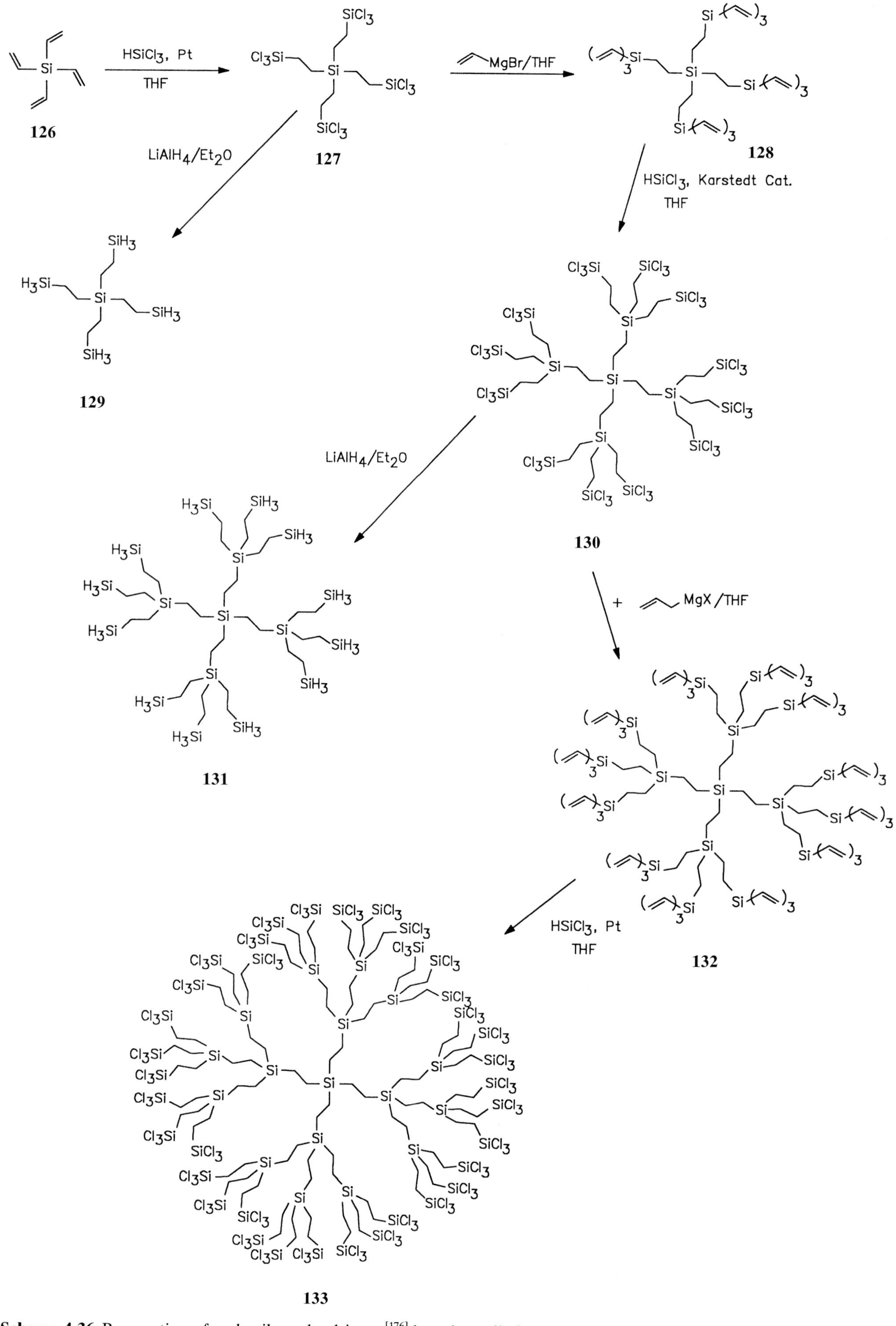

Scheme 4.36 Preparation of carbosilane dendrimers[176] based on alkyl-saturated silicon chemistry.

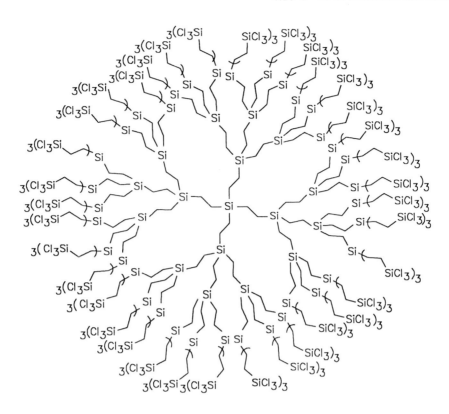

134

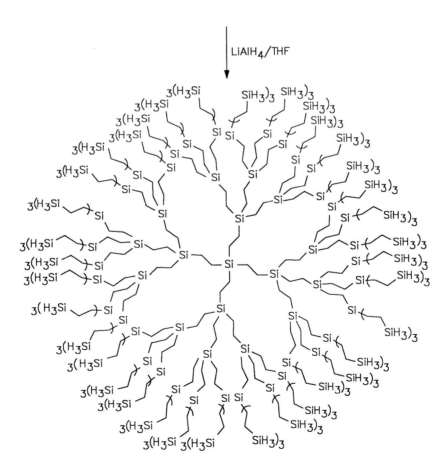

135

Scheme 4.37 Reduction of terminal trichlorosilane moieties to trihydridosilane units.[176]

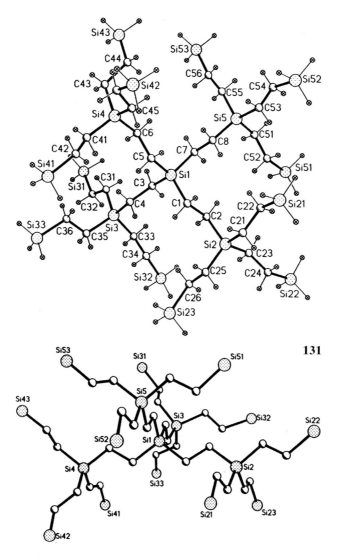

131

Figure 4.4 X-ray structure of a dendrimer possessing silicon-based superstructure **131**. [Reproduced by permission of the American Chemical Society[176]]

However, the use of a Zr-cross-linking agent led to products that were insoluble in most common organic solvents. X-ray crystallographic data and the corresponding structure were reported for the 1st generation polyhydrido **131** (Figure 4.4). Frey et al. prepared analogous carbosilane dendrimers bearing alcohol[182] and rigid cyanobiphenyl mesogenic termini.[183]

Lambert, Pflug, and Stern[184, 185] reported the preparation of the first dendritic polysilane consisting of an all-silicon framework. The small dendrimer can be visualized by considering the following formula: MeSi[SiMe$_2$Si(SiMe$_3$)$_3$]$_3$ {methyl[tris(permethylneopentasilyl)]silane} (**136**). The impetus for the construction of this material stemmed from the electronic, optical, and chemical properties of oligo- and polymeric silanes [i.e., (–SiR$_2$–)$_n$]. However, Si–Si bond lability can, under certain conditions, adversely affect these properties. Branched silane structures might inhibit internal Si–Si bond scission and thereby maintain bulk properties. Lambert et al.[186] delineated the first use of 2-D ^{29}Si-^{29}Si INADEQUATE NMR for unequivocal structural verification of the small Si-based [(Me$_3$Si)$_2$SiMeSiMe$_2$]$_3$SiMe construct.[185] The unavailability of crystals, preventing structure confirmation by X-ray analysis, provided the impetus for the development of this technique. The related tris[2,2,5,5-tetrakis(trimethylsilyl)hexasilyl]methylsilane has also been prepared[187] and characterized by its ^{29}Si-^{29}Si INADEQUATE spectrum, in which the entire Si–Si connectivity pattern was assigned.[188]

Preparation of the branched silane **136** began with the reaction of tris(trimethylsilyl)silane with CHCl$_3$ (CCl$_4$) and MeLi to afford the peralkylated methyl[tris(trimethylsilyl)]silane. Subsequent treatment with AlCl$_3$ and ClSiMe$_3$ gave the trichlorosilane, which was then reacted with tris(trimethylsilyl)silyllithium yielding the final silane dendrimer.

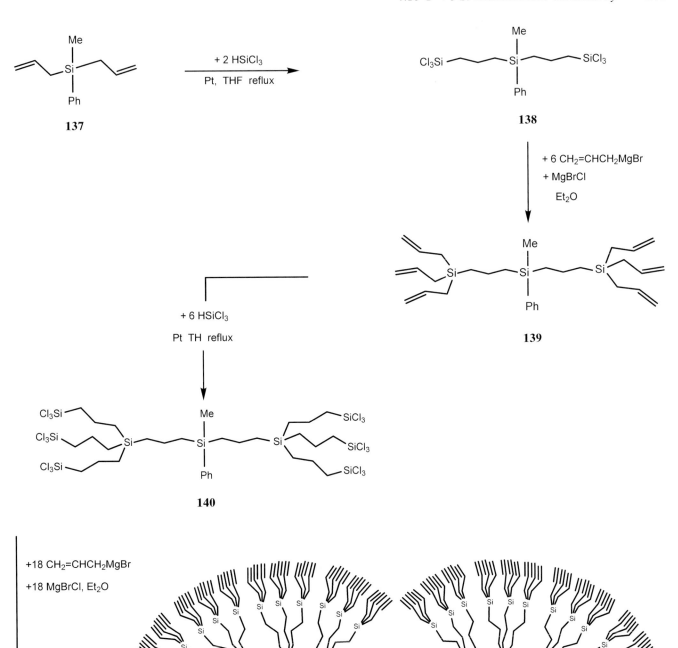

Scheme 4.38 Preparation of a carbosilane dendrimer with a purported 162 allylic termini.[190]

Seven silicon nuclei comprise the silane chain, which is repeated 27 times. X-ray crystallography confirmed a threefold axis of symmetry with respect to the core Si–C bond. Suzuki et al.[189] reported another slightly modified synthesis of silane **136**.

Kim et al.[190] prepared a carbosilane dendrimer possessing a purported 162 allyl end groups (Scheme 4.38). Starting from bis(allyl)methylphenylsilane (**137**), the bis(trichlorosilane) (**138**) was obtained by treatment with 2 equivalents of Cl_3SiH (Pt, THF, heat), and then this was treated with 6 equivalents of $CH_2=CHCH_2MgBr$ to give the hexaalkene **139**. Hydrosilylation afforded the corresponding poly(trichlorosilane) **140**. Repetition of the sequence afforded the 4th generation dendrimer **141**. The poly(trichlorosilane) precursor was reduced (LAH) to yield the poly(silane) possessing 54 SiH_3 terminal moieties. Attempts to construct the 5th generation were unsuccessful, presumably due to dense packing limitations. The authors also noted that the viscosity of the dendrimers increased with increasing generation. A modified procedure, whereby the nucleophilic addition to the $SiCl_3$ moiety is accomplished using allyl alcohol, has been reported.[191] A new class of amphiphilic dendritic diblock copolymers based on hydrophilic linear PEO and hydrophobic dendritic carbosilane possessing allyl termini has been reported.[192, 192a] Friedmann et al.[193] employed similar technology for the construction of the 1st and 2nd generation Si-based dendrimers terminated with bulky triphenylsilane groups; the crystal structure of the smaller dendrimer was obtained. Later, the crystal structure of the dendrimer/THF inclusion complex was reported.[194] Méry and coworkers[195] reported the formation of worm-like dendrimers starting from a poly(methylhydrosiloxane) core and attaching short propylsilane trees by the above technology; due to the premature onset of steric congestion, only the 2nd generation was reached.

Carbosilane dendrons have been focally modified with pyrene and investigated using time-correlated single-photon counting and steady-state fluorescence spectroscopy.[196] Similarities were observed for pyrenyl excimer formation on dendrons of different generations. A similarly prepared carbosilane possessing hydroxyl termini was studied with respect to its "wetting" of different substrates using tapping force microscopy.[197]

Frey et al.[198] coated this type of branched scaffolding possessing terminal vinyl groups by free-radical addition of 3,3,4,4,5,5,6,6,7,7,8,8,8-tridecafluoro-*n*-octylmercaptan. The generation 1 product exhibited a highly ordered smectic mesophase in the –15 to –30 °C temperature range, while the 2nd and 3rd generations were observed to form hexagonally ordered columnar arrays. Increased dense packing effects were postulated to account for the generation-dependent thermal properties.

A segmental dynamics study of these end group perfluorinated ($-C_6F_{13}$) carbosilanes using quasielastic neutron scattering; X-ray scattering has also been undertaken.[199] As a result of end group and carbosilane microphase separation, generation-dependent superstructures were observed, whereby helical end chains formed layers between branched framework domains. Dielectric relaxation of these materials has also been examined;[200] fast β relaxation was observed, conforming to Arrhenius behavior, while the dominant α-process was found to be comprised of fast and slow components.

Frey et al. undertook a molecular force field study concerning the host properties of carbosilane dendrimers.[201] Core structural variation was examined, as was outer-shell density. The inner cavity dimensions were determined to be of the order of 5–15 Å, while higher generation constructs were found to possess peripheral holes of dimensions 2–3 Å. A surface fractal dimension of 2.1 was calculated.

These silane-based architectures have been constructed on polyhedral silsesquioxane (S_8O_{12}) cores;[202] a single-crystal X-ray structure was obtained for a 24-vinyl-terminated dendrimer, which revealed disorder in the vinyl moieties.

An interesting assembly of tri- or tetravalent carbosilane cores functionalized with three and four β-cyclodextrin moieties, respectively, has been reported;[203] the synthesis involved a single-pot Birch reduction, followed by an S_N2 displacement reaction.

Kriesel and Tilley[204, 205] reported the preparation of dendrimer-based xerogels using these 2nd and 3rd generation triethoxysilyl-terminated polysilanes. Extended network generation, and thus gel formation, was achieved by acid-catalyzed (HCl) hydrolysis followed by solvent (H_2O, THF) removal. The observation of small pore volumes suggested a denser xerogel structure than that obtained using hard spheres of comparable size. Additional reports on the preparation of similar "stargels"[206] based on the corresponding 1st generation polysilane, and of hybrid xerogels incorporating these carbosilane dendrimers and arborols,[207] are available. These xerogels have also been examined with regard to their use as new catalyst supports.[208]

Kuzuhara et al.[209] capped these architectures with trisaccharide groups [globotriaosyl ceramide; Gb$_3$] and examined the products as host receptors for verotoxins.

A novel organosilane 4-triallylsilylphenol monomer that can be used convergently or divergently was developed by van Koten et al.[210] Facile core attachment (Et$_3$N) was demonstrated using 1,3,5-tris(chlorocarbonyl)benzene. Van Koten and others have recently used this 1 → 3 Si core to build reactive organometallic reagents,[210a–e] as well as tertiary phosphine catalysts.[210f]

4.11   1 → 3 *Adamantane*-Branched, *Ester* Connectivity

Chapman et al.[211] reported the construction of "polycules" (**144**), which were generated from "*quasi*-atoms", in this case substituted 1,3,5,7-tetraphenyladamantanes, as well as the related diadamantanes. This assembly process can be demonstrated (Scheme 4.39) by the treatment of core **142**[212–214] with building block **143**.

Tetrakis(4-iodophenyl)methane[215] has been subjected to both Pd-catalyzed Heck coupling with styrene, pentafluorophenyl styrene, and 4,4'-*tert*-butylvinylstilbene to yield the corresponding polyalkenes,[216] and to Jeffery's PTC conditions[134] with excess allyl alcohol to produce the sensitive tetrakis(2-formylethyl) derivative.[217] Recently, stilbenoid moieties have been similarly attached to 4-substituted tetraphenylmethane, tetraphenyladamantane, and tetraphenylsilane cores under Pd-catalyzed Heck conditions; the structural and optical properties of the products were evaluated.[218]

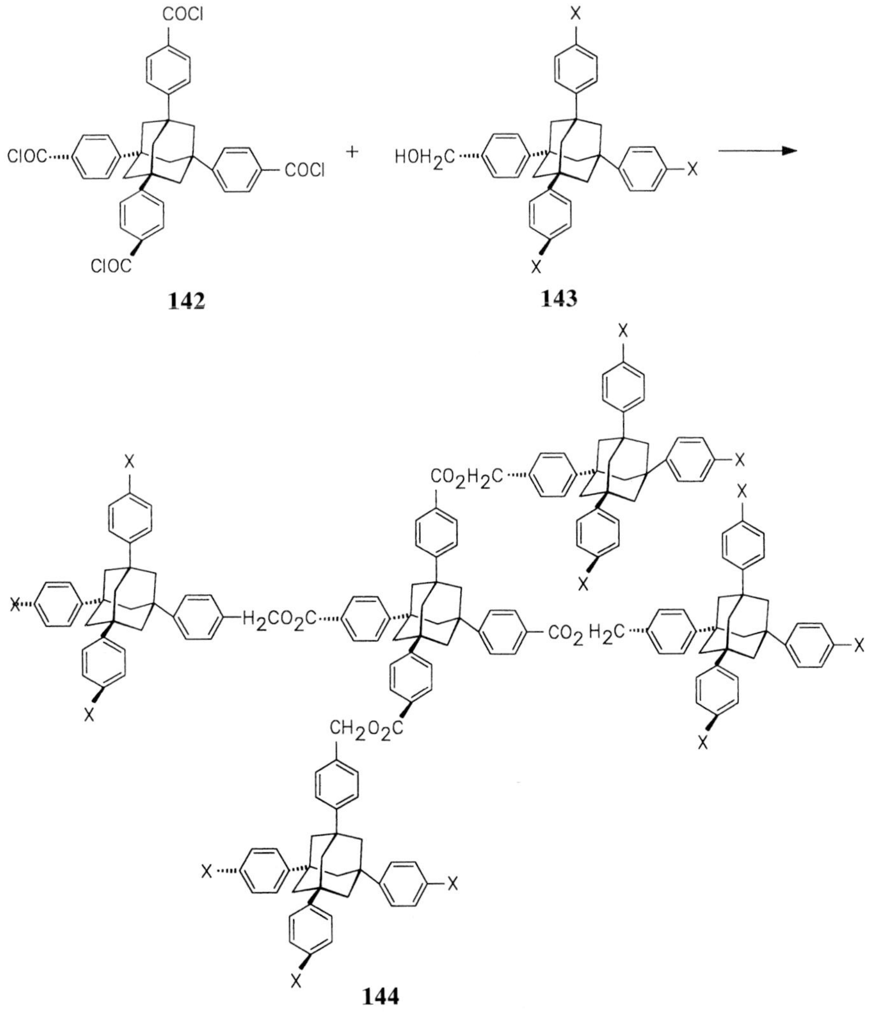

Scheme 4.39 Use of rigid monomers for the construction of adamantane-based dendrimers.[211]

4.12   1 → 4 *N,C*- (1 → 2 *N* & 1 → 2 *C*) Branched, *Ether & Amide* Connectivity

Bis(1,3-dihydroxyisopropyl)amine {[(HOCH₂)₂C]₂NH} has been utilized as a 1 → 4 symmetrical AB₄ monomer[219] in the construction of multifunctional 1st generation dendrimers; conversion [ketalization; ClCH₂CHO, NaBH(OAc)₃; NaN₃; H₂, Pd] of the secondary amine to the extended and more reactive primary amine was readily accomplished.[220] Treatment of the hydroxy moieties with CH₂=CHCO₂*t*Bu afforded the tetraester; subsequent extension to the primary amine gave the desired monomer, which was coupled with 1,3,5-tris(chlorocarbonyl)benzene to generate the dendrimer.

4.13   1 → 5 *N₃P₃*-Branched, *Alkyl* Connectivity

Labarre et al.[221] prepared a series of "dandelion"[222] dendrimers based on the repetitive addition of alkyl diamines with a hexavalent cyclotriphosphazene monomer (**145**; Scheme 4.40); eight generations were purported. The strategy commenced with the reaction of 6 equivalents of 1,6-diaminohexane with N₃P₃Cl₆ to afford hexaamine **146**, which, in turn, was reacted with more cyclotriphosphazene monomer **145** to yield the 1st generation polychloride **147**. This cascade process was repeated through 8 generations to afford the purported dandelion possessing 2,343,750 termini (P–Cl moieties) and a calculated molecular weight of 228,977,179 amu. Asymmetrical dandelion dendrimers were prepared by regiospecific aminolysis of N₃P₃Cl₆ with amino alcohols.[223]

Cyclotriphosphazene cores[224] have also been employed as scaffolds for star and cross-linked polymers, the chemical and thermal properties of which were investigated with regard to flame retardancy and oil repellence.

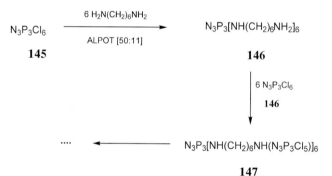

Scheme 4.40 Strategy for the synthesis of "dandelions."[221]

4.14   References

[1]   G. R. Newkome, Z. Yao, G. R. Baker, V. K. Gupta, "Cascade Molecules: A New Approach to Micelles. A [27]-Arborol", *J. Org. Chem.* **1985**, *50*, 2003–2004.

[2]   F. Hallé, R. A. A. Oldeman, *Essai sur l'Architecture et la Dynamique de Croissance des Arbres Tropicaux*, Masson, Paris, **1970**.

[3]   F. Hallé, R. A. A. Oldeman, P. B. Tomlinson, *Tropical Trees and Forests: An Architectural Analysis*, Springer, Berlin, **1982**.

[4]   P. B. Tomlinson, "Tree Architecture", *Am. Sci.* **1983**, *71*, 141–149.

[5]   G. R. Newkome, G. R. Baker, "The Chemistry of Methanetricarboxylic Esters. A Review", *Org. Prep. Proced. Int.* **1986**, *18*, 117–144.

[6]   J. Skarzewski, "Carbon-Acylations in the Presence of Magnesium Oxide. A Simple Synthesis of Methanetricarboxylic Esters", *Tetrahedron* **1989**, *45*, 4593–4598.

[7]   J. Skarzewski, "The Michael Reaction of Methanetricarboxylic Esters. A Simple Method of Two-Carbon Chain Elongation", *Synthesis* **1990**, 1125–1127.

[8] G. R. Newkome, C. N. Moorefield, G. R. Baker, "Building Blocks for Dendritic Macromolecules", *Aldrichim. Acta* **1992**, *25*, 31–38.

[9] G. R. Newkome, G. R. Baker, M. J. Saunders, P. S. Russo, V. K. Gupta, Z. Yao, J. E. Miller, K. Bouillion, "Two-Directional Cascade Molecules: Synthesis and Characterization of [9]-*n*-[9] Arborols", *J. Chem. Soc., Chem. Commun.* **1986**, 752–753.

[10] G. R. Newkome, G. R. Baker, S. Arai, M. J. Saunders, P. S. Russo, K. J. Theriot, C. N. Moorefield, L. E. Rogers, J. E. Miller, T. R. Lieux, M. E. Murray, B. Phillips, L. Pascal, "Synthesis and Characterization of Two-Directional Cascade Molecules and Formation of Aqueous Gels", *J. Am. Chem. Soc.* **1990**, *112*, 8458–8465.

[11] R. M. Fuoss, D. J. Edelson, "Bolaform Electrolytes; I. Di-(β-trimethylammonium Ethyl) Succinate Dibromide and Related Compounds", *J. Am. Chem. Soc.* **1951**, *73*, 269–273.

[12] J.-H. Fuhrhop, J. Mathieu, "Routes to Functional Vesicle Membranes without Proteins", *Angew. Chem.* **1984**, *96*, 124–137; *Angew. Chem.* **1984**, *96*, 124–137; *Int. Ed. Engl.* **1984**, *23*, 100–113.

[13] G. H. Escamilla, G. R. Newkome, "Bolaamphiphiles: From Golf Balls to Fibers", *Angew. Chem.* **1994**, *106*, 2013–2016; *Angew. Chem.* **1994**, *106*, 2013–2096; *Int. Ed. Engl.* **1994**, *33*, 1937–1940.

[14] G. H. Escamilla, G. R. Newkome, "Bolaamphiphiles: Golf Balls to Fibers", in *Organic Synthesis Highlights III* (Eds.: J. Mulzer, H. Waldmann), Wiley-VCH, Weinheim, **1998**, pp. 382–390.

[15] G. H. Escamilla, "Dendritic Bolaamphiphiles and Related Molecules", in *Advances in Dendritic Macromolecules* (Ed.: G. R. Newkome), JAI Press, Inc., Greenwich, Conn., **1995**, pp. 157–190.

[16] G. R. Newkome, C. N. Moorefield, G. R. Baker, R. K. Behera, G. H. Escamilla, M. J. Saunders, "Supramolecular Self-Assemblies of Two-Directional Cascade Molecules: Automorphogenesis", *Angew. Chem.* **1992**, *104*, 901–903; *Angew. Chem.* **1992**, *104*, 901–903; *Int. Ed. Engl.* **1992**, *31*, 917–919.

[17] J.-M. Lehn, "Perspectives in Supramolecular Chemistry – From Molecular Recognition Toward Molecular Information Processing and Self Organization", *Angew. Chem.* **1990**, *102*, 1347–1362; Int. Ed. Engl. **1990**, *29*, 1304–1319.

[18] G. R. Newkome, X. Lin, Y. Chen, G. H. Escamilla, "Two-Directional Cascade Polymer Synthesis: Effects of Core Variation", *J. Org. Chem.* **1993**, *58*, 3123–3129.

[19] K. H. Yu, P. S. Russo, L. Younger, W. G. Henk, D.-W. Hua, G. R. Newkome, G. R. Baker, "Observations on the Thermoreversible Gelation of Two-Directional Arborols in Water-Methanol Mixtures", *J. Polym. Sci., Polym. Phys. Ed.* **1997**, *35*, 2787–2793.

[20] S. Bhattacharya, S. N. G. Acharya, A. R. Raju, "Exceptional Adhesive and Gelling Properties of Fibrous Nanoscopic Tapes of Self-Assembled Bipolar Urethane Amides of L-Phenylalanine", *Chem. Commun.* **1996**, 2101–2102.

[21] F. Hentrich, C. Tschierske, H. Zaschke, "Bolaamphiphilic Polyols: A Novel Class of Amphotropic Liquid Crystals", *Angew. Chem.* **1991**, *103*, 429–431; Int. Ed. Engl. **1991**, *30*, 440–441.

[22] J. S. Choi, D. K. Joo, C. H. Kim, K. Kim, J. S. Park, "Synthesis of a Barbell-like Triblock Copolymer, Poly(L-lysine) Dendrimer–*block*–Poly(ethylene glycol)–*block*–Poly(L-lysine) Dendrimer, and Its Self-Assembly with Plasmid DNA", *J. Am. Chem. Soc.* **2000**, *122*, 474–480.

[23] M. Masuda, T. Hanada, K. Yase, T. Shimizu, "Polymerization of Bolaform Butadiyne 1-Glucosamide in Self-Assembled Nanoscale-Fiber Morphology", *Macromolecules* **1998**, *31*, 9403–9405.

[24] Y. Rivaux, N. Noiret, H. Patin, "Synthesis, Aqueous Solution Behaviour, and Amphotropic Liquid Crystalline Properties of Glycerol Based Bolaamphiphiles", *New J. Chem.* **1998**, 857–863.

[25] T. W. Davey, W. A. Ducker, A. R. Hayman, "Aggregation of ω-Hydroxy Quaternary Ammonium Bolaform Surfactants", *Langmuir* **2000**, *16*, 2430–2435.

[26] F. M. Menger, V. A. Migulin, "Synthesis and Properties of Multi-armed Geminis", *J. Org. Chem.* **1999**, *64*, 8916–8921.

[27] I. Nakazawa, M. Masuda, Y. Okada, T. Hanada, K. Yase, M. Asai, T. Shimizu, "Spontaneous Formation of Helically Twisted Fibers from 2-Glucosamide Bolaamphiphiles: Energy-Filtering Transmission Electron Microscopic Observation and Even-Odd Effect of Connecting Bridge", *Langmuir* **1999**, *15*, 4757–4764.

[28] C. Prata, N. Mora, A. Polidori, J.-M. Lacombe, B. Pucci, "Synthesis and Molecular Aggregation of New Sugar Bola-Amphiphiles", *Carbohydr. Res.* **1999**, *321*, 15–23.

[29] I. Nakazawa, S. Suda, M. Masuda, M. Asai, T. Shimizu, "pH-Dependent Reversible Polymers Formed from Cyclic Sugar- and Aromatic Boronic Acid-Base Bolaamphiphiles", *Chem. Commun.* **2000**, 881–882.

[30] W. E. Lindsell, P. N. Preston, J. M. Seddon, G. M. Rosair, A. J. Woodman, "Macroscopic Helical and Cylindrical Morphologies from Achiral 1,3-Diynes", *Chem. Mater.* **2000**, *12*, 1572–1576.

[31] D. Zhao, Q. Huo, J. Feng, J. Kim, Y. Han, G. D. Stucky, "Novel Mesoporous Silicates with Two-Dimensional Mesostructure Direction Using Rigid Bolaform Surfactants", *Chem. Mater.* **1999**, *11*, 2668–2672.

[32] C. D. Meglio, S. B. Rananavare, S. Svenson, D. H. Thompson, "Bolaamphiphilic Phospho-cholines: Structure and Phase Behavior in Aqueous Media", *Langmuir* **2000**, *16*, 128–133.

[33] P. Camilleri, A. Kremer, A. J. Edwards, K. H. Jennings, O. Jenkins, I. Marshall, C. McGre-gor, W. Neville, S. Q. Rice, R. J. Smith, M. J. Wilkinson, A. J. Kirby, "A Novel Class of Cat-ionic *Gemini* Surfactants Showing Efficient *in Vitro* Gene Transfection Properties", *Chem. Commun.* **2000**, 1253–1254.

[34] S. Gao, B. Zou, L. Chi, H. Fuchs, J. Sun, X. Zhang, J. Shen, "Nano-Size Stripes of Self-Assembled Bolaform Amphiphiles", *Chem. Commun.* **2000**, 1273–1274.

[35] A. A. Shukla, S. S. Bae, J. A. Moore, K. A. Barnthouse, S. M. Cramer, "Synthesis and Char-acterization of High-Affinity, Low Molecular Weight Displacers for Cation-Exchange Chro-matography", *Ind. Eng. Chem. Res.* **1998**, *37*, 4090–4098.

[36] F. M. Menger, J. S. Keiper, "Gemini Surfactants", *Angew. Chem.* **2000**, *112*, 1980–1996; *Int. Ed.* **2000**, *39*, 1906–1920.

[36a] M. Masuda, V. Vill, T. Shimizu, "Conformational and Thermal Phase Behavior of Oligome-thylene Chains Constrained by Carbohydrate Hydrogen-Bond Network", *J. Am. Chem. Soc.* **2000**, *122*, 12327–12333.

[36b] J. Song, Q. Cheng, S. Kopta, S. C. Stevens, "Modulating Artifical Membrane Morphology: pH-Induced Chromatic Transition and Nanostructural Transformation of a Bolaamphiphilic Conjugated Polymer from Blue Helical Ribbons to Red Nanofibers", *J. Am. Chem. Soc.* **2001**, *123*, 3205–3213.

[36c] A. Lubineau, A. Malleron, C. Le Narvor, "Chemo-enzymatic synthesis of oligosaccharides using a dendritic soluble support", *Tetrahedron Lett.* **2000**, *41*, 8887–8891.

[37] G. R. Newkome, Z. Yao, G. R. Baker, V. K. Gupta, P. S. Russo, M. J. Saunders, "Cascade Molecules: Synthesis and Characterization of a Benzene[9]^3-arborol", *J. Am. Chem. Soc.* **1986**, *108*, 849–850.

[38] T.-P. Engelhardt, L. Belkoura, D. Woermann, W. Grimme, "Dynamic Light Scattering Exper-iments with Aqueous Solutions of a Cascade Molecule: Benzene[9]^3-arborol", *Ber. Bunsen-Ges. Phys. Chem.* **1993**, *97*, 33–35.

[39] G. R. Newkome, Y. Hu, M. J. Saunders, F. R. Fronczek, "Silvanols: Water-Soluble Calixare-nes", *Tetrahedron Lett.* **1991**, *32*, 1133–1136.

[40] C. D. Gutsche, *Calixarenes*, Royal Society of Chemistry, London, **1989**.

[41] C. D. Gutsche, *Calixarenes Revisited*, Royal Society of Chemistry, London, **1998**.

[42] G. R. Newkome, V. K. Gupta, G. R. Baker, National Meeting of the American Chemical Society, Miami, FL, *ORGN-166*, 1985.

[43] M. Sawamoto, "Recent Advances in Topologically Well-Defined Polymers", *Kagaku (Kyoto)* **1990**, *45*, 537–539.

[44] P. Lhotak, S. Shinkai, "Synthesis and Metal-Binding Properties of Oligo-Calixarenes. An Approach towards the Calix[4]arene-Based Dendrimers", *Tetrahedron* **1995**, *51*, 7681–7696.

[45] T. Sugawara, T. Matsuda, "Synthesis and Photochemical Surface Fixation of Hydroxylated Cascade Molecules", *J. Polym. Sci., Part A: Polym. Chem.* **1997**, *35*, 137–142.

[46] R. Esfand, D. A. Tomalia, A. E. Beezer, J. C. Mitchell, M. Hardy, C. Orford, "Dendripore and Dendrilock Concepts – New Controlled Delivery Strategies", *Polym. Prepr.* **2000**, *41*, 1324–1325.

[47] C. I. Alvarez, M. C. Strumia, "Different Synthetic Routes to Obtain Hydrophilic Matrices", *J. Polym. Sci., Part A: Polym. Chem.* **2000**, *38*, 489–497.

[48] M. Jørgensen, K. Bechgaard, T. Bjørnholm, P. Sommer-Larsen, L. G. Hansen, K. Schaum-burg, "Synthesis and Structural Characterization of a Bis-arborol–Tetrathiafulvalene Gel: Toward a Self-Assembling 'Molecular' Wire", *J. Org. Chem.* **1994**, *59*, 5877–5882.

[49] M. B. Nielsen, C. Lomholt, J. Becher, "Tetrathiafulvalenes as Building Blocks in Supramo-lecular Chemistry II", *Chem. Soc. Rev.* **2000**, *29*, 153–164.

[50] N. Saito, T. Sugawara, T. Matsuda, "Synthesis and Hydrophilicity of Multifunctionally Hydroxylated Poly(acrylamides)", *Macromolecules* **1996**, *29*, 313–319.

[51] T. Matsuda, T. Sugawara, "Synthesis of Multifunctional, Nonionic Vinyl Polymers and their ^{13}C Spin-Lattice Relaxation Times in Deuterium Oxide Solutions", *Macromolecules* **1996**, *29*, 5375–5383.

[52] S. M. Ngola, P. C. Kearney, S. Mecozzi, K. Russell, D. A. Dougherty, "A Selective Receptor for Arginine Derivatives in Aqueous Media. Energetic Consequences of Salt Bridges that are Highly Exposed to Water", *J. Am. Chem. Soc.* **1999**, *121*, 1192–1201.

[53] G. E. DuBois, B. Zhi, G. M. Roy, S. Y. Stevens, M. Yalpani, "New Non-ionic Polyol Deriva-tives with Sucrose Mimetic Properties", *J. Chem. Soc., Chem. Commun.* **1992**, 1604–1605.

[54] D. Patel, B. D. McKinley, T. P. Davis, F. Porreca, H. I. Yamamura, V. J. Hruby, "Peptide Targeting and Delivery across the Blood–Brain Barrier Utilizing Synthetic Triglyceride Esters: Design, Synthesis, and Bioactivity", *Bioconj. Chem.* **1997**, *8*, 434–441.

[55] M. B. Smith, J. March, *March's Advanced Organic Chemistry*, 5th ed., Wiley-Interscience, New York, 2001.

[56] G. R. Newkome, C. N. Moorefield, K. J. Theriot, "A Convenient Synthesis of 'Bishomotris': 4-Amino-4-[1-(3-hydroxypropyl)]-1,7-heptanediol and 1-Azoniapropellane", *J. Org. Chem.* **1988**, *53*, 5552–5554.

[57] G. R. Newkome, C. N. Moorefield, "Multifunctional Synthons as Used in the Preparation of Cascade Polymers or Unimolecular Micelles", **1992**, *U. S. Pat.*, 5, 136, 096.

[58] G. R. Newkome, C. N. Moorefield, "Multifunctional Synthons as Used in the Preparation of Cascade Polymers or Unimolecular Micelles", **1993**, *U. S. Pat.*, 5, 206, 410.

[59] G. R. Newkome, C. N. Moorefield, "Multifunctional Synthons as Used in the Preparation of Cascade Polymers or Unimolecular Micelles", **1993**, *U. S. Pat.*, 5, 210, 309.

[60] J. K. Whitesell, H. K. Chang, "Directionally Aligned Helical Peptides on Surfaces", *Science* **1993**, *261*, 73–76.

[61] J. G. Tirrell, M. J. Fournier, T. L. Mason, D. A. Tirrell, "Biomolecular Materials", *Chem. Eng. News* **1994**, 40–51.

[62] M. Scheffler, A. Dorenbeck, S. Jordan, M. Wüstefeld, G. von Kiedrowski, "Self-Assembly of Trisoligonucleotidyls: The Case for Nano-Acetylene and Nano-Cyclobutadiene", *Angew. Chem.* **1999**, *111*, 3514–3518; *Int. Ed.* **1999**, *38*, 3312–3315.

[63] G. R. Newkome, C. D. Weis, "Di-*tert*-butyl 4-[2-(*tert*-butoxycarbonyl)ethyl]-4-aminoheptane-dicarboxylate", *Org. Prep. Proced. Int.* **1996**, *28*, 485–488.

[64] H. A. Bruson, T. W. Riener, "The Chemistry of Acrylonitrile; IV. Cyanoethylation of Active Hydrogen Groups", *J. Am. Chem. Soc.* **1943**, *65*, 23–27.

[65] C. D. Weis, G. R. Newkome, "Reduction of Nitro-Substituted Tertiary Alkanes", *Synthesis* **1995**, 1053–1065.

[66] D. E. Butler, "5-Oxo-2,2-pyrrolidinedipropanoic Acid and Ester Derivatives Thereof", **1984**, *U. S. Pat.*, 4, 454, 327.

[67] G. R. Newkome, R. K. Behera, G. R. Baker, F. R. Fronczek, "Di-*tert*-butyl 4-[2-(*tert*-butoxycarbonyl)ethyl]-4-nitroheptanedioate, $C_{22}H_{39}NO_8$", *Acta Cryst.* **1994**, *C50*, 120–122.

[68] G. R. Newkome, A. Nayak, R. K. Behera, C. N. Moorefield, G. R. Baker, "Cascade Polymers: Synthesis and Characterization of Four-Directional Spherical Dendritic Macromolecules Based on Adamantane", *J. Org. Chem.* **1992**, *57*, 358–362.

[69] G. R. Newkome, R. K. Behera, C. N. Moorefield, G. R. Baker, "Cascade Polymers: Syntheses and Characterization of One-Directional Arborols Based on Adamantane", *J. Org. Chem.* **1991**, *56*, 7162–7167.

[70] G. R. Newkome, V. V. Narayanan, A. K. Patri, J. Groß, C. N. Moorefield, G. R. Baker, "Cascade Infrastructure Modification via Integration of Application-Based Monomers" *Polym. Mater. Sci. Eng.* **1995**, *73*, 222–223.

[71] D. Braun, C.-C. Keller, M. D. Roth, B. Schartel, M. Voigt, J. H. Wendorff, "Tetrahedral Adamantane Derivatives: Glass Formation and Melting Behaviour", *J. prakt. Chem. Chem.-Zeit.* **1997**, *339*, 708–713.

[72] A. Bashir-Hashemi, J. Li, N. Gelber, "Photochemical Carbonylation of Adamantanes; Simple Synthesis of 1,3,5,7-Tetracarbomethoxyadamantane", *Tetrahedron Lett.* **1995**, *36*, 1233–1236.

[73] Y. S. Klausner, M. Bodansky, "Coupling Reagents in Peptide Synthesis", *Synthesis* **1972**, 453–463.

[74] G. R. Newkome, J. K. Young, G. R. Baker, R. L. Potter, L. Audoly, D. Cooper, C. D. Weis, K. F. Morris, C. S. Johnson, Jr., "Cascade Polymers. pH Dependence of Hydrodynamic Radii of Acid Terminated Dendrimers", *Macromolecules* **1993**, *26*, 2394–2396.

[75] G. R. Newkome, C. D. Weis, "6,6-Bis(carboxy-2-oxabutyl)-4,8-dioxaundecane-1,11-dicarboxylic Acid", *Org. Prep. Proced. Int.* **1996**, *28*, 242–246.

[76] H. A. Bruson, "Cyanoalkyl Ethers of Polyhydric Alcohols", **1946**, *U. S. Pat.*, 2, 401, 607.

[77] G. R. Newkome, C. D. Weis, X. Lin, F. R. Fronczek, "5,5-Bis(4-cyano-2-oxabutyl)-1,9-dicyano-3,7-dioxanonane", *Acta Cryst.* **1993**, *C49*, 998–1000.

[78] K. F. Morris, C. S. Johnson, Jr., "Resolution of Discrete and Continuous Molecular Size Distributions by means of Diffusion-Ordered 2D NMR Spectroscopy", *J. Am. Chem. Soc.* **1993**, *115*, 4291–4299.

[78a] W. Chen, D. A. Tomalia, J. L. Thomas, "Unusual pH-Dependent Polarity Changes in PAMAM Dendrimers: Evidence for pH-Responsive Conformational Changes", *Macromolecules* **2000**, *33*, 9169–9172.

[79] S. A. Kuzdzal, C. A. Monnig, G. R. Newkome, C. N. Moorefield, "A Study of Dendrimer–Solute Interactions via Electrokinetic Chromatography", *J. Am. Chem. Soc.* **1997**, *119*, 2255–2261.

[80] S. A. Kuzdzal, C. A. Monnig, G. R. Newkome, C. N. Moorefield, "Dendrimer Electrokinetic Capillary Chromatography: Unimolecular Micellar Behaviour of Carboxylic Acid Terminated Cascade Macromolecules", *J. Chem. Soc., Chem. Commun.* **1994**, 2139–2140.

[81] K. Otsuka, S. Terabe, "Micellar Electrokinetic Chromatography", *Bull. Chem. Soc. Jpn.* **1998**, *71*, 2465–2481.

[82] J. P. Harmon, S. K. Emran, H. Gao, B. Wang, G. Newkome, G. R. Baker, C. N. Moorefield, "Molecular Relaxations in Amide-Based Dendrimers", *Polym. Prepr.* **1996**, *37*, 421–422.

[82a] M. C. Strumia, A. Halabi, P. A. Pucci, G. R. Newkome, C. N. Moorefield, J. D. Epperson, "Surface Modifications of Activated Polymeric Matrices by Dendritic Attachments", *J. Polym. Sci. , Part A: Polym. Chem.* **2000**, *38*, 2779–2786.

[83] H. Zhang, P. L. Dubin, J. Kaplan, C. N. Moorefield, G. R. Newkome, "Dissociation of Carboxyl-Terminated Cascade Polymers: Comparison with Theory", *J. Phys. Chem. B* **1997**, *101*, 3494–3497.

[83a] Q. R. Huang, P. L. Dubin, C. N. Moorefield, G. R. Newkome, "Counterion Binding on Charged Spheres: Effect of pH and Ionic Strength on the Mobility of Carboxyl-Terminated Dendrimers", *J. Phys. Chem. B.* **2000**, *104*, 898–904.

[84] L. L. Miller, Y. Kunugi, A. Canavesi, S. Rigaut, C. N. Moorefield, G. R. Newkome, "'Vapo-conductivity'. Sorption of Organic Vapors Causes Large Increases in the Conductivity of a Dendrimer", *Chem. Mater.* **1998**, *10*, 1751–1754.

[85] R. G. Duan, L. L. Miller, D. A. Tomalia, "An Electrically Conducting Dendrimer", *J. Am. Chem. Soc.* **1995**, *117*, 10783–10784.

[86] L. L. Miller, R. G. Duan, D. C. Tully, D. A. Tomalia, "Electrically Conducting Dendrimers", *J. Am. Chem. Soc.* **1997**, *119*, 1005–1010.

[87] D. S. Tarbell, Y. Yamamoto, B. M. Pope, "New Method to Prepare *N*-*t*-Butoxycarbonyl Derivatives and the Corresponding Sulfur Analogs from Di-*t*-butyl Dicarbonate or Di-*t*-butyl Dithiol Dicarbonates and Amino Acids", *Proc. Natl. Acad. Sci. U. S. A.* **1972**, *69*, 730–732.

[88] E. Ponnusamy, U. Fotadar, A. Spisni, D. Fiat, "A Novel Method for the Rapid, Non-Aqueous *t*-Butoxycarbonylation of Some ^{17}O-Labeled Amino Acids and ^{17}O NMR Parameters of the Products", *Synthesis* **1986**, 48–49.

[89] X. A. Dominguez, I. C. Lopez, R. Franco, "Simple Preparation of a Very Active Raney Nickel Catalyst", *J. Org. Chem.* **1961**, *26*, 1625.

[90] J. K. Young, G. R. Baker, G. R. Newkome, K. F. Morris, C. S. Johnson, Jr., "'Smart' Cascade Polymers. Modular Syntheses of Four-Directional Dendritic Macromolecules with Acidic, Neutral, or Basic Terminal Groups and the Effect of pH Changes on their Hydrodynamic Radii", *Macromolecules* **1994**, *27*, 3464–3471.

[91] D. F. DeTar, R. Silverstein, F. F. Rogers, Jr., "Reactions of Carbodiimides; III. The Reactions of Carbodiimides with Peptide Acids", *J. Am. Chem. Soc.* **1966**, *88*, 1024–1030.

[92] W. König, R. Geiger, "Eine neue Methode zur Synthese von Peptiden: Aktivierung der Carboxylgruppe mit Dicyclohexylcarbodiimid unter Zusatz von 1-Hydroxy-benztrizolen", *Chem. Ber.* **1970**, *103*, 788–798.

[93] G. R. Newkome, B. D. Woosley, E. He, C. N. Moorefield, R. Güther, G. R. Baker, G. H. Escamilla, J. Merrill, H. Luftmann, "Supramolecular Chemistry of Flexible, Dendritic-Based Structures Employing Molecular Recognition", *Chem. Commun.* **1996**, 2737–2738.

[94] G. R. Newkome, C. N. Moorefield, G. R. Baker, "Lock and Key Micelles and Monomer Building Blocks Thereof", **1999**, *U. S. Pat.* 5, 863, 919.

[95] M. C. Strumia, A. Halabi, P. A. Pucci, G. R. Newkome, C. N. Moorefield, J. D. Epperson, "Surface Modifications of Activated Polymeric Matrices by Dendritic Attachments", *J. Polym. Sci., Part A: Polym. Chem.* **2000**, *38*, 2779–2786.

[95a] A. Halabi, M. C. Strumia, "Synthesis and Characterization of a Novel Dendritic Acrylic Monomer", *J. Org. Chem.* **2000**, *65*, 9210–9213.

[96] V. V. Narayanan, G. R. Newkome, L. Echegoyen, E. Pérez-Cordero, "Novel Dendrimers Possessing Internal Electroactive Quinonoid Moieties" *Polym. Prepr.* **1996**, *37*, 419–420.

[97] G. R. Newkome, V. V. Narayanan, L. Echegoyen, E. Pèrez-Cordero, H. Luftmann, "Synthesis and Chemistry of Novel Dendritic Macromolecules Possessing Internal Electroactive Anthraquinonoid Moieties", *Macromolecules* **1997**, *30*, 5187–5191.

[97a] G. R. Newkome, V. V. Narayanan, L. A. Godínez, "Anthraquinoid-based extended dendritic monomers: electrochemical comparisons", *Designed Monomers and Polymers* **2000**, *3*, 17–24.

[98] G. R. Newkome, V. V. Narayanan, L. A. Godínez, E. Pérez-Cordero, L. Echegoyen, "A Tailored Approach to the Syntheses of Electroactive Dendrimers Based on Diaminoanthraquinones", *Macromolecules* **1999**, *32*, 6782–6791.

[99] G. R. Newkome, V. V. Narayanan, L. A. Godínez, "Electroactive, Internal Anthraquinonoid Dendritic Cores", *J. Org. Chem.* **2000**, *65*, 1643–1649.

[100] G. R. Newkome, V. V. Narayanan, L. A. Godínez, "Anthraquinoid-based Extended Dendritic Monomers: Electrochemical Comparisons", *Designed Monomers and Polymers* **2000**, *3*, 17–24.

[101] G. R. Newkome, J. Groß, C. N. Moorefield, B. D. Woosley, "Approaches Towards Specifically Functionalized Cascade Macromolecules: Dendrimers with Incorporated Metal Binding Sites and their Palladium(II) and Copper(II) Complexes", *Chem. Commun.* **1997**, 515–516.

[102] G. R. Newkome, C. D. Weis, B. J. Childs, "Synthesis of 1 → 3 Branched Isocyanate Monomers for Dendritic Construction", *Designed Monomers and Polymers* **1997**, *1*, 3–14.

[103] G. R. Newkome, L. A. Godínez, C. N. Moorefield, "Molecular Recognition using β-Cyclodextrin Modified Dendrimers: Novel Building Blocks for Convergent Self-Assembly", *Chem. Commun.* **1998**, 1821–1822.

[104] J. Boger, R. J. Corcoran, J.-M. Lehn, "Cyclodextrin Chemistry: Selective Modification of All Primary Hydroxyl Groups of α- and β-Cyclodextrins", *Helv. Chim. Acta* **1978**, *61*, 2190–2218.

[104a] E. Alvarez-Parrilla, P. R. Cabrer, W. Al-Soufi, F. Meijide, E. R. Núñez, J. T. Tato, "Dendritic Growth of a Supramolecular Complex", *Angew. Chem. Int. Ed.* **2000**, *39*, 2856–2858.

[105] G. R. Newkome, B. J. Childs, M. J. Rourk, G. R. Baker, C. N. Moorefield, "Dendrimer Construction and Macromolecular Property Modification via Combinatorial Methods", *Biotechnol. Bioeng.* **1999**, *61*, 243–253.

[106] G. R. Newkome, C. D. Weis, C. N. Moorefield, G. R. Baker, B. J. Childs, J. Epperson, "Isocyanate-Based Dendritic Building Blocks: Combinatorial Tier Construction and Macromolecular Property Modification", *Angew. Chem.* **1998**, *110*, 318–321; *Int. Ed.* **1998**, *37*, 307–310.

[107] G. R. Newkome, G. R. Baker, C. N. Moorefield, E. He, J. Epperson, C. D. Weis, "Modular Approaches to the Construction of Branched Macromolecules" *Polym. Mater. Sci. Eng.* **1997**, *77*, 65–66.

[108] G. R. Newkome, C. N. Moorefield, "Combinatorial Method of Forming Cascade Polymer Surfaces", **1999**, *U.S. Pat.*, 5, 886, 127.

[109] G. R. Newkome, C. N. Moorefield, "Combinatorial Method of Forming Cascade Polymer Surfaces", **1999**, *U.S. Pat.*, 5, 886, 126.

[110] K. Feldrapp, W. Brütting, M. Schwoerer, M. Brettreich, A. Hirsch, "Photovoltaic Effect in Blend Systems and Heterostructures of Poly(*p*-phenylenevinylene) and C_{60}", *Synth. Metals* **1999**, *101*, 156–157.

[111] M. Brettreich, S. Burghardt, C. Böttcher, T. Bayerl, S. Bayerl, A. Hirsch, "Globular Amphiphiles: Membrane-Forming Hexaadducts of C_{60}", *Angew. Chem.* **2000**, *112*, 1915–1918; Int. Ed. **2000**, *39*, 1845–1848.

[112] V. V. Narayanan, E. C. Wiener, "Metal-Directed Self-Assembly of Ethylenediamine-Based Dendrons", *Macromolecules* **2000**, *33*, 3944–3946.

[112a] C. M. Cardona, J. Alvarez, A. E. Kaifer, T. D. McCarley, S. Pandey, G. A. Baker, N. J. Bonzagni, F. V. Bright, "Dendrimers Functionalized with a Single Fluorescent Dansyl Group Attached "Off Center": Synthesis and Photophysical Studies", *J. Am. Chem. Soc.* **2000**, *122*, 6139–6144.

[113] G. R. Newkome, C. D. Weis, C. N. Moorefield, I. Weis, "Detection and Functionalization of Dendrimers Possessing Free Carboxylic Acid Moieties", *Macromolecules* **1997**, *30*, 2300–2304.

[114] G. Shah, P. L. Dubin, J. I. Kaplan, G. R. Newkome, C. N. Moorefield, G. R. Baker, "Size-Exclusion Chromatography of Carboxyl-terminated Dendrimers as a Model for Permeation of Charged Particles into Like-Charged Cavities", *J. Colloid Interface Sci.* **1996**, *183*, 397–407.

[115] F. G. Smith, III, W. M. Deen, "Electrostatic Effects on the Partitioning of Spherical Colloids between Dilute Bulk Solution and Cylindrical Pores", *J. Colloid Interface Sci.* **1983**, *91*, 571–590.

[116] H. Zhang, P. L. Dubin, J. Ray, G. S. Manning, C. N. Moorefield, G. R. Newkome, "Interaction of a Polycation with Small Oppositely Charged Dendrimers", *J. Phys. Chem. B* **1999**, *103*, 2347–2354.

[117] N. Miura, P. L. Dubin, C. N. Moorefield, G. R. Newkome, "Complex Formation by Electrostatic Interaction between Carboxyl-Terminated Dendrimers and Oppositely Charged Polyelectrolytes", *Langmuir* **1999**, *15*, 4245–4250.

[118] S. K. Emran, G. R. Newkome, C. D. Weis, J. P. Harmon, "Molecular Relaxations in Ester-Terminated, Amide-Based Dendrimers", *J. Polym. Sci., Part B: Polym. Phys.* **1998**, *37*, 2025–2038.

[118a] S. K. Emran, G. R. Newkome, J. P. Harmon, "Viscoelastic Properties and Phase Behavior of 12-*tert*-Butyl Ester Dendrimer/Poly(Methyl Methacrylate) Blends", *J. Polym. Sci., Part B: Polym. Phys.* **2001**, *39*, 1381–1393.

[119] Q. R. Huang, P. L. Dubin, C. N. Moorefield, G. R. Newkome, "Counterion Binding on Charged Spheres: Effect of pH and Ionic Strength on the Mobility of Carboxyl-Terminated Dendrimers", *Langmuir* **2000**, *in press*.

[120] G. R. Newkome, C. N. Moorefield, G. R. Baker, A. L. Johnson, R. K. Behera, "Alkane Cascade Polymers Possessing Micellar Topology: Micellanoic Acid Derivatives", *Angew. Chem.* **1991**, *103*, 1205–1207; *Int. Ed. Engl.* **1991**, *30*, 1176–1178.

[121] G. R. Newkome, G. R. Baker, C. N. Moorefield, M. J. Saunders, "I Think I Shall Never See a Polymer Lovely as a Tree", *Polym. Prepr.* **1991**, *32*, 625–626.

[122] G. R. Newkome, C. N. Moorefield, "Unimolecular Micelles and Method of Making the Same", **1992**, *U. S. Pat.*, 5, 154, 853.

[123] N. Ono, H. Miyake, A. Kamimura, I. Hamamoto, R. Tamura, A. Kaji, "Denitrohydrogenation of Aliphatic Nitro Compounds and a New Use of Aliphatic Nitro Compounds as Radical Precursors", *Tetrahedron* **1985**, *41*, 4013–4023.

[124] C. K. Ingold, L. C. Nickolls, "Experiments on the Synthesis of the Polyacetic Acids of Methane; Part VII. *iso*-Butylene-α,γ,γ'-Tricarboxylic Acid and Methanetetraacetic Acid", *J. Chem. Soc.* **1922**, *121*, 1638–1648.

[125] L. M. Rice, B. S. Sheth, T. B. Zalucky, "New Compounds: Symmetrically Substituted Tetraalkyl Methanes", *J. Pharm. Chem.* **1971**, *60*, 1760.

[126] G. R. Newkome, V. K. Gupta, R. W. Griffin, S. Arai, "A Convenient Synthesis of Tetrakis(2-bromoethyl)methane", *J. Org. Chem.* **1987**, *52*, 5480–5482.

[127] G. R. Newkome, S. Arai, F. R. Fronczek, C. N. Moorefield, X. Lin, C. D. Weis, "Synthesis of Functionalized Cascade Cores: Tetrakis(ς-bromoalkyl)methanes", *J. Org. Chem.* **1993**, *58*, 898–903.

[128] K. S. Feldman, K. M. Masters, "Facile Preparation of Tetra(2-aminoethyl)methane and Tetra(3-aminopropyl)methane: Novel Tetravalent Monomers for Materials Synthesis", *J. Org. Chem.* **1999**, *64*, 8945–8947.

[129] G. R. Newkome, C. N. Moorefield, G. R. Baker, M. J. Saunders, S. H. Grossman, "Unimolecular Micelles", *Angew. Chem.* **1991**, *103*, 1207–1209; *Int. Ed. Engl.* **1991**, *30*, 1178–1180.

[130] G. R. Newkome, C. N. Moorefield, "Chemistry within a Unimolecular Micelle: Metallomicellanoic Acids", *Polym. Prepr.* **1993**, *34*, 75–76.

[131] G. R. Newkome, C. N. Moorefield, J. M. Keith, G. R. Baker, G. H. Escamilla, "Chemistry within a Unimolecular Micelle Precursor: Boron Superclusters by Site- and Depth-Specific Transformations of Dendrimers", *Angew. Chem.* **1994**, *106*, 701–703; *Int. Ed. Engl.* **1994**, *33*, 666–668.

[132] G. R. Newkome, C. N. Moorefield, "Metallo- and Metalloido-Micellane™ Derivatives: Incorporation of Metals and Nonmetals within Unimolecular Superstructures", in *International Symposium on New Macromolecular Architectures and Supramolecular Polymers* (Eds.: V. Percec, D. A. Tirrell), Hüthig & Wepf Verlag, Basel, **1994**, pp. 63–71.

[133] S. Sengupta, S. K. Sadhukhan, "Synthetic Studies on Tetraphenylmethane Dendrimers", *Tetrahedron Lett.* **1999**, *40*, 9157–9161.

[134] T. Jeffery, "On the Efficiency of Tetraalkylammonium Salts in Heck-Type Reactions", *Tetrahedron* **1996**, *52*, 10113–10130.

[135] A. B. Padias, H. K. Hall, Jr., D. A. Tomalia, J. R. McConnell, "Starburst Polyether Dendrimers", *J. Org. Chem.* **1987**, *52*, 5305–5312.

[136] A. B. Padias, H. K. Hall, Jr., D. A. Tomalia, "Starburst Polyether and Polythioether Dendrimers", *Polym. Prepr.* **1989**, *30*, 119–120.

[137] S. Rustad, R. Stølevik, "Conformational Analysis; XI. The Molecular Structure, Torsional Oscillations, and Conformational Equilibria of Gaseous Tetrakis(bromomethyl)methane, $C(CH_2Br)_4$, as Determined by Electron Diffraction and Compared with Molecular Mechanics Calculation", *Acta Chem. Scand. A* **1976**, *30*, 209–218.

[138] D. A. Tomalia, A. M. Naylor, W. A. Goddard, III, "Starburst Dendrimers: Molecular-Level Control of Size, Shape, Surface Chemistry, Topology, and Flexibility in the Conversion of Atoms to Macroscopic Materials", *Angew. Chem.* **1990**, *102*, 119–126; *Int. Ed. Engl.* **1990**, *29*, 113–163.

[139] D. A. Tomalia, D. M. Hedstrand, L. R. Wilson, "Dendritic Polymers", in *Encyclopedia of Polymer Science and Engineering*, Wiley & Sons, Inc., New York, **1990**, p. 46.

[140] J.-J. Lee, W. T. Ford, J. A. Moore, Y. Li, "Reactivity of Organic Anions Promoted by a Quaternary Ammonium Ion Dendrimer", *Macromolecules* **1994**, *27*, 4632–4634.

[141] J. A. Moore, Personal communication, **1994**.

[142] G. Jayaraman, Y.-F. Li, J. A. Moore, S. M. Cramer, "Ion-Exchange Displacement Chromatography of Proteins. Dendritic Polymers as Novel Displacers", *J. Chromatogr. A* **1995**, *702*, 143–155.

[143] A. F. Bochkov, B. E. Kalganov, V. N. Chernetskii, "Synthesis of Cascadol – A Highly Branched, Functionalized Polyether", *Izv. Akad. Nauk SSSR, Ser. Khim.* **1989**, 2394–2395.

[144] J. Chang, O. Oyelaran, C. K. Esser, G. S. Kath, G. W. King, B. G. Uhrig, Z. Konteatis, R. M. Kim, K. T. Chapman, "Synthesis of Di- and Trisubstituted Guanidines on Multivalent Soluble Supports", *Tetrahedron Lett.* **1999**, *40*, 4477–4480.

[145] R. A. Cormier, B. A. Gregg, "Synthesis and Characterization of Liquid-Crystalline Perylene Diimides", *Chem. Mater.* **1998**, *10*, 1309–1319.

[146] D. Astruc, J.-C. Blais, E. Cloutet, L. Djakovitch, S. Rigaut, J. Ruiz, V. Sartor, C. Valério, "The First Organometallic Dendrimers: Design and Redox Functions," *Top. Curr. Chem.* **2000**, *210*, 229–259.

[147] G. R. Newkome, X. Lin, "Symmetrical, Four-Directional, Poly(ether-amide) Cascade Polymers", *Macromolecules* **1991**, *24*, 1443–1444.

[148] P. J. Dandliker, F. Diederich, M. Gross, C. B. Knobler, A. Louati, E. M. Sanford, "Dendritic Porphyrins: Modulating Redox Potentials of Electroactive Chromophores with Pendant Multifunctionality", *Angew. Chem.* **1994**, *106*, 1821–1824; Int. Ed. Engl. **1994**, *33*, 1739–1742.

[149] N. Tirelli, F. Cardullo, T. Habicher, U. W. Suter, F. Diederich, "Thermotropic Behaviour of Covalent Fullerene Adducts Displaying 4-Cyano-4'-oxybiphenyl Mesogens", *J. Chem. Soc., Perkin Trans. 2* **2000**, 193–198.

[150] S. Mattei, P. Seiler, F. Diederich, V. Gramlich, "Dendrophanes: Water-Soluble Dendritic Receptors", *Helv. Chim. Acta* **1995**, *78*, 1904–1912.

[151] P. Wallimann, P. Seiler, F. Diederich, "Dendrophanes: Novel Steroid-Recognizing Dendritic Receptors", *Helv. Chim. Acta* **1996**, *79*, 779–788.

[152] S. Mattei, P. Wallimann, B. Kenda, W. Amrein, F. Diederich, "Dendrophanes: Water-Soluble Dendritic Receptors as Models for Buried Recognition Sites in Globular Proteins", *Helv. Chim. Acta* **1997**, *80*, 2391–2417.

[153] P. Wallimann, S. Mattei, P. Seiler, F. Diederich, "New Cyclophanes as Initiator Cores for the Construction of Dendritic Receptors: Host–Guest Complexation in Aqueous Solutions and Structures of Solid-State Inclusion Compounds", *Helv. Chim. Acta* **1997**, *80*, 2368–2390.

[154] J.-F. Nierengarten, T. Habicher, R. Kessinger, F. Cardullo, F. Diederich, V. Gramlich, J.-P. Gisselbrecht, C. Boudon, M. Gross, "Macrocyclization on the Fullerene Core: Direct Regio- and Diastereoselective Multi-Functionalization of [60]Fullerene, and Synthesis of Fullerene-Dendrimer Derivatives", *Helv. Chim. Acta* **1997**, *80*, 2238–2276.

[155] B. Kayser, J. Altman, W. Beck, "Benzene-Bridged Hexaalkynylphenylalanines and First-Generation Dendrimers Thereof", *Chem. Eur. J.* **1999**, *5*, 754–758.

[156] R. G. Denkewalter, J. F. Kolc, W. J. Lukasavage, "Preparation of Lysine-Based Macromolecular Highly Branched Homogeneous Compound", **1979**, *U. S. Pat.*, 4, 360, 646.

[157] J. Shao, J. P. Tam, "Unprotected Peptides as Building Blocks for the Synthesis of Peptide Dendrimers with Oxime, Hydrazone, and Thiazolidine Linkages", *J. Am. Chem. Soc.* **1995**, *117*, 3893–3899.

[158] J.-E. S. Sohna, F. Fages, "Sensitized Luminescence Emission of the Europium(III) Ion Bound to a Pyrene-Containing Triacid Ligand", *Tetrahedron Lett.* **1997**, *38*, 1381–1384.

[158a] S. Koenig, L. Müller, D. K. Smith, "Dendritic Biomimicry: Microenvironmental Hydrogen-Bonding Effects on Tryptophan Fluorescence", *Chem. Eur. J.* **2001**, *7*, 979–986.

[158b] H. Wei, H. Kou, W. Shi, H. Yuan, Y. Chen, "Photopolymerization kinetics of dendritic poly(ether-amide)s", *Polymer* **2001**, *42*, 6741–6746.

[159] I. Bodnár, A. S. Silva, Y. H. Kim, N. J. Wagner, "Structure and Rheology of Hyperbranched and Dendritic Polymers; II. Effects of Blending Acetylated and Hydroxy-Terminated Poly (propyleneimine) Dendrimers with Aqueous Poly(ethylene oxide) Solutions", *J. Polym. Sci., Part B: Polym. Phys.* **2000**, *38*, 874–882.

[160] C. Fromont, M. Bradley, "High-Loading Resin Beads for Solid-Phase Synthesis Using Triple Branching Symmetrical Dendrimers", *Chem. Commun.* **2000**, 283–284.

[161] N. A. J. M. Sommerdijk, J. D. Wright, "Matrix Effects on Selective Chemical Sensing by Sol-Gel Entrapped Complexing Agents", *J. Sol-Gel Sci. Technol.* **1998**, *13*, 565–568.

[162] K. Rengan, R. Engel, "Ammonium Cascade Molecules", *J. Chem. Soc., Chem. Commun.* **1992**, 757–758.

[163] R. Engel, "Ionic Dendrimers and Related Materials", in *Advances in Dendritic Macromolecules* (Ed.: G. R. Newkome), JAI, Greenwich, Conn., **1995**, pp. 73–99.

[164] R. Engel, "Cascade Molecules", *Polymer News* **1992**, *17*, 301–305.

[165] A. Cherestes, R. Engel, "Dendrimeric Ion-Exchange Materials", *Polymer* **1994**, *35*, 3343–3344.

[166] E.-R. Kenawy, "Synthesis and Modifications of Dendrimers on Polymer System Supported on Montmorillonite and their use in Organic Synthesis", *J. Macromol. Sci. – Pure Appld. Chem.* **1998**, *A35*, 657–672.

[167] R. Engel, K. Rengan, C. Milne, "Cationic Cascade Molecules", *Polym. Prepr.* **1991**, *32*, 601.

[168] K. Rengan, R. Engel, "Phosphonium Cascade Molecules", *J. Chem. Soc., Chem. Commun.* **1990**, 1084–1085.

[169] K. Rengan, R. Engel, "The Synthesis of Phosphonium Cascade Molecules", *J. Chem. Soc., Perkin Trans. 1* **1991**, 987–990.

[170] R. Engel, K. Rengan, C.-S. Chan, "New Cascade Molecules Centered about Phosphorus", *Heteroat. Chem.* **1993**, *4*, 181–184.

[171] A. W. van der Made, P. W. N. M. van Leeuwen, "Silane Dendrimers", *J. Chem. Soc., Chem. Commun.* **1992**, 1400–1401.

[172] A. W. van der Made, P. W. N. M. van Leeuwen, J. C. de Wilde, R. A. C. Brandes, "Dendrimeric Silanes", *Adv. Mater. (Weinheim, Fed. Repub. Ger.)* **1993**, *5*, 466–468.

[173] R. van Heerbeek, J. N. H. Reek, P. C. J. Kamer, P. W. N. M. van Leeuwen, "Divergent Synthesis of Carbosilane Wedges as Dendritic Building Blocks: A New Strategy Towards Core Functionalized Carbosilane Dendrimers", *Tetrahedron Lett.* **1999**, *40*, 7127–7130.

[174] M. C. Coen, K. Lorenz, J. Kressler, H. Frey, R. Mülhaupt, "Mono- and Multilayers of Mesogen-Substituted Carbosilane Dendrimers on Mica", *Macromolecules* **1996**, *29*, 8069–8076.

[175] C. Schlenk, H. Frey, "Carbosilane Dendrimers – Synthesis, Functionalization, Application", *Monatsh. Chem.* **1999**, *130*, 3–14.

[176] D. Seyferth, D. Y. Son, A. L. Rheingold, R. L. Ostrander, "Synthesis of an Organosilicon Dendrimer Containing 324 Si–H Bonds", *Organometallics* **1994**, *13*, 2682–2690.

[177] L.-L. Zhou, J. Roovers, "Synthesis of Novel Carbosilane Dendritic Macromolecules", *Macromolecules* **1993**, *26*, 963–968.

[178] H. Frey, C. Schlenk, "Silicon-Based Dendrimers", *Top. Curr. Chem.* **2000**, *210*, 69–129.

[179] H. Frey, K. Lorenz, R. Mülhaupt, U. Rapp, F. J. Mayer-Posner, "Dendritic Polyols Based on Carbosilanes – Lipophilic Dendrimers with Hydrophilic Skin", *Macromol. Symp.* **1996**, *102*, 19–26.

[180] D. Seyferth, "Carbosilane Dendrimers: Molecules with Many Possibilities", Proceedings of the 50th Anniversary Conference of the Korean Chemical Society, Seoul, Korea, **1996**.

[181] B. D. Karstedt, "Platinum Complexes of Unsaturated Siloxanes and Platinum-Containing Organopolysiloxanes", *U. S. Pat.* **1973**, 3,775,452.

[182] K. Lorenz, R. Mülhaupt, H. Frey, U. Rapp, F. J. Mayer-Posner, "Carbosilane-Based Dendritic Polyols", *Macromolecules* **1995**, *28*, 6657–6661.

[183] H. Frey, K. Lorenz, D. Hölter, R. Mülhaupt, "Mesogen-Functionalized Carbosilane Dendrimers – Dendritic Liquid-Crystalline Polymers", *Polym. Prepr.* **1996**, *37*, 758–759.

[184] J. B. Lambert, J. L. Pflug, C. L. Stern, "Synthesis and Structure of a Dendritic Polysilane", *Angew. Chem.* **1995**, *107*, 106–107; *Int. Ed. Engl.* **1995**, *34*, 98–99.

[185] J. B. Lambert, J. L. Pflug, J. M. Denari, "First-Generation Dendritic Polysilanes", *Organometallics* **1996**, *15*, 615–625.

[186] J. B. Lambert, E. Basso, N. Qing, S. H. Lim, J. L. Pflug, "Two-Dimensional Silicon-29 INADEQUATE as a Structural Tool for Branched and Dendritic Polysilanes", *J. Organomet. Chem.* **1998**, *554*, 113–116.

[187] J. B. Lambert, H. Wu, "Synthesis and Crystal Structure of a Nanometer-Scale Dendritic Polysilane", *Organometallics* **1998**, *17*, 4904–4909.

[188] J. B. Lambert, H. Wu, "Atom Connectivity and Spectral Assignments from the ^{29}Si-^{29}Si INADEQUATE Experiment on a Nanometer Scale Dendritic Polysilane", *Magn. Reson. Chem.* **2000**, *38*, 389.

[189] H. Suzuki, Y. Kimata, S. Satoh, A. Kuriyama, "Polysilane Dendrimer. Synthesis and Characterization of [2,2-(Me$_3$Si)$_2$Si$_3$Me$_5$]$_3$SiMe", *Chem. Lett.* **1995**, 293–294.

[190] C. Kim, E. Park, E. Kang, "Preparation of Organosilane Dendrimer Containing Allyl End Groups", *Bull. Korean Chem. Soc.* **1996**, *17*, 419–424.

[191] C. Kim, A. Kwon, "Silane Arborols; XV: A Dendritic Carbosilane Based on Siloxane Tetramers", *Synthesis* **1998**, 105–108.

[192] Y. Chang, Y. C. Kwon, S. C. Lee, C. Kim, "Amphiphilic Linear PEO-Dendritic Carbosilane Block Copolymers", *Macromolecules* **2000**, *33*, 4496–4500.

[192a] Y. Chang, C. Kim, "Synthesis and Photophysical Characterization of Amphiphilic Dendritic-Linear-Dendritic Block Copolymers", *J. Polym. Sci., Part A: Polym. Chem.* **2001**, *39*, 918–926.

[193] G. Friedmann, Y. Guilbert, J.-C. Wittmann, "Dendrimères à Structure Superficielle Aromatique: Synthèse, Caractérisation, Propriétés: Mise en Évidence d'une Structure Cristalline", *Eur. Polym. J.* **1997**, *33*, 419–426.

[194] G. Friedmann, Y. Guilbert, J. C. Wittmann, "Crystalline Dendritic Arylalkylsilane/Tetrahydrofuran Inclusion Complex", *Eur. Polym. J.* **1999**, *35*, 1097–1105.

[195] N. Ouali, S. Méry, A. Skoulios, L. Noirez, "Backbone Stretching of Worm-like Carbosilane Dendrimers", *Macromolecules* **2000**, *33*, 6185–6193.

[196] A. G. Vitukhnovsky, M. I. Sluch, V. G. Krasovskii, A. M. Muzafarov, "Study of Polyallylcarbosilane Dendrimer Structure by Pyrenyl Excimer Formation", *Synth. Metals* **1997**, *91*, 375–377.

[197] S. S. Sheiko, A. M. Muzafarov, R. G. Winkler, E. V. Getmanova, G. Eckert, P. Reineker, "Contact Angle Microscopy on a Carbosilane Dendrimer with Hydroxy End Groups: Method for Mesoscopic Characterization of the Surface Structure", *Langmuir* **1997**, *13*, 4172–4181.

[198] K. Lorenz, H. Frey, B. Stühn, R. Mülhaupt, "Carbosilane Dendrimers with Perfluoroalkyl End Groups. Core-Shell Macromolecules with Generation-Dependent Order", *Macromolecules* **1997**, *30*, 6860–6868.

[199] B. Stark, B. Stühn, H. Frey, C. Lach, K. Lorenz, B. Frick, "Segmental Dynamics in Dendrimers with Perfluorinated End Groups: A Study Using Quasielastic Neutron Scattering", *Macromolecules* **1998**, *31*, 5415–5423.

[200] B. Trahasch, B. Stühn, H. Frey, K. Lorenz, "Dielectric Relaxation in Carbosilane Dendrimers with Perfluorinated End Groups", *Macromolecules* **1999**, *32*, 1962–1966.

[201] C. Lach, D. Brizzolara, H. Frey, "Molecular Force Field Study Concerning the Host Properties of Carbosilane Dendrimers", *Macromol. Theory Simul.* **1997**, *6*, 371–380.

[202] P.-A. Jaffrès, R. E. Morris, "Synthesis of Highly Functionalised Dendrimers Based on Polyhedral Silsesquioxane Cores", *J. Chem. Soc., Dalton Trans.* **1998**, 2767–2770.

[203] K. Matsuoka, M. Terabatake, Y. Saito, C. Hagihara, Y. Esumi, D. Terunuma, H. Kuzuhara, "Synthesis of Carbosilane Compounds Functionalized with Three or Four β-Cyclodextrin Moieties. Use of a One-Pot Reaction in Liquid Ammonia for Birch Reduction and the Subsequent S_N2 Replacement", *Bull. Chem. Soc. Jpn.* **1998**, *71*, 2709–2713.

[204] J. W. Kriesel, T. D. Tilley, "Dendrimers as Building Blocks for Nanostructured Materials: Micro- and Mesoporosity in Dendrimer-Based Xerogels", *Chem. Mater.* **1999**, *11*, 1190–1193.

[205] J. W. Kriesel, T. D. Tilley, "Hybrid Organic–Inorganic Gels Based on Dendrimeric Building Blocks: High Surface Area Xerogels as Catalyst Supports", *Polym. Prepr.* **2000**, *41*, 566–567.

[206] K. G. Sharp, M. J. Michalczyk, "Star Gels: New Hybrid Network Materials from Polyfunctional Single Component Precursors", *J. Sol-Gel Sci. Technol.* **1997**, *8*, 541–546.

[207] B. Boury, R. J. P. Corriu, R. Nuñez, "Hybrid Xerogels from Dendrimers and Arborols", *Chem. Mater.* **1998**, *10*, 1795–1804.

[208] J. W. Kreisel, T. D. Tilley, "Synthesis and Chemical Functionalization of High Surface Dendrimer-Based Xerogels and Their Use as New Catalyst Supports", *Chem. Mater.* **2000**, *12*, 1171–1179.

[209] K. Matsuoka, M. Terabatake, Y. Esumi, D. Terunuma, H. Kuzuhara, "Synthetic Assembly of Trisaccharide Moieties of Globtriaosyl Ceramide using Carbosilane Dendrimers as Cores. A New Type of Functional Glyco-Material", *Tetrahedron Lett.* **1999**, *40*, 7839–7842.

[210] R. A. Gossage, E. Muñoz-Martínez, G. van Koten, "Synthesis of a Key Reactive Unit for Use in the Divergent or Convergent Synthesis of Carbosilane Dendrimers", *Tetrahedron Lett.* **1998**, *39*, 2397–2400.

[210a] N. J. Hovestad, A. Ford, J. T. B. H. Jastrzebski, G. van Koten, "Functionalized Carbosilane Dendritic Species as Soluble Supports in Organic Synthesis", *J. Org. Chem.* **2000**, *65*, 6338–6344.

[210b] E. B. Eggeling, N. J. Hovestad, J. T. B. H. Jastrzebski, D. Vogt, G. van Koten, "Phosphino Carboxylic Acid Ester Functionalized Carbosilane Dendrimers: Nanoscale Ligands for the Pd-Catalyzed Hydrovinylation Reaction in a Membrane Reactor", *J. Org. Chem.* **2000**, *65*, 8857–8865.

[210c] A. W. Kleij, R. van de Coevering, R. J. M. K. Gebbink, A.-M. Noordman, A. L. Spek, G. van Koten, "Polycationic (Mixed) Core – Shell Dendrimers for Binding and Delivery of Inorganic/Organic Substrates", *Chem. Eur. J.* **2001**, *7*, 181–192.

[210d] C. Schlenk, A. W. Kleij, H. Frey, G. van Koten, "Macromolecular-Multisite Catalysts Obtained by Grafting Diaminoaryl Palladium(II) Compllexes onto a Hyperbranched-Polytriallylsilane Support", *Angew. Chem. Int. Ed.* **2000**, *39*, 3445–3447.

[210e] A. W. Kleij, R. A. Gossage, R. J. M. K. Gebbink, N. Brinkmann, E. J. Reijerse, U. Kragl, M. Lutz, A. L. Spek, G. van Koten, "A "Dendritic Effect" in Homogeneous Catalysis with Carbosilane-Supported Arylnickel(II) Catalysts: Observation of Active-Site Proximity Effects in Atom-Transfer Radical Addition", *J. Am. Chem. Soc.* **2000**, *122*, 12112–12124.

[210f] L. Ropartz, R. E. Morris, G. P. Schwarz, D. F. Foster, D. J. Cole-Hamilton, "Dendrimer-bound tertiary phosphines for alkene hydroformylation", *Inorg. Chem. Commun.* **2000**, *3*, 714–717.

[211] O. L. Chapman, J. Magner, R. Ortiz, "Polycules", *Polym. Prepr.* **1995**, *36*, 739–740.

[212] M. Dotrong, M. H. Dotrong, G. J. Moore, R. C. Evers, "Three-Dimensional Benzobisoxazole Rigid-Rod Polymers", *Polym. Prepr.* **1994**, *35*, 673–674.

[213] V. R. Reichert, L. J. Mathias, "Rigid-Expanded Tetrahedral Cores for Four-Armed Branched Structures: 1,3,5,7-Tetrakis(4-iodophenyl)adamantane and its Derivatives", *Polym. Prepr.* **1993**, *34*, 495–496.

[214] V. R. Reichert, J. P. Mathias, "Expanded Tetrahedral Molecules from 1,3,5,7-Tetraphenyl-adamantane", *Macromolecules* **1994**, *27*, 7015–7023.

[215] O. Mongin, A. Gossauer, "Synthesis of Nanometer-sized Homo- and Heteroorganometallic Tripodaphyrins", *Tetrahedron* **1997**, *53*, 6835–6846.

[216] W. J. Oldham, Jr., R. J. Lachicotte, G. C. Bazan, "Synthesis, Spectroscopy, and Morphology of Tetrastilbenoidmethanes", *J. Am. Chem. Soc.* **1998**, *120*, 2987–2988.

[217] S. Sengupta, S. K. Sadhukhan, "Four-fold Heck Reactions on the Tetraphenylmethane Tripod: Model Studies Towards Construction of Centrally Based Three Dimensional Networks", *Tetrahedron Lett.* **1998**, *39*, 1237–1238.

[218] S. Wang, W. J. Oldham, Jr., R. A. Hudack, Jr., G. C. Bazan, "Synthesis, Morphology, and Optical Properties of Tetrahedral Oligo(phenylenevinylene) Materials", *J. Am. Chem. Soc.* **2000**, *122*, 5695–5709.

[219] D. A. Scott, T. M. Krülle, M. Finn, R. J. Nash, A. L. Winters, N. Asano, T. D. Butters, G. W. J. Fleet, "Synthesis of Bis(1,3-dihydroxy-isopropyl)amine by Reductive Amination of Dihydroxyacetone: Open Chain Equivalent of DMDP and a Potential AB$_4$ Dendritic Monomer", *Tetrahedron Lett.* **1999**, *40*, 7581–7584.

[220] D. A. Scott, T. M. Krülle, M. Finn, G. W. J. Fleet, "Bis(1,3-dihydroxy-isopropyl)amine (BDI) as an AB$_4$ Dendritic Building Block: Rapid Synthesis of a Second Generation Dendrimer", *Tetrahedron Lett.* **2000**, *41*, 3959–3962.

[221] F. Sournies, F. Crasnier, C. Vidal, M.-C. Labarre, J.-F. Labarre, "On the Scent of Dandelion Dendrimers; Part IV. Molecular Modeling of Cyclophosphazenic Dandelion Dendrimers Up to the Fourth Generation", *Main Group Chemistry* **1996**, *1*, 207–218.

[222] J.-F. Labarre, F. Sournies, F. Crasnier, M.-C. Labarre, C. Vidal, J.-P. Faucher, M. Graffeuil, "On the Scent of Spherical Dendrimers: Cyclophosphazenic Dandelion Dendrimers up to the Eighth Generation", *Phosphorus, Sulfur Silicon Relat. Elem.* **1996**, *109–110*, 525–528.

[223] F. Sournies, K. Zai, K. Vercruysse, M.-C. Labarre, J.-F. Labarre, "On the Scent of Asymmetrical Dandelion Dendrimers; Part 1. Starting Materials from the Regiospecific Aminolysis of N$_3$P$_3$Cl$_6$ by Amino-Alcohols", *J. Mol. Struct.* **1997**, *412*, 19–26.

[224] J. Y. Chang, M. J. Han, "Star-Branched and Cross-linked Polymers (Cyclotriphosphazene Cores)", in *Polymeric Materials Encyclopedia* (Ed.: J. C. Solamone), CRC Press, Boca Raton, FL, **1996**, pp. 7873–7880.

5 Convergent Methodologies

5.1 General Concepts

The "convergent" mode of dendritic construction is an alternative strategy to that of the divergent procedure whereby branched polymeric arms (dendrons) are synthesized from the "outside-in". This concept (Scheme 5.1), initially described by Fréchet and his coworkers[1] and shortly thereafter by Miller and Neenan,[2] can best be illustrated, as well as reviewed,[3–8] by envisaging the attachment of two terminal units (**1**) containing one reactive group X to one monomer (**2**) possessing protected functionality Z resulting in the preparation of the 1st generation or tier (e.g., **3**). Transformation of the unique focal site (Z → X), followed by treatment with ½ equivalent of the masked monomer **2** affords the next higher generation (e.g., **4**). A notable advantage of this procedure is the requirement of a minimum number of transformations for tier construction. Thus, using a three-directional building block, only two reactions (of Y at X) need to be effected in order to attach each consecutive generation. In general, the focal site bonding the wedge is engulfed in the infrastructure with each successive tier attachment; thus, at some point of wedge development, chemical connectivity to a core will become impossible, or very difficult, due to the juxtaposed steric interference. On the other hand, purification becomes less problematic than in the divergent case due to larger byproducts and the smaller number of reactions required.

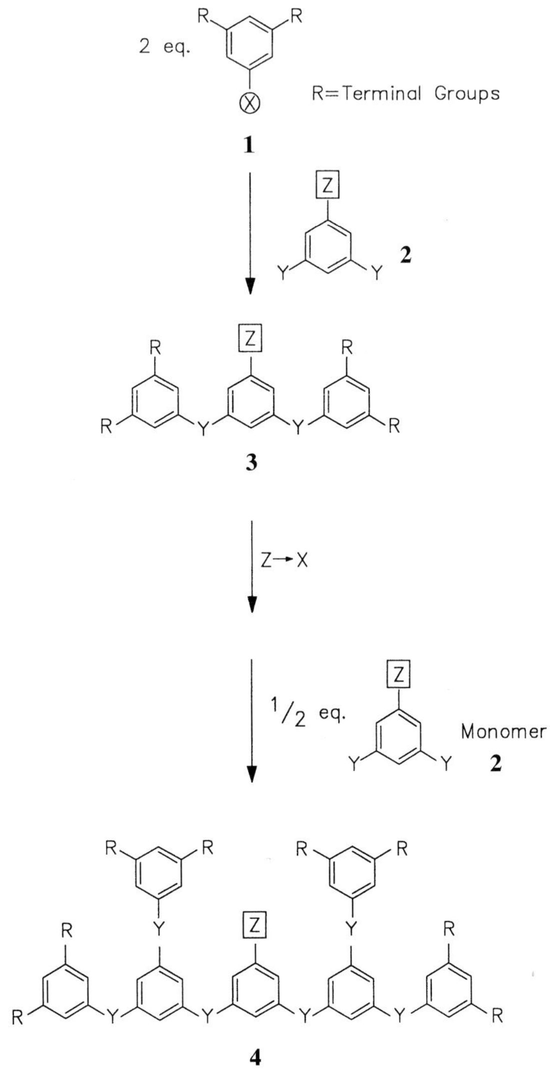

Scheme 5.1 Illustration of the general concept of convergent synthesis.

5.2   1 → 2 Branched

5.2.1   1 → 2 *C*-Branched and Connectivity

Rajca[9–12] reported the construction of dendritic triphenylmethyl polyradicals and polyanions; his work was aimed at the examination of high-spin organomagnetic materials,[13–18] and it has been reviewed.[19–21] An example of this 1 → 2 *C*-branched iterative methodology (Scheme 5.2) starts with the monolithiation of 1,3-dibromobenzene (**5**), followed by the addition of 4,4'-di-*tert*-butylbenzophenone (**6**); subsequent quenching [EtOC(O)Cl, EtOH] afforded the terminal unit **7**. Subsequent metal–halogen exchange and reaction with 3-bromo-4'-*tert*-butylbenzophenone (**8**), performed twice, followed by ethoxylation generated the desired oligomeric, linear arms. Further metallation of hepta-

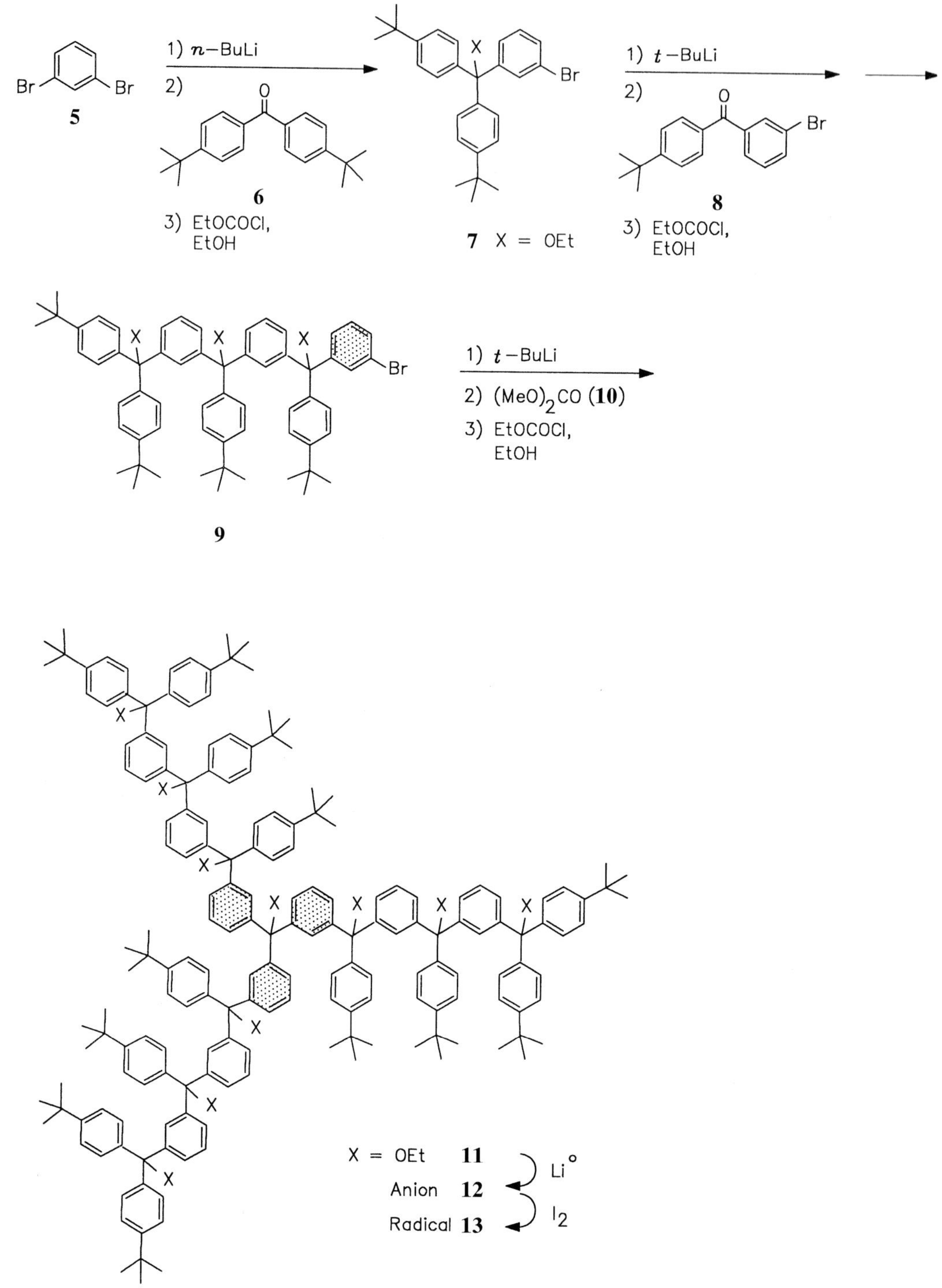

Scheme 5.2 Preparation of poly(arylmethyl) macromolecules.[9–12]

phenylenebromide **9** and addition of the core **10** was followed by treatment with EtOC(O)Cl and EtOH to give poly(arylmethyl)decaether **11**. The corresponding deca-anion **12** was generated (Li°, THF) from **11** and its subsequent oxidation (I$_2$) provided the related decaradical **13**, which was supported by ESR and magnetic susceptibility data.[22] Further studies[23] of these macromolecules with respect to the "topological control of electron localization in π-conjugated carbopolyanions and radical anions" demonstrated the "relationship between high spin in polyradicals, polyanion charge distribution, and electron localization in the radical anions". 1,3-Bridging benzene units result in weaker coupling between arylmethyl moieties than with 1,4-benzenoid coupling; thus, controlled molecular construction effects electron localization. Magnetic interactions of the corresponding polyarylmethyl triplet diradicals were also studied.[24–27] Inoue et al.[28] have studied trinitroxide radicals stabilized by Mn coordination that behave as molecule-based magnets (T_c = 46 K).

Rajca and Utamapanya extended[29] their iterative procedures to include the synthesis (Scheme 5.3) and characterization of these dendrimers possessing 15 and 31 centers of unpaired electrons. Their convergent synthesis relied on the iterative application of aryl lithiation via metal–halogen exchange (*tert*-BuLi), addition of an alkyl benzoate (**14**; methyl 5-bromo-2,4-dimethylbenzoate), and etherification (MeI). Thus, pentadecaether **16** (X = Y = OMe) was prepared by the reaction of two equivalents of heptamethyl ether **15** (X = OMe) with methyl 4-*tert*-butylbenzoate and subsequent methylation of the

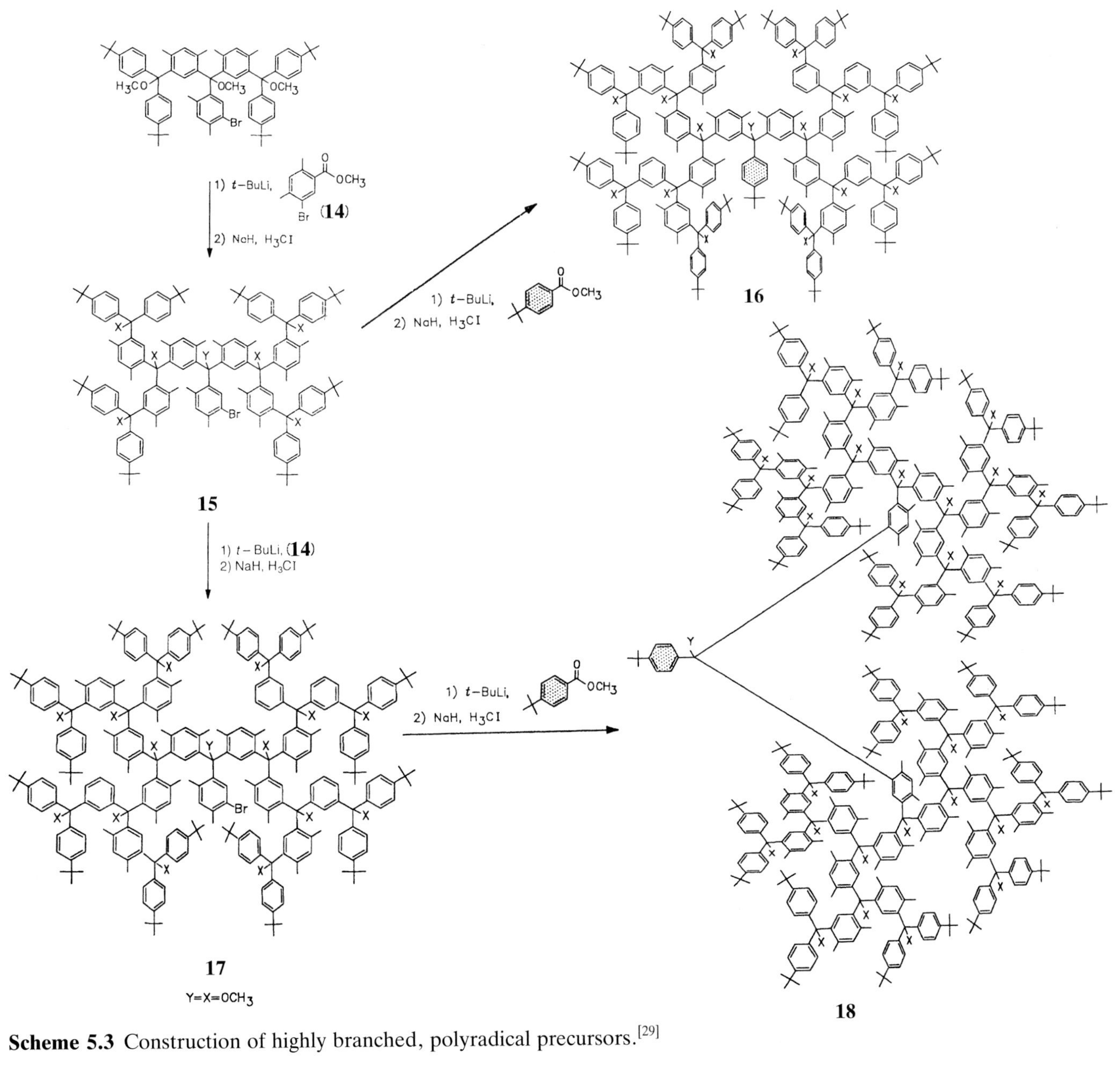

Scheme 5.3 Construction of highly branched, polyradical precursors.[29]

resulting central hydroxyl moiety. Treatment of two equivalents of metallated aryl bromide **15** with benzoate **14** led to monobromopentadecaether **17**, which was then lithiated and added to methyl *tert*-butylbenzoate to afford the 31-methyl ether **18**. Syntheses, based on repetitive addition of aryllithiums to carbonyl precursors, of a related series of sterically hindered 1,3-connected tri-, tetra-, and heptaarylmethanes have been reported.[30, 31]

Synthesis of the closely related acyclic and macrocyclic polyradicals has been described.[32] The π-conjugated carbanions (e.g., the calix[4]arene-based tetraanion and the related calix[3]arene-based trianion) were synthesized and studied.[33, 34] Oxidation of these tetra- and tri-anions gave the corresponding tetra- and tri-radicals, respectively. It has been shown[35] in closely related systems that it is not the shape or overall geometric symmetry of the molecules, but rather the juxtaposition of the carbenic centers within the π-cross-conjugated structure that is most important in determining the spin multiplicity of the alternate hydrocarbon molecule. Matsuda and Iwamura[36] have shown that the [3-(*N*-ylooxy-*tert*-butylamino)-5-*tert*-butylphenyl]phenylcarbene monomer has potential in syntheses of related carbene networks. Rajca et al.[37] subsequently repeated the synthesis of another macrocyclic-based organic polyradical possessing a high spin $S = 10$, which was still lower than the $S = 12$ value predicted for 24 ferromagnetically-coupled unpaired electrons; defects resulting from less than 100 % radical generation accounted for the discrepancy.

The preparation of hexaterphenylyl- and hexaquaterphenylylbenzene has been reported;[38] a Pd-catalyzed cyclotrimerization and multiple Suzuki couplings were key reactions in route to these interesting molecules. A single-pot Heck reaction of 2-bromo-4-(acetoxyphenyl)styrene with either 1,3,5-triiodobenzene or 1,3,5-tris(3',5'-diiodophenyl)benzene has been employed[39] to produce star-shaped polyphenoxyl radicals; a hydrolysis, phenolate formation, and oxidation sequence afforded the desired high-spin polyradical. A novel phenylazomethine is also known.[39a] [39b]

5.2.2   1 → 2 *C*-Branched, *Alkyne* Connectivity

Fomina et al.[40] prepared a series of rigid, fully conjugated branched molecules predicated on silylated β,β-dibromo-4-ethynylstyrene[41] (**19**; Scheme 5.4); pertinent synthetic aspects included a tandem Heck reaction of the terminal aryl alkyne to give tetraaldehyde **21** and silyl group removal allowing repetition of the sequence. Quantum-mechanical calculations[42] have been conducted on these β,β-dibromo-4-ethynylstyrene oligomers, which led to the conclusion that the dendritic structure contributes little to the instability and conjugation disruption compared to 1 → 2 branched polyacetylene.[43]

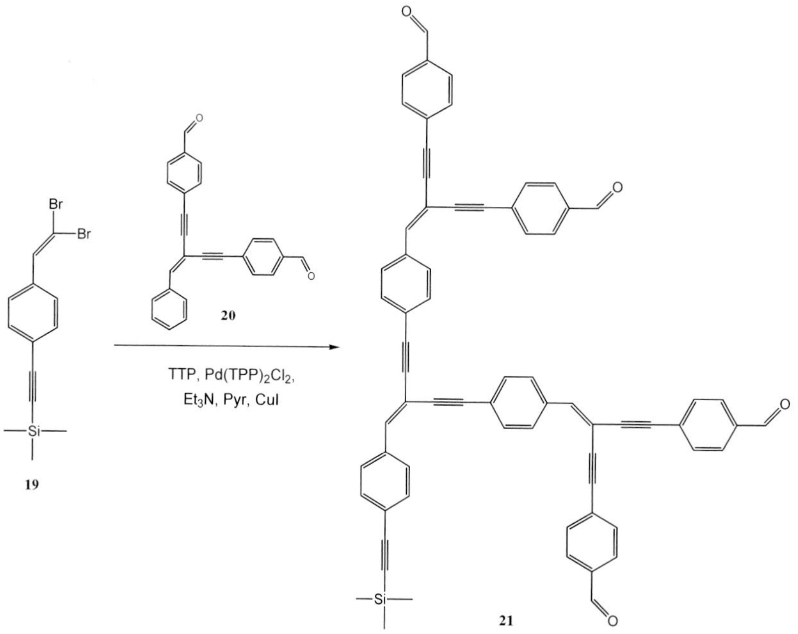

Scheme 5.4 Fomina's Heck-based monomer addition.[40]

Molecular modeling of these branched and the related linear polyacetylenes showed the dendrimers to be less stable than *trans*-polyacetylene oligomers and that the instability increases with molecular mass.[41] Solid support synthesis of hyperbranched analogs has also been studied.[41a]

5.2.3 1 → 2 *C*-Branched, *Ether* Connectivity

Wooley, Hawker, and Fréchet[44, 45] described a 'double-stage convergent growth' approach that led to dendrimers comprised of different monomers at different generations. Dendrons (e.g., **25**) were convergently prepared (Scheme 5.5) using 4,4-bis(4'-hydroxyphenyl)pentanol (**23**), which was obtained from the corresponding pentanoic acid (**22**) by four simple steps. Reaction of two equivalents of bis(benzyl)-protected

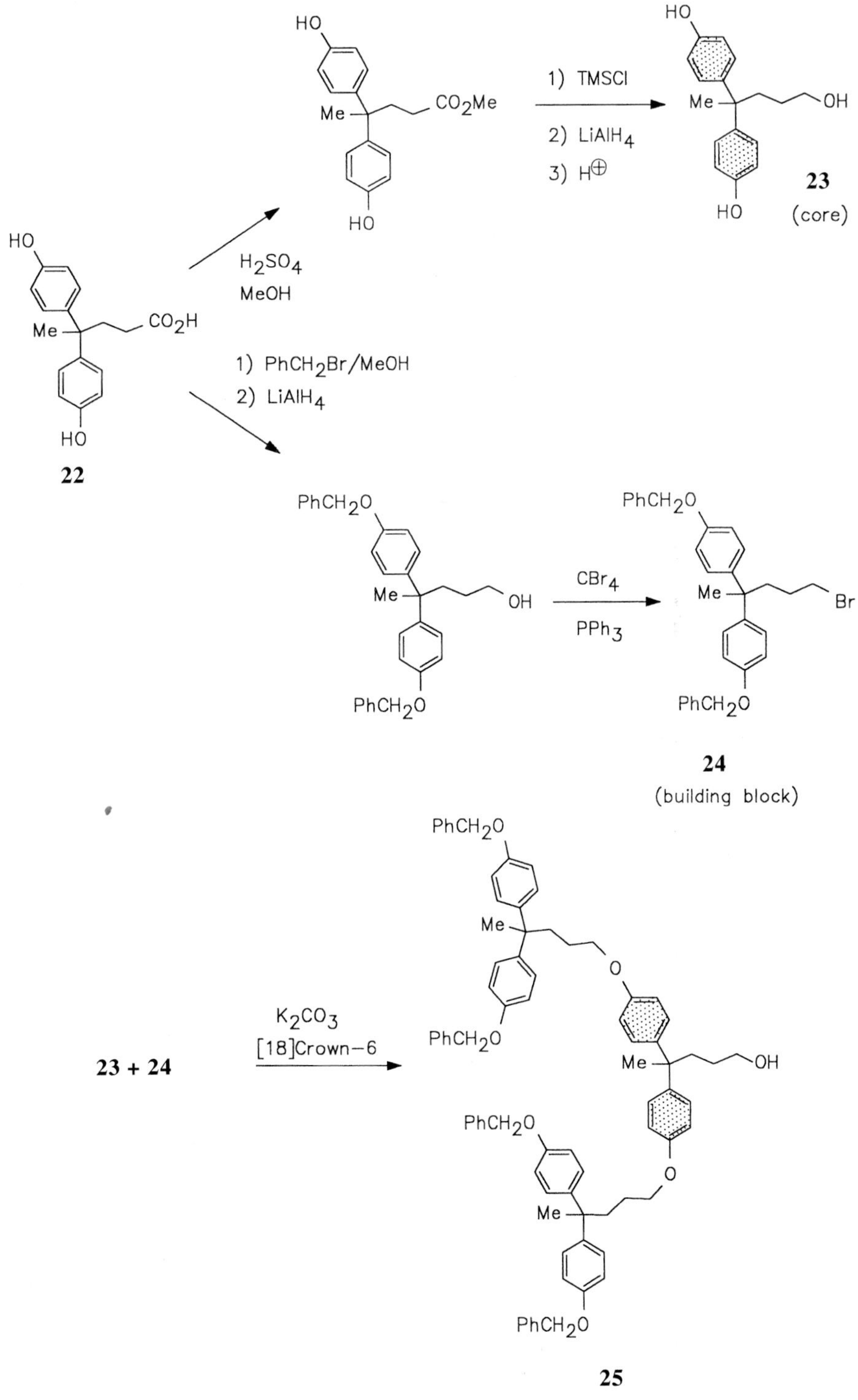

Scheme 5.5 Synthesis of dendritic "wedges" (**25**)[44] used in "double-stage convergent growth".

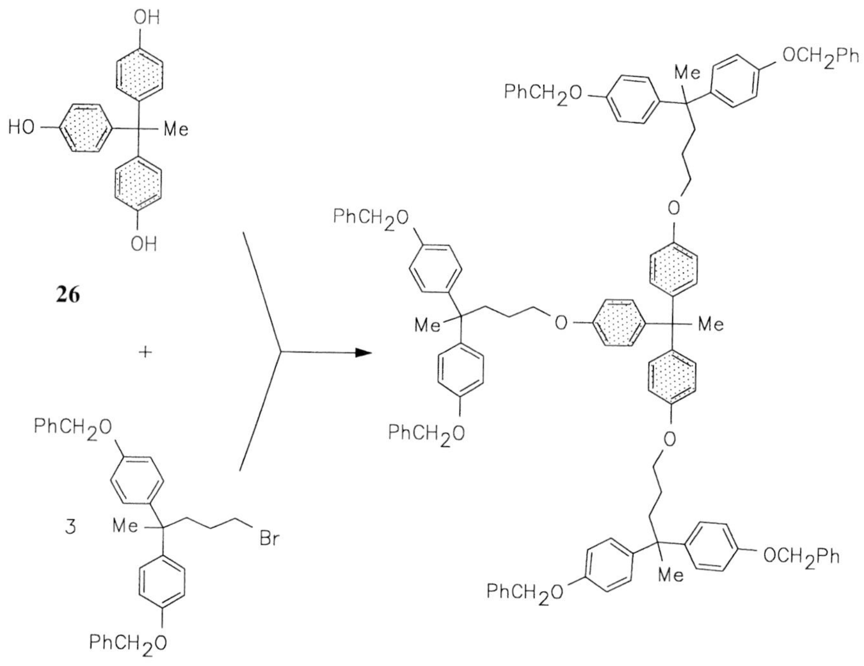

Scheme 5.6 Construction of flexible "hypercore" precursor **27**.[44]

monomer **24**, prepared from the same common acid **22**, with bis(hydroxyphenyl)penta-nol **23** afforded the new 1 → 2 wedge **25**. Using these monomers, 'hypercores' up to the 3rd generation were constructed. The simplest flexible core, **27**, was prepared (Scheme 5.6) by treatment of alkyl bromide **24** with the three-directional, tris(phenolic) core **26**. After debenzylation (Pd/C, cyclohexene) of **27**, the hexaphenolic core **28** was treated (Scheme 5.7) with aryl-branched, benzyl ether dendrons such as **29**, that were prepared by similar iterative transformations,[46] i.e., benzylic bromination and phenolic *O*-alkylation (see Section 5.3.2), to give, in this case, the benzyloxy-terminated dendrimer **30**. Key features of these dendrimers include cores with flexible alkyl spacers and a three-directional, quaternary carbon branching center. Such use of two different monomers in this convergent approach to a species possessing a nominal M_w of *ca.* 84,200 amu has been realized. Variations of T_g with molecular weight and chain-end composition for related dendritic polyethers (as well as polyesters) were studied.[47] Substituted unimole-cular micelles with a hydrophobic core surrounded by a hydrophilic shell were prepared by coupling the appropriate hypercore with PEG mesylate;[48] encapsulation of indo-methacin was achieved and supports their potential as drug delivery systems.

Chen and Gorman[49] described the synthesis of focally-modified aryl thiol-based dendrons (e.g., **31**; Figure 5.1) based on this cascade series. Wedge construction was facilitated by modification of the original[44] protocol (Scheme 5.5) whereby an (alkyl) mesylate, instead of a benzylic bromide, was reacted with a phenol in the presence of 18-crown-6, Me$_2$CO, and K$_2$CO$_3$. Formation of the desired thiol monomers was effected by conversion of the focal alcohols (for generations 1–4) to mesylate groups, reaction (DMSO, 60–70 °C) with H$_2$NMe, acid chloride coupling of a protected aryl thiol, and thiol deprotection (NaOH). Aryl thiol formation employed, as a key step, rearrange-ment of a dimethylthiocarbamate. These materials were envisaged as liquids for hybrid organic–inorganic clusters, as surface coatings (i.e., for gold-thiol coordination), and as appendages for FeS clusters.[50]

Yamamoto et al.[51] reported the convergent preparation (Scheme 5.8) of a simple one-directional, two-tier cascade (**35**) for the purpose of solubilizing a terminal *o*-carborane moiety (1,2-RHC$_2$B$_{10}$H$_{10}$) in an aqueous environment. The dibenzyl ether **33** was reacted with epichlorohydrin (**32**) to afford the 2nd tier tetrabenzyl ether (**34**) possessing 1 → 2 *C*-branching centers. Subsequent treatment with propargyl bromide gave the terminal acet-ylene, which was then treated with decaborane (B$_{10}$H$_{14}$/MeCN) and deprotected [Pd(OH)$_2$] to provide the water-soluble, terminal *o*-carborane cascade **35**. Later modifi-

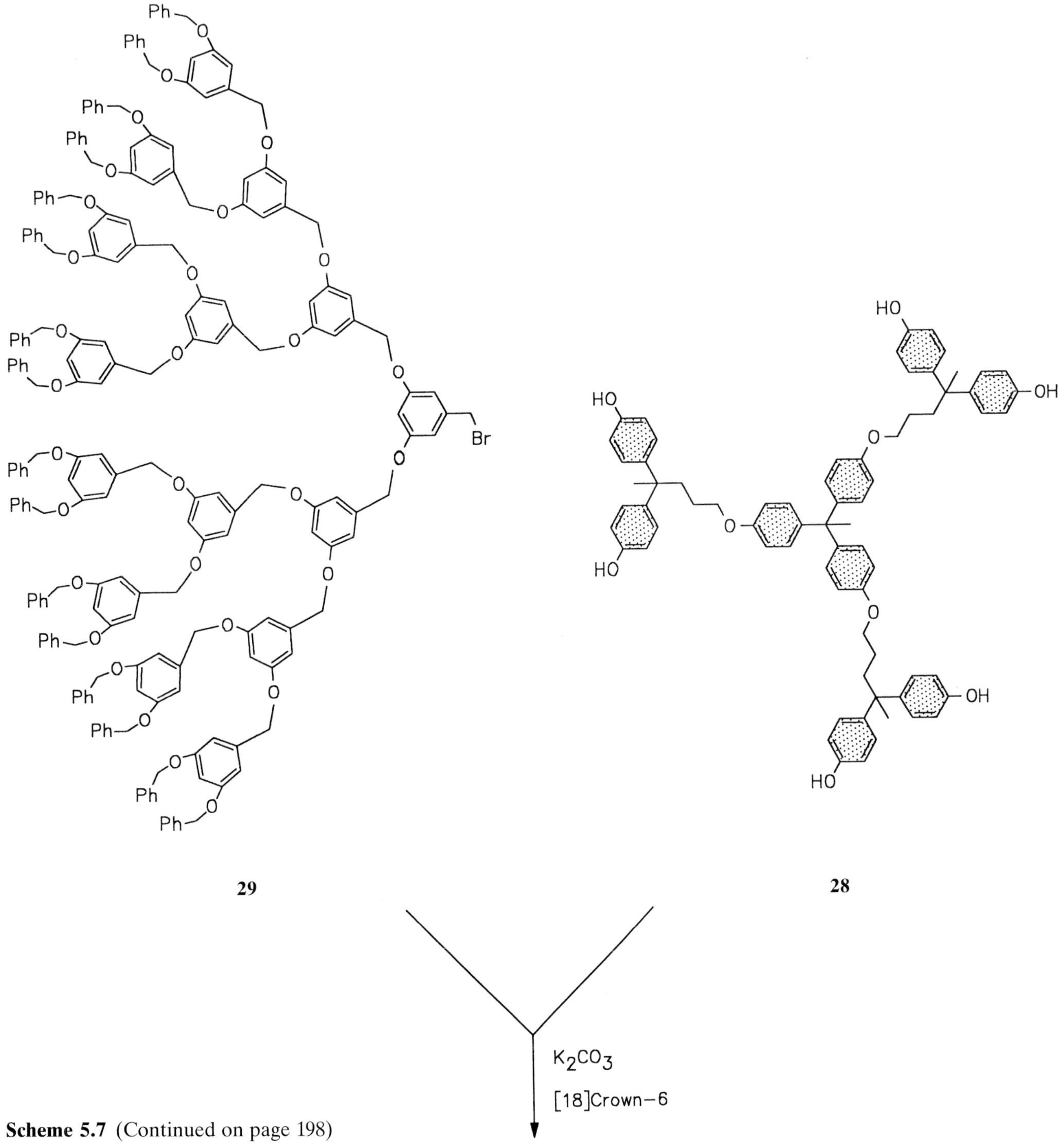

29

28

K₂CO₃

[18]Crown−6

Scheme 5.7 (Continued on page 198)

cations of the water-soluble carborane included the attachment of a tumor-seeking uracil moiety (Scheme 5.9)[52] as well as netropsin and distamycin A analogues. Key transformations allowing the synthesis of this carborane **39** included construction of masked uracil allyl carbonate **36** and its subsequent connection {Pd(dba)₂}, dppe to the benzyl protected *o*-carborane cascade **37**, derived from **35**, and finally catalytic debenzylation of **38**.

Percec et al.[53] employed 1 → 2 *C*-branching and ethereal-type connectivity for the preparation of nonspherical, thermotropic liquid-crystalline dendrimers. Their observed thermotropic behavior is predicated on mesogenic monomers that are capable of conformational isomerism. Monomer preparation (Scheme 5.10) began with a Suzuki cross-coupling of aryl bromide **40** with boronic acid **41** to give the bis(methoxy) ketone **42**; reduction (LAH) followed by phenol deprotection then afforded the trihydroxyl mesogenic monomer **43**.

Thus, the selective elaboration (Scheme 5.11) of the phenolic moieties of **43** with 1-bromo-10-undecene (10 N NaOH, TBAH, *o*-DCB) gave the diene **44**, which with CBr₄

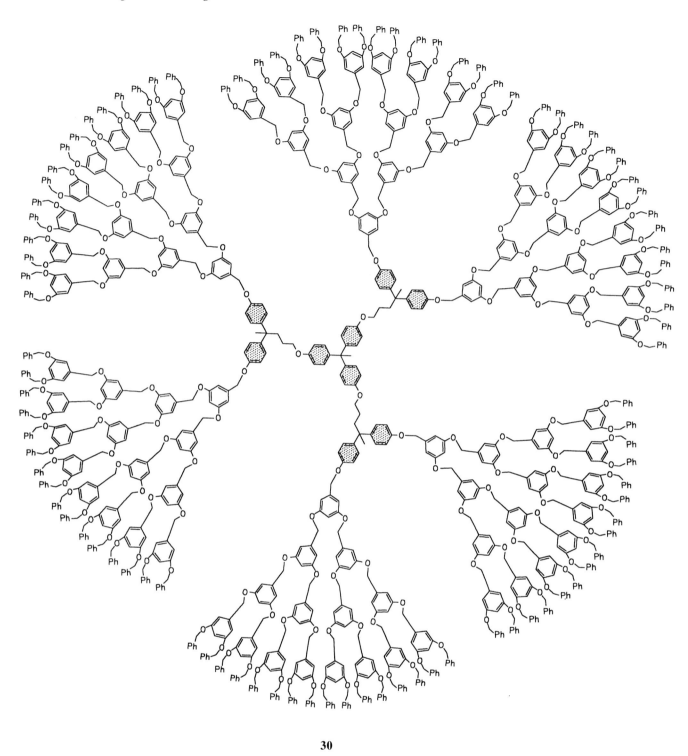

30

Scheme 5.7 Attachment of (polybenzyl ether) dendrons[46] to flexible hypercores (**28**) results in "dendritic block copolymers" possessing monomer differentiated generations.

and PPh₃ afforded the focally-substituted bromide **45**. Treatment of **45** with 0.5 equivalent of triol **44** and subsequent bromination, again at the focal position, gave the 2nd generation monomer **47**. This sequence was repeated to afford the larger tiered **48**. Dendrimers were then prepared from each generation of monomer in a manner similar to that shown in Scheme 5.12, in which monomer **48** was treated (DMAP) with 1,3,5-tris(chlorocarbonyl)benzene (**49**) to afford alkene-terminated dendrimer **50**.

Each monodendron, as well as each dendrimer, was characterized by DSC and optical polarized microscopy for thermal property analysis. Both dendrons and dendrimers were found to display thermotropic cybotactic nematic, smectic, and crystalline phases as a direct result of the *anti*-conformation of the monomeric units. Trends in transition temperature dependence on molecular weight or generation number were found to be analo-

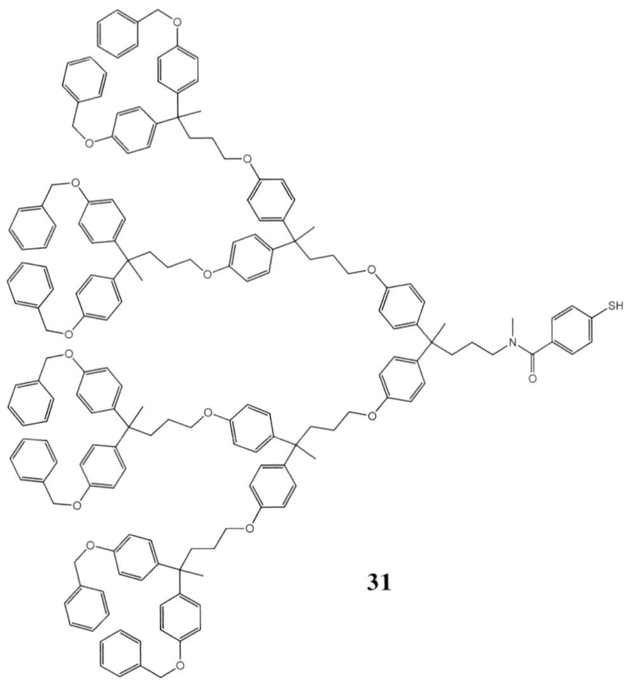

Figure 5.1 Thiol focal group on an ethereal dendron.[49]

gous to those for linear mesogenic liquid-crystalline polymers; however, the neumatic phase viscosity for the dendrimers was found to be lower in comparison to the corresponding hyperbranched[54] and linear series.[53] These were the first examples of branched macromolecules exhibiting conventional calorimetric cybotactic nematic and smectic crystalline phases.

The isotropic-nematic pretransitional behavior of these liquid crystals has been examined.[55] Nematic wetting and fluctuations were probed using Kerr measurements, ellipsometry, and quasielastic light scattering. Pretransitional behavior was qualitatively observed to be similar to that of liquid crystals possessing low molecular weights. Monodendron transition temperatures were generally found to be elevated 0.5 K above the supercooling limit of the isotropic phase; orientational viscosities were fitted to an M_n power law for an exponent ≤ 1.0. For both the dendrons and dendrimers in the isotropic phase, nematic fluctuations showed a Landau-like mean field behavior.

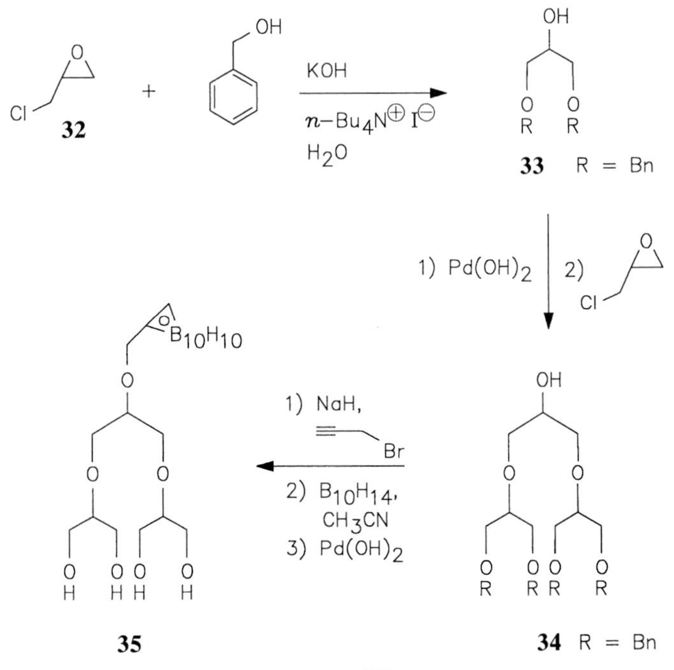

Scheme 5.8 Synthesis of water-soluble *o*-carboranes (**35**)[51] for Boron Neutron Capture Therapy reagents.

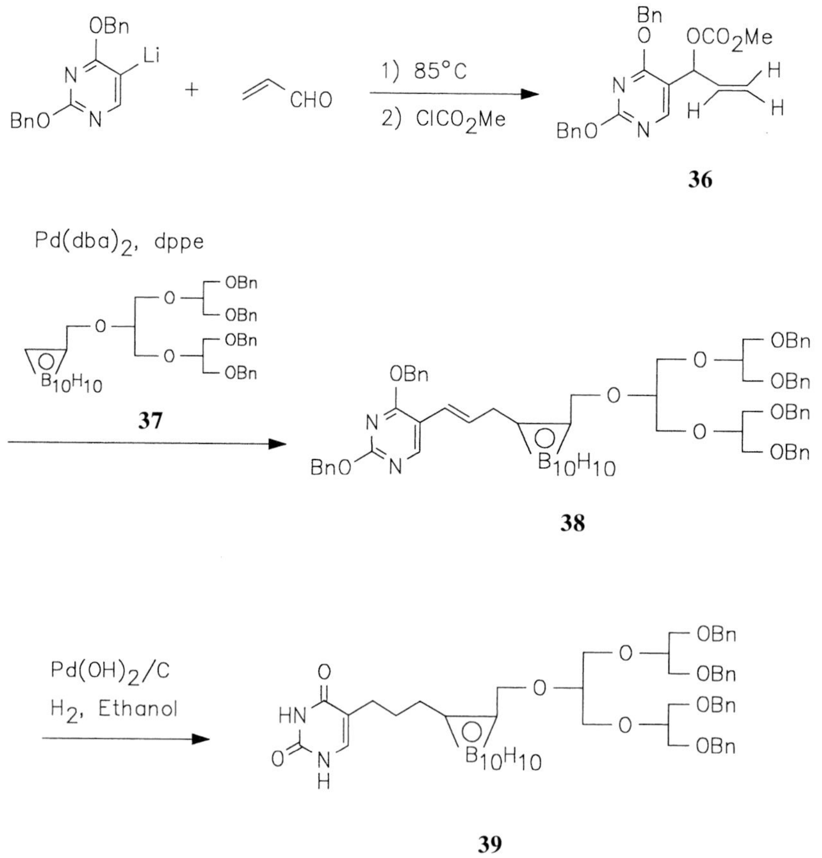

Scheme 5.9 Introduction of a tumor-seeking uracil moiety[52] onto a water-soluble *o*-carborane.

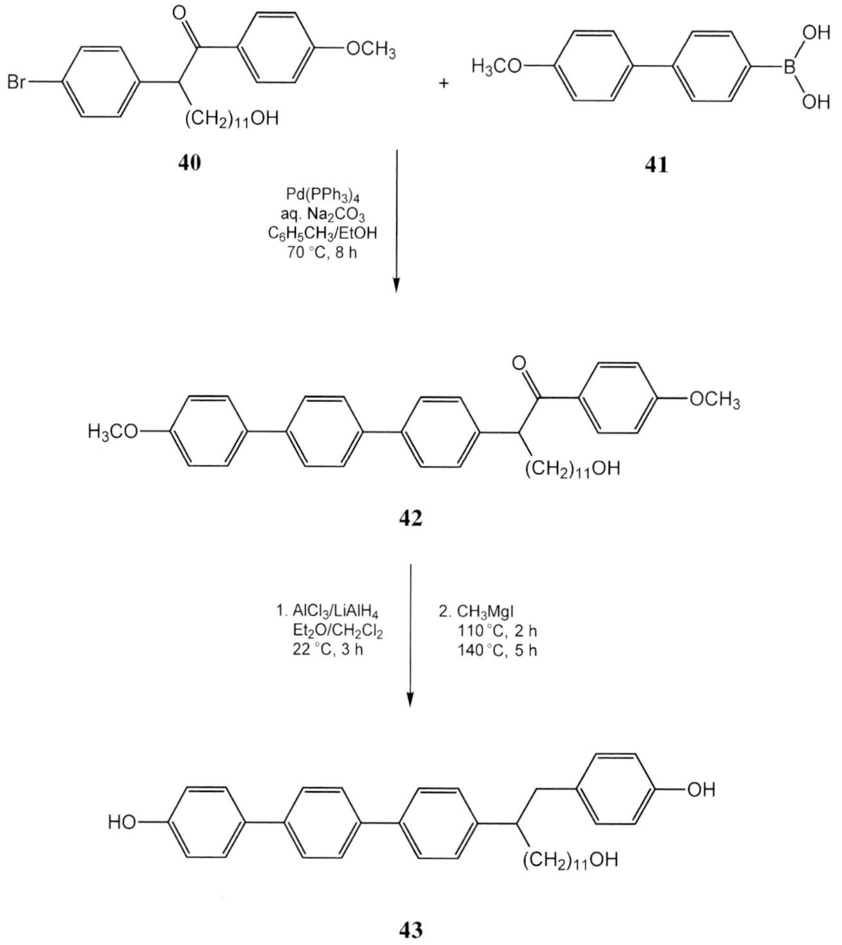

Scheme 5.10 Synthesis of an AB$_2$ polyaryl monomer (**43**).[53]

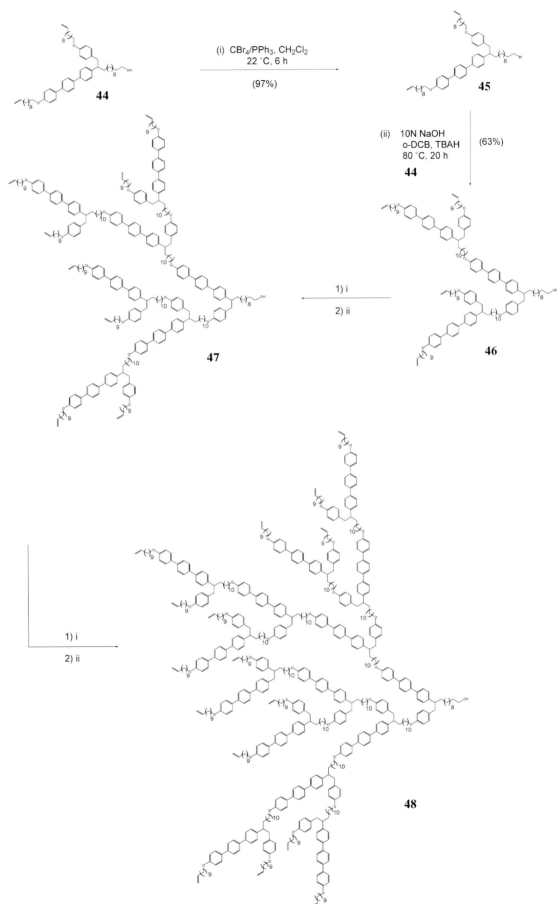

Scheme 5.11 Construction of the polyaryl dendrons **46–48**.[53]

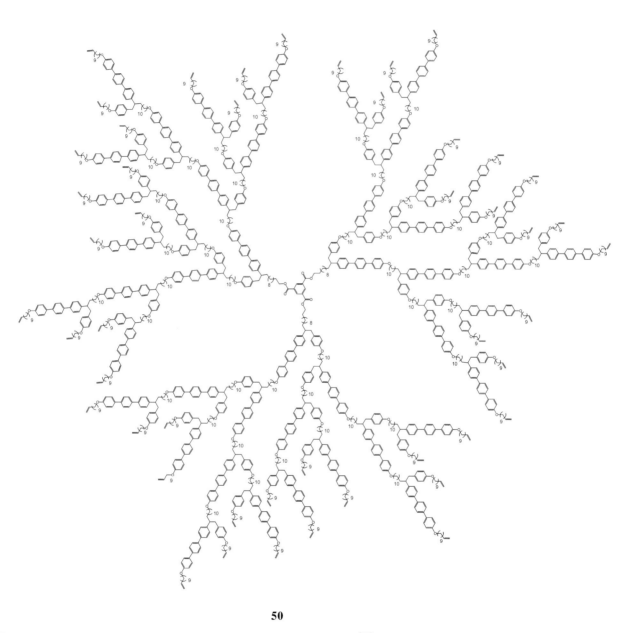

50

Scheme 5.12 Subsequent assembly of the aryl-cored dendrimer **50**.[53]

Allcock et al.[56] synthesized poly(ethereal) dendrimers for use as an electrolyte in solid-state energy storage devices (Scheme 5.13). Preparation of these small dendrimers began by reaction of diol **51**, generated by benzoylation of the free hydroxy moiety of acetonide-protected 1,2,3-trihydroxypropane and subsequent acid-catalyzed deprotection, with tosylate **52** to give benzyl ether **53**. Hydrogenation of **53** gave alcohol **54**, which was attached to the hexacore hexachlorocyclotriphosphazene (**55**) to afford the desired dendrimer **56**.

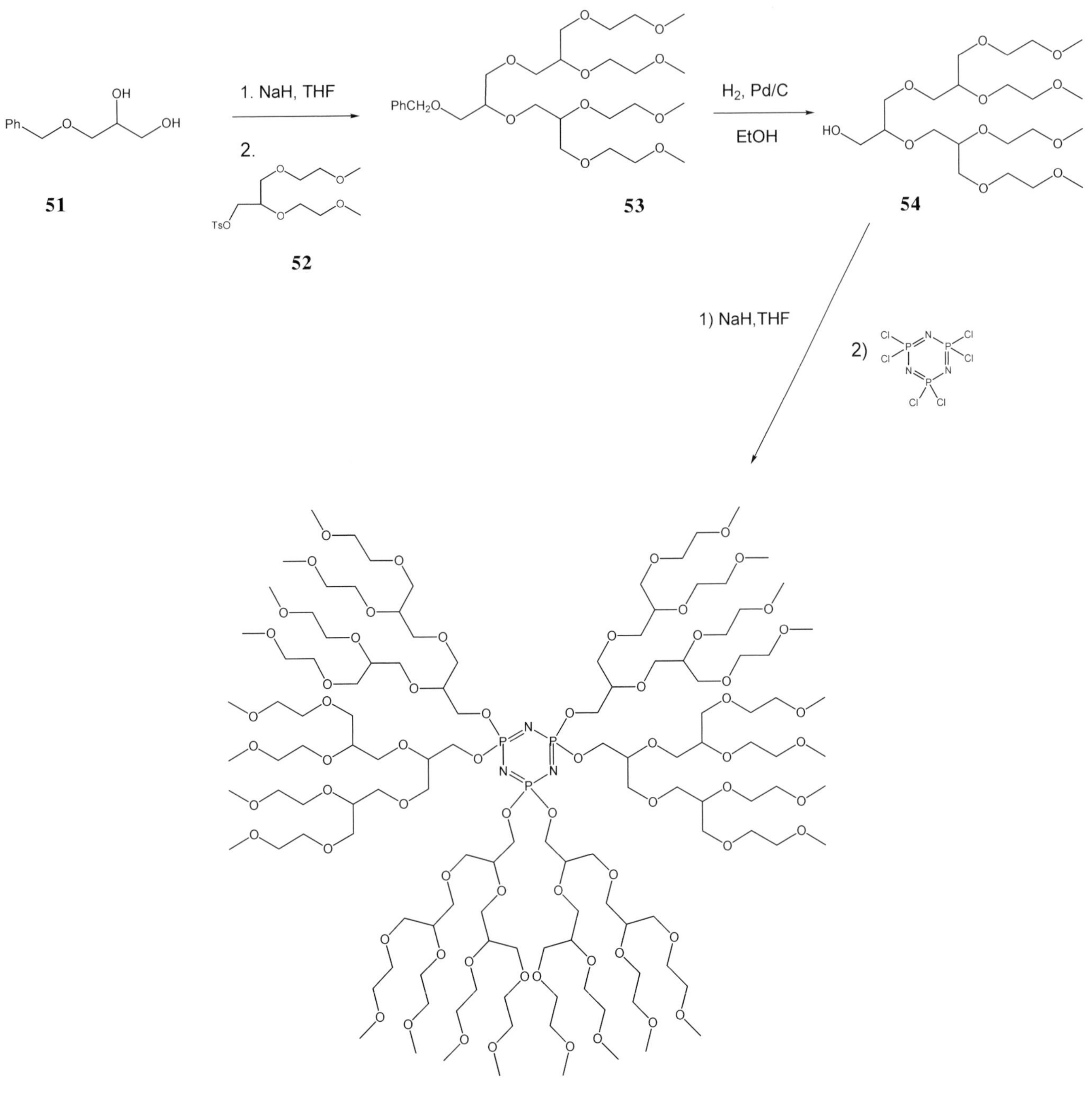

56

Scheme 5.13 Allcock et al.'s poly(ethyleneoxy)cyclotriphosphazene **56** constructed as solid polymer electrolytes.[56]

Jayaraman and Fréchet[57] employed methylallyl dichloride **58**[58] (Scheme 5.14), obtained using pentaerythritol, as the key monomer in the synthesis of branched polyethers. Iteration consisted of oxyanion chloride displacement followed by regiospecific alkene hydroboration–oxidation. The 4th and 5th generation dendrons (i.e., **62** and **63**) were readily obtained in high yields using this protocol; MALDI-TOF MS analysis supported the defect-free architectures.[52a] Grayson, Jayaraman, and Fréchet[59, 60] reported other poly(ether)-based dendrimers, for which the initial monomers included the monoalcohol derived from the di-*n*-butyl ketal of 1,1,1-tris(hydroxymethyl)ethane. Reaction (NaH, KI, THF, 18-crown-6) of these protected monomers with **58** afforded the desired focal alkene, which could be hydroborated and subsequently treated with dichloride **58** to give dendrons possessing different surface components.

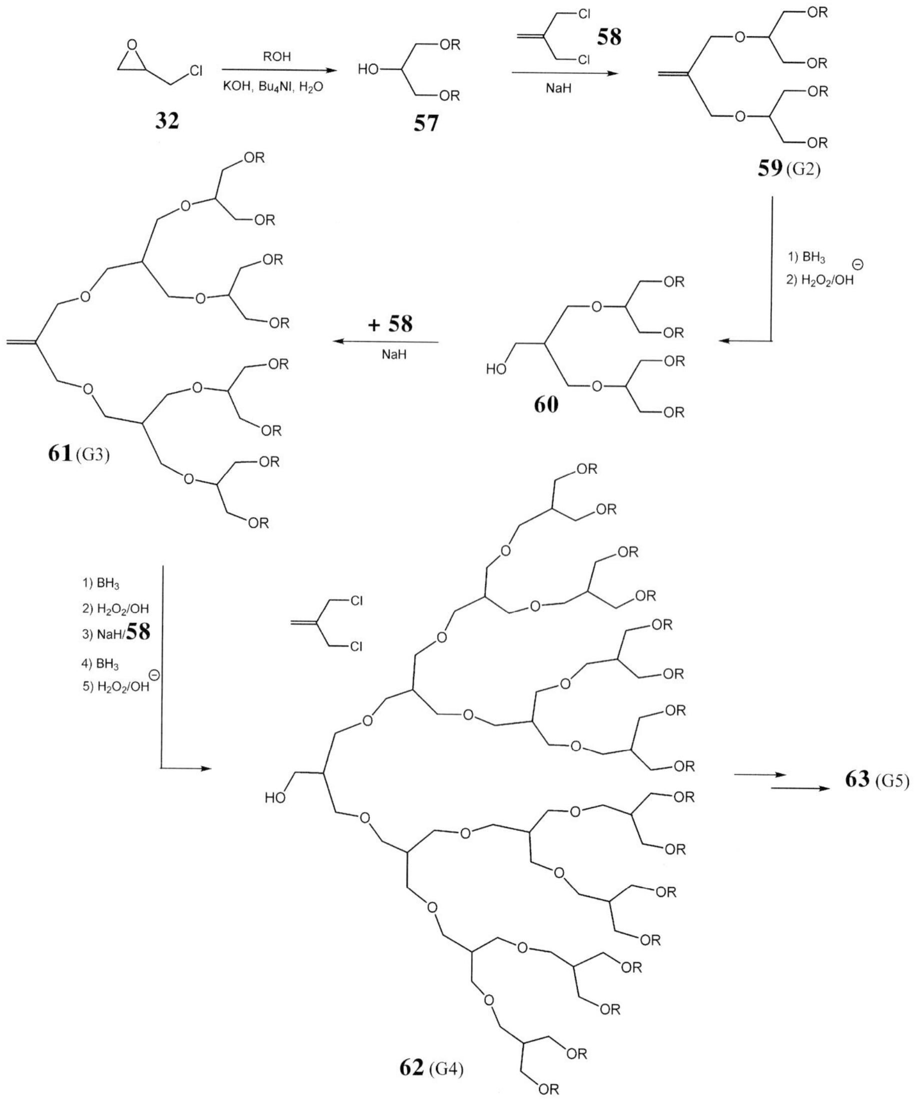

Scheme 5.14 Polyethereal dendron family (**57**, **60–63**) prepared by a hydroboration–substitution protocol.[57]

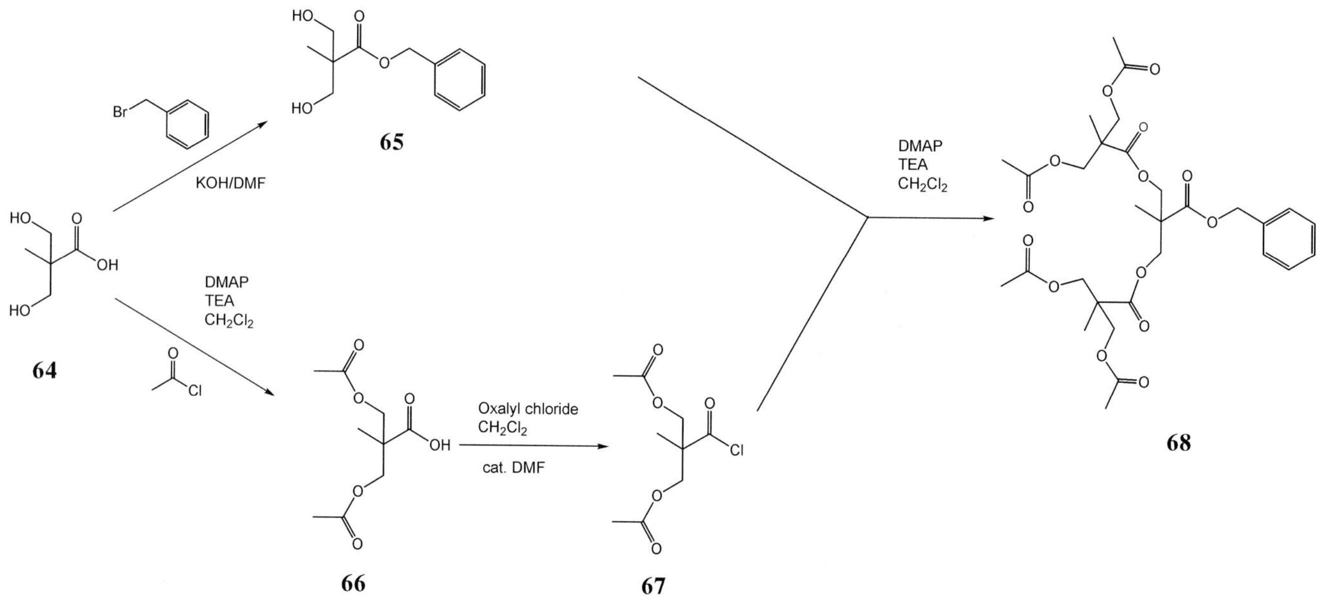

Scheme 5.15 Preparation of monomers (**65** and **67**) for the construction of dendritic polyester.[61]

5.2.4 1 → 2 *C*-Branched, *Ester* Connectivity

Hult et al.[61] reported the preparation of aliphatic polyester dendrimers relying on key monomers synthesized from 2,2-bis(hydroxymethyl)propionic acid (bis-MPA). Scheme 5.15 illustrates the synthetic protocol whereby diol **65** is reacted with two equivalents of acid chloride **67** to yield tetraester **68**. Diol **65** was accessed by treatment (KOH/DMF) of the substituted propionic acid **64** with benzyl bromide, while acid halide **67** was prepared by diol protection (DMAP, Et₃N, CH₂Cl₂) of **64** with MeCOCl, followed by conversion

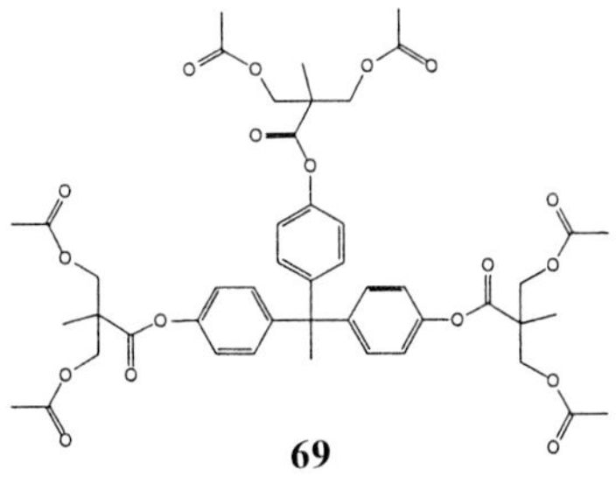

69
Mᵥᵥ : 906.31 g/mol
C₄₇H₅₄O₁₈

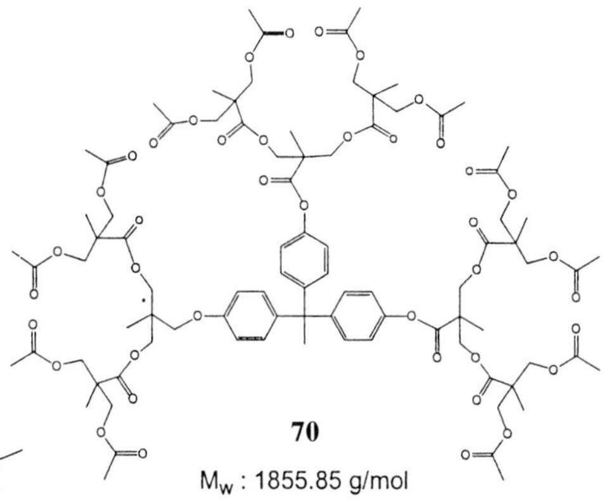

70
Mᵥᵥ : 1855.85 g/mol
C₈₉H₁₁₄O₄₂

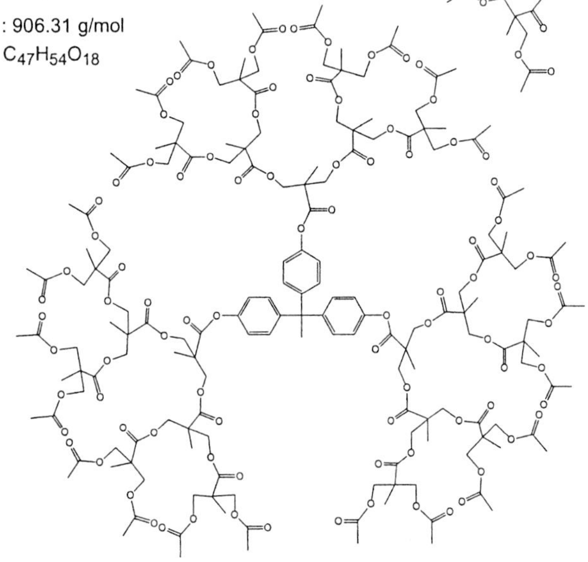

71
Mᵥᵥ : 3753.69 g/mol
C₁₇₃H₂₃₄O₉₀

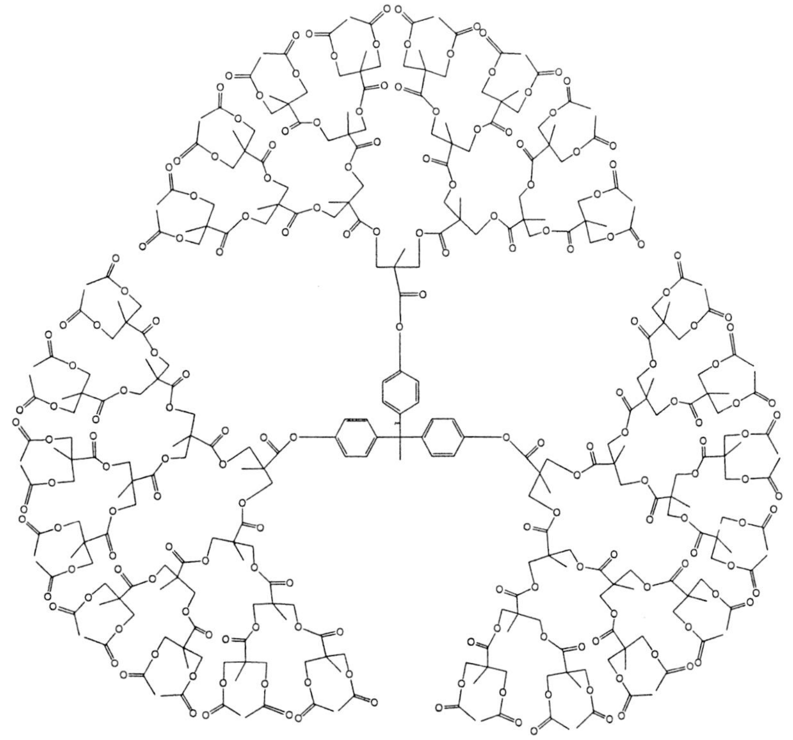

72
Mᵥᵥ : 7549.36 g/mol
C₃₄₁H₄₇₄O₁₈₆

Figure 5.2 Hult et al.'s polyester dendrimers (**69–72**).[61]

[(COCl)$_2$, CH$_2$Cl$_2$, DMF] to give the acyl chloride. Formation of higher generation dendrons was accomplished by focal hydrogenation [10% Pd/C] to give the corresponding carboxylic acid, which was reacted with first (COCl)$_2$ and then 0.5 equivalent of diol **65**. Four generations were prepared, which, in turn, were each reacted with the core, 1,1,1-tris(hydroxyphenyl)ethane (**27**), affording dendrimers (**69–72**; Figure 5.2). Generations 1 to 4 were obtained in 93, 89, 74, and 30% yield, respectively. Using pulsed-field spin-echo ^{1}H NMR spectroscopy, diffusion coefficients were calculated by means of the Stejskal–Tanner equation. Assuming spherical geometries, the dendrimer's effective radii were determined to be 7.8, 10.3, 12.6, and 17.1 Å for the 1st to the 4th tier structures, respectively. A double-stage convergent synthesis of these materials has subsequently been reported;[62] thus, stable hydroxy-terminated dendrimers were prepared and several surface modifications were undertaken including synthesis of a wax-like palmitoate ester. Iversen et al.[63] used small dendrimers, with cores based on the dihydroxypropionic acid monomer, for the ring-opening polymerization of ε-caprolactone promoted by Lipase B. A series of star poly(lactide)s with well-defined structures was synthesized from several simple bis-MPA derivatives by this "living" polymerization process;[64] ferroelectric properties in this series are known.[64a]

5.2.5   1 → 2 *C*-Branched, *Ether & Imide* Connectivity

Shu et al.[65] [65a] described the synthesis and characterization of a series of poly(ether imide)s using 1-(4-aminophenyl)-1,1-bis(4-hydroxyphenyl)ethane as the monomer, predicated on the nucleophilic displacement of an activated aromatic nitro group with 3-nitro-*N*-phenylphthalimide. This procedure led to a dendron possessing an aminophenyl group at the focal point, which was treated with 3-nitrophthalic anhydride; the procedure was continued to produce the next generation.

5.3   1 → 2 *Aryl*-Branched

5.3.1   1 → 2 *Aryl*-Branched and Connectivity

At about the same time as Hawker and Fréchet's initial publication,[1] Miller and Neenan reported[2] the synthesis of monodisperse molecular spheres based on 1,3,5-trisubstituted benzene (Scheme 5.16). The completely aromatic hydrocarbon **77** was prepared using 1,3-dibromo-5-(trimethylsilyl)benzene (**73**) as the key monomer. Two equivalents of phenylboronic acid were catalytically [Pd(PPh$_3$)$_4$] coupled to dibromobenzene **73** yielding terphenylsilane **74**, the silyl group of which was subsequently transformed to a boronic acid moiety and further coupled to monomer **73**. The resultant heptaphenyl trimethylsilyl wedge **75** was then boronated and attached to the three-directional core **76** to generate hydrocarbon **77**, termed a 12-Cascade:benzene[3–1,3,5]:(1,3,5-phenylene)^2: benzene.

Miller et al.[66, 67] described (Scheme 5.17) the preparation of the related 46-phenylene dendrimer **80**. The Suzuki reaction[68] was again used for dendrimer assembly.[67] Since the smaller 2nd generation 7-phenylene boronic acid monomer (**79**) was readily accessible (from trimethylsilyl precursor **75**), an extended hexabromotetraphenyl core **78** was prepared by acid-catalyzed trimerization of the dibromoacetophenone **77**. Subsequent coupling of core **78** with heptaphenylboronic acid (**79**) afforded the desired hydrocarbon **80**. The fluorinated analogs were prepared[66, 69] by analogous technology, except that the 1st tier derivative was obtained in poor yields (<5%) from the treatment (Cu°, 190°C) of bromopentafluorobenzene with tribromobenzene **76**. Higher generations were derived in greater yields by the coupling of 3,5-bis(pentafluorophenyl)phenylboronic acid with the tri- and hexabromoarene cores. The 2nd generation fluorinated dendrimer was reported to possess very limited solubility in most organic solvents; it was rationalized "that it can adopt a pancake-like conformation in which all of the benzene rings are

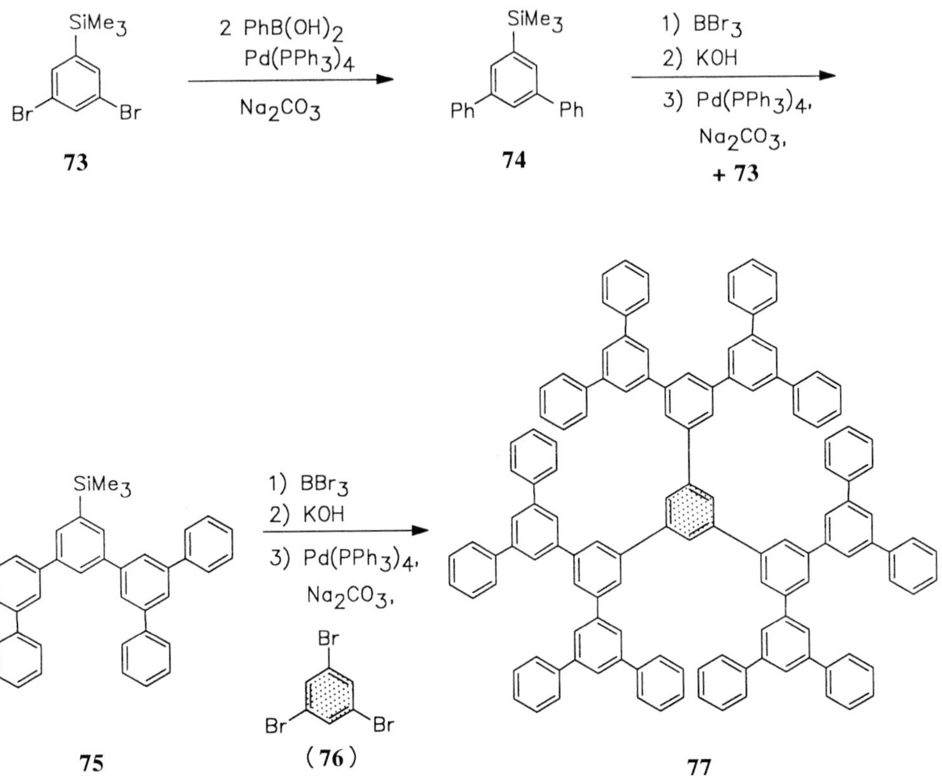

Scheme 5.16 Employment of the Suzuki coupling leads to the construction of all-aromatic dendrimers.[2]

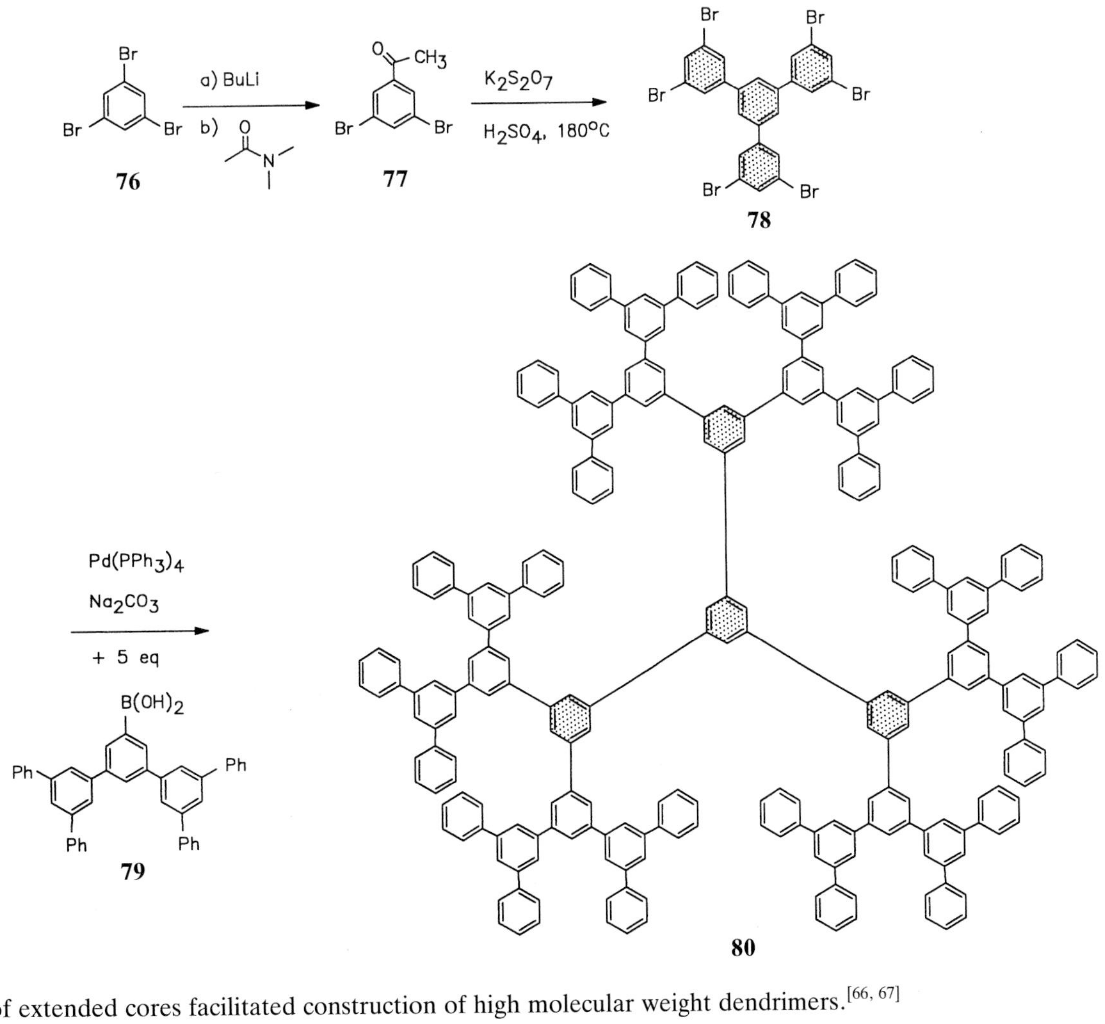

Scheme 5.17 Use of extended cores facilitated construction of high molecular weight dendrimers.[66, 67]

approximately in a plane". Although divergently constructed, this example fits here in this context. A dodecabromo analogue of **78**, in which the terminal phenyl groups are replaced by bromo moieties. This was treated [Pd(PPh$_3$)$_4$] with 2-chlorozincthiophene to afford the dodecathienyl derivative;[70] this material was subsequently brominated with NBS in the α-position of the twelve thiophene units in high yield.

Wiesler and Müllen[71] have expanded their Diels–Alder and Knoevenagel protocol to include a convergent approach (Scheme 5.18). Addition of an excess of tetraphenylcyclopentadienone (**81**) to diyne **82** produced the 1st generation dione **83**, which, upon reaction with **84**, afforded the 2nd generation cyclopentadienone **85**, treatment (Ph$_2$O, 200 °C, 7 d) of which with tetrayne **86** gave the polyphenylene **87**. Notably, these materials exhibited high chemical and thermal stability, remaining unmodified in boiling HCl, 30 % KOH

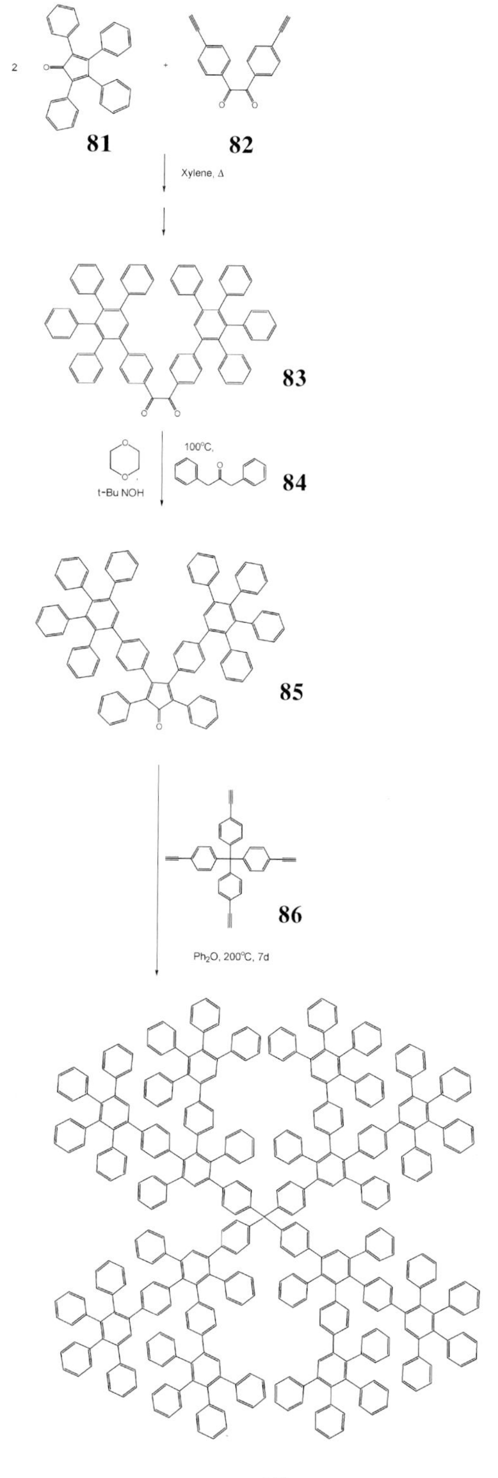

Scheme 5.18 Müllen's convergent [2+4] cycloaddition protocol.[71]

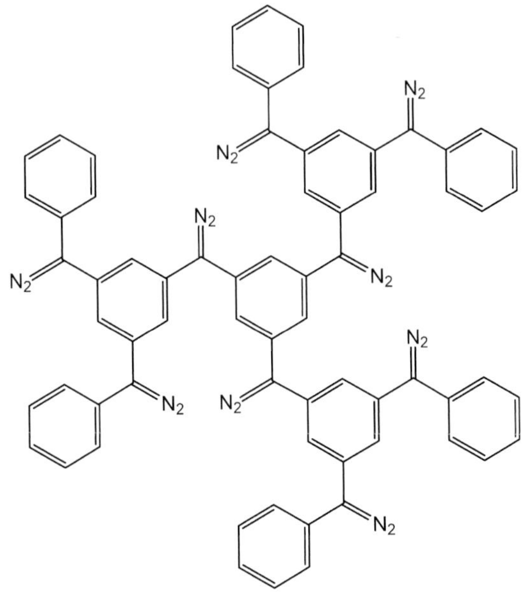

88

Figure 5.3 A starburst-type nonadiazo poly(*m*-phenylene).[78]

(7 d), or on heating to 550 °C. Surface modification could, however, be effected with strong electrophiles (e.g., H_2SO_4). An interesting tetraethynylbiphenyl core was also reported. De Schryver, Müllen et al.[72–74] [74a] modified the surface of these polyphenylenes with peryleneimide chromophores and examined the time-dependent fluorescence spectra. Results indicated the presence of a decay channel resulting in excited-state chromophore interaction in a single dendrimer. Surface-alkylated polyphenylene dendrimers possessing a tetrahedral or disc-like shape have been demonstrated[75] to form supramolecular structures, e.g., parallel rods or two-dimensional crystals on graphite. Due to their interesting material properties, theoretical calculations on the geometric and electronic properties of these dendrimers addressing the effects of chemical doping have appeared.[76, 77]

Iwamura et al.[78] reported the construction of a "starburst"-type, nonadiazo polyphenylene (**88**; Figure 5.3), in which a cyclotrimerization of 3,5-dibenzyl ethynyl ketone was pivotal to the synthesis. The magnetic data from this photoproduct suggested a pentadecet ($S = 7$) ground state in contrast to the predicted nonadecet ($S = 9$) state.

5.3.2 1 → 2 *Aryl*-Branched, *Ether* Connectivity

In 1990, the first, and now best known, convergent dendritic synthesis, based on the combination of 3,5-aryl branching and ethereal connectivity, was reported[1, 46, 79] by Hawker and Fréchet. Two simple synthetic transformations were used: (1) selective alkylation of phenolic hydroxyl groups, and (2) conversion of a benzylic alcohol to a benzylic bromide to generate a reactive focal moiety (Scheme 5.19). Thus, addition of two equivalents of the previously reported[80] benzylic bromide **89** to the initial monomer **90** yielded the heptaphenyl ethereal wedge, which was converted (CBr_4, PPh_3) to the corresponding bromide **91**. This sequence was successfully repeated to prepare dendritic wedges through to the 6th generation. These wedges were subsequently attached to the three-directional core **27** to afford a benzylic ether-based dendritic series represented by **92**, the largest of which possessed 192 terminal phenyl rings and a M_w of 40,689 amu. Many research groups have initiated their entrance to the field of dendrimers by using these easily accessible, versatile wedges and their facile attachment to diverse cores.

Tyler and Hanson[81] developed a synthesis of poly(aryl ether)s whereby monomer connectivity is based on focal phenolate displacement of benzylic bromides (Scheme 5.20) in

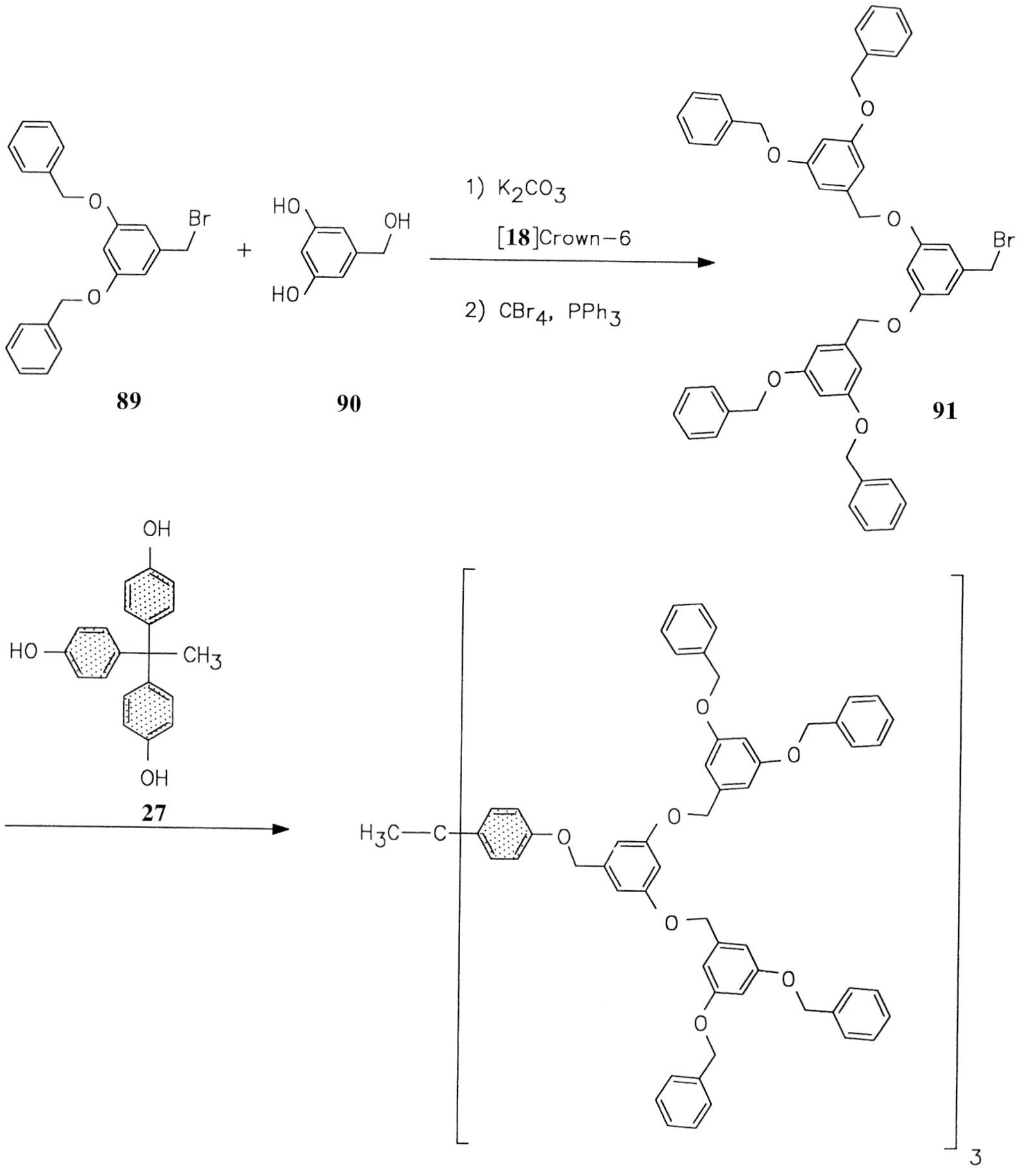

Scheme 5.19 Fréchet et al.'s[1, 44] original convergent methodology.

a protocol that is essentially a "reversed" Fréchet method.[82] Beginning with 3,5-bis(bromomethyl)phenyl hexadecanesulfonate (**93**), prepared in five steps from dimethyl 5-hydroxyisophthalate, phenolic dendrons were successively attached (K₂CO₃, Me₂CO) and deprotected (NaOH, EtOH) to access higher generations (e.g., **95**). Up to six generations (i.e., **96**) were prepared; attachment to a trivalent core was reported. These are essentially constitutional isomers of the original Fréchet dendrons. Weintraub and Parquette[83] have employed similar technology, albeit incorporating an aldehyde moiety as a latent benzylic bromide (**99**; Scheme 5.21), for the synthesis of unsymmetrical architectures. Upon coupling of the phenolic units, reduction (NaBH₄, MeOH) of the aldehyde to corresponding alcohol permits access (PBr₃, CH₂Cl₂) to a benzyl bromide thus allowing repetition of the sequence; up to the 4th generation dendron **100** was reported.

Polyethereal dendrons possessing -(CH₂)₃- connectors, thereby eliminating reactive benzyl positions, have also been reported as monomers for functional dendrimers.[76, 84–87] Building blocks were prepared by reacting 5-benzyloxyresorcinol and a monophenol-substituted 1,3-dibromopropane to provide the starting peripheral units, or an unmodified 1,3-dibromopropane to access the key bis(alkyl halide) used for repetitive coupling. Focal modifications with terpyridine and a bis(oxazoline)[86, 88] unit were published.[85] Fréchet and Gitsov[89] reported the preparation of hybrid "star-like" macromolecules possessing poly(aryl ether) dendrons attached to a four-directional PEG molecule. These unique "dendritic stars" form conformationally differing unimolecular

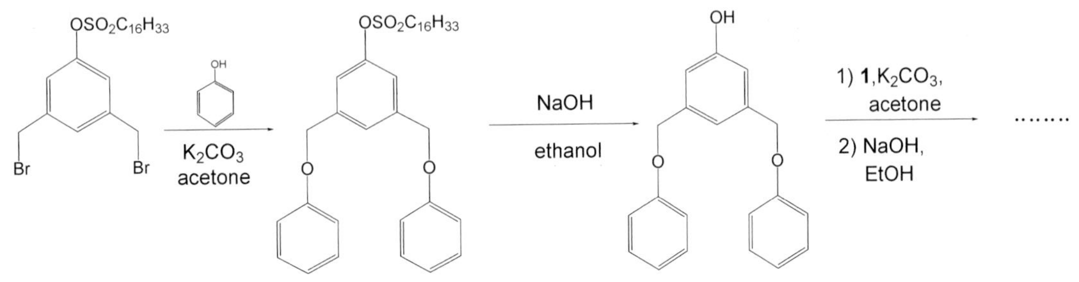

93 **94** **95**

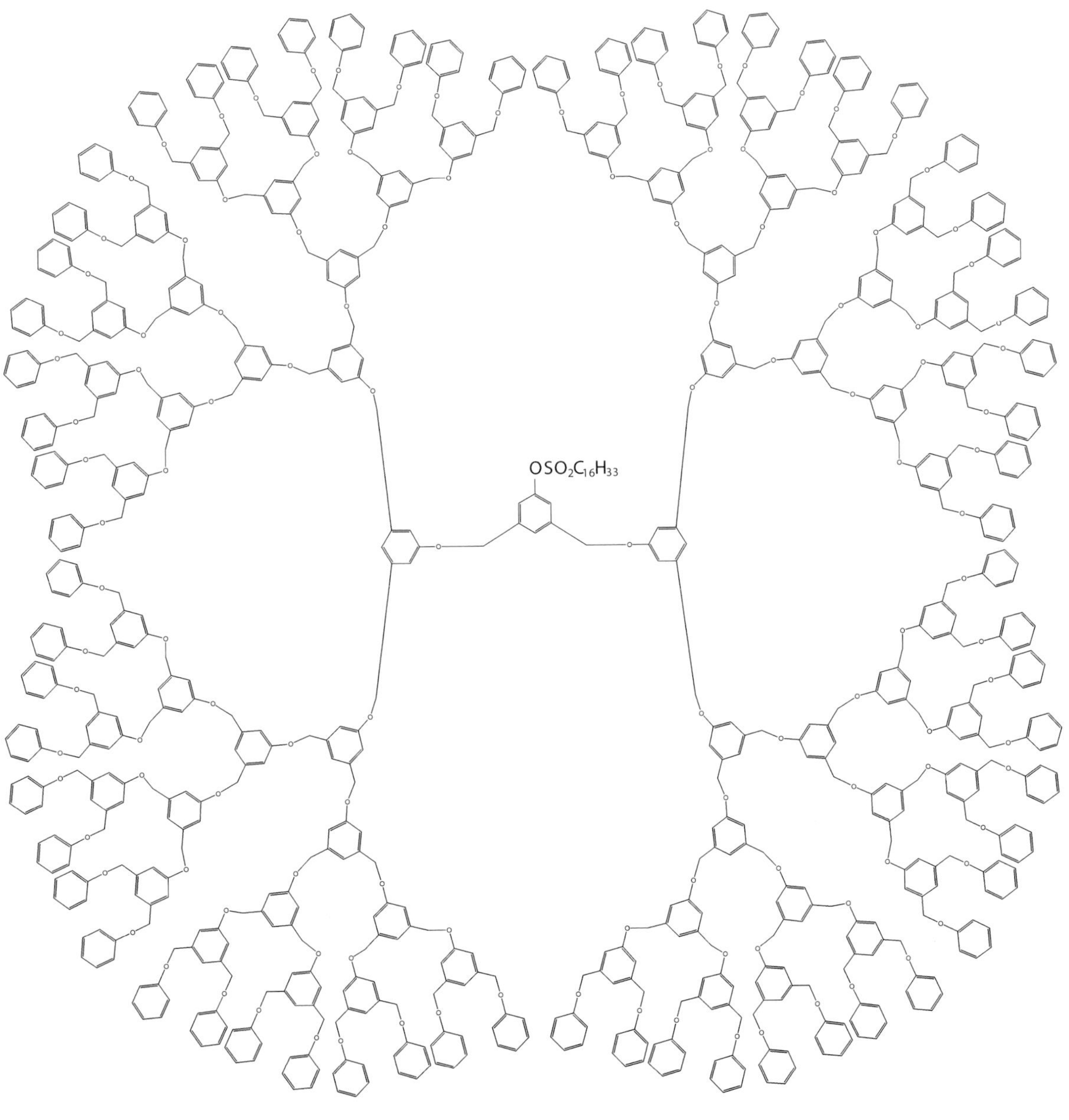

96

Scheme 5.20 Hanson's isomeric Fréchet-type dendrons.[81]

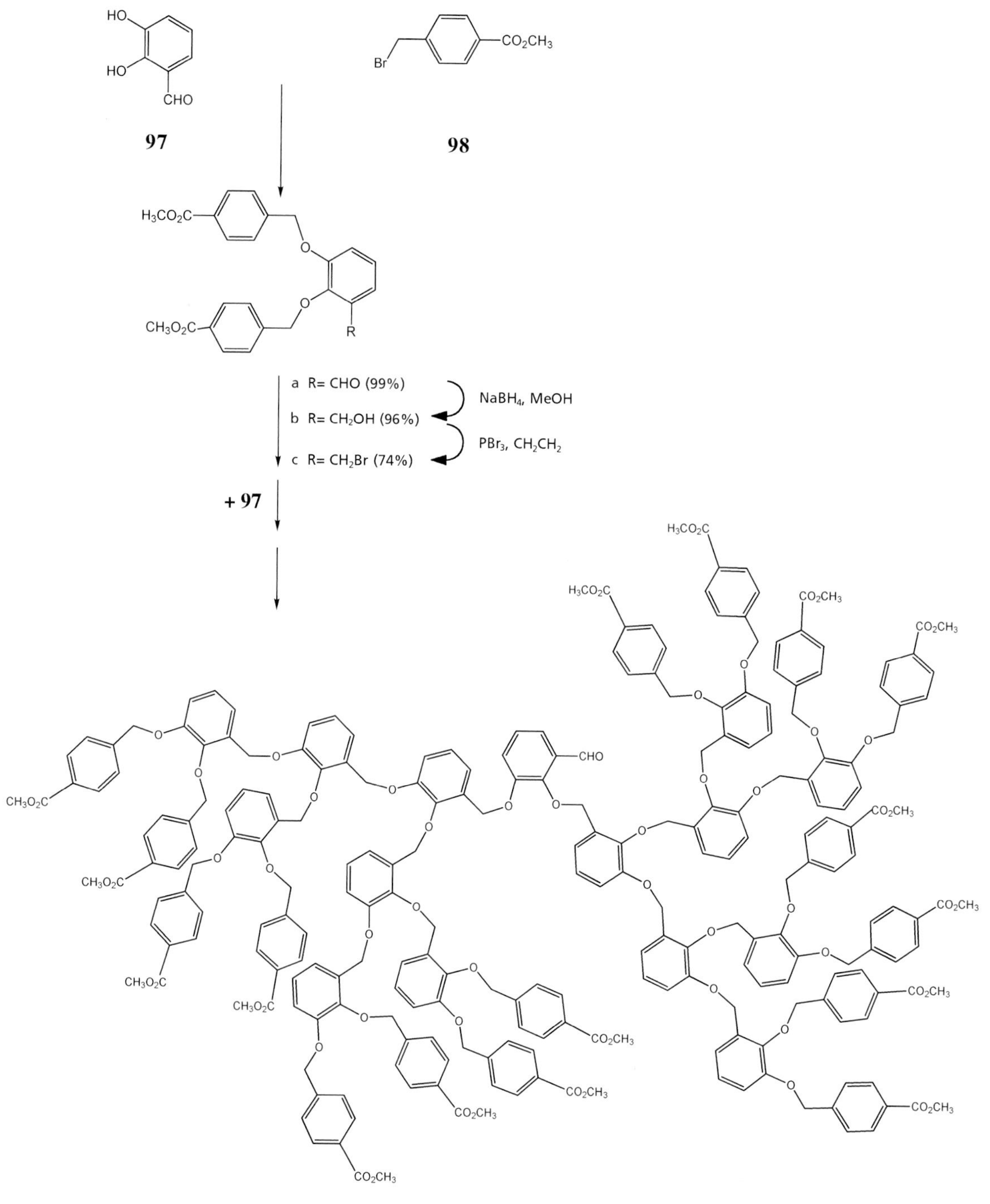

Scheme 5.21 Parquette's "unsymmetrical" poly(aryl ether)s.[83]

micelles depending on the polarity of the surrounding environment. Poly(aryl ether) dendrons have been attached to 1,10-phenanthroline cores and complexed with ruthenium.[90] Other branched star polymers are known.[90a]

Recently, one core (an alkyne) was transformed to another (a benzene); Hecht and Fréchet[91] thus demonstrated the possibility of assembling a dendrimer by $Co_2(CO)_8$-promoted cyclotrimerization of poly(aryl ether) disubstituted alkynes.

5.3.2.1   Focal Functionalization

The creation and use of activated focal points have been interesting attributes of the convergently generated dendritic macromolecules. Researchers can thus probe a specific or specifically created microenvironment; in one early study[92] a solvatochromic probe [4-(*N*-methylamino)-1-nitrobenzene] was attached at the inner focal position of a comb (R = N(Me)—〈 〉—NO₂)

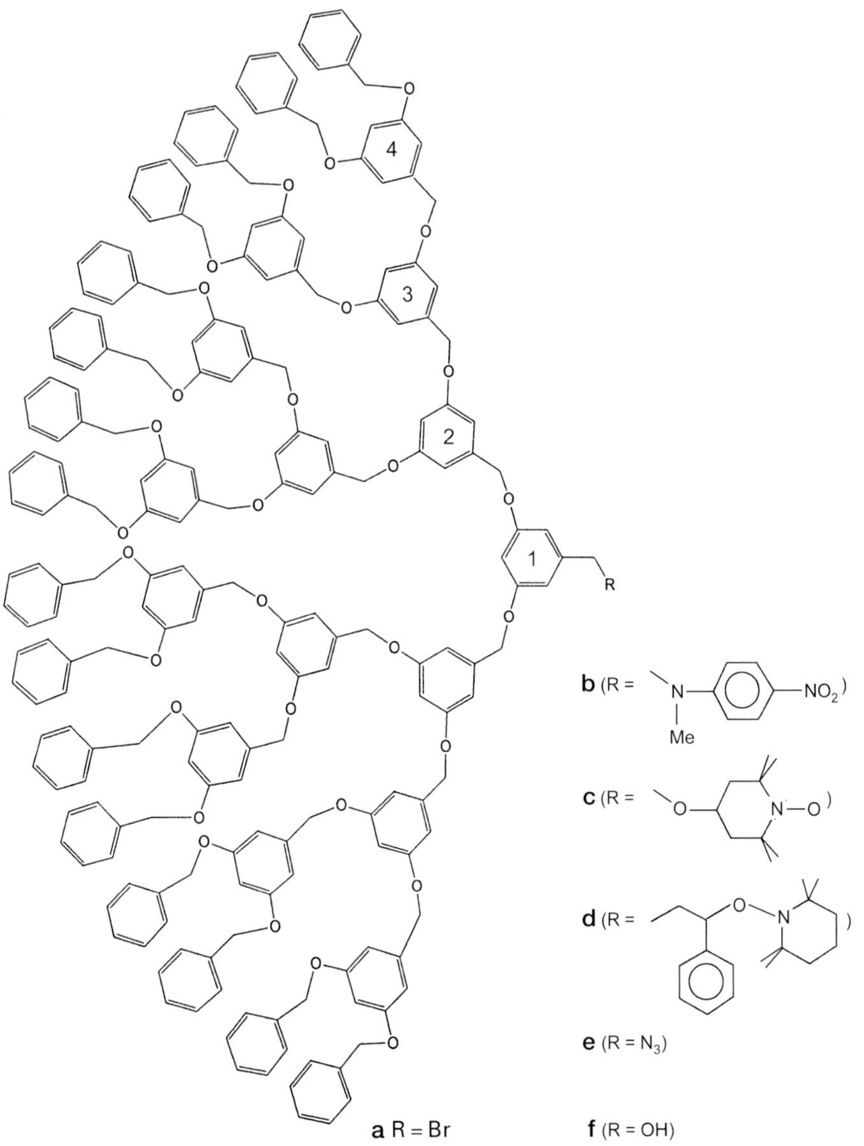

c (R = —O—〈piperidinyl〉N—O)

d (R = —CH(Ph)—O—N〈piperidinyl〉)

e (R = N₃)

a R = Br   f (R = OH)

101

Figure 5.4 The microenvironment of solvatochromic probes,[92] TEMPO-modified,[94] benzylic TEMPO,[95] and azide.

plete series of related ethereal dendrimers [Figure 5.4; **101b** (G4)]. Variations in the absorption maximum with increasing generations and different solvents demonstrated that the microenvironment near the focal point (core) had a high polarity, especially with the more spherical dendrimers. Fréchet-type dendrons attached to aminotriethanol or phenylchalcogeno group[92a] have been reported[93] to be efficient catalysts of the nitroaldol (Henry) or H_2O_2 oxidation reaction, respectively.

Matyjaszewski et al.[94] modified the focal sites of low generation Fréchet-type dendrimers with TEMPO-based free radicals (**101c**) for the initiation and control of "living" radical polymerizations. Styrene, vinyl acetate, methyl methacrylate, and *n*-butyl acrylate were found to be polymerized in a similar manner as when TEMPO alone was employed or polymerization control was maintained only during early stages of growth. Xi et al.[96] reported the preparation of the 1st and 2nd generation dendrons possessing acrylate moieties at the focal site; polymerization was accomplished by using AIBN. Dumbbell-shaped ABA triblock copolymers composed of poly(benzyl ether) wedges and polystyrene were prepared using a TEMPO-mediated "living" free-radical polymerization;[97] a new bis-dendritic unimolecular initiator was utilized. Poly(aryl ether) dendrimers have been exploited[95] for the controlled free radical polymerization of vinyl monomers to give hybrid dendritic-linear block copolymers possessing low polydispersities. Focally substituted nitroxide dendrons (**101d**) were prepared by reaction (NaH) of a hydroxybenzylic TEMPO reagent with a series of focally-functionalized benzylic bromide dendrons. Each class of dendron, nitroxide- or halide-based, was used for nitroxide-promoted or atom-transfer free radical polymerization, respectively. Single T_gs for all of the hybrid polymers prepared suggested compatible miscibility of both linear and branched polymer segments.

Isophthalate ester-terminated and benzyl chloride focally-modified dendrons have been employed for metal-catalyzed "living" radical polymerizations of styrene to produce dendritic-linear block copolymers.[98] Copolymer ester moieties were transformed to carboxylic acids, butyl amides, benzyl alcohols, and halides, as well as higher generations through dendron transesterification. Controlled M_ws of ca. 30,000 amu were achieved with associated polydispersities of less than 1.2. Simple hydroxymethylaryl ether dendrons were also employed as "macroinitiators" for the "living" ring-opening polymerization (ROP) of lactones and lactides to afford dendritic-linear block copolymers.[99]

Attached preformed polyethereal dendrons, functionalized at the focal point by reaction with *p*-(chloromethyl)styrene, have been shown[100] to undergo free radical copolymerization with styrene giving rise to polystyrene with appended benzyl ether dendritic wedges. Schlüter et al.[101, 102] reported the polymerization of focally-modified, styrene-based dendrons of generations 1–3 (Scheme 5.22). Introduction of the styrene component was effected by transforming *p*-bromostyrene to *p*-(2-hydroxyethyl)styrene via generation of a Grignard reagent and reaction with ethylene oxide. Substitution of the hydroxy moiety at the focal bromide **101a** afforded monomers **102** for polymerization (initiated with either DBPO or *t*-BPB) to give the cylindrical dendrimers **103**. Light-scattering experiments using fractionated samples (M_w = 82,000 and 600,000 amu, respectively) obtained from the 2nd generation wedge showed that the polymer can best be described as a Gaussian, as opposed to a worm-like chain, and thus should not be considered as a rigid rod. Formation of a focal-sited isocyanate moiety on the Fréchet-type dendrons (**105**) for attachment to hydroxylated poly(*p*-phenylene) (**104**; Scheme 5.23) generated the cylindrical dendrimer **106**.[103] Reaction of the isocyanate dendron **105**, promoted by added DABCO, was driven to completion using the 2nd generation wedge and to ca. 95 % completion with that of the 3rd generation.

L'Abbé, Haelterman, and Dehean[104] reported the functionalization of the focal site with 4-(*p*-hydroxyphenyl)-1,2,3-thiadizole. Treatment with strong base (*t*-BuOK) transformed the thiadiazole units to alkynethiolate moieties, which were subsequently reacted

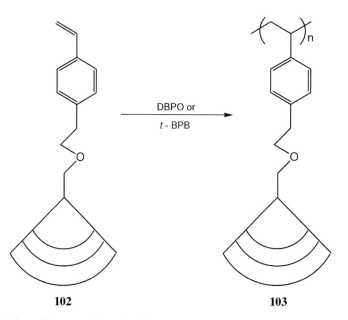

102   **103**

Scheme 5.22 Polymerization of styrene-functionalized dendrons leading to cylindrical dendrimers.[101]

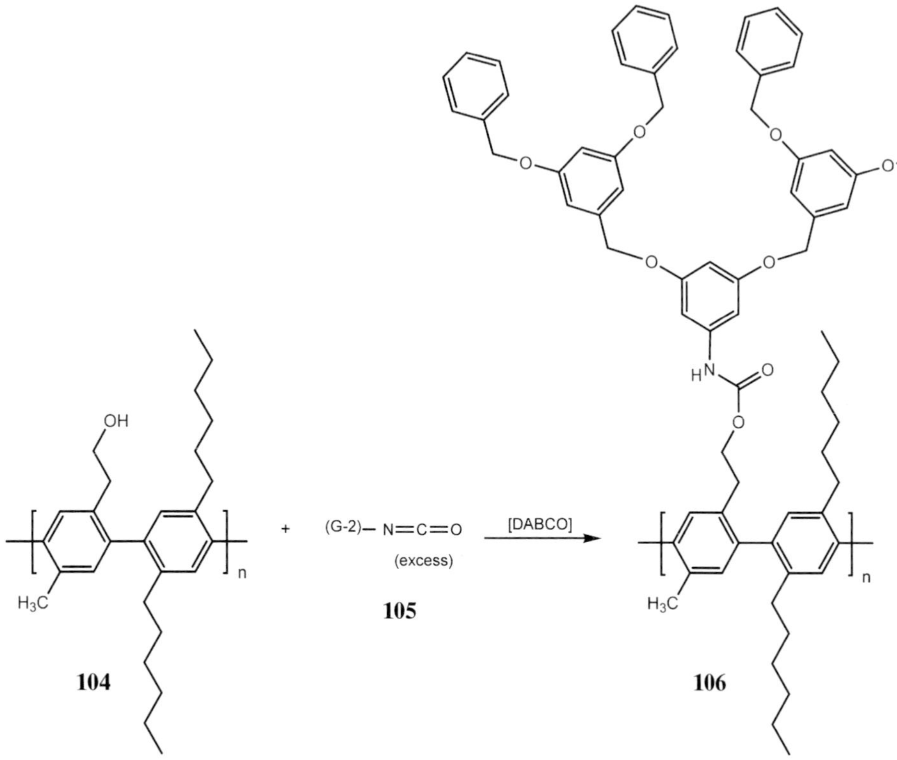

Scheme 5.23 Schlüter's isocyanate reaction with a functionalized polymer core[103]

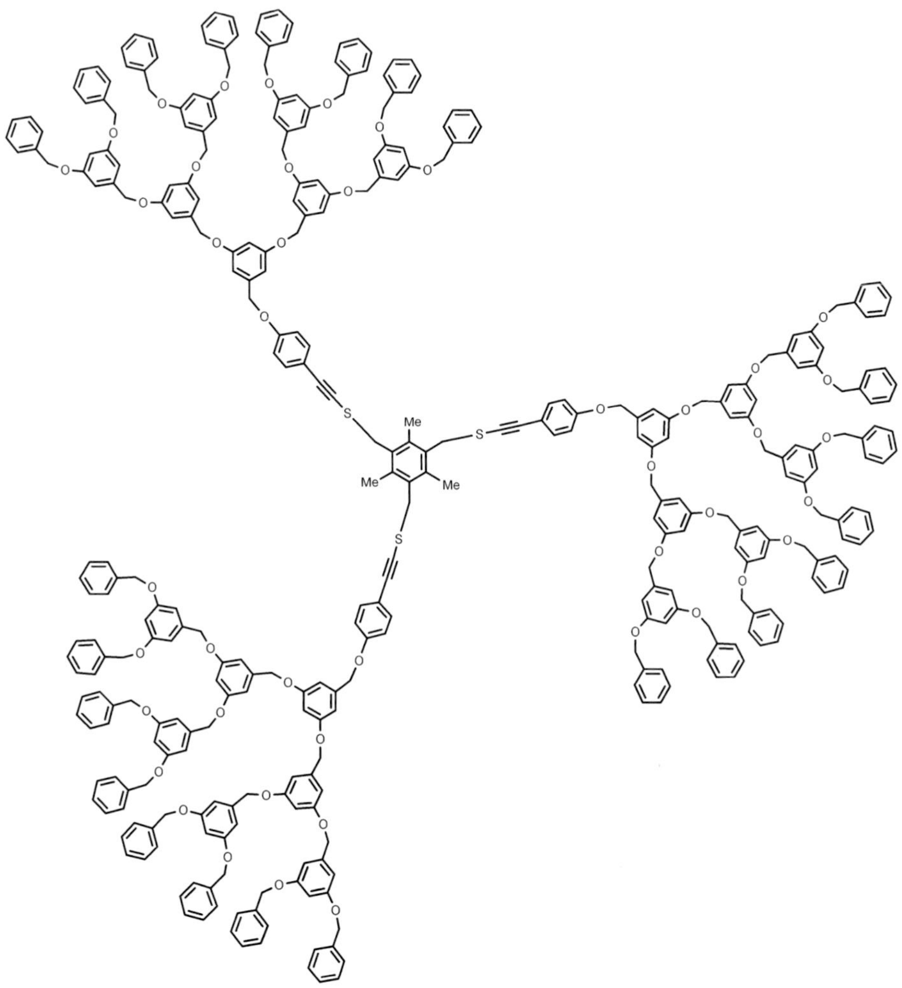

107

Figure 5.5 Base-promoted decomposition of focal site thiadiazole units.[104]

with either hexakis(bromomethyl)benzene or 1,3,5-tris(bromomethyl)-2,4,6-trimethyl-benzene to yield the corresponding dendrimer (e.g., **107**; Figure 5.5).

Reaction of C_{60} with bis(*p*-methoxyphenyl)diazomethane **108** and subsequent cleavage of the methyl ethers afforded a 6–6 methano-bridged fullerene **109** (Scheme 5.24) possessing two phenolic moieties as the major product.[105–107] Treatment of the bisphenolic fullerene **109** with 2.7 equivalents of the activated dendron **101a** afforded the desired substituted fullerene **110** possessing two dendritic arms. Similarly, treatment (C_6H_5Cl, Δ) of C_{60} with wedge **101e** possessing an azide focal point, prepared (NaN$_3$, DMSO) from the corresponding bromide (**101a**), gave rise (69 %) to a controlled, one-step cycloaddition affording the amine-substituted fullerene **111**.[108]

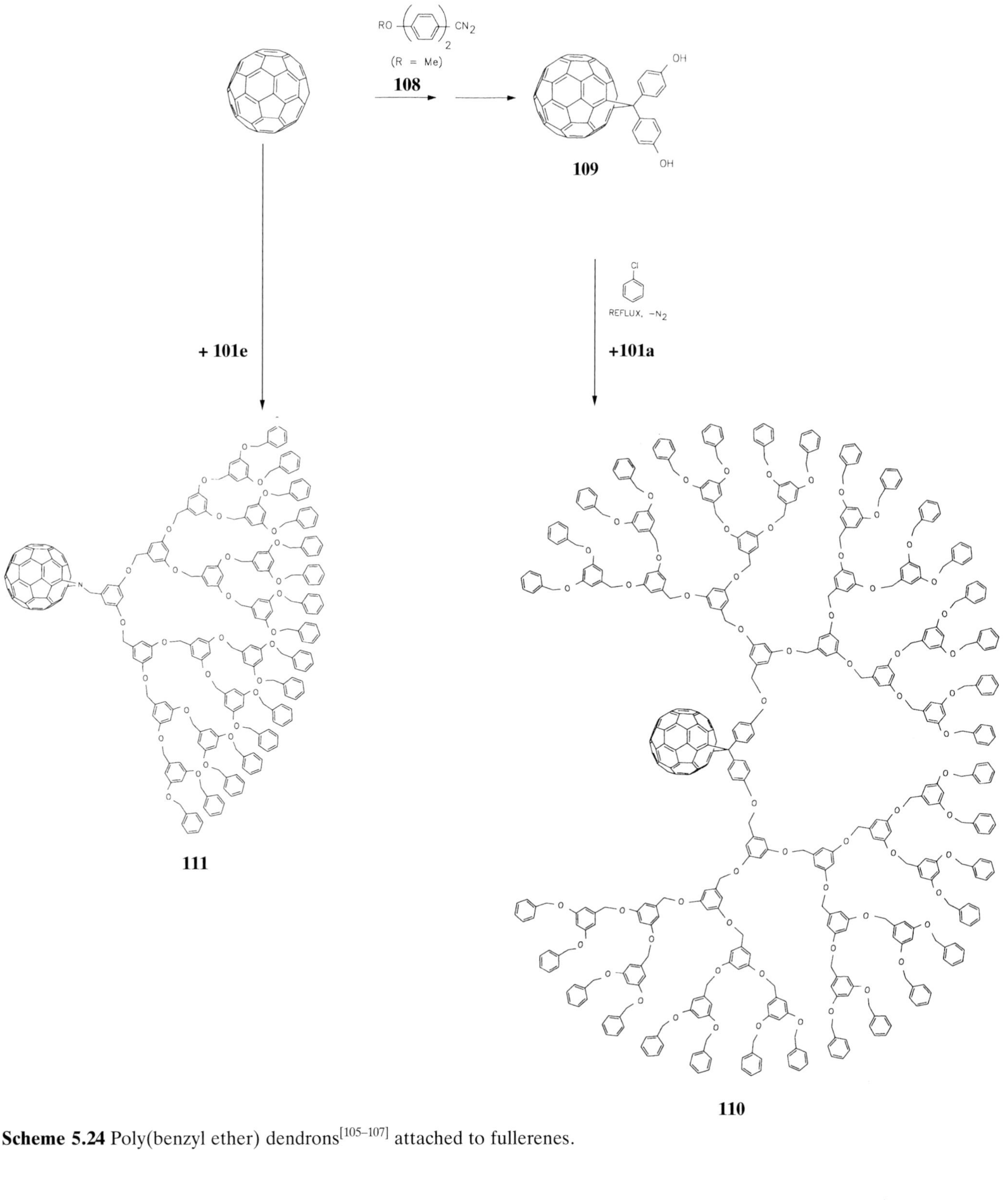

Scheme 5.24 Poly(benzyl ether) dendrons[105–107] attached to fullerenes.

Camps, Schönberger, and Hirsch[109] described the preparation of poly(aryl ether) dendrimers using functionalized C_{60} cores. Dendrons of generations 1 to 3 were modified by reaction of the focal benzyl alcohol moieties with malonoyl dichloride to give the corresponding dendritic malonates, which were then brominated (DBU, CBr_4) and subsequently treated with C_{60} and NaH affording the desired [6,6]-bridged adducts. Using the 1st generation bis(dendron) malonate and a template activation protocol,[110] the penta- and hexa-bridged malonates were prepared possessing 10 and 12 core branching multiplicities, respectively. Molecular dynamics simulations were included in the characterization and analysis of these novel materials. A later report described mixed hexakis adducts of C_{60} that possessed an additional pattern with T_h symmetry.[111] In a related report, repetitive Diels–Alder reactions based on *o*-quinodimethane chemistry for the synthesis of C_{60}-containing main-chain polymers with M_w up to 80,000 amu have been described.[112] In order to enhance the unimolecular micellar character of the hexakis adducts, short PEGed esters were capped with a 1 → 2 branched hydrophobic surface;[113] the resultant structures were used to prepare nanocrystalline silver particles as confirmed by energy-dispersion X-ray analysis.

Zimmerman et al.[114] reported the remarkable self-assembly of these dendritic wedges into ordered hexameric arrays based on focal modification to incorporate rigidly juxtaposed tetracarboxylic acid moieties facilitating *H*-bonding (see Section 9.4.2.3). SANS data were supportive of the hexameric assembly.[115, 116] Zimmerman et al.[117] also attached different wedges to an anthyridine focal moiety to produce strongly *H*-bonding, or appropriately described as "sticky", dendrons [**112** (G2); Figure 5.6]. A focally-substituted benzamidinium dendron was used for a binding study while a bis(benzamidinium) connector further demonstrated the potential for self-assembly. With a related heteroaryl core, naphthyridine, Zimmerman et al.[118] analyzed amidinium guest complexation; fast "on-off" rates, high stability, and relative insensitivity to solvent polarity or dendrimer size and flexibility/rigidity suggested a highly porous host in this supramolecular process. Other closely aligned heterocycles at the focal point of these Fréchet dendrons have been demonstrated[119] to self-assemble in an unambiguous DDA·AAD (where, D = donor and A = acceptor) *H*-bonding motif.[119a]

Lehn, Zimmerman, and coworkers[120] have connected these poly(ether) wedges to phthalhydrazide [6,7-bis(alkyloxy)-2,2-dihydrophthalazine-1,4-dione], which exists as an equilibrium mixture of three tautomers with the lactim-lactam form being the most stable in ethanolic solution. The self-assembly of this modified tautomer has resulted in dendrimer formation through *H*-bonded core creation (**113**; Figure 5.7). Characteriza-

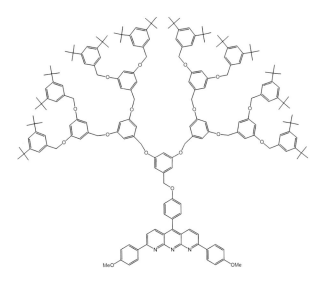

112

Figure 5.6 Zimmerman's anthyridine-modified G3 dendrons.[117]

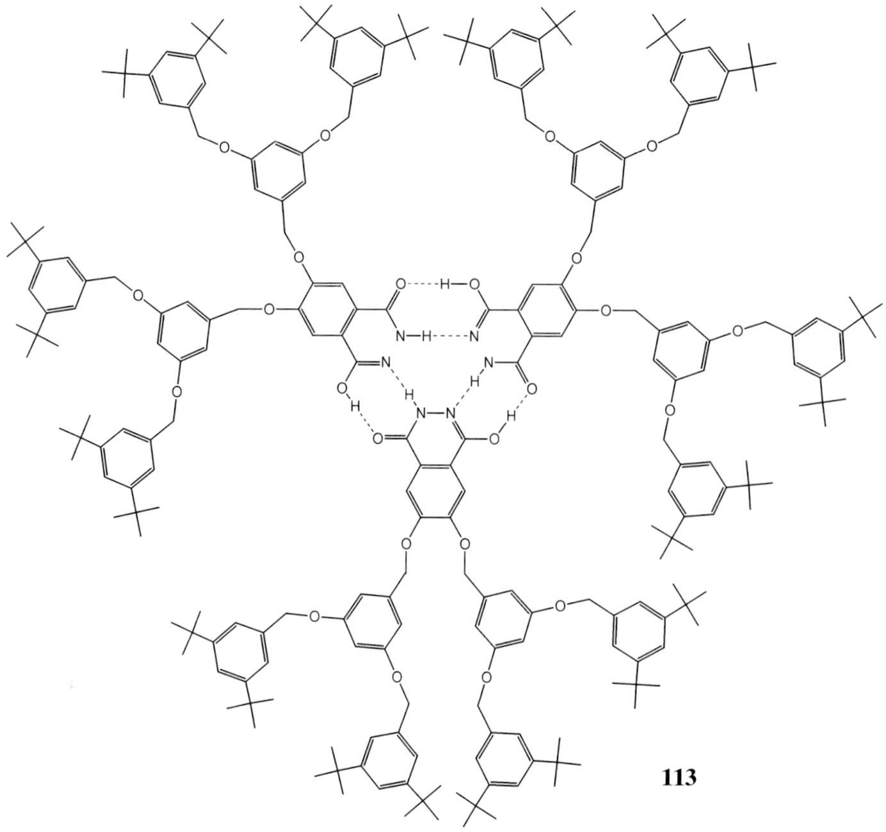

113

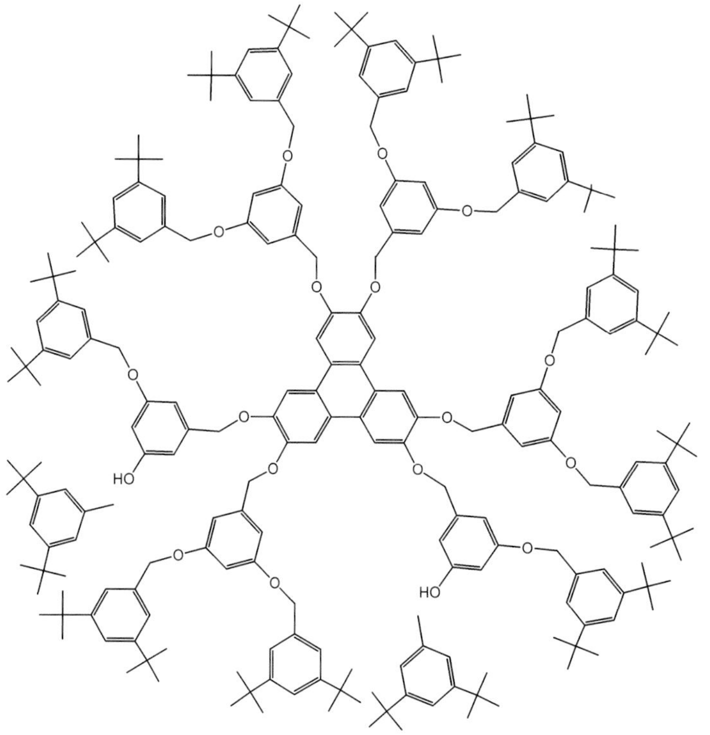

114

Figure 5.7 Dendrimer self-assembly based on a trimeric *H*-bonded core and a covalently connected counterpart.[120]

tion of the trimeric aggregate included the synthesis of the covalent-based core molecule **114**, which exhibited a nearly identical SEC retention time to that observed for the trimer.

Kawa and Fréchet[121] focally modified their poly(benzyl ether)s to include carboxylate anions amenable to the coordination of the trivalent lanthanide cations Er^{3+}, Tb^{3+}, and Eu^{3+}.[121a, b] Self-assembly of three of these dendron-carboxylates with a single lanthanide produced luminescent molecular balls whereby luminescence activity was found to be dependent on dendritic shell size. This dependence was attributed to the antenna effect provided by the branched framework as well as the site-isolation of the individual lanthanide cations prohibiting self-quenching. Use as energy-harvesting devices or fiber optic amplifiers was suggested. An alternative approach to complexation was recently reported by Hannon, Mayers, and Taylor,[122] who incorporated a tripyridinylmethyl moiety at the focal center for metal binding; in order to ensure water solubility, the periphery of the wedges was coated with small PEG units. Poly(benzyl ether)s with a carboxylate focal moiety have been used as acid/base systems in nonpolar organic solvents;[123] the effect of different ratios of the two buffer forms on the catalytic activity of subtilisin Carlsberg and chymotrypsin was evaluated in toluene.

Focal modification of these dendrons (G1–3) with 4-carboxydibenzo[24]crown-8, accessed through the high-dilution reaction of tris(ethylene glycol) dichloride, methyl 3,4-dihydroxybenzoate, and 1,2-dihydroxybenzene, via esterification (DEAD, Ph_3P) afforded the corresponding macrocycles **115** (Scheme 5.25).[124] The self-assembly of three equivalents of **115** with tris(ammonium) core **116** generated pseudorotaxanes **117**, which were characterized by ^1H NMR and MALDI-TOF MS. Notably, the formation of **117** was slow relative to that of its smaller analogs due to steric congestion and decreased solubility in $CDCl_3$. Gitsov and Ivanova[125] focally modified these dendrons, through to the 4th generation, with crown ethers in order to study the self-assembly and binding properties.

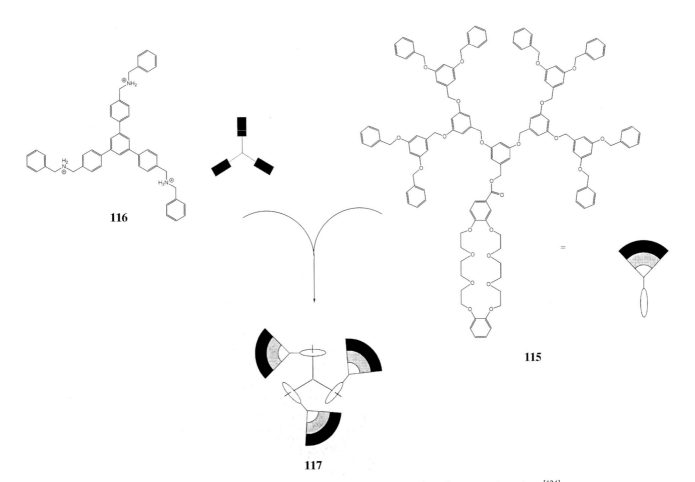

116

115

117

Scheme 5.25 Gibson's pseudorotaxanes obtained by crown ether ammonium ion complexation.[124]

Hawker et al.[126] modified the focal site of a 4th generation dendron with di-, tri-, and tetraethylene glycol to examine the properties of the products at the air–water interface. Later, focally-modified dendrons with linear oligo(ethylene glycol) "tails" were prepared and Langmuir stability studies thereof were conducted by Hawker, Frank, et al.[127] The 3rd and 4th generation constructs were observed to form stable films, while the 5th tier dendron films were not stable at 20 °C. Film stability increased with increasing oligomer length (e.g., 4.1 mN/m per ethylene oxide repeat unit for the 4th generation material). Li and coworkers[128] have also modified these dendrons with focal chains and carboxylic termini to examine the Langmuir characteristics. Highly compressible and stable monolayers were observed at the air–water interface; AFM data also supported the highly ordered aggregation.

The 4th generation dendritic wedge, possessing a hydroxymethyl focal group that was functionalized first with BuLi and potassium *t*-pentoxide and then with Me_3SiX or dodecyl electrophiles, has been analyzed by reversed-phase HPLC.[129] In contrast to the rough picture of composition obtained using size-exclusion techniques, reversed-phase HPLC facilitated component separation and analysis.

Stewart and Fox[130] focally modified naphthyl-terminated dendrons with a norbornenediol ketal moiety then subjected it to a ring-opening metathesis polymerization producing hybrid polymers. Notable polymer aggregation and chromophore interaction in dilute solution was suggested by substantial excimer emission.

Zhang et al.[131] and others[131a] prepared thio-based, self-assembled monolayers with these ubiquitous dendrons on silver and gold films and studied them by FT-surface enhanced Raman scattering spectroscopy and STM. Ionically bound dendron monolayers have been prepared for use as positive resists in scanning probe lithography; these new poly(benzyl ether) dendrimers possess either focal or peripheral carboxylic acid moieties[132, 133] as well as other functionalities.[134, 135] Hilborn, Hedrick, and coworkers[136] developed a mild and versatile synthesis of thiol-modified polymers; the method is based on Sanger's reagent, i.e., 2,4-dinitrofluorobenzene.

Aida et al.[137] prepared a series of *tert*-Boc-(L)-tyrosinyl-(L)-alanine moieties attached at the focal site of the poly(aryl ether) wedges. Gelation occurred upon standing overnight; it was proposed that ca. 20,000 solvent molecules were gelled to a single dendrimer and that this transparent gel was induced by *H*-bonding interactions of the dipeptide moiety.

5.3.2.2   Attachments

The attachment of the poly(aryl ethereal) Fréchet-type wedges to other materials or molecules to change their chemophysical properties has by far dominated their common use.

Gitsov et al.[138–141] reported the syntheses (Scheme 5.26) of dendritic polyethereal copolymers (e.g., **119**) by the treatment of the benzyl ether wedges (e.g., **101a**) with polyethylene glycol (PEG) and polyethylene oxide (PEO).[142] It was determined that the rate constants for the reaction of PEG or PEO with bromodendrons of various sizes increased as the dendritic generation and linear block length increased. Synthesis of the related linear–dendritic 'triblock' copolymers (i.e., **120**)[143–146] was accomplished by anionic polymerization of styrene, initiated by potassium naphthalide, modification of the resulting "living" polystyrene with 1,1-diphenylethylene, and finally quenching with a reactive dendron, such as bromide **101a**. MALDI-TOF MS has been utilized for the characterization of the dendrimer–PEG copolymers.[147] Similar linear-dendritic architectures, using "well-defined blocks" of PEO, PEG, polyester, as well as polystyrene, have been reported.[148] These hybrid linear–dendritic block copolymers consisting of poly-(ethylene ether) dendrons attached to PEGs have also been investigated by static and dynamic surface tension measurements as well as by adsorption experiments on polymeric substrates thereby enhancing the substrate hydrophilicity.[149] Use of focal reactive sites for polymerization initiation leads to the sterically induced avoidance of unwanted side reactions, thereby demonstrating the potential of the dendrimers to perform as nanoscale reactors. Cozzi et al.[150] reported the use of poly(ethylene glycol)s (4,600

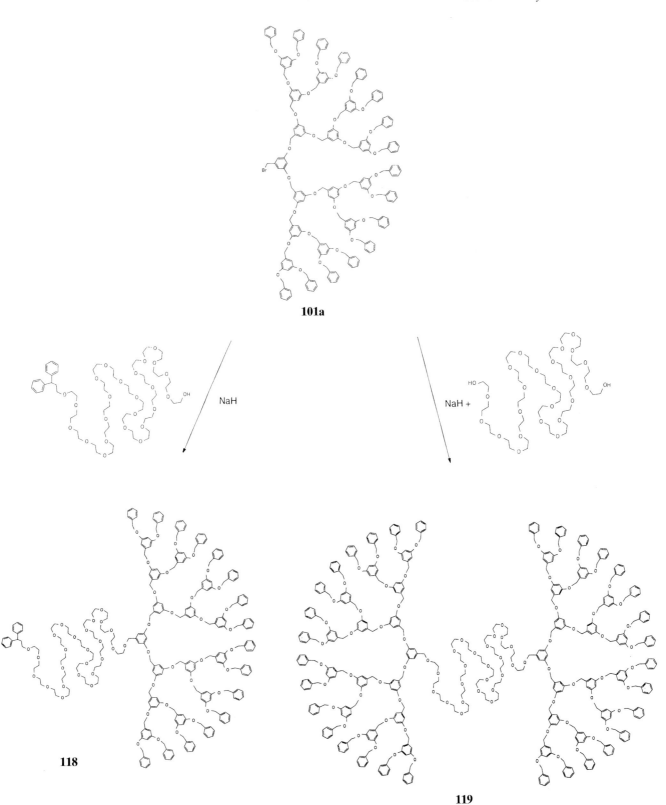

Scheme 5.26 Fréchet-type dendritic wedges with PEO and PEG.[138–140]

amu) functionalized with low generation poly(aryl ether)s that were terminally modified to facilitate "soluble, solid-supported synthesis". The solution behavior of these ABA block copolymers that self-assemble in aqueous media to form micelles with highly branched nanoporous cores has recently been reported by Gitsov et al.[151]

Gitsov and Fréchet[89] also reported the attachment of poly(aryl ether) dendrons, constructed through to the 3rd generation, to the ends of a four-arm, PEG star polymer (**121**) possessing a nominal M_w of 15,000 amu. Terminal hydroxy star attachment to the focal benzyl bromides was accomplished under classical THF/NaH conditions. Due to the contrast in polarity of the connected dendrons and the PEG chains, these hybrid macromol-

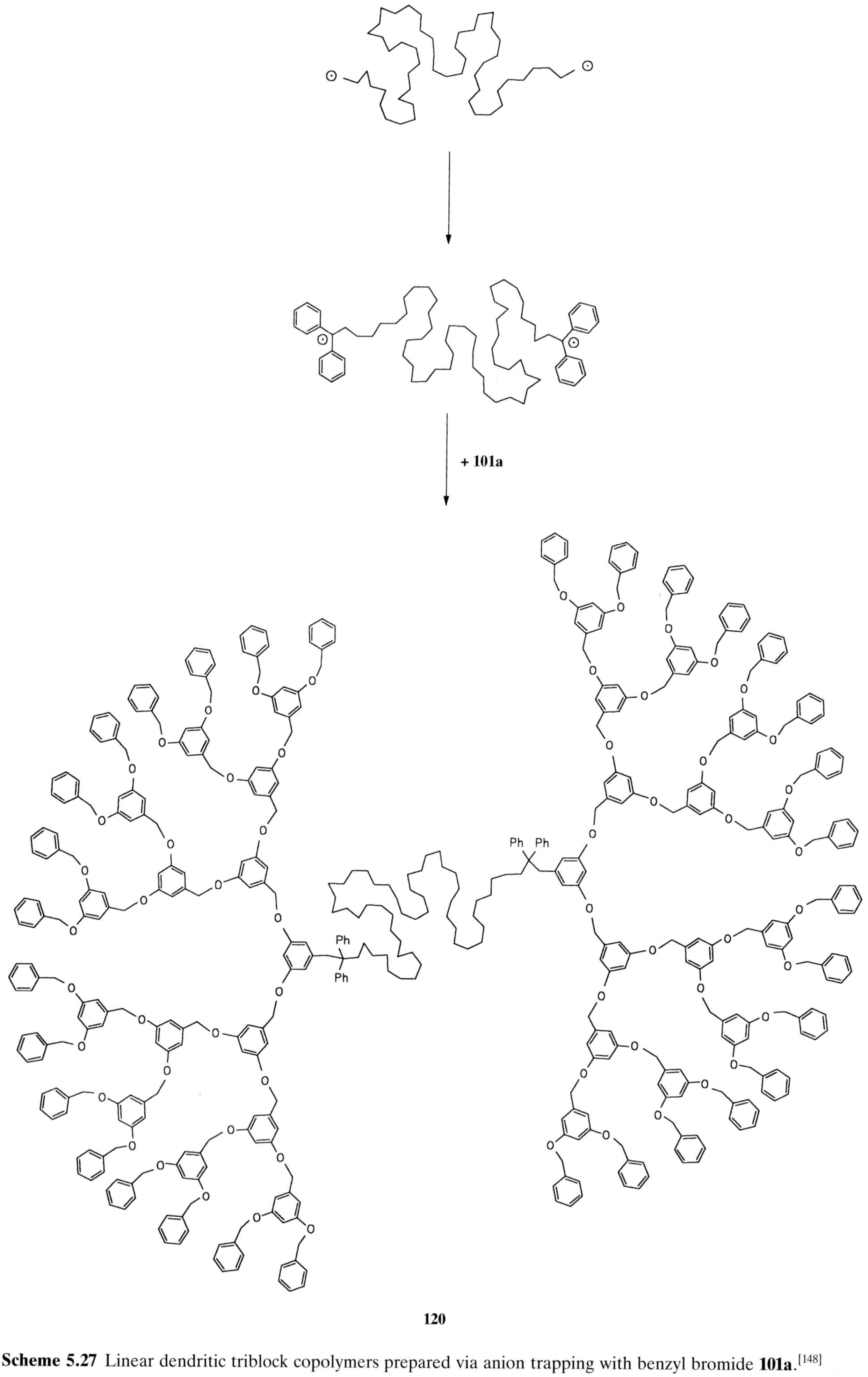

120

Scheme 5.27 Linear dendritic triblock copolymers prepared via anion trapping with benzyl bromide **101a**.[148]

Figure 5.8 Gitsov and Fréchet's star PEGs (**121**).[89]

ecules adjust conformationally to accommodate changes in solvent polarity; this affects the molecular size as well as the geometry. Studies of the hybrids using SEC and NMR suggested a conformational equilibrium in response to solvent polarity changes as depicted in Figure 5.8. Progressing from THF to $CHCl_3$, the compacted star portion (**A**) becomes more open (**B**); a change to the more polar MeOH forces shielding of the hydrophobic dendrons. Two forms of the MeOH solvated structure were noted, in which the dendritic portion is tightly compacted near the core (**C**) or loosely wrapped by the PEG chains (**D**).

Inoue et al.[152] reported the attachment of preconstructed polyethereal dendrons, prepared by the method of Fréchet,[46] to an eight-directional porphyrin core (**122**) to form dendritic porphyrin **124** (Scheme 5.28). Attachment of the wedges afforded a series of photochemically active porphyrins buried within an increasingly crowded environment. Characterization of these macromolecules was accomplished by SEC and FAB MS, as well as by a photochemical investigation of the corresponding zinc complexes. For the larger dendrimers, access to the core by relatively small quenchers, such as vitamin K_3, was facile, whereas with larger quenchers, such as the 1st tier dendrimer, core access was effectively denied. Aida et al. have also investigated fluorescence quenching of these dendrimers using methyl viologen.[153] Tang and Nikles reported similar dendrimerized porphyrins.[154]

McKeown et al.[155] [155a] attached these arylether dendrons to phthalocyanines to produce glass-forming columnar mesogens, which were demonstrated to produce clear, crack-free solid films upon cooling from the melt phase or spin-coating on a glass surface. They examined[156] [156a] the effects of axial substitution by dendrons on phthalocyanine-coordinated silicon. Tunable optical and electrical properties were envisaged. These materials were accessed through bound Si–Cl substitution with focal benzyl alcohols. An X-ray structure was obtained for the 2nd generation construct, which showed the symmetrical conformation of the ethereal arms as well as the crystal lattice packing in the unit cell.

Ferguson et al.[157] reported the preparation of Fréchet-type dendrimers employing calix[4]- and -[5]-arenes as polyfunctional core moieties. Using the 1st to 3rd generation benzyl-capped, bromobenzyl focally-substituted dendrons, the corresponding calix[*n*]arene-based dendrimers were formed; reaction (NaH, THF, DMF) of the 3rd tier wedge with each core produced the fully substituted calix[4]- and -[5]-arenes, in 21 and 22% yield, respectively. These authors investigated partial substitution of the calix[4]arene through modification of the reaction conditions; thus, reaction (Me_2CO, K_2CO_3) with an ester-terminated 3rd generation dendron produced hexadecaester **125a** (Figure 5.9), which was converted (NaOH, THF, H_2O, MeOH) to the corresponding polyacid **125b**. A wedged calixarene core was prepared by a "double-stage" approach using silyl-protected hypermonomers.[158] Small-angle X-ray scattering was then used for structure confirmation, the scattering data suggesting an ellipsoidal shape with a uniform electron density distribution.

L'Abbé et al.[159, 160] combined the novel trithiolate core (**126**), prepared in three-steps from triacylbenzene, with Fréchet's (1st and) 2nd generation ethereal benzyl bromides to yield the rigid, trialkyne core dendrimers **128** (Scheme 5.29); NMR and MS techniques structurally supported these symmetrical products.

Nierengarten et al.[161] [161a] attached these dendrons to a cyclotriveratrylene core, which is known to form host–guest complexes with C_{60}. Based on UV/vis titrations in CH_2Cl_2 and C_6H_6, C_{60}–dendrimer binding constants were observed to significantly increase with generation rising from a value of 85 ± 20 dm^3 mol^{-1} for generation zero to 340 ± 20 dm^3 mol^{-1} for the 3rd generation construct. Shinkai et al.[162] also demonstrated the ability of Fréchet-type dendrimers, possessing a phloroglucinol core, to host the C_{60} guest. Later, Nierengarten et al. reported[163] monomers[163a, b] bearing C_{60} so as to allow the construction of dendrons with methanofullerene moieties at each successive generation. Other C_{60}-hybrids are known.[163c]

De Schryver et al.[164] examined the solvent dependence of the hydrodynamic volume of these poly(aryl ether)s attached to a rubicene core. The volume, determined by a time-resolved fluorescence depolarization technique, was found to be substantially smaller in MeCN (a poor solvent) than in toluene or Me_2CO (good and medium sol-

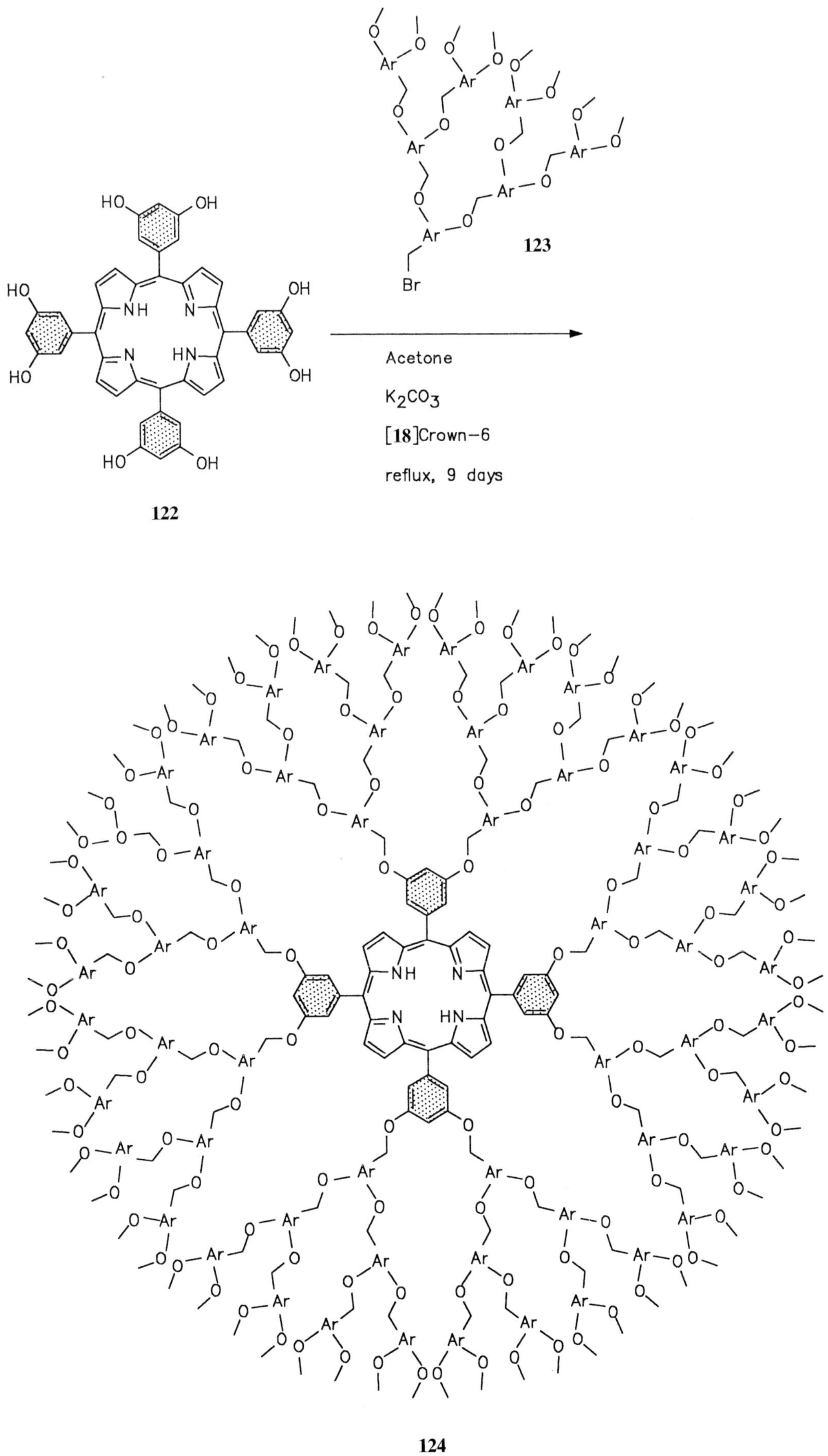

Scheme 5.28 Porphyrin modification[152] by the attachment of Fréchet-type wedges.

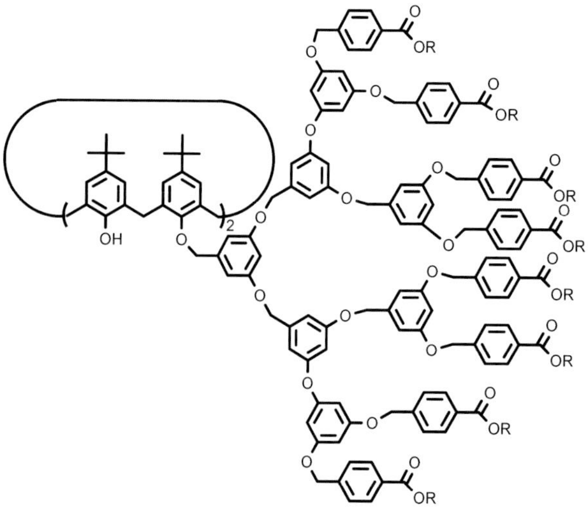

125a (R = Me)
  b (R = H)

Figure 5.9 Dendritic calixarenes (**125**)[157] based on poly(aryl ether) dendron attachment.

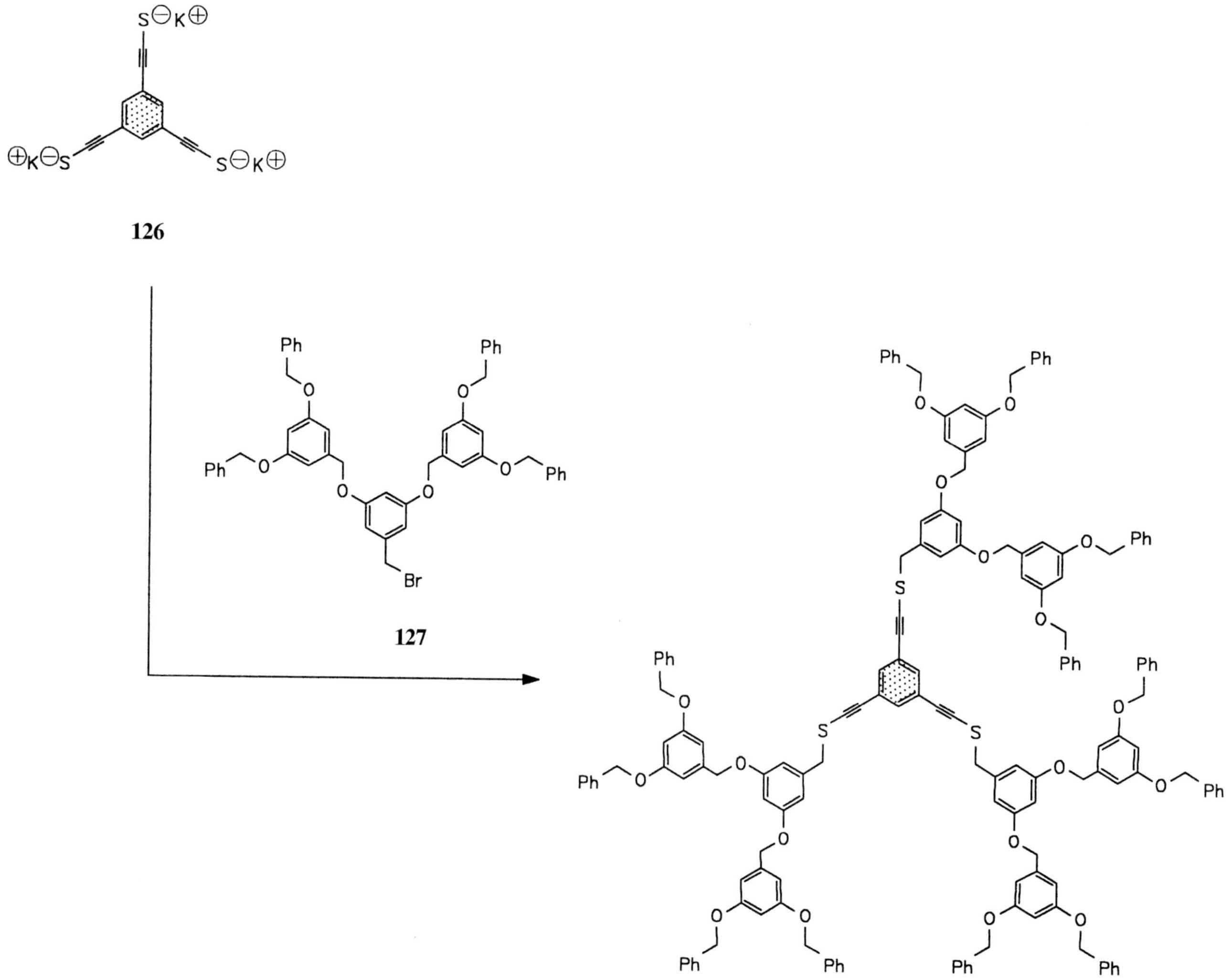

126

127

128

Scheme 5.29 Introduction of rigid, trialkyne cores[160] by the convergent methodology.

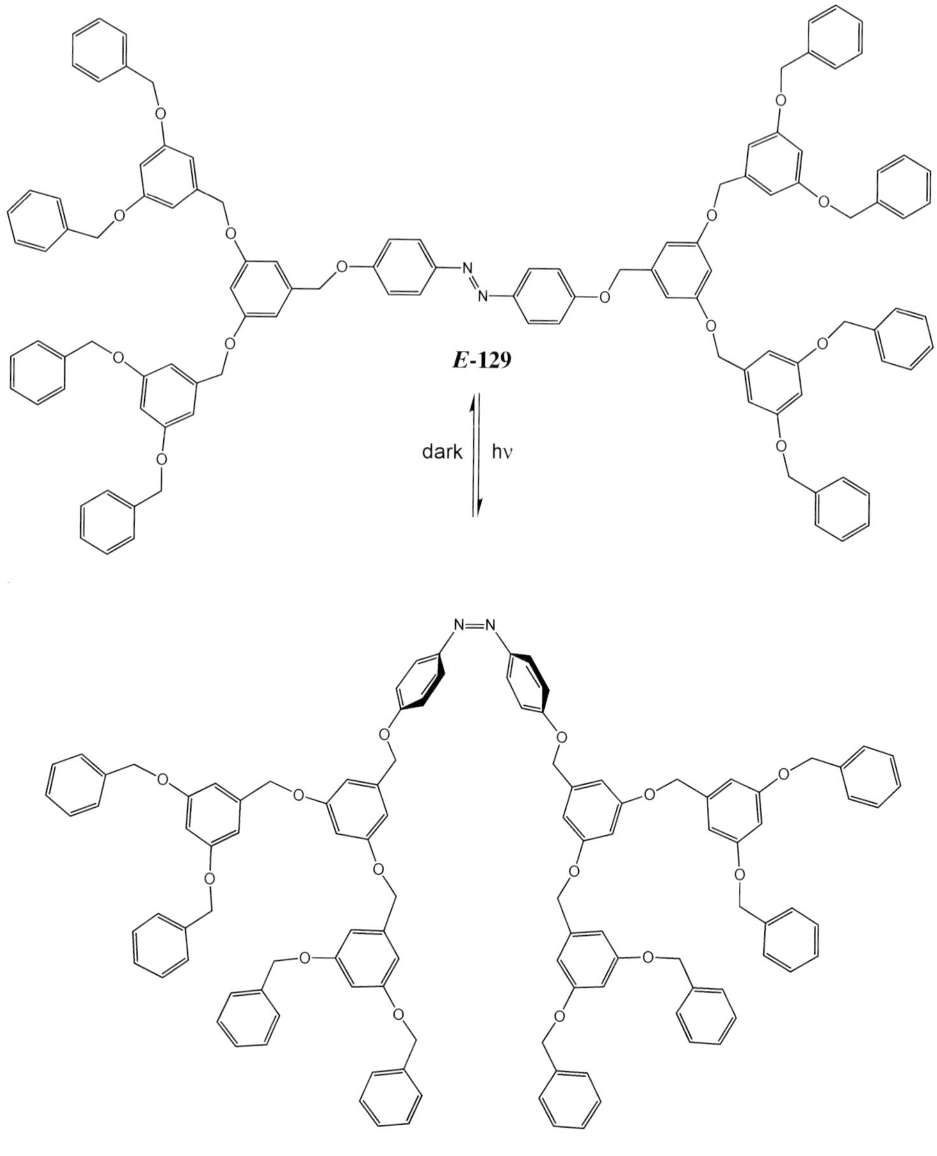

E-129

dark ↕ hν

Z-129

Figure 5.10 Azobenzene cores possessing (aryl ether) dendrons exhibit photoisomerization.[166]

vents, respectively). The attachment of these dendrons, up to the 3rd generation, to fluorescent dihydropyrrolopyrroledione has been accomplished[165] and subsequently used to probe localized information on compatibility and microviscosity in polymer films. Spin coating a solution of these unique chromophores and polystyrene on glass permitted the imaging of single dendritic molecules using a confocal microscope. Use of an electrooptical modulator, two avalanche photodiodes, and a polarizing beam splitter facilitated differentiation of dendrimer clusters and single species.

Poly(aryl ether) dendrons have been attached[166] to a bis(phenolic)azobenzene core to produce photoresponsive dendrimers (**129**; Figure 5.10). The 1st and 2nd generation dendrimers were prepared, as well as analogous monomethyl monodendrons. Isomerization was evidenced by standard UV absorption data; exposure to 350 nm light produced an increase in UV absorbances at 314 and 450 nm indicating the *cis* isomer, while prolonged absence of light maximized the 360 nm absorbance indicative of the *trans* isomer. 1H NMR data obtained for the monomethyl constructs further supported the isomerization process. An analogous PAMAM-based architecture has been reported.[166a] Other photoresponsive azobenzene-modified poly(aryl ether)s (Figure 5.11) have been created,[167–169] in which dendrons, focally transformed to possess a hydroxybenzylazobenzene moiety, were subsequently reacted with a suitable core to afford the tris(azobenzene) architecture (e.g., **130**). Due to the potential for *E/Z* isomerism about each azo

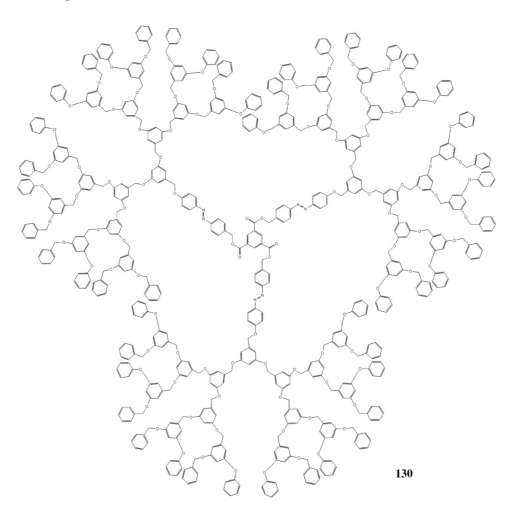

130

Figure 5.11 Photoresponsive dendrimers.[167]

moiety, four discrete conformations can exist for each dendrimer (i.e., *EEE, EEZ, EZZ,* and *ZZZ*). Notably, all four states were characterized and each possessed different physical properties. Li and McGrath[170] proceeded to evaluate the placement of the photochromic groups by synthesizing isomeric structures bearing the stimuli-responsive groups on their surfaces; it was concluded that the relative placement of this azobenzene moiety within the macromolecule can dramatically influence the extent of photomodulation of the dendrimer's properties.

Jiang and Aida[171] attached these dendrons to an azobenzene core such that channeling of absorbed energy at the core was shielded by the branched matrix; similar methoxy-terminated poly(aryl ether) antennae based on a diazobenzene core have also been reported.[172] For the 4th and higher generations in the series, *cis* to *trans* isomerization of the compartmentalized moiety was observed following exposure to infrared excitation (1597 cm^{-1}). Smaller dendrimers (i.e., generation $\leq$ 3) did not exhibit this phenomenon. A dependence of the isomerization ratio constant on the distance of the radiation source indicated an "unusual multiphotonic" process. For the L5AZO spherical macromolecule, the number of IR photons absorbed was estimated to be only 10^{-3} (photons/sec), suggesting that this large dendrimer does not absorb five photons simultaneously but rather sequentially;[173] thus, there is potential for long-term intramolecular energy storage.

McGrath et al.[174] attached these dendrons to a photolabile core prepared by focal modification with an α-substituted-*o*-nitrobenzyloxy group, which was subsequently attached via esterification to 1,3,5-tris(chlorocarbonyl)benzene; notable site-specific photodegradation (i.e., sequential cleavage of the dendrons from the core) was demonstrated upon UV irradiation.

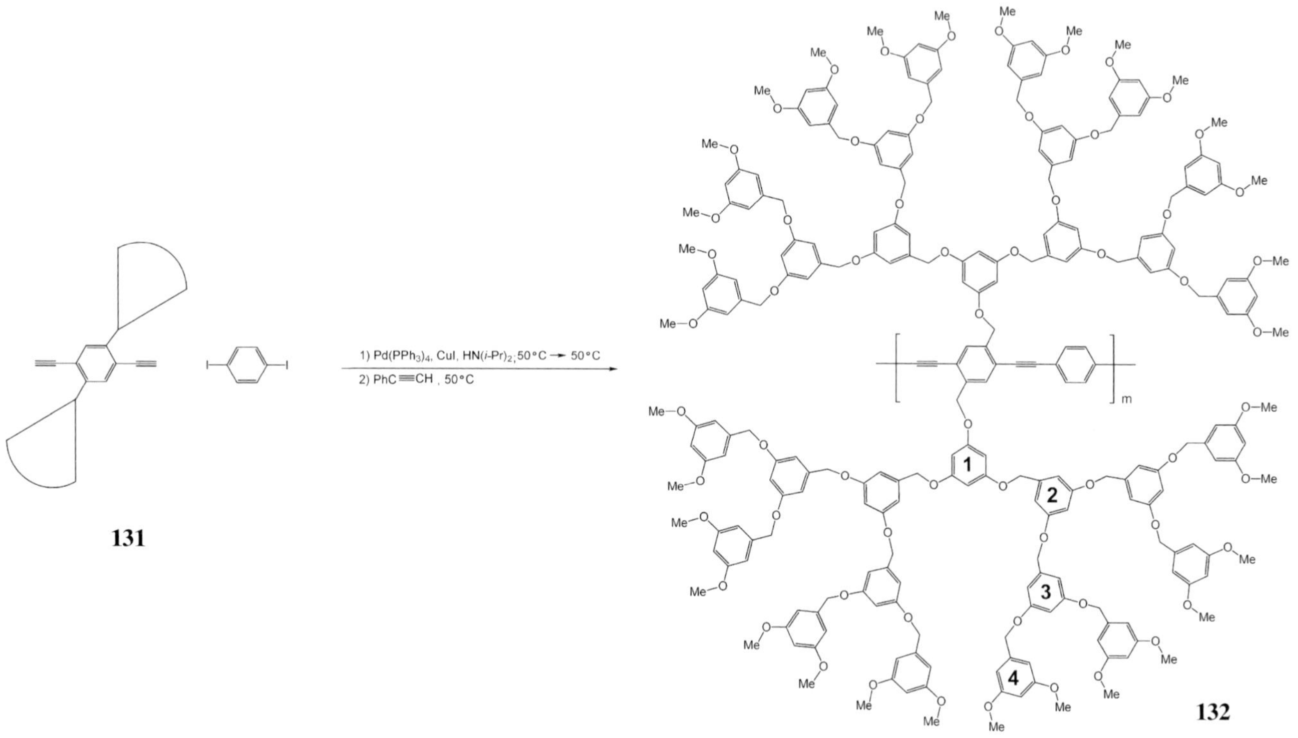

Scheme 5.30 A simple route to blue-luminescent molecular rods (**132**).[176]

Diederich et al.[175] [175a] coated phenylacetylene end-capped poly(triacetylene) oligomers with these G1–3 dendrons. Distortion from planarity due to steric effects of the branched fragments did not affect the π-electron conjugation; the dendritic units constituted an insulating layer.

By attaching these dendrons to the diyne precursor of poly(phenylenethynylene) (**131**; Scheme 5.30), Sato, Jiang, and Aida[176] [176a] created blue-luminescent dendritic rods **132**. The key precursor was accessed by reaction of 2,5-diethynylhydroquinone with benzyl bromide dendrons (G2–4). Evidence of dendritic light harvesting was provided by 100 % intramolecular singlet energy transfer quantum efficiency.

Stoddart et al.[177] incorporated Fréchet's dendrons into the self-assembly of [2]-, [3]-, and [4]-rotaxanes. The general strategy for assembly of these unique macromolecules is illustrated in Scheme 5.31; two equivalents of bromodendron **133** are reacted with the bis(bipyridinium) rod **134** in the presence of 8 molar equivalents of bis(*p*-phenylene)-34-crown-10 (**135**) under ultra high pressure conditions (12 Kbar, 25 °C, 72 h) to afford the [3]- (**136**) and [2]- (**137**) rotaxanes. An added benefit of the dendritic capping units is enhanced solubility of the polycationic bipyridinium rods. Analogously, the higher order [*n*]rotaxanes were constructed using a tris(bipyridinium) rod and 12 equivalents of crown ether (Figure 5.12). Isolated yields of the [4]- (**138**), [3]- (**139**), and [2]- (**140**) rotaxanes were reported to be 25, 15, and 4 %, respectively. Rotaxane **140** was investigated[177] with respect to the *molecular shuttle* component of the assembly by means of variable-temperature ^{1}H NMR spectroscopy. The movement of the crown ether between the bipyridinium units was found to depend on solvent polarity. For example, the rate constant increased from ca. 200 to 33,000 times per second on switching from CDCl$_3$ to (CD$_3$)$_2$CO. Molecular modeling indicated lengths for the rotaxanes ranging from 40 to 60 nm. These dendritic rotaxanes were compared to naturally occurring chemical systems possessing an active component. Perhaps as notable as the reaction-dependent selective substitution was the confirmation by X-ray diffraction analysis of the first generation, fully-substituted calixarene dendrimer. A core cone conformation was evidenced, whereby opposing core aromatic rings were seen to be nearly normal [A and C] or parallel to one another [B and D], along with a random, branched arm distribution. Vögtle et al.[178] [178a] used tailored dendrons as stoppers to quantify their spatial demands through deslipping of the rotaxanes.

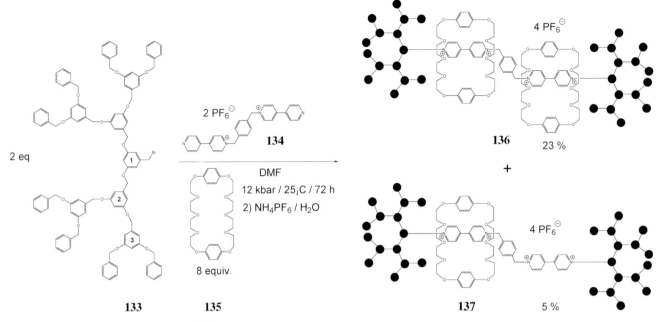

2 eq

133   **134**   **135**

DMF

12 kbar / 25¡C / 72 h

2) NH₄PF₆ / H₂O

8 equiv.

136   23 %

+

137   5 %

Scheme 5.31 Dendritic rotaxanes.[177]

Tezuka et al.[179] demonstrated the fascinating use of focal carboxylate dendrons as bulky counterions for ionic *N*-methylpyrrolidinium end groups on poly(tetrahydrofuran). Upon heating, the carboxylate moiety reacted with the activated heterocycle to produce covalently attached chain end stoppers in polyrotaxane formation.

Miller et al.[180] built nanoscale molecular dumbbells by focal attachment of poly(aryl ether) dendrons to a rigid oligoimide measuring 4.4 nm in length. Connection of the 4th generation wedges to each rod end afforded a dumbbell (**141**) of 9.6 nm when fully expanded (Figure 5.13). The well-known potential to form diimide anions and dianions, demonstrated in this case by the coulometric generation of a tetraanion, combined with the electrical and optical properties of functionalized dendrimers dramatically enhance the potential for the creation of application-oriented materials.

Electroactive oligothienylenevinylene rods have been terminally modified with trifluoromethylaryl surface-coated Fréchet-type dendrons through to the 3rd generation.[181] Dendron attachment was facilitated by a Wittig–Horner aldehyde olefination. These extended π-conjugated oligomers gave identical cyclic voltammograms showing two reversible one-electron oxidation waves attributable to cation radical and dication states. Fréchet and coworkers[182] also prepared "lengthy" oligothiophene cores (e.g., 8 and 17 units in length) and attached poly(benzyl ether) and poly(triarylamine)-terminated[182a] dendrons to the ends. These constructs have been employed as solubilizing agents for oligothiophene-based conducting polymers.[183]

Schlüter et al.[184, 185] used rod-shaped polymers,[186] such as poly(*p*-phenylene)s and poly([1.1.1]propellane)s, as cores for the attachment of convergently generated dendrons.[187, 188] These new branched macromolecules possess a rigid backbone wrapped with structural wedges that become increasingly dense toward the outer cylindrical surface. Pd-catalyzed copolymerization of dibromobiphenyl derivatives with Fréchet-type ethereal dendrons bearing substituted aryl diboronic acid moieties afforded the dendritic coated poly(*p*-phenylene) rod.

Jahromi et al.[189] [189a] focally modified these dendrons, through to the 4th generation, with the bis(ethanol)amine moiety, which, in turn, was reacted with four different diisocyanates [1,6-hexamethylene diisocyanate, 1,12-diisocyanatododecane, 2,4-di(isocyanato)-toluene, bisphenol-F diisocyanate]. The resulting side-chain dendritic polyenes did not show liquid-crystalline behavior, although X-ray diffraction experiments (G4-bisphenol-based polymer) suggested the occurrence of supramolecular aggregation into body-centered cubic lattices. Intrinsic viscosity experiments as well as molecular modeling facilitated conformational analysis in solution.

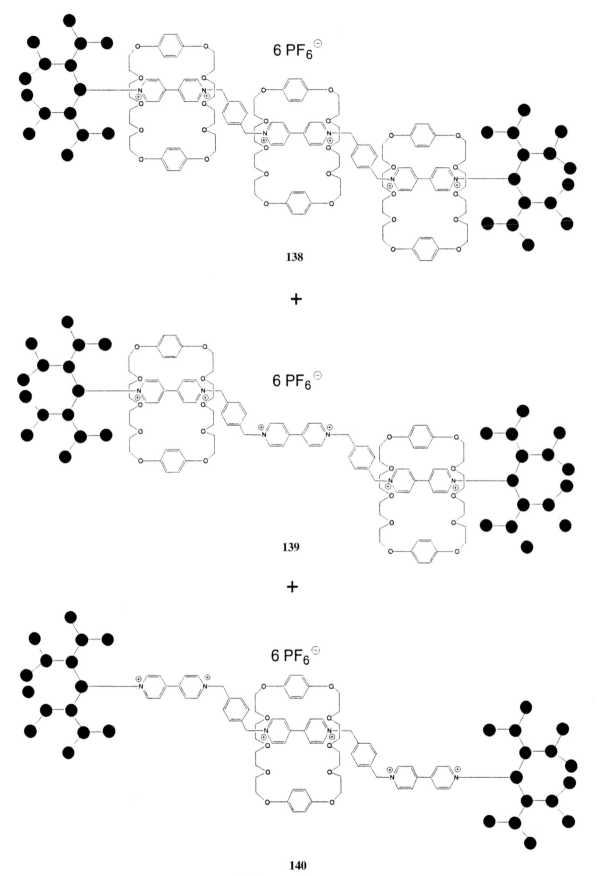

138

+

139

+

140

Figure 5.12 Dendritic-capped rotaxanes.[177]

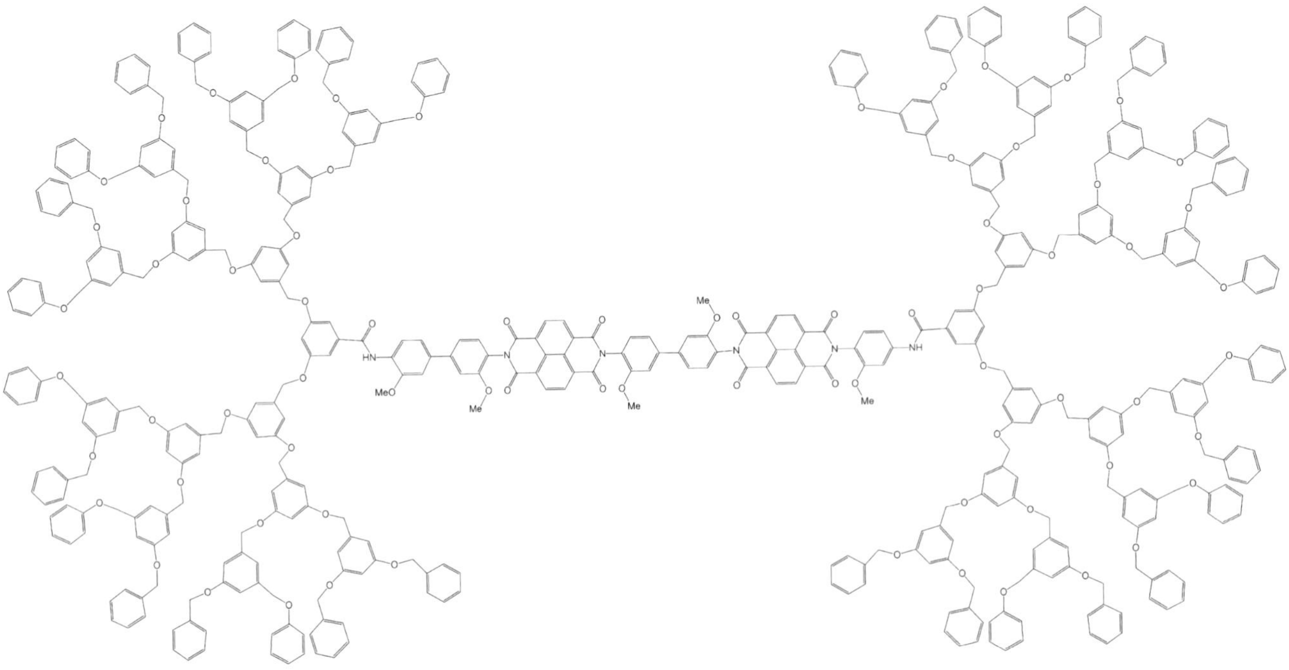

141

Figure 5.13 Miller's nanoscale molecular dumbbell (**141**).[180]

Shimomura et al.[190] described the preparation of a model system for an isolated chromophore in an amorphous matrix based on chromophoric focal modification of Fréchet-type dendrons. The novel diarylethylene chromophore, depicted in Figure 5.14, has been shown to undergo thermally irreversible, fatigue-resistant photochromic reactions. AFM was used for structural study of monolayers formed using the 2nd and 3rd generation chromophoric dendrimers (e.g., **142**). Dilute dendrimer/benzene solutions were spread on clean water surfaces and, on the basis of observed pressure/area data, G2 formed 3-D aggregates while G3 formed monolayer films. Conventional Langmuir–Blodgett dipping facilitated form transfer to solid supports. In support of G2 aggregation and G3 monolayer formation, AFM images revealed rough and smooth surfaces for mica-transferred films of G2 and G3, respectively.

Bargon, Vögtle et al.[191] reported the use of these poly(benzyl) ethers attached to a benzophenone core as selective sensor materials for the detection of carbonyl compounds. The attachment of these poly(benzyl ether) dendrons to the 9,9-positions of poly-2,7-fluorene has also recently appeared.[192]

Bo et al.[193] internally functionalized each generation of a poly(aryl ether) dendrimer with a single aryl bromide and showed them to react by Suzuki coupling with phenyl boronic acid.

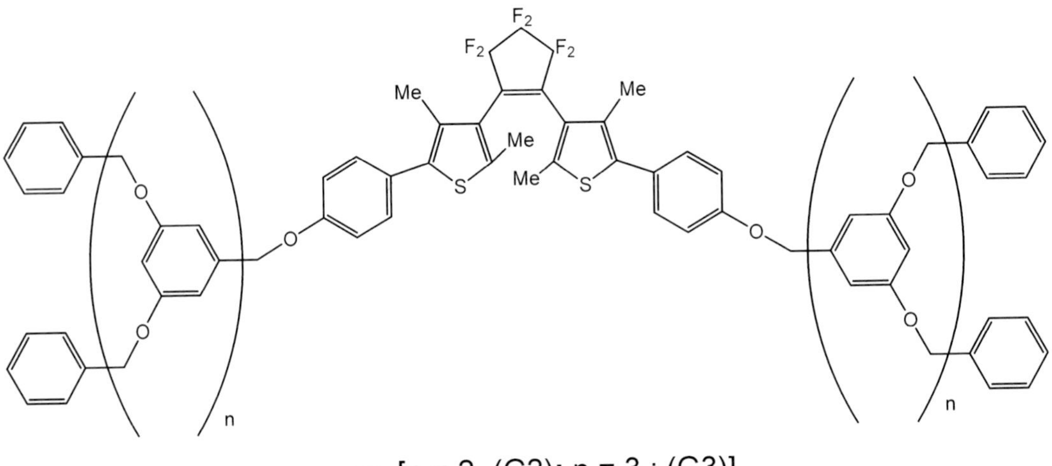

142 [n = 2 (G2); n = 3 : (G3)]

Figure 5.14 Photochromic dendrimer that forms monolayers and aggregates at the air/water interface.[190]

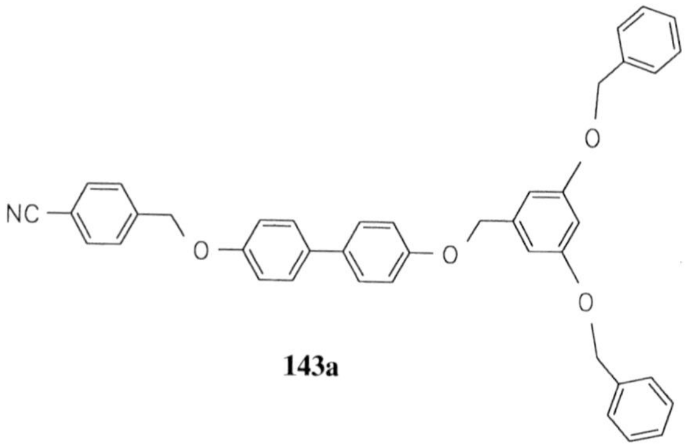

143a

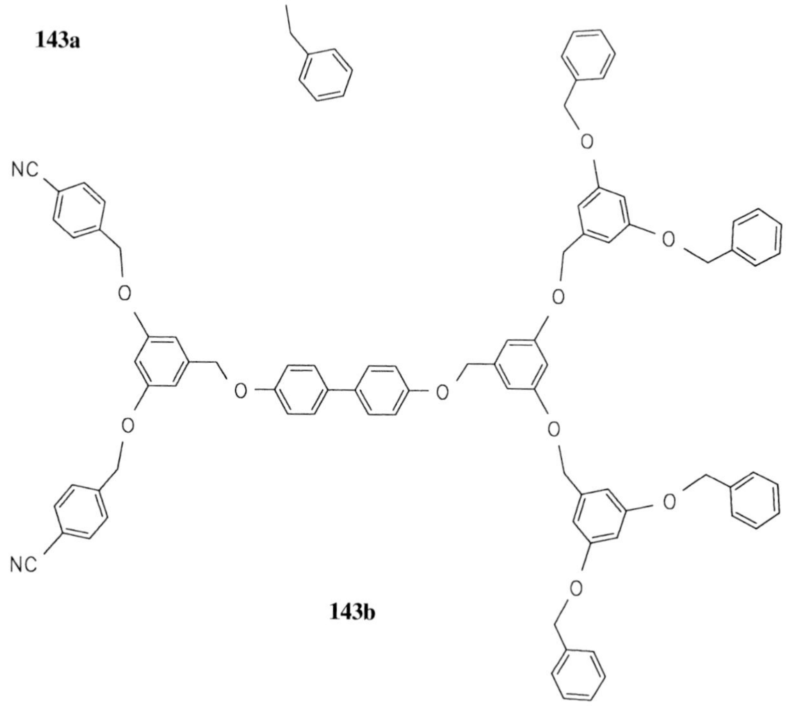

143b

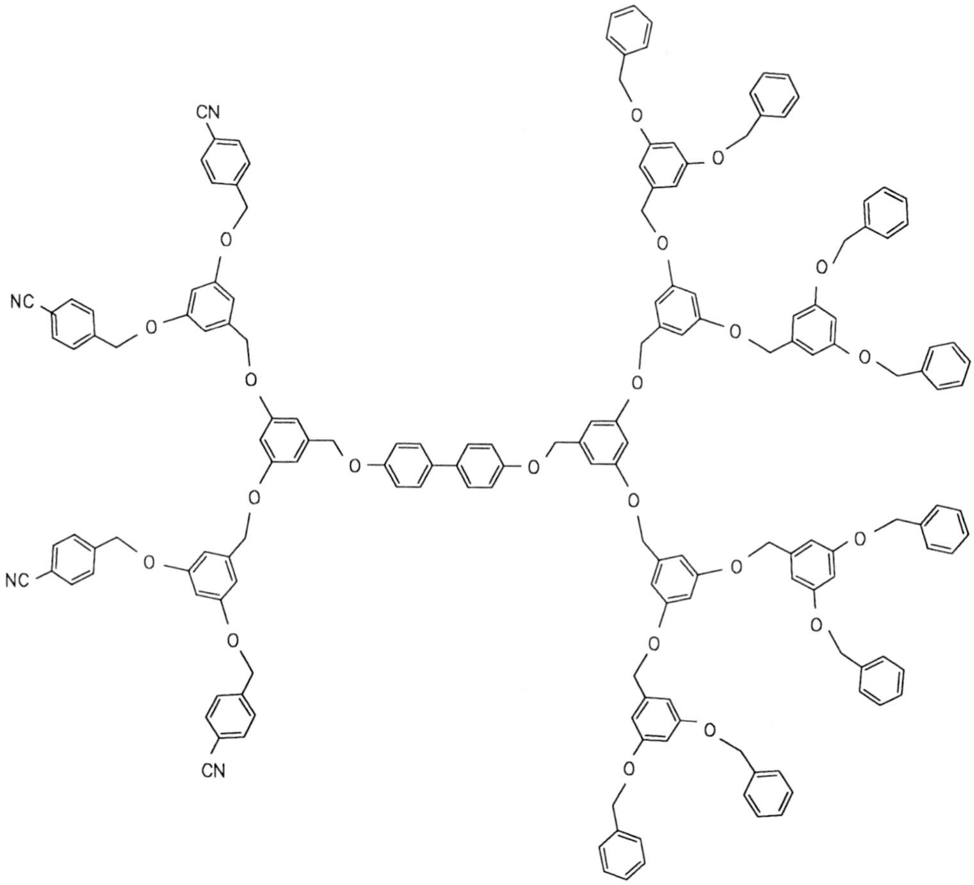

143c

Figure 5.15 (Continued on pages 232 and 233)

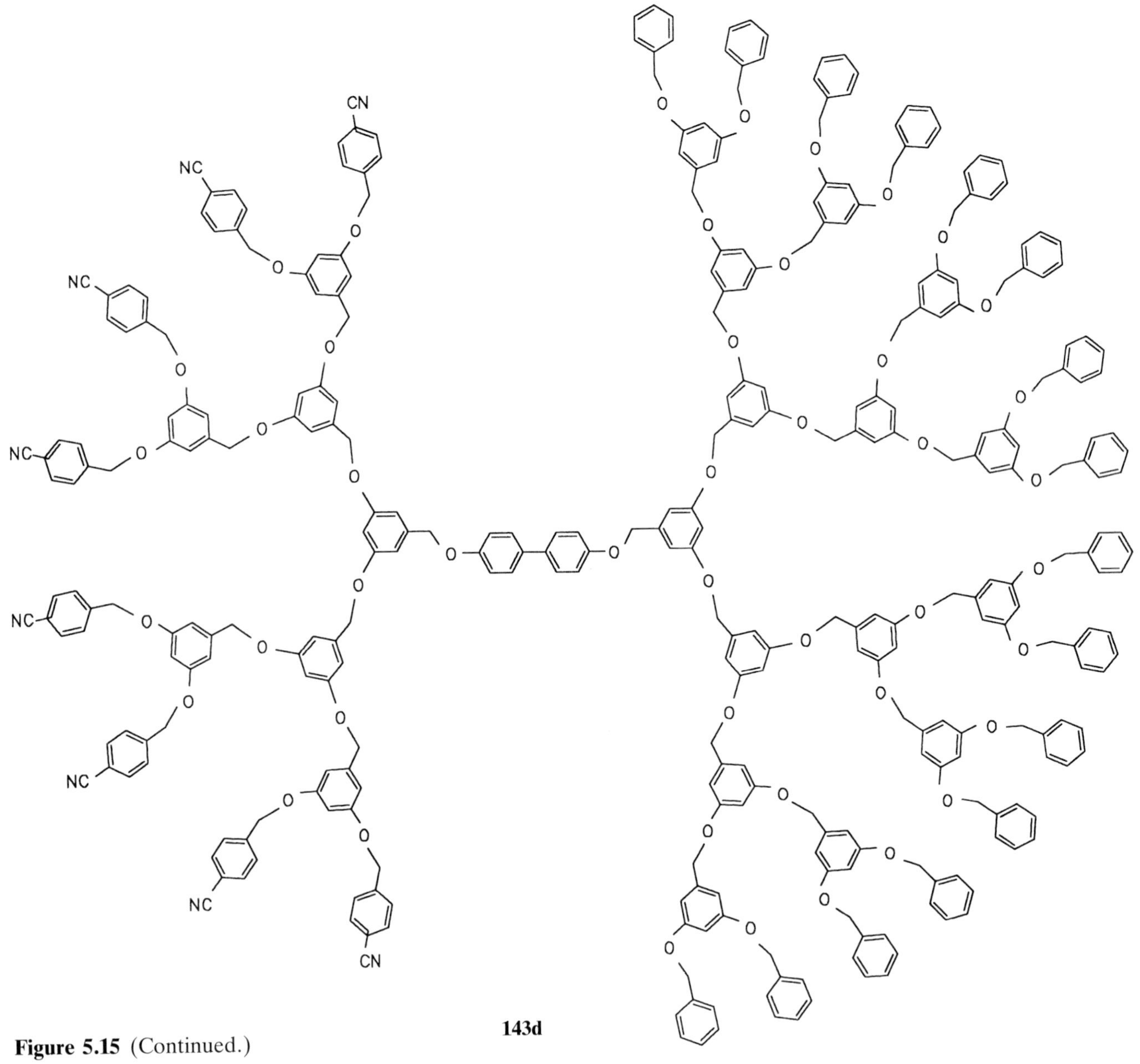

143d

Figure 5.15 (Continued.)

5.3.2.3   Terminal Modifications

The coupling of two different (e.g., one coated with cyano and the other coated with benzyl ether moieties) convergently built (aryl ether) wedges has been demonstrated[194, 195] to generate monodisperse dendritic macromolecules possessing enhanced dipole moments (e.g., **143**; Figure 5.15). Dendritic block copolymers possessing amphiphilic character, such as macromolecule **149**, have also been reported (Scheme 5.32).[196–198] Construction was based on the attachment of hemispherical wedges possessing lipophilic (**147**) or hydrophilic (precursor **144**) surface groups to various cores (e.g., **145**). Surfactant studies employing these dendritic amphiphiles as micellar substrates indicated a significant increase in the aqueous saturation concentration of polycyclic aromatics.

Fréchet et al.[199, 200] reported the control of surface functionality with respect to the number and placement of terminal groups using the convergent method and ethereal connectivity. Dendrons were constructed with single cyano, bromo, or methoxycarboxy moieties at the surface. The preparative strategy for mono-, di-, and tri-substituted cascades is illustrated in Scheme 5.33, where c, f_p, f_r, S, and X represent coupling sites, protected functionalities, reactive functionalities, surface groups, and surface functionalities, respectively. An example of this methodology utilizing termini-differentiated building blocks is given in Scheme 5.34. Thus, beginning with cyano-substituted benzyl bromide **152** and phenoxybenzyl alcohol **151**, the 4th tier cyanohalide cascade **153** was constructed in five steps.

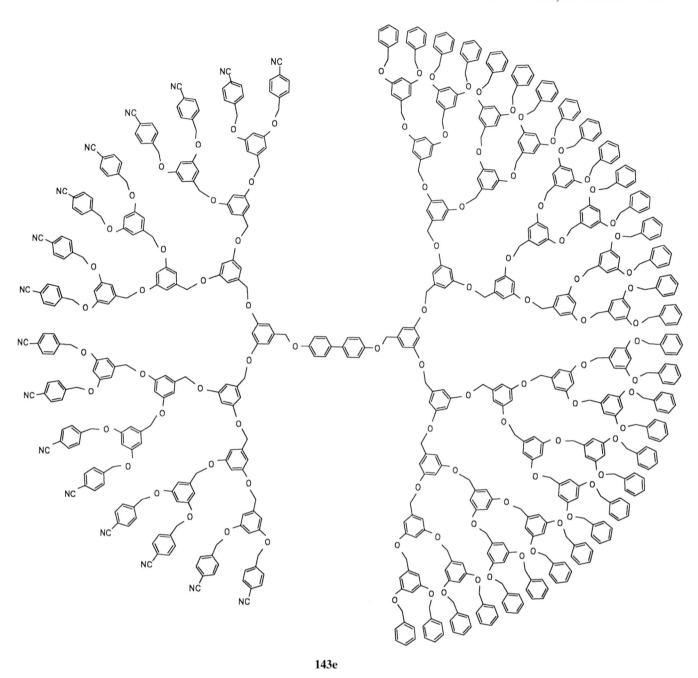

143e

Figure 5.15 Surface-modified dendrimers[194, 195] with enhanced macromolecular dipole moments.

Scheme 5.35 illustrates the preparation of tri-surface modified dendrimers (**157**). Hence, using the 4th tier, cyano-substituted wedge **153** and the non-substituted analog (not depicted) for attachment to the trihydroxy core **27**, the desired substituted dendrimers (**155–157**) were obtained by careful stepwise addition of the appropriate substituted wedge(s).

Fréchet et al.[201] further investigated the functionalization of the poly(benzyl ether) dendritic superstructure (Scheme 5.36); thus treatment of monoalcohol **101f** with a 1:3 mixture of n-BuLi and potassium $tert$-pentoxide (THF, hexane, –85 °C) gave the polypotassium salt **158**. Quenching with electrophiles, including D_2O, Me_3SiCl, $C_{18}H_{37}Br$, and CO_2, provided the functionalized **159**; however, electrophilic addition was not site-specific and multisite-substituted dendrimers, including surface-functionalized materials, were produced.

Fox and Stewart[202] constructed Fréchet-type dendrons coated with pyrenyl, naphthyl, or methyl moieties for examination as potential "light-harvesting antennae". Based on steady-state fluorescence measurements, intramolecular excimer emission in the case of naphthalene-modified dendrons was absent, while excimer emission was predominate

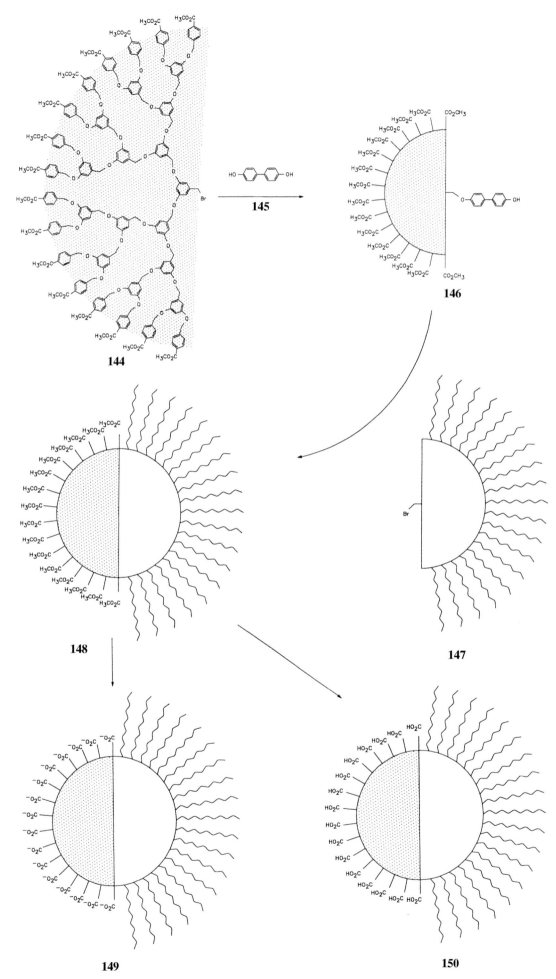

Scheme 5.32 Convergent methodology facilitated the construction of amphiphilic micellar dendrimers.[196–198]

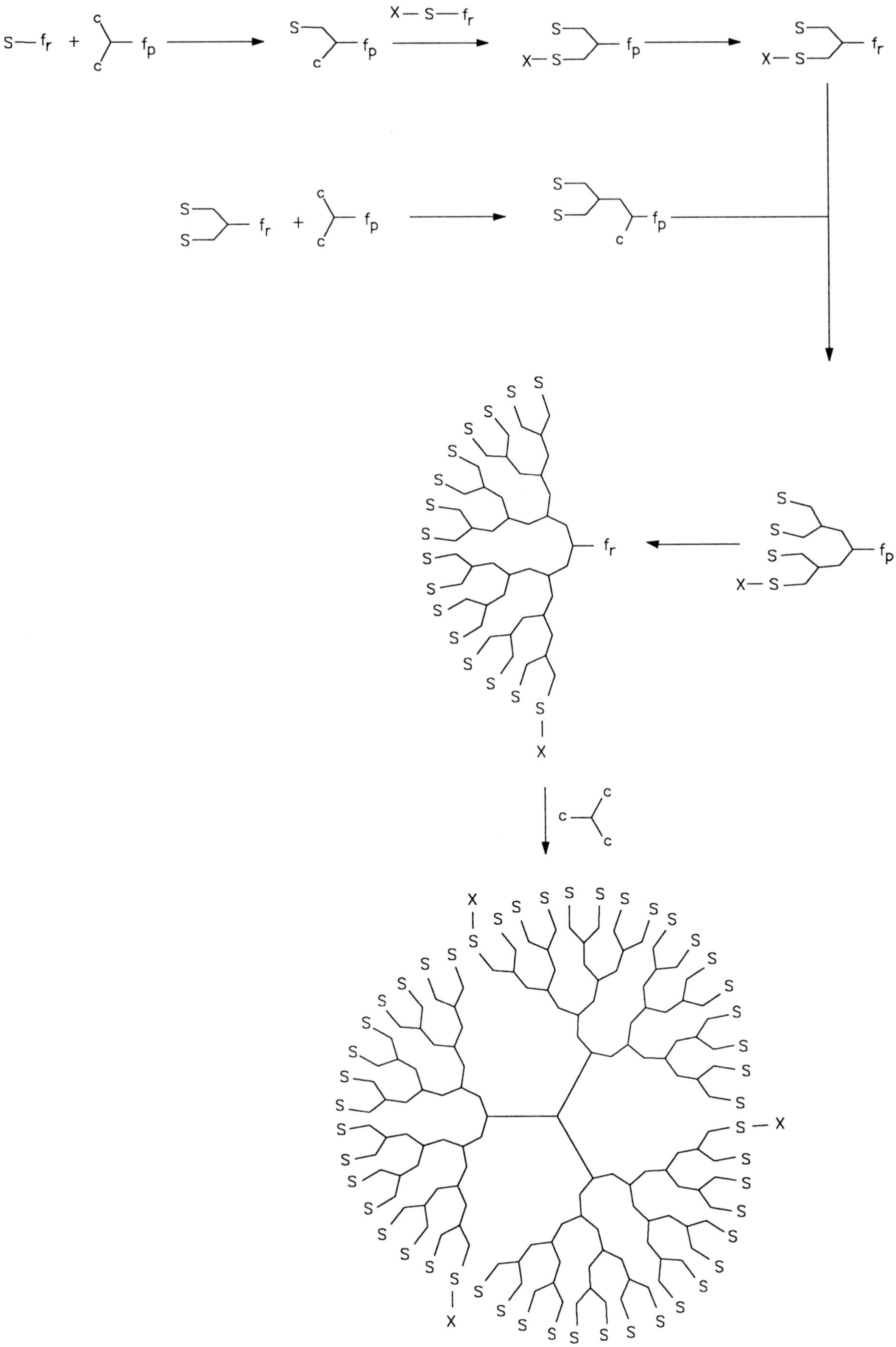

Scheme 5.33 General strategy for the construction of dendritic wedges[199, 200] possessing single, differentiated surface groups.

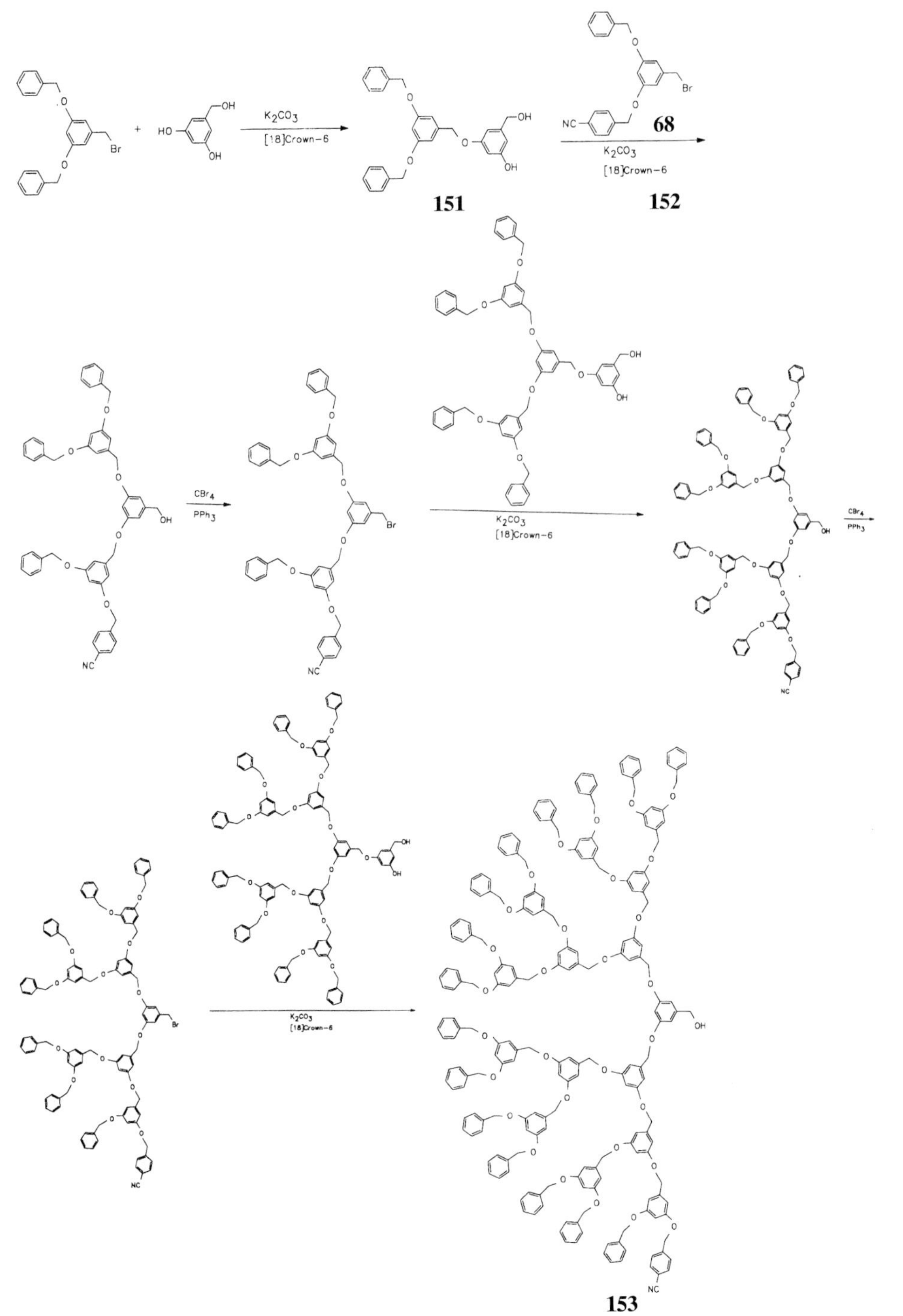

Scheme 5.34 Preparation of a poly(aryl ether) dendron (**153**)[199, 200] possessing a single nitrile terminus.

with the pyrene-capped monomers; focal modification of these "antennae" with an electron donor such as a 3-(dimethylamino)phenoxy moiety quenched the fluorescence. Fluorescence quantum yields as well as lifetimes were investigated; a 2-naphthyl-terminated 1st tier dendron and a 1-pyrenyl-terminated 2nd tier dendron exhibited lifetimes (τ) of 62 and 60 ns, respectively.

Fréchet et al.[203] created ester-terminated poly(aryl ether) dendrimers through the synthesis of a novel 1 → 2 aryl branched diethyl 5-(bromomethyl)isophthalate. Scheme 5.37 shows a "bidendron" G3 dendrimer **161**, employing 4,4'-biphenol as the core, which was reacted with alcohol **162** in the presence of dibutyltin dilaurate **160** to afford the transesterified higher generation dendrimer **163**. Analogously, the 3rd and 4th generation polyesters (see Section 5.3.7) were in turn treated with benzyl amide and carboxylic acid substructures, respectively. Notably, purification of the polyester dendrimers relied primarily on recrystallization, facilitating large-scale preparation.

Klopsch, Frank, and Schlüter[204] developed an exponential growth method (Scheme 5.38) for the construction of Fréchet-type dendrons possessing peripheral functionality with the goal of attachment to polyhydroxy biphenyl polymers. The method was used to prepare gram quantities of the 1st, 2nd, and 4th generation wedges. Construction of the 4th tier dendron began with the preparation of bis(THP ether) **164**; half was transformed to the acyl azide **165** by saponification (KOH) followed by activated anhydride formation [ClCO$_2$Et] and reaction with NaN$_3$, while the other half was converted to the diol **166** by deprotection (HCl). Two equivalents of the isocyanate, obtained by heating the azide **165**, were reacted with diol **166** to yield the 2nd generation, urethane-linked **167**. Repetition of this sequence afforded the desired hexadecaTHP wedge **168**. Preliminary experiments involving ethereal dendron attachment to the rigid polymer have been reported[205] and have led to new types of polymeric amphiphiles.[188, 206] The gram-scale preparation of a single 4th generation ethereal dendron α-attached to 2,5-dibromo-*p*-xylene has been reported; this product was then subjected to a Suzuki oligomerization.[207, 208] The dendrimerization of polystyrene with different surface-coated dendrons has recently appeared.[209] Additional notable and pioneering references, in which these authors examine rigid rod or cylindrical dendrimers, are available.[210–215]

L'Abbé, Forier, and Dehaen[76, 216] constructed two synthons described as facilitating "double-stage" convergent construction (Figure 5.16). Essentially, these building blocks can be considered as AB$_4$ (**169**) and AB$_8$ (**170**) monomers whereby concomitant deprotection of the *t*-BuPh$_2$Si protecting groups and alkylation of the resulting phenoxide anion with an appropriate dendritic wedge [e.g., (G$_n$)-CH$_2$Br] allows rapid access to higher generation materials. This was demonstrated by treatment of monomer YY with the well-known 1st to 3rd generation dendrons. These synthons were used to synthesize dendrons with functional groups at the periphery.[76]

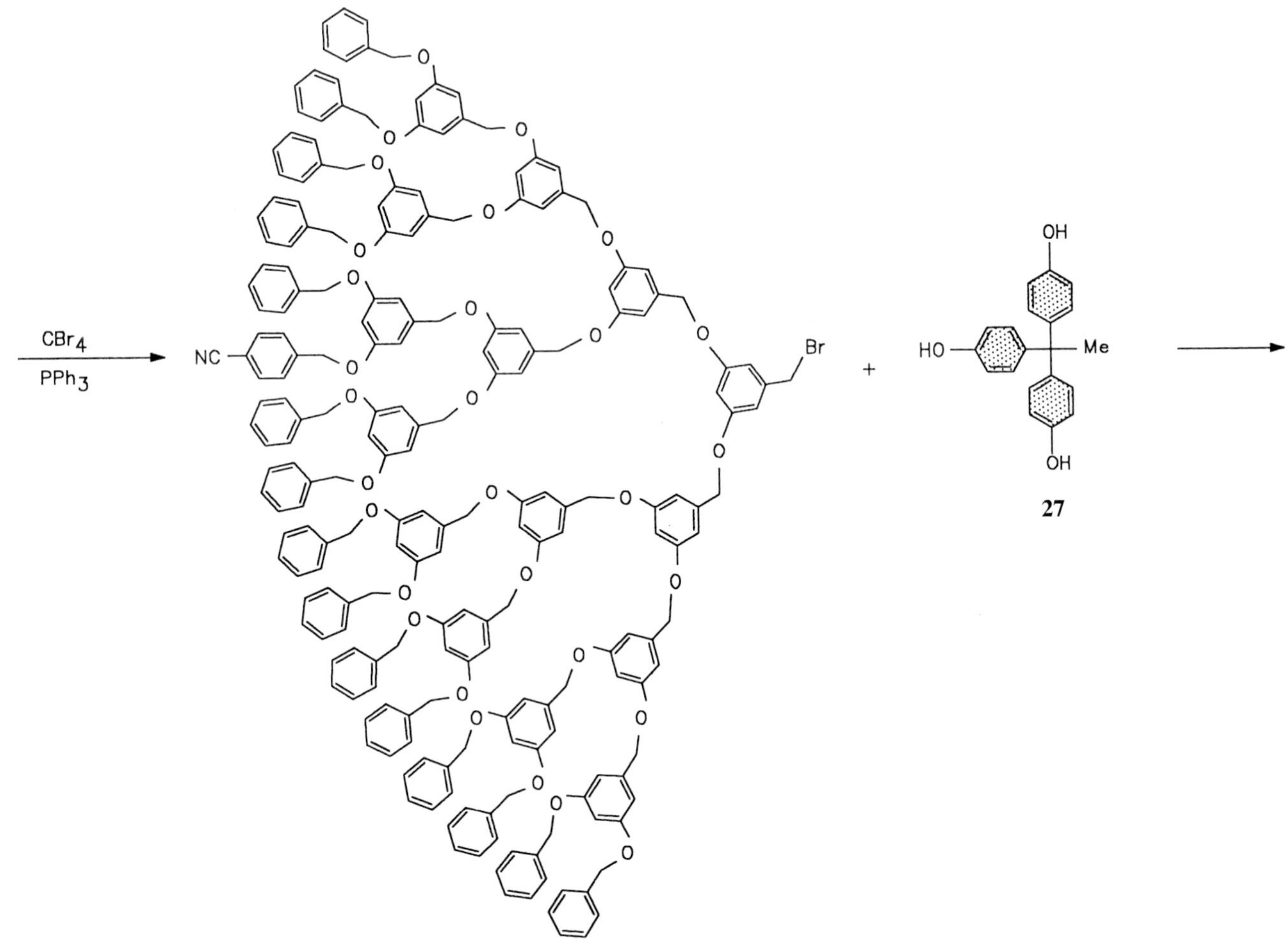

Scheme 5.35 (Continued on page 238)

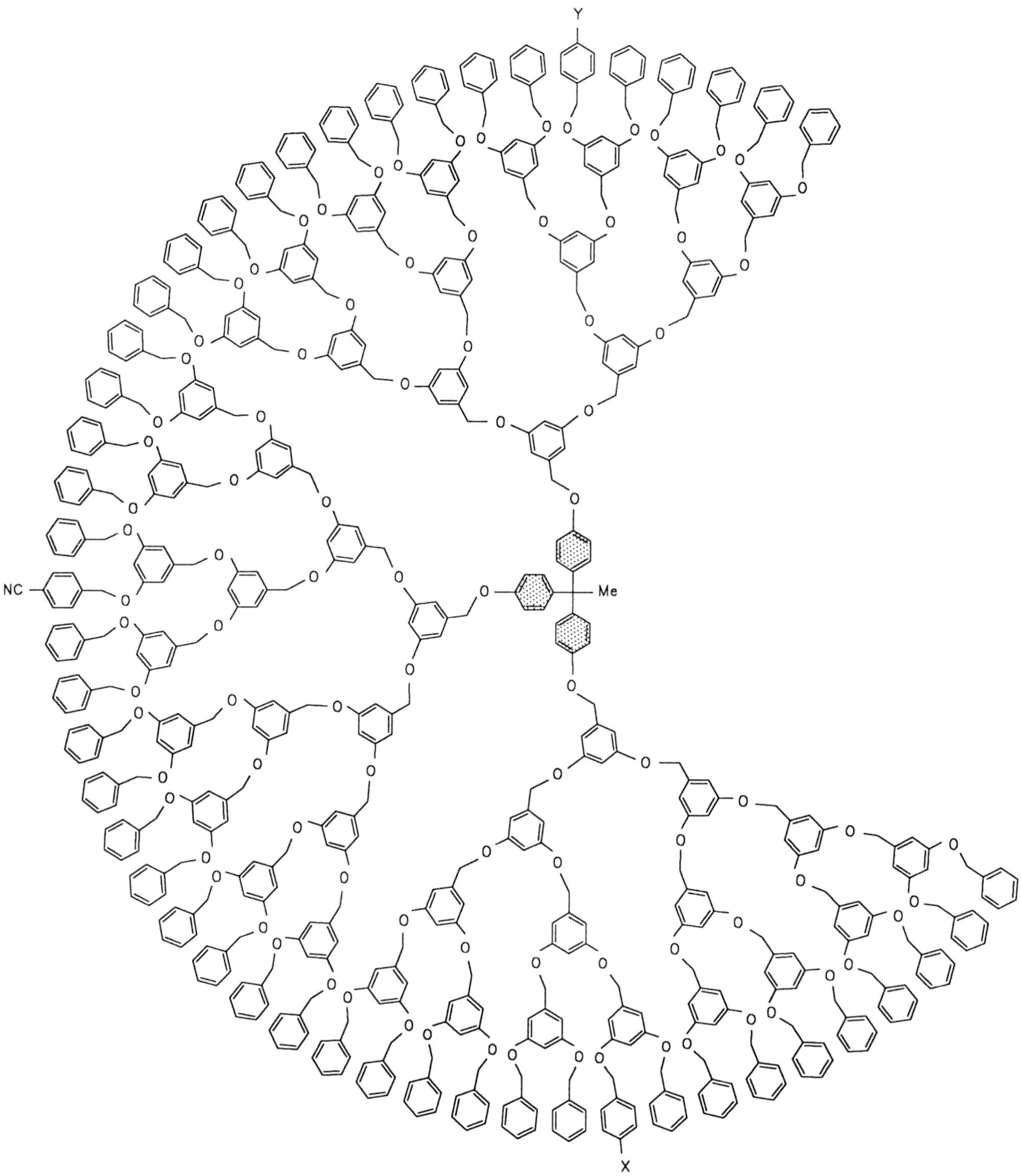

155   X = Y = H

156   X = H, Y = CN

157   X = Y = CN

Scheme 5.35 Statistical preparation of dendrimers possessing 1, 2, or 3 nitrile surface groups.

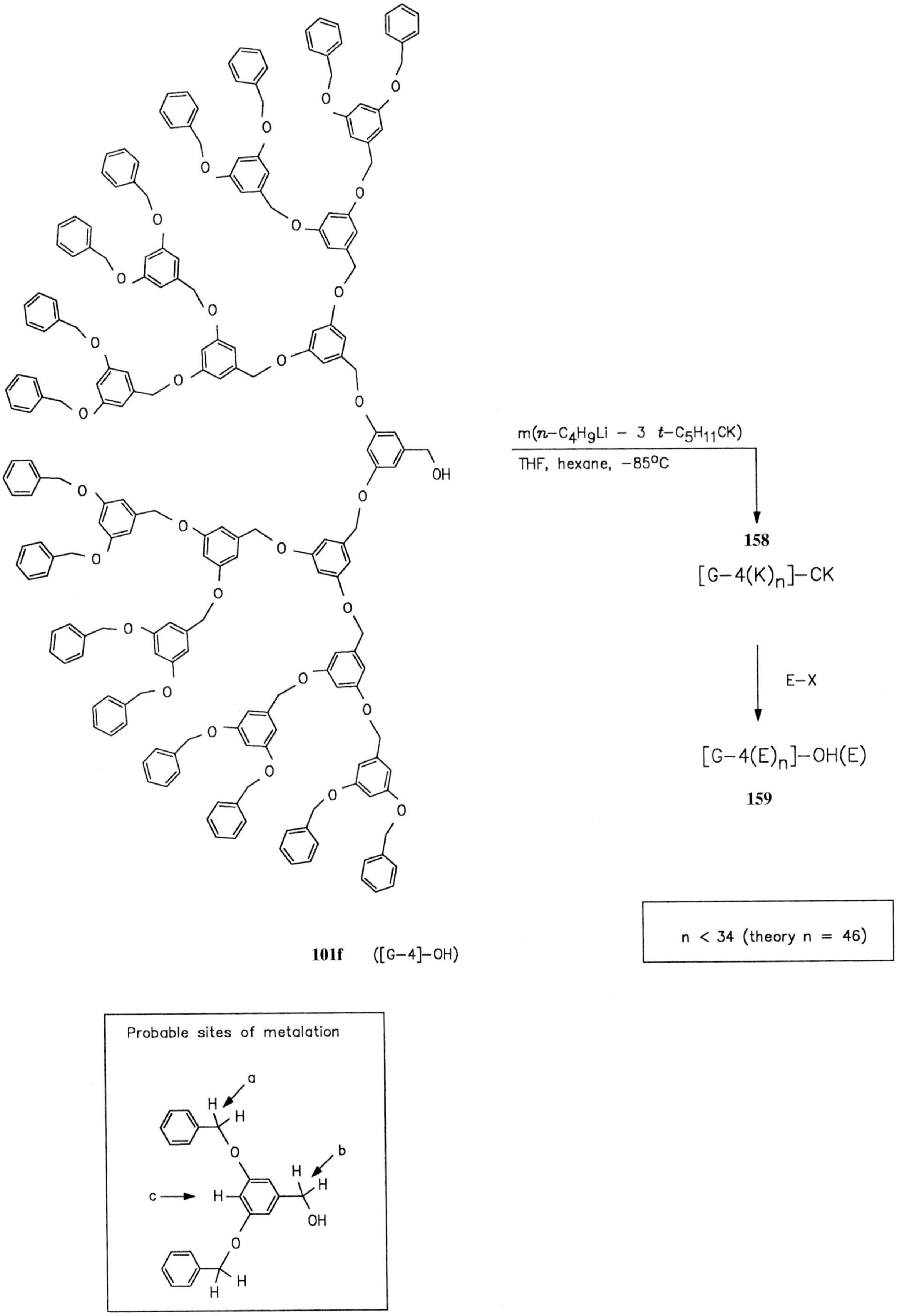

$$\dfrac{m(n\text{–}C_4H_9Li \; - \; 3 \; t\text{–}C_5H_{11}CK)}{THF, \; hexane, \; -85^oC}$$

158

$$[G\text{–}4(K)_n]\text{–}CK$$

E–X

$$[G\text{–}4(E)_n]\text{–}OH(E)$$

159

n < 34 (theory n = 46)

101f ([G–4]–OH)

Probable sites of metalation

Scheme 5.36 Superstructure modification of poly(benzyl ether) dendrimers[201] via metallation followed by electrophilic addition.

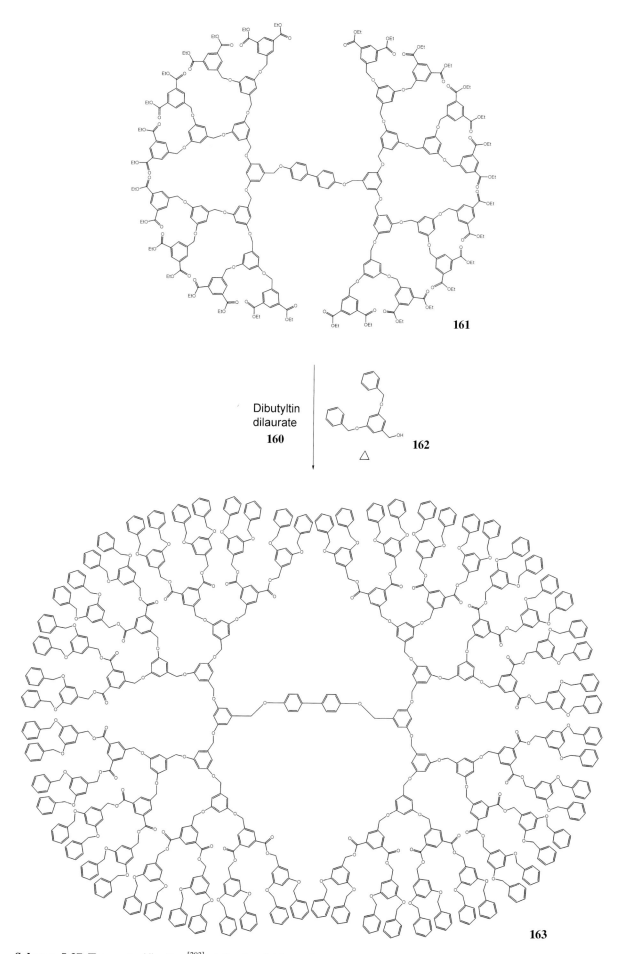

161

Dibutyltin
dilaurate
160

162

163

Scheme 5.37 Transesterification[203] at the dendritic surface.

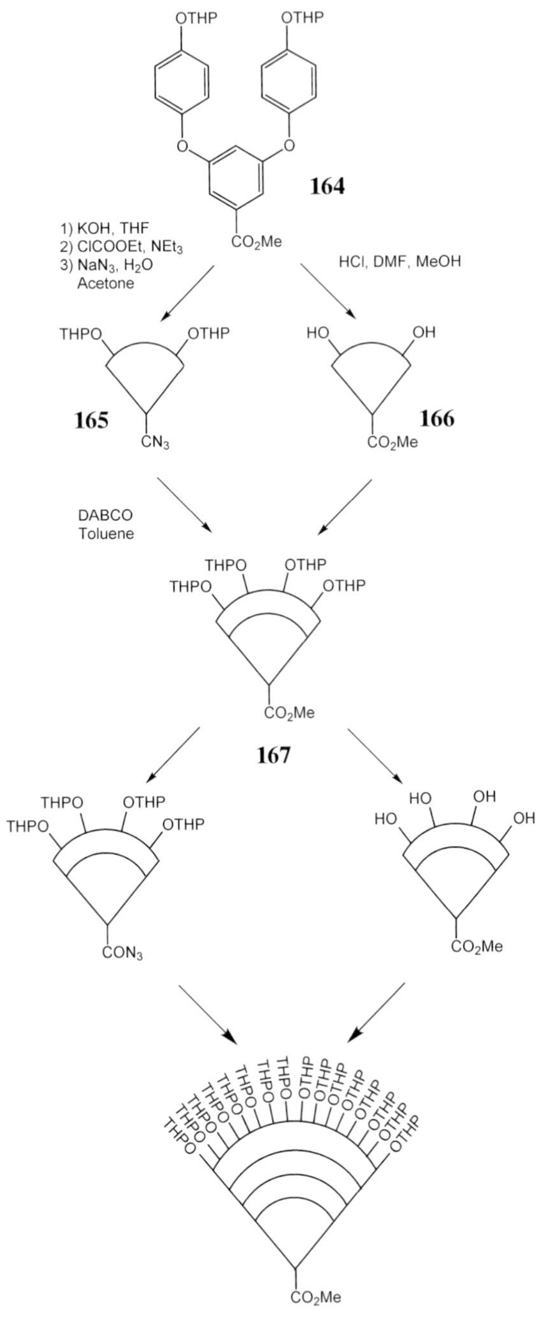

Scheme 5.38 Schlüter's approach to a polymer core functionalized with poly(ethereal) wedges.[204]

Chen et al.[217] prepared the 1st to 3rd generations of aryl ether dendrimers possessing 6, 12, and 24 (**171**) terminal allyl moieties, respectively (Figure 5.17). The key terminal building block was prepared by the addition of allyl bromide to methyl dihydroxybenzo-ate to afford the bis(allyloxy)benzoate. Reduction (LAH) of the ester moiety followed by treatment with PBr₃ gave the benzyl bromide group, which was used for standard attachment to dihydroxybenzyl alcohol for the construction of higher generation dend-rons. In each case, the core used was tris(4'-hydroxyphenyl)methane.

Höger[218] reported the facile preparation of ester-terminated, poly(aryl ether) dend-rons that could be employed for hyperbranched construction as well as for iteration appli-cations. A key transformation to these dendrons utilized the Mitsunobu reaction,[219] whereby bis(benzyl alcohol) moieties are converted (PPh₃, DEAD) to suitable leaving groups and then displaced with phenol. Protection of dendron focal phenolic alcohols was effected by the use of *tert*-butyldimethylsilyl groups. Notably, employing the Mitsunobu conditions facilitates single-pot leaving group generation and subsequent displacement.

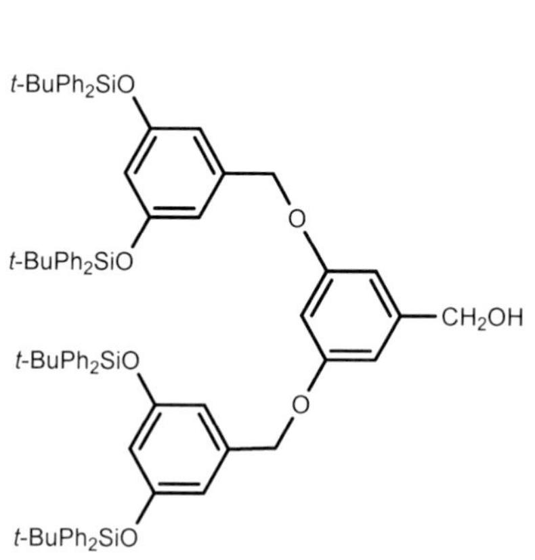

169

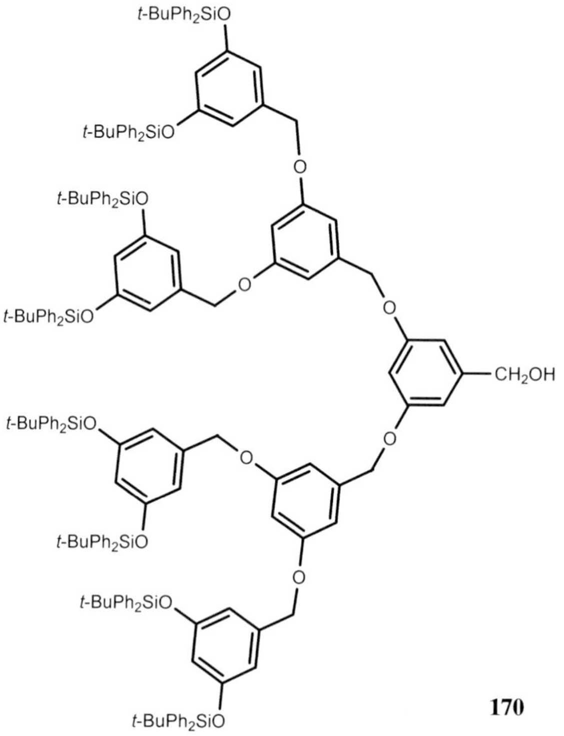

170

Figure 5.16 L'Abbé's dendrons for "double-stage" growth.[216]

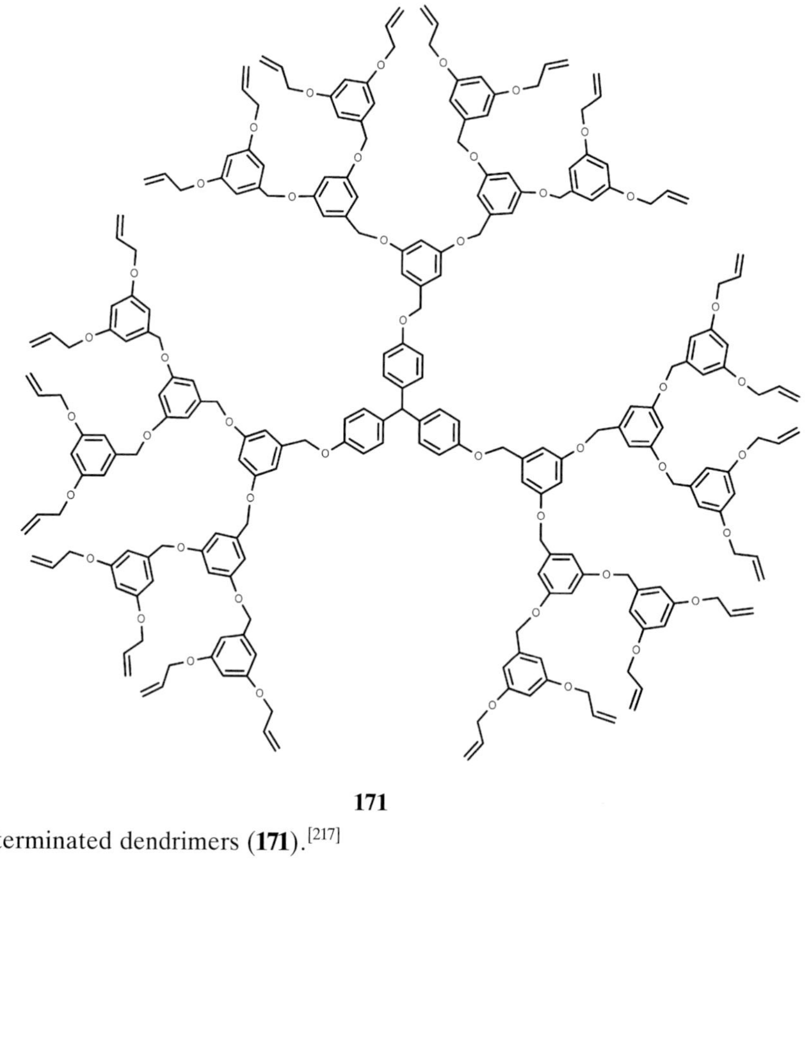

171

Figure 5.17 Bis(allyl)-terminated dendrimers (**171**).[217]

172   **173**

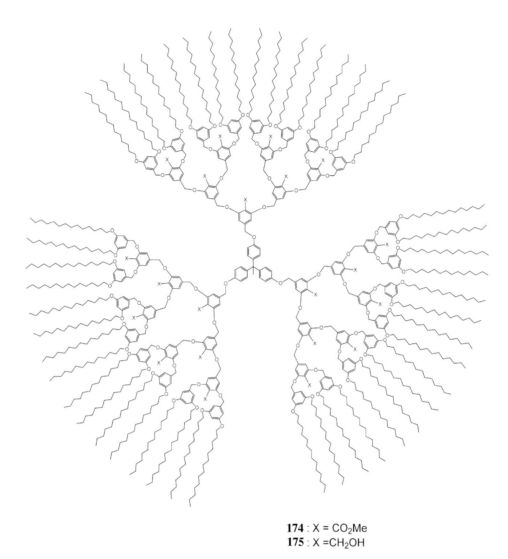

174 : X = CO₂Me
175 : X =CH₂OH

Figure 5.18 A reverse unimolecular micelle possessing internal functionality.[220]

Employing building blocks **172** (inner) and **173** (terminal) (Figure 5.18) and the standard etherification–bromination protocol, a reverse unimolecular micelle possessing specifically juxtaposed internal methyl ester moieties (**174**) has been prepared.[220, 221] [221a] Methyl ester reduction (LAH) was readily accomplished to afford the corresponding polyol **175**. Notably, the catalytic activity of the 4th generation polyol led to 99 % conversions for substrate-dendrimer ratios of 176,000 to 1 in the base-catalyzed E₁ elimination reaction of a tertiary alkyl iodide in cyclohexane; in contrast, little or no reaction was observed without added dendrimer.

Wendland and Zimmerman[222] reported the unique synthesis of "cored" dendrimers using these dendrons (Scheme 5.39) terminally modified with homoallyl moieties and anchored to a trivalent, ester-based core. (e.g., **177**). Treatment of the allyl groups with 4 mol % of Grubbs' Ru catalyst per dendrimer afforded the intramolecular polymerized product **178**. The core unit was cleaved (KOH, EtOH) to give the focally disconnected dendrimer **180**. Potential uses in molecular imprinting and moulding were suggested. The use of isomeric ester-based cores demonstrated the flexibility of this strategy for the preparation of unique molecular structures.

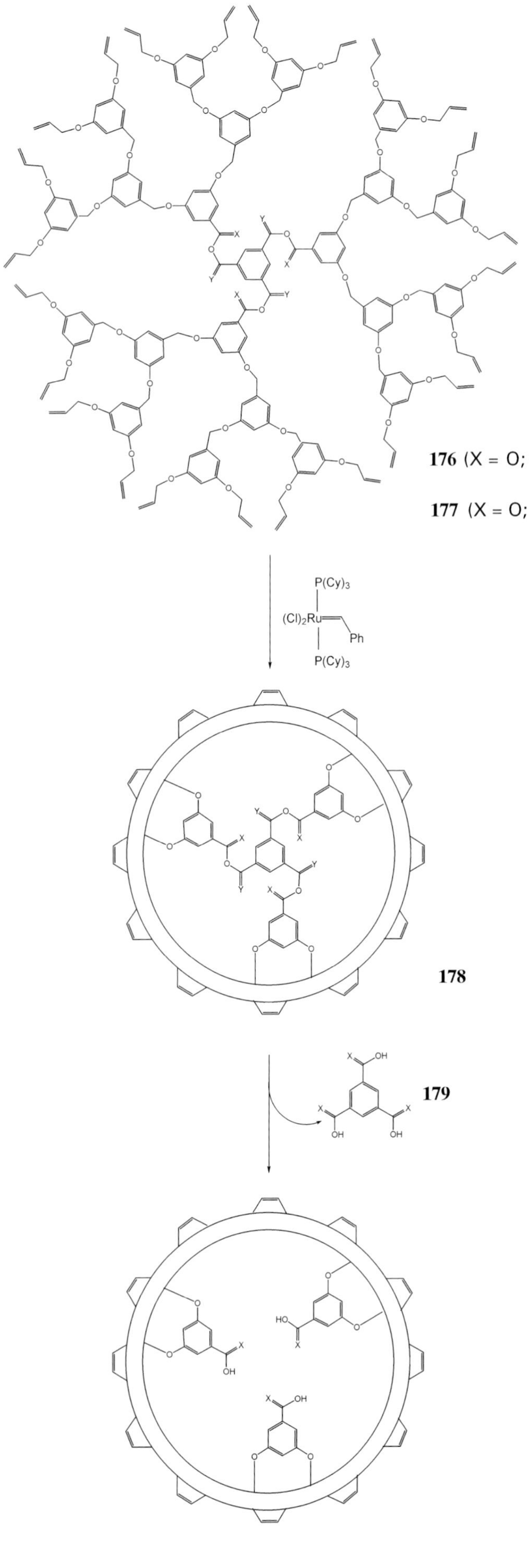

176 (X = O; Caps = O)

177 (X = O; Caps = H)

P(Cy)₃
|
(Cl)₂Ru══Ph
|
P(Cy)₃

178

179

180

Scheme 5.39 Access to "cored" dendrimers.[222]

Gilat, Adronov, and Fréchet[82] reported the synthesis and energy-transfer potential[223–226] of laser dye-modified poly(aryl ether)s[226a] (Figure 5.19). Construction of these dendritic antennae was achieved by a "reversed" focal modification strategy, whereby dendron nucleophilic attack of the focal phenolate at an electrophilic benzyl bromide monomer group effects generational growth; this is in contrast to the normal convergent protocol, in which a phenolate moiety is generated on the monomer and then attacks a dendron focal benzyl bromide. Use of a 2nd generation, tetrabenzyl bromide "hyper-monomer" further facilitated the rapid synthesis of the higher generation materials. Labeled dendrimers (**181–183**) were each terminally modified with the terminal donor chromophore coumarin and the focal acceptor chromophore coumarin-343. Favorable characteristics such as chromophoric solubility, emission overlap, high fluorescence quantum yields, high purity, and commercial availability supported the choice of the donor–acceptor pair. Energy absorption and transfer to the core was readily demon-

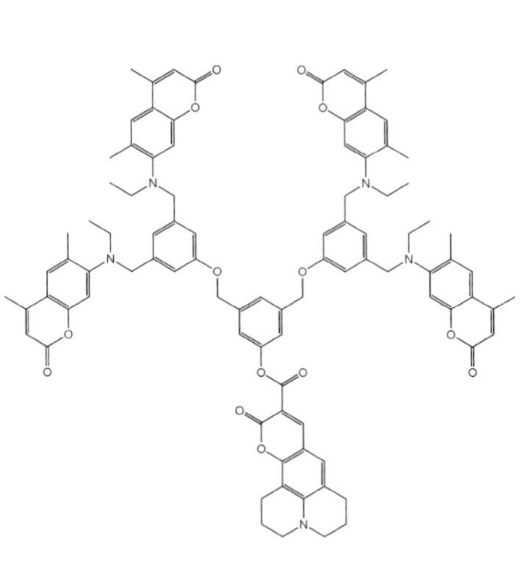

181

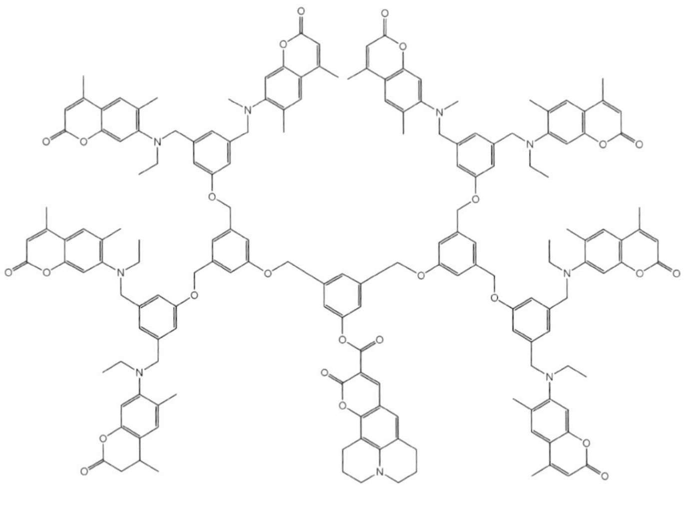

182

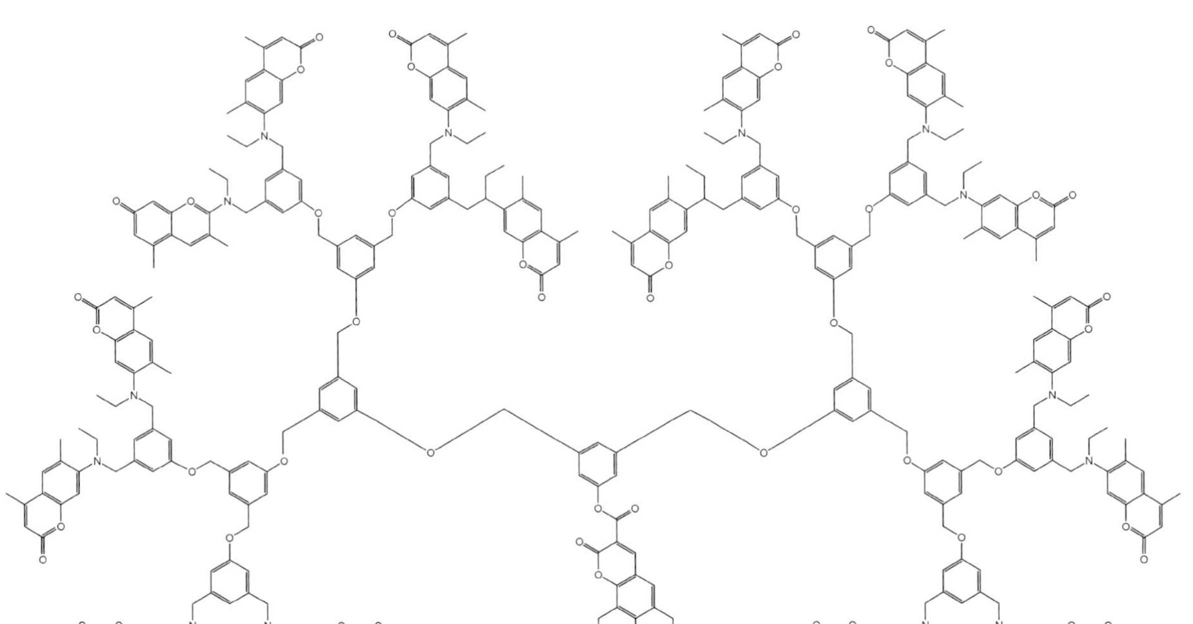

183

Figure 5.19 Laser dye-labeled dendrons for light harvesting and energy transfer.[82, 223]

strated; when the 3rd generation dendrimer was irradiated at 343 nm, fluorescence at 470 nm from the focal dye was observed. Thus, the focal coumarin functions as an energy "sink" or "concentrator". Steady-state photophysical properties of the dye-based dendrons, which were focally modified with oligothiophenes, have also been reported.[227] Recently, donor coumarin-2 chromophores and acceptor coumarin-343 chromophores have been coated on a silicon wafer; efficient energy transfer within the *xy* plane of these self-assembled adsorbates has been demonstrated.[228] Ru-based analogs are known.[228a]

Marquardt and Lüning[229, 230] reported the synthesis of dendrimers with "concave" pyridine-based end groups (**184**; Figure 5.20) that impart a surface reminiscent of a dimpled (molecular) golf ball.[231] The selectivity of base-catalyzed acylation of alcohols (by ketenes) was found to be similar to that achieved with the free monopyridine, but better than that with the base fixed to either a Merrifield resin or a linear polymer.

Hong et al.[232] prepared small terpyridine-terminated polyethers and attached them via phosphonium bromide formation to an octavalent silsesquioxane core.

Using Pd-mediated cross-coupling reactions (Suzuki and Stille couplings), Groenendaal and Fréchet[233] transformed these dendron peripheries from *p*-bromophenyl groups to phenyl, 2-thienyl, and 2-pyridinyl moieties. Gingras et al.[234] reported the synthesis of interesting star-shaped molecules termed "molecular asterisks" through the MacNicol reaction[235] (NaH, 1,3-dimethyl-2-imidazolidinone) of poly(phenylene sulfide) oligomers with perchlorinated benzene and coronene cores.

White et al.[236] prepared methyl ester and cyano-terminated 4th generation dendrons and studied the surface pressure/surface area isotherms to assess their properties at the air-water interface. In contrast to the unmodified dendrons, monolayer formation was observed for the substituted materials. Films at high compression were found to move into the subphase instead of forming multilayer structures.

Selective protection of terminal monomers has allowed access to water-soluble PEG conjugates (**185**; Figure 5.21) that have potential as dendrimer-based drug carriers.[237] Attached model drugs include cholesterol, phenylalanine, and tryptophan moieties. Cell targetable cationic amphiphiles capable of gene delivery and possessing glycosyl residues have also been reported.[238]

Percec et al.[239, 240] reported the attachment of 3,4-bis(*n*-dodecan-1-yloxy)benzyl ether to the periphery of the typical 3,5-disubstituted benzyl ether series to produce a family of wedges that exhibited novel supramolecular dendron shapes; the 4th generation wedge was shown to possess a disc-like shape.

Polyether dendrimers having a surface bearing folate residues have been prepared as model drug carriers with a view to tumor cell specificity.[241] In essence, the ester-terminated dendrimers were readily converted (H₂NNH₂) to the corresponding polyhy-

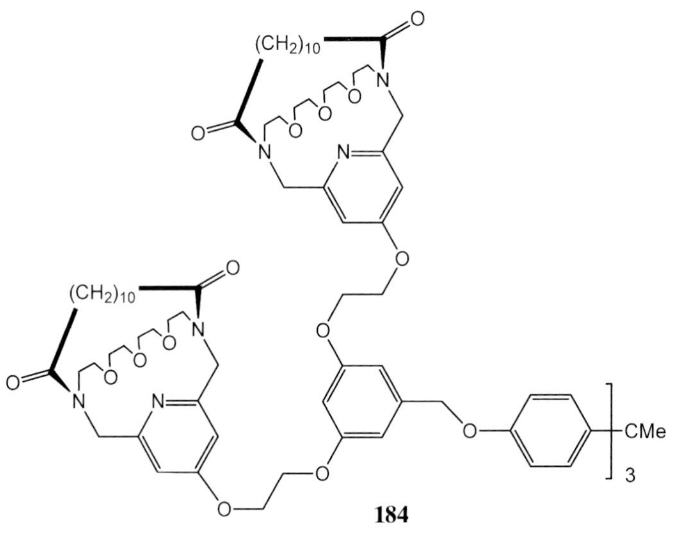

184

Figure 5.20 Terminal concave pyridinyl units reminiscent of a golf balls' surface.[229]

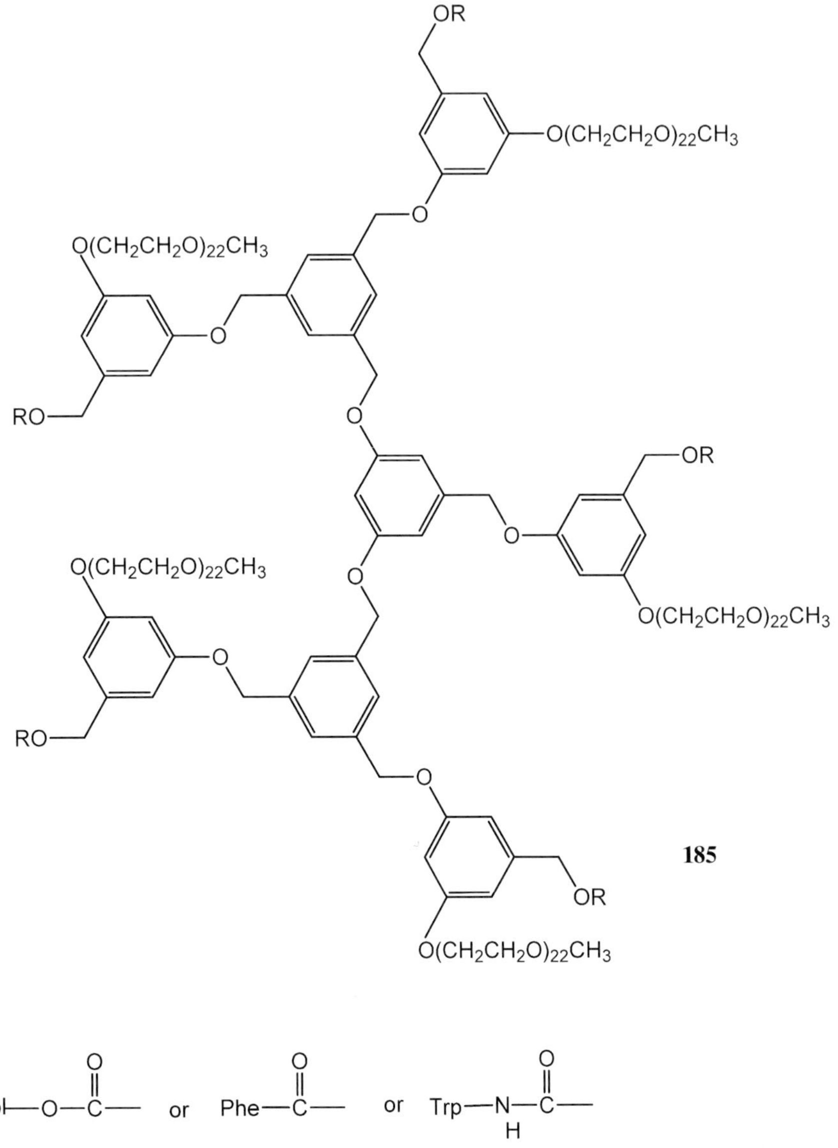

Figure 5.21 Star-like conjugates for potential drug carriers.[237]

drazides, which were then successfully treated with an active ester derivative of folic acid or methotrexate.

5.3.2.4 Physical Properties

Mourey et al.[242] evaluated the viscosities of these polybenzyl ether dendrimers[1, 44] by means of size-exclusion chromatography coupled with differential viscometry. For the 0–6th generations, the intrinsic viscosity for these three-directional family (e.g., **92**) was found to pass through a maximum at the 3rd generation, whereas the refractive index was found to pass through a minimum at ca. the 2nd generation. These data suggest a monotonic decrease in density from the center of these polyethers. The results support Lescanec and Muthukumar's[243] theoretical model of dendrimers, which indicates higher internal density relative to surface density, but are at variance with the de Gennes and Hervet model,[244] which suggests greater surface congestion.

Saville et al.[245, 246] studied the surface pressure/area isotherms at the water-air interface at 25 °C for monomolecular films of the ethereal dendrimer series (e.g., **92**). They reported a strong dependence of the isotherms on molecular weight, which compared well with similar findings for hydroxyl-terminated polystyrene.

The melt viscosity behavior for this family of dendrimers generated from 3,5-dihydroxybenzyl alcohol has been investigated.[247] In general, evaluation of the melt viscosity for this series revealed a profile with no critical molecular weight for molecules as large as 85,000 amu. At high molecular weights, branching and surface congestion in this family prevent appreciable intermolecular entanglement affording "ball-bearing-like" macromolecules, which are only capable of interdigitation. The zero shear melt viscosity (η_o) has been measured for a variety of these poly(benzyl ether)s.[248] For higher generation materials, a direct correlation with molecular mass was observed, while smaller generations exhibited only a "stronger" dependence. The viscosity was also found to scale with molecular mass for mono- and tridendrons rather than with generation number. On re-examination of the literature data, the authors determined that intrinsic viscosity and T_gs scale as a function of molecular mass.

Wooley et al.[249–251] examined the shapes, sizes, and intermolecular packings of the 1st through to the 5th generations of these poly(benzyl ether) dendrons by means of REDOR NMR and molecular dynamics simulations. The REDOR experiments revealed dipolar coupling between the ^{19}F-labeled focal site and the ^{13}C-labeled benzylic methylene termini. ^{13}C–^{19}F intramolecular distances were, on average, determined to be ca. 12 Å for generations 3–5 suggesting branch chain backfolding, while intermolecular dipolar coupling decreased as a function of increasing generation, suggesting less interdigitation at higher generations. These results support mathematical models[243, 252, 253] of flexible dendrimers that point to structure distributed, inward folding, chain ends.

Hawker et al.[254] reported linear analogs of the classic Fréchet dendrons. The protocol used for their construction involved a "convergent dual exponential" strategy, whereby two linear components, one possessing 2^n repeat units and the other possessing 2^{n-1} repeat units, were synthesized and subsequently coupled. This strategy yielded identical numbers of repeat units in the linear constructs as in the corresponding dendrons. It was found[254] that for generations 4 or less, the hydrodynamic volumes of the dendrimers compared well with those of the linear models. A significant increase, however, was observed for the hydrodynamic volumes of the 5th and 6th generation linear models as compared to their dendritic counterparts. Differences were also observed in solubilities and crystallinity; the linear materials displayed higher crystallinity compared to the dendrimers.

Rheological studies of side chain dendritic polymers, i.e., rod-like polymers based on Fréchet-type dendrons attached to the linear backbone, have been reported.[255] Similar architectures have been described by Hawker et al.,[256] who employed a coupling approach; the copolymers were accessed by coupling the dendrons to the preformed linear rod. Due to steric crowding and thus incomplete reaction, globular hybrids were generated rather than extended rods.

Fréchet-type dendrons have been analyzed by MALDI-TOF MS.[257] Structures surveyed include standard dendrons terminally modified with ester or perdeuterophenyl moieties, bis(dendron)s connected by PEG, and polystyrene focally modified materials. Indoleacrylic acid was employed as a matrix for the regular dendrons, while *t*-retinoic acid and 2-nitrophenyl octyl ether were used for polystyrene and PEG-based architectures, respectively.

5.3.3   1 → 2 *Aryl*-Branched, *Amide* Connectivity

Miller and Neenan[2] reported the preparation (Scheme 5.40) of a polyamido cascade **189**. The bis(acid halide) **186** of 5-nitroisophthalic acid was treated with aniline to generate bis(amide) **187**, which was catalytically reduced to give the disubstituted aniline **188**. Addition of three equivalents of amine **188** to **49** then yielded the polyaromatic amido cascade **189**. Feast et al.[258–260] published the series of aromatic polyamide dendrimers exactly analogous to the amide-based macromolecules.[2] The synthetic transformations employed for wedge construction were also similar except that reduction of the aromatic nitro group was effected by catalytic reduction (NaBH$_4$, SnCl$_2$·H$_2$O). Analysis of the ^{1}H NMR spectra of the dendrimers suggested that, due to sterically restricted branch rotation, the interior secondary amide protons, which are equidistant from the core, reside in different chemical environments. The solubilities of the 1st, 2nd, and 3rd tier cascades were measured as 24, 298, and 40 g/L, respectively, in THF at 25 °C. Recently, Ishida, Jikei,

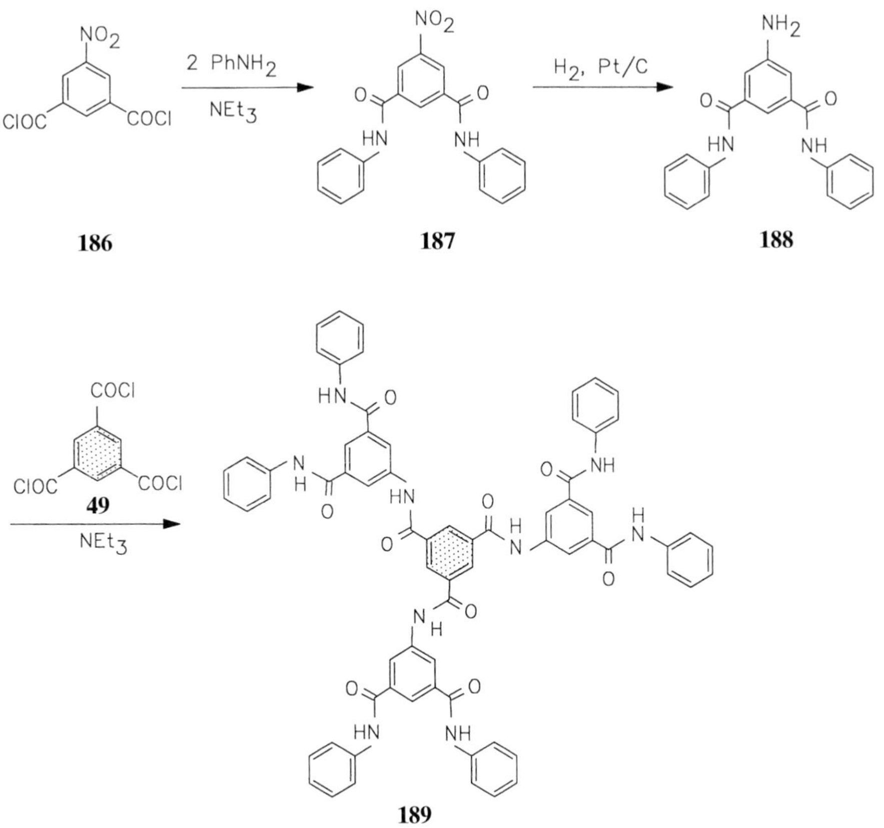

Scheme 5.40 Miller and Neenan's[2] arylamide-based sequence.

and Kakimoto[261] reported a rapid orthogonal and double-stage convergent approach to the construction of this family of dendrimers.

Voit and Wolf[262] [262a, b] developed a convergent protocol based on 5-(2-aminoethoxy)-isophthalic acid affording "perfectly" branched dendrons up to the 4th generation (Scheme 5.41). Pertinent steps in the sequence involved the conversion of the Boc-protected amine **190**, derived in four steps from dimethyl 3-hydroxyisophthalate, to the corresponding diimidazole **191** and reaction with amine **192** to yield tetramethyl ester **193**. Deprotection (HCl/MeOH) of the primary amine affording the 2nd generation dendron **194** facilitated repetition of the sequence with more diimidazole **191**. Huwe et al.[263] later used broadband dielectric spectroscopy to study molecular dynamics of these materials up to the 3rd generation.

Ritzén and Frejd[264] reported the construction of similar motifs albeit with 1 → 2 branched chiral amino acid building blocks; the complementary protection–deprotection scheme was founded on Boc-protected terminal amines and focal benzylic esters.

Liskamp et al.[265] developed amino acid based dendrimers using methyl 3,5-bis(2-*tert*-butyloxycarbonylaminoethoxy)benzoate (**195**; Scheme 5.42). Treatment of this mono-methyl ester **195** with base (NaOH) or acid (HCl) afforded the corresponding monocar-boxylic acid **196** or complementary diamine **197**, respectively, required for this conver-gent strategy. Thus, use of a "BOP" coupling reaction[266] afforded dendron **198**; saponifi-cation of the focal methyl ester and reaction with more diamine generated the next tier (or generation); up to the 5th generation was reported. Liskamp et al.[267] later extended this protocol to include a number of 1 → 2- and 1 → 3-branched aryl amino acids, thereby demonstrating the "diversity" of these monomers. Similar β-alanine constructs are known.[267a]

Schlüter et al. published a series of papers dedicated to the functionalization of 1 → 2 aryl branched architectures whereby terminal modification with glucose,[268] hydroxyl,[269] and amino[270] moieties was described. The glucose derivatives were focally modified with a subsequently polymerized acrylate ester affording a cylindrical motif.

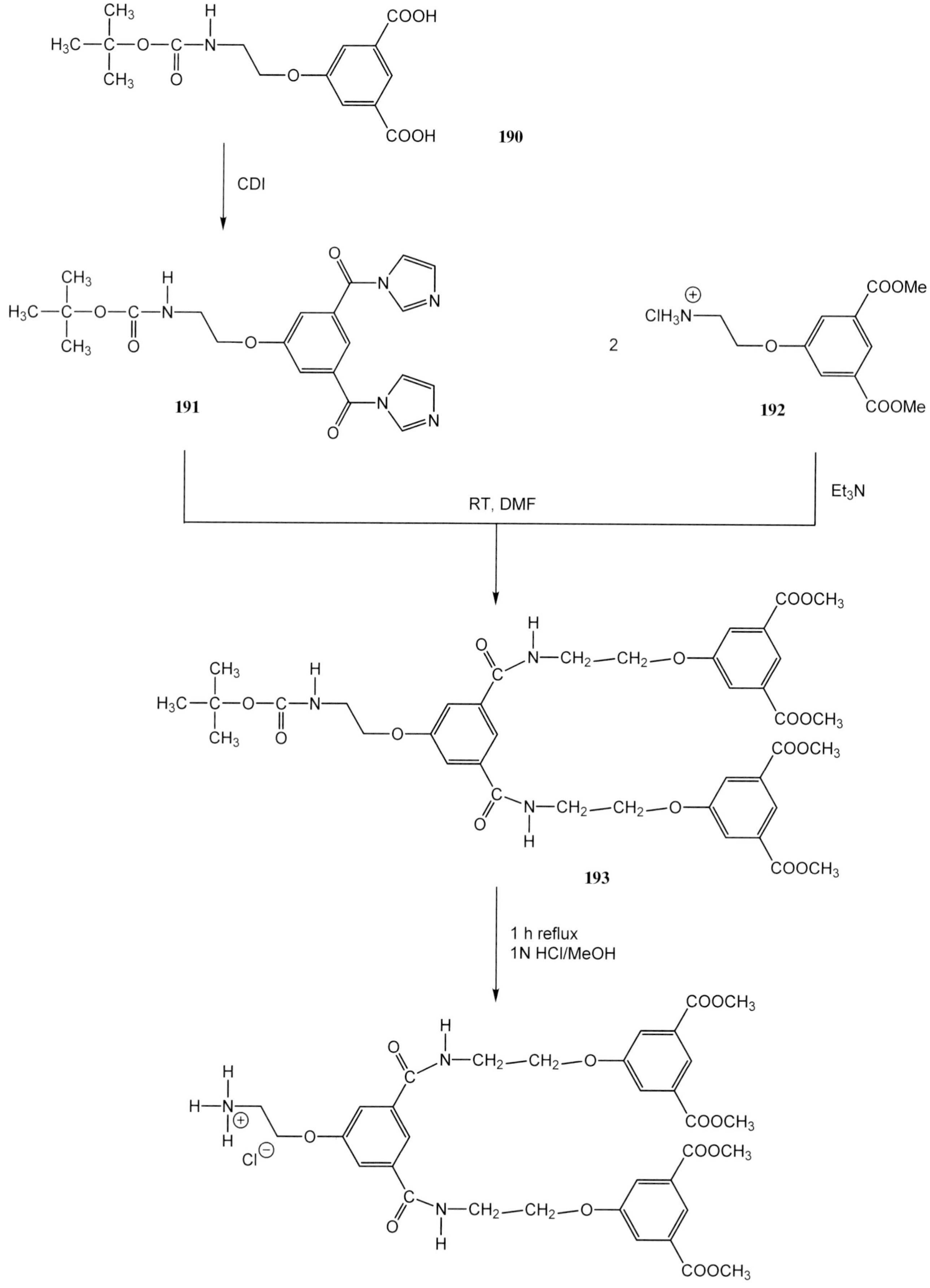

Scheme 5.41 Voit's protocol for perfect dendron construction.[262]

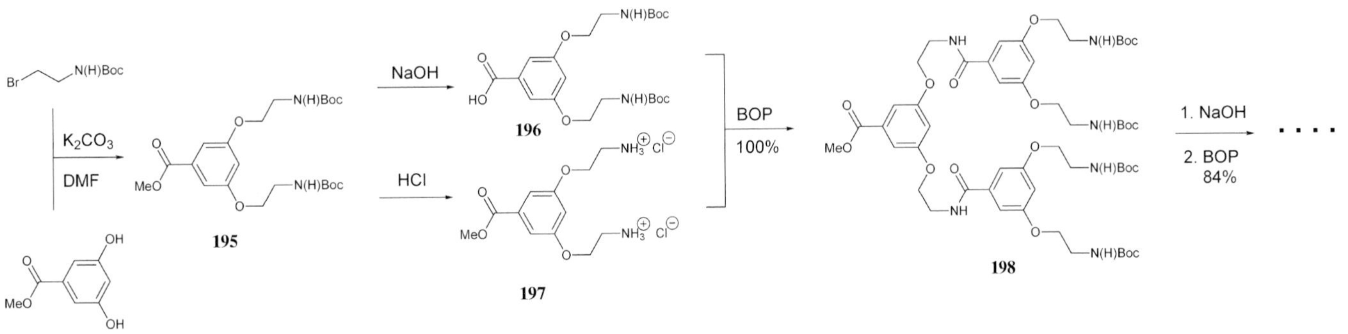

Scheme 5.42 Liskamp's "amino acid"-based monomers.[265]

5.3.4 1 → 2 *Aryl*-Branched, *Ammonium* Connectivity

Stoddart et al.[271] created polycationic dendrimers based on the key benzylic alcohol monomer **199** (Scheme 5.43) prepared by bromination (2 equiv. Ph₃P, CBr₄) of 1,3,5-tris(hydroxymethyl)benzene followed by treatment excess Et₂NH. Peripheral dendron **200**, prepared by reaction of the precursor to **199** with excess Et₃N, followed by bromination (HBr/AcOH) of the single hydroxy group and counterion exchange (NH₄PF₆), was then attached to diamine **199** via halide displacement (Me₂CO, Δ). Following simple procedures, the 2ⁿᵈ generation dendron **201** was obtained in high overall yield. Repetition of this sequence provided the 3ʳᵈ generation wedge **202**, which was treated with the triamine core **203**, prepared by reaction of Et₂NH with 1,3,5-tris(bromomethyl)benzene, to give polycationic dendrimer **204**.

5.3.5 1 → 2 *Aryl*-Branched & 1 → 3 *C*-Branched, *Ether & Amide* Connectivity

Newkome et al.[272] reported the convergent construction of an aromatic cascade based on a building block developed for use in a divergent synthetic method (Scheme 5.44), utilizing the Miller and Neenan thought process.[2] Treatment of 5-nitroisophthaloyl dichloride (**206**) with two equivalents of the amine building block[273] **205** gave the hexaester dendron, which was catalytically reduced (PtO₂, H₂) to yield aromatic amine **207**. Monomer **207** was treated with triacyl core **49** affording the three-directional dendrimer **208**, which was hydrolyzed to the corresponding polyacid **209**. Related two-directional cascades (**211**) were also reported[272] employing terephthaloyl dichloride (**210**) as the core.

5.3.6 1 → 2 *Aryl*-Branched, *Ether & Amide/Amine* Connectivity

Shinkai et al.[274, 275] described the construction of 'crowned arborols' that incorporated diaza crown ethers as spacers, and three-directional, aromatic branching centers (Scheme 5.45). In this case, the convergent construction was found to be more effective than the divergent approach. Tetraoxadiaza crown monomer **215** was prepared from *N*-benzyloxycarbonyldiaza crown ether (**212**) and a monoacyl chloride **213**, affording the diprotected intermediate **214**, which was catalytically debenzylated (Pd/C, H₂) to give diester **215** or hydrolyzed (aq. base) to afford diacid **216**. Monomer **216** was subsequently transformed to tetraester **217**, by means of the mixed anhydride method, with two equivalents of the corresponding amine building block **215**. The resultant dendron **217** was attached to **49** affording the 2ⁿᵈ generation polyethereal crown cascade **218**.

A related reduced series was also reported,[275] in which the selectivity of these crowned arborols towards alkali metal cation binding was examined and allosteric as well as conformational binding effects were studied. The 'sterically less crowded' arborols (e.g., **219**; Figure 5.22) were shown to enhance the dissolution of myoglobin in organic solvents, e.g., DMF; this example affords an interesting entrance to internal metal ion complexation at specific loci; see Section 8.3.

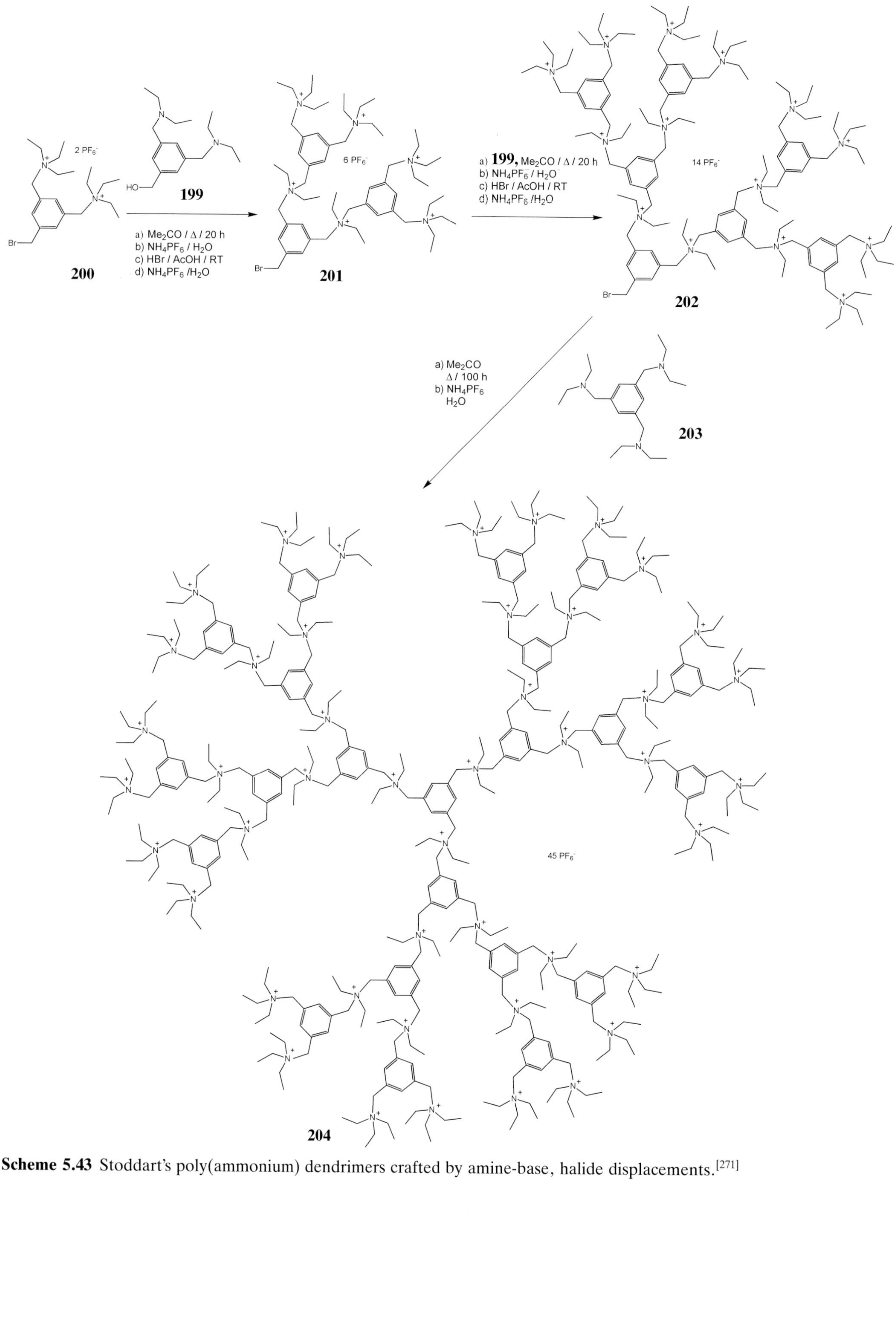

Scheme 5.43 Stoddart's poly(ammonium) dendrimers crafted by amine-base, halide displacements.[271]

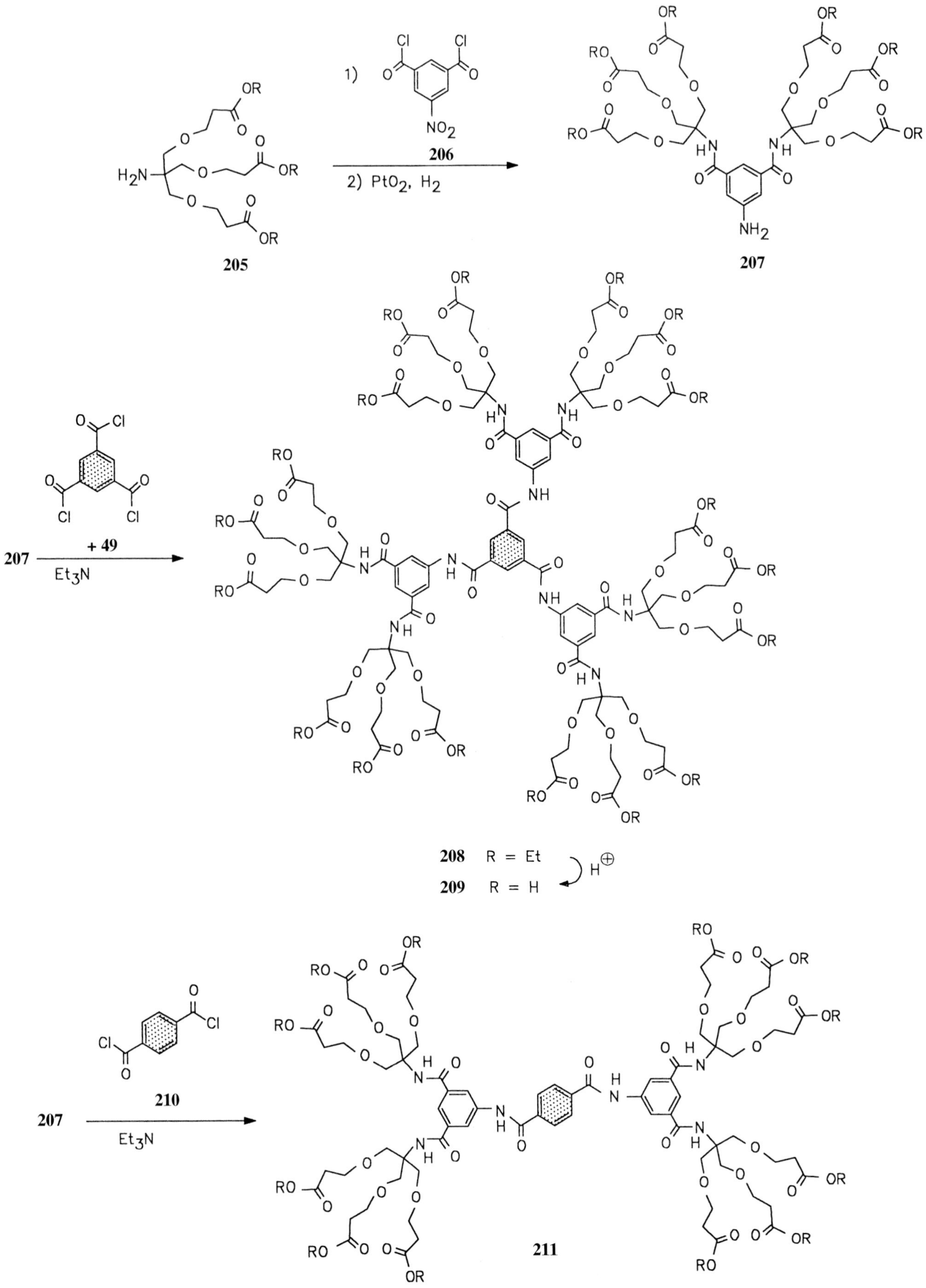

Scheme 5.44 Employment of a flexible aminotriester for the construction of dendrimers[272] possessing arylamide-based cores.

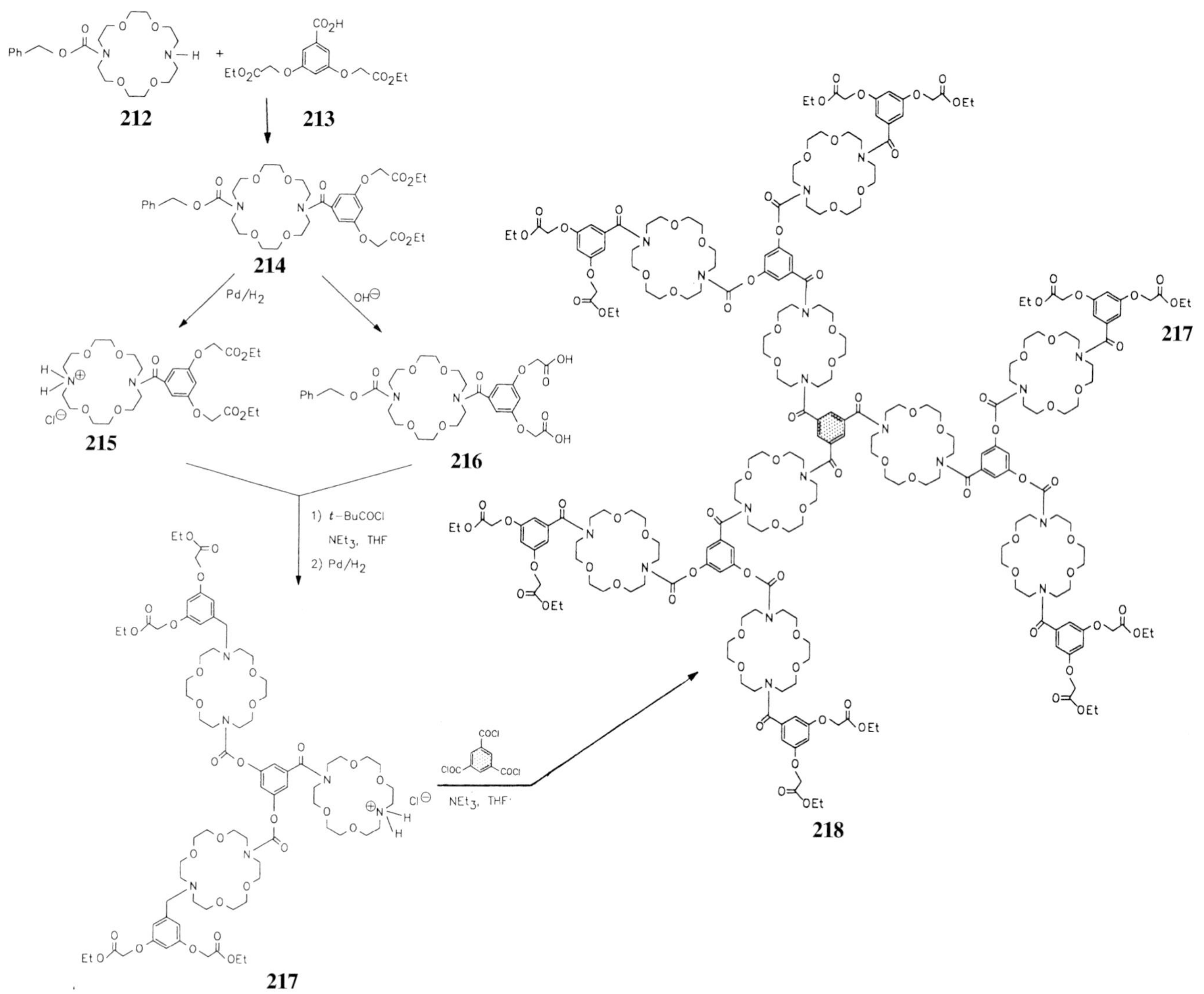

Scheme 5.45 Shinkai et al.'s[274] procedure for the synthesis of "crowned arborols".

Although strictly not a dendritic system, Agar et al.[276] have reported the preparation of Cu(II) phthalocyaninate substituted with eight 12-membered tetraaza macrocycles as well as its Ni(II), Cu(II), Co(II), and Zn(II) complexes. Thus, the use of 1,4,7-tritosyl-1,4,7,10-tetraazacyclododecane offers a novel approach to the $1 \rightarrow 3$ branching pattern and a locus for metal ion encapsulation.

Pan and Ford[277, 278] created architectures possessing alternating amine and ether functionalities with the goal of studying their potential as catalysts in aqueous media. Construction (Scheme 5.46) proceeded via conversion of the bis(acid chloride) **220**, derived from the corresponding phenolic diacid, to the bis(amide) **221**, which was subsequently reacted with benzyl bromide **222** to give the focal amide **223**. After reduction (LAH), the pentaamino polyether dendron **224** was finally reacted with a trivalent core to afford dendrimer **225** after reduction of the carbonyl moieties. Di- and tetravalent cores were also employed; synthetic details as well as spectral characterizations have appeared.[279] These materials were shown to enhance pyrene solubility in water by a factor of 30.

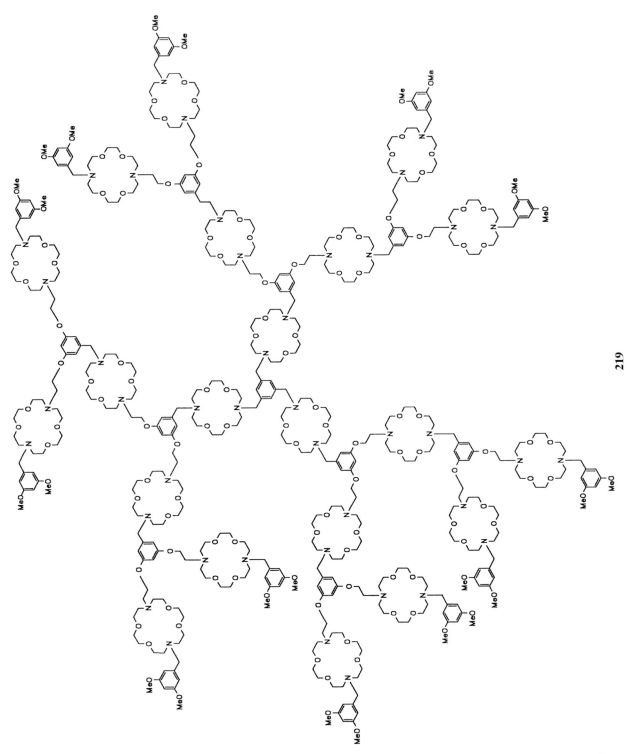

219

Figure 5.22 Reduction of the carbonyl moieties of the "crowned arborols" (Scheme 5.45) generated enhanced internal lipophilicity.[275]

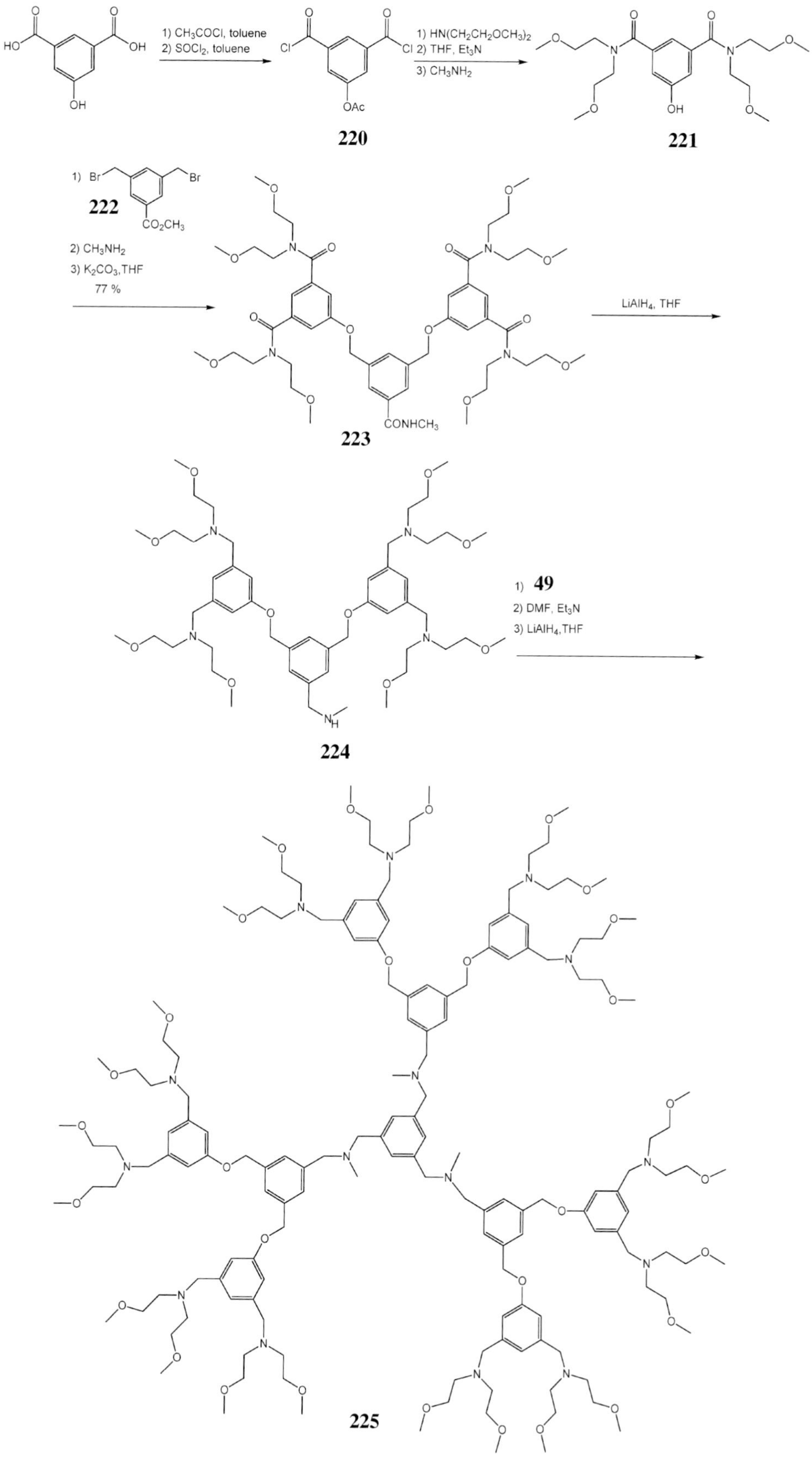

225

Scheme 5.46 A protocol for poly(amine-ether) convergent construction.[277]

5.3.7 1 → 2 *Aryl*-Branched, *Ether & Urethane* Connectivity

Spindler and Fréchet[280] demonstrated the construction of cascade polymers possessing both ethereal and urethane bonds as well as the growth of two generations in a single synthetic operation (Scheme 5.47). Their one-pot, multistep approach utilizes the preformed 1st generation alcohol **226**, which is reacted with the 3,5-diisocyanatobenzyl chloride (**227**) to build the 2nd generation bis(carbamate) **228** possessing a benzyl chloride focal group. Without further purification, 3,5-dihydroxybenzyl alcohol is added giving rise to the 3rd generation alcohol **229**. As expected, the judicious selection and use of specific monomers is critical to the successful application of this accelerated approach to dendritic materials.

Taylor and Puapaiboon[281] developed a novel urethane-based protocol for branched construction (Scheme 5.48). Treatment of a siloxane diacid, such as **230**, with an alcohol and diphenylphosphoryl azide (DPPA) generated the bis(urethane), which was desilylated (TBAF) to afford the branched alcohol **231**, which could be similarly added to more diacid. Four generations were thus prepared and subsequently attached to trimesic acid (**49**). Post source decay MALDI-TOF MS has been employed for structural analysis and confirmation of polyurethane-based dendrimers.[282]

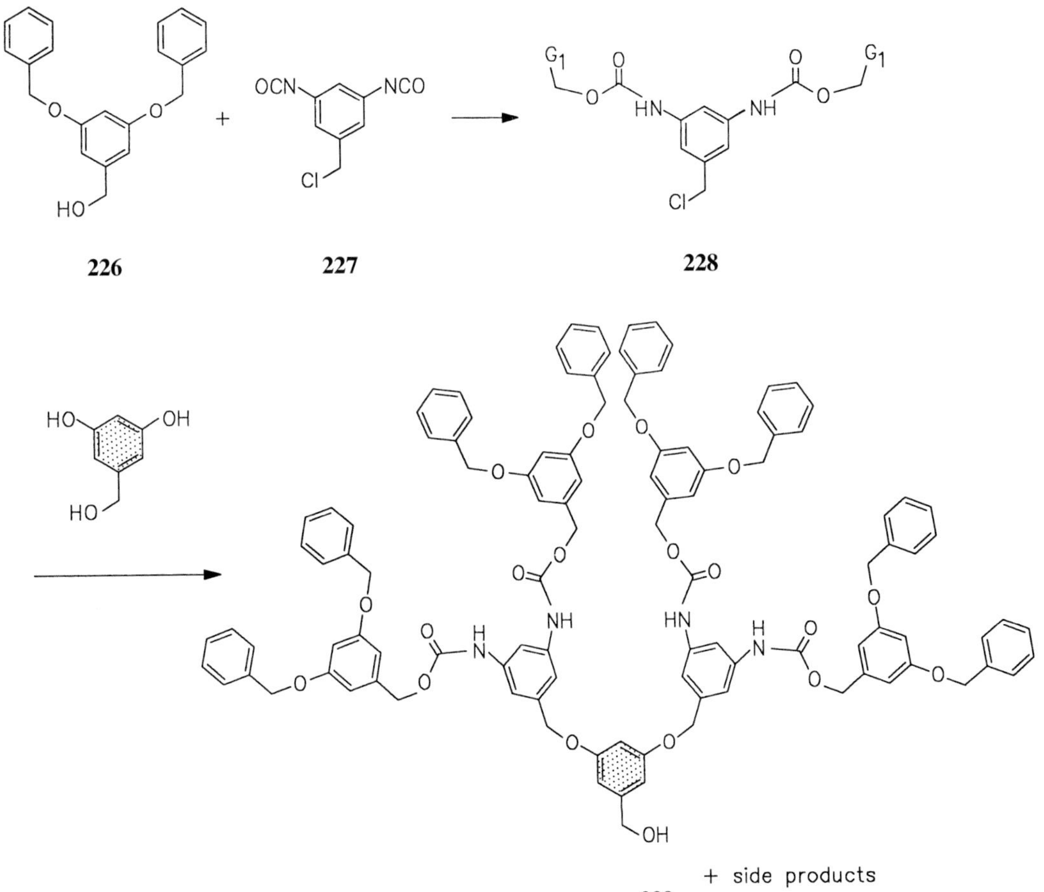

226   **227**   **228**

+ side products
229

Scheme 5.47 Construction of two generations using a single-pot, multistep approach.[280]

5.3.8 1 → 2 *Aryl*-Branched, *Ester* Connectivity

Ester-based cascades (e.g., **236**) have been prepared[283–285] by using 5-(*tert*-butyldimethylsiloxy)isophthaloyl dichloride (**233**), which was synthesized in high yield from 5-hydroxyisophthalic acid (Scheme 5.49). The dendron wedges were prepared by treatment of siloxane **233** with phenol to give bis(aryl ester) **234**, which was hydrolyzed, or desilylated (HCl, Me₂CO), to generate a new phenolic terminus. Treatment of this

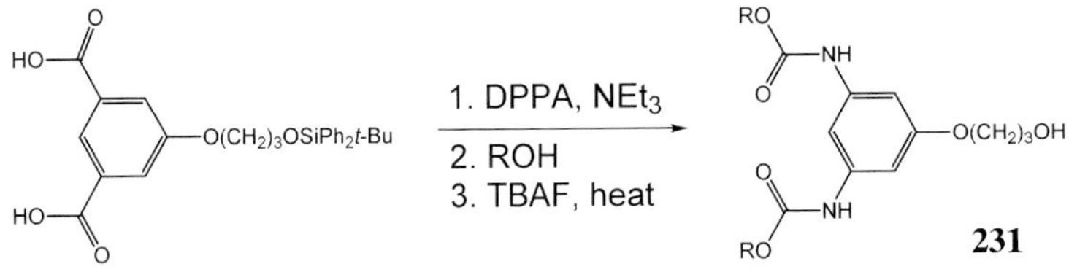

230

1. DPPA, NEt₃
2. ROH
3. TBAF, heat

231

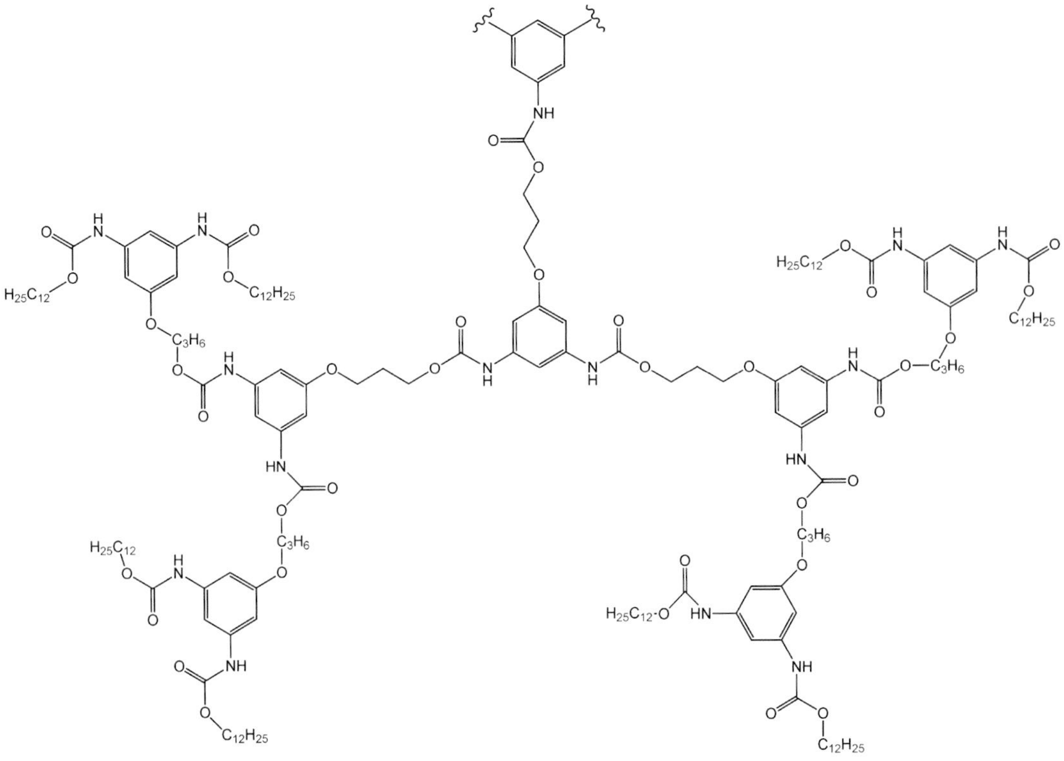

232

Scheme 5.48 Taylor's polyurethanes based on the Curtius rearrangement.[281]

free phenolic moiety with monomer **234**, followed by hydrolysis, afforded the next tier (**235**). Repetition of the sequence followed by reaction of the free focal phenols with a triacyl chloride core (e.g., **49**) afforded the 4th tier dendrimer **236** of the polyester aryl series; the choice of base (*N,N*-dimethylaniline) used in the final esterification was critical since with pyridine or DMAP facile transesterification resulting in branch fragmentation occurred. Characterization of the series was achieved by standard ^{13}C and ^{1}H NMR spectroscopy and gel permeation chromatography. The melting point of the 1st tier ester (176–178 °C) was found not to differ greatly from that of the 4th tier ester (192–203 °C) even though the molecular weights differ by a factor of nearly 10. Haddleton et al.[286] demonstrated the utility of MALDI-MS for these aromatic polyester dendrimers.

Bryce et al.[287, 288] [288a] created highly functionalized tetrathiafulvene (TTF) derivatives; several reviews by Bryce et al.[287, 289, 290] and Becher et al.[291] have appeared. Cascade peripheral modification was envisaged as providing redox-active materials for areas such as molecular self-assembly and molecular electronics. Tetrathiafulvalene is of interest due to numerous factors that include: (1) sequential and reversible oxidation to the monocation and dication radical species, (2) the ability to effect oxidation potential change through substituent choice, (3) radical thermodynamic stability, and (4) stacking propensity. The preparation (Scheme 5.50) of the TTF-capped dendrons initially utilized 5-(*tert*-butyldimethylsiloxy)isophthaloyl chloride (**233**),[284] which upon treatment with the TTF–MeOH[292] derivative (**237**) gave diester **238**. Deprotection of the phenol moiety (*n*-Bu₄N⁺F⁻, THF) of diester **238**, followed by reaction with 0.5 equivalents of bis(acid chloride) **233** and further desilylation afforded the tetra-TTF dendron **239**.[293] Reaction of three equivalents of hexaester **239** with **49** gave the dodeca-TTF dendrimer **240**.

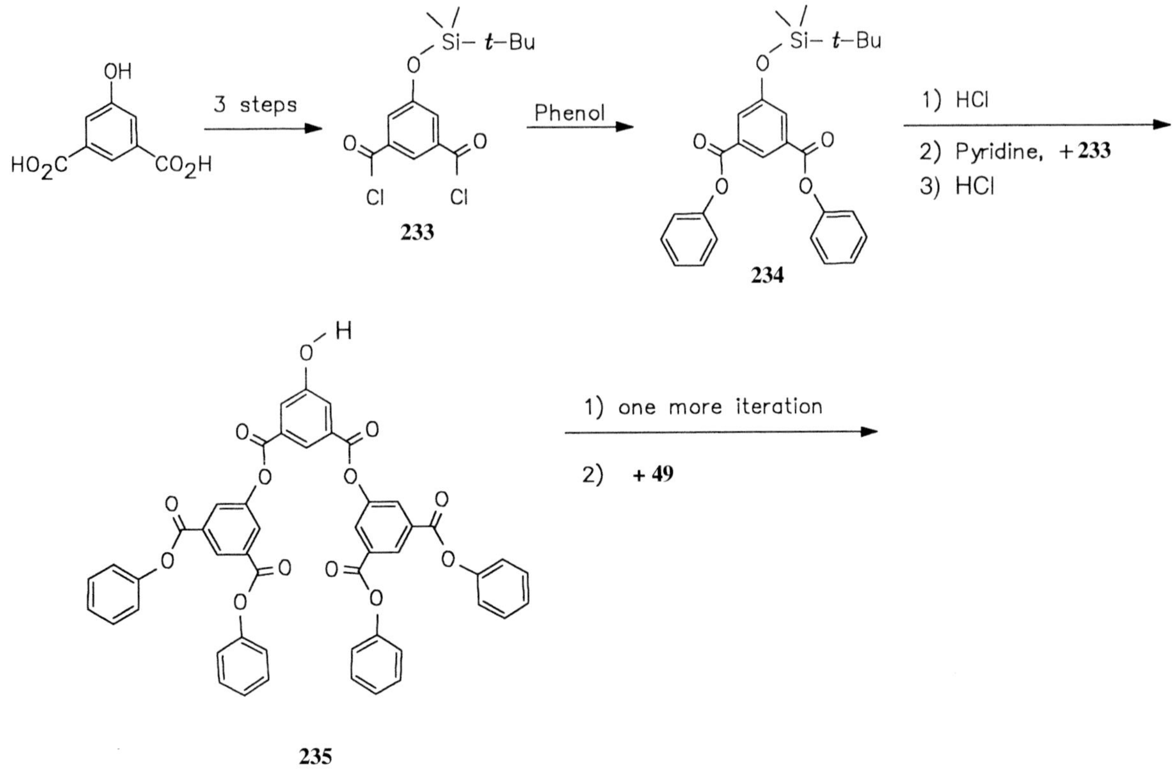

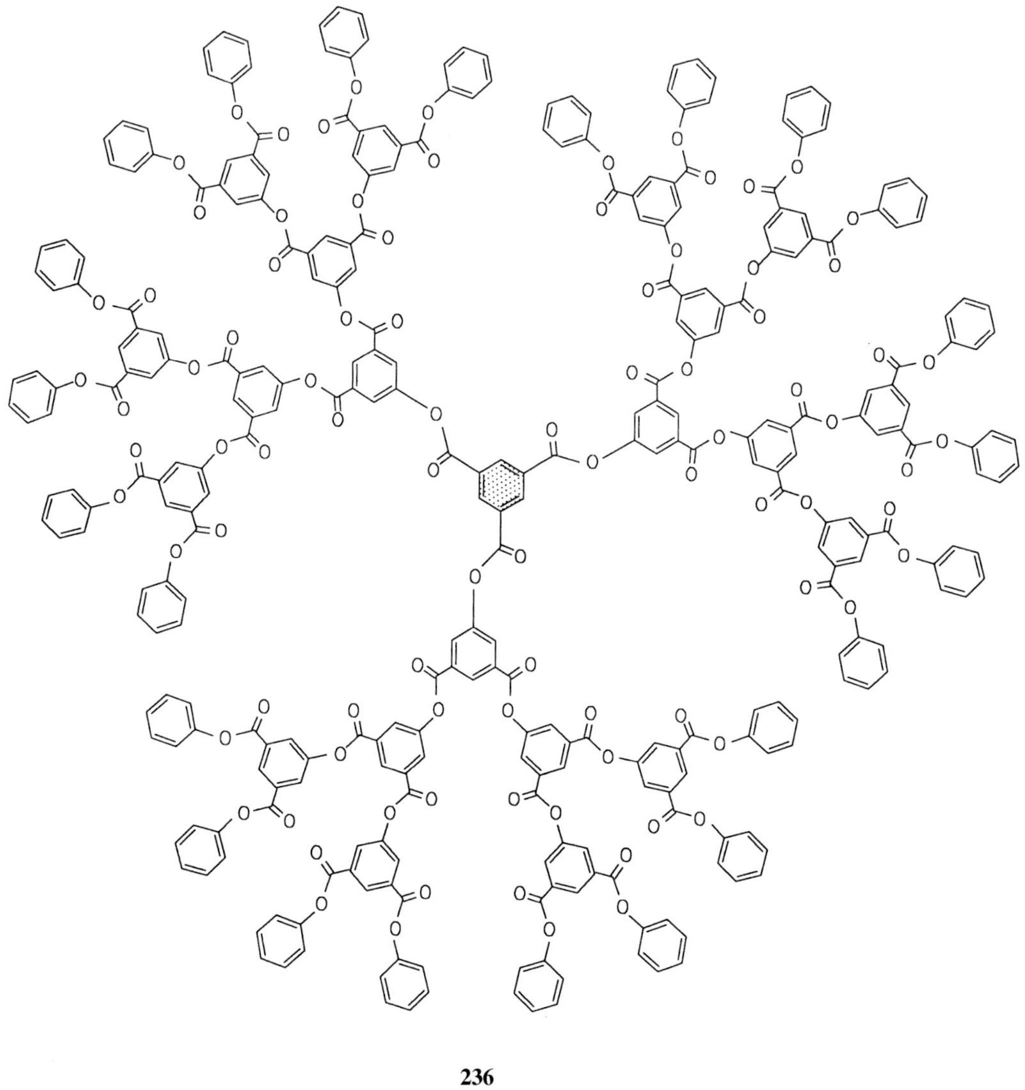

236

Scheme 5.49 Employment of silane protective groups facilitated the preparation of polyester-based dendrimers.[283–285]

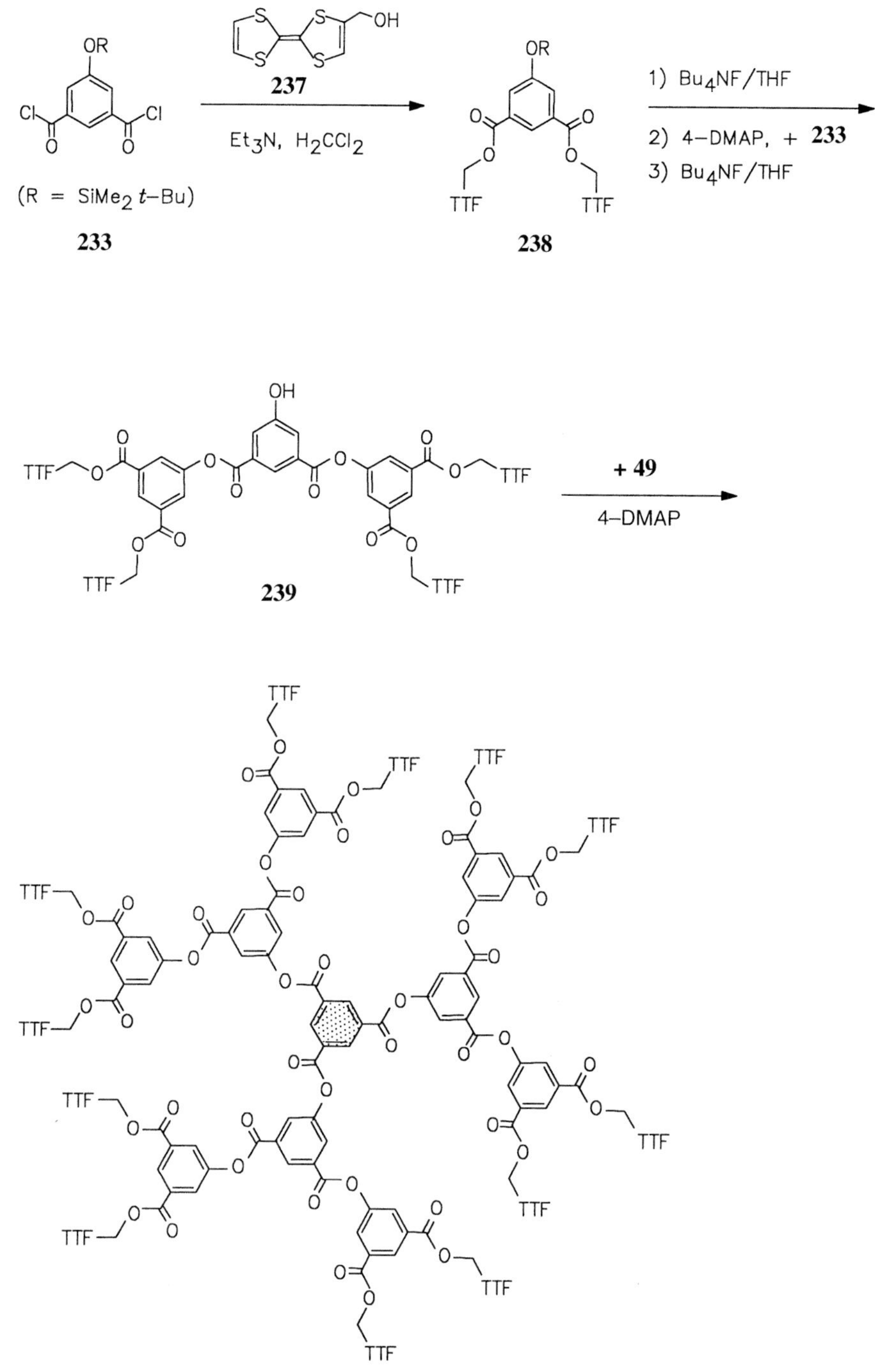

Scheme 5.50 Tetrathiafulvalene-terminated dendrimers[287, 288] prepared using ester connectivity.

These TTF dendrimers were reportedly stable when stored below 0 °C; however, within seven days at 25 °C in the absence of light and under an inert atmosphere (argon), notable decomposition was observed. The redox behavior of dodeca-TTF dendrimers **240**, as well as the corresponding lower generations, gave rise to two well-defined, quasi-reversible couples at $E_1^{1/2} \approx +0.45$ and $E_1^{1/2} \approx +0.85$ V (*vs.* Ag/AgCl). Treatment (I$_2$, CH$_2$Cl$_2$) of the TTF dendrimers oxidized the fulvalene groups as indicated by the appearance of a broad absorption in the UV spectrum (λ_{max} = 590 nm). This signal is consistent with TTF radical cation formation. From their studies, these investigators concluded that each couple corresponds to a multi-electron transfer process and that with respect to the charged TTF moieties on the dendrimers there are no significant interactions (see also[294]).

Reacting 2.1 equivalents of a TTF-terminated, phenol-based dendron with 1,3,5-tris(chlorocarbonyl)benzene allowed Bryce et al.[295] to introduce a thiol component pos-

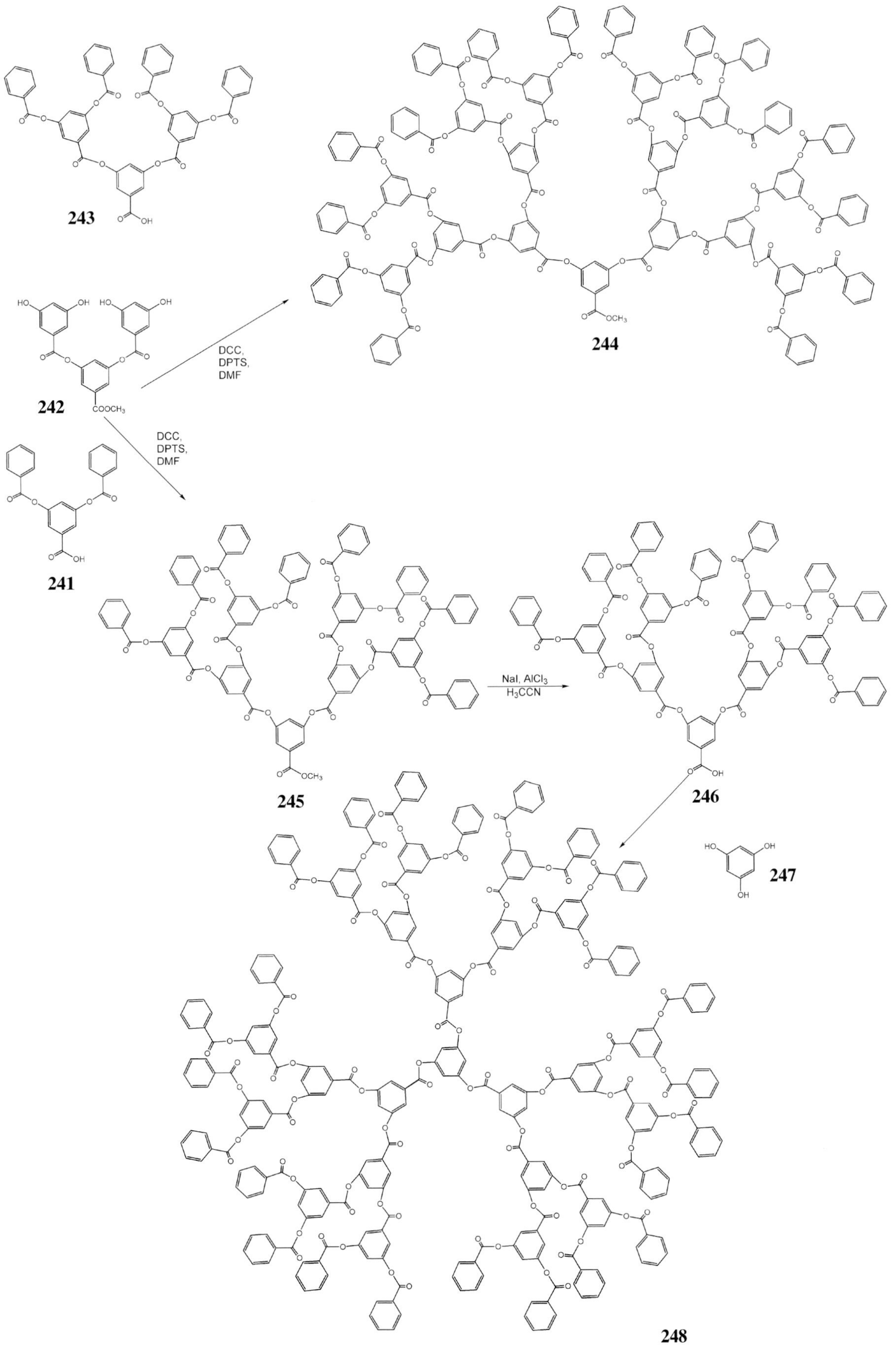

Scheme 5.51 The branched monomer strategy for construction of polyaryl esters.[297]

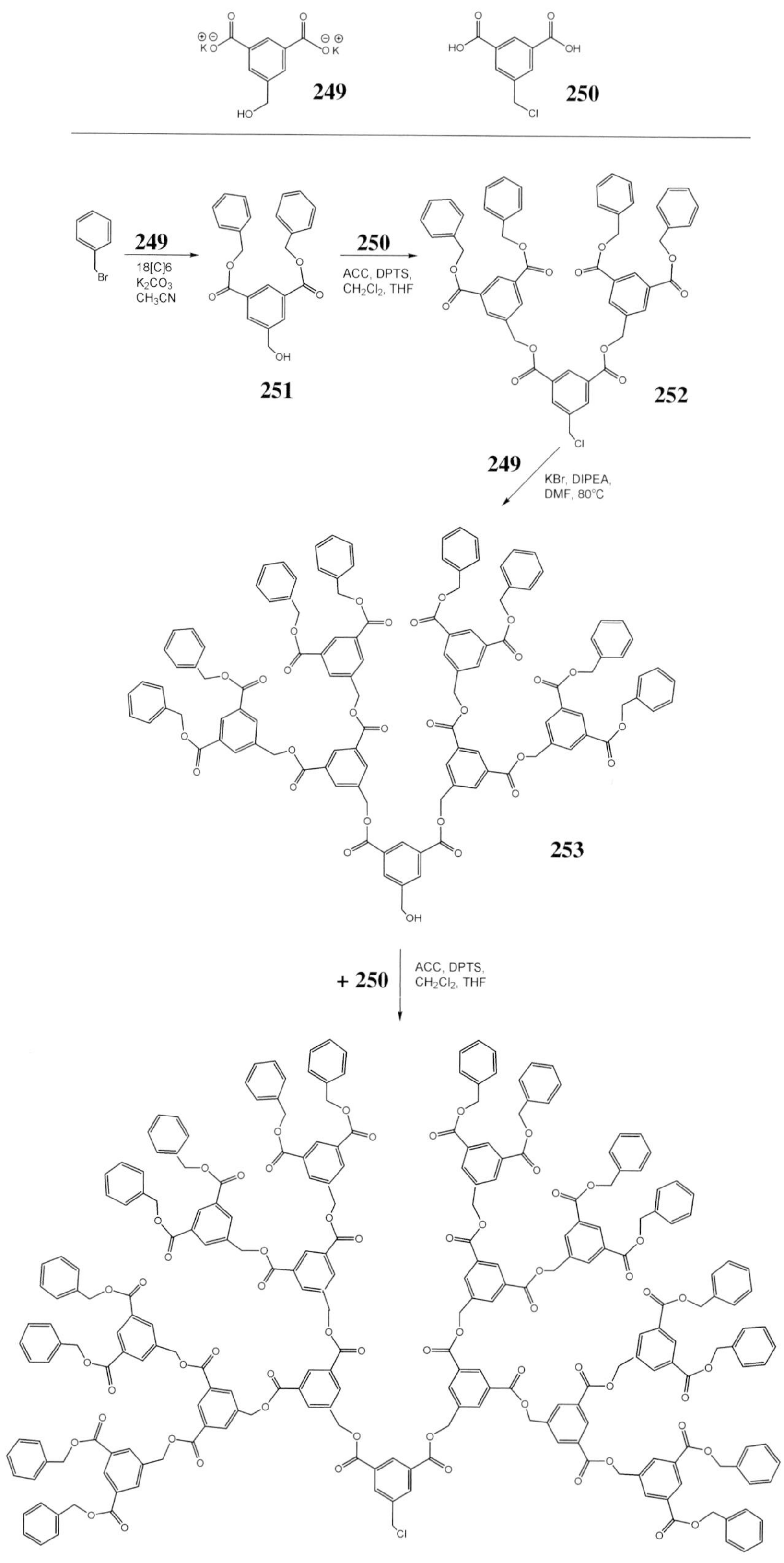

254

Scheme 5.52 Fréchet's activated monomer approach to dendritic construction.[301]

sessing anthraquinone units [i.e., $(TTF)_4(AQ)_2$ and $(TTF)_8(AQ)_4$]. Amphoteric redox behavior was observed with reversible cationic and anionic switching; for the higher generation construct, intradendrimer charge transfer was observed.

Feast et al.[296] investigated miscible blends of two small poly(aryl ether) dendrimers with poly(ethylene terephthalate) (PET). Dendrimer blends produced a noticeable effect on material processing characteristics, as evidenced by dielectric measurements and tensile analysis. The smaller of the two constructs (G0) reduced polymer chain interaction acting as a plasticizer, whereas the larger molecule (G1) acted in an antiplasticizer capacity increasing intermolecular interactions. This was rationalized by considering that the more highly branched structure can entangle and knot the polymer chain so as to increase chain contact probability, while the minimally branched material was incapable of inducing entanglements and served only to separate and restrict chain interactions.

A branched monomer strategy (Scheme 5.51) was reported by Zhang, Shen, and coworkers[297] utilizing the key monomers **242** and **243**, which when coupled afforded the 4th generation wedge **244**. Treatment with the 1st tier dendron **241** produced the 3rd generation monoester **245**, which was transformed to the corresponding focal acid **246** and subsequently reacted with the trisphenolic core **247** to afford dendrimer **248**. These poly(ester)s have been peripherally derivatized with carbazole groups[298] for examination of the resultant electro-optical properties.[299] Fluorescence studies revealed strong carbazole-based intramolecular interactions. Goodson et al.[300] have also constructed carbazole-stilbenoid branched structures for studying their fast electronic nonlinear optical properties.

Freeman and Fréchet[301] described and defined the "activated" monomer approach to dendrimer construction using two functionally orthogonal monomers whereby one is the "activated" analog of the other (Scheme 5.52). The protocol can be envisaged by considering the repetitive use of monomer **249** (activated) and **250**. Treatment of bis(carboxylate) **249** with benzyl bromide followed by coupling with diacid **250** afforded benzyl chloride **252**, which was subsequently treated with more dicarboxylate [*via in situ* bromination (KBr, DMF, DIPEA)] and then coupled again with further diacid to give the 4th generation dendron **254**. This procedure allows for the rapid construction of perfect dendrimers up to a focally hindered limit.

Feast and Stainton[302] prepared related but extended aryl ester dendrimers employing a phenolic acetate–phenol protection–deprotection scheme whereby dendron formation was based on bis(phenolic ester) formation followed by focal acetate deprotection and subsequent repetition.

5.3.9 1 → 2 *Aryl*-Branched, *Ether & Ester* Connectivity

Fréchet et al.[303] reported a branched-monomer approach to dendritic macromolecules; this approach permits an accelerated growth through the replacement of the simplest repeat unit with a larger repeat unit of the next generation. In essence, the traditional AB_2 monomer is replaced with an AB_4 unit. Scheme 5.53 depicts the transformation [$CF_3CONMe(SiMe_3)$] of the AB_4 unit (**255**) to the trimethylsilyl protected tetraester **256** as well as to the benzyloxy-terminated building block **257** by treatment ([18]-crown-6, K_2CO_3, Me_2CO) with bromide **89**. Activation (CBr_4, PPh_3) of the alcoholic moiety in **257** by conversion to the corresponding bromide **258** and treatment with 0.25 equivalents of tetrakis(trimethylsiloxane) **256** generated the desired dendron **259**. The incorporation of the internal hydrolysable ester linkages affords entry to internal hydrolytic sites.

The use of a terminal 3,5-bis(benzyloxy)benzoic acid in the convergent process with internal ester connectivity permitted the catalytic deprotection (H_2/Pd-C) of the benzyl groups. The resultant spherical macromolecule possessed the reactive phenolic functionality, which facilitated aqueous solubility and functionalization.[304] Oxadiazole, photon absorbing structures are known.[304a]

Hawker, Wooley, and Fréchet[305, 306] created novel "mix and match" combinations utilizing dendritic polyesters and polyethers. The resultant globular block architectures were generated by the controlled location of different ester and ether groups affording either radial or concentric patterns. Thus, a dendritic "segment"-block copolymer is prepared by the attachment of radially alternating dendritic segments from the core,

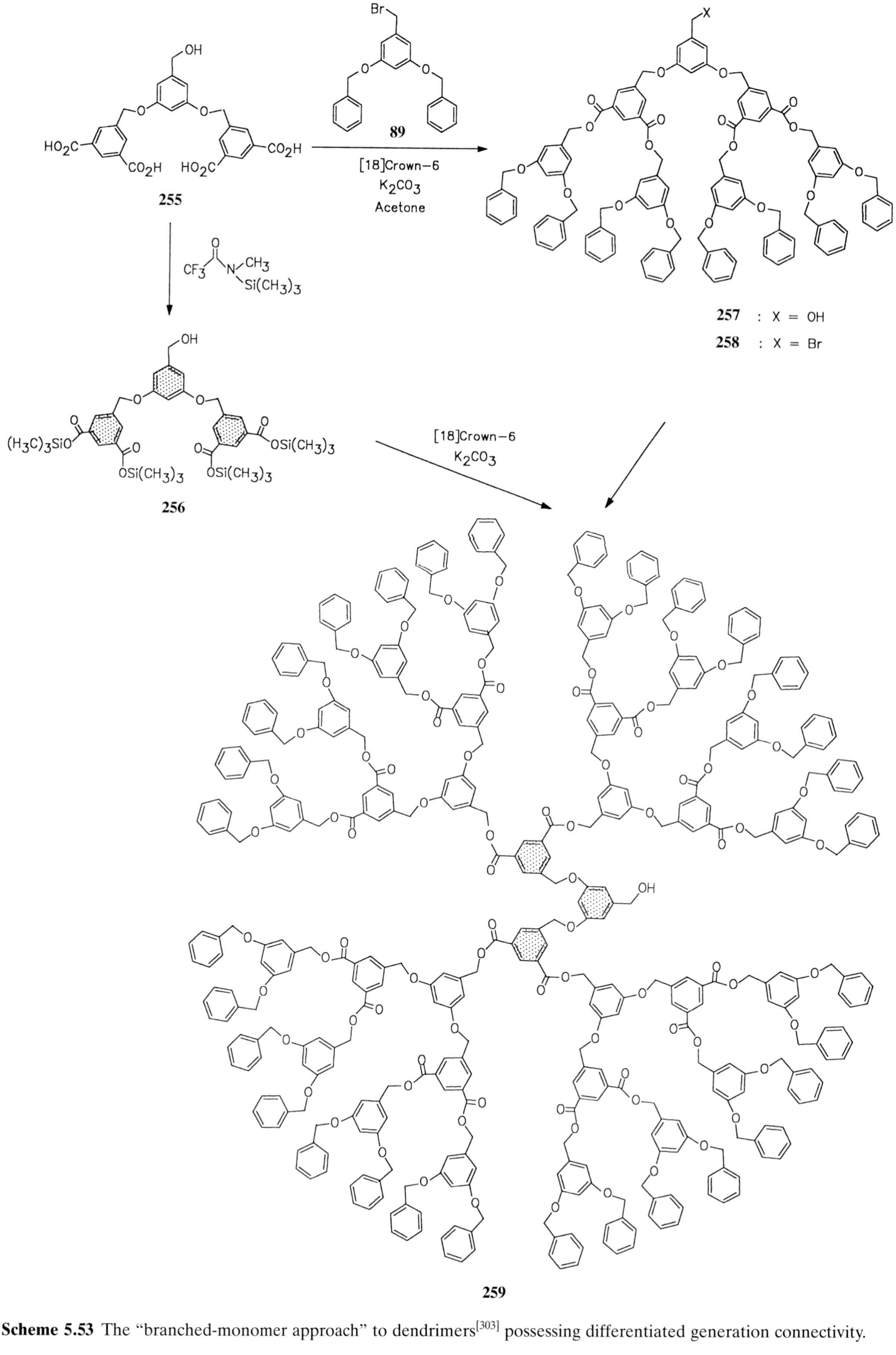

259

Scheme 5.53 The "branched-monomer approach" to dendrimers[303] possessing differentiated generation connectivity.

whereas a dendritic "layer"-block is created by concentric alternation of ester and ether layers from the core. Since the termini and core are similar in each case, the relative proportion of ether and ester building blocks rather than their precise geometries controlled the glass transition temperatures of the various block copolymers.

Nierengarten, Felder, and Nicoud created poly(aryl ester) architectures possessing peripheral fullerene units (Scheme 5.54).[307] The starting terminal monomer **260** was accessed (C$_{60}$, I$_2$, DBU) by bis(malonate) macrocyclization.[308] *tert*-Butyl ester deprotection followed by reaction (DCC, DMAP) with diol **261** yielded the 1st generation bis(C$_{60}$)

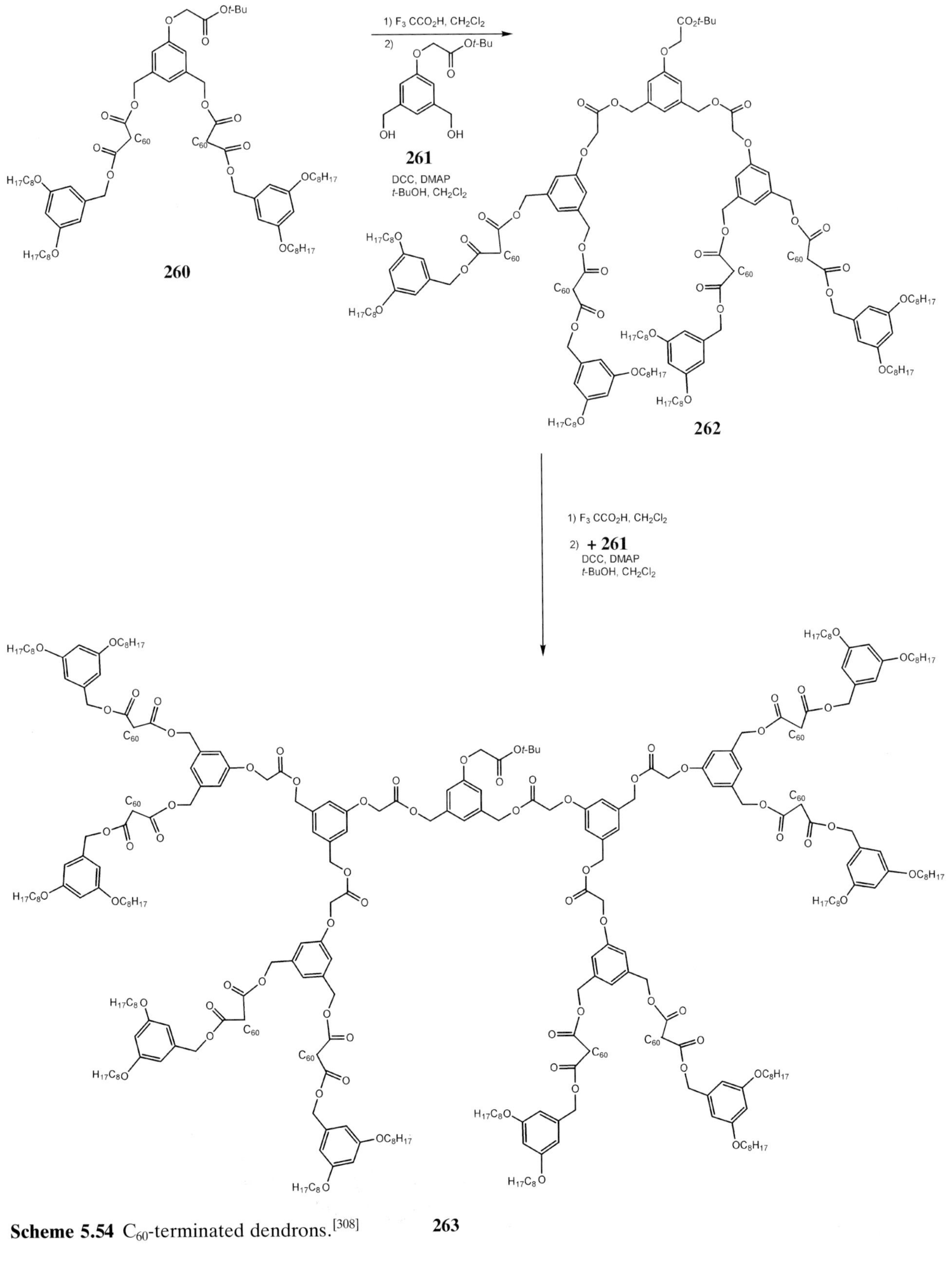

Scheme 5.54 C$_{60}$-terminated dendrons.[308]

262. Iteration led to the corresponding tetrakis(C$_{60}$) dendron **263**. These unique materials were subsequently attached to a phenanthroline diol and complexed with Cu(I). The interaction of similar amphiphilic fullerene-terminated dendrons with thin films has been examined. High quality Langmuir films were shown to form at the air–water interface, which could be readily transferred to solid substrates.

5.3.10   1 → 2 *Aryl*-Branched, *Ether* and *Ketone* Connectivity

Morikawa et al.[309] prepared a series of poly(ether ketone) dendrimers by a convergent approach utilizing 3,5-bis(4-fluorobenzoyl)anisole. Initial conversion of the fluoro groups to phenyl ether moieties was followed by deprotection (AlCl$_3$) of the methyl ether moiety; nucleophilic aromatic substitution using the fluoro building block completed the general scheme. Four generations were easily constructed by repetition of this procedure; the cascade polymers each possessed a narrow molecular distribution.

5.3.11   1 → 2 *Aryl*-Branched, *Alkyne* Connectivity

Moore and Xu[310, 311] reported the convergent preparation of phenylacetylene dendritic wedges (Scheme 5.55) by a repetitive strategy involving a Pd-catalyzed coupling of a terminal alkyne to an aryl halide possessing a trimethylsilyl-capped alkyne. The concept is illustrated by considering the following transformations: two equivalents of phenylacetylene (**264**) were attached to a silyl-protected 3,5-dibromophenylacetylene[312, 313] (**265**) by a cross-coupling [Pd(dba)$_2$] procedure. After deprotection (MeOH, M$_2$CO$_3$), the process was either repeated to give the next higher generation (path a) or the terminal alkyne was removed from the growth sequence and subjected to core attachment (path b). Dendritic wedge synthesis proceeded as expected through to the 2nd genera-

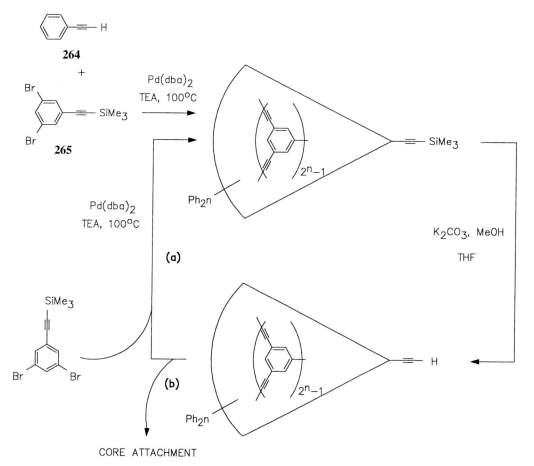

Scheme 5.55 Moore et al.'s[311] iterative procedure for preparing poly(aryl alkyne) dendrimers.

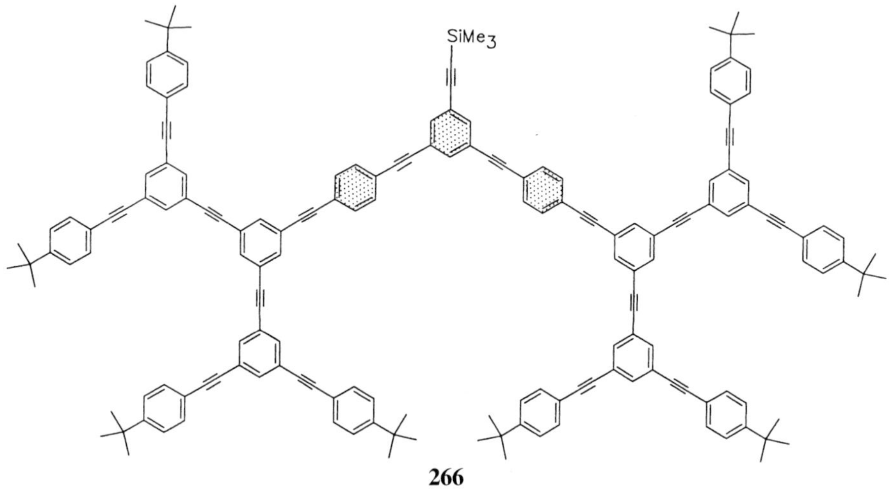

Figure 5.23 Use of extended monomers facilitated construction of large, rigid dendrons.[310, 311]

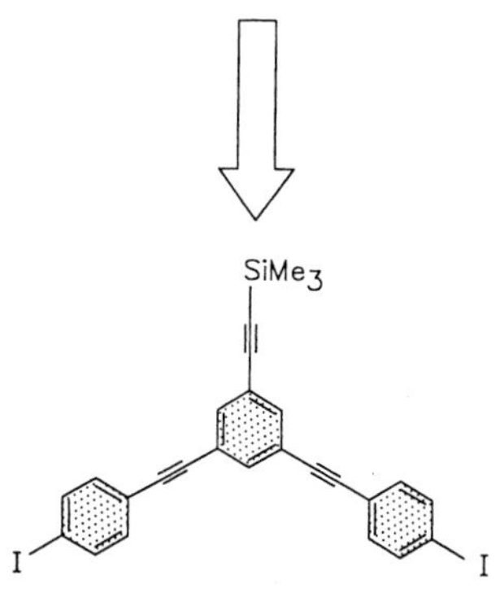

tion; however, two factors, dendron stability and steric inhibition to reaction at the focal point, were operative. Modest gains in the yield of the 3rd tier dendron were realized by using *p*-substituted peripheral phenyl units, such as 4-MeOC$_6$H$_5$ or 4-(*tert*-Bu)C$_6$H$_5$. Even higher yields of the 3rd generation wedge **266** were achieved by the use of an elongated monomer **267** (Figure 5.23) indicating that increased spacer length should allow facile construction of large, rigid dendrimers.

Moore and Xu[314, 315] later reported improvements in phenylacetylene dendron and dendrimer preparation. Essentially, a dendritic construction set (**268–279**; Figure 5.24) was prepared by using the previously described[311, 314] aryl halide–alkyne coupling combined with trimethylsilyl alkyne masking as well as an aryldiethyltriazene to aryl iodide transformation.[316] The aryldiethyltriazene moiety served as a protected aryl iodide that could be introduced by triazene treatment with MeI at 110 °C in a sealed tube. Specifically, the use of peripheral 3,5-di-*tert*-butylphenyl units, rigorous exclusion of molecular oxygen from the Pd-mediated alkyne–aryl halide cross-coupling reaction, and particularly, the employment of optimized conditions (35–40 °C, 2 d) allowed the preparation (37 %) of the '94-mer' dendrimer. With the increased alkyl surface (i.e., aryl 3,5-di-*tert*-butyl groups), all of the dendrimers were readily soluble in pentane at 25 °C. Characterization included a discussion of the COSY-45 ^{1}H NMR spectrum of the hydrocarbon dendrimer. Mass spectrometry was used to confirm[317] purity and structural composition. Terminal functionalization of the 1st–3rd generations of these rigid dendrimers with bis-{2-[2-(2-methoxyethoxy)ethoxy]ethyl} ester units facilitated the formation of columnar discotic liquid crystalline phases.[318]

Moore et al.[319, 320] subsequently employed the concept of dendritic spacer elongation for the construction of a series of phenylacetylene cascades (e.g., **280**; Scheme 5.56) with the longest being 12.5 nm in diameter. Hence, reaction of an extended acetylene dendron **280** with 1,3,5-triiodobenzene afforded the 4th generation '127-mer' **281** (Scheme 5.56). The method of avoidance of steric inhibition to dendritic construction through the use of increasingly longer spacer moieties gave rise to the acronym *SYNDROME* (*SYN*thesis of

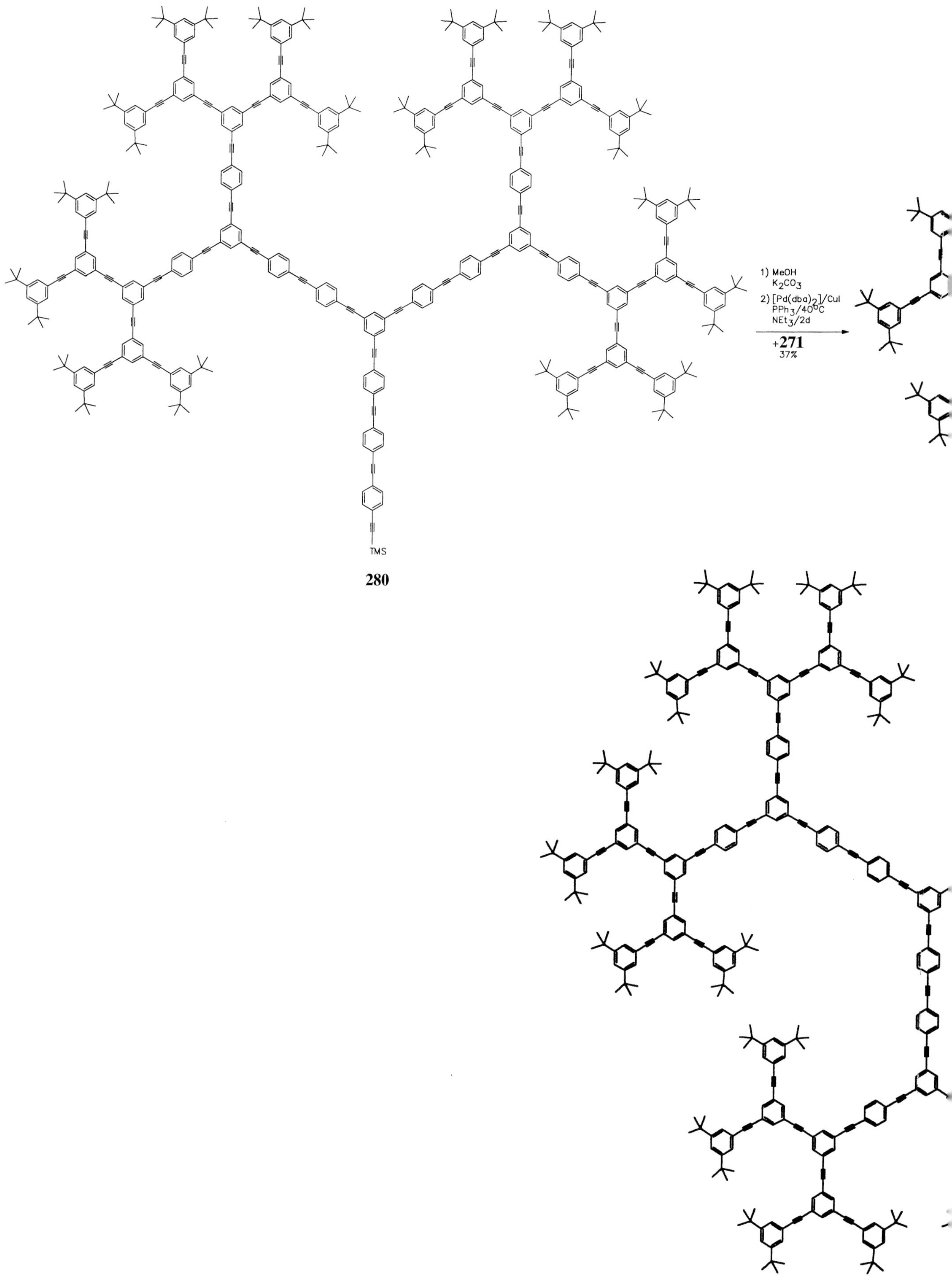

280

1) MeOH
   K$_2$CO$_3$
2) [Pd(dba)$_2$]/CuI
   PPh$_3$/40°C
   NEt$_3$/2d

+271
37%

Scheme 5.56 Synthesis by the *SYNDROME* procedure.[319, 320]

281

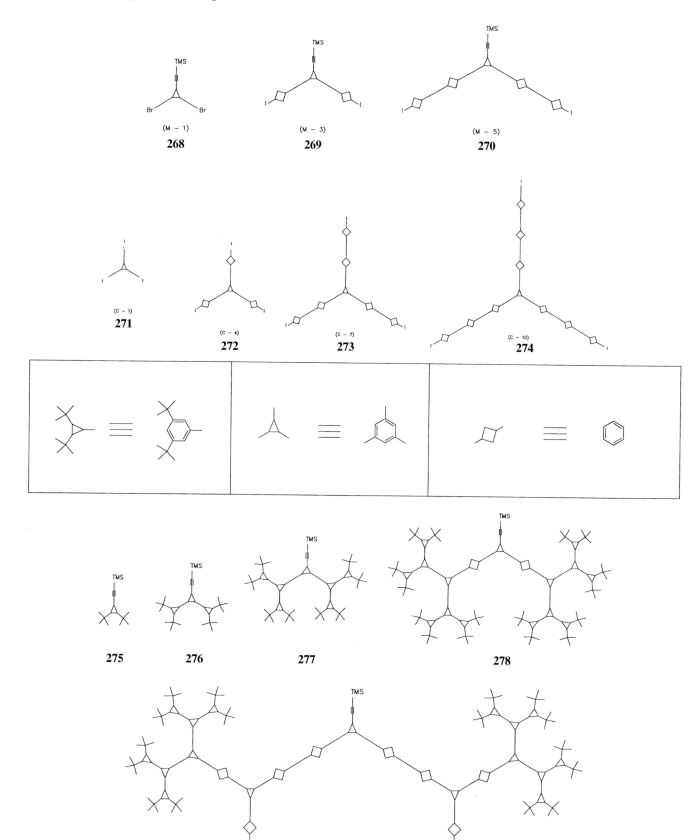

Figure 5.24 Dendritic construction set based on Moore et al.'s[321] phenylalkyne chemistry (aryl connectors consist of –C°C≡ bonds).

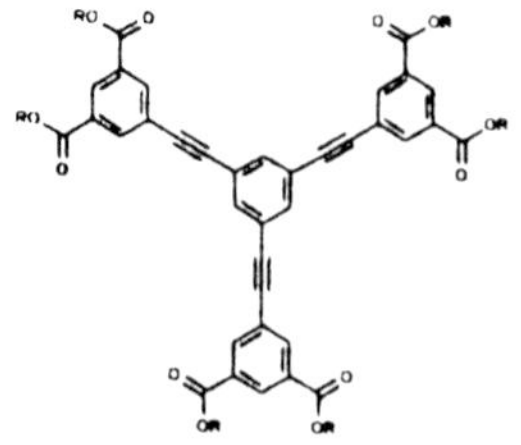

a R = *t*-Bu
282 **b** R = H
      c R =

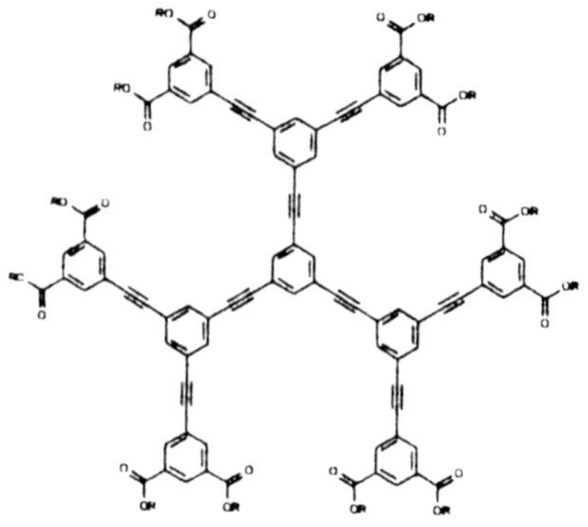

a R = *t*-Bu
283 **b** R = H
      c R =

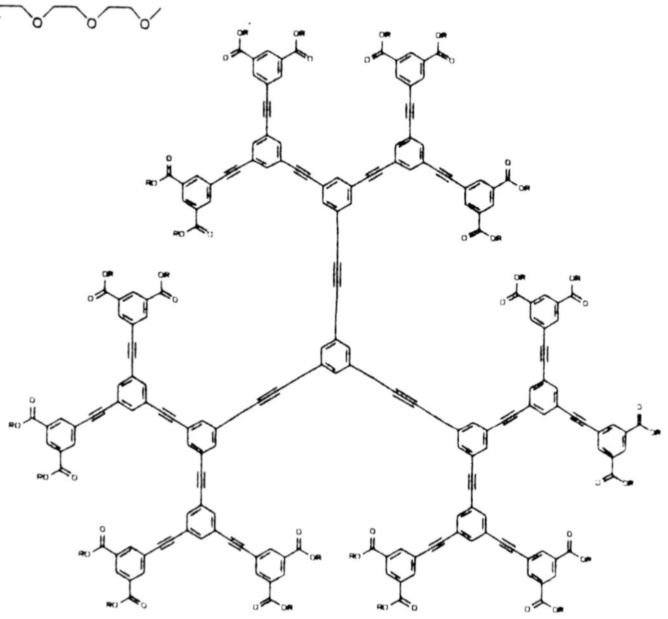

a R = *t*-Bu
284 **b** R = H
      c R =

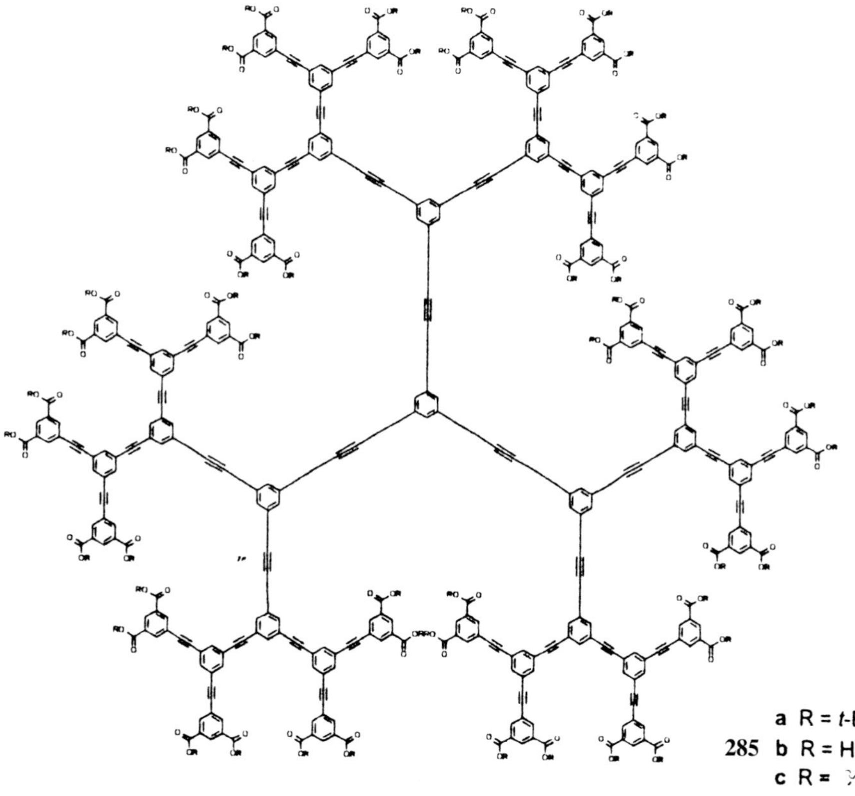

a R = *t*-Bu
285 **b** R = H
      c R =

Figure 5.25 Moore et al.'s water-soluble, shape-persistent dendrimers.[325]

*Dendrimers by *Repetition Of Monomer Enlargement*). These dendrimers[319, 322, 323] are 'shape-persistent' and 'dimension-persistent', which can play an important role in the development of molecular frameworks[324] that require rigorous control of functional group juxtaposition.

Using a slightly modified synthetic protocol, Moore et al.[325] prepared and studied water-soluble analogues of the classic poly(phenylacetylene) dendrimers; for additional references concerning related acetylene chemistry, see.[326–338] [338a] Water-soluble dendrimers, up to the 5th generation (Figure 5.25; **282–285**), were accessed by coating the periphery with 3,5-di-*tert*-butyl ester benzene moieties. Transformation of the ester groups to carboxylic acids by solid-state thermolysis imparted aqueous solubility; however, thermolysis was suspected to cause a small amount of "cross-linking". A second series of esters was prepared terminating with 2-[2-(2-methoxyethoxy)ethoxy]ethyl ester moieties, thereby enabling aqueous solution through ester hydrolysis.

Pesak and Moore[339] further created a family of amphiphilic poly(phenylacetylene)s by focal modification of hydrocarbon-terminated dendrons with carboxylic acid terminated dendrons to examine polar and non-polar promoted dendrimer organization. Aggregation in lipophilic solvents was consistent with a proposed higher-order assembly, whereby the polar regions orient towards each other in the same manner as in an inverted micelle.

Kawaguchi and Moore[340] described the construction of 'ball & chain' dendritic polymers for comparative purposes with 'monodendron-based' copolymers (Figure 5.26; **286** and **287**, respectively). Architecturally, the 'ball & chain' design features a spherical or

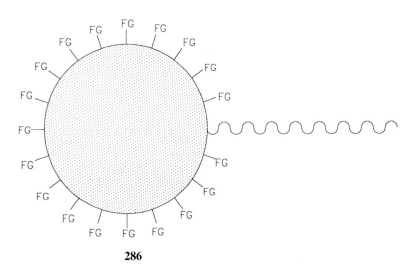

286

"Ball and chain" copolymer architecture

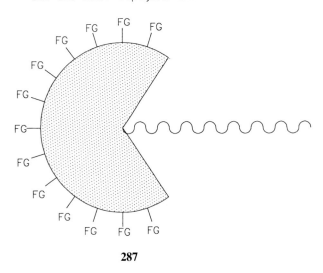

287

Monodendron—based copolymer

Figure 5.26 "Ball and Chain"-type dendrimers.[340]

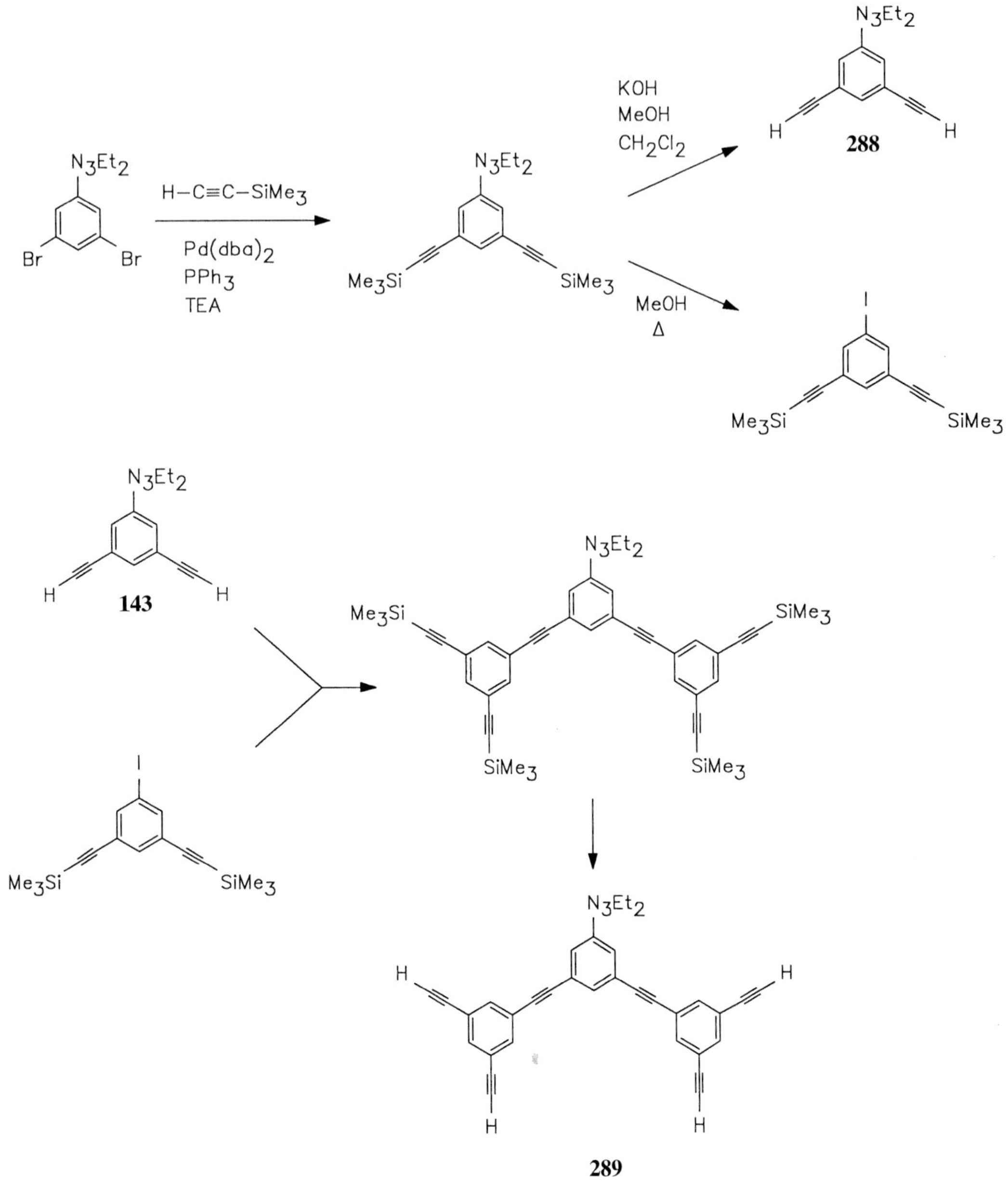

Scheme 5.57 Preparation of polyalkyne-terminated monomers.[316]

globular dendrimer, possessing a single functionally differentiated terminal group connected to a linear polymeric chain. This is in contrast to the monodendron topology that arises from connection of the linear polymer to the focal point of a dendritic wedge. The convergent strategy to these 'site-specified' phenylacetylene dendrimers included construction of (1) dendritic wedges with 3,5-di-*tert*-butyl phenyl terminated groups, (2) site-specified wedges containing one aldehyde moiety at the periphery, and (3) spherical dendrimers with one functionally differentiated terminal group. The strategy is similar to Fréchet's procedure for the control of surface functionality.[200]

Bis- and tetrakis-terminated alkyne cores (**288** and **289**, respectively) were prepared by standard iterative transformations[316] (Scheme 5.57). These hydrido-terminated cores were used for the preparation of *tert*-butyl dendritic wedges (e. g., **292**; Scheme 5.58) containing up to 63 aryl 1 → 2-branching centers.

Scheme 5.59 shows the transformations used for the synthesis of dendritic wedges (**297**) that are *di*functionalized with aryl iodide and trimethylsilyl-protected arylalkyne moieties at the focal region. Reaction of 3,5-dibromobenzotriazene (**293**) with one equivalent of isopropoxy-protected acetylene **294**, followed by treatment with base, gave the 3-bromo-5-ethynylaryltriazene (**295**). Higher selectivity of Pd-catalyzed alkynylation of aryl iodides (e.g., **291**) afforded the difunctionalized wedges (**296** and **297**).

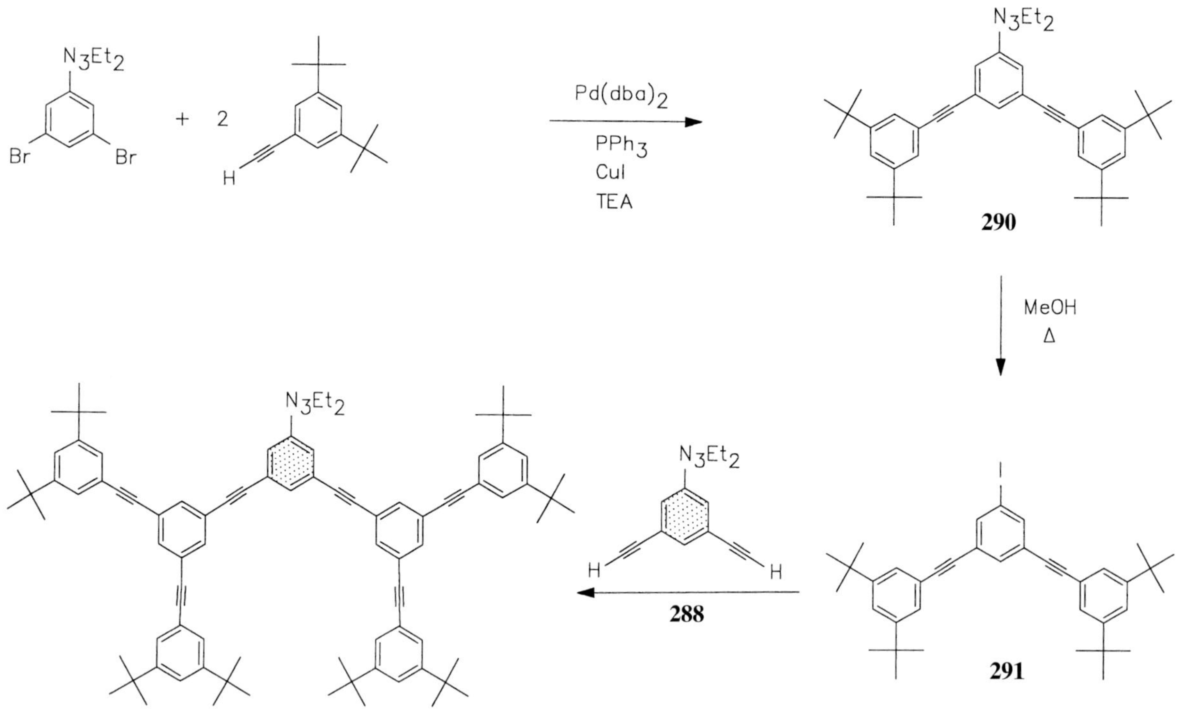

292

Scheme 5.58 Construction of dendrons[316] possessing diethyltriazine focal groups.

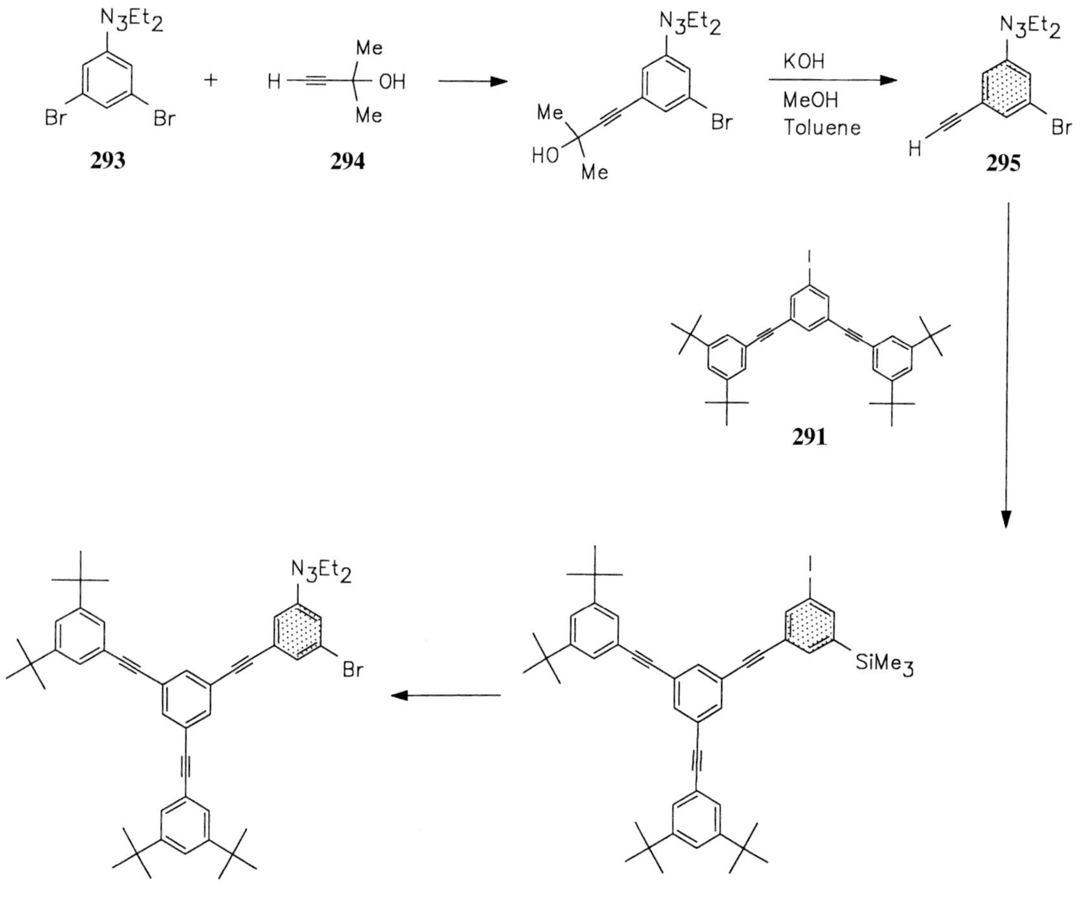

297 296

Scheme 5.59 Preparation of dendrons[316] possessing dual functionality at the focal region.

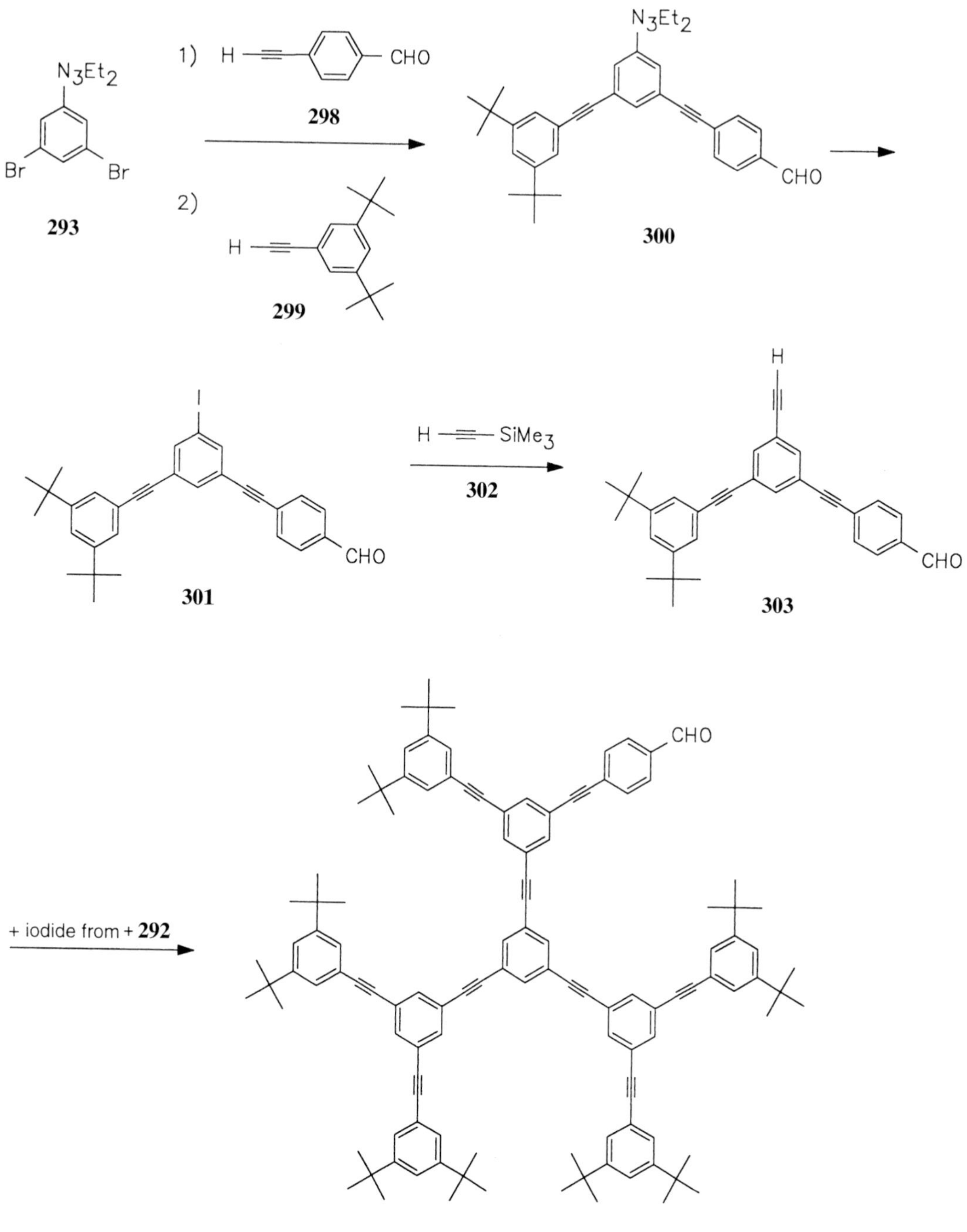

Scheme 5.60 Sequence for the construction of "site-specific" phenylalkyne dendrimers.[316]

Monocarboxaldehyde wedges (**304**) were prepared analogously (Scheme 5.60) by coupling one equivalent each of 4-ethynylbenzenecarboxaldehyde (**298**) and 1,3-di-*tert*-butyl-5-ethynylbenzene (**299**) to aryl dibromide **293** to afford the monoaldehyde diazene building block (**300**). Conversion of the diazene moiety to an aryl iodide and treatment with trimethylsilylacetylene, followed by silane removal, allowed the preparation of ynealdehyde **303**. Reaction of bifunctional building block **300** with the corresponding aryl iodide derived from **292** afforded the monofunctionalized phenylacetylene dendrimer **304**.

Using this technology, the preparation of the 5th tier, monoaldehyde dendrimer was reported.[340] Subsequently, "living" poly(methyl methacrylate) [PMMA, prepared by group transfer polymerization] was treated with the site-specific cascades. Polydispersities determined for the copolymer were similar to that recorded for the living polymer when smaller dendrimers were used. Using larger dendrimers for copolymer formation led to a dependence of polydispersity on the dendrimer.

Xu and Moore[341] proposed the use of their dendritic arylacetylenes as directional molecular antennae. The directional transduction of energy through a convergent pathway is conveniently incorporated in the macromolecular framework of **305**. The three-directional branching centers are nodes in the molecular electronic wavefunctions and the phenylacetylene connectors are localized regions of extended π-conjugation. A luminescent probe, perylene, was appended at the dendron's core in order to test the hypothesis of molecular harvesting of light energy. Energy is collected at the outer surface and funneled through the convergent cascade until it reaches the single fluorescent emitter. Figure 5.27 depicts the collection and transmission route through the dendron. Photophysical properties of dendrimer **305** indicate that perylene emission after excitation at a wavelength (312 nm) corresponding to the absorption maximum of the peripheral aryl groups occurs with enhanced intensity when compared to that of 3-ethynylperylene itself; spectroscopic evidence indicated a transfer efficiency of 98 % from the dendrimer framework to the focally located perylene.[342] Enhanced energy transfer in extended as opposed to compacted molecules has also been considered.[342a]

Electronic coupling in dendrimers has been considered, including the role that disorder plays in the determination of electronic communication pathways.[343]

Bar-Haim, Klafter, and Kopelman[344] mathematically examined the potential for dendrimers to act as controlled artificial energy antennae. An exact solution for the

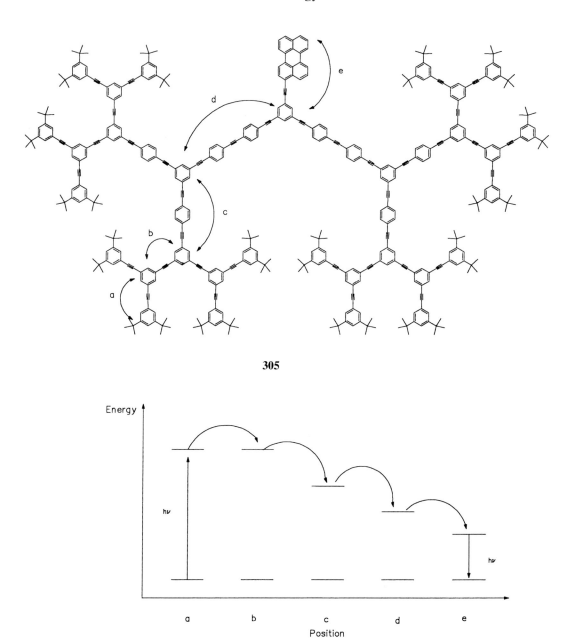

305

Figure 5.27 Rigid phenylalkyne dendrimers[341] employed as molecular antennae for the directional transduction of energy.

"mean first passage time" (MFPT; τ) or the "mean time for an excitation that starts "at the periphery to reach the center." Essentially, the MFPT, which is a measure of energy trapping efficiency, can be derived by the solution of a set of differential equations. Local rates of energy migration k_{up} (towards the periphery) or k_{down} (towards the core) directly influence overall energy migration to the surface or core (k_1 and k_2, respectively) along with coordination number. Notably, the authors point out that "symmetric dendrimers are characterized by an *inherent geometrically-induced bias* toward the periphery, a property unique to these supermolecules." It was suggested that control of dendrimer geometric bias and energetics should facilitate artificial antenna design.

Moore et al.[331, 345] reported a 'double exponential dendrimer growth', which constituted an accelerated convergent process for the synthesis of monodendrons by means of a bidirectional procedure, where the degree of polymerization (DP) follows Equation 5.1.

$$dp = s^{2n} - 1 \qquad (5.1)$$

This represents a dramatic increase in the average polymer growth considering that for a $1 \rightarrow 2$ branching pattern and using a convergent procedure, the DP follows Equation 5.2.

$$dp = 2^{(n+1)} - 1 \qquad (5.2)$$

This double-exponential growth scheme is represented by the reaction–deprotection cyclic diagram shown in Figure 5.28. The significance of the strategy is reinforced by a comparison of dendritic growth through examination of the DP using this procedure *versus* that achieved by simple convergent growth. Theoretically, the DP using the double-exponential method could potentially reach 250 after three generations (or after only nine synthetic steps), while by a simple convergent route this DP would take 21 synthetic steps, i.e. after seven generations. The advantages are analogous to those of the "double-stage convergent method",[44] but here the hypercore also grows concomitantly with peripheral monodendrons; the disadvantages are the need for a pair of orthogonal protecting groups and the rapid onset of steric crowding.

A series of rigid monodendrons focally substituted with perylene (Figure 5.29; **305–311**) was prepared in order to investigate intramolecular energy transfer using time-resolved and steady-state fluorescence spectroscopies.[346] A corresponding family, focally substituted with bis(*tert*-butyl)phenyl moieties, was synthesized as reference dendrons. It was found that the light-harvesting ability of dendrons **306–311** increased with generation, primarily due to an increase in the molar extinction coefficients, while energy-transfer efficiency was found to decrease as the generation increased. Notably, the branched-chain extended construct **305** exhibited energy-transfer rate constants approximately two orders of magnitude greater than those for the rest of the series. It was postulated that monomer variation at successive generations develops an "energy

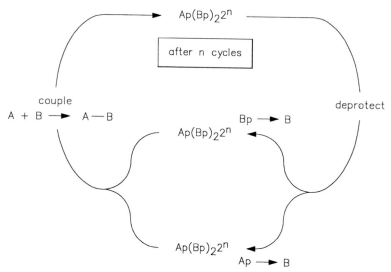

Figure 5.28 Moore et al.'s[345] scheme for "double-exponential dendrimer growth".

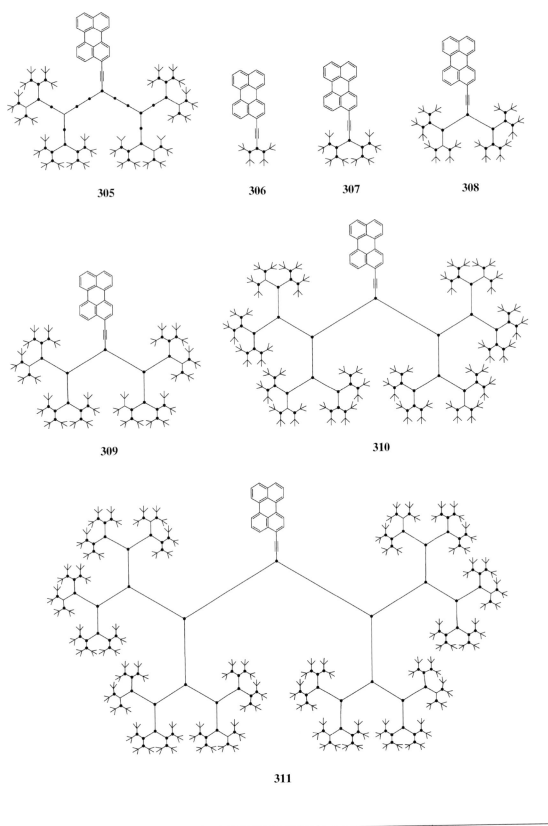

Figure 5.29 A series of molecular antennae[346] with a perylene moiety at the focal point.

funnel" created by the presence of an energy gradient. A modified energy funnel was later prepared that incorporated amine-branching centers for enhanced redox activity (see Section 5.6.5).[347]

These authors[348] further described the results of a fluorescence quenching study on the same series of monodendrons that had been focally modified with di-*tert*-butyl benzene. Stern–Volmer quenching constants in the case of higher generations being quenched by DABCO were found to be unusually large, and a significant static quenching contribution for higher generations was also observed. DABCO–dendrimer exciplex formation was found in THF.

Martin et al.[349] employed low-dose techniques to carry out selected area electron diffraction analysis and high-resolution electron microscopy on these phenylacetylene dendrimers.

Xavier et al.[350] used "hyperbolic" modeling to investigate these poly(aryl alkyne)s, whereby a density profile was obtained that possessed characteristics of all the prevailing theories. It was observed to initially decrease with radius and then gradually increase to a maximum. Kopelman et al.[351] studied the dynamics of directed, multistep energy transport in these phenylacetylene-based dendrimers.[327, 341] Molecules investigated included a symmetrical dendrimer (D127, **281**; see Scheme 5.56) and the 4th tier dendron focally substituted with an energy-absorbing perylene unit therein described as a "nanostar". Steady-state as well as time-dependent lifetime data have been collected[352] for these novel materials, which demonstrated their potential as nanoscopic LEDs, photovoltaic sensitizers, and optical probes or sensors. Correlated excimer formation was observed in certain solvents, which was found to depend strongly upon initial excited-state placement within the dendritic structure;[353] models of excimer formation dynamics and relaxation kinetics were considered. These molecules were further described as "extended" or "compact" Bethe dendrimers owing to the theoretically demonstrated[341, 344, 354, 355] optimal energy channels of Bethe lattices.[356] For example, Bar-Haim and Klafter[357] studied the light-harvesting geometric *versus* energetic competition in dendrimers. These authors proposed that for ideal dendrimers, control of the site of excitation should be possible through monitoring of the temperature with respect to the energy funnel. Thus, the potential for controlled photochemical processes might be realized. Extended Bethe dendrimers possessed increasingly long branch chain connectors (spacer units) progressing from the periphery to the center or focal point, while the compact Bethe dendrimers were constructed with equal length chains between branching sites. Based on equal chain length, disruption of π excitation conjugation because of *meta*-aromatic branching and the resultant localized excitation, "compact" dendrimers do not exhibit energy funneling. However, localized excitations in the "extended" dendrimers are comprised of varying energies that correlate inversely with chain length; extended Bethe dendrimers thus serve as energy funnels directing transfer towards the center or focal sites of dendrimers and dendrons, respectively. Thus, energy transfer efficiencies of 98 % to the "perylene trap" were recorded using the nanostar dendrimer.

These rigid energy funnels have been further studied[358] with respect to their electronic absorption and energy transfer. Discrete exciton localization on isolated units within the branched framework was observed. A linear-chain oscillation model was used for band-edge energy calculations, which show good agreement with experimental data, thereby supporting tight excitation localization.

Harigaya developed[359, 360] a new theoretical model composed of coupled exciton states with off-diagonal disorder to characterize the optical excitation in extended dendrimers. This work supports the evidence that these extended dendrimers can harvest light energy and thereby function as "fractal antennae". A simple tight-binding model of phenylene molecules possessing a *meta*-substitution pattern has been shown to describe the essential optical properties of these phenylacetylene dendrimers.[361]

para-Dimethoxybenzene focally modified phenylacetylene has also been reported.[362] Steady-state and time-resolved fluorescence spectroscopy of higher generation dendrons revealed an anomalous spectral shift in non-polar hydrocarbon solvents that was subtly dependent on solvent architecture. For example, the 5th generation dendron exhibited a fluorescence maximum shift from 380 to 421 nm on changing the solvent from cyclohexane to pentane. Pu et al.[363] attached these dendrons to the optically pure diacetate of

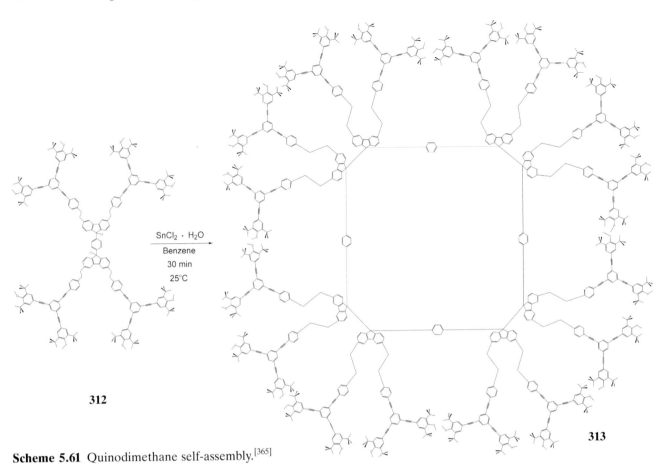

312

SnCl$_2$ · H$_2$O
Benzene
30 min
25°C

313

Scheme 5.61 Quinodimethane self-assembly.[365]

4,4',6,6'-tetrabromo-1,1'-bi-2-naphthol and demonstrated efficient energy transfer to the central core by carrying out UV and fluorescence spectroscopic studies.[363a, b] The 2nd generation construct was used in the asymmetric catalysis of the reaction of diethylzinc with aldehydes in the presence of Ti(O-*i*Pr)$_4$. Using an alternative core, 1,2-dihydro-1,2-methano[60]fullerenes possessing 1st and 2nd generation 3,5-di-*tert*-butylphenylacetylene wedges have been prepared and fully characterized;[364] cyclic voltammetry studies indicated that no significant interaction occurred in the ground state between the wedges and the core.

Ipaktschi et al.[365] attached Moore-type dendrons (Scheme 5.61) to bis(fluorenone) derivatives (e.g., **312**), which then underwent high yield tetramerizations (e.g., to give **313**). These tetramers also exhibit a reversible color change to blue-violet upon energy addition due to the reversible quinodimethane-based reaction.

Using a slightly modified procedure for access to Moore's phenylacetylene dendrons, Oikawa et al.[366] used a rhodium catalyst [Rh(C$_7$H$_8$)Cl]$_2$ to polymerize the focal alkyne moieties and obtained polymer chains described as "polydendrons". The columnar polymers produced were estimated to have diameters of 17.6, 23.2, and 31.9 Å when polymerized from the 0th, 1st, and 2nd generation dendrons, respectively.

Peng et al.[367] recently reported the synthesis of a 1 → 2 (3,4-disubstituted) aryl branching pattern connected by alkyne linkers; the first unsymmetrical conjugated dendrimers have been prepared from one synthon, namely 4,5-diethynyl-2-methoxyphenol, using two sets of reactions.

5.3.12   1 → 2 *Aryl*-Branched, *Ethyne* & *Ester* Connectivity

Zimmerman and Zeng[368] developed an "orthogonal coupling strategy" for the rapid construction of poly(alkyne ester) dendrimers. Scheme 5.62 delineates the protocol whereby key transformations include the Mitsunobu esterification[219,369] and the Sonogashira aryl alkynylation.[370, 371] Thus, coupling (PPh$_3$, DEAD) of alcohol **314** with diacid

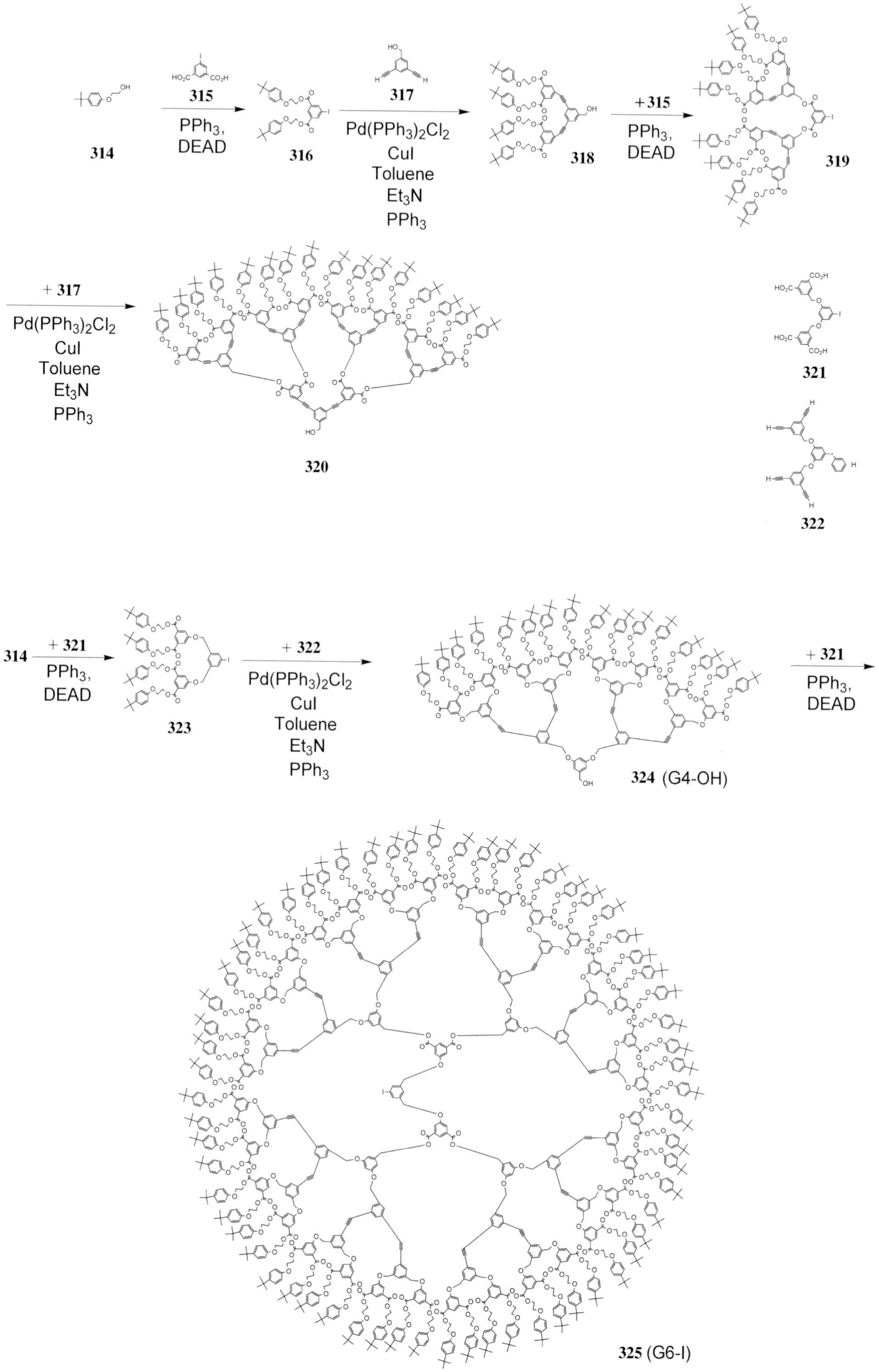

Scheme 5.62 Zimmerman's polyfunctional synthesis.[368]

315 afforded terminal unit **316**, which was then reacted [Pd(PPh$_3$)$_2$Cl$_2$, CuI or Pd$_2$(dba)$_3$, CuI, PPh$_3$, NEt$_3$, toluene] with bis(alkyne) **317** to yield the mono(benzyl alcohol) bis(al-kyne) **318**. Repetition of the Mitsunobu esterification using alcohol **318** and diacid **315** gave aryl iodide **319**, which was then coupled to bis(alkyne) **317** to give the dendron **320** possessing 10 alkyne moieties. Up to a 4th generation wedge was prepared by this route.

Incorporating Fréchet's "branched monomer approach"[303] improved the efficiency of orthogonal strategy. From tetraacid **327** (Scheme 5.63), which was prepared by initial reduction of 5-iodoisophthalic acid to give the corresponding diol followed by coupling (PPh$_3$, DEAD) to dimethyl 5-hydroxyisophthalate and hydrolysis (KOH, H$_2$O), and tet-raalkyne **328**, synthesized by initial bromination (PPh$_3$, CBr$_4$) of 3,5-bis(ethynyl)benzyl alcohol and coupling to 3,5-dihydroxybenzyl alcohol, larger building blocks were accessed. Capping of the tetraacid **327** with a modified phenol afforded the correspond-ing 2nd generation aryl iodide, which was subsequently coupled with tetraalkyne **328** to give the 4th tier dendron **329**; repetition yielded the 6th generation aryl iodide **330** possess-ing a molecular weight of 20,896 amu and a formula of C$_{1292}$H$_{1369}$IO$_{242}$. Notably, this material was obtained in just three steps from monomers **327** and **328**. Esterification of these dendrons with trimesic acid afforded the corresponding dendrimers.

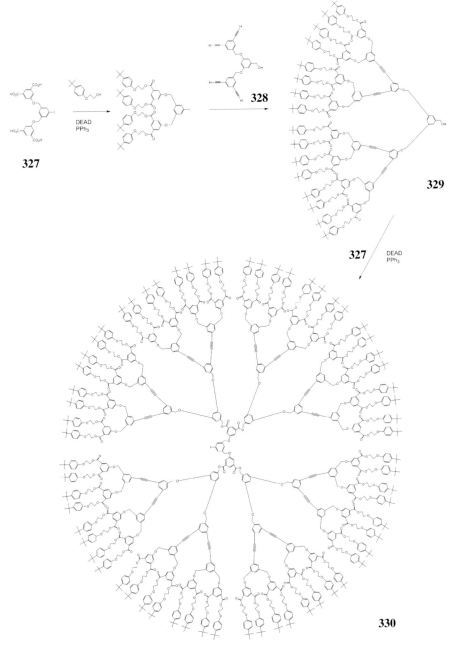

Scheme 5.63 Zimmerman's orthogonal coupling strategy.[368]

5.3.13   1 → 2 *Aryl*-Branched, *Oxadiazole* Connectivity

Kraft[372–374] described the incorporation of the oxadiazole unit (Scheme 5.64), where the key transformations included Pd-catalyzed carbonylation of an aryl iodide, followed by attachment to a hydrazide moiety and chlorosulfonic acid dehydration giving rise to the desired oxadiazole ring. Thus, three equivalents of bis(oxadiazole) **331** were reacted with tris(hydrazide) **332** to afford the hexa(oxadiazole) **333**, which was then dehydrated to give nona(oxadiazole) **334**. Self-association of the small dendrimer was postulated on the basis of upfield shifts and signal broadening in the ^{1}H NMR spectrum; this association was confirmed by vapor-pressure osmometry. Kraft[375] later modified access to oxadiazole connectors to include the Pd(0)-mediated coupling of aryl iodides to aryl tetrazo-

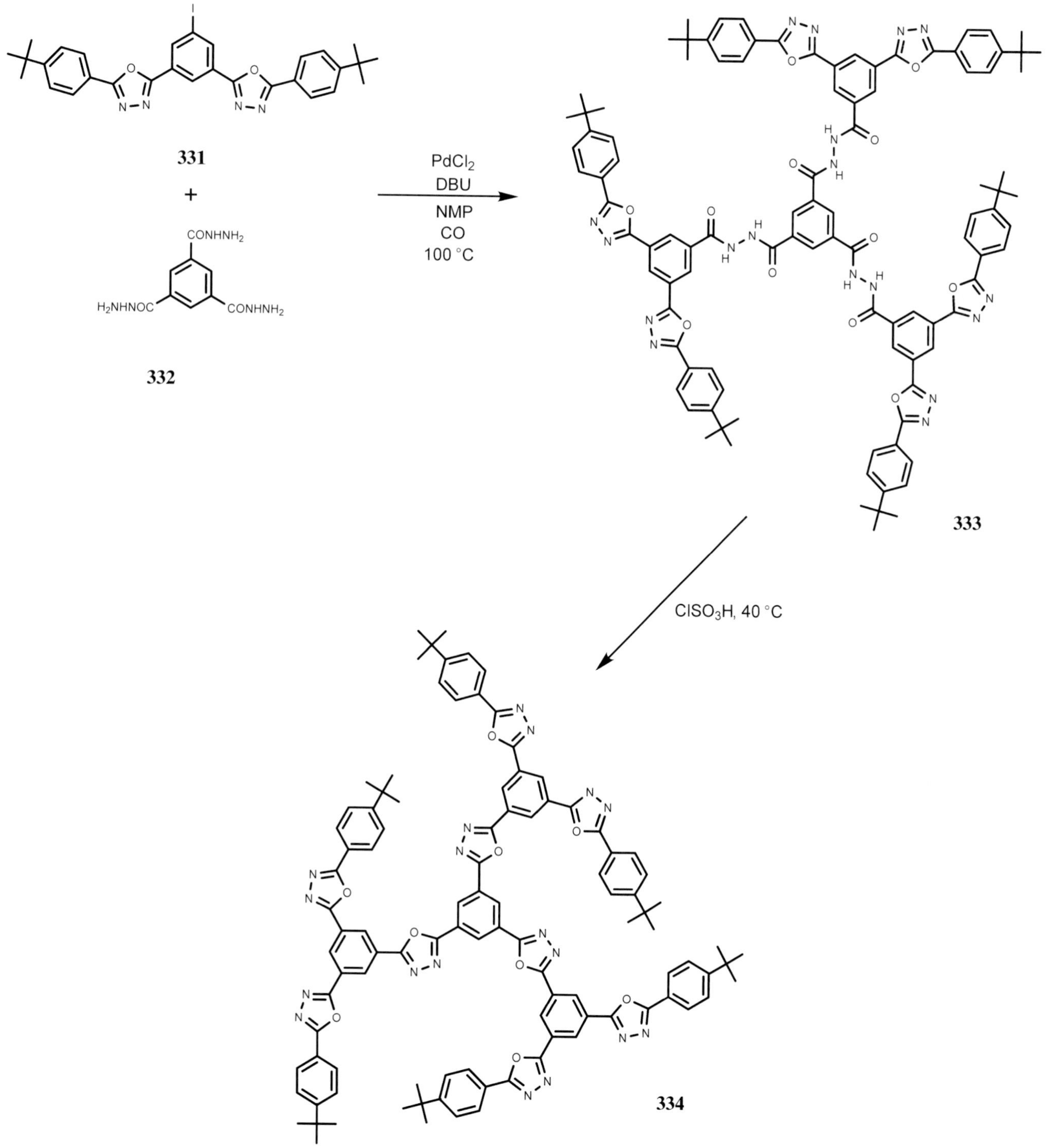

Scheme 5.64 Branched oxadiazole dendrimers, which show a propensity to stack in solution.[372–375]

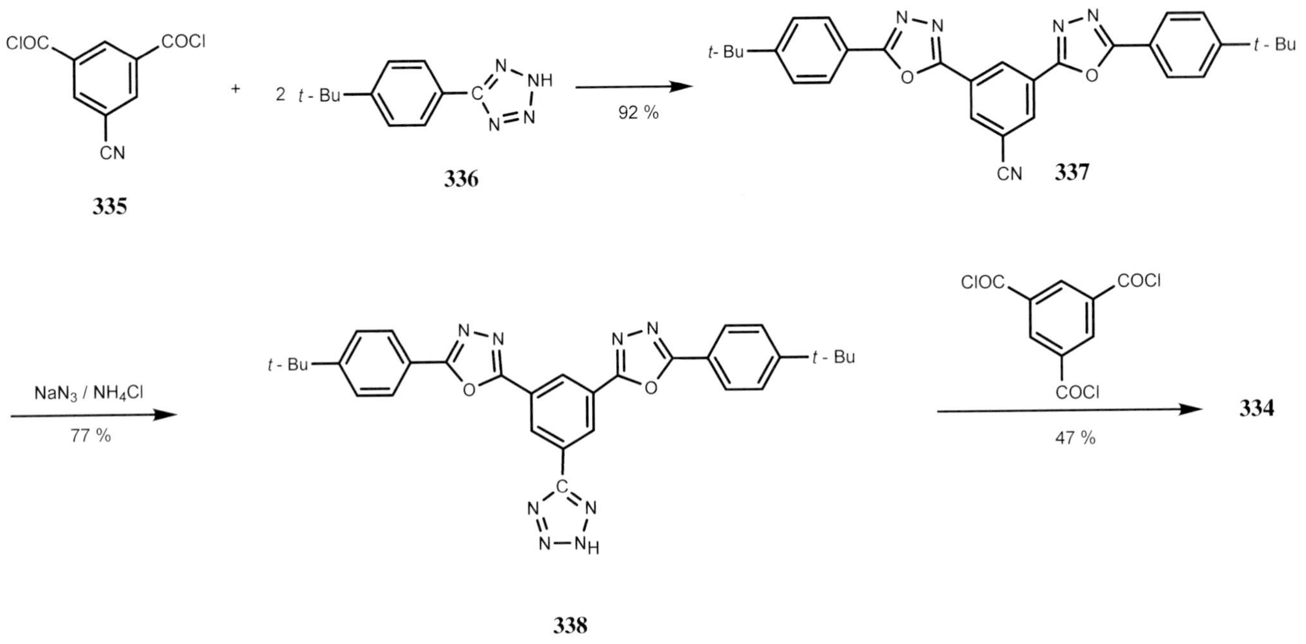

Scheme 5.65 Branched oxadiazole connectivity.[376]

les. ^{1}H NMR data indicated π-stacking of these materials through significant upfield shifts of the signals attributed to centrally located protons.

Bettenhausen and Strohriegl[376] reported an alternative route for the construction of the 1st generation **334** (Scheme 5.65) and evaluated its potential in organic light-emitting diodes. Convergent construction relied on the key reaction of acyl chlorides with substituted tetrazole, followed by intramolecular ring-contraction to yield the oxadiazole.[377] The starting tetrazoles are accessible by treatment of aromatic nitriles with NaN$_3$ and NH$_4$Cl. Thus, reaction of bis(acid halide) **335** with tetrazole **336** afforded nitrile **337**, which was then converted to the corresponding tetrazole (**338**) and reacted with tris(acid chloride) **49** to give the 1st generation nonaoxadiazole dendrimer **334**. Stable glass formation was observed for **334**. Following an observed melting point of 407 °C during the first DSC heating cycle, a T_g of 222 °C is found during subsequent cycles. A preliminary report of the synthesis and optical properties of poly(*p*-phenylene vinylene)s coated with substituted oxadiazole moieties has appeared.[378]

5.3.14   1 → 2 *Aryl* Branched, *Alkene* Connectivity

An orthogonal approach to the preparation of poly(phenylenevinylene) dendrimers has been described[379] and is predicated on the Horner–Wadsworth–Emmons modification of the Wittig and Heck reactions. Due to tolerance of the respective functionalities employed under the different reaction conditions, these protocols are essentially self-protecting. The strategy can be envisaged as follows: 2 equivalents of aryl bromide **340** (Scheme 5.66), generated from the corresponding diethyl phosphite and the bis(*tert*-butyl) aryl aldehyde, are coupled [Pd(acac)$_2$, *n*-Bu$_4$NBr, K$_2$CO$_3$] with 3,5-divinylbenzaldehyde (**341**), and then two equivalents of the resultant 2nd generation aryl aldehyde **342** are reacted (NaH, NMP) with the key bis(phosphite) monomer. The 4th generation dendrimer (**344**) was obtained (9 % due to solubility problems) by coupling the 3.5 generation dendron (focal aldehyde) to a tris(phosphite)-substituted benzene core. The yellowish dendrimer exhibiting a blue fluorescence was easily isolated and its identity was confirmed by NMR and MALDI MS data.

Meier and Lehmann[380–382] reported the use of the Wittig–Horner reaction as a basis for the preparation of poly(*trans*-alkene) architectures. Relying on the high *trans*-selectivity of this now classical transformation, the rigid dendrimers illustrated in Scheme 5.67 (**345–347**) were prepared. Precursors of the Wittig-based monomers, such as core

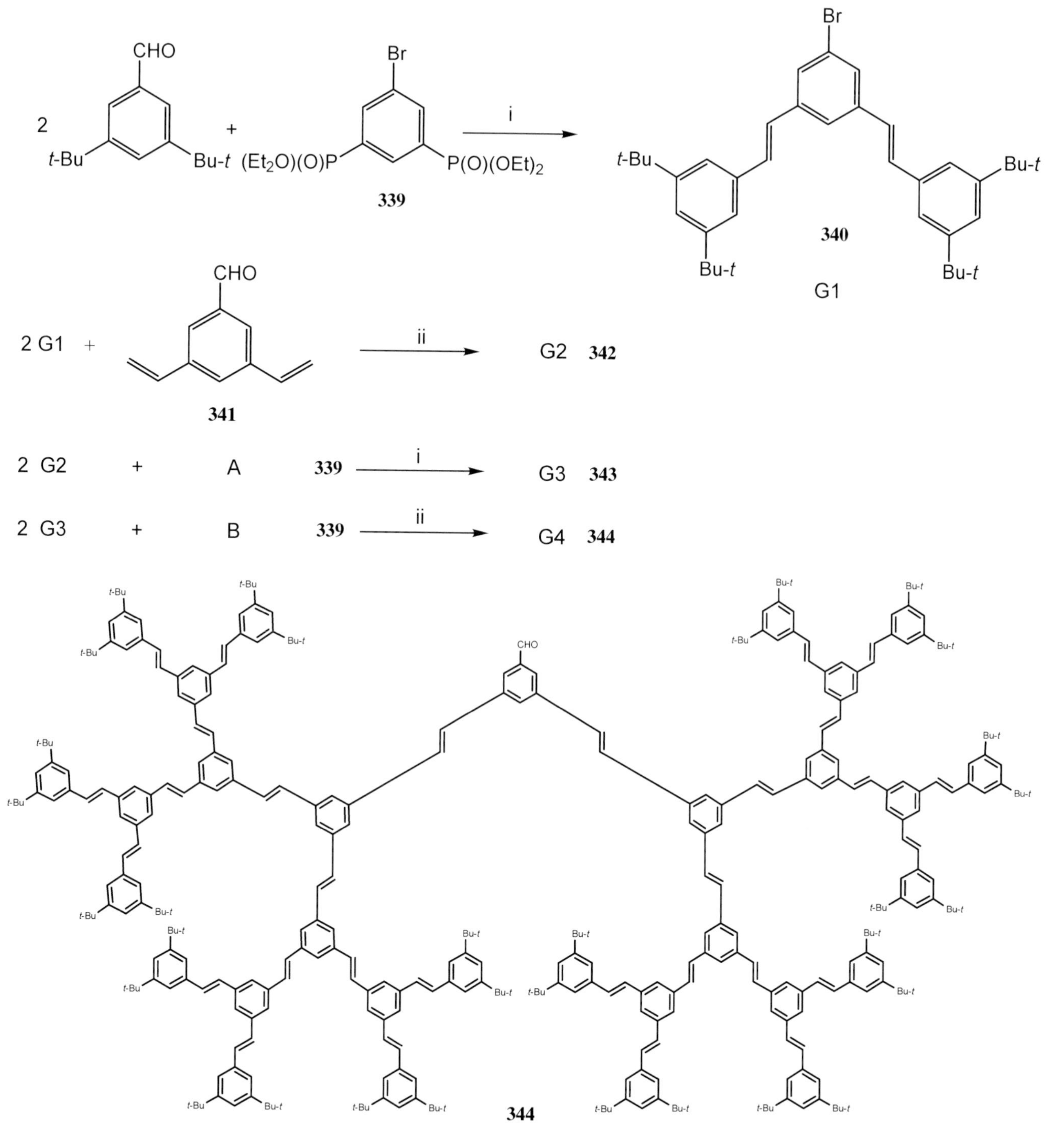

Reaction conditions: i, NaH/NMP ; ii, Pd(OAc)$_2$/(n-Bu)$_4$NBr/K$_2$CO$_3$

Scheme 5.66 Orthogonal coupling strategy for the construction of poly(phenylenevinylene) dendrimers.[379]

348, were derived from 1,3,5-tris(bromomethyl)benzene or 1,3-bis(bromomethyl)-5-methylbenzene, which were oxidized and protected as their acetates, and then treated with the phosphonate to form the desired monomers (not shown). Treatment of the 0th, 1st, or 2nd generation dendrons (i.e., **349–351**) with the core **348** afforded the desired "stilbenoid" dendrimers. Neat **345** and **346** each exhibited two liquid crystalline mesophases and the tendency to aggregate increased with increasing generation, as shown by NMR studies.

Rodrígues-López et al.[383] [383a] also reported the use of the Horner–Wadsworth–Emmons transformation for the construction of poly(phenylene vinylene)s. Dipolar and non-dipolar constructs were accessed. Essentially, the benzyl alcohol moieties on growing

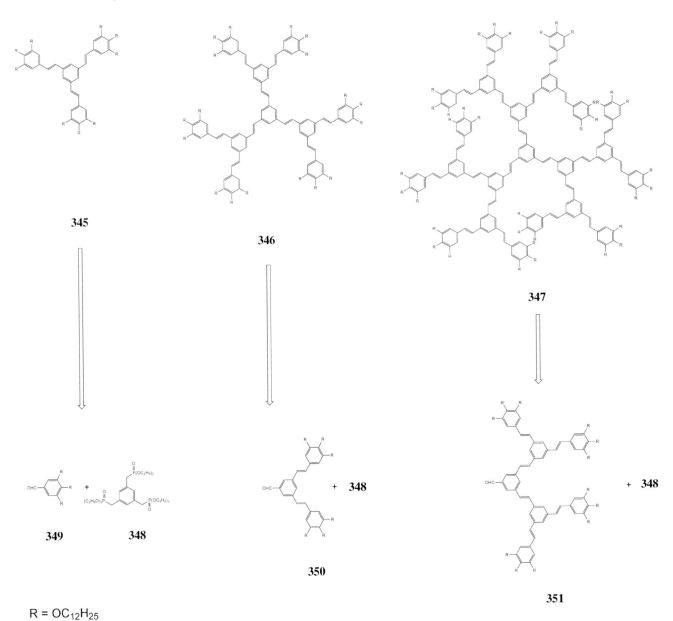

Scheme 5.67 A route to "stilbenoid" dendrimers.[380]

dendrons were oxidized (PCC) to benzaldehyde functions and convergently reacted with a bis(phosphonate) to produce two new vinylene-connected units. Focal attachment to fullerenes has been reported.[383b]

Similar poly(alkene)s have been reported[384] [384a] (Scheme 5.68); construction involved a two-step procedure whereby a focal aldehyde group, i.e., **352**, was converted to the corresponding methylene analogue by a Wittig reaction (Ph$_3$P$^+$–Me I$^-$, base, THF) and coupled [*trans*-di(μ-acetato)bis{*o*-(di-*o*-tolylphosphonyl)benzyl}dipalladium(II)] to dibromobenzaldehyde in a Heck-type reaction to give the next higher generation dendron **353**. Attachment of these dendrons, prepared through to the 3rd generation, to distyrylbenzene, distyrylanthracene, or *meso*-tetraphenylporphyrin cores afforded photoluminescent architectures exhibiting blue, yellow-green, and red light emission, respectively, that were shown to form good quality thin films upon spin-coating.[385]

5.3.15   1 → 2 *Aryl* Branched, *Alkene* & *Ether* Connectivity

An oligophenylenevinylene-fullerene conjugate has recently been shown to exhibit photoinduced energy transfer to the C$_{60}$ moiety.[386]

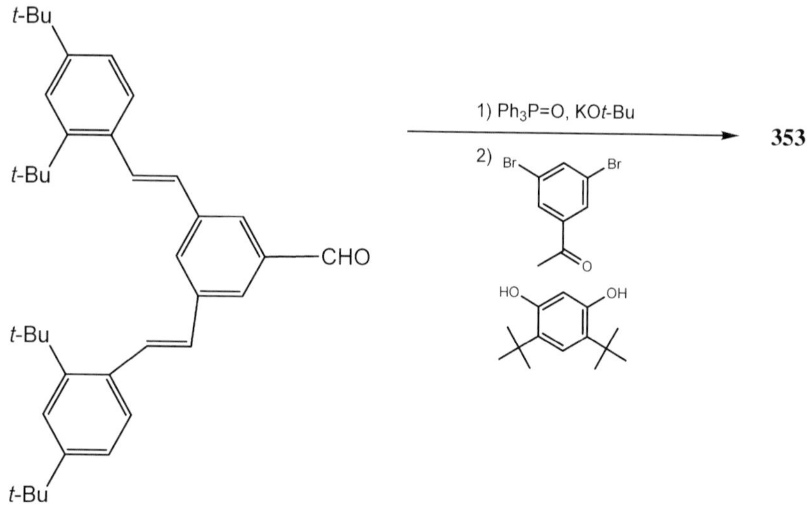

Scheme 5.68 Construction of poly(stilbene)-based dendrons.[384]

5.4 1 → 2 *Pyridine*-Branched, *Ether* Connectivity

Chessa and Scrivanti[387] described the synthesis of dendritic polypyridines possessing 2,6-bis(ethoxycarbonyl)pyridine moieties. Key monomers (Scheme 5.69) used for construction of these materials included diethyl 4-hydroxypyridine-2,6-dicarboxylate (**354**), prepared by esterification of the corresponding diacid, and 4-benzyloxy-2,6-bis(chloromethyl)pyridine (**355**), synthesized by initial esterification of the hydroxy diacid (MeOH, SOCl$_2$) followed by benzylation (C$_6$H$_5$CH$_2$Br, K$_2$CO$_3$), reduction (NaBH$_4$), and subsequent chlorination (SOCl$_2$). Dendrimer construction proceeded smoothly upon reaction of 2 equivalents of pyridinol **354** with derivative **355** to give the 1st generation tetraester **356**. Debenzylation (H$_2$, Pd/C) of monomer **356** afforded the focal hydroxy moiety, which was treated with further monomer **355** to give the 2nd tier dendron **357** possessing eight ester groups. Following deprotection of the polyester, reaction with 1,3,5-tris(bromomethyl)benzene yielded the 3rd generation dendrimer **358**. These polypyridines were analyzed by laser desorption and MALDI MS;[388] in the latter a strong matrix effect was seen, leading to fragments related to the original skeleton.

5.5 1 → 2 *Other Heteroaryl*-Branched and *N*-Connectivity

Konishi, Aida, et al.[389] created polyuracil architectures (Scheme 5.70) employing the nucleobase monomers **359** and **360**. Conversion (HBr, HOAc) of benzyl ether **360** to a bromomethyl group and reaction (K$_2$CO$_3$, DMF) with the starting bis(amide) **359** afforded the 1st generation dendron **361**; iteration facilitated access to higher generations. La^{3+} coordination (1:1) was demonstrated by MALDI-TOF MS. Locked dendrimers were then reported[390] based on the notable ability of these materials to undergo intramolecular [2+2] photodimerization.

Zhang and Simanek[391, 392] reported the convergent as well as divergent synthesis of poly(heteroaryl amine)s possessing specifically the polymelamine architecture. Notably, the divergent procedure furnished less than 1 % yield, whereas the 3rd generation wedge was prepared in high yield and purity by the convergent protocol. The synthesis was predicated on amine displacement of a chloro group from cyanuric chloride.

Marsh et al.[392a] developed a diverse protocol for preparing triazine-based dendrimers.

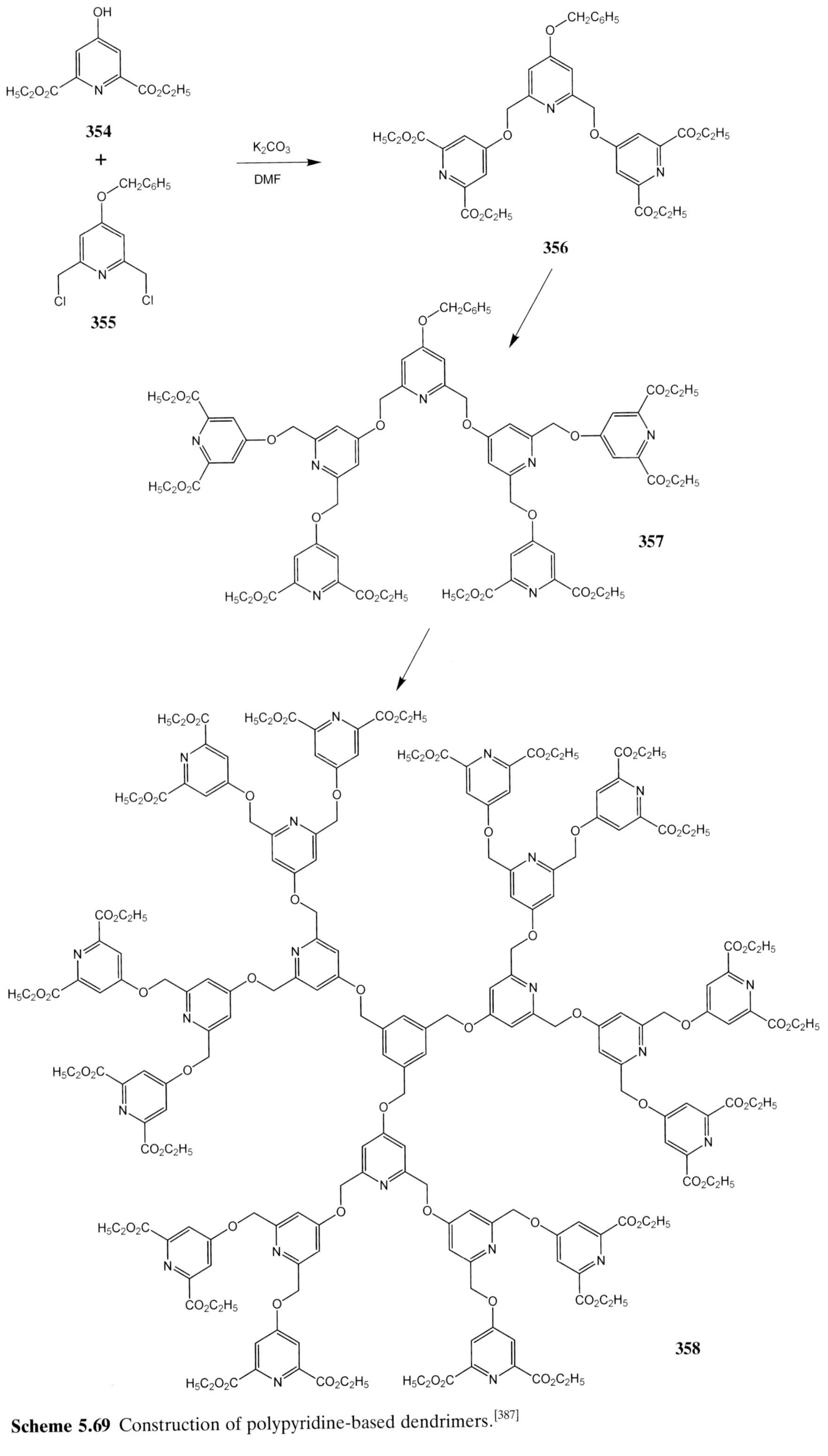

Scheme 5.69 Construction of polypyridine-based dendrimers.[387]

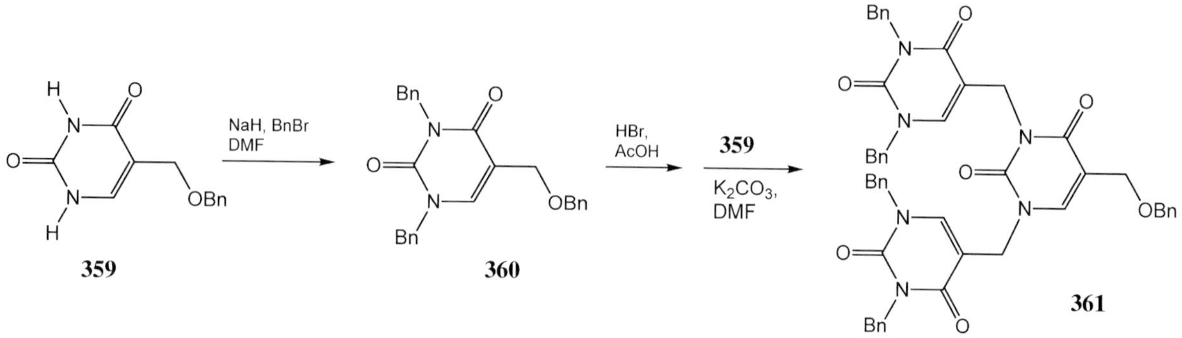

Scheme 5.70 Architectures based on uracil monomers serve as multidentate ligands for rare-earth ions.[389]

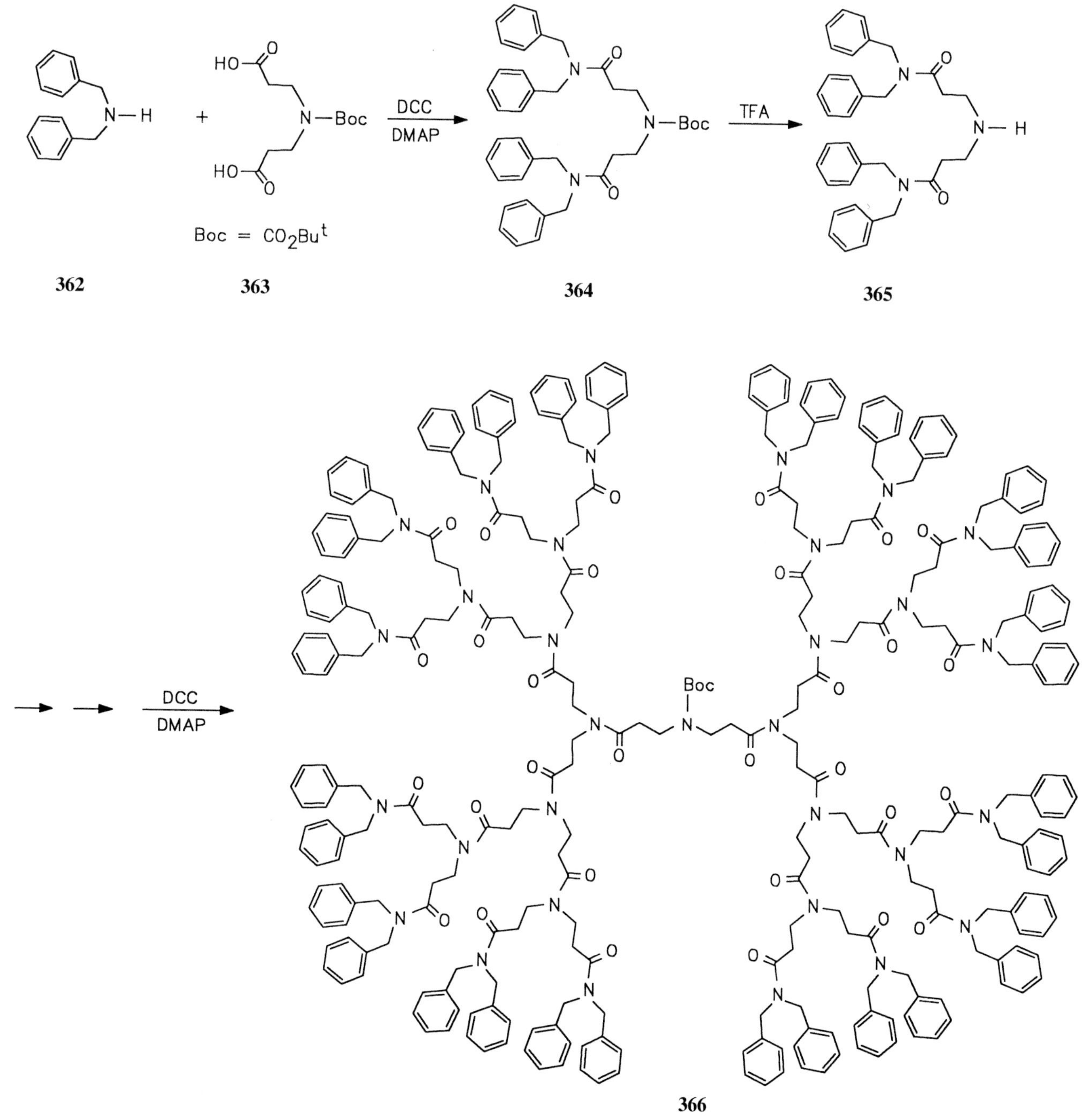

Scheme 5.71 Construction of *N*-benzyl-terminated amide-based dendrimers.[393]

5.6   1 → 2 *N*-Branched

5.6.1   1 → 2 *N*-Branched, *Amide* Connectivity

Uhrich and Fréchet[393] reported an approach to polyamido cascades (Scheme 5.71). The iterative synthesis was accomplished by coupling two equivalents of a secondary amine **362** to the *N*-Boc-protected aminodicarboxylic acid (**363**). Subsequent deprotection (TFA) of bis(amide) **364** afforded a new secondary amine **365**, which, in turn, was acylated with 0.5 equivalents of the imino-protected diacid **363**. The procedure was repeated three times affording the 3rd generation polyamide **366**. Higher generations were not prepared due to decreasing reactivity at the focal amine, which was attributed to steric inhibition, as well as difficulties with purification.

The preparation of PAMAMs[394] (see Scheme 4.3.2) by the construction of dendritic wedges possessing a reactive focal group for a convergent-type construction has been reported.[395] The resultant branched macromolecules were prepared by an iterative treatment of an amino alcohol, first with methyl acrylate and then with ethylenediamine, which allowed the multigram synthesis of 1 → 2 amino-branched dendrimers. Dendritic termini were transformed to diverse groups (–CO$_2$H, –CN, –CONH$_2$, –OH, or –SH) by standard methods from either a CO$_2$Me or NH$_2$ moiety. Implementation of these dendrons in dendrimer syntheses as building blocks with three-directional cores allows the preparation of polytermini-differentiated dendrimers. Peripheral differentiation results from statistical reactivity and hence product distribution relies on collision frequency.

Adamczyk et al.[396] connected chemiluminescent acridinium moieties to 1st and 2nd generation amide-based dendrons to examine their efficacy as conjugate labels; these new labels were termed "TRACERMERS™" (Scheme 5.72). Branched framework construction began with the known monoacid **367**, which was deprotected (HBr) to afford diammonium bromide **368**. Reaction (DMF/buffer) of the *N*-succinimidyl activated ester of **367** with diamine **368** gave the 2nd generation wedge **369**. Following deprotection to give acid **370**,[397] reaction with the acridinium activated ester **371** afforded the corresponding monoacid **372**, which was then converted (*N*-hydroxysuccinimidyl trifluoroacetate) to its activated ester **373** for subsequent attachment to bovine serum albumin (BSA). Comparison of conjugates obtained from acridinium sulfate alone and the 1st and 2nd tier TRACERMER™ labels revealed chemiluminescent emissions corresponding to $0.85 \pm 0.1 \, ' \, 10^{20}$, $2.0 \pm 0.2 \, ' \, 10^{20}$, and $5.3 \pm 0.7 \, ' \, 10^{20}$ counts/mol/label, respectively.

5.6.2   1 → 2 *N*-Branched, *Amide & Azo* Connectivity

Photochromic dendrimers possessing azobenzene branch chain connectors and a novel calix[4]arene core were reported by Nagasaki et al.[398] The key monomers **378** and **379** (Scheme 5.73) were each accessed by initial coupling of diazocarboxylic acid **375** and secondary amine **374**. Following phthalimide removal (NH$_2$NH$_2$, H$_2$O) and amine *t*-Boc protection [(Boc)$_2$O], alcohol **376** was extended with methyl bromoacetate yielding ester **377**. Basic hydrolysis afforded acid **378**, while treatment with trifluoroacetic acid afforded diamine **379**. Dendron construction was then effected by reaction of the mixed anhydride of acid **378** (*i*BuOCOCl) and 0.5 equiv. of diamine **379** affording the 2nd generation, tetravalent, tris(azobenzene) **380**. Ester hydrolysis facilitated construction of the 3rd and 4th generation dendrons, which were attached to a tetraamine calix[4]arene core **381** (prepared by phthalimide amination of the corresponding tetrachloro derivative) to give dendrimers **382** and **383**. The smaller dendrimer was obtained as a 1:4 mixture of *trans/cis* conformers following exposure to UV radiation, whereas the larger dendrimer acquired only a 35:65 *trans/cis* ratio. This was attributed to increased steric congestion.

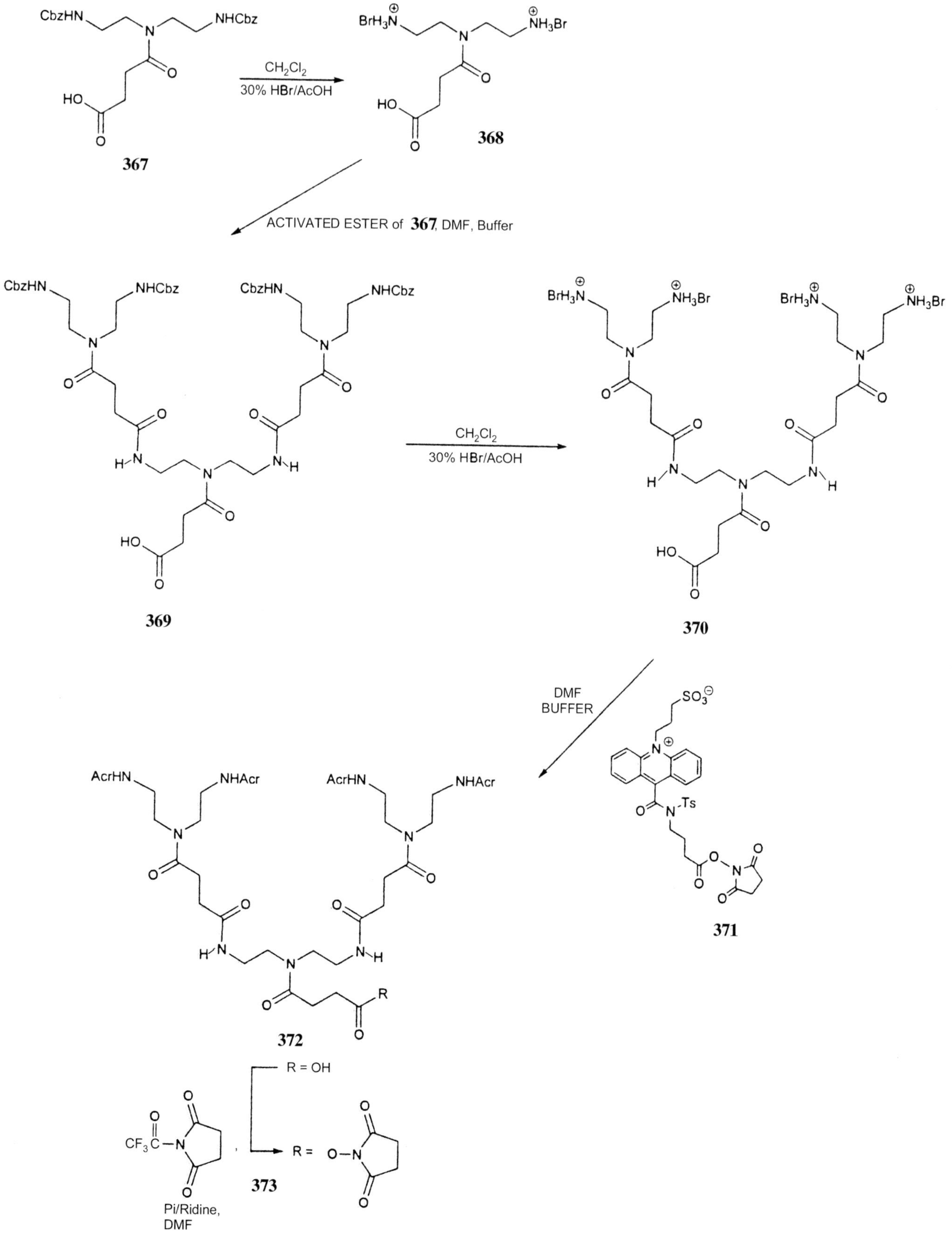

Scheme 5.72 Route to TRACERMER™ signal generators by incorporation of multiple chemiluminescent acridinium labels.[396]

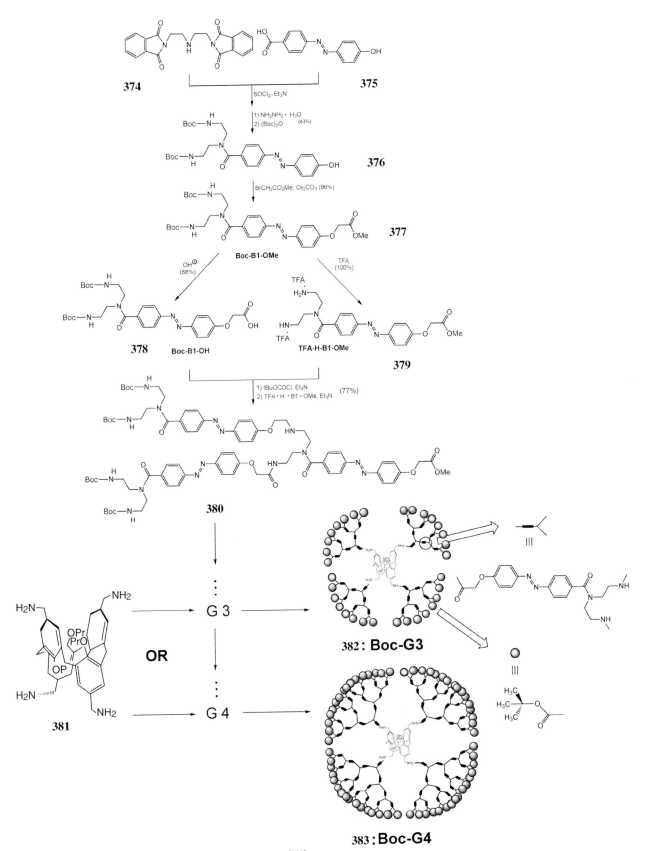

Scheme 5.73 Azobenzene dendrimer construction.[398]

5.6.3 1 → 2 *N*-Branched, *Ester & Azo* Connectivity

A series of nonlinear optical macromolecules comprised of polyfunctional azobenzene wedges have been synthesized and their conformational properties and molecular nonlinear optical properties have been evaluated by second-order optical measurements.[399] The first-order molecular hyperpolarizability of the azobenzene dendron having 15 chromophoric moieties was measured as $3{,}010 \times 10^{-30}$ esu using the hyper-Rayleigh scattering method. A single-pot route to polycarbonates is known.[399a]

5.6.4 1 → 2 *N*-Branched and Connectivity

Hartwig et al.[400, 401] employed Pd-catalyzed amination to gain access to triarylamine dendrimers that are of interest with respect to conducting polymers and ferromagnetic coupling as a result of their potential to form high-spin polyradical species. The strategy for their construction is illustrated in Scheme 5.74, whereby reaction [Pd{P(*o*-tolyl)$_3$}$_2$; 2 mol %] of two equivalents of the lithiobis(aryl bromide) **384** with the bis(aryl bromide) **385** afforded triamine **386**. Hydrogenation (Pd/C, H$_2$) of the benzyl moiety followed by *N*-lithiation (*n*BuLi) and reaction with a tris(aryl bromide)amine yielded decaamine **389**. Reaction of dendron **388** with bromide **385** gave the higher generation dendron **390**, which, upon deprotection, treatment with NaO*t*-Bu, and reaction with biphenyl dibromide gave the dendrimer **393**. Chromatography of the crude materials gave analytically pure, bright-yellow samples, which were fluorescent in solution. Hole-transporting hydrazone derivatives are also known.[401a]

5.6.5 1 → 2 *N*-Branched, *Alkyne* Connectivity

Moore et al.[347] incorporated amine-based, redox-active branching into their poly(phenyl acetylene) dendrimers for examination as photoelectrocatalysts. Synthesis (Scheme 5.75) of these *N*-branched constructs was achieved based on the trigonal building block unit N(C$_6$H$_4$)$_3$-. From this synthon, dendron **394** was prepared, deprotected (K$_2$CO$_3$, MeOH, CH$_2$Cl$_2$), and reacted [Pd(dba)$_2$, CuI, PPh$_3$, Et$_3$N] with *p*-BrC$_6$H$_4$-*t*Bu to afford **395**. Subsequent reaction with *N*-(*p*-BrC$_6$H$_4$)$_3$ gave the tetraamine **396**. Elaboration of **395** to the corresponding 2nd generation with dibromide **397** then allowed access to the dendron **398**. Initial photo- and electrochemical studies on dendrimer **396** by steady-state fluorescence measurements and cyclic voltammetry, respectively, revealed efficient fluorescence and reversible oxidation.

Zhu and Moore[402] used alkyne-based protection and coupling chemistries, as well as incorporated a 9-phenylcarbazole moiety as the 1 → 2 branching unit, to create a unique family of dendrimers possessing well-organized arrays of redox sites. Having only been able to realize up to the 3rd generation due to purification problems, monodendrons possessing terminal 3,5-di-*tert*-butyl-4-(2-methoxyethoxy)phenylene or 1,1,3,3-tetramethylbutyl groups were prepared.[403] These surface units increased solubility and facilitated purification such that a 4th generation construct became accessible.

5.7 1 → 2 *C*- & *N*-Branched, *Ester* Connectivity

Mitchell et al.[404] reported the use of 1,3-diaminopropan-2-ol (**399**) as a building block that incorporates both 1 → 2 *C*- as well as 1 → 2 *N*-branching. Michael-type addition of alkyl acrylates (Scheme 5.76) to the diamine provided the unique alcohol **400**, which was reacted with the acyl chloride **49** to give dodecaester **401**. Using this technology and appropriate α,β-unsaturated carbonyl compounds, i.e., simple alkyl or phenyl acrylates, a series of ester-terminated, medium-sized dendrimers was prepared. Size-exclusion chromatography did not give reasonable relative molecular masses for the higher molecular weight members in this series; however, FABMS correlated precisely with the calculated masses of the dendrimers.

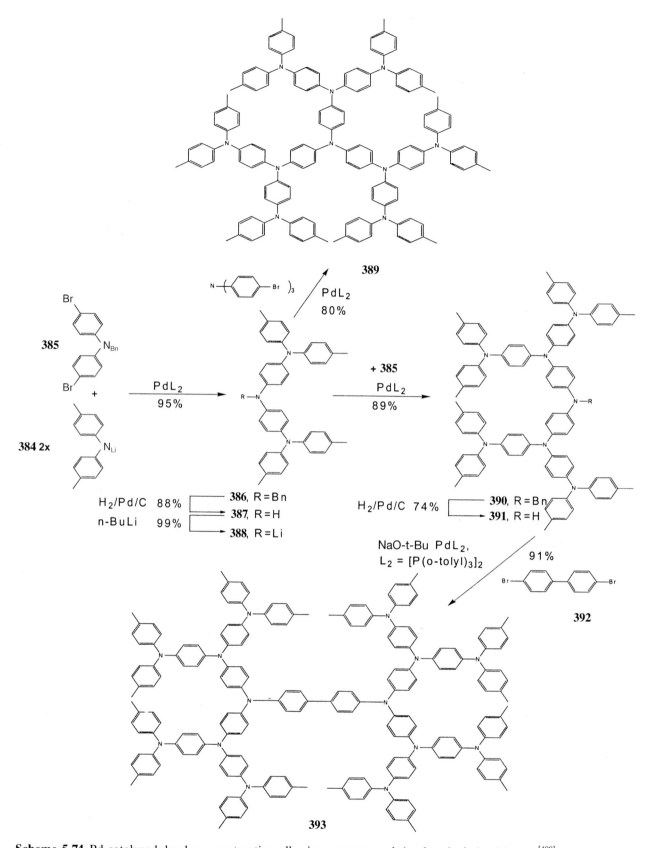

Scheme 5.74 Pd-catalyzed dendron construction allowing access to poly(aryl amine) dendrimers.[400]

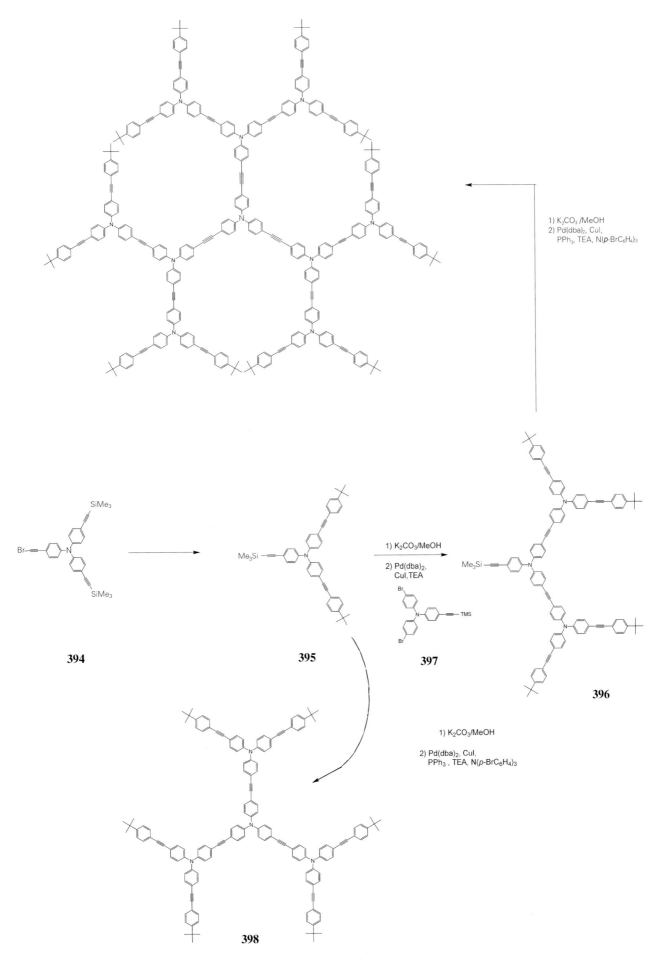

Scheme 5.75 Moore et al.'s[347] *N*-branched phenylacetylenes designed for photoelectrocatalysis.

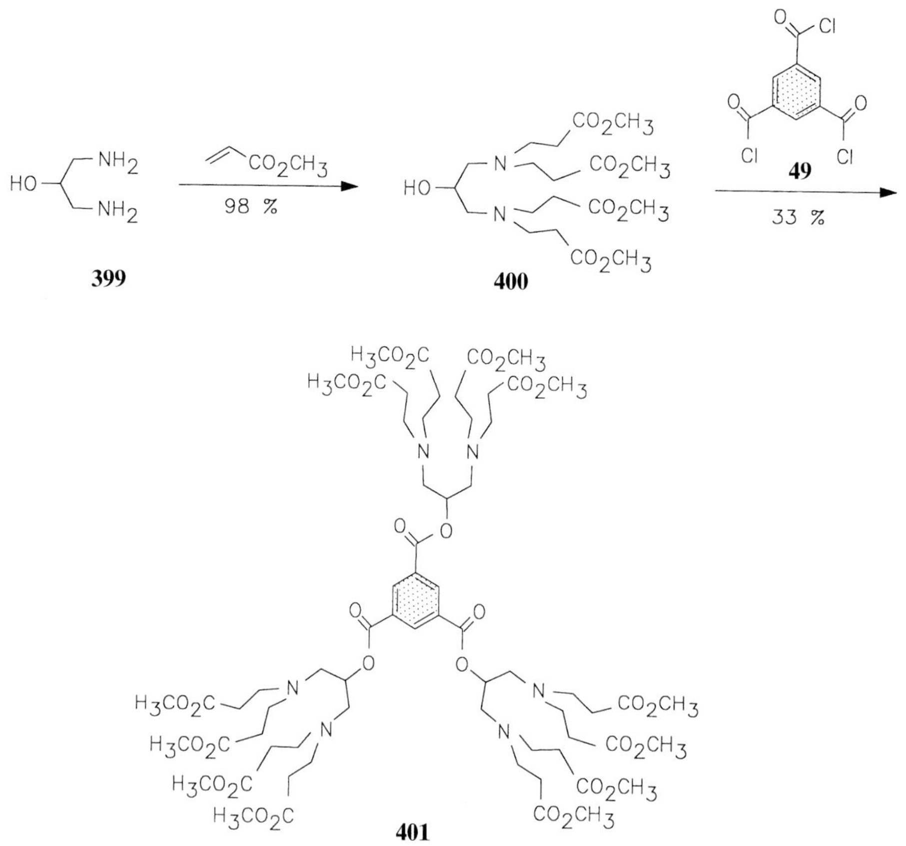

Scheme 5.76 Preparation of small, amino ester-based dendrimers.[404]

5.8 1 → 2 *Si*-Branched, *Silyloxy* Connectivity

Imai et al.[405, 406] synthesized an initial series of branched polysiloxane polymers based on tris[(phenyldimethylsiloxy)dimethylsiloxy]methylsilane as the core and bis[(phenyldimethylsiloxy)methylsiloxy]dimethylsilanol as the building block. The construction of this series was based on allylbis[4-(hydroxydimethylsilyl)phenyl]methylsilane (**405**) as the key building block.[407] The general procedure is depicted in Scheme 5.77. Conversion of an allyl moiety to an (*N,N*-diethylamino)silane (**404**) was accomplished by Pt-mediated hydrosilylation, followed by treatment with HNEt$_2$. Treatment of the monomer **405** with two equivalents of silylamine **404** afforded the 2nd generation tetranitrile **406**. Analogously, the polysiloxane "wedges" possessing 8 and 16 terminal groups, corresponding to 3rd and 4th generation dendrons, respectively, were prepared. The attachment of the 3rd generation wedge to a tris(hydroxysilyl)silane core afforded a three-directional cascade (not shown). The larger polymers showed higher glass transition temperatures (T_gs) than the smaller homologues in a T_g range that spanned from –61 to 36 °C.

Sheiko et al.[408] reported the investigation of the absorption and aggregation properties of these carbosilane dendrimers on mica and graphite by means of SFM. Aggregation phenomena were observed to lead first to clusters and then to fluid droplets starting from single molecules and finally to layer formation; particle diameters of 25 Å were reported. Droplets of the dendrimer were observed to spread slowly to two molecular layer thick polygonal lamellae; a regular packing of the globular molecules was indicated.

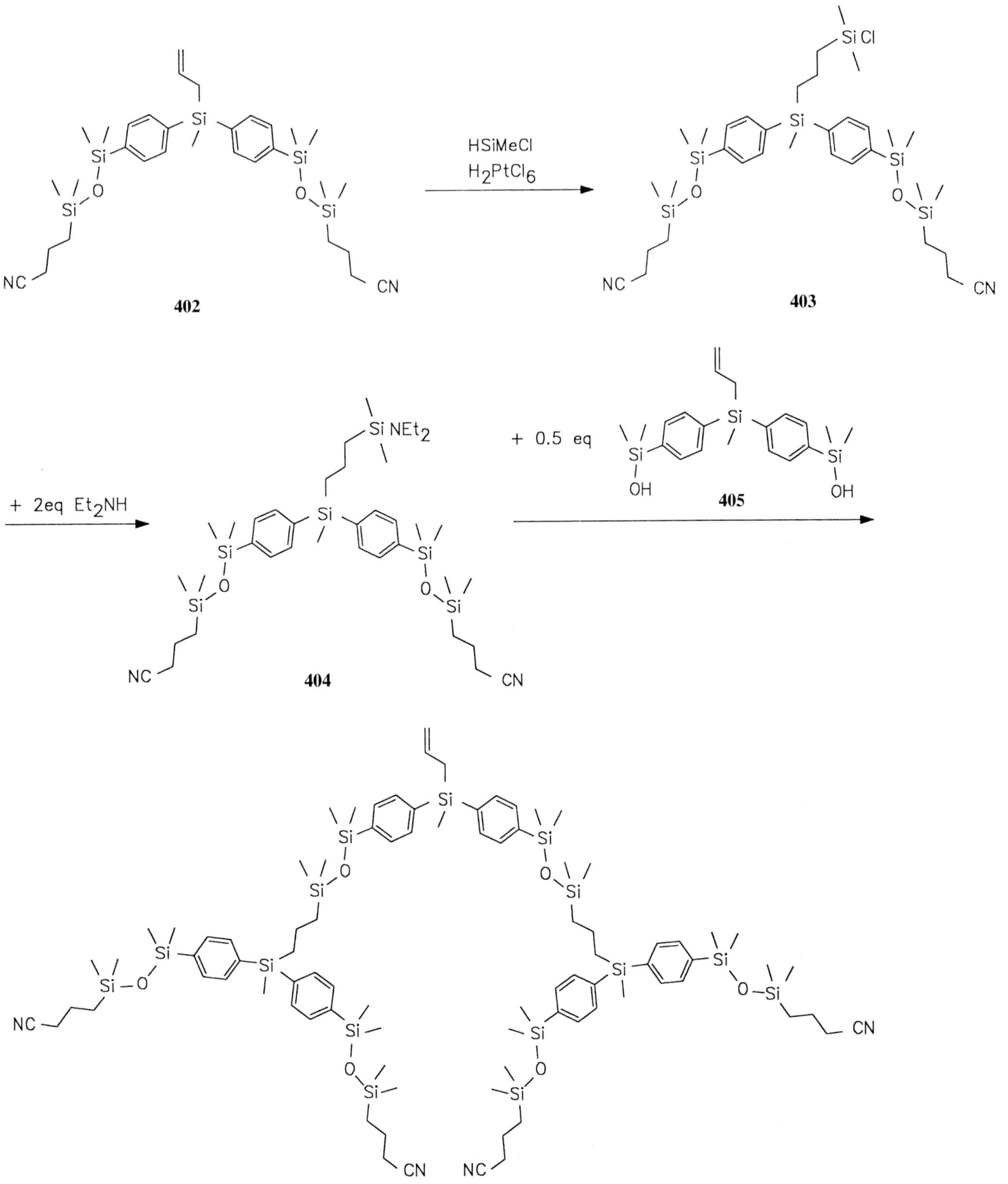

406

Scheme 5.77 Convergent preparation of siloxane dendrimers[407] employing silicon as a branching center.

5.9   1 → 2 *B*-Branched, *Aryl* Connectivity

The coupling of 9-(dimesitylboryl)-10-lithioanthracene, prepared (*n*-BuLi, Et$_2$O) from the 10-bromo derivative, with BF$_3$ · Et$_2$O afforded (49 %) the desired trigonally branched species, in which the core possesses three anthracene spacers attached to a 1 → 2 branching boron moiety.[409]

5.10   1 → 2 *TTF*-Branched and Connectivity

Bryce and coworkers[410] reported the synthesis of tetrathiafulvalene (TTF) dendrimers, whereby TTF units were employed as branching centers (Scheme 5.78). Staring with bis(nitrile) **407**, the monothiomethyl derivative **408** was obtained (CsOH·H$_2$O, DMF, MeOH, MeI) in high yield[411] and then converted [Hg(OAc)$_2$] to the corresponding ketone **409**, reaction of which with bis(benzyl chloride) **410** gave the mononitrile TTF unit **411**. The branched monomer **412** was obtained by CsOH·H$_2$O-promoted thiobenzylation using the methyl-terminated monomer **413**. Finally, treatment of monomer **412** with the tetravalent core **414** afforded the desired poly-TTF dendrimer through analogous cesium-promoted thiolate generation. Thin-layer cyclic voltammetry of **415**, as well as of its smaller homologue, showed that two single-electron oxidations were possible for each TTF moiety. Redox-active, 1,3-dithiole-functionalized [3]- and [4]dendralenes have also been reported.[412]

Using this same protocol employing TTF-branching centers, Bryce et al.[413] incorporated triethylene glycol spacers into their poly(TTF)s. CV studies on a 21 TTF construct revealed two single-electron oxidations at each TTF unit, affording the +42 redox state. It was further demonstrated that intradendrimer interactions occur in partially oxidized materials.

5.11   1 → 3 *Aryl*-Branched, *Ether* Connectivity

An interesting branched architecture (Scheme 5.79) has been prepared by Malthête.[414] The mesogenic monomer **416** was coupled (DCC, DMAP) to pentaerythritol to afford aryloxy mesogen **417**, which exhibits columnar mesophase formation despite possessing a tetrahedral core component; dimeric and trimeric constructs were also reported.

Percec et al.[415] also constructed simple fluorinated monodendrons that, based on a fluorophobic effect, self-assemble into columnar dendrimers; these dendrimers further exhibit a homotropic hexagonal columnar liquid-crystalline phase. Building blocks for the rod-like dendrimers are depicted in Figure 5.30. Each monomer (i.e., **418–421**) is comprised of essentially three components – the fluorinated alkyl chains, a 1 → 3 branching monocarboxylic acid, and a 15-crown-5 ether connected via an ester moiety. In all, six monomers were constructed, including crown ethers possessing differing ratios of fluorinated to non-fluorinated methylene groups in the alkyl chain. Monomer self-assembly and the resultant cylindrical architecture (**423**) are related to complexation of the crown ethers with Li, Na, or K trifluoromethanesulfonates and the monodendron's tapered-shape (**422**), respectively. The columns were reportedly stable in solution, in melts, and in the solid phase. A fluorophobic effect was deemed responsible for the hexagonal thermotropic columnar (Φ_h) liquid crystalline (LC) phase (**424**). These novel structures were probed using DSC, X-ray diffraction, and thermal optical polarized microscopy.

Similar tapered monodendrons have been attached to a polymerizable 7-oxanorbornene monomer for mechanistic studies on the formation of supramolecular cylindrically shaped polymers and oligomers.[416]

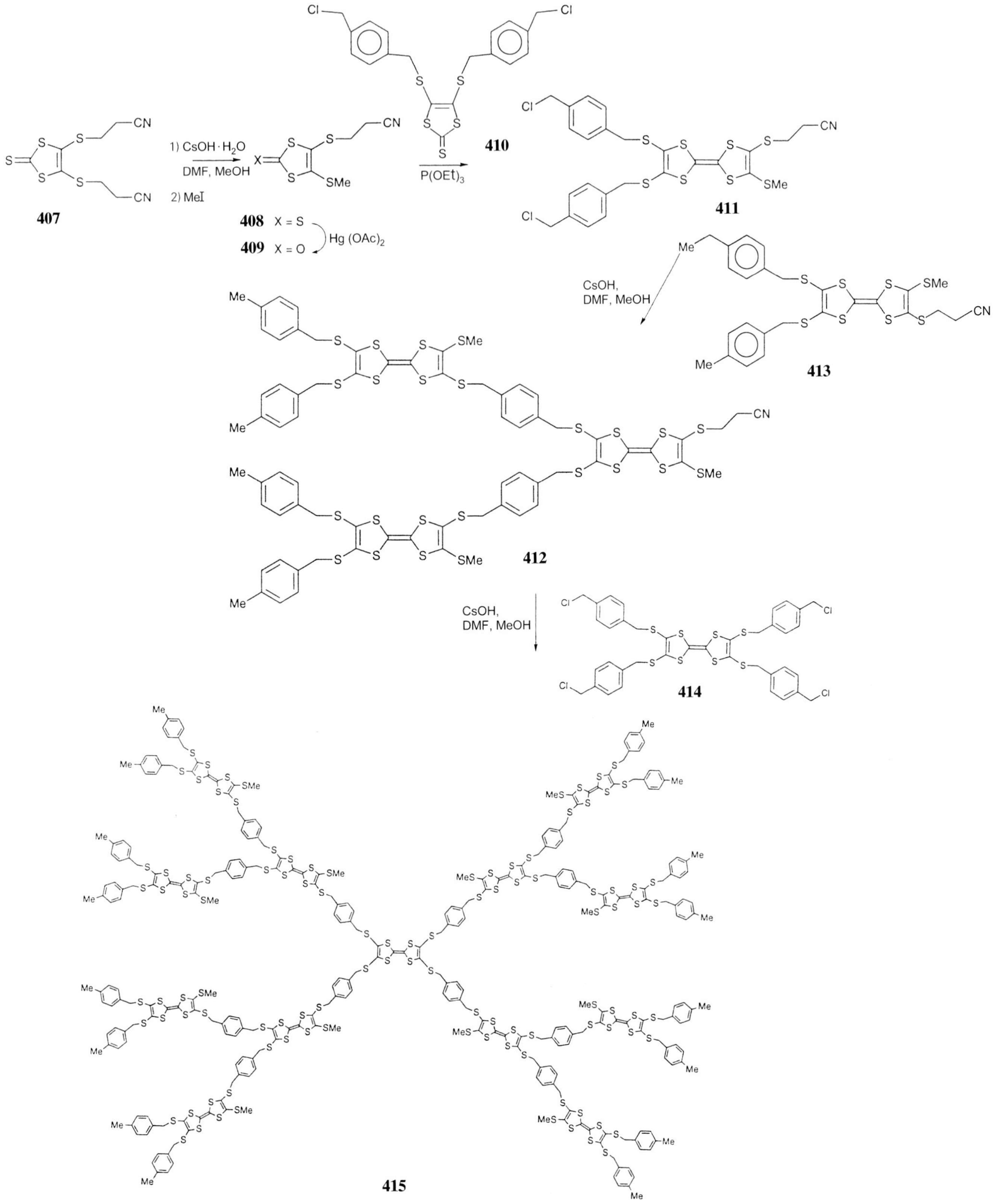

Figure 5.78 Preparation of dendrimers using TTF units as branching centers.[410]

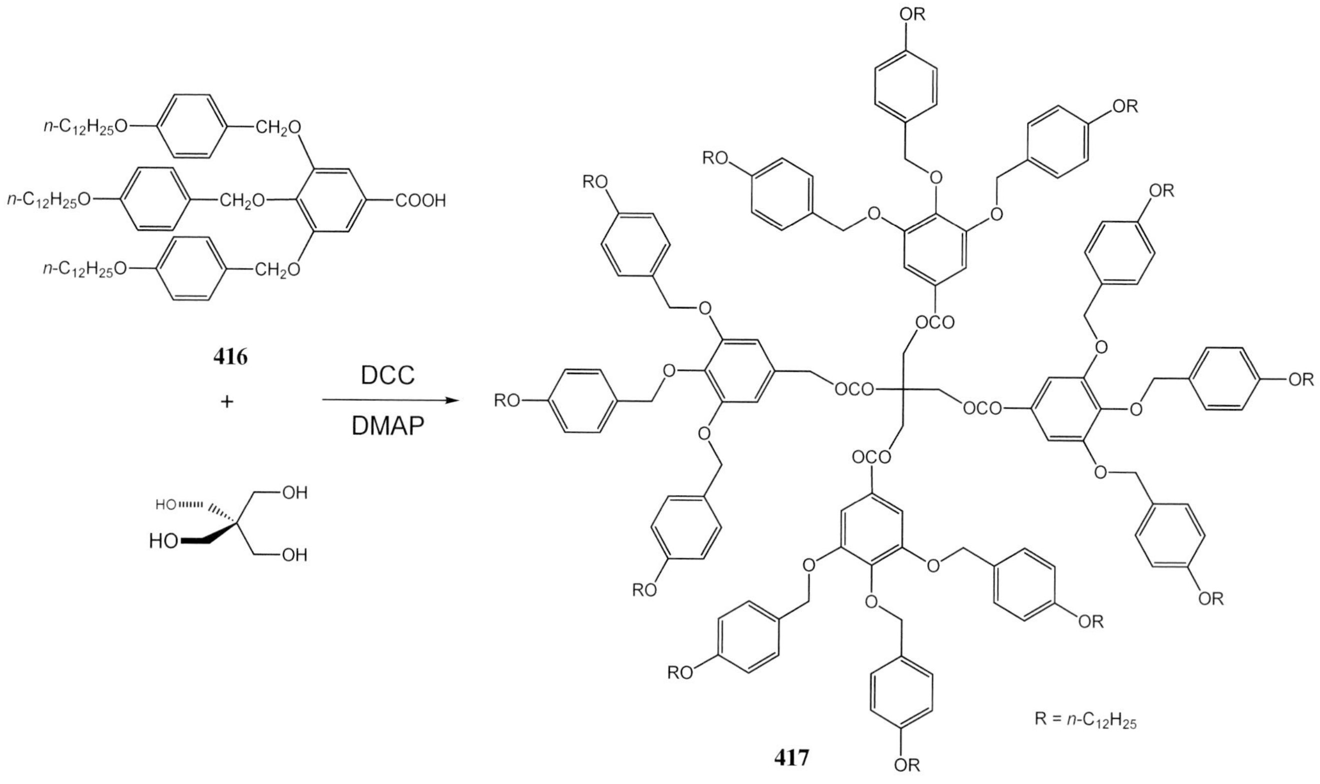

Scheme 5.79 Branched architecture that forms a columnar mesophase.[414]

Dendrimers that self-assemble into supramolecular aggregates forming the basis of a unique thermotropic cubic liquid-crystalline phase (Figure 5.31; **425**) have been reported.[417] Synthesis of these materials (Scheme 5.80) was predicated on methyl 3,4,5-trihydroxybenzoate (**426**) possessing a 1 → 3-branched multiplicity. Alkylation of triol **426** with a dodecyl halide gave the tris(alkylated) ester **427**, which was then reduced (LAH, Et$_2$O) and transformed (SOCl$_2$, CH$_2$Cl$_2$) to the corresponding benzyl chloride **428**. Reaction (K$_2$CO$_3$, DMF) of three equivalents of chloride **428** with triol **426** then afforded the 1st generation ester **429**. Reduction (LAH, Et$_2$O) and halogenation [SOCl$_2$, 2,6-di(*tert*-butyl)pyridine, DMF (cat.), CH$_2$Cl$_2$] led to chloride **430**, which was then treated (NMP, K$_2$CO$_3$) with triol **426** to give the 2nd generation ester **431**. Repetition of the sequence afforded chloride **432** and the 3rd generation ester **433**. Conversion (saponification) of each generation monodendron ester to the corresponding carboxylic acid afforded the self-assembling precursors that exhibited crystalline and isomorphic cubic liquid-crystalline (LC) phases ($Pm\bar{3}n$ space group), as determined by X-ray analysis.[418] This was in contrast to the 1st generation material, which only showed a crystalline phase. For generations 2 through to 4, cubic LC phases were determined to possess 12, 6, and 2 monodendrons per cubic unit cell, respectively. Focally substituted carboxylic acids were found to form the most stable complexes based on *H*-bonding, followed by the corresponding benzylic alcohols. This report constituted the first quantitative confirmation of overall spherical architecture for dendrimers in an ordered liquid state.

These fascinating dendrons have been used for acceleration and control in radical polymerizations of dendritic monomers by self-assembly.[419] X-ray analysis has confirmed the existence of spherical and cylindrical architectures that form highly ordered liquid-crystalline phases.[420] [420a, b] Libraries of these types of self-assembling dendrons have been examined with regard to aggregate rigidity[421] as well as their combined potential to form flat tapers, discs, cones, cylinders, and spheres.[239, 422] Monodendron shape control with respect to generation number and structural analysis of supramolecular cylindrical and spherical dendrimers has also been reported.[423] More recently, similar heat-shrinking spherical and columnar supramolecular constructs, with structures depending on the shape and molecular angle of the dendritic wedges, have been reported.[424] Investigations into smaller, albeit branched and tapered dendrons showed

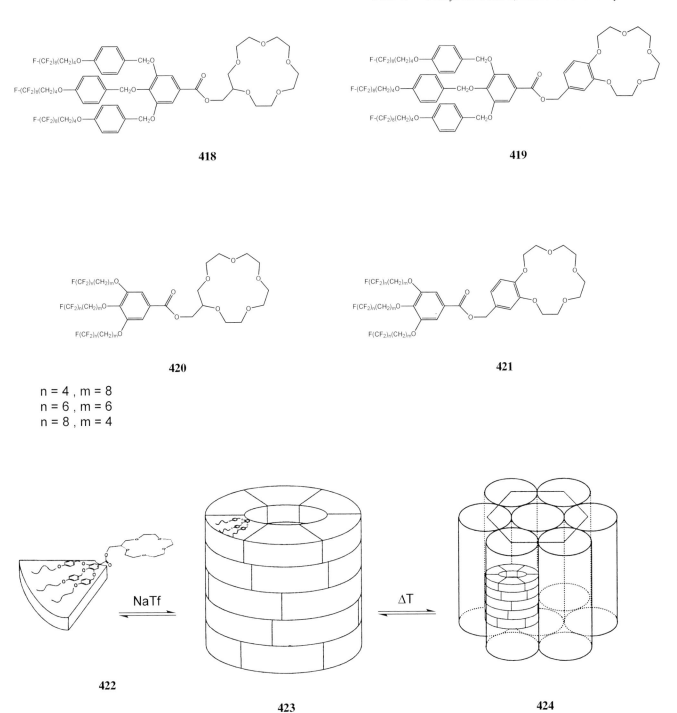

418

419

420

421

n = 4 , m = 8
n = 6 , m = 6
n = 8 , m = 4

422

423

424

Figure 5.30 Percec's monodendrons and their resultant self-assembly into columnar superstructures. Reproduced by permission of the American Chemical Society.[415]

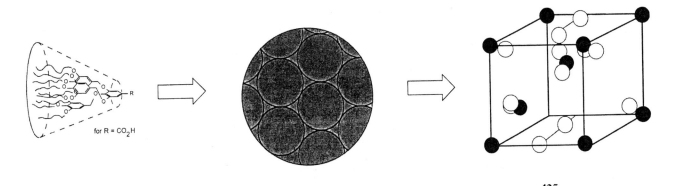

425

Figure 5.31 Self-assembly of alkyl-terminated dendrons into spherical clusters that exhibit a thermotropic cubic liquid-crystalline phase. Reproduced by permission of the American Chemical Society.[417]

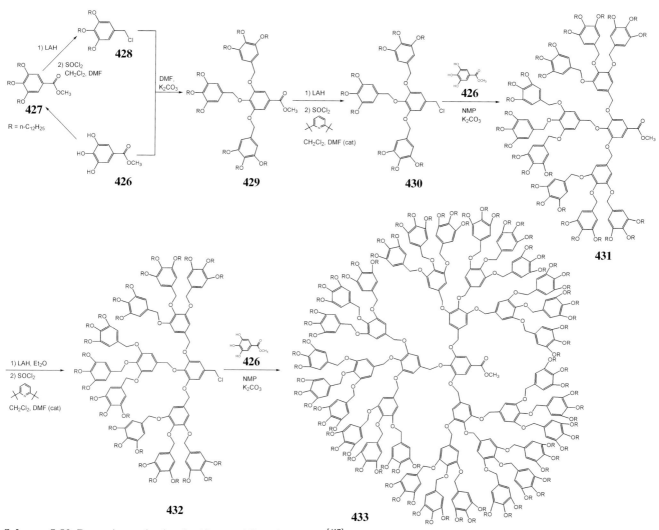

Scheme 5.80 Percec's synthesis of self-assembling dendrons.[417]

that these also formed hexagonal columnar liquid-crystalline superlattices[425] as well as worm-like structures.[426]

Poly(phenylenevinylene) coated with C_6H_{13} hydrocarbon chains attached via 1st or 2nd generation 1 → 3 branched poly(aryl ether) dendrons have been reported.[427] These rod-like polymers, with M_Ws approaching 10,000 amu, exhibit solid-state self-ordering and produce thermotropic nematic phases.

Polystyrene and polymethacrylate coated with either 3,4,5-tris[4-(tetradecyloxy)ben-zyloxy]benzoic acid or 3,4,5-tris[3,4,5-tris(dodecyloxy)benzyloxy]benzyl alcohol Percec-type dendrons have been prepared[428] and imaged by SFM, which allowed quantitative assessment of their length distribution. For the more sterically demanding dendrons, an almost fully extended backbone was observed due to the steric constraints of the side chains. Reduced length polymers were postulated to result from disordered helical con-formations.

Percec-type dendrons, up to the 2nd generation, have been attached to biphenyl diamines and the resulting adducts were subsequently condensed with bis(anhydrides) leading to dendron-jacketed polyimides with enhanced solubility characteristics.[429]

5.12   1 → 3 *C*-Branched, *Amide* Connectivity

Brettreich and Hirsch[430] communicated the convergent synthesis of dendrons up to the 3rd generation, based on Lin's amine monomer[273] (Scheme 5.81; **434–436**). These polyester dendrons were subsequently used to aid the aqueous solubilization of C_{60}; bis(dendron) attachment (DCC, 1-HBT, THF) to an extended malonic acid, obtained by

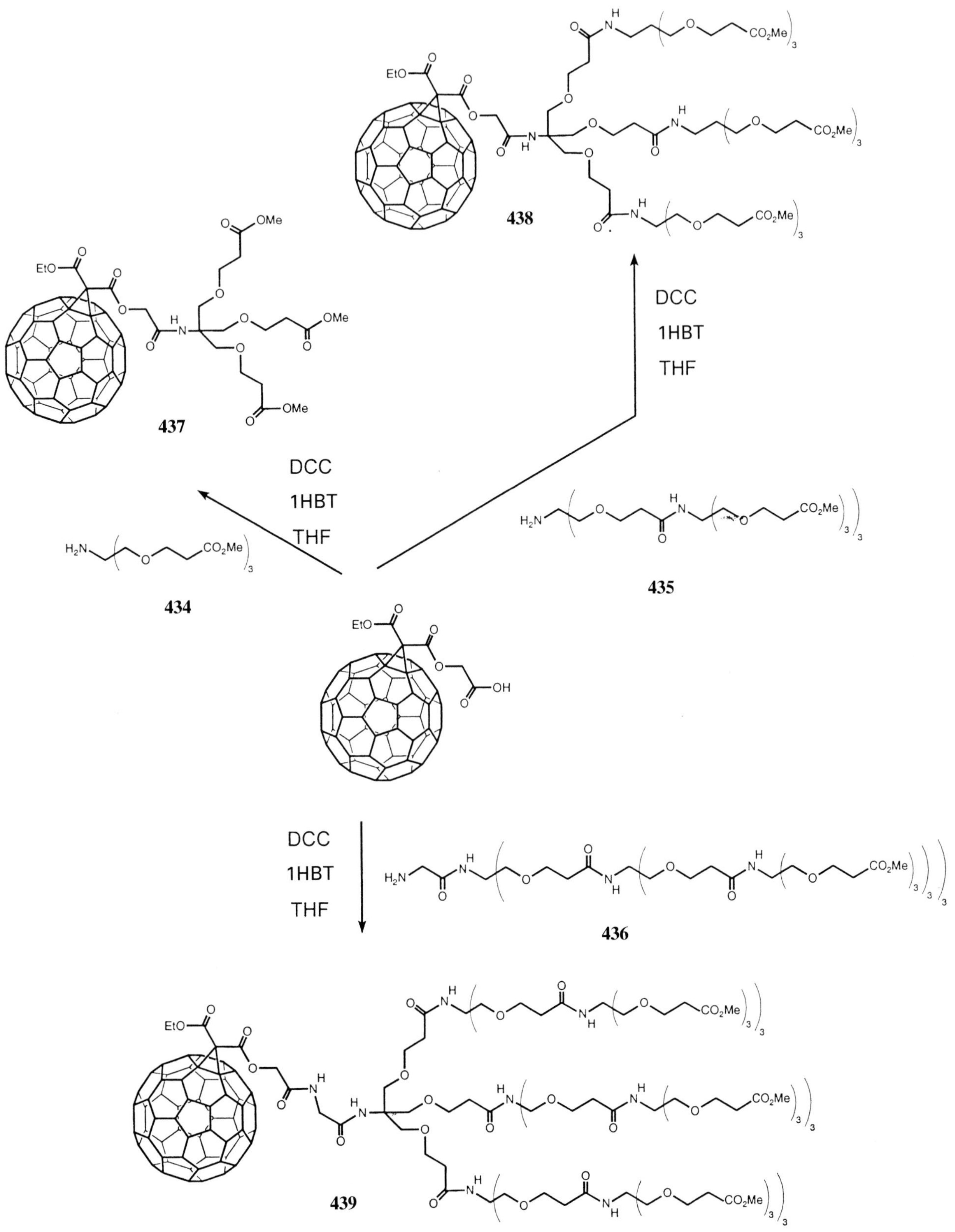

Scheme 5.81 A C_{60} dendritic family. [430]

cyclopropanation of C_{60}, furnished highly water-soluble dendro[60]fullerenes (**437–439**) after ester hydrolysis.[431] These 2nd and 3rd tiered wedges have each been focally func-tionalized with either a ferrocene[432] or dansyl moiety;[433] the latter has been evaluated as regards host–guest interactions through the dansyl unit with β-cyclodextrin and poly-clonal anti-dansyl antibodies. The anti-dansyl antibodies were found to bind to the dan-syl residue in all three examples with "remarkably large binding affinities", whereas the cyclodextrin was found not to associate with the two larger species.[433]

5.13   1 → 3 *C*-Branched, *Ether & Amide* Connectivity

Diederich et al.[434] examined the macrocyclization of C_{60} based on a modified double "Bingel" reaction[435] that furnishes bis(malonato) adducts of C_{60}. These adducts were subsequently converted to dendrimers by treatment with divergently synthesized dendrons using Newkome's tris-based aminotriester.[273]

An example of the fullerene macrocyclization and the use of these novel cores is shown in Scheme 5.82. Reaction of diol **440** with malonic acid afforded the bis(malonic acid) **441**, while treatment of triol **442** with $BrCH_2CO_2tBu$ afforded monoalcohol **443**. Coupling (DCC, DMAP, THF) then gave the bis(malonate) tetraester **444**, reaction (DBU, I_2, toluene) of which with C_{60} afforded (22 %) the tetravalent core **445**. After transformation of the ester moieties to carboxylic acids (CF_3CO_2H, CH_2Cl_2), reaction with the glycine-modified 2nd generation dendron **446** (DCC, 1-HBT, THF) yielded the 36-ester dendrimer **447**. Analogously, the 1st through to the 3rd generation dendrimers were prepared starting with a monovalent core. The 3rd generation dendrimer was only accessible by amine homologation of the dendron with glycine. Other divalent C_{60} cores were also transformed to branched architectures. Redox potential evaluation of the core fullerenes showed no influence of the attached dendrons regardless of their size or density. Amphiphilic cyclic fullerene bis-adducts possessing multiple alkyl chains ($C_{12}H_{25}$) have also been reported[436] and their air–water interface (Langmuir) properties have been studied.

Kenda and Diederich[437] reported the supramolecular aggregation of a rigid phenylacetylene rod terminated at one or both ends by cholesteric testosterone with the cyclophane core of their dendrophanes; generations 0 through to 3 were examined. The driving force behind complex formation included apolar bond formation and hydrophobic desolvation. Association constants and binding enthalpies for the 1:1 complexes ranged from 900 to 91,000 L mol^{-1} and –4.0 to –6.8 Kcal mol^{-1}, respectively, while the corresponding values determined for 2:1 complexes ranged from 700 to 30,000 L mol^{-1} and –3.9 to –6.1 Kcal mol^{-1}, respectively.

Similar dendrophanes possessing catalytic thiazolomethyl-substituted cyclophane cores and either methyl or monomethyltriethylene glycol ester termini have been created, albeit convergently as functional mimics of the thiaminediphosphate-dependent enzyme pyruvate oxidase.[438] Initial catalytic studies of the oxidation of naphthalene-2-carboxaldehyde to the corresponding acid showed only weak activity, which was attributed to inhibition to reaction transition states by the branched appendages.

Smith and Diederich[439] employed these alkyl-based monomers to create a PEG-coated, water-soluble branched scaffolding attached to an enantiomerically pure 9,9'-spirobis[9*H*-fluorene] core possessing diamidopyridine complexing moieties (see Section 7.0. Enantio- and diastereoselectivity for glucopyranoside complexation were "modulated" by the dendritic component.

5.14   1 → 3(2) *Silatrane*-Branched, *Amine* Connectivity

Dendritic silatrane arrays have been prepared[440] by a synthetic strategy (Scheme 5.83) employing as key steps the reaction of an alkanol amine (i.e., **448**) with the trimethoxy-(glycidoxypropyl)silane (**449**) to give silatrane **450**, and subsequent reaction of the latter with ethanolamine or NH_3 to afford the bis- (**451**) and tris(silatrane) **452**, respectively. Repeating the reaction sequence allowed access to the dendrimers **453** and **454**. A notable feature of this sequence is that it has the potential, through variation of the amine component (i.e., NH_3, $H_2NCH_2CH_2OH$, etc.), to incorporate different levels of branching multiplicity at different generations (e.g., **453**). Furthermore, due to the rigidity of the silatrane moiety, stereoisomeric forms are generated.

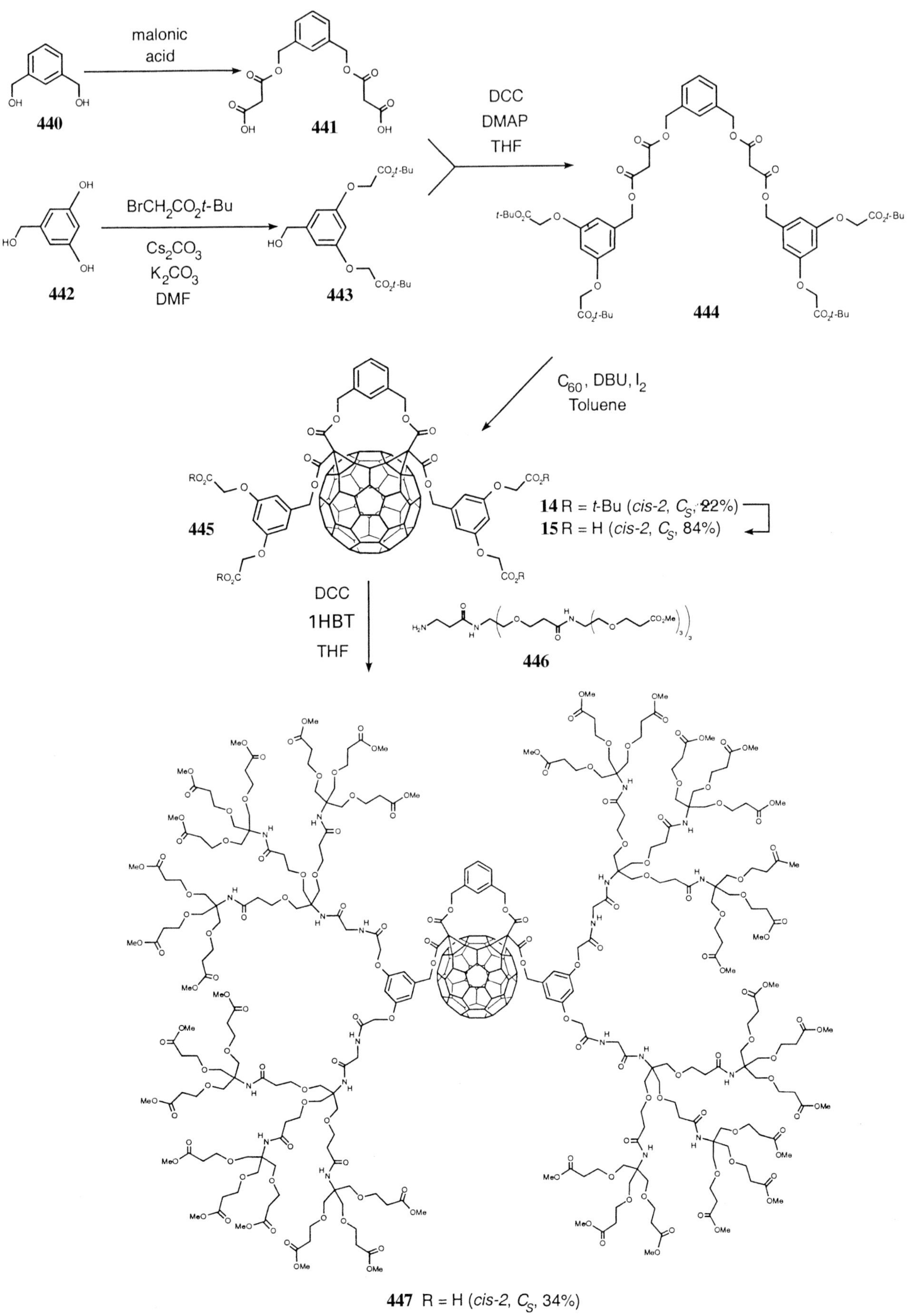

14 R = t-Bu (*cis-2*, C_S, 22%)
15 R = H (*cis-2*, C_S, 84%)

447 R = H (*cis-2*, C_S, 34%)

Scheme 5.82 C_{60} macrocycles provide unique cores for dendrimer formation.[434]

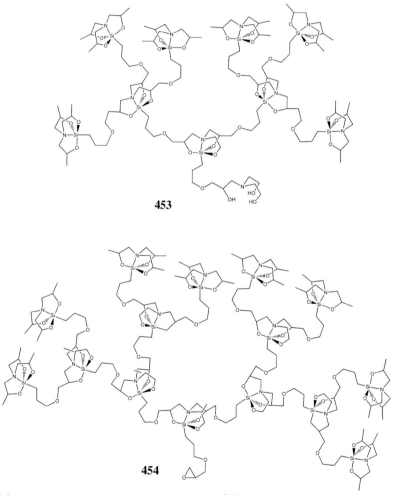

Scheme 5.83 Protocol for access to silatrane-based dendrimers.[440]

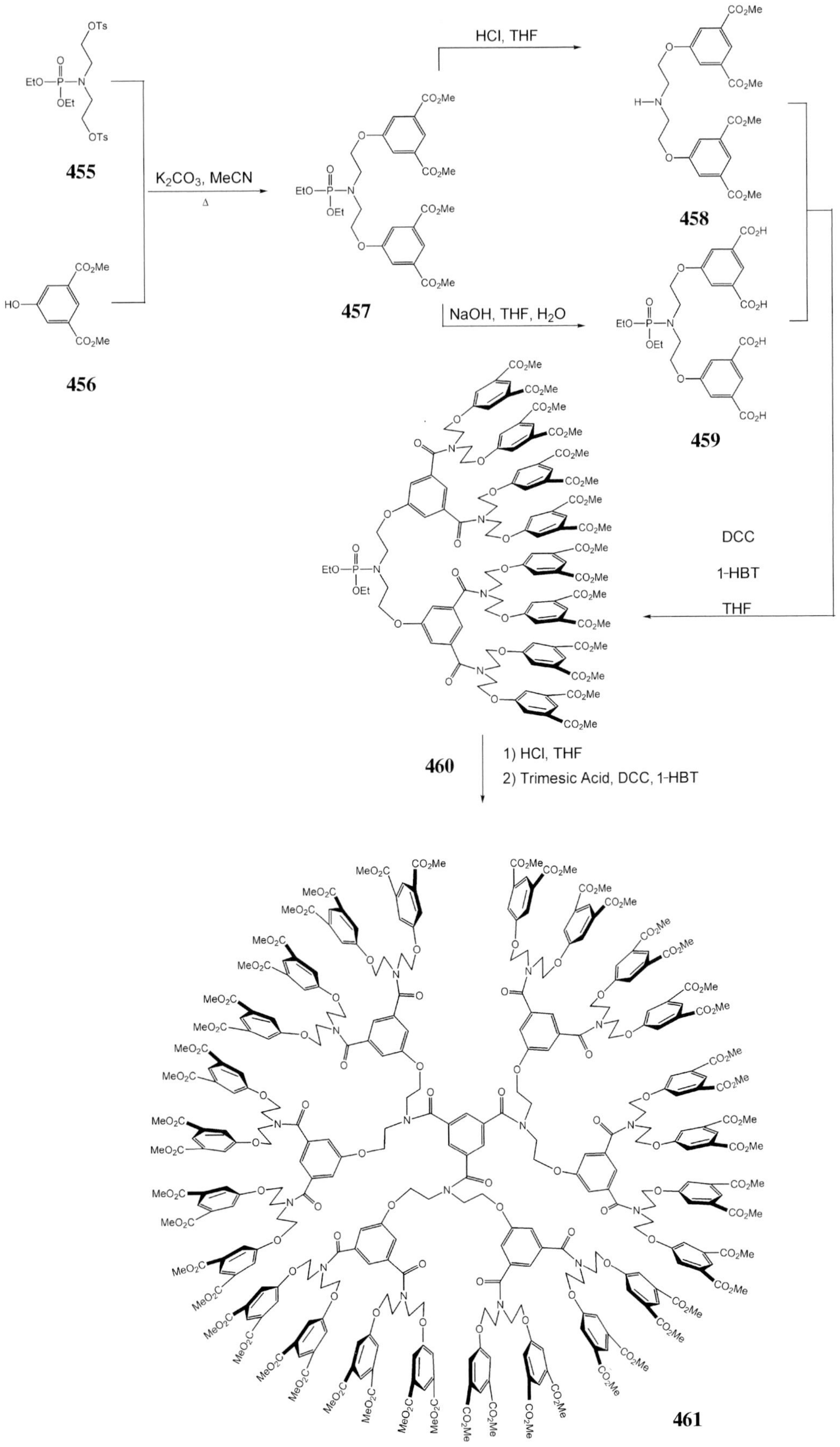

Scheme 5.84 Stoddart's poly(amide) dendrimers based on diethanolamine and 5-hydroxyisophthalic acid.[441]

5.15 1 → 4 *Amine* and *Aryl* Branched, *Amide* Connectivity

Stoddart et al.[441] reported the creation of a family of polyamide dendrimers predicated on $HN(CH_2CH_2OH)_2$ and 5-hydroxyisophthalic acid. Elaboration of these materials to provide the requisite complementary protection–deprotection functionality included the creation of bis(tosylate) **455** (Scheme 5.84) by treatment with diethyl phosphite followed by standard tosylation. Esterification provided phenol **456**. Reaction of these two elements generated monomer **457** (reminiscent of the double-exponential type monomers introduced by Moore[345]) that could, in turn, be selectively deprotected both focally (HCl, THF) and terminally (NaOH, THF, H_2O) to give **458** and **459**, which could be recombined (DCC, 1-HBT, THF) to give the 4th generation dendron **460**. Following focal deprotection and reaction (1-HBT, DCC, THF) with trimesic acid, dendrimer **461** was produced (13 %) along with ca. 32 % of the bis(dendron). These monomers were also employed divergently for dendrimer construction.

5.16 References

[1] C. Hawker, J. M. J. Fréchet, "A New Convergent Approach to Monodisperse Dendritic Macromolecules", *J. Chem. Soc., Chem. Commun.* **1990**, **1010**–1013.

[2] T. M. Miller, T. X. Neenan, "Convergent Synthesis of Monodisperse Dendrimers Based upon 1,3,5-Trisubstituted Benzenes", *Chem. Mater.* **1990**, *2*, 346–349.

[3] J. M. J. Fréchet, "Functional Polymers and Dendrimers: Reactivity, Molecular Architecture, and Interfacial Energy", *Science* **1994**, *263*, 1710–1715.

[4] J. M. J. Fréchet, C. J. Hawker, K. L. Wooley, "The Convergent Route to Globular Dendritic Macromolecules: A Versatile Approach to Precisely Functionalized Three-Dimensional Polymers and Novel Block Copolymers", *J. Macromol. Sci., Pure Appl. Chem.* **1994**, *A31*, 1627–1645.

[5] C. J. Hawker, K. L. Wooley, "The Convergent-Growth Approach to Dendritic Macromolecules" in *Advances in Dendritic Macromolecules* (Ed.: G. R. Newkome), JAI Press, Greenwich, Conn., **1995**, pp. 1–39.

[6] J. M. J. Fréchet, C. J. Hawker, "Synthesis and Properties of Dendrimers and Hyperbranched Polymers" in *Comprehensive Polymer Chemistry, 2nd Supplement* (Eds.: S. L. Aggarwal, S. Russo), Elsevier, Oxford, U.K., **1996**, pp. 71–132.

[7] J.-L. Six, Y. Gnanou, "Dendritic Architectures by the Convergent Method" in *Star and Hyperbranched Polymers* (Eds.: M. K. Mishra, S. Kobayashi), Marcel Dekker, Inc., New York, **1999**, pp. 239–266.

[8] C. J. Hawker, "Dendritic and Hyperbranched Macromolecules – Precisely Controlled Macromolecular Architectures", *Adv. Polym. Sci.* **1999**, *147*, 114–160.

[9] A. Rajca, "A Polyarylmethyl Carbotetraanion", *J. Am. Chem. Soc.* **1990**, *112*, 5889–5890.

[10] A. Rajca, "A Polyarylmethyl Quintet Tetraradical", *J. Am. Chem. Soc.* **1990**, *112*, 5890–5892.

[11] A. Rajca, "Synthesis of 1,3-Connected Polyarylmethanes", *J. Org. Chem.* **1991**, *56*, 2557–2563.

[12] A. Rajca, S. Utamapanya, "Poly(arylmethyl) Quartet Triradicals and Quintet Tetraradicals", *J. Am. Chem. Soc.* **1993**, *115*, 2396–2401.

[13] K. Nakatani, J. Y. Carriat, Y. Journaux, O. Kahn, F. Lloret, J. P. Renard, Y. Pei, J. Sletten, M. Verdaguer, "Chemistry and Physics of the Novel Molecular-Based Compound Exhibiting a Spontaneous Magnetization Below T_c = 14 K, MnCu(obbz)· $1H_2O$ [obbz = oxamidobis(-benzoato)]. Comparison with the Antiferromagnet MnCu(obbz)· $5H_2O$. Crystal Structure and Magnetic Properties of NiCu(obbz)· $6H_2O$", *J. Am. Chem. Soc.* **1989**, *111*, 5739–5748.

[14] A. Caneschi, D. Gatteschi, R. Sessoli, P. Rey, "Toward Molecular Magnets: The Metal-Radical Approach", *Acc. Chem. Res.* **1989**, *22*, 392–398.

[15] H. Iwamura, "High-spin Organic Molecules and Spin Alignment in Organic Molecular Assemblies" in *Advances in Physical Organic Chemistry* (Ed.: D. Bethell), Academic Press, New York, **1990**, pp. 179–253.

[16] N. Nakamura, K. Inoue, H. Iwamura, T. Fujioka, Y. Sawaki, "Synthesis and Characterization of a Branched-Chain Hexacarbene in a Tridecet Ground State. An Approach to Superparamagnetic Polycarbenes", *J. Am. Chem. Soc.* **1992**, *114*, 1484–1485.

[17] D. A. Dougherty, "Spin Control in Organic Molecules", *Acc. Chem. Res.* **1991**, *24*, 88–94.

[18] J. S. Miller, A. J. Epstein, "Organic and Organometallic Molecular Magnetic Materials – Designer Magnets", *Angew. Chem.* **1994**, *106*, 399–432; *Int. Ed. Engl.* **1994**, *33*, 385–415.

[19] A. Rajca, "Toward Organic Synthesis of a Nanometer-Size Magnetic Particle", *Adv. Mater. (Weinheim, Fed. Repub. Ger.)* **1994**, *6*, 605–607.

[20] A. Rajca, "High-spin Polyarylmethyl Polyradicals" in *Advances in Dendritic Macromolecules* (Ed.: G. R. Newkome), JAI Press, Inc., Greenwich, Conn., **1994**, pp. 133–168.

[21] A. Rajca, "Organic Diradicals and Polyradicals: From Spin Coupling to Magnetism?", *Chem. Rev.* **1994**, *94*, 871–893.

[22] A. Rajca, S. Utamapanya, S. Thayumanavan, "Poly(arylmethyl) Octet ($S = 7/2$) Heptaradical and Undecet ($S = 5$) Decaradical", *J. Am. Chem. Soc.* **1992**, *114*, 1884–1885.

[23] S. Utamapanya, A. Rajca, "Topological Control of Electron Localization in π-Conjugated Polyarylmethyl Carbopolyanions and Radical Anions", *J. Am. Chem. Soc.* **1991**, *113*, 9242–9251.

[24] A. Rajca, S. Utamapanya, J. Xu, "Control of Magnetic Interactions in Polyarylmethyl Triplet Diradicals Using Steric Hindrance", *J. Am. Chem. Soc.* **1991**, *113*, 9235–9241.

[25] A. Rajca, S. Utamapanya, "π-Conjugated Systems with Unique Electronic Structure: A Case of 'Planarized' 1,3-Connected Polyarylmethyl Carbodianion and Stable Triplet Hydrocarbon Diradical", *J. Org. Chem.* **1992**, *57*, 1760–1767.

[26] A. Rajca, S. Utamapanya, D. J. Smithhisler, "Near-Degeneracy Between the Low- and High-Spin States in an Alternant Hydrocarbon Diradical: Topology and Geometry", *J. Org. Chem.* **1993**, *58*, 5650–5652.

[27] A. Rajca, J. Wongsriratanakul, S. Rajca, "Organic Spin Clusters: Ferromagnetic Spin Coupling through a Biphenyl Unit in Polyarylmethyl Tri-, Penta-, Hepta-, and Hexadecaradicals", *J. Am. Chem. Soc.* **1997**, *119*, 11674–11686.

[28] K. Inoue, T. Hayamizu, H. Iwamura, D. Hashizume, Y. Ohashi, "Assemblage and Alignment of the Spins of the Organic Trinitroxide Radical with a Quartet Ground State by Means of Complexation with Magnetic Metal Ions. A Molecule-Based Magnet with Three-Dimensional Structure and High T_c of 46 K", *J. Am. Chem. Soc.* **1996**, *118*, 1803–1804.

[29] A. Rajca, S. Utamapanya, "Toward Organic Synthesis of a Magnetic Particle: Dendritic Polyradicals with 15 and 31 Centers for Unpaired Electrons", *J. Am. Chem. Soc.* **1993**, *115*, 10688–10694.

[30] A. Rajca, S. Janicki, "Synthesis of Sterically Hindered 1,3-Connected Polyarylmethanes", *J. Org. Chem.* **1994**, *59*, 7099–7107.

[31] A. Rajca, S. Utamapanya, "Spin Balls and Spin Barbells. Preparation and Magnetic Studies of $S = 7/2$ Dendritic Heptaradical and Progress Toward Very High Spin Dendrimers", *Mol. Cryst. Liq. Cryst.* **1993**, *232*, 305–312.

[32] A. Rajca, S. Rajca, R. Padmakumar, "Calixarene-Based Macrocyclic Nonet ($S = 4$) Octaradical and its Acyclic Sextet ($S = 5/2$) Pentaradical Analogue", *Angew. Chem.* **1994**, *106*, 2193–2195; *Int. Ed. Engl.* **1994**, *33*, 2091–2093.

[33] A. Rajca, S. Rajca, S. R. Desai, "Macrocyclic π-Conjugated Carbopolyanions and Polyradicals Based upon Calix[4]arene and Calix[3]arene Rings", *J. Am. Chem. Soc.* **1995**, *117*, 806–816.

[34] A. Rajca, K. Lu, S. Rajca, "High-Spin Polyarylmethyl Polyradical: Fragment of a Macrocyclic 2-Strand Based upon Calix[4]arene Rings", *J. Am. Chem. Soc.* **1997**, *119*, 10335–10345.

[35] K. Matsuda, N. Nakamura, K. Takahashi, K. Inoue, N. Koga, H. Iwamura, "Design, Synthesis, and Characterization of Three Kinds of π-Cross-Conjugated Hexacarbenes with High-Spin ($S = 6$) Ground States", *J. Am. Chem. Soc.* **1995**, *117*, 5550–5560.

[36] K. Matsuda, H. Iwamura, "Synthesis and EPR Characterisation of [3-(*N*-ylooxy-*tert*-butylamino)-5-*tert*-butylphenyl]phenylcarbene with a Quartet Ground State", *Chem. Commun.* **1996**, 1131–1132.

[37] A. Rajca, J. Wongsriratanakul, S. Rajca, R. Cerny, "A Dendritic Macrocyclic Organic Polyradical with a Very High Spin of $S = 10$", *Angew. Chem.* **1998**, *110*, 1284–1288; *Int. Ed.* **1998**, *37*, 1229–1232.

[38] M. A. Keegstra, S. De Feyter, F. C. De Schryver, K. Müllen, "Hexaterphenylyl- Hexaquaterphenylylbenzene: The Behavior of Chromophores and Electrophores in a Restricted Space", *Angew. Chem.* **1996**, *108*, 830–833; *Int. Ed. Engl.* **1996**, *35*, 774–776.

[39] H. Nishide, M. Miyasaka, E. Tsuchida, "High-Spin Polyphenoxyls Attached to Star-Shaped Poly(phenylenevinylene)s", *J. Org. Chem.* **1998**, *63*, 7399–7407.

[39a] M. Higuchi, S. Shiki, K. Ariga, K. Yamamoto, "Novel Phenylazomethine Dendrimers: Synthesis and Structural Properties", *Org. Lett.* **2000**, *2*, 3079–3082.

[39b] M. Higuchi, S. Shiki, K. Ariga, K. Yamamoto, "First Synthesis of Phenylazomethine Dendrimer ligands and structural Studies", *J. Am. Chem. Soc.* **2001**, *123*, 4414–4420.

[40] L. Fomina, P. Guadarrama, S. Fomine, R. Salcedo, T. Ogawa, "Synthesis and Characterization of Well-Defined Fully Conjugated Hyperbranched Oligomers of β,β-Dibromo-4-ethynylstyrene", *Polymer* **1998**, *39*, 2629–2635.

[41] L. Fomina, R. Salcedo, "Synthesis and Polymerization of β,β-Dibromo-4-ethynylstyrene; Preparation of a New Polyconjugated, Hyperbranched Polymer", *Polymer* **2000**, *37*, 1723–1728.

[41a] P. Guadarrama, L. Fomina, S. Fomine, "Solid-supported synthesis of hyperbranched polymer with β,β-diethynylstyryl units", *Polym. Int.* **2001**, *50*, 76–83. (hyperbranched title)

[42] L. Fomina, P. Ponce, P. Guadarrama, S. Fomine, "Effect of Terminal Groups on the Electronic Structure of Hyperbranched Polyacetylene", *Macromol. Theory Simul.* **1999**, *8*, 403–408.

[43] S. Fomine, L. Fomina, P. Guadarrama, "Electronic Structure of Fully Conjugated Dendritic Oligomers of β,β-Dibromo-4-ethynylstyrene", *THEOCHEM* **1999**, *488*, 216.

[44] K. L. Wooley, C. J. Hawker, J. M. J. Fréchet, "Hyperbranched Macromolecules via a Novel Double-Stage Convergent Growth Approach", *J. Am. Chem. Soc.* **1991**, *113*, 4252–4261.

[45] K. L. Wooley, C. J. Hawker, J. M. J. Fréchet, "Hyperbranched Macromolecules via a Novel Double-Stage Convergent Growth Approach" *Polym. Mater. Sci. Eng.* **1991**, 235–236.

[46] C. J. Hawker, J. M. J. Fréchet, "Preparation of Polymers with Controlled Molecular Architecture. A New Convergent Approach to Dendritic Macromolecules", *J. Am. Chem. Soc.* **1990**, *112*, 7638–7647.

[47] K. L. Wooley, C. J. Hawker, J. M. Pochan, J. M. J. Fréchet, "Physical Properties of Dendritic Macromolecules: A Study of Glass Transition Temperature", *Macromolecules* **1993**, *26*, 1514–1519.

[48] M. Liu, K. Kono, J. M. J. Fréchet, "Water-Soluble Dendritic Unimolecular Micelles: Their Potential as Drug Delivery Agents", *J. Controlled Release* **2000**, *65*, 121–131.

[49] K.-Y. Chen, C. B. Gorman, "Synthesis of a Series of Focally-Substituted Organothiol Dendrons", *J. Org. Chem.* **1996**, *61*, 9229–9235.

[50] C. B. Gorman, B. L. Parkhurst, W. Y. Su, K.-Y. Chen, "Encapsulated Electroactive Molecule Based upon an Inorganic Cluster Surrounded by Dendron Ligands", *J. Am. Chem. Soc.* **1997**, *119*, 1141–1142.

[51] H. Nemoto, J. G. Wilson, H. Nakamura, Y. Yamamoto, "Polyols of a Cascade Type as a Water-Solublizing Element of Carborane Derivatives for Boron Neutron Capture Therapy", *J. Org. Chem.* **1992**, *57*, 435.

[52] H. Nemoto, J. Cai, Y. Yamamoto, "Synthesis of a Water-Soluble *o*-Carbaborane bearing a Uracil Moiety *via* a Palladium-catalysed Reaction under Essentially Neutral Conditions", *J. Chem. Soc., Chem. Commun.* **1994**, 577–578.

[53] V. Percec, P. Chu, G. Ungar, J. Zhou, "Rational Design of the First Nonspherical Dendrimer which Displays Calamitic Nematic and Smectic Thermotropic Liquid Crystalline Phases", *J. Am. Chem. Soc.* **1995**, *117*, 11441–11454.

[54] V. Percec, C. G. Cho, C. Pugh, D. Tomazos, "Synthesis and Characterization of Branched Liquid-Crystalline Polyethers Containing Cyclotetraveratrylene-based Disk-like Mesogens", *Macromolecules* **1992**, *25*, 1164–1176.

[55] J. Li, K. A. Crandall, P. Chu, V. Percec, R. G. Petschek, C. Rosenblatt, "Dendrimeric Liquid Crystals: Isotropic–Nematic Pretransitional Behavior", *Macromolecules* **1996**, *29*, 7813–7819.

[56] H. R. Allcock, R. Ravikiran, S. J. M. O'Connor, "Effect of Oligo(ethyleneoxy)cyclotriphosphazenes, Tetraglyme, and Other Small Molecules on the Ionic Conductivity of the Poly[bis(methoxyethoxyethoxy)phosphazene] (MEEP)/Lithium Triflate System", *Macromolecules* **1997**, *30*, 3184–3190.

[57] M. Jayaraman, J. M. J. Fréchet, "A Convergent Route to Novel Aliphatic Polyether Dendrimers", *J. Am. Chem. Soc.* **1998**, *120*, 12996–12997.

[57a] S. M. Grayson, J. M. J. Fréchet, "Synthesis and Surface Functionalization of Aliphatic Polyether Dendrons", *J. Am. Chem. Soc.* **2000**, *122*, 10335–10344.

[58] K. M. Lynch, W. P. Dailey, "Improved Preparations of 3-Chloro-2-(chloromethyl)-1-propene and 1,1-Dibromo-2,2-bis(chloromethyl)cyclopropane: Intermediates in the Synthesis of [1.1.1]Propellane", *J. Org. Chem.* **1995**, *60*, 4666–4668.

[59] S. M. Grayson, M. Jayaraman, J. M. J. Fréchet, "Convergent Synthesis and 'Surface' Functionalization of a Dendritic Analog of Poly(ethylene glycol)", *Chem. Commun.* **1999**, 1329–1330.

[60] S. M. Grayson, M. Jayaraman, J. M. J. Fréchet, "Selective Surface Modification of Orthogonally Protected Aliphatic Polyether Dendrons" *Polym. Prepr.* **2000**, *41*, 167–168.

[61] H. Ihre, A. Hult, E. Söderlind, "Synthesis, Characterization, and ¹H NMR Self-Diffusion Studies of Dendritic Aliphatic Polyesters Based on 2,2- Bis(hydroxymethyl)propionic acid and 1,1,1-Tris(hydroxyphenyl)ethane", *J. Am. Chem. Soc.* **1996**, *118*, 6388–6395.

[62] H. Ihre, A. Hult, J. M. J. Fréchet, I. Gitsov, "Double-Stage Convergent Approach for the Synthesis of Functionalized Dendritic Aliphatic Polyesters Based on 2,2-Bis(hydroxymethyl) propionic Acid", *Macromolecules* **1998**, *31*, 4061–4068.

[63] A. Córdova, A. Hult, K. Hult, H. Ihre, T. Iversen, E. Malmström, "Synthesis of a Poly(ε-caprolactone) Monosubstituted First Generation Dendrimer by Lipase Catalysis", *J. Am. Chem. Soc.* **1998**, *120*, 13521–13522.

[64]   B. Atthoff, M. Trollsås, H. Claesson, J. L. Hedrick, "Poly(lactides) with Controlled Molecular Architecture Initiated from Hydroxy Functional Dendrimers and the Effect on the Hydrodynamic Volume", *Macromol. Chem. Phys.* **1999**, *200*, 1333–1339.

[64a] P. Busson, J. Örtegren, H. Ihre, U. W. Gedde, A. Hult, G. Andersson, "Ferroelectric Liquid Crystalline Dendrimers: Synthesis, Thermal Behavior, and Electrooptical Characterization", *Macromolecules* **2001**, *34*, 1221–1229.

[65]   C.-M. Leu, Y.-T. Chang, C.-F. Shu, C.-F. Teng, J. Shiea, "Synthesis and Characterization of Dendritic Poly(ether imide)s", *Macromolecules* **2000**, *33*, 2855–2861.

[65a] C.-M. Leu, C.-F. Shu, C.-F. Teng, J. Shiea, "Dendritic poly(ether-imide)s: synthesis, characterization, and modification", *Polymer* **2001**, *42*, 2339–2348.

[66]   T. M. Miller, T. X. Neenan, R. Zayas, H. E. Bair, "Synthesis and Characterization of a Series of Monodisperse, 1,3,5-Phenylene-Based Hydrocarbon Dendrimers Including $C_{276}H_{186}$ and their Fluorinated Analogues", *J. Am. Chem. Soc.* **1992**, *114*, 1018–1025.

[67]   T. M. Miller, T. X. Neenan, H. E. Bair, "The Synthesis and Characterization of a Series of Monodisperse, 1,3,5-Phenylene Based Hydrocarbon Dendrimers Including $C_{276}H_{186}$" *Polym. Prepr.* **1991**, *32*, 627–628.

[68]   N. Miyaura, T. Yanagi, A. Suzuki, "The Palladium-Catalyzed Cross-Coupling Reaction of Phenylboronic Acid with Haloarenes in the Presence of Bases", *Synth. Commun.* **1981**, *11*, 513–519.

[69]   T. M. Miller, T. X. Neenan, R. Zayas, H. E. Bair, "The Synthesis and Thermal Characterization of a Series of Monodisperse, 1,3,5-Phenylene-Based, Fluorinated Dendrimers" *Polym. Prepr.* **1991**, *32*, 599–600.

[70]   F. Wang, A. B. Kon, R. D. Rauh, "Synthesis of a Terminally Functionalized Bromothiophene Polyphenylene Dendrimer by a Divergent Method", *Macromolecules* **2000**, *33*, 5300–5302.

[71]   U.-M. Wiesler, K. Müllen, "Polyphenylene Dendrimers *via* Diels–Alder Reactions: The Convergent Approach", *Chem. Commun.* **1999**, 2293–2294.

[72]   T. Gensch, J. Hofkens, A. Herrmann, K. Tsuda, W. Verheijen, T. Vosch, T. Christ, T. Basché, K. Müllen, F. C. De Schryver, "Fluorescence Detection from Single Dendrimers with Multiple Chromophores", *Angew. Chem.* **1999**, *111*, 3970–3974; *Int. Ed.* **1999**, *38*, 3752–3756.

[73]   Y. Karni, S. Jordens, G. De Belder, G. Schweitzer, J. Hofkens, T. Gensch, M. Maus, F. C. De Schryver, A. Hermann, K. Müllen, "Intramolecular Evolution from a Locally Excited State to an Excimer-like State in a Multichromophoric Dendrimer Evidenced by a Femtosecond Fluorescence Upconversion Study", *Chem. Phys. Lett.* **1999**, *310*, 73–78.

[74]   J. Hofkens, L. Latterini, G. De Belder, T. Gensch, M. Maus, T. Vosch, Y. Karni, G. Schweitzer, F. C. De Schryver, A. Hermann, K. Müllen, "Photophysical Study of a Multichromophoric Dendrimer by Time-Resolved Fluorescence and Femtosecond Transient Absorption Spectroscopy", *Chem. Phys. Lett.* **1999**, *304*, 1–9.

[74a] G. De Belder, G. Schweitzer, S. Jordens, M. Lor, S. Mitra, J. Hofkens, S. De Feyter, M. V. der Auweraer, A. Herrmann, T. Weil, K. Müllen, F. C. De Schryver, "Singlet Singlet Annihilation in Multichromophoric Peryleneimide Dendrimers, Determined by Fluoresence Upconversion", *Chem. Phys. Chem.* **2001**, 49–55.

[75]   S. Loi, U.-M. Wiesler, H.-J. Butt, K. Müllen, "Formation of Nanorods by Self-Assembly of Alkyl-Substituted Polyphenylene Dendrimers on Graphite", *Chem. Commun.* **2000**, 1169–1170.

[76]   B. Forier, W. Dehaen, "Alternative Convergent and Accelerated Double-Stage Convergent Approaches Towards Functionalized Dendritic Polyethers", *Tetrahedron* **2000**, *55*, 9829–9846.

[77]   P. Brocorens, E. Zojer, J. Cornil, Z. Shuai, G. Leising, K. Müllen, J.-L. Brédas, "Theoretical Characterization of Phenylene-Based Oligomers, Polymers, and Dendrimers", *Syn. Metals* **1999**, *100*, 141–162.

[78]   K. Matsuda, N. Nakamura, K. Inoue, N. Koga, H. Iwamura, "Design and Synthesis of a "Starburst"-Type Nonadiazo Compound and Magnetic Characterization of its Photoproduct", *Chem. Eur. J.* **1996**, *2*, 259–264.

[79]   J. M. J. Fréchet, C. J. Hawker, A. E. Philippides, "Dendritic Molecules and Method of Production", **1991**, *U. S. Pat.*, 5, 041, 516.

[80]   E. Reimann, "Natürliche Stilbene. Synthese von Polyhydroxystilbenäthern durch Wittig-Reaktion", *Chem. Ber.* **1969**, *102*, 2881–2888.

[81]   T. L. Tyler, J. E. Hanson, "An Efficient Synthesis of Poly(aryl ether) Monodendrons and Dendrimers based on 3,5-Bis(hydroxymethyl)phenol", *Chem. Mater.* **1999**, *11*, 3452–3459.

[82]   S. L. Gilat, A. Adronov, J. M. J. Fréchet, "Modular Approach to the Accelerated Convergent Growth of Laser Dye-Labeled Poly(aryl ether) Dendrimers Using a Novel Hypermonomer", *J. Org. Chem.* **1999**, *64*, 7474–7484.

[83]   J. G. Weintraub, J. R. Parquette, "Synthesis of Unsymmetrically Branched Dendrimeric Wedges up to the Fourth Generation Based on 2,3-Dihydroxybenzyl Alcohol", *J. Org. Chem.* **1999**, *64*, 3796–3797.

[84]   H.-F. Chow, I. Y. K. Chan, C. C. Mak, M.-K. Ng, "Synthesis and Properties of a New Class of Polyether Dendritic Fragments: Useful Building Blocks for Functional Dendrimers", *Tetrahedron* **1996**, *52*, 4277–4290.

[85]   H.-F. Chow, I. Y. K. Chan, C. C. Mak, "Facile Construction of Acid–Base and Redox-Stable Polyether-based Dendritic Fragments", *Tetrahedron Lett.* **1995**, *36*, 8633–8636.

[86]   H.-F. Chow, C. C. Mak, "Dendritic Bis(oxazoline)copper(II) Catalysts. 2. Synthesis, Reactivity, and Substrate Selectivity", *J. Org. Chem.* **1997**, *62*, 5116–5127.

[87]   H.-F. Chow, Z.-Y. Wang, Y.-F. Lau, "An Accelerated, Improved Synthetic Route for the Preparation of Polyether-based Dendritic Fragments", *Tetrahedron* **1998**, *54*, 13813–13824.

[88]   C. C. Mak, H.-F. Chow, "Dendritic Catalysts: Reactivity and Mechanism of the Dendritic Bis(oxazoline)metal Complex Catalyzed Diels–Alder Reaction", *Macromolecules* **1997**, *30*, 1228–1230.

[89]   I. Gitsov, J. M. J. Fréchet, "Stimuli-Responsive Hybrid Macromolecules: Novel Amphiphilic Star Copolymers With Dendritic Groups at the Periphery", *J. Am. Chem. Soc.* **1996**, *118*, 3785–3786.

[90]   S. Serroni, S. Campagna, A. Juris, M. Venturi, V. Balzani, G. Denti, "Polyether Arborols Mounted on a Luminescent and Redox-Active Ruthenium(II)-Polypyridine Core", *Gazz. Chim. Ital.* **1994**, *124*, 423–427.

[90a] S. Hecht, N. Vladimirov, J. M. J. Fréchet, "Encapsulation of Functional Moieties within Branched Star Polymers: Effect of Chain Length and Solvent on Site Isolation", *J. Am. Chem. Soc.* **2001**, *123*, 18–25.

[91]   S. Hecht, J. M. J. Fréchet, "An Alternative Synthetic Approach toward Dendritic Macromolecules: Novel Benzene-Core Dendrimers via Alkyne Cyclotrimerization", *J. Am. Chem. Soc.* **1999**, *121*, 4084–4085.

[92]   C. J. Hawker, K. L. Wooley, J. M. J. Fréchet, "Solvatochromism as a Probe of the Microenvironment in Dendritic Polyethers: Transition from an Extended to a Globular Structure", *J. Am. Chem. Soc.* **1993**, *115*, 4375–4376.

[92a] C. Francavilla, M. D. Drake, F. V. Bright, M. R. Detty, "Dendrimeric Organochalcogen Catalysts for the Activation of Hydrogen Peroxide: Improved Catalytic Activity through Statistical Effects and Coorperativity in Successive Generations", *J. Am. Chem. Soc.* **2001**, *123*, 57–67.

[93]   I. Morao, F. P. Cossío, "Dendritic Catalysts for the Nitroaldol (Henry) Reaction", *Tetrahedron Lett.* **1997**, *38*, 6461–6464.

[94]   K. Matyjaszewski, T. Shigemoto, J. M. J. Fréchet, M. Leduc, "Controlled/'Living' Radical Polymerization with Dendrimers Containing Stable Radicals", *Macromolecules* **1996**, *29*, 4167–4171.

[95]   M. R. Leduc, C. J. Hawker, J. Dao, J. M. J. Fréchet, "Dendritic Initiators for 'Living' Radical Polymerizations: A Versatile Approach to the Synthesis of Dendritic-Linear Block Copolymers", *J. Am. Chem. Soc.* **1996**, *118*, 11111–11118.

[96]   Y.-M. Chen, C.-F. Chen, W.-H. Liu, Y.-F. Li, F. Xi, "Poly(methacrylates) with Different Kinds of Dendritic Moieties as Side Groups", *Macromol. Rapid Commun.* **1996**, *17*, 401–407.

[97]   T. Emrick, W. Hayes, J. M. J. Fréchet, "A TEMPO-Mediated 'Living' Free-Radical Approach to ABA Triblock Dendritic Linear Hybrid Copolymers", *J. Polym. Sci., Part A: Polym. Chem.* **1999**, *37*, 3748–3755.

[98]   M. R. Leduc, W. Hayes, J. M. J. Fréchet, "Controlling Surface and Interfaces with Functional Polymers: Preparation and Functionalization of Dendritic-Linear Block Copolymers via Metal-Catalyzed 'Living' Free Radical Polymerization", *J. Polym. Sci., Part A: Polym. Chem.* **1998**, *36*, 1–10.

[99]   D. Mecerreyes, Ph. Dubois, R. Jérôme, J. L. Hedrick, C. J. Hawker, "Synthesis of Dendritic-Linear Block Copolymers by 'Living' Ring-Opening Polymerization of Lactones and Lactides Using Dendritic Initiators", *J. Polym. Sci., Part A: Polym. Chem.* **1999**, *37*, 1923–1930.

[100] C. J. Hawker, J. M. J. Fréchet, "The Synthesis and Polymerization of a Hyperbranched Polyether Macromonomer", *Polymer* **1992**, *33*, 1507–1511.

[101] I. Neubert, E. Amoulong-Kirstein, A.-D. Schlüter, H. Dautzenberg, "Polymerization of Styrenes Carrying Dendrons of the First, Second, and Third Generation", *Macromol. Rapid Commun.* **1996**, *17*, 517–527.

[102] I. Neubert, R. Klopsch, W. Claussen, A.-D. Schlüter, "Polymerization of Styrenes and Acrylates Carrying Dendrons of the First and Second Generation", *Acta Polym.* **1996**, *47*, 455–459.

[103] B. Karakaya, W. Claussen, A. Schaefer, A. Lehmann, A.-D. Schlüter, "Full Coverage of a Hydroxy-Substituted Poly(*p*-phenylene) with First- and Second-Generation Dendritic Wedges having Isocyanate Focal Points", *Acta Polym.* **1996**, *47*, 79–84.

[104] G. L'Abbé, B. Haelterman, W. Dehaen, "The Use of 1,2,3-Thiadiazoles in the Convergent Synthesis of Dendrimers", *Bull. Chim. Soc. Belg.* **1996**, *105*, 419–420.

[105] K. L. Wooley, C. J. Hawker, J. M. J. Fréchet, F. Wudl, G. Srdanov, S. Shi, C. Li, M. Kao, "Fullerene-Bound Dendrimers: Soluble, Isolated Carbon Clusters", *J. Am. Chem. Soc.* **1993**, *115*, 9836–9837.

[106] F. Diederich, L. Isaacs, D. Philp, "Syntheses, Structures, and Properties of Methanofullerenes", *Chem. Soc. Rev.* **1994**, *23*, 243–255.

[107] F. Diederich, M. Gómez-López, "Supramolecular Fullerene Chemistry", *Chem. Soc. Rev.* **1999**, 263–277.

[108] C. J. Hawker, K. L. Wooley, J. M. J. Fréchet, "Dendritic Fullerenes: A New Approach to Polymer Modification of C_{60}", *J. Chem. Soc., Chem. Commun.* **1994**, 925–926.

[109] X. Camps, H. Schönberger, A. Hirsch, "The C_{60} Core: A Versatile Tecton for Dendrimer Chemistry", *Chem. Eur. J.* **1997**, *3*, 561–567.

[110] I. Lamparth, A. Herzog, A. Hirsch, "Synthesis of [60]Fullerene Derivatives with Octahedral Addition Pattern", *Tetrahedron* **1996**, *52*, 5065–5075.

[111] A. Herzog, A. Hirsch, O. Vostrowsky, "Dendritic Mixed Hexakis Adducts of C_{60} with a T_h Symmetrical Addition Pattern", *Eur. J. Org. Chem.* **2000**, 171–180.

[112] A. Gügel, P. Belik, M. Walter, A. Kraus, E. Harth, M. Wagner, J. Spickermann, K. Müllen, "The Repetitive Diels–Alder Reaction: A New Approach to [60]Fullerene Main-Chain Polymers", *Tetrahedron* **1996**, *52*, 5007–5014.

[113] K. Fu, A. Kitaygorodskiy, Y.-P. Sun, "Fullerene-Centered Macromolecules as Unimolecular Micellar Structures", *Chem. Mater.* **2000**, *12*, 2073–2075.

[114] S. C. Zimmerman, F. Zeng, D. E. C. Reichert, S. V. Kolotuchin, "Self-Assembling Dendrimers", *Science* **1996**, *271*, 1095–1098.

[115] K. R. Stickley, T. D. Selby, S. C. Blackstock, "Isolable Polyradical Cations of Polyphenylenediamines with Populated High-Spin States", *J. Org. Chem.* **1997**, *62*, 448–449.

[116] P. Thiyagarajan, F. Zeng, C. Y. Ku, S. C. Zimmerman, "SANS Investigation of Self-Assembling Dendrimers in Organic Solvents", *J. Mater. Chem.* **1997**, *7*, 1221–1226.

[117] Y. Wang, F. Zeng, S. C. Zimmerman, "Dendrimers with Anthyridine-Based Hydrogen-Bonding Units at their Cores: Synthesis, Complexation and Self-Assembly Studies", *Tetrahedron Lett.* **1997**, *38*, 5459–5462.

[118] S. C. Zimmerman, Y. Wang, P. Bharathi, J. S. Moore, "Analysis of Amidinium Guest Complexation by Comparison of Two Classes of Dendrimer Hosts Containing a Hydrogen Bonding Unit at the Core", *J. Am. Chem. Soc.* **1998**, *120*, 2172–2173.

[119] S. V. Kolotuchin, S. C. Zimmerman, "Self-Assembly Mediated by the Donor–Donor–Acceptor·Acceptor–Acceptor–Donor (DDA·AAD) Hydrogen-Bonding Motif: Formation of a Robust Hexameric Aggregate", *J. Am. Chem. Soc.* **1998**, *120*, 9092–9093.

[119a] E. Mertz, S. Mattei, S. C. Zimmerman, "Synthetic Receptors for CG Base Pairs", *Org. Lett.* **2000**, *2*, 2931–2934.

[120] M. Suárez, J.-M. Lehn, S. C. Zimmerman, A. Skoulios, B. Heinrich, "Supramolecular Liquid Crystals. Self-Assembly of a Trimeric Supramolecular Disk and Its Self-Organization into a Columnar Discotic Mesophase", *J. Am. Chem. Soc.* **1998**, *120*, 9526–9532.

[121] M. Kawa, J. M. J. Fréchet, "Self-Assembed Lanthanide-Cored Dendrimer Complexes: Enhancement of the Luminescence Properties of Lanthanide Ions through Site-Isolation and Antenna Effects", *Chem. Mater.* **1998**, *10*, 286–296.

[121a] T. Seto, M. Kawa, K. Sugiyama, M. Nomura, "XAFS studies of Tb or Eu cored dendrimer complexes with various properties of luminescence", *J. Synch. Rad.* **2001**, *8*, 710–712.

[121b] M. Kawa, K. Motoda, "An antenna effect influenced by the focal structure of Tb^{3+}-cored dendrimer complexes", *Kobunshi Ronbunshu* **2000**, *57*, 855–858.

[122] M. J. Hannon, P. C. Mayers, P. C. Taylor, "Synthesis and Characterization of Water-Soluble Poly(aryl ether) Dendrimers for Encapsulaiton of Biomimetic Active Site Analogues", *J. Chem. Soc., Perkin Trans. 1* **2000**, 1881–1889.

[123] M. Dolman, P. J. Halling, B. D. Moore, "Functionalized Dendritic Polybenzylethers as Acid/Base Buffers for Biocatalysis in Nonpolar solvents", *Biotechnol. Bioeng.* **1997**, *55*, 278–282.

[124] N. Yamaguchi, L. M. Hamilton, H. W. Gibson, "Dendritic Pseudorotaxanes", *Angew. Chem.* **1998**, *110*, 3463–3466; *Int. Ed.* **1998**, *37*, 3275–3279.

[125] I. Gitsov, P. T. Ivanova, "Synthesis of New Hybrid Macromolecules with Cyclo-dendritic Architecture", *Chem. Commun.* **2000**, 269–270.

[126] J. P. Kampf, C. W. Frank, C. J. Hawker, "Physical Properties of Poly(benzylether) Dendrimers at the Air–Water Interface" *Polym. Prepr.* **1997**, *38*, 908–909.

[127] J. P. Kampf, C. W. Frank, E. Malmström, C. J. Hawker, "Stability and Molecular Conformation of Poly(benzyl ether) Monodendrons with Oligo(ethylene glycol) Tails at the Air–Water Interface", *Langmuir* **1999**, *15*, 227–233.

[128] G. Cui, Y. Xu, M. Liu, F. Fang, T. Ji, Y. Chen, Y. Li, "Highly Ordered Assemblies of Dendritic Molecules Bearing Multihydrophilic Head Groups", *Macromol. Rapid Commun.* **1999**, *20*, 71–76.

[129] J. M. J. Fréchet, L. Lochmann, V. Smigol, F. Svec, "Reversed-Phase High-Performance Liquid Chromatography of Functionalized Dendritic Macromolecules", *J. Chromatogr. A* **1994**, *667*, 284–289.

[130] G. M. Stewart, M. A. Fox, "Dendrimer-Linear Polymer Hybrids through ROMP", *Chem. Mater.* **1998**, *10*, 860–863.

[131] Z. Bo, L. Zhang, B. Zhao, X. Zhang, J. Shen, S. Höppener, L. Chi, H. Fuchs, "Self-Assembled Monolayers of Dendron-Thiol on Solid Substrate", *Chem. Lett.* **1998**, 1197–1198.

[131a] A. Friggeri, H. Schönherr, H.-J. van Manen, B.-H. Huisman, G. J. Vancso, J. Huskens, F. C. J. M. van Veggel, D. N. Reinhoudt, "Insertion of Individual Dendrimer Molecules into Self-Assembled Monolayers on Gold: A Mechanistic Study", *Langmuir* **2000**, *16*, 7757–7763.

[132] D. C. Tully, A. R. Trimble, J. M. J. Fréchet, K. Wilder, C. F. Quate, "Synthesis and Preparation of Ionically Bound Dendrimer Monolayers and Application toward Scanning Probe Lithography", *Chem. Mater.* **1999**, *11*, 2892–2898.

[133] D. C. Tully, A. R. Trimble, J. M. J. Fréchet, "Dendrimer-Based Chemically Amplified Resist Materials" *Polym. Prepr.* **2000**, *41*, 142–143.

[134] D. C. Tully, K. Wilder, J. M. J. Fréchet, A. R. Trimble, C. F. Quate, "Dendrimer-Based Self-Assembled Monolayers as Resists for Scanning Probe Lithography", *Adv. Mater.* **1999**, *11*, 314–318.

[135] D. C. Tully, A. R. Trimble, J. M. J. Fréchet, "Dendrimers with Thermally Labile End Groups: An Alternative Approach to Chemically Amplified Resist Materials Designed for Sub-100 nm Lithography", *Adv. Mater.* **2000**, *12*, 1118–1122.

[136] M. Trollsås, C. J. Hawker, J. L. Hedrick, G. Carrot, J. Hilborn, "A Mild and Versatile Synthesis for the Preparation of Thiol-Functionalized Polymers", *Macromolecules* **1998**, *31*, 5960–5963.

[137] W.-D. Jang, D.-L. Jiang, T. Aida, "Dendritic Physical Gel: Hierarchical Self-Organization of a Peptide-Core Dendrimer to Form a Micrometer-Scale Fibrous Assembly", *J. Am. Chem. Soc.* **2000**, *122*, 3232–3233.

[138] I. Gitsov, K. L. Wooley, J. M. J. Fréchet, "Novel Polyether Copolymers Consisting of Linear and Dendritic Blocks", *Angew. Chem.* **1992**, *104*, 1282–1285; *Int. Ed. Engl.* **1992**, *31*, 1200–1202.

[139] I. Gitsov, K. L. Wooley, C. J. Hawker, P. T. Ivanova, J. M. J. Fréchet, "Synthesis and Properties of Novel Linear-Dendritic Block Copolymers. Reactivity of Dendritic Macromolecules toward Linear Polymers", *Macromolecules* **1993**, *26*, 5621–5627.

[140] I. Gitsov, J. M. J. Fréchet, "Solution and Solid-State Properties of Hybrid Linear-Dendritic Block Copolymers", *Macromolecules* **1993**, *26*, 6536–6546.

[141] I. Gitsov, J. M. J. Fréchet, "Stimuli Responsive Dendritic Macromolecules. Novel Amphiphilic Star Copolymers with Dendritic Groups at the Periphery" *Polym. Mater. Sci. Eng.* **1995**, *73*, 129–130.

[142] K. W. Pollak, J. M. J. Fréchet, "Nonionic Unimolecular Micelles: Poly(ethylene oxide)-Coated Dendrimers" *Polym. Mater. Sci. Eng.* **1996**, *75*, 273–274.

[143] I. Gitsov, J. M. J. Fréchet, "Novel Nanoscopic Architectures. Linear–Globular ABA Copolymers with Polyether Dendrimers as A Blocks and Polystyrene as B Block", *Macromolecules* **1994**, *27*, 7309–7315.

[144] I. Schipor, J. M. J. Fréchet, "Use of Anionic Polymerization in the Preparation of Hybrid Linear-Dendritic Block Copolymers" *Polym. Prepr.* **1994**, *35*, 480–481.

[145] I. Gitsov, P. T. Ivanova, J. M. J. Fréchet, "Dendrimers as Macroinitiators for Anionic Ring-Opening Polymerization. Polymerization of ε-Caprolactone", *Macromol. Rapid Commun.* **1994**, *15*, 387–393.

[146] I. Gitsov, K. L. Wooley, C. J. Hawker, J. M. J. Fréchet, "Synthesis and Solution Properties of Polystyrenes with Dendritic End Groups" *Polym. Prepr.* **1991**, *32*, 631–632.

[147] D. Yu, N. Vladimirov, J. M. J. Fréchet, "MALDI-TOF in the Characterization of Dendritic-Linear Block Copolymers and Stars", *Macromolecules* **1999**, *32*, 5186–5192.

[148] J. M. J. Fréchet, I. Gitsov, "Nanoscopic Supermolecules with Linear-Dendritic Architecture: Their Preparation and Their Supramolecular Behavior", *Macromol. Symp.* **1995**, *98*, 441–465.

[149] J. M. J. Fréchet, I. Gitsov, T. Monteil, S. Rochat, J.-F. Sassi, C. Vergelati, D. Yu, "Modification of Surfaces and Interfaces by Non-Covalent Assembly of Hybrid Linear-Dendritic Block Copolymers: Poly(benzyl ether) Dendrons as Anchors for Poly(ethylene glycol) Chains on Cellulose or Polyester", *Chem. Mater.* **1999**, *11*, 1267–1274.

[150] M. Benaglia, R. Annunziata, M. Cinquini, F. Cozzi, S. Ressel, "Synthesis of New Poly(ethyleneglycol)s with a High Loading Capacity", *J. Org. Chem.* **1998**, *63*, 8628–8629.

[151] I. Gitsov, K. R. Lambrych, V. A. Remnant, R. Pracitto, "Micelles with Highly Branched Nanoporous Interior: Solution Properties and Binding Capabilities of Amphiphilic Copolymers with Linear Dendritic Architecture", *J. Polym. Sci., Part A: Polym. Chem.* **2000**, *38*, 2711–2727.

[152] R.-H. Jin, T. Aida, S. Inoue, "'Caged' Porphyrin: The First Dendritic Molecule having a Core Photochemical Functionality", *J. Chem. Soc., Chem. Commun.* **1993**, 1260–1262.

[153] R. Sadamoto, N. Tomioka, T. Aida, "Photoinduced Electron-Transfer Reactions *through* Dendrimer Architecture", *J. Am. Chem. Soc.* **1996**, *118*, 3978–3979.

[154] H. Tang, D. E. Nikles, "Preparation of Porphyrin Dendrimers with Ester Linkages" *Polym. Prepr.* **1999**, *40*, 847–848.

[155] M. Brewis, G. J. Clarkson, A. M. Holder, N. B. McKeown, "Phthalocyanines Substituted with Dendritic Wedges: Glass-Forming Columnar Mesogens", *Chem. Commun.* **1998**, 969–970.

[155a] M. Brewis, G. J. Clarkson, M. Helliwell, A. M. Holder, N. B. McKeown, "The Synthesis and Glass-Forming Properties of Phthalocyanine-Containing Poly(aryl ether) Dendrimers", *Chem. Eur. J.* **2000**, *6*, 4630–4636.

[156] M. Brewis, G. J. Clarkson, V. Goddard, M. Helliwell, A. M. Holder, N. B. McKeown, "Silicon Phthalocyanines with Axial Dendritic Substituents", *Angew. Chem.* **1998**, *110*, 1185–1187; *Int. Ed.* **1998**, *37*, 1092–1094.

[156a] M. Brewis, M. Helliwell, N. B. McKeown, S. Reynolds, A. Shawcross, "Phthalocyanine-centred aryl ether dendrimers with oligo(ethyleneoxy) surface groups", *Tetrahedron Lett.* **2001**, *42*, 813–816.

[157] G. Ferguson, J. F. Gallagher, M. A. McKervey, E. Madigan, "Calixarene-Bound Dendritic Macromolecules", *J. Chem. Soc., Perkin Trans. 1* **1996**, 599–602.

[158] R. Kleppinger, H. Reynaers, K. Desmedt, B. Forier, W. Dehaen, M. Koch, P. Verhaert, "A Small-Angle X-ray Scattering Study of Sizes and Shapes of Poly(benzyl ether) Dendrimer Molecules", *Macromol. Rapid Commun.* **1998**, *19*, 111–114.

[159] G. L'Abbé, B. Haelterman, W. Dehaen, "Potassium Benzene-1,3,5-Triyltris(ethynethiolate): A New Core Reagent for Dendrimer Synthesis", *J. Chem. Soc., Perkin Trans. 1* **1994**, 2203–2204.

[160] G. L'Abbé, W. Dehaen, B. Haelterman, D. Vangeneugden, "1,3,5-Tris(1,2,3-thiadiazol-4-yl)benzene. A New Core Synthon for Dendrimers", *Acros Organics Acta* **1995**, *1*, 61–62.

[161] J.-F. Nierengarten, L. Oswald, J.-F. Eckert, J.-F. Nicoud, N. Armaroli, "Complexation of Fullerenes with Dendritic Cyclotriveratrylene Derivatives", *Tetrahedron Lett.* **1999**, *40*, 5681–5684.

[161a] J.-F. Eckert, D. Byrne, J.-F. Nicoud, L. Oswald, J.-F. Nierengarten, M. Numata, A. Ikeda, S. Shinkai, N. Armaroli, "Polybenzyl ether dendrimers for the complexation of [60]fullerenes", *New J. Chem.* **2000**, *24*, 749–758.

[162] M. Numata, A. Ikeda, C. Fukuhara, S. Shinkai, "Dendrimers can act as a Host for [60]Fullerene", *Tetrahedron Lett.* **1999**, *40*, 6945–6948.

[163] J.-F. Nierengarten, D. Felder, J.-F. Nicoud, "Methanofullerene-Functionalized Dendritic Branches", *Tetrahedron Lett.* **2000**, *41*, 41–44.

[163a] D. Felder, H. Nierengarten, J.-P. Gisselbrecht, C. Boudon, E. Leize, J.-F. Nicoud, M. Gross, A. V. Dorsselaer, J.-F. Nierengarten, "Fullerodendrons. Synthesis, electrochemistry and reduction in the electrospray source for mass spectrometry analysis", *New J. Chem.* **2000**, *24*, 687–695.

[163b] Y. Rio, J.-F. Nicoud, J.-L. Rehspringer, J.-F. Nierengarten, "Fullerodendrimers with peripheral triethyleneglycol chains", *Tetrahedron Lett.* **2000**, *41*, 10207–10210.

[163c] T. Nishioka, K. Tashiro, T. Aida, J.-Y. Zheng, K. Kinbara, K. Saigo, S. Sakamoto, K. Yamaguchi, "Molecular Design of a Novel Dendrimer Porphyrin for Supramolecular Fullerene/Dendrimer Hybridization", *Macromolecules* **2000**, *33*, 9182–9184.

[164] S. De Backer, Y. Prinzie, W. Verheijen, M. Smet, K. Desmedt, W. Dehaen, F. C. De Schryver, "Solvent Dependence of the Hydrodynamic Volume of Dendrimers with a Rubicene Core", *J. Phys. Chem. A.* **1998**, *102*, 5451–5455.

[165] J. Hofkens, W. Verheijen, R. Shukla, W. Dehaen, F. C. De Schryver, "Detection of a Single Dendrimer Macromolecule with a Fluorescent Dihydropyrrolopyrroledione (DPP) Core Embedded in a Thin Polystyrene Polymer Film", *Macromolecules* **1998**, *31*, 4493–4497.

[166] D. M. Junge, D. V. McGrath, "Photoresponsive Dendrimers", *Chem. Commun.* **1997**, 857–858.

[166a] S. Ghosh, A. K. Banthia, "Synthesis of photoresponsive polyamidoamine (PAMAM) dendritic architecture", *Tetrahedron Lett.* **2001**, *42*, 501–503.

[167] D. M. Junge, D. V. McGrath, "Photoresponsive Azobenzene-Containing Dendrimers with Multiple Discrete States", *J. Am. Chem. Soc.* **1999**, *121*, 4912–4913.

[168] A. Sidorenko, C. Houphouet-Boigny, A. C. Greco, O. Villavicencio, M. Hashemzadeh, D. V. McGrath, V. V. Tsukruk, "Langmuir Monolayers from Azobenzene-Containing Dendrons", *Polym. Prepr.* **2000**, *41*, 1487–1488.

[169] S. Li, S. Sikder, D. V. McGrath, "Synthesis of Amphiphilic Photoresponsive Dendrons" *Polym. Prepr.* **1999**, *40*, 267–268.

[170] S. Li, D. V. McGrath, "Effect of Macromolecular Isomerism on the Photomodulation of Dendrimer Properties", *J. Am. Chem. Soc.* **2000**, *122*, 6795–6796.

[171] D.-L. Jiang, T. Aida, "Photoisomerization in Dendrimers by Harvesting of Low-Energy Photons", *Nature* **1997**, *388*, 454–456.

[172] T. Aida, D.-L. Jiang, E. Yashima, Y. Okamoto, "A New Approach to Light-Harvesting with Dendritic Antenna", *Thin Solid Films* **1998**, *331*, 254–258.

[173] Y. Wakabayashi, M. Tokeshi, D.-L. Jiang, T. Aida, T. Kitamori, "Long-Term Energy Storage of Dendrimers", *J. Lumin.* **1999**, *83–84*, 313–315.

[174] M. Smet, L.-X. Liao, W. Dehaen, D. V. McGrath, "Photolabile Dendrimers Using *o*-Nitrobenzyl Ether Linkages", *Org. Lett.* **2000**, *2*, 511–513.

[175] A. P. H. J. Schenning, R. E. Martin, M. Ito, F. Diederich, C. Boudon, J.-P. Gisselbrecht, M. Gross, "Dendritic Rods with a Poly(triacetylene) Backbone: Insulated Molecular Wires", *Chem. Commun.* **1998**, 1013–1014.

[175a] A. P. H. J. Schenning, J.-D. Arndt, M. Ito, A. Stoddart, M. Schreiber, P. Siemsen, R. E. Martin, C. Boudon, J.-P. Gisselbrecht, M. Gross, V. Gramlich, F. Diederich, "Insulated Molecular Wires: Dendritic Encapsulation of Poly(triacetylene) Oligomers, Attempted Dendritic Stabilization of Novel Poly(pentaacetylene) Oligomers, and an Organometallic Approach to Dendritic Rods", *Helv. Chim. Acta* **2001**, *84*, 296–334.

[176] T. Sato, D.-L. Jiang, T. Aida, "A Blue-Luminescent Dendritic Rod: Poly(phenylene ethynylene) within a Light-Harvesting Dendritic Envelope", *J. Am. Chem. Soc.* **1999**, *121*, 10658–10659.

[176a] D. L. Jiang, T. Sato, T. Aida, "Design and photofunctions of dendrimer-encapsulated poly(phenyleneethynylene)s", *Chin. J. Polym. Sci.* **2001**, *19*, 161–166.

[177] D. B. Amabilino, P. R. Ashton, V. Balzani, C. L. Brown, A. Credi, J. M. J. Fréchet, J. W. Leon, F. M. Raymo, N. Spencer, J. F. Stoddart, M. Venturi, "Self-Assembly of [*n*]Rotaxanes Bearing Dendritic Stoppers", *J. Am. Chem. Soc.* **1996**, *118*, 12012–12020.

[178] G. M. Hübner, G. Nachtsheim, Q. Y. Li, C. Seel, F. Vögtle, "The Spacial Demand of Dendrimers: Deslipping of Rotaxanes", *Angew. Chem.* **2000**, *112*, 1315–1318; *Int. Ed.* **2000**, *39*, 1269–1272.

[178a] F. Osswald, E. Vogel, O. Safarowsky, F. Schwanke, F. Vögtle, "Rotaxane Assemblies with Dendritic Architecture", *Adv. Synth. Catal.* **2001**, *343*, 303–309.

[179] H. Oike, F. Kobayashi, Y. Tezuka, S. Hashimoto, T. Shiomi, "Covalent Conversion of Cyclic Onium Salt End Groups of Poly(tetrahydrofuran) by Bulky Counterions in the Absence and Presence of Macrocyclic Compounds", *Macromolecules* **1999**, *32*, 2876–2882.

[180] L. L. Miller, B. Zinger, J. S. Schlechte, "Synthesis and Electrochemistry of Nanometer-Scaled Molecular Dumbbells", *Chem. Mater.* **1999**, *11*, 2313–2315.

[181] I. Jestin, E. Levillain, J. Roncali, "Electroactive Dendritic π-Conjugated Oligothienylenevinylenes", *Chem. Commun.* **1998**, 2655–2656.

[182] P. R. L. Malenfant, L. Groenendaal, J. M. J. Fréchet, "Well-Defined Triblock Hybrid Dendrimers Based on Lengthy Oligothiophene Cores and Poly(benzyl ether) Dendrons", *J. Am. Chem. Soc.* **1998**, *120*, 10990–10991.

[182a] J. J. Apperloo, R. A. J. Janssen, P. R. L. Malenfant, J. M. J. Fréchet, "Concentration-Dependent Thermochromism and Supramolecular Aggregation in Solution of Triblock Copolymers Based on Lengthy Oligothiophene Cores and Poly(benzyl ether) Dendrons", *Macromolecules* **2000**, *33*, 7038–7043.

[182b] A. W. Freeman, S. C. Koene, P. R. L. Malenfant, M. E. Thompson, J. M. J. Fréchet, "Dendrimer-Containing Light-Emitting Diodes: Toward Site-Isolation of Chromophores", *J. Am. Chem. Soc.* **2000**, *122*, 12385–12386.

[183] P. R. L. Malenfant, J. M. J. Fréchet, "Dendrimers as Solubilizing Groups for Conducting Polymers: Preparation and Characterization of Polythiophene Functionalized Exclusively with Aliphatic Ether Convergent Dendrons", *Macromolecules* **2000**, *33*, 3634–3640.

[184] A.-D. Schlüter, "Dendritic Structures with Polyfunctional Cores" *Polym. Prepr.* **1995**, *36*, 745–746.

[185] W. Claussen, N. Schulte, A.-D. Schlüter, "A Poly(*p*-phenylene) Decorated with Fréchet-type Dendritic Fragments of the First Generation", *Macromol. Rapid Commun.* **1995**, *16*, 89–94.

[186] R. Freudenberger, W. Claussen, A.-D. Schlüter, H. Wallmeier, "Functionalized Rod-Like Polymers: One-Dimensional Rigid Matrices", *Polymer* **1994**, *35*, 4496–4501.

[187] A.-D. Schlüter, "Dendrimers with Polymeric Cores: Towards Nanocylinders", *Top. Curr. Chem.*, **1998**, *197*, 165–191.

[188] A. D. Schlüter, J. P. Rabe, "Dendronized Polymers: Synthesis, Characterization, Assembly at Interfaces, and Manipulation", *Angew. Chem.* **2000**, *112*, 860–880; *Int. Ed.* **2000**, *39*, 864–883.

[189] S. Jahromi, B. Coussens, N. Meijerink, A. W. M. Braam, "Side Chain Dendritic Polymers: Synthesis and Physical Properties", *J. Am. Chem. Soc.* **1998**, *120*, 9753–9762.

[189a] S. Jahromi, V. Litvinov, B. Coussens, "Polyurethane Networks Bearing Dendritic Wedges: Synthesis and Some Properties", *Macromolecules* **2001**, *34*, 1013–1017.

[190] O. Karthaus, K. Ijiro, M. Shimomura, J. Hellmann, M. Irie, "Monomolecular Layers of Diarylethene-Containing Dendrimers", *Langmuir* **1996**, *12*, 6714–6716.

[191] C. Heil, G. R. Windscheif, S. Braschohs, F. Flörke, J. Gläser, M. Lopez, J. Müller-Albrecht, U. Schramm, J. Bargon, F. Vögtle, "Highly Selective Sensor Materials for Discriminating Carbonyl Compounds in the Gas Phase Using Quartz Microbalances", *Sens. Actuators, B* **1999**, *61*, 51–58.

[192] D. Marsitzky, R. Vestberg, C. J. Hawker, K. R. Carter, "Self-Encapsulation of Poly-2,7-fluorenes in a Dendrimer Matrix", *Polym. Prepr.* **2000**, *41*, 1344–1345.

[193] Z. Bo, A. Schäfer, P. Franke, A. D. Schlüter, "A Facile Synthetic Route to a Third-Generation Dendrimer with Generation-Specific Functional Aryl Bromides", *Org. Lett.* **2000**, *2*, 1645–1648.

[194] C. J. Hawker, K. L. Wooley, J. M. J. Fréchet, "Novel Dendritic Macromolecules by the Convergent Growth Approach: Hyperbranched Block Copolymers and Monodisperse Polyesters" *Polym. Prepr.* **1991**, *32*, 623–624.

[195] K. L. Wooley, C. J. Hawker, J. M. J. Fréchet, "Unsymmetrical Three-Dimensional Macromolecules: Preparation and Characterization of Strongly Dipolar Dendritic Macromolecules", *J. Am. Chem. Soc.* **1993**, *115*, 11496–11505.

[196] C. J. Hawker, K. L. Wooley, J. M. J. Fréchet, "Unimolecular Micelles and Globular Amphiphiles: Dendritic Macromolecules as Novel Recyclable Solubilization Agents", *J. Chem. Soc., Perkin Trans. 1* **1993**, 1287–1297.

[197] C. J. Hawker, K. L. Wooley, J. M. J. Fréchet, "Synthesis and Properties of Covalent Micelle-Like Structures Based on Dendritic Polyethers" *Polym. Prepr.* **1993**, *34*, 54–55.

[198] E. M. Sanford, J. M. J. Fréchet, K. L. Wooley, C. J. Hawker, "Amphiphilic Dendritic Block Copolymers, and Approaches to Their Accelerated Synthesis" *Polym. Prepr.* **1993**, *34*, 654–655.

[199] C. J. Hawker, J. M. J. Fréchet, "Control of Surface Functionality in the Synthesis of Dendritic Macromolecules Using the Convergent-Growth Approach", *Macromolecules* **1990**, *23*, 4726–4729.

[200] K. L. Wooley, C. J. Hawker, J. M. J. Fréchet, "Polymers with Controlled Molecular Architecture: Control of Surface Functionality in the Synthesis of Dendritic Hyperbranched Macromolecules Using the Convergent Approach", *J. Chem. Soc., Perkin Trans. 1* **1991**, 1059–1076.

[201] L. Lochmann, K. L. Wooley, P. T. Ivanova, J. M. J. Fréchet, "Multisite Functionalized Dendritic Macromolecules Prepared via Metalation by Superbases and Reaction with Electrophiles", *J. Am. Chem. Soc.* **1993**, *115*, 7043–7044.

[202] G. M. Stewart, M. A. Fox, "Chromophore-Labeled Dendrons as Light-Harvesting Antennae", *J. Am. Chem. Soc.* **1996**, *118*, 4354–4360.

[203] J. W. Leon, M. Kawa, J. M. J. Fréchet, "Isophthalate Ester-Terminated Dendrimers: Versatile Nanoscopic Building Blocks with Readily Modifiable Surface Functionalities", *J. Am. Chem. Soc.* **1996**, *118*, 8847–8859.

[204] R. Klopsch, P. Franke, A.-D. Schlüter, "Repetitive Strategy for Exponential Growth of Hydroxy-Functionalized Dendrons", *Chem. Eur. J.* **1996**, *2*, 1330–1334.

[205] A.-D. Schlüter, W. Claussen, R. Freudenberger, "Cylindrically-Shaped Dendritic Structures", *Macromol. Symp.* **1995**, *98*, 475–482.

[206] Z. Bo, C. Zhang, N. Severin, J. P. Rabe, A. D. Schlüter, "Synthesis of Amphiphilic Poly(*p*-phenylene)s with Pendant Dendrons and Linear Chains", *Macromolecules* **2000**, *33*, 2688–2694.

[207] Z. Bo, A. D. Schlüter, "Progress Toward the Polymerization of a Fourth Generation Dendritic Macromonomer", *Macromol. Rapid Commun.* **1999**, *20*, 21–25.

[208] Z. Bo, A. D. Schlüter, "Entering a New Level of Use for Suzuki Cross-Coupling: Poly(*para*-phenylene)s with Fourth-Generation Dendrons", *Chem. Eur. J.* **2000**, *6*, 3235–3241.

[209] L. Shu, A. Schäfer, A. D. Schlüter, "Dendronized Polymers: Increasing of Dendron Generation by the Attach-to Approach", *Macromolecules* **2000**, *33*, 4321–4328.

[210] W. Stocker, B. L. Schürmann, J. P. Rabe, S. Förster, P. Lindner, I. Neubert, A.-D. Schlüter, "A Dendritic Nanocylinder: Shape Control Through Implementation of Steric Strain", *Adv. Mater.* **1998**, *10*, 793–797.

[211] Z. Bo, J. P. Rabe, A. D. Schlüter, "A Poly(*para*-phenylene) with Hydrophobic and Hydrophilic Dendrons: Prototype of an Amphiphilic Cylinder with the Potential to Segregate Lengthwise", *Angew. Chem.* **1999**, *111*, 2540–2542; *Int. Ed.* **1999**, *38*, 2370–2372.

[212] I. Neubert, A. D. Schlüter, "Dendronized Polystyrenes with Hydroxy and Amino Groups in the Periphery", *Macromolecules* **1998**, *31*, 9372–9378.

[213] W. Stocker, B. Karakaya, B. L. Schürmann, J. P. Rabe, A. D. Schlüter, "Ordered Dendritic Nanorods with a Poly(*p*-phenylene) Backbone", *J. Am. Chem. Soc.* **1998**, *120*, 7691–7695.

[214] B. Karakaya, W. Claussen, K. Gessler, W. Saenger, A.-D. Schlüter, "Toward Dendrimers with Cylindrical Shape in Solution", *J. Am. Chem. Soc.* **1997**, *119*, 3296–3301.

[215] B. Karakaya, W. Claussen, A. D. Schlüter, "Toward Macrocylinders: Complete Coverage of Rigid-Rod Polymers with Dendritic Fragments" *Polym. Prepr.* **1996**, *37*, 216–217.

[216] G. L'Abbé, B. Forier, W. Dehaen, "A Fast Double-Stage Convergent Synthesis of Dendritic Polyethers", *Chem. Commun.* **1996**, 2143–2144.

[217] Y. Chen, D. Xia, W. Liu, C. Chen, F. Xi, "Syntheses of Dendritic Polymers with Branch Ends Modified by Allyl Groups", *Gaofenzi Xuebao* **1996**, 456–461.

[218] S. Höger, "Methoxycarbonyl-Terminated Dendrons via the Mitsunobu Reaction: An Easy Way to Functionalized Hyperbranched Building Blocks", *Synthesis* **1997**, 20–22.

[219] O. Mitsunobu, "The Use of Diethyl Azodicarboxylate and Triphenylphosphine in Synthesis and Transformation of Natural Products", *Synthesis* **1981**, 1–28.

[220] M. E. Piotti, F. Rivera, Jr., R. Bond, C. J. Hawker, J. M. J. Fréchet, "Synthesis and Catalytic Activity of Unimolecular Dendritic Reverse Micelles with 'Internal' Functional Groups", *J. Am. Chem. Soc.* **1999**, *121*, 9471–9472.

[221] M. E. Piotti, C. Hawker, J. M. J. Fréchet, F. Rivera, J. Dao, R. Bond, "Synthesis and Catalytic Activity of Unimolecular Dendritic Reverse Micelles" *Polym. Prepr.* **1999**, *40*, 410–411.

[221a] A. W. Freeman, L. A. J. Chrisstoffels, J. M. J. Fréchet, "A Simple Method for Controlling Dendritic Architecture and Diversity: A Parallel Monomer Combination Approach", *J. Org. Chem.* **2000**, *65*, 7612–7617.

[222] M. S. Wendland, S. C. Zimmerman, "Synthesis of Cored Dendrimers", *J. Am. Chem. Soc.* **1999**, *121*, 1389–1390.

[223] S. L. Gilat, A. Adronov, J. M. J. Fréchet, "Light Harvesting and Energy Transfer in Novel Convergently Constructed Dendrimers", *Angew. Chem.* **1999**, *111*, 1519–1524; *Int. Ed.* **1999**, *38*, 1422–1427.

[224] J. M. J. Fréchet, A. Adronov, D. R. Robello, "Dendritic and Polymeric Light-Harvesting Systems" *Polym. Prepr.* **2000**, *41*, 851–852.

[225] L. A. J. Chrisstoffels, A. Adronov, J. M. J. Fréchet, "Light Harvesting and Energy Transfer Within Chromophore-Labeled Monolayers" *Polym. Prepr.* **2000**, *41*, 793–794.

[226] A. Adronov, S. L. Gilat, J. M. J. Fréchet, K. Ohta, F. V. R. Neuwahl, G. R. Fleming, "Light Harvesting and Energy Transfer in Laser-Dye-Labeled Poly(aryl ether) Dendrimers", *J. Am. Chem. Soc.* **2000**, *122*, 1175–1185.

[226a] J. M. Riley, S. Alkan, A. Chen, M. Shapiro, W. A. Khan, W. R. Murphy, Jr., J. E. Hanson, "Pyrene-Labeled Poly(aryl ether) Monodendrons: Synthesis, Characterization, Diffusion Coefficients, and Photophysical Studies", *Macromolecules* **2001**, *34*, 1797–1809.

[227] A. Adronov, P. R. L. Malenfant, J. M. J. Fréchet, "Synthesis and Steady-State Photophysical Properties of Dye-Labeled Dendrimers Having Novel Oligothiophene Core: A Comparative Study", *Chem. Mater.* **2000**, *12*, 1463–1472.

[228] L. A. J. Chrisstoffels, A. Adronov, J. M. J. Fréchet, "Surface-Confined Light Harvesting, Energy Transfer, and Amplification of Fluorescence Emission in Chromophore-Labeled Self-Assembled Monolayers", *Angew. Chem.* **2000**, *112*, 2247–2251; *Int. Ed.* **2000**, *39*, 2163–2167.

[228a] X. Zhou, D. S. Tyson, F. N. Castellano, "First Generation Light-Harvesting Dendrimers with a [Ru(bpy)$_3$]$^{2+}$ Core and Aryl Ether Ligands Functionalized with Coumarin 450", *Angew. Chem. Int. Ed.* **2000**, *39*, 4301–4305.

[229] T. Marquardt, U. Lüning, "Towards Golf-Ball Shaped Reagents: Dendrimer-Fixed Concave Pyridines", *Chem. Commun.* **1997**, 1681–1682.

[230] U. Lüning, T. Marquardt, "Dendrimer-Fixed Concave Pyridine", *J. Prakt. Chem.* **1999**, *341*, 222–227.

[231] U. Lüning, M. Hagen, F. Löffler, T. Marquardt, B. Meynhardt, "Molecular Lamps and Molecular Golf Balls", *J. Inclusion Phenom. Macrocycl. Chem.* **1999**, *35*, 381–387.

[232] B. Hong, T. P. S. Thoms, H. J. Murfee, M. J. Lebrun, "Highly Branched Dendritic Macromolecules with Core Polyhedral Silsesquioxane Functionalities", *Inorg. Chem.* **1997**, *36*, 6146–6147.

[233] L. Groenendaal, J. M. J. Fréchet, "Surface Functionalization of Polyether Dendrimers Using Palladium-Catalyzed Cross-Coupling Reactions", *J. Org. Chem.* **1998**, *63*, 5675–5679.

[234] M. Gingras, A. Pinchart, C. Dallaire, "Synthesis of *p*-Phenylene Sulfide Molecular Asterisks", *Angew. Chem.* **1998**, *110*, 3338–3341; *Int. Ed.* **1998**, *37*, 3149–3151.

[235] D. D. MacNicol, "Structure and Design of Inclusion Compounds: The Hexa-Hosts and Symmetry Considerations" in *Inclusion Compounds* (Eds.: J. L. Atwood, J. E. Davies, D. D. MacNicol), Academic Press, London, **1984**, pp. 123–168.

[236] G. F. Kirton, A. S. Brown, C. J. Hawker, P. A. Reynolds, J. W. White, "Surface Activity of Modified Dendrimers at High Conversion", *Physica B (Amsterdam)* **1998**, *248*, 184–190.

[237] M. Liu, K. Kono, J. M. J. Fréchet, "Water-Soluble Dendrimer-Poly(ethylene glycol) Star-like Conjugates as Potential Drug Carriers", *J. Polym. Sci., Part A: Polym. Chem.* **1999**, *37*, 3492–3503.

[238] T. Ren, D. Liu, "Synthesis of Targetable Cationic Amphiphiles", *Tetrahedron Lett.* **1999**, *40*, 7621–7625.

[239] V. Percec, W.-D. Cho, G. Ungar, D. J. P. Yeardley, "From Molecular Flat Tapers, Discs, and Cones to Supramolecular Cylinders and Spheres using Fréchet-Type Monodendrons Modified on their Periphery", *Angew. Chem.* **2000**, *112*, 1662–1666; *Int. Ed.* **2000**, *39*, 1598–1602.

[240] D. J. P. Yeardley, G. Ungar, V. Percec, M. N. Holerca, G. Johansson, "Spherical Supramolecular Minidendrimers Self-Organized in an 'Inverse Micellar'-like Thermotropic Body-Centered Cubic Liquid-Crystalline Phase", *J. Am. Chem. Soc.* **2000**, *122*, 1648–1689.

[241] K. Kono, M. Liu, J. M. J. Fréchet, "Design of Dendritic Macromolecules Containing Folate or Methotrexate Residues", *Bioconj. Chem.* **1999**, *10*, 1115–1121.

[242] T. H. Mourey, S. R. Turner, M. Rubinstein, J. M. J. Fréchet, C. J. Hawker, K. L. Wooley, "Unique Behavior of Dendritic Macromolecules: Intrinsic Viscosity of Polyether Dendrimers", *Macromolecules* **1992**, *25*, 2401–2406.

[243] R. L. Lescanec, M. Muthukumar, "Configurational Characteristics and Scaling Behavior of Starburst Molecules: A Computational Study", *Macromolecules* **1990**, *23*, 2280–2288.

[244] P. G. de Gennes, H. Hervet, "Statistics of <<Starburst>> Polymers", *J. Phys. Lett.* **1983**, *44*, L351–L360.

[245] P. M. Saville, J. W. White, C. J. Hawker, K. L. Wooley, J. M. J. Fréchet, "Dendrimer and Polystyrene Surfactant Structure at the Air–Water Interface", *J. Phys. Chem.* **1993**, *97*, 293–294.

[246] P. M. Saville, P. A. Reynolds, J. W. White, C. J. Hawker, J. M. J. Fréchet, K. L. Wooley, J. Penfold, J. R. P. Webster, "Neutron Reflectivity and Structure of Polyether Dendrimers as Langmuir Films", *J. Phys. Chem.* **1995**, *99*, 8283–8289.

[247] C. J. Hawker, P. J. Farrington, M. E. Mackay, K. L. Wooley, J. M. J. Fréchet, "Molecular Ball Bearings: The Unusual Melt Viscosity Behavior of Dendritic Macromolecules", *J. Am. Chem. Soc.* **1995**, *117*, 4409–4410.

[248] P. J. Farrington, C. J. Hawker, J. M. J. Fréchet, M. E. Mackay, "The Melt Viscosity of Dendritic Poly(benzyl ether) Macromolecules", *Macromolecules* **1998**, *31*, 5043–5050.

[249] K. L. Wooley, C. A. Klug, K. Tasaki, J. Schaefer, "Shapes of Dendrimers from Rotational-Echo Double-Resonance NMR", *J. Am. Chem. Soc.* **1997**, *119*, 53–58.

[250] K. Wooley, C. Klug, T. Kowalewski, J. Schaefer, "Conformational Studies of Dendritic Macromolecules by Rotational-Echo Double Resonance (REDOR) Solid-State NMR" *Polym. Mater. Sci. Eng.* **1995**, *73*, 230–231.

[251] H.-M. Kao, A. D. Stefanescu, K. L. Wooley, J. Schaefer, "Location of Terminal Groups of Dendrimers in the Solid State by Rotational-Echo Double-Resonance NMR", *Macromolecules* **2000**, *33*, 6214–6216.

[252] M. L. Mansfield, L. I. Klushin, "Monte Carlo Studies of Dendrimer Macromolecules", *Macromolecules* **1993**, *26*, 4262–4268.

[253] M. L. Mansfield, "Dendron Segregation in Model Dendrimers", *Polymer* **1994**, *35*, 1827–1830.

[254] C. J. Hawker, E. E. Malmström, C. W. Frank, J. P. Kampf, "Exact Linear Analogs of Dendritic Polyether Macromolecules: Design, Synthesis, and Unique Properties", *J. Am. Chem. Soc.* **1997**, *119*, 9903–9904.

[255] S. Jahromi, J. H. M. Palmen, P. A. M. Steeman, "Rheology of Side Chain Dendritic Polymers", *Macromolecules* **2000**, *33*, 577–581.

[256] A. Desai, N. Atkinson, F. Rivera, Jr., W. Devonport, I. Rees, S. E. Branz, C. J. Hawker, "Hybrid Dendritic-Linear Graft Copolymers: Steric Considerations in 'Coupling-to' Approach", *J. Polym. Sci., Part A: Polym. Chem.* **2000**, *38*, 1033–1044.

[257] J. W. Leon, J. M. J. Fréchet, "Analysis of Aromatic Polyether Dendrimers and Dendrimer-Linear Block Copolymers by Matrix-Assisted Laser Desorption Ionization Mass Spectrometry", *Polym. Bull.* **1995**, *35*, 449–455.

[258] P. M. Bayliff, W. J. Feast, D. Parker, "The Synthesis and Properties of a Series of Aromatic Dendritic Polyamides", *Polym. Bull.* **1992**, *29*, 265–270.

[259] S. C. E. Backson, P. M. Bayliff, W. J. Feast, A. M. Kenwright, D. Parker, R. W. Richards, "Synthesis and Properties of Aramid Dendrimers" *Polym. Prepr.* **1993**, *34*, 50–51.

[260] S. C. E. Backson, P. M. Bayliff, W. J. Feast, A. M. Kenwright, D. Parker, R. W. Richards, "Synthesis and Properties of Aramid Dendrimers", *Makromol. Chem., Macromol. Symp.* **1994**, *77*, 1–10.

[261] Y. Ishida, M. Jikei, M. Kakimoto, "Rapid Synthesis of Aromatic Polyamide Dendrimers by an Orthogonal and a Double-Stage Convergent Approach", *Macromolecules* **2000**, *33*, 3202–3211.

[262] B. I. Voit, D. Wolf, "Perfectly Branched Polyamide Dendrons Based on 5-(2-Amino-ethoxy)isophthalic Acid", *Tetrahedron* **1997**, *53*, 15535–15551.

[262a] S. Wong, D. Applehans, B. Voit, U. Scheler, "Effect of Branching on the Scaling Behavior of Poly(ether amide) Dendrons and Dendrimers", *Macromolecules* **2001**, *34*, 678–680.

[262b] D. Appelhans, H. Komber, D. Voigt, L. Häussler, B. I. Voit, "Synthesis and Characterization of Poly(ether amide) Dendrimers Containing Different Core Molecules", *Macromolecules* **2000**, *33*, 9503.

[263] A. Huwe, D. Appelhans, J. Prigann, B. I. Voit, F. Kremer, "Broadband Dielectric Spectroscopy on the Molecular Dynamics in Dendritic Model Systems", *Macromolecules* **2000**, *33*, 3762–3766.

[264] A. Ritzén, T. Frejd, "Synthesis of a Chiral Dendrimer Based on Polyfunctional Amino Acids", *Chem. Commun.* **1999**, 207–208.

[265] S. J. E. Mulders, A. J. Brouwer, P. G. J. van der Meer, R. M. J. Liskamp, "Synthesis of a Novel Amino Acid Based Dendrimer", *Tetrahedron Lett.* **1997**, *38*, 631–634.

[266] B. Castro, J. R. Dormoy, G. Evin, C. Selve, "Reactifs de Couplage Peptidique IV: L'Hexafluorophosphate de Benzotriazolyl *N*-ooytridimethylamino Phosphonium (B.O.P.)", *Tetrahedron Lett.* **1975**, *14*, 1219–1222.

[267] S. J. E. Mulders, A. J. Brouwer, R. M. J. Liskamp, "Molecular Diversity of Novel Amino Acid Based Dendrimers", *Tetrahedron Lett.* **1997**, *38*, 3085–3088.

[267a] T. K. K. Mong, A. Niu, H.-F. Chow, C. Wu, L. Li, R. Chen, "β-Alanine-Based Dendritic β-Peptides: Dendrimers Possessing Unusually Strong Binding Ability Towards Protic Solvents and Their Self-Assembly into Nanoscale Aggregates through Hydrogen-Bond Interactions", *Chem. Eur. J.* **2001**, *7*, 686–699.

[268] A. Zistler, S. Koch, A. D. Schlüter, "Dendronized Polyacrylates with Glucose Units in the Periphery", *J. Chem. Soc., Perkin Trans. 1* **1999**, 501–508.

[269] A. Ingerl, I. Neubert, R. Klopsch, A. D. Schlüter, "Hydroxy-Functionalized Dendritic Building Blocks", *Eur. J. Org. Chem.* **1998**, 2551–2556.

[270] R. Klopsch, S. Koch, A. D. Schlüter, "Amino-Functionalized, Second-Generation Dendritic Building Blocks", *Eur. J. Org. Chem.* **1998**, 1275–1283.

[271] P. R. Ashton, K. Shibata, A. N. Shipway, J. F. Stoddart, "Polycationic Dendrimers", *Angew. Chem.* **1997**, *109*, 2902–2905; *Int. Ed. Engl.* **1997**, *36*, 2781–2783.

[272] G. R. Newkome, X. Lin, J. K. Young, "Syntheses of Amine Building Blocks for Dendritic Macromolecule Construction", *Synlett* **1992**, 53–54.

[273] G. R. Newkome, X. Lin, "Symmetrical, Four-Directional, Poly(ether amide) Cascade Polymers", *Macromolecules* **1991**, *24*, 1443–1444.

[274] T. Nagasaki, M. Ukon, S. Arimori, S. Shinkai, "'Crowned' Arborols", *J. Chem. Soc., Chem. Commun.* **1992**, 608–610.

[275] T. Nagasaki, O. Kimura, M. Ukon, S. Arimori, I. Hamachi, S. Shinkai, "Synthesis, Metal-Binding Properties and Polypeptide Solubilization of 'Crowned' Arborols", *J. Chem. Soc., Perkin Trans. 1* **1994**, 75–81.

[276] E. Agar, B. Bati, E. Erdem, M. Özdemir, "Synthesis and Characterization of a Novel Phthalocyaninate Substituted with Eight Tetraazamacrocycles and its Nonanuclear Complexes", *J. Chem. Res., Synop.* **1995**, 16–17.

[277] Y. Pan, W. T. Ford, "Dendrimers with Alternating Amine and Ether Generations", *J. Org. Chem.* **1999**, *64*, 8588–8593.

[278] Y. Pan, W. T. Ford, "Synthesis of Polyamine Dendrimers with Alternating Amine and Ether Generations" *Polym. Prepr.* **1999**, *40*, 135–136.

[279] Y. Pan, W. T. Ford, "Ester- and Amide-Terminated Dendrimers with Alternating Amide and Ether Generations", *J. Polym. Sci., Part A: Polym. Chem.* **2000**, *38*, 1533–1543.

[280] R. Spindler, J. M. J. Fréchet, "Two-Step Approach towards the Accelerated Synthesis of Dendritic Macromolecules", *J. Chem. Soc., Perkin Trans. 1* **1993**, 913–918.

[281] R. T. Taylor, U. Puapaiboon, "Polyurethane Dendrimers via Curtius Reaction", *Tetrahedron Lett.* **1998**, *39*, 8005–8008.

[282] U. Puapaiboon, R. T. Taylor, J. Jai-nhuknan, "Structural Configuration of Polyurethane Dendritic Wedges and Dendrimers Using Post Source Decay Matrix-assisted Laser Desorption/Ionization Time-of-flight Mass Spectrometry", *Rapid Commun. Mass Spectrom.* **1999**, *13*, 516–520.

[283] E. W. Kwock, T. X. Neenan, T. M. Miller, "The Convergent Synthesis of Four Generations of Monodisperse Aryl Ester Dendrimers" *Polym. Prepr.* **1991**, *32*, 635–636.

[284] E. W. Kwock, T. X. Neenan, T. M. Miller, "Convergent Synthesis of Monodisperse Aryl Ester Dendrimers", *Chem. Mater.* **1991**, *3*, 775–777.

[285] T. M. Miller, E. W. Kwock, T. X. Neenan, "Synthesis of Four Generations of Monodisperse Aryl Ester Dendrimers Based on 1,3,5-Benzenetricarboxylic Acid", *Macromolecules* **1992**, *25*, 3143–3148.

[286] H. S. Sahota, P. M. Lloyd, S. G. Yeates, P. J. Derrick, P. C. Taylor, D. M. Haddleton, "Characterisation of Aromatic Polyester Dendrimers by Matrix-assisted Laser Desorption Ionisation Mass Spectroscopy", *J. Chem. Soc., Chem. Commun.* **1994**, 2445–2446.

[287] M. R. Bryce, A. S. Batsanov, W. Devonport, J. N. Heaton, J. A. K. Howard, G. J. Marshallsay, A. J. Moore, P. J. Skabara, S. Wegener, "New Materials Based on Highly-Functionalised Tetrathiafulvalene Derivatives" in *Molecular Engineering for Advanced Materials* (Ed.: J. Becher), Kluwer, Dordrecht, The Netherlands, **1994**, pp. 235–250.

[288] G. J. Marshallsay, T. K. Hansen, A. J. Moore, M. R. Bryce, J. Becher, "Synthesis of 4,5-Bis- and 4,4',5,5'-Tetrakis(2-hydroxyethylthio)tetrathiafulvalene: The Assembly of Tris- and Pentakis(tetrathiafulvalene) Macromolecules", *Synthesis* **1994**, 926–930.

[288a] C. A. Christensen, M. R. Bryce, J. Becher, "New Multi(tetrathiafulvalene) Dendrimers", *Synthesis* **2000**, 1695–1704.

[289] M. R. Bryce, W. Devonport, "Redox-Active Dendrimers, Related Building Blocks, and Oligomers" in *Advances in Dendritic Macromolecules* (Ed.: G. R. Newkome), JAI, Greenwich, Conn., **1996**, pp. 115–149.

[290] M. R. Bryce, W. Devonport, L. M. Goldenberg, C. Wang, "Macromolecular Tetrathiafulvalene Chemistry", *Chem. Commun.* **1998**, 945–951.

[291] M. B. Nielsen, C. Lomholt, J. Becher, "Tetrathiafulvalenes as Building Blocks in Supramolecular Chemistry II", *Chem. Soc. Rev.* **2000**, *29*, 153–164.

[292] J. Garín, J. Orduna, S. Uriel, A. J. Moore, M. R. Bryce, S. Wegener, D. S. Yufit, J. A. K. Howard, "Improved Syntheses of Carboxytetrathiafulvalene and (Hydroxymethyl)tetrathiafulvalene: Versatile Building Blocks for New Functionalised Tetrathiafulvalene Derivatives", *Synthesis* **1994**, 489–493.

[293] M. R. Bryce, W. Devonport, A. J. Moore, "Dendritic Macromolecules Incorporating Tetrathiafulvalene Units", *Angew. Chem.* **1994**, *106*, 1862–1864; *Int. Ed. Engl.* **1994**, *33*, 1761–1763.

[294] W. Devonport, M. R. Bryce, G. J. Marshallsay, A. J. Moore, L. M. Goldenberg, "Aryl Ester Dendrimers Incorporating Tetrathiafulvalene Units: Convergent Synthesis, Electrochemistry and Charge-Transfer Properties", *J. Mater. Chem.* **1998**, *8*, 1361–1372.

[295] M. R. Bryce, P. de Miguel, W. Devonport, "Redox-Switchable Polyester Dendrimers Incorporating Both π-Donor (Tetrathiafulvalene) and π-Acceptor (Anthraquinone) Groups", *Chem. Commun.* **1998**, 2565–2566.

[296] P. L. Carr, G. R. Davies, W. J. Feast, N. M. Stainton, I. M. Ward, "Dielectric and Mechanical Characterization of Aryl Ester Dendrimer/PET Blends", *Polymer* **1996**, *37*, 2395–2401.

[297] Z. Bo, X. Zhang, C. Zhang, Z. Wang, M. Yang, J. Shen, Y. Ji, "Rapid Synthesis of Polyester Dendrimers", *J. Chem. Soc., Perkin Trans. 1* **1997**, 2931–2935.

[298] Y. Zhang, T. Wada, H. Sasabe, "Carbazole Photorefractive Materials", *J. Mater. Chem.* **1998**, *8*, 809–828.

[299] Z. Bo, W. Zhang, X. Zhang, C. Zhang, J. Shen, "Synthesis and Properties of Polyester Dendrimers Bearing Carbazole Groups in their Periphery", *Macromol. Chem. Phys.* **1998**, *199*, 1323–1327.

[300] O. Varnavski, A. Leanov, L. Liu, J. Takacs, T. Goodson, III, "Large Nonlinear Refraction and Higher Order Effects in a Novel Organic Dendrimer", *J. Phys. Chem. B.* **2000**, *104*, 179–188.

[301] A. W. Freeman, J. M. J. Fréchet, "A Rapid, Orthogonal Synthesis of Poly(benzyl ester) Dendrimers via an 'Activated' Monomer Approach", *Org. Lett.* **1999**, *1*, 685–687.

[302] W. J. Feast, N. M. Stainton, "A Convergent Synthesis of Extended Aryl Ester Dendrimers", *J. Mater. Chem.* **1994**, *4*, 1159–1165.

[303] K. L. Wooley, C. J. Hawker, J. M. J. Fréchet, "A 'Branched-Monomer Approach' for the Rapid Synthesis of Dendrimers", *Angew. Chem.* **1994**, *106*, 123–126; *Int. Ed. Engl.* **1994**, *33*, 82–85.

[304] C. J. Hawker, J. M. J. Fréchet, "Monodispersed Dendritic Polyesters with Removable Chain Ends: A Versatile Approach to Globular Macromolecules with Chemically Reversible Polarities", *J. Chem. Soc., Perkin Trans. 1* **1992**, 2459–2469.

[304a] A. Adronov, J. M. J. Fréchet, G. S. He, K.-S. Kim, S.-J. Chung, J. Swiatkiewicz, P. N. Prasad, "Novel Two-Photon Absorbing Dendritic Structure", *Chem. Mater.* **2000**, *12*, 2838–2841.

[305] C. J. Hawker, J. M. J. Fréchet, "Unusual Macromolecular Architectures: The Convergent Growth Approach to Dendritic Polyesters and Novel Block Copolymers", *J. Am. Chem. Soc.* **1992**, *114*, 8405–8413.

[306] C. J. Hawker, K. L. Wooley, J. M. J. Fréchet, "Novel Macromolecular Architectures: Globular Block Copolymers Containing Dendritic Components", *Makromol. Chem. , Macromol. Symp.* **1994**, *77*, 11–20.

[307] J.-F. Nierengarten, D. Felder, J.-F. Nicoud, "Preparation of Dendrons with Peripheral Fullerene Units", *Tetrahedron Lett.* **1999**, *40*, 269–272.

[308] J.-F. Nierengarten, V. Gramlich, F. Cardullo, F. Diederich, "Regio- and Diastereoselective Bisfunctionalization of C_{60} and Enantioselective Synthesis of a C_{60} Derivative with a Chiral Addition Pattern", *Angew. Chem.* **1996**, *108*, 2242–2244; *Int. Ed. Engl.* **1996**, *35*, 2101–2103.

[309] A. Morikawa, M. Kakimoto, Y. Imai, "Convergent Synthesis of Starburst Poly(ether ketone) Dendrons", *Macromolecules* **1993**, *26*, 6324–6329.

[310] J. S. Moore, Z. Xu, "Synthesis of Rigid Dendrimers that overcome Steric Inhibition" *Polym. Prepr.* **1991**, *32*, 629–630.

[311] J. S. Moore, Z. Xu, "Synthesis of Rigid Dendritic Macromolecules: Enlarging the Repeat Unit Size as a Function of Generation Permits Growth to Continue", *Macromolecules* **1991**, *24*, 5893–5894.

[312] D. L. Trumbo, C. S. Marvel, "Polymerization using Palladium(II) Salts: Homopolymers and Copolymers from Phenylethynyl Compounds and Aromatic Bromides", *J. Polym. Sci.* **1986**, *24*, 2311–2326.

[313] N. A. Bumagin, A. B. Ponomarev, I. P. Beletskaya, "Cross-Coupling of Terminal Acetylenes with Organic Halides in the R_3N–Cu–Pd Catalytic System", *Izv. Akad. Nauk SSSR, Ser. Khim.* **1984**, 1561–1566.

[314] Z. Xu, J. S. Moore, "Synthesis and Characterization of a Stiff Dendrimer of High Molecular Weight", *Angew. Chem.* **1993**, *105*, 261–264; *Int. Ed. Engl.* **1993**, *32*, 246–248.

[315] Z. Xu, M. Kahr, K. L. Walker, C. L. Wilkins, J. S. Moore, "Phenylacetylene Dendrimers by the Divergent, Convergent, and Double-Stage Convergent Methods", *J. Am. Chem. Soc.* **1994**, *116*, 4537–4550.

[316] J. S. Moore, E. J. Weinstein, Z. Wu, "A Convenient Masking Group for Aryl Iodides", *Tetrahedron Lett.* **1991**, *32*, 2465–2466.

[317] K. L. Walker, M. S. Kahr, C. L. Wilkins, Z. Xu, J. S. Moore, "Analysis of Hydrocarbon Dendrimers by Laser Desorption Time-of-Flight and Fourier-Transform Mass Spectrometry", *J. Am. Soc. Mass Spectrom.* **1994**, *5*, 731–739.

[318] D. J. Pesak, J. S. Moore, "Columnar Liquid Crystals from Shape-Persistent Dendritic Molecules", *Angew. Chem.* **1997**, *109*, 1709–1712; *Int. Ed. Engl.* **1997**, *36*, 1636–1639.

[319] Z. Xu, J. S. Moore, "Rapid Construction of Large-size Phenylacetylene Dendrimers up to 12.5 Nanometers in Molecular Diameter", *Angew. Chem.* **1993**, *105*, 1394–1396; *Int. Ed. Engl.* **1993**, *32*, 1354–1357.

[320] Z. Xu, Z.-Y. Shi, W. Tan, R. Kopelman, J. S. Moore, "Phenylacetylene Dendrimers with Extended π-Conjugation" *Polym. Prepr.* **1993**, *34*, 130–131.

[321] Z. Xu, J. S. Moore, "Synthesis and Characterization of a Stiff Dendrimer of High Molecular Weight", *Angew. Chem.* **1993**, *105*, 261–264; *Int. Ed. Engl.* **1993**, *32*, 246–248.

[322] T. Kawaguchi, Z. Xu, J. S. Moore, "Architectures of Stiff Dendritic Macromolecules and Linear Flexible Chains" *Polym. Prepr.* **1993**, *34*, 124–125.

[323] Z. Xu, J. S. Moore, "Stiff Dendritic Macromolecules: Extending Small Organic Chemistry to the Nanoscale Regime" *Polym. Prepr.* **1993**, *34*, 128–129.

[324] J. S. Moore, "Shape-Persistent Molecular Architectures of Nanoscale Dimension", *Acc. Chem. Res.* **1997**, *30*, 402–413.

[325] D. J. Pesak, J. S. Moore, T. E. Wheat, "Synthesis and Characterization of Water-Soluble Dendritic Macromolecules with a Stiff, Hydrocarbon Interior", *Macromolecules* **1997**, *30*, 6467–6482.

[326] J. Zhang, J. S. Moore, Z. Xu, R. A. Aguirre, "Nanoarchitectures; 1. Controlled Synthesis of Phenylacetylene Sequences", *J. Am. Chem. Soc.* **1992**, *114*, 2273–2274.

[327] P. Bharathi, U. Patel, T. Kawaguchi, D. J. Pesak, J. S. Moore, "Improvements in the Synthesis of Phenylacetylene Monodendrons Including a Solid-Phase Convergent Method", *Macromolecules* **1995**, *28*, 5955–5963.

[328] J. C. Nelson, J. K. Young, J. S. Moore, "Solid-Phase Synthesis of Phenylacetylene Oligomers Utilizing a Novel 3-Propyl-3-(benzyl-supported) Triazene Linkage", *J. Org. Chem.* **1996**, *61*, 8160–8168.

[329] P. Bharathi, J. S. Moore, "Solid-Supported Hyperbranched Polymerization: Evidence for Self-Limited Growth", *J. Am. Chem. Soc.* **1997**, *119*, 3391–3392.

[330] J. Zhang, D. J. Pesak, J. L. Ludwick, J. S. Moore, "Geometrically-Controlled and Site-Specifically-Functionalized Phenylacetylene Macrocycles", *J. Am. Chem. Soc.* **1994**, *116*, 4227–4239.

[331] T. Kawaguchi, J. S. Moore, "Towards Double Exponential Dendrimer Growth" *Polym. Prepr.* **1994**, *35*, 669–670.

[332] L. Jones, II, D. L. Pearson, J. S. Schumm, J. M. Tour, "Synthesis of Well-Defined Conjugated Oligomers for Molecular Electronics", *Pure Appl. Chem.* **1996**, *68*, 145–148.

[333] Z. Wu, J. S. Moore, "A Freely Hinged Macrotricycle with a Molecular Cavity", *Angew. Chem.* **1996**, *108*, 320–322; *Int. Ed. Engl.* **1996**, *35*, 297–299.

[334] M. Brake, V. Enkelmann, U. H. F. Bunz, "Synthesis and Characterization of Oxygen-Substituted Pericyclynes", *J. Org. Chem.* **1996**, *61*, 1190–1191.

[335] M. Mayor, J.-M. Lehn, K. M. Fromm, D. Fenske, "Reducible Nanoscale Molecular Rods Based on Diacetylene-Linked Poly(phenylthio)-Substituted Benzenes", *Angew. Chem.* **1997**, *109*, 2468–2471; *Int. Ed. Engl.* **1997**, *36*, 2370–2372.

[336] U. H. F. Bunz, G. Roidl, M. Altmann, V. Enkelmann, K. D. Shimizu, "Synthesis and Structural Characterization of Novel Organometallic Dehydroannulenes with Fused CpCo-Cyclobutadiene and Ferrocene Units Including a Cyclic Fullerenyne Segment", *J. Am. Chem. Soc.* **1999**, *121*, 10719–10726.

[337] S. Höger, A.-D. Meckenstock, "Template-Directed Synthesis of Shape-Persistent Macrocyclic Amphiphiles with Convergently Arranged Functionalities", *Chem. Eur. J.* **1999**, *5*, 1686–1691.

[338] Y. Yao, J. M. Tour, "Facile Convergent Route to Molecular Caltrops", *J. Org. Chem.* **1999**, *64*, 1968–1971.

[338a] C. Chi, J. Wu, X. Wang, X. Zhao, J. Li, F. Wang, "A new solid-supported iterative divergent/convergent strategy for the synthesis of dendrimers", *Tetrahedron Lett.* **2001**, *42*, 2181–2184.

[339] D. J. Pesak, J. S. Moore, "Polar Domains on Globular Macromolecules: Shape-Persistent, Amphiphile Tridendrons", *Tetrahedron* **1997**, *53*, 15331–15347.

[340] T. Kawaguchi, J. S. Moore, "'Ball and Chain' Copolymer Architecture" *Polym. Prepr.* **1994**, *35*, 872–873.

[341] Z. Xu, J. S. Moore, "Design and Synthesis of a Convergent and Directional Molecular Antenna", *Acta Polym.* **1994**, *45*, 83–87.

[342] M. R. Shortreed, S. F. Swallen, Z.-Y. Shi, W. Tan, Z. Xu, C. Devadoss, J. S. Moore, R. Kopelman, "Directed Energy Transfer Funnels in Dendrimeric Antenna Supermolecules", *J. Phys. Chem. B* **1997**, *101*, 6318–6322.

[342a] D. Rana, G. Gangopadhyay, "Steady-state spectral properties of dendrimer supermolecule as a light harvesting system", *Chem. Phys. Lett.* **2001**, *334*, 314–324.

[343] S. M. Risser, D. N. Beratan, J. N. Onuchic, "Electronic Coupling in Starburst Dendrimers: Connectivity, Disorder, and Finite Size Effects in Macromolecular Bethe Lattices", *J. Phys. Chem.* **1993**, *97*, 4523–4527.

[344] A. Bar-Haim, J. Klafter, R. Kopelman, "Dendrimers as Controlled Artificial Energy Antennae", *J. Am. Chem. Soc.* **1997**, *119*, 6197–6198.

[345] T. Kawaguchi, K. L. Walker, C. L. Wilkins, J. S. Moore, "Double Exponential Dendrimer Growth", *J. Am. Chem. Soc.* **1995**, *117*, 2159–2165.

[346] C. Devadoss, P. Bharathi, J. S. Moore, "Energy Transfer in Dendritic Macromolecules: Molecular Size Effects and the Role of an Energy Gradient", *J. Am. Chem. Soc.* **1996**, *118*, 9635–9644.

[347] J. K. Young, C. Devadoss, Z. Zhu, P.-W. Wang, J. S. Moore, "Ordered Arrays of Chromophores and Redox Centers in Dendritic Macromolecules" *Polym. Mater. Sci. Eng.* **1995**, *73*, 224–225.

[348] C. Devadoss, P. Bharathi, J. S. Moore, "Photoinduced Electron Transfer in Dendritic Macromolecules; 1. Intermolecular Electron Transfer", *Macromolecules* **1998**, *31*, 8091–8099.

[349] C. J. Buchko, P. M. Wilson, Z. Xu, J. Zhang, J. S. Moore, D. C. Martin, "Electron Microscopy and Diffraction of Crystalline Dendrimers and Macrocycles", *Polymer* **1995**, *36*, 1817–1825.

[350] I. M. Xavier, Jr., A. De Pádua, F. Moraes, J. A. De Miranda-Neto, "Hyperbolic Modeling of Starburst Dendrimers", *Mol. Eng.* **2000**, *7*, 283–291.

[351] R. Kopelman, M. Shortreed, Z.-Y. Shi, W. Tan, Z. Xu, J. S. Moore, A. Bar-Haim, J. Klafter, "Spectroscopic Evidence for Excitonic Localization in Fractal Antenna Supermolecules", *Phys. Rev. Lett.* **1998**, *78*, 1239–1242.

[352] S. F. Swallen, R. Kopelman, J. S. Moore, C. Devadoss, "Dendrimer Photoantenna Supermolecules: Energetic Funnels, Exciton Hopping and Correlated Excimer Formation", *J. Mol. Struct.* **1999**, *485–486*, 585–597.

[353] S. F. Swallen, Z. Zhu, J. S. Moore, R. Kopelman, "Correlated Excimer Formation and Molecular Rotational Dynamics in Phenylacetylene Dendrimers", *J. Phys. Chem. B.* **2000**, *104*, 3988–3995.

[354] A. Bar-Haim, J. Klafter, "Dendrimers as Light-Harvesting Antennae", *J. Lumin.* **1998**, *76/77*, 197–200.

[355] R. Kopelman, M. Shortreed, Z.-Y. Shi, W. Tan, Z. Xu, J. S. Moore, A. Bar-Haim, J. Klafter, "Spectroscopic Evidence for Excitonic Localization in Fractal Antenna Supermolecules", *Phys. Rev. Lett.* **1997**, *78*, 1239–1242.

[356] B. B. Mandelbrot, *The Fractal Geometry of Nature*, Freeman, San Francisco, **1982**.

[357] A. Bar-Haim, J. Klafter, "Geometric versus Energetic Competition in Light Harvesting by Dendrimers", *J. Phys. Chem. B.* **1998**, *102*, 1662–1664.

[358] S. F. Swallen, Z.-Y. Shi, W. Tan, Z. Xu, J. S. Moore, R. Kopelman, "Exciton Localization Hierarchy and Directed Energy Transfer in Conjugated Linear Aromatic Chains and Dendrimeric Supermolecules", *J. Lumin.* **1998**, *76/77*, 193–196.

[359] K. Harigaya, "Coupled Exciton Model with Off-Diagonal Disorder for Optical Excitations in Extended Dendrimers", *Phys. Chem. Chem. Phys.* **1999**, *1*, 1687–1689.

[360] K. Harigaya, "Optical Excitations in Fractal Antenna Supramolecules with Conjugated Electrons: Extended Dendrimers", *Chem. Phys. Lett.* **1999**, *300*, 33–36.

[361] Y. Shimoi, B. A. Friedman, "A Tight-Binding Model of Phenylene Molecules with *meta*-Connections – Implications for Phenylacetylene Dendrimers", *Chem. Phys.* **1999**, *250*, 13–22.

[362] C. Devadoss, P. Bharathi, J. S. Moore, "Anomalous Shift in the Fluorescence Spectra of a High-Generation Dendrimer in Nonpolar Solvents", *Angew. Chem.* **1997**, *109*, 1706–1709; *Int. Ed. Engl.* **1997**, *36*, 1633–1635.

[363] Q.-S. Hu, V. Pugh, M. Sabat, L. Pu, "Structurally Rigid and Optically Active Dendrimers", *J. Org. Chem.* **1999**, *64*, 7528–7536.

[363a] V. J. Pugh, Q.-S. Hu, L. Pu, "The First Dendrimer-Based Enantioselective Fluorescent Sensor for the Recognition of Chiral Amino Alcohols", *Angew. Chem. Int. Ed.* **2000**, *39*, 3638–3641.

[363b] L.-Z. Gong, Q.-S. Hu, L. Pu, "Optically Active Dendrimers with a Binaphthyl Core and Phenylene Dendrons: Light Harvesting and Enantioselective Fluorescent Sensing", *J. Org. Chem.* **2001**, *66*, 2358–2367.

[364] A. G. Avent, P. R. Birkett, F. Paolucci, S. Roffia, R. Taylor, N. K. Wachter, "Synthesis and Electrochemical Behaviour of [60]Fullerene Possessing Poly(arylacetylene) Dendrimer Addends", *J. Chem. Soc., Perkin Trans. 2* **2000**, 1409–1414.

[365] J. Ipaktschi, R. Hosseinzadeh, P. Schlaf, "Self-Assembly of Quinodimethanes through Covalent Bonds: A Novel Principle for the Synthesis of Functional Macrocycles", *Angew. Chem.* **1999**, *111*, 1765–1768; *Int. Ed.* **1999**, *38*, 1658–1660.

[366] T. Kaneko, T. Horie, M. Asano, T. Aoki, E. Oikawa, "Polydendron: Polymerization of Dendritic Phenylacetylene Monomers", *Macromolecules* **1997**, *30*, 3118–3121.

[367] Z. Peng, Y. Pan, B. Xu, J. Zhang, "Synthesis and Optical Properties of Novel Unsymmetrical Conjugated Dendrimers", *J. Am. Chem. Soc.* **2000**, *122*, 6619–6623.

[368] F. Zeng, S. C. Zimmerman, "Rapid Synthesis of Dendrimers by an Orthogonal Coupling Strategy", *J. Am. Chem. Soc.* **1996**, *118*, 5326–5327.

[369] D. L. Hughes, "The Mitsunobu Reaction", *Org. React. (N. Y.)* **1992**, *42*, 335–656.

[370] K. Sonogashira, Y. Tohda, N. Hagihara, "A Convenient Synthesis of Acetylenes: Catalytic Substitutions of Acetylenic Hydrogen with Bromoalkenes, Iodoarenes, and Bromopyridines", *Tetrahedron Lett.* **1975**, 4467–4470.

[371] T. Hundertmark, A. F. Littke, S. L. Buchwald, G. C. Fu, "Pd(PhCN)$_2$Cl$_2$/P(*t*-Bu)$_3$: A Versatile Catalyst for Sonogashira Reactions of Aryl Bromides at Room Temperature", *Chem. Lett.* **2000**, *2*, 1729–1731.

[372] A. Kraft, "Self-Association of a 1,3,4-Oxadiazole-Containing Dendrimer. [Erratum to *Chem. Commun.* **1996**, 77–79]", *Chem. Commun.* **1996**, 2103.

[373] A. Kraft, "Self-Association of a 1,3,4-Oxadiazole-Containing Dendrimer", *Chem. Commun.* **1996**, 77–79.

[374] A. Kraft, F. Osterod, "Self-Association of Branched and Dendritic Aromatic Amides", *J. Chem. Soc., Perkin Trans. 1* **1998**, 1019–1024.

[375] A. Kraft, "Synthesis and Self-Association of First-Generation 1,3,4-Oxadiazole-Containing Dendrimers", *Liebigs Ann. Chem.* **1997**, 1463–1471.

[376] J. Bettenhausen, P. Strohriegl, "Dendrimers with 1,3,4-Oxadiazole Units; Part 1. Synthesis and Characterization", *Macromol. Rapid Commun.* **1996**, *17*, 623–631.

[377] R. Huisgen, J. Sauer, H. J. Sturm, J. H. Markgraf, "Die Bildung von 1,3,4-Oxdiazolen bei der Acylierung 5-substituierter Tetrazole", *Chem. Ber.* **1960**, *93*, 2106–2124.

[378] Z. Peng, J. Zhang, B. Xu, "Synthesis and Unusual Optical Properties of Conjugated Polymers Containing Multibranched Oxadizole Substituents" *Polym. Prepr.* **2000**, *41*, 881–882.

[379] S. K. Deb, T. M. Maddux, L. Yu, "A Simple Orthogonal Approach to Poly(phenylenevinylene) Dendrimers", *J. Am. Chem. Soc.* **1997**, *119*, 9079–9080.

[380] H. Meier, M. Lehmann, "Stilbenoid Dendrimers", *Angew. Chem.* **1998**, *110*, 666–669; *Int. Ed.* **1998**, *37*, 643–645.

[381] M. Lehmann, B. Schartel, M. Hennecke, H. Meier, "Dendrimers Consisting of Stilbene or Distyrylbenzene Building Blocks: Synthesis and Stability", *Tetrahedron* **1999**, *55*, 13377–13394.

[382] H. Meier, M. Lehmann, U. Kolb, "Stilbenoid Dendrimers", *Chem. Eur. J.* **2000**, *6*, 2462–2469.

[383] E. Díez-Barra, J. C. García-Martínez, J. Rodrígues-López, "A Horner–Wadsworth–Emmons Approach to Dipolar and Non-Dipolar Poly(phenylenevinylene)dendrimers", *Tetrahedron Lett.* **1999**, *40*, 8181–8184.

[383a] E. Díez-Barra, J. C. García-Martínez, J. Rodriguez-Lópis, R. Gómez, J. L. Segura, N. Martín, "Synthesis of New 1,1'-Binaphthyl-Based Chiral Phenylenevinylene Dendrimers", *Org. Lett.* **2000**, *2*, 3651–3653.

[383b] F. Langa, M. J. Gómez-Escalonilla, E. Díez-Barra, J. C. García-Martínez, A. de la Hoz, J. Rodrígues-López, A. González-Cortés, V. López-Arza, "Synthesis, electrochemistry and photophysical properties of phenylenevinylene fullerodendrimers", *Tetrahedron Lett.* **2001**, *42*, 3435–3438.

[384] J. N. G. Pillow, M. Halim, J. M. Lupton, P. L. Burn, I. D. W. Samuel, "A Facile Iterative Procedure for the Preparation of Dendrimers Containing Luminescent Cores and Stilbene Dendrons", *Macromolecules* **1999**, *32*, 5987–5993.

[384a] J. M. Lupton, I. D. W. Samuel, R. Beavington, P. L. Burn, H. Bässler, "Nanoengineering of organic semiconductors for light-emitting diodes: control of charge transport", *Syn. Metals* **2001**, *116*, 357–362.

[385] M. Halim, J. N. G. Pillow, I. D. W. Samuel, P. L. Burn, "Conjugated Dendrimers for Light-Emitting Diodes: Effect of Generation", *Adv. Mater.* **1999**, *11*, 371–374.

[386] N. Armaroli, F. Barigelletti, P. Ceroni, J.-F. Eckert, J.-F. Nicoud, J.-F. Nierengarten, "Photoinduced Energy Transfer in a Fullerene-Oligophenylenevinylene Conjugate", *Chem. Commun.* **2000**, 599–600.

[387] G. Chessa, A. Scrivanti, "Synthesis of Dendritic Polypyridines with Ethoxycarbonyl Groups as Surface Functionality", *J. Chem. Soc., Perkin Trans. 1* **1996**, 307–311.

[388] G. Chessa, A. Scrivanti, R. Seraglia, P. Traldi, "Matrix Effects on Matrix-Assisted Laser Desorption/Ionization Mass Spectrometry Analysis of Dendrimers with a Pyridine-Based Skeleton", *Rapid Commun. Mass Spectrom.* **1998**, *12*, 1533–1537.

[389] M. Tominaga, J. Hosoggi, K. Konishi, T. Aida, "Nucleobase Dendrimer as a Multidentate Ligand for a Rare-Earth Metal Ion", *Chem. Commun.* **2000**, 719–720.

[390] M. Tominaga, K. Konishi, T. Aida, "A Photocrosslinkable Dendrimer Consisting of a Nucleobase", *Chem. Lett.* **2000**, 374–375.

[391] W. Zhang, E. E. Simanek, "Dendrimers Based on Melamine. Divergent and Orthogonal, Convergent Synthesis of G3 Dendrimer", *Org. Lett.* **2000**, *2*, 843–845.

[392] W. Zhang, E. E. Simanek, "Dendrimers Based on Melamine", *Polym. Prepr.* **2000**, *41*, 1579.

[392a] A. Marsh, S. J. Carlisle, S. C. Smith, "High-loading scavenger resins for combinatorial chemistry", *Tetrahedron Lett.* **2001**, *42*, 493–496.

[393] K. E. Uhrich, J. M. J. Fréchet, "Synthesis of Dendritic Polyamides *via* a Convergent Growth Approach", *J. Chem. Soc., Perkin Trans. 1* **1992**, 1623–1630.

[394] D. A. Tomalia, H. Baker, J. Dewald, M. Hall, G. Kallos, S. Martin, J. Roeck, J. Ryder, P. Smith, "A New Class of Polymers: Starburst-Dendritic Macromolecules", *Polym. J. (Tokyo)* **1985**, *17*, 117–132.

[395] D. A. Tomalia, D. R. Swanson, J. W. Klimash, H. M. Brothers, III, "Cascade (Starburst™) Dendrimer Synthesis by the Divergent Dendron/Divergent Core Anchoring Methods" *Polym. Prepr.* **1993**, *34*, 52–53.

[396] M. Adamczyk, J. Fishpaugh, P. G. Mattingly, K. Shreder, "Tracermer™ Signal Generators: An Arborescent Approach to the Incorporation of Multiple Chemiluminescent Labels", *Bioorg. Med. Chem. Lett.* **1998**, *8*, 3595–3598.

[397] M. Adamczyk, J. R. Fishpaugh, K. J. Hauser, "Chemoselective Synthesis of Protected Polyamines and Facile Synthesis of Polyamine Derivatives using *o*-Alkyl-*o'-N* -succinimidyl)carbonates", *Org. Prep. Proced. Int.* **1998**, *30*, 339–348.

[398] T. Nagasaki, S. Tamagaki, K. Ogino, "Syntheses and Characterization of Photochromic Dendrimers Including a 1,3-Alternate Calix[4]arene as a Core and Azobenzene Moieties as Branches", *Chem. Lett.* **1997**, 717–718.

[399] S. Yokoyama, T. Nakahama, A. Otomo, S. Mashiko, "Intermolecular Coupling Enhancement of the Molecular Hyperpolarizability in Multichromophoric Dipolar Dendrons", *J. Am. Chem. Soc.* **2000**, *122*, 3174–3181.

[399a] S. P. Rannard, N. J. Davis, "A Highly Selective, One-Pot Multiple-Addition Convergent Synthesis of Polycarbonate Dendrimers", *J. Am. Chem. Soc.* **2000**, *122*, 11729–11730.

[400] J. Louie, J. F. Hartwig, A. J. Fry, "Discrete High Molecular Weight Triarylamine Dendrimers Prepared by Palladium-Catalyzed Amination", *J. Am. Chem. Soc.* **1997**, *119*, 11695–11696.

[401] J. F. Hartwig, "Palladium-Catalyzed Synthesis of Triarylamine Macromolecules" *Polym. Prepr.* **2000**, *41*, 420–421.

[401a] H. Nam, D. H. Kang, J. K. Kim, S. Y. Park, "Synthesis of Hole-Transporting Hydrazone Dendrimers", *Chem. Lett.* **2000**, 1298–1299.

[402] Z. Zhu, J. S. Moore, "Synthesis and Characterization of Monodendrons Based on 9-Phenylcarbazole", *J. Org. Chem.* **2000**, *65*, 116–123.

[403] Z. Zhu, J. S. Moore, "Synthesis and Characterization of 9-Phenylcarbazole Monodendrons: An Exploration of Peripheral Groups To Facilate Purification", *Macromolecules* **2000**, *33*, 801–807.

[404] L. J. Twyman, A. E. Beezer, J. C. Mitchell, "An Approach for the Rapid Synthesis of Moderately Sized Dendritic Macromolecules", *J. Chem. Soc., Perkin Trans. 1* **1994**, 407–411.

[405] Y. Imai, "Synthesis of New Silicon-Based Condensation Polymers", *Kagaku (Kyoto)* **1991**, *46*, 280–281.

[406] Y. Imai, "Synthesis of New Functional Silicon-Based Condensation Polymers", *J. Macromol. Sci., Chem.* **1991**, *A28*, 1115–1135.

[407] A. Morikawa, M. Kakimoto, Y. Imai, "Convergent Synthesis of Siloxane Starburst Dendrons and Dendrimers via Hydrosilylation", *Macromolecules* **1992**, *25*, 3247–3253.

[408] S. S. Sheiko, G. Eckert, G. Ignat'eva, A. M. Muzafarov, J. Spickermann, H. J. Rädar, M. Möller, "Solid-Like States of a Dendrimer Liquid Displayed by Scanning Force Microscopy", *Macromol. Rapid Commun.* **1996**, *17*, 283–297.

[409] S. Yamaguchi, S. Akiyama, K. Tamao, "Tri-9-anthrylborane and its Derivatives: New Boron-Containing π-Electron Systems with Divergently Extended π-Conjugated through Boron", *J. Am. Chem. Soc.* **2000**, *122*, 6335–6336.

[410] C. Wang, M. R. Bryce, A. S. Batsanov, L. M. Goldenberg, J. A. K. Howard, "Synthesis and Electrochemistry of New Tetrathiafulvalene (TTF) Dendrimers: X-ray Crystal Structure of a Tetrafunctionalised TTF Core Unit", *J. Mater. Chem.* **1997**, *7*, 1189–1197.

[411] J. Becher, J. Lau, P. Leriche, P. Mørk, N. Svenstrup, "Caesium Tetrathiafulvalene-Thiolates: Key Synthetic Intermediates", *J. Chem. Soc., Chem. Commun.* **1994**, 2715–2716.

[412] M. A. Coffin, M. R. Bryce, A. S. Batsanov, J. A. K. Howard, "Redox-Active, Functionalised [3]- and [4]-Dendralenes", *J. Chem. Soc., Chem. Commun.* **1993**, 552–554.

[413] C. A. Christensen, L. M. Goldenberg, M. R. Bryce, J. Becher, "Synthesis and Electrochemistry of a Tetrathiafulvalene (TTF)$_{21}$-Glycol Dendrimer: Intradendrimer Aggregation of TTF Cation Radicals", *Chem. Commun.* **1998**, 509–510.

[414] J. Malthête, "Une Mésophase Colonnaire constituée de Molécules à Cœur Tétraédrique", *New J. Chem.* **1996**, *20*, 925–928.

[415] V. Percec, G. Johansson, G. Ungar, J. Zhou, "Fluorophobic Effect Induces the Self-Assembly of Semifluorinated Tapered Monodendrons Containing Crown Ethers into Supramolecular Columnar Dendrimers Which Exhibit a Homeotropic Hexagonal Columnar Liquid Crystalline Phase", *J. Am. Chem. Soc.* **1996**, *118*, 9855–9866.

[416] V. Percec, D. Schlueter, "Mechanistic Investigations on the Formation of Supramolecular Cylindrical Shaped Oligomers and Polymers by Living Ring Opening Metathesis Polymerization of a 7-Oxanorbornene Monomer Substituted with Two Tapered Monodendrons", *Macromolecules* **1997**, *30*, 5783–5790.

[417] V. S. K. Balagurusamy, G. Ungar, V. Percec, G. Johansson, "Rational Design of the First Spherical Supramolecular Dendrimers Self-Organized in a Novel Thermotropic Cubic Liquid-Crystalline Phase and the Determination of Their Shape by X-ray Analysis", *J. Am. Chem. Soc.* **1997**, *119*, 1539–1555.

[418] V. Percec, W.-D. Cho, M. Möller, S. A. Prokhorova, G. Ungar, D. J. P. Yeardley, "Design and Structural Analysis of the First Spherical Monodendron Self-Organizable in a Cubic Lattice", *J. Am. Chem. Soc.* **2000**, *122*, 4249–4250.

[419] V. Percec, C.-H. Ahn, B. Barboiu, "Self-Encapsulation, Acceleration and Control in the Radical Polymerization of Monodendritic Monomers via Self-Assembly", *J. Am. Chem. Soc.* **1997**, *119*, 12978–12979.

[420] S. D. Hudson, H.-T. Jung, V. Percec, W.-D. Cho, G. Johansson, G. Ungar, V. S. K. Balagurusamy, "Direct Visualization of Individual Cylindrical and Spherical Supramolecular Dendrimers", *Science* **1997**, *278*, 449–452.

[420a] V. Percec, W.-D. Cho, G. Ungar, D. J. P. Yeardley, "Synthesis and Structural Analysis of Two Constitutional Isomeric Libraries of AB$_2$-Based Monodendrons and Supramolecular Dendrimers", *J. Am. Chem. Soc.* **2001**, *123*, 1302–1315.

[420b] V. Percec, W.-D. Cho, G. Ungar, "Increasing the Diameter of Cylindrical and Spherical Supramolecular Dendrimers by Decreasing the Solid Angle of Their Monodendrons via Periphery Functionalization", *J. Am. Chem. Soc.* **2000**, *122*, 10273–10281.

[421] V. Percec, C.-H. Ahn, W.-D. Cho, A. M. Jamieson, J. Kim, T. Leman, M. Schmidt, M. Gerle, M. Möller, S. A. Prokhorova, S. S. Sheiko, S. Z. D. Cheng, A. Zhang, G. Ungar, D. J. P. Yeardley, "Visualizable Cylindrical Macromolecules with Controlled Stiffness from Backbones Containing Libraries of Self-Assembling Dendritic Side Groups", *J. Am. Chem. Soc.* **1998**, *120*, 8619–8631.

[422] V. Percec, C.-H. Ahn, G. Ungar, D. J. P. Yeardley, M. Möller, S. S. Sheiko, "Controlling Polymer Shape Through the Self-Assembly of Dendritic Side-Groups", *Nature* **1998**, *391*, 161–164.

[423] V. Percec, W.-D. Cho, P. E. Mosier, G. Ungar, D. J. P. Yeardley, "Structural Analysis of Cylindrical and Spherical Supramolecular Dendrimers Quantifies the Concept of Mondendron Shape Control by Generation Number", *J. Am. Chem. Soc.* **1998**, *120*, 11061–11070.

[424] G. Ungar, V. Percec, M. N. Holerca, G. Johansson, J. A. Heck, "Heat-Shrinking Spherical and Columnar Supramolecular Dendrimers: Their Interconversion and Dependence of Their Shape on Molecular Taper Angle", *Chem. Eur. J.* **2000**, *6*, 1258–1266.

[425] V. Percec, C.-H. Ahn, T. K. Bera, G. Ungar, D. J. P. Yeardley, "Coassembly of a Hexagonal Columnar Liquid Crystalline Superlattice from Polymer(s) Coated with a Three-Cylindrical Bundle Supramolecular Dendrimer", *Chem. Eur. J.* **1999**, *5*, 1070–1083.

[426] S. A. Prokhorova, S. S. Sheiko, C.-H. Ahn, V. Percec, M. Möller, "Molecular Conformations of Monodendron-Jacketed Polymers by Scanning Force Microscopy", *Macromolecules* **1999**, *32*, 2653–2660.

[427] Z. Bao, K. R. Amundson, A. J. Lovinger, "Poly(phenylenevinylene)s with Dendritic Side Chains: Synthesis, Self-Ordering, and Liquid Crystalline Properties", *Macromolecules* **1998**, *31*, 8647–8649.

[428] S. A. Prokhorova, S. S. Sheiko, M. Möller, C.-H. Ahn, V. Percec, "Molecular Imaging of Monodendron Jacketed Linear Polymers by Scanning Force Microscopy", *Macromol. Rapid Commun.* **1998**, *19*, 359–366.

[429] Y. Li, T. Ji, J. Zhang, "Preparation and Properties of Novel Aromatic Polyamides and Polyimides Jacked with Dendritic Fragments", *J. Polym. Sci., Part A: Polym. Chem.* **2000**, *38*, 189–197.

[430] M. Brettreich, A. Hirsch, "Convergent Synthesis of 1 → 3 C-Branched Polyamide Dendrons", *Synlett* **1998**, 1396–1398.

[431] M. Brettreich, A. Hirsch, "A Highly Water-Soluble Dendro[60]fullerene", *Tetrahedron Lett.* **1998**, *39*, 2731–2734.

[432] C. M. Cardona, T. D. McCarley, A. E. Kaifer, "Synthesis, Electrochemistry, and Interactions with β-Cyclodextrin of Dendrimers Containing a Single Ferrocene Subunit Located 'Off-Center'", *J. Org. Chem.* **2000**, *65*, 1857–1864.

[433] C. M. Caradona, J. Alvarez, A. E. Kaifer, T. D. McCarley, S. Pandey, G. A. Baker, N. J. Bonzagni, F. V. Bright, "Dendrimers Functionalized with a Single Fluorescent Dansyl Group Attached 'Off-Center': Synthesis and Photophysical Studies", *J. Am. Chem. Soc.* **2000**, *122*, 6139–6144.

[434] J.-F. Nierengarten, T. Habicher, R. Kessinger, F. Cardullo, F. Diederich, V. Gramlich, J.-P. Gisselbrecht, C. Boudon, M. Gross, "Macrocyclization on the Fullerene Core: Direct Regio- and Diastereoselective Multi-Functionalization of [60]Fullerene, and Synthesis of Fullerene-Dendrimer Derivatives", *Helv. Chim. Acta* **1997**, *80*, 2238–2276.

[435] C. Bingel, "Cyclopropanierung von Fullerenen", *Chem. Ber.* **1993**, *126*, 1957–1959.

[436] J.-F. Nierengarten, C. Schall, J.-F. Nicoud, B. Heinrich, D. Guillon, "Amphiphilic Cyclic Fullerene Bisadducts: Synthesis and Langmuir Films at the Air–Water Interface", *Tetrahedron Lett.* **1998**, *39*, 5747–5750.

[437] B. Kenda, F. Diederich, "Supramolecular Aggregates of Dendritic Cyclophanes (Dendrophanes) Threaded on Molecular Rods with Steroid Termini", *Angew. Chem.* **1998**, *110*, 3357–3361; *Int. Ed.* **1998**, *37*, 3154–3158.

[438] T. Habicher, F. Diederich, V. Gramlich, "Catalytic Dendrophanes as Enzyme Mimics: Synthesis, Binding Properties, Micropolarity Effect, and Catalytic Activity of Dendritic Thiazoliocyclophanes", *Helv. Chim. Acta* **1999**, *82*, 1066–1095.

[439] D. K. Smith, F. Diederich, "Dendritic Hydrogen Bonding Receptors: Enantiomerically Pure Dendroclefts for the Selective Recognition of Monosaccharides", *Chem. Commun.* **1998**, 2501–2502.

[440] T. Kemmitt, W. Henderson, "Dendrimeric Silatrane Wedges", *J. Chem. Soc., Perkin Trans. 1* **1997**, 729–739.

[441] P. R. Ashton, D. W. Anderson, C. L. Brown, A. N. Shipway, J. F. Stoddart, M. S. Tolley, "The Synthesis and Characterization of a New Family of Polyamide Dendrimers", *Chem. Eur. J.* **1998**, *4*, 781–795.

6 Hyperbranched Materials

6.1 General Concepts

Whereas the well-characterized, perfect (or nearly so) structures of dendritic macromolecules constructed by discrete stepwise procedures have been illustrated in the preceding chapters, this Chapter describes the related, less-than-perfect, hyperbranched polymers, which can be synthesized by means of a direct, one-step polycondensation of AB_x monomers where $x \geq 2$. Flory's prediction and subsequent demonstration[1, 2] that AB_x monomers can generate highly branched polymers heralded advances in the creation of idealized dendritic systems; thus, the desire for simpler, and in most cases more economical (one-step), procedures for obtaining the hyperbranched relatives became more attractive. However, it was not until three decades later, led by the efforts of Professor H. R. Kricheldorf,[3] that synthetic protocols aimed at the investigation of these unique polymers began to evolve.

Such one-step polycondensations afford products possessing a high degree of branching (DB), but which are not as faultless as the stepwise constructed dendrimers; for reviews, see refs.[4–9] The supramolecular assemblies and micellar properties of these hyperbranched polymers offer synthetic and physical insights as well as noteworthy comparative relationships to the monomolecular dendritic analogues. The DB of these hyperbranched polymers generally ranges from 55–70 % and is independent of their molecular weights. For an extensive presentation of DB and average number of branches (ANB) per non-terminal monomer moiety, Frey and Hölter[10] considered the copolymerization of AB_x monomers with AB and AB_y monomers.

Synthetic high molecular weight polymers with spherical symmetry have also been created[11] by a graft-on-graft procedure (chloromethylation, followed by anionic grafting), which results in tree-like structures, analogous to these hyperbranched materials. These have been termed "arborescent graft polymers".[12–14] Control of structural rigidity in these materials has been investigated.[15] In general, grafting side chains of comparable molecular weight on a linear core forms "comb-branched" structures, ultimately leading to materials with increasing globular or spherical shape as generations increase. Such polymers are obtained with molecular weights ideally increasing geometrically as expressed by:

$$M = M_b + M_b f + M_b f^2 + \ldots = \sum_{x=0}^{G+1} M_b f^\infty, \tag{6.1}$$

where M_b is the molecular weight per branch, and f is the branching functionality, which remains constant for each generation G. For the graft-on-graft procedure, high molecular weights ($> 10^6$) were realized after three graftings with $M_w/M_N \approx 1.1$–1.3 at each generation. Monolayer films of arborescent polystyrene-based grafts have been created,[16] and SANS studies of arborescent graft polystyrenes have been published.[17, 18] The synthesis[19] of poly(ethylene imine)-based comb-burst dendrimers, their Monte Carlo simulations,[20] and density profiles from simulated comb-burst molecules[21] have appeared. A similar all-hydrocarbon, hyperbranched polyethene has also been reported.[22] Schultz and Wilks[23] addressed registration of hyperbranched polymers possessing symmetrical I-shaped structural repeat units with the Chemical Abstracts Registry Service as well as nomenclature and structural representation of asymmetrical I-shaped hyperbranched polymers.[24] Möller et al.[25] considered and discussed the conversion dependence of the polycondensation branching density for AB_x monomers. Polycondensation kinetic formulae were derived, thereby affording the time-dependent degree of polymerization and conversion. Other Monte Carlo simulations of hyperbranched reactions are known.[25a]

Mansfield[26] examined molecular weight distributions of imperfect dendrimers. Fourier analysis of the products revealed that essentially monodisperse molecular weight

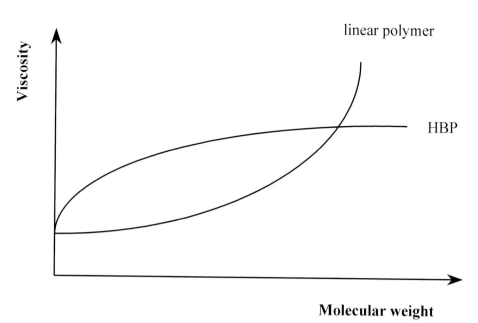

Figure 6.1 Viscosity behavior as related to molecular weight of linear and hyperbranched polymers. Reproduced by permission of Elsevier Science.[32]

ranges could be obtained (for divergent growth) if, in the early stages of growth, perfection was maintained, or nearly so, while later generations inevitably possess arbitrary amounts of defects. Hanselmann, Hölter, and Frey[27, 28] used computer simulation to model the kinetics of a novel core dilution/slow addition protocol for hyperbranched polymer preparation. Essentially, the technique consists of the slow addition of AB_x monomers to a B_f core with rapid and quantitative reaction ($x = 2$ or 3; $f = 2$–12). This method was shown to control resultant molecular weights, lower polydispersity, and enhance DB. McCoy[29] later presented an analytical solution for the growth of hyperbranched polymers; distribution kinetics of branched growth was addressed. A report on the kinetics of reactions with non-uniform reaction rate constants leading to hyperbranched materials has recently appeared,[30] in which three situations for AB_2 systems and the kinetics of self-condensing vinyl polymerization for the ABB' system are considered. It was found[30] that the influence of the first B substituent on the reactivity with the second B group has the greatest effect on structural outcome. The competition between hyperbranched growth and cyclization that can occur with flexible AB_x ($x = 2$ or 4) monomers in a step growth process has been simulated with a three-dimensional lattice model.[31] Two mean-field kinetic models have also been considered.[31a] Phase behaviour has been addressed.[31b]

In general, the comparative viscosity behavior for linear *vs.* hyperbranched macromolecules plotted against M_w is as shown in Figure 6.1.[32] Spherical dendritic materials also exhibit[33] a similar property for equivalent molecular weights; their mechanical properties are, for the most part, delineated by the core, whereas the chemical properties are determined by the shell, thus controlling the phase-separation processes. Thus, hyperbranched polymers have been shown to instill outstanding performance as tougheners in epoxy resins and do not lead to decreased resin stiffness or T_gs.

6.2   1 → 2 *Aryl*-Branched

6.2.1   1 → 2 *Aryl*-Branched and Connectivity

Kim and Webster reported,[34–36] patented,[37] and reviewed[38–42] the facile one-step conversion[38] of 3,5-dibromophenylboronic acid[43] (**1**) in the presence of a catalyst [Pd(PPh₃)₄] under reflux conditions in aqueous carbonate to give the hyperbranched polyphenylene **3** (Scheme 6.1).[44, 45] An alternative route to **3** utilized the mono-Grignard[34] (**2**), prepared from 1,3,5-tribromobenzene with activated magnesium,[46] and Ni(PPh₃)₂Cl₂; this method proved advantageous in large-scale runs. Based on ¹³C NMR

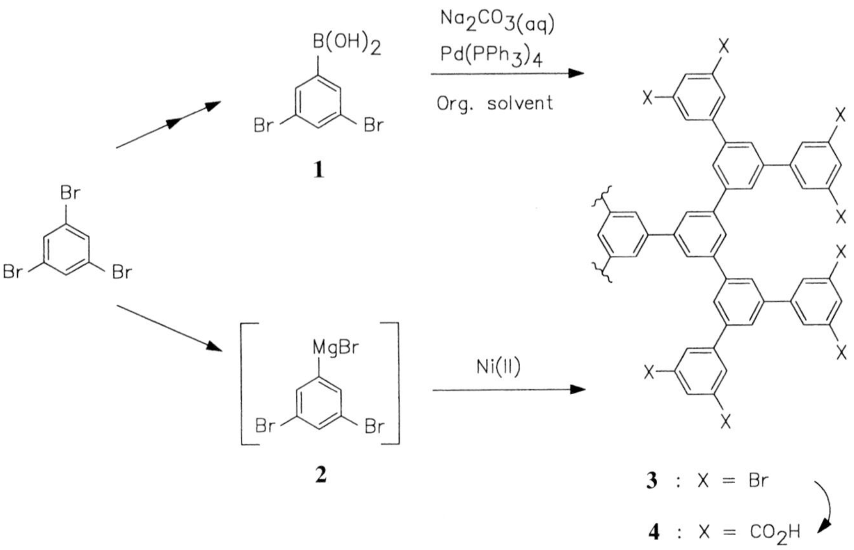

Scheme 6.1 Transition metal mediated preparation of polyphenylene macromolecules.[44, 45]

spectroscopy, the DB was estimated to be ca. 70 %, while the molecular weight of **3** was qualitatively found to be a function of the organic solvent employed. Polymerization with Ni(II) gave polymers of $\overline{M}_n$ in the 2,000 to 4,000 amu range, often with greater polydispersity than those obtained by the boronic acid route. Molecular weight limitations may result from steric hindrance at the organometallic center and/or intramolecular cyclization(s). The effect of different terminal groups on the T_g of these polyphenylenes and triphenylbenzenes has been studied.[47]

Diverse derivatives have been prepared from polylithio-polyphenylene, which, due to instability, was generated in less than quantitative yields by a metal–halogen exchange process. Electrophiles used to quench the polylithiated polyphenylene included CO_2, CH_3OCH_2Br, Me_2CO, DMF, $C_6H_5C(O)Me$, MeOH, Me_2SO_4, and Me_3SiCl.

Reaction of the polybromide **3** with the anion of 2-methyl-3-butyn-2-ol was examined. In this case, "polymer reactivity seems to be enhanced when a small amount of the bromide groups had reacted with the reagent." It was speculated[35] that "accelerated reactivity of a partially converted polymer could be one characteristic of highly branched materials." Wooley et al.[48] supported this supposition with the finding that the convergent coupling of dendritic wedges to a hexavalent core proceeded to completion with no evidence of partially substituted cores based on GPC experiments, thus indicating an accelerated reactivity of partially transformed branched macromolecules. The phenomenon was attributed to "localized polarity or microenvironmental effects" favoring monomer connectivity. These observations lend support to the assertion of complete surface transformation for many *divergently* constructed dendrimers.

When biphenyl ether or 1-methylnaphthalene was used as solvent, the polyphenylenes were determined to possess higher M_n values than in xylene solution. No molecular weight increase was obtained by the addition of further monomer at or toward the end of the polymerization. The highest molecular mass was obtained using nitrobenzene as solvent. In order to investigate the unimolecular micellar behavior of the water-soluble lithium salt of polycarboxylic acid **4**, its ${}^1H$ NMR spectrum in the presence of an NaOAc solution of *p*-toluidine was recorded. Signals associated with the guest molecule(s) were shifted upfield and dramatically broadened.

Webster et al.[49] expanded on the preparation of the polyphenylenes to develop the one-pot synthesis of hyper-cross-linked poly(triphenylcarbinol); thus, reaction (–80 → 25 °C, THF) of 4,4'-dilithiobiphenyl with Me_2CO_3 afforded trityl alcohol based polymer. The absence of carbonyl or methoxycarbonyl NMR resonances led to speculation that the polymer grows via a branched convergent process.

6.2.2   1 → 2 *Aryl*-Branched, *Ester* Connectivity

Kricheldorf et al.[3] laid the very early synthetic foundation for their, as well as others', hyperbranched efforts. Branched polycondensations were achieved by the copolymerization of 3-acetoxybenzoic acid and 3,5-bis(acetoxy)benzoic acid (**5**) or [3-(trimethylsiloxy)benzoyl chloride] and 3,5-bis(trimethylsiloxy)benzoyl chloride (**6**). Formulae for the most probable branching units (**a**–**c**) derived from the copolymerizations (Scheme 6.2) are depicted in Figure 6.2. The $\overline{M}_n$ using the trimethylsiloxy-based monomers ranged from 10,000 to 29,000 amu, while that for the acetoxy-based building blocks varied slightly from 3,300 to 3,700 amu.

In this series, a relatively clean condensation process was demonstrated[50–55] by reactions of the silylated carboxylic acids with acetylated phenol moieties. This procedure avoids acidic protons as well as reduces the effects of acid-catalyzed side reactions.

Fréchet et al.[56] reported the high-yielding, reproducible preparation of the hyperbranched aromatic polyester (**7**) with controllable molecular weight by the self-condensation of **6**, which was synthesized from 3,5-dihydroxybenzoic acid by silylation (Me$_3$SiCl, Et$_3$N) followed by treatment with SOCl$_2$ and a catalyst (Me$_4$NCl). Polymerization[56, 57] of acid chloride **6** (Scheme 6.2) was effected thermally; a polystyrene-equivalent M_w of 184,000 amu for the polyester was obtained using a reaction temperature of 200 °C and a catalyst (DMF or Me$_3$N·HCl). The thermal stability was reported to be analogous to that of similar linear materials, whereas solubility was found to be enhanced. The DB was determined to be between 55 and 60 %, the branches bearing a large number of both internal and external free phenolic groups, functionalization of which was possible. The molecular weight was found to increase with higher reaction temperatures, longer reaction times, and increased quantities of catalyst.

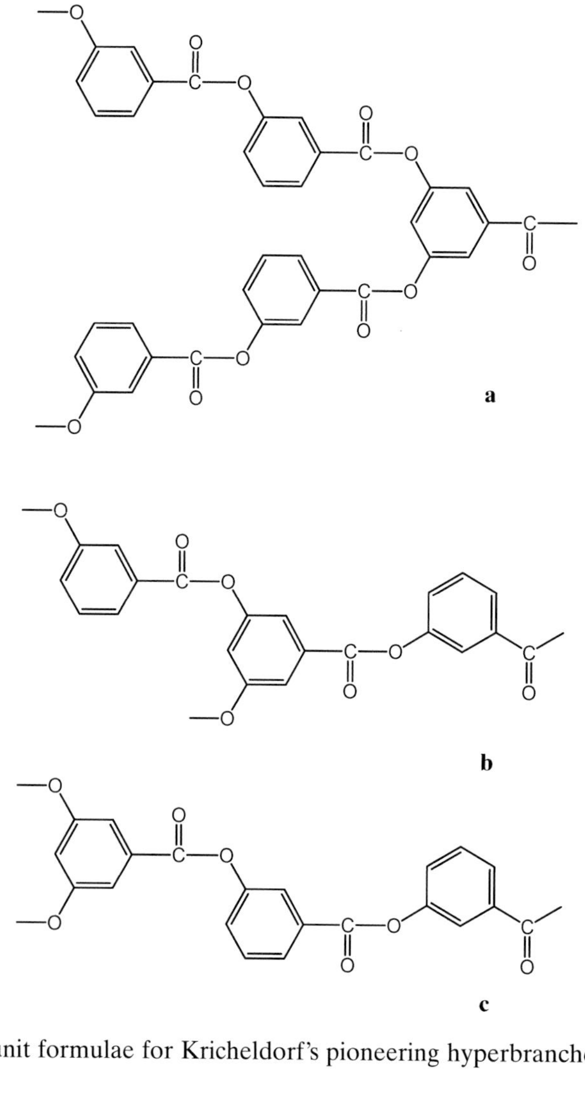

Figure 6.2 Branching unit formulae for Kricheldorf's pioneering hyperbranched polymers.[3]

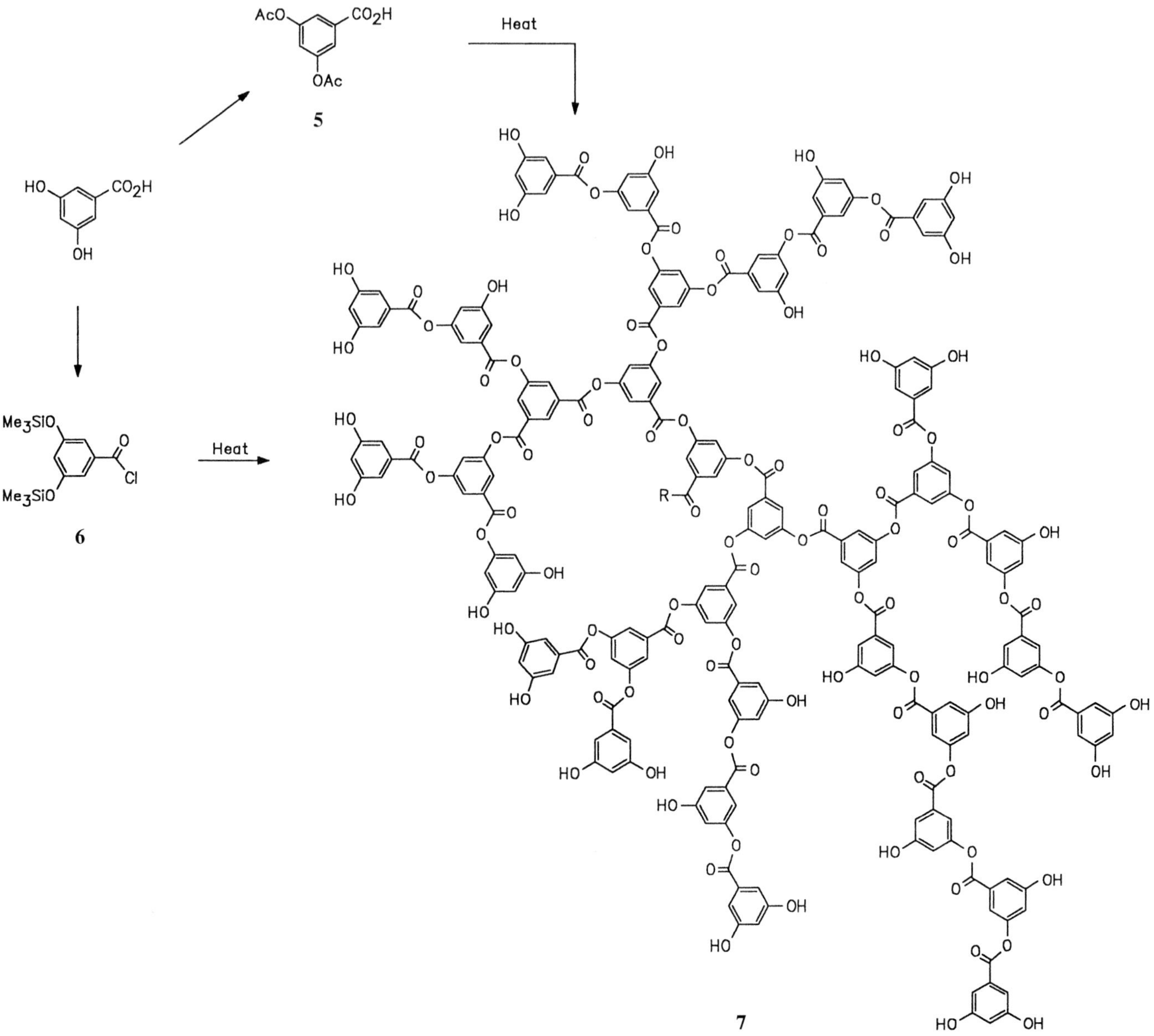

Scheme 6.2 Thermolysis of AB$_2$-type monomers affords high molecular weight polyesters.[57]

Fréchet and Hawker[58] employed a step-growth polymerization for the preparation of hyperbranched aromatic polyesters based on 3,5-dihydroxybenzoic acid derived monomers. The physical properties of these dendrimers were compared and contrasted to those of Kim and Webster's polyphenylenes[44] and those of Kambouris and Hawker's partially aliphatic polyesters.[59] Interesting comparisons have been made[33] between dendritic and hyperbranched structures; the thermal properties (T_gs and thermogravimetric analysis) were found to be independent of architecture, while their solubilities proved comparable, but greater than those of their linear counterparts. Analogous hyperbranched poly(silyl esters) have been prepared.[59a]

Turner, Voit, and Mourey[60, 61] later reported analogous hyperbranched polyester macromolecules (**7**) obtained by the thermal polymerization (Scheme 6.2) of AB$_2$ diacetate **5**, prepared (Ac$_2$O) from 3,5-dihydroxybenzoic acid. Polymer structures derived from **5** were supported by similar spectral characterization as that described for polymers obtained from **6**. Condensation of **5** below 170 °C was found to be "slow", whereas at 250 °C the rate was substantially increased. Products (possessing $\overline{M}_w > 1,000,000$ amu) were much less sensitive to purity of the starting material (i.e., **5**) than those synthesized employing TMS monomer **6**. When monomer **5** was heated *in vacuo*, it was the quality and duration of the vacuum that had the largest effect on the resultant weight; the $\overline{M}_w$ ranged from 5,000 to 800,000 amu depending on the temperature, time, and vacuum.

Catalysts (Mg° or *p*-MeC₆H₄SO₃H) did not improve the reaction course. These materials[60, 62] have been characterized by their molecular weights, viscosities, and rheology; the influence on the properties of different terminal groups as well as blend behavior have also been discussed.[63]

Massa, Voit, et al.[64] conducted a survey of the phase behavior of blends of these polyester hyperbranched polymers with linear polymers. Blend miscibility of a hydroxy-terminated polyester was found to be comparable to that of poly(vinylphenol), indicating strong *H*-bonding interactions, whereas miscibility of an acetoxy-terminated analogue decreased relative to the hydroxy derivative.

Kumar and Ramakrishnan[65] prepared hyperbranched polyesters by subjecting 3,5-dihydroxybenzoic acid and its derivatives to standard self-condensation conditions (i.e., transesterification). Spacer lengths between branching centers were varied using meso-genic segments and the resulting products were studied by DSC. These hyperbranched materials were found to be amorphous with no liquid-crystalline phases. This was postulated as being attributable to a "random distribution of the mesogenic segments."

A related series of hyperbranched polymers possessing high molecular weights (20,000–50,000 amu) was created[62, 66] by melt condensation of either 5-acetoxy- (**8**) or 5-(2-hydroxyethoxy)isophthalic acid (**9**) (Scheme 6.3). Polymerization of diacid **8** was effected in two stages: (1) melting at 250 °C combined with removal of AcOH with the aid of an inert gas, and (2) application of a vacuum at the onset of solid-state formation. Refluxing the resultant acid-terminated, ester-linked polymer **10** in THF/H₂O decomposed the labile anhydride cross-links that were generated under the reaction conditions. The DB[56] was determined to be ca. 50 %. The acetoxy monomer was also copolymerized with various AB-type monomers, e.g., 3-(4-acetoxyphenyl)propionic acid. Isophthalic acid **9**, prepared from 5-hydroxyisophthalic acid with ethylene oxide, was polymerized at 190 °C using a catalyst [Bu₂Sn(OAc)₂]. The resulting carboxylic acid terminated hyperbranched polymer (not shown) proved to be readily soluble in common organic solvents. Due to a lower condensation temperature than that employed for the polymerization of diacid **8**, evidence of anhydride bond formation was not observed.

Feast et al.,[67, 67a] in their continuing study of the polymerization of dimethyl hydroxy-alkyloxyisophthalates,[68] reported step-growth polymerization of these AB₂ monomers

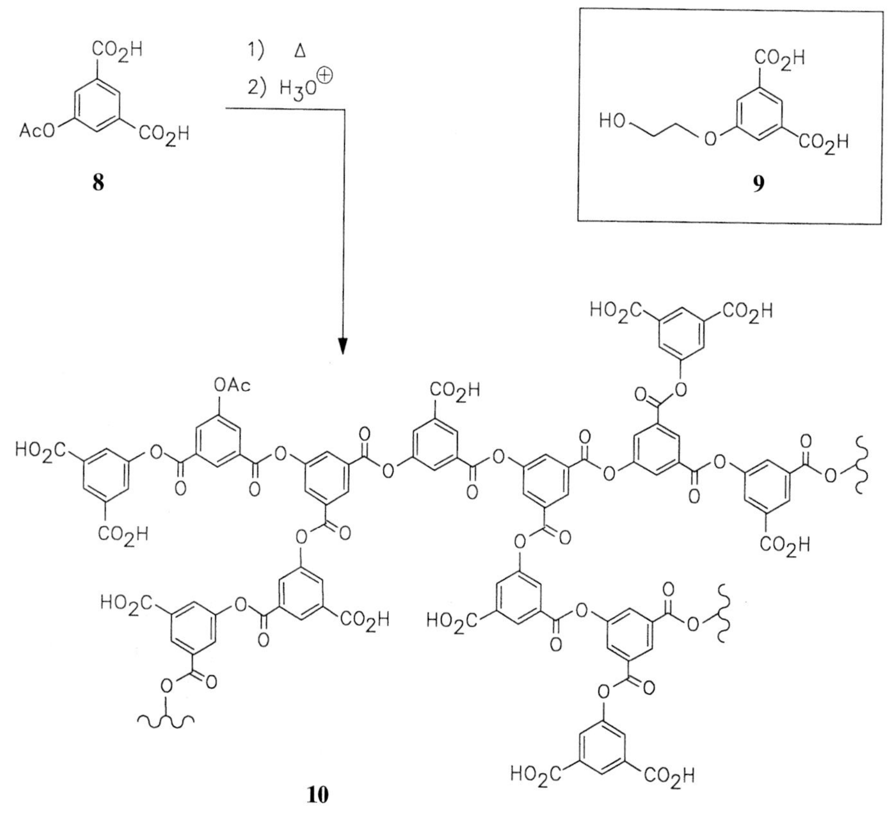

Scheme 6.3 Melt condensation of isophthalic acid derivatives affording polyesters.[62, 66]

possessing alkyloxy methylene chain lengths of 2 to 6 units. Polymerization was effected by heating (210 to 240 °C) under nitrogen in the presence of a transesterification catalyst [Mn(OAc)$_2$/Sb$_2$O$_3$] and a thermal degradation suppressant [(PhO)$_3$PO]. For polymers obtained using the monomer with the shortest chain length, molecular weight and polydispersity increased with increasing reaction time up to a maximum molecular weight of 107,000 amu and an M_w/M_n of 15. Similar behavior for the higher chain length polymeric homologues was not observed, although an "odd-even effect" was noted in that alkene lengths of 2, 4, or 6 methylene units produced higher molecular weight polymers than the monomers possessing an odd number (3 or 5) methylenes.

Kricheldorf et al.[69] prepared two classes of liquid-crystalline hyperbranched copolyesters by copolymerization of silated **5** with difunctional mesogenic monomers. One class was prepared from a binary mesogen mixture comprised of silylated derivatives of β-(4-acetoxyphenyl)propionic acid and 4-acetoxybenzoic acid, while the other class was obtained from a ternary mixture that also incorporated silated 6-acetoxy-2-naphthoic acid. It was found that, in this case, at least a six-difunctional monomer segment length

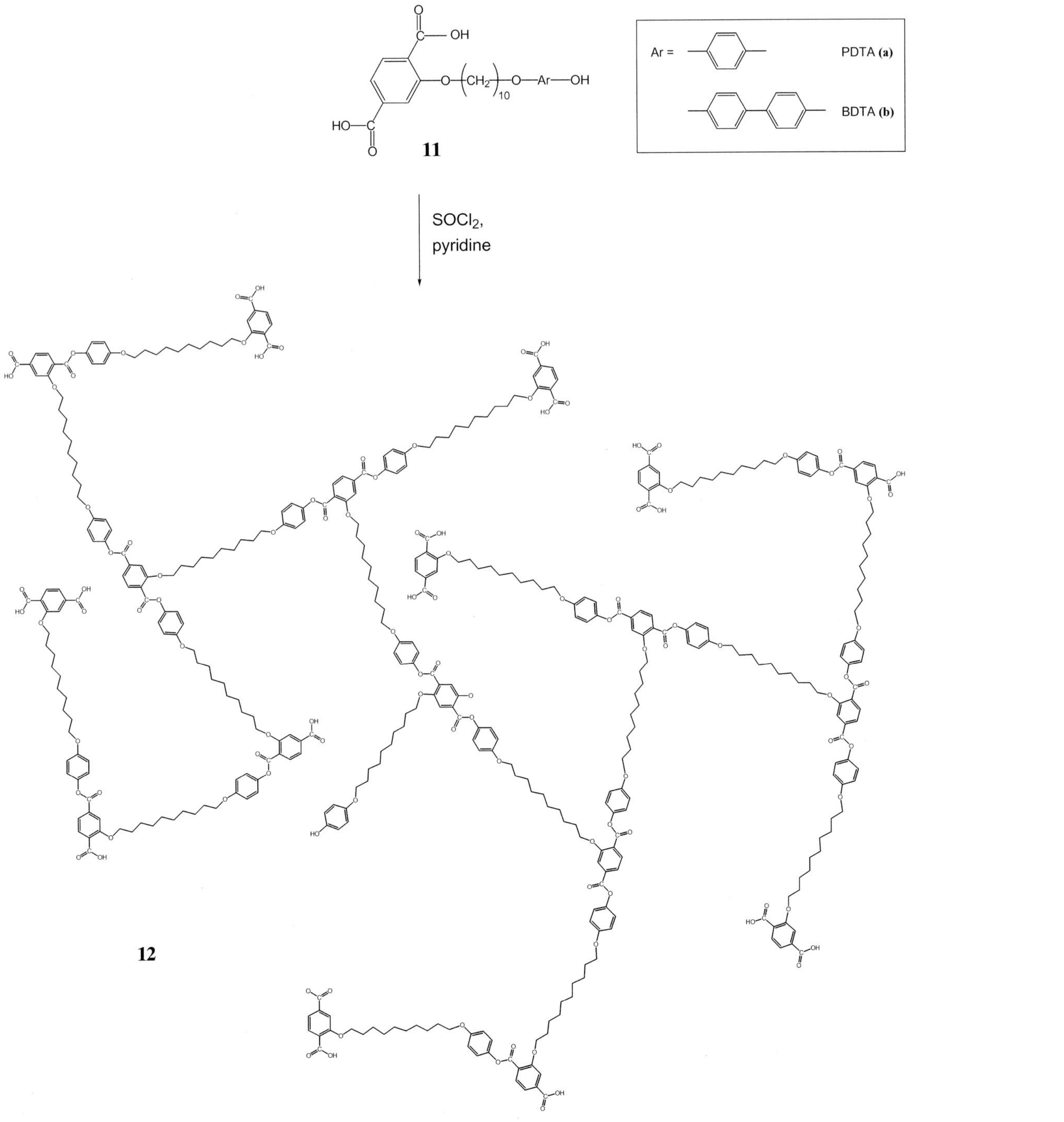

Scheme 6.4 Thermotropic, liquid-crystalline hyperbranched polymers based on terephthalate derivatives.[72]

between branch junctures was required in order to obtain a nematic phase. Melt rheology was used to further characterize the binary-based polymers. Additional poly(ester)s prepared by Kricheldorf et al. included liquid-crystalline poly(ester)s based on β-(4-hydroxyphenyl)propionic acid and either 4-hydroxybenzoic[70] or gallic[71] acids.

Jin et al.[72] reported the preparation of thermotropic, liquid-crystalline hyperbranched polymers (e.g., **12** in Scheme 6.4) based on terephthalic acid derivatives **11** possessing either phenyl (a) or biphenyl (b) moieties. Monomer preparation was effected by methoxy-*para*-xylene oxidation (KMnO$_4$) to give the diacid, phenol liberation (HBr), esterification, phenol alkylation [Br(CH$_2$)$_{10}$Br], aryl diol (or biphenyl diol) chain attachment, and finally ester hydrolysis. Polymerization was brought about by polycondensation in the presence of SOCl$_2$ in pyridine. The $\overline{M}_n$s were found to be 6,970 and 15,050 amu for the phenyl- and biphenyl-based polymers, respectively. Each was observed to form nematic liquid crystals, although this property was not exhibited by the corresponding methyl ester terminated polymers; the DB for both materials was determined to be ca. 40 %.

Voit et al.[73] examined blends of amphiphilic polyesters and polyolefins. Polyesters, prepared by polycondensation of 3,5-dihydroxybenzoyl chloride, in turn obtained by hydrolysis of the corresponding bis(TMS) derivative, were surface-modified with dodecanoyl chloride to afford amphiphilic globular polymers. The ability to incorporate organic dyes was demonstrated. Blends with polypropylene or polyethylene were prepared, with the polymer content ranging from 0.05 to 20 wt %. The dye-hosting potential facilitated homogeneous dye distribution in the polyolefin matrices.

Kricheldorf, Bolender, and Wollheim[74] also reported the polycondensation of bissilylated acid chloride **6** (see Scheme 6.2) as well as its co-polycondensation with 3-trimethylsiloxybenzoyl chloride or *N,N'*-bis(trimethylsilyl)bis(aniline-P). Terminal group modification with acetyl, chloroacetyl, undecanoyl, stearoyl, 4-methoxycinnamoyl, and perfluorooctanoyl acid chlorides was also reported; solubilities and T_gs were found to vary greatly with end group. Films cast from 4-methoxycinnamoyl-coated polymers were cross-linked by UV irradiation.

6.2.3   1 → 2 *Aryl*-Branched, *Ether* Connectivity

A one-step synthesis (Scheme 6.5) of the hydroxy-terminated, dendritic polyether **15** from 5-(bromomethyl)-1,3-dihydroxybenzene (**14**), prepared (PPh$_3$, CBr$_4$) from 1,3-dihydroxy-5-(hydroxymethyl)benzene (**13**), has been reported.[75, 76] Bromomethyl monomer **14** was polymerized upon addition to an acetone suspension of K$_2$CO$_3$ and 18-crown-6 (an ubiquitous method of connectivity in convergent, iterative sequences).[77] The rate of monomer addition did not significantly affect the polymer characteristics and $\overline{M}_w$ exceeded 10^5. ^{1}H NMR spectroscopic analysis indicated that *C*- as well as *O*-alkylation occurred during polymerization, to varying degrees ranging from 11 to 32 %, when parameters such as reaction time, concentration, and solvent were varied; the least amount of *O*-alkylation occurred utilizing prolonged reaction times (92 h) with acetone as the solvent. Polymer characterization also included molecular weight determinations by SEC and low-angle laser light scattering.

Wooley and coworkers[78] prepared similar perfluorinated materials using a pentafluorinated aryl-terminated 1st generation Fréchet-type monomer (**16**; Scheme 6.6). Treatment of monomer **16** with Na metal afforded the hyperbranched material **17** through benzyloxy anion substitution of the terminal *p*-fluoro moieties. These polymers were derivatized with lithium trifluoroethoxide, lithium perfluorodecanoxide, and *p*-iodophenol. AFM revealed a significant decrease (2-fold) in the friction coefficient of films of the perfluoroalkyl-substituted material relative to the unsubstituted polymers. Analysis (MALDI-TOF MS) of the cyclization that occurred during this hyperbranched synthesis has been reported.[79]

Miller, Neenan, et al.[80, 81] reported a general single-pot method (Scheme 6.7) for the preparation of poly(aryl ether) hyperbranched macromolecules (**20a–d**) that were functionally analogous to linear poly(aryl ether) engineering plastics.[80–82] These hyper-

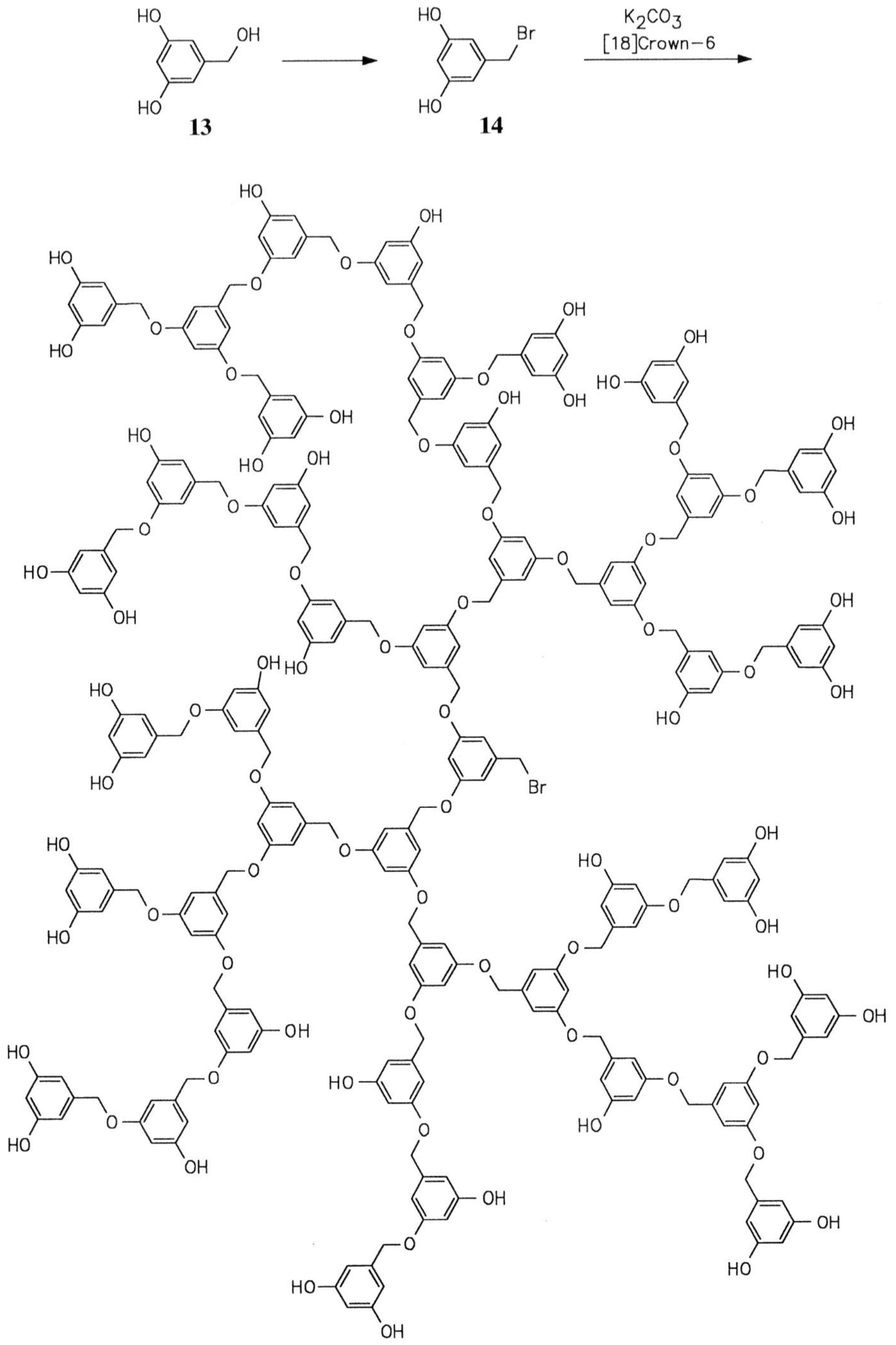

Scheme 6.5 Preparation of hyperbranched analogs (**15**) of Fréchet's convergently constructed, (benzyl ether)-based dendrimers.[75]

branched polymers were generated from phenolic A_2B-type monomers (**18a–d**), which were converted (NaH, THF) to the corresponding sodium phenoxides (**19a–d**) and subjected to polymerization. The procedure utilized phenoxide monomers possessing two carbonyl-, sulfonyl-, or tetrafluorophenyl-activated aryl fluoride moieties, which were displaced during monomer connection. Reaction conditions included a short duration (0.5–2 h) and moderate temperatures (100–180 °C). Polymerizations involving monomers **18a,b** were found to be insensitive to the reaction temperature, whereas the self-assembly of monomers **18c,d** at 140 °C afforded insoluble gels, presumably due to acetylene cross-linking, although at 100 °C completely soluble polymers were realized. A related procedure (bromide displacement by phenoxide ion) using 2,4-dibromophenol or 2,4,6-tribromophenol as the starting material has also been reported.[36]

Polyhydroxylated, oxygen-organosolv lignin (also referred to as oxygen-acetone lignin) possessing a highly branched architecture has been employed as a micromonomer

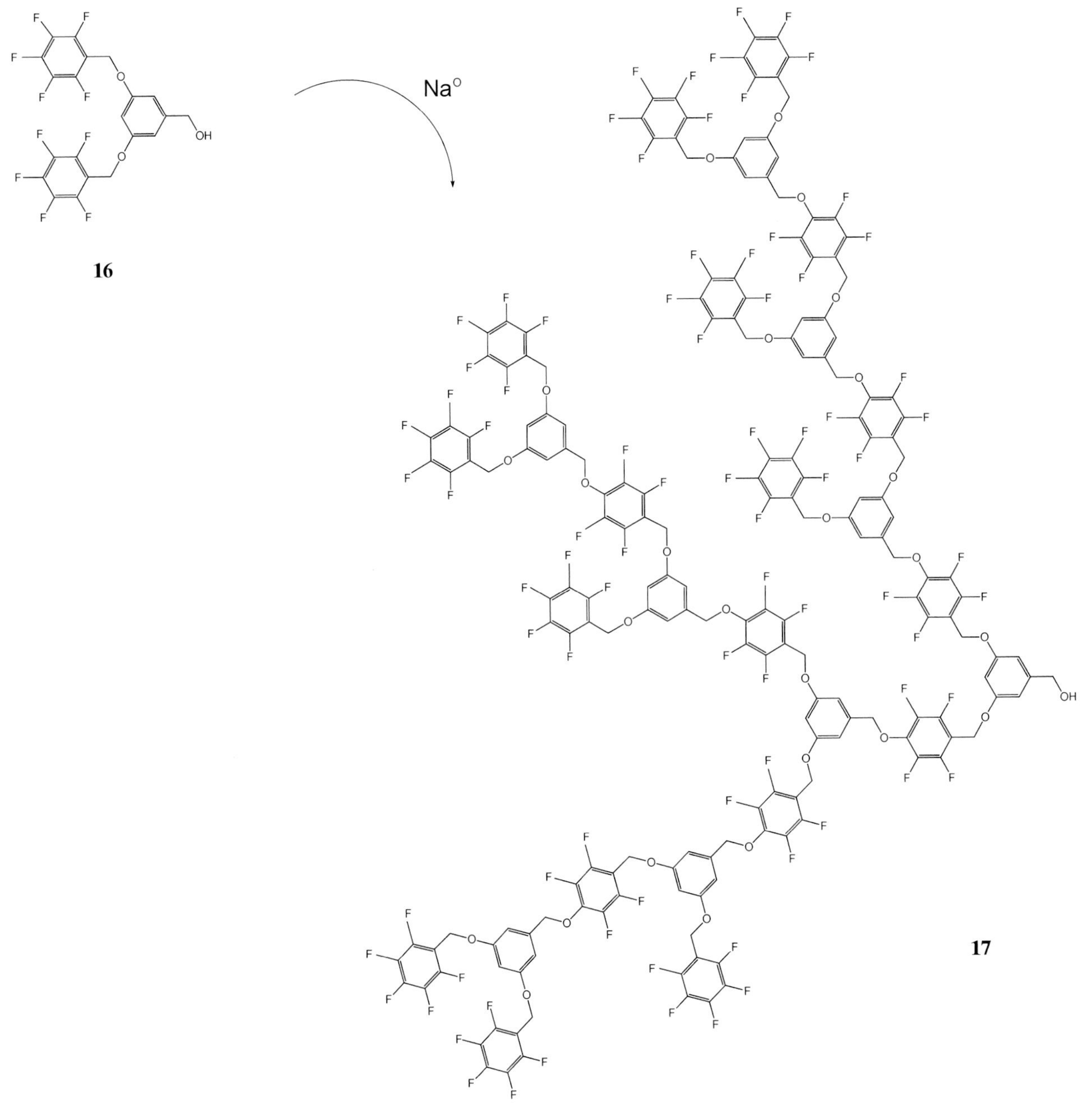

16

Na°

17

Scheme 6.6 Wooley's hyperbranched perfluorinated construct.[78]

(**21**; Figure 6.3) for polyesterification using diacyl chlorides and oligoethylene oxide glycol by Evtugin and Gandini.[83] Notably, lignin use in the production of hyperbranched architectures is of practical concern in view of the ready availability of these abundant and renewable biomass resources.

Chang and Fréchet[84] applied their proton-transfer polymerization approach to the construction of hyperbranched poly(aryl ether)s. Using epoxy-modified benzyl ether **22** (Scheme 6.8) under basic conditions, repetitive phenolic proton abstraction, epoxide ring-opening, and proton transfer facilitated formation of the polymer (e.g., **23**). Chemo- and regioselectivity were maintained due to the greater nucleophilicity of the phenolate compared to the secondary alkoxide; an M_w up to 206,000 amu was reported and the polymerization kinetics were also addressed.

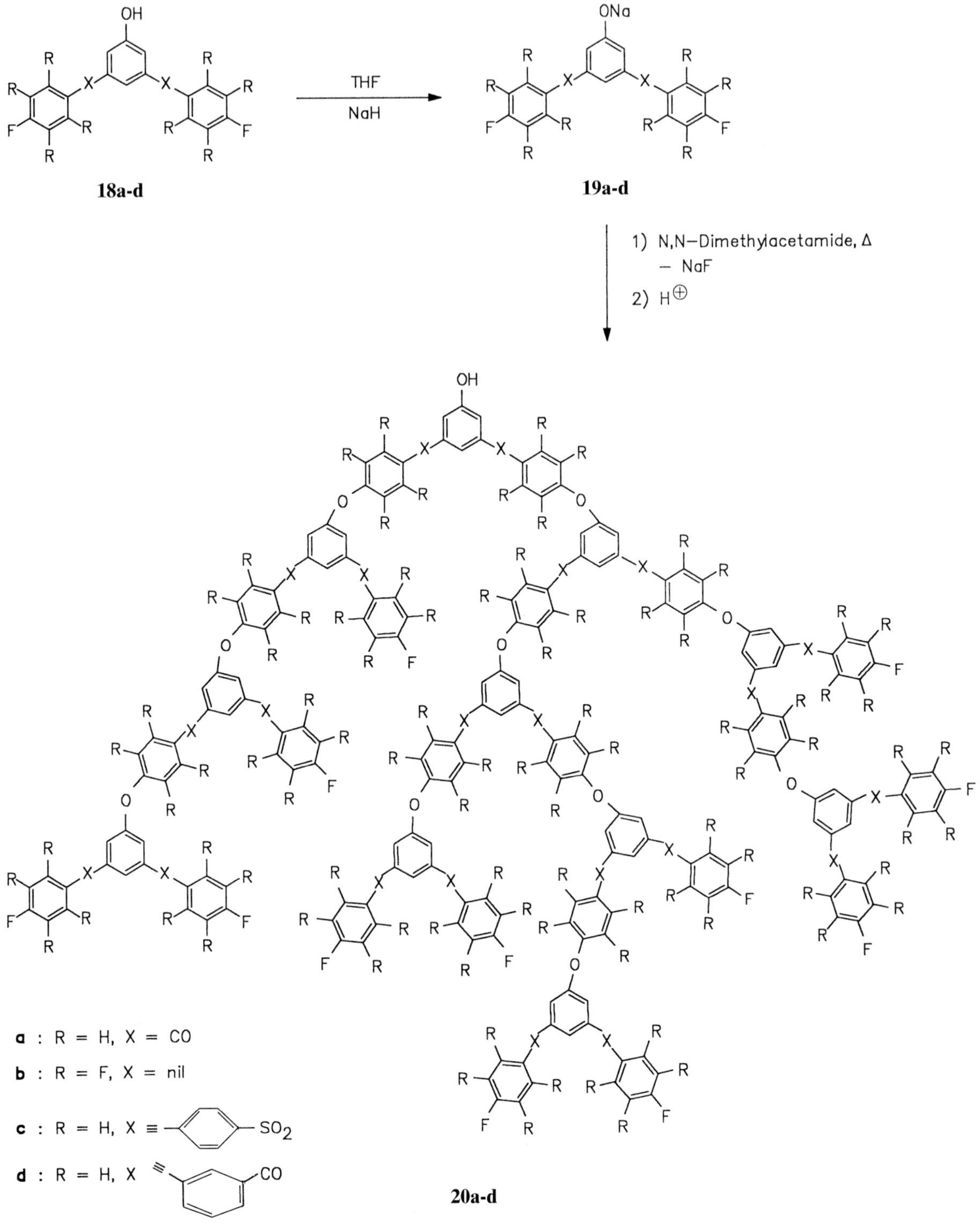

a : R = H, X = CO

b : R = F, X = nil

c : R = H, X ≡ —⟨benzene⟩—SO₂

d : R = H, X ≡ ⟨benzene⟩—CO

20a-d

Scheme 6.7 Preparation[80, 81] of branched analogs of (aryl ether) engineering plastics.

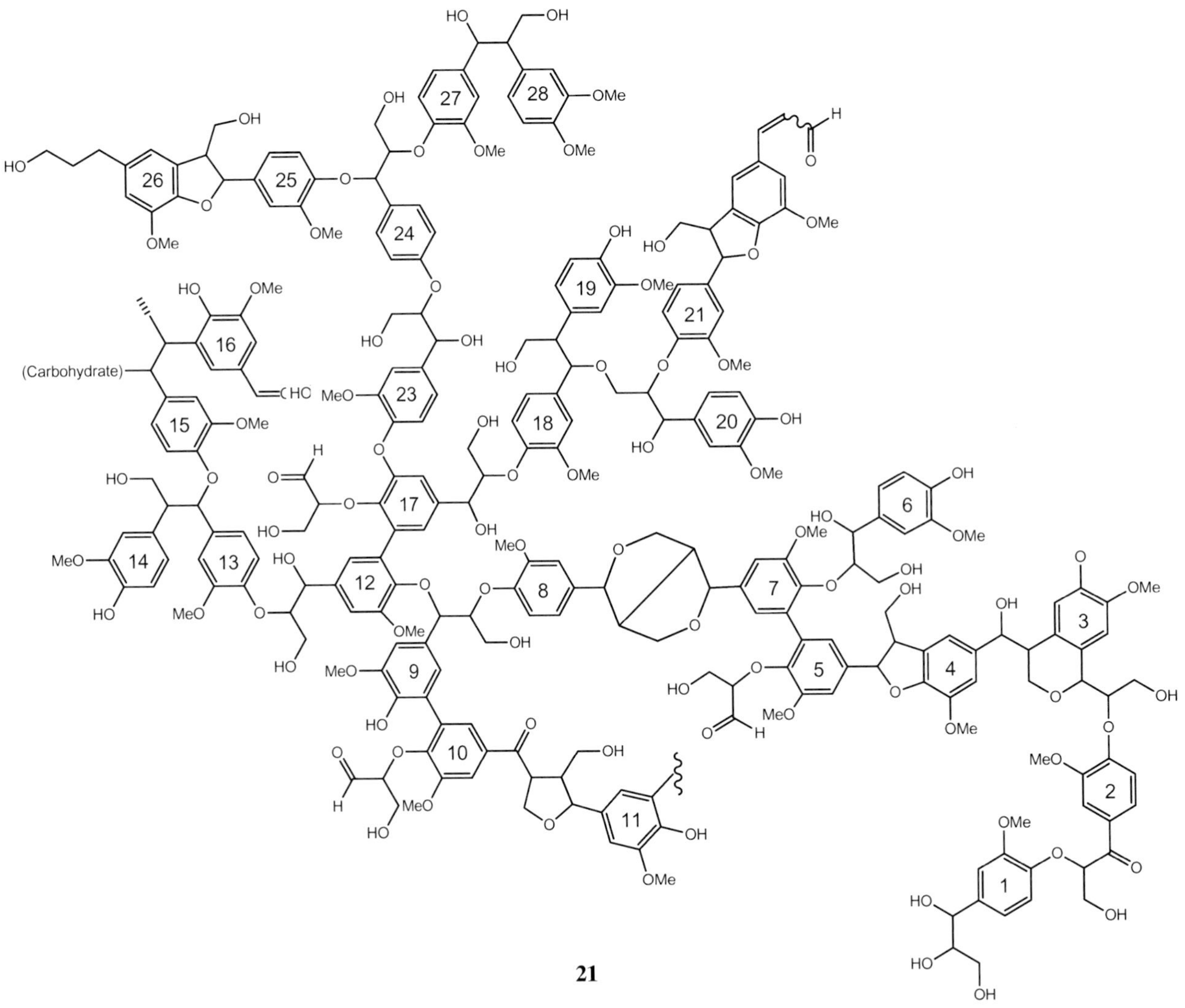

21

Figure 6.3 Branched lignins (**21**) used as "micromonomers" for hyperbranched polymers.[83]

6.2.4 1 → 2 *Aryl*-Branched, *Amine* Connectivity

Hyperbranched *m*-polyanilines (**25**; Scheme 6.9), prepared for the study of organic magnetic systems, have been reported.[85] Pd-catalyzed (Pd$_2$dba$_3$/3 equiv. BINAP) condensation of 3,5-dibromoaniline (**24**) yielded branched polymer **25**; analogous linear materials have also been reported. Molecular weights determined for these materials ranged from 7,000 to 22,000 amu, with a DP of 50 as determined by GPC.

6.2.5 1 → 2 *Aryl*-Branched, *Ether & Ketone* Connectivity

Hawker et al.[86, 87] reported the structural characterization of hyperbranched macromolecules (Scheme 6.10; **28** and **29**), obtained from the polymerization of monomers related to the AB$_2$ systems such as 3,5-difluoro-4'-hydroxybenzophenone (**26**) and 3,5-dihydroxy-4'-fluorobenzophenone (**27**); also see.[87a] These new related families of poly(ether ketone)s possessed the same internal linkages as well as terminal groups, but differed in the DB. Interestingly, as DB increased, the solubilities of the polymers increased, whereas their viscosities decreased; no change in the thermal characteristics was observed.

Hawker and Chu[88] later expanded on their poly(ether ketone) hyperbranched polymers to include AB$_3$- and AB$_4$-type monomers. The thermal properties of these materials were shown not to depend on molecular architecture, but rather on chain end functional-

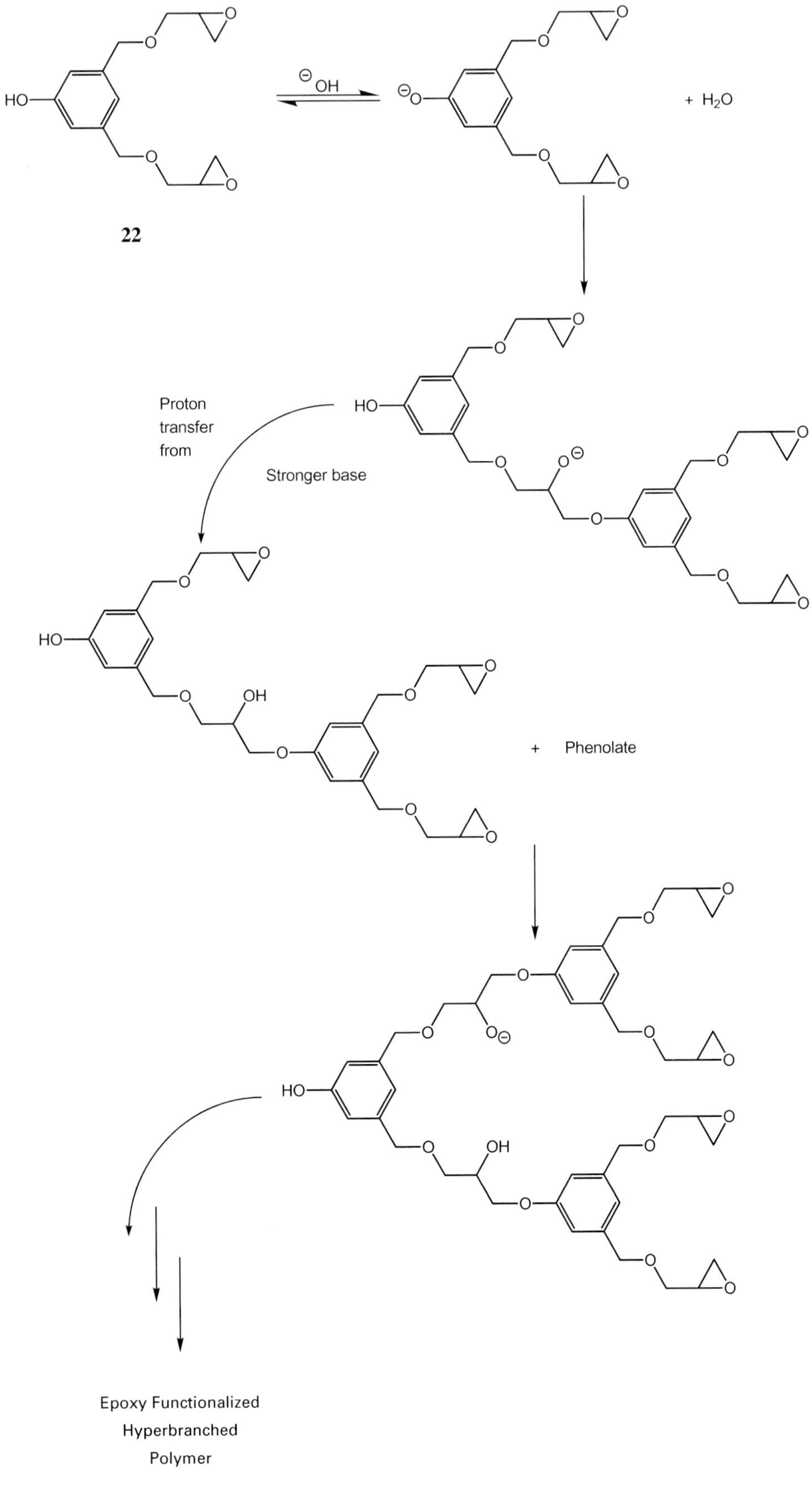

22

23

Scheme 6.8 Fréchet's poly(aryl ether)s obtained by proton-transfer polymerization.[84]

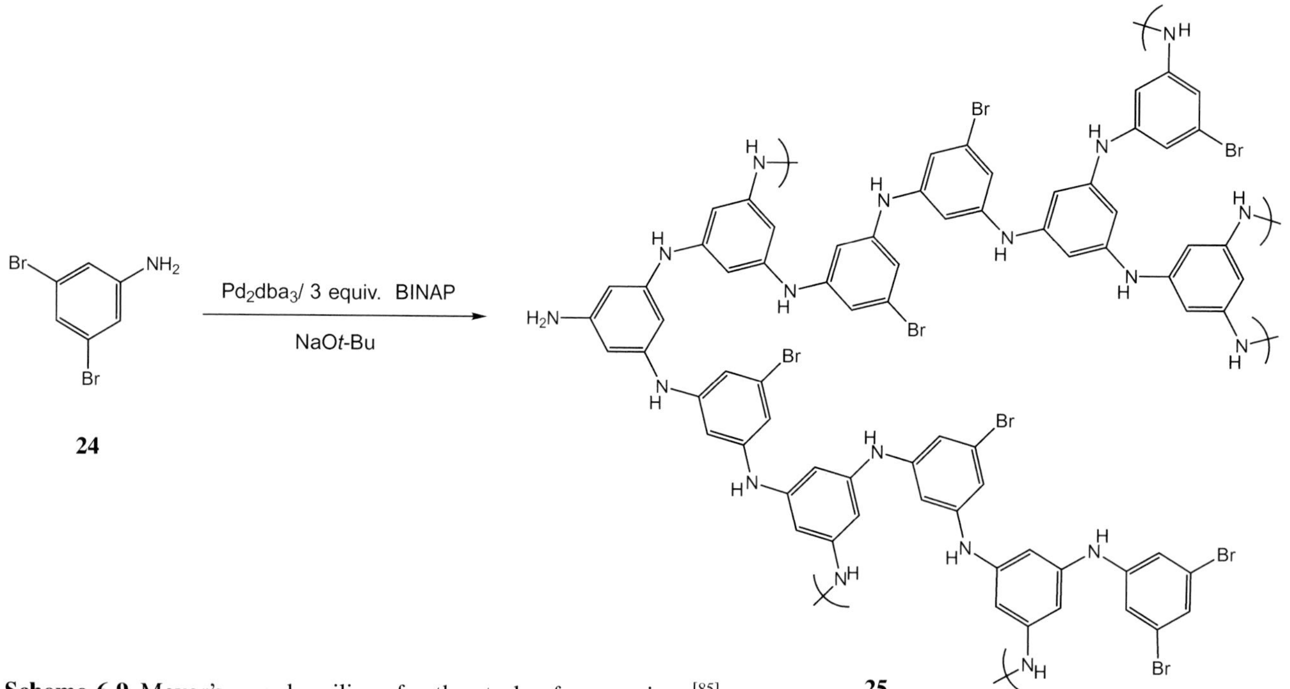

Scheme 6.9 Meyer's *m*-polyanilines for the study of magnetism.[85]

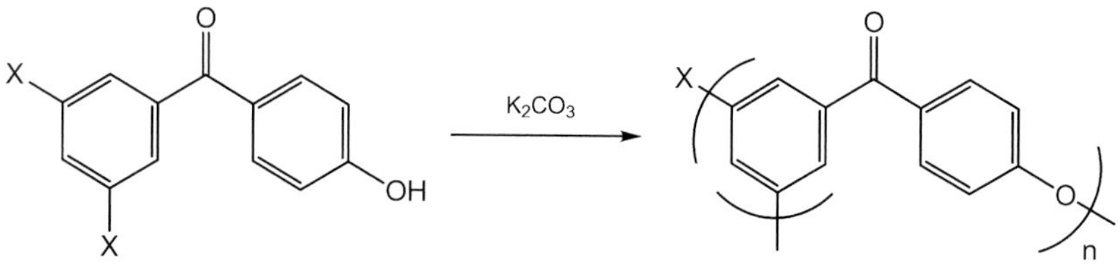

26 (X = F)
27 (X = OH)

28 (X = F)
29 (X = OH)

Scheme 6.10 Preparation of hyperbranched keto aryl ether polymers.[86,87]

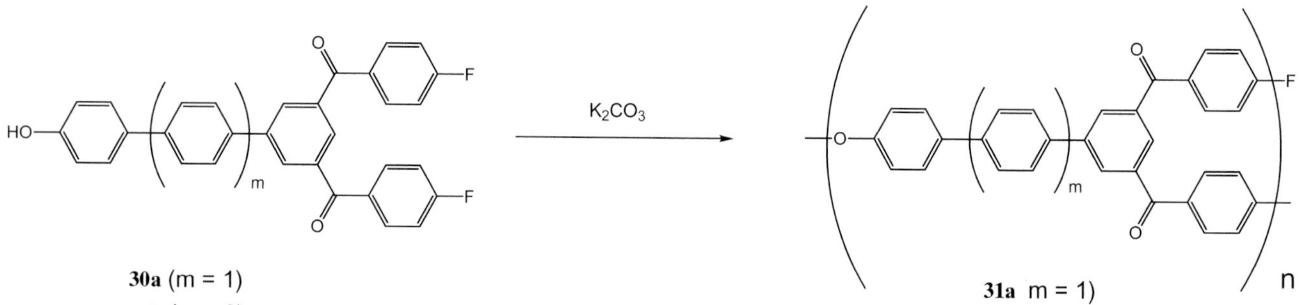

30a (m = 1)
b (m = 2)
c (m = 3)

31a m = 1)
b (m = 2)
c (m = 3)

Scheme 6.11 Morikawa's polyphenylene polymers.[89]

ity; the T_gs ranged from 97 to 290 °C. Imparting aqueous solubility by capping with carboxylate moieties led to the creation of unimolecular micelles, as demonstrated by the solubilization of 1,4-diaminoanthraquinone in water (40-fold increase in solubility compared to that in neat water).

Morikawa[89] reported the use of phenolic displacement of aryl fluoro substituents to achieve hyperbranched monomer connectivity. A series of phenyl-homologated phenolic monomers (**30**; Scheme 6.11) was thus polymerized (K_2CO_3) to afford a new family of polymers (**31**) possessing thermal properties that varied in relation to the phenylene spacer length. For example, T_gs increased with increasing spacer length. The thermal properties were also compared with those of other poly(ether ketone)s. Different fluoroterminated hyperbranched poly(ether ketone)s with varying DB and their linear counterparts have been synthesized and the effects of the DB have been evaluated.[90]

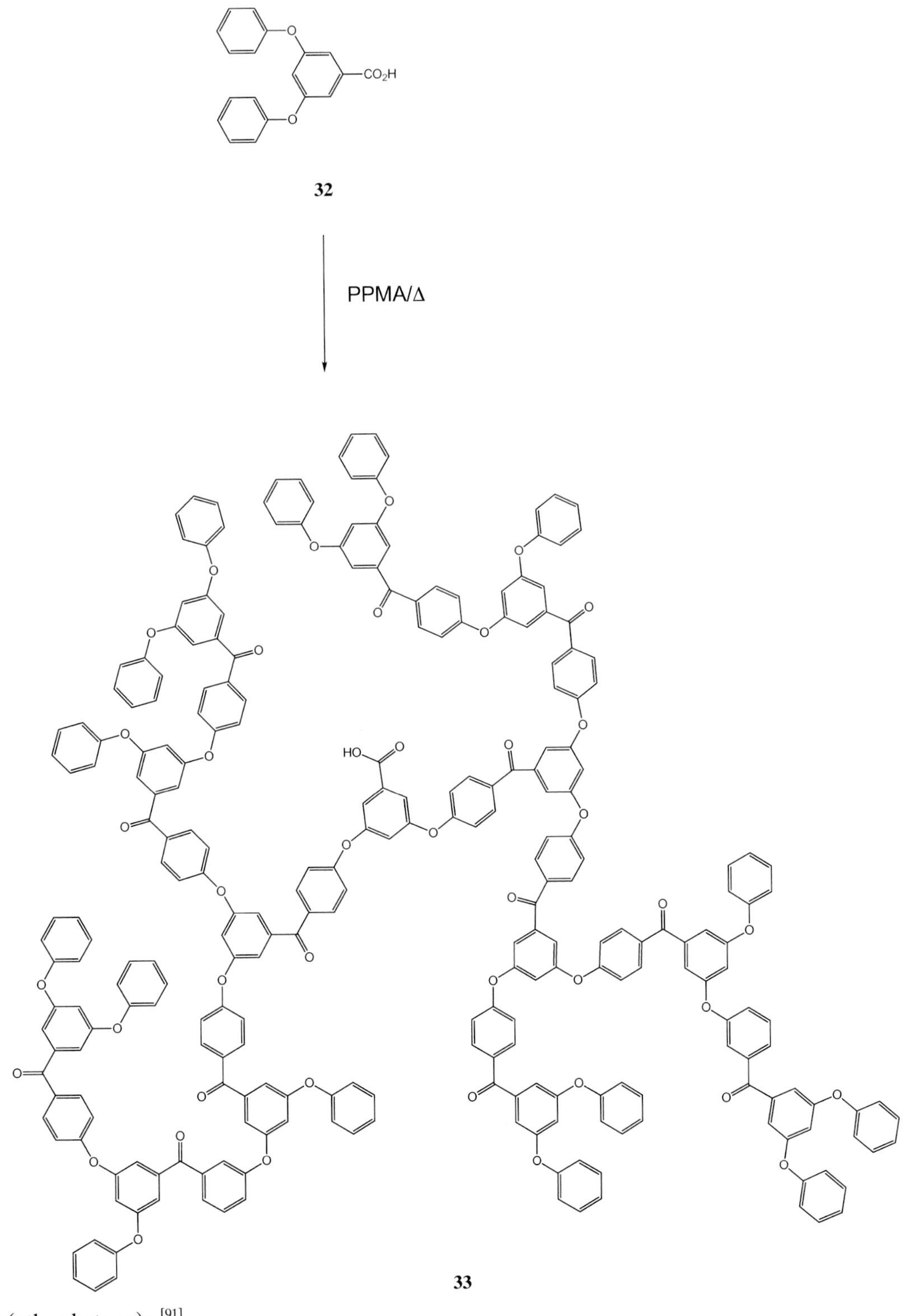

32

PPMA/Δ

33

Scheme 6.12 Shu's poly(ether ketone)s.[91]

Shu, Leu, and Huang[91] prepared poly(ether ketone)s (**33**) by the polycondensation of 3,5-di(phenoxy)benzoic acid (**32**; Scheme 6.12) using $P_2O_5/MeSO_3H$ as both the condensing agent and solvent. These materials (e.g., **33**) were further modified with *p*-toluic, 4-*n*-octylbenzoic, or stearic acid, effectively changing the thermal and solution properties of the polymers. For example, as the length of the terminal alkyl chains was increased, the T_g of the polymer was observed to decrease. Similar poly(ether ketone)s have been constructed[92] using 5-phenyloxyphthalic acid; the resulting polymer, possessing carboxylic acid termini, was modified with a variety of capping groups including MeOH, diphenyl ether, toluene, and NH_3, and was found to have DBs of ca. 55 %. 1,4-Diaminoanthraquinone was used to demonstrate the micellar character of the ammonium carboxylate derivative. Poly(aryl ether oxazole)s are also known.[92a]

Telechelic oligo(ether ketone)s possessing two trimethylsiloxy end groups and one methyl moiety per repeat group were accessed by polycondensation of 4-fluoro-2'-methyl-4'-(trimethylsiloxy)benzophenone.[93] Condensation of a small amount of bis(phenol-P) imparted the telechelic character. The oligomers were condensed with silated 3,5-bis(acetoxy)benzoic acid to produce A–B–A triblock copolymers.

6.2.6 1 → 2 *Aryl*-Branched, *Amide* Connectivity

Kim[36, 94, 95] reported two related types of hyperbranched aromatic polyamides; each employed amide bond formation for monomer connection and was based on three-directional, aryl branching centers (Scheme 6.13). Polymerization occurred upon neu-

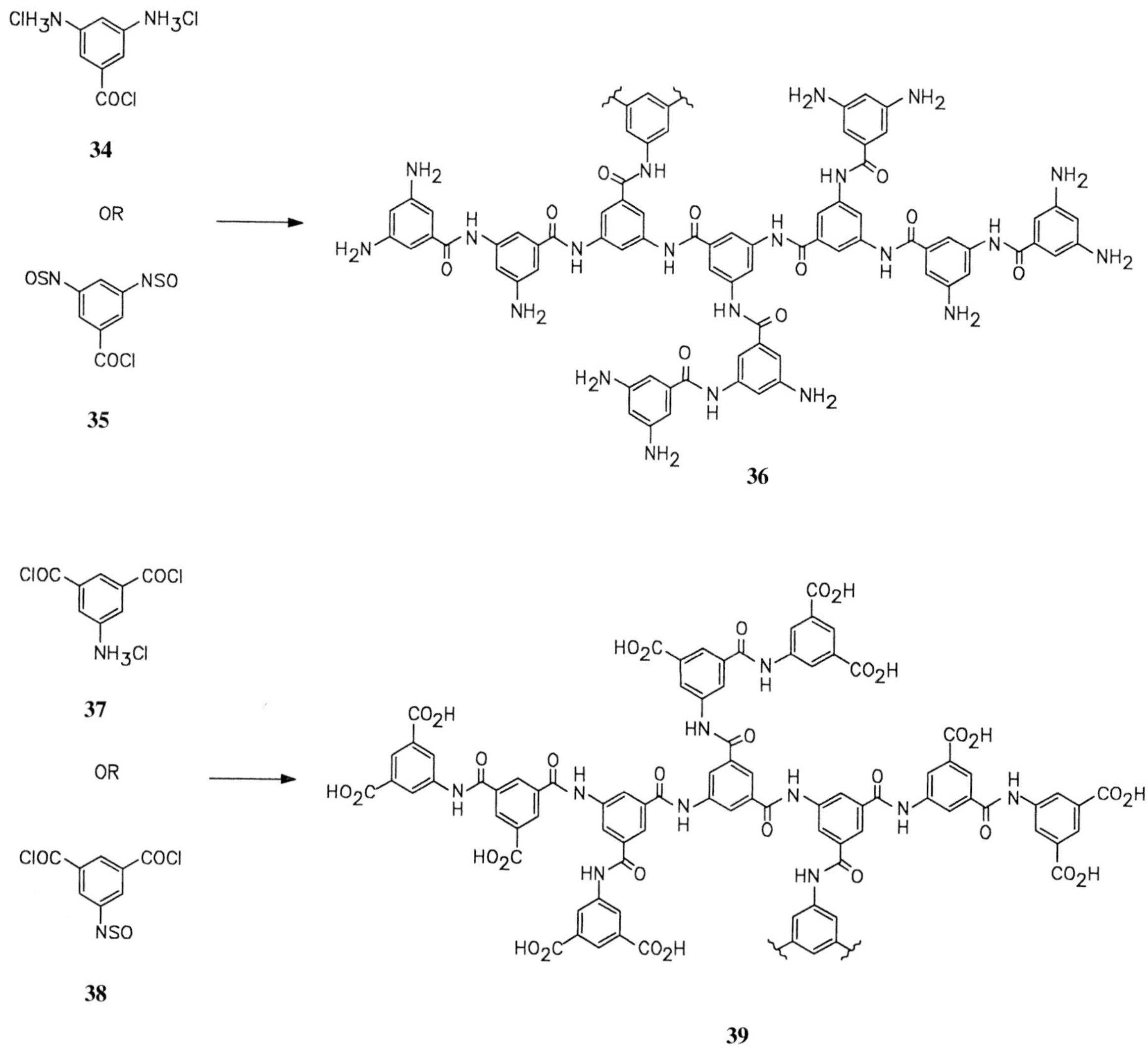

Scheme 6.13 Procedures for the preparation of poly(aryl amide) polymers.[36, 94] The carboxyl-terminated polymers exhibit lyotropic liquid-crystalline behavior.

tralization of either of the amino acyl chloride hydrochlorides **34** or **37** to afford the amine or carboxylic acid terminated materials **36** and **39**, respectively. Similar polymerizations, utilizing the sulfinyl-masked amino acid chlorides **35** and **38**, were effected by hydrolysis of the amine-protecting group. Lyotropic liquid-crystalline behavior was manifested in birefringence exhibited by the acid-terminated polymer **39** in either NMP or DMF at a concentration > 40 wt %. The corresponding poly(methyl ester) also displayed birefringence. Molecular weights (GPC) in the range 24,000 to 46,000 amu were indicated when NMP/LiBr/H$_3$PO$_4$/THF was used as eluent; however, elution with pure DMF indicated aggregates of 700,000 to 1,000,000 amu.

Reichert and Mathias[96] prepared branched aramids akin to those of Kim[36, 94] from 3,5-dibromoaniline (**40**) under Pd-catalyzed carbonylation conditions (Scheme 6.14). These brominated hyperbranched materials **41** were found to be insoluble in solvents such as DMF, DMAc, and NMP, in contrast to the soluble polyamine and polycarboxylic acid terminated polymers synthesized by Kim. This supports the observation that surface functionality plays a major role in determining the physical properties of hyperbranched as well as dendritic macromolecules.[34, 97] A high degree of cross-linking was also found to significantly affect solubility. When a four-directional core was incorporated into the polymerization through the use of tetrakis(4-iodophenyl)adamantane (**42**),[98, 99] the resultant hyperbranched polybromide (e.g., **43**) exhibited enhanced solubility in the aforementioned solvents, possibly as a result of the disruption of crystallinity and increased porosity. Adamantane derivative **42** has been transformed to the corresponding tetraacetylene derivatives,[99, 99a] which were subsequently polymerized in a stepwise process to afford highly cross-linked, hyperbranched polymers.

The solid-phase preparation of hyperbranched dendritic polyamides based on 3,5-diaminobenzoic acid was attempted, but was deemed to "have severe limitations", such as that the critical condensation reaction could not be forced to completion.[100] The inaccessibility of chain ends within the solid support was suggested[35] to be the major obstacle to high yield conversions.

Ueda et al.[101] employed a "direct polycondensation method" for the synthesis of poly(aryl amide)s, in which polymerization of 5-aminoisophthalic acid was effected by

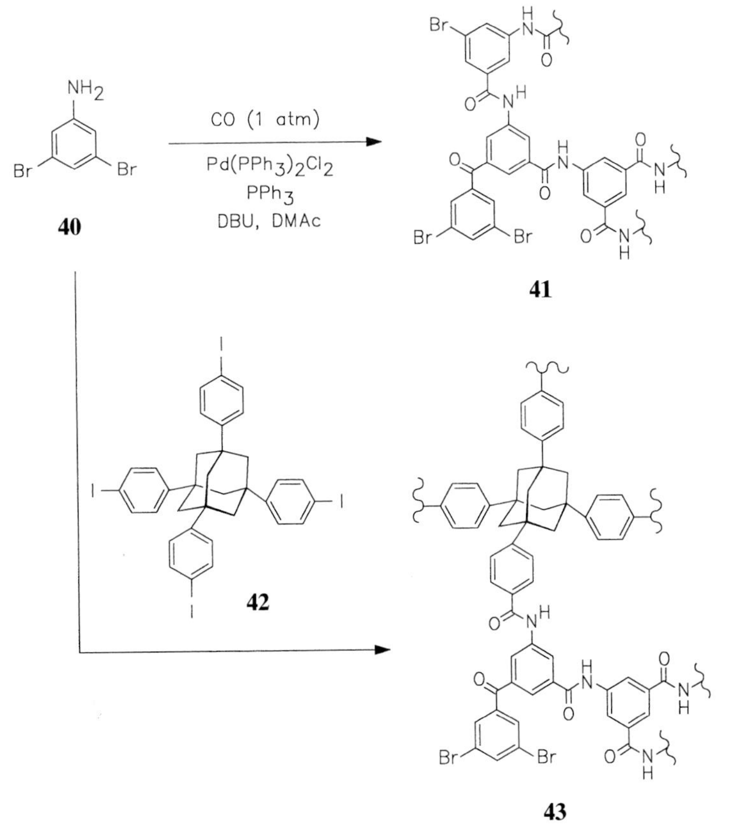

Scheme 6.14 Concomitant carbonylation and hyperbranched polyaramide formation,[96] where a rigid, tetrahedral adamantane-based core imparted enhanced solubility characteristics.

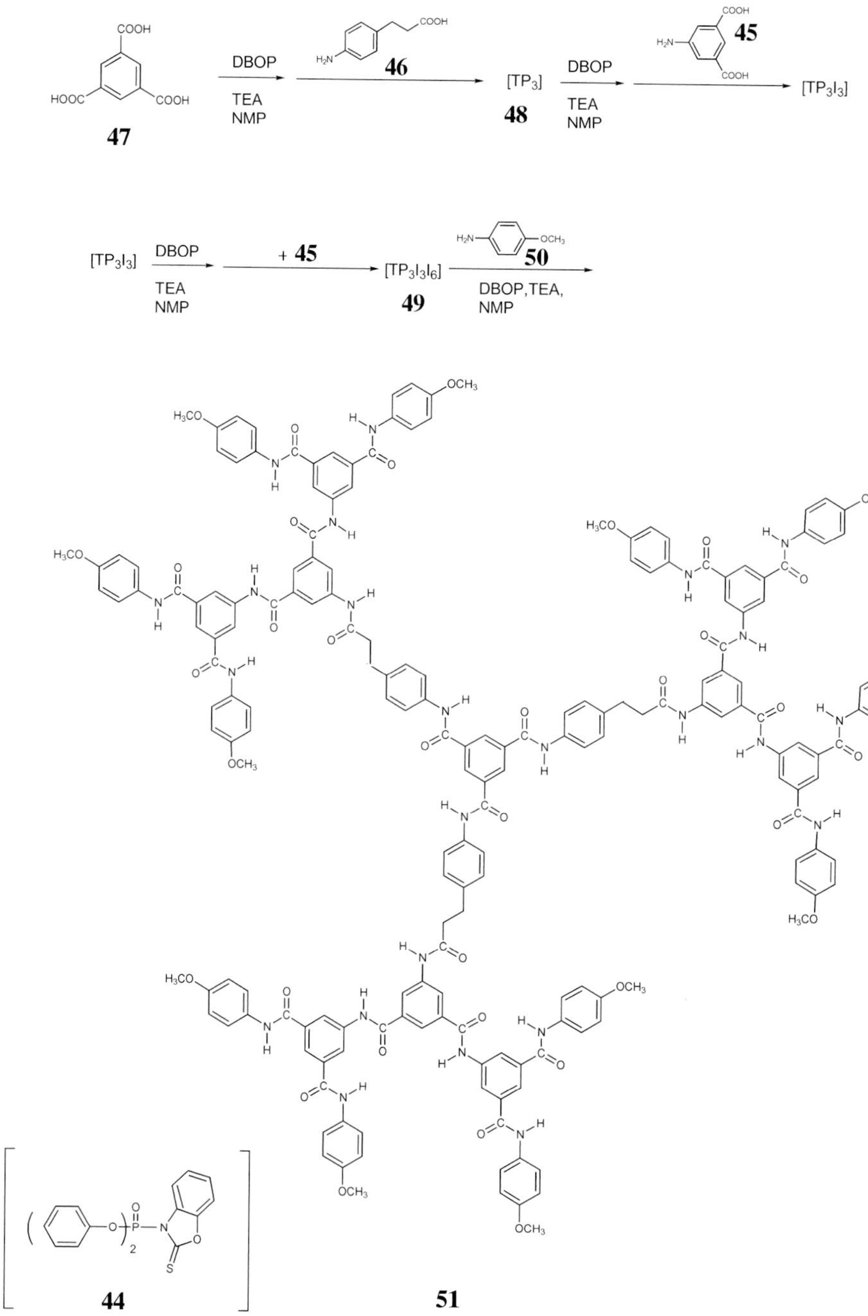

Scheme 6.15 Sequential, single-pot activation and addition procedure leading to hyperbranched materials.[102, 103]

the generation of activated esters using diphenyl (2,3-dihydro-2-thioxo-3-benzoxazolyl)-phosphonate (**44**; DBOP). The polymer thus obtained was subsequently capped with *para*-anisidine; it possessed an $\overline{M}_w$ of 30,000 amu, as determined by GPC (polystyrene).

Yamakawa et al.[102–104] developed an interesting single-pot protocol (Scheme 6.15) based on repetitive carboxylic acid activation by treatment with DBOP followed by amidation with an amine such as 5-aminoisophthalic acid (**45**) or 4-aminophenylpropionic acid (**46**). Starting with trimesic acid (**47**), DBOP activation followed by spacer (**46**) addition furnished the homologated triacid **48**. Three more activation–addition steps afforded polyaramide **49**, which was capped with *p*-anisidine (**50**) to give **51**; using this procedure, a 10% defect level was realized.

Kakimoto et al.[105] reported the thermal polycondensation of a diarylaminomonocarboxylic acid monomer to give polyaramide polymers. Thermal polymerization was effected at 235 °C for 1 hour to give a material possessing an M_w of 74,600 amu and a T_g of 200 °C. The authors evaluated[106] the effect of monomer multiplicity on the DB using related polyamide AB_2, AB_4, and AB_8 wedges to access poly(arylamide)s.

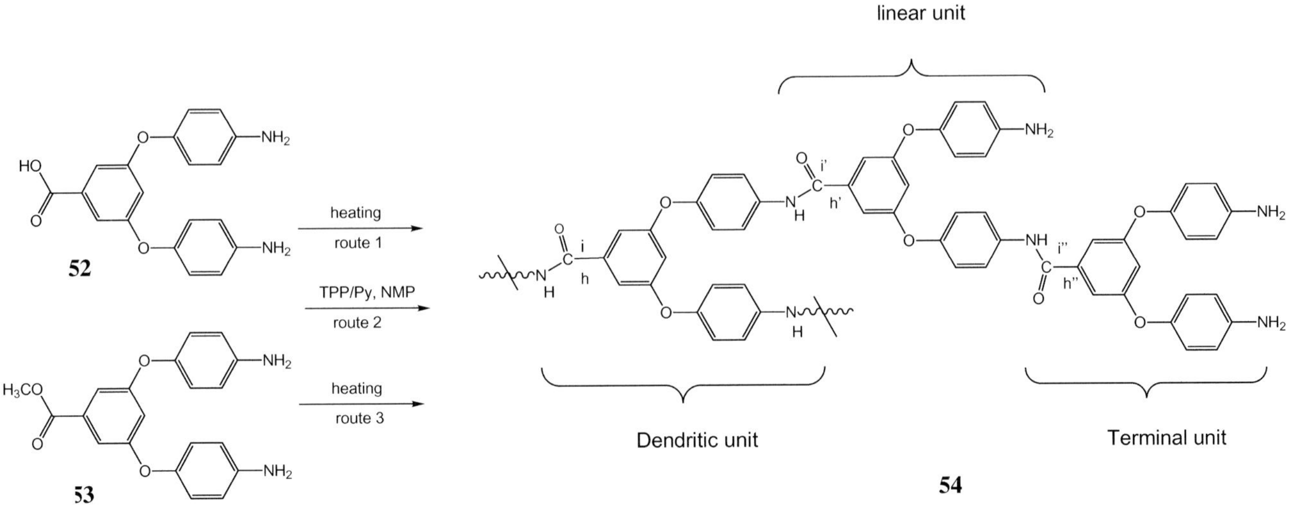

Scheme 6.16 Various routes to hyperbranched species.[107]

Kakimoto, Jikei, and Yang[107] reported the preparation of poly(amide) polymers (**54**) via numerous routes (Scheme 6.16). Thermal condensation of monomer **52**,[105] as well as of the corresponding ester **53**, led to similar materials. Treatment of amino acid **52** with triphenylphosphite and pyridine in NMP also led to the same architecture; M_ws obtained following routes 1, 2, and 3 were determined as 74,600, 36,800, and 47,800 amu, respectively, with corresponding T_gs of 200, 200, and 180 °C. The synthesis of these polyamides has been extended to incorporate aromatic diamine spacer units.[108]

6.2.7 1 → 2 *Aryl*-Branched, *Urethane* Connectivity

Spindler and Fréchet[109] prepared hyperbranched polyurethanes (**57**) by step-growth polymerization (Scheme 6.17) of protected or "blocked" isocyanate AB₂ monomers. The method is dependent on the thermal dissociation of a carbamate unit into the corresponding isocyanate and alcohol moieties.[110, 111] Decomposition temperatures range from ca. 250 °C for alkyl carbamates to ca. 120 °C for aryl carbamates.[110] Thus, 3,5-bis[(benzoxycarbonyl)imino]benzyl alcohol (**55**; R = H) was refluxed in THF to give the bis(isocyanate) benzyl alcohol (**56**) and two equivalents of phenol (Scheme 6.17a). Self-

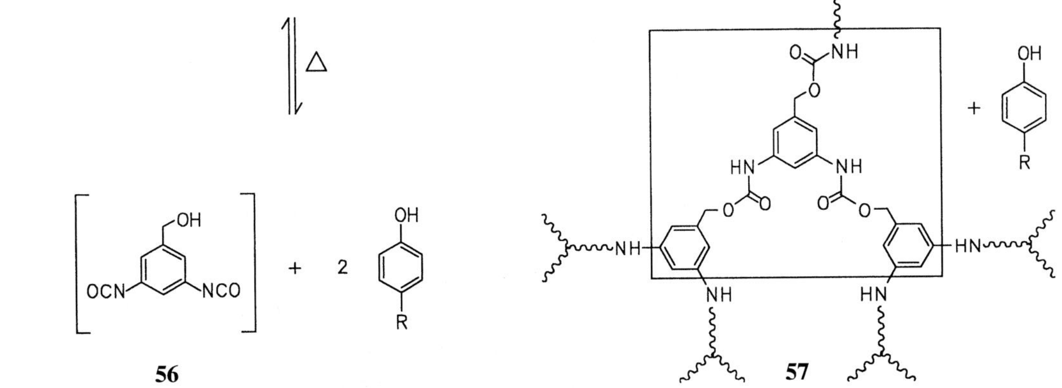

Scheme 6.17 Synthesis of polyurethanes[109] based on the thermal decomposition of a "blocked' diisocyanate monomer.

addition of the isocyanate moieties with the alkyl alcohol group subsequently gave the hyperbranched carbamate **57** (Scheme 6.17b). Addition of an alcohol [i.e., *p*-MeC$_6$H$_4$CH$_2$OH, *p*-NO$_2$C$_6$H$_4$CH$_2$OH, CH$_3$OCH$_2$CH$_2$OCH$_2$CH$_2$OH, or CH$_3$(CH$_2$)$_9$OH] in the early stages of polymerization reduced the amount of cross-linking side reactions that are characteristic of isocyanate transformations through the trapping of reactive surface groups. This method of "end-capping" of the hyperbranched polymers yielded polyurethanes that were soluble in common solvents such as THF, DMSO, and DMF, with conversions in the range 28–83 %.

Kumar and Ramakrishnan[112] demonstrated that the thermal decomposition (107 °C) of 3,5-dihydroxybenzoyl azide gave rise to the labile 3,5-dihydroxyphenylisocyanate, which afforded (110 °C, dry DMSO, catalytic dibutyltin dilaurate) the polyurethane in 95 % yield. A polydispersity greater than two was indicative of the poor mobility of the monomers; the polymers were found to be soluble in aqueous base, confirming the presence of free phenolic termini.

6.2.8 1 → 2 *Aryl*-Branched, *Urea* Connectivity

Kumar and Meijer[113, 113a] developed a urea-based hyperbranched architecture (**61**; Scheme 6.18) predicated on the reaction of 3,5-diaminophenylisocyanate (**59**), which was generated *in situ* by thermal decomposition of the corresponding diaminoacylazide **60**, prepared in several steps from the substituted benzoic acid **58**. The DB for these poly-(urea)s was determined to be 0.55, with the M_w approaching 20,000 amu and polydispersities of 1.56.

6.2.9 1 → 2 *Aryl*-Branched, *Ether & Ester* Connectivity

Ringsdorf et al.[114] reported the single-pot preparation of liquid-crystalline materials with terminal chiral groups. The hyperbranched macromolecule **65** (Scheme 6.19) was prepared using two different building blocks: (1) an AB$_2$-type monomer **64** possessing a mixed anhydride and two mesogenic biphenyl acetate moieties, and (2) a terminal unit **63** possessing a 3,4-disubstituted benzoyl chloride moiety. Both monomers were synthesized from methyl 3,4-dihydroxybenzoate (**62**). Dendron synthesis was effected by treatment of the diacetate monomer **64** with *p*-toluenesulfonic acid and heating (240 °C, 8 h), and then adding more acid catalyst followed by the capping agent **63**. Thermal analysis of the CHCl$_3$-soluble **65** by polarization microscopy, DSC, and X-ray scattering was included in the characterization of this asymmetric polymer; polarimetric experiments substantiated the macromolecular chirality, indicating an optical rotation of $[\alpha]_D = +6.8°$ in CH$_2$Cl$_2$.

Itoh et al.[115] incorporated diethylene and triethylene glycol units as spacers in the construction of 1 → 2 aryl-branched polyesters and complexed the resulting materials with lithium metal salts, such as lithium triflate and lithium trifluoromethanesulfonimide [(CF$_3$SO$_2$)$_2$NLi], in order to examine their ionic conductivities.

6.2.10 1 → 2(3) *Aryl*-Branched, *Amide & Ester* Connectivity

Kricheldorf et al.[116] prepared a novel pentavalent monomer for the construction of hyperbranched poly(ester-amide)s. The monomer was obtained by acylation of silylated 3,5-diaminobenzoic acid (DABA) with 3,5-bis(acetoxy)benzoyl chloride. Polycondensation using this monomer afforded a polymer possessing a nearly uniform sequence of alternating DABA and two 3,5-dihydroxybenzoic acid units. It was determined that approximately 15–20 % of the DABA segments cease to function as branching units due to ester–amide exchange. This monomer was also copolycondensed with 3-acetoxy-benzoic acid or its trimethylsilyl ester to produce copoly(ester-amide)s.

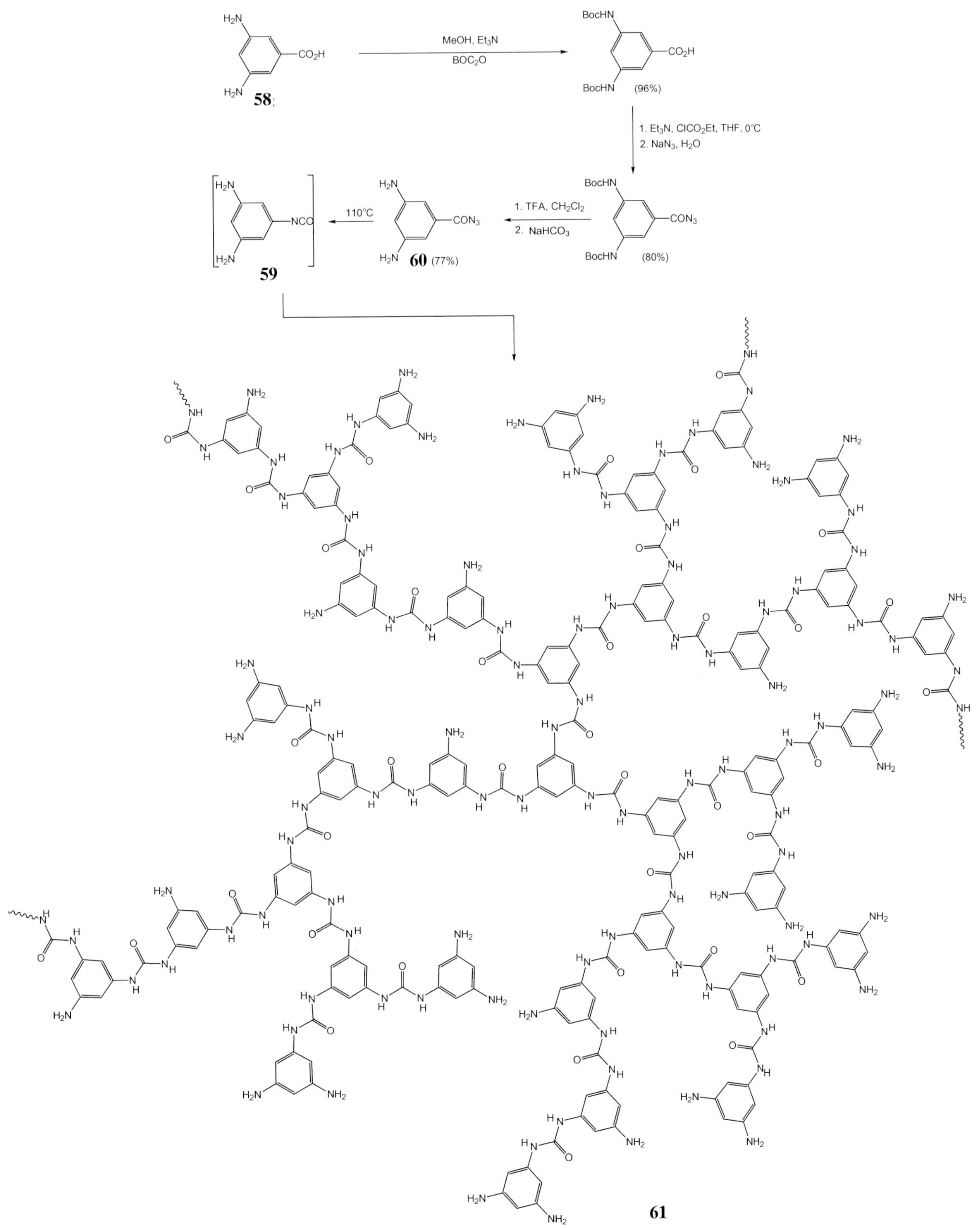

Scheme 6.18 Hyperbranched materials derived from the *in situ* generation of 3,5-diaminophenylisocyanate (**59**).[113]

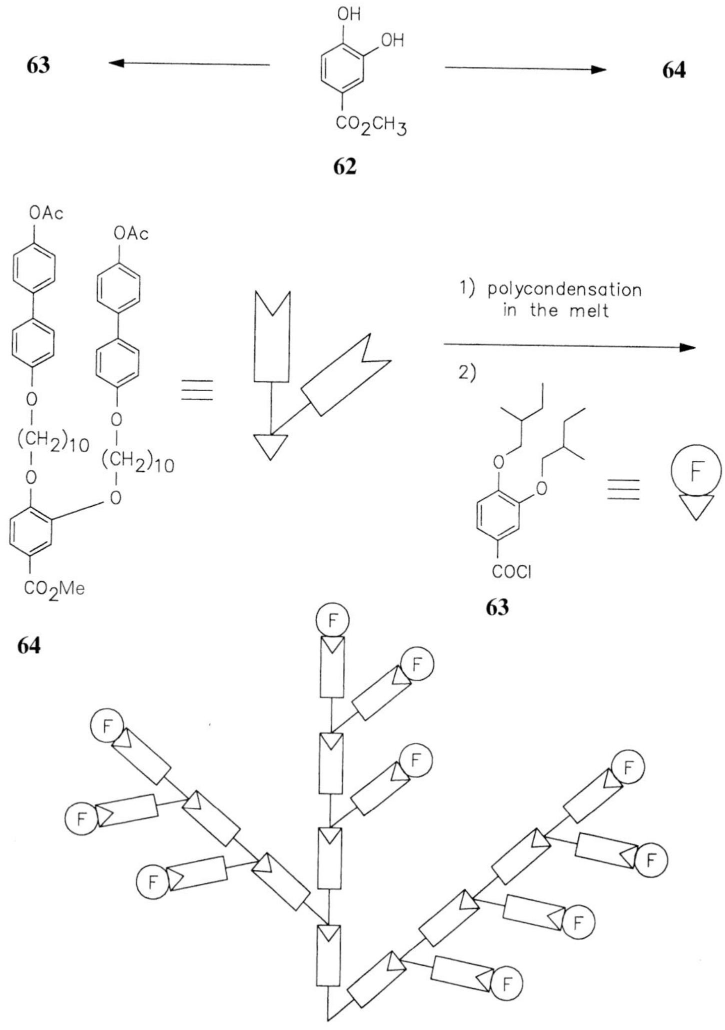

Scheme 6.19 Ringsdorf et al.'s[114] preparation of hyperbranched liquid-crystalline polymers.

New poly(ester-amide)s were prepared by condensation of silylated 4-amino- or 3,5-diaminobenzoic acid with acetylated naturally occurring vanillic or phloretic acids.[117] These amorphous polymers possessed T_gs in the range 170–230 °C; the T_gs were observed to increase with the molar fraction of amide.

6.2.11   1 → 2 *Aryl*-Branched, *Ether & Ester* Connectivity

A series of hyperbranched aromatic polyamide copolymers was obtained[118] by treatment of 3-(4-aminophenyloxy)benzoic with 3,5-bis(4-aminophenyloxy)benzoic acid in the presence of triphenylphosphite and pyridine.

6.2.12   1 → 2 *Aryl*-Branched, *S*-Connectivity

Kakimoto et al.[119, 120] reported the preparation of the hyperbranched engineering plastic poly(phenylene sulfide)s (**68**; Scheme 6.20) by the acid-catalyzed polymerization of methyl 3,5-bis(phenylthio)phenyl sulfoxide (**66**) to give the polymethylated polymer **67**, which was subsequently treated with pyridine to afford the corresponding polysulfide. The degree of polymerization and M_w were determined to be 80 and 25,700 amu, respectively, while the T_g was found to increase (from 102 to 124 °C) in proportion to the molecular weight.

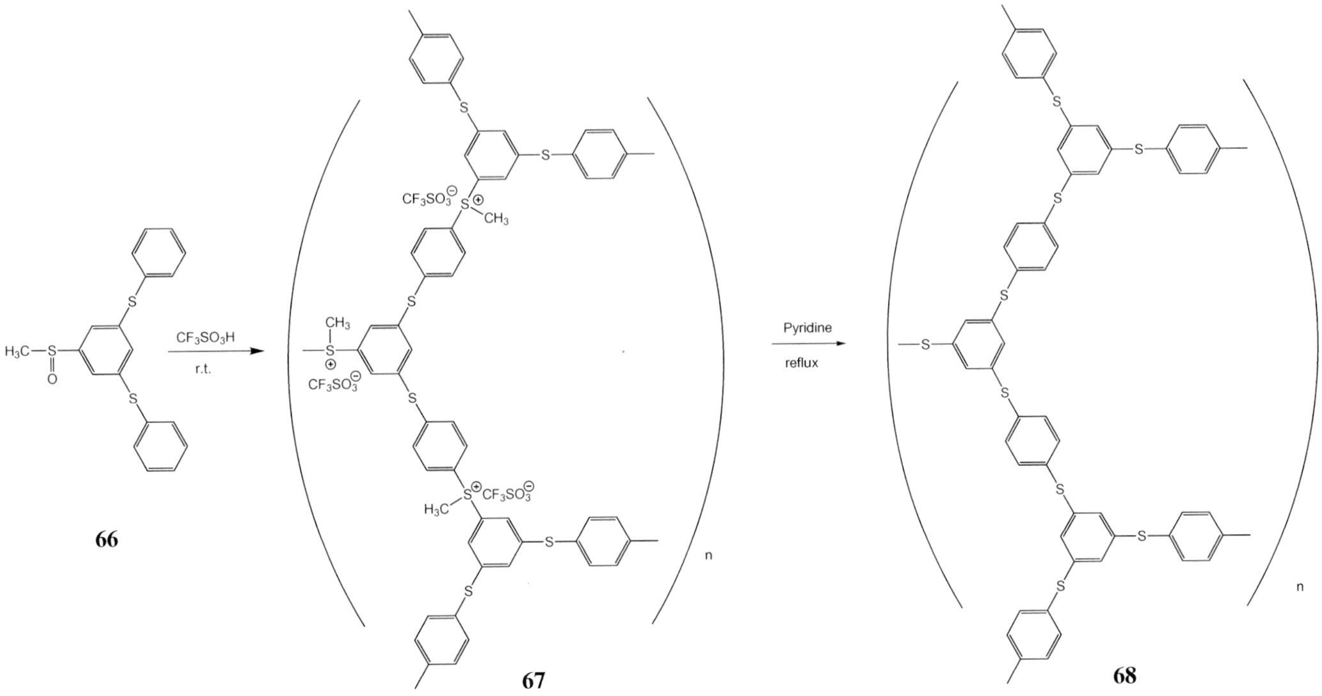

Scheme 6.20 Acid-catalyzed polymerization of methyl 3,5-bis(phenylthio)phenyl sulfoxide (**66**).[120]

6.2.13 1 → 2 *Aryl*-Branched, *C*-Connectivity

Fréchet et al.[121] reported the preparation of branched materials employing a self-condensing vinyl polymerization of 3-(1-chloroethyl)ethenylbenzene, which undergoes a "living" vinyl polymerization resulting in increasing numbers of reactive sites and branches on the growing chains. Polymers synthesized by this method possess M_w greater than 100,000 amu. A similar hyperbranched polystyrene has been reported[122] and this methodology has been extended to monomers, e.g., 4-(chlorodimethylsilyl)styrene, possessing a polymerizable vinyl moiety and a group capable of undergoing quantitative S_N2 chemistry.[123] Weimer, Fréchet, and Gitsov[124] discussed the importance of active-site reactivity and reaction conditions for self-condensing vinyl polymerizations in relation to hyperbranched materials. Comparison of linear *versus* branched poly[4-(chloromethyl)-styrene] was considered.

6.2.14 1 → 2 *Aryl*-Branched, *Alkyne* Connectivity

Bharathi and Moore[125] described a solid-supported hyperbranched polymerization leading to poly(phenylalkyne)s that exhibit a "self-limited" growth. Polymerization (Scheme 6.21) was effected using a triazene-modified support **69**[126] and the key monomer **70**, affording aryl iodide-terminated structures such as polymer **71**. These were capped using the corresponding bis(*tert*-butyl)phenyl alkyne, making the product more soluble (i.e., **72**) prior to support cleavage and isolation. For all cases examined, low polydispersities were obtained in the range 1.28–1.89 depending on monomer concentration relative to support focal point loading. The molecular weights of these materials were controlled over a range of ca. 5,000–25,000 amu. Limitations included "impingement of adjacent dendrimers" and "confinement of the growing polymer within the support boundaries". Later studies of these materials included the continuous, slow addition of AB$_2$ monomer to a multifunctional core to produce narrow polydispersity, high molecular weight poly(phenylacetylene)s.[127]

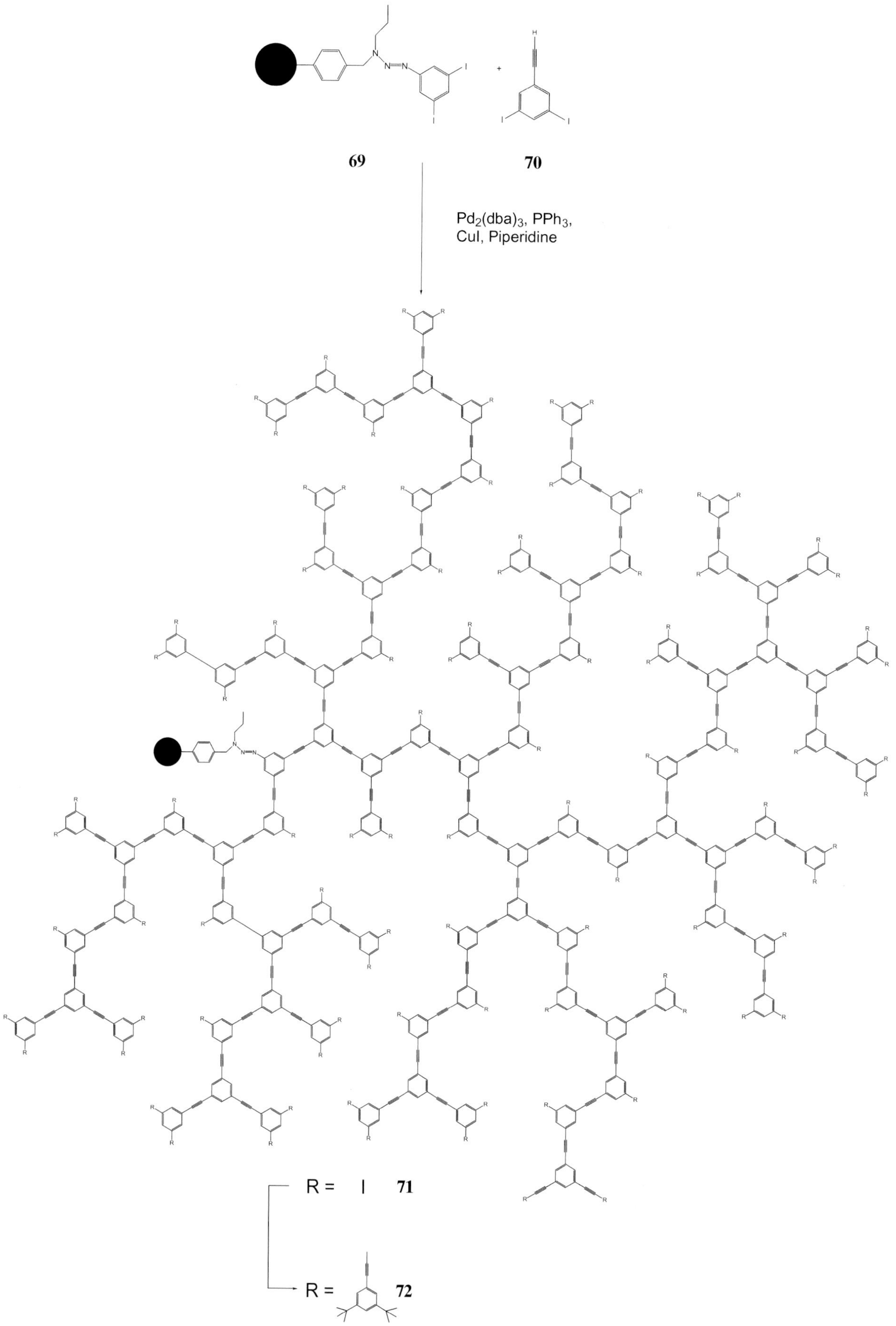

Scheme 6.21 Moore's "self-limited" growth polymerization producing low polydispersity poly(phenylacetylene)s.[126]

6.2.15 1 → 2 Aryl-Branched, *Alkene* Connectivity

One-step preparations of poly[3,5-bis(vinyl)benzene] via two routes have been reported:[128] (a) the treatment of 1,3,5-tris(bromomethyl)benzene with *t*-BuOK (THF), and (b) a Wittig reaction (PPh₃, KOEt) on 3,5-bis(formyl)(bromomethyl)benzene. Both routes gave only poor yields and low molecular weights.

6.2.16 1 → 2 Aryl-Branched, *Imide & Ether* Connectivity

The preparation of soluble hyperbranched aromatic poly(imide)s was accomplished[129, 129a] by the self-polycondensation of an isomeric mixture of 3,5-bis(4-aminophenyloxy)-diphenyl ether-3',4'-dicarboxylic monomethyl ester in the presence of DBOP. The resultant polymer was found to have no cross-linked structure and exhibited good solubility in NMP, DMF, and DMSO. It also retained good thermal stability.

Thompson et al.[130, 131, 131a] published the rapid (ca. 2.5 min), single-pot synthesis of high molecular weight aromatic poly(ether imide)s (**76**; Scheme 6.22). Monomer preparation began with treatment of aminodiol **73** with 4-fluorophthalic anhydride, followed by protection (*t*-BuMe₂SiCl) of the resulting diol **74** to afford imide **75**. CsF-catalyzed polymerization of the latter in diphenylsulfone at 240 °C afforded the desired polymer **76** possessing a DB of 0.66 and an M_w up to 100,400 amu. These materials were found to be thermally stable, showing only 10 % weight loss at 530 °C. The use of these aromatic ether imide copolymers to tune molecular architectures with regard to the optimization of physical properties and the inherent processability has been reported.[132]

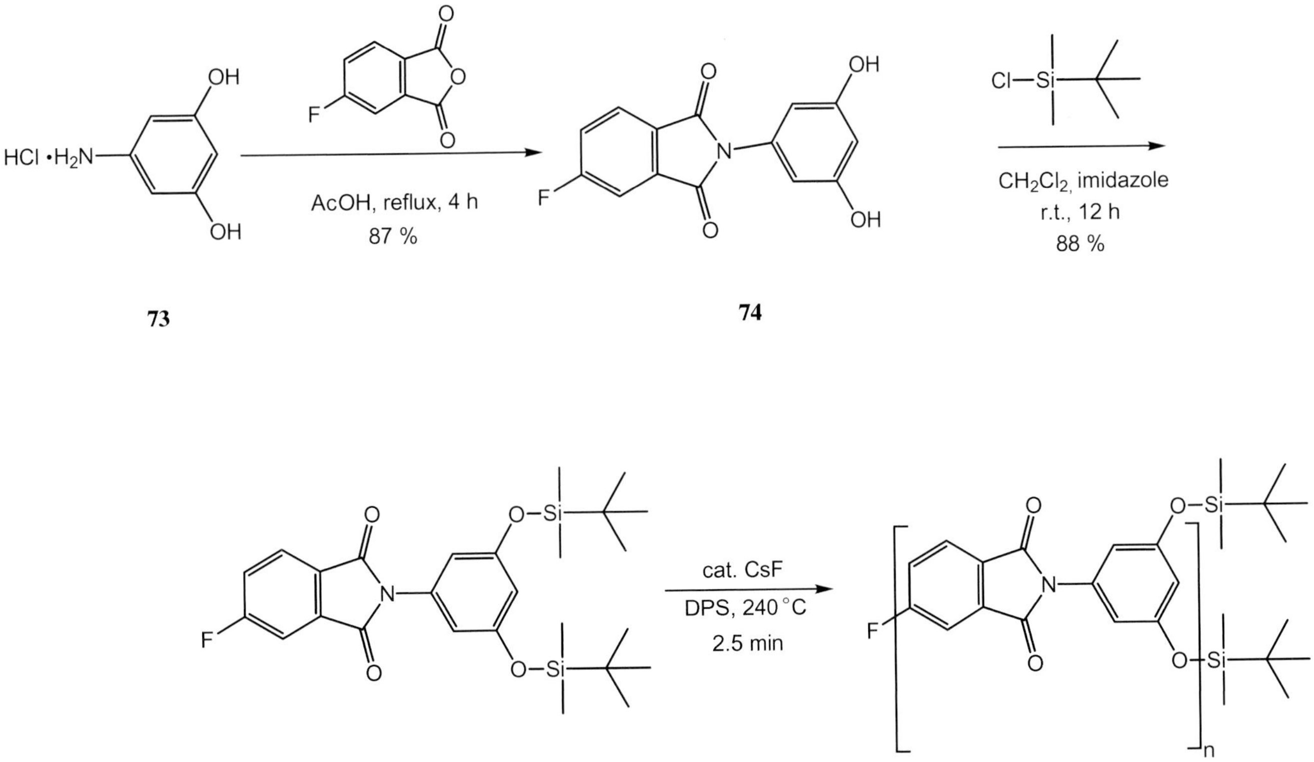

Scheme 6.22 Moore's synthesis of aromatic poly(ether imide)s (**76**).[130, 131]

6.3   1 → 2 *Heteroaryl*-Branched

6.3.1   1 → 2 *Heteroaryl*-Branched, *Ether* Connectivity

Hedrick and coworkers[133, 134] reported the synthesis of quinoxaline-containing hyper-branched materials for examination as engineering thermoplastics; these were found to possess melt and solution-phase processability, high transition temperatures, and favorable mechanical properties. Polymer formation (Scheme 6.23) was effected by phenoxide displacement of heterocycle-activated aryl fluoride moieties. Thus, AB$_2$-type monomers **78**, accessed by reaction of 1,2-diamino-4-fluorobenzene with the respective diketones **77**, were polymerized (NMP[135], K$_2$CO$_3$) to give polymers **79**. Other solvents, such as a 1:1 mixture of NMP/N-cyclohexyl-2-pyrrolidone (CHP) or DMPU, could also be successfully used; on the basis of the dramatic increases in the viscosities of the reaction mixtures, high molecular weight polymers were obtained in all cases. As indicated by MALDI-TOF and ^{1}H NMR data, the primary (or starting) focal aryl fluoride was no lon-

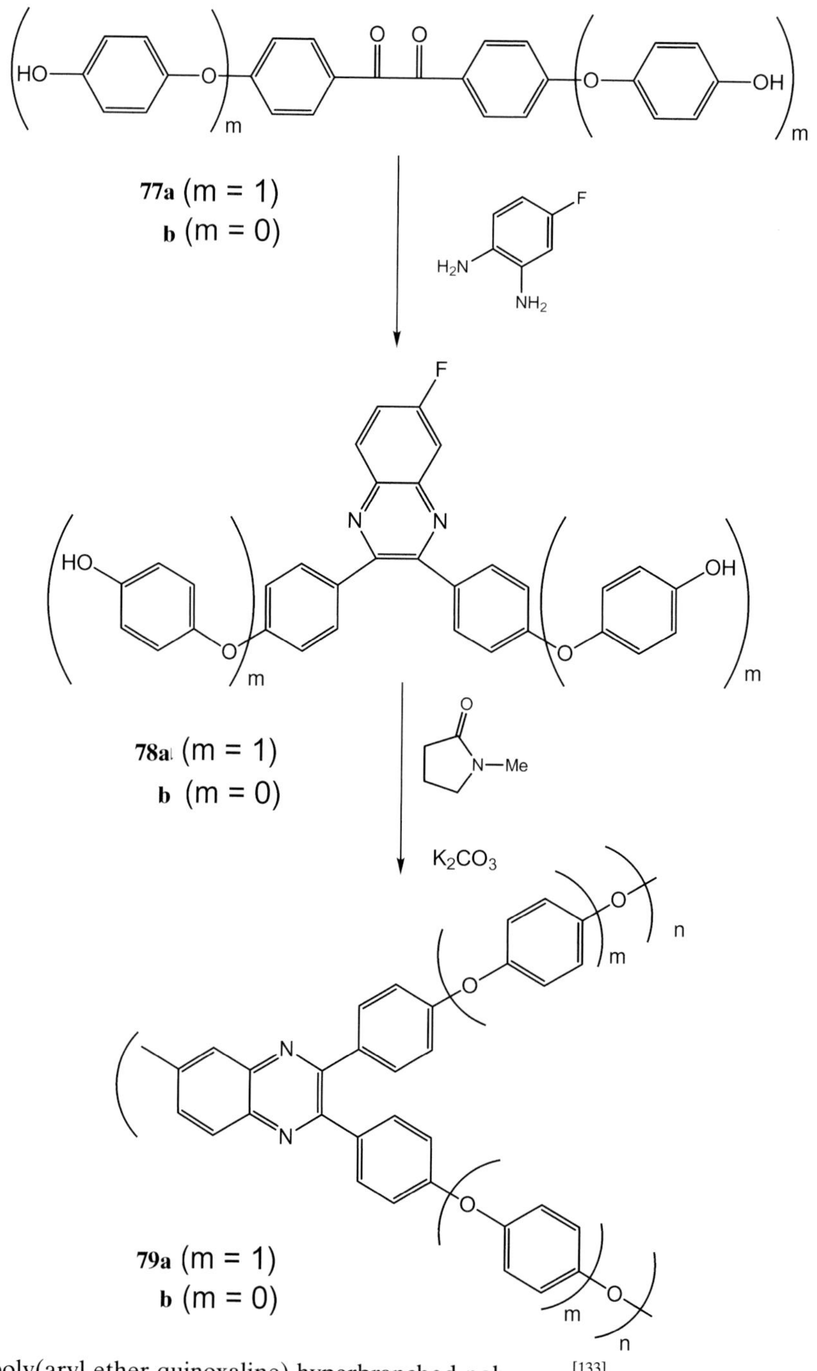

77a (m = 1)
  b (m = 0)

78a (m = 1)
  b (m = 0)

79a (m = 1)
  b (m = 0)

Scheme 6.23 Hedrick's poly(aryl ether quinoxaline) hyperbranched polymers.[133]

ger present in the hyperbranched polymers due to intramolecular cyclization. The phenol-terminated material (**79**) was capped with 3-(isocyanatopropyl)triethoxysilane to give the triethoxysilyl-terminated polymer, which was then treated with poly(silsesquioxane) and cured (410 °C, 2 h). The thermal stabilities of the hybrid composites were unaffected by the incorporation of the hyperbranched materials, although modification of the chain ends showed a dramatic effect. The triethoxysilyl-terminated dendrimer exhibited nanophase separation, while the phenol-terminated construct showed micron-sized phase separation.

6.3.2 1 → 2 *Heteroaryl*-Branched, *Ethyne* and *Ether* Connectivity

Kim, Chang, and Kim[136, 136a] described the use of cyanuric chloride (**80**; Scheme 6.24) in the construction of unique monomers by reaction with 4-hydroxyiodobenzene to give bis(aryl iodide) (**81**), followed by treatment with 3-ethynylphenol to generate the key monomer **82**. Polymerization of **82** to give polymer **83** was effected employing Pd(PPh$_3$)$_2$Cl$_2$. The M_w of **83**, as determined by GPC, ranged from 6,000 to 10,000 amu, while polydispersity values varied from 1.6 to 3.0. Notably, sequential chloride displacement from cyanuric chloride can be expected to facilitate the preparation of a variety of AB$_2$-type monomers. Other triazene constructs are known.[136b]

6.4   1 → 2 *C*-Branched

6.4.1 1 → 2 *C*-Branched, *Ester* Connectivity

Kambouris and Hawker[59] described the one-step polymerization of methyl 4,4-bis(4'-hydroxyphenyl)pentanoate (**84**) and the structural elucidation of the resultant hyperbranched polymer **85** (Scheme 6.25). Neat methyl ester monomer **84** was melted (120 °C) *in vacuo* in the presence of a catalyst [Co(OAc)$_2$] to give the phenolic-terminated, hyperbranched polyester possessing an $\overline{M}_w$ of 47,000 amu. Chu and coworkers[137] investigated the intramolecular cyclization of these types of polyesters. Application of a previously described method[56] was not helpful in ascertaining the DB, hence a degradative technique was devised and demonstrated.[59] Exhaustive methylation (MeI, Ag$_2$O) of phenolic polymer **85** gave the methyl ether **86**, which was formed in high yield with no evidence of side reactions. This ethereal ester **86** was then saponified (KOH, THF, H$_2$O) to afford only three building block products in yields of 24 % (**87**), 51 % (**88**), and 25 % (**89**), as determined by capillary gas chromatography and HPLC. From the product distribution, the DB was determined as 49 %.

A hyperbranched aliphatic polyester (**92**) based on 2,2-bis(hydroxymethyl)propionic acid (**91**, bis-MPA) as the AB$_2$ monomer, and 2-ethyl-2-(hydroxymethyl)-1,3-propanediol (**90**) as the core, has been reported.[138–144] Hult et al.[145, 146] described the preparation of thermally stable polymers exhibiting second-order nonlinear optical activity. In general, the one-step esterification (Scheme 6.26) was performed in the bulk using an acid catalyst, involved no purification steps, and resulted in theoretically calculated molar masses of 1200–44,300 amu. The DB was determined to be ca. 80 %, suggesting that the polymers are highly branched. These materials, e.g., **92**, exhibited good thermal stability when analyzed by TGA in an inert atmosphere. The relaxation processes associated with these hyperbranched macromolecules have been reported,[147] as well as the transport properties.[148] Jang and Oh[149] examined the *in situ* FT-IR spectra of these materials so as to investigate their microstructural parameters by varying the stoichiometry of the monomer-to-core ratio in the 1st, 2nd, and 3rd generations. Later, the influence of the terminal groups on the relaxation processes was investigated by dielectric spectroscopy, DSC, and dynamic mechanical analysis.[150] The effect on interfacial tension of the addition of these polyesters to polypropylene–polyamide blends has been reported.[151] The kinetics of formation of hyperbranched polyesters accessed by polymerization of bis-

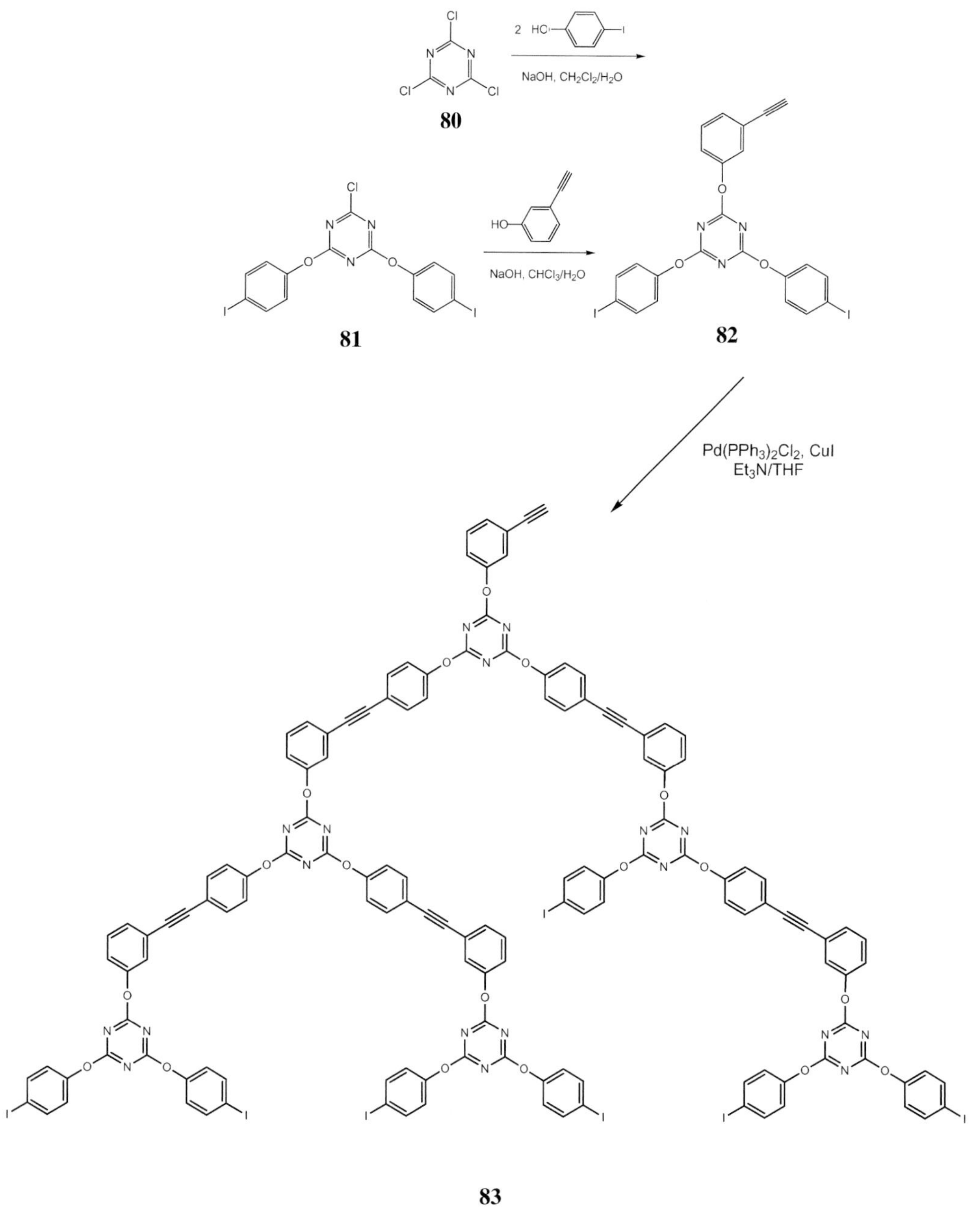

Scheme 6.24 Construction of triazene-branched polymers.[136]

MPA have been investigated.[152] When these hyperbranched polyesters were maintained at the polymerization temperature in order to ascertain the effects of heat treatment on composition, it was found that:[153] (1) repeating units changed to a large extent with conversion; (2) slow monomer addition to a trifunctional core afforded a product with a DB of 47%; and (3) heat treatment caused small changes in the fractions of the different repeating units. The thermal cationic ring-opening polymerization of 3-ethyl-3-(hydroxymethyl)oxetane in the presence of benzyltetramethylenesulfonium hexafluoroantimonate[122] gave rise to hyperbranched aliphatic polyethers with a DB of 0.4 and capable of initiating ε-caprolactone.[154] The role of intramolecular cyclization during the preparation of these polyesters has been investigated;[155] in stark contrast to previous reports,[140] the molecular weights were observed to be significantly lower due to this process. The synthesis and solution properties of six constitutional isomers of star polymers, based on high molecular weight poly(ε-caprolactone) with branching units based on bis-MPA generated on a triarylmethane core, have been described.[156]

Hult and coworkers[157] later surface-modified these hyperbranched materials with alkyl chains. Preparative details were analogous to those for polyol **92**, except that a tet-

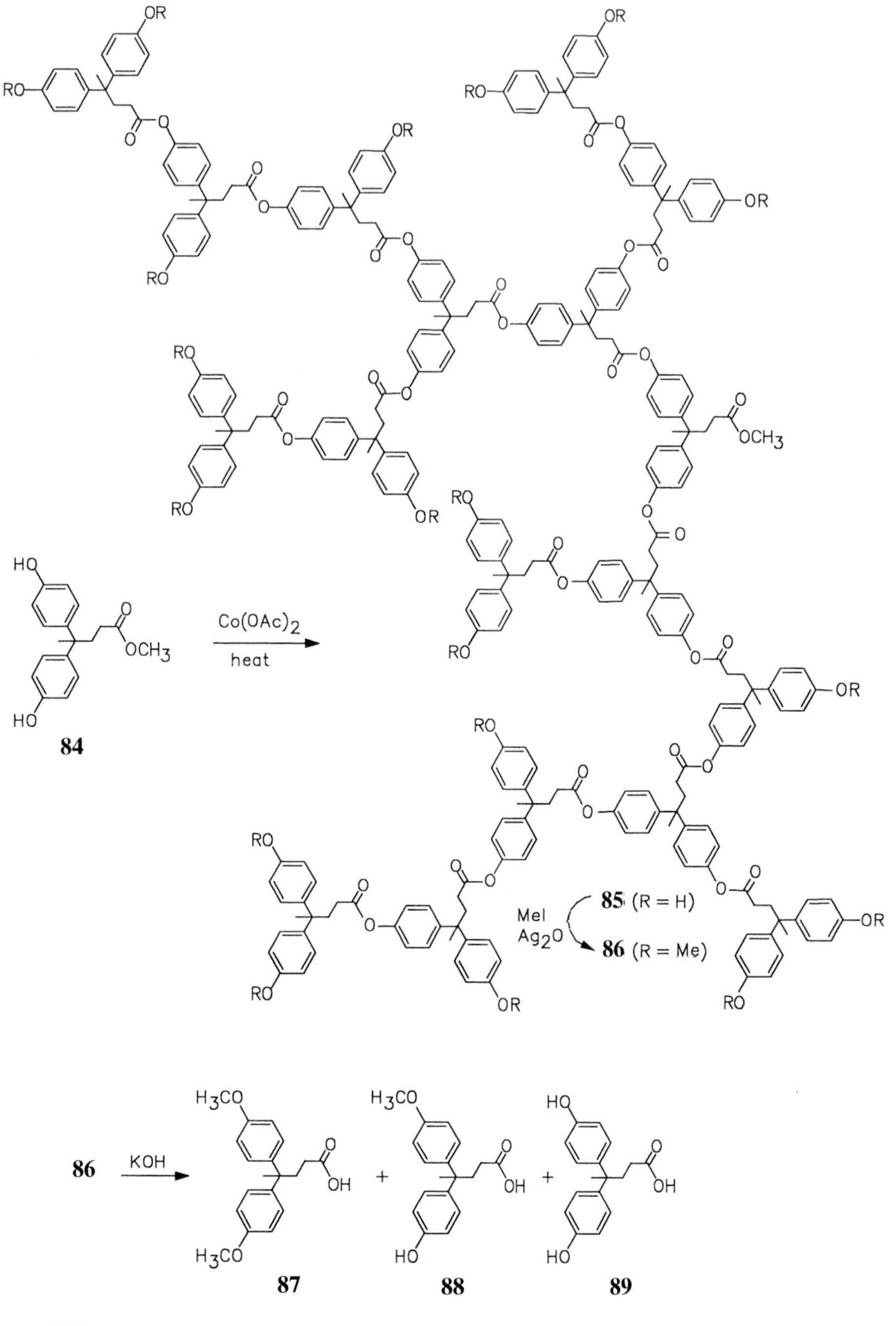

Scheme 6.25 Hawker et al.'s[59] Co-catalyzed polymerization affording polyester macromolecules.

ravalent core (tetraethoxylated pentaerythritol) was employed; these 4- or 2-tiered (average 64 or 16 hydroxy termini, respectively) polyols are now commercially available. Surface modification was readily accomplished by treatment with various alkyl acyl chlorides. Shorter alkyl termini (3–6 carbons) were found to decrease T_gs to well below zero, whereas much longer chains (12–16 carbons) led to crystalline materials. On the basis of X-ray data for the C_{16}-modified materials, it was suggested that crystallization occurred in a "non-interpenetrating" manner; for the shorter chain species, however, solidification occurred with interdigitation of the end groups. A series of additives known as Bolton®, consisting of hydroxy-functionalized polyesters based on bis-MPA and ethoxylated pentaerythritol, was partially functionalized with a mixture of eicosanoic and docosanoic acids. The resultant derivatives were used as processing aids for linear low-density polyethylene in the tubular film blowing process.[158] Surface properties of these polyesters have been reported.[158a]

Similar, although architecturally more complex dendrigraft and graft polymers, have been prepared by a "tandem" approach[159] based on "living" free radical polymerization.

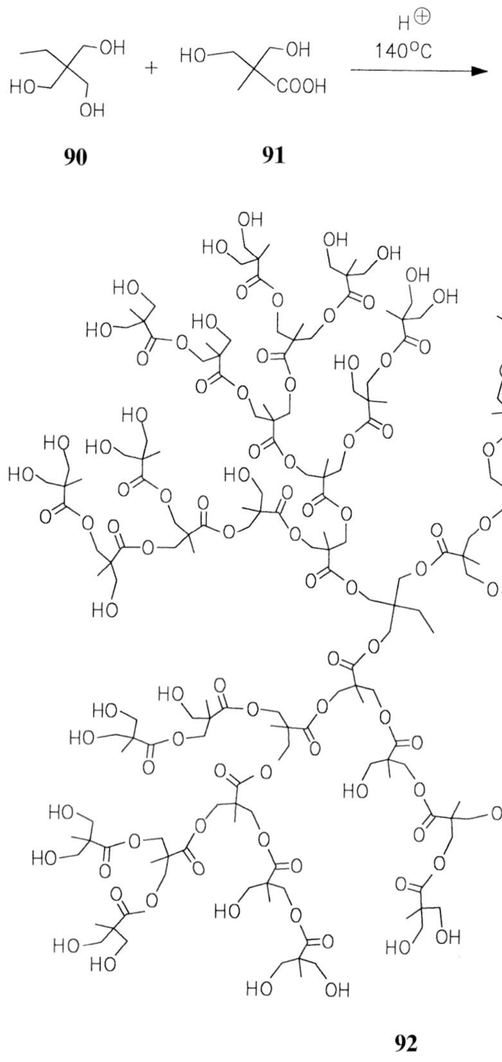

Scheme 6.26 Preparation of hyperbranched aliphatic polyesters.[146]

Other investigated star and graft polymers have been prepared analogously.[160] Molecular weights were modulated by controlling the stoichiometry of monomer addition; low polydispersities were also realized.

Sakamoto, Aimiya, and Kira[161] described the preparation of hyperbranched polymethacrylates by group-transfer self-condensation polymerization using 2-(2-methyl-1-triethylsiloxy-1-propenyloxy)ethyl methacrylate (**93**; Scheme 6.27), which was accessed by ethyleneglycol dimethylacrylate monohydrosilylation using Et₃SiH under promotion by RhCl(PPh₃)₃. Polymerization was effected (THF, 0 °C) with either (trimethylsilyl)difluoride or tetrabutylammonium dibenzoate, affording the desired branched material (e.g. **94**). The extent of branching, as determined by inverse-gated-decoupled ¹³C NMR, was found to be dependent on the catalyst used; the tetrabutylammonium salt gave fewer defects and led to lower polydispersities.

Feast, Hamilton, and Rannard[162] reported the preparation of oligomeric poly(diethyl 3-hydroxyglutarate) by step-growth condensation promoted by titanium(IV) butoxide. These hyperbranched products were characterized by MALDI-TOF MS; high laser power was found to fracture the polymer, whereas with lower incident laser power a representative spectrum proved difficult to obtain.[163]

Hedrick et al.[164] employed small branched polyols for the controlled synthesis of star polymers and branched poly(caprolactone)s[165–169] by ring-opening polymerizations. Later, these authors reported[170, 171] the construction of "dendrimer-like star block copolymers". This new architecture was characterized as possessing "radial geometry with different generations of high molecular weight polymer emanating from a central branched core". The synthetic protocols employed were "living" ROP and atom-transfer radical

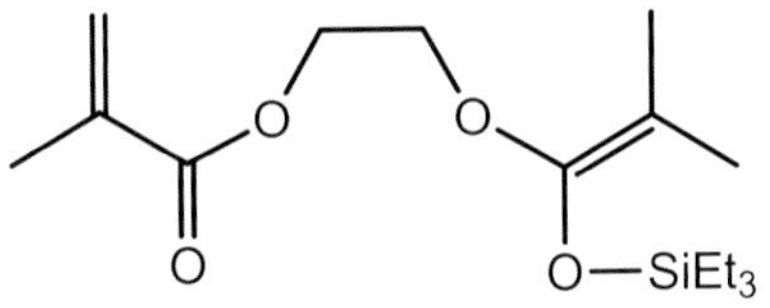

93

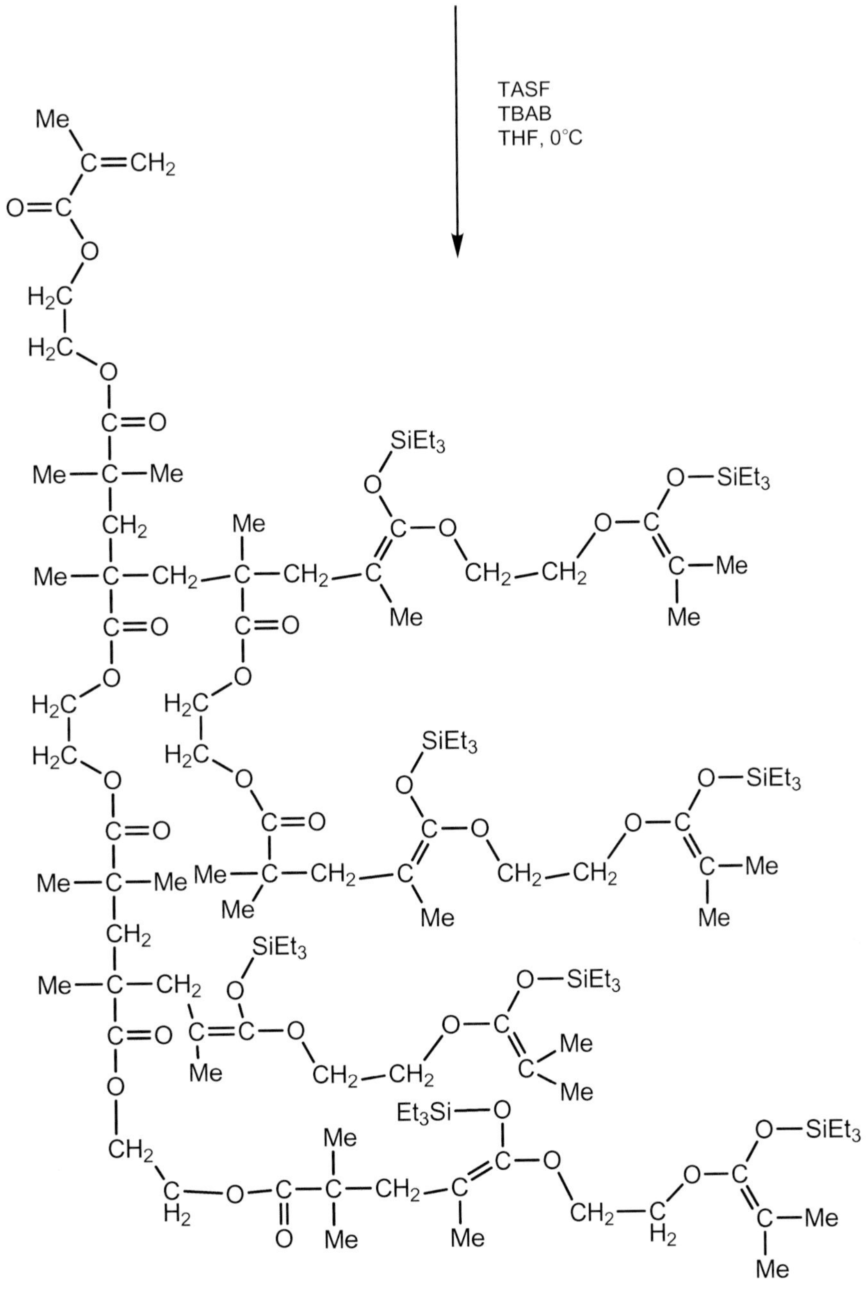

TASF
TBAB
THF, 0°C

94

Scheme 6.27 Dimethacrylate group-transfer polymerization proceeding through active and latent functionalities.[161]

polymerizations. Similar shell cross-linked nanoparticles have been reported by Wooley et al.[172]

Related reports include those by Hedrick, Miller, and coworkers,[173] who have used the 1st and 2nd generations of these polyesters[168, 174] possessing terminal tertiary alkyl bromides as "microinitiators" for the atom-transfer radical polymerization of *tert*-butyl acrylate and methyl methacrylate.[175] The resulting star-like block copolymers exhibited unimolecular micellar character.

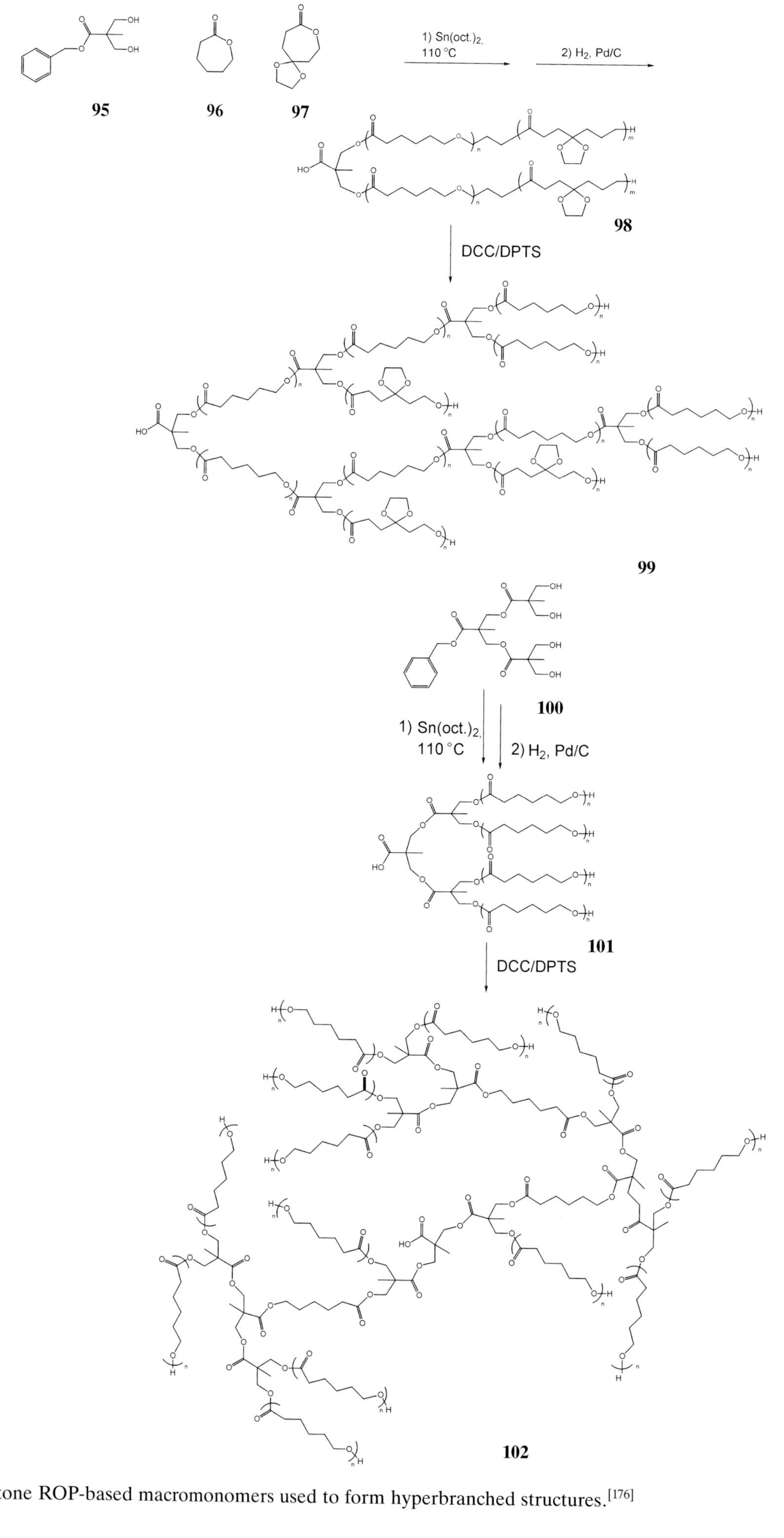

Scheme 6.28 Caprolactone ROP-based macromonomers used to form hyperbranched structures.[176]

Similar, albeit internally functionalized, polyesters have been reported,[176] as shown in Scheme 6.28. Starting with the benzyl-protected diol **95**, caprolactone **96**, and 1,4,9-trioxaspiro[4.6]-9-undecanone (**97**), macromonomers with the general structure of poly-(ester) **98** were obtained by ring-opening polymerization, as promoted by the $Sn(oct)_2$ catalyst, followed by facile deprotection (Pd/C, H_2). These flexible monomers, possessing M_ns (GPC) ranging from 21,000 to 93,000 amu, were then polymerized (DCC, DPTS) to effect esterification. Analogously, an AB_4-type monomer **101**, prepared using only caprolactone, was found to have an M_n ranging from 19,000 to 32,000 amu. Following polymerization and coupling, flexible architectures, e.g., **99** and **102**, were obtained. Similar layered block copolymers have been reported[177] that integrate alkyl polyester dendrons based on bis-MPA with Sn-based ring-opening polymerization of caprolactone. Nanoporous organosilicate films prepared from analogous star-shaped hydroxyl-terminated poly(ε-caprolactone) silsesquioxane-based materials have also been reported.[178, 179]

Subsequently, a double-stage convergent approach to these architectures was developed.[180] Thereafter, Hult, Ihre, and Busson[181] attached chiral mesogens to the first fer-roelectric dendritic liquid-crystalline polymer, which exhibited chiral SmA* and SmC* phases.

Hedrick and coworkers polymerized substituted caprolactones so as to generate branched and dendrigrafted aliphatic polyesters,[182] branched block copolymers,[183, 184] and hyperbranched polyesters.[185] For example (Scheme 6.29), DCC-based coupling of macromonomers **103** and **104**, each prepared by ring-opening polymerization of either L-lactide or a phenyl-substituted caprolactone as promoted by $Sn(oct)_2$, afforded the desired copolymer **105**. Unique layered copolymers were prepared by sequential lactone–lactide polymerizations. Hyperbranched polyesters (Scheme 6.30) were also accessed by the self-condensing cyclic ester polymerization of bis(hydroxymethyl)-substituted ε-caprolactone **106**. Polymers such as **107**, possessing M_ws of 3,000 to 8,000 amu, were observed only after high conversions. The caprolactone monomer was prepared in four steps starting from 1,4-cyclohexanediol.

A new model describing the phase behavior of binary hyperbranched polymer systems has been developed, based on lattice cluster theory.[186] Thermo-optical analysis was used for cloud-point determinations of these polyols. Kantchev and Parquette[187] employed these polyols as scaffolds for the construction of disaccharides; see Chapter 7.

The ring-opening polymerization of 4-(2-hydroxyethyl)-ε-caprolactone (**108**; Scheme 6.31) to produce alcohol-terminated polyesters (i.e., **109**) has been reported.[188] Polymerization was promoted by a catalyst [$Sn(oct)_2$] at 110 °C; molecular weights in the range of 65,000 to 85,000 amu and polydispersities of 3.2 were obtained with a DB of 0.5, as determined by NMR.

Mulkern and Tan[189] reported on the processing and characterization of polystyrene and hyperbranched polyester blends. The hyperbranched materials were shown to act as lubricating agents during processing and as toughening agents in final formulations.

The solution rheology of these hyperbranched polyesters has been presented,[190] as well as the crystallization behavior[191] of blends of poly(ethylene terephthalate) with these polyesters, which were end-capped with hydroxy, acetate, and benzoate groups.

A new hyperbranched material has been synthesized through the free radical alternating copolymerization of an allyloxymaleic acid/maleic anhydride combination; this procedure provides a general methodology for highly branched as well as highly functionalized polymers.[192] Maleic anhydride functionalized polyesters have also been used as radical crosslinkers.[192a]

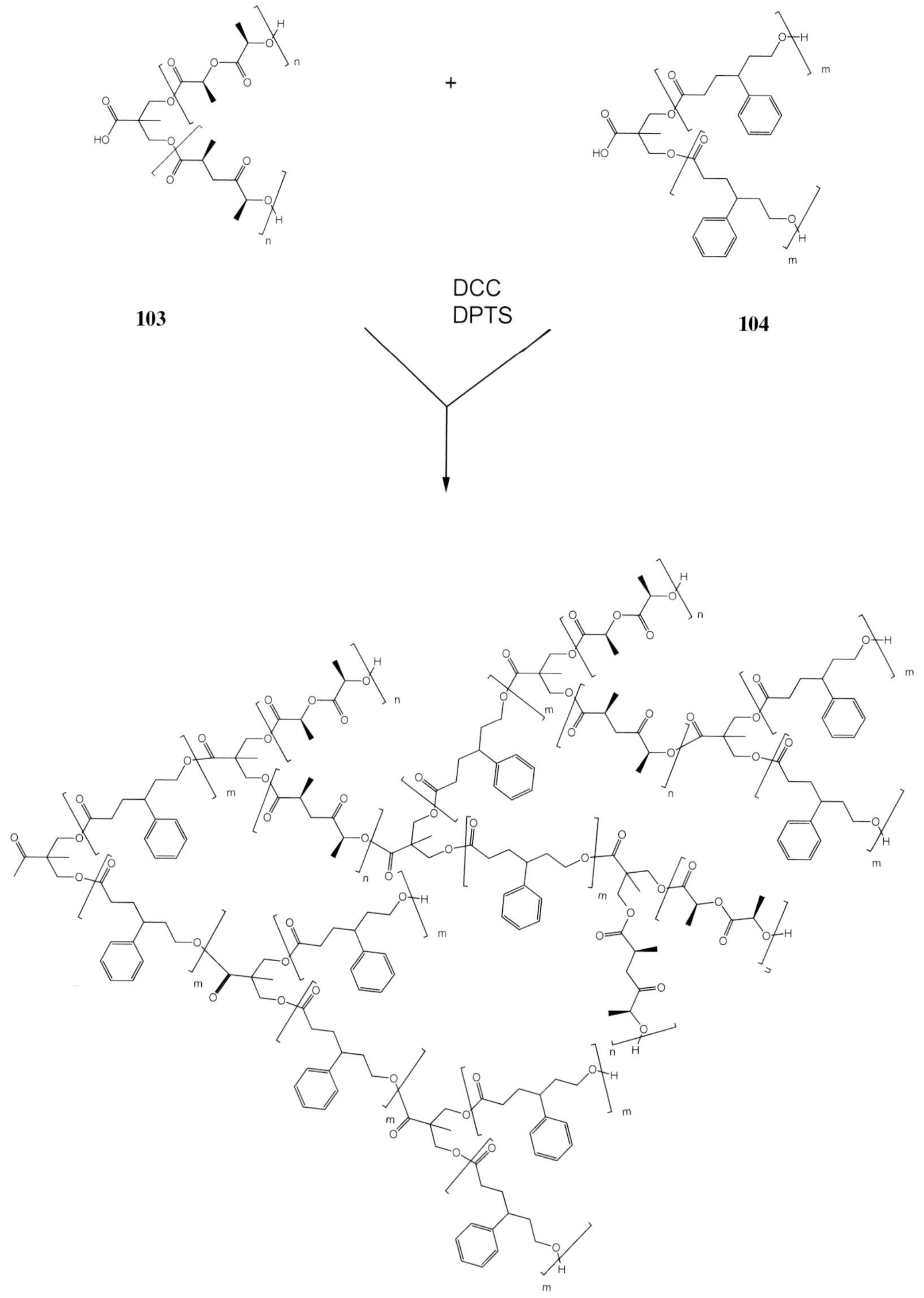

Scheme 6.29 Hedrick's block copolymers based on macromonomer coupling.[184]

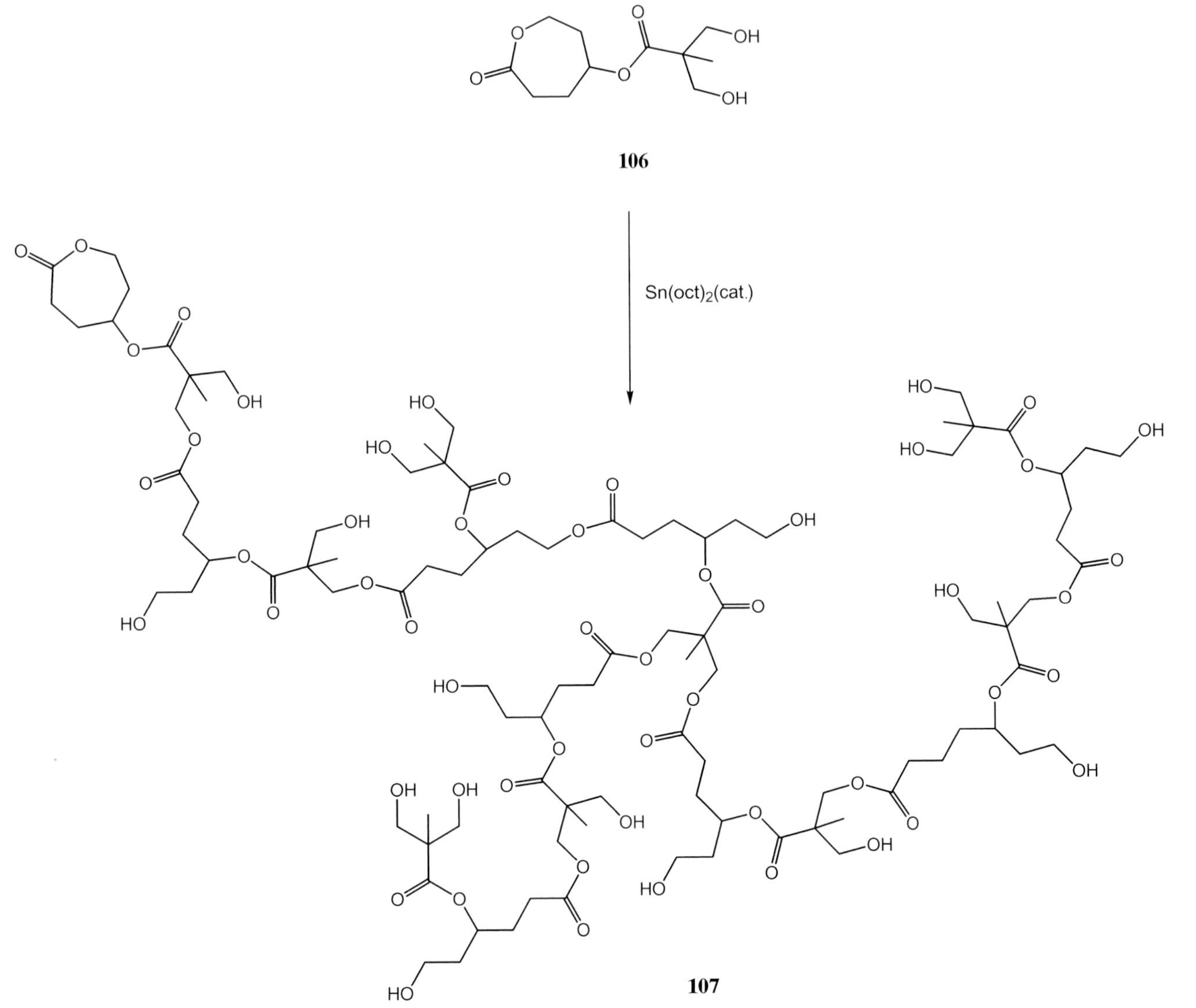

106

Sn(oct)$_2$(cat.)

107

Scheme 6.30 Hedrick's self-condensing polymerization.[185]

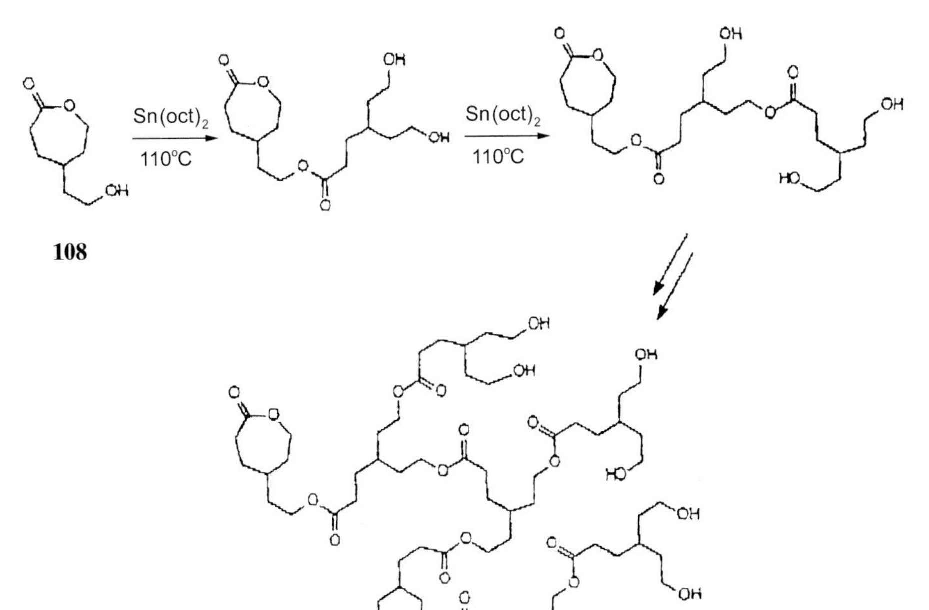

108

Sn(oct)$_2$
110°C

Sn(oct)$_2$
110°C

109

Scheme 6.31 Ring-opening polymerization to produce polyesters.[188]

6.4.2   1 → 2 *C*-Branched, *Ether* Connectivity

Percec and Kawasumi[193, 194] initially reported a preliminary insight into the preparation of a fascinating series of thermotropic, liquid-crystalline polymers possessing tertiary *C*-branching centers. Percec et al.[195–198] later published more complete summaries of their branching systems. Since dendritic design was initially derived from the branching architecture of trees, the authors compared the synthetic approach to these novel cascades to the nature of the willow tree, due to conformationally flexible branching centers inherent in the monomers. Under conditions where the monomer units adopt '*gauche*' configurations, the arms will be relatively extended, thus imparting typical cascade morphology. Conditions favoring monomer '*anti*' configurations will cause a minimization of the interior void volume through the formation of a layered, nematic mesophase.

"Willow-like" cascade construction was facilitated by the preparation of the AB$_2$-type building blocks, e.g., 6-bromo-1-(4-hydroxy-4'-biphenylyl)-2-(4-hydroxyphenyl)hexane (**114**), 13-bromo-1-(4-hydroxyphenyl)-2-[4-(6-hydroxy-2-naphthalenylyl)phenyl]tridecane (**115**), and 13-bromo-1-(4-hydroxyphenyl)-2-(4-hydroxy-4''-*p*-terphenylyl)tridecane (**116**). Scheme 6.32 illustrates the preparative strategy for the incorporation of the monomeric mesogenic moieties predicated on conformational (*gauche versus anti*) isomerism. Thus, one equivalent of 4-hydroxybiphenyl was reacted with 1,4-dibromobutane to yield monobromide **110**, which was converted under Finkelstein conditions to the corresponding iodide **111**. Treatment of this iodide with ketone **112** under phase-transfer conditions afforded the α-substituted ketone **113**. Decarbonylation, followed by concomitant demethylation and bromination, gave the desired monomer **114**.

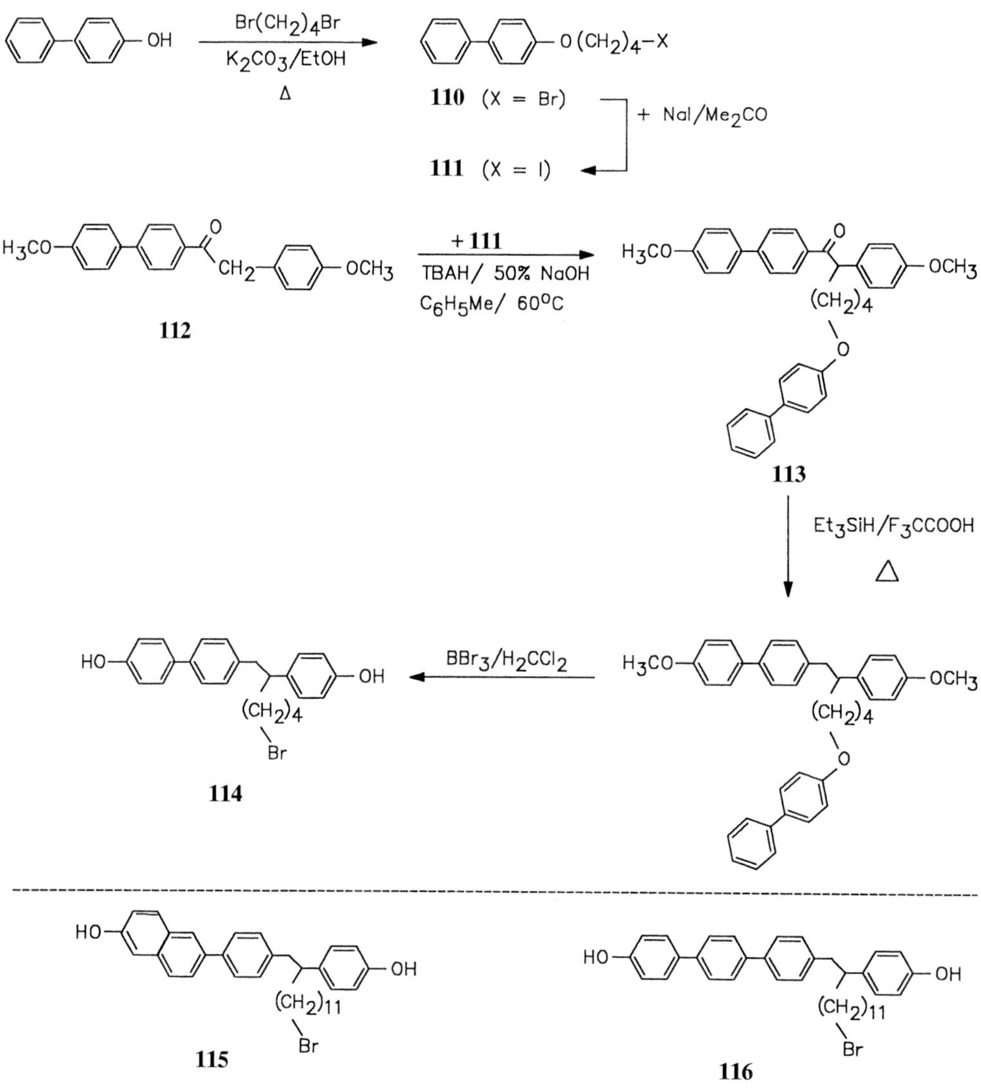

Scheme 6.32 Monomer preparation for the construction of liquid-crystalline polymers.[193, 194]

The synthesis of the phenolic bromide monomer **114** was similar to that of a previously reported monomer,[194] but the alkyl chain spacer unit was shortened in an attempt to increase the nematic mesophase isotropization temperature.[196] Construction of the related building blocks **115** and **116** followed similar pathways; these were prepared with the goal of creating hyperbranched macromolecules with a broader mesophase.[196]

Single-pot construction[199] of the polyethereal polymers (**117**; Scheme 6.33) utilized standard phenoxide alkylation methods [Bu₄NHSO₄, 10 M NaOH]. Capping of the phenolic terminal groups with either alkyl or benzyl groups gave hyperbranched macromolecule **118**. Scheme 6.33 also shows idealized representations of the isotropic (**117a**) and nematic (**117b**) states. All such polymers were found to be soluble in *o*-dichlorobenzene and other more typical solvents when the counterion associated with the phenolic termini was the tetrabutylammonium ion. Polymers possessing alkali metal phenolate termini were water-soluble. The characteristic ¹H NMR spectra of the materials were extensively discussed, and comments were made on diverse side reactions such as alkyl halide hydrolysis and cyclization. Other uses for these novel building blocks, such as the preparation of macrocyclic liquid-crystalline oligopolyethers,[199–202] and convergently generated, nonspherical dendrimers that also exhibit thermotropic liquid-crystalline phases,[203] have been reported. Percec's hyperbranched liquid-crystalline polymers,[196] in low molar mass nematic solvents, were also examined by dynamic light scattering for twist and bend viscosity analysis by Jamieson.[204] Monodisperse, linear analogs of these polymers possessing a DP of 33 and $M_n = 15,090$ have been prepared and examined.[205]

Hyperbranched architectures accessed by polymerization of 4-(4'-chloromethylbenzyloxy)phenylacetonitrile have been reported;[206] the reaction was controlled by the addition of a chloromethylarene and tetrabutylammonium chloride. Number average molecular weights ranged from 2,900 to 9,100 amu.

Zhang and Ruckenstein[207] reported the preparation of vinyl ether-based hyperbranched polymers. Polymerization was effected by Lewis acid (e.g., ZnCl₂) activated, cationic self-condensation of 1-[(2-vinyloxy)ethoxy]ethyl acetate, which was accessed by the reaction of ethylene glycol divinyl ether with acetic acid.

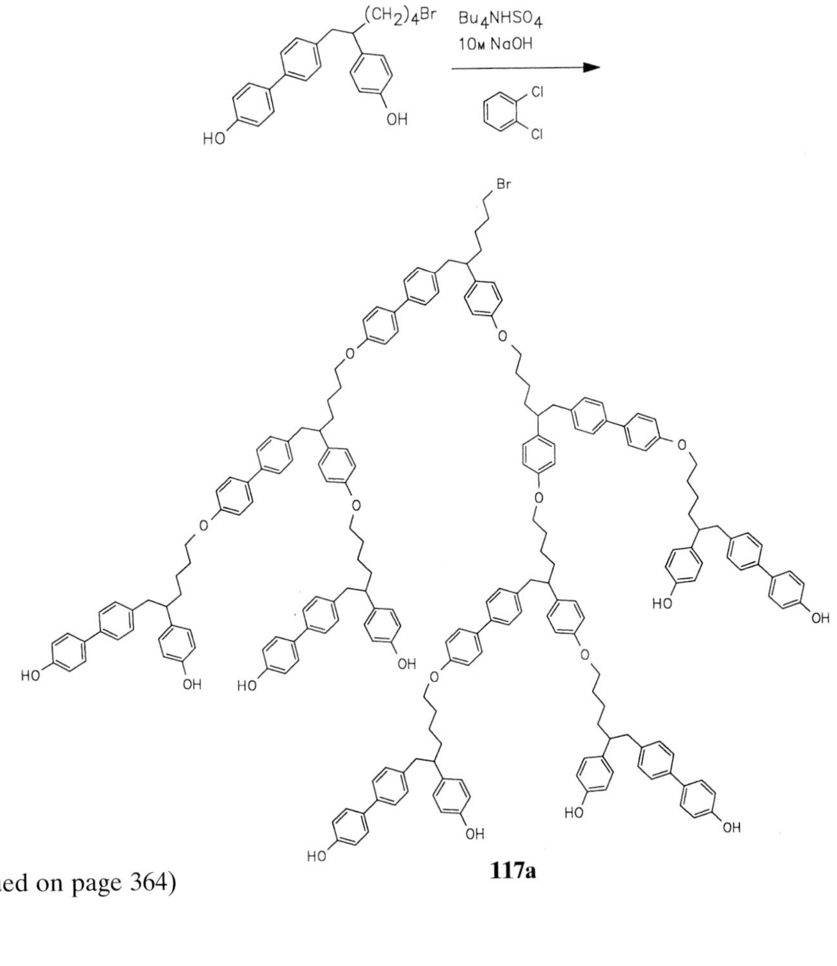

Scheme 6.33 (Continued on page 364)   **117a**

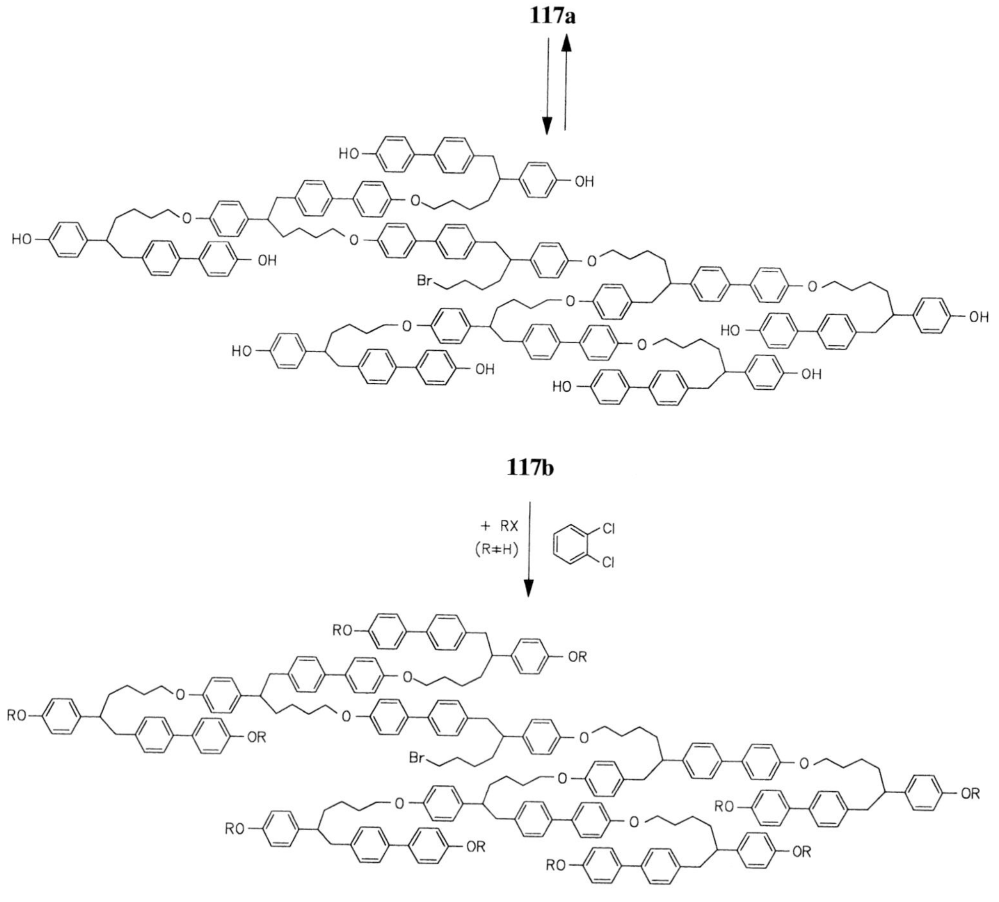

117a

117b

+ RX
(R≠H)

118

Scheme 6.33 Percec et al.'s[199] application of mesogenic monomers for the preparation of hyperbranched polymers.

Frey et al.[208, 208a] reported the creation of "molecular nanocapsules" based on the partial esterification of polyethereal polyol polymers (Scheme 6.34). Anionic ring-opening polymerization of glycidol[209] afforded the amphiphilic alcohol **119**, which was esterified to various degrees using palmitoyl (C_{16}) or caprylic (C_8) acid chlorides to give hydrophobic shells **120**. These materials were demonstrated to quantitatively extract Congo red dye from an aqueous phase into a $CHCl_3$ phase. Release of the encapsulated dye could only be effected by cleavage of the alkyl chains. Other dyes that were examined showed similar behavior. These polyols have also been coated with mesogenic end groups (cyanobiphenyl groups with C_5 or C_{11} alkyl spacers) resulting in low-viscosity, nematic liquid-crystalline materials,[210] and they have been employed as initiator cores for the formation of star block copolymers based on Sn-promoted ε-caprolactone polymerization.[211] A core-first method of PEO multi-armed star polymer formation using these polyglycerols as initiators has also been reported.[212] These hyperbranched amphiphilic materials were used to prepare nano-sized metal colloids, which were purported to be the first examples of such species to find application as transition metal catalysts.[213] Similar base-catalyzed ring-opening polymerization using either enantiomer of the chiral glycidols has led to the creation of chiral polyols possessing M_ws in the range of 1,000 to 10,000 amu.[214] Block copolymerization with propylene oxide allows access to poly(glycerol)s, such that the polarity of these highly hydrophilic polyols can be varied.[215] The thermal behavior of the corresponding esterified polyols has been reported.[216] Similar poly(glycerol)s have also been reported by Haag et al.[217, 217a] (see Scheme 3.30).

Watanabe et al.[218] reported the preparation of solvent-free polymer electrolytes possessing high ionic conductivities. Essentially, a cross-linked polymer network, accessed by UV irradiation of an acrylic acid modified poly(ethylene oxide)-based macromonomer with hyperbranched side chains, was doped with such electrolytes as lithium bis(trifluoromethylsulfonyl)imide (LiTFSI) to give conductivities of 1×10^{-4} S cm^{-1} at 30 °C.

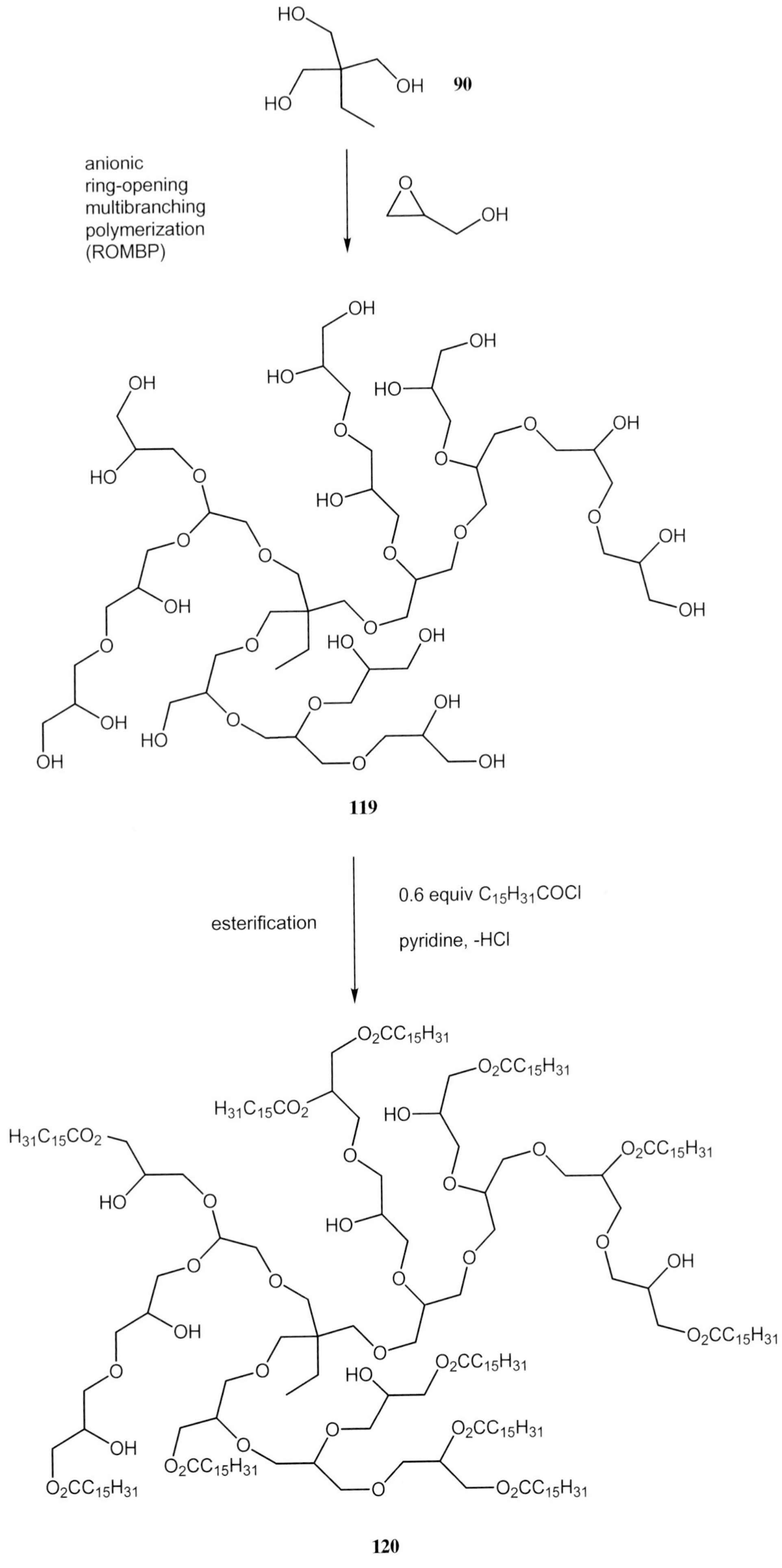

Scheme 6.34 Construction of "Molecular Nanocapsules".[208]

6.4.3   1 → 2 *C*-Branched, *Amide* Connectivity

Bergbreiter, Crooks et al.[219] described the synthesis of hyperbranched polymer films grafted onto self-assembled monomers. Branching within the polymeric superstructure was effected by the use of an α,ω-diamino-terminated poly(*tert*-butyl acrylate) monomer. Advantages of this procedure include: (1) the production of thick polymer layers even when reactions proceed in low yield, and (2) a resultant polymer film possessing a high density of modifiable functional groups. The formation of films of fluorinated polymers on gold and silica surfaces by means of layered, hyperbranched poly(acrylic acid) surface-grafts using a fluorinated octylamine has also been reported.[220] The synthesis and characterization of surface-grafted hyperbranched films possessing fluorescent, hydrophobic, ion-binding, biocompatible, and electroactive groups have been described.[221] Poly(acrylic acid) (PAA) films were prepared (3 to 100 nm thickness) and derivatized with functionalities such as pyrene and ferrocene. Inhibition of electrochemical reactions was observed at gold surfaces grafted with highly fluorinated hyperbranched PAA films.[222] Further, a β-cyclodextrin-functionalized hyperbranched poly-(acrylic acid) film possessing an ultrathin polyamine capping layer has been coated on Au electrodes,[223] thus giving rise to a new "molecular filter" approach for selective chemical sensors. A similar ultrathin film was employed as a host matrix for electrostatic enzyme entrapment; amine-modified glucose peroxidase, horseradish peroxidase, and PAMAM-modified horseradish peroxidase were shown[224] to be reversibly absorbed into hyperbranched poly(sodium acrylate) films. PPA micropatterned surfaces on gold are also known.[224a]

Novel architectures described as "amphiphilic core-shell nanospheres" have been prepared[225, 226] by intramicellar polymer chain cross-linking of polystyrene–poly(acrylic acid) micelles with bis(amino-terminated) poly(ethylene glycol). These shell cross-linked Knedel-like structures[227–229] possessed hydrodynamic radii of 50 nm and were subject to a ca. five-fold volume increase upon aqueous dissolution from the solid state. Their sizes, structures, and stabilities make these materials attractive drug delivery vehicles. An excellent review has recently appeared covering these Knedel-like nanostructures.[230]

Very recently, Bergbreiter et al.[231, 231a] described the use of polyethylene powder supports bearing useful "loadings" of functional groups, which were accessed by hyperbranched grafting procedures. The available carboxylic acid surface moieties were functionalized covalently or ionically to generate powders exhibiting physical durability and solvent resistance.

6.4.4   1 → 2 *C*-Branched, *Alkane* Connectivity

A hyperbranched polystyrene has been reported,[122] which was obtained by atom-transfer radical polymerization of commercially available *p*-(chloromethyl)styrene in the presence of Cu(I) and bipyridine. Molecular weights of 31,600 (M_n) and 164,500 (M_w) amu were determined on the basis of viscosity and light-scattering data, respectively; this was in contrast to values of 13,400 (M_n) and 75,000 (M_w) determined by calibration with linear polystyrene. Smaller hydrodynamic volumes are generally obtained for branched polymers as compared to linear polymers of similar molecular weight. Similar poly(isobutylene)s have been shown.[231b]

Ru-catalyzed step-growth polymerization of 4-acetylstyrene has been reported by Weber et al.[232] The AB$_2$ monomer was treated with [(PPh$_3$)$_3$Ru(H)$_2$CO] to afford the poly(4-acetylstyrene) branched architecture. Aryl C–H bonds adjacent to the activating acetyl moieties were essentially added across the vinyl double bond; addition proceeded with both Markovnikov and *anti*-Markovnikov regioselectivities.

A series of papers has appeared on "self-condensing vinyl polymerizations" (SCVP) of acrylate monomers, examining: (1) molecular weight distribution,[233] (2) the degree of branching,[234] (3) the kinetics and mechanism of chain growth,[235] and (4) the effect of reaction conditions.[236] Synthesis of these polyacrylates by atom-transfer radical polymerization has also been described.[237] The authors examined polymers obtained by "living" radical polymerization of acrylic AB* monomers, where A is a vinyl group and B* equates to a functional group that can be transformed to an active center for the initiation of double-bond polymerization.

Müller et al.[238] further studied the effect of "core-forming" molecules (i.e., a multifunctional initiator) on the DB and M_w distribution in hyperbranched polymers. A significant decrease in the polydispersity of self-condensing vinyl polymerization was noted, which was found to be dependent on the initiation functionality, while the DB for a semibatch process was found to be 2/3. Molecular weight averages, average DB, and the kinetics[238a] of self-condensing vinyl copolymerizations of a vinyl monomer with an "inimer" (a molecule possessing both a vinyl moiety and an initiating group) have also been addressed.[239] Yan, Zhou, and Müller[240] derived the molecular weight distributions for these SCVP reactions of vinyl monomers and "inimers".

Fomina et al.[240a] polymerized a dibromostyrene monomer via solid-supported Pd-catalyzed cross-coupling of vinyl halides and terminal acetylenes.

6.4.5   1 → 2 *C*-Branched, *Carbonate* Connectivity

Bolton and Wooley[241] reported the preparation of polycarbonates (**122**; Scheme 6.35). Using a bis(acylimidazole) AB₂ monomer (**121**), rigid architectures such as **122** were accessed by treatment with AgF should be (20 % CH₃CN/THF, 70 °C, 2 d). The DB was determined to be 53 %, while M_ws were found to correspond to 16,000, 77,000, and 82,000 amu based on GPC (polystyrene standard), or 23,000, 180,000, and 83,000 amu based on GPC with LALLS for the carbonylimidazole, phenol, and *t*-butyldimethylsilyl ether-terminated materials, respectively.

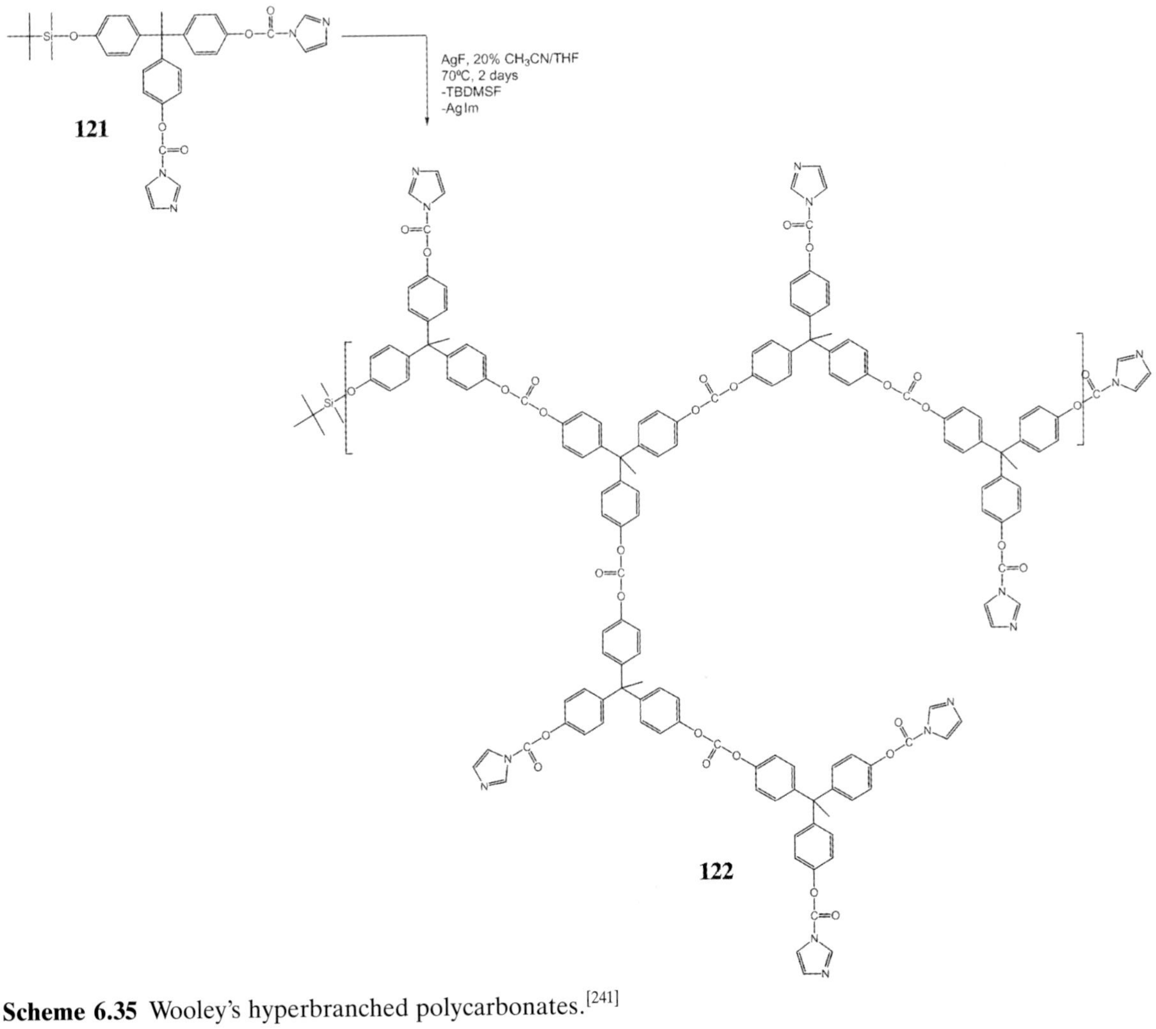

Scheme 6.35 Wooley's hyperbranched polycarbonates.[241]

6.4.6 1 → 2 *C*-Branched, *Ether & Ester* Connectivity

Gong and Fréchet[242] transformed commercial 4,4-bis(hydroxyphenyl)pentanoic acid (**89**) into 4,4-bis(oxiranylmethoxyphenyl)pentanoic acid, which was polymerized (Et$_4$N$^+$Br$^-$) by a proton-transfer polymerization process[84] affording a product possessing an M_w of 44,000 amu and a PDI of 5.9 within 27 h at 65 °C. The use of chloride ion in place of bromide led to a slower polymerization and a lower M_w of 24,700 amu under similar conditions.

These poly(glycerol)s have also been used as cores for methyl acrylate polymerization to produce poly(acrylate) star polymers.[243] The hyperbranched polyol was converted to a macroinitiator by surface modification with 2-bromoisobutyryl bromide. Atom-transfer radical polymerization was then conducted in the presence of CuBr and pentamethyl-diethylenetriamine to produce 45- to 55-armed star polymers with $\overline{M}_n$ ranging from 16,500 to 184,500 amu based on GPC data.

Kadokawa et al.[244] employed proton-transfer polymerization of an acrylate diol monomer [i.e., 2,2-bis(hydroxymethyl)propyl acrylate]. Using PPh$_3$ to initiate the reaction, polymers possessing $\overline{M}_n$ (VPO) approaching 3,000 amu and DBs ranging from 0.45 to 0.60 (^{1}H NMR) were obtained. Two pathways leading to product formation were discussed, including a "macrozwitterion" mechanism and a "gegenion" mechanism, whereby repetitive oxyanion Michael-type addition and proton-transfer occur concomitantly.

6.5 1 → 2 *N*-Branched

6.5.1 1 → 2 *N*-Branched and Connectivity

Suzuki et al.[245] explored the Pd-catalyzed ring-opening of the monomer 5,5-dimethyl-6-ethenylperhydro-1,3-oxazin-2-one to give a hyperbranched polyamine. This polymerization was conducted at ambient temperatures in THF under catalysis by [Pd$_2$(dba)$_3$·CHCl$_3$· 2dppe] to afford, after the evolution of CO$_2$, the desired hyperbranched polyamine. NMR spectroscopy confirmed the high yield conversion of the monomer. The DB was ascertained (NMR) as the ratio of the number of tertiary amine units to the total number of secondary and tertiary moieties.

Copoly(sulfone-amine)s have also been reported.[245a–c]

6.5.2 1 → 2 *N*-Branched, *Amide* Connectivity

Feast and coworkers[246, 247] in a remarkable experiment, created hyperbranched analogs of PAMAMs.[248] The reaction is depicted in Scheme 6.36; *N*-acryloyl-1,2-diaminoethane hydrochloride (**123**),[249] prepared in three simple steps, undergoes melt polymerization (210 °C) to afford the HCl salt-terminated polymer **124** possessing a branching ratio very close to one. Methylene spacer chain lengths of 2 to 7 units were examined; monomers with (CH$_2$)$_2$ chains led to faster reaction times than those with (CH$_2$)$_6$ chains. Following the polymerization, complete retention of HCl in the product was observed; an amine to carbonyl HCl equilibrium was thus proposed to account for the requisite free amine attack on the alkene. Characterization included MALDI-TOF MS, T_g, DSC, and solution viscosity measurements. A decrease in intrinsic viscosity was observed as the molecular weight increased, supporting the dendritic structure. The postgrafting of hyperbranched PAMAM wedges initiated from a surface amine of a polymer chain grafted to the surface of silica has recently appeared,[249a] as have similar cationic, DNA condensing poly(amino ester)s.[250, 250a] As well, PAMAMs have been used for the creation of a chitosan-sialic acid hybrid.[250b, c] The UV curing behavior of similar materials was reported.[250d]

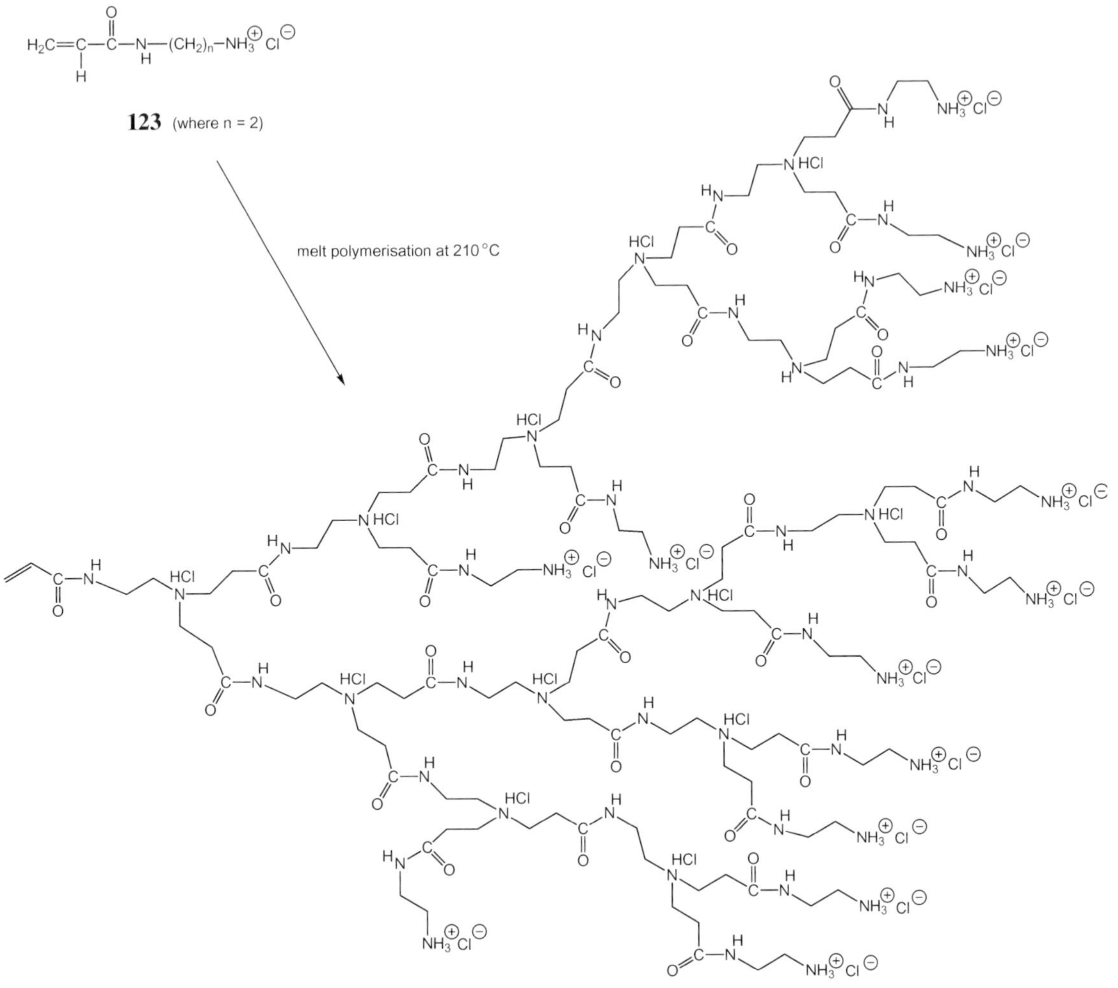

123 (where n = 2)

melt polymerisation at 210 °C

124

Scheme 6.36 Feast's "one-pot" preparation of the PAMAMs.[246]

6.5.3   1 → 2 *N*-Branched and Connectivity

Recently, families of aromatic hyperbranched polyimides were prepared[251] by polymerization of tris(4-aminophenyl)amine with a series of commercial dianhydrides such as 2,2-bis(3,4-dicarboxyphenyl)hexafluoropropane anhydride, 3,3',4,4'-diphenylsulfone-tetracarboxylic dianhydride, and pyromellitic anhydride. It was suggested that the initial formation of the hyperbranched polyamic acid precursors was followed by thermal or chemical imidization. Gas permeation studies are known.[251a]

6.5.4   1 → 2 *N*-Branched, *Aryl* Connectivity

Tanaka, Iso, and Doke[252] reported the preparation of hyperbranched poly(triphenyl-amine) dendrimers (Scheme 6.37). Reaction of one equivalent of *n*-BuLi (–78 °C, THF) with tris(aryl bromide) **125** followed by treatment with MgBr$_2$·Et$_2$O gave the Grignard dihalide **126**, which was subsequently polymerized [Ni(acac)$_2$] to afford the desired poly-(aryl amine) **127**. The $\overline{M}_w$ was determined to be 4.0×10^3 and good solubility in organic media was maintained. On the basis of GPC, the DP was determined to be ca. 12, while bromide content suggested a DP of ca. 10. Conductivity was examined by means of vis/near IR spectroscopy. Following doping of polybromide **127** and applying a potential dif-

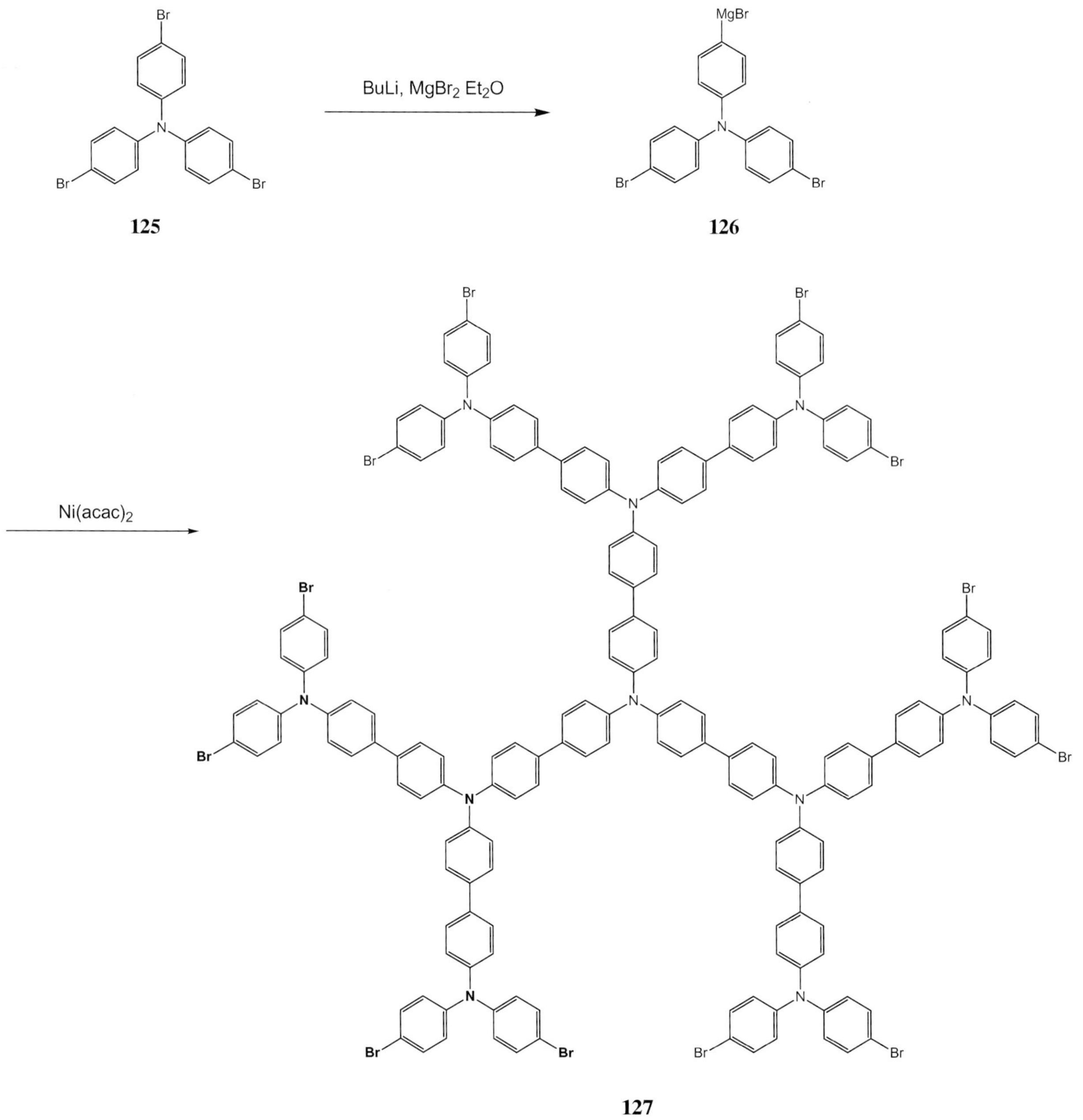

Scheme 6.37 Poly(triphenylamine) polymers as potential molecular electronic devices.[252]

ference of 1.0 or 1.2 V, three new transition bands were observed at 500 and 1500 nm, attributable to valence band transitions to the higher and lower polaron bands, respectively, and at 750 nm, attributable to a low to high polaron transition. In a related report,[253] these poly(triarylamine)s were used as cores for the attachment of poly(3-hexylthiophene)s, resulting in electroactive materials that were shown to possess high conductivities (100 S cm^{-1}) when doped.

An interesting ring-opening polymerization (Scheme 6.38) to prepare a poly(alkene amine) polymer was reported,[254] in which the monomer (5-methyleneperhydro-1,3-oxazin-2-one; **128**), prepared from 1,3-dichloro-2-methylenepropane, was polymerized [Pd(dba)$_2$/PPh$_3$] to afford polyamine **129**, which was subsequently capped (*n*-butyl isocyanate) to give the corresponding urea-terminated material **130**. The DB ranged from 61 to 81 depending on the solvent used, while $\overline{M}_n$ values ranged from 1,800 to 3,000 depending on the method of determination.

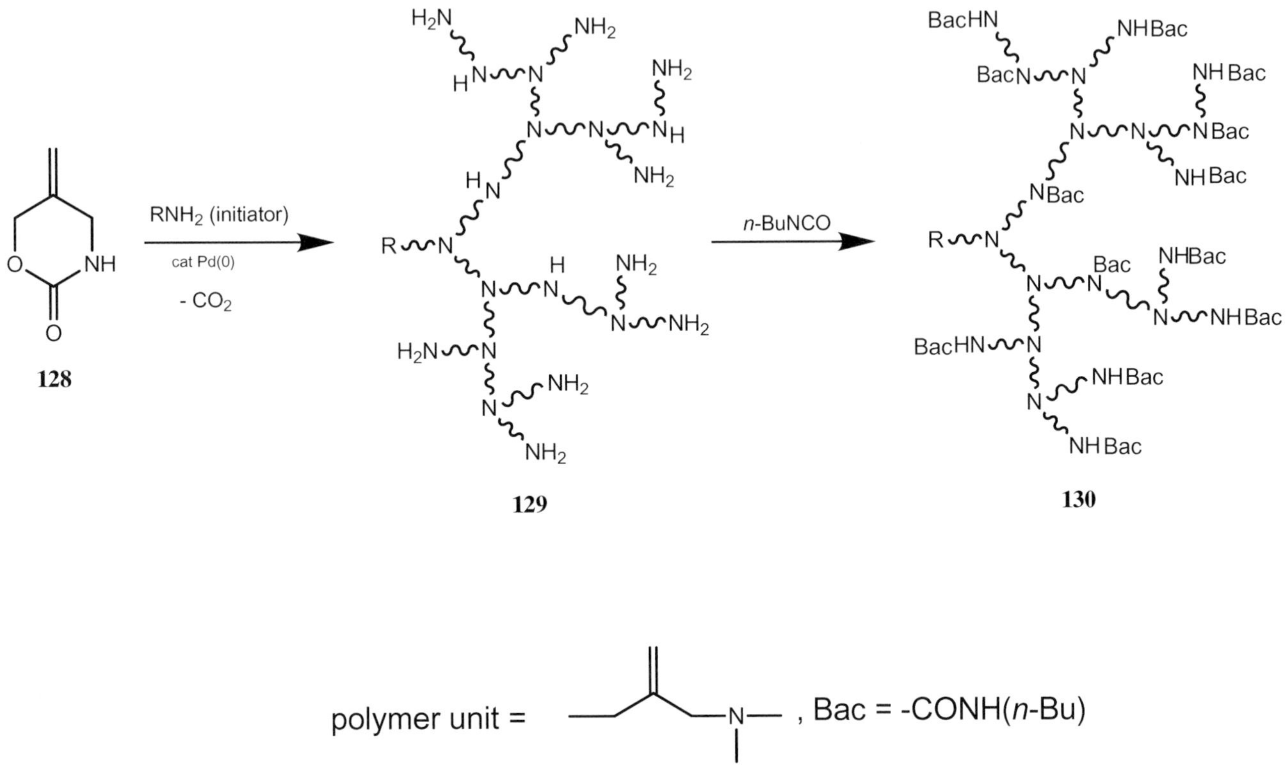

Scheme 6.38 Suzuki's ring-opening polymerization leading to hyperbranched polyalkenes.[254]

6.6 1 → 3 *C*-Branched, *Ester* Connectivity

Wu et al.[255] condensed phthalic anhydride with pentaerythritol to produce "dendritic-like" polyelectrolytes, which were described as highly branched polymer clusters. Light-scattering experiments revealed a hydrodynamic radius of 4.16 nm and an $\overline{M}_w$ of 1.1×10^4 g/mol. Aggregation and complex formation of these hyperbranched polymers with cetyltrimethylammonium bromide (CTAB) micelles was examined.

6.7 1 → 3 *C*-Branched, *Ether* Connectivity

An A_2 + B_3 approach to hyperbranched material (Scheme 6.39) has been developed,[256] whereby an epoxide ring-opening gave the corresponding oxyanion (e.g., A_2, **131b**) providing the base for deprotection (proton transfer) of a B_3 monomer to give anion **132**, which could then react with more epoxide and continue the propagation. Polymers of this type, e.g., **133**, were obtained with $\overline{M}_w$'s (GPC) of 5,000 to 15,000 amu, while polydispersities were observed to increase with increasing molecular weight.

6.8 1 → 3 *Ge*-Branched and Connectivity

The anionic polymerization of tris(pentafluorophenyl)germanium halide (**134**) was reported by Bochkarev et al.[257] to give Ge-based polymers, specifically a 9-Cascade: germane[3]:(2,3,5,6-tetrafluorogermanylidyne):pentafluorobenzene (**135**; Scheme 6.40). These polymers possessed M_ws ranging from 100,000 to 170,000 amu. Polymerization was attributed to the formation of a metal anion, leading to aromatic substitution of the *p*-fluoro substituent. Unique phosphine oxide based materials have also been reported.[257a]

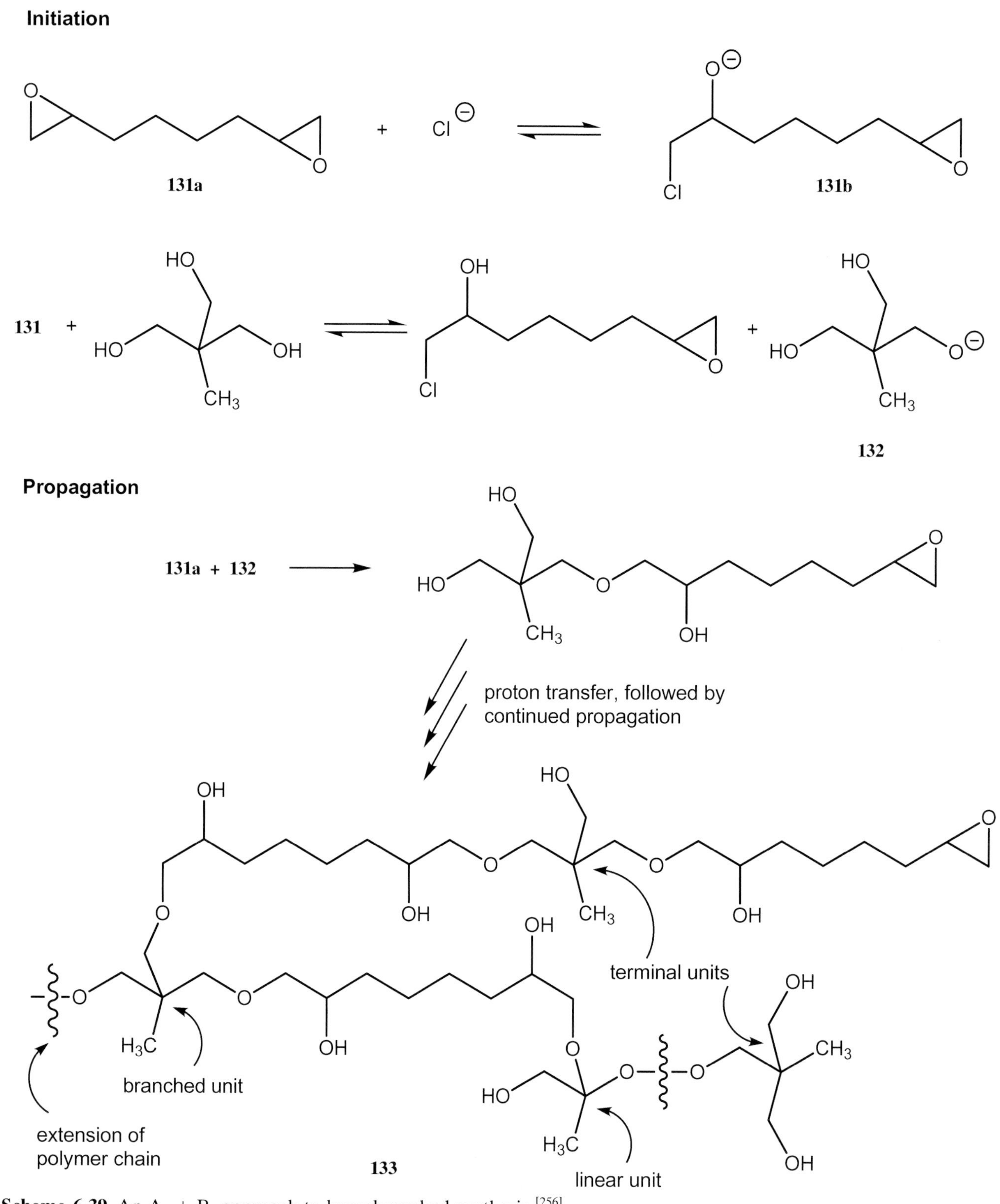
Initiation
131a
131b
131
132
Propagation
131a + 132
proton transfer, followed by
continued propagation
terminal units
branched unit
extension of
polymer chain
linear unit
133

Scheme 6.39 An $A_2 + B_3$ approach to hyperbranched synthesis.[256]

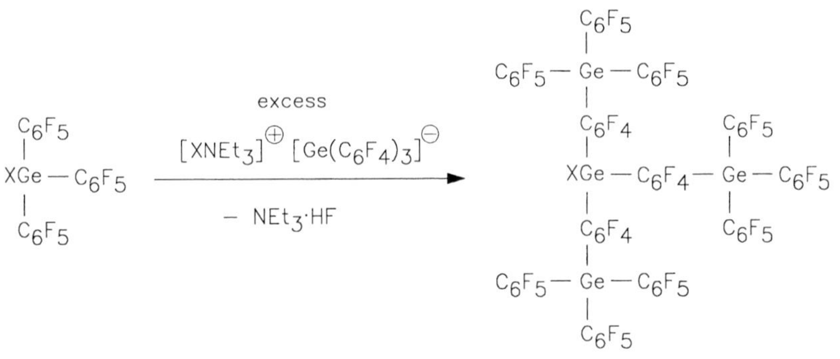

Scheme 6.40 Synthesis of Ge-hyperbranched polymers.[257]

6.9   1 → 3(2) *Si*-Branched and Connectivity

Mathias and Carothers[258–261] reported the classical polymerization (Scheme 6.41) of an A$_3$B-type, Si-based monomer (**136**) employing Pt-catalyzed alkene hydrosilylation to afford the hyperbranched poly(siloxysilane)s, e.g., **137**. ^{1}H NMR integration suggested that polymer growth continued through to the 3rd or 4th generation. The addition of more catalyst to the reaction did not increase the molecular weight (ca. 19,000 amu), as indicated by a single, narrow SEC peak, probably due to the onset of dense packing. Gradual coupling of the unreacted Si–H moieties to form Si–O–Si bonds was circumvented by termination with allyl phenyl ether and by the formation of allyl-terminated oxyethylene oligomers.

Mathias et al.[262] also reported the preparation of a rigid, four-directional core for the construction of dendrimers possessing "sterically dispersed" functionalities. Pathways for divergent cascade synthesis beginning with core **42** were described. Conventional polymerization of a 1 → 3 branched, polysiloxane monomer by hydrosilylation–polyaddition was also reported. Terminal functionalization of the siloxane-based polymers and potential applications of these novel materials, e.g. as controlled interphases

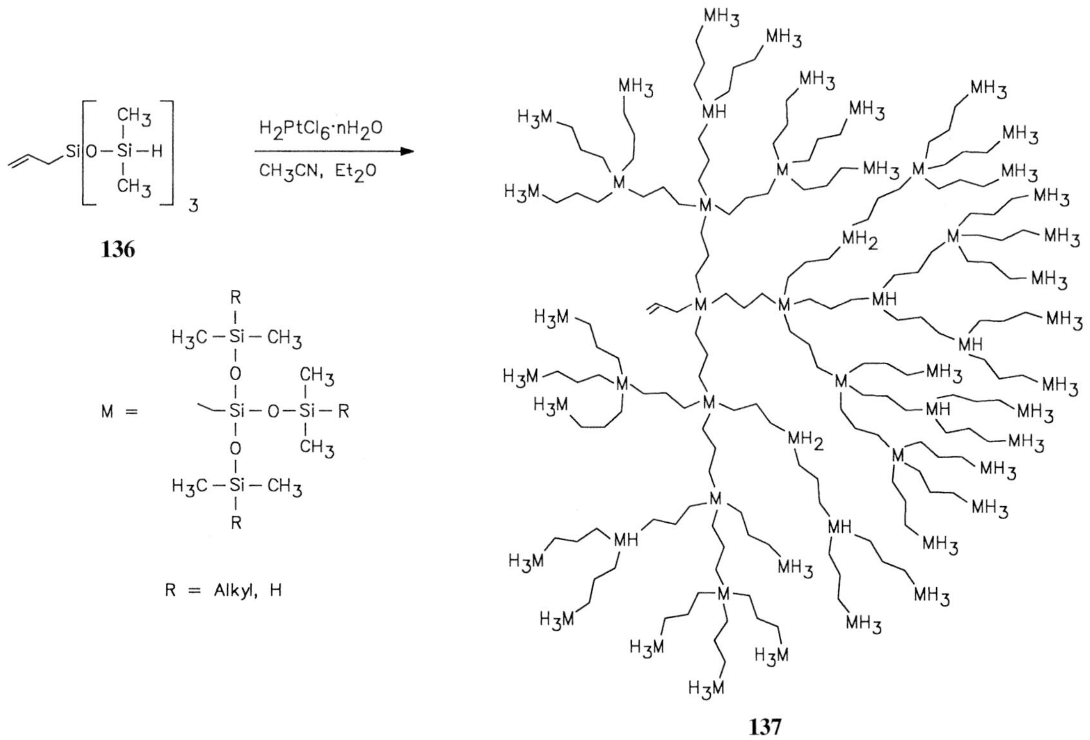

Scheme 6.41 Hyperbranched poly(siloxysilane)s[258–261] obtained by polymerization of Si-based monomers.

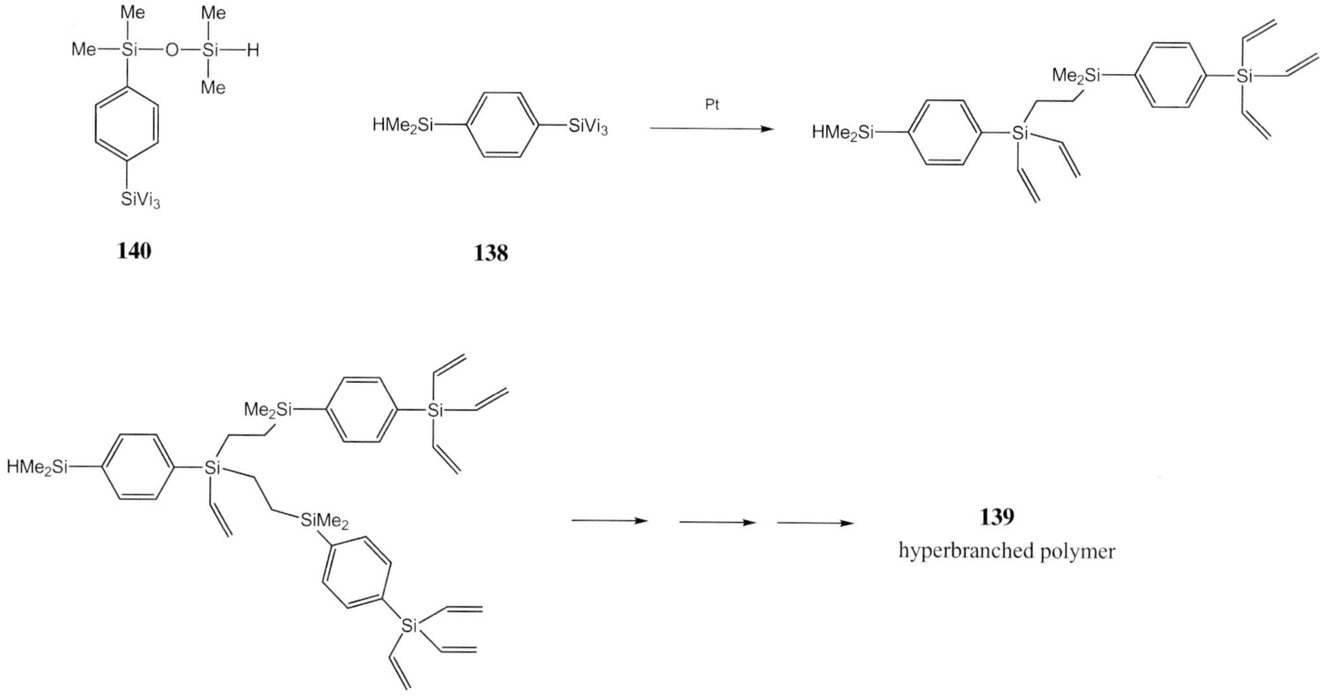

Scheme 6.42 Son's synthesis of poly(carbosilarylene)s.[265]

for composites, as artificial blood substitutes, and as "microreactors" possessing terminally bound catalytic moieties, were noted. Muzafarov, Golly, and Möller[263] prepared similar poly(alkyloxysilane)s that are easily hydrolyzed under acidic conditions and are thus biodegradable. Additional branched organosilicon polymers of interest include those of Ishikawa et al.,[264] whereby regiospecific 1,2-hydrosilylation reactions of silylene-based diethynylene polymers with 1,4-bis(methylphenylsilyl)benzene, as promoted by $Rh_6(CO)_{16}$, yielded the corresponding branched polymers.

Yoon and Son[265] described the preparation of poly(carbosilarylene)s (Scheme 6.42). Beginning with the trivinylbis(silane) **138**, accessed from 1,4-dibromobenzene, polymerization was effected by Pt-mediated hydrosilylation to give polymer **139**. Use of the siloxane-based monomer **140** afforded a more flexible polymer, albeit with reduced thermal stability. T_gs for the materials obtained from monomers **138** and **140** were determined to be –45 and 12 °C, respectively, with $\overline{M}_n$s ranging from 2,560 to 5,600 amu; the DB was found to be ca. 0.44. Poly(silylene vinylene)s with alkyne termini are also known.[265a]

Möller et al.[266] prepared degradable dendrimers for templated cavity incorporation into resins. The dendrimer termini were envisaged as remaining attached to the resulting pore surface after degradation through the cross-linking of alkene moieties into the matrix during polymerization. Hyperbranched organosilicon polymers were employed, possessing molecular weights in the 100,000 amu range. Silyl ether hydrolysis was effected by methanolysis, while hyperbranched materials aggregated during resin formation resulting in larger pore formation than expected. Pt-mediated hydrosilylation of methyldivinylsilane, methyldiallylsilane, triallylsilane, and methyldiundecenylsilane has also led to the creation of polyalkenylsilanes.[267]

Matyjaszewski et al.[268] prepared and analyzed polysilanes of various architectures including copolymers prepared from $RSiCl_3$ and $R'R''SiCl_2$, as well as hyperbranched polymers incorporating $[(SiMe_2)_4Si]_n$ and $(Si)_m(Me_2Si)_n$ units. The electronic properties of these materials were examined (e.g., by emission and UV spectroscopies) and related to those of porous silica.

Hyperbranched poly(siloxysilanes) possessing either vinyl or Si–H peripheral groups, prepared from tris(dimethylvinylsiloxy)silane and vinyltris(dimethylsiloxy)silane, respectively, have been reported by Rubinsztajn.[269]

Oligomeric and polymeric silanes containing branched Si_4 units have been reported.[270] Approaches allowing control over the molecular weight and polydispersity,[271] as well as the functionalization of such poly(siloxysilanes) with epoxy, amine, and hydroxyl groups,[272] have also been discussed.

Base-catalyzed proton-transfer polymerization has been reported as a means of synthesizing polysiloxanes; comparisons were made between the properties and microstructures of these systems and those of their linear analogs.[273]

Oxazoline, focally-modified carbosilane hyperbranched macromolecules[274] have been polymerized and copolymerized with 2-phenyl-1,3-oxazoline to give hyperbranched analogs of comb-burst polymers.[275] Hyperbranched trimers based on trimeric acid cores were also noted to form superstructures either in solution or in bulk. Lach and Frey[276] have also described the enhancement of the DB of these unique macromonomers by reaction of the unreacted alkane moieties in the hyperbranched product with trichlorosilane (Pt-mediated), followed by treatment with allylmagnesium bromide, effectively combining the hyperbranched and standard dendritic protocols. They also introduced an improved DB definition,[28,277,278] along with the concept of the *average number of branches* (ANB), which quantifies the average number of polymer chains attached to non-terminal branch junctures including the linear direction. A measure of structural "branch density", allowing comparisons to be made, was thus delineated.

Fréchet and Miravet[279] examined the variation of branching motif and end-functionalization on hyperbranched poly(siloxysilane)s; studies included the use of AB_2, AB_4, and AB_6 monomers to obtain weight-averaged molecular weights in the range 5,000–10,000 amu and polydispersities approaching 2.

6.10   1 → 3 *Cyclotetraveratrylene*-Branched, *Ether* Connectivity

Percec et al.[280] described the preparation of the first hyperbranched polymers possessing disc-shaped mesogens as the branching centers (Scheme 6.43). Construction of these architectures was predicated on the electrophilic cyclotetramerization of 3,4-bis(*n*-alkyloxy)benzyl alcohol units to give octaalkyloxy-substituted cyclotetraveratrylene derivatives (CTTV-*n*; where *n* is the length of the alkyloxy side chain); alkyl chain connectivity of the tetrameric monomer was introduced by the use of a dimeric bis(benzyl alcohol). Thus, reaction of 3,4-bis(heptaalkyloxy)benzyl alcohol (**141**) with the hexadecaalkyloxy-connected benzyl alcohol dimer (**142**) in the presence of a catalyst (CF_3CO_2H) afforded the poly(CTTV-7)-based-polymer **143**. Polymers possessing shorter pendant side chains (C_6) and shorter connectors (C_{12}) were also examined. A key by-product obtained from these polymerizations was the simple dimer (CTTV), which could be completely removed by precipitation; adjustment of the starting monomer concentrations controlled its formation.

Notably, these materials were observed to be "completely soluble", suggesting that it was unlikely that any of the CTTV mesogen was attached via spacers to four adjacent discs, while there was a high probability of a 1 → 3 branching motif. The poly(CTTV-7) polymer was subjected to DSC, GPC, and polarized optical micrograph analysis. Columnar mesophase stacking, obtained by isotropic melt cooling (115 °C) of the polymer, was observed in the optical micrographs.

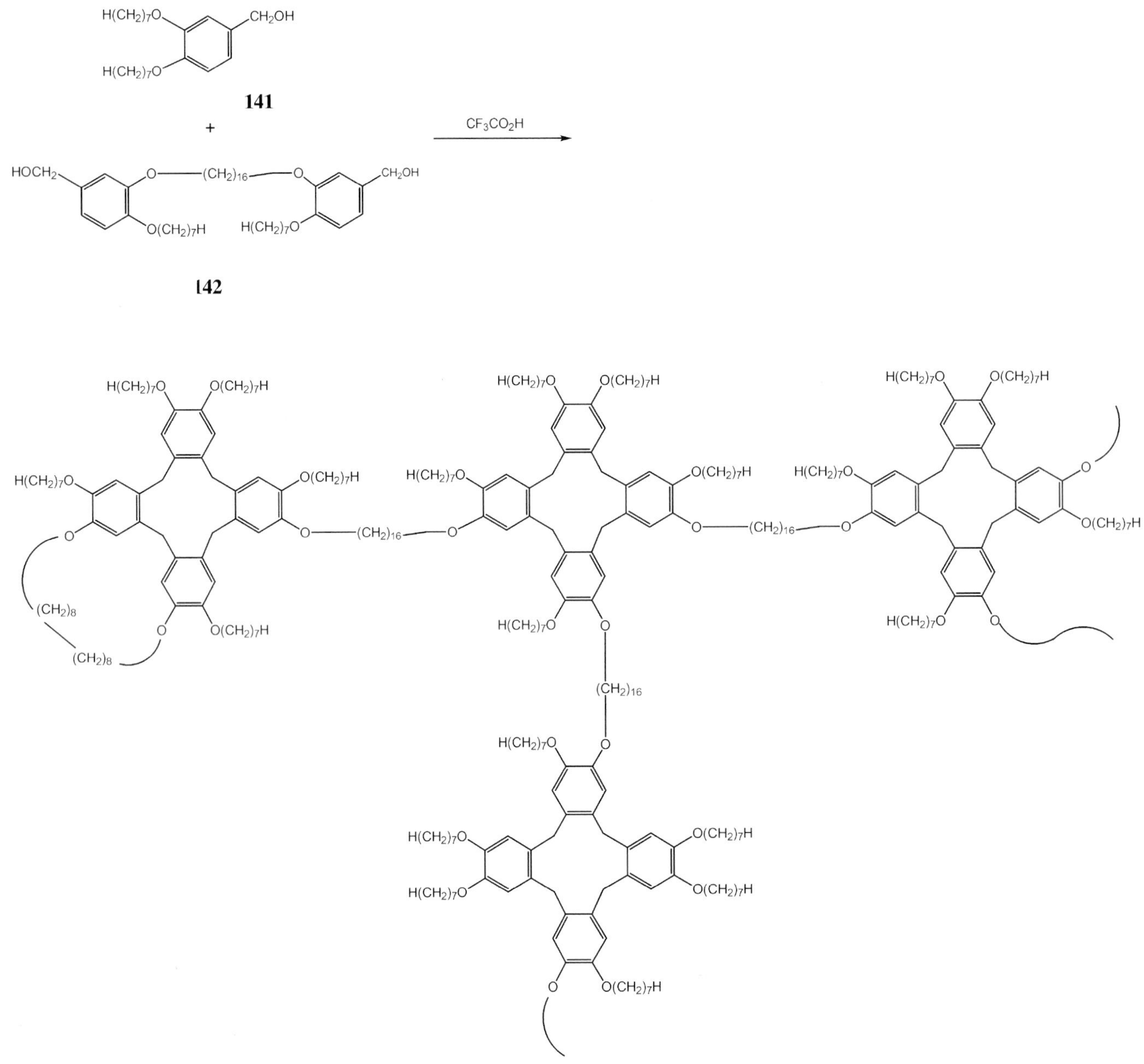

Scheme 6.43 Synthesis of hyperbranched polymers containing disc-shaped mesogens.[280]

6.11   1 → 3 *Aryl*-Branched, *Ester* Connectivity

Kricheldorf and Stukenbrock[281] reported the synthesis of hyperbranched materials employing the 1 → 3 branching motif derived from gallic acid (3,4,5-trihydroxybenzoic acid). Homopolyesters were accessed by bulk condensation of acetylated gallic acid, while copolyesters were obtained by the addition of silylated derivatives of 3-hydroxybenzoic acid or β-(4-hydroxyphenyl)propionic acid. Analogs incorporating equal quantities of the propionic acid moiety and 2-, 3-, or 4-hydroxybenzoic acid, vanillic acid, or 4-hydroxycinnamic acid were also described. Model reactions point to the reaction of all three hydroxyl groups on the gallic acid, while ^{13}C NMR spectra recorded during homopolycondensation suggested otherwise. These amorphous materials were found to possess T_gs much lower than that of the parent poly(gallic acid) architecture.

6.12   1 → 2 Heteroaromatic Branched

The synthesis of dendrimers and hyperbranched polymers based on the cycloaddition of cyanoguanidine to nitriles followed by cyanoethylation has recently been described.[282] These procedures are conducted in the presence of basic catalysts (MOH or Triton B, respectively). This guanamine dendrimerization process, giving materials terminated with cyano or amine groups has been conducted through to the 4th generation; diverse side products were isolated and characterized.

6.13   References

[1]   P. J. Flory, "Molecular Size Distribution in Three-Dimensional Polymers; IV. Branched Polymers Containing A–R–B$_{f-1}$ Type Units", *J. Am. Chem. Soc.* **1952**, *74*, 2718–2723.

[2]   P. J. Flory, *Principles of Polymer Chemistry*, Cornell University Press, Ithaca, New York, **1953**.

[3]   H. R. Kricheldorf, Q.-Z. Zang, G. Schwarz, "New Polymer Syntheses; 6. Linear and Branched Poly(3-hydroxybenzoates)", *Polymer* **1982**, *23*, 1821–1829.

[4]   M. N. Bochkarev, M. A. Katkova, "Dendritic Polymers Obtained by a Single-Stage Synthesis", *Russ. Chem. Rev.* **1995**, *64*, 1035–1048.

[5]   J. M. J. Fréchet, C. J. Hawker, "Synthesis and Properties of Dendrimers and Hyperbranched Polymers", in *Comprehensive Polymer Chemistry, 2nd Supplement* (Eds.: S. L. Aggarwal, S. Russo), Elsevier, Oxford, U.K., **1996**, pp. 71–132.

[6]   J. M. J. Fréchet, C. J. Hawker, I. Gitsov, J. W. Leon, "Dendrimers and Hyperbranched Polymers: Two Families of Three-Dimensional Macromolecules with Similar but Clearly Distinct Properties", *J. Macromol. Sci. – Pure Appl. Chem.* **1996**, *A33*, 1399–1425.

[7]   A. Hult, M. Johansson, E. Malmström, "Hyperbranched Polymers", in *Advances in Polymer Science: Branched Polymers* (Ed.: J. Roovers), Springer-Verlag, Berlin, Heidelberg, New York, **1999**, pp. 2–34.

[8]   B. Voit, "New Developments in Hyperbranched Polymers", *J. Polym. Sci., Part A: Polym. Chem.* **2000**, *38*, 2505–2525.

[9]   A. Sunder, J. Heinemann, H. Frey, "Controlling the Growth of Polymer Trees: Concepts and Perspectives For Hyperbranched Polymers", *Chem. Eur. J.* **2000**, *6*, 2499–2506.

[10]   H. Frey, D. Hölter, "Degree of Branching in Hyperbranched Polymers; 3. Copolymerization of AB$_m$ Monomers with AB and AB$_n$ Monomers", *Acta Polym.* **1999**, *50*, 67–76.

[11]   M. Gauthier, M. Möller, "Uniform Highly Branched Polymers by Anionic Grafting: Arborescent Graft Polymers", *Macromolecules* **1991**, *24*, 4548–4553.

[12]   M. Gauthier, L. Tichagwa, J. S. Downey, S. Gao, "Arborescent Graft Copolymers: Highly Branched Macromolecules with a Core-Shell Morphology", *Macromolecules* **1996**, *29*, 519–527.

[13]   R. S. Frank, G. Merkle, M. Gauthier, "Characterization of Pyrene-Labeled Arborescent Polystyrenes Using Fluorescence Quenching Techniques", *Macromolecules* **1997**, *30*, 5397–5402.

[14]   M. A. Hempenius, W. Michelberger, M. Möller, "Arborescent Graft Polybutadienes", *Macromolecules* **1997**, *30*, 5602–5605.

[15]   M. Gauthier, M. Möller, W. Burchard, "Structural Rigidity Control in Arborescent Graft Polymers", *Makromol. Chem., Macromol. Symp.* **1994**, *77*, 43–49.

[16]   S. S. Sheiko, M. Gauthier, M. Möller, "Monomolecular Films of Arborescent Graft Polystyrenes", *Macromolecules* **1997**, *30*, 2343–2349.

[17]   S. Choi, R. M. Briber, B. J. Bauer, A. Topp, M. Gauthier, L. Tichagwa, "Small-Angle Neutron Scattering of Solutions of Arborescent Graft Polystyrenes", *Macromolecules* **1999**, *32*, 7879–7886.

[18]   S. Choi, R. M. Briber, B. J. Bauer, D.-W. Liu, M. Gauthier, "Small-Angle Neutron Scattering of Blends of Arborescent Polystyrenes", *Macromolecules* **2000**, *33*, 6495–6501.

[19]   D. A. Tomalia, D. M. Hedstrand, M. S. Ferritto, "Comb-Burst Dendrimer Topology. New Macromolecular Architecture Derived from Dendritic Grafting", *Macromolecules* **1991**, *24*, 1435–1438.

[20]   Y. Rouault, O. V. Borisov, "Comb-Branched Polymers: Monte Carlo Simulation and Scaling", *Macromolecules* **1996**, *29*, 2605–2611.

[21]   R. L. Lescanec, M. Muthukumar, "Density Profiles of Simulated Comb-Burst Molecules", *Macromolecules* **1991**, *24*, 4892–4897.

[22] S. Murtuza, S. B. Harkins, G. S. Long, A. Sen, "Tantalum- and Titanium-Based Catalytic Systems for the Synthesis of Hyperbranched Polyethene", *J. Am. Chem. Soc.* **2000**, *122*, 1867–1872.

[23] J. L. Schultz, E. S. Wilks, "Symmetrical "'I'"-Shaped Hyperbranched Structural-Repeating-Unit (SRU) Polymers: Converting Unregistrable SRUs to SRUs Registrable by Chemical Abstracts Service's Registry System", *J. Chem. Inf. Comput. Sci.* **1996**, *36*, 967–972.

[24] J. L. Schultz, E. S. Wilks, "A Nomenclature and Structural Representation System for Asymmetrical "'I'"-Shaped Hyperbranched Polymers", *J. Chem. Inf. Comput. Sci.* **1996**, *36*, 1109–1117.

[25] U. Beginn, C. Drohmann, M. Möller, "Conversion Dependence of the Branching Density for the Polycondensation of AB_n Monomers", *Macromolecules* **1997**, *30*, 4112–4116.

[25a] X. He, H. Liang, C. Pan, "Monte Carlo Simulation of Hyperbranched Copolymerizations in the Presence of a Multifunctional Initiator", *Macromol. Theory Simul.* **2001**, *10*, 196–203.

[26] M. L. Mansfield, "Molecular Weight Distributions of Imperfect Dendrimers", *Macromolecules* **1993**, *26*, 3811–3814.

[27] R. Hanselmann, D. Hölter, H. Frey, "Hyperbranched Polymers Prepared via the Core-Dilution/Slow Addition Technique: Computer Simulation of Molecular Weight Distribution and Degree of Branching", *Macromolecules* **1998**, *31*, 3790–3801.

[28] D. Hölter, H. Frey, "Degree of Branching in Hyperbranched Polymers; 2. Enhancement of the DB: Scope and Limitations", *Acta Polym.* **1997**, *48*, 298–309.

[29] B. J. McCoy, "Hyperbranched Polymers and Aggregates: Distribution Kinetics of Dendrimer Growth", *J. Colloid Interface Sci.* **1999**, *216*, 235–241.

[30] D. Schmaljohann, J. G. Barratt, H. Komber, B. I. Voit, "Kinetics of Nonideal Hyperbranched Polymerizations; 1. Numeric Modeling of the Structural Units and the Diads", *Macromolecules* **2000**, *33*, 6284–6294.

[31] C. Cameron, A. H. Fawcett, C. R. Hetherington, R. A. W. Mee, F. V. McBride, "Step Growth of Two Flexible $A–B_f$ Monomers: The Self-Return of Random Branching Walks Eventually Frustrates Fractal Formation", *Macromolecules* **2000**, *33*, 6551–6568.

[31a] H. Galina, J. B. Lechowicz, K. Kaczmarski, "Kinetic Models of the Polymerization of an AB_2 Monomer", *Macromol. Theory Simul.* **2001**, *10*, 174–178.

[31b] J. G. Jang, Y. C. Bae, "Phase behavior of hyperbranched polymer solutions with specific interctions", *J. Chem. Phys.* **2001**, *114*, 5034–5042.

[32] L. Boogh, B. Pettersson, J.-A. E. Månson, "Dendritic Hyperbranched Polymers as Tougheners for Epoxy Resins", *Polymer* **1999**, *40*, 2249–2261.

[33] K. L. Wooley, J. M. J. Fréchet, C. J. Hawker, "Influence of Shape on the Reactivity and Properties of Dendritic, Hyperbranched, and Linear Aromatic Polyesters", *Polymer* **1994**, *35*, 4489–4495.

[34] Y. H. Kim, O. W. Webster, "Water-Soluble Hyperbranched Polyphenylene: 'A Unimolecular Micelle'?", *J. Am. Chem. Soc.* **1990**, *112*, 4592–4593.

[35] Y. H. Kim, O. W. Webster, "Hyperbranched Polyphenylenes", *Macromolecules* **1992**, *25*, 5561–5572.

[36] Y. H. Kim, "Highly Branched Aromatic Polymers Prepared by Single-Step Syntheses", *Polym. Prepr.* **1993**, *34*, 56–57.

[37] Y. H. Kim, "Hyperbranched Polyarylene", **1989**, *U. S. Pat.*, 4, 857, 630.

[38] Y. H. Kim, "Highly Branched Polymers", *Adv. Mater. (Weinheim, Fed. Repub. Ger.)* **1992**, *4*, 764–766.

[39] Y. H. Kim, "Highly Branched Aromatic Polymers: Their Preparation and Applications", in *Advances in Dendritic Macromolecules (Ed.: G. R. Newkome), JAI Press, Greenwich, Conn.,* **1995**, pp. 123–156.

[40] Y. H. Kim, "Highly Branched Polymers: Dendrimers and Hyperbranched Polymers", *Plastics Engineering (New York)* **1997**, *40 (Macromolecular Design of Polymeric Materials)*, 365–378.

[41] Y. H. Kim, "Hyperbranched Polymers 10 Years After", *J. Polym. Sci., Part A: Polym. Chem.* **1998**, *36*, 1685–1698.

[42] Y. H. Kim, O. W. Webster, "Hyperbranched Polymers", in *Star and Hyperbranched Polymers* (Eds.: M. K. Mishra, S. Kobayashi), Marcel Dekker, Inc., New York, **1999**, pp. 201–238.

[43] N. Miyaura, T. Yanagi, A. Suzuki, "The Palladium-Catalyzed Cross-Coupling Reaction of Phenylboronic Acid with Haloarenes in the Presence of Bases", *Synth. Commun.* **1981**, *11*, 513–519.

[44] Y. H. Kim, O. W. Webster, "Hyperbranched Polyphenylenes", *Polym. Prepr.* **1988**, *29*, 310–311.

[45] Y. H. Kim, "Highly Branched Aromatic Polymers Prepared by Single-Step Syntheses", *Makromol. Chem., Macromol. Symp.* **1994**, *77*, 21–33.

[46] Y.-H. Lai, "Grignard Reagents from Chemically Activated Magnesium", *Synthesis* **1981**, 585–604.

[47] Y. H. Kim, R. Beckerbauer, "Role of End Groups on the Glass Transition of Hyperbranched Polyphenylene and Triphenylbenzene Derivatives", *Macromolecules* **1994**, *27*, 1968–1971.

[48] K. L. Wooley, C. J. Hawker, J. M. J. Fréchet, "Hyperbranched Macromolecules via a Novel Double-Stage Convergent Growth Approach", *J. Am. Chem. Soc.* **1991**, *113*, 4252–4261.

[49] O. W. Webster, Y. H. Kim, F. P. Gentry, R. D. Farlee, B. E. Smart, "Hyperbranched and Hypercrosslinked Rigid Polymers", *Polym. Prepr.* **1992**, *33*, 186–187.

[50] H. R. Kricheldorf, O. Stöber, D. Lübbers, "New Polymer Syntheses; 78. Star-Shaped and Hyperbranched Polyesters by Polycondensation of Trimethylsilyl 3,5-Diacetoxybenzoate", *Macromolecules* **1995**, *28*, 2118–2123.

[51] H. R. Kricheldorf, G. Löhden, "New Polymer Syntheses; 80. Linear, Star-Shaped, and Hyperbranched Poly(ester-amide)s from Silicon-Mediated One-Pot Condensations of 3-Acetoxy-, 3,5-Bisacetoxybenzoic Acid, and 3-Aminobenzoic Acid", *J. Macromol. Sci. – Pure Appl. Chem.* **1995**, *A32*, 1915–1930.

[52] H. R. Kricheldorf, G. Löhden, "New Polymer Syntheses; 79. Hyperbranched Poly(ester-amide)s Based on 3-Hydroxybenzoic Acid and 3,5-Diaminobenzoic Acid", *Macromol. Chem. Phys.* **1995**, *196*, 1839–1854.

[53] H. R. Kricheldorf, O. Stöber, D. Lubbers, "New Polymer Syntheses; 81. Poly(3-oxybenzoate) Randomly Branched with 3,5-Dihydroxybenzoic Acid or 5-Hydroxyisophthalic Acid", *Macromol. Chem. Phys.* **1995**, *196*, 3549–3562.

[54] H. R. Kricheldorf, G. Schwarz, F. Ruhser, "New Polymer Syntheses; 50. Whiskers of Poly(4-hydroxybenzoate) by Condensation of Trimethylsilyl 4-Acetoxybenzoate", *Macromolecules* **1991**, *24*, 3485–3488.

[55] H. R. Kricheldorf, D. Lübbers, "New Polymer Synthesis; 56. Synthesis of Aromatic Polyesters by Condensation of Silylated Aromatic Dicarboxylic Acids and Acetylated Diphenols", *Makromol. Chem., Rapid Commun.* **1991**, *12*, 691–696.

[56] C. J. Hawker, R. Lee, J. M. J. Fréchet, "One-Step Synthesis of Hyperbranched Dendritic Polyesters", *J. Am. Chem. Soc.* **1991**, *113*, 4583–4588.

[57] K. L. Wooley, C. J. Hawker, R. Lee, J. M. J. Fréchet, "One-Step Synthesis of Hyperbranched Polyesters. Molecular Weight Control and Chain End Functionalization", *Polym. J. (Tokyo)* **1994**, *26*, 187–197.

[58] J. M. J. Fréchet, C. J. Hawker, "Hyperbranched Polyphenylene and Hyperbranched Polyesters: New Soluble, Three-Dimensional, Reactive Polymers", *React. Funct. Polym.* **1995**, *26*, 127–136.

[59] P. Kambouris, C. J. Hawker, "A Versatile New Method for Structure Determination in Hyperbranched Macromolecules", *J. Chem. Soc., Perkin Trans. 1* **1993**, 2717–2721.

[59a] M. Wang, D. Gan, K. L. Wooley, "Linear and Hyperbranched Poly(silyl ester)s: Synthesis via Cross-Dehydrocoupling-Based Polymerization, Hydrolytic Degradation Properties, and Morphological Analysis by Atomic Force Microscopy", *Macromolecules* **2001**, *34*, 3215–3223.

[60] S. R. Turner, B. I. Voit, T. H. Mourey, "All-Aromatic Hyperbranched Polyesters with Phenol and Acetate End Groups: Synthesis and Characterization", *Macromolecules* **1993**, *26*, 4617–4623.

[61] B. I. Voit, S. R. Turner, "Synthesis and Characterization of High-Temperature Hyperbranched Polyesters", *Polym. Prepr.* **1992**, *33*, 184–185.

[62] S. R. Turner, F. Walter, B. I. Voit, T. H. Mourey, "Hyperbranched Aromatic Polyesters with Carboxylic Acid Terminal Groups", *Macromolecules* **1994**, *27*, 1611–1616.

[63] A. R. Brenner, B. I. Voit, D. J. Massa, S. R. Turner, "Hyperbranched Polyesters: End Group Modification and Properties", *Macromol. Symp.* **1996**, *102*, 47–54.

[64] D. J. Massa, K. A. Shriner, S. R. Turner, B. I. Voit, "Novel Blends of Hyperbranched Polyesters and Linear Polymers", *Macromolecules* **1995**, *28*, 3214–3220.

[65] A. Kumar, S. Ramakrishnan, "Structural Variants of Hyperbranched Polyesters", *Macromolecules* **1996**, *29*, 2524–2530.

[66] F. Walter, S. R. Turner, B. I. Voit, "Hyperbranched Polyesters with Carboxylic Acid End Groups", *Polym. Prepr.* **1993**, *34*, 79–80.

[67] W. J. Feast, A. J. Keeney, A. M. Kenwright, D. Parker, "Synthesis, Structure, and Cyclics Content of Hyperbranched Polyesters", *Chem. Commun.* **1997**, 1749–1750.

[67a] D. Parker, W. J. Feast, "Synthesis, Structure, and Properties of Hyperbranched Polyesters Based on Dimethyl 5-(2-Hydroxyethoxy)isophthalate", *Macromolecules* **2001**, *34*, 2048–2059.

[68] W. J. Feast, N. M. Stainton, "Synthesis, Structure, and Properties of Some Hyperbranched Polyesters", *J. Mater. Chem.* **1995**, *5*, 405–411.

[69] H. R. Kricheldorf, T. Stukenbrock, C. Friedrich, "New Polymer Syntheses; 97. Hyperbranched LC-Polyesters Based on β-(4-Hydroxyphenyl)propionic Acid and 4-Hydroxybenzoic Acid", *J. Polym. Sci., Part A: Polym. Chem.* **1998**, *36*, 1397–1405.

[70]   H. R. Kricheldorf, T. Stukenbrock, C. Friedrich, "New Polymer Syntheses; 97. Hyperbranched LC-Polyester Based on β-(4-Hydroxyphenyl)propionic Acid and 4-Hydroxybenzoic Acid", *J. Polym. Sci., Part A: Polym. Chem.* **1998**, *36*, 1397–1405.

[71]   H. R. Kricheldorf, T. Stukenbrock, "New Polymer Syntheses; 93. Hyperbranched Homo- and Copolymers Derived from Gallic Acid and β-(4-Hydroxyphenyl)propanoic Acid", *J. Polym. Sci., Part A: Polym. Chem.* **1998**, *36*, 2347–2357.

[72]   S.-W. Hahn, Y.-K. Yun, J.-I. Jin, O. H. Han, "Thermotropic Hyperbranched Polyesters Prepared from 2-{[10-(4-Hydroxyphenoxy)decyl]oxy}terephthalic Acid and 2-({10-[(4'-Hydroxy-1,1'-biphenyl-4-yl)oxy]decyl}oxy)terephthalic Acid", *Macromolecules* **1998**, *31*, 6417–6425.

[73]   D. Schmaljohann, P. Pötschke, R. Hässler, B. I. Voit, P. E. Froehling, B. Mostert, J. A. Loontjens, "Blends of Amphiphilic, Hyperbranched Polyesters and Different Polyolefins", *Macromolecules* **1999**, *32*, 6333–6339.

[74]   H. R. Kricheldorf, O. Bolender, T. Wollheim, "New Polymer Syntheses; 103. In Situ End Group Modification of Hyperbranched Poly(3,5-dihydroxybenzoate)", *Macromolecules* **1999**, *32*, 3878–3882.

[75]   K. E. Uhrich, C. J. Hawker, J. M. J. Fréchet, S. R. Turner, "One-Pot Synthesis of Hyperbranched Polyethers", *Macromolecules* **1992**, *25*, 4583–4587.

[76]   K. E. Uhrich, C. J. Hawker, J. M. J. Fréchet, S. R. Turner, "'One-Step' Synthesis of Hyperbranched Macromolecules", *Polym. Mater. Sci. Eng.* **1991**, *64*, 237–238.

[77]   C. J. Hawker, J. M. J. Fréchet, "Preparation of Polymers with Controlled Molecular Architecture. A New Convergent Approach to Dendritic Macromolecules", *J. Am. Chem. Soc.* **1990**, *112*, 7638–7647.

[78]   A. Mueller, T. Kowalewski, K. L. Wooley, "Synthesis, Characterization, and Derivatization of Hyperbranched Polyfluorinated Polymers", *Macromolecules* **1998**, *31*, 776–786.

[79]   J. K. Gooden, M. L. Gross, A. Mueller, A. D. Stefanescu, K. L. Wooley, "Cyclization in Hyperbranched Polymer Syntheses: Characterization by MALDI-TOF Mass Spectrometry", *J. Am. Chem. Soc.* **1998**, *120*, 10180–10186.

[80]   T. M. Miller, T. X. Neenan, E. W. Kwock, S. M. Stein, "Dendritic Analogues of Engineering Plastics: A General One-Step Synthesis of Dendritic Polyaryl Ethers", *J. Am. Chem. Soc.* **1993**, *115*, 356–357.

[81]   T. M. Miller, T. X. Neenan, E. W. Kwock, S. M. Stein, "Dendritic Analogues of Engineering Plastics: A General One-Step Synthesis of Dendritic Polyaryl Ethers", *Polym. Prepr.* **1993**, *34*, 58–59.

[82]   T. M. Miller, T. X. Neenan, E. W. Kwock, S. M. Stein, "Dendritic Analogs of Engineering Plastics: A General One-Step Synthesis of Dendritic Polyaryl Ethers", *Macromolecular Symposia* **1994**, *77*, 35–42.

[83]   D. V. Evtugin, A. Gandini, "Polyesters Based on Oxygen-Organosolv Lignin", *Acta Polym.* **1996**, *47*, 344–350.

[84]   H.-T. Chang, J. M. J. Fréchet, "Proton-Transfer Polymerization: A New Approach to Hyperbranched Polymers", *J. Am. Chem. Soc.* **1999**, *121*, 2313–2314.

[85]   N. Spetseris, R. E. Ward, T. Y. Meyer, "Linear and Hyperbranched *m*-Polyaniline: Synthesis of Polymers for the Study of Magnetism in Organic Systems", *Macromolecules* **1998**, *31*, 3158–3161.

[86]   F. Chu, C. J. Hawker, "A Versatile Synthesis of Isomeric Hyperbranched Polyetherketones", *Polym. Bull.* **1993**, *30*, 265–272.

[87]   C. J. Hawker, F. Chu, P. Kambouris, "Architectural Control in Hyperbranched Macromolecules", *Polym. Prepr.* **1995**, *36*, 747–748.

[87a]   S.-Y. Kwak, D. U. Ahn, "Processability of Hyperbranched Poly(ether ketone)s with Different Degrees of Branching from Viewpoints of Molecular Mobility and Comparison with Their Linear Analogue", *Macromolecules* **2000**, *33*, 7557–7563.

[88]   C. J. Hawker, F. Chu, "Hyperbranched Poly(ether ketones): Manipulation of Structure and Physical Properties", *Macromolecules* **1996**, *29*, 4370–4380.

[89]   A. Morikawa, "Preparation and Properties of Hyperbranched Poly(ether ketones) with a Various Number of Phenylene Units", *Macromolecules* **1998**, *31*, 5999–6009.

[90]   S.-Y. Kwak, H. Y. Lee, "Molecular Relaxation and Local Motion of Hyperbranched Poly(ether ketone)s with Reference to Their Linear Counterparts; 1. Effect of Degrees of Branching", *Macromolecules* **2000**, *33*, 5536–5543.

[91]   C.-F. Shu, C.-M. Leu, F.-Y. Huang, "Synthesis, Modification, and Characterization of Hyperbranched Poly(ether ketones)", *Polymer* **1999**, *40*, 6591–6596.

[92]   C.-F. Shu, C.-M. Leu, "Hyperbranched Poly(ether ketone) with Carboxylic Acid Terminal Groups: Synthesis, Characterization, and Derivatization", *Macromolecules* **1999**, *32*, 100–105.

[92a]   Z.-H. Gong, C.-M. Leu, F.-I. Wu, C.-F. Shu, "Hyperbranched Poly(aryl ether oxazole)s: Synthesis, Characterization, and Modification", *Macromolecules* **2000**, *33*, 8527–8533.

[93] H. R. Kricheldorf, T. Stukenbrock, "New Polymer Syntheses, 90. A–B–A Triblock Copolymers with Hyperbranched Polyester A-Blocks", *J. Polym. Sci., Part A: Polym. Chem.* **1998**, *36*, 31–38.

[94] Y. H. Kim, "Lyotropic Liquid Crystalline Hyperbranched Aromatic Polyamides", *J. Am. Chem. Soc.* **1992**, *114*, 4947–4948.

[95] Y. H. Kim, "Rigid Chain Hyperbranched Polymer", *Polym. Mater. Sci. Eng.* **1995**, *73*, 175.

[96] V. R. Reichert, L. J. Mathias, "Tetrahedrally-Oriented Four-Armed Star Branched Aramids", *Macromolecules* **1994**, *27*, 7024–7029.

[97] K. L. Wooley, C. J. Hawker, J. M. Pochan, J. M. J. Fréchet, "Physical Properties of Dendritic Macromolecules: A Study of Glass Transition Temperature", *Macromolecules* **1993**, *26*, 1514–1519.

[98] L. J. Mathias, V. R. Reichert, A. V. G. Muir, "Synthesis of Rigid Tetrahedral Tetrafunctional Molecules from 1,3,5,7-Tetrakis(4-iodophenyl)adamantane", *Chem. Mater.* **1993**, *5*, 4–5.

[99] V. R. Reichert, J. P. Mathias, "Expanded Tetrahedral Molecules from 1,3,5,7-Tetraphenyladamantane", *Macromolecules* **1994**, *27*, 7015–7023.

[99a] V. R. Reichert, L. J. Mathius "Highly Cross-linked Polymers Based on Acetylene Derivatives of Tetraphenyladamantane", *Macromolecules* **1994**, *27*, 7030–7034.

[100] K. E. Uhrich, S. Boegeman, J. M. J. Fréchet, S. R. Turner, "The Solid-Phase Synthesis of Dendritic Polyamides", *Polym. Bull.* **1991**, *25*, 551–558.

[101] O. Haba, H. Tajima, M. Ueda, R. Nagahata, "Synthesis of Hyperbranched Aromatic Polyamide by Direct Polycondensation Method", *Chem. Lett.* **1998**, 333–334.

[102] Y. Yamakawa, M. Ueda, K. Takeuchi, M. Asai, "One-Pot Synthesis of Dendritic Polyamide", *J. Polym. Sci., Part A: Polym. Chem.* **1999**, *37*, 3638–3645.

[103] Y. Yamakawa, M. Ueda, K. Takeuchi, M. Asai, "One-Pot Synthesis of Dendritic Polyamide; 2. Dendritic Polyamide from 5-[3-(4-Aminophenyl)propionylamino]isophthalic Acid Hydrochloride", *Macromolecules* **1999**, *32*, 8363–8369.

[104] Y. Yamakawa, M. Ueda, K. Takeuchi, M. Asai, "One-Pot Synthesis of Polyamide Dendrimers", *Polym. Prepr.* **1999**, *40*, 91–92.

[105] G. Yang, M. Jikei, M. Kakimoto, "Successful Thermal Self-Polycondensation of AB$_2$ Monomer to Form Hyperbranched Aromatic Polyamide", *Macromolecules* **1998**, *31*, 5964–5966.

[106] Y. Ishida, A. C. F. Sun, M. Jikei, M. Kakimoto, "Synthesis of Hyperbranched Aromatic Polyamides Starting from Dendrons as AB$_x$ Monomers: Effect of Monomer Multiplicity on the Degree of Branching", *Macromolecules* **2000**, *33*, 2832–2838.

[107] G. Yang, M. Jikei, M. Kakimoto, "Synthesis and Properties of Hyperbranched Aromatic Polyamide", *Macromolecules* **1999**, *32*, 2215–2220.

[108] M. Jikei, S.-H. Chon, M. Kakimoto, S. Kawauchi, T. Imase, J. Watanebe, "Synthesis of Hyperbranched Aromatic Polyamide from Aromatic Diamines and Trimesic Acid", *Macromolecules* **1999**, *32*, 2061–2064.

[109] R. Spindler, J. M. J. Fréchet, "Synthesis and Characterization of Hyperbranched Polyurethanes Prepared from Blocked Isocyanate Monomers by Step-Growth Polymerization", *Macromolecules* **1993**, *26*, 4809–4813.

[110] Z. W. Wicks, Jr., "Blocked Isocyanates", *Prog. Org. Coat.* **1975**, *3*, 73–99.

[111] J. H. Saunders, K. C. Frisch, "Polyurethanes: Chemistry and Technology", Wiley, New York, **1962**, pp. 118–121.

[112] A. Kumar, S. Ramakrishnan, "A Novel One-Pot Synthesis of Hyperbranched Polyurethanes", *J. Chem. Soc., Chem. Commun.* **1993**, 1453–1454.

[113] A. Kumar, E. W. Meijer, "Novel Hyperbranched Polymer Based on Urea Linkages", *Chem. Commun.* **1998**, 1629–1630.

[113a] A. V. Abade, A. Kumar, "An Efficient Route for the Synthesis of Hyperbranched Polymers and Dendritic Building Blocks Based on Urea Linkages", *J. Polym. Sci., Part A: Polym. Chem.* **2001**, *39*, 1295–1304.

[114] S. Bauer, H. Fischer, H. Ringsdorf, "Highly Branched Liquid Crystalline Polymers with Chiral Terminal Groups", *Angew. Chem.* **1993**, *105*, 1658–1660; *Angew. Chem. Int. Ed. Engl.* **1993**, *32*, 1589–1592.

[115] T. Itoh, M. Ikeda, N. Hirata, Y. Moriya, M. Kubo, O. Yamamoto, "Ionic Conductivity of the Hyperbranched Polymer–Lithium Metal Salt Systems", *J. Power Sources* **1999**, *81–82*, 824–829.

[116] H. R. Kricheldorf, O. Bolender, T. Stukenbrock, "New Polymer Syntheses; 91. Hyperbranched Poly(ester-amide)s Derived from 3,5-Dihydroxybenzoic Acid and 3,5-Diaminobenzoic Acid", *Macromol. Chem. Phys.* **1997**, *198*, 2651–2666.

[117] H. R. Kricheldorf, O. Bolender, "New Polymer Syntheses; 98. Hyperbranched Poly(ester-amide)s Derived from Naturally Occurring Monomers", *J. Macromol. Sci. – Pure Appl. Chem.* **1998**, *A35*, 903–918.

[118] M. Jikei, K. Fujii, G. Yang, M. Kakimoto, "Synthesis and Properties of Hyperbranched Aromatic Polyamide Copolymers from AB and AB$_2$ Monomers by Direct Polycondensation", *Macromolecules* **2000**, *33*, 6228–6234.

[119] M. Jikei, Z. Hu, M. Kakimoto, Y. Imai, "Synthesis of Poly(phenylene sulfide) Dendron via Poly(sulfonium cation) as a Precursor", *Polym. Mater. Sci. Eng.* **1995**, *73*, 135–136.

[120] M. Jikei, Z. Hu, M. Kakimoto, Y. Imai, "Synthesis of Hyperbranched Poly(phenylene sulfide) via a Poly(sulfonium cation) Precursor", *Macromolecules* **1996**, *29*, 1062–1064.

[121] J. M. J. Fréchet, M. Henmi, I. Gitsov, S. Aoshima, M. R. Leduc, R. B. Grubbs, "Self-Condensing Vinyl Polymerization: An Approach to Dendritic Materials", *Science* **1995**, *269*, 1080–1083.

[122] S. G. Gaynor, S. Edelman, K. Matyjaszewski, "Synthesis of Branched and Hyperbranched Polystyrenes", *Macromolecules* **1996**, *29*, 1079–1081.

[123] D. M. Knauss, H. A. Al-Muallem, T. Huang, D. T. Wu, "Polystyrene with Dendritic Branching by Convergent Living Anionic Polymerization", *Macromolecules* **2000**, *33*, 3557–3568.

[124] M. W. Weimer, J. M. J. Fréchet, I. Gitsov, "Importance of Active-Site Reactivity and Reaction Conditions in the Preparation of Hyperbranched Polymers by Self-Condensing Vinyl Polymerization: Highly Branched *vs.* Linear Poly[4-(chloromethyl)styrene] by Metal-Catalyzed 'Living' Radical Polymerization", *J. Polym. Sci., Part A: Polym. Chem.* **1998**, *36*, 955–970.

[125] P. Bharathi, J. S. Moore, "Solid-Supported Hyperbranched Polymerization: Evidence for Self-Limited Growth", *J. Am. Chem. Soc.* **1997**, *119*, 3391–3392.

[126] J. K. Young, J. C. Nelson, J. S. Moore, "Synthesis of Sequence-Specific Phenylacetylene Oligomers on an Insoluble Solid Support", *J. Am. Chem. Soc.* **1994**, *116*, 10841–10842.

[127] P. Bharathi, J. S. Moore, "Controlled Synthesis of Hyperbranched Polymers by Slow Monomer Addition to a Core", *Macromolecules* **2000**, *33*, 3212–3218.

[128] T. Lin, Q. He, F. Bai, L. Dai, "Design, Synthesis and Photophysical Properties of a Hyperbranched Conjugated Polymer", *Thin Solid Films* **2000**, *363*, 122–125.

[129] K. Yamanaka, M. Jikei, M. Kakimoto, "Synthesis of Hyperbranched Aromatic Polyimides via Polyamic Acid Methyl Ester Precursor", *Macromolecules* **2000**, *33*, 1111–1114.

[129a] K. Yamanaka, M. Jikei, M. Kakimoto, "Preparation and Properties of Hyperbranched Aromatic Polyimides via Polyamic Acid Methyl Ester Precursors", *Macromolecules* **2000**, *33*, 6937–6944.

[130] D. S. Thompson, L. J. Markoski, J. S. Moore, "Rapid Synthesis of Hyperbranched Aromatic Polyetherimides", *Macromolecules* **1999**, *32*, 4764–4768.

[131] D. S. Thompson, L. J. Markoski, J. S. Moore, I. Sendijarevic, A. Lee, A. J. McHugh, "Synthesis and Characterization of Hyperbranched Aromatic Poly(ether imide)s with Varying Degrees of Branching", *Macromolecules* **2000**, *33*, 6412–6415.

[131a] L. J. Markoski, J. S. Moore, I. Sendijarevic, A. J. McHugh, "Effect of Linear Sequence Length on the Properties of Branched Aromatic Etherimide Copolymers", *Macromolecules* **2001**, *34*, 2695–2701.

[132] L. J. Markoski, J. L. Thompson, J. S. Moore, "Synthesis and Characterization of Linear-Dendritic Aromatic Etherimide Copolymers: Tuning Molecular Architecture to Optimize Properties and Processability", *Macromolecules* **2000**, *33*, 5315–5317.

[133] S. Srinivasan, R. Twieg, J. L. Hedrick, C. J. Hawker, "Heterocycle-Activated Aromatic Nucleophilic Substitution of AB$_2$ Poly(aryl ether phenylquinoxaline) Monomers, 3", *Macromolecules* **1996**, *29*, 8543–8545.

[134] J. L. Hedrick, C. J. Hawker, R. D. Miller, R. Twieg, S. A. Srinivasan, M. Trollsås, "Structure Control in Organic–Inorganic Hybrids Using Hyperbranched High-Temperature Polymers", *Macromolecules* **1997**, *30*, 7607–7610.

[135] For a versatile solvent for the synthesis of poly(aryl ether)s, see: J. W. Labadie, J. L. Hedrick, K. Carter, R. Twieg, *Polym. Bull.* **1993**, *30*, 25.

[136] C. Kim, Y. Chang, J. S. Kim, "Dendritic Hyperbranched Polyethynylenes with the 1,3,5-*s*-Triazine Moiety", *Macromolecules* **1996**, *29*, 6353–6355.

[136a] S. Y. Cho, Y. Chang, J. S. Kim, S. C. Lee, C. Kim, "Hyperbranched poly(ether ketone) analogues with heterocyclic triazine moiety: Synthesis and peripheral functionalization", *Macromol. Chem. Phys.* **2001**, *202*, 263–269.

[136b] M. Eigner, H. Komber, B. Voit, "Hyperbranched poly(triazene ester)s as novel globular photolabile and thermolabile polymers", *Macromol. Chem. Phys.* **2001**, *202*, 245–256.

[137] F. Chu, C. J. Hawker, P. J. Pomery, D. J. T. Hill, "Intramolecular Cyclization in Hyperbranched Polyesters", *J. Polym. Sci., Part A: Polym. Chem.* **1997**, *35*, 1627–1633.

[138] A. Hult, M. Johansson, E. Malmström, "Dendritic Resins for Coating Applications", *Macromol. Symp.* **1995**, *98*, 1159–1161.

[139] M. Johansson, E. Malmström, A. Hult, "Synthesis, Characterization, and Curing of Hyperbranched Allyl Ether–Maleate Functional Ester Resins", *J. Polym. Sci., Part A: Polym. Chem.* **1993**, *31*, 619–624.

[140] E. Malmström, M. Johansson, A. Hult, "Hyperbranched Aliphatic Polyesters", *Macromolecules* **1995**, *28*, 1698–1703.

[141] A. Hult, E. Malmström, M. Johansson, K. Sorensen, "Dendritic Macromolecule and Process for Preparation Thereof", *U. S. Pat.* **1995**, *5,418,301*.

[142] M. Johansson, A. Hult, "Synthesis, Characterization and Curing of Resins Based on Hyperbranched Structures", *Polym. Mater. Sci. Eng.* **1995**, *73*, 178–179.

[143] H. Ihre, E. Söderlind, A. Hult, "Synthesis and Characterization of Dendritic Aliphatic Polyesters based on 2,2-Bis(hydroxymethyl)propionic Acid", *Polym. Mater. Sci. Eng.* **1995**, *73*, 351–352.

[144] M. Johansson, E. Malmström, A. Hult, "The Synthesis and Properties of Hyperbranched Polyesters", *Trends in Polymer Science* **1996**, *4*, 398–403.

[145] A. Hult, E. Malmström, M. Johansson, "Synthesis and Properties of Aliphatic Hyperbranched Polyesters", *Polym. Mater. Sci. Eng.* **1995**, *73*, 173–174.

[146] M. Trollsås, F. Sahlén, U. W. Gedde, A. Hult, D. Hermann, P. Rudquist, L. Komitov, S. T. Lagerwall, B. Stebler, J. Lindström, O. Rydlund, "Novel Thermally Stable Polymer Materials for Second-Order Nonlinear Optics", *Macromolecules* **1996**, *29*, 2590–2598.

[147] E. Malmström, F. Liu, R. H. Boyd, A. Hult, U. W. Gedde, "Relaxation Processes in Hyperbranched Polyesters", *Polym. Bull.* **1994**, *32*, 679–685.

[148] M. S. Hedenqvist, H. Yousefi, E. Malmström, M. Johansson, A. Hult, U. W. Gedde, M. Trollsås, J. L. Hedrick, "Transport Properties of Hyperbranched and Dendrimer-Like Star Polymers", *Polymer* **2000**, *41*, 1827–1840.

[149] J. Jang, J. H. Oh, "In situ FT-IR Spectroscopic Investigation on the Microstructure of Hyperbranched Aliphatic Polyesters", *Polymer* **1999**, *40*, 5985–5992.

[150] E. Malmström, A. Hult, U. W. Gedde, F. Liu, R. H. Boyd, "Relaxation Processes in Hyperbranched Polyesters: Influence of Terminal Groups", *Polymer* **1997**, *38*, 4873–4879.

[151] G. Jannerfeldt, L. Boogh, J.-A. E. Månson, "Influence of Hyperbranched Polymers on the Interfacial Tension of Polypropylene/Polyamide-6 Blends", *J. Polym. Sci., Part B: Polym. Phys.* **1999**, *37*, 2069–2077.

[152] E. Malmström, A. Hult, "Kinetics of Formation of Hyperbranched Polyesters based on 2,2-Bis(methylol)propionic Acid", *Macromolecules* **1996**, *29*, 1222–1228.

[153] G. Magnusson, E. Malmström, A. Hult, "Structure Build-up in Hyperbranched Polymers from 2,2-Bis(hydroxymethyl)propionic Acid", *Macromolecules* **2000**, *33*, 3099–3104.

[154] H. Magnusson, E. Malmström, A. Hult, "Synthesis of Hyperbranched Aliphatic Polyethers via Cationic Ring-Opening Polymerization of 3-Ethyl-3-(hydroxymethyl)oxetane", *Macromol. Rapid Commun.* **1999**, *20*, 453–457.

[155] A. Burgath, A. Sunder, H. Frey, "Role of Cyclization in the Synthesis of Hyperbranched Aliphatic Polyesters", *Macromol. Chem. Phys.* **2000**, *201*, 782–791.

[156] M. Trollsås, B. Atthoff, A. Würsch, J. L. Hedrick, J. A. Pople, A. P. Gast, "Constitutional Isomers of Dendrimer-like Star Polymers: Design, Synthesis, and Conformational and Structural Properties", *Macromolecules* **2000**, *33*, 6423–6438.

[157] E. Malmström, M. Johansson, A. Hult, "The Effect of Terminal Alkyl Chains on Hyperbranched Polyesters based on 2,2-Bis(hydroxymethyl)propionic Acid", *Macromol. Chem. Phys.* **1996**, *197*, 3199–3207.

[158] Y. Hong, S. J. Coombs, J. J. Cooper-White, M. E. Mackay, C. J. Hawker, E. Malmström, N. Rehnberg, "Film Blowing of Linear Low-Density Polyethylene Blended with a Novel Hyperbranched Polymer Processing Aid", *Polymer* **2000**, *41*, 7705–7713.

[158a] M. E. Mackay, G. Carmezini, B. B. Sauer, W. Kampert, "On the Surface Properties of Hyperbranched Polymers", *Langmuir* **2001**, *17*, 1708–1712.

[159] R. B. Grubbs, C. J. Hawker, J. Dao, J. M. J. Fréchet, "A Tandem Approach to Graft and Dendritic Graft Copolymers Based on 'Living' Free Radical Polymerizations", *Angew. Chem.* **1997**, *109*, 261–164; *Angew. Chem. Int. Ed. Engl.* **1997**, *36*, 270–272.

[160] C. J. Hawker, "Architectural Control in 'Living' Free Radical Polymerizations: Preparation of Star and Graft Polymers", *Angew. Chem.* **1995**, *107*, 1623–1627; *Angew. Chem. Int. Ed. Engl.* **1995**, *34*, 1456–1459.

[161] K. Sakamoto, T. Aimiya, M. Kira, "Preparation of Hyperbranched Polymethacrylates by Self-Condensing Group Transfer Polymerization", *Chem. Lett.* **1997**, 1245–1246.

[162] W. J. Feast, L. M. Hamilton, L. J. Hobson, S. P. Rannard, "The Synthesis and Characterization of Hyperbranched Poly(diethyl 3-hydroxyglutarate)", *J. Mater. Chem.* **1997**, *8*, 1121–1125.

[163] W. J. Feast, L. M. Hamilton, S. Rannard, "The Influence of Laser Power on the Observed MALDI-TOF Mass Spectra of Poly(diethyl 3-hydroxyglutarate), an AB_2 Hyperbranched Aliphatic Polyester; M_n from MALDI-TOF MS? – Caveat Emptor", *Polym. Bull.* **1997**, *39*, 347–352.

[164] M. Trollsås, J. L. Hedrick, D. Mecerreyes, Ph. Dubois, R. Jérôme, H. Ihre, A. Hult, "Versatile and Controlled Synthesis of Star and Branched Macromolecules by Dendritic Initiation", *Macromolecules* **1997**, *30*, 8508–8511.

[165] M. Trollsås, B. Atthoff, H. Claesson, J. L. Hedrick, "Hyperbranched Poly(ε-caprolactone)s", *Macromolecules* **1998**, *31*, 3439–3445.

[166] M. Trollsås, J. L. Hedrick, "Hyperbranched Poly(ε-caprolactone) Derived from Intrinsically Branched AB$_2$ Macromonomers", *Macromolecules* **1998**, *31*, 4390–4395.

[167] J. L. Hedrick, M. Trollsås, C. Nguyen, J. Remenar, C. J. Hawker, K. R. Carter, W. Volksen, D. Yoon, R. D. Miller, "Templating Nanoporosity in Organosilicates with Well-Defined Branched Macromolecules and Block Copolymers", *Polym. Prepr.* **1999**, *40*, 978–979.

[168] J. L. Hedrick, M. Trollsås, D. Mecerreyes, "ABC/BCD Polymerization: A Self-Condensing Vinyl and Cyclic Ester Polymerization by Combination Free Radical and Ring Opening Techniques for the Preparation of Novel Graft Copolymers", *Polym. Prepr.* **1999**, *40*, 97–98.

[169] A. Heise, M. Trollsås, T. Magbitang, J. L. Hedrick, C. W. Frank, R. D. Miller, "Star Polymers with Perfectly Alternating Arms from Dendritic Initiators", *Polym. Prepr.* **1999**, *40*, 376–377.

[170] J. L. Hedrick, M. Trollsås, C. J. Hawker, B. Atthoff, H. Claesson, A. Heise, R. D. Miller, D. Mecerreyes, R. Jérôme, Ph. Dubois, "Dendrimer-like Star Block and Amphiphilic Copolymers by Combination of Ring Opening and Atom Transfer Radical Polymerization", *Macromolecules* **1998**, *31*, 8691–8705.

[171] M. Trollsås, J. L. Hedrick, "Dendrimer-like Star Polymers", *J. Am. Chem. Soc.* **1998**, *120*, 4644–4651.

[172] Q. Zhang, E. E. Remsen, K. L. Wooley, "Shell Cross-Linked Nanoparticles Containing Hydrolytically Degradable, Crystalline Core Domains", *J. Am. Chem. Soc.* **2000**, *122*, 3642–3651.

[173] A. Heise, J. L. Hedrick, C. W. Frank, R. D. Miller, "Star-like Block Copolymers with Amphiphilic Arms as Models for Unimolecular Micelles", *J. Am. Chem. Soc.* **1999**, *121*, 8647–8648.

[174] A. Heise, J. L. Hedrick, M. Trollsås, R. D. Miller, C. W. Frank, "Novel Star-like Poly(methyl methacrylate)s by Controlled Dendritic Free Radical Initiation", *Macromolecules* **1999**, *32*, 231–234.

[175] R. D. Miller, A. Heise, S. Diamanti, "Initiation Behavior a of Dendritic Multi-Arm Initiator in Atom Transfer Radical Polymerization (ATRP)", *Polym. Prepr.* **2000**, *41*, 48–49.

[176] M. Trollsås, J. L. Hedrick, D. Mecerreyes, R. Jérôme, Ph. Dubois, "Internal Functionalization in Hyperbranched Polyesters", *J. Polym. Sci., Part A: Polym. Chem.* **1998**, *36*, 3187–3192.

[177] M. Trollsås, H. Claesson, B. Atthoff, J. L. Hedrick, "Layered Dendritic Block Copolymers", *Angew. Chem.* **1998**, *110*, 3308–3312; *Angew. Chem. Int. Ed.* **1998**, *37*, 3132–3136.

[178] C. V. Nguyen, K. R. Carter, C. J. Hawker, J. L. Hedrick, R. L. Jaffe, R. D. Miller, J. F. Remenar, H.-W. Rhee, P. M. Rice, M. F. Toney, M. Trollsås, D. Y. Yoon, "Low-Dielectric, Nanoporous Organosilicate Films Prepared via Inorganic/Organic Polymer Hybrid Templates", *Chem. Mater.* **1999**, *11*, 3080–3085.

[179] C. Nguyen, C. J. Hawker, R. D. Miller, E. Huang, J. L. Hedrick, R. Gauderon, J. G. Hilborn, "Hyperbranched Polyesters as Nanoporosity Templating Agents for Organosilicates", *Macromolecules* **2000**, *33*, 4281–4284.

[180] H. Ihre, A. Hult, J. M. J. Fréchet, I. Gitsov, "Double-Stage Convergent Approach for the Synthesis of Functionalized Dendritic Aliphatic Polyesters Based on 2,2-Bis(hydroxymethyl)-propionic Acid", *Macromolecules* **1998**, *31*, 4061–4068.

[181] P. Busson, H. Ihre, A. Hult, "Synthesis of a Novel Dendritic Liquid Crystalline Polymer Showing a Ferroelectric SmC* Phase", *J. Am. Chem. Soc.* **1998**, *120*, 9070–9071.

[182] M. Trollsås, J. L. Hedrick, D. Mecerreyes, Ph. Dubois, R. Jérôme, H. Ihre, A. Hult, "Highly Functional Branched and Dendri-Graft Aliphatic Polyesters through Ring Opening Polymerization", *Macromolecules* **1998**, *31*, 2756–2763.

[183] M. Trollsås, C. J. Hawker, J. F. Remenar, J. L. Hedrick, M. Johansson, H. Ihre, A. Hult, "Highly Branched Radial Block Copolymers via Dendritic Initiation of Alphatic Polyesters", *J. Polym. Sci., Part A: Polym. Chem.* **1998**, *36*, 2793–2798.

[184] M. Trollsås, M. A. Kelly, H. Claesson, R. Siemens, J. L. Hedrick, "Highly Branched Block Copolymers: Design, Synthesis, and Morphology", *Macromolecules* **1999**, *32*, 4917–4924.

[185] M. Trollsås, P. Löwenhielm, V. Y. Lee, M. Möller, R. D. Miller, J. L. Hedrick, "New Approach to Hyperbranched Polyesters: Self-Condensing Cyclic Ester Polymerization of Bis(hydroxymethyl)-Substituted ε-Caprolactone", *Macromolecules* **1999**, *32*, 9062–9066.

[186] J. G. Jang, Y. C. Bae, "Phase Behaviors of Hyperbranched Polymer Solutions", *Polymer* **1999**, *40*, 6761–6768.

[187] A. B. Kantchev, J. R. Parquette, "Disaccharide Synthesis on a Soluble Hyperbranched Polymer", *Tetrahedron Lett.* **1999**, *40*, 8049–8053.

[188] M. Liu, N. Vladimirov, J. M. J. Fréchet, "A New Approach to Hyperbranched Polymers by Ring-Opening Polymerization of an AB Monomer: 4-(2-Hydroxyethyl)-ε-caprolactone", *Macromolecules* **1999**, *32*, 6881–6884.

[189] T. J. Mulkern, N. C. B. Tan, "Processing and Characterization of Reactive Polystyrene/ Hyperbranched Polyester Blends", *Polymer* **2000**, *41*, 3193–3203.

[190] C. M. Nunez, B.-S. Chiou, A. L. Andrady, S. A. Khan, "Solution Rheology of Hyperbranched Polyesters and Their Blends with Linear Polymers", *Macromolecules* **2000**, *33*, 1720–1726.

[191] J. Jang, J. H. Oh, S. I. Moon, "Crystalline Behavior of Poly(ethylene terephthalate) Blended with Hyperbranched Polymers: The Effect of Terminal Groups and Composition of Hyperbranched Polymers", *Macromolecules* **2000**, *33*, 1864–1870.

[192] H. Liu, J. H. Näsman, M. Skrifvars, "Radical Alternating Copolymerization: A Strategy for Hyperbranched Materials", *J. Polym. Sci., Part A: Polym. Chem.* **2000**, *38*, 3074–3085.

[192a] H. Liu, C.-E. Wilén, "Hyperbranched Polymers with Maleic Functional Groups As Radical Crosslinkers", *J. Polym. Sci. , Part A: Polym. Chem.* **2001**, *39*, 964–972.

[193] V. Percec, M. Kawasumi, "Synthesis and Characterization of a Thermotropic Nematic Liquid Crystalline Dendrimeric Polymer", *Macromolecules* **1992**, *25*, 3843–3850.

[194] V. Percec, M. Kawasumi, "Synthesis and Characterization of a Thermotropic Liquid Crystalline Dendrimer", *Polym. Prepr.* **1992**, *33*, 221–222.

[195] V. Percec, J. Heck, G. Johansson, D. Tomazos, M. Kawasumi, P. Chu, G. Ungar, "Molecular Recognition Directed Self-Assembly of Supramolecular Architectures", *J. Macromol. Sci. – Pure Appl. Chem.* **1994**, *A31*, 1719–1758.

[196] V. Percec, P. Chu, M. Kawasumi, "Toward 'Willow like' Thermotropic Dendrimers", *Macromolecules* **1994**, *27*, 4441–4453.

[197] V. Percec, J. Heck, G. Johansson, D. Tomazos, M. Kawasumi, G. Ungar, "Molecular-Recognition-Directed Self-Assembly of Supramolecular Polymers", *J. Macromol. Sci. – Pure Appl. Chem.* **1994**, *A31*, 1031–1070.

[198] V. Percec, "Molecular Design of Novel Liquid Crystalline Polymers with Complex Architecture: Macrocyclics and Dendrimers", *Pure Appl. Chem.* **1995**, *67*, 2031–2038.

[199] V. Percec, M. Kawasumi, "Synthesis and Characterization of Thermotropic Liquid Crystalline Dendrimers", *Polym. Prepr.* **1993**, *34*, 158–159.

[200] V. Percec, M. Kawasumi, "Synthesis and Characterization of Macrocyclic Liquid Crystalline Oligopolyethers", *Polym. Prepr.* **1993**, *34*, 154–155.

[201] V. Percec, M. Kawasumi, "Synthesis and Characterization of Chiral Linear And Macrocyclic Liquid Crystalline Polyethers", *Polym. Prepr.* **1993**, *34*, 156–157.

[202] V. Percec, P. Chu, "Willow-like Hyperbranched and Dendrimeric Liquid Crystals", *Polym. Prepr.* **1995**, *36*, 743–744.

[203] V. Percec, P. Chu, G. Ungar, J. Zhou, "Rational Design of the First Nonspherical Dendrimer which Displays Calamitic Nematic and Smectic Thermotropic Liquid Crystalline Phases", *J. Am. Chem. Soc.* **1995**, *117*, 11441–11454.

[204] F.-L. Chen, A. M. Jamieson, M. Kawasumi, V. Percec, "Viscoelastic Properties of Dilute Nematic Mixtures Containing Cyclic and Hyperbranched Liquid Crystal Polymers Dissolved in a Nematic Solvent", *J. Polym. Sci., Part B: Polym. Phys.* **1995**, *33*, 1213–1223.

[205] V. Percec, A. D. Asandei, "Monodisperse Linear Liquid Crystalline Polyethers *via* a Repetitive 2^n Geometric Growth Algorithm", *Macromolecules* **1997**, *30*, 7701–7720.

[206] R.-H. Jin, S. Motokucho, Y. Andou, T. Nishikubo, "Controlled Polymerization of an AB_2 Monomer using a Chloromethylarene as Comonomer: Branched Polymers from Activated Methylene Compounds", *Macromol. Rapid Commun.* **1998**, *19*, 41–46.

[207] H. Zhang, E. Ruckenstein, "Dendritic Polymers from Vinyl Ether", *Polym. Bull.* **1997**, *39*, 399–406.

[208] A. Sunder, M. Krämer, R. Hanselmann, R. Mülhaupt, H. Frey, "Molecular Nanocapsules Based on Amphiphilic Hyperbranched Polyglycerols", *Angew. Chem.* **1999**, *111*, 3758–3761; *Angew. Chem. Int. Ed.* **1999**, *38*, 3552–3555.

[208a] A. Sunder, H. Türk, R. Haag, H. Frey, "Copolymers of Glycidol and Glycidyl Ethers: Design of Branched Polyether Polyols by Combination of Latent Cyclic AB_2 and ABR Monomers", *Macromolecules* **2000**, *33*, 7682–7692.

[209] A. Sunder, R. Hanselmann, H. Frey, R. Mülhaupt, "Controlled Synthesis of Hyperbranched Polyglycerols by Ring-Opening Multibranching Polymerization", *Macromolecules* **1999**, *32*, 4240–4246.

[210] A. Sunder, M.-F. Quincy, R. Mülhaupt, H. Frey, "Hyperbranched Polyether Polyols with Liquid Crystalline Properties", *Angew. Chem.* **1999**, *111*, 3107–3110; *Angew. Chem. Int. Ed.* **1999**, *38*, 2928–2930.

[211] A. Burgath, A. Sunder, I. Neuner, R. Mülhaupt, H. Frey, "Multi-Arm Star Block Copolymers based on ε-Caprolactone with Hyperbranched Polyglycerol Core", *Macromol. Chem. Phys.* **2000**, *201*, 792–797.

[212] R. Knischka, P. J. Lutz, A. Sunder, R. Mülhaupt, H. Frey, "Functional Poly(ethylene oxide) Multi-Arm Star Polymers: Core-First Synthesis Using Hyperbranched Polyglycerol Initiators", *Macromolecules* **2000**, *33*, 315–320.

[213] S. Macking, R. Thomann, H. Frey, A. Sunder, "Preparation of Catalytically Active Palladium Nanoclusters in Compartments of Amphiphilic Hyperbranched polyglycerols", *Macromolecules* **2000**, *33*, 3958–3960.

[214] A. Sunder, R. Mülhaupt, R. Haag, H. Frey, "Chiral Hyperbranched Dendron Analogues", *Macromolecules* **2000**, *33*, 253–254.

[215] A. Sunder, R. Mülhaupt, H. Frey, "Hyperbranched Polyether-Polyols Based on Polyglycerol: Polarity Design by Block Copolymerization with Propylene Oxide", *Macromolecules* **2000**, *33*, 309–314.

[216] A. Sunder, T. Bauer, R. Mülhaupt, H. Frey, "Synthesis and Thermal Behavior of Esterified Aliphatic Hyperbranched Polyether Polyols", *Macromolecules* **2000**, *33*, 1330–1337.

[217] R. Haag, A. Sunder, J.-F. Stumbé, "An Approach to Glycerol Dendrimer and Pseudo-Dendritic Polyglycerols", *J. Am. Chem. Soc.* **2000**, *122*, 2954–2955.

[217a] R. Haag, J.-F. Stumbé, A. Sunder, H. Frey, A. Hebel, "An Approach to Core – Shell -Type Architectures in Hyperbranched Polyglycerols by Selective Chemical Differentiation", *Macromolecules* **2000**, *33*, 8158–8166.

[218] A. Nishimoto, K. Agehara, N. Furuya, T. Watanabe, M. Watanabe, "High Ionic Conductivity of Polyether-Based Network Polymer Electrolytes with Hyperbranched Side Chains", *Macromolecules* **1999**, *32*, 1541–1548.

[219] Y. Zhou, M. L. Bruening, D. E. Bergbreiter, R. M. Crooks, M. Wells, "Preparation of Hyperbranched Polymer Films Grafted on Self-Assembled Monolayers", *J. Am. Chem. Soc.* **1996**, *118*, 3773–3774.

[220] Y. Zhou, M. L. Bruening, Y. Liu, R. M. Crooks, D. E. Bergbreiter, "Synthesis of Hyperbranched, Hydrophilic Fluorinated Surface Grafts", *Langmuir* **1996**, *12*, 5519–5521.

[221] M. L. Bruening, Y. Zhou, G. Aguilar, R. Agee, D. E. Bergbreiter, R. M. Crooks, "Synthesis and Characterization of Surface-Grafted, Hyperbranched Polymer Films Containing Fluorescent, Hydrophobic, Ion-Binding, Biocompatible, and Electroactive Groups", *Langmuir* **1997**, *13*, 770–778.

[222] M. Zhao, Y. Zhou, M. L. Bruening, D. E. Bergbreiter, R. M. Crooks, "Inhibition of Electrochemical Reactions at Gold Surfaces by Grafted, Highly Fluorinated, Hyperbranched Polymer Films", *Langmuir* **1997**, *13*, 1388–1391.

[223] D. L. Dermody, R. F. Peez, D. E. Bergbreiter, R. M. Crooks, "Chemically Grafted Polymeric Filters for Chemical Sensors: Hyperbranched Poly(acrylic acid) Films Incorporating β-Cyclodextrin Receptors and Amine-Functionalized Filter Layers", *Langmuir* **1999**, *15*, 885–890.

[224] J. G. Franchina, W. M. Lackowski, D. L. Dermody, R. M. Crooks, D. E. Bergbreiter, K. Sirkar, R. J. Russell, M. V. Pishko, "Electrostatic Immobilization of Glucose Oxidase in a Weak Acid, Polyelectrolyte Hyperbranched Ultrathin Film on Gold: Fabrication, Characterization, and Enzymatic Activity", *Anal. Chem.* **1999**, *71*, 3133–3139.

[224a] M. L. Amirpour, P. Ghosh, W. M. Lackowski, R. M. Crooks, M. V. Pishko, "Mammalian Cell Cultures on Micropatterned Surfaces of Weak-Acid, Polyelectrolyte Hyperbranched Thin Films on Gold", *Anal. Chem.* **2001**, *73*, 1560–1566.

[225] H. Huang, E. E. Remsen, K. L. Wooley, "Amphiphilic Core-Shell Nanospheres Obtained by Intramicellar Shell Cross-Linking of Polymer Micelles with Poly(ethylene oxide) Linkers", *Chem. Commun.* **1998**, 1415–1416.

[226] H. Huang, T. Kowalewski, E. E. Remsen, R. Gertzmann, K. L. Wooley, "Hydrogel-Coated Glassy Nanospheres: A Novel Method for the Synthesis of Shell Cross-Linked Knedels", *J. Am. Chem. Soc.* **1997**, *119*, 11653–11659.

[227] K. L. Wooley, "From Dendrimers to Knedel-like Structures", *Chem. Eur. J.* **1997**, *3*, 1397–1399.

[228] K. B. Thurmond, II, K. L. Wooley, "Shell Cross-linked Knedels: Amphiphilic Core-Shell Nanospheres and their Potential Applications", *Polym. Prepr.* **1998**, *39*, 303.

[229] Q. Ma, E. E. Remsen, T. Kowalewski, K. L. Wooley, "Entirely Hydrophilic Shell Cross-Linked Knedel-like (SCK) Nanoparticles", *Polym. Prepr.* **2000**, *41*, 1571–1572.

[230] K. L. Wooley, "Shell Cross-linked Polymer Assemblies: Nanoscale Constructs Inspired from Biological Systems", *J. Polym. Sci., Part A: Polym. Chem.* **2000**, *38*, 1397–1407.

[231] D. E. Bergbreiter, G. Tao, A. M. Kippenberger, "Functionalized Hyperbranched Polyethylene Powder Supports", *Org. Lett.* **2000**, *2*, 2853–2855.

[231a] D. E. Bergbreiter, G. Tao, "Chemical Modification of Hyperbranched Ultrathin Films on Gold and Polyethylene", *J. Polym. Sci., Part A: Polym. Chem.* **2000**, *38*, 3944–3953.

[231b] C. Paulo, J. E. Puskas, "Synthesis of Hyperbranched Polyisobutylene by Inimer-Type Living Polymerization. 1. Investigation of the Effect of Reaction Conditions", *Macromolecules* **2001**, *34*, 734–739.

[232] P. Lu, J. K. Paulasaari, W. P. Weber, "Hyperbranched Poly(4-acetylstyrene) by Ruthenium-Catalyzed Step-Growth Polymerization of 4-Acetylstyrene", *Macromolecules* **1996**, *29*, 8583–8586.

[233] A. H. E. Müller, D. Yan, M. Wulkow, "Molecular Parameters of Hyperbranched Polymers Made by Self-Condensing Vinyl Polymerization; 1. Molecular Weight Distribution", *Macromolecules* **1997**, *30*, 7015–7023.

[234] D. Yan, A. H. E. Müller, K. Matyjaszewski, "Molecular Parameters of Hyperbranched Polymers Made by Self-Condensing Vinyl Polymerization; 2. Degree of Branching", *Macromolecules* **1997**, *30*, 7024–7033.

[235] K. Matyjaszewski, S. G. Gaynor, A. H. Müller, "Preparation of Hyperbranched Polyacrylates by Atom Transfer Radical Polymerization; 2. Kinetics and Mechanism of Chain Growth for the Self-Condensing Vinyl Polymerization of 2-[(2-Bromopropionyl)oxy]ethyl Acrylate", *Macromolecules* **1997**, *30*, 7034–7041.

[236] K. Matyjaszewski, S. G. Gaynor, "Preparation of Hyperbranched Polyacrylates by Atom Transfer Radical Polymerization; 3. Effect of Reaction Conditions on the Self-Condensing Vinyl Polymerization of 2-[(2-Bromopropionyl)oxy]ethyl Acrylate", *Macromolecules* **1997**, *30*, 7042–7049.

[237] K. Matyjaszewski, S. G. Gaynor, A. Kulfan, M. Podwika, "Preparation of Hyperbranched Polyacrylates by Atom Transfer Radical Polymerization; 1. Acrylate AB* Monomers in 'Living' Radical Polymerization", *Macromolecules* **1997**, *30*, 5192–5194.

[238] W. Radke, G. Litvinenko, A. H. E. Müller, "Effect of Core-Forming Molecules on Molecular Weight Distribution and Degree of Branching in the Synthesis of Hyperbranched Polymers", *Macromolecules* **1998**, *31*, 239–248.

[238a] G. I. Litvinenko, P. F. W. Simon, A. H. E. Müller, "Molecular Parameters of Hyperbranched Copolymers Obtained by Self-Condendsing Vinyl Copolymerization, 2. Non-Equal Rate Constants", *Macromolecules* **2001**, *34*, 2418–2426.

[239] G. I. Litvinenko, P. F. W. Simon, A. H. E. Müller, "Molecular Parameters of Hyperbranched Copolymers Obtained by Self-Condensing Vinyl Copolymerization; 1. Equal Rate Constants", *Macromolecules* **1999**, *32*, 2410–2419.

[240] D. Yan, Z. Zhou, A. H. E. Müller, "Molecular Weight Distribution of Hyperbranched Polymers Generated by Self-Condensing Vinyl Polymerization in Presence of a Multifunctional Initiator", *Macromolecules* **1999**, *32*, 245–250.

[240a] P. Guadarrama, L. Fomina, S. Fomine, "Solid-supported synthesis of hyperbranched polymer with β,β-diethynylstyryl units", *Polym. Int.* **2001**, *50*, 76–83.

[241] D. H. Bolton, K. L. Wooley, "Synthesis and Characterization of Hyperbranched Polycarbonates", *Macromolecules* **1997**, *30*, 1890–1896.

[242] C. Gong, J. M. J. Fréchet, "Proton Transfer Polymerization in the Preparation of Hyperbranched Polyesters with Epoxide Chain-Ends and Internal Hydroxy Functionalities", *Macromolecules* **2000**, *33*, 4997–4999.

[243] S. Maier, A. Sunder, H. Frey, R. Mülhaupt, "Synthesis of Poly(glycerol)–*block*–poly(methyl acrylate) Multi-Arm Star Polymers", *Macromol. Rapid Commun.* **2000**, *21*, 226–230.

[244] J. Kadokawa, Y. Kaneko, S. Yamada, K. Ikuma, H. Tagaya, K. Chiba, "Synthesis of Hyperbranched Polymers via Proton-Transfer Polymerization of Acrylate Monomer Containing Two Hydroxy Groups", *Macromol. Rapid Commun.* **2000**, *21*, 362–368.

[245] M. Suzuki, A. Ii, T. Saegusa, "Multibranching Polymerization: Palladium-Catalyzed Ring-Opening Polymerization of Cyclic Carbamate to Produce Hyperbranched Dendritic Polyamine", *Macromolecules* **1992**, *25*, 7071–7072.

[245a] C. Gao, D. Yan, "Polyaddition of B_2 and BB'_2 Type Monomers to A_2 Type Monomer. I Synthesis of Highly Branched Copoly(sulfone-amine)s", *Macromolecules* **2001**, *34*, 156–161.

[245b] D. Yan, C. Gao, "Hyperbranched Polymers Made from A_2 and BB'_2 Type Monomers. 1. Polyaddition of 1-(2-Aminoethyl)piperazine to Divinyl Sulfone", *Macromolecules* **2000**, *33*, 7693–7699.

[245c] C. Gao, W. Tang, D. Yan, P. Zhu, P. Tao, "Hyperbranched polymers made from A_2, B_2, and BB'_2 type monomers, 2. Preparation of hyperbranched copoly(sulfone-amine)s by polyaddition of *N*-ethylethylenediamine and piperazine to divinylsulfone", *Polymer* **2001**, *42*, 3437–3443.

[246] L. J. Hobson, A. M. Kenwright, W. J. Feast, "A Simple 'One-Pot' Route to the Hyperbranched Analogues of Tomalia's Poly(amidoamine) Dendrimers", *Chem. Commun.* **1997**, 1877–1878.

[247] L. J. Hobson, W. J. Feast, "Poly(amidoamine) Hyperbranched Systems: Synthesis, Structure, and Characterization", *Polymer* **1999**, *40*, 1279–1297.

[248] D. A. Tomalia, H. Baker, J. Dewald, M. Hall, G. Kallos, S. Martin, J. Roeck, J. Ryder, P. Smith, "A New Class of Polymers: Starburst-Dendritic Macromolecules", *Polym. J. (Tokyo)* **1985**, *17*, 117–132.

[249] G. L. Stahl, R. Walter, C. W. Smith, "General Procedure for the Synthesis of Mono-*N*-acylated 1,6-Diaminohexanes", *J. Org. Chem.* **1978**, *43*, 2285–2286.

[249a] M. Okazaki, M. Murota, Y. Kawaguchi, N. Tsubokawa, "Curing of Epoxy Resin by Ultrafine Silica Modified by Grafting of Hyperbranched Polyamidoamine Using Dendrimer Synthesis Methodology", *J. Appl. Polym. Sci.* **2001**, *80*, 573–579.

[250] K. Fujiki, M. Sakamoto, T. Sato, N. Tsubokawa, "Postgrafting of Hyperbranched Dendritic Polymer from Terminal Amino Groups of Polymer Chains Grafted onto Silica Surface", *J. Macromol. Sci. – Pure Appl. Chem.* **2000**, *A37*, 357–377.

[250a] Y. Lim, S.-M. Kim, Y. Lee, W. Lee, T. Yang, M. Lee, H. Suh, J. Park, "Cationic Hyperbranched Poly(amino ester): A Novel Class of DNA Condensing Molecule with Cationic Surface, Biodegradable Three-Dimensional Structure, and Tertiary Amine Groups in the Interior", *J. Am. Chem. Soc.* **2001**, *123*, 2460–2461.

[250b] H. Sashiwa, Y. Shigamasa, R. Roy, "Chemical Modification of Chitosan. 3. Hyperbranched Chitosan-Sialic Acid Dendrimer Hybrid with Tetraethylene Glycol Spacer", *Macromolecules* **2000**, *33*, 6913–6915.

[250c] H. Sashiwa, Y. Shigemasa, R. Roy, "Highly Convergent Synthesis of Dendrimerized Chitosan-Sialic Acid Hybrid", *Macromolecules* **2001**, *34*, 3211–3214.

[250d] H. Wei, Y. Lu, W. Shi, H. Yuan, Y. Chen, "UV Curing Behavior of Methacrylated Hyperbranched Poly(amine-ester)s", *J. Appl. Polym. Sci.* **2001**, *80*, 51–57.

[251] J. Fang, H. Kita, K. Okamoto, "Hyperbranched Polyimides for Gas Separation Applications; 1. Synthesis and Characterization", *Macromolecules* **2000**, *33*, 4639–4646.

[251a] J. Fang, H. Kita, K. Okamoto, "Gas permeation properties of hyperbranched polyimide membranes", *J. Membr. Sci.* **2001**, *182*, 245–256.

[252] S. Tanaka, T. Iso, Y. Doke, "Preparation of New Branched Poly(triphenylamine)", *Chem. Commun.* **1997**, 2063–2064.

[253] F. Wang, M. S. Wilson, R. D. Rauh, P. Schottland, J. R. Reynolds, "Electroactive and Conducting Star-Branched Poly(3-hexylthiophene)s with a Conjugated Core", *Macromolecules* **1999**, *32*, 4272–4278.

[254] M. Suzuki, S. Yoshida, K. Shiraga, T. Saegusa, "New Ring-Opening Polymerization via a π-Allylpalladium Complex; 5. Multibranching Polymerization of Cyclic Carbamate to Produce Hyperbranched Dendritic Polyamine", *Macromolecules* **1998**, *31*, 1716–1719.

[255] C. Wu, R. Ma, B. Zhou, J. Shen, K. K. Chan, K. F. Woo, "Laser Light-Scattering Study of the Dendritic-like Polyelectrolytes and CTAB Complex Formation", *Macromolecules* **1996**, *29*, 228–232.

[256] T. Emrick, H.-T. Chang, J. M. J. Fréchet, "An A$_2$ + B$_3$ Approach to Hyperbranched Aliphatic Polyethers Containing Chain End Epoxy Substituents", *Macromolecules* **1999**, *32*, 6380–6382.

[257] M. N. Bochkarev, V. B. Cilkin, L. P. Mayorova, G. A. Razuvaev, U. D. Cemchkov, V. E. Sherstyanux, "Polyphenylenegermane – A New Type of Polymeric Material", *Metalloorg. Khim.* **1988**, *1*, 196–200.

[257a] H. L. Lee, M. Takeuchi, M. Kakimoto, S. Y. Kim, "Hyperbranched poly(arylene ether phosphine oxide)s", *Polym. Bull.* **2000**, *45*, 319–326.

[258] T. W. Carothers, L. J. Mathias, "Divergent and Convergent Synthesis of Hyperbranched Poly(siloxysilanes)", *Polym. Prepr.* **1993**, *34*, 503–504.

[259] L. J. Mathias, T. W. Carothers, "Hyperbranched Poly(siloxysilanes)", *J. Am. Chem. Soc.* **1991**, *113*, 4043–4044.

[260] L. J. Mathias, T. W. Carothers, "Terminal-Group Modifications of Hyperbranched Poly(siloxy-silanes)", *Polym. Prepr.* **1991**, *32*, 633–634.

[261] L. J. Mathias, R. M. Bozen, "Linear and Star-Branched Siloxy-Silane Polymers: One-Pot A–B Polymerization and End-Capping", *Polym. Prepr.* **1992**, *33*, 146–147.

[262] L. J. Mathias, V. R. Reichert, T. W. Carothers, R. M. Bozen, "Stars, Dendrimers, and Hyperbranched Polymers: Towards Understanding Structure–Property Relationships for Single Molecule Constructs", *Polym. Prepr.* **1993**, *34*, 77–78.

[263] A. M. Muzafarov, M. Golly, M. Möller, "Degradable Hyperbranched Poly[bis(undecenyloxy)methylsilane]s", *Macromolecules* **1995**, *28*, 8444–8446.

[264] A. Kunai, E. Toyoda, I. Nagamoto, T. Horio, M. Ishikawa, "Polymeric Organosilicon Systems; 25. Preparation of Branched Polymers by Regiospecific Hydrosilylation of Poly[(silylene)diethynylenes] and their Properties", *Organometallics* **1996**, *15*, 75–83.

[265] K. Yoon, D. Y. Son, "Syntheses of Hyperbranched Poly(carbosilarylenes)", *Macromolecules* **1999**, *32*, 5210–5216.

[265a] Y. Xiao, R. A. Wong, D. Y. Son, "Synthesis of a New Hyperbranched Poly(silylenevinylene) with Ethynyl Functionalization", *Macromolecules* **2000**, *33*, 7232–7234.

[266] A. M. Muzafarov, E. A. Rebrov, O. B. Gorbacevich, M. Golly, H. Gankema, M. Möller, "Degradable Dendritic Polymers – A Template for Functional Pores and Nanocavities", *Macromol. Symp.* **1996**, *102*, 35–46.

[267] C. Drohmann, M. Möller, O. B. Gorbatsevich, A. M. Muzafarov, "Hyperbranched Polyalkenylsilanes by Hydrosilylation with Platinum Catalysts; I. Polymerization", *J. Polym. Sci., Part A: Polym. Chem.* **2000**, *38*, 744–751.

[268] J. Maxka, J. Chrusciel, M. Sasaki, K. Matyjaszewski, "Polysilanes with Various Architectures", *Makromol. Chem., Macromol. Symp.* **1994**, *77*, 79–92.

[269] S. Rubinsztajn, "Synthesis and Characterization of New Poly(siloxysilanes)", *J. Inorg. Organomet. Polym.* **1994**, *4*, 61–77.

[270] U. Herzog, G. Roewer, A. Moll, W. Habel, T. Windmann, P. Sartori, "Darstellung und Charakterisierung oligomerer und polymerer Ethinylsilane mit Si_2-Si_4-Einheiten", *Journal für praktische Chemie Chemiker-Zeitung* **1997**, *339*, 603–606.

[271] C. Gong, J. Miravet, J. M. J. Fréchet, "Intramolecular Cyclization in the Polymerization of AB_x Monomers: Aproaches to the Control of Molecular Weight and Polydispersity in Hyperbranched Poly(Siloxysilane)", *J. Polym. Sci., Part A: Polym. Chem.* **1999**, *37*, 3193–3201.

[272] C. Gong, J. M. J. Fréchet, "End Functionalization of Hyperbranched Poly(siloxysilane): Novel Cross-linking Agents and Hyperbranched-Linear Star Block Copolymers", *J. Polym. Sci., Part A: Polym. Chem.* **2000**, *38*, 2970–2978.

[273] J. K. Paulasaari, W. P. Weber, "Synthesis of Hyperbranched Polysiloxanes by Based-Catalyzed Proton-Transfer Polymerization. Comparison by Hyperbranched Polymer Microstructure and Properties to those of Linear Analogues Prepared by Cationic or Anionic Ring-Opening Polymerization", *Macromolecules* **2000**, *33*, 2005–2010.

[274] C. Lach, P. Müller, H. Frey, R. Mülhaupt, "Hyperbranched Polycarbosilane Macromonomers Bearing Oxazoline Functionalities", *Macromol. Rapid Commun.* **1997**, *18*, 253–260.

[275] C. Lach, R. Hanselmann, H. Frey, R. Mülhaupt, "Hyperbranched Carbosilane Oxazoline-Macromonomers: Polymerization and Coupling to a Trimesic Acid Core", *Macromol. Rapid Commun.* **1998**, *19*, 461–465.

[276] C. Lach, H. Frey, "Enhancing the Degree of Branching of Hyperbranched Polymers by Post-synthetic Modification", *Macromolecules* **1998**, *31*, 2381–2383.

[277] D. Hölter, A. Burgath, H. Frey, "Degree of Branching in Hyperbranched Polymers", *Acta Polym.* **1997**, *48*, 30–35.

[278] H. Frey, C. Lach, C. Schlenk, T. Pusel, "Hyperbranched Polycarbosilanes: A Versatile Class of Inorganic Polymers", *Polym. Prepr.* **2000**, *41*, 568–569.

[279] J. F. Miravet, J. M. J. Fréchet, "New Hyperbranched Poly(siloxysilanes): Variations of the Branching Pattern and End-Functionalization", *Macromolecules* **1998**, *31*, 3461–3468.

[280] V. Percec, C. G. Cho, C. Pugh, D. Tomazos, "Synthesis and Characterization of Branched Liquid-Crystalline Polyethers Containing Cyclotetraveratrylene-based Disk-like Mesogens", *Macromolecules* **1992**, *25*, 1164–1176.

[281] H. R. Kricheldorf, T. Stukenbrock, "New Polymer Syntheses 93; Hyperbranched Homo- and Copolymer Derived from Gallic Acid and β-(4-Hydroxyphenyl)propionic Acid", *J. Polym. Sci., Part A: Polym. Chem.* **1998**, *36*, 2347–2357.

[282] M. Maciejewski, E. Bednarek, J. Janiszewska, J. Janiszewska, G. Szczygiel, M. Zapora, "Guanamine Dendrimerization", *J. Macromol. Sci. – Pure Appl. Chem.* **2000**, *A37*, 753–783.

7 Chiral Dendritic Macromolecules

In the preceding Chapters, the syntheses and properties of highly symmetrical dendrimers and of their less perfect hyperbranched counterparts were presented. Although aspects of chirality have appeared, this Chapter will deal more directly with the combination of chirality and molecular symmetry. Several excellent reviews[1-4] have recently appeared and have started to provide a better understanding of this area. As different types of chirality are incorporated at specific levels or positions within these molecular spheres, new and potentially novel properties should appear, especially in relation to catalytic pockets akin to those found in biological systems.

Interestingly, Avnir et al.[5] examined the classical definitions and terminology of chirality and subsequently concluded that they are too restrictive to describe complex objects such as large random supermolecular structures and spiral diffusion-limited aggregates. Architecturally, these structures resemble chiral (and fractal) dendrimers; therefore, new insights into chiral concepts and nomenclature were introduced that have a direct bearing on the nature of dendritic macromolecular assemblies, e.g., "continuous chirality measure" and "virtual enantiomers". A new category of graph-theoretical trees has been modeled for asymmetrical dendrimers and analyzed using the Cayley scheme.[6]

As noted in Section 3.2, Denkewalter et al.[7-9] were the first (1979) to report a divergent chiral procedure for obtaining high molecular weight cascade polymers, which utilized a protected, naturally occurring amino acid monomer, N,N'-bis(tert-butyloxycarbonyl)-L-lysine. These poly(α,ε-L-lysine) dendrimers, with purported molecular weights up to 233,600 amu (at the 10th generation), were described as having utility as "surface modifying agents, metal chelating agents, and substrates for the preparation of pharmaceutical dosages".[7] In these patents, as well as in a related study of physical properties,[10] discussion of the dendrimers' chirality or lack thereof was omitted.

Chirality at different locations within a dendrimer can be categorized as: (a) surface: Newkome et al.[11] utilized an amino acid as a capping mode of chirality in 1991; (b) connectors: Chow et al.[12] used chiral connectors between the branching centers in 1994; (c) core: Seebach et al.[13] used a chiral core, and (d) asymmetric core: Meijer et al.[14] described the first attempt in 1994. Other examples and combinations of site chirality are presented according to their structural composition.

7.1 Divergently Generated Chiral Dendrimers

7.1.1 1 → 2 Branched

7.1.1.1 1 → 2 C-Branched, *Amide* Connectivity

In 1979, Denkewalter et al.[7-9] reported a divergent procedure using the chiral, protected amino acid monomer, N,N'-bis(tert-butyloxycarbonyl)-L-lysine to prepare poly(α,ε-L-lysine) dendrimers (Figure 7.1; **1**) through to the 10th generation (details can be found in Section 3.2).

Tam et al.[15-20] coupled amino acids by means of a Merrifield-type peptide synthesizer using similar lysine-based amine acylation technology[8] to produce octopus-immunogens (**2**) and scaffolding, or core matrices, for a *multiple antigen peptide* (Figure 7.2; MAP core, **3**); Tam has also reviewed this topic.[21] The synthesis of peptide dendrimers utilized a novel tetravalent glyoxylyl-lysinyl core peptide, [(OHCCO)$_2$Lys]$_2$Lys–AlaOH, for the formation of oxime, hydrazone, or thiazolidine linkages.[22] In general, a branched lysine core matrix bearing aldehyde functions ligated four copies of the unprotected peptides possessing bases such as aminooxy, hydrazide, or cysteine 1,2-aminothiol groups at their N-termini, thereby forming the synthetic proteins. The use of the mutually reactive weak

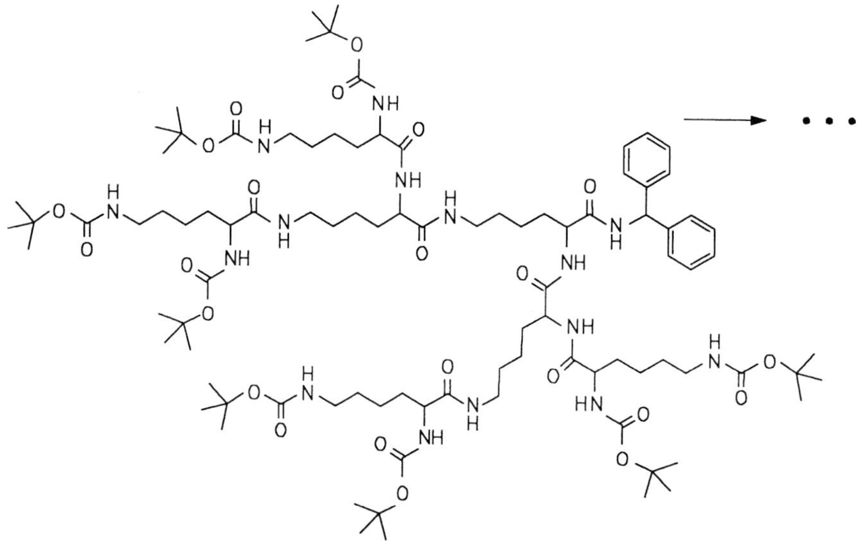

1

Figure 7.1 A representative of Denkewalter et al.'s polylysine dendrimers.[7]

base and aldehyde pair provides a convenient route to peptide dendrimers and artificial proteins.[23] Synthesis of the 1st generation peptide dendrimers by thiol-based Michael addition to pentaerythritol tetraacrylate has been described.[24] Hamilton et al.[25] reported the preparation of a tetracarboxylic acid-capped calix[4]arene, which was subsequently coupled to four equivalents of a "constrained" hexapeptide loop; the resulting "antibody" mimicked the binding of natural proteins to cytochrome *c*. A series of lipidic polylysine dendrimers has been synthesized, of which the 4th generation was evaluated for its absorption after oral administration to female rats.[26]

Roy et al.[27, 28, 28a] described the solid-state preparation of four generations of dendritic sialoside inhibitors of influenza A virus haemagglutinin (Scheme 7.1(a); e.g., **4**). They applied a similar backbone as that used in the Denkewalter series,[8] by employing N^α,N^ε-di(fluorene-9-ylmethoxycarbonyl)-L-lysine benzotriazole (Scheme 7.1(b); **5**) as the activated ester monomer for tier construction. After acylation and subsequent deprotection, the two-step process was repeated. Each tier was terminally functionalized with the preformed benzotriazole ester of chloroacetylglycylglycine, and then treated with

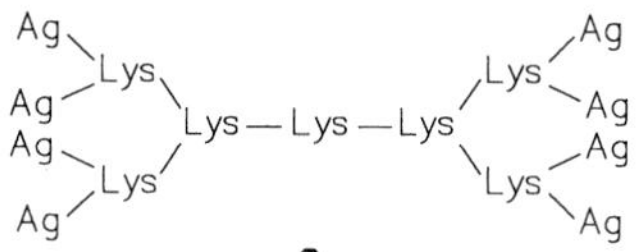

2

An Octopus—Immunogen

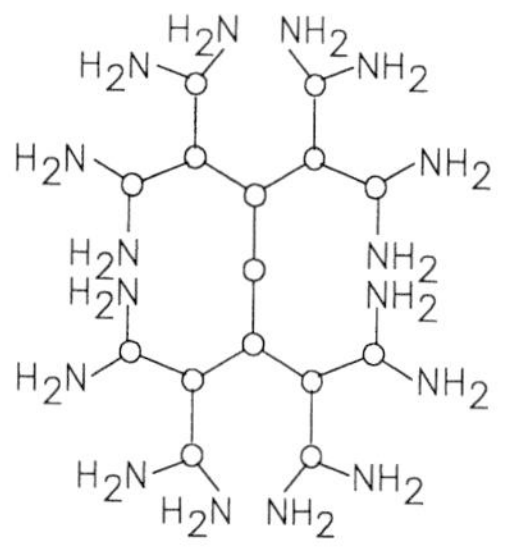

3   (○ ≡ Lys)

A Lysine—base core matrix used
for the preparation of a
Multiple—Antigen—Peptide

Figure 7.2 Lysine-based dendritic "scaffolding" used for the preparation of polyimmunogens.[15–20]

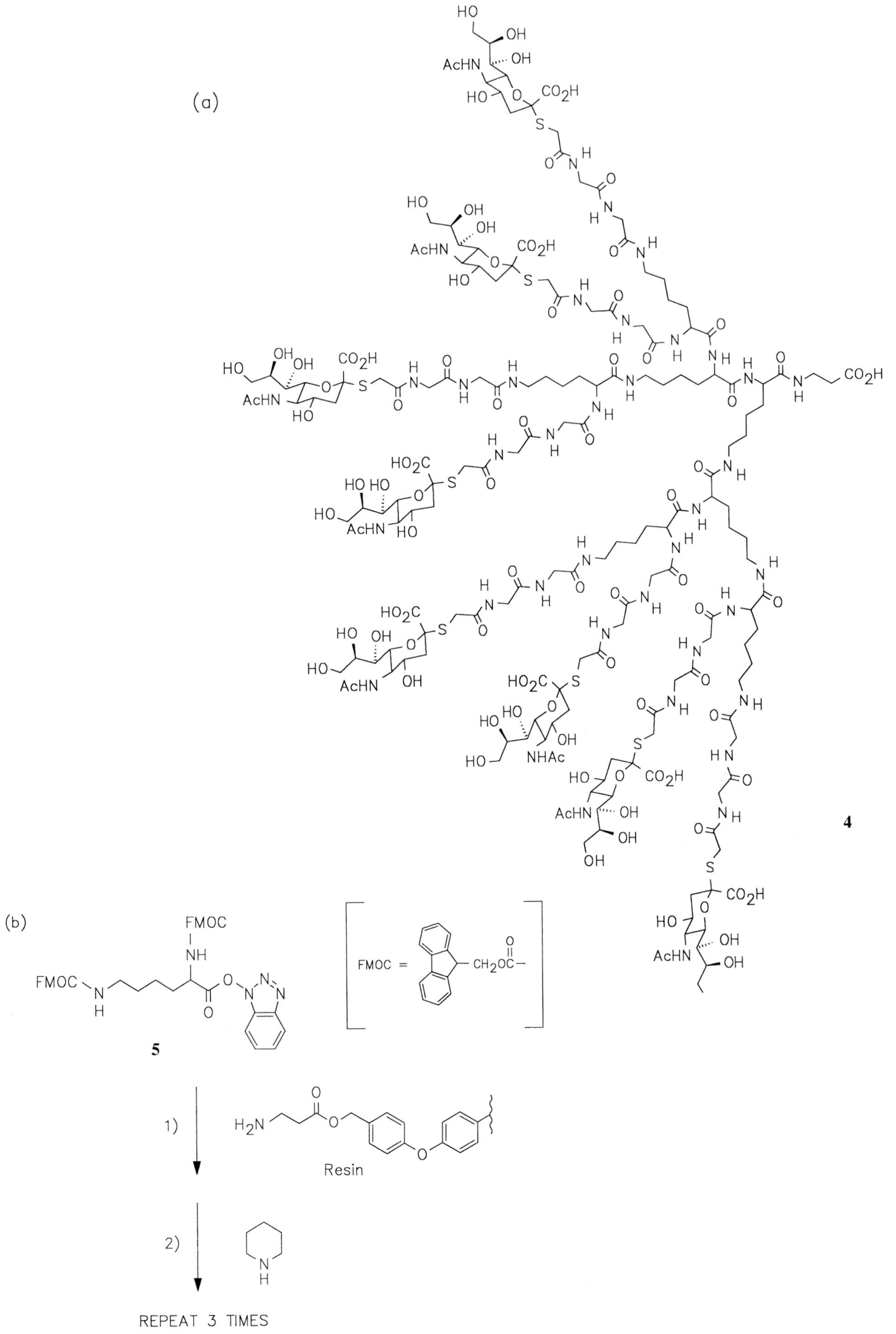

(a)

4

(b)

5

FMOC

FMOC =

1)

H₂N

Resin

2)

REPEAT 3 TIMES

Scheme 7.1 Dendritic sialoside inhibitors of influenza A virus haemagglutinin constructed by the solid-phase synthesis of a lysine-based superstructure.[27, 28]

peracetylated 2-thiosialic acid. Removal from the solid support and subsequent hydrolysis afforded the desired inhibitor in nearly quantitative yield.

Zanini and Roy[29] later employed this lysine-based scaffolding, prepared by a solid-phase Fmoc and HBT protocol, for the synthesis and study of *N*-acetylglucosamine- and *N*-acetyllactosamine-terminated dendrimers; inhibition of lectin–porcine stomach mucin interactions was examined. For the 1st to 3rd generations, the glucosamine dendrimers exhibited a 20-fold inhibitory increase using wheatgerm lectin as a binder, whereas no significant inhibition was observed using the lactosamine dendrimers and *Erythrina cristagalli* lectin. Hyperbranched L-lysine cores with *N*-acetylglucosamine residues were synthesized and enzymatically transformed into sialyl Lewis^x-containing dendrimers;[30, 30a] these dendrimers are being evaluated as selectin antagonists. A polymeric version has recently appeared, whereby glycoconjugates possessing *N*-acetyllactosamine or sialyl Lewis^X moieties were synthesized;[31] these monomers possessed an acrylamide group as well as different spacer groups, and were subsequently polymerized. Lactosylated PAMAM hydrodynamic properties have been reported.[31a, b]

Chapman et al.[32] reported the synthesis (Scheme 7.2) of poly(ethylene oxide)-supported dendritic *t*-Boc-poly(α,ε-L-lysines). Methoxy-terminated poly(ethylene oxide)

$$H_3CO-PEO-OH \;+\; N-Boc-L-glycine \xrightarrow{\text{dicyclohexylcarbodiimide}}$$

6

$$H_3CO-PEO-Gly-Boc \xrightarrow[\text{2) xs } 7 \text{ xs DPEA}]{\text{1) TFA/CH}_2\text{Cl}_2} H_3CO-PEO \sim Lys \overset{Boc}{\underset{Boc}{<}} \xrightarrow[\text{2) xs } 7 \text{ xs DPEA}]{\text{1) TFA/CH}_2\text{Cl}_2}$$

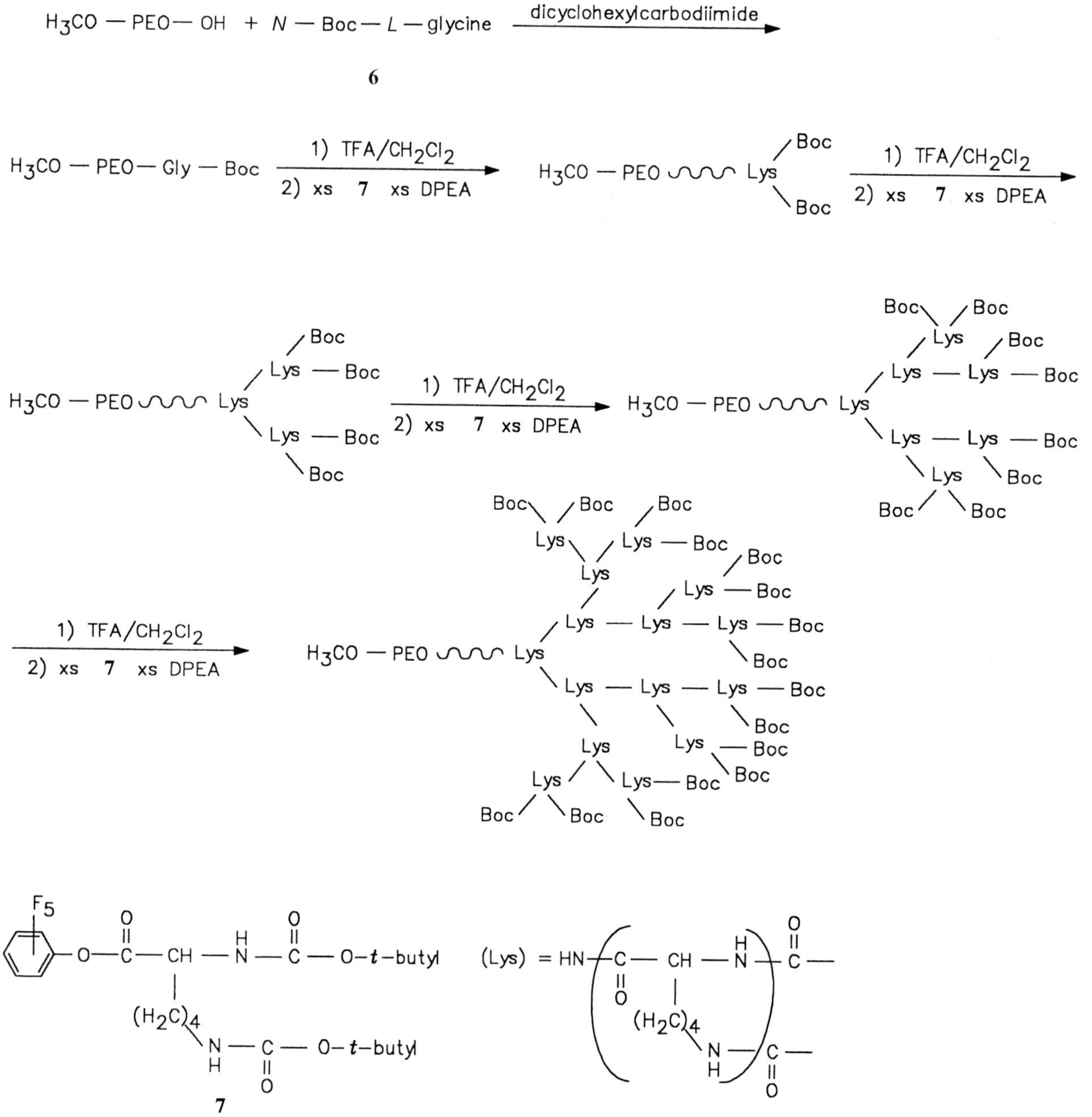

Scheme 7.2 Lysine-based, dendritic-terminated poly(ethylene oxide), described as a "hydroamphiphile" due to the formation of stable foams in water.[32]

7

was esterified with Boc-protected glycine (**6**), then deprotected (TFA) to liberate the free amine, and subsequently used as a support for the divergent construction of the poly(α,ε-L-lysine) scaffolding. Employing a procedure akin to that of Denkewalter,[8] pentafluorophenyl N^α,N^ε-di(*t*-Boc)-L-lysinate (**7**) was repetitively coupled and then deprotected. These 'chimerical' molecules with multiple terminal hydrophobic *tert*-butyl groups were termed "hydroamphiphiles", since in water they were observed to form foams possessing good temporal stability.

Tam et al. reported[33] and reviewed[34] the synthesis of an MAP system[16, 18–20] utilizing the divergently constructed lysine-based scaffolding, which was accomplished using a Merrifield-type, solid-phase method, thus eliminating the traditional step of peptide to carrier conjugation. Six different antigen peptide sequences were attached to the cascade imine periphery through triglycyl linkers.

Rao and Tam[17] connected a 24-residue peptide to a small, lysine-based dendrimer (Scheme 7.3). The peptide connection was effected by treatment of the octameric, glyoxylyl-terminated dendrimer **8** with the 1,2-aminothiol polypeptide **9** to afford the thiazolidinyl peptide dendrimer **10**. The thiazolidine linkage also introduces an additional asymmetric center.

Terminally-boronated lysine-based dendrimers have been reported,[35] which were shown to possess free thiol and amine sites for the attachment of carrier proteins, such as an antibody fragment and a fluorescent probe, respectively. Aqueous solubility was accomplished by the use of an ester-connected PEG unit. These materials were envisaged as being potentially useful in boron neutron capture therapy (BNCT), electron energy loss spectroscopy (EELS), and electron spectroscopic imaging (ESI); the latter has been examined.[36] Tam and Shao[22] extended the usefulness of terminal glyoxylyl moieties to prepare (Scheme 7.4) peptide dendrimers possessing additional modes of peptide ligation (i.e., terminal peptide connectivity). Thus, the tetrakis(glyoxylyl) cascade **12**, prepared from tetrakis(acetal) **11**, was reacted with peptides containing terminal hydrazide, 1,2-aminothiol, and oxime moieties to give the hydrazone, thiazolidine, and oxime peptide ligated dendrimers **13**, **14**, and **15**, respectively.

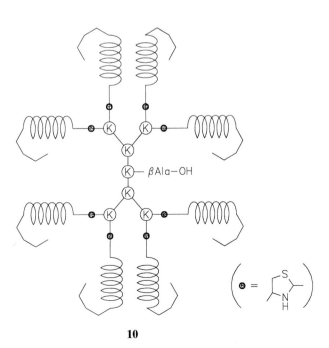

Scheme 7.3 Use of a lysine-based scaffolding for the ordering of 24-residue peptides.[17]

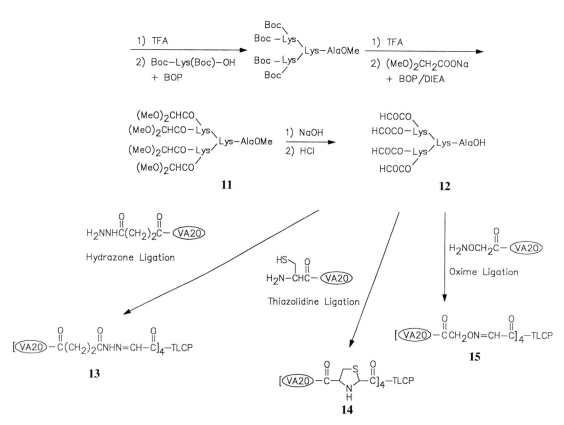

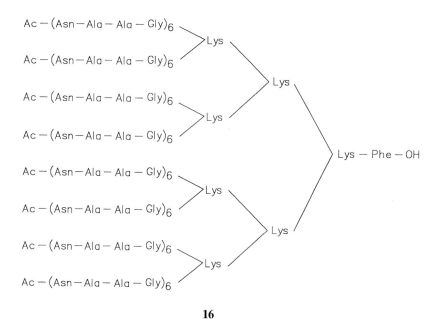

Scheme 7.4 Employment of terminal glyoxylyl groups to effect different modes of peptide attachment.[22]

Zhang and Tam[37] further described an efficient, regiospecific protocol for the macrocyclization of unprotected peptide segments predicated on intramolecular transthioesterification and their subsequent use as terminal groups on lysine-based scaffolds. The resultant dendrimers possessed enhanced, ring-based rigidity and offered a route to new surface antigen mimics. Unprotected peptides were also employed as building blocks for dendrimers.[38]

Figure 7.3 Octaantigen polypeptide **16**, prepared by solid-phase peptide synthesis.[39]

Pessi et al.[39] prepared a different multiple antigen peptide employing the continuous-flow polypeptide procedure[40] for solid-phase peptide synthesis. The resultant one-directional, octaantigen polypeptide cascade **16** is depicted in Figure 7.3; it was characterized by gel permeation chromatography, FAB MS, and amino acid ratio analysis.

Pallin and Tam[41] later employed a small, tetravalent lysine-based scaffold for the attachment of cyclized polypeptides and thus created a new type of multiple antigen peptide (MAP). Cyclic antigens were chosen due to their structural similarity to native protein, as a result of which the induced antibodies could be expected to show an enhanced affinity for the native protein as compared to those induced by linear analogs. Hamilton et al.[25] also constructed four peptide loops on the rim of a calix[4]arene, thereby creating an antibody mimic for protein surface recognition.

Grandjean et al.[42] used solid-phase synthesis to prepare L-lysinyl-based structures possessing 4, 8, and 16 termini, which were subsequently terminally modified with D-(–)-quinic and (–)-shikimic acids as potential *C*-lectin ligands. Similar fluorescein-labeled "cluster glycosides" have been reported.[43] Boons et al.[44] recently delineated a highly convergent, chemoselective glycosylation strategy for the assembly of phytoalexin-elicitor-active branched oligosaccharides.

A sulfonic acid-terminated, lysine-based manifold has been developed[45] for use as a protic dopant in polyaniline emeraldine base conducting films. Doping with these dendrimers afforded films with conductivities reportedly up to 10 S cm^{-1}, while a sulfonic acid-coated C_{60} dopant led to remarkable ambient temperature conductivities up to 100 S cm^{-1}.

Poly(lysine)-based MAPs have been employed for the preparation of biomineralization templates with the ultimate goal of developing antimalarial treatments.[46] Analogous poly(L-lysine) cascades,[47] also accessed via Fmoc-based chemistry, have been used to create poly(ethylene glycol)block–poly(L-lysine) dendrimers that form globular, water-soluble polyionic complexes with DNA.

The poly(lysine) wedge has been attached to the ends of a PEG chain to produce a triblock copolymer,[48] self-assembly of which with plasmid DNA was demonstrated. The complexes were observed by AFM to be nearly spherical, with diameters in the range 50–150 nm; *in vitro* cytotoxicity[48a] was examined.

Grandjean, Melnyk et al.[49] described the surface modification of a lysine framework attached to an antigen; a single-pot, double-orthogonal reaction was reported to give the desired glycodendrimers in respectable yields.

The recent creation of a "loligomer 4", a peptide-based intracellular vehicle that is able to penetrate eukaryotic cells and self-localize in the nucleus, has been reported.[50, 51] The *C*-terminal region consists of a five amino acid moiety and the *N*-terminal arms are attached through a branched lysine dendrimer. The term "loligomer" was coined for these multi-tasking agents.[52]

Birchall and North[53] reported lysine monomers as branch sites for the synthesis of block copolymers possessing enantiomerically pure amino acids; 4th generation materials were reported.

Ito et al.[54] generated novel dendritic "mastoparans", a wasp venom toxin of the sequence Ile–Asn–Leu–Lys–Ala–Leu–Ala–Ala–Leu–Ala–Lys–Lys–Ile–Leu–NH$_2$, based on a lysine–aminobutyric acid branched framework. The most potent construct, an octameric mastoparan, was found to be 8,000 times more active than mastoparan alone.

Smith[55] prepared lysine-based dendrons with carboxylic acid focal groups and used them to solubilize a polybasic dye (proflavine hydrochloride). The optical properties were found to be regulated by dendron generation; UV absorption and fluorescence intensity were observed to increase with increasing size.

Nishino and coworkers synthesized[56,57] and evaluated systems incorporating crowded (8 to 32) porphyrin rings attached to a dendritic poly(L-lysine) scaffold. Their synthetic endeavors coupled with CD and NMR studies paint an interesting picture of the novel interactions.[58]

Other reports involving the use of related lysine building blocks for dendritic scaffold construction include the preparation of (1) a lipid–core–peptide system,[59] (2) an octameric synthetic HIV-1 antigen,[60] and (3) glucodendrimers such as thiolated glycosides.[61, 62]

Scrimin et al.[63] reported the preparation of a small, three-directional polypeptide that proved useful for the modulation of membrane permeability. Decapeptide fragments were attached to a tren [tris(2-aminoethyl)amine] core.

Sharpless et al.[64] constructed a series of chiral dendrimers by employing the "double-exponential" synthetic method.[65] Chirality was induced through the use of aryl acetonide monomers, prepared by asymmetric dihydroxylation of the corresponding prochiral alkenes. Each monomeric unit possessed two asymmetric carbons. Examples described are C_3-symmetric and contain up to 45 chiral building blocks with 24 acetonide termini.

7.1.1.2   1 → 2 *C*-Branched, *Amide & Ester* Connectivity

Kress, Rosner, and Hirsch[66] undertook the synthesis of dendrimers resembling depsipeptides; branching groups consisted of (*R,R*)-, (*S,S*)-, or *meso*-tartaric acid, while spacer moieties were derived from di- and tripeptides of either glycine, L-alanine, or L-leucine. Numerous combinations of the chiral components afforded different dendrons up to the 3rd generation.

7.1.1.3   1 → 2 *N*-Branched & Connectivity

Meijer et al.[67] constructed a series of novel "dendritic boxes" by surface modification of divergently prepared, amine-terminated PPI dendrimers[68] (Scheme 7.5; e.g., **17**) by the attachment of *N*-protected chiral amino acid caps (see Section 3.3.1.1). Thus, treatment of polyamine **17**, possessing 64 terminal amino groups, with the *N*-hydroxysuccinimide activated ester of *N*-(*tert*-butyloxycarbonyl)phenylalanine **18** afforded the capped dendrimer **19**.[69] Using this procedure, the amino-terminated cascade family from generations 1 through to 5 was converted to the corresponding protected amino acid derivatives, e.g., L-alanine, L-leucine, L- and D-phenylalanines, L-tyrosine, and L-tryptophan. The 5th generation dendrimer, possessing 64 amino groups, was also coated with other chiral amino acid moieties (e.g., L-*t*-Bu-serine, L-Tyr-cysteine, L-*t*-Bu aspartic ester) by similar technology.[70–72]

Although the optical properties of the dendrimers, such as specific rotations, were not initially discussed,[70, 71] the possibility of racemization during amino acid acylation was investigated. Amino acid isolation and HPLC analysis following acidic hydrolysis of the asymmetric dendrimers revealed a > 96 % enantiomeric excess. Other aspects of these macromolecules that were discussed included T_gs, spin-spin (T_2) and spin-lattice (T_1) relaxation times, and molecular inclusion through the dense-packed entrapment of guest molecules. A molecular dynamics study of generations 1 through to 5, modified at the periphery with *N*-*t*-Boc-L-phenylalanine, has been conducted.[73] The shape was found to be dependent on generation level, with higher generations being more spherical, which corresponds well with findings concerning the architecture of the PAMAM series. Results further suggested end-group self-inclusion within the inner shell, while an observed decrease in optical activity was attributed to conformational disorder at higher generations.

Vögtle et al.[74] prepared the 1st through to the 3rd generation PPIs (Figure 7.4; **20–22**) possessing planar chiral 4-hydroxy[2.2]paracyclophane termini. Peripheral modification was effected by reaction of the 5-formylparacyclophane with the amine termini to afford the corresponding imine-connected cyclophane unit. These materials were constructed for further investigation as multi(cobalt-salen)-type derivatives,[75] which are known to be efficient catalysts of oxygen-transfer reactions, and as asymmetric poly(manganese-salen) complexes, which are known to promote high enantioselectivity.[76–78] Furthermore, since non-dendritic manganese-salen complexes have been reported to split water,[79] the potential use of such dendrimer-supported catalysts in this context was suggested. Circular dichromism spectra were discussed depending on generation number.

Based on the reaction (BF$_3$·OEt$_2$, CH$_2$Cl$_2$) of β-thiopropionic acid with D-galactose β-pentaacetate and lactose octaacetate, followed by transformation of the resulting saccharide-substituted carboxylic acids to their corresponding *N*-hydroxysuccinimide activated esters, Meijer, Stoddart, and coworkers[80] prepared a series of sugar-coated

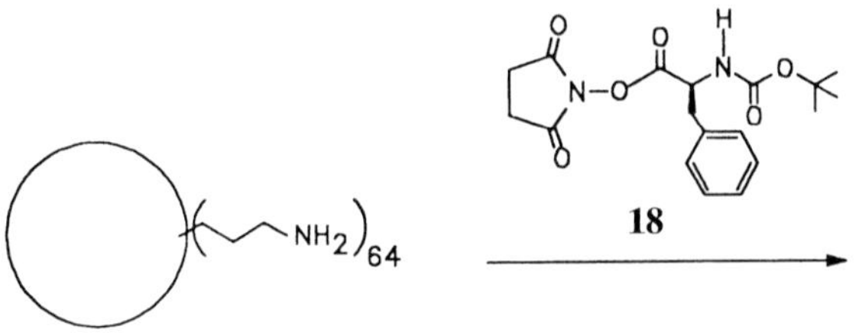

Scheme 7.5 Construction of a "dendritic box" by peripheral modification through reaction with chiral amino-protected activated esters[67] (R = benzyl; hydrogen atoms have been omitted for clarity).

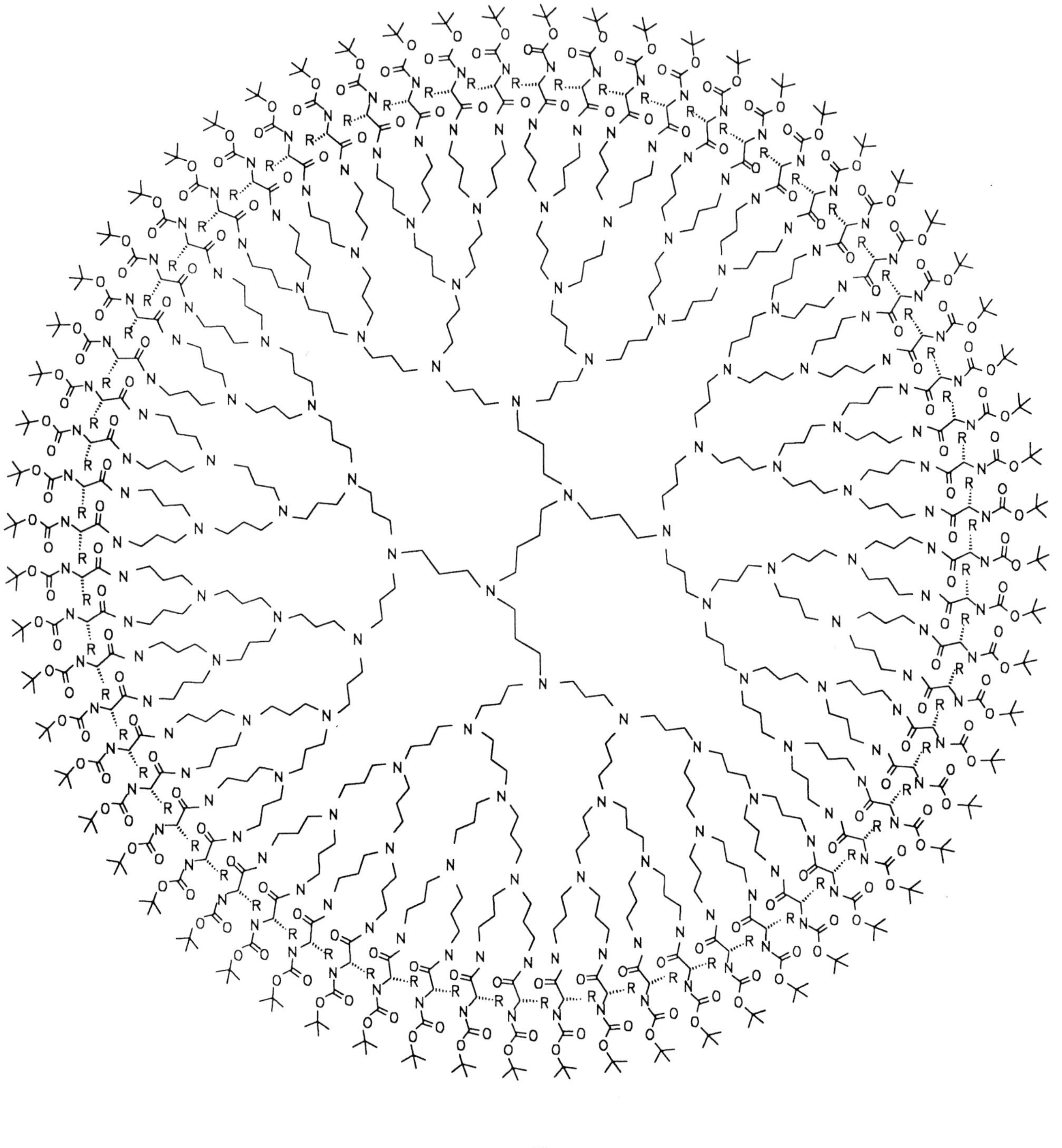

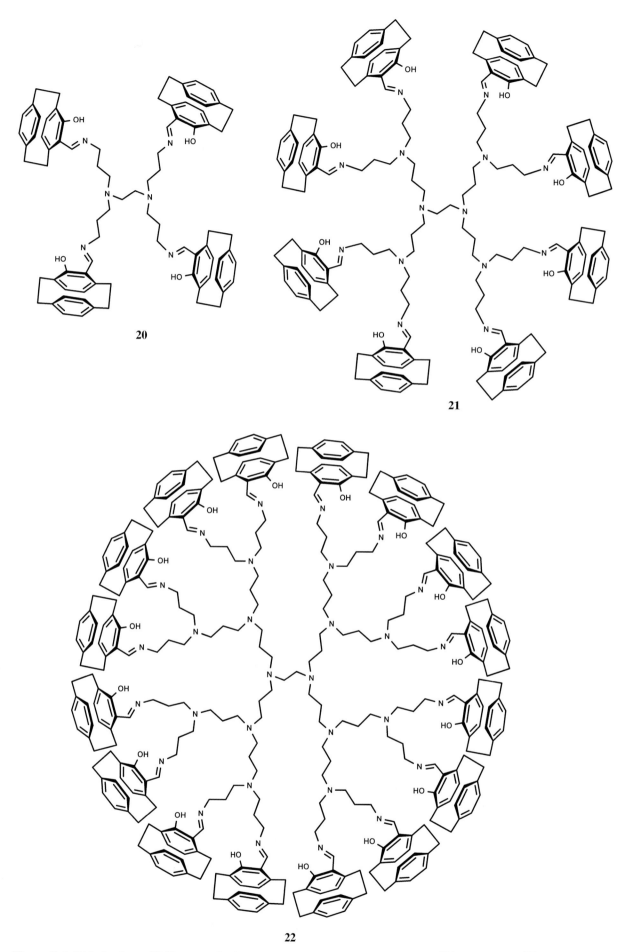

20

21

22

Figure 7.4 Chiral, planar [2.2]paracyclophane termini modification of Astramol™ dendrimers.[74]

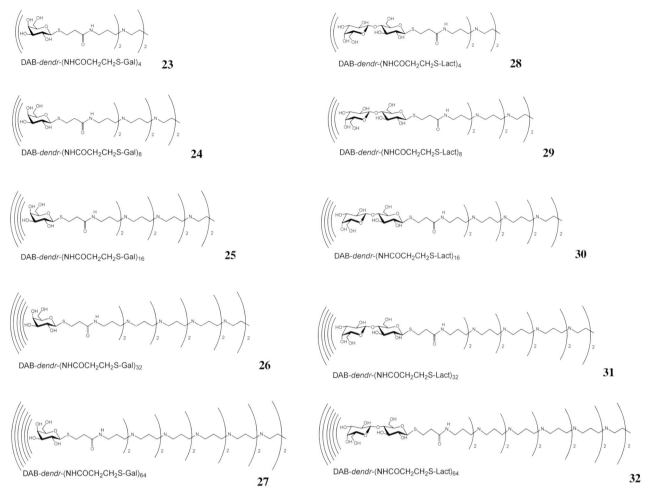

Figure 7.5 Glycodendrimers based on the PPI framework.[80]

dendrimers.[81] Treatment of the mono- and disaccharide activated esters with successive 1st–5th generations of PPIs followed by deacetylation of the terminal sugar moieties afforded the corresponding glycodendrimers (Figure 7.5; **23–32**). Specific optical rotations were determined for each acetylated and deprotected member [**23–27** (acetylated) –4.5, –4.8, –4.9, –5.8, –5.5; (deacetylated) +8.2, +5.1, +6.3, +5.6, +7.5; **28–32** (acetylated) –5.5, –7.2, –8.0, –8.5, –7.7; (deacetylated) +0.5, 0, +0.2, 0, +1]. Calculated carbon/sulfur ratios were compared with those determined microanalytically as a check of terminal reaction completeness. Saccharide residues have been attached to generations 1–5 of PPIs possessing connective spacer lengths of 1, 5, and 10 methylene units in order to evaluate local saccharide surface concentration effects.[82] The hydrodynamic properties of PPIs coated with β-thiolactosyl residues have been evaluated by velocity sedimentation, translational diffusion, and viscosity measurements in saline media.[83]

Oligosaccharide-derivatized PPIs have been shown to inhibit the adherence of cholera toxin and *E. coli* enterotoxin to cell surface GM1 {monosialoganglioside Gal(β1 → 3)GalNAc(β1 → 4)[sialic acid (α2 → 3)]Gal(β1 → 4)Glcβ1-ceramide}. This result was attributed to the radial distribution of the clustered oligosaccharides acting as an artificial micelle.

Magnusson et al.[84] reported small di-, tri-, and tetravalent dendritic galabiosides that inhibit *Streptococcus suis* hemagglutination at nanomolar concentrations. These 1st generation dendrimers were found to be several hundred orders of magnitude more efficient than the corresponding monomeric materials. A combined chemical and enzymatic preparation of a branched *C*-glycopeptide has been reported,[85] along with an analysis of its glycoamidase inhibitory activity. Fructose modified constructs are also known[85a] as well as cell targeting galactoside-based PPI_9.[85b]

7.1.1.4   1 → 2 *N*-Branched, *Amide* Connectivity

As a result of the commercial availability of the PAMAM family, numerous chiral applications involving its use as molecular scaffolding have appeared.

Okada et al.[86] reported the persubstitution of PAMAMs with either lactose or maltose derivatives. Treatment (DMSO, 27–40 °C, under N_2) of PAMAM (e.g. **33**; Scheme 7.6) with an excess of either *O*-β-D-galactopyranosyl-(1 → 4)-D-glucono-1,5-lactone (**34**) or the related *O*-α-D-glucopyranosyl-(1 → 4)-D-glucono-1,5-lactone afforded the surface-coated cascade **35**, which was ascertained to be a single component by SEC and characterized by IR and ^{13}C NMR. According to ^1H NMR and vapor-pressure osmometry measurements, the surface of the PAMAM was *"almost quantitatively"* coated with the sugar residue. These "sugar balls" were found to be water-soluble, but insoluble in EtOH and $CHCl_3$. The molecular recognition properties of the sugar groups were investigated; Concanavalin A showed a strong interaction with the gluco-coated sugar ball, but showed little or no interaction with the galacto-coated surface. Apparent aggregation of the lactose-sugar ball with peanut agglutinin (*Arachis hypogaea*) was demonstrated, whereas no interaction was observed with the maltose derivative.

In a complementary paper, Aoi[87] reported the preparation of "sugar balls" by the attachment of saccharides to the surface of amine-terminated PAMAM dendrimers. Aoi, Itoh, and Okada[88] further described the synthesis of PAMAM-based dendrons with focally-protected amines (Scheme 7.7). Reaction of the terminal amines with either maltono lactone **36** or anhydride **37** afforded the corresponding "hydrophilic block" **38** and "hydrophobic block" **39**, respectively. Following tandem focal deprotection (H_2, Pd/C) and reaction of the liberated hydrophobic amine with trichloromethyl chloroformate to generate the corresponding acid chloride focal group **40**, the two blocks were connected to give the AB-type surface block dendrimer **41**. The ability of the sugar-coated portion to bind Concanavalin A lectin demonstrated the potential utility of these materials as cell-recognizable biomedical tools.

Okada et al.[89] later reported the preparation of additional sugar balls by the reactions of *O*-(tetra-*O*-acetyl-β-D-glucopyranosyl)-L-serine α-amino acid *N*-carboxyanhydride (NCA) and *O*-(2-acetamido-3,4,6-tri-*O*-acetyl-2-deoxy-β-D-glucopyranosyl)-L-serine α-amino acid (NCA) with 1st–3rd generations PAMAMs. Deacetylation to give the final products was accomplished using hydrazine hydrate. Similar sugar balls, prepared by rapid growth polymerization of sugar-modified α-amino acid *N*-carboxyanhydrides on the surface of PAMAMs, have also been reported by Aoi et al.[90]

In a related communication, Lindhorst and Kieburg[91] described "glyco" amine-terminated PAMAMs (0th and 1st generations) constructed with a divalent diamino-ethane core, as well as from the trivalent tris(2'-aminoethyl)amine with six termini. Isothiocyanate-functionalized *glyco-*, *manno-*, *galacto-*, *celobio-*, and *lacto-*glycosides were reacted with the amine termini to give the corresponding thiourea-attached clusters. Notably, these materials are reminiscent of Hanessian's[92] "megacaloric cluster compounds". Deprotection of the acetyl groups followed by lyophilization afforded the dendrimers, as white powders.[93] Kieburg and Lindhorst[94] also reported the reaction of water-soluble, unprotected saccharides bearing –NCS moieties with PAMAMs in aqueous solution, thereby eliminating the need for final deprotection. Recently, orthogonally protected glucose-based AB_2 monomers have been reported,[95] which allow the assembly of branched glycopeptidomimetics under peptide coupling conditions.

Pagé and Roy[96] reported the synthesis of large-scale quantities of mannopyranoside-terminated PAMAMs and then evaluated their biological properties. Realization of the 1st–4th generations was achieved by reaction of *p*-isothiocyanatophenyl 2,3,4,6-tetra-*O*-acetyl-α-D-mannopyranoside with the peripheral amines. Referenced against monosaccharides, the binding properties were evaluated using yeast mannan and peroxidase-labeled lectins; the PAMAM-based systems were found to be potent inhibitors (IC_{50} values > 400 times greater than that of monomeric methyl α-D-mannopyranoside). The binding of lectin led to insoluble complexes and selective binding of the mannose-binding protein Concanavalin A in a lectin mixture was demonstrated. Methods for oligosaccharide assembly based on cyclohexane-1,2-diacetal chemistry, including a single-pot procedure, have also been reported by Grice et al.;[97] cluster lactoside telomers have been generated.[98, 98a, b]

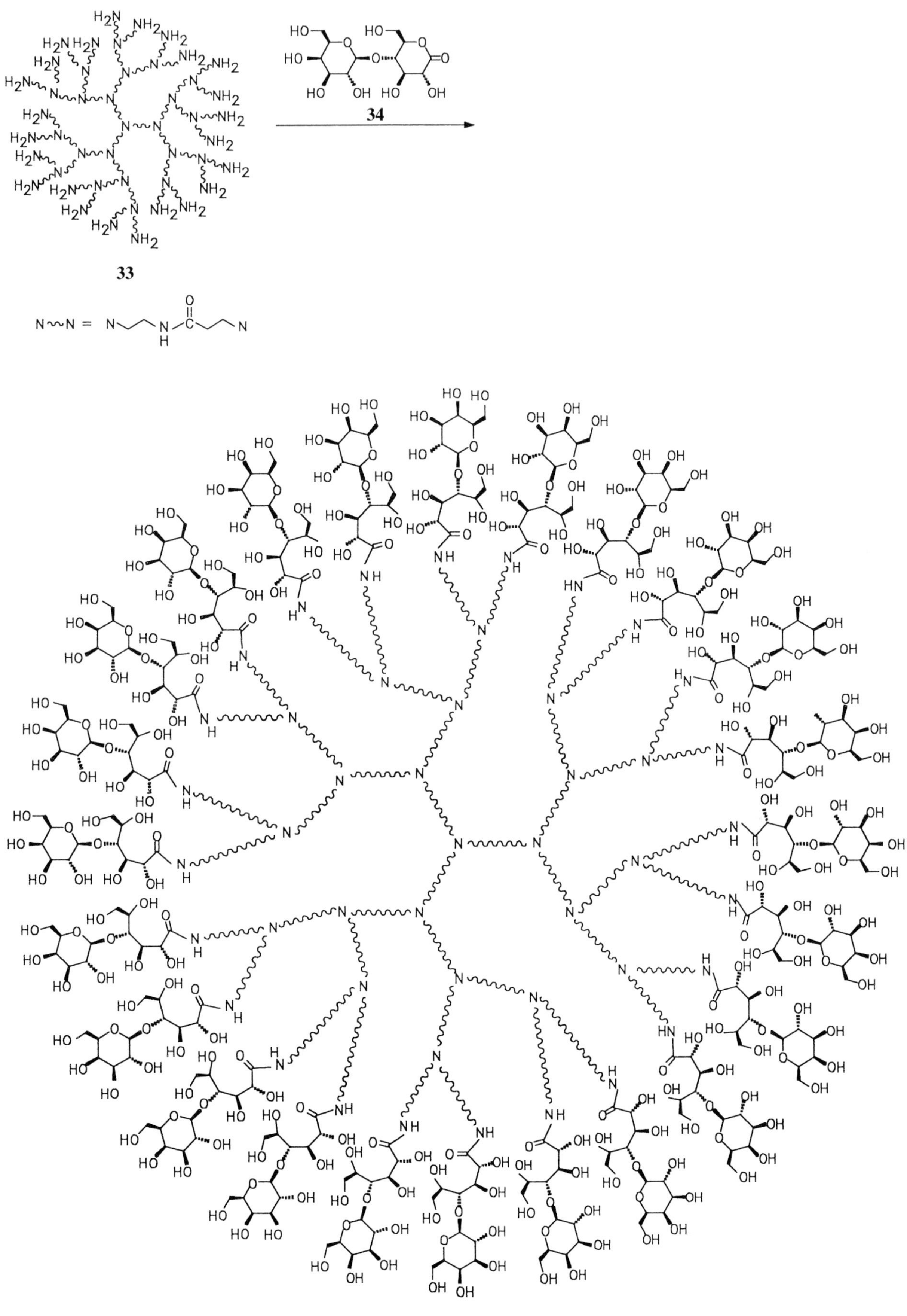

Scheme 7.6 Synthesis of "sugar balls" **35** by treatment of PAMAMs with galacto- or gluco-pyranosyl lactone derivatives.[86]

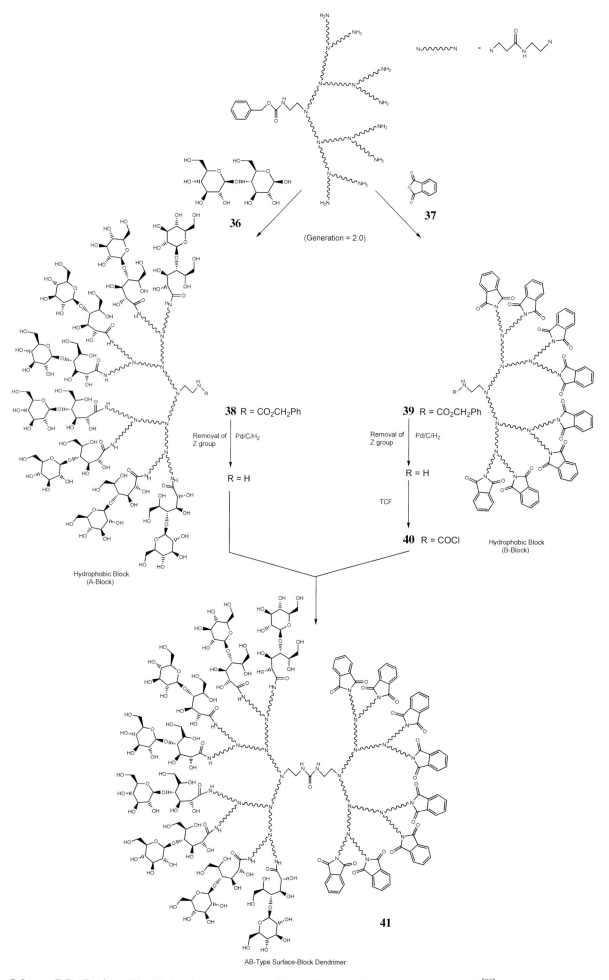

Scheme 7.7 "Surface-Block" dendrimers prepared by a divergent/convergent approach.[88]

Rico-Lattes et al.[99] presented an example of high asymmetric induction at the "pseudo-micellar" interface of a glucose-modified PAMAM. Chiral alcohols were obtained (NaBH₄) from their corresponding dendrimer-coordinated ketones in high yields (> 90 %) and with high *ee* (99.4 %) in the case of acetophenone reduction in THF. Moreover, this illustrates the potential of "heterogeneous" dendrimer-based asymmetric induction allowing filtration, recovery, and reuse of the catalyst. Rico-Lattes et al.[100] further demonstrated the micellar potential of their systems by solubilizing pyrene and aromatic ketones in water. Aggregates of these polysaccharides were observed by means of electron microscopy and light scattering. These chiral-coated polymers were also shown to effect asymmetric induction (NaBH₄ reduction of benzoylcyclohexane; 95 % yield; 50 % *ee*).

Lambert et al.[101] developed methods for the modification of amine-bearing surfaces with either peptides or carbohydrates; thus, coupling of the amine moieties with levulinic acid afforded terminal ketone groups available for peptidyl amine condensation, while end-capping with a Boc-aminooxyacetic acid followed by deprotection gave oxyamine termini capable of reacting with carbohydrates. Higashi et al.[102, 102a] modified a PAMAM surface with helical oligopeptides comprised of γ-benzyl-L-glutamic acid and demonstrated a significant enhancement of the helicity, which was postulated to arise from templated close assembly resulting in a hydrophobic effect that stabilizes the helices.

Llinares and Roy[103] demonstrated the solid-phase preparation of the 3rd generation *N*-linked, α-sialo-based dendrimers using Wang resin. Roy and Kim[104] also constructed carbohydrate-coated dendrons atop the hydroxylated rim of a calix[4]arene for use as coating antigens on a hydrophobic surface (e.g., polystyrene) in solid-phase immunoassays.

Soai et al.[105] modified PAMAM surfaces with chiral amino alcohols; thus (1*R*,2*S*)-*N*-(4-formylbenzyl)ephedrine was condensed with the terminal amines of generations 0 and 1, then reduced (NaBH₄) to afford the desired chiral constructs. Use of these poly-(asymmetric ligands) for enantioselective Et₂Zn addition to *N*-diphenylphosphinylimines resulted in *ee*'s of up to 93 %.

Through a sequence of perallylation, ozonolysis, bis(*N*-benzylamination), and benzyl reduction, Dubber and Lindhorst[106] transformed D-glucose to the pentaamino-modified sugar. This was subsequently subjected to the PAMAM protocol to produce dendrimers with chiral glucose cores. Specific rotation values were found to decrease with increasing generation, while molar rotation values remained in a narrow range (ca. 400). The recent preparation[107,108] of glycopeptide dendrons through the coupling of AB₂ carbohydrate moieties led to a family of carbohydrate-containing dendrons prepared in either a divergent or convergent manner; this procedure can be easily adapted to either automated techniques or combinatorial approaches.

Zimmerman et al.[109] reported the creation of optically active dendrons based on the generation of a chiral monomer (Scheme 7.8), which could be accessed on a 100 g scale via a five-step process that included cyanoethylation followed by hydrogenation of either L-valine or L-leucine. Dendron synthesis was then realized starting with a glycine-modified poly(ethylene glycol) monomethyl ether resin **42**, which was deprotected (TFA, CH₂Cl₂) and treated (1-HBT, DIEA, DMF, CH₂Cl₂) with the key building block **43** to afford the protected diamine **44**. Repetition of the sequence yielded the higher generation materials **45** and **46**, as well as a 4th generation wedge. At each stage, any unreacted amine moieties were capped with Ac₂O; cleavage from the resin was effected with dilute methanolic NaOH. Characterization of these poly(peptide)s included optical microscopy, DSC, and measurements of specific as well as molar optical rotations. The protocol was envisaged as potentially leading to novel biomaterials based on the combinatorial use of additional amino acids.

Similar "caged" lencyl leucine methylesters have been attached to the PAMAM surface.[109a]

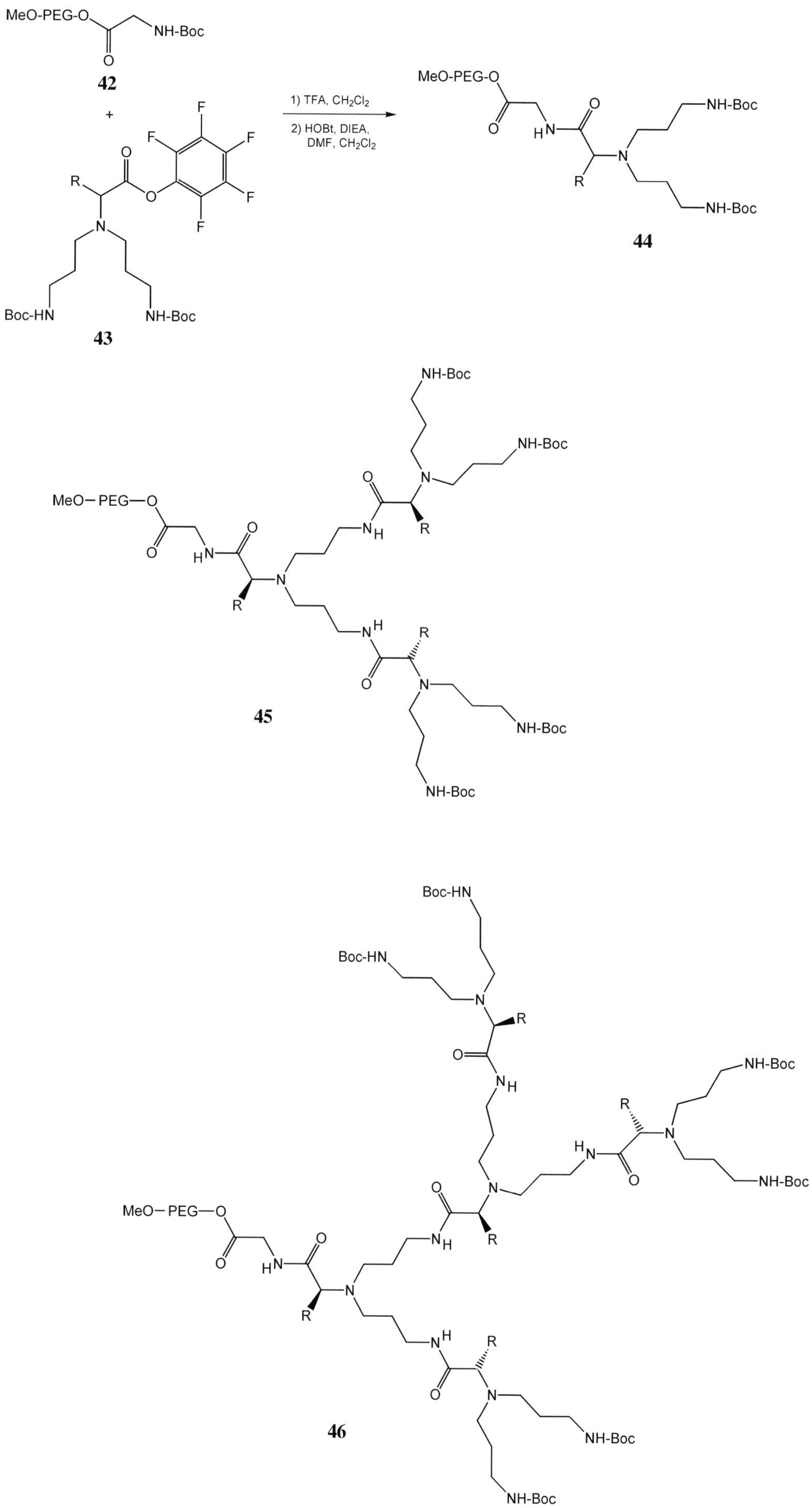

Scheme 7.8 Zimmerman's optically active dendrimers.[109]

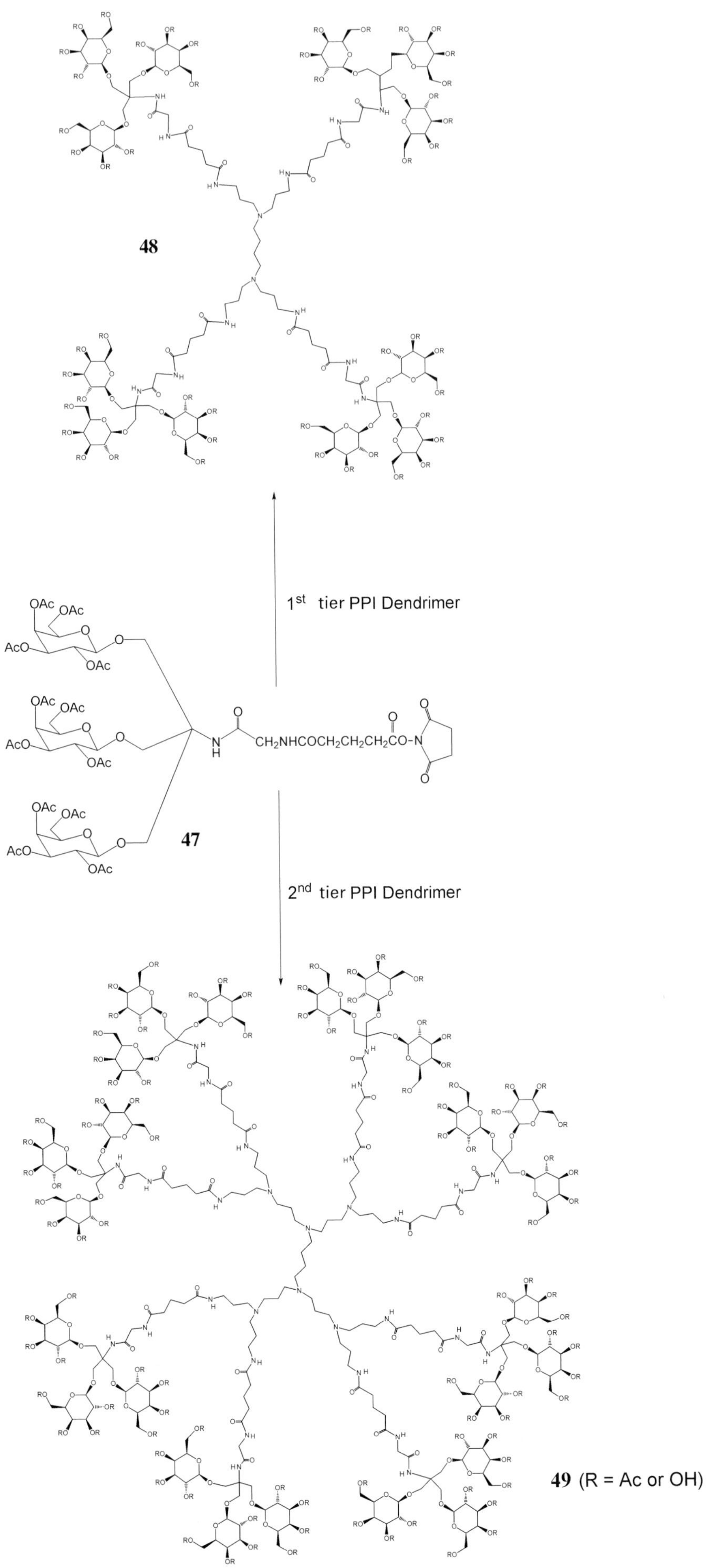

48

1st tier PPI Dendrimer

47

2nd tier PPI Dendrimer

49 (R = Ac or OH)

Scheme 7.9 Tris-based terminated gluco-functionalization of PPI.[80]

7.1.1.5   1 → 2 *N*- & 1 → 3 *C*-Branched, *Amide* Connectivity

The use of tris in the preparation of a (1 → 3) trisaccharide monomer[110] for attachment to a PPI dendrimer has appeared.[80] Following modification of the tris(galactoside) monomer with glutaric anhydride and then *N*-hydroxysuccinimide to give the homologated monomer **47** (Scheme 7.9), reaction with the 1st and 2nd tier PPIs afforded the desired 12- and 24-galactoside-terminated dendrimers **48** and **49**, respectively, which were also subjected to deacetylation.

7.1.1.6   1 → 2 *Aryl*-Branched, *Ether* Connectivity

Seebach et al.[111] prepared a series of small dendrimers terminated with α,α,α',α'-tetraaryl-1,3-dioxolane-4,5-dimethanol units (TADDOL's) for Ti and Al coordination (Figure 7.6). Metal complexes of these dendrimers (**50–53**) were employed as chiral aux-

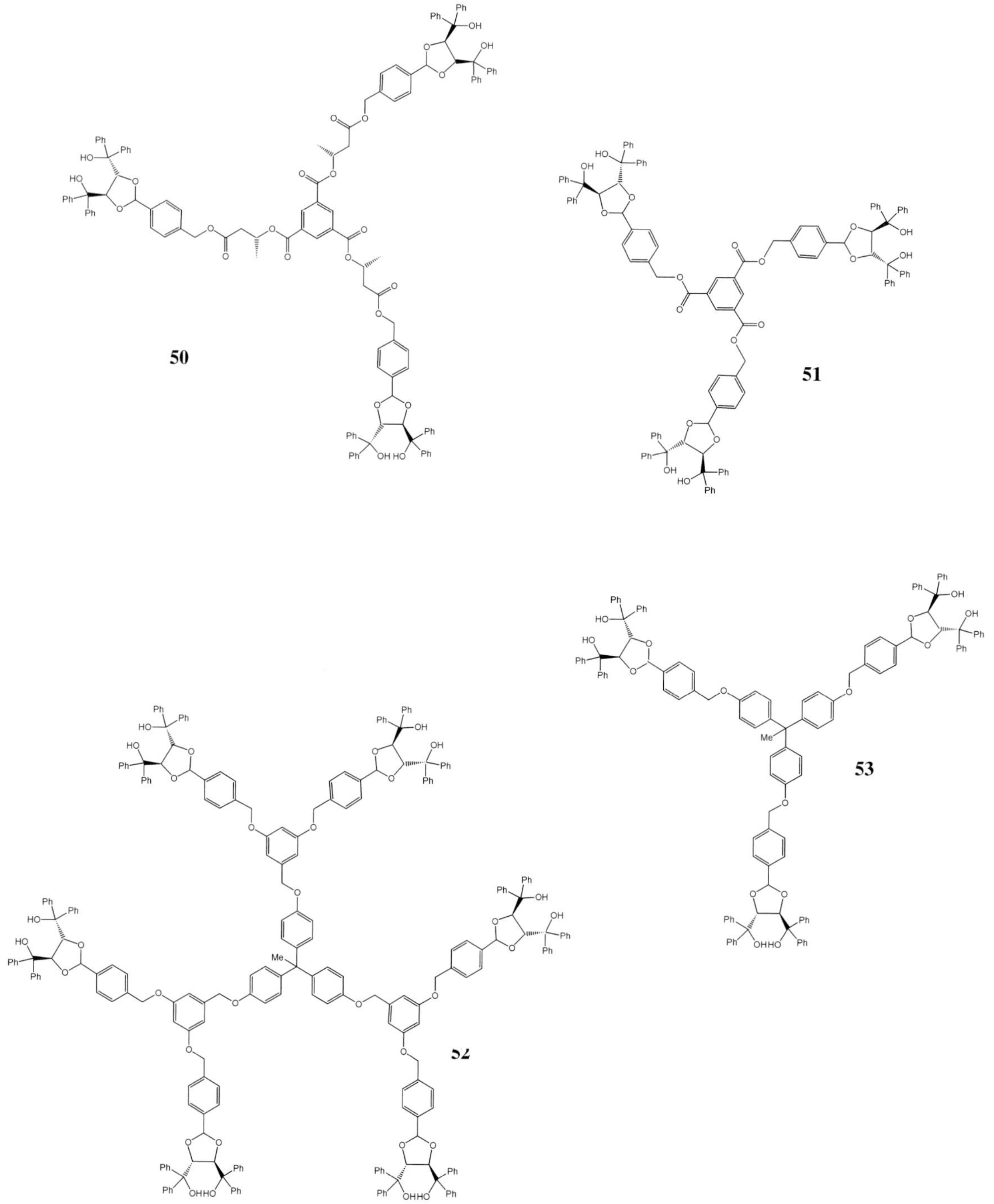

Figure 7.6 Seebach's tetraaryl dioxolane dimethanols (TADDOLs) that can coordinate Ti and Al.[111]

iliaries in enantioselective nucleophilic additions to aldehydes and ketones, and in the ring-opening of *meso*-anhydrides. Enantioselectivities were observed to correspond well with those achieved using the soluble analogs. The ease of recovery and reuse of the catalyst was described. Similar styryl-terminated TADDOLs have been constructed as polymer cross-linkers.[112] Copolymerization with styrene afforded polymer beads, which were treated with Ti[OCH(CH$_3$)$_2$]$_4$ and used for enantioselective Et$_2$Zn additions to benzaldehyde. Similar studies on catalytic TADDOLs were conducted using a poly(aryl ether) support,[113] while dendritically cross-linked polystyrene-bound moieties afforded Ti-TADDOLate catalysts.[114] Higher catalytic activities were observed compared to those seen for classical, rather insoluble, polymer-based Ti-TADDOLates, which approached the activity of freely soluble catalysts. A brief overview[115a] of TADDOLs as enantioselective catalysts, dendritic cross-linkers, and cholesteric liquid crystals and dendritic binaphthols[115a] have appeared.

7.1.1.7   1 → 2 *Aryl*-Branched, *Ester & Amide* Connectivity

Seebach et al.[116, 117] prepared a series of chiral aryl dendrimers (Scheme 7.10) from chiral triol cores by treatment with 3,5-dinitrobenzoyl chloride 56 to afford the 1st generation hexanitro cascade 54, which was catalytically reduced to give hexaamine 55. Reaction of this hexaamine with the same aroyl chloride (56) gave rise to the dodecanitro dendrimer 57 (83 % with R = isopropyl). A notable change in the molar rotation was observed on proceeding from the 1st tier hexanitro dendrimer 54 [$\Phi_D^{rt} - 124°$] to the 2nd tier dodecanitro dendrimer 57 [$\Phi_D^{rt} + 155°$], suggesting that the chromophoric conformational asymmetry at the dendritic surface makes a significant contribution to the optical activity. These amino- and nitro-terminated dendrimers were further shown to act as hosts in the formation of host–guest complexes by the inclusion of EtOAc, MeOH, and Me$_2$CO.

7.1.1.8   1 → 2 *Aryl*-Branched, *Ether & Ester* Connectivity

Ringsdorf et al.[118] reported the single-pot preparation of liquid-crystalline hyperbranched polymers possessing terminal chiral groups. Preparation of the branched polymer (Scheme 7.11; 58) utilized two different monomers: an AB$_2$-type monomer 59, incorporating a mixed anhydride and two mesogenic biphenyl acetate moieties, and an end-capping 3,4-disubstituted benzoyl chloride (60). Both of these were prepared from methyl 3,4-dihydroxybenzoate, in five and three steps, respectively (see Section 6.2.9).

The chiral polymer showed good solubility in typical organic solvents, e.g., CHCl$_3$, toluene, and THF; it was further determined by gel permeation chromatography to pos-

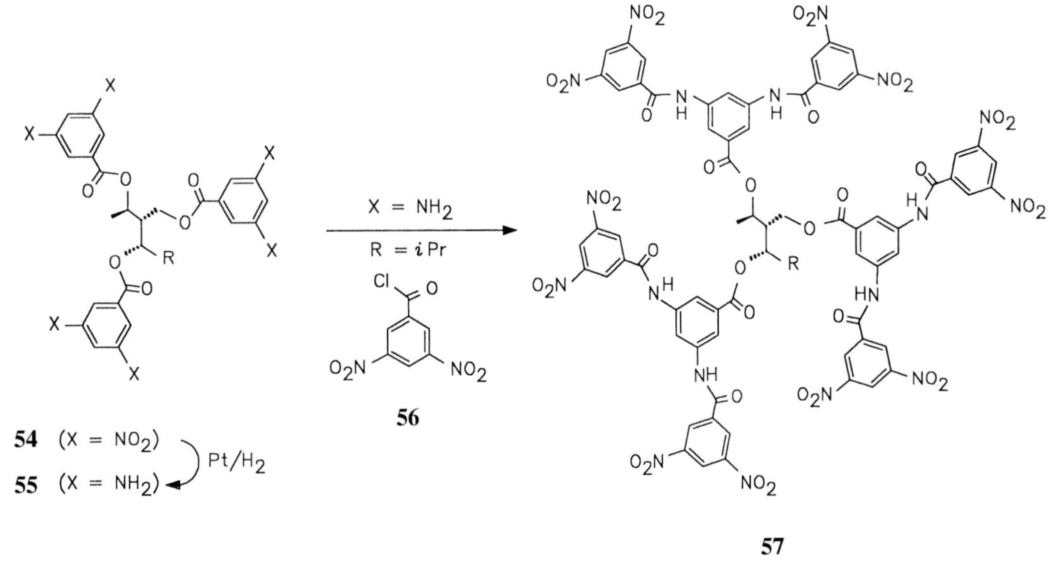

Scheme 7.10 Divergent construction of polyaramide dendrimers possessing a chiral, ester-based core.[116, 117]

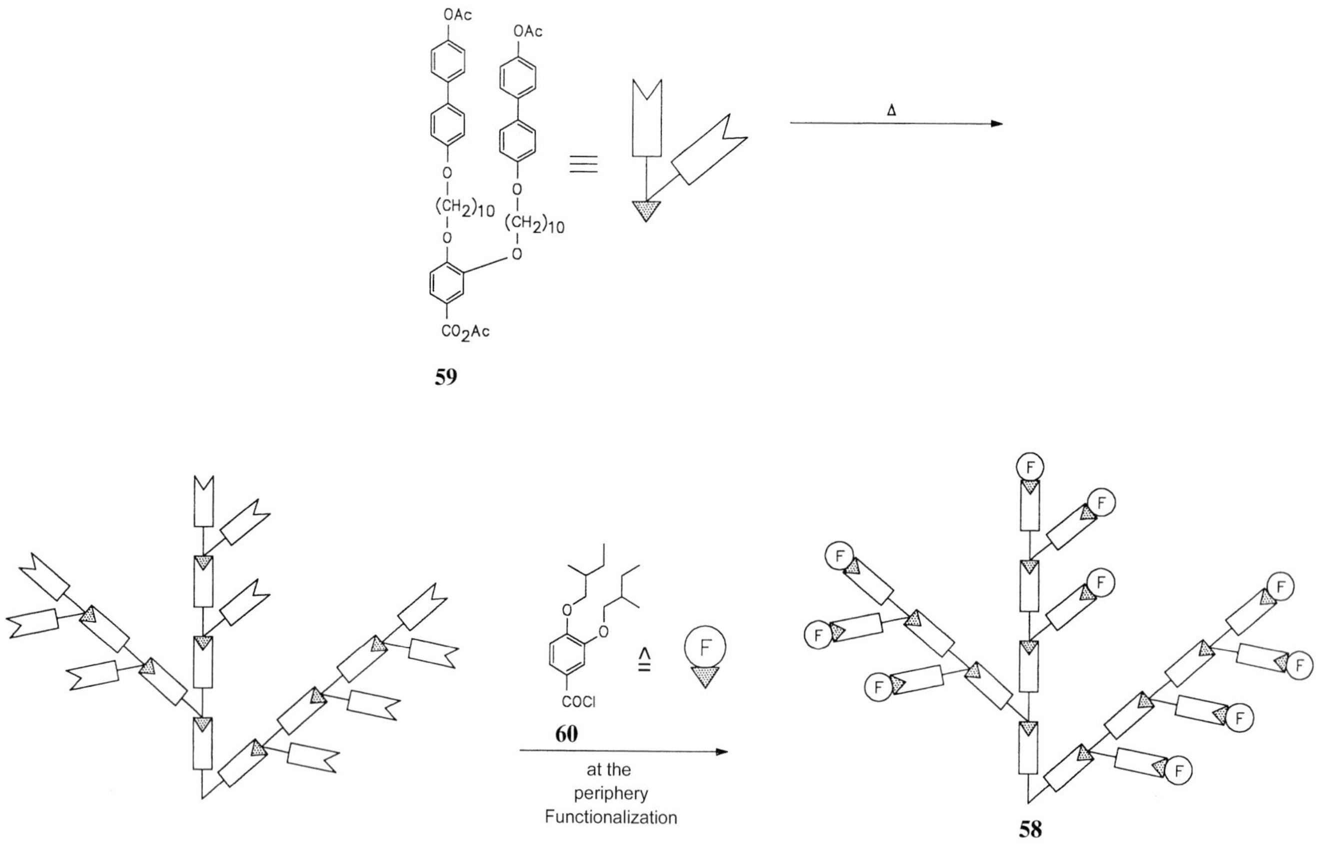

59

60

at the
periphery
Functionalization

58

Scheme 7.11 Preparation of a hyperbranched liquid crystalline polymer possessing mesogenic monomers and chiral termini.[118]

sess an intermediate degree of branching ($P_n = 6$). Its optical rotation was determined by polarographic measurements in CH_2Cl_2 to be $[\alpha]_D^{20} = +6.8°$. Wide-angle X-ray scattering patterns indicated an average mesogen–mesogen distance of 4.4 Å, while small-angle reflections suggested a spacing between the cholesteric phase layers of 28 Å.

7.1.1.9   1 → 2 *Aryl*-Branched, *Ether & Amide* Connectivity

Roy et al.[119] described the construction and lectin-binding properties of a branched mannopyranoside (Scheme 7.12). The core scaffolding was prepared by coupling of methyl 3,5-dihydroxybenzoate with two equivalents of $TsOCH_2CH_2OCH_2CH_2N_3$ followed by treatment with tris(2-aminoethyl)amine to give the tris(amide) core, and the terminal azide groups were reduced to amines. Reaction of thioisocyanate isocyanate **61** with the hexavalent core **62**, followed by deacetylation (NaOMe), gave the desired polysugar (**63**). The mannopyranoside-terminated dendrimer exhibited a sharp precipitin band in a double immunodiffusion assay using lectin Concanavalin A; enzyme-linked lectin assays were also performed. Compared to the monosaccharide analog, inhibition of the binding of Concanavalin A to mannan was increased by a factor of 1.7.

7.1.1.10   1 → 2 *Ethano*-Branched, *Ether* Connectivity

Wong et al.[120] reported the solid-phase preparation of branched glycopeptides. A key step in the synthesis involved the connection of β-1,4-galactosyl monomeric units by treatment with β-1,4-galactosyltransferase. Enzymatic fucosylation after α-chymotrypsin-mediated cleavage of the linear glycopeptide from the silica-based solid phase afforded the sialyl Lewis^X [121] glycopeptide (Figure 7.7; **64**).

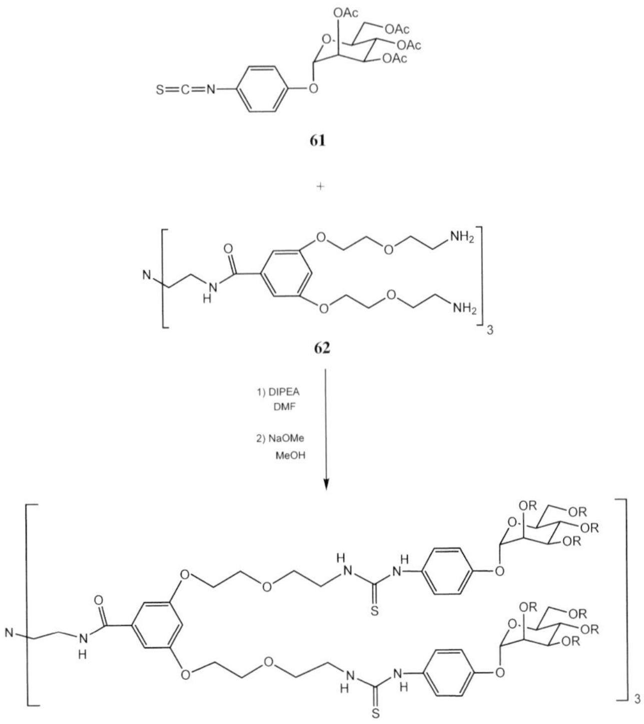

Scheme 7.12 Roy's branched mannopyranosides.[119]

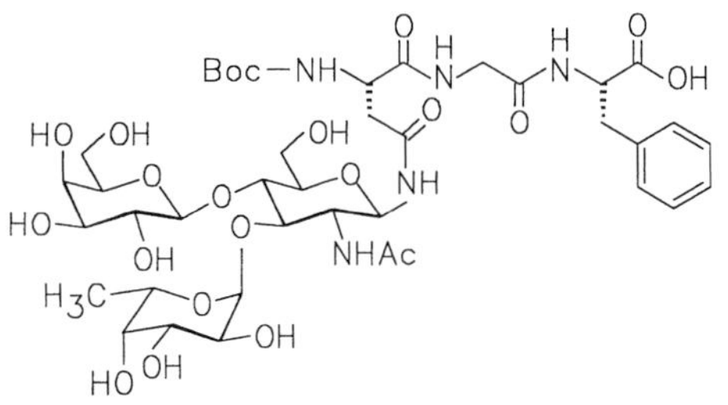

Figure 7.7 Solid-phase synthesis[120] afforded "sialyl Lewis^X" glycopeptide **64**.

7.1.2 1 → 3 Branched

7.1.2.1 1 → 3 *C*-Branched, *Ether & Amide* **Connectivity**

In 1991, Newkome and Lin[122] reported a series of chiral poly(ether-amido) dendrimers (e.g. **65**) based on a tetraacid core and terminated with the amino acid monomer **66** (Scheme 7.13). The resultant series was examined by ORD/CD; preliminary data indicated a linear relationship between optical rotation and the number of surface tryptophan moieties. Tryptophane focally-modified dendrimers are also known.[122a]

In a quest for novel amphiphilic dendritic galactosides[123] for the selective targeting of liposomes to the hepatic asialoglycoprotein receptor, Lin's amine (Chapter 4; **96**) was

C[CH₂OCH₂CH₂CONHC(CH₂OCH₂CH₂CONHC(CH₂OCH₂CH₂CO₂H)₃)₃]₄

i: ; DCC/1–HBT/DMF

66

R =

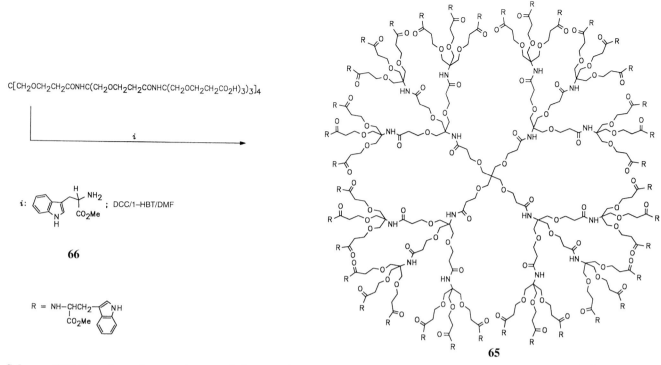

65

Scheme 7.13 Reaction of tryptophan methyl ester with a series of carboxylic acid terminated dendrimers.[122]

extended with *Z*-Gly and its ester units were saponified. This was followed by extension of the branched portion with *N*-Boc-1,3-diaminopropane. Treatment with a galactoside and sequential removal of the protecting groups gave the desired synthon, which was coupled to a functionalized steroid moiety.

7.1.2.2   1 → 3 *C*-Branched, *Ether* Connectivity

In 1993, Seebach et al.[124] reported the utilization of their chiral tris(hydroxymethyl)-methane derivatives[116] as building blocks for divergent dendrimer construction (Scheme 7.14). Analogs of polyol **67** were constructed by means of aldol additions of the enolate of (2R,6R)-2-*tert*-butyl-6-methyl-1,3-dioxan-4-one to various aldehydes, followed by reduction (LAH) and derivatization with functionalized halides bearing allyl, 4-(silyloxy) but-2-en-1-yl, or 4-substituted benzyl moieties, as well as pent-4-enoic and 3,5-dinitrobenzoic acid chlorides. Homologation of alcohol **67** was accomplished by treatment with allyl bromide (NaH, THF) to give triene **68**, which was followed by hydroboration (BH₃, THF) to afford triol **69**. Subsequent conversion to the corresponding trimesylate allowed formation of nonaester **70** by reaction (110 °C) with triethyl sodiomethanetricarboxylate.[125] However, the purification of ester **70** proved troublesome. The asymmetric triol **67** was also subjected to a Michael-type addition with acrylonitrile affording, albeit in low yield, trinitrile **71**. Hydrolysis and esterification smoothly furnished the tris-(methyl ester) **72**. Murer and Seebach[126] extended the utility of their asymmetric monomers to include the preparation of dendrimers (through to the 3rd tier) with doubly- and triply-branched architectures. Chirality was incorporated into the core as well as the superstructure. Greiveldinger and Seebach[127] also labeled these chiral monomers with trifluoromethyl substituents for use as NMR probes; ¹⁹F NMR showed "constitutional heterotopicity" of substituents separated by a 1,1'-biphenyl-4,4'-diyl spacer moiety. Simple pentaerythritol-based "cluster" α-D-mannosides were prepared;[128] the problems associated with glycosyl orthoester formation were solved. These clusters blocked the binding of *E. coli* to yeast mannan *in vitro* and were found to be in excess of 200 times more potent in inhibiting mannose-specific adhesion than simple α-D-mannoside. Le Narvor et al.[128a] has demonstrated chemenzymatic synthesis of oligosaccharides employing soluble dendritic support.

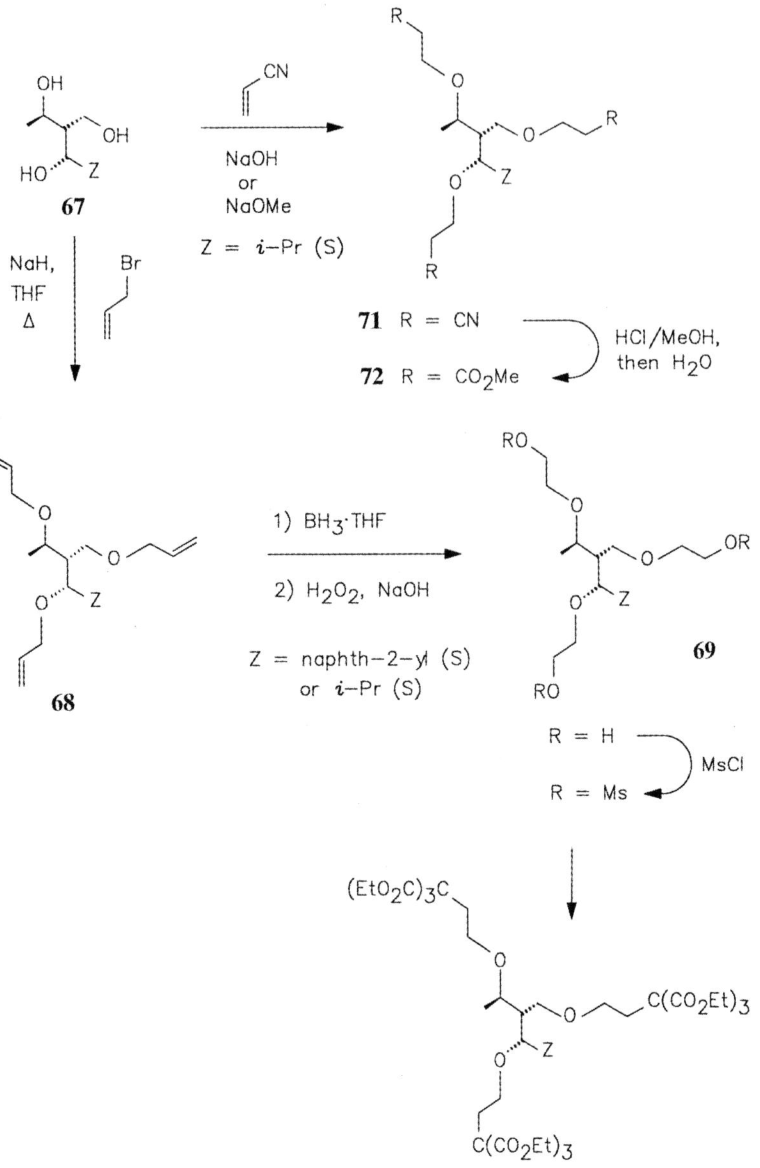

Scheme 7.14 Seebach et al.'s[124] chiral core preparation.

7.1.2.3   1 → 3 *Aryl*-Branched, *Ether & Amide* Connectivity

Roy et al.[129] reported the synthesis of hyperbranched dendritic lactosides based on gallic acid as the 1 → 3 branching center, and described the use of tetraethylene glycol as a hydrophilic spacer moiety. Roy et al.[130] later reported the synthesis of similar poly(α-thiosialoside)s, although convergent construction was employed.

7.1.2.4   1 → 3 *Si*-Branched, *Thio & Alkyl* Connectivity

Carbosilanes[131, 131a] bearing the trisaccharide moieties of globotriaosyl ceramide at their surfaces have been prepared. The more dendritic materials exhibited potent activity against the cytoxicity of Verotoxins 1 and 2.

7.1.2.5   1 → 3 *P*-Branched and *Phosphate* Connectivity

A novel approach to the synthesis of oligonucleotide dendrimers was presented,[132] in which the phosphoramidite-based synthon tris-2,2,2-[3-(4,4-dimethoxytrityloxy)propyl-oxymethyl]ethyl-*N,N*-diisopropylamino-cyanoethoxyphosphoramidite (Figure 7.8; **73**), prepared in seven steps starting from pentaerythritol, was used to prepare oligonucleo-

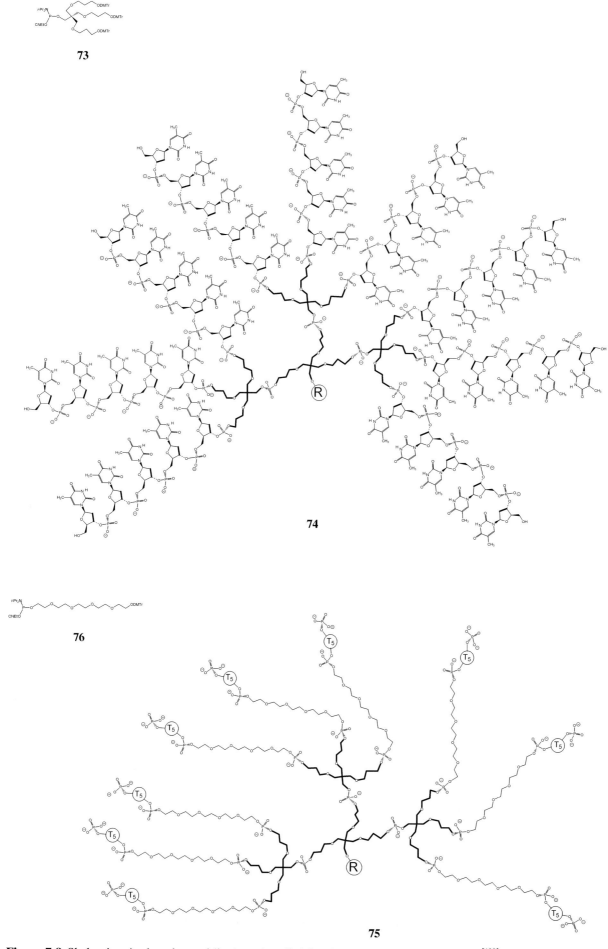

73

74

76

75

Figure 7.8 Shchepinov's phosphoramidite-based scaffolding for oligonucleotide support.[132]

tide-terminated dendron **74**. The phosphate-terminated analog **75** was then accessed by incorporation of spacer **76**. These materials exhibited signal enhancement when employed as probes in oligonucleotide (DNA) assays. The authors further demonstrated that these dendrimers could be incorporated into PCR (polymerase chain reaction) products, thereby significantly affecting the electrophoretic mobility of the latter. Shchepinov and Southern later reported[133] the synthesis of similar dendritic oligonucleotides incorporating the concept of "mixed oligonucleotides".

Shchepinov et al.[134, 135] synthesized seven generations of these poly(oligonucleotide)s on derivatized 250 nm controlled pore-size glass supports possessing either two, three, six, nine, or twenty-seven branches. The solution association properties between the solid-supported oligonucleotides and their complementary bases revealed enhanced stability relative to unbranched analogs of equal length. Their use in "bottom up" nanoassembly was described. Matsuda et al.[136] reported the synthesis of similar branched oligonucleotides using pentaerythritol as a branching junction; their thermal stabilization of triplex formation was studied.

Additional sugar-terminated branched structures of interest include a high mannose content nonamannan residue,[137] clustered glycopolymers,[138] and tris(sialyl Lewis *N*-glycopeptides).[139] Binder and Schmid[140] have also reported the synthesis of small carbohydrate-bearing molecules.

Di- and tetramannosylated tetrathymidylic acid homologues have been synthesized on a solid-support.[141] Pentaerythritol-based tetraols have been prepared[142] by allylation–hydroboration and attached to acetylated mannosides through phenylselenide displacement; trisaccharide glyco-clusters have also been prepared and connected to the ethereal core.

7.2 Convergently Generated Chiral Dendrimers

7.2.1 1 → 2 Branched

7.2.1.1 1 → 2 *C*-Branched, *Amide* Connectivity

Mitchell et al.[143] reported the synthesis of chiral dendrimers using L-glutamic acid as the monomer. Construction (Scheme 7.15) was accomplished by starting from *N*-benzyloxycarbonyl-protected L-glutamic acid, which was converted to the bis(activated ester) (**77**) by reaction (DCC, DMAP) with *N*-hydroxysuccinimide. Treatment of the activated ester **77** with L-glutamic acid diethyl ester (**78**) yielded tetraethyl ester **79**, which was debenzylated (ISiMe$_3$) to give the free amine **80**. Acylation of two equivalents of amino tetraester **80** with one equivalent of bis(succinimide ester) **77** resulted in the 2nd generation, i.e. the octaethyl ester, which was subsequently catalytically deprotected (H$_2$/Pd/C) to give the free amine **81**. This alternative method was employed when attempted cleavage (ISiMe$_3$) failed, presumably due to the increased steric hindrance caused by the growing dendrimer. Iteration then afforded the *N*-protected dendrimer **82** possessing 16 terminal ester moieties and 15 chiral centers, all with the identical L-configuration. No optical activity data were reported. These studies were later expanded to include poly-L-aspartic acid and poly-L-lysine motifs.[144]

Ranganathan and Kurur[146] also employed aspartic and glutamic acids as building blocks for branched scaffolding; a bifunctional adamantane core or 1,3,5-benzenetricarboxylic acid[147] was used. Interestingly, the 1st generation of glutamic acid-based dendrons incorporating the latter benzenetriacid core formed a cylindrical assembly; X-ray diffraction analysis showed the resultant benzene π-π stack to be held together by vertical *H*-bonding.[147] Polyglutamic dendritic porphyrins[148] (free-base *meso*-tetra-4-carboxyphenylporphyrin) were synthesized and characterized by NMR and MALDI-TOF MS; these dendrimerized porphyrins acted as fluorescent pH sensors in the biological pH range.

Nilsen et al.[149] described the use of a heterodimer of two single-stranded nucleic acid oligomers, comprised of a central double-stranded "waist" and four single-stranded

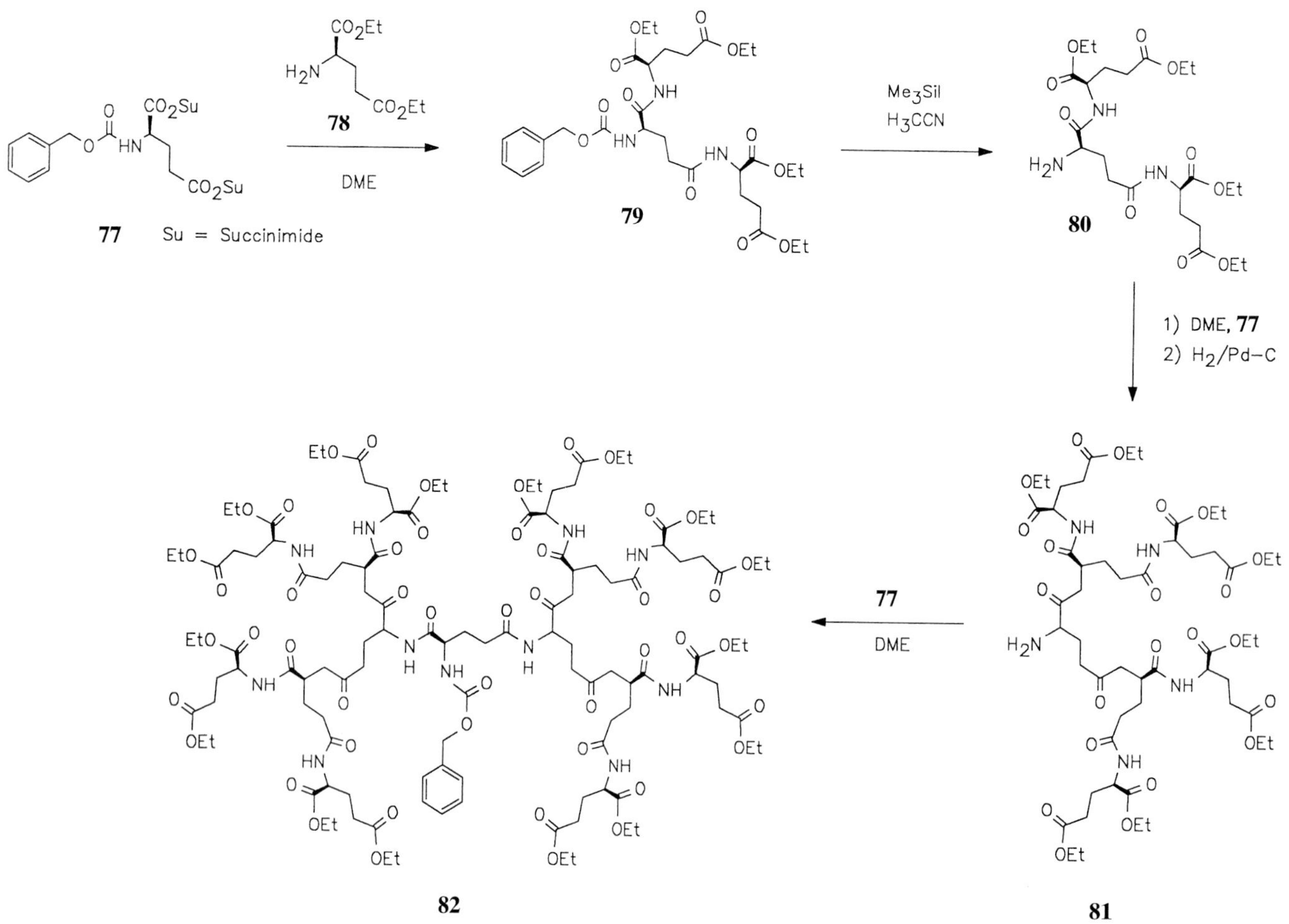

Scheme 7.15 Employment of L-glutamic acid monomers for the construction of chiral superstructures.[145]

binding arms, as a monomer. It was suggested that the resulting dendrimers might find use in the development of semipermeable nucleic acid membranes. Their use as signal amplifiers in nucleic acid diagnostics has been described.[150] These novel nucleic acid dendrimers (4th generation) have also been used as nucleic acid probes in DNA biosensors.[151] Mass-sensitive, piezoelectric transducers coated with these dendrimers were exposed to oligonucleotide targets that bind to the dendrimer and thereby produce a transducer response. The degree of loading was shown to be greater with the dendrimers than with "linear" binding materials. Orentas et al.[152] further used these DNA architectures for detection of the Epstein–Barr virus EBER sequence in post-transplant lymphoma patients.

7.2.1.2 1 → 2 *Aryl*-Branched, *Ether* Connectivity

Seebach et al.[13] successfully utilized a convergent approach for the preparation of chiral dendrimers. They wished to address three basic questions: (1) will the incorporation of chiral cores impart asymmetry to dendritic structures? (2) can enantioselective host–guest interactions occur near the core? (3) can chiral recognition be transposed to the dendritic surface?

Employing the asymmetric tris(hydroxymethyl)methane derivatives **86** and **87** as cores, cascades **83** and **84** were prepared (Scheme 7.16) by the attachment of preconstructed benzyl ether wedges (e.g., **85**).[153] The optical activities exhibited by these dendrimers were diminished (**83**, $[\alpha]_D^{25} = +3.7°$; **84**, $[\alpha]_D^{25} = -0.2°$) compared to those of the cores (**86**, $[\alpha]_D^{25} = +12.6°$; **87**, $[\alpha]_D^{25} = -14.2°$), suggesting a chiral dilution effect. The authors also noted that on going from the tris(allyl ether) precursors ($[\sigma]_D^{25} = +10.8°$) to the homologated triol core **87**, the sign of the optical rotation is reversed. Polyether **83** was shown to form a clathrate with CCl4; extrusion of CCl4 was possible only by heating at 100°C (0.5 mm Hg) for several hours.

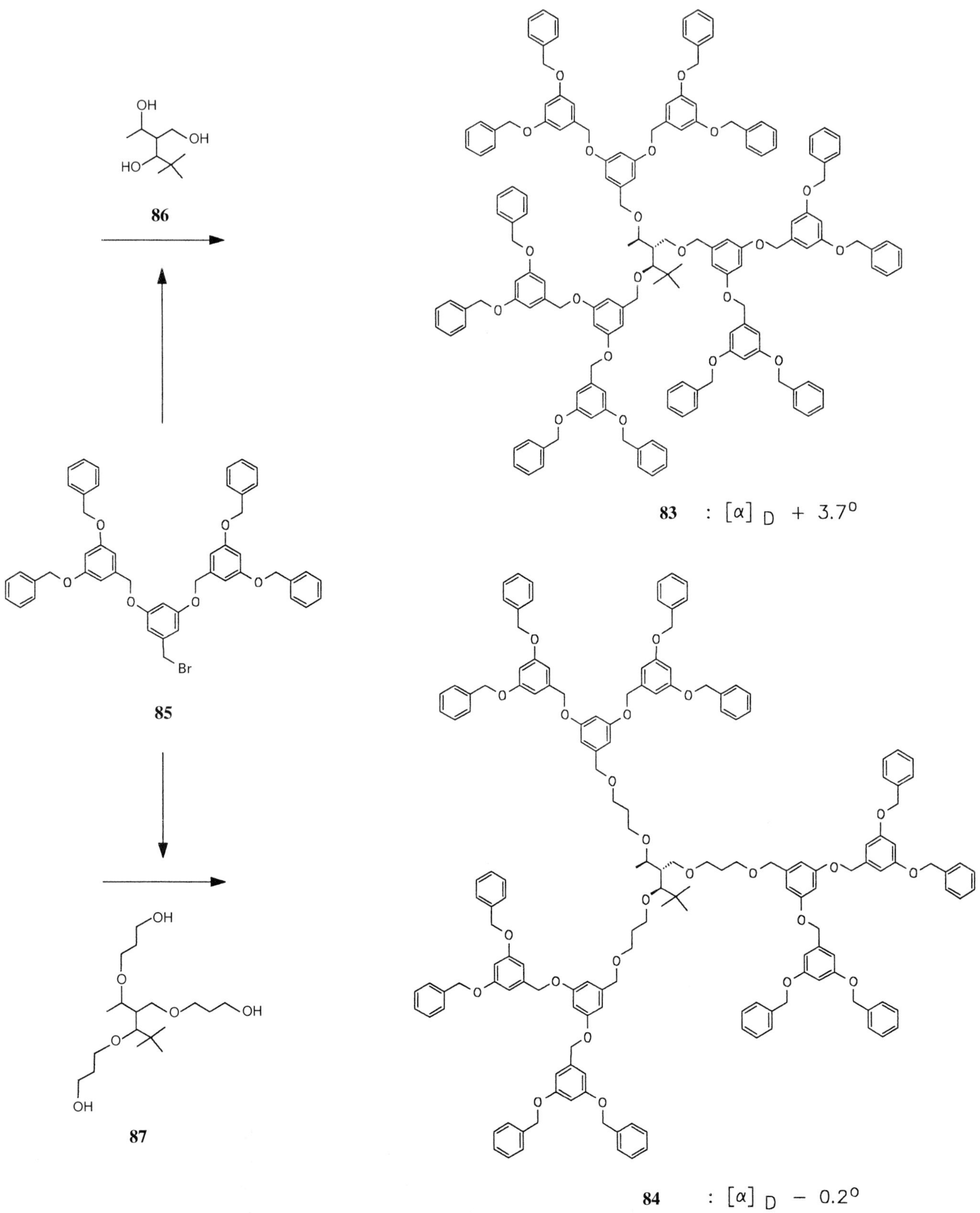

83 : $[\alpha]_D + 3.7^0$

84 : $[\alpha]_D - 0.2^0$

Scheme 7.16 Preparation of chiral dendrimers by attachment of benzyl ether dendrons to asymmetric trigonal cores.[13]

Seebach et al.[154] modified their chiral triol building blocks for use as monomers in the construction of 2nd generation dendrimers and dendrons possessing either all *S*, all *R*, or mixed combinations. Figure 7.9 shows the architectures of the triols used as core components (**88–90**), the (*R*)- or (*S*)-branching units (**91** and **92**), and the chiral dendrons (**93–96**) that were employed for branched assembly (**97–106**). Attempted preparation of the full (triply-substituted core) dendrimers resulted only in the disubstituted dendrons (i.e., **99**, **100**, **103**, and **104**), revealing a remarkable diastereodifferentiation or molecular

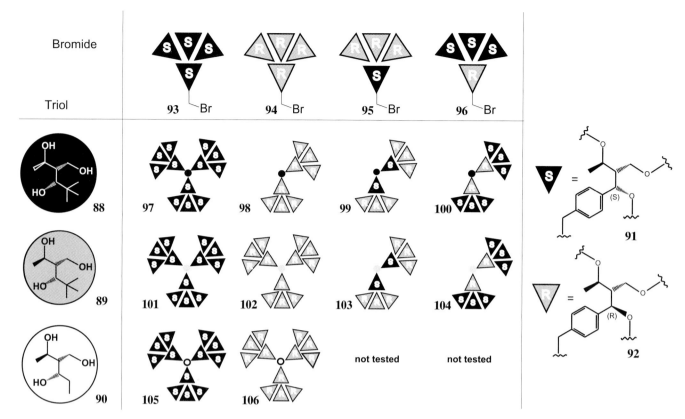

Figure 7.9 Seebach's family of chiral dendrimers.[154] Reproduced with permission from Wiley-VCH.

recognition phenomenon. MALDI-TOF MS studies suggested that these structures were defect-free. A novel and unambiguous nomenclature for these materials was proposed. Similar architectures have been reported[155] that show specific rotations as an average of all chiral units, which are independent of solvent, whereas circular dichroism regularity is lost upon variation of the solvent.

Kremers and Meijer reported a novel approach to chiral dendrimer synthesis,[14, 156] while Meijer reported an interesting relationship between cryptochirality and dendrimers.[157] Conceptionally, this molecular cascade can be envisaged as a tetrasubstituted pentaerythritol, where each core oxygen is substituted by a different generation dendritic wedge, i.e. bromides **108**, **109**, and **110** corresponding to the 1st through to the 3rd tier wedges, respectively. Construction of the poly(benzyl ether) **107** proved slightly more problematic than its conception (Scheme 7.17). Starting from diethyl (ethoxymethylene)-malonate, diol **111** was prepared in five steps. Bis-silylation of diol **111** followed by alde-hyde liberation, reduction, and reaction with benzyl bromide gave the bis(*tert*-butyldimethylsilane) **112**. The key step in the introduction of chirality through core manipulation was the removal (ZnBr$_2$, CH$_2$Cl$_2$) of a single t-BuMe$_2$Si- (TBDMS) group from polyether **112**, in the presence of a similar TBDMS group as well as an MEM ether moiety, to afford monoalcohol **113**. Subsequent reaction with the 1st tier bromide **108**, desilylation (Bu$_4$NF), and treatment with the 2nd tier bromide **109** afforded the MEM-protected alcohol **114**. MEM removal (β-chlorocatecholborane) and final attachment of the 3rd tier wedge **110** gave the desired racemic asymmetric ether **107**. The use of chiral auxiliaries to induce enantioselectivity was unsuccessful, as was attempted racemic separation on chiral HPLC stationary phases. Nevertheless, this constitutes an excellent example of the utility of the convergent process coupled with controlled deprotection of a core unit.

In view of the absence of observable optical activity for the polyether (*S*)-**115**[158] (Scheme 7.18) by means of ORD, CD, and optical rotation methods (i.e., its "cryptochirality"), a more conformationally rigid constitutional isomer was designed and prepared by Meijer et al.,[159] and others,[160] in order to verify the effect of flexibility (or lack thereof) on chirality. The strategy to access such a molecule involved the construction of two dendrons, each possessing a branching center with conformationally restricting back-folded arms (i.e., **116** and **117**). Access to bromide **116** was predicated on 2,6-

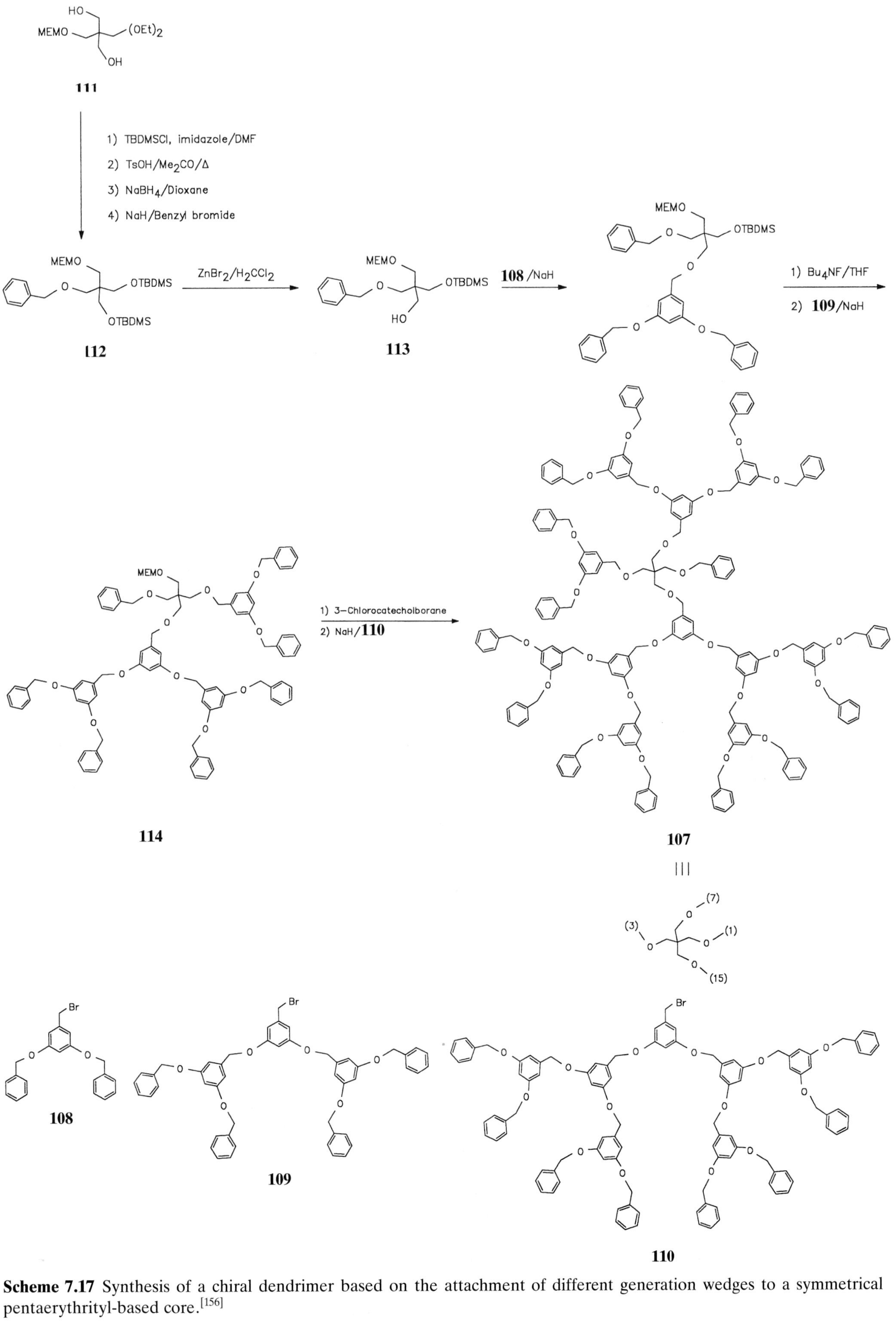

Scheme 7.17 Synthesis of a chiral dendrimer based on the attachment of different generation wedges to a symmetrical pentaerythrityl-based core.[156]

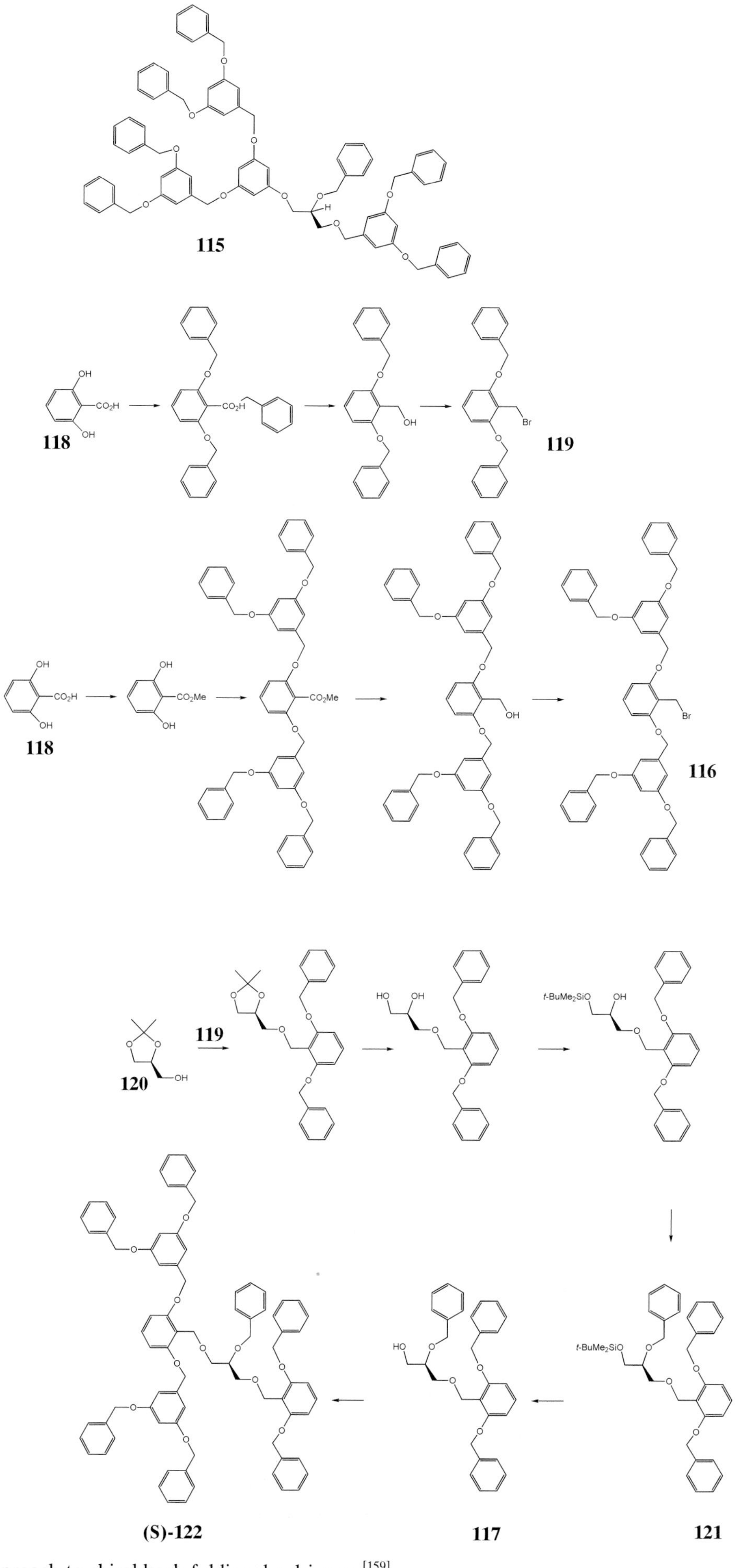

Scheme 7.18 Meijer's approach to chiral back-folding dendrimers.[159]

dihydroxybenzoic acid **118**, which was also central to the preparation of the smaller analog. Addition of dendron **119** to the protected diol **120** followed by deprotection and treatment of the primary alcohol with *t*-BuMe$_2$SiCl and reaction of the secondary alcohol with benzyl bromide gave dendron **121**. Subsequent silyl deprotection and reaction with benzyl bromide **117** afforded the desired polyether (*S*)-**122**, which exhibited an optical activity of $[\alpha]^{20}_D = +0.8°$ (*c* = 2.2, CH$_2$Cl$_2$). Notably, an induced chiral effect was observed in the CD spectrum at 15 °C that vanished at 30 °C, suggesting temperature-dependent chirality based on flexibility.

Axially chiral dendrimers have been synthesized[161] using enantiomerically pure (*S*)-1,1'-bi-2-naphthol as a core; generations zero through to 4 were examined. Increasing molar optical activity was observed for higher generations, an effect attributed to greater naphthyl torsion angles caused by increasing dendron steric interactions. However, this effect was described as marginal due to the inherent flexibility of the poly(aryl ether)s. Fan et al.[162] prepared similar BINAP derivatives possessing a *P*-binding site for use in asymmetric hydrogenation; Ru complexes were formed, the utility of which as catalysts was demonstrated since they could be readily prepared, separated, and recycled.

Ester-terminated Fréchet-type dendrons have been attached to an enantiomerically pure (1*R*,2*S*)-2-amino-1-phenyl-1,3-propanediol core by Parquette et al.[163] The molar rotations of these dendrimers were found to decrease with increasing dendrimer generation. Comparison of methyl ester-terminated with benzyl ester-terminated materials suggested that this decrease in rotation resulted from a sterically induced conformational effect of the chiral core, which increased with higher generation. Rohde and Parquette[164] attached allyl ester-terminated dendrons to a tetravalent, chiral 2,5-anhydro-D-mannitol core; up to three generations were prepared. Dendrimer specific molar rotations increased with generation, but were still more negative than that of the unsubstituted core. Parguette et al.[164a, b] also demonstrated chiral confirmational order in intramolecularly N-bonded dendrons.

A new series of chiral dendrimers possessing planar, chiral cycloenantiomeric and topologically chiral cores has been created from racemic 4-hydroxy[2.2]paracyclophane, a [2]rotaxane with sulfonamide groups in the "wheel and axle" positions, and a [2]catenane with a sulfonamide moiety on both of the macrocycles.[57, 165] Chirality was found to be dependent not only on the chiral elements, but also on the dendritic wedges and their size.

Chow and coworkers[12, 166] reported the synthesis and characterization of homochiral dendrimers (Scheme 7.19) through to the 1st generation using (2*R*,3*R*)-tartaric acid as the chiral building block. The mode of preparation was based on three constituent parts: the three-directional core, 1,3,5-trihydroxybenzene (**123**); the chiral connector, 1-*O*-tosyl-4-*tert*-butylphenyloxy-2,3-*O*-isopropylidene-L-threitol (**124**), and a branching unit, 1-benzyloxy-3,5-dihydroxybenzene (**125**). Construction began with the treatment of (2*S*,3*S*)-(–)-1,4-di-*O*-tosyl-2,3-*O*-isopropylidene-L-threitol with half an equivalent of 4-*tert*-butylphenol under basic conditions to yield the mono-*O*-arylation product (**124**). Subsequent reaction of the chiral aryl ether **124** with half an equivalent of dihydroxybenzene[167] **125**, followed by hydrogenolysis (Pd/C), gave the dendron **126**. Attempted preparation of the disubstituted phenol **126** by reaction of the tosylate **124** with one-third of an equivalent of core **123** resulted in low yields (ca. 26 %); poor solubility of the tris(phenolic) trianion was suggested as a rationale for the low conversion.

The 0th generation cascade **127** was prepared from core **123** and chiral block **124**, whereas the 1st generation dendrimer **129** was synthesized by extension of dendron **126** with the bis(tosylate) **128**, followed by the addition of the same core **123**. The specific rotations ($[\alpha]_D$) of the 0th and 1st generation dendrimers (**127** and **129**) were measured as –59.6° and –69.7°, respectively. The molar rotations were reported to be –569° (**127**) and –1769° (**129**), with molar rotations per tartaric acid unit of –190° (**127**) and –197° (**129**), respectively. The use of bis(aryl ether) **126** allowed the construction of tetrakis(aryl ether) **130**, treatment of which with the bis(tosylate) **128** afforded the corresponding dendron for envisaged use in the preparation of the 2nd generation dendrimers. However, subsequent reaction with tris(phenol) **123** failed to give the desired material, presumably due to steric and solubility problems.

Chow and Mak[168,169] later reported the preparation of homochiral, layered-block dendrimers based on the key monomers D- and L-tartaric acid (Figure 7.10). Construc-

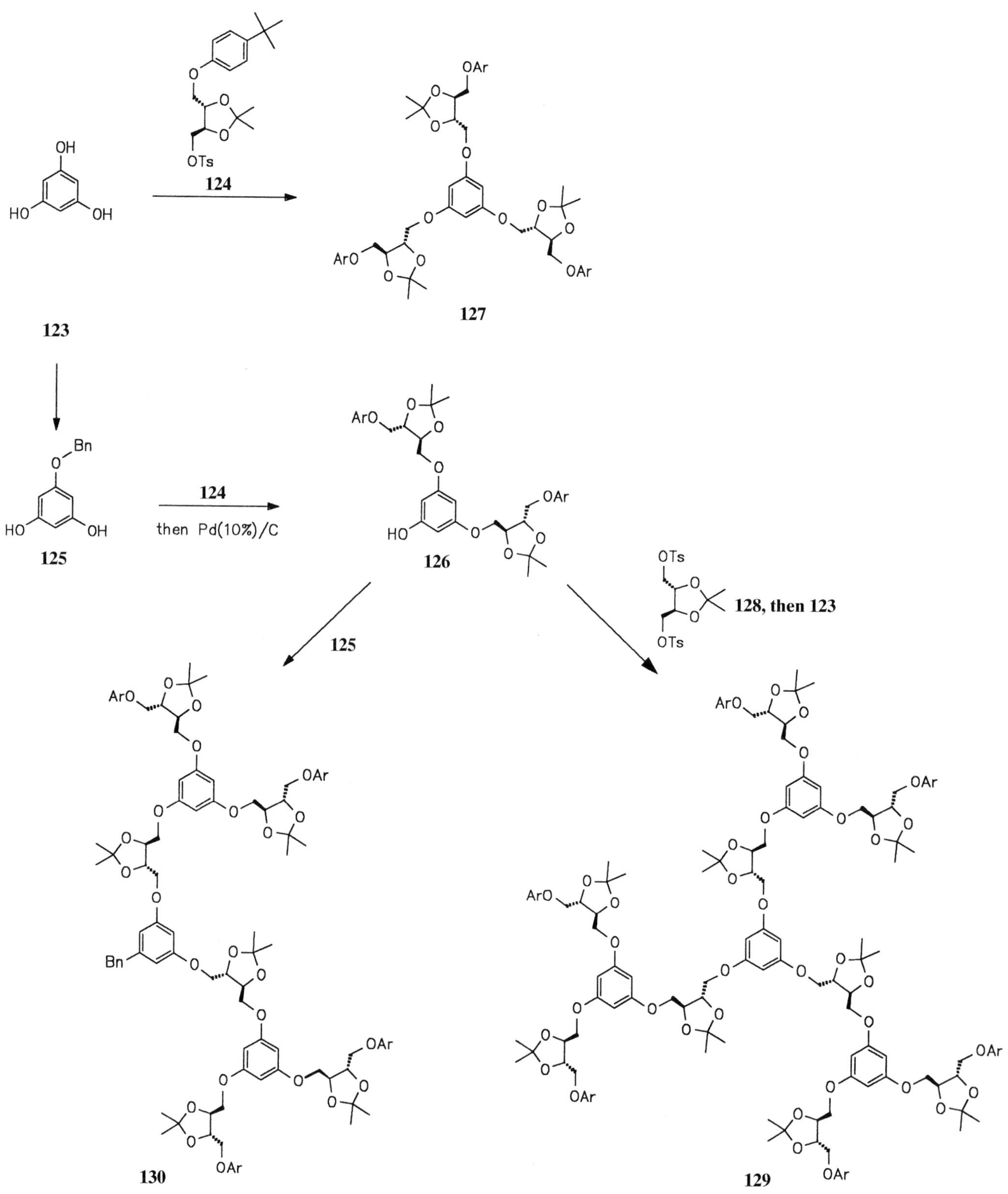

Scheme 7.19 Construction of homochiral dendrimers effected using (2*R*,3*R*)-tartaric acid based monomers.[12,166]

tion of the two small dendrimers (**131**, **132**) was achieved in a similar manner as that of the previously described materials.[12, 166] Dendrimer **131** consisted of a triad of D-tartaric acid units around an aryl core surrounded by a hexameric layer of L-tartaric acid, while the outer layer of **132** consisted of alternating D- and L-moieties around the D-series triad. On a one-to-one basis, the optical rotation due to a D-unit was found to cancel that due to an L-unit, although CD experiments suggested that this effect was more clear-cut when the chiral units were located within the same generation. Thus, the observed molar rotations were dependent on the number of chiral units in excess. The measured values for molar rotation, specific rotation [α], and molar rotation per tartaric moiety for **131** and

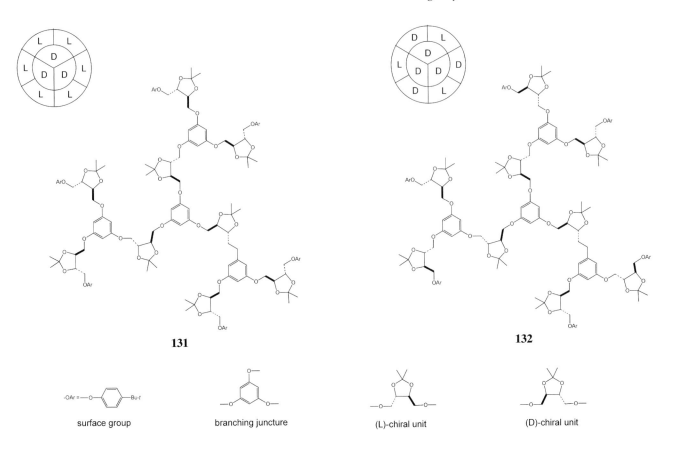

surface group   branching juncture   (L)-chiral unit   (D)-chiral unit

-OAr = —O—⟨ ⟩—Bu-*t*

131   **132**

Figure 7.10 Chiral layer-block dendrimers with diverse subunits.[168, 169]

132 were $-29_{(1)}°$ and $+23_{(2)}°$, $-748_{(1)}°$ and $+579_{(2)}°$, and $-249_{(1)}°$ and $+193_{(2)}°$, respectively. Additional chiroptical data have been published.[170]

Chow and Mak[171] also terminally functionalized poly(aryl ether) wedges possessing –(CH$_2$)$_3$– spacer moieties with L-tartaric units to give a series of optically active, homochiral dendrons (Figure 7.11; **133** and **134**). The chiroptical properties were found to be directly proportional to the number of attached chiral moieties.

McGrath et al.[172, 173] described the convergent synthesis of chiral monomers[174] possessing hydroxylated frameworks up to the 2nd generation. Schemes 7.20 and 7.21 illustrate the key aspects of the strategy. Primary monomer construction was predicated on two key reactions: (1) the Horner–Emmons Wittig modification resulting in alkene formation, and (2) Os-catalyzed asymmetric dihydroxylation[175] (AD) for the introduction of a latent functionality following acetonide protection. Reduction (LAH) of the focal

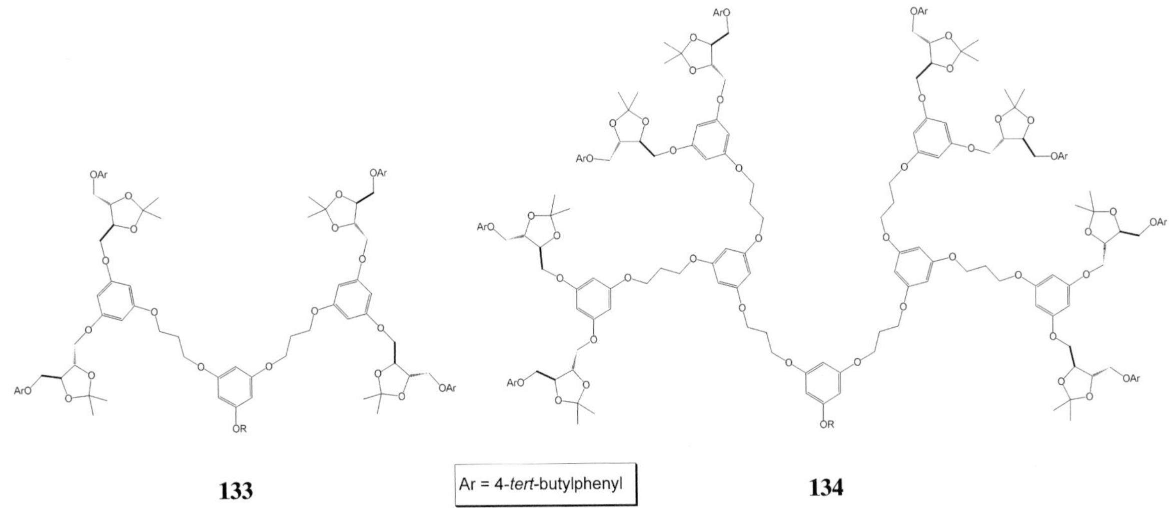

133   Ar = 4-*tert*-butylphenyl   **134**

Figure 7.11 Chow's homochiral poly(aryl ether)s.[171]

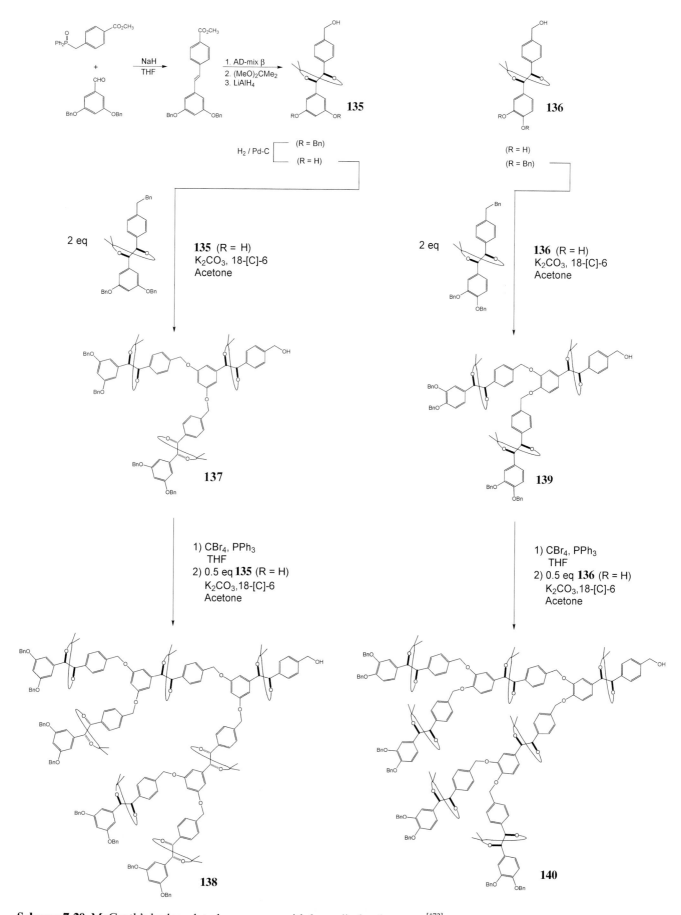

Scheme 7.20 McGrath's hydroxylated monomers with benzylic focal groups.[172]

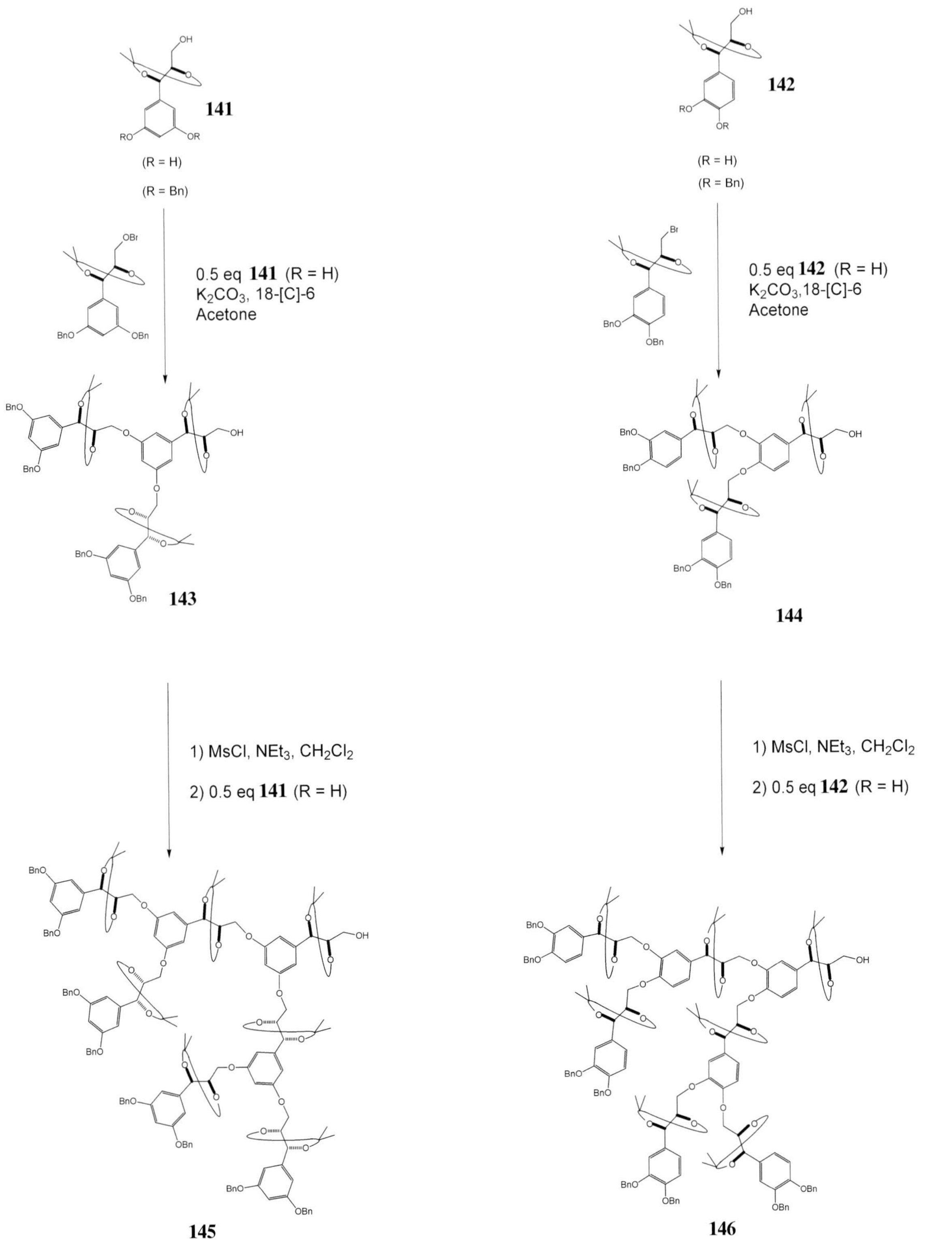

Scheme 7.21 Polyhydroxylated dendrons possessing hydroxyalkyl focal sites.[172]

ester produced the starting benzyl alcohol (e.g., **135**) for use in the subsequent Fréchet-type construction.

Four analogous starting monomers were prepared, benzylic alcohols **135a** and **136**, possessing 1,3,5- and 1,3,4-benzenoid branching, respectively, and the corresponding hydroxyalkyl ethers **141** and **142**, characterized by similar motifs. Monomer **135** (R = H) was reacted (K$_2$CO$_3$, 18-crown-6) with two equivalents of its benzyl-protected, brominated analog to afford the 1st generation tris(acetonide) **137**, which was then focally brominated (CBr$_4$, PPh$_3$) and treated with further **135** to give the 2nd generation dendron **138**. Monomer **136** was treated analogously to give dendrons **139** and **140**. While the monomer attachment transformations of dendrons **143** and **144** proved sluggish, conver-

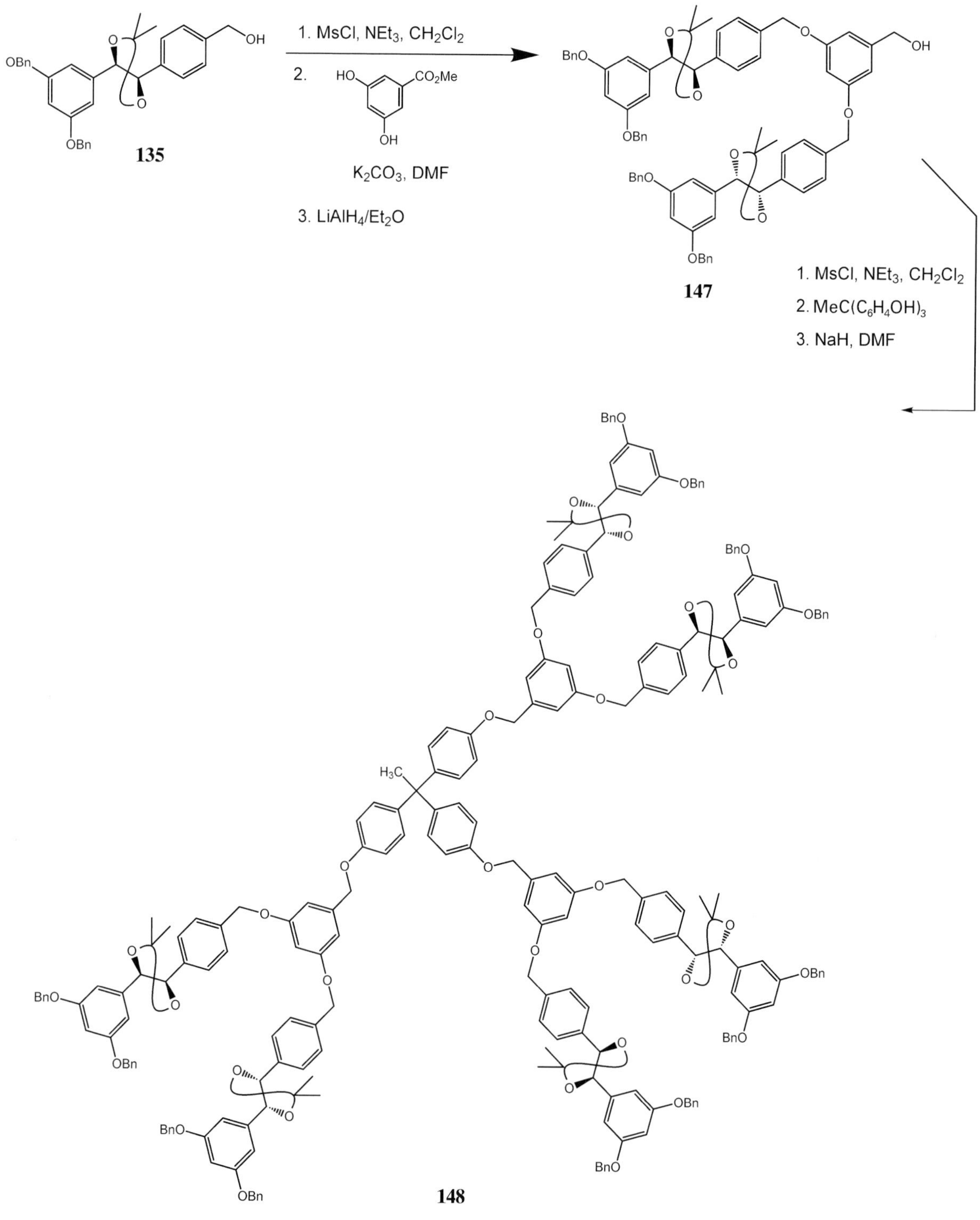

Scheme 7.22 Dendrimers possessing single layers of chirality.[176]

sion (MsCl, NEt₃) to the mesylates facilitated sequence repetition to furnish the higher generations **145** and **146**.

McGrath et al.[176,177] also reported the use of their key monomer (**135**; Scheme 7.22) for the preparation of dendrimers possessing single layers of chirality. Thus, mesylation of benzyl alcohol **135** and reaction with methyl 3,5-dihydroxybenzoate followed by reduction (LAH) afforded the building block **147**; this was subsequently attached to a tris(phenol) core to give dendrimer **148**.

Use of these novel hydrobenzoin-based chiral monomers facilitated the synthesis of "optically active shell dendrons", whereby Junge and McGrath[178] prepared the 4th generation wedges positioning chiral monomers at increasing distances from the focal site. Conformational order, as suggested by chiroptical data, was not observed in solution. The constitutions, configurations, and optical activities of the chiral architectures have been reported.[179] Similar polyhydroxylated dendrimers and dendrons have been constructed and their chiroptical properties have been examined.[180, 181]

Bolm et al.[182] reported the focal modification of *n*-propyl-terminated poly(aryl ether) dendrons with a chiral (*S*)-hydroxymethylene pyridyl moiety. These new asymmetric dendrons have been investigated with regard to their catalytic enantioselectivity in the alkylation of benzaldehyde with Et$_2$Zn to afford a preponderance of the (*S*)-enantiomer. The 1st and 2nd generation dendrons afforded *ee*'s of 92.5 and 93 % (5 mol %) and 90 and 89.5 % (1 mol %), respectively. Compared with catalysis by the "non-appended" or free pyridyl alcohol, *ee*'s were found to be only 2–3 % less with the modified dendrons. Further focal modification of these poly(aryl ether)s with (*S*)-2-amino-3-(*p*-hydroxyphenyl)-1,1-diphenylpropan-1-ol has been reported.[183] These chiral amino alcohols were then used to catalyze enantioselective borane ketone reductions; using 10 mol % of the dendritic catalyst, *ee*'s ranging from 82 to 96 % were realized.

Shinkai, Ikeda, and Numata[184] modified saccharide moieties for investigation and control of *H*-bonding-based self-assembly. Larger dendrons are less sterically demanding, while smaller aggregates were observed with higher generation dendrons. Aggregates were obtained from these saccharide dendrons by immobilization through *in situ* cross-linking with 1,3-phenylene diisocyanate.[185]

Hirsch et al.[186] modified chiral, C_3-symmetrical tris[bis(4-phenyl-2-oxazoline)methano]-substituted fullerenes with these poly(aryl ether) dendrons via a three-fold cyclopropanation. Use of the resulting hexakis-adducts as catalysts for the stereoselective cyclopropanation of styrene with ethyl diazoacetate was demonstrated.

Lacking other connectivity, phenylene dendrons have been constructed onto chiral binapthol cores.[186a, b] Polyphenylenevinylene binaphthyl constructs are also known.[186c]

7.2.1.3   1 → 2 *Aryl*-Branched, *Ester* Connectivity

Seebach et al.[187] constructed a series of biodegradable dendrimers (Figure 7.12; **149**) using (*R*)-3-hydroxyglutanoic acid and trimesic acid as monomers. Following attachment of the dimer or tetramer of butanoic acid benzyl ester to a TBDPS-protected 5-hydroxymethyl-1,3-benzenedicarboxylic acid dichloride, terminal debenzylation, and acid halide generation, the focally deprotected counterpart was connected. Reaction of the resultant 2nd generation wedge with trimesic acid trichloride afforded the desired dendrimers. Biodegradability in the presence of various hydrolases, such as PHB-depolymerase, was examined in the context of the envisaged use of these materials as templates for defined nanocavities. These chiral, branched poly(β-hydroxyalkanoate)s have been examined by MALDI-TOF MS,[188] which confirmed them to be essentially unimolecular species. The technique of post-source decay or fragment analysis by standard time-of-flight MS permits the observation of metastable fragments, representing a new tool for structural oligomer analysis.

7.2.1.4   1 → 2 *Aryl*-Branched, *Amide* Connectivity

Draheim and Ritter[189] prepared an acrylate-based dendron bearing polymerizable focal units. Synthesis began with the amine-protected, bis(carboxylic acid) **150** (Scheme 7.23), which was converted (ClCO$_2$Et) to the activated anhydride and treated (Et$_3$N) with L-aspartic acid dimethyl ester hydrochloride to afford tetraester **151**. Following transfer hydrogenation to liberate the aryl amine **152** and transformation to the corresponding silylamine **153**, a 2nd generation dendron **155** was accessed by reaction with modified diacid **154**. Polymerization was then effected (AIBN, DMSO, N$_2$) to generate polymers adopting helical conformations as a result of the relief of steric strain.

7.2.1.5   1 → 2 *Aryl*-Branched, *Alkyne* Connectivity

Soai and coworkers[190] modified the termini of Moore's poly(phenylalkyne)s (Section 5.3.11) with chiral amino alcohols and examined the utility of the products in the catalytic and enantioselective alkylation of aldehydes. Use of chiral dendrimers[191] possessing

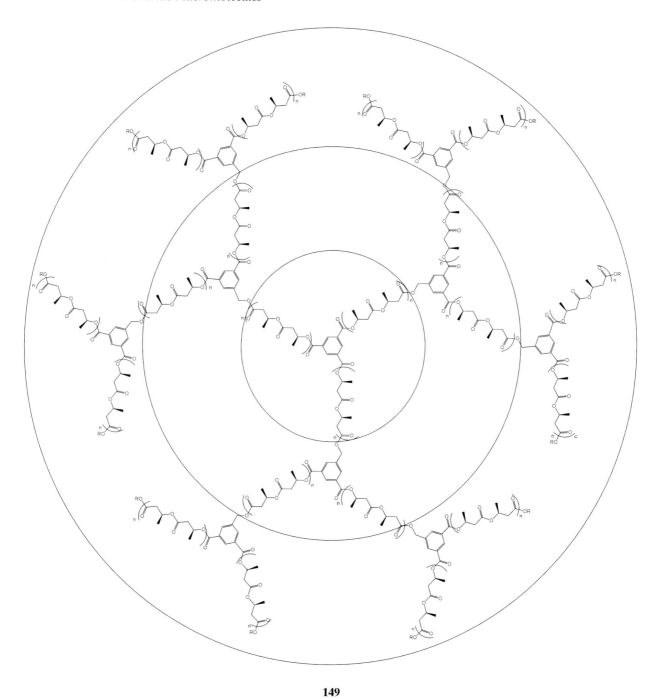

149

Figure 7.12 Seebach's biodegradable dendrimers.[187]

three or six chiral terminal β-amino alcohols in the enantioselective addition of Et$_2$Zn to *N*-diphenylphosphinylimines afforded enantiomerically enriched *N*-diphenylphosphinyl-amines with up to 94 % *ee*.

7.2.1.6   1 → 2 *Ethano*-Branched, *Phosphate* Connectivity

Damha, and coworkers described[192–194] the synthesis of dendritic polymers based on nucleic acids using an automated DNA synthesizer and employing thymidine and adenosine as building blocks. Convergent construction began by anchoring a thymidine moiety (Scheme 7.24) to a long-chain-alkylamine controlled-pore glass (LCAA-CPG) support possessing long alkyl spacers (18 Å) that enhance bound reagent accessibility and a 500 Å pore size that facilitates high molecular weight oligonucleotide synthesis. Repetitive treatment of the thymidine-bound units (**156**) with deoxythymidine phosphoramidite in a "chain extension" procedure afforded pentathymidine chains (**157**). The

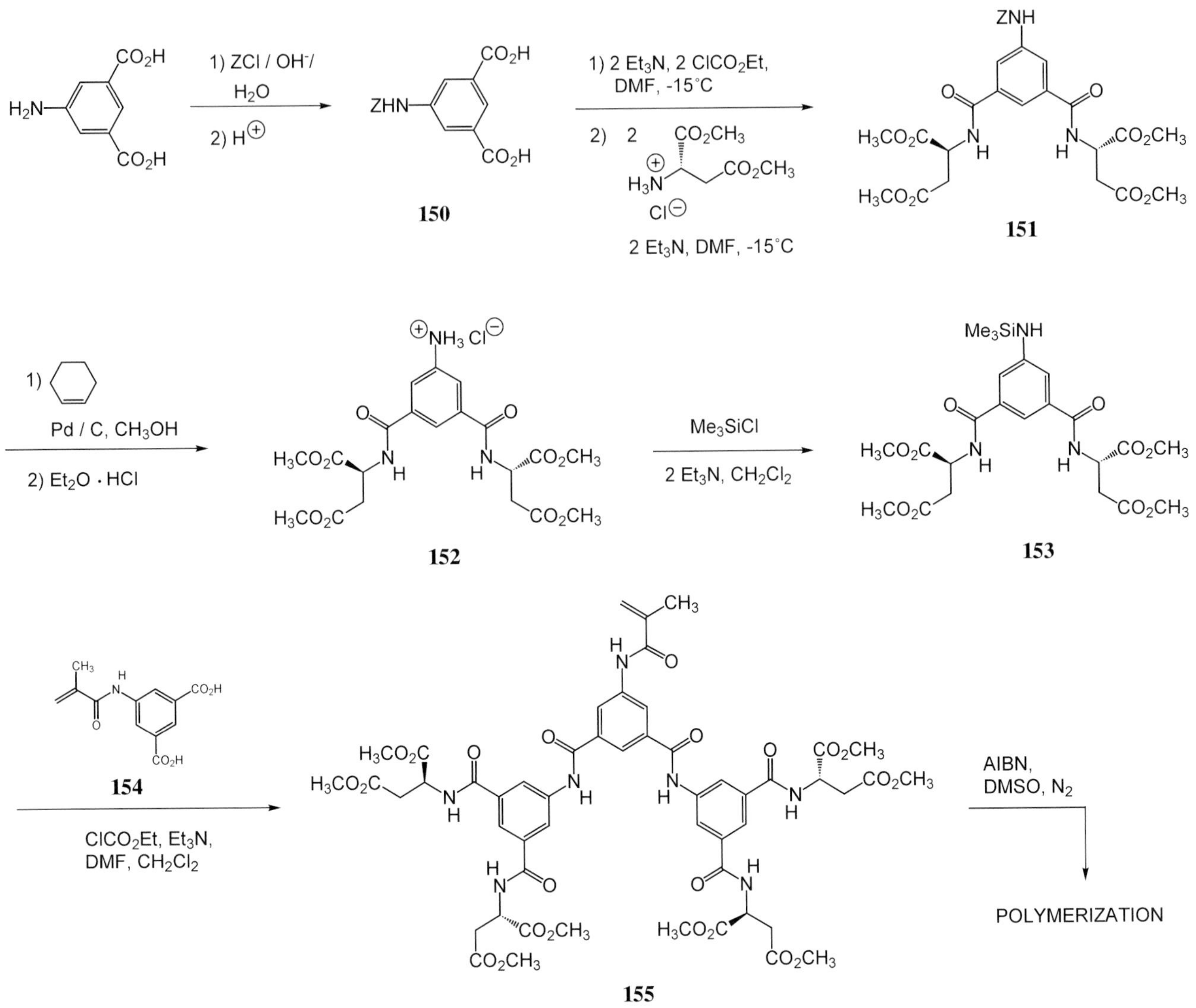

Scheme 7.23 Drahein and Ritter's procedure for generating polymerizable dendrimers.[189]

polymer-bound chains were then "branched", or coupled, by reaction with a tetrazole-activated adenosine 2',3'-bis(phosphoramidito) reagent **158**. Continued "chain extension" and "branching" allowed the preparation of an 87-mer (**159**), which was simultaneously cleaved from the support and deprotected (NH$_4$OH). Attempted coupling of two equivalents of 38-mer **160** was unsuccessful, presumably due to the increased distances between the reactive sites. Further chain extension to give the 43-mer (**161**) subsequently allowed the preparation of the two-directional dendrimer by standard coupling techniques.

Characterization of the dendrimers included composition analysis by snake venom phosphodiesterase and alkaline phosphatase enzymatic digestion to afford inosine, derived from the deamination of adenosine and thymidine, which was found in the expected quantities by HPLC. Polyacrylamide gel electrophoresis was also employed to monitor digestion. The authors noted an increasing occurrence of branch defects, described as "default" oligomers, and suggested that divergent methodology might be better suited for the construction of such macromolecules. Indeed, Hudson, Robidoux, and Damha[195] later modified their procedure for access to nucleic acid dendrimers to include a divergent solid-phase protocol. The surface of a controlled pore-size glass support was used for chain assembly; comparable or better yields of various materials were realized as compared with the convergent approach.

The construction of an asymmetric diamino acid has also been reported,[196] which was envisaged as a replacement for the *C*-terminal half of a class of DNA binding proteins known as "leucine zipper" proteins.

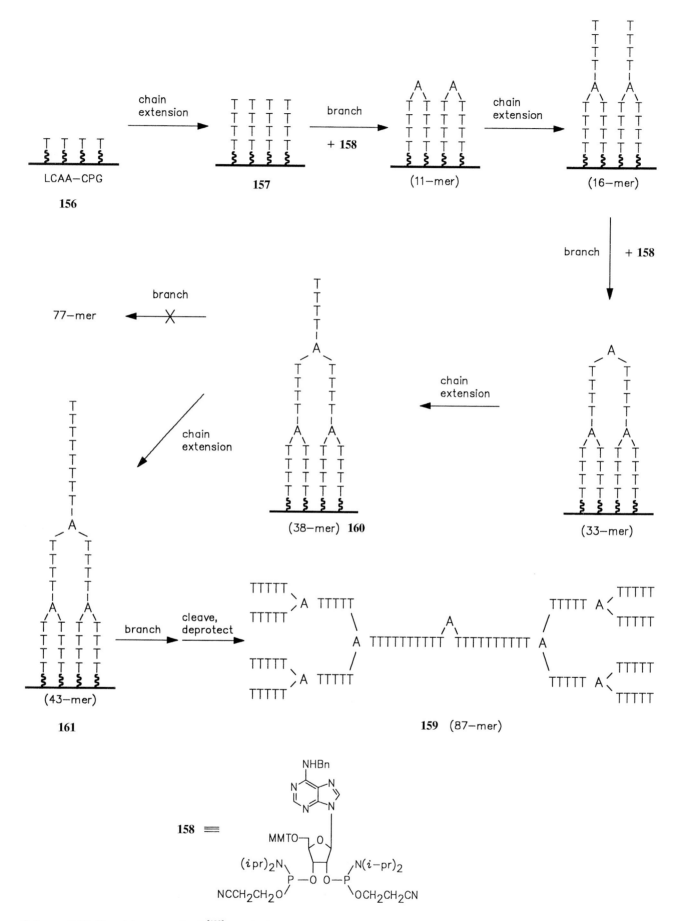

Scheme 7.24 Dendrimer synthesis[192] employing thymidine and adenosine monomers.

7.2.1.7 1 → 2 *N*-, 1 → 3 *C*-Branched *Amide* Connectivity

Stoddart et al.[197] reported the synthesis of carbohydrate-terminated dendrimers based on the preparation of a tris(glucosylated) tris(hydroxymethyl)aminomethane (**162**; Scheme 7.25). Glucosylate triad **162** was prepared by a sequence of amine benzoylation, hydroxyl glucosylation promoted by AgOTf, transformation of the saccharide benzyl-protecting groups to acetyl moieties (NaOMe, then Ac₂O; this was necessary because unacceptable mixtures were obtained when acetylated glucosyl bromide was employed), and finally removal (10 % Pd/C) of the *N*-benzoyloxy group. Monomer **162** was then sub-jected to numerous reactions to form dendrimers as well as larger dendrons (Scheme 7.26). Since amine **162** failed to react with benzenetriacyl chloride, it was homologated by treatment with N^α-Boc-Gly-perfluorophenol (**163**) to give the glycine-extended mono-mer **164**, which was deprotected and reacted with bis-activated ester **165** to afford the protected dendron **166**. Homologated focal amine hexasaccharide **167** was prepared by treating **162** with **168** to circumvent the sluggish reactivity, but was also found to give mix-tures, as ascertained by MALDI-TOF analysis.

Since DCC-type coupling proved to be quite successful for dendrimer formation, amine **162** was coupled to the bis(glycine)-extended benzenedicarboxylic acid **169** to quantitatively yield the "6-mer" dendrimer **170**. Correspondingly, amine **162** was reacted with the glycine-extended benzenetricarboxylic acid **171** to afford **172** (Scheme 7.27). A 2ⁿᵈ generation dendrimer **173** was constructed in a similar manner from the free amine focally homologated dendron (**74**; Scheme 7.28).

Optical rotations of the acylated dendrimers and dendrons were measured in CHCl₃ and H₂O, respectively, and were found to be proportional to the number of asymmetric moieties attached to the surface. This is in agreement with other reported[11,12] data, pro-vided that the dense-packing limits are not reached.[1] Molecular modeling indicated dendritic radii of ca. 12.5 and 18 Å for dendrimers **172** and **173**, respectively.

Using a benzenetricarboxylic acid core, these constructs[198] have been evaluated for their efficacy in the inhibition of Concanavalin A lectin binding to purified yeast by Stod-dart et al.[199] Compared to a dendritic wedge possessing three terminal mannose moie-ties, inhibition efficacy was found to be ca. four times greater on a molar basis for dendri-mers possessing 9, 18, or 36 mannose termini. Other poly- and oligosaccharides have been prepared by these authors by the attachment of carbohydrates to the surfaces of PAMAMs, PPIs, and lysine cores, and have also been used for framework construc-tion.[200]

Stoddart et al.[110] further prepared carbohydrate-terminated dendrimers employing two similar strategies (Scheme 7.28). Construction of a 36-D-glucopyranoside-terminated structure was initially attempted by coupling (DCC, 1-HBT) of "6-mer" dendron **174** with the extended hypercore **175** in a "6 + 6" reaction, which afforded a 24-monosaccharide dendrimer as the primary product, as characterized by MALDI-TOF MS. Forcing conditions resulted only in the 30-termini structure. Thus, the approach was modified to incorporate a "12 + 3" protocol. Reaction of amine **162** with tetraacid **176** followed by focal deprotection (Pd/C, H₂) gave the "12-mer" dendron **177**, which was

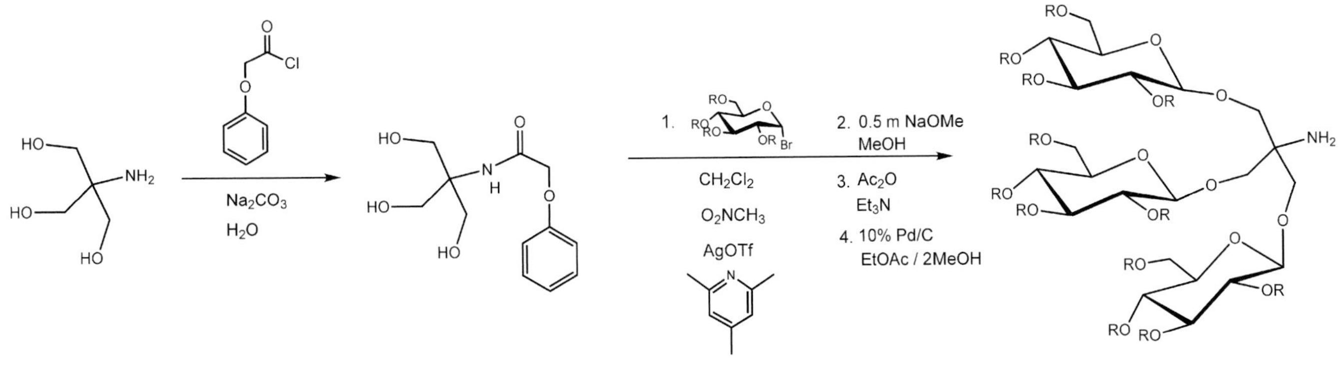

Scheme 7.25 Construction of glucoside-based monomers.[197]   **162**

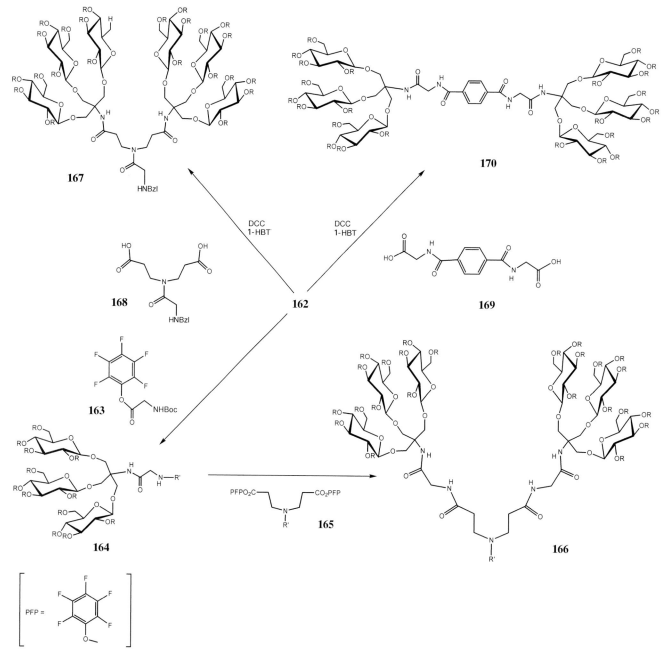

Scheme 7.26 Transformation of tris-based glucoside monomers.[197]

coupled (DCC, 1-HBT) to the trivalent core **178** to yield the desired 36-monosaccharide dendrimer **179**, as evidenced by mass spectrometry. Building block components were accessed employing triglycosyl, amino-protected tris(hydroxymethyl)aminomethane (tris), and a protected 3,3'-glycinamido dipropionic acid. The core units were derived via a 3,3'-iminodipropionic acid extended tricarbonylbenzene of a glycine-elongated analog.

Diederich et al.[201] attached these tris-based glycodendrons (e.g. **162**) to carboxylic acid functionalized fullerenes. These materials were found to form stable, ordered air–water Langmuir monolayers with little C_{60} aggregation due to the bulk of the dendron. Transfer of the Langmuir–Blodgett films onto quartz slides suggested potential optical and biosensor applications.

Enantiomerically pure dendrimers (Figure 7.13) based on a *trans*-3,4-dihydroxypyrrolidine unit have been reported.[202] Convergent synthesis of the chiral dendrons involved the use of *tert*-butyl dimethylsilyl and *N*-benzyl orthogonal protecting groups. "Linear" as well as "radial" derivatives were prepared (e.g., **180** and **181**); the propensity for complete reaction of the secondary amine with acyl chlorides was demonstrated by the preparation of the hexaamide **182**. Self-organization of the chiral moieties was suggested based on analysis of the chiroptical properties (i.e., $[\alpha]_D$ and CD).

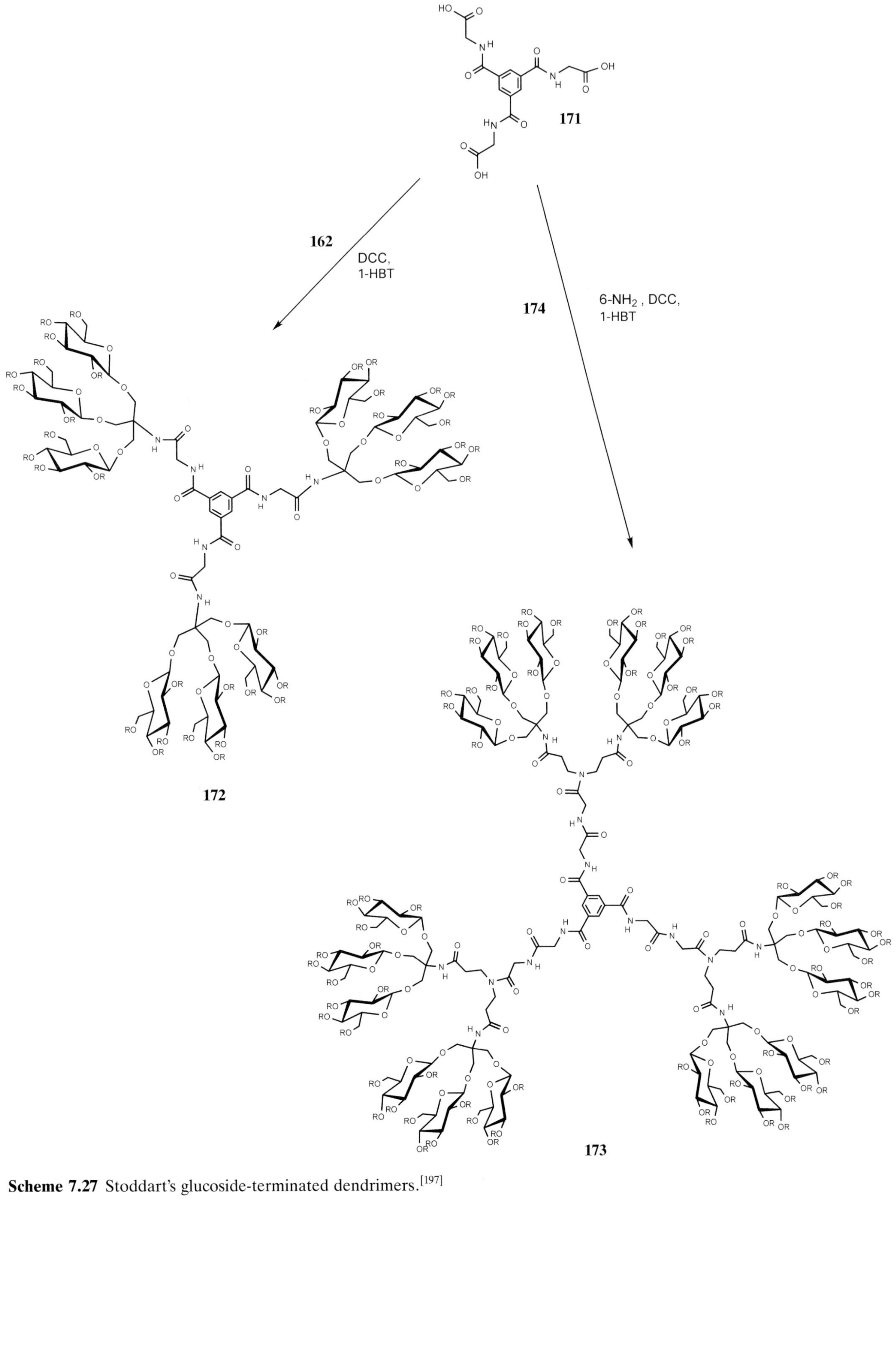

Scheme 7.27 Stoddart's glucoside-terminated dendrimers.[197]

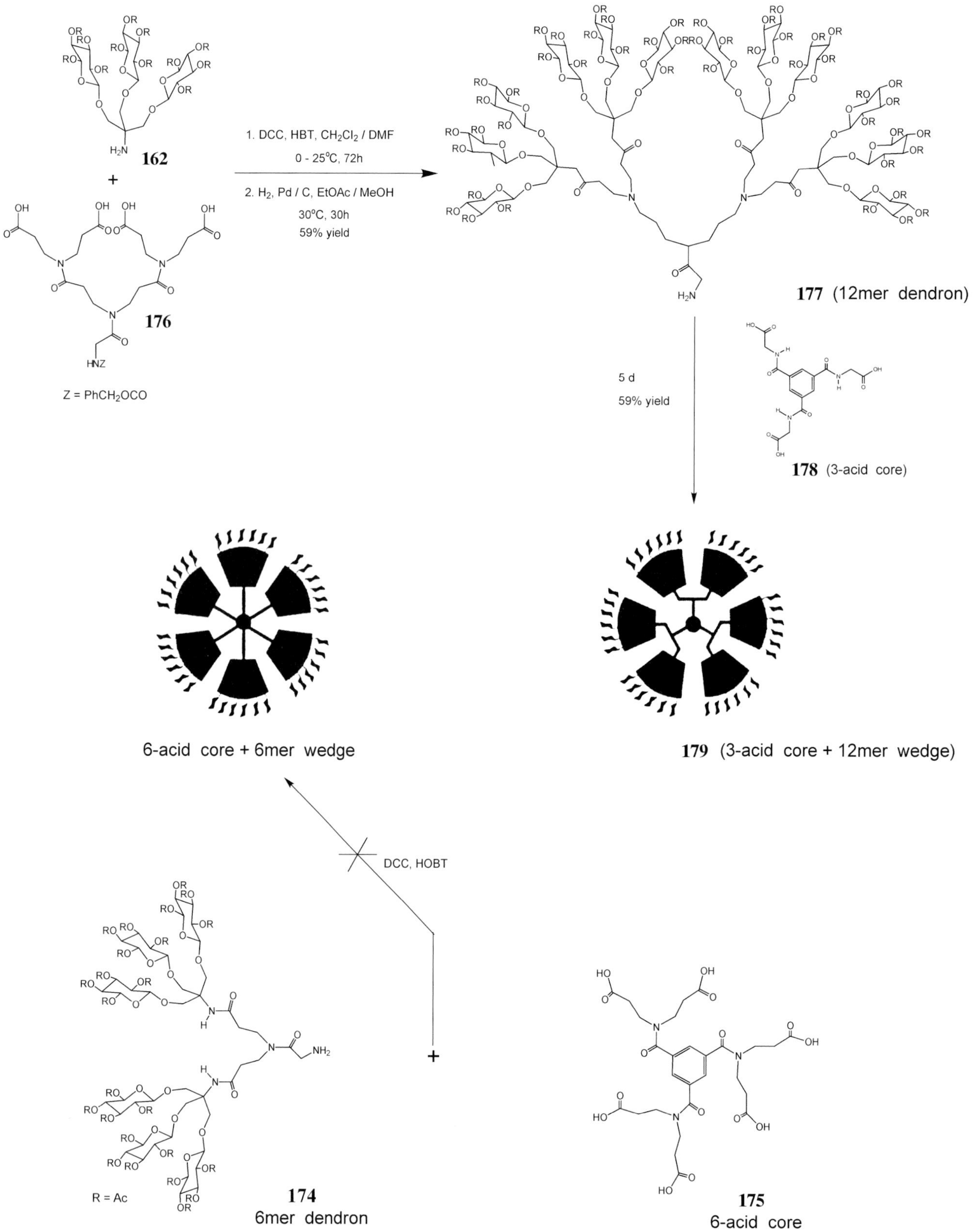

Scheme 7.28 Differing strategies towards similar architectures.[110]

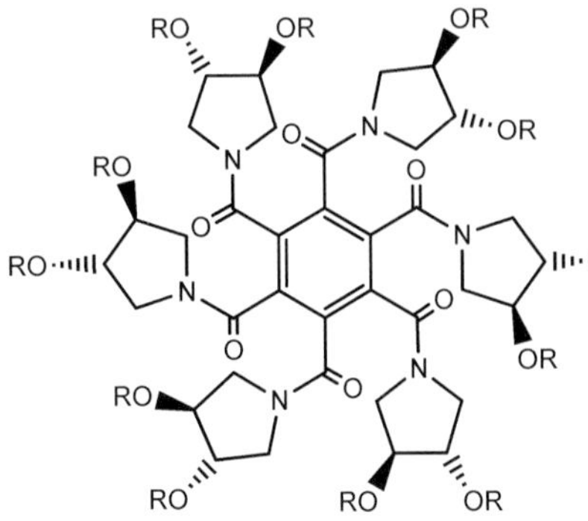

182

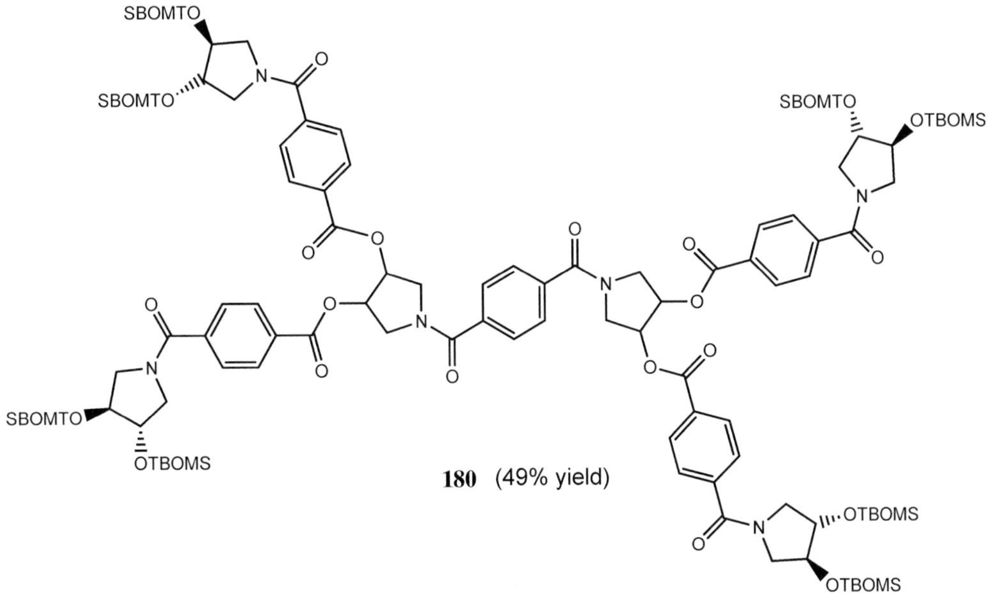

180 (49% yield)

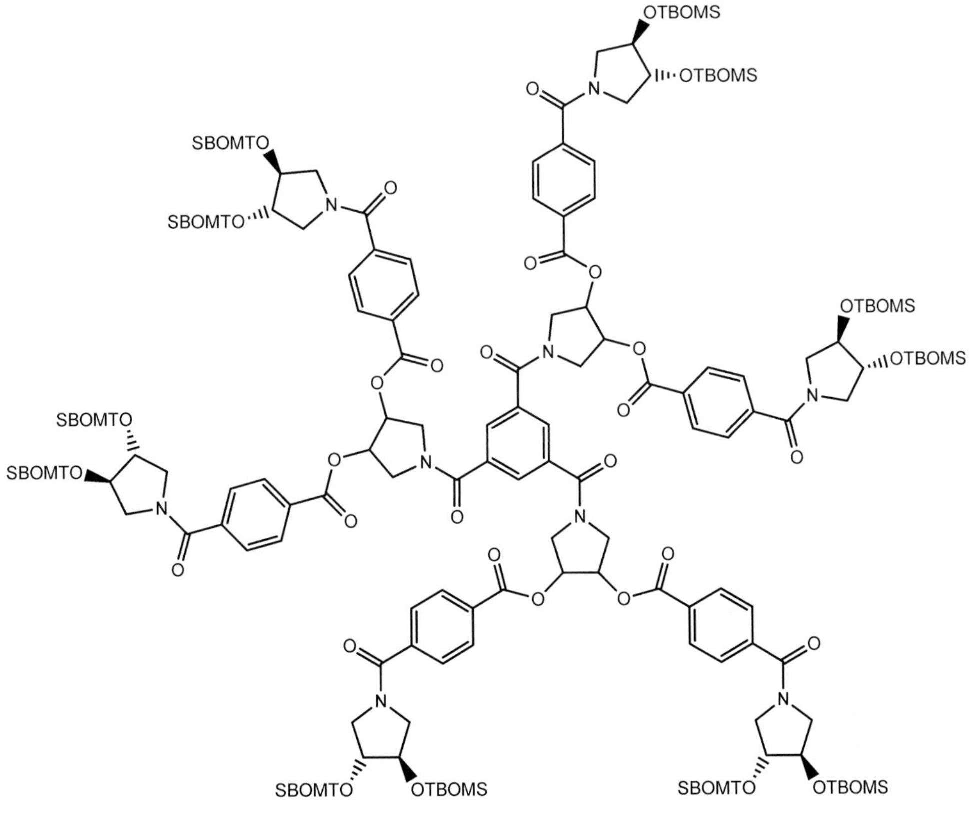

181 (42% yield)

Figure 7.13 Enantiomerically pure dendrimers based on *trans*-3,4-dihydroxypyrrolidine.[202]

7.2.1.8   1 → 2 & 3 *C*-Branched *Amide & Thiourea* (& *Ether*) Connectivity

In an attempt to prepare drug delivery systems, Fernández, Defaye, et al. have recently reported the thiourea attachment of various wedges comprised of mannosyl-coated dendritic branches to a single locus on the upper surface of β-cyclodextrins.[203] These multivalent adducts exhibited high Concanavalin A lectin binding ability and intact inclusion capabilities.

7.2.1.9   1 → 2-*Saccharide*-Branched, *Ether & Amine* Connectivity

Turnbull, Pease, and Stoddart[204] recently described the synthesis of large oligosaccharide-dendrimers generated in a convergent sequence from peracetylated α-maltosyl trichloroacetimidate and 1,2:3,4-di-*O*-isopropylidene α-D-galactopyranose through a linear trisaccharide and the first AB$_2$ trisaccharide. Molecular modeling studies indicated that the 1st generation 9-mer wedges attached to a three-directional benzene core had diameters of 8 nm, while the 2nd generation construct had a diameter of 11–12 nm.

7.2.2   1 → 3 Branched

7.2.2.1   1 → 3 *C*-Branched, *Amide* Connectivity

Diederich et al.[205] prepared a series of "Dendroclefts", which are optically active, dendrimer-based receptors constructed for *H*-binding complexation of monosaccharides.[206] These unique dendrimers (Figure 7.14; **183** and **184**), possessing recognition sites surrounded by branched PEG-based architectures, can be viewed as structural mimics of bacterial sugar-binding and -transporting proteins.[55, 207, 208]

Core construction was accomplished by reaction of diaminopyridine with the bis(acid chloride) of (–)-(*R*)-spirobifluorene, followed by treatment of the free arylamines with *N*-[(*tert*-butyloxy)carbonyl]glycine. Subsequent *t*-Boc removal yielded the chiral diamine dendritic core (**185**). The PEG-based dendrons were constructed (Scheme 7.29) starting from nitromethanetris(propanol), which was converted to the trimesylate and reacted with monomethyltriethylene glycol to afford the nitrotris(PEG) intermediate (**186**). Central to the development of these monomers was the treatment (Bu$_3$SnH, AIBN) of **186** with acrylonitrile to effect cyanoethylation at the quaternary position yielding nitrile **187**. Hydrolysis then afforded monoacid **188**. The 2nd generation monomer was prepared by coupling of the acid chloride of **188** to triol **189** to give tetraester **190**, which was converted to the monocarboxylic acid for attachment to the core. Two new series of optically active dendroclefts for carbohydrate recognition were prepared by attachment of Fréchet-type dendrons through an ethynediyl linker to the core comprised of one or two 1,1'-binaphthalene-2,2'-diyl phosphate units;[209] binding selectivities were weak in all cases and small changes in association strengths were observed in relation to wedge size.

Molar optical rotation values [M_D^{298}] were found to be similar for generations zero through to two, which is in contrast to the observations of Meijer and Peerlings,[161] who found optical rotation to increase with generation. Enhanced flexibility of the 1,1'-binaphthalene (Meijer's core) and the attendant change in the dihedral angle of the chiral axis with increasing generation were postulated to account for the differences observed in [M_D^{298}].

Circular dichroism spectra, however, did change with generation. The maxima of the positive bands were shifted to shorter wavelengths, while major positive and negative band intensity (*I*) decreased with increasing generation. This was postulated to arise from shell–core *H*-bonding or core–pyridine torsional changes. Binding studies were performed using 1-*O*-octyl glucopyranosides, where the formation of 1:1 complexes was confirmed by Job plot analysis. Diastereoselectivity was found to increase on progressing from the 1st to the 2nd generation. This represents the first instance of dendritic branching being involved in the modulation of molecular recognition stereoselectivity. Additional data reported include binding constants, free enthalpies of complexation, and pyridine-carboxamide proton chemical shift changes following complex formation.

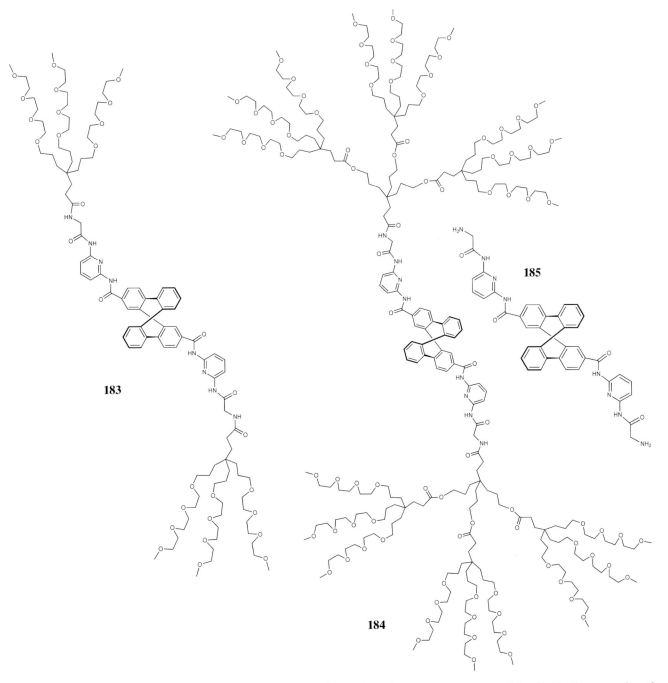

Figure 7.14 Dendroclefts possessing optically active, dendrimer-based receptors constructed for H-binding complexation of monosaccharides.[206]

The chiroptical properties of 1,1'-binaphthyl focally-modified poly(aryl ether) dendrons, i.e., normal Fréchet-type as well as "back-folded" types, have been reported.[210] CD spectroscopy was employed to examine the geometrical conformation of the torsional angle (θ) between the two naphthalene planes. Torsional angles were found to fall in the range 95–110°.

7.2.2.2 1 → 3 *P*- & *Aryl*-Branched, *P*- & *Ether* Connectivity

Brunner and Fürst[211, 212] reported the synthesis of a series of optically active, extended chelate phosphines, one of which was derived from the reaction of 5-bromo-1,3-di(borneoxymethyl)benzene (**191**) and bis(dichlorophosphanyl)methane (Scheme 7.30). Technically, the product (–)-1,1-bis{3',5'-di[(borneoxymethyl)phenyl]phosphanyl}methane (**192**) is both a 1st tier *P*- and 1,3-aryl-branched dendrimer. Excluding distortions, the maximum distance between the *P* atom and the most distant H atom(s) is 11.9 Å for

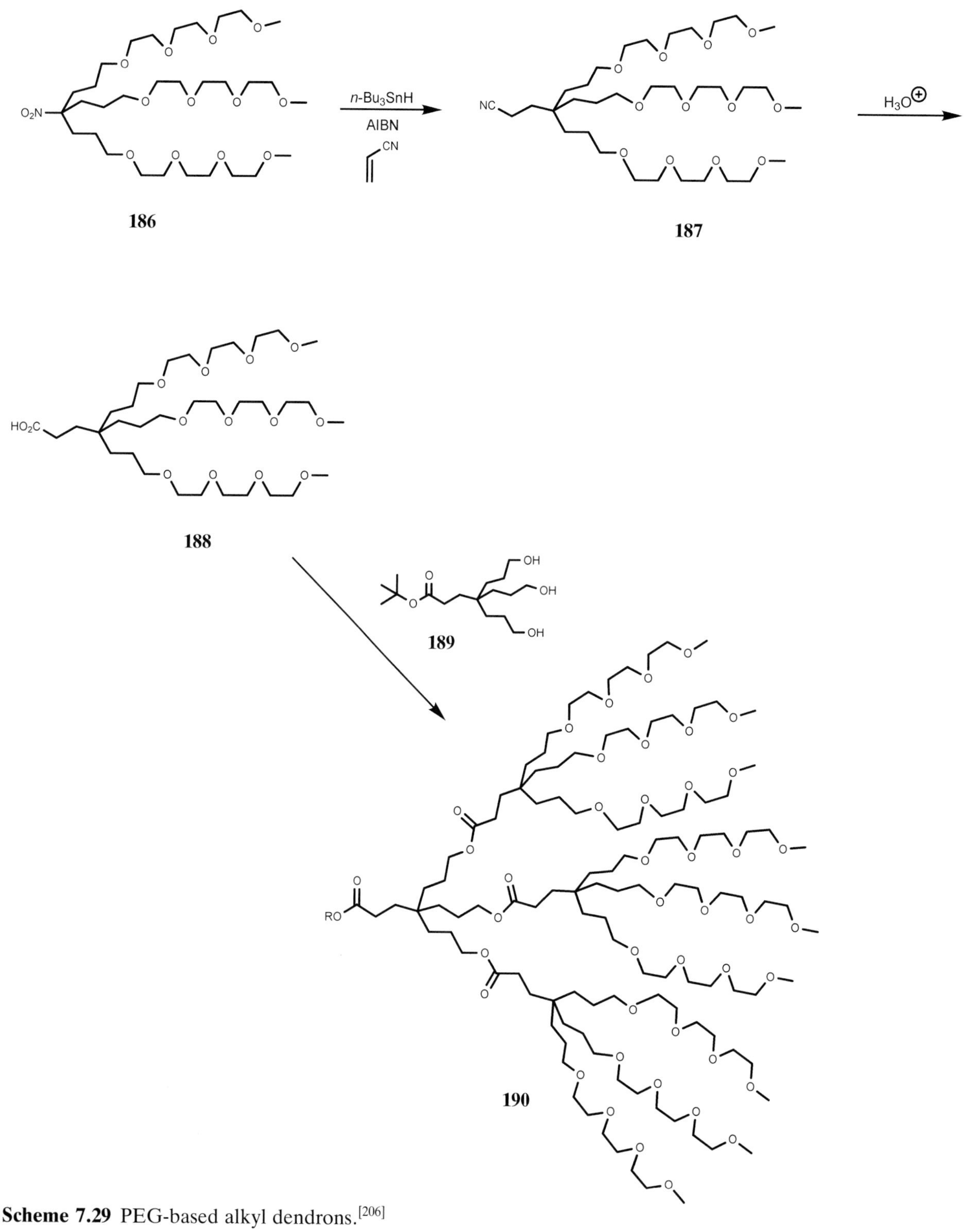

Scheme 7.29 PEG-based alkyl dendrons.[206]

the two-layer ligand **192**. The Rh-catalyzed {[Rh(cod)Cl]$_2$} hydrogenation of α-N-acetamidocinnamic acid in the presence of **192** afforded a low [5.2 % *ee* (*R*)] optical induction. Related expanded bis(phosphane) ligands possessing a chiral *trans*-1,2-substituted cyclopentane core have recently been reported in Brunner's quest for improved enantioselective catalysts.[213]

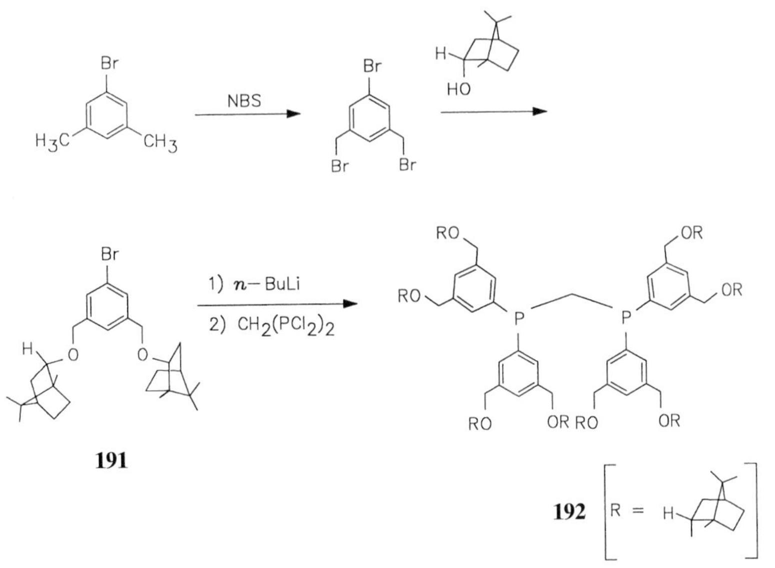

Scheme 7.30 First generation phosphine-core dendrimers possessing optically active (–)-borneol-terminated moieties.[211]

7.2.2.3 1 → 3(2) *Cholic Acid*-Branched, *Ester* Connectivity

Maitra et al.[214, 214a] described the construction of novel architectures (Scheme 7.31) based on the bile acids deoxycholic acid and cholic acid (AB₂ and AB₃ monomers, respectively). Complementary protective groups included hydroxyacetates and 1-naphthylmethyl ester moieties. Thus, reaction of two equivalents of the denaphthylated analog of **193** with the zeroth generation deacetylated diol gave the 2nd generation construct **194**. Similar transformations afforded dendrons incorporating 1 → 3 branching monomers. In each generation dendron, the reported molar rotation values followed a roughly linear relationship with the bile acid number, as well as with the molecular weight.

7.2.2.4 1 → 3 *Aryl*-Branched, *Amide & Ether* Connectivity

Hoping to gain a greater understanding of multivalency and the role it plays in carbohydrate–protein interactions, Roy et al.[130] described the synthesis of dendrimers possessing PEG spacer units and α-thiosialoside termini. The key monomers **195** and **196** (Scheme 7.32) were each obtained from methyl ester **197**, which was prepared by coupling three equivalents of a functionally differentiated tetraethylene glycol [$N_3(CH_2CH_2O)_3CH_2CH_2OTs$] to gallic acid methyl ester (K_2CO_3, DMF). Pd-mediated hydrogenolysis gave the aryl acid **196**. Coupling (EDC, 1-HBT, DIPEA, EtOH, MeCN) of triamine **195** to three equivalents of aryl acid **196** afforded dendron **198** after azide reduction and reaction with chloroacetic anhydride. Sialoside incorporation was then effected by terminal halide displacement using sialic acid thiol (**199**, 11 equiv.) to give, after deacetylation and methyl ester hydrolysis, dendrimer **200**. The 1st generation, tris-(sialoside) dendrimer was also prepared along with the corresponding triethylene glycol homologues. Initial results from binding studies using dimeric wheatgerm agglutinin and slug *Limax flavus* lectin showed that the nonavalent dendrons possessed greater affinities than their trivalent counterparts.

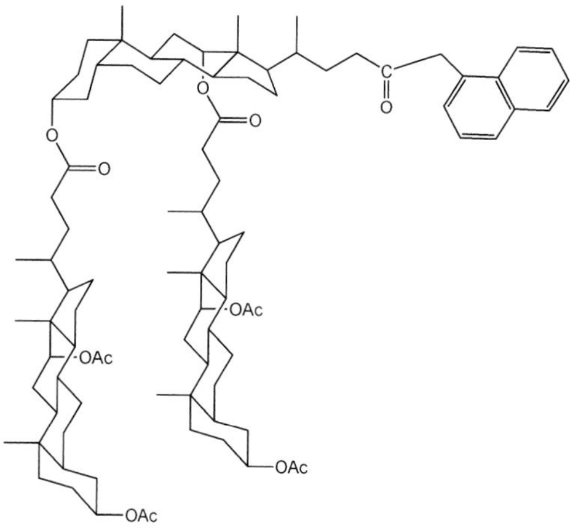

193

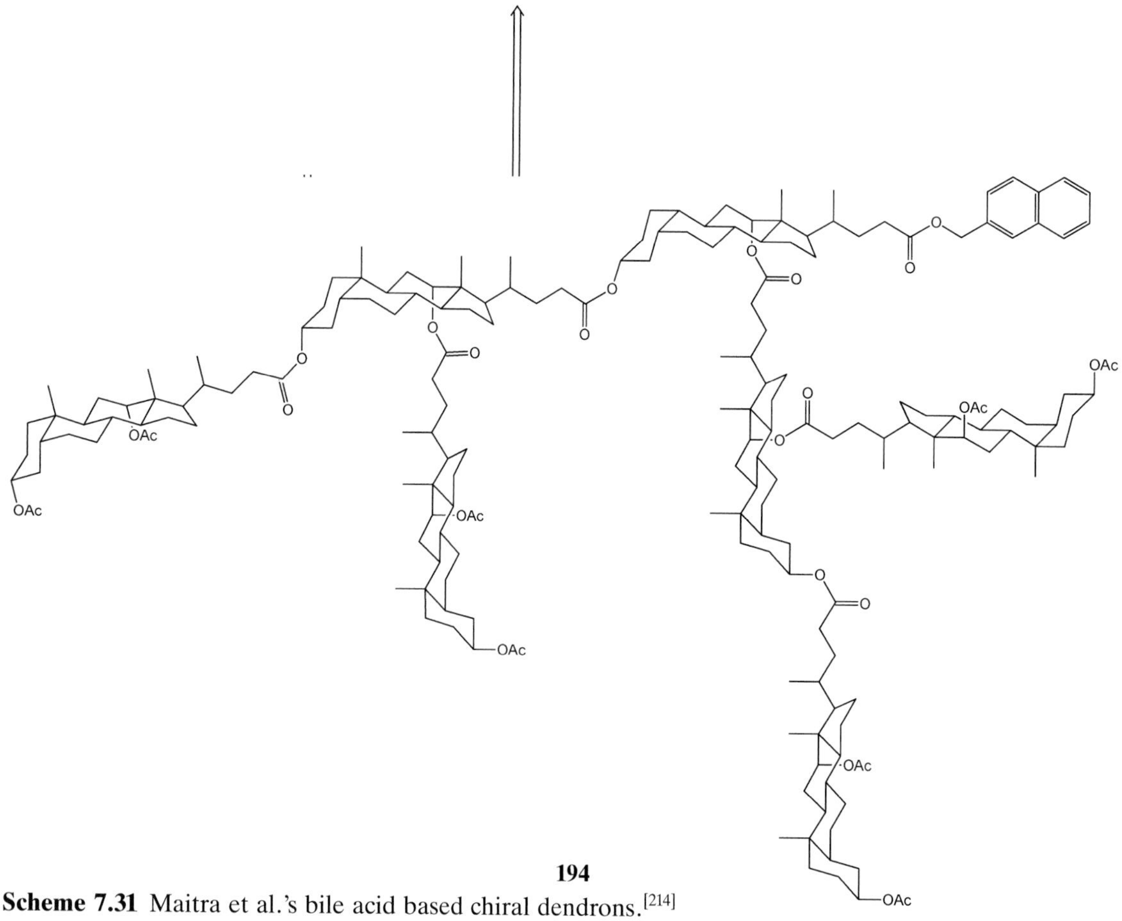

194

Scheme 7.31 Maitra et al.'s bile acid based chiral dendrons.[214]

7.3 Hyperbranched Derived Motifs

7.3.1 *Saccharide*-Branched, *Ether*-Connectivity

Kadokawa et al.[215] reported the synthesis of architectures based on an aminopolysaccharide monomer; construction was facilitated by acid-catalyzed polymerization of 2-methyl-(6-*O*-tosyl-1,2-dideoxy-α-D-glucopyrano)[2,1-*d*]-4,5-dihydrooxazole, which acts as an AB$_2$ monomer. Molecular weights increased with reaction times and ranged from 5,500 to 6,600 amu, while polydispersities were observed in the range 1.44 to 1.93.

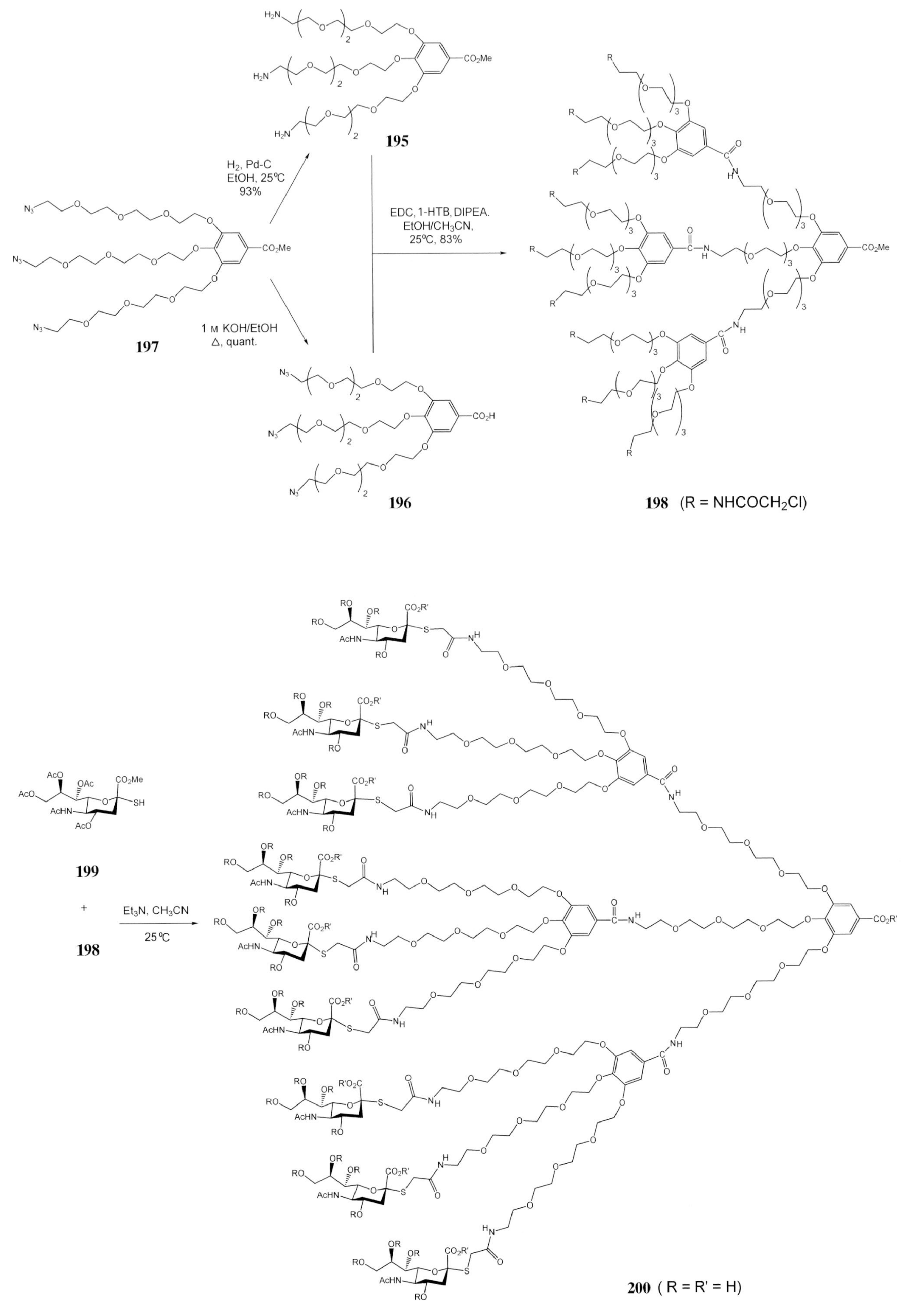

Scheme 7.32 Roy's 1 → 3 branched glycodendrimers.[130]

7.4 References

[1] H. W. I. Peerlings, E. W. Meijer, "Chirality in Dendritic Architectures", *Chem. Eur. J.* **1997**, *3*, 1563–1570.

[2] D. Seebach, P. B. Rheiner, G. Greiveldinger, T. Butz, H. Sellner, "Chiral Dendrimers", *Top. Curr. Chem.* **1998**, *197*, 125–164.

[3] H.-F. Chow, T. K. K. Mong, C.-W. Wan, Z.-Y. Wang, "Chiral Dendrimers", in *Advances in Dendritic Macromolecules* (Ed.: G. R. Newkome), JAI Press Inc., Greenwich, CN, **1999**, pp. 107–133.

[4] C. W. Thomas, Y. Tor, "Dendrimers and Chirality", *Chirality* **1998**, *10*, 59.

[5] O. Katzenelson, H. Z. Hel-Or, D. Avnir, "Chirality of Large Random Supramolecular Structures", *Chem. Eur. J.* **1996**, *2*, 174–181.

[6] C. Yeh, "Isomerism of Asymmetric Dendrimers and Stereoisomerism of Alkanes", *J. Mol. Struct.* **1998**, *432*, 153–159.

[7] R. G. Denkewalter, J. F. Kolc, W. J. Lukasavage, "Preparation of Lysine-Based Macromolecular Highly Branched Homogeneous Compound", **1979**, *U. S. Pat.*, 4, 360, 646.

[8] R. G. Denkewalter, J. F. Kolc, W. J. Lukasavage, "Macromolecular Highly Branched Homogeneous Compound Based on Lysine Units", **1981**, *U. S. Pat.*, 4, 289, 872.

[9] R. G. Denkewalter, J. F. Kolc, W. J. Lukasavage, "Macromolecular Highly Branched Homogeneous Compound", **1983**, *U. S. Pat.*, 4, 410, 688.

[10] S. M. Aharoni, C. R. Crosby, III, E. K. Walsh, "Size and Solution Properties of Globular *tert*-Butyloxycarbonyl-poly(α,ε-L-lysine)", *Macromolecules* **1982**, *15*, 1093–1098.

[11] G. R. Newkome, X. Lin, C. D. Weis, "Polytryptophan-Terminated Dendritic Macromolecules", *Tetrahedron: Asymmetry* **1991**, *2*, 957–960.

[12] H.-F. Chow, C. C. Mak, "Synthesis and Structure–Optical Rotation Relationships of Homochiral, Monodisperse, Tartaric Acid-based Dendrimers", *J. Chem. Soc., Perkin Trans. 1* **1994**, 2223–2228.

[13] D. Seebach, J.-M. Lapierre, K. Skobridis, G. Greiveldinger, "Chiral Dendrimers from Tris(hydroxymethyl)methane Derivatives", *Angew. Chem. Int. Ed. Engl.* **1994**, *33*, 440–442.

[14] J. A. Kremers, E. W. Meijer, "Synthesis and Characterization of a Chiral Dendrimer Derived from Pentaerythritol", *J. Org. Chem.* **1994**, *59*, 4262–4266.

[15] J.-P. Defoort, B. Nardelli, W. Huang, D. D. Ho, J. P. Tam, "Macromolecular Assemblage in the Design of a Synthetic AIDS Vaccine", *Proc. Natl. Acad. Sci. U.S.A.* **1992**, *89*, 3879–3883.

[16] D. N. Posnett, H. McGrath, J. P. Tam, "A Novel Method for Producing Anti-Peptide Antibodies. Production of Site-Specific Antibodies to the T-Cell Antigen Receptor β-Chain", *J. Biol. Chem.* **1988**, *263*, 1719–1725.

[17] C. Rao, J. P. Tam, "Synthesis of Peptide Dendrimer", *J. Am. Chem. Soc.* **1994**, *116*, 6975–6976.

[18] J. P. Tam, "Synthetic Peptide Vaccine Design: Synthesis and Properties of a High-Density Multiple Antigenic Peptide System", *Proc. Natl. Acad. Sci. U.S.A.* **1988**, *85*, 5409–5413.

[19] J. P. Tam, Y.-A. Lu, "Vaccine Engineering: Enhancement of Immunogenicity of Synthetic Peptide Vaccines Related to Hepatitis in Chemically Defined Models Consisting of T- and B-Cell Epitopes", *Proc. Natl. Acad. Sci. U.S.A.* **1989**, *86*, 9084–9088.

[20] K.-J. Chang, W. Pugh, S. G. Blanchard, J. McDermed, J. P. Tam, "Antibody Specific to the α-Subunit of the Guanine Nucleotide-Binding Regulatory Protein G$_o$: Developmental Appearance and Immunocytochemical Localization in the Brain", *Proc. Natl. Acad. Sci. U.S.A.* **1988**, *85*, 4929–4933.

[21] J. P. Tam, "Recent Advances in Multiple Antigen Peptides", *J. Immunol. Methods* **1996**, *196*, 17–32.

[22] J. Shao, J. P. Tam, "Unprotected Peptides as Building Blocks for the Synthesis of Peptide Dendrimers with Oxime, Hydrazone, and Thiazolidine Linkages", *J. Am. Chem. Soc.* **1995**, *117*, 3893–3899.

[23] C.-F. Liu, J. P. Tam, "Chemical Ligation Approach to Form a Peptide Bond Between Unprotected Peptide Segments. Concept and Model Study", *J. Am. Chem. Soc.* **1994**, *116*, 4149–4153.

[24] C.-F. Liu, J. P. Tam, "Synthesis of a Symmetric Branched Peptide. Assembly of a Cyclic Peptide on a Small Tetraacetate Template", *Chem. Commun.* **1997**, 1619–1620.

[25] Y. Hamuro, M. C. Calama, H. S. Park, A. D. Hamilton, "A Calixarene with Four Peptide Loops: An Antibody Mimic for Recognition of Protein Surfaces", *Angew. Chem.* **1997**, *109*, 2475–2476; *Int. Ed. Engl.* **1997**, *36*, 2680–2683.

[26] A. T. Florence, T. Sakthival, I. Toth, "Oral Uptake and Translocation of a Polylysine Dendrimer with a Lipid Surface", *J. Controlled Release* **2000**, *65*, 253–259.

[27] R. Roy, D. Zanini, S. J. Meunier, A. Romanowska, "Solid-Phase Synthesis of Dendritic Sialoside Inhibitors of Influenza A Virus Haemagglutinin", *J. Chem. Soc., Chem. Commun.* **1993**, 1869–1872.

[28] R. Roy, D. Zanini, S. J. Meunier, A. Romanowska, "Synthesis and Antigenic Properties of Sialic Acid Based Dendrimers" *ACS Symposium Series*, American Chemical Society, Washington DC, **1994**, pp. 104–119.

[28a] J. D. Reuter, A. Myc, M. M. Hayes, Z. Gan, R. Roy, D. Qin, R. Yin, L. T. Piehler, R. Esfand, D. A. Tomalia, J. R. Baker, Jr., "Inhibition of Viral Adhesion and Infection by Sialic-Acid-Conjugated Dendritic Polymers", *Bioconj. Chem.* **1999**, *10*, 271–278.

[29] D. Zanini, R. Roy, "Chemoenzymatic Synthesis and Lectin Binding Properties of Dendritic *N*-Acetyllactosamine", *Bioconj. Chem.* **1997**, *8*, 187–192.

[30] M. M. Palcic, H. Li, D. Zanini, R. S. Bhella, R. Roy, "Chemoenzymatic Synthesis of Dendritic Sialyl Lewis^X", *Carbohydr. Res.* **1998**, *305*, 433–442.

[30a] R. Roy, M.-G. Baek, K. Rittenhouse-Olson, "Synthesis of *N,N'*-bis(Acrylamido)acetic Acid-Based T-Antigen Glycodendrimers and Their Mouse Monoclonal IgG Antibody Binding Properties", *J. Am. Chem. Soc.* **2001**, *123*, 1809–1816.

[31] F. Sallas, S.-I. Nishimura, "Chemoenzymatic Synthesis of Glycoconjugates and Sequential Glycopeptides bearing Lactosamine and Sialyl Lewis^x Unit Pendant Chains", *J. Chem. Soc., Perkin Trans. 1* **2000**, 2091–2103.

[32] T. M. Chapman, G. L. Hillyer, E. J. Mahan, K. A. Shaffer, "Hydraamphiphiles: Novel Linear Dendritic Block Copolymer Surfactants", *J. Am. Chem. Soc.* **1994**, *116*, 11195–11196.

[33] J. P. Tam, Y.-A. Lu, "Coupling Difficulty Associated with Interchain Clustering and Phase Transition in Solid-Phase Peptide Synthesis", *J. Am. Chem. Soc.* **1995**, *117*, 12058–12063.

[34] J. P. Tam, J. C. Spetzler, "Chemoselective Approaches to the Preparation of Peptide Dendrimers and Branched Artificial Proteins using Unprotected Peptides as Building Blocks", *Biomed. Pept., Proteins, Nucleic Acids* **1995**, *1*, 123–132.

[35] B. Qualmann, M. M. Kessels, H.-J. Musiol, W. D. Sierralta, P. W. Jungblut, L. Moroder, "Synthesis of Boron-Rich Lysine Dendrimers as Protein Labels in Electron Microscopy", *Angew. Chem.* **1996**, *108*, 970–973; *Int. Ed. Engl.* **1996**, *35*, 909–911.

[36] B. Qualmann, M. M. Kessels, F. Klobasa, P. W. Jungblut, W. D. Sierralta, "Electron Spectroscopic Imaging of Antigens by Reaction with Boronated Antibodies", *J. Microscopy (Oxford)* **1996**, *183*, 69–77.

[37] L. Zhang, J. P. Tam, "Synthesis and Application of Unprotected Cyclic Peptides as Building Blocks for Peptide Dendrimers", *J. Am. Chem. Soc.* **1997**, *119*, 2363–2370.

[38] J. C. Spetzler, J. P. Tam, "Unprotected Peptides as Building Blocks for Branched Peptides and Peptide Dendrimers", *Int. J. Pept. Protein Res.* **1995**, *45*, 78–85.

[39] A. Pessi, E. Bianchi, F. Bonelli, L. Chiappinelli, "Application of the Continuous-Flow Polyamide Method to the Solid-Phase Synthesis of a Multiple Antigen Peptide (MAP) based on the Sequence of a Malaria Epitope", *J. Chem. Soc., Chem. Commun.* **1990**, 8–9.

[40] A. Dryland, R. C. Sheppard, "Peptide Synthesis; Part 8. A System for Solid-Phase Synthesis Under Low-Pressure Continuous Flow Conditions", *J. Chem. Soc., Perkin Trans. 1* **1986**, 125–137.

[41] T. D. Pallin, J. P. Tam, "Assembly of Cyclic Peptide Dendrimers from Unprotected Linear Building Blocks in Aqueous Solution", *Chem. Commun.* **1996**, 1345–1346.

[42] C. Grandjean, C. Rommens, H. Gras-Masse, O. Melnyk, "Convergent Synthesis of D-(–)-Quinic and Shikimic Acid-Containing Dendrimers as Potential C-Lectin Ligands by Sulfide Ligation of Unprotected Fragments", *J. Chem. Soc., Perkin Trans. 1* **1999**, 2967–2975.

[43] C. Grandjean, C. Rommens, H. Gras-Masse, O. Melnyk, "Convergent Synthesis of Fluorescein-labelled Lysine-based Cluster Glycosides", *Tetrahedron Lett.* **1999**, *40*, 7235–7238.

[44] R. Geurtsen, F. Côté, M. G. Hahn, G.-J. Boons, "Chemoselective Glycosylation Strategy for the Convergent Assembly of Phytoalexin-Elicitor Active Oligosaccharides and their Photoreactive Derivatives", *J. Org. Chem.* **1999**, *64*, 7828–7835.

[45] L. Dai, J. Lu, B. Matthews, A. W. H. Mau, "Doping of Conducting Polymers by Sulfonated Fullerene Derivatives and Dendrimers", *J. Phys. Chem. B* **1998**, *102*, 4049–4053.

[46] J. Ziegler, R. T. Chang, D. W. Wright, "Multiple-Antigenic Peptides of Histidine-Rich Protein II of *Plasmodium falciparum*: Dendrimeric Biomineralization Templates", *J. Am. Chem. Soc.* **1999**, *121*, 2395–2400.

[47] J. S. Choi, E. J. Lee, Y. H. Choi, Y. J. Jeong, J. S. Park, "Poly(ethylene glycol)–*block*–Poly(L-lysine) Dendrimer: Novel Linear Polymer/Dendrimer Block Copolymer Forming a Spherical Water-Soluble Polyionic Complex with DNA", *Bioconj. Chem.* **1999**, *9*, 62–65.

[48] J. S. Choi, D. K. Joo, C. H. Kim, K. Kim, J. S. Park, "Synthesis of a Barbell-like Triblock Copolymer, Poly(L-lysine) Dendrimer–*block*–Poly(ethylene glycol)–*block*–Poly(L-Lysine)

Dendrimer, and its Self-Assembly with Plasmid DNA", *J. Am. Chem. Soc.* **2000**, *122*, 474–480.

[48a] C. Grandjean, H. Gras-Masse, O. Melnyk, "Synthesis of Clustered Glycoside – Antigen Conjugates by Two One-Pot, Orthogonal, Chemoselective Ligation Reactions: Scope and Limitations", *Chem. Eur. J.* **2001**, *7*, 230–239.

[49] C. Grandjean, C. Rommens, H. Gras-Masse, O. Melnyk, "One-Pot Synthesis of Antigen-Bearing, Lysine-Based Cluster Mannosides Using Two Orthogonal Chemoselective Ligation Reactions", *Angew. Chem.* **2000**, *112*, 1110–1114; *ED.* **2000**, *39*, 1068–1072.

[50] D. Singh, R. Kiarash, K. Kawamura, E. C. Lacasse, J. Gariépy, "Penetration and Intracellular Routing of Nucleus-Directed Peptide-Based Shuttles (Lolitomers) in Eukarytic Cells", *Biochem.* **1998**, *37*, 5798–5809.

[51] D. Singh, S. K. Bisland, K. Kawamura, J. Gariépy, "Peptide-Based Intracellular Shuttle able to Facilitate Gene Transfer in Mammalian Cells", *Bioconj. Chem.* **1999**, *10*, 745–754.

[52] K. Sheldon, D. Liu, J. Ferguson, J. Gariépy, "Loligomers: Design of *de novo* Peptide-Based Intracellular Vehicles", *Proc. Natl. Acad. Sci. U.S.A.* **1995**, *92*, 2056–2060.

[53] A. C. Birchall, M. North, "Synthesis of Highly Branched Block Copolymers of Enantiomerically Pure Amino Acids", *Chem. Commun.* **1998**, 1335–1336.

[54] T. Kurita, Y. Kosemura, K. Kumakura, H. Kasai, H. Ito, "Syntheses and Biological Activities of Dendrimeric Mastoparans", *Chem. Lett.* **1999**, 193–194.

[55] D. K. Smith, "Supramolecular Dendritic Solubilisation of a Hydrophilic Dye and Tuning of its Optical Properties", *Chem. Commun.* **1999**, 1685–1686.

[56] T. Kato, M. Uchiyama, N. Maruo, T. Arai, N. Nishino, "Fluorescence Energy Transfer in Dendritic Poly(L-lysine)s Combining Thirty-two Free Base- and Zinc(II)-porphyrins in Scrambling Fashion", *Chem. Lett.* **2000**, 144–145.

[57] N. Maruo, M. Uchiyama, T. Kato, T. Arai, H. Akisada, N. Nishino, "Hemispherical Synthesis of Dendritic Poly(L-lysine) Containing Sixteen Free-Base Porphyrins and Sixteen Zinc Porphyrins", *Chem. Commun.* **1999**, 2057–2058.

[58] T. Kato, N. Maruo, H. Akisada, T. Arai, N. Nishino, "The NMR Spectroscopic Evaluation of Immobility of a Crowd of Porphyrin Rings Combined with Dendritic Poly(L-lysine)s", *Chem. Lett.* **2000**, 890–891.

[59] I. Toth, M. Danton, N. Flinn, W. A. Gibbons, "A Combined Adjuvant and Carrier System for Enhancing Synthetic Peptides Immunogenicity Utilizing Lipidic Amino Acids", *Tetrahedron Lett.* **1993**, *34*, 3925–3928.

[60] C. Y. Wang, D. J. Looney, M. L. Li, A. M. Walfield, J. Ye, B. Hosein, J. P. Tam, F. Wong-Staal, "Long-Term High-Titer Neutralizing Activity Induced by Octameric Synthetic HIV-1 Antigen", *Science* **1991**, *254*, 285–288.

[61] D. Zanini, W. K. C. Park, R. Roy, "Synthesis of Novel Dendritic Glycosides", *Tetrahedron Lett.* **1995**, *36*, 7383–7386.

[62] D. Zanini, W. K. C. Park, S. J. Meunier, Q. Wu, S. Aravind, B. Kratzer, R. Roy, "Synthesis and Biological Properties of Glycodendrimers" *Polym. Mater. Sci. Eng.* **1995**, *73*, 82–83.

[63] P. Scrimin, A. Veronese, P. Tecilla, U. Tonellato, V. Monaco, F. Formaggio, M. Crisma, C. Toniolo, "Metal Ion Modulation of Membrane Permeability Induced by a Polypeptide Template", *J. Am. Chem. Soc.* **1996**, *118*, 2505–2506.

[64] H.-T. Chang, C.-T. Chen, T. Kondo, G. Siuzdak, K. B. Sharpless, "Asymmetric Dihydroxylation Enables Rapid Construction of Chiral Dendrimers Based on 1,2-Diols", *Angew. Chem.* **1996**, *108*, 202–206; *Int. Ed. Engl.* **1996**, *35*, 182–186.

[65] T. Kawaguchi, K. L. Walker, C. L. Wilkins, J. S. Moore, "Double-Exponential Dendrimer Growth", *J. Am. Chem. Soc.* **1995**, *117*, 2159–2165.

[66] J. Kress, A. Rosner, A. Hirsch, "Depsipeptide Dendrimers", *Chem. Eur. J.* **2000**, *6*, 247–257.

[67] J. F. G. A. Jansen, E. M. M. de Brabander-van den Berg, E. W. Meijer, "Encapsulation of Guest Molecules into a Dendritic Box", *Science* **1994**, *266*, 1226–1229.

[68] E. M. M. de Brabander-van den Berg, E. W. Meijer, "Poly(propylene imine) Dendrimers: Large-Scale Synthesis via Heterogeneously Catalyzed Hydrogenation", *Angew. Chem.* **1993**, *105*, 1370–1373; *Int. Ed. Engl.* **1993**, *32*, 1308–1311.

[69] J. F. G. A. Jansen, E. M. M. de Brabander-van den Berg, E. W. Meijer, "Induced Chirality of Guest Molecules Encapsulated into a Dendritic Box", *Recl. Trav. Chim. Pays-Bas* **1995**, *114*, 225–230.

[70] J. F. G. A. Jansen, H. W. I. Peerlings, E. M. M. de Brabander-van den Berg, E. W. Meijer, "Optical Activity of Chiral Dendritic Surfaces", *Angew. Chem.* **1995**, *107*, 1321–1324; *Int. Ed. Engl.* **1995**, *34*, 1206–1209.

[71] J. F. G. A. Jansen, E. W. Meijer, E. M. M. de Brabander-van den Berg, "The Dendritic Box: Shape-Selective Liberation of Encapsulated Guests", *J. Am. Chem. Soc.* **1995**, *117*, 4417–4418.

[72]   H. W. I. Peerlings, J. F. G. A. Jansen, E. M. M. de Brabander-van den Berg, E. W. Meijer, "Optical Activity of Dendrimers with Chiral End Groups" *Polym. Mater. Sci. Eng.* **1995**, *73*, 342–343.

[73]   L. Cavallo, F. Fraternali, "A Molecular Dynamics Study of the First Five Generations of Poly(propylene imine) Dendrimers Modified with *N-t*-Boc-L-phenylalanines", *Chem. Eur. J.* **1998**, *4*, 927–934.

[74]   J. Issberner, M. Böhme, S. Grimme, M. Nieger, W. Paulus, F. Vögtle, "Poly(amine/imine) Dendrimers Bearing Planar Chiral Terminal Groups – Synthesis and Chiroptical Properties", *Tetrahedron: Asymmetry* **1996**, *7*, 2223–2232.

[75]   R. Moors, F. Vögtle, "Cascade Molecules: Building Blocks, Multiple Functionalization, Complexing Units, Photoswitches", in *Advances in Dendritic Macromolecules* (Ed.: G. R. Newkome), JAI, Greenwich, Conn., **1995**, pp. 41–71.

[76]   S. Chang, J. M. Galvin, E. N. Jacobsen, "Effect of Chiral Quaternary Ammonium Salts on (salen)Mn-Catalyzed Epoxidation of *cis*-Olefins. A Highly Enantioselective, Catalytic Route to *trans*-Epoxides", *J. Am. Chem. Soc.* **1994**, *116*, 6937–6938.

[77]   P. A. Ganeshpure, S. Satish, "Oxidation of (*E*)-4-Stilbenols Catalysed by Cobalt(II) Schiff Base Chelates", *Tetrahedron Lett.* **1988**, *29*, 6629–6632.

[78]   H. Sasaki, R. Irie, T. Katsuki, "Construction of Highly Efficient Mn-salen Catalyst for Asymmetric Epoxidation of Conjugated *cis*-Olefins", *Synlett* **1994**, 356–358.

[79]   M. Watkinson, A. Whiting, C. A. McAuliffe, "Synthesis of a Bis-manganese Water-Splitting Complex", *J. Chem. Soc., Chem. Commun.* **1994**, 2141–2143.

[80]   P. R. Ashton, S. E. Boyd, C. L. Brown, S. A. Nepogodiev, E. W. Meijer, H. W. I. Peerlings, J. F. Stoddart, "Synthesis of Glycodendrimers by Modification of Poly(propylene imine) Dendrimers", *Chem. Eur. J.* **1997**, *3*, 974–984.

[81]   N. Jayaraman, S. A. Nepogodiev, J. F. Stoddart, "Synthetic Carbohydrate-Containing Dendrimers", *Chem. Eur. J.* **1997**, *3*, 1193–1199.

[82]   H. W. I. Peerlings, S. A. Nepogodiev, J. F. Stoddart, E. W. Meijer, "Synthesis of Spacer-Armed Glucodendrimers Based on the Modification of Poly(propylene Imine) Dendrimers", *Eur. J. Org. Chem.* **1998**, 1879–1886.

[83]   G. M. Pavlov, E. V. Korneeva, K. Jumel, S. E. Harding, E. W. Meijer, H. W. I. Peerlings, J. F. Stoddart, S. A. Nepogodiev, "Hydrodynamic Properties of Carbohydrate-Coated Dendrimers", *Carbohydr. Polym.* **1999**, *38*, 195–202.

[84]   H. C. Hansen, S. Haataja, J. Finne, G. Magnusson, "Di-, Tri-, and Tetravalent Dendritic Galabiosides that Inhibit Hemagglutination by *Streptococcus suis* at Nanomolar Concentration", *J. Am. Chem. Soc.* **1997**, *119*, 6974–6979.

[85]   L.-X. Wang, M. Tang, T. Suzuki, K. Kitajima, Y. Inoue, S. Inoue, J.-Q. Fan, Y. C. Lee, "Combined Chemical and Enzymatic Synthesis of a *C*-Glycopeptide and its Inhibitory Activity toward Glycoamidases", *J. Am. Chem. Soc.* **1997**, *119*, 11137–11146.

[85a] M. Kawase, T. Shiomi, H. Matsui, Y. Ouji, S. Higashiyama, T. Tsutsui, K. Yagi, "Suppression of apoptosis in heptocytes by fructose-modified dendrimers", *J. Biomed. Mater. Res.* **2001**, *54*, 519–524.

[85b] T. Ren, G. Zhang, D. Liu, "Synthesis of bifunctional cationic compounds for gene delivery", *Tetrahedron Lett.* **2001**, *42*, 1007–1010.

[86]   K. Aoi, K. Itoh, M. Okada, "Globular Carbohydrate Macromolecule '"Sugar Balls"'; 1. Synthesis of Novel Sugar-Persubstituted Poly(amido amine) Dendrimers", *Macromolecules* **1995**, *28*, 5391–5393.

[87]   K. Aoi, "Sugar balls", *Kobunshi* **1996**, *45*, 260.

[88]   K. Aoi, K. Itoh, M. Okada, "Divergent/Convergent Joint Approach with a Half-Protected Initiator Core to Synthesize Surface-Block Dendrimers", *Macromolecules* **1997**, *30*, 8072–8074.

[89]   K. Aoi, K. Tsutsumiuchi, A. Yamamoto, M. Okada, "Globular Carbohydrate Macromolecules 'Sugar Balls'; 2. Synthesis of Mono(glycopeptide)-persubstituted Dendrimers by Polymer Reaction with Sugar-substituted α-Amino Acid *N*-Carboxyanhydrides (glycoNCAs)", *Macromol. Rapid Commun.* **1998**, *19*, 5–9.

[90]   K. Aoi, K. Tsutsumiuchi, A. Yamamoto, M. Okabe, "Globular Carbohydrate Macromolecules 'Sugar Balls'; 3. 'Radial-Growth Polymerization' of Sugar-substituted α-Amino *N*-Carboxyanhydrides (glycoNCAs) with a Dendritic Initiator", *Tetrahedron* **1997**, *53*, 15415–15427.

[91]   T. K. Lindhorst, C. Kieburg, "Glyco-Coating of Oligovalent Amines: Synthesis of Thiourea-Bridged Cluster Glycosides from Glycosyl Isothiocyanates", *Angew. Chem.* **1996**, *108*, 2083–2086; *Int. Ed. Engl.* **1996**, *35*, 1953–1956.

[92]   S. Hanessian, C. Hoornaert, A. G. Pernet, A. M. Nadzan, "Design and Synthesis of Potential Megacaloric Parenteral Nutrients", *Carbohydr. Res.* **1985**, *137*, C14–C16.

[93]   T. K. Lindhorst, "Glycodendrimers", *Nachrichten aus Chemie, Technik und Laboratorium* **1996**, *44*, 1073–1079.

[94]   C. Kieburg, T. K. Lindhorst, "Glycodendrimer Synthesis without using Protecting Groups", *Tetrahedron Lett.* **1997**, *38*, 3885–3888.

[95]   C. Kieburg, K. Sadalapure, T. K. Lindhorst, "Glucose-based AB₂-Building Blocks for the Construction of Branched Glycopeptidomimetics", *Eur. J. Org. Chem.* **2000**, 2035–2040.

[96]   D. Pagé, R. Roy, "Synthesis and Biological Properties of Mannosylated Starburst Poly(amidoamine) Dendrimers", *Bioconj. Chem.* **1997**, *8*, 714–723.

[97]   P. Grice, S. V. Ley, J. Pietruszka, H. M. I. Osborn, H. W. M. Priepke, S. L. Warriner, "A New Strategy for Oligosaccharide Assembly Exploiting Cyclohexane-1,2-diacetal Methodology: An Efficient Synthesis of a High Mannose Type Nonasaccharide", *Chem. Eur. J.* **1997**, *3*, 431–440.

[98]   W. K. C. Park, S. Aravind, A. Romanowska, J. Renaud, R. Roy, "Syntheses of Clustered Lactosides by Telomerization", *Methods Enzymol.* **1994**, *242*, 294–304.

[98a] G. M. Pavlov, N. Errington, S. E. Harding, E. V. Korneeva, R. Roy, "Dilute solution properties of lactosylated polyamidoamine dendrimers and their structural characteristics", *Polymer* **2001**, *42*, 3671–3678.

[98b] L. J. Twyman, "Post synthetic modification of the hydrophobic interior of a water-soluble dendrimer", *Tetrahedron Lett.* **2000**, *41*, 6875–6878.

[99]   A. Schmitzer, E. Perez, I. Rico-Lattes, A. Lattes, "First Example of High Asymmetric Induction at the 'Pseudo-Micellar' Interface of a Chiral Amphiphilic Dendrimer", *Tetrahedron Lett.* **1999**, *40*, 2947–2950.

[100] A. Schmitzer, E. Perez, I. Rico-Lattes, A. Lattes, S. Rosca, "First Example of Supramolecular Assemblies in Water of New Amphiphilic Glucose-persubstituted Poly(amidoamine) Dendrimers", *Langmuir* **1999**, *15*, 4397–4403.

[101] J. P. Mitchell, K. D. Roberts, J. Langley, F. Koentgen, J. N. Lambert, "A Direct Method for the Formation of Peptide and Carbohydrate Dendrimers", *Bioorg. Med. Chem. Lett.* **1999**, *9*, 2785–2788.

[102] N. Higashi, T. Koga, N. Niwa, M. Niwa, "Enhancement in Helicity of an Oligopeptide by its Organization onto a Dendrimer Template", *Chem. Commun.* **2000**, 361–362.

[102a] N. Higashi, T. Koga, M. Niwa, "Dendrimers with Attached Helical Peptides", *Adv. Mater.* **2000**, *12*, 1373–1375.

[103] M. Llinares, R. Roy, "Multivalent Neoglycoconjugates: Solid-Phase Synthesis of *N*-Linked α-Sialodendrimers", *Chem. Commun.* **1997**, 2119–2120.

[104] R. Roy, J. M. Kim, "Amphiphilic *p-tert*-Butylcalix[4]arene Scaffolds Containing Exposed Carbohydrate Dendrons", *Angew. Chem.* **1999**, *111*, 380–384; *Int. Ed.* **1999**, *38*, 369–372.

[105] T. Suzuki, Y. Hirokawa, K. Ohtake, T. Shibata, K. Soai, "Chiral Amino Alcohols bound to Diimines, Diamines, and Dendrimers as Chiral Ligands for the Enantioselective Ethylation of *N*-Diphenylphosphinylimines", *Tetrahedron: Asymmetry* **1997**, *8*, 4033–4040.

[106] M. Dubber, T. K. Lindhorst, "Synthesis of Chiral Carbohydrate-Centered Dendrimers", *Chem. Commun.* **1998**, 1265–1266.

[107] K. Sadalapure, T. K. Lindhorst, "A General Entry into Glycopeptide 'Dendrons'", *Angew. Chem. Int. Ed.* **2000**, *39*, 2010–2013.

[108] M. Dubber, T. K. Lindhorst, "Synthesis of Carbohydrate-Centered Oligosaccharide Mimetics Equipped with a Functionalized Tether", *J. Org. Chem.* **2000**, *65*, 5275–5281.

[109] Y. Kim, F. Zeng, S. C. Zimmerman, "Peptide Dendrimers from Natural Amino Acids", *Chem. Eur. J.* **1999**, *5*, 2133–2138.

[109a] S. Watanabe, M. Sato, S. Sakamoto, K. Yamaguchi, M. Iwamura, "New Dendritic Caged Compounds: Synthesis, Mass Spectrometric Characterization, and Photochemical Properties of Dendrimers with α-Carboxy-2-nitrobenzyl Caged Compounds at Their Periphery", *J. Am. Chem. Soc.* **2000**, *122*, 12588–12589.

[110] P. R. Ashton, S. E. Boyd, C. L. Brown, N. Jayaraman, J. F. Stoddart, "A Convergent Synthesis of a Carbohydrate-Containing Dendrimer", *Angew. Chem.* **1997**, *109*, 756–759; *Int. Ed. Engl.* **1997**, *36*, 732–735.

[111] D. Seebach, R. E. Marti, T. Hintermann, "146. Polymer- and Dendrimer-Bound Ti-TADDOLates in Catalytic (and Stoichiometric) Enantioselective Reactions: Are Pentacoordinate Cationic Ti Complexes the Catalytically Active Species?", *Helv. Chim. Acta* **1996**, *79*, 1710–1740.

[112] P. B. Rheiner, H. Sellner, D. Seebach, "Dendritic Styryl TADDOLs as Novel Polymer Cross-Linkers: First Application in an Enantioselective Et₂Zn Addition Mediated by a Polymer-Incorporated Titanate", *Helv. Chim. Acta* **1997**, *80*, 2027–2032.

[113] P. B. Rheiner, D. Seebach, "Dendritic TADDOLs: Synthesis, Characterization and Use in the Catalytic Enantioselective Addition of Et₂Zn to Benzaldehyde", *Chem. Eur. J.* **1999**, *5*, 3221–3236.

[114] H. Sellner, D. Seebach, "Dendritically Cross-Linking Chiral Ligands: High Stability of a Polystyrene-Bound Ti-TADDOLate Catalyst with Diffusion Control", *Angew. Chem.* **1999**, *111*, 2039–2041; *Int. Ed.* **1999**, *38*, 1918–1920.

[115] D. Seebach, "TADDOLs – from Enantioselective Catalysis to Dendritic Cross-Linkers to Cholesteric Liquid Crystals", *Chimia* **2000**, *54*, 60–62.

[115a] H. Sellner, C. Faber, P. B. Rheiner, D. Seebach, "Immobilization of BINOL by Cross-Linking Copolymerization of Styryl Derivatives with Styrene, and Applications in Enantioselectivity Ti and Al Lewis Acid Mediated Additions of Et₂Zn and Me₃SiCN to Aldehydes and of Diphenyl Nitrone to Enol Ethers", *Chem. Eur. J.* **2000**, *6*, 3692–3705.

[116] D. Seebach, J.-M. Lapierre, W. Jaworek, P. Seiler, "29. A Simple Procedure for the Preparation of Chiral 'Tris(hydroxymethyl)aminomethane' Derivatives", *Helv. Chim. Acta* **1993**, *76*, 459–475.

[117] D. Seebach, J.-M. Lapierre, G. Greiveldinger, K. Skobridis, "151. Synthesis of Chiral Starburst Dendrimers from PHB-Derived Triols as Central Cores", *Helv. Chim. Acta* **1994**, *77*, 1673–1688.

[118] S. Bauer, H. Fischer, H. Ringsdorf, "Highly Branched Liquid Crystalline Polymers with Chiral Terminal Groups", *Angew. Chem.* **1993**, *105*, 1658–1661; *Int. Ed. Engl.* **1993**, *32*, 1589–1592.

[119] D. Pagé, S. Aravind, R. Roy, "Synthesis and Lectin Binding Properties of Dendritic Mannopyranoside", *Chem. Commun.* **1996**, 1913–1914.

[120] M. Schuster, P. Wang, J. C. Paulson, C. H. Wong, "Solid-Phase Chemical-Enzymatic Synthesis of Glycopeptides and Oligosaccharides", *J. Am. Chem. Soc.* **1994**, *116*, 1135–1136.

[121] L. A. Lasky, "Selectins: Interpreters of Cell-Specific Carbohydrate Information During Inflammation", *Science* **1992**, *258*, 964–969.

[122] G. R. Newkome, X. Lin, "Symmetrical, Four-Directional, Poly(ether-amide) Cascade Polymers", *Macromolecules* **1991**, *24*, 1443–1444.

[122a] S. Koenig, L. Müller, D. K. Smith, "Dendritic Biomimicry: Microenvironmental Hydrogen-Bonding Effects on Tryptophan Fluorescence", *Chem. Eur. J.* **2001**, *7*, 979–986.

[123] L. A. J. M. Sleidregt, P. C. N. Rensen, E. T. Rump, P. J. van Santbrink, M. K. Bijsterbosch, A. R. P. M. Valentijn, G. A. van der Marel, J. H. van Boom, T. J. C. van Berkel, E. A. L. Biessen, "Design and Synthesis of Novel Amphiphilic Dendritic Galactosides for Selective Targeting of Liposomes to the Hepatic Asialoglycoprotein Receptor", *J. Med. Chem.* **1999**, *42*, 609–618.

[124] J.-M. Lapierre, K. Skobridis, D. Seebach, "Preparation of Chiral Building Blocks for Starburst Dendrimer Synthesis", *Helv. Chim. Acta* **1993**, *76*, 2419–2432.

[125] G. R. Newkome, G. R. Baker, "The Chemistry of Methanetricarboxylic Esters. A Review", *Org. Prep. Proced. Int.* **1986**, *18*, 117–144.

[126] P. Murer, D. Seebach, "Synthesis and Properties of First to Third Generation Dendrimers with Doubly and Triply Branched Chiral Building Blocks", *Angew. Chem.* **1995**, *107*, 2297–2300; *Int. Ed. Engl.* **1995**, *34*, 2116–2119.

[127] G. Greiveldinger, D. Seebach, "Second-Generation Trifluoromethyl-Substituted Chiral Dendrimers Containing Triply Branched Building Blocks: CF₃ as Sensitive NMR Probe for 'Remote' Diastereotopicity", *Helv. Chim. Acta* **1998**, *81*, 1003–1022.

[128] T. K. Lindhorst, M. Dubber, U. Krallmann-Wenzel, S. Ehlers, "Cluster Mannosides as Inhibitors of Type 1 Fimbriae-mediated Adhesion of *Escherichia coli*: Pentaerythritol Derivatives as Scaffolds", *Eur. J. Org. Chem.* **2000**, 2027–2034.

[128a] A. Lubineau, A. Malleron, C. Le Narvor, "Chemo-enzymatic synthesis of oligosaccharides using a dendritic soluble support", *Tetrahedron Lett.* **2000**, *41*, 8887–8891.

[129] R. Roy, W. K. C. Park, Q. Wu, S.-N. Wang, "Synthesis of Hyperbranched Dendritic Lactosides", *Tetrahedron Lett.* **1995**, *36*, 4377–4380.

[130] S. J. Meunier, Q. Wu, S.-N. Wang, R. Roy, "Synthesis of Hyperbranched Glycodendrimers Incorporating α-Thiosialosides based on a Gallic Acid Core", *Can. J. Chem.* **1997**, *75*, 1472–1482.

[131] K. Matsuoka, M. Terbatake, Y. Esumi, D. Terunuma, H. Kuzuhara, "Synthetic Assembly of Trisaccharide Moieties of Globotriaosyl Ceramide using Carbosilane Dendrimers as Cores. A New Type of Functional Glyco-Material", *Tetrahedron Lett.* **1999**, *40*, 7839–7842.

[131a] K. Matsuoka, H. Oka, T. Koyama, Y. Esumi, D. Terunuma, "An alternative route for the construction of carbosilane dendrimers uniformly functionalized with lactose or sialyllactose moieties", *Tetrahedron Lett.* **2001**, *42*, 3327–3330.

[132] M. S. Shchepinov, I. A. Udalova, A. J. Bridgman, E. M. Southern, "Oligonucleotide Dendrimers: Synthesis and Use as Polylabelled DNA Probes", *Nucleic Acids Res.* **1997**, *25*, 4447–4454.

[133] M. S. Shchepinov, E. M. Southern, "The Synthesis of Branched Oligonucleotide Structures", *Russ. J. Bioorg. Chem.* **1998**, *24*, 794–797.

[134] M. S. Shchepinov, K. U. Mir, J. K. Elder, M. D. Frank-Kamenetskii, E. M. Southern, "Oligonucleotide Dendrimers: Stable Nano-Structures", *Nucleic Acids Res.* **1999**, *27*, 3035–3041.

[135] M. S. Shchepinov, "Oligonucleotide Dendrimers: From Poly-Labelled DNA Probes to Stable Nano-Structures", *The Glenn Report* **1999**, *12*, 1–4.

[136] Y. Ueno, M. Takeba, M. Mikawa, A. Matsuda, "Nucleosides and Nucleotides; 182. Synthesis of Branched Oligodeoxynucleotides with Pentaerythritol at the Branch Point and their Thermal Stabilization of Triplex Formation", *J. Org. Chem.* **1999**, *64*, 1211–1217.

[137] P. Grice, S. V. Ley, J. Pietruszka, H. W. M. Priepke, "Synthesis of the Nonamannan Residue of a Glycoprotein with High Mannose Content", *Angew. Chem.* **1996**, *108*, 206–208; *Int. Ed. Engl.* **1996**, *35*, 197–200.

[138] T. Furuike, N. Nishi, S. Tokura, S.-I. Nishimura, "Synthesis of Novel Clustered Glycopolymers Containing Triantennary Glycosides of *N*-Acetyllactosamine", *Chem. Lett.* **1995**, 823–824.

[139] U. Sprengard, M. Schudok, W. Schmidt, G. Kretzschmar, H. Kunz, "Multiple Sialyl Lewis^x *N*-Glycopeptides: Effective Ligands for *E*-Selectin", *Angew. Chem.* **1996**, *108*, 359–362; *Int. Ed. Engl.* **1996**, *35*, 321–324.

[140] W. H. Binder, W. Schmid, "Synthesis of a Symmetric Multivalent Molecule Containing Four Carbohydrate Substituents", *Monatsh. Chem.* **1995**, *126*, 923–931.

[141] E. R. Wijsman, D. Filippov, A. R. P. M. Valentijn, G. A. van der Marel, J. H. van Boom, "Solid-Support Synthesis of Di- and Tetramannosylated Tetrathymidylic Acid", *Recl. Trav. Chim. Pays-Bas* **1996**, *115*, 397–401.

[142] P. Langer, S. J. Ince, S. V. Ley, "Assembly of Dendritic Glycoclusters from Monomeric Mannose Building Blocks", *J. Chem. Soc., Perkin Trans. 1* **1998**, 3915.

[143] L. J. Twyman, A. E. Beezer, J. C. Mitchell, "The Synthesis of Chiral Dendritic Molecules Based on the Repeat Unit L-Glutamic Acid", *Tetrahedron Lett.* **1994**, *35*, 4423–4424.

[144] L. J. Twyman, A. E. Beezer, R. Esfand, B. T. Mathews, J. C. Mitchell, "The Synthesis of Chiral Dendrimeric Molecules Based on Amino Acid Repeat Units", *J. Chem. Res., Synop.* **1998**, 759.

[145] L. J. Twyman, A. E. Beezer, J. C. Mitchell, "An Approach for the Rapid Synthesis of Moderately Sized Dendritic Macromolecules", *J. Chem. Soc., Perkin Trans. 1* **1994**, 407–411.

[146] D. Ranganathan, S. Kurur, "Synthesis of Totally Chiral, Multiple Armed, Poly Glu and Poly Asp Scaffoldings on Bifunctional Adamantane Core", *Tetrahedron Lett.* **1997**, *38*, 1265–1268.

[147] D. Ranganathan, S. Kurur, R. Gilardi, I. L. Karle, "Design and Synthesis of AB$_3$-Type (A = 1,3,5-Benzenetricarbonyl Unit; B = Glu diOMe or Glu7 OctaOMe) Peptide Dendrimers: Crystal Structure of the First Generation", *Biopolym.* **2000**, *54*, 289–295.

[148] S. A. Vinogradov, D. F. Wilson, "Electrostatic Core Shielding in Dendritic Polyglutamic Porphyrins", *Chem. Eur. J.* **2000**, *6*, 2456–2461.

[149] T. W. Nilsen, J. Grayzel, W. Prensky, "Dendritic Nucleic Acid Structures", *J. Theor. Biol.* **1997**, *187*, 273–284.

[150] H. H. Vogelbacker, R. C. Getts, N. Tian, R. Labaczewski, T. W. Nilsen, "DNA Dendrimers: Assembly and Signal Amplification", *Polym. Mater. Sci. Eng.* **1997**, *76*, 458–460.

[151] J. Wang, M. Jiang, T. W. Nilsen, R. C. Getts, "Dendritic Nucleic Acid Probes for DNA Biosensors", *J. Am. Chem. Soc.* **1998**, *120*, 8281–8282.

[152] R. J. Orentas, S. J. Rospkopf, J. T. Casper, R. C. Getts, T. W. Nilsen, "Detection of Epstein--Barr Virus EBER Sequence in Post-Transplant Lymphoma Patients with DNA Dendrimers", *J. Virol. Methods* **1999**, *77*, 153–163.

[153] C. Hawker, J. M. J. Fréchet, "A New Convergent Approach to Monodisperse Dendritic Macromolecules", *J. Chem. Soc., Chem. Commun.* **1990**, 1010–1013.

[154] P. K. Murer, J.-M. Lapierre, G. Greiveldinger, D. Seebach, "119. Synthesis and Properties of First and Second Generation Chiral Dendrimers with Triply Branched Units: A Spectacular Case of Diastereoselectivity", *Helv. Chim. Acta* **1997**, *80*, 1648–1681.

[155] P. Murer, D. Seebach, "Synthesis and Properties of Monodisperse Chiral Dendrimers (up to Fourth Generation) with Doubly Branched Building Blocks: An Intriguing Solvent Effect", *Helv. Chim. Acta* **1998**, *81*, 603–631.

[156] J. A. Kremers, E. W. Meijer, "Chirality and Dendrimers: The Issue of Chiral Recognition at the Nanoscopic Level", *Macromol. Symp.* **1995**, *98*, 491–499.

[157] M. P. Struijk, H. W. I. Peerlings, E. W. Meijer, "Cryptochirality and Dendrimers" *Polym. Prepr.* **1996**, *37*, 497–498.

[158] H. W. I. Peerlings, M. P. Struijk, E. W. Meijer, "Chiral Objects with a Dendritic Architecture", *Chirality* **1998**, *10*, 46–52.

[159] H. W. I. Peerlings, D. C. Trimbach, E. W. Meijer, "Chiral Dendrimers with Backfolding Wedges", *Chem. Commun.* **1998**, 497–498.

[160] H. W. I. Peerlings, "Chirality in Dendritic Architectures", Ph.D. Dissertation, Technische Universiteit Eindhoven, **1998**.

[161] H. W. I. Peerlings, E. W. Meijer, "Synthesis and Characterization of Axially Chiral Molecules Containing Dendritic Substituents", *Eur. J. Org. Chem.* **1998**, 573–577.

[162] Q.-H. Fan, Y.-M. Chen, X.-M. Chen, D.-Z. Jiang, F. Xi, A. S. C. Chan, "Highly Effective and Recyclable Dendritic BINAP Ligands for Asymmetric Hydrogenation", *Chem. Commun.* **2000**, 790.

[163] J.-L. Chaumette, M. J. Laufersweiler, J. R. Parquette, "Synthesis and Chiroptical Properties of Dendrimers Elaborated from a Chiral, Nonracemic Central Core", *J. Org. Chem.* **1998**, *63*, 9399–9405.

[164] J. M. Rohde, J. R. Parquette, "Synthesis of Dendrimers Containing 2,5-Anhydro-2,5-D-mannitol as a Chiral, Tetrafunctional Central Core with C_2-Symmetry", *Tetrahedron Lett.* **1998**, *39*, 9161–9164.

[164a] J. Recker, D. J. Tomcik, J. R. Parquette, "Folding Dendrons: The Development of Solvent-, Temperature-, and Generation-Dependent Chiral Conformational Order in Intramolecularly Hydrogen-Bonded Dendrons", *J. Am. Chem. Soc.* **2000**, *122*, 10298–10307.

[164b] B. Huang, J. R. Parquette, "Effect of an Internal Anthranilamide Turn Unit on the Structure and Conformational Stability of Helically Biased Intramolecularly Hydrogen-Bonded Dendron", *J. Am. Chem. Soc.* **2001**, *123*, 2689–2690.

[165] C. Reuter, G. Pawlitzki, U. Wörsdörfer, M. Plevoets, A. Mohry, T. Kubota, Y. Okamoto, F. Vögtle, "Chiral Dendrophanes, Dendro[2]rotaxanes, and Dendro[2]catenanes: Synthesis and Chiroptical Phenomena", *Eur. J. Org. Chem.* **2000**, 3059–3067.

[166] H.-F. Chow, L. F. Fok, C. C. Mak, "Synthesis and Characterization of Optically Active, Homochiral Dendrimers", *Tetrahedron Lett.* **1994**, *35*, 3547–3550.

[167] W. D. Curtis, J. F. Stoddart, G. H. Jones, "1,6,13,18,25,30-Hexaoxa[6.6.6](1,3,5)cyclophane. Attempted Synthesis of a [4]Cryptand", *J. Chem. Soc., Perkin Trans. 1* **1977**, 785–788.

[168] C. C. Mak, H.-F. Chow, "Synthesis and Chiroptical Properties of Optically Active Layer-Block Dendrimers", *Chem. Commun.* **1996**, 1185–1186.

[169] H.-F. Chow, C. C. Mak, "Preparation and Structure-Chiroptical Relationships of Tartaric Acid-Based Layer-Block Chiral Dendrimers", *J. Chem. Soc., Perkin Trans. 1* **1997**, 91–95.

[170] H.-F. Chow, C. C. Mak, "Synthesis and Chiroptical Properties of Layer-Block Dendrimers", *Pure Appl. Chem.* **1997**, *69*, 483–488.

[171] H.-F. Chow, C. C. Mak, "Facile Preparation of Optically Active Dendritic Fragments Containing Multiple Tartrate-derived Chiral Units", *Tetrahedron Lett.* **1996**, *37*, 5935–5938.

[172] J. R. McElhanon, M.-J. Wu, M. Escobar, D. V. McGrath, "Toward Chiral Dendrimers with Highly Functionalized Interiors. Dendrons from Synthetic AB_2 Monomers", *Macromolecules* **1996**, *29*, 8979–8982.

[173] J. R. McElhanon, D. V. McGrath, "Constitution, Configuration, and the Optical Activity of Chiral Dendrimers" *Polym. Prepr.* **1997**, *38*, 278–279.

[174] J. R. McElhanon, M.-J. Wu, M. Escobar, U. Chaudhry, C.-L. Hu, D. V. McGrath, "Asymmetric Synthesis of a Series of Chiral AB_2 Monomers for Dendrimer Construction", *J. Org. Chem.* **1997**, *62*, 908–915.

[175] H. C. Kolb, M. S. van Nieuwenhuize, K. B. Sharpless, "Catalytic Asymmetric Dihydroxylation", *Chem. Rev.* **1994**, *94*, 2483–2547.

[176] D. V. McGrath, M.-J. Wu, U. Chaudhry, "An Approach to Highly Functionalized Dendrimers from Chiral, Non-Racemic Synthetic Monomers", *Tetrahedron Lett.* **1996**, *37*, 6077–6080.

[177] J. McElhanon, M.-J. Wu, M. Escobar, D. V. McGrath, "Synthesis of Optically Active Chiral Dendrimers with 1,2-Diol Linkages" *Polym. Prepr.* **1996**, *37*, 495–496.

[178] D. M. Junge, D. V. McGrath, "Synthesis of Optically Active Chiral Shell Dendrons", *Tetrahedron Lett.* **1998**, *39*, 1701–1704.

[179] J. R. McElhanon, D. V. McGrath, "Constitution, Configuration, and the Optical Activity of Chiral Dendrimers", *J. Am. Chem. Soc.* **1998**, *120*, 1647–1656.

[180] J. R. McElhanon, D. V. McGrath, "Toward Chiral Polyhydroxylated Dendrimers. Preparation and Chiroptical Properties", *J. Org. Chem.* **2000**, *65*, 3525–3529.

[181] D. M. Junge, M.-J. Wu, J. R. McElhanon, D. V. McGrath, "Synthesis and Chiroptical Analysis of Optically Active Chiral Shell Dendrons", *J. Org. Chem.* **2000**, *65*, 5306–5314.

[182] C. Bolm, N. Derrien, A. Seger, "Hyperbranched Macromolecules in Asymmetric Catalysis", *Synlett* **1996**, 387–388.

[183] C. Bolm, N. Derrien, A. Seger, "Hyperbranched Chiral Catalysts for the Asymmetric Reduction of Ketones with Borane", *Chem. Commun.* **1999**, 2087–2088.

[184] A. Ikeda, M. Numata, S. Shinkai, "A Novel Attempt to Control the Aggregation Number of Dendrons with a Saccharide", *Chem. Lett.* **1999**, 929–930.

[185] M. Numata, A. Ikeda, S. Shinkai, "Properly Assembled Dendrons can be Immobilized into Dendrimers by *in situ* Cross-Link", *Chem. Lett.* **2000**, 370–371.

[186] F. Djojo, E. Ravanelli, O. Vostrowsky, A. Hirsch, "Fullerene Dendrimers and Lipofullerenes with an Inherently Chiral Hexaaddition Pattern", *Eur. J. Org. Chem.* **2000**, 1051–1059.

[186a] L.-Z. Gong, Q.-S. Hu, L. Pu, "Optically Active Dendrimers with a Binaphthyl Core and Phenylene Dendrons: Light Harvesting and Enantioselective Fluorescent Sensing", *J. Org. Chem.* **2001**, *66*, 2358–2367.

[186b] V. J. Pugh, Q.-S. Hu, L. Pu, "The First Dendrimer-Based Enantioselective Fluorescent Sensor for the Recognition of Chiral Amino Alcohols", *Angew. Chem. Int. Ed.* **2000**, *39*, 3638–3641.

[186c] E. Díez-Barra, J. C. García-Martínez, J. Rodriguez-Lópis, R. Gómez, J. L. Segura, N. Martín, "Synthesis of New 1,1'-Binaphthyl-Based Chiral Phenylenevinylene Dendrimers", *Org. Lett.* **2000**, *2*, 3651–3653.

[187] D. Seebach, G. F. Herrmann, U. D. Lengweiler, B. M. Bachmann, W. Amrein, "Synthesis and Enzymatic Degradation of Dendrimers from (*R*)-3-Hydroxybutanoic Acid and Trimesic Acid", *Angew. Chem.* **1996**, *108*, 2969–2972; *Int. Ed. Engl.* **1996**, *35*, 2795–2797.

[188] D. Seebach, G. F. Herrmann, U. D. Lengweiler, W. Amrein, "76. Synthesis of Monodisperse Macromolecular Bicyclic and Dendritic Compounds from (*R*)-3-Hydroxybutanoic Acid and Benzene-1,3,5-tricarboxylic Acid and Analysis by Fragmenting MALDI-TOF Mass Spectroscopy", *Helv. Chim. Acta* **1997**, *80*, 989–1026.

[189] G. Draheim, H. Ritter, "Polymerizable Dendrimers. Synthesis of a Symmetrically Branched Methacryl Derivative Bearing Eight Ester Groups", *Macromol. Chem. Phys.* **1995**, *196*, 2211–2222.

[190] I. Sato, T. Shibata, K. Ohtake, R. Kodaka, Y. Hirokawa, N. Shirai, K. Soai, "Synthesis of Chiral Dendrimers with a Hydrocarbon Backbone and Application to the Catalytic Enantioselective Addition of Dialkylzincs to Aldehydes", *Tetrahedron Lett.* **2000**, *41*, 3123–3126.

[191] I. Sato, R. Kodaka, T. Shibata, Y. Hirokawa, N. Shirai, K. Ohtake, K. Soai, "Highly Enantioselective Addition of Diethylzinc to *N*-Diphenylphosphinylimines using Dendritic Chiral Ligands with Hydrocarbon Backbones", *Tetrahedron: Asymmetry* **2000**, *11*, 2271–2275.

[192] R. H. E. Hudson, M. J. Damha, "Nucleic Acid Dendrimers: Novel Biopolymer Structures", *J. Am. Chem. Soc.* **1993**, *115*, 2119–2124.

[193] M. J. Damha, K. Ganeshan, R. H. E. Hudson, S. V. Zabarylo, "Solid-Phase Synthesis of Branched Oligoribonucleotides Related to Messenger RNA Splicing Intermediates", *Nucleic Acids Res.* **1992**, *20*, 6565–6573.

[194] M. J. Damha, S. Zabarylo, "Automated Solid-Phase Synthesis of Branched Oligonucleotides", *Tetrahedron Lett.* **1989**, *30*, 6295–6298.

[195] R. H. E. Hudson, S. Robidoux, M. J. Damha, "Divergent Solid-Phase Synthesis of Nucleic Acid Derivatives", *Tetrahedron Lett.* **1998**, *39*, 1299–1302.

[196] F. Bambino, R. T. Brownlee, F. C. Chiu, "Synthesis of a Symmetrically Branched Template for Parallel α-Helix Dimers", *Tetrahedron Lett.* **1994**, *35*, 4619–4622.

[197] P. R. Ashton, S. E. Boyd, C. L. Brown, N. Jayaraman, S. A. Nepogodiev, J. F. Stoddart, "A Convergent Synthesis of Carbohydrate-Containing Dendrimers", *Chem. Eur. J.* **1996**, *2*, 1115–1128.

[198] N. Jayaraman, J. F. Stoddart, "Synthesis of Carbohydrate-Containing Dendrimers; 5. Preparation of Dendrimers Using Unprotected Carbohydrates", *Tetrahedron Lett.* **1997**, *38*, 6767–6770.

[199] P. R. Ashton, E. F. Hounsell, N. Jayaraman, T. M. Nilsen, N. Spencer, J. F. Stoddart, M. Young, "Synthesis and Biological Evaluation of α-D-Mannopyranoside-Containing Dendrimers", *J. Org. Chem.* **1998**, *63*, 3429–3437.

[200] B. Colonna, V. D. Harding, S. A. Nepogodiev, F. M. Raymo, N. Spencer, J. F. Stoddart, "Synthesis of Oligosaccharide Dendrimers", *Chem. Eur. J.* **1998**, *4*, 1244–1254.

[200a] K.-S. Kim, K. Katsuraya, Y. Yachi, K. Hatanaka, T. Uryu, "Synthesis of Lysine-Core Dendrimer Containing Long Pyrrole-Terminated Alkylene Derivative", *Sen'i Gakkaishi* **2000**, *56*, 584–591.

[201] F. Cardullo, F. Diederich, L. Echegoyen, T. Habicher, N. Jayaraman, R. M. Leblanc, J. F. Stoddart, S. Wang, "Stable Langmuir and Langmuir–Blodgett Films of Fullerene-Glycodendron Conjugates", *Langmuir* **1998**, *14*, 1955–1959.

[202] S. Cicchi, A. Goti, C. Rosini, A. Brandi, "Enantiomerically Pure Dendrimers Based on a *trans*-3,4-Dihydroxypyrrolidine", *Eur. J. Org. Chem.* **1998**, 2591–2597.

[203] I. Baussanne, J. M. Benito, C. O. Mellet, J. M. G. Fernández, H. Law, J. Defaye, "Synthesis and Comparative Lectin-Binding Affinity of Mannosyl-Coated β-Cyclodextrin-Dendrimer Constructs", *Chem. Commun.* **2000**, 1489–1490.

[204] W. B. Turnbull, A. R. Pease, J. F. Stoddart, "Synthetic Carbohydrate Dendrimers, Part 8. – Toward the Synthesis of Large Oligosaccharide-Based Dendrimers", *Chem. Eur. J. Chem. Bio.* **2000**, *1*, 70–74.

[205] D. K. Smith, A. Zingg, F. Diederich, "Dendroclefts: Optically Active Dendritic Receptors for the Selective Recognition and Chiroptical Sensing of Monosaccharide Guests", *Helv. Chim. Acta* **1999**, *82*, 1225–1241.

[206] D. K. Smith, F. Diederich, "Functional Dendrimers: Unique Biological Mimics", *Chem. Eur. J.* **1998**, *4*, 1353–1361.

[207] D. K. Smith, F. Diederich, "Dendritic Hydrogen Bonding Receptors: Enantiomerically Pure Dendroclefts for the Selective Recognition of Monosaccharides", *Chem. Commun.* **1998**, 2501–2502.

[208] D. K. Smith, L. Müller, "Dendritic Biomimicry: Microenvironmental Effects on Tryptophan Fluorescence", *Chem. Commun.* **1999**, 1915–1916.

[209] A. Bähr, B. Felber, K. Schneider, F. Diederich, "Dendritic 1,1'-Binaphthalene-Derived Cleft-Type Receptors (*Dendroclefts*) for the Molecular Recognition of Pyranosides", *Helv. Chim. Acta* **2000**, *83*, 1346–1376.

[210] C. Rosini, S. Superchi, H. W. I. Peerlings, E. W. Meijer, "Enantiopure Dendrimers Derived from the 1,1'-Binaphthyl Moiety: A Correlation Between Chiroptical Properties and Conformation of the 1,1'-Binaphthyl Template", *Eur. J. Org. Chem.* **2000**, 61–71.

[211] H. Brunner, J. Fürst, "Enantioselective Catalysis, 85 [1]. Optically Active Expanded Chelate Phosphines Derived from 1,5-Bis(dichlorophosphino)alkanes", *Tetrahedron* **1994**, *50*, 4303–4310.

[212] H. Brunner, "Dendrizymes: Expanded Ligands for Enantioselective Catalysis", *J. Organomet. Chem.* **1995**, *500*, 39–46.

[213] H. Brunner, S. Stefaniak, M. Zabel, "Enantioselective Catalysis; 130. Optically Active Expanded Ligands Based on the *trans*-1,2-Substituted Cyclopentane Skeleton", *Synthesis* **1999**, 1775–1784.

[214] S. Balasubramanian, P. Rao, U. Maitra, "First Bile Acid-Derived Chiral Dendritic Species with Nanometric Dimensions", *Chem. Commun.* **1999**, 2353–2354.

[214a] R. Balasubramanian, U. Maitra, "Design and Synthesis of Novel Chiral dendritic Species Derived from Bile Acids", *J. Org. Chem.* **2001**, *66*, 3035–3040.

[215] J. Kadokawa, M. Sato, M. Karasu, H. Tagaya, K. Chiba, "Synthesis of Hyperbranched Aminopolysaccharides", *Angew. Chem.* **1998**, *110*, 2540–2543; *Int. Ed. Engl.* **1998**, *37*, 2373–2376.

8 Metallodendrimers

8.1 Introduction

The introduction of metal centers on the surface or within dendrimers has been a recent trend reflecting a shift from the initial synthetic directions to a more applied emphasis. In general, surface modification is the easier mode of attachment as compared to the incorporation of metal centers within the interior of the dendrimers or hyperbranched polymers, due to a lack of appropriately designed building blocks or the internal availability of binding loci. The organometallic dendrimers offer attractive advantages over their polymeric counterparts in that they possess a precise molecular architecture as well as a predetermined chemical composition. Metal centers have been shown to act as connectors, branching points, or terminal (surface) centers; moreover, metal sites can be located at either specific or random loci within the dendritic infrastructure. For reviews, see refs.[1–18]

The merger of supramolecular chemistry, defined as "chemistry beyond the molecule" by Lehn,[19–22] and the chemistry of dendrimers, leads directly to "suprasupermolecular" chemistry[23] and the creation of the metallodendritic regime. The incorporation of metal ions into dendritic frameworks was initiated by Balzani's[24–26] and Newkome's[27] research groups in the early 1990's through the use of metal branching centers or metal complexation, respectively. Commonly, metals have served as *branching centers, building block connectors* (including core as well as monomer connection), *terminal groups*, and *structural auxiliaries* (whereby metals are introduced into a framework after dendritic construction) (Figure 8.1). The latter category allows for *site-specific* versus *random* internal inclusion. These main categories shall serve as the basis for metallodendrimer classification in this review.

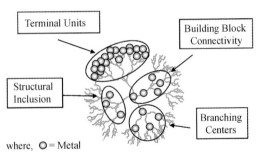

Figure 8.1 Key positions of metals within the dendritic framework. Reprinted with permission of the American Chemical Society.[3]

8.2 Metals as Branching Centers

The first synthesis of metallodendrimers possessing homo- and hetero-metallic branching centers (ruthenium and osmium) was reported by Balzani et al.[24–26, 28, 29] (Scheme 8.1). The decanuclear polypyridine complex **3** was prepared [MeOH/H$_2$O/(CH$_2$OH)$_2$/ AgNO$_3$] from a trigonal core, M(BL)$_3^{2+}$ (**1**), containing a metal coordinated to three 2,4-bis(2-pyridyl)pyrazine (BL) ligands,[30] and the preconstructed building block Ru[(BL)M(L)$_2$]$_2$Cl(PF$_6$)$_4$ (**2**), where L is either bipyridine (bpy) or biquinoline (biq).[31] Balzani et al.[32] further reported the examination of analogous metallodendrimers derived from transition metal complexes[33, 34] and subsequently studied their luminescence properties, electrochemical and redox behavior, intercomponent energies, and electron-transfer abilities. Purely Ru-containing complexes exhibit luminescence bands at around 770 nm, whereas all the Ru/Os mixed-metal complexes display bands in the 850–1000 nm spectral region, which can be assigned to triplet metal-to-ligand charge-transfer transitions localized on the Os-bearing components. For such Ru/Os mixed-metal constructs, only an Os-based luminescence is observed. Electrochemical results indicate that each redox step in these metal complexes is essentially localized on a specific ligand. Assignments of the redox sites have been proposed, and their mutual interactions have been discussed.[33] The electrochemistry of these dendrimers in liquid SO$_2$ has appeared.[35] Furthermore, the electrochemical behaviour of a family of these materials has been studied;[36] an extensive ligand-centered redox series comprised of up to 26 reversible reduction processes [for a hexanuclear, Ru(II)-based complex] was reported.

The synthetic strategy developed by Balzani's group allows the preparation of dendrimers possessing two or more different metals and/or ligands. Each dendrimer can be viewed as an ordered ensemble of weakly interacting [M(L)$_n$(BL)$_{3-n}$]$^{2+}$ units [M = Ru(II) or Os(II); L = bpy or biq; BL = 2,3- or 2,5-dpp; n = 0 or 2]. Absorption bands and the electrochemical properties of each unit are only slightly perturbed in the assembled structure, so that the absorption spectrum and redox pattern of a dendrimer resemble the "sum" of the spectra and redox patterns of the constituent units. Electronic interactions, however, are sufficiently strong to allow very fast (exergonic) energy transfer between adjacent units. Thus, with their inherent maximized degree of control of the electrochemical behavior, light absorption properties, and direction of energy transfer, these dendrimers have been touted as being ideal for use in light-harvesting devices.[37, 38]

Campagna and coworkers[39, 40] pursued the investigation of higher generations of this series (Scheme 8.2), whereas earlier reports were focused on tetra-[41] or decanuclear[42] complexes. Some core 2D-COSY experiments have also been performed.[43] The largest member (i.e., **6**, possessing 22 metal centers) was prepared by treatment of the tetrakis-Ru(II) core **4** with the tris-Ru(II) dendron **5**. These metallodendrimers exhibit[39] strong absorption in the UV/vis spectral region, a moderate red luminescence, as well as a number of metal-based oxidation and ligand-based reduction processes. For instance, **6** shows a UV absorption maximum at 542 nm with ε = 2.02 × 10^5 (M^{-1}cm^{-1}), a luminescence band at 786 nm (τ = 45 ns), and a peak at +1.52 V (*vs.* SCE) corresponding to the simul-

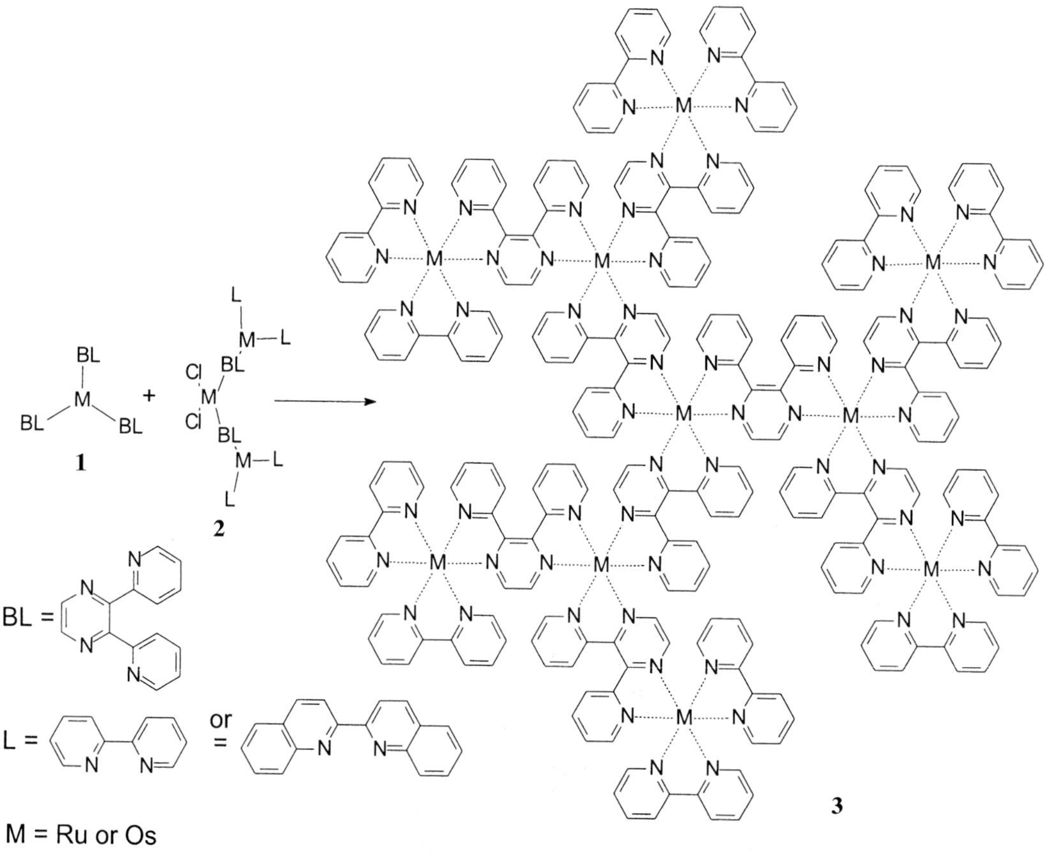

Scheme 8.1 Balzani's pyridyl-based metallodendrimers.[24–26, 28, 29]

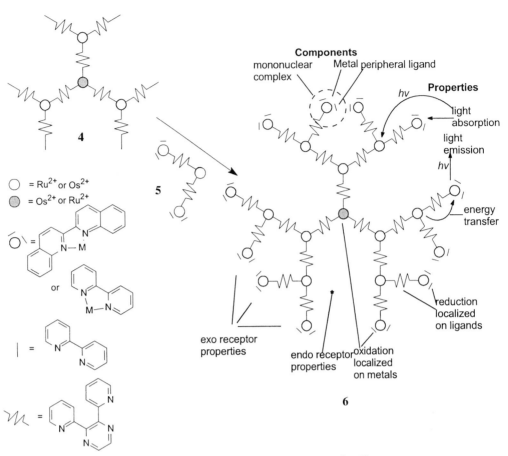

Components

mononuclear   Metal peripheral ligand
complex

Properties

light
absorption

light
emission

energy
transfer

reduction
localized
on ligands

exo receptor
properties

endo receptor
properties

oxidation
localized
on metals

○ = Ru²⁺ or Os²⁺
◉ = Os²⁺ or Ru²⁺

Scheme 8.2 Ru- and Os-containing macromolecules possessing up to 22 metal centers.[39, 40]

taneous oxidation of the twelve weakly interacting metal ions. The absorption spectra, luminescence properties, and electrochemical behavior of the various generations of these metallodendrimers have been reported.[40] Some electrochemical interactions between neighboring units were observed, although as noted above, each metal-based building block in the polymetallic dendrimers gave rise to discrete absorption properties.

The analytical determination of the amount of ruthenium and osmium in these types of dendrimers was reported by Taddia, Lucano, and Juris.[44] Their procedure involved electrothermal atomic absorption spectrometry and was also applied to the characterization of polynuclear complexes containing both metals. A heptanuclear Ru(II) dendrimer based on 1,4,5,8,9,12-hexaazatriphenylene (HAT) has been imaged by means of STM as a mono-add layer on graphite.[45]

A tetranuclear ruthenium complex {Ru[(tpphz)Ru(bpy)$_2$]$_3$}$^{8+}$, where tpphz is tetrapyridino[3,2-*a*:2',3'-*c*:3'', 2''-*h*:2'',3''-*j*]phenazine, has been constructed[46] both divergently and convergently. Despite it being a mixture of eight stereoisomers, full ^{1}H NMR characterization was possible. Campagna et al.[47] also investigated branched Ru(II) complexes using tpphz as a spacer moiety. Absorption spectra, photophysical properties, and redox behavior of the stereochemically pure dendritic tetramers was reported.

Arakawa, Haga, et al.[48] prepared a related tetrakis-Ru(II) complex, which was characterized by analysis of multiply-charged ions generated by electrospray ionization mass spectrometry. The proton-induced switching of electron-transfer pathways in the analogous mono-Ru(II)–tris-Os(II) complex was also evaluated.[49] A spectroelectrochemical analysis by the "flow-through" method was subsequently reported for these mixed valence di- and tetranuclear Ru building blocks.[50] Kirsch-De Mesmaeker et al.[51] reported the synthesis and electrospray mass spectrometric characterization of a larger hepta-Ru(II) construct based on HAT; its mass spectra featured four sets of peaks attributed to 3–6 charges.

The first synthesis of "dendritic bismuthanes," reported by Suzuki and coworkers[52] (Scheme 8.3), involved directed tris-*ortho*-lithiation (*t*-BuLi) of bismuthane **7**, followed by treatment with three equivalents of bis[2-(diethylaminosulfonyl)phenyl]bismuth

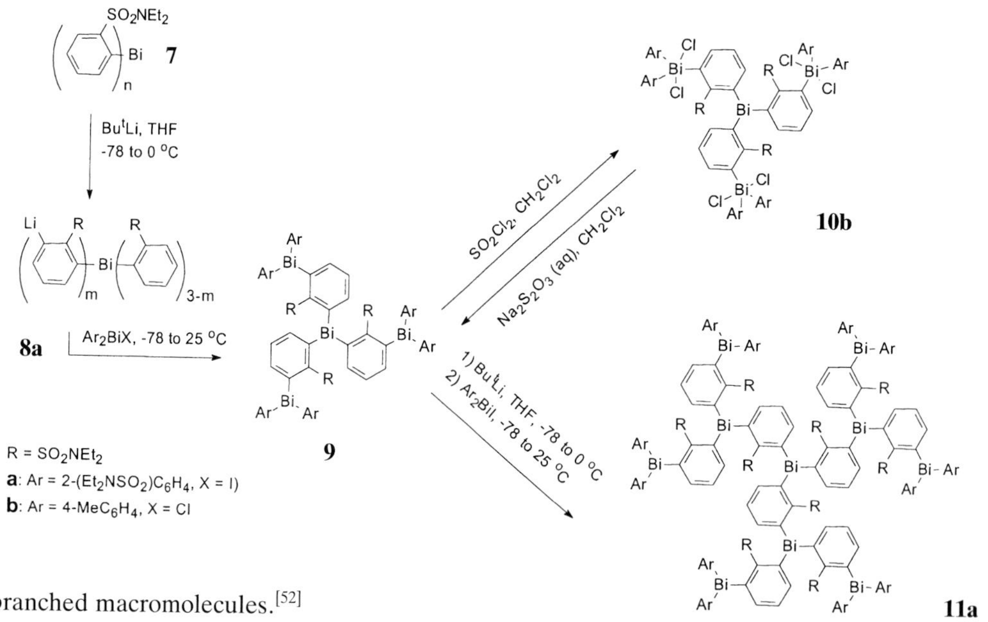

SO₂NEt₂

$$\left(\begin{array}{c}\text{Bi}\end{array}\right)_n \quad \mathbf{7}$$

Bu^{t}Li, THF
-78 to 0 °C

$$\left(\begin{array}{c}\text{Li} \quad R \quad R\\ \text{Bi}\end{array}\right)_{m}\left(\right)_{3-m}$$

8a   Ar₂BiX, -78 to 25 °C

R = SO₂NEt₂
a: Ar = 2-(Et₂NSO₂)C₆H₄, X = I
b: Ar = 4-MeC₆H₄, X = Cl

9

SO₂Cl₂, CH₂Cl₂

Na₂S₂O₃ (aq), CH₂Cl₂

10b

1) Bu^{t}Li, THF, -78 to 0 °C
2) Ar₂BiI, -78 to 25 °C

11a

Scheme 8.3 Suzuki's Bi-branched macromolecules.[52]

iodide (**8**) to give the symmetrically branched Bi₄-bismuthane **9** as the main product. Similarly, tetrakis-Bi **9** was converted, in a one-pot procedure, to Bi₁₀-bismuthane **11**. Reaction of tetrakis-Bi **9** with SO₂Cl₂ gave the corresponding chloride **10**, which could be reduced (Na₂S₂O₃) with reversion to **9**. Later, a similar Bi₁₀-bismuthane was successfully prepared by Matano et al.,[53] along with related phenylene-bridged oligomers.

Bochkarev et al.[54] constructed hyperbranched, Ge-based macromolecules by polymerization of tris(pentafluorophenyl)germanium anions (Scheme 8.4). Building block assembly was accomplished by *Ge*-substitution of the para-fluoro substituents in the branched monomers such as **12**, affording the hyperbranched polymers (e.g., **13**) possessing M_ws in the range of 100,000 to 170,000 amu. The rheological properties of these poly(fluorophenylene germane) materials were studied.[55]

Mazerolles et al.[56] reported the synthesis of Ge-branched, alkane-based metallodendrimers (Scheme 8.5). The 1st generation dodecaalkene **15** was prepared by both divergent and convergent procedures from core **14**, whereas the 2nd generation 36-alkene **16** was obtained only by the divergent approach. The authors further noted that differentiation of different generations by IR and NMR was not feasible, although mass spectrometric analysis was facilitated by terminal functionalization using methyl thioglycolate affording lipophilic interiors and hydrophilic shells.

Nanjo and Sekiguchi[57] reported the synthesis of metallodendrimers possessing alternating Si-branching centers and Ge-branched chain connectors (Scheme 8.6). Employing divergent construction, initial reaction of chlorodimethylphenylgermane with lithiosilane monomer **17** gave core **18**; this was followed by phenyl cleavage (TrfOH) and subsequent reaction with more silane monomer **17** to afford (12 %) the 1st generation nonagermane **19**. Sequential reaction (TrfOH, NH₄$^+$Cl$^-$, and MeMgI) of the dendritic hybrid **19** gave the corresponding hexatriflate **20**, the hexachloride **21**, and the permethylated species **22**. The molecular structure of this permethylated analog of **19** was confirmed by X-ray dif-

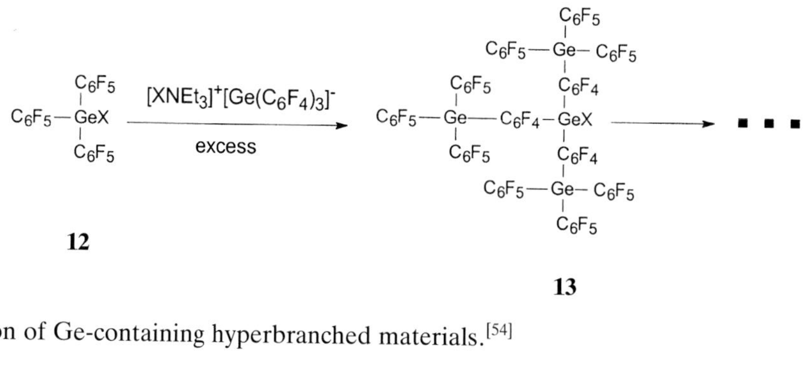

C₆F₅

$$\text{C}_6\text{F}_5-\overset{\overset{\displaystyle \text{C}_6\text{F}_5}{|}}{\underset{\underset{\displaystyle \text{C}_6\text{F}_5}{|}}{\text{GeX}}}$$

[XNEt₃]$^+$[Ge(C₆F₄)₃]$^-$
—————————→
excess

C₆F₅
|
C₆F₅—Ge—C₆F₄
|
C₆F₅

C₆F₅
|
C₆F₅—Ge—C₆F₅
|
C₆F₄
|
—C₆F₄—GeX
|
C₆F₄
|
C₆F₅—Ge—C₆F₅
|
C₆F₅

■ ■ ■

12

13

Scheme 8.4 Construction of Ge-containing hyperbranched materials.[54]

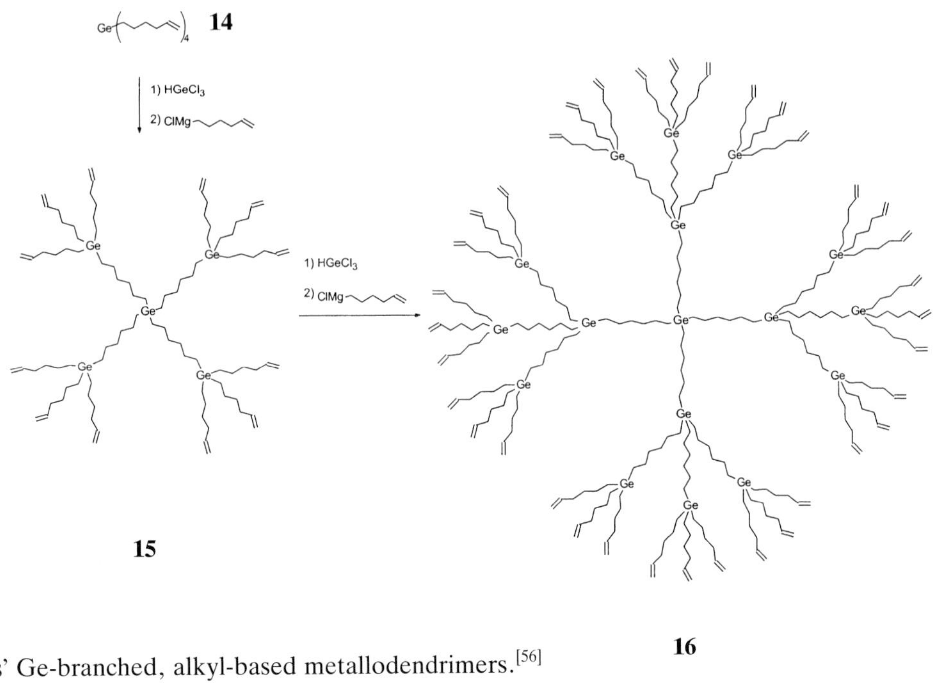

Scheme 8.5 Mazerolles' Ge-branched, alkyl-based metallodendrimers.[56]

fraction analysis, while its UV spectrum in 3-methylpentane at 300 K exhibited an absorption (λ = 271 nm; ε = 5.07 × 10^4 M^{-1}cm^{-1}) attributable to the σ-σ^* band of the Si–Ge chains.

A porphyrin–bis(fullerene) conjugate was prepared by Nierengarten et al.[58] using benzyl ether based dendrons; as demonstrated by variable-temperature NMR, the construct was obtained as a mixture of two conformers.

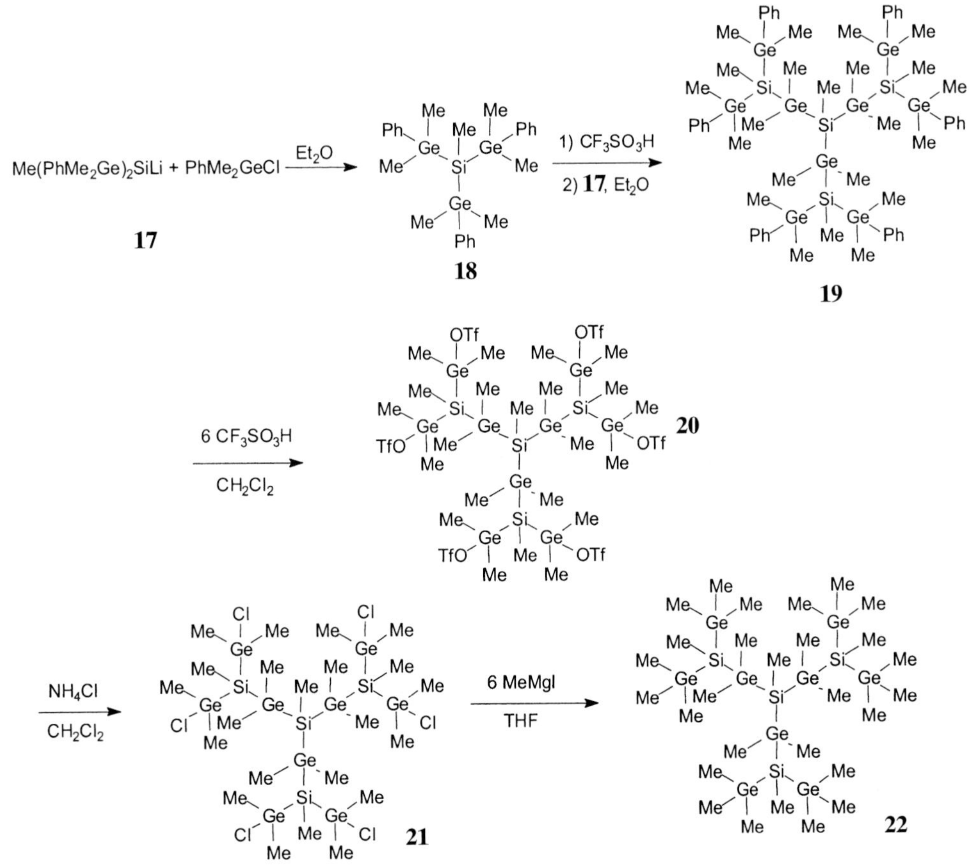

Scheme 8.6 Metallomacromolecules possessing both Ge and Si branching centers.[57]

MacDonnell and coworkers[59] prepared 1st generation chiral metallodendrimers based on condensation of enantiopure Δ-[Ru(1,10-phenanthroline-5,6-dione)₃][PF₆]₂ with the corresponding diamine. The diastereomerically and enantiomerically pure products obtained were fully characterized, including by MALDI-TOF MS. Larger structures, along with a discussion of their global chirality, were subsequently reported.[60]

8.3   Metals as Building Block Connectors

Newkome et al.[61] first reported the preparation of metallodendrimers incorporating convenient ligand–metal–ligand (denoted herein as –<M>–, where '>–' ≡ 4'-substituted terpy) connectivity for component attachment (Scheme 8.7). The resulting dendrimer **27** possessed twelve pseudooctahedral Ru(II) centers,[62, 63] where each metal center was coordinated by two different orthogonal 4'-substituted 2,2':6',2"-terpyridine units.

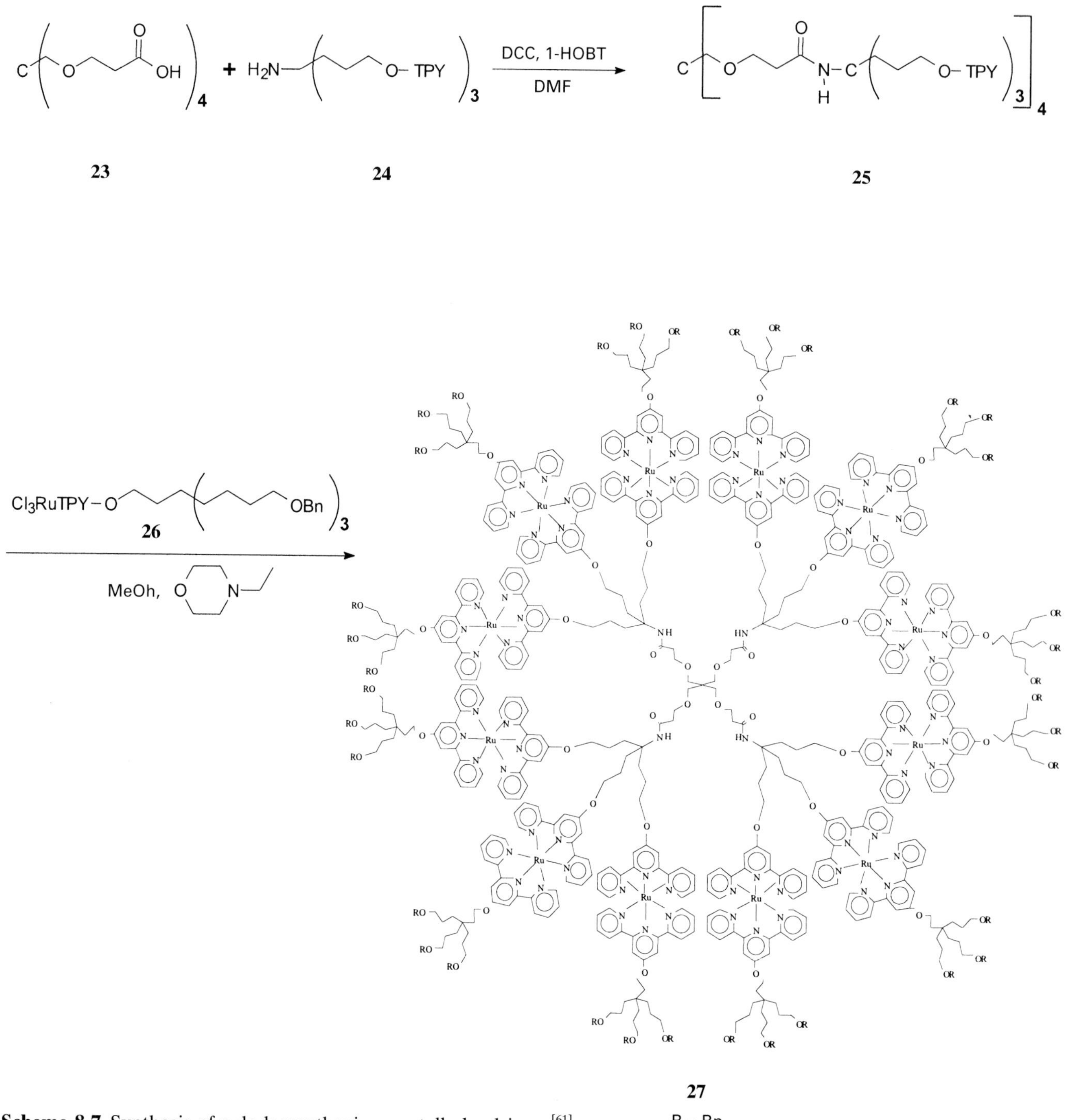

Scheme 8.7 Synthesis of a dodecaruthenium metallodendrimer.[61]

Metallodendrimer construction was achieved by coupling tetracarboxylic acid[64] **23** with **24** (accessed by base-promoted alkoxylation of 4'-chloroterpyridine[65] with the corresponding aminotriol) to give dodecaterpyridine **25**, reaction (MeOH, *N*-ethylmorpholine) of which with the Ru(III) adduct **26** gave the desired diamagnetic metallodendrimer **27**.

This strategy of –<Ru>– connectivity has been applied to the development of metal complexes capable of connecting multiple, preconstructed branched polymers, thereby resulting in the creation of dendritic assemblies or networks.[66, 67] Thus, dendrimers **30** was subsequently prepared (Scheme 8.8) through the formation of a single –<Ru>– connection between two independently constructed dendrons (i.e., **28** and **29**).[68] Preconstructed dendron synthesis was facilitated by 4'-chloroterpyridine alkoxylation using the appropriate ω-hydroxycarboxylic acid to provide the requisite functionalized cores. Generational, amide-based dendrimer construction was then accomplished by iterative DCC coupling and ester deprotection using "Behera's amine"[69, 70] (as described in Section 4.1.4). Cyclic voltammetry experiments conducted on this series of complexes revealed electrochemically reversible redox behavior in cases of connected lower-generation components, whereas irreversible behavior was observed for complexes comprised of locks of generation 3 or higher and keys of generation 2 or higher. It was postulated that this resulted from slowed electron transfer due to branch chain steric hindrance.

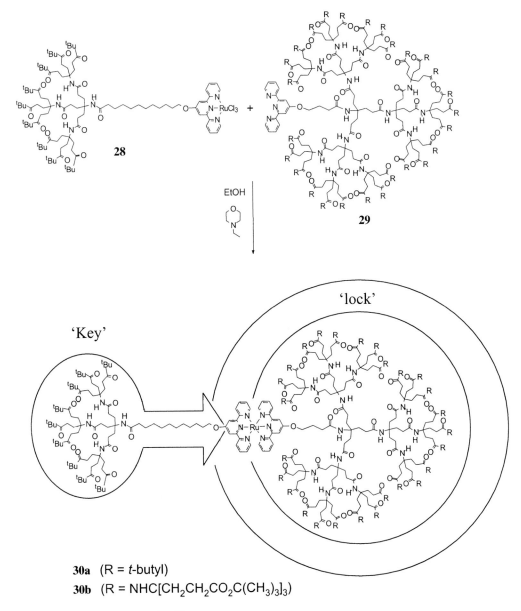

30a (R = *t*-butyl)
30b (R = NHC[CH₂CH₂CO₂C(CH₃)₃]₃)

Scheme 8.8 Newkome's 'key' and 'lock' assembly.[66, 67]

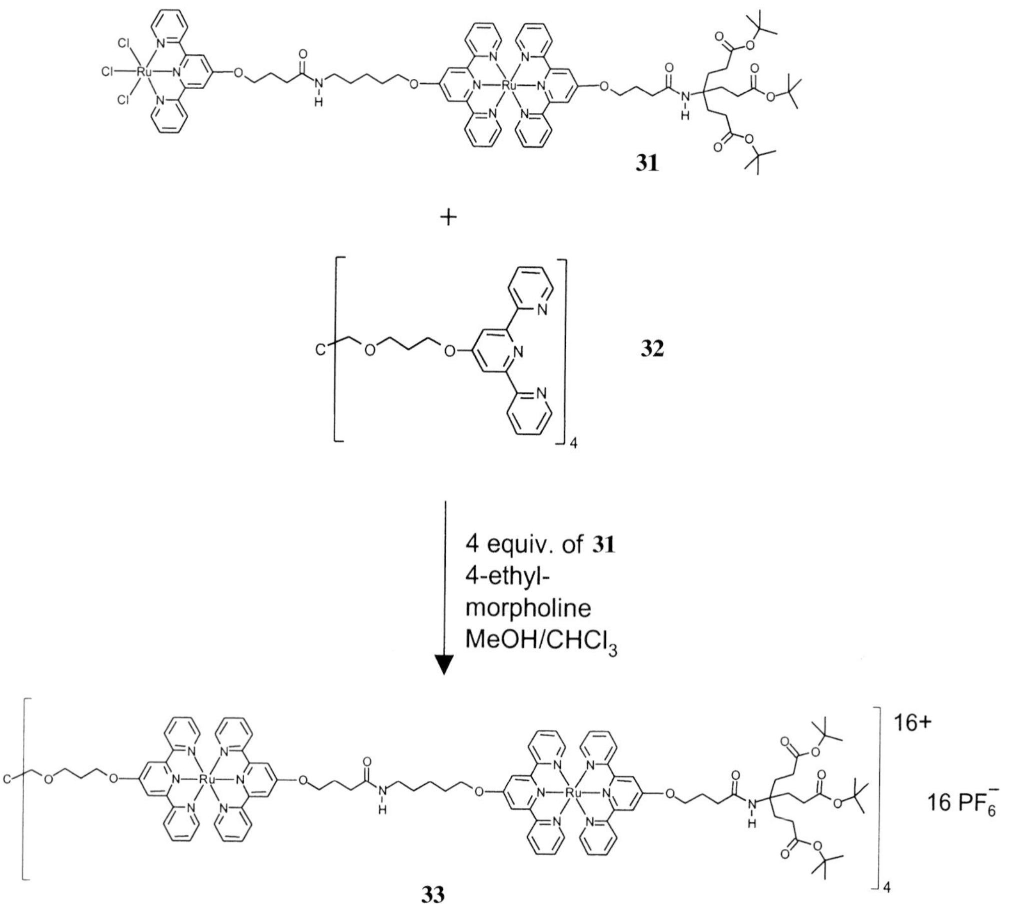

+

32

4 equiv. of **31**
4-ethyl-
morpholine
MeOH/CHCl$_3$

Scheme 8.9 Newkome and He's macromolecule possessing double –<Ru>–(x)–<Ru>– connectivities.[71]

Newkome and He[71] expanded the –<Ru>– single connectivity to include the incorporation of linearly connected complexes (i.e., –<Ru>–(x)–<Ru>–) in each arm of a tetrahedral "hypercore" (Scheme 8.9). Construction of the octa-Ru(II) core **33** was predicated on the reaction of the two complementary fragments, the paramagnetic Ru(III) triester **31** and the tetrakis(terpyridyl ether) **32**. Key transformations for building block synthesis included the treatment of 4'-chloroterpyridine with the potassium alkoxide of either 4-hydroxybutyric acid, 5-aminopentanol, or a tetrahydroxyl-terminated core derived from reduction of the corresponding tetracarboxylic acid[72] (**23**; Scheme 8.7). Characterization of the 16$^+$ cationic metallocore **33** included UV/vis spectrophotometric analysis, which revealed an eight-fold absorbance increase as compared to the spectrum of a single Ru(II)-terpyridine-based complex. Similar but branched and more rigid constructs have been reported.[73] Terpyridine–ruthenium connectivity was incorporated into rigid phenylacetylene monomers focally attached to bipyridines, allowing self-assembly through Ru(II) cores.

Newkome and He[74] exploited the '–<Ru>–' connectivity to examine the potential to create new metallodendrimer architectures comprised of more than two preconstructed branched macromolecules (Scheme 8.10). Based on the inherent tetrahedral architectures of dendrimers prepared with saturated, sp^3-based *C*-foundations, nanoscale dendritic networks were obtained possessing *methane-based* geometries. Complementary components, used for network assembly, were prepared by initial acylation of amine **35** (**a** and **b**; synthesized by treatment of 5-nitroisophthalic acid monoacid halide with the aminotriester, Behera's amine, followed by DCC coupling of an aminoterpyridine) with tetraacyl halide **34** to afford the 1st and 2nd generation "*carbon*" fragment **36a,b**. The "*hydrogen-type*" components **37** and **38** were synthesized as described for the "lock and key" assemblies in Scheme 8.8. Final network construction was achieved by treatment of the cores **36a,b** with four equivalents each of the 1st and 2nd generation Ru(III) dendrons to yield the desired metallodendrimers **39** and **40** (Figure 8.2).

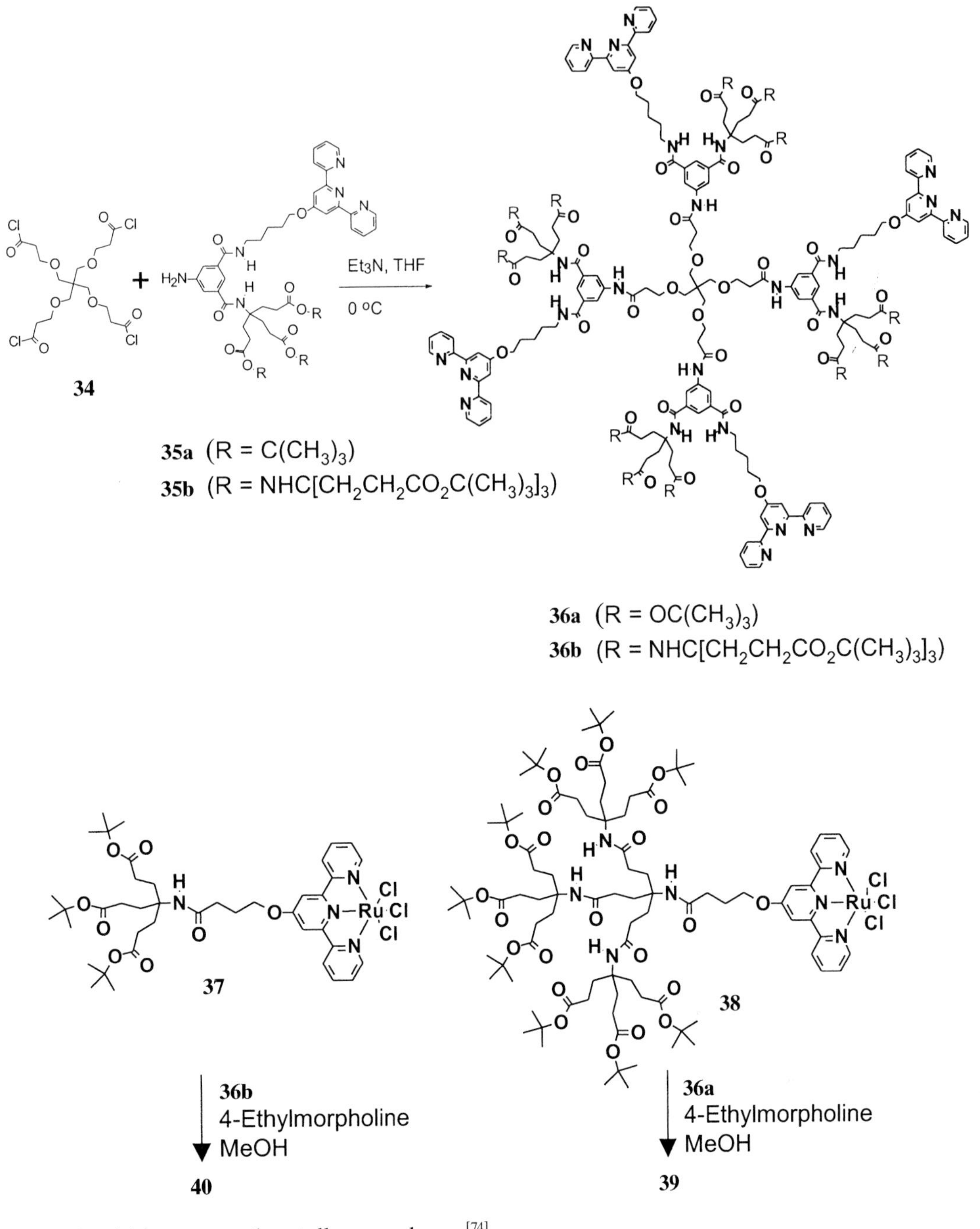

34

35a (R = C(CH₃)₃)

35b (R = NHC[CH₂CH₂CO₂C(CH₃)₃]₃)

36a (R = OC(CH₃)₃)

36b (R = NHC[CH₂CH₂CO₂C(CH₃)₃]₃)

37

38

36b
4-Ethylmorpholine
MeOH

40

36a
4-Ethylmorpholine
MeOH

39

Scheme 8.10 Construction of dendritic cores and metalloappendages.[74]

Notably, the mixing of 1ˢᵗ and 2ⁿᵈ generation components has led to the creation of the first dendritic constitutional isomers.[74] Each isomer possesses a nominal molecular weight of 12,526 amu and displays nearly identical IR, UV, NMR, and MALDI-TOF mass spectra. Other physical properties, such as decomposition temperature and solubility, are also similar. In contrast, electrochemical studies suggest that the internal densities and void regions differ. The cyclic voltammograms exhibit a single, quasi-reversible pattern for the Ru centers of isomer **40** and two waves for the metallic centers of isomer **39**. This suggests that the internal dendritic scaffolding of such metallodendrimers can be tailored to facilitate or negate internal electrochemical communication.[75] The use of this –<Ru>– mode of dendritic construction has recently been extended to the isomeric 5,5″-disubstituted terpyridine ligands,[76] and has been applied in the construction of metallo-supramolecular fullerene assemblies and polymers.[77]

Constable et al. developed a number of metal chelating agents useful as bridging ligands[78] that were subsequently incorporated into related metallosupramolecules. Constable and Harverson[79] reported a convergent strategy for the construction of –<Ru>––based, branched macromolecules, in which arm construction involved oxyanion displace-

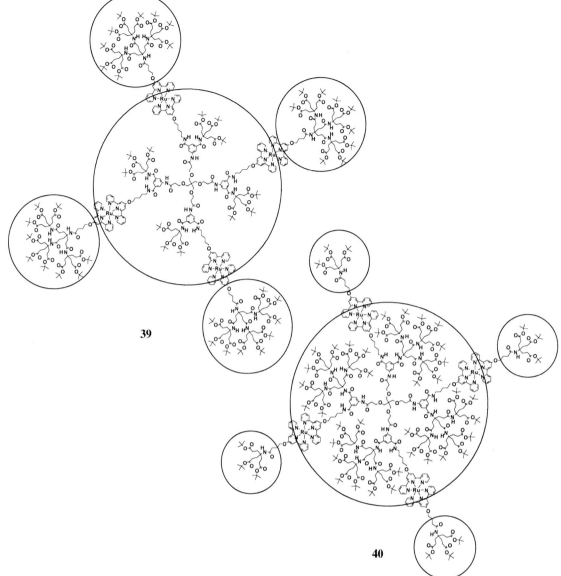

Figure 8.2 First two isomeric metallodendrimers mimicking CR$_4$.[74]

ment of a chloroterpyridyl substituent resulting in 'back-to-back' terpyridine connectivity. The resulting linear arms were then attached to a benzene core. Adsorption of the simple nona-Ru species onto silica/titania surfaces, as studied by optical wave-guide light mode spectroscopy[80] (OWLS), revealed that the deposition mode strongly depends on the bulk concentration. A similar approach allowed the convergent synthesis[81] of an octadecanuclear complex, as well as of another non-branching "high-nuclearity" material, employing pentaerythritol as the four-directional core.[82] These authors' synthesis[83] of a hexadeca-Ru(II) dendrimer is depicted in Scheme 8.11. Pentaerythritol-modified terpyridine **41** was coupled to core **42** under reductive conditions to give the 12-polyol **44**. In turn, this was reacted (KOH, DMSO) with the monochloroterpyridine complex **43** to afford the desired polyruthenium complex **45**.

Holmstron and Cox[83a] developed a conducting composite electrode incorporating these pentaerythritol-based Ru(II) dendrimers for electrocatalytic studies on the oxidation of methionine, cysteine, and As^{3+}; detection limits for methionine and cysteine were 0.6 and 0.5 μM, respectively. Constable et al.[84] also used 4,4'-dihydroxy-2,2'-bipyridine coupled with two terpyridyl Ru(II) moieties and coordination to transition metals, e.g., Fe(II) and Co(II), to create a non-branched heptanuclear metallomacromolecule. Other polypyridinyl-based molecules have also appeared in the literature, e.g. hexakis(4-pyridinyl)benzene,[85] 1,3,5-tris(4'-terpyridinyl)benzene,[86] as well as poly(4-vinylpyridino)-benzenes and poly(4-ethylpyridino)benzenes,[87] which could lead to the synthesis of large metallodendrimers.

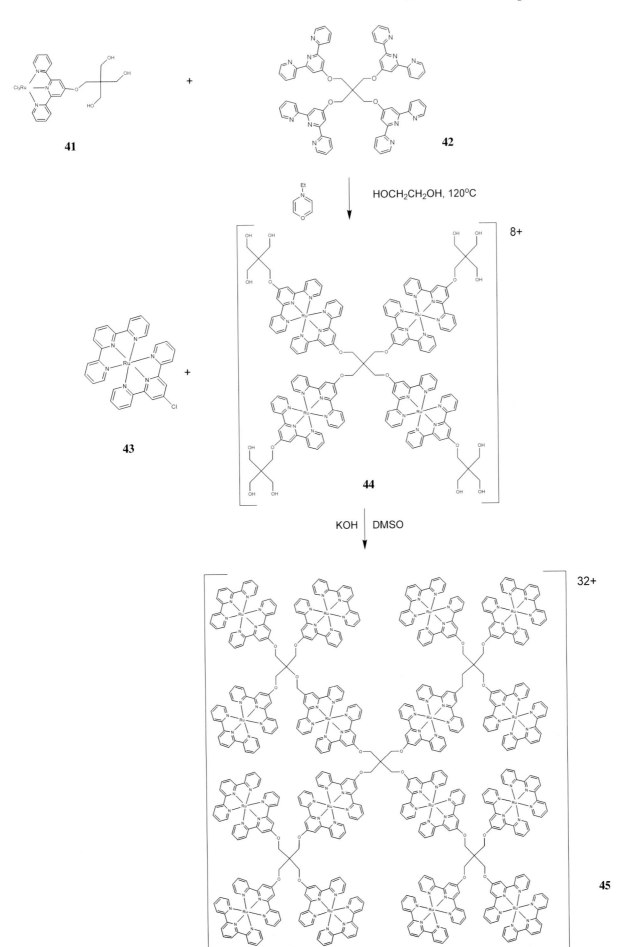

Scheme 8.11 Constable's pentaerythritol-based Ru constructs.[83]

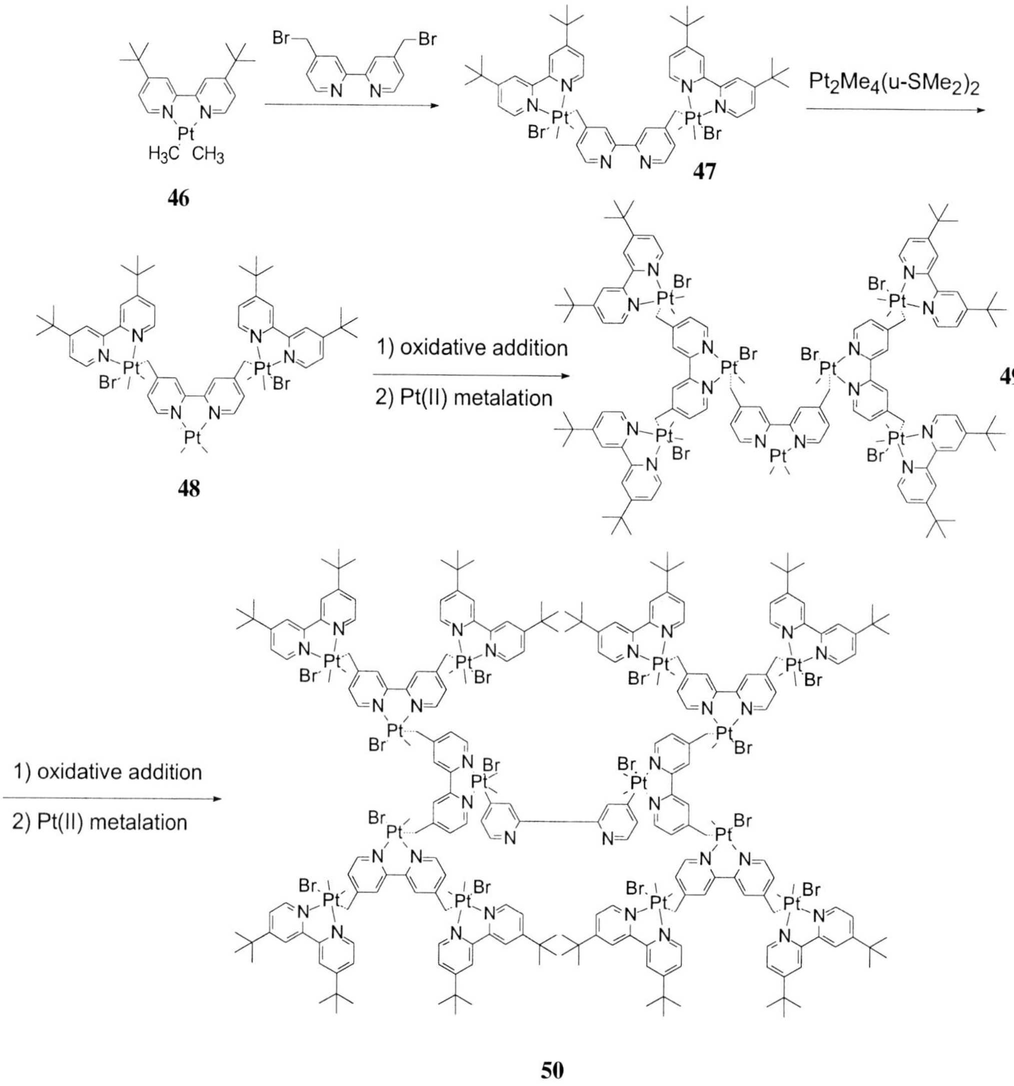

Scheme 8.12 Convergent approach to Pt-containing metallodendrimers.[88]

Achar and Puddephatt[88] created a series of convergently prepared Pt dendrimers (Scheme 8.12) using two repeating reactions: (1) oxidative addition of PtMe$_2$(t-Bu$_2$bpy) through CH$_2$Br insertion generating stable Pt(IV) centers, and (2) SMe$_2$ ligand displacement from Pt$_2$Me$_4$(μ-SMe$_2$)$_2$ by a free bipyridine. Thus, treatment of two equivalents of **46** with one equivalent of 4,4'-bis(bromomethyl)-2,2'-bipyridine afforded complex **47**, which was subsequently treated with a platinum dimethylsulfide reagent to afford the mixed valence triplatinum complex **48**. Oxidative addition and Pt(II) metallation of **48** gave the heptaplatinum **49**. Repetition of this synthetic sequence afforded the 3rd generation **50** with 14 platinum centers. The 4th generation metallodendrimer with 28 platinum centers has also been reported.[89]

Puddephatt et al.[90] also reported oxidative addition reactions of four equivalents each of the Pt-complex precursors **48** and **49** to 1,2,4,5-tetrakis(bromomethyl)benzene (Scheme 8.13) to form Pt dendrimers possessing 12 (**51**) and 28 (**52**) metal centers, respectively. These authors[91] further reported the preparation of oligomers and a polymer containing organoplatinum centers.

Recently, Liu and Puddephatt[92] explored routes to the synthesis of Pt/Pd-containing dendrimers using their oxidation–addition protocol. Although synthesis of the desired metallodendrimer was not successful, the building blocks may prove very useful in future approaches. These authors[93] successfully developed a divergent route to Pt- and Pt/Pd-containing dendrimers (**53** and **54**) comprised of inner generation layers incorporating platinum centers and outer layers incorporating either platinum or palladium complexes (Figure 8.3).

Scheme 8.13 Pt dendrimers possessing up to 22 metal centers.[90]

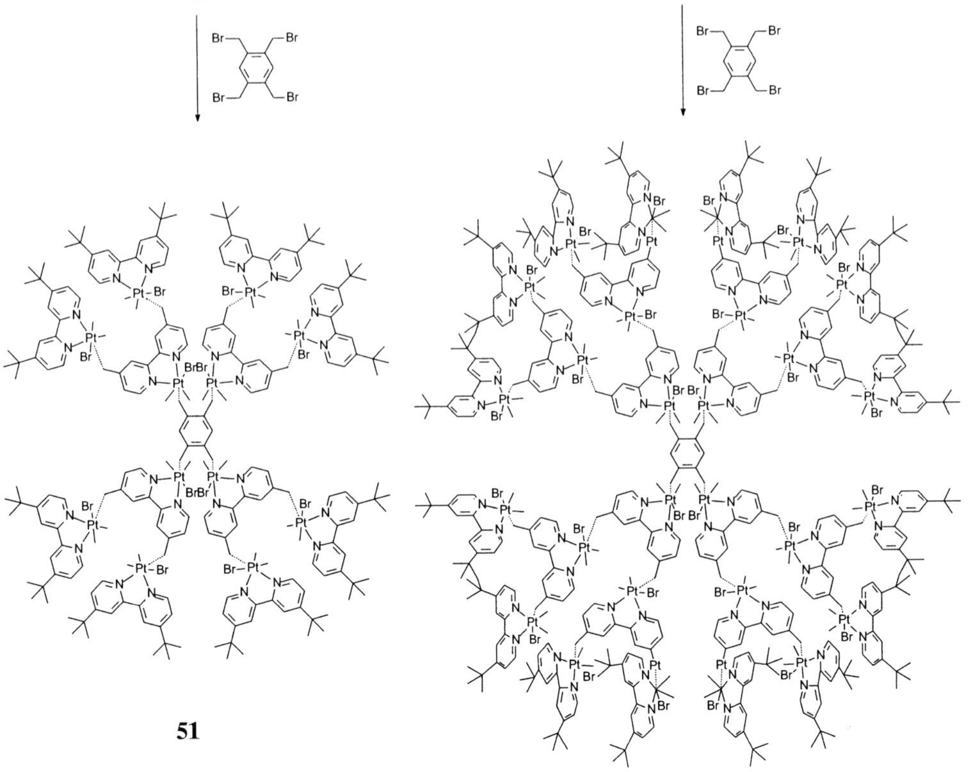

Reinhoudt and coworkers[94, 95] reported the assembly of hyperbranched materials by the polymerization of AB$_2$ monomer **55** (Scheme 8.14). Polymerization was induced by MeCN displacement promoted by heating under low pressure, which allowed benzyl nitrile coordination leading to "self-assembled metallospheres". Monomers **56a–g** possessing different ligating groups (*S-tert*-butyl, *S*-naphthyl, *S*-ethyl,[96] and *P*-Ph$_2$[97]) as well as different metals [e.g., Ni(II) and Pt(II)] were synthesized. Reinhoudt and coworkers[98, 99] later reported Pd dendrimers up to the 3rd generation with 47 Pd connections, which were prepared by divergent and convergent approaches (Scheme 8.14). This

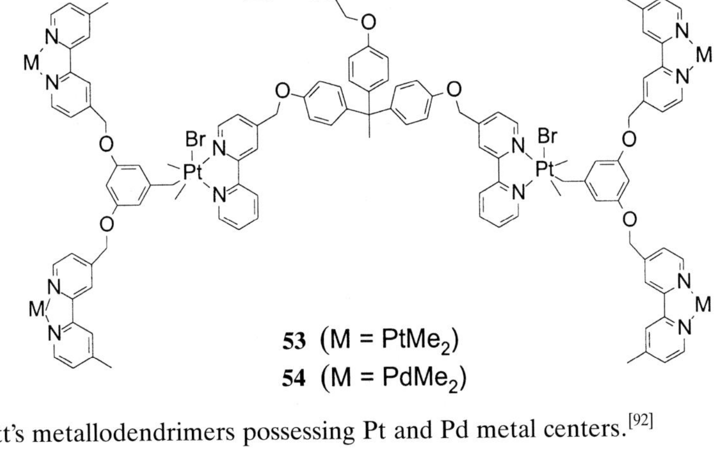

53 (M = PtMe$_2$)
54 (M = PdMe$_2$)

Figure 8.3 Puddephatt's metallodendrimers possessing Pt and Pd metal centers.[92]

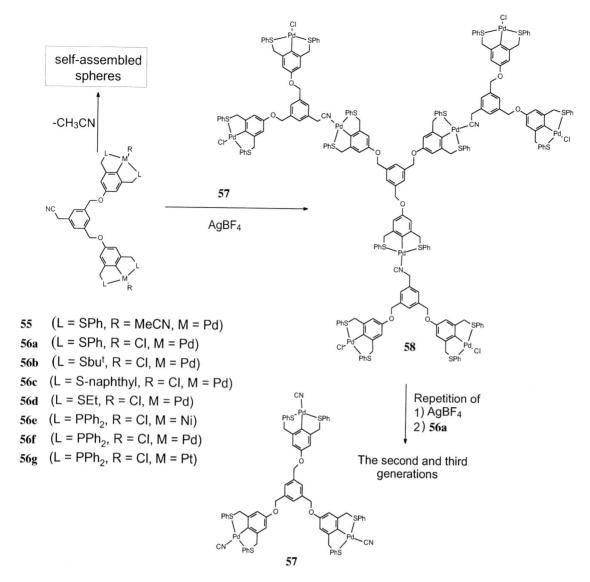

Scheme 8.14 Reinhoudt's hyperbranched spheres and Pd macromolecules.[94, 95]

assembly relies on a kinetically inert tridentate (S–C–S) Pd complex that is labile toward substitution when the "external" ligand is chloride. Substitution of the halide ligand by a nitrile or pyridyl moiety effects monomer connectivity. This type of non-covalent self-assembly has been employed for the synthesis of multiporphyrin systems.[100] Essentially, a monopyridine-bearing porphyrin was used to cap or terminate a 2nd generation poly-Pd metallodendrimer resulting in a 12-porphyrin array. A tetrakis(Pd–Cl) porphyrin was also used as a core. Spherical assemblies based on these unique building blocks, with diameters ranging from 100 to 400 nm, have been reported.[101] The 1st generation metallodendrimer **58** was assembled by addition to dipalladium chloride **56a** using a mixture of tris (palladium chloride) **57** and AgBF$_4$. By repetition of this procedure, 2nd and 3rd generation metallodendrimers were prepared, which were subsequently characterized by NMR and MALDI-TOF mass spectrometry. These authors also modified[102] the focal sites of dendrons prepared in this manner with diallyl sulfide chains. Incorporation of these dendrimer-modified chains into gold-supported alkanethiol monolayers afforded "surface-confined" metallodendrimers.

Reinhoudt, van Veggel, and coworkers[103] developed a synthesis of metallodendrimers based on a combination of coordination chemistry and *H*-bonding (Scheme 8.15). "Rosettes" were constructed by addition of triazine **59** to the metallodendrons G$_{1-3}$ (**60**) to afford up to the 3rd generation *H*-bonded, mixed-metal dendrimer **61**.

Ohshiro et al.[104] reported Pt-dendrimers with a backbone composed of platinum--acetylide units (Scheme 8.16). Similar metal–acetylide units with iridium and rhodium

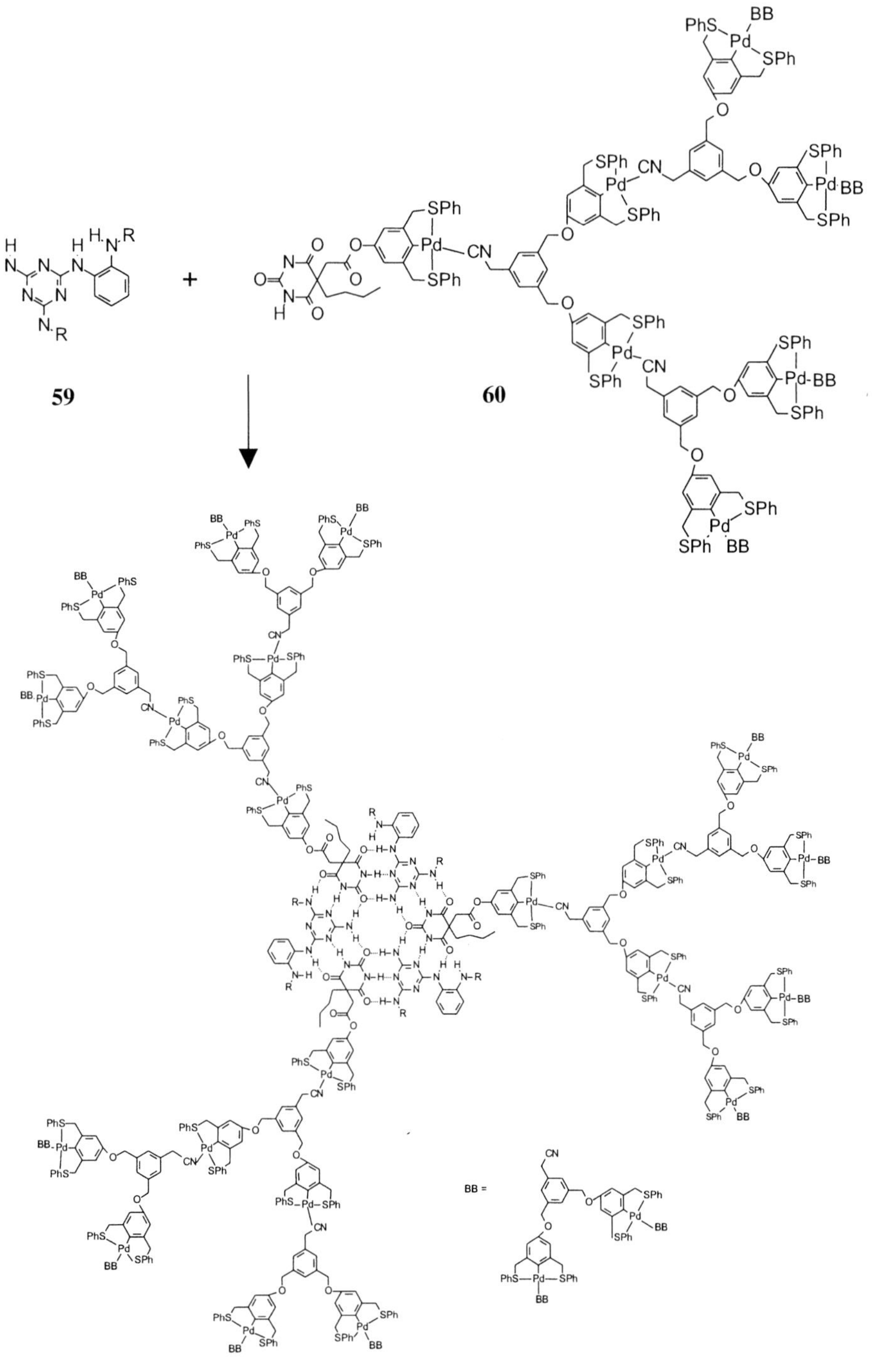

Scheme 8.15 "Rosettes" based on metal coordination and *H*-bonding.[103]

had previously been reported by Tykwinski and Stang,[105] while similar organoplatinum dendrimers have also been reported.[106] Treatment of 1,3,5-triethynyl-2,4,6-trimethyl-benzene (**62**) with dichlorobis(tri-*n*-butylphosphine)platinum or dichlorobis(triethyl-phosphine)platinum in the presence of CuCl as a catalyst in piperidine afforded the tri-nuclear complex **63**, which was subsequently treated with PhC ≡ CH to give the triphe-nyl analog **64**. A larger construct **65** was convergently synthesized by coupling **62** with **63**, which was ultimately converted to its corresponding phenylacetylene derivative **66**. Higher generation constructs (e.g., a heneicosanuclear complex), along with an X-ray structure of the core unit, were later reported.[107] The synthesis of these materials has also been examined from a completely convergent perspective. Three generations were prepared[108] in this manner and subsequently characterized.

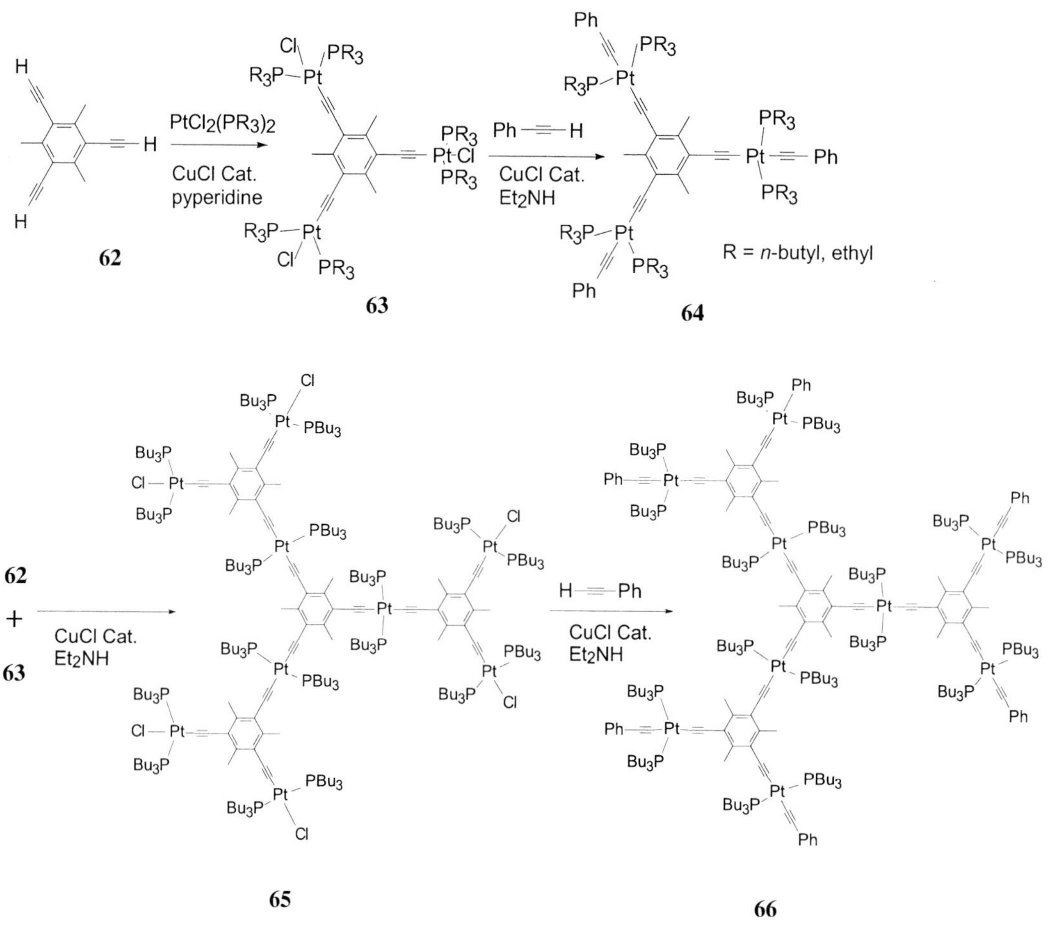

Scheme 8.16 Metallodendrimers possessing Pt atoms in the backbone.[104]

8.4   Metals as Cores

Vögtle, Balzani, et al.[109] prepared architectures with a photoactive and redox-active [Ru(bpy)$_3$]$^{2+}$-based core (**67–69**; Figure 8.4). Using Fréchet-type dendrons focally attached at the 4,4'-positions of a bipyridine moiety, the dendrimers were constructed by supramolecular reaction with RuCl$_3$, or Ru(bpy)$_2$Cl$_2$ in the case of a singly complexed bipyridinyl dendron. Photophysical properties, electrochemical behavior, and excited-state electron-transfer reactions were discussed. For example, the dendritic shell was demonstrated to protect the core's luminescent excited state from dioxygen quenching. Similar heteroheptanuclear star complexes have been reported by Constable and Haverson.[110] Small Ru(bpy)$_3$-based polycarboxylates have been reported[111] that act as selective cytochrome *c* receptors; an asymmetric derivative was demonstrated to be a photo-mediated modulator of the cytochrome *c* redox state. Nierengarten, Felder, and Nicoud[112] focally attached a phenanthroline ligand to their ester-based C$_{60}$-terminated dendrons (see Section 5.13) facilitating Cu(I) complexation of two units (**70**; Figure 8.5). Complex formation was effected using Cu(MeCN)$_4$BF$_4$.

Catalano and Parodi[113] reported reversible C$_{60}$ binding to metallodendrimer core complexes [i.e., Ir(CO)Cl(PPh$_2$R)$_2$; Scheme 8.17]. Dendritic wedges possessing focal *P*-ligands were constructed by treatment (NaPPh$_2$) of the Fréchet-type dendritic fragments[114] **71** and **72**. Under reductive conditions (CO), addition of Na$_3$IrCl$_6$· 6H$_2$O to the Ph$_2$P-based wedges yielded complexes **73** (*cis* and *trans*) and **74** (*trans* only). Each was subsequently bound to C$_{60}$ to afford metallodendrimers **75** and **76**, the binding in which was found to be thermally reversible. Thermodynamic data on the reversible binding were obtained in chlorobenzene solution by line-width analysis of the ^{31}P{^{1}H} NMR spectra, while rates of reaction with O$_2$ for both **75** and **76** indicated that the steric bulk

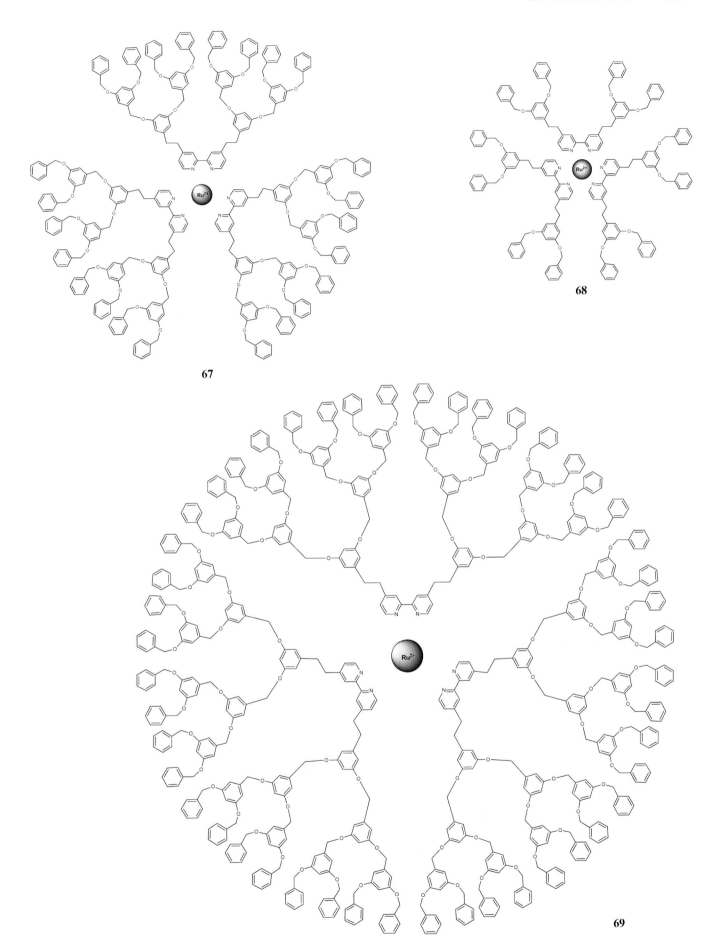

67

68

69

Figure 8.4 Vögtle's photo- and redox-active ruthenium core dendrimers (drawn flat, but octahedral inreality).[109]

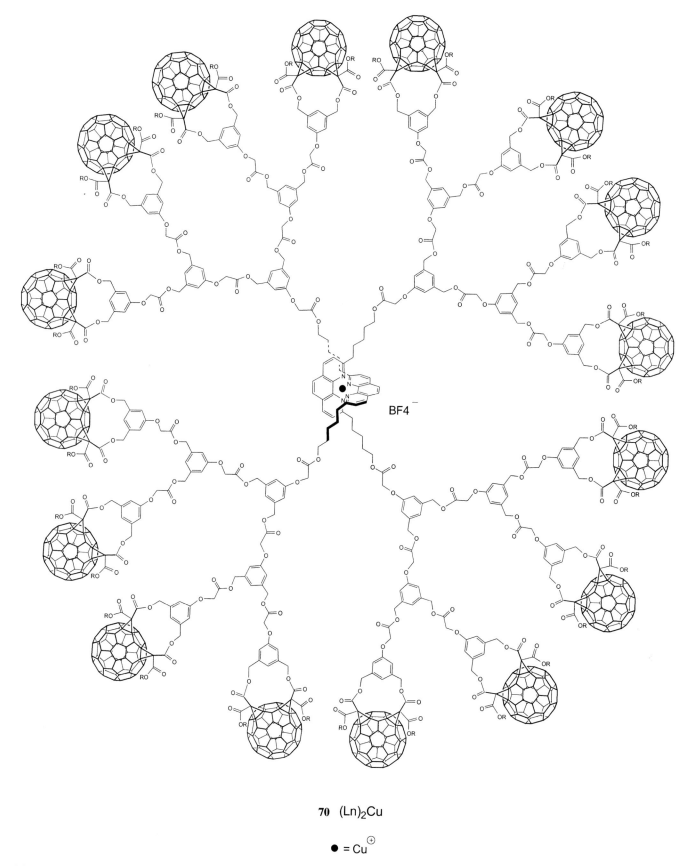

70   (Ln)₂Cu

● = Cu⊕

Figure 8.5 Nierengarten's phenanthroline-based assembly of C₆₀-terminated dendrimers.[112]

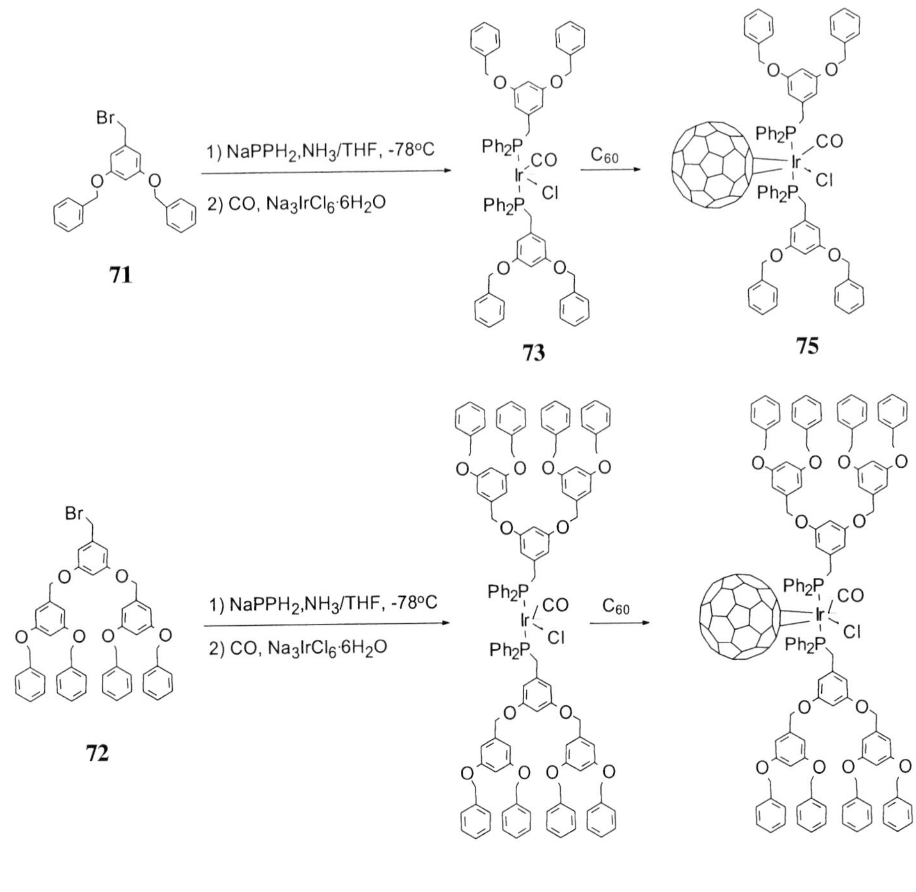

Scheme 8.17 C_{60}-containing complexes possessing an Ir branching center.[113]

of these low generation dendritic branches does not significantly perturb or obstruct the metal center.

Inoue and coworkers[115] reported a dendritic 'caged' porphyrin–zinc complex with photochemical functionality. Metallodendrimers up to the 4th generation (**77**; Figure 8.6) were synthesized employing a convergent method.[114] Fluorescence quenching studies using the 4th generation metallodendrimer revealed that access to the porphyrin core by small-sized quenchers (i.e., vitamin K$_3$) was facile, whereas larger quenchers (e.g., the 1st generation dendrimer) were denied an approach. These authors suggested that dendritic branches in higher tiers behaved as "shielding barriers" to the metal core, while acting as traps for small molecules.[116]

Jiang and Aida[118, 119] synthesized similar Fe-porphyrin metallodendrimers up to the 4th generation (**79**; Figure 8.6), which displayed reversible dioxygen-binding activity. On the other hand, the O$_2$ adduct was found to have a lifetime of several months, even in the presence of water, and a half-life of 50 h on exposure to CO. Aida and coworkers[120] prepared the 5th generation zinc porphyrin dendrimer **78**, which was subsequently used as a spectroscopic probe. By the 5th generation, the λ_{max} of the Soret band became solvent independent, suggesting that the metallo-core was almost completely shielded by the dendritic surroundings. Aida and coworkers[121, 122] surface-modified the same zinc porphyrin metallodendrimers (e.g., **80**) with carboxylate groups. Using higher generation dendrimers, the addition of methyl viologen (MV^{2+}) or the more negatively charged naphthalenesulfonate did not lead to noticeable solvatochromic changes, whereas a significant change was observed with the 1st generation. This result further illustrated the shielding effect of the large dendritic shell, corroborating the earlier observations of these authors.[115] Aida and coworkers[123] later examined the electrostatic assembly of 4th generation porphyrin-based dendrimers coated with CO$_2$Me, CO$_2$H, CO$_2^-$K$^+$, CONH-(CH$_2$)$_2$NMe$_2$, or CONH(CH$_2$)$_2$NMe$_3^+$Cl$^-$ groups. By combining metallated and non-metallated core constructs in protic media, organized assemblies were observed, which

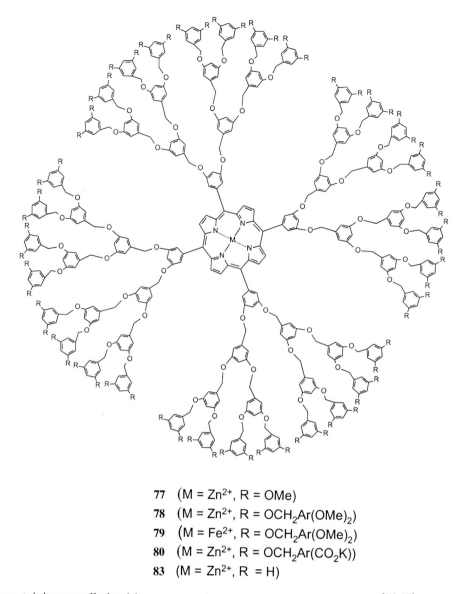

77 (M = Zn²⁺, R = OMe)
78 (M = Zn²⁺, R = OCH₂Ar(OMe)₂)
79 (M = Fe²⁺, R = OCH₂Ar(OMe)₂)
80 (M = Zn²⁺, R = OCH₂Ar(CO₂K))
83 (M = Zn²⁺, R = H)

Figure 8.6 Porphyrin-containing metallodendrimers possessing certain biological properties.[115, 117]

formed the basis of a strategy allowing nanoscopic control over macromolecular spatial arrangements.

Morphology-dependent photochemical events have been reported[124] for these poly-(aryl ether) dendritic porphyrins. A higher energy-transfer quantum yield [Φ_{ent} = 80.3 %] was observed for fully substituted porphyrins, i.e., $(L5)_4P$, where $L5$ represents a 5th generation dendron and P represents the core, relative to partially substituted materials, i.e., $(L5)_1P$ through to $(L5)_3P$ [$\Phi_{ent} \approx$ 10.1–31.6 %].

Enomoto and Aida[117, 125] reported the "oxygen-promoted" self-assembly of copper-ligating dendrimers in order to examine dendritic effects on a non-heme metalloprotein mimic (e.g., **82**; Scheme 8.18). Construction of the "copper-ligating" metallodendrimers was achieved by attachment of methyl-terminated, Fréchet-type dendrons to 1,4,7-triazacyclononane by amine displacement of benzyl bromide focal groups (e.g., **81**), followed by treatment with [Cu(MeCN)₄]PF₆ and O₂ to give the bis(μ-oxo)-bridged dicopper core moiety. Oxidative decomposition studies on each member of the series suggested a greater stability for the higher generation metallodendrimer on the basis of the entropy loss associated with the reaction. Steric effects were postulated to account of this phenomenon. Analogous N₃-based Zn(II)-Coordination is known.[125a]

Fréchet and coworkers[126–128] prepared zinc porphyrin metallodendrimers based on benzyl ether dendrons up to the 4th generation (**83**; see Figure 8.6). They concluded that the dendritic shell did not significantly affect the electrochemical or photophysical nature

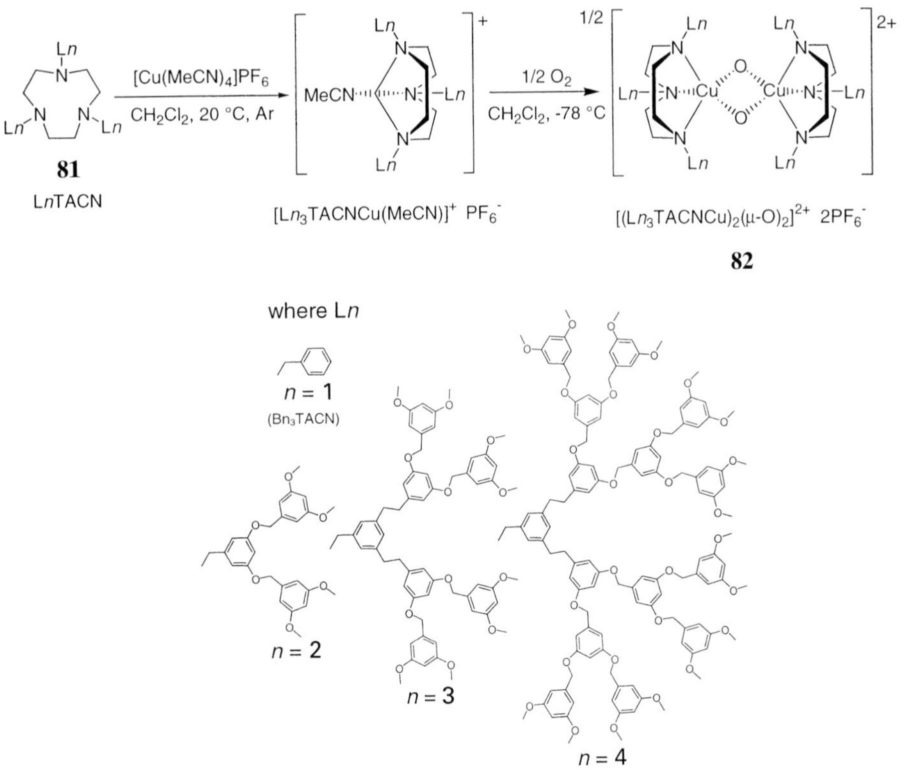

Scheme 8.18 Copper-ligating mimic of a non-heme metalloprotein.[117]

of the metalloporphyrin core. However, even with lower generation dendrimers, the rate of interfacial electron transfer was noticeably reduced, presumably due to the greater separation of the porphyrin core from the electrode surface in cyclic voltammetry experiments. No hindrance to penetration of the dendritic shell by small molecules was noticed, which is in accord with Inoue's and Aida's observations.[115] Only the 4th generation metallodendrimer **83** showed a 33 % rate enhancement of the viologen fluorescence quenching of the zinc porphyrin. Suslick et al.[129, 130] reported the "shape-selective control of ligation" of dendritic metalloporphyrins. Pollak, Sanford, and Fréchet[131] later compared two different routes for the construction of porphyrin-cored dendrimers using the ubiquitous poly(aryl ether) dendrons (see Section 5.3.2). It was determined that aldehyde focal modification of dendrons up to the 4th generation allowed rapid and smooth porphyrin assembly through condensation with pyrrole; higher generation dendrons were subject to reduced yields due to steric limitations. Attachment of dendrons to a preformed porphyrin core also proved useful, although this method necessitated more rigorous purification. MALDI-TOF MS was used to monitor the progress of the reaction. Micellar complexes have been reported.[131a]

Yamago et al.[132] prepared binaphthol-based dendrimers substituted at the 6,6'-positions with poly(aryl ether) dendrons. Titanium complexes of these dendrimers were demonstrated to effect asymmetric allylations with yields in the range 18–36 % and *ee*'s in the range 92–88 %. These results compared favorably to those achieved employing (*R*)-binaphthol (i.e., yield 31 %, *ee* 87 %).

Fréchet and coworkers[133] reported on the effect of core structure on the photophysical and hydrodynamic properties of these porphyrin dendrimers. The intrinsic viscosity of the zinc adducts was found to pass through a maximum as a function of generation and thus supports the internal density profile of dendrimers as decreasing monotonically outward from the molecular center.

Moore, Suslick and coworkers[134, 135] synthesized Mn-containing dendrimers as selective oxidation catalysts (Figure 8.7). Four ester-based dendritic wedges were appended to the *meta* positions of a 5,10,15,20-tetrakis(3',5'-hydroxyphenyl)porphyrinatomanganese(III) chloride to obtain the 3rd and 4th generation metallodendrimers (e.g., **84**). Both dendrimers have been examined as regioselective oxidation catalysts for both intra- and intermolecular cases. Epoxidation of non-conjugated dienes and of 1:1 mixtures of

84

Figure 8.7 A representative of Moore and Suslick's Mn-containing metallodendrimers that act as selective oxidation catalysts.[134, 135]

various alkenes of different shapes and sizes was examined. The dendritic catalysts showed superior regioselectivity at less hindered double-bond positions when compared to the corresponding parent 5,10,15,20-tetrakis(3',5'-hydroxyphenyl)porphyrinatomanganese(III) cation. A dimolybdenum construct is known.[135a]

Diederich and coworkers[136] prepared zinc porphyrin metallodendrimers (e.g. **85** and **86**; Figure 8.8) by a divergent strategy, employing a 1 → 3 *C*-branched amino triester building block[137] to construct the dendritic shell. These molecules were regarded as models for electron-transfer proteins such as cytochrome *c*, in which the dendritic polypeptide branches can affect the metalloporphyrin core redox potential. The first reduc-

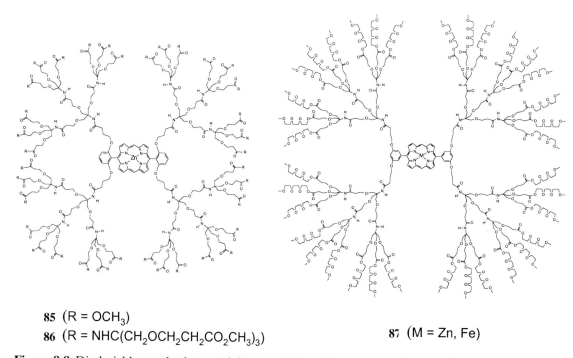

85 (R = OCH$_3$)

86 (R = NHC(CH$_2$OCH$_2$CH$_2$CO$_2$CH$_3$)$_3$)

87 (M = Zn, Fe)

Figure 8.8 Diederich's porphyrin-containing metallodendrimers possessing novel electrochemical and biological properties.[136]

tion potential of the zinc porphyrin unit becomes more negative with increasing dendrimer generation by 90 to 300 mV, which can be rationalized in terms of the increasingly electron-rich microenvironment around the porphyrin core. Meanwhile, the first oxidation potential becomes less positive by up to 300 mV compared to the tetraester porphyrin core alone. Both reduction and oxidation processes become totally irreversible at the 3rd generation (**86**).

Diederich, Gross and coworkers[138, 139] later synthesized similar Fe(II) porphyrin metallodendrimers (e.g. **87**; Figure 8.8) bearing triethylene glycol monomethyl ether groups on their surfaces, thereby facilitating solubility in solvents with polarities ranging from non-polar (*p*-xylene) to polar (H$_2$O). Electrochemical studies indicated that solvent polarity strongly affects the redox potential of electrochemical reactions that occur at the heme metal center. In organic solvents, e.g., CH$_2$Cl$_2$, the iron porphyrins in both the 1st and 2nd generations are exposed to similar microenvironments so that similar potentials for the Fe^{III}/Fe^{II} couple are measured. In water, however, the loosely packed, 1st generation dendrimer does not impede aqueous solvation of the iron porphyrin, whereas the more densely packed dendritic shell of metallodendrimer **87** significantly reduces contact between the heme and external H$_2$O molecules, thereby destabilizing the oxidized form, increasing the charged state, and shifting the redox potential to a more positive value. The large potential shift in water is comparable to the difference observed between cyto-

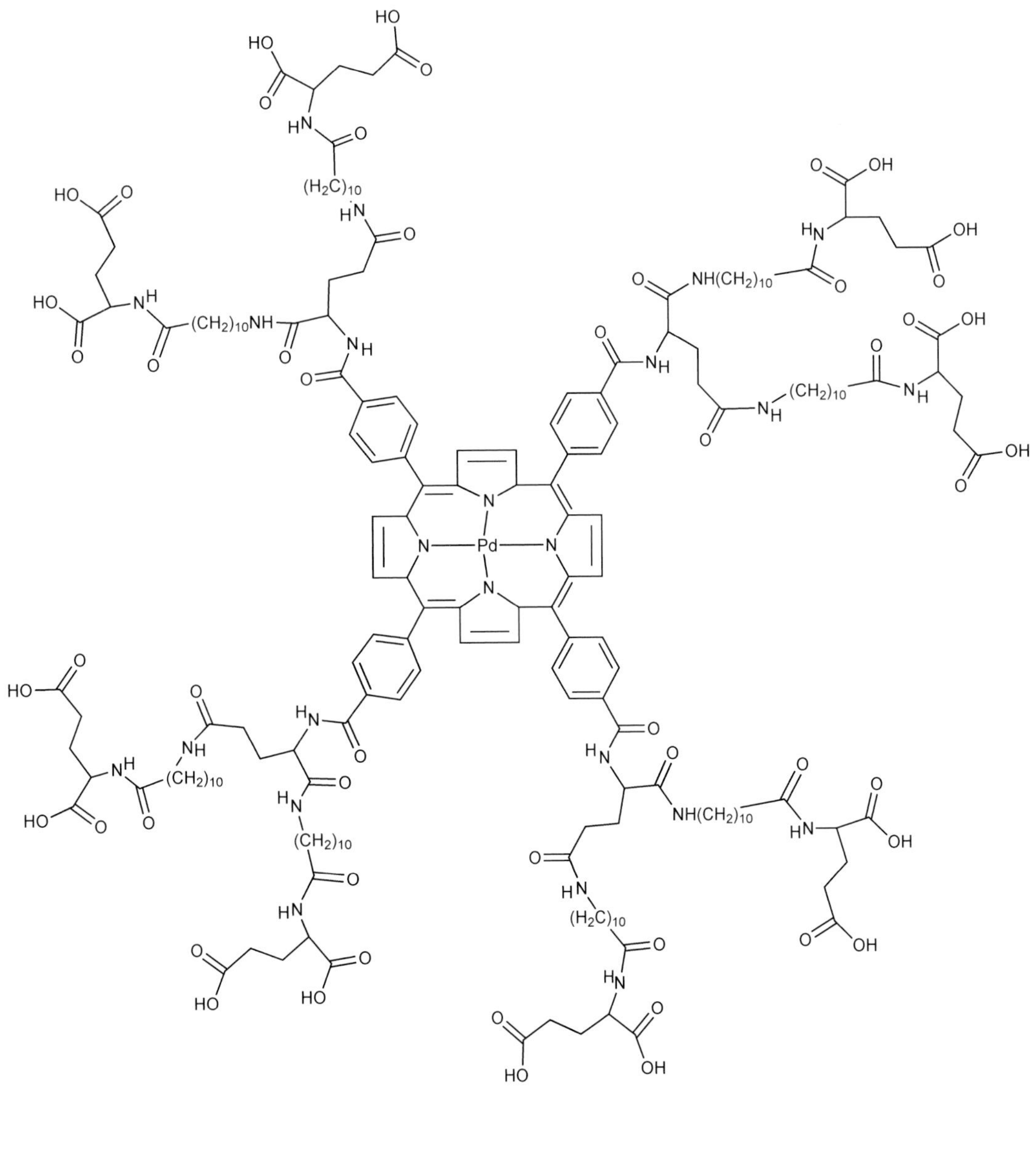

88

Figure 8.9 Vinogradov's polyglutamic acid porphyrins.[141]

chrome *c* and a cytochrome *c* heme model possessing a more open structure. Collman, Diederich, et al.[140] reported the O₂ affinities of the dendritic porphyrins to be about 1500 times greater than those of T-state hemoglobins, whereas the CO affinities of the dendritic porphyrins are lower than those of the 'picket fence' model, but close to the value of hemoglobin in the T-state.

Glutamic acid based Pd porphyrins have been created by Vinogradov, Lo, and Wilson (**88**; Figure 8.9).[141] Glutamic monomers up to the 4th generation were constructed by standard DCC coupling and ester deprotection protocols, starting from a tetraacid palladium porphyrin complex. Phosphorescence quenching constants for generations 1–4 using O₂ in DMF were similar, whereas in water the values decreased significantly as the generation increased. Less open and more compact conformations in aqueous media were postulated to alter the barrier to oxygen diffusion to the core.

Diederich and coworkers[142] later prepared similar poly(ethylene glycol)-terminated dendrimers, although the iron porphyrin core was ligated in a fixed manner by two tethered imidazole moieties in the axial positions. Thus, this dendrimer was designed as a cytochrome mimic. The redox properties of these materials were said to "firmly establish well-designed dendrimers as powerful mimics for globular proteins."

Kimura et al.[143] synthesized zinc phthalocyanine metallodendrimers based on a divergent-growth strategy using the 1 → 3 *C*-branched amino ester[144] (Figure 8.10). The UV/vis spectra of lipophilic metallodendrimers **89** and **91** in CHCl₃ showed a sharp signal at 621 nm ($\varepsilon = 2.95 \times 10^5$ dm³mol⁻¹cm⁻¹) with a shoulder at 621 nm, and were found to be essentially independent of the dendrimer generation. However, the spectra of the hydrophilic metallodendrimers **90** and **92** were quite different, owing to the non-aggregation of the zinc phthalocyanine of **91**, presumably because of the bulky dendritic branches. The higher generation **92** also displayed a strong fluorescence peak at 674 nm in aqueous media, while lower generations do not fluoresce due to zinc phthalocyanine aggregation. A report on the electrochemical properties and catalytic activities of these materials has appeared.[145]

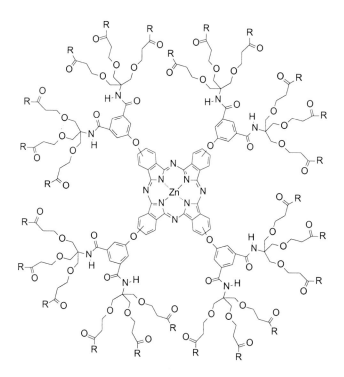

89 (R = OEt)
90 (R = OLi)
91 (R = NHC(CH₂OCH₂CH₂CO₂Et)₃)
92 (R = NHC(CH₂OCH₂CH₂CO₂Li)₃)

Figure 8.10 Divergent construction of phthalocyanine-containing metallodendrimers.[143]

Ng et al.[146] synthesized a series of phthalocyanines bearing poly(aryl ether) dendrons with terminal ester or carboxylate groups. A fluorescence quenching method was used to study the photoinduced electron-transfer; quenching rates were found to be dependent on the size of the attached dendrons. Later, the influence of surfactants on the aggregation of these water-soluble phthalocyanines was reported.[147]

Kimura et al.[148] later designed and synthesized 1,3,5-phenylene-based dendritic porphyrins for the study of intramolecular energy-transfer. Steady-state fluorescence spectra supported highly efficient energy-transfer from the dendron units to the porphyrin core. Similar phenanthroline-based, Ru(II)-bis(bipyridine) core constructs have also been reported;[149] preliminary energy-transfer studies were described. Kimura et al.[150] later focally modified their rigid dendrons with terpyridine to facilitate bis-dendron coupling through Ru(II) complexation, thereby producing a redox-active core.

Kawa and Fréchet[151] reported lanthanide-containing metallodendrimers (Scheme 8.19) up to the 5th generation (e.g., **94**), obtained by the self-assembly of three convergent polybenzyl ether dendrons (e.g. **93**) around a central trivalent cation (Er^{3+}, Tb^{3+}, Eu^{3+}) through carboxylate anion coordination. The luminescence properties of these metallodendrimers, measured both in solution and in the bulk, were enhanced as the size of the dendritic surroundings increased.[152] This was attributed to both a large "antenna effect" from the non-conjugated phenyl benzyl ether dendritic cluster and a shielding effect arising from the site-isolation of the lanthanide cations, which prevents their interaction and hence decreases the rate of self-quenching. Zhu et al.[152a] have also studied this phenomena. Antenna effects[152b] and XAFS studies[152c] have been reported for these materials.

Chow et al.[153, 154] reported the synthesis and characterization of –<Fe>– dendrimers up to the fourth generation (**95**; Figure 8.11) using benzyl ether-based wedges with propylene spacer moieties.[155, 156] Cyclic voltammetry revealed similar behavior as that reported[67] whereby the reversibility of the processes occurring at the metal redox centers decreases with increasing generation. These results were rationalized in terms of the steric hindrance stemming from the bulky dendritic shell, which isolates the redox center from the electrode surface.

Several examples are given here involving the use of bidentate ligands, e.g., 2,2'-bipyridine and 1,10-phenanthroline, as the dendritic core. Tzalis and Tor[157] synthesized the metallodendrimers **96** by incorporating Cu^+ and Fe^{2+} ions within symmetrically substituted 1,10-phenanthroline ligands. Branching was effected using pentaerythritol-based

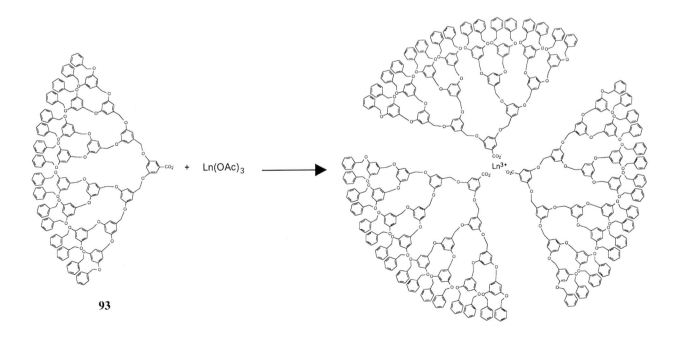

93

94 (Ln = Er, Eu, Tb)

Scheme 8.19 Kawa and Fréchet's lanthanide-containing metallodendrimers.[151]

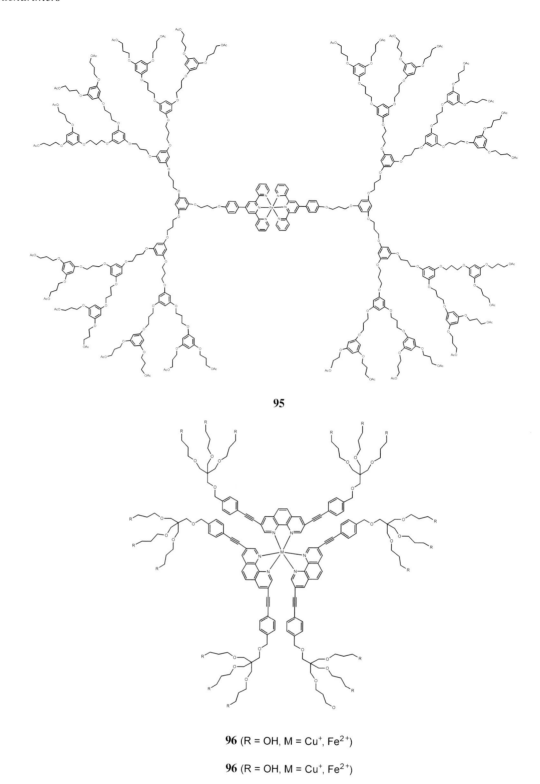

95

96 (R = OH, M = Cu⁺, Fe²⁺)

96 (R = OH, M = Cu⁺, Fe²⁺)

Figure 8.11 Chow's Fe(II)-terpyridine macromolecules[153, 154] and Tzalis and Tor's Cu- and Fe-containing complexes.[157]

building blocks (Figure 8.11). All structures were subjected to ESI-MS, which revealed singly- or doubly-charged molecular peaks.

Issberner et al.[158] reported metallodendrimers constructed up to the 3rd generation with an [Ru(bpy)₃]²⁺ complex as the core (**97**; Figure 8.12); the largest generation possessed 54 peripheral methyl ester groups. Branching was introduced by using an amino triester monomer prepared by addition of methyl acrylate to "tris".[159, 160] In aerated solutions, the larger metallomacromolecule exhibited a more intense emission spectrum and a longer excited-state lifetime than the parent Ru(II) cation. This was rationalized by considering the shielding effect of the dendritic scaffolding. Notably, the 3rd generation metallodendrimer had an excited-state lifetime longer than 1 μs. The quenching constant

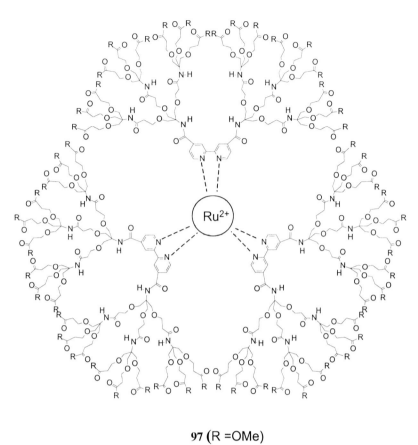

97 (R = OMe)

Figure 8.12 Vögtle and Balzani's metallodendrimers possessing and Ru(II)-bipyridine complex as the core.[158]

of the luminescent Ru(II)-bipyridine MLCT level decreases on going from the 1st to the 3rd generation metallodendrimer and thereafter attains a value of one-twelfth of that of the original Ru(II)-bipyridine core. Vögtle, Balzani, and coworkers[161] later expanded this work to include naphthyl units at the dendron periphery. Efficient energy-transfer processes were shown to take place from the aromatic fluorescence excited states to the core, constituting an antenna effect. Other accounts of poly(aryl ether) dendrons attached to luminescent and redox-active Ru(II)-bipyridyl cores had been presented earlier.[162] Photophysical and electrochemical properties were reported therein.

Narayanan and Wiener[163] prepared a protected ethylenediamine core, which, after divergent construction by established procedures using Behera's amine, is deprotected and subsequently self-assembles around a Co(III) center in a convergent manner (to give, e.g., **98** and **99**; Figure 8.13).

Gorman et al.[164] reported the synthesis of metallodendrimers **100** based on an Fe–S cluster core up to the 4th generation (Figure 8.14). Focally-substituted thiol-based dendrons[165] were treated with 1/4 equivalent of $(n\text{-Bu}_4\text{N})_2[\text{Fe}_4\text{S}_4(\text{S-}t\text{-Bu})]$ to give the desired product. Cyclic voltammetry experiments revealed an increasingly negative potential for one-electron reduction of the metal center with increasing dendrimer generation. The largest molecule also displayed an irreversible oxidation wave. These observations were rationalized by considering that the increased steric hindrance of the dendritic shell would render the redox centers, both kinetically and thermodynamically, more difficult to reduce or oxidize. The molecular weights of these metallomolecules were determined by VPO. The paramagnetic FeS core has been used as an NMR probe to study the conformation of these metallodendrimers.[166] The T_1 relaxation time of the protons in these dendrimers was found to be shorter than that of those in a similar dendrimer possessing a diamagnetic (tetraphenylmethane) core. Gorman et al.[167] also created dendrimers employing an Mo_6Cl_8 core. Later, they studied structure–property relationships in the context of electron-transfer processes in their redox-active core dendrimers.[168] Two series of dendrimers were investigated, one with flexible, aryl ether-based dendrons and the other with rigid, phenylacetylene dendrons. Electron-transfer rate constants supported the "rigid" series as being more effective for electron-transfer rate attenuation.

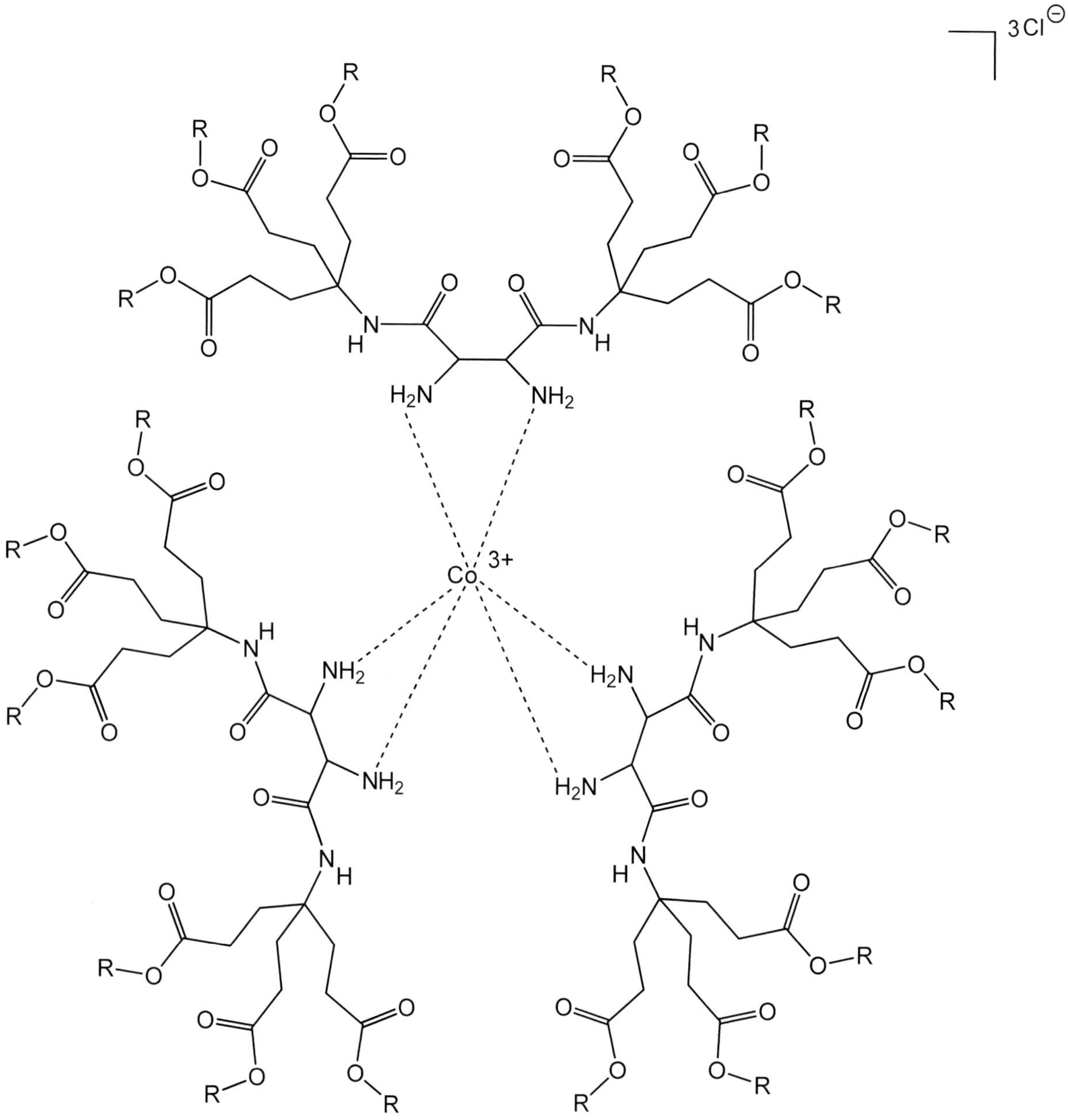

Figure 8.13 Wiener's Co-bound ethylenediamine architecture.[163]

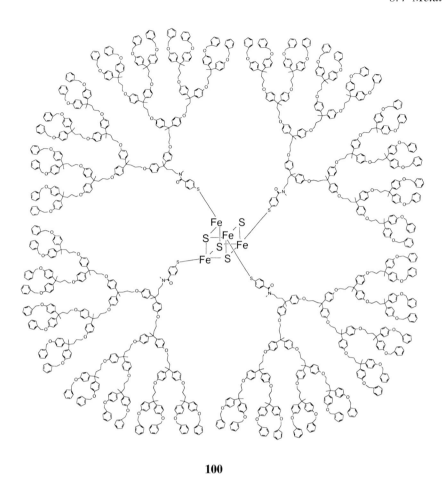

100

Figure 8.14 Gorman's novel Fe-containing metallodendrimers.[164]

Additionally, the electrochemical behavior of these series of dendrimers in films has been compared to that in solution.[169]

Wang and Zheng[170] reported similar architectures using $[Re_6Se_8]^{2+}$ as the metal cluster core. The 1st generation aryl ether $1 \rightarrow 3$ branched dendrons were focally-modified with pyridine to facilitate hexadendron adduct formation. Modification of the dendron structural components was shown to produce dramatic color changes in these materials.

Chow and Mak[171, 172] prepared metallodendrimers up to the 4th generation containing bis(oxazoline)copper(II) **101** and reported their evaluation as catalysts for the Diels--Alder reaction of cyclopentadiene with crotonyl imide (Figure 8.15). Dendritic synthetic details were analogous to these authors' previously described examples, whereby phenolic-based dendrons were prepared convergently.[156, 173] The binding constant ($K_c = k_1/k_{-1}$) of the bis(oxazoline)copper(II)–dienophile complex was found to decrease from 9.8 M^{-1} [G_1·Cu(OTf)$_2$] to 5.7 M^{-1} [G_3·Cu(OTf)$_2$]. The rate constant (k_2) for the Diels--Alder reaction was found to be almost the same ($3.3 \rightarrow 10^3\ M^{-1}s^{-1}$) for the 1st and 2nd generation catalysts, but showed a sharp decrease at the 3rd generation ($1.9 \rightarrow 10^3\ M^{-1}s^{-1}$). These observations were attributed to the loosely packed branches having little effect upon the metal catalytic center, with the increased steric hindrance caused by the larger dendritic and more densely packed shell impeding both the reactivity and binding profiles of the central metal.

Seebach and coworkers[174] developed dendritic TADDOLs to serve as polymer cross-linkers that incorporate titanium metal in the core upon copolymerization with styrene (Scheme 8.20). The dioxolane core moiety was tetrasubstituted with styryl-terminated dendrons prepared by the convergent protocol[114] to afford the 1st and 2nd generation dendrimers (e.g., **102**). These dendrimers were individually copolymerized with styrene to give beads ranging in size from 100 to 600 μm. Treatment of these dried beads with Ti(OCHMe$_2$)$_4$ in toluene gave the Ti-loaded copolymers **103**. Both dendritic copolymers exhibited comparable catalytic activities as the non-dendritic Ti-TADDOLates in the addition of Et$_2$Zn to PhCHO.

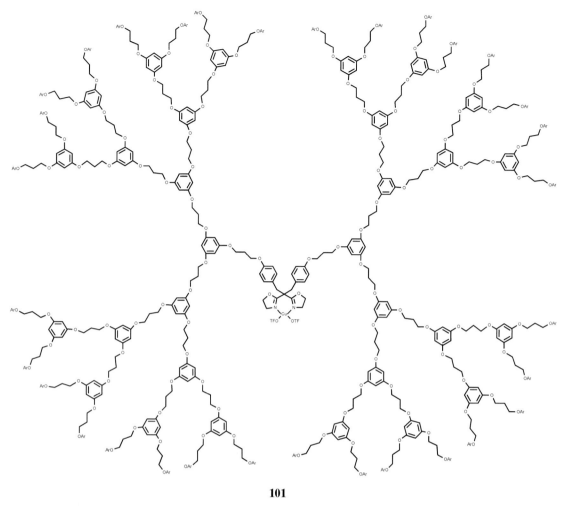

101

Figure 8.15 An oxazoline-based copper(II) catalyst for Diels–Alder reactions.[171]

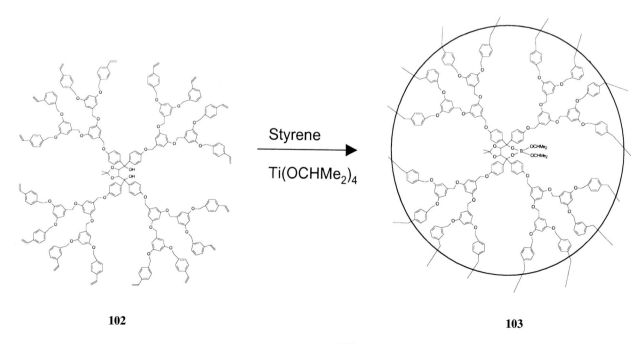

102   **103**

Scheme 8.20 Seebach's Ti-containing polymer cross-linkers.[174]

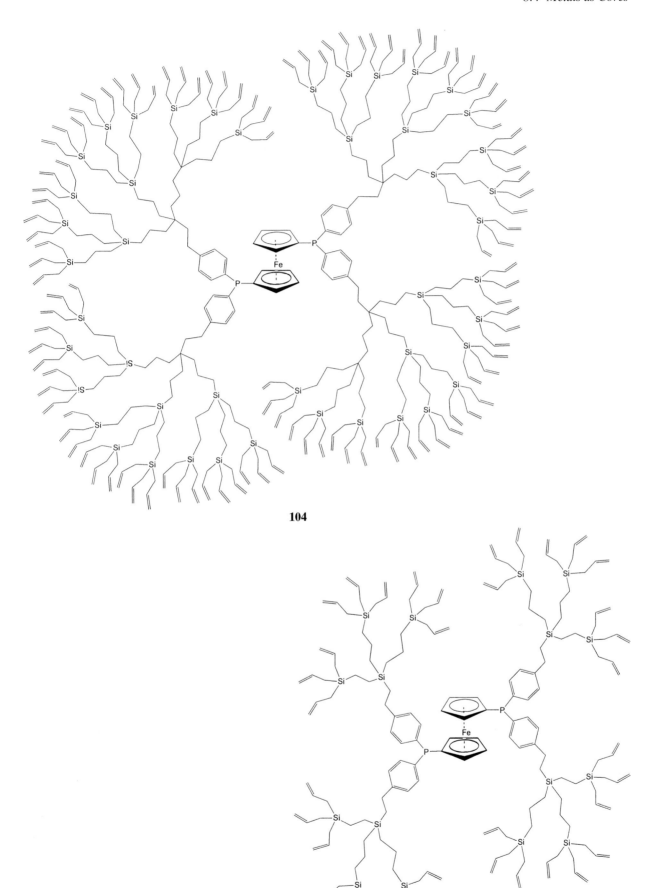

104

105

Figure 8.16 Van Leeuwen's allylation catalysts.[177]

Engel et al.[175] reported an unusual metallodendrimer, with gold chloride attached to the trivalent, neutral phosphine core of a polyphosphorium construct (**119**; Scheme 4.33).

Armaroli, Gross, Nierengarten and coworkers[176] connected their C_{60}-terminated polyester dendrons (see Section 5.3.9) to a tetrahedral bis(phenanthroline) copper(I) complex to create a "fullerene-functionalized black box". Studies showed the C_{60} units to behave as independent redox centers, while the central electroactive site appeared to be totally inaccessible.

Recently, van Leeuwen et al.[177] described the construction of a $1 \rightarrow 3$ branched carbosilane architecture starting from $-POEt_2$-bearing ferrocene rings (**104** and **105**; Figure 8.16). Focal lithio-aryl displacement of each ethoxide ligand allowed the attachment of preconstructed dendrons through to the 3rd generation. Reaction of each dendrimer with $Pd(MeCN)_2Cl_2$ then afforded palladium chloride complexes bound in a *cis* bidentate manner at the core. Their catalytic activities in the allylic alkylation of malonate anion were examined; regioselectives for the "*trans*" addition product ranged from 88–79% with conversions of 17–76%.

Cardona and Kaifer[178] created metallodendrimers based on ferrocene cores by both divergent and convergent procedures utilizing Behera's amine.[70] The partially "buried" redox-active core was found to be similar to that observed in numerous redox-active proteins. The potential for orientation-dependent electron-transfer rates was postulated. Later, molecular orientation effects on the rates of heterogeneous electron-transfer in these unsymmetrical dendrimers were explored;[179] charged dendrimers interacting with oppositely charged surfaces were found to hinder molecular rotation and slowed electron-transfer from the surface to the dendrimer.

Smith[180] created ferrocene core dendrimers (generations 1 and 2) possessing poly(ethylene glycol) termini attached to the core through aryl ethers. The electrochemical properties of these constructs suggested that the dendritic branches are associated with critical parameters in controlling the molecular properties. The electrochemistry and interaction with β-cyclodextrin of these dendrimers possessing an "off-center" ferrocene moiety has been described; the dendrons were found to hamper the formation of the host–guest complex.[181]

8.5   Metals as Termination Groups (Surface Functionalization)

In 1994, van Koten and coworkers[182] reported the development of novel Ni dendrimers up to the 2nd generation to serve as homogeneous catalysts for the Kharasch addition[183] of polyhaloalkanes to simple C=C double bonds (Scheme 8.21). Dodeca(aryl bromide) **106** was functionalized at the periphery by oxidative addition $[Ni(PPh_3)_4]$ to afford the metallodendrimer **107**. The catalytic activity of the poly-Ni dendrimers was found to be comparable, in kinetic terms, to that observed for the monomeric organometallic complex. A notable advantage of these metallodendritic catalysts, or "dendrocatalysts", is their facile removal from the reaction mixture, thus making them recyclable.

Scheme 8.21 van Koten's novel Ni-containing macromolecular catalysts for the Kharasch addition of polyhalogenoalkanes to simple alkenes.[182]

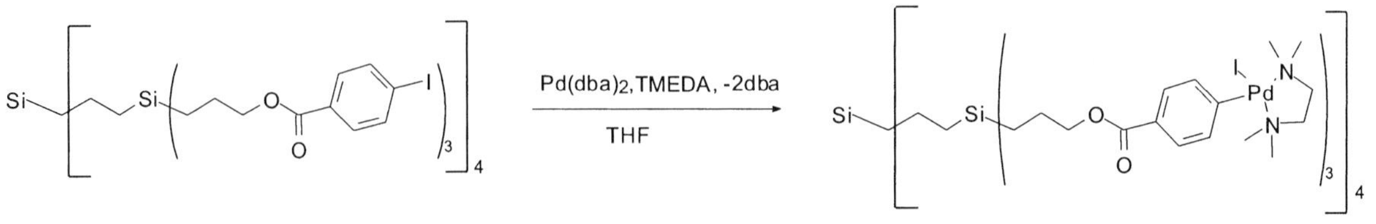

108 **109**

Scheme 8.22 Metallomacromolecule with Pd σ-bonded to the surface.[189]

Soluble polysiloxane polymers have been grafted with similar aryldiamine nickel catalysts[184] and the products showed good catalytic activity in Kharasch reactions[185] (i.e., polyhaloalkane addition to alkenes).[184] The attachment of similar nickel complexes to silica surfaces[186] and the use of amino acid-based scaffolding for these poly-Ni complexes have also been reported;[187] the respective products proved to be effective catalysts for Kharasch additions. Pincer-type ligands have also been grafted onto rigid polyphenylacetylene frameworks and subsequently converted to the Pd complexes.[188]

Additionally, van Koten et al.[189] prepared a small metallodendrimer possessing 12 peripheral Pd centers (Scheme 8.22) attached to an Si-based framework. Dodecaaryliodo-terminated dendrimer **108** was treated with Pd(dba)$_2$ and TMEDA to yield metallodendrimer **109**: this was the first example of Pd being σ-bonded to a dendritic scaffolding. The use of similar Pt-terminated dendrimers as SO$_2$ gas detectors acting through reversible η^1-SO$_2$–Pt bond formation has appeared.[190, 191, 191a] Zinc-

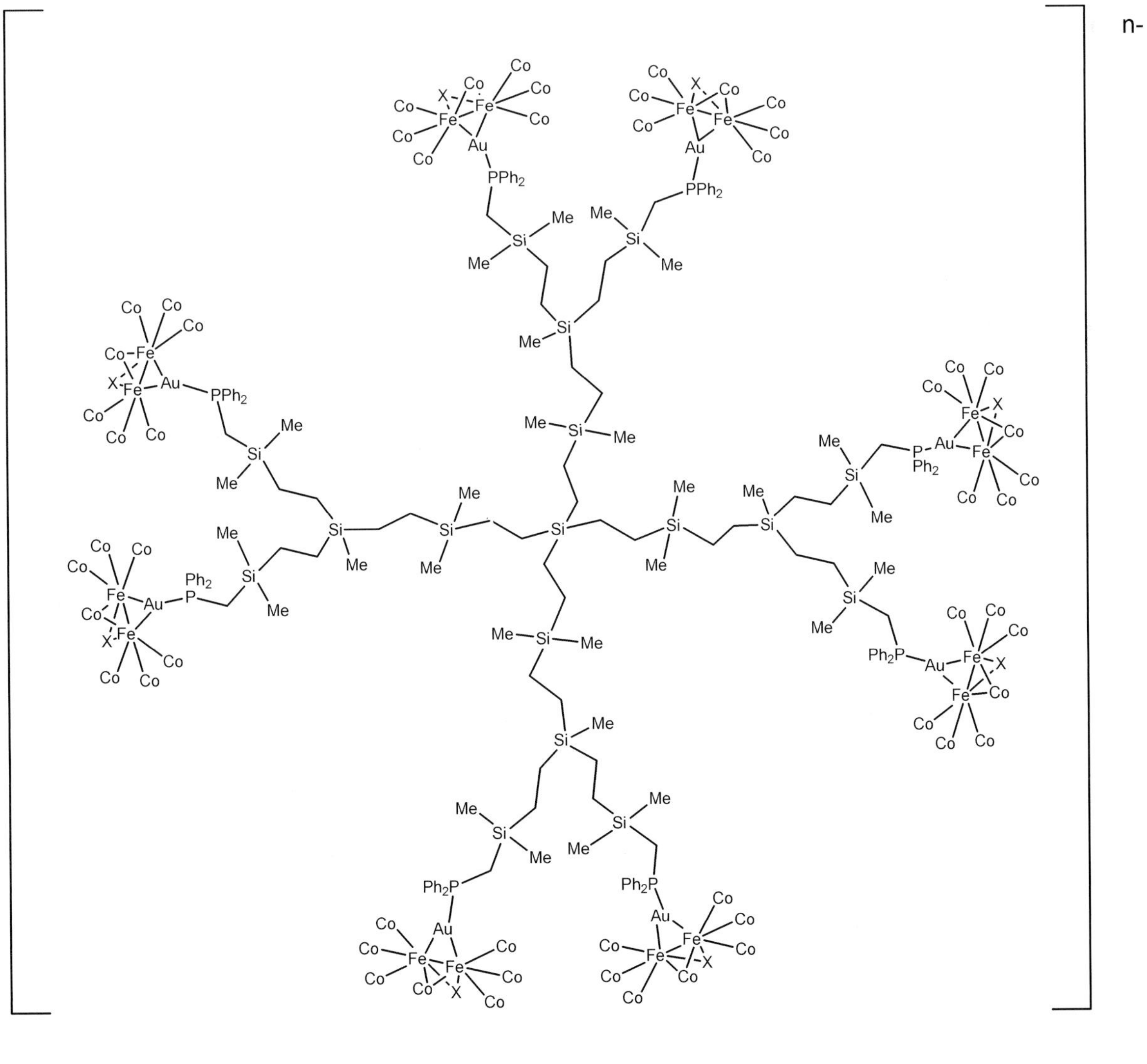

110 (X = Fe (CO)$_4$)

Figure 8.17 Rossell's AuFe$_3$ carbonyl clusters.[192, 193]

terminated Si-based analogs are also known[190b] along with a Pd-terminated, hyper-branched construct.[190c]

Rossell and coworkers[192, 193] terminally functionalized carbosilanes with AuFe$_2$ and AuFe$_3$ carbonyl clusters as well as AuCl.[193a] Chlorosilane termini were transformed (LiCH$_2$PPh$_3$·TMEDA, THF) to the methyldiphenylphosphine analogs, which were treated with ClAu(THT) and then with [Fe$_3$(CO)$_{11}$(PPh$_4$)$_2$] to yield the desired polyclusters (e.g., **110**; Figure 8.17). Notably, these dendritic materials showed good solubility in common organic solvents. Analogous tris(bipyridine)Ru(II)terminated carbosilanes are known.[193b]

Later, van Koten and coworkers[194] prepared periphery-palladated carbosilanes (**111**; Figure 8.18). The palladium terminal groups were accessed by transformation of benzylic alcohol end groups to the corresponding bromide and treatment with 4-iodophenol. Metallation was effected by reaction with Pd(dba)$_2$ and TMEDA to give the Pd(IV) complex, followed by treatment with MeLi and bipyridine to afford the Pd(II) complex. Other reports related to carbosilanes possessing Si-bonded 1-[C$_6$H$_2$(CH$_2$NMe$_2$)$_2$-3,5-Li-4] and 1-[C$_6$H$_3$(CH$_2$NMe$_2$)-4-Li-3] mono- and bis(amino)aryllithium termini have appeared.[195] An examination of the utility of similar Ni "pincer" complexes in homogeneous catalysis has been described;[196] catalytic activity was shown to decrease as terminal steric hindrance increased, which was probably associated with an intramolecular redox process. Similar Pd-based dendrimers with hemilabile *P* and *O* ligands constructed for the selective hydrovinylation of styrene in a membrane reactor have been described,[197, 197a] along with "molecular-multisite catalysts" accessed through the attachment of diamino-aryl Pd(II) complexes to a hyperbranched triallylsilane-based support.[198]

A "dendritic effect" in homogeneous catalysis using arylnickel(II)-terminated carbosilanes has been reported by van Koten and coworkers;[199, 199a] a notable decrease in cata-

111

Figure 8.18 van Koten's organopalladium carbosilanes.[194]

lytic activity with increasing generation number was observed. This deactivation process was ascribed to the formation of mixed-valence intermediates at the surfaces of more highly branched structures. Alper and coworkers[200] constructed related PAMAM-type Rh complexes on a silica surface to address a similar catalytic retardation effect; incorporation of diamine spacers 4 to 6 methylene units in length facilitated increased activity. However, on going to longer spacers with 12 methylene units, no significant change was observed.

Another related series of Rh-based catalysts for hydroformylations has been reported by Arya et al.,[201, 201a] where a divergent, solid-phase approach to framework construction was employed for synthesis on polystyrene beads. Standard Fmoc-benzyl ester protection–deprotection schemes allowed access to the 3rd generation materials following the introduction of bis(diphenylphosphanyl)methyl termini and their subsequent complexation. Hydroformylations of olefins proceeded with remarkable (>99%) conversions.

Hoveyda and colleagues[202] created efficient and recyclable poly(organoruthenium) metathesis catalysts (Figure 8.19) The benzylidene Ru catalysts **112** and **113** were isolated as air-stable, dark-green and brown solids, respectively. The activity of dendrimer **113** in promoting ring-closing metathesis of a bis(allyl)-*N*-tosylamide to the corresponding cyclopentane consistently led to yields of 87% or higher, with a slight loss of Ru upon recovery. Dendrimer **112** reportedly exhibited even higher catalytic activity in promoting the formation of a cyclic trisubstituted allylic alcohol. Dendrimer recovery after silica gel chromatography amounted to 90%, with a concomitant 8% Ru loss based on ^{1}H NMR.

Kriesel et al.[203] demonstrated the construction of aryl-terminated carbosilanes capped with CpRu^{+} moieties resulting in 12^{+}-, 24^{+}-, 36^{+}-, and 72^{+}-polycations. These materials were characterized by detailed electrospray ionization (ESI) mass spectrometry. A full resolution of the isotopic distributions was achieved for the first time. The 1st generation construct was also characterized by X-ray diffraction.

Liao and Moss[204] prepared metallodendrimers up to the 4th generation[205, 206] containing 48 peripheral ruthenium centers on a dendritic framework constructed by a conver-

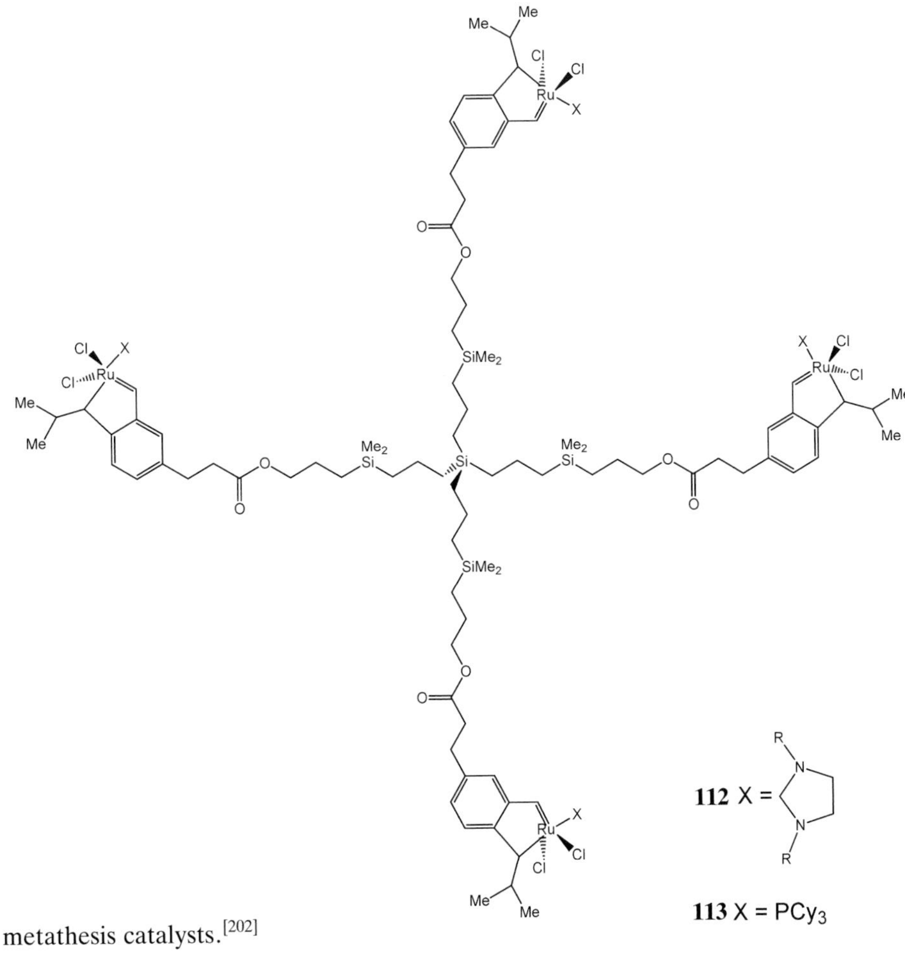

Figure 8.19 Hoveyda's metathesis catalysts.[202]

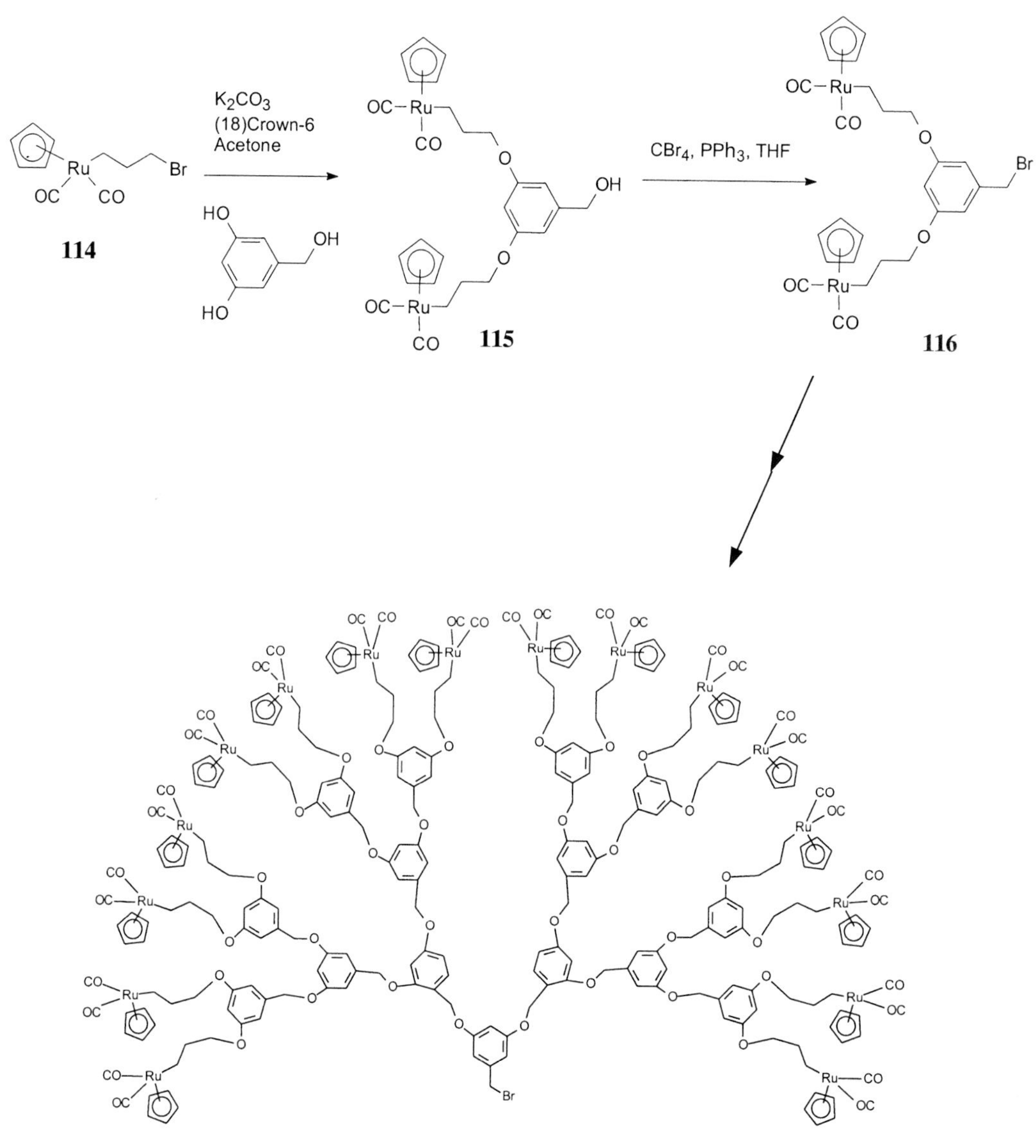

Scheme 8.23 Convergent construction of building blocks possessing 16 Ru centers at the periphery.[204]

gent approach.[114] The preparation of the 1st generation metallodendritic building blocks **115** and **116** commenced with the reaction of $(\eta^5\text{-}C_5H_5)Ru(CO)_2[(CH_2)_3Br]$ (**114**) with 3,5-dihydroxybenzyl alcohol (Scheme 8.23). Transformation of the focal hydroxyl moiety to the corresponding bromide permitted repetition of the sequence to produce the 4th generation wedge **117** with 16 ruthenium centers. Three equivalents of each generation dendron were then assembled through nucleophilic substitution with 1,1,1-tris(4'-hydroxyphenyl)ethane to afford the corresponding larger metallodendrimers (e.g., **118**; Scheme 8.24), the characterization of which included standard spectroscopic methods and elemental analysis, as well as an examination of the thermal properties such as T_g. A computational study on the use of these dendrimers as supports for organometallic catalysts has appeared.[207] A 1st generation organorhenium dendrimer (six termini) has also been reported[208] along with a silsesguioxane-cored construct.[208a]

Takada et al.[209] studied the thermodynamics and kinetics of adsorption of PAMAMs terminally modified with either bis(terpyridyl)Ru(II) or tris(bipyridyl)Ru(II) complexes. These materials were found to be readily adsorbed onto Pt electrodes (+0.8 V *vs.* Ag/AgCl) with the adsorption thermodynamics being well-characterized by the Langmuir adsorption isotherm. The kinetics of adsorption were found to be activation controlled, with the rate constant decreasing with decreasing generation. Electrochemical quartz crystal microbalance studies on the materials have also appeared.[210] Kimura et al.[211] reported the synthesis and characterization of bis(terpyridine)Fe(II)-terminated

117 + HO—⟨⟩—C(CH₃)—⟨⟩—OH → (K₂CO₃, (18)Crown-6, Acetone)

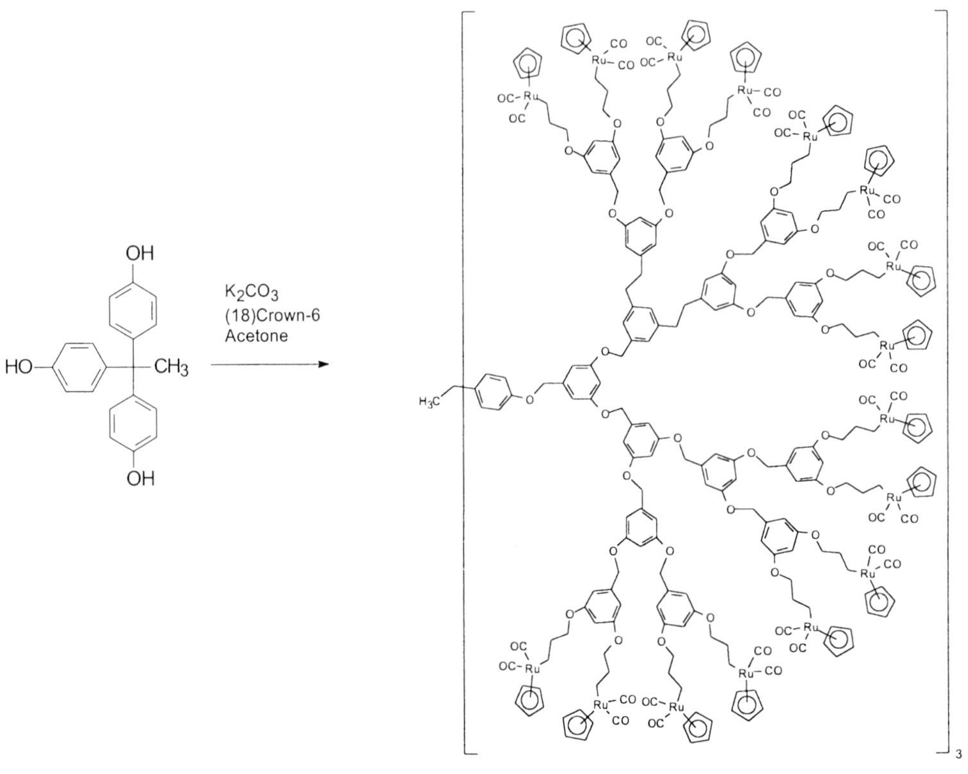

118

Scheme 8.24 Liao and Moss's Ru-containing metallodendrimer.[205, 206]

PAMAMs. Cyclic voltammetry suggested that the external iron centers were electrochemically equivalent.

Astruc and coworkers[212, 213] developed a strategy for the construction of metallodendrimers possessing iron sandwich complexes at the periphery (Scheme 8.25). Nonaene **119** (prepared by exhaustive addition of allyl bromide to an iron-activated mesitylene complex) was hydroborated, oxidized (H_2O_2), and treated with [FeCp(η^6-p-MeC$_6$H$_4$F)]PF$_6$ to give the branched polyferrocene **120**. Astruc and colleagues[214] synthesized amidoferrocene dendrimers with up to 18 ferrocene moieties on the surface (Scheme 8.26). The amino dendrimer **121** was treated (Et$_3$N) with FcCOCl (Fc = ferrocenyl) to give the 18-ferrocenyl construct **122**. Similarly, 36- and 72-amine-terminated dendrimers were converted to metallodendrimers, which were found to be insoluble in all solvents tested. It was speculated that for these constructs, peripheral steric hindrance reached a limiting,

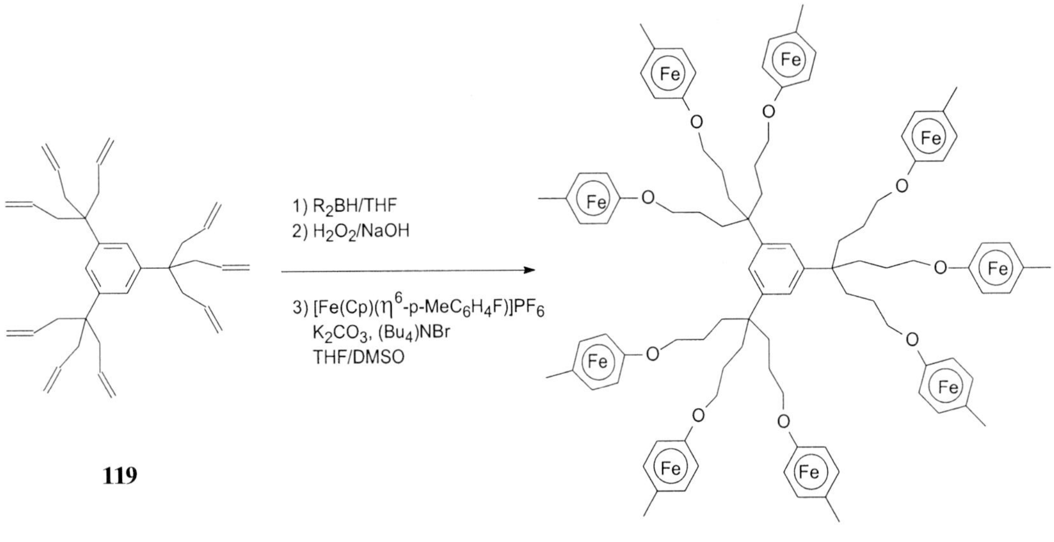

119

1) R₂BH/THF
2) H₂O₂/NaOH
3) [Fe(Cp)(η^6-p-MeC₆H₄F)]PF₆ K₂CO₃, (Bu₄)NBr THF/DMSO

120

Scheme 8.25 Astruc's iron sandwich bearing species.[212, 213]

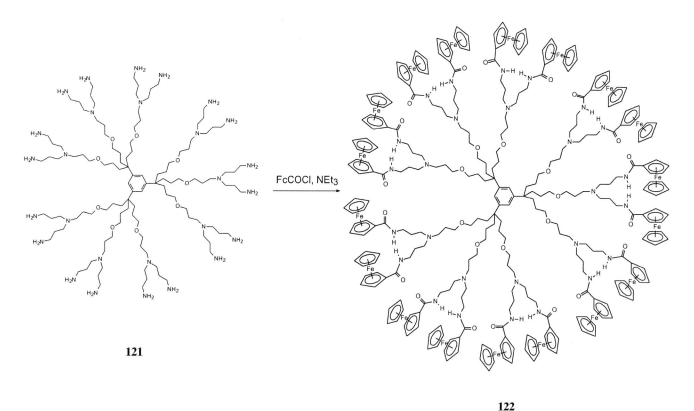

121   FcCOCl, NEt₃ →   **122**

Scheme 8.26 Amido-ferrocene dendrimer with 18 ferrocenes on the surface.[214]

dense-packed point between the 18-ferrocenyl and 36-ferrocenyl metallodendrimers. ΔE° values were found to increase with each generation, with all the Fe^{II}/Fe^{III} redox centers behaving independently. These authors have also reported the syntheses of several related branched macromolecules[215–217] and the heterobifunctionalization of a hexaferrocenyl 19-electron branched redox catalyst.[218] Hexa-arm polyisobutylene star polymers have also been crafted using a "core-first" protocol.[219] Chiral ferrocenyl diphosphorus ligands attached to the surface of small dendrimers have been described by Köllner, Pugin, and Togni[220] for use in asymmetric catalysis (Figure 8.20). A ferrocenyl "Josiphos" derivative[221] **123** was coupled to a 3,5-bis(aryl halide) TBDMS-protected phenol to afford the requisite monomer, which was subsequently attached to tri- and tetravalent cores to yield the 6 and 8 metal-terminated dendrimers (i.e., **124** and **125**, respectively). Characterization of these materials by MALDI-TOF MS and NMR has been discussed[222] and cyclophosphazene cores have also been used[223] to assemble a 16-ferrocenyl construct. Organometallic "triskelia" possessing triruthenium-triferrocenyl complexes at the periphery have also been published.[224]

Astruc and coworkers[225] also reported an "organoiron route" to a monomer, i.e., 4-$HOC_6H_4C(CH_2CH=CH_2)_3$, for use in rapid divergent or convergent protocols. A polycationic metallodendrimer possessing $[Fe(\eta^5\text{-}C_5Me_5)(\eta^6\text{-}N\text{-alkylaniline})]^+$ termini has been described;[226] these materials recognize Cl⁻ and Br⁻ anions. Other inorganic anions ($H_2PO_4^-$, HSO_4^-, and Cl⁻) have been shown to be efficiently recognized by a polycationic nona-amido-cobalticinium dendrimer.[227]

Astruc and coworkers[228] employed their 4-hydroxytris(allyl)methylbenzene monomer in the synthesis of phenol dendrons and arene-centered polyolefin dendrimers. Building block transformations included phenoxide substitution of alkyl iodides or alkyl mesylates and alkene oxidation to give terminal alcohols, which could be smoothly converted to better leaving groups (e.g., mesylate). The largest member of the family, a 243-allyl-terminated construct, was obtained in 11 % yield after chromatography. Divergent as well as convergent protocols were described. This series has been terminated with ferrocenylsilane moieties through hydrosilylation of the alkene groups.[229] It was found that these materials could be reduced from the ferrocenium form back to the ferrocenyl form, supporting their use as molecular batteries. Other ferrocene-based architectures incorporating Newkome's aminotrinitrile[144] have been reported.[230]

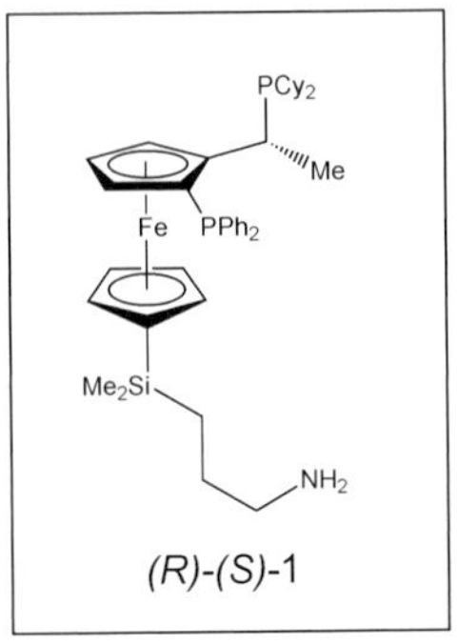

Figure 8.20 Togni's chiral ferrocenyl-based dendrimers for asymmetric catalysis.[220]

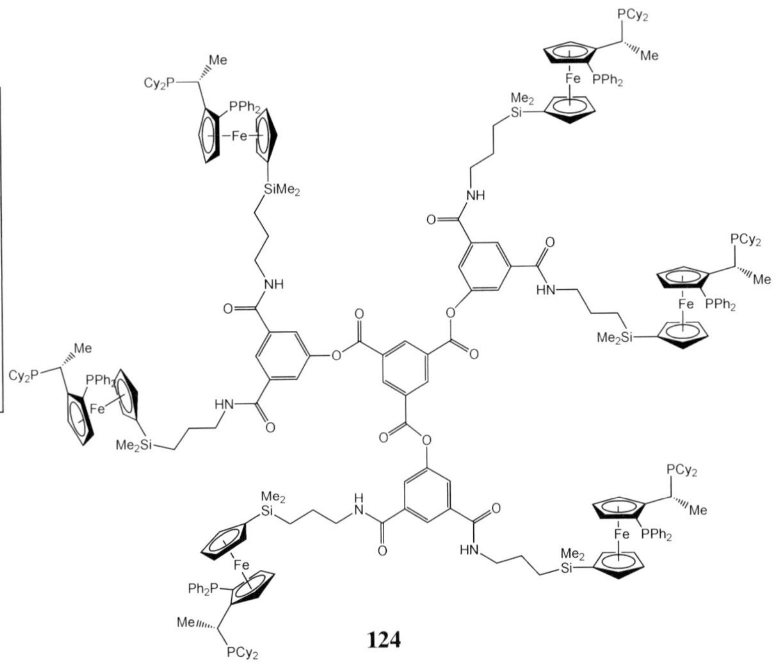

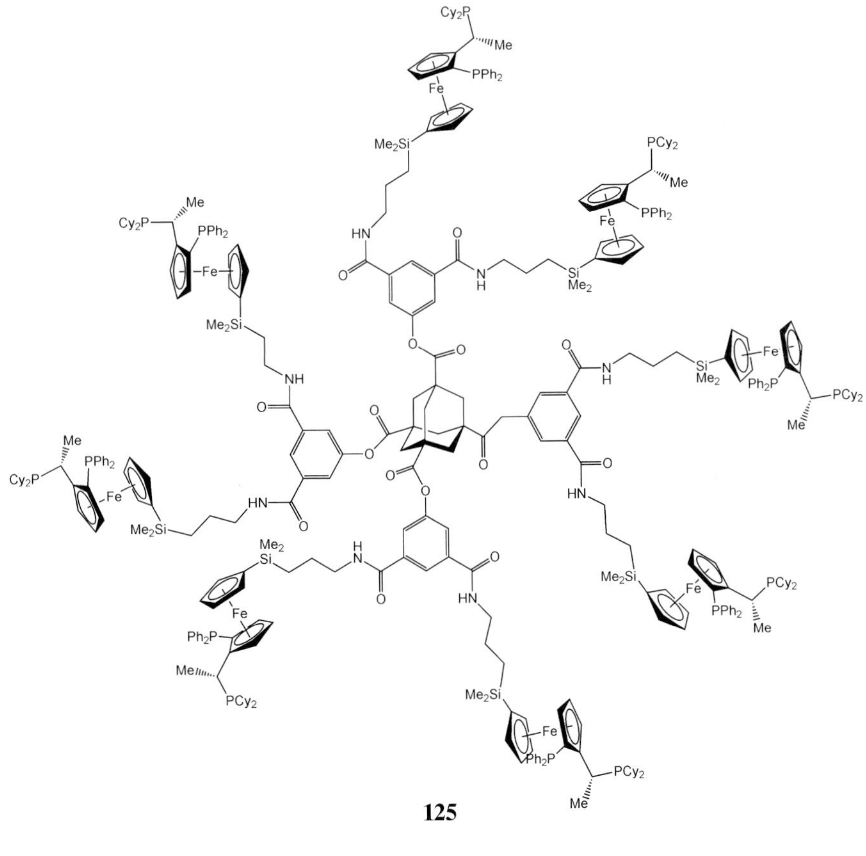

Alonso and Astruc[231] introduced $Ru_3(CO)_{11}$ clusters at the periphery of bis(diphenyl-phosphanylmethyl)amine-terminated PPIs. Mixed cluster formation through ligand-exchange reactions was avoided by using electron-transfer-chain catalysis with an $[FeCp(C_6Me_6)]_{64}$ complex acting as an electron-reservoir. Up to the 5th generation $[Ru_3(CO)_{11}]_{64}$ was constructed; notably, the bulk of the surface enabled solvent retention (e.g., THF was retained even after several days in vacuo). Dendritic multielectron reservoirs showing activity as anion sensors have been described.[1]

Ipaktschi, Hosseinzadeh, and Schlaf[232] reported the novel self-assembly of quinodimethanes through covalent bonds to produce unique functional macrocycles. Reduction

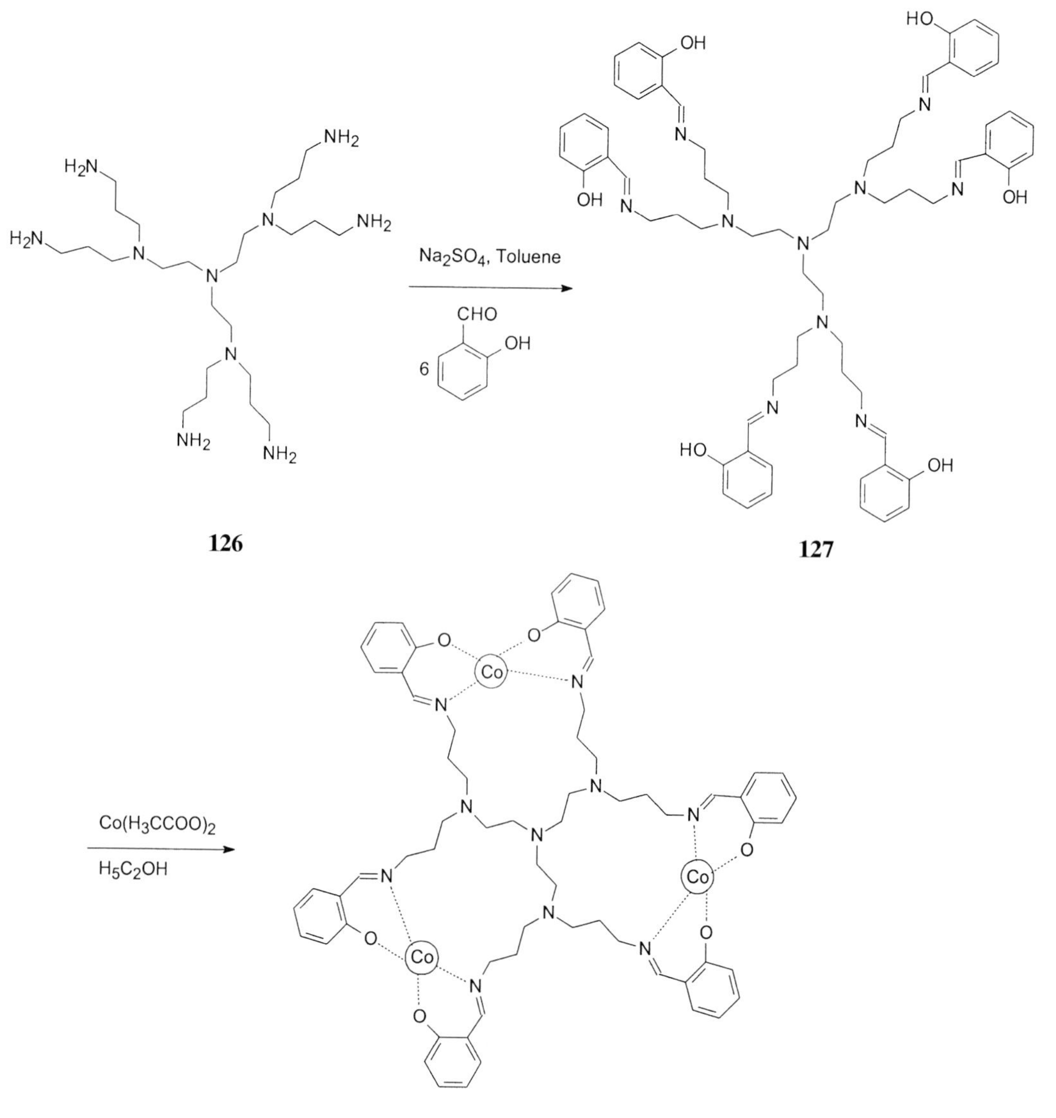

Scheme 8.27 Moors and Vögtle's oligocobalt complex.[233]   **128**

of a fluorenone-derived, ferrocene-modified diol with Sn/HCl led to a quinodimethane, which spontaneously self-assembled to give a tetrameric product possessing 16 ferrocene moieties (see also Scheme 5.61)

Moors and Vögtle[233] prepared hexaamine dendron **126**, which was treated with 2-hydroxybenzaldehyde to give the corresponding hexaimine **127** (Scheme 8.27). Reaction of hexaimine **127** with Co(II) acetate afforded the tricobalt complex **128** in quantitative yield. Seyferth et al.[234] and Jacobsen et al.[234a] described Co-capped metallodendrimers up to the 2nd and 3rd generations (e.g., **130**), respectively. Formation of the terminal bis-Co complexes was effected by treatment of the alkyne moieties on the carbosilane dendrimer **129** with $Co_2(CO)_8$ (Scheme 8.28). Absorption and electrochemical properties have been reported.[234b]

Morán et al. have made numerous contributions to the literature by preparing Co-, Cr-, and ferrocene-containing metallodendrimers, as well as hyperbranched polymers. For example, Cuadrado, Morán, and coworkers[235] functionalized the surfaces of carbosilane dendrimers with cobalt or iron complexes (Scheme 8.29). Tetrachlorosilane core **131** was treated with $Na[\eta^5\text{-}C_5H_5Fe(CO)_2]$ to give the small Fe-containing dendrimer **134**, while treatment with sodium cyclopentadienide followed by $Co_2(CO)_8$ gave the cobaltocene-terminated construct **135**. Reduction of the Si–Cl moieties of the core afforded carbosilane **132**, which was further treated with $Co_2(CO)_8$ to generate the Co-terminated dendrimer **133**.

Morán and coworkers[236, 237] also modified the termini of phenylsilane dendrimers (Scheme 8.30). The 0th generation phenylsilane dendrimer (not shown) was treated with

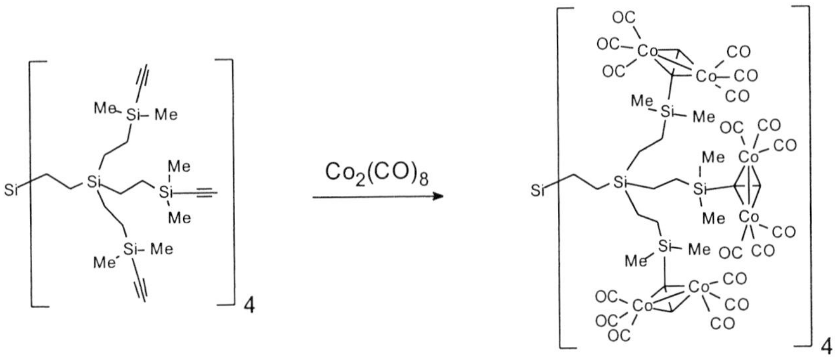

Scheme 8.28 Seyferth's Co-containing dendrimers derived from alkyne-terminated carbosilanes.[234]

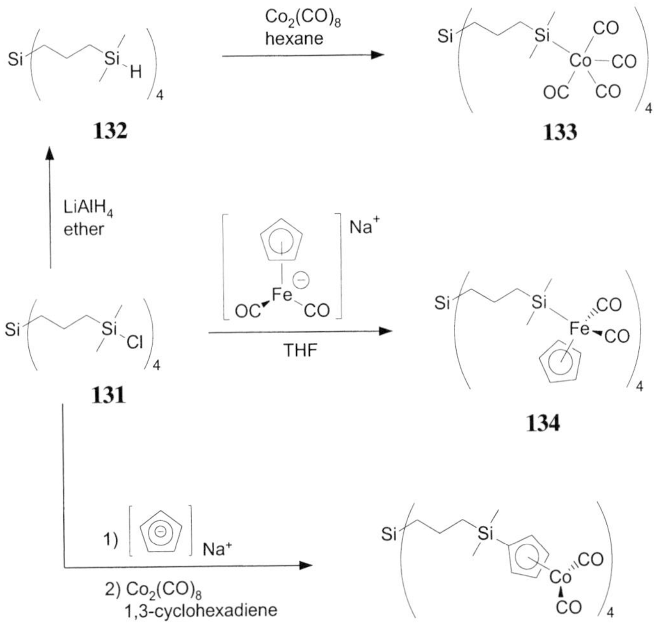

Scheme 8.29 Silane-based branched molecules used as "scaffolding" for ferrocenyl units.[235]

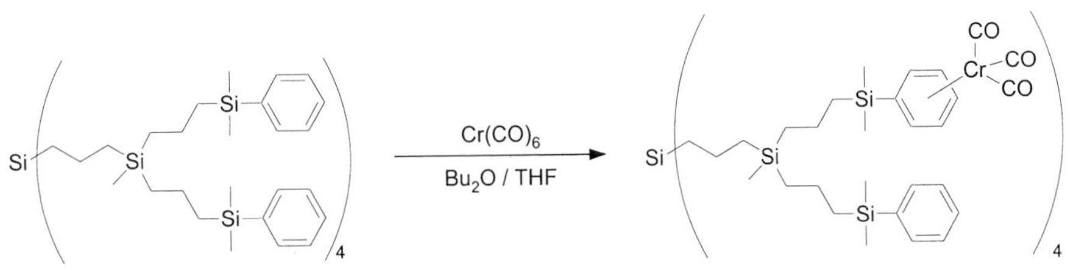

Scheme 8.30 Morán and Cuadrado's Cr-containing macromolecules.[236, 237]

excess Cr(CO)$_6$ to afford the tetra-Cr(CO)$_3$ complex, whereas the 1st generation octa-phenyl construct **136** afforded only partially metallated metallodendrimers (e.g., **137**). The Si-based infrastructures were prepared as noted in Section 3.5. Employing similar Si-based architectures, these authors[238] have created a series of ferrocenyl-terminated metallodendrimers (Scheme 8.31). Treatment of octachlorosilane **138** with (η^5-C$_5$H$_5$)Fe(η^5-C$_5$H$_4$CH$_2$NH$_2$) or (η^5-C$_5$H$_5$)Fe(η^5-C$_5$H$_4$Li) afforded the corresponding poly-ferrocenes **139** and **140**, respectively. Electrochemical studies on these polyferrocenes revealed an eight-electron oxidation process, while the 0th generation analog, possessing

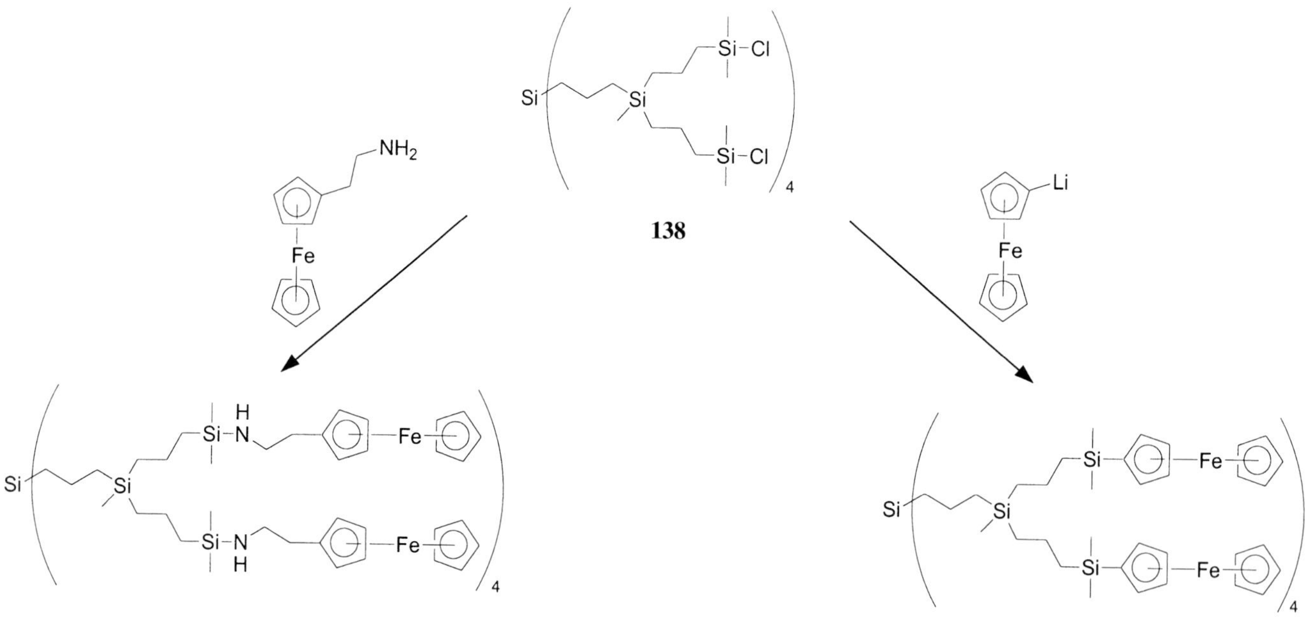

Scheme 8.31 Ferrocenyl-terminated complexes constructed on Si–Cl surface-functionalized carbosilanes.[238]

only four ferrocenyl units, showed a four-electron oxidation process. It was concluded that the ferrocenyl moieties were essentially *non-interacting* redox centers. The Fe-modified carbosilanes have recently been employed as electroactive films on electrode surfaces.[239] Reductive substitution (Scheme 8.32) of an octachlorosilane afforded the corresponding hydrosilane, which was subsequently treated with $(\eta^5\text{-}C_5H_5)Fe(\eta^5\text{-}C_5H_4CH=CH_2)$ in the presence of the Karstedt catalyst[240] to give metallodendrimer **141** (Figure 8.21).[241, 242]

Losada et al.[243] reported the novel application of these metallodendrimers as mediators in amperometric biosensors by incorporating them into carbon paste electrodes. Electrochemical data suggest that ferrocene **141** possesses better mediating capabilities than its related homolog with a shorter Si–Fe bridge. It was reasoned that chain length and the spacing between the ferrocene moieties play important roles in facilitating electron-transfer from the reduced enzyme. A smaller energy barrier to bond rotation for longer, less rigid ferrocene connectors renders the interaction between the dendritic redox centers and enzyme surfaces more efficient.

Cuadrado et al.[75] observed the electronic interactions between the transition metal centers in some surface-functionalized, convergently synthesized metallodendrimers (e.g., **142**; Figure 8.22).[244, 245] The dendritic wedges, as well as the corresponding metallodendrimers, exhibit two well-separated and reversible oxidation waves of equal intensity in cyclic voltammetry experiments. This behavior is in agreement with that observed for two ferrocenyl units linked through a bridging silicon atom.[246] It was postulated that the first oxidation occurs on non-adjacent ferrocene centers, making the adjacent ferrocenyl centers more difficult to oxidize. Thus, there is an 'electronic communication' between the Si-bridged ferrocenyl centers. This is in accord with similar observations made for some Ru-containing metallodendrimers.[74] Electrochemical investigations[247] of similar silicon-based ferrocenyl-terminated dendrimers possessing Si–NH ferrocenyl connectivity revealed that electrodes modified with these materials recognize and sense anionic guests. For example, the dendrimer exhibits selective preferences for HSO_4^- and $H_2PO_4^-$. The presence of the Si–NH moiety was deemed a critical parameter in the recognition of anions.

Cuadrado, Morán, and coworkers[248] reported other ferrocenyl metallodendrimers based on the PPI series (Scheme 8.32). Construction of the series through to the 5th generation relied on treatment of the requisite polyamine (e.g., **143**) with $(\eta^5\text{-}C_5H_5)Fe(\eta^5\text{-}C_5H_4COCl)$ to afford the corresponding polyferrocenes (e.g., **144** and **145**). Each member of the series was seen to exhibit a single, reversible oxidation process, suggesting that

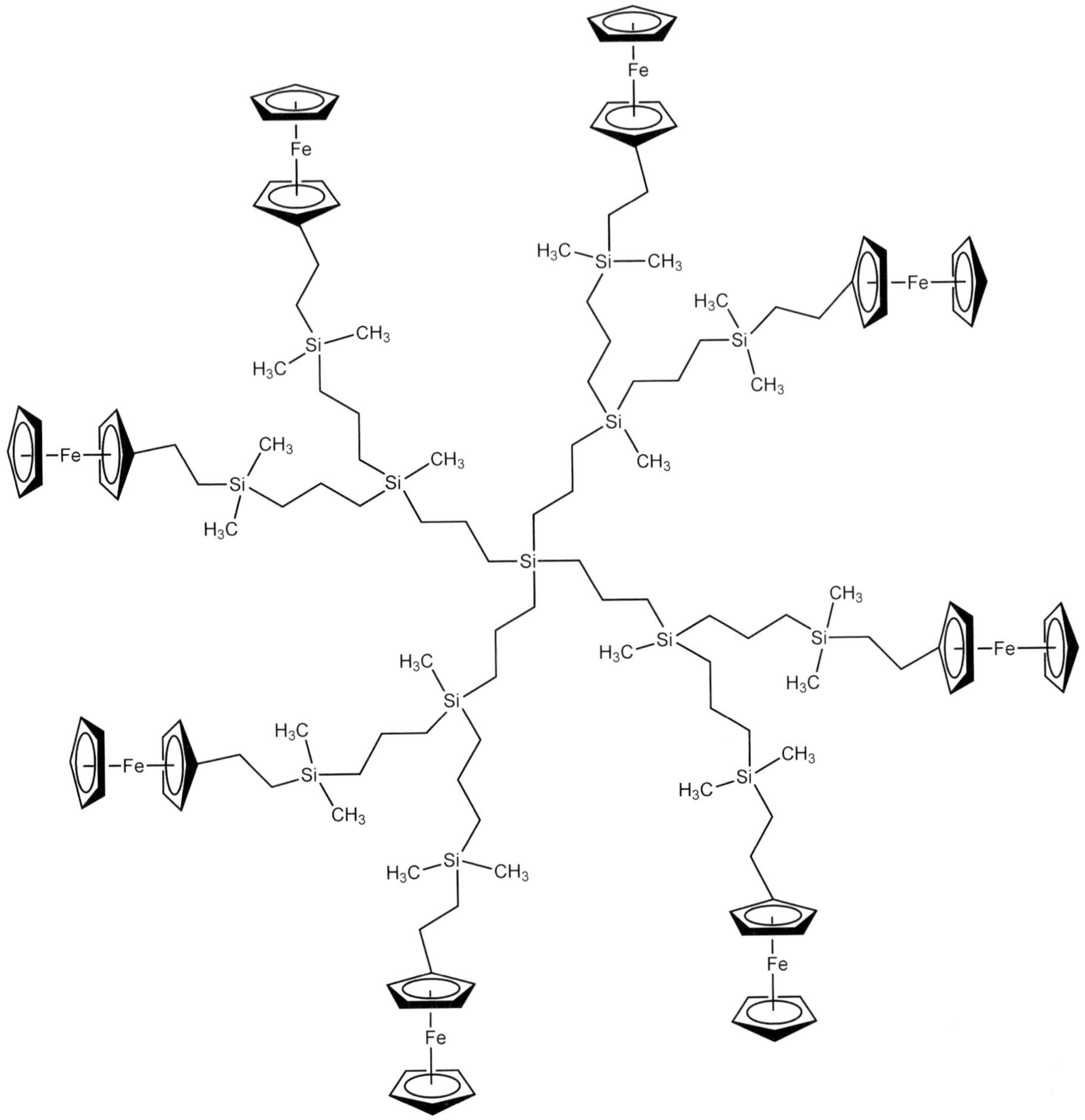

141

Figure 8.21 Cuadrado and Morán's mediators in amperometric biosensors.[241, 242]

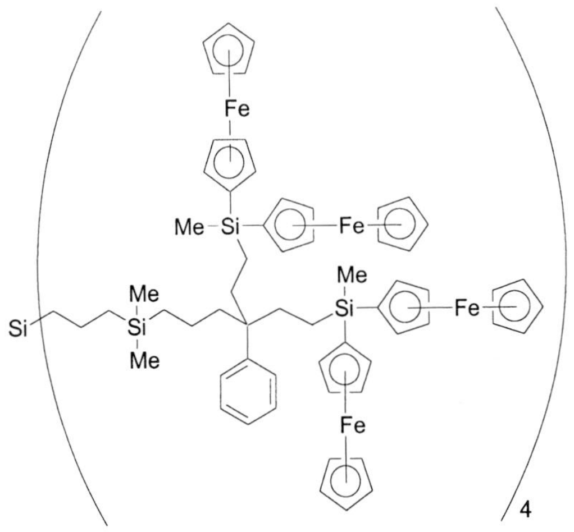

142

Figure 8.22 Cuadrado's ferrocenyl-complexes exhibiting electronic interaction between the transition metal centers.[75]

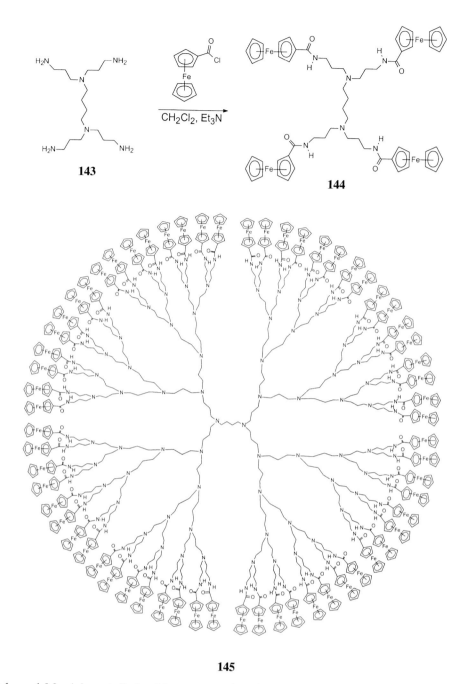

Scheme 8.32 Cuadrado and Morán's metallodendrimer possessing 64 amide-linked ferrocenyl metal centers on the surface.[248]

all of the metal centers within a single molecule are electrochemically equivalent. Takada et al.[249] investigated these redox-active cascades to determine the thermodynamics and kinetics of their adsorption onto a platinum electrode surface. Rate constants for the kinetics of absorption were found to be largely dependent on generation and less dependent on the concentration. Similar metallodendrimers, up to the 3rd generation, have been used to form large supramolecular polycomplexes by the coordination of cyclodextrin moieties at their ferrocenium termini by Kaifer and colleagues[250] A later report by Kaifer, Morán, Cuadrado, and coworkers[251, 252] described the terminal functionalization of the PPIs through to generation four, to yield a series of polycobaltocene macromolecules (e.g., **146**; Figure 8.23). β-Cyclodextrin derivatives were shown to facilitate aqueous solubilization of the reduced [Cp$_2$Co(0)] form through encapsulation of the termini. Recently, a combinatorial approach to the surface-functionalization of PPIs has appeared[253] where the parent polyamine was treated with an equimolar mixture of chlorocarbonylferrocene and the PF$_6^-$ salt of chlorocarbonylcobaltocenium.

Shu and Shen[254] reported the preparation of ferrocene-coated metallodendrimers up to the 3rd generation (e.g., **147**) based on a convergent approach[114] (Figure 8.24). The

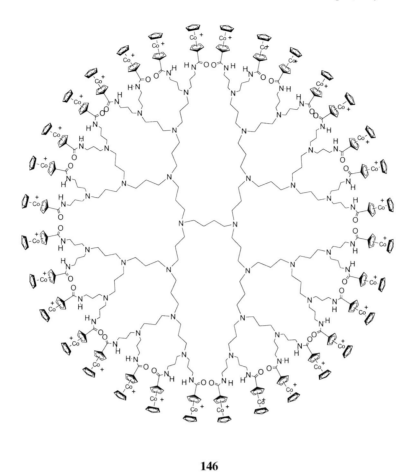

146

Figure 8.23 Polycobaltocene dendrimer shown to coordinate β-cyclodextrin.[251]

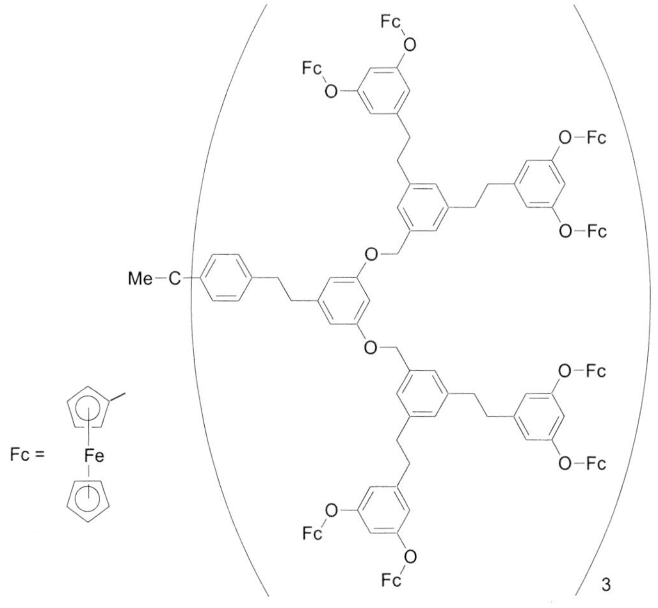

147

Figure 8.24 Shu and Shen's Fc-containing macromolecules[254] and highly branched macromolecules incorporating Pt, Pd, and Rh complexes.[255]

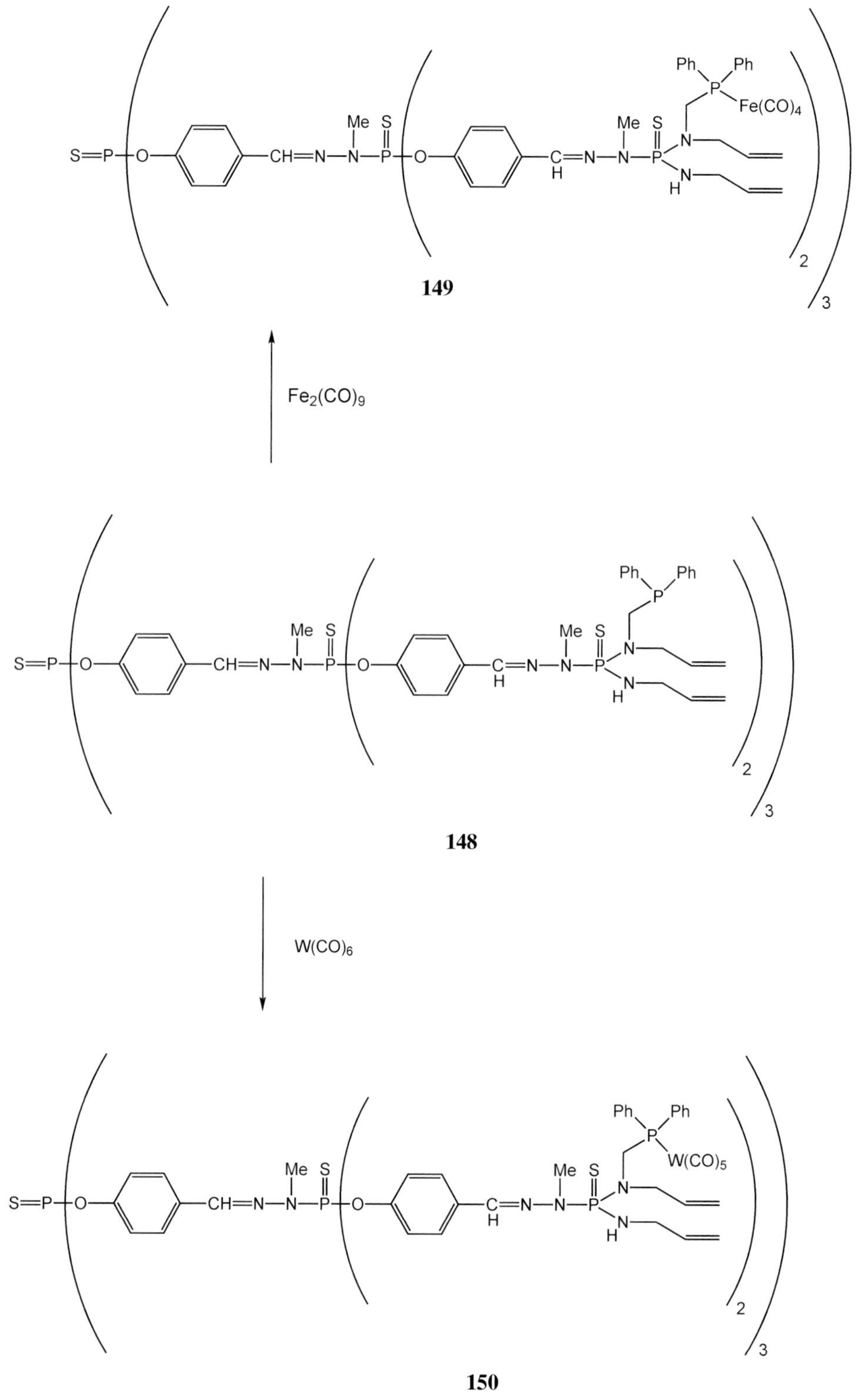

Scheme 8.33 Construction of phosphorus-containing metallodendrimers and surface-functionalization with W and Fe complexes.[256]

peripheral metal centers were shown to be electrochemically equivalent, which is consistent with other examples of ferrocene-containing metallodendrimers.

Majoral et al.[256] have synthesized a variety of *P*-based macromolecules (see Section 3.6) and have functionalized their cascades with numerous transition metals, e.g., iron and tungsten[257] (Scheme 8.33). The entire series of –P(S)Cl$_2$-terminated dendrimers was treated with H$_2$NCH$_2$CH=CH$_2$ followed by Ph$_2$PCH$_2$OH to generate phosphanyl-alkene-terminated cascades (e.g., **148**). This example was converted to the hexa-Fe(CO)$_3$ complex **149** or the hexa-W(CO)$_3$ complex **150** by treatment with Fe$_2$(CO)$_9$ and W(CO)$_6$, respectively. Caminade, Majoral, Chaudret, and coworkers[255] also functionalized the

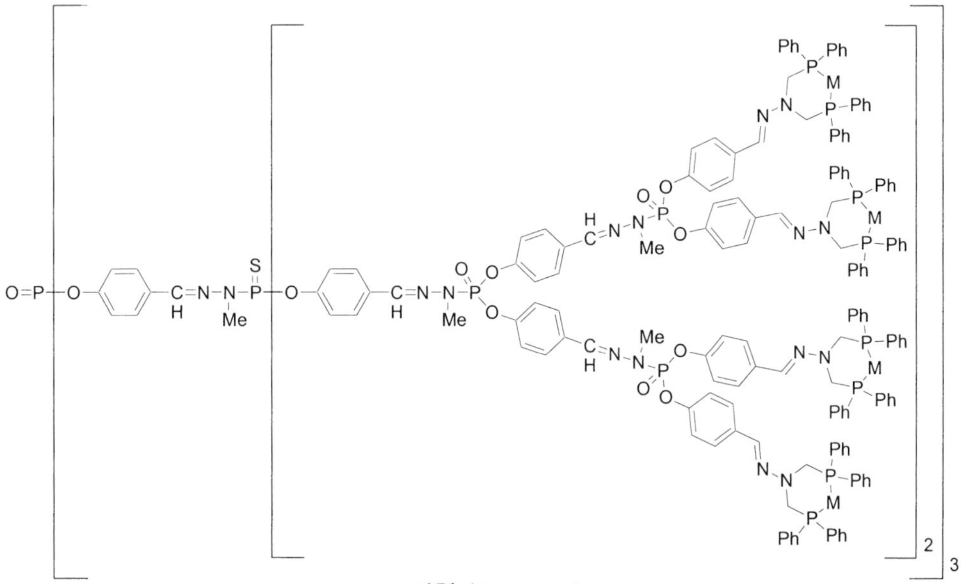

151 (M = no metal)
152 (M = PtR$_2$)
153 (M = PdR$_2$)
154 (M = RhR$_2$)

Figure 8.25 Surface modification with Ru complexes.[258]

surfaces of their *P*-branched dendrimers with platinum, palladium, and rhodium cations (Figure 8.25). Polycomplexes **152–154** were prepared by treatment of the Ph$_2$P-capped dendrimer **151** with the reagents PtX$_2$(COD), PdX$_2$(COD), and Rh(acac)(COD), respectively.

Majoral, Chaudret, et al.[258] also modified the surfaces of the same *P*-dendrimers with ruthenium moieties (Figure 8.26). The metallodendrimer **155** was prepared by treatment of the third generation *P*-dendrimer with RuH$_2$(PPh$_3$)$_4$. Notably, the novel complexes were found to exist as mixtures of different stereoisomers with different arrangements of

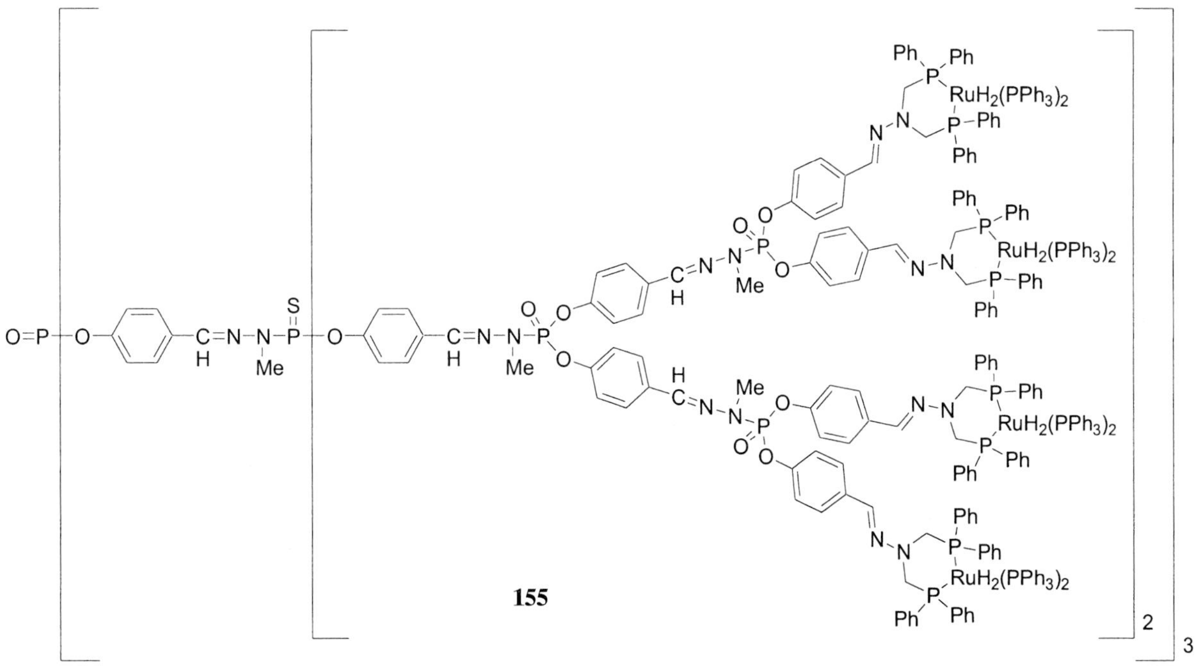

155

M = Fe(CO)$_4$, W(CO)$_5$, AuCl, Au-Me, Rh(acac)(CO), RhCl(COD)

Figure 8.26 Surface functionalization with Fe, W, Au, and Rh complexes.[259]

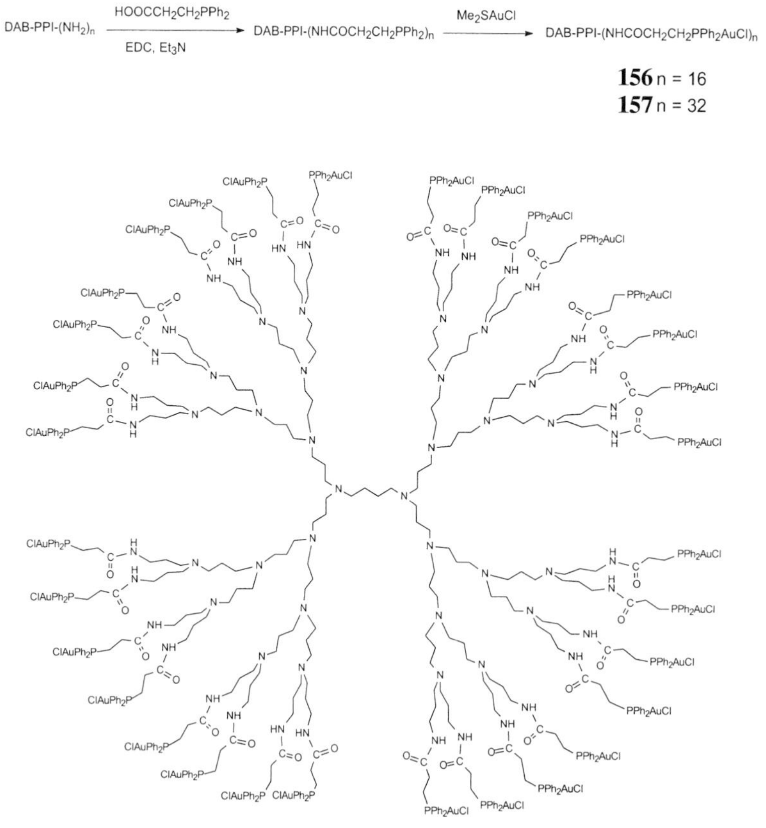

$$\text{DAB-PPI-(NH}_2)_n \xrightarrow[\text{EDC, Et}_3\text{N}]{\text{HOOCCH}_2\text{CH}_2\text{PPh}_2} \text{DAB-PPI-(NHCOCH}_2\text{CH}_2\text{PPh}_2)_n \xrightarrow{\text{Me}_2\text{SAuCl}} \text{DAB-PPI-(NHCOCH}_2\text{CH}_2\text{PPh}_2\text{AuCl})_n$$

156 n = 16
157 n = 32

157

Scheme 8.34 Schmidbaur's gold-coated metallodendrimers.[261, 262]

the coordinated ligands. These authors[259] applied the surface-functionalization method to a wide variety of metals (i.e., iron, tungsten, gold, and rhodium), which were also coordinated to monophosphine-based ligands. These constructs have also been terminated with zwitterionic complexes;[260] phosphonium anionic zirconocene(IV) species were generated by [3+3] cycloadditions between surface aldehydes and 2-phosphanyl-1-zirconaindene.

Schmidbaur and coworkers[261, 262] reported Au-containing metallodendrimers based on amino-terminated cascades (Scheme 8.34), where the 3rd and 4th generation diaminobutane-poly(trimethyleneamine)s [DAB–PPI–(NH$_2$)$_n$, n = 16 or 32] were treated with β-(diphenylphosphanyl)propionic acid to give the Ph$_2$P-terminated dendrimers, subsequent treatment of which with (Me$_2$S)AuCl afforded the corresponding Au-substituted dendrimers **156** and **157**.

Several other research groups have also utilized this family of PPI dendrimers in surface-functionalization studies. Reetz et al.[263] started with the 3rd generation dendrimer **158**, which was treated with Ph$_2$PCH$_2$OH to afford phosphanyl-terminated dendrimer **159** (Scheme 8.35), which was then converted to the related metallodendrimers **160a–e** by treatment with [PdCl$_2$(PhCN)$_2$], [Pd(CH$_3$)$_2$(TMEDA)], [Ir(cod)$_2$BF$_4$], [Rh(cod)$_2$BF$_4$], and a 50:50 mixture of [Pd(CH$_3$)$_2$(TMEDA)] and [Ni(CH$_3$)$_2$(TMEDA)], respectively. The catalytic activity of **160a** was tested in the Heck reaction of bromobenzene and styrene to form stilbene. The conversion to stilbene was 89 %, with 11 % of 1,1-diphenylethylene as a by-product. The macromolecular catalyst showed a significantly higher activity than the parent material [RN(CH$_2$PPh$_2$)$_2$PdMe$_2$], typically giving turnover numbers of 50 as opposed to 16 for the mononuclear complex. When the *recycled* dendrimer catalyst **160a** was subsequently reused for the same reaction, it displayed comparable catalytic activity. Feeder et al.[264] surface-functionalized the 3rd generation PPI-type core with (–CH$_2$PPh$_2$)$_2$ groups, which were subsequently transformed to either the [Ru$_5$C(CO)$_{12}$] or [Au$_2$Ru$_6$C(CO)$_{16}$] clusters; these materials can be envisaged as spheres comprised of an organic core coated with an outer layer of conducting metal particles. High-resolution transmission electron microscopy coupled with molecular modeling was used to ascertain the morphology of these nanoparticles.

DAB-PPI-(NH$_2$)$_{16}$

158

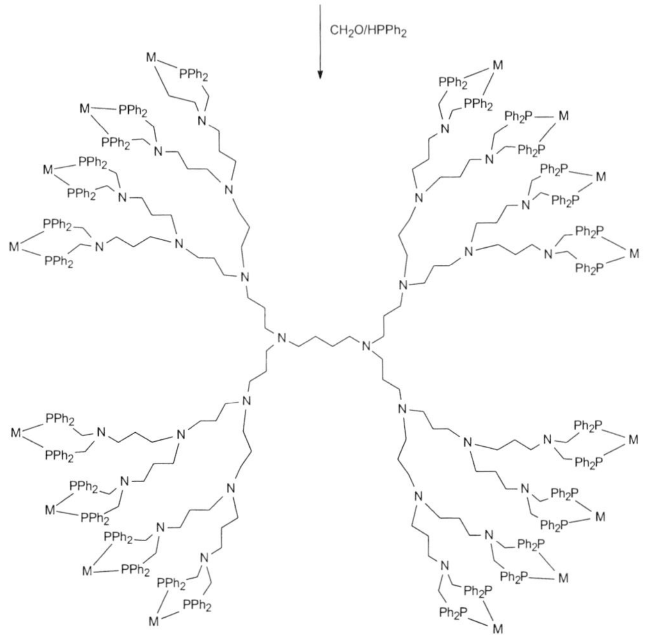

CH$_2$O/HPPh$_2$

159 (M = no metal)
160a (M = Pd(Cl)$_2$)
160b (M = Pd(CH$_3$)$_2$)
160c (M = Ir(cod)$_2$)
160d (M = Rh(cod)$_2$)
160e (M = Pd(CH$_3$)$_2$ and Ni(CH$_3$)$_2$)

Scheme 8.35 Reetz's Pd, Ir, Rh, and Ni complexes based on the PPI family.[263]

Similar bis(diphenylphosphanylmethyl)amine ligands at the termini of siloxane surface-bound PAMAMs have been complexed with palladium and used in the styrene–bromobenzene Heck reaction by Alper et al.[265] Stilbene conversions amounted to 67, 10, and 17 % for generations 0–2, respectively.

Meijer and coworkers[266] treated unmodified PPI dendrimers (up to the 5th generation) with NiCl$_2$, CuCl$_2$, and ZnCl$_2$. These complexes were studied by various spectroscopic methods, which confirmed their structural assignment. For example, ^{1}H NMR studies on the Zn(II) complexes indicated that metal binding occurs exclusively at the surface and involves the terminal primary amines and the outermost tertiary amines. UV/vis extinction coefficients of the fully Cu(II) complexed dendrimers were found to increase linearly on progressing from low to high generations.

PPIs have been functionalized with 2-vinylpyridine to generate peripheral bis[2-(2-pyridyl)ethyl]amine ligands, treatment of which with Zn(ClO$_4$)$_2$ or Cu(ClO$_4$)$_2$ afforded the corresponding complexes.[267] Low-temperature UV/vis spectrophotometry of a complex formed between [Cu(MeCN)$_4$(ClO$_4$)] functions and dioxygen revealed that ca. 60–70 % of the copper sites were bound by the diatomic molecule; this metallodendrimer was thus suggested as a possible synthetic analog of hemocyanin (i.e., an oxygen-transport protein found in molluscs and arthropods). Similar Pd(II) complexes based on the PPI series terminated with bis(diphenylphosphanylmethyl)amine ligands have been reported;[268] these materials promoted selective hydrogenation of conjugated dienes to monoenes.

A 1st generation PPI has been terminally modified with *trans*-diamminochloroplatinum by reaction with cisplatin in an effort to overcome cisplatin resistance.[269] Cytotoxicity towards mouse leukemia cell lines and seven human tumor cell lines was low; high charge at physiological pH and the branched architecture were postulated to retard cell membrane crossing.

Wiener et al.[270] prepared PAMAM-based[271] Gd-coordinating metallodendrimers,[271a] which were used as magnetic resonance imaging (MRI) contrast agents (Figure 8.27).

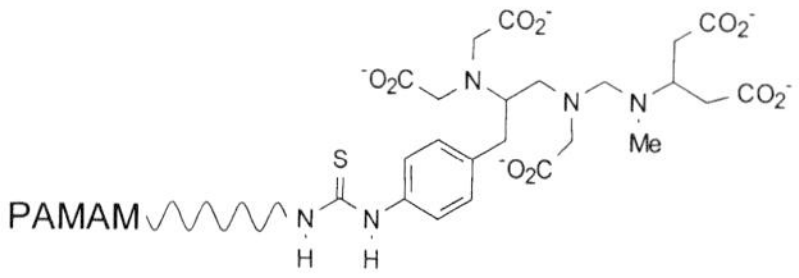

161 2nd generation, 92% substituted, 11-12 site Gd(III) ligand

162 6nd generation, 88% subsituted, 170-192 site Gd(III) ligand

Figure 8.27 Gd-coordinating PAMAM dendrimers used as MRI contrast agents.[270]

The 2nd and 6th generation PAMAM dendrimers were treated with diethylenetriamine-pentaacetic acid and then subjected to Gd(III) complexation to give partially complexed **161** and **162**, respectively. Their applications in medical imaging have recently been reviewed.[272] The molecular dynamics of these ion-chelate complexes attached to dendrimers have been studied.[273] Data were consistent with the hypothesis that the relaxivities of these MRI agents respond to the molecular weight partially as a result of differing rotational correlation times with increasing molecular weight.

A zeroth generation PAMAM was treated with pentaamino-triflato-Cr(III) and -Co(III), giving rise to the formation of five-membered chelate rings possessing amine and amide nitrogens.[274]

DuBois and coworkers[275] reported Pd-containing metallomolecules (e.g., **163** and **164**; Figure 8.28) that catalyze the electrochemical reduction of CO_2 to CO. Beer et al.[276] constructed dodecaferrocene **165** for a study of multiple redox centers. Deschenaux et al.[277] prepared a "ferrocene-containing liquid-crystalline" dendrimer **166** (Figure 8.29), which was shown to exhibit a "broad enantiotropic smectic A phase" along with suitable thermal properties. The mesogenic group was comprised of a cholesterol-substituted ferrocene, which was anchored to a small aryl-based dendrimer.

A mixed C_{60}/ferrocene liquid-crystalline dendrimer[277a] incorporating an ester-based framework (**167**; Figure 8.30) has been formed by the convergent construction of dend-

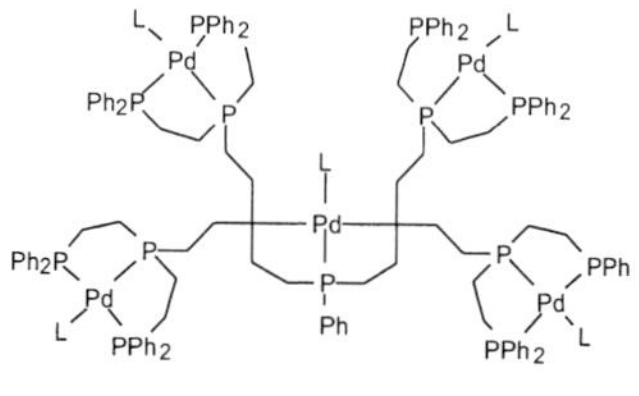

163

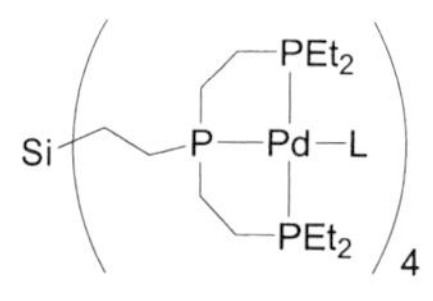

L = MeCN or Pet₃

164

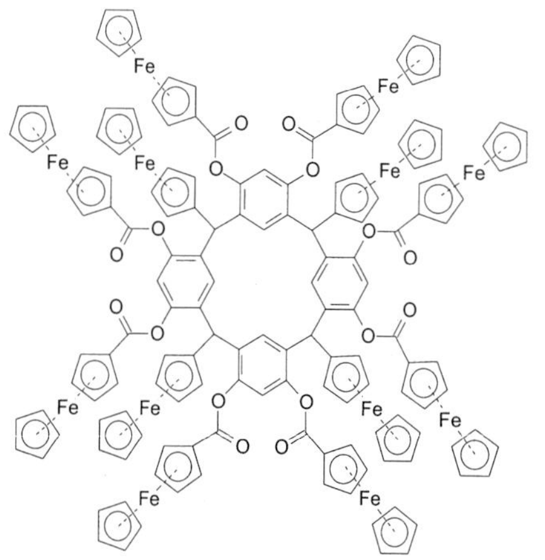

165

Figure 8.28 DuBois's Pd complexes used as catalysts for electrochemical reduction of CO_2 to CO[275] and Beer's dodecaferrocene.[276]

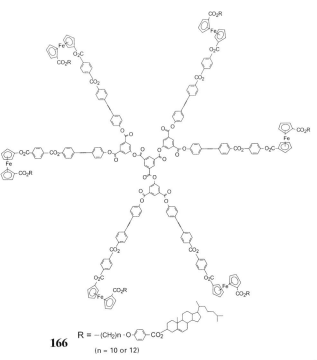

166   R = —(CH2)n-O-⬡-CO2-

(n = 10 or 12)

Figure 8.29 Deschenaux's ferrocene-containing liquid-crystalline macromolecules.[277]

rons possessing terminal cholesteric moieties attached to ferrocene groups via an alkyl chain connected to a two-directional malonate-based core.[292] Fullerene cyclopropanation (DBU, I₂, C₆₀) after dendrimer preparation allowed C₆₀ incorporation. Mesomorphic properties included the formation of an enantiotropic smectic A phase.

Zeng, Newkome, and Hill[278] have described the preparation of prototype polyoxometallates (POMS) based on a "tris"-terminated tetravalent dendritic core, which exhibit catalytic properties. The termini consisted of $[H_4P_2V_3W_{15}O_{62}]^{5-}$ moieties and the relative rates of tetrahydrothiophene oxidation to the corresponding oxide were investigated.

The first phthalocyanine-terminated architecture was reported by Kraus and Louw,[279] where a hydroxysilylated phthalocyanine was employed as a nucleophile for substitution of the halide substituents on dichlorotriazine end groups. The 1st generation ester-based construct purportedly possessed six phthalocyanine units. Analogously, poly(lysine) scaffolding (up to the 4th generation) has been created,[280] where each hemisphere of the amine-terminated dendrimers was protected (t-Boc or Fmoc) to allow for selective removal. Thus, at the 4th generation, 16 free-base porphyrins and 16 Zn-porphyrins were attached to the same dendrimer. This unique material showed intramolecular fluorescence energy transfer in DMF. Later, a 64 terminal poly(lysine) scaffold was similarly capped with 32 free base and 32 Zn(II) porphyrins in a scrambled fashion.[281] These materials exhibited highly efficient (85 %) fluorescence energy-transfer from the metallated to the non-metallated ligands.

Kim and Jung[282] created carbosilane dendrimers (see Section 4.9) possessing internal alkene moieties and external alkyne groups (Figure 8.31). In a later report,[283] these authors described the preparation of standard 1 → 2 branched carbosilane frameworks possessing ethynyl termini and their subsequent treatment with dicobalt octacarbonyl to form the corresponding hexacarbonyl peripheral clusters (e.g., **168**). Core units consisted of 2,4,6,8-tetramethyl-2,4,6,8-tetrasila-1,3,5,7-tetraoxacyclooctane $(CH_2=CHMeSiO)_4$ and 1,2-bis(triallylsilyl)ethane $[(CH_2=CHCH_2)_3SiCH_2]_2$. The phenylethynyl surface groups were introduced by reaction of the chlorosilane termini with lithium phenylacetylide.

Low-generation PAMAMs have been functionalized with dithiocarbamate termini and reacted with either $[Ru(\mu-Cl)_2Cl_2(p\text{-cymene})_2]$ or $[RuCl(PPh_3)_2(\eta^5\text{-}C_5H_5)]$ to afford the corresponding Ru-terminated constructs.[284]

Brüning and Lang[285] described the synthesis of a 2nd generation carbosiloxane terminated with $(\eta^5\text{-}C_5H_4SiMe_2)(\eta\text{-}C_5H_5)TiCl_2$ moieties. Dendrimer construction was

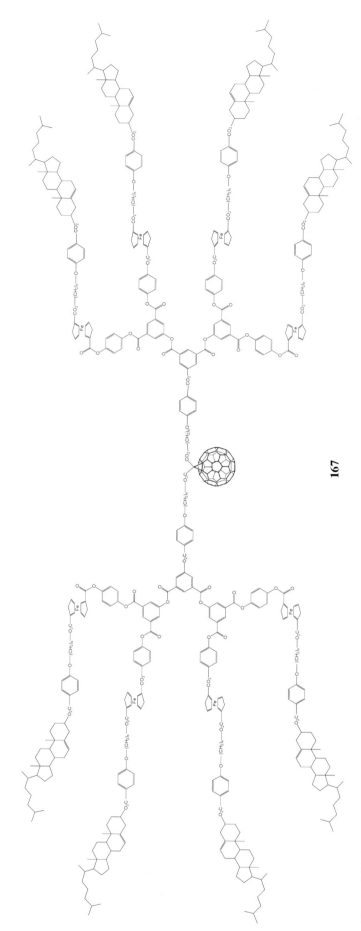

Figure 8.30 Deschenaux's fullerene–ferrocene liquid-crystalline dendrimers.[292]

167

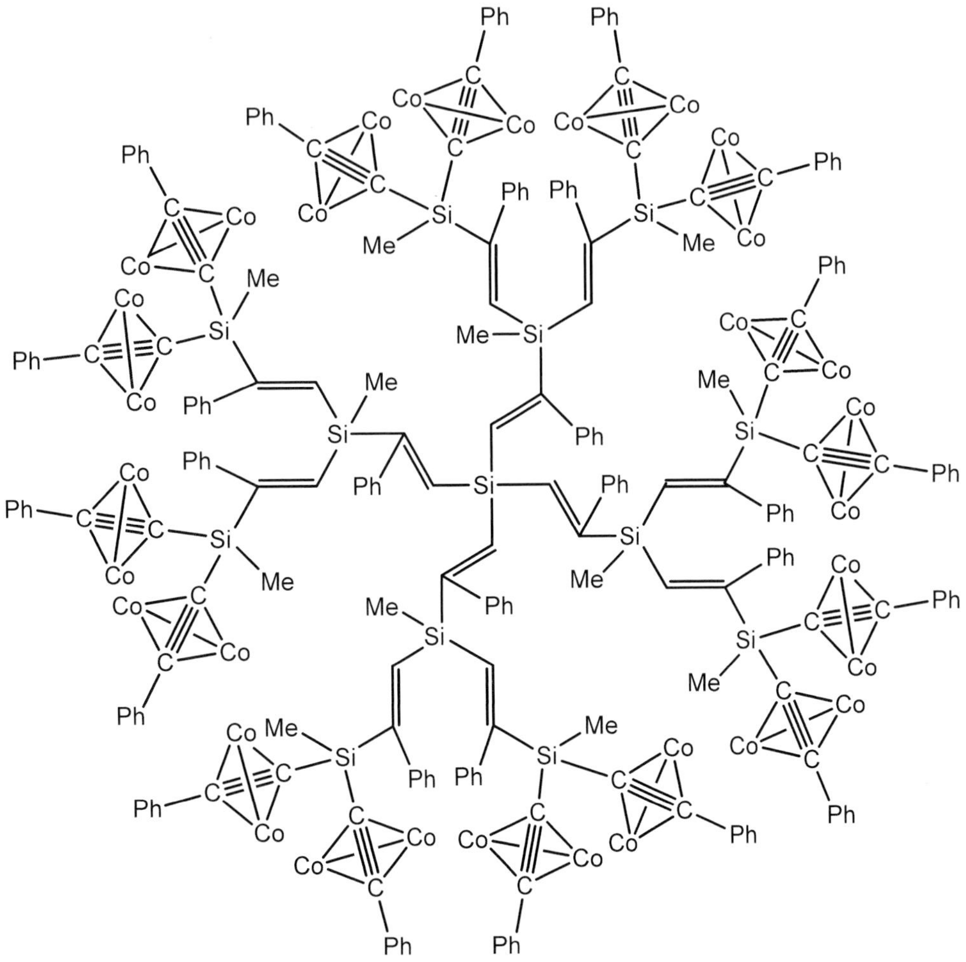

168

Figure 8.31 Kim's carbosilanes with peripheral ethynyldicobalt hexacarbonyl complexes.[282]

effected by repetitive standard hydrosilylation (H_2PtCl_6) and subsequent treatment with allylic alcohol.

Hughes et al.[286] examined the functionalization of the peripheral aromatic rings of poly(aryl ether)s with tricarbonylchromium(0) complexes. A force field suitable for molecular mechanics and dynamics calculations of these materials was delineated.

Brinkmann et al.[287] created Pd-coated PPIs for use as allylic substitution catalysts in a continuously operating membrane reactor. Two types of catalyst surface were prepared, one by treatment with dimethylpalladium chloride and the other generated *in situ* by reaction with allylpalladium chloride. The dendritic polyligands consisted of bis(diphenylphosphanylmethyl)amine units. The model reaction studied was the conversion of methyl 3-phenyl-2-propenylcarbonate to *N*-(3-phenyl-2-propenyl)morpholine. The allylpalladium-based catalysts were found to be superior, initially giving complete conversion, which then gradually decreased to 80 %.

Later, van Leeuwen and coworkers[288] prepared mono- and bidentate phosphine-terminated carbosilane dendrimers,[289] also for use as catalysts in a continuous-flow membrane reactor. *P*-ligands on the dendritic surface were treated with $[(\eta^3\text{-}C_3H_7)PdCl]_2$ to form the corresponding allylpalladium complexes. These complexes were employed as catalysts for the allylic alkylation of allyl trifluoroacetate and diethyl sodiomethylmalonate to afford diethyl allylmethylmalonate. Batch as well as continuous-flow catalytic processes were explored using these constructs; high activity was observed in each case, although some catalyst decomposition was noted in the continuous flow mode.

PAMAMs have been constructed on a silica surface, phosphonated [Ph_2PCH_2OH] at the periphery, and terminated [$Rh(CO)_2Cl_2$] with rhodium.[290] Excellent selectivities in hydroformylation reactions, favoring branched vs. linear products, were found with these catalysts.

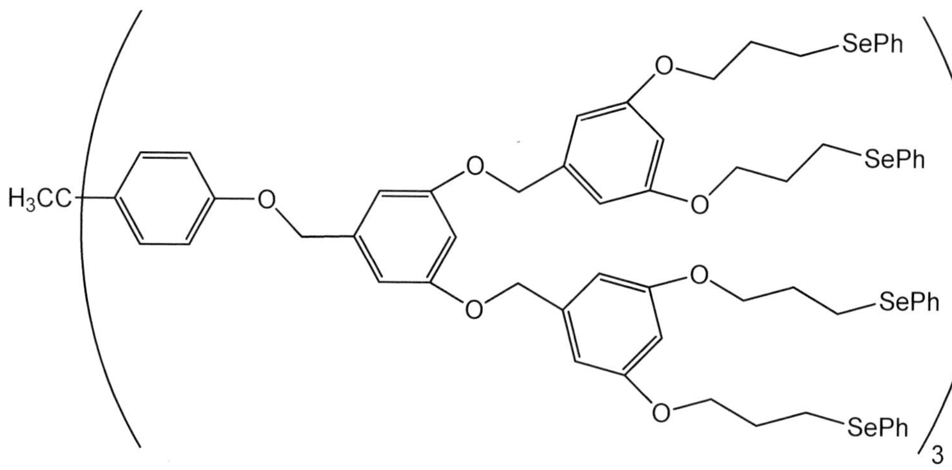

169

Figure 8.32 Detty's catalyst for H$_2$O$_2$ activation.[293]

Vassilev and Ford[291] generated PPI complexes with Cu(II), Zn(II), and Co(III) for study as catalysts in the hydrolysis of *p*-nitrophenyl diphenylphosphate. The diamino-[Cu(II) and Zn(II)] and tetraamino- [Co(III)] base complexes were characterized spectrophotometrically and titrimetrically. Hydrolysis rates were found to be dependent on ionic strength. Rates for buffered NaCl solutions of dendrimers and 1.0 mM Cu(II) were found to be 1.3 to 6.3 times faster relative to those in the absence of Cu(II).

Francavilla, Bright, and Detty[293] created poly(phenylselenide)s (**169**; Figure 8.32) that catalyze the oxidation (H$_2$O$_2$) of bromine to give positive bromine species, which, in turn, can react with cyclohexene in a two-phase system. Catalytic activity was found to increase per phenylseleno group (i.e., beyond statistical contributions) with increasing generation.

A family of CpW(CO)$_3$Me-terminated PPIs has been created and successfully employed as organometallic photonucleases.[294] The smaller di- and tetra-metallated constructs were found to cleave plasmid DNA in a double-stranded manner. In contrast, the larger 8- and 16-terminated complexes led to DNA aggregation and precipitation as a dominant competing process.

The use of low-generation carbosilanes as cores in star polymer formation employing a Ru-ROMP catalyst has been reported.[295]

8.6   Metals as Structural Auxiliaries

8.6.1   Site-Specific Inclusion

The first organometallic dendrimer possessing a site-specific attachment of a metal center within the structure was reported by Newkome et al.[296] During the construction of an all-carbon unimolecular micelle,[27] the intermediate dodecaalkyne **170** was treated with dicobalt octacarbonyl to afford the corresponding Co-containing cascade **171**, which was termed a "cobaltomicellane" (Scheme 8.36). The potential to perform reactions on a branched framework[297, 298] was thus realized, along with the precedent to construct reaction sites within a highly branched framework. Marx et al.[299] prepared a 1st generation cobalt complex from its hexaalkyne precursor; this has been followed by numerous other related examples.[300–304]

Newkome et al.[305] described a series of macromolecules possessing multiple piperazine moieties within the dendritic framework (Figure 8.33). Construction of these flexible architectures was facilitated by a three-component, single-pot reaction to produce a homologated aminotriester, which was reacted with a tetraacyl chloride (see Section 4.1.4). Both the 1st and 2nd generation polyligands were treated with Pd(MeCN)$_2$Cl$_2$ and CuCl$_2$ to afford the corresponding metallodendrimers **172** and **173**, respectively. Complexation with Cu was followed by ^{1}H NMR titration.[306–309]

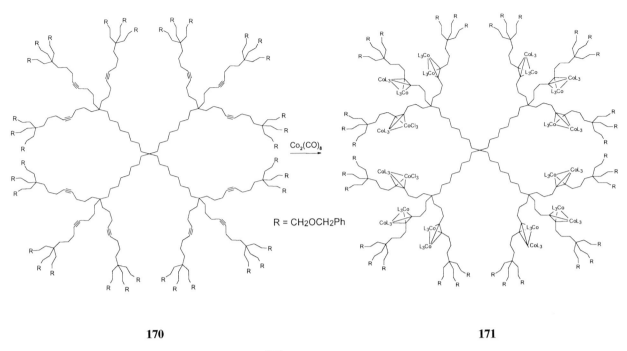

170   171

Scheme 8.36 Site-specific internal metallation.[296]

Epperson et al.[70, 310] utilized Co(II) as a paramagnetic ¹H NMR probe to investigate the internal cavities of dendrimers possessing specifically located binding sites (**174**; Figure 8.34). Addition of < 1/4 equivalent of Co(II) to the diaminopyridinyl-containing macromolecules (see Section 4.1.4)[311] led to the formation of a 1:1 Co(II)–dendrimer complex. Using EXSY, TOCSY, and COSY techniques, it was observed that protons close to the paramagnetic complex exhibited hyperfine shifted signals and possessed shortened relaxation times (T_1 and T_2). Analysis of the data thus provided information pertaining to the geometry of the inner dendritic region.

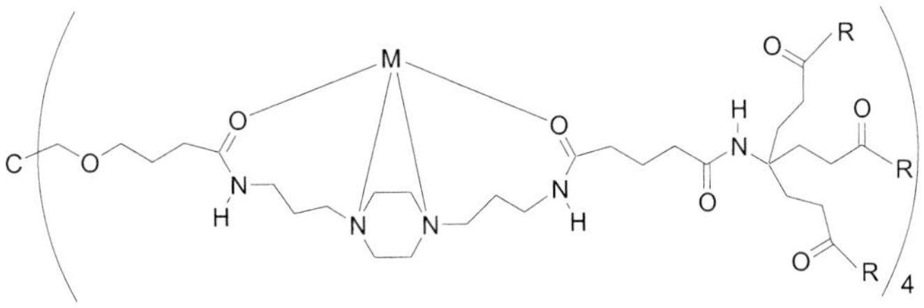

172a (R = O-*t*-Bu, M = Pd²⁺)
172b (R = O-*t*-Bu, M = Cu²⁺)
173a (R = NHC(CH₂CH₂CO₂-*t*-Bu)₃
M = Pd²⁺)
173b (R = NHC(CH₂CH₂CO₂-*t*-Bu)₃
M = Cu²⁺)

Figure 8.33 Examples of site-specific metal inclusion.[305]

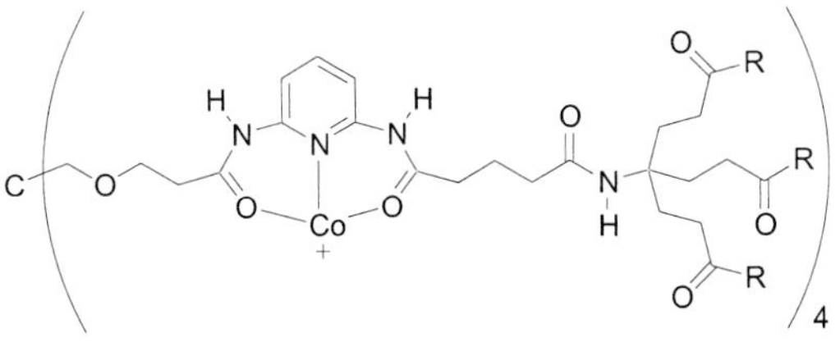

174a (R = O-*t*-Bu)
174b (R = NHC(CH₂CH₂CO₂H)₃)

Figure 8.34 Pyridinyl site specific locus for metal inclusion.[70, 310]

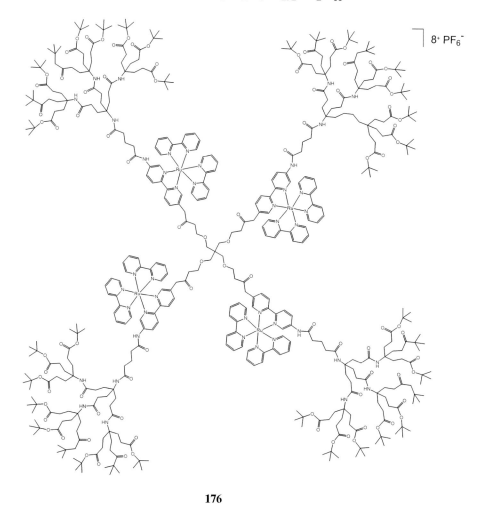

175

$$R' = (CH_2)_3CONHC[CH_2O(CH_2)_2CO_2Et]_3$$

176

Figure 8.35 Newkome's neutral (**175**)[312] and charged (**176**)[314] branched architectures.

Newkome and coworkers expanded their work (Figure 8.35) with Ru-based architectures to include the highly branched "neutral" tetrakis complex **175** possessing internal carboxylate counterions[312, 313] and the tetrakis[tris(bipyridine) ruthenium(II)] dendrimer[314] **176**. Cyclic voltammetry and absorption spectra were employed to aid in structural confirmation. Notably, the solubility of the neutral complex **175** was seen to decrease in common organic solvents such as DMF, but this could be readily reversed by the addition of external counterions, i.e., BF_4^- or PF_6^-.

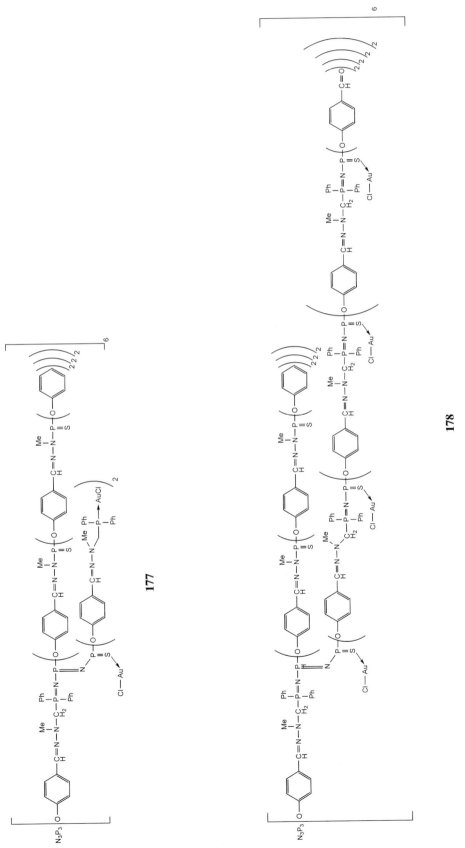

Figure 8.36 Incorporation of Au–Cl groups through complexation with –P=S and –PPh₂ moieties.[315]

Larré et al.[315] described regioselective gold complexation on the infrastructure of Majoral's *P*-dendrimers (see Section 3.6). Site-specific construction of –P=N–P=N–P=S, –P=N–P=S, and –PPh$_2$ groups facilitated Au–Cl adduct formation. Constructs bearing 18 and 90 gold centers (**177** and **178**, respectively) were reported. An X-ray crystal structure of the 1st generation dinuclear complex was presented.

Mak, Bampos, and Sanders[316] demonstrated the formation of "site-specific" metallodendrimers that exhibit controlled folding of the framework (Scheme 8.37). Following preparation of the requisite porphyrin monomers [**179** and **182** were accessed by condensation of the appropriate aldehydes and bis(pyrrole)], construction began with Pd-mediated coupling of monoalkyne **179** to a diiodobenzyl alcohol to yield the bis(porphyrin) **180**, which was then connected to an iododiacid under Mitsunobu conditions (PPh$_3$, DIAD, THF) to afford tetraporphyrin **181**. Coupling of aryl iodide **181** to the bis(alkyne) core **182** yielded dendrimer **183**. UV/vis titrations of **181** and **183** with DABCO revealed a controlled folding of the porphyrin units such that an "induced stacking" of parallel planar rings occurs (e.g., **184**). Sanders and coworkers[317] later constructed similar porphyrin dendrimers using a more flexible tetraacid-based, metal porphyrin core. Photophysical studies indicated energy transfer from the periphery to the core to occur with an efficiency of approximately 60 %. Norsten and Branda[318] functionally differentiated planar porphyrins and connected them through phenolic displacement of benzyl chloride units to yield a 1st generation pentaporphyrin. The pentametallo Zn adduct was prepared and its identity was verified by MALDI-TOF MS.

Humphrey and colleagues[319] created similar Ru-based phenylacetylenes for nonlinear optical applications. The phenylacetylene architecture was constructed using standard Pd-mediated alkyne–aryl iodide coupling sequences. Similarly, alkynide displacement of a chloride ligand from an Ru(dppe)$_2$ moiety allowed metal incorporation into the framework. The resulting nona-Ru(II) construct showed no loss of optical transparency, an increase in the second hyperpolarizability (γ), and a dramatic enhancement of two-photon absorption as compared to less branched analogs and intermediates.

The attachment of a single porphyrin unit to a C$_{60}$ moiety by cyclopropanation, followed by similar attachment of generation one and two poly(aryl ether) dendrons at the remaining C$_{60}$ octahedral positions, has been described by Camps et al.[320] These constructs were found to exhibit fluorescence and singlet-oxygen formation properties reminiscent of the porphyrin–C$_{60}$ dyad. Other branched porphyrin-based architectures have been reported,[58, 321–332] although they are not strictly dendritic due to minimal branching or the potential to repeat the coupling reactions.

A divergent route to the preparation of organophosphine metallodendrimers using simple acid-base chemistry was delineated by Kakkar and coworkers[333] (Scheme 8.38). Their protocol relied on the hydrolysis of aminosilanes by alkylhydroxy groups. Thus, diaminosilane **186** was reacted with triol **185** to give tris(aminosilane) **187**, which was then treated with further triol to afford hexaalcohol **188**. Subsequent reaction with [(μ-Cl)(1,5-C$_8$H$_{12}$)Rh] yielded the tetrarhodium complex **189**. Dendritic construction could also be achieved using the metallated reagents (e.g., **190** and **191**). Construction was continued through to the 4th generation, providing the 4, 10, 22, and 46 Rh metallated complexes.

Chessa et al.[334] incorporated 2,6-di(thiomethyl)pyridine subunits into pyridine-based dendrons; these subunits readily formed Pd(II) complexes when reacted with Pd(PhCN)$_2$Cl$_2$.

8.6.2   Random Metal Inclusion

Ottaviani et al.[335] characterized 'half-generation' (carboxylate-terminated) PAMAMs with randomly located Cu(II) complexes in aqueous solution by means of EPR (e.g., **192**; Figure 8.37). Three types of signal were observed, indicating three distinct types of metal complexes. Coordination sites were shown to be comprised of carboxylate-amine, amido-amine, and bis(carboxylate) ligand groups. Ottaviani et al.[336] also employed Mn(II) as a probe for the characterization of PAMAMs. This probe showed no interactions with full-generation dendrimers, although half-generation (CO$_2^-$)-terminated shells did give interactions. Complexation was favored at higher generations (i.e., > 4.5), by

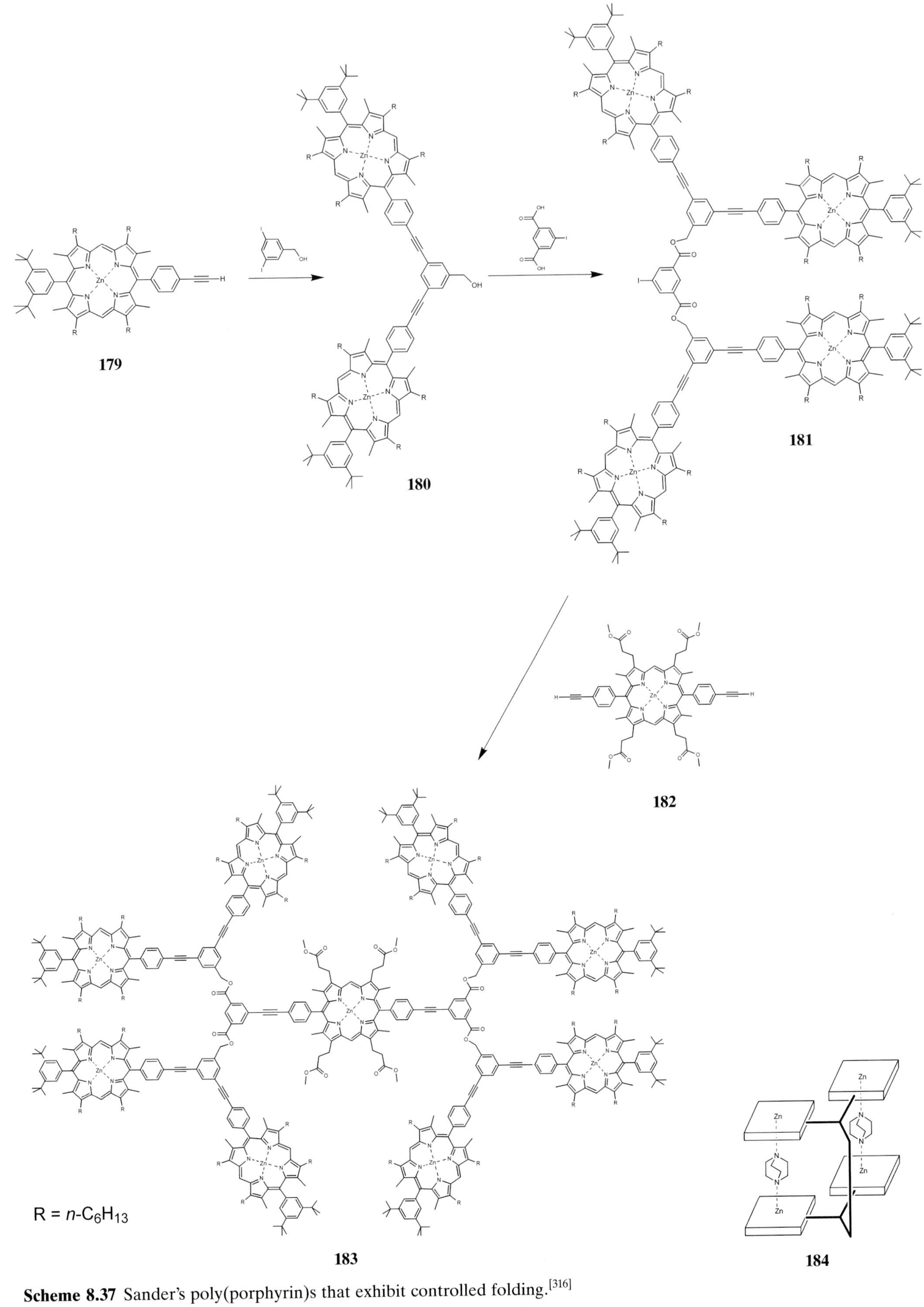

179

180

181

182

R = *n*-C₆H₁₃

183

184

Scheme 8.37 Sander's poly(porphyrin)s that exhibit controlled folding.[316]

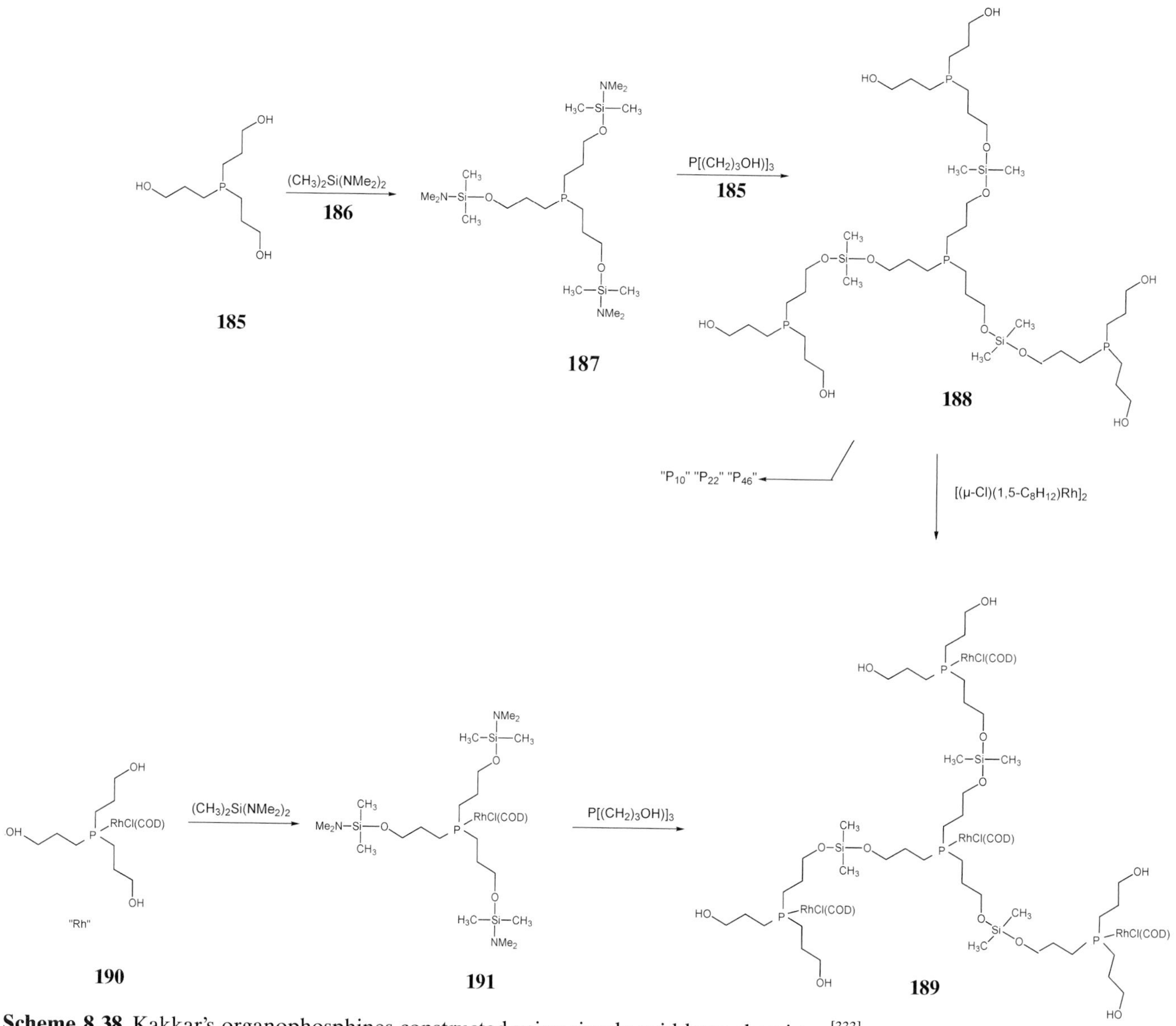

Scheme 8.38 Kakkar's organophosphines constructed using simple acid-base chemistry.[333]

higher pH, and by low Mn(II) concentrations. Ottaviani et al.[337] also characterized the 'full-generation' (amino-terminated) series of PAMAM dendrimers (e.q., **193**) incorporating Cu(II) complexes by EPR spectroscopy. One to four different types of signals were observed depending on the generation and the pH of the solution. At pH 4–5, the EPR spectra showed all four signals (A–D) at low generation, while signal D disappeared at higher generations; at pH 6, only signal B persisted in the Cu(II) solution.

Perhaps one of the most innovative applications of dendrimers is their use as templates for the preparation of metal nanoclusters within a branched framework or "nanoreactor". The first published account of this was by Crooks, Zhao, and Sun.[338] A verification of the concept was demonstrated by the synthesis of Cu clusters, although the authors noted that "this protocol was amenable to any transition metal that can be encapsulated by a dendrimer and subsequently reduced". Essentially, Cu^{2+} was extracted into a 4th generation PAMAM through coordination by interior tertiary amines, and then chemically reduced with excess $NaBH_4$ to produce intradendrimer copper clusters. Cluster size was controlled by the dendrimer nanoreactor size. Support for cluster formation included EPR, absorbance spectra, and TEM evidence. It was later demonstrated[339] that colloids could be formed using dendrimers to surround, encapsulate, and stabilize the 2–3 nm sized particles (**194**; Figure 8.38). *In situ* reduction of $HAuCl_4$ in the presence of poly(amidoamine)s led to such products. Notably, lower generation dendrimers afforded larger clusters. Crooks and Zhao[340] subsequently reported the formation of Pt clusters within dendritic hosts (Scheme 8.39). Platinum(II) coordination (**195**) followed

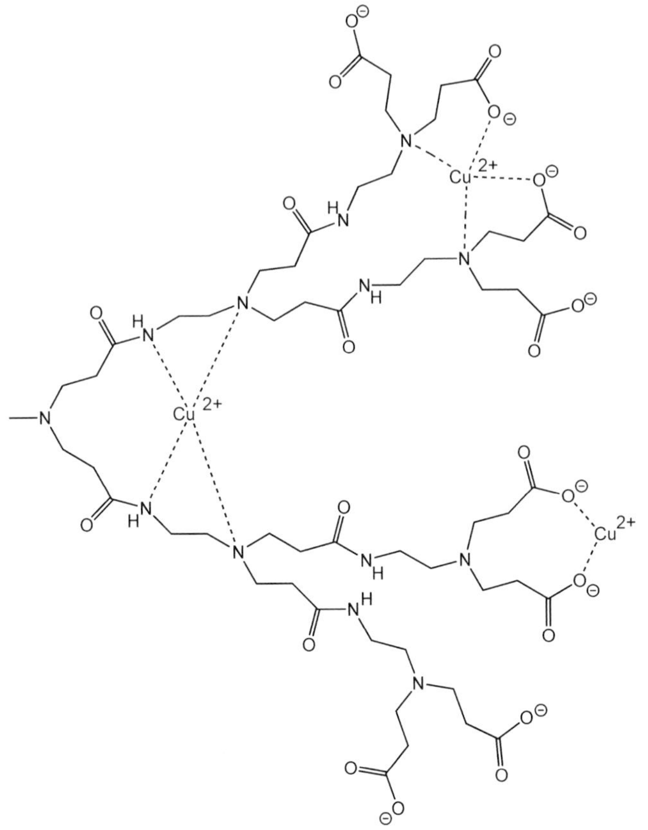

192

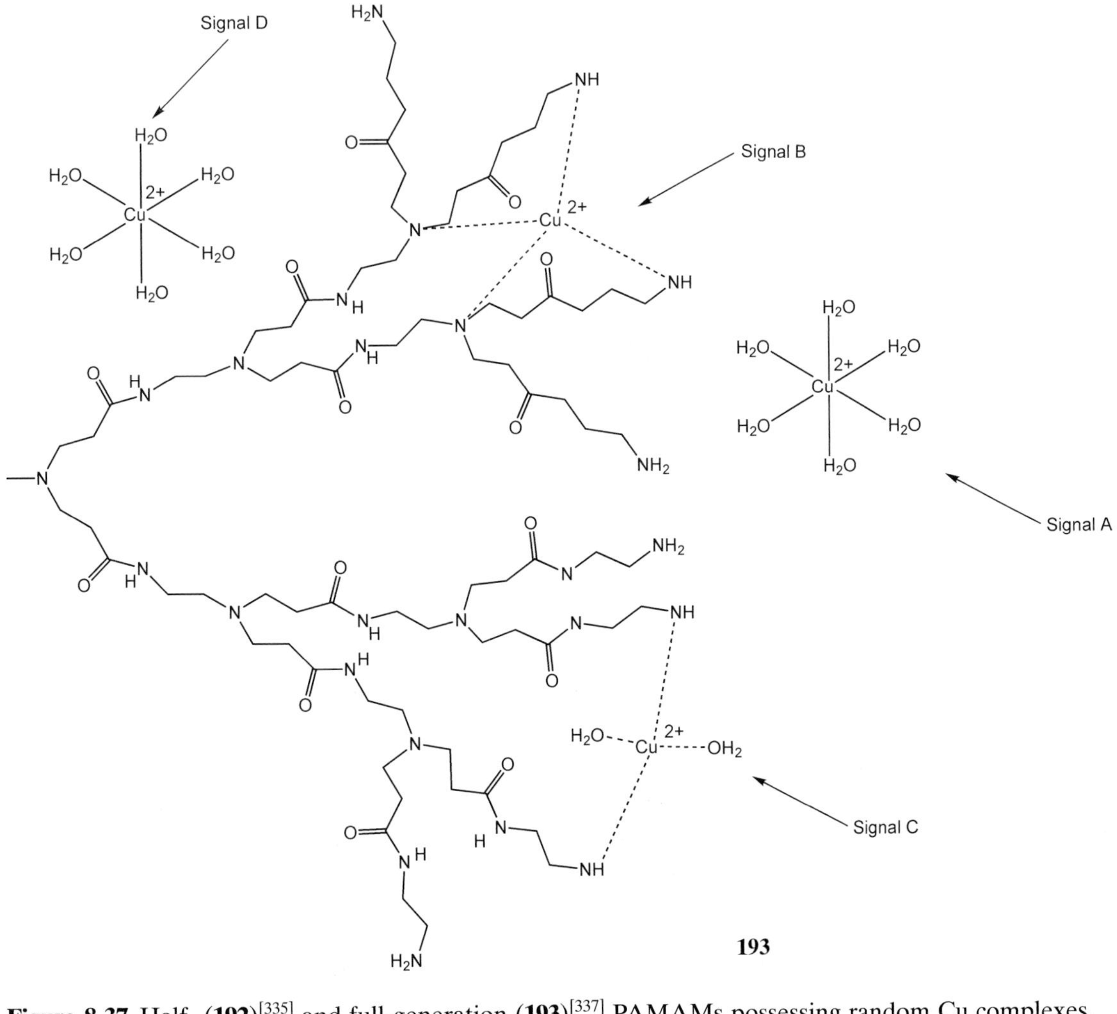

193

Figure 8.37 Half- (**192**)[335] and full-generation (**193**)[337] PAMAMs possessing random Cu complexes.

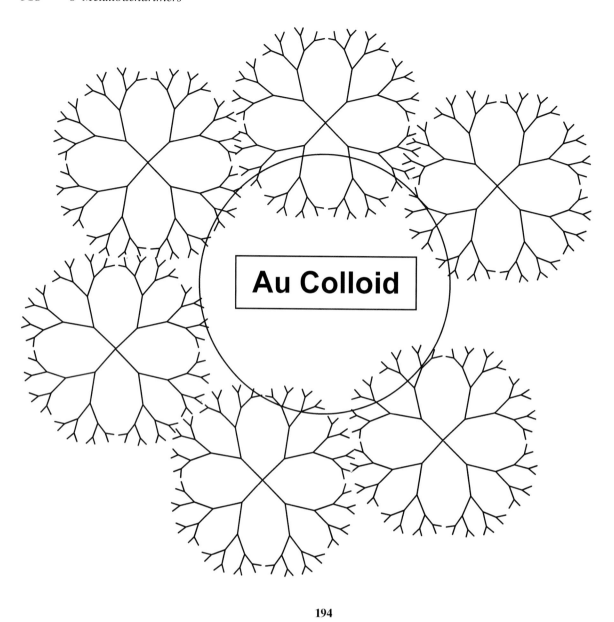

194

Figure 8.38 Crooks' colloidal nanocomposites.[339]

by reduction afforded the clusters (**196**), which were shown to be electrocatalytically active, reducing O_2 at a dendrimer-modified gold electrode (**197**). It was further shown that these zero-valent nanoparticles could be displaced within the dendrimer by other more noble transition metal ions.[341] Displacements were described as being rapid and complete, resulting in relatively monodisperse, 1–3 nm, stable particles. Homogeneous hydrogenation catalysis has been performed with dendrimer-encapsulated Pd and Pt particles.[342] Dendrimer–gold colloids have also been prepared by Esumi et al.[343] by UV irradiation of the metal salts. Dendrimer-encapsulated Pd nanoclusters were reported as useful fluorous phase-soluble catalysts by Chechik and Crooks.[344]

About two months after the initial report by Crooks and coworkers,[338] Balogh and Tomalia[345] reported the preparation of stable metallic Cu(II) PAMAM complexes **198** and Cu(0) PAMAM composites **199** (Scheme 8.40). A generation 4 or 5 PAMAM dendrimer was added to a Cu(II) acetate solution to give ca. 15 to 32 Cu(II) complex sites within the dendrimers, as detected by spectrophotometric titration. Dropwise addition of aqueous N_2H_4 afforded the reduced Cu(0) composites. Strong UV/vis absorption at very short wavelengths (250, 310, 350 nm) suggested the presence of individual copper atoms or very small copper domains. A small-angle X-ray scattering study of these dendrimer–copper sulfide nanocomposites has also been reported.[346] Earlier, Rubin et al.[347] described the use of 4th generation amine-terminated PPI's as glass, silicon, or ITO surface modifiers for Au and Au colloid monolayer formation. These rough noble metal

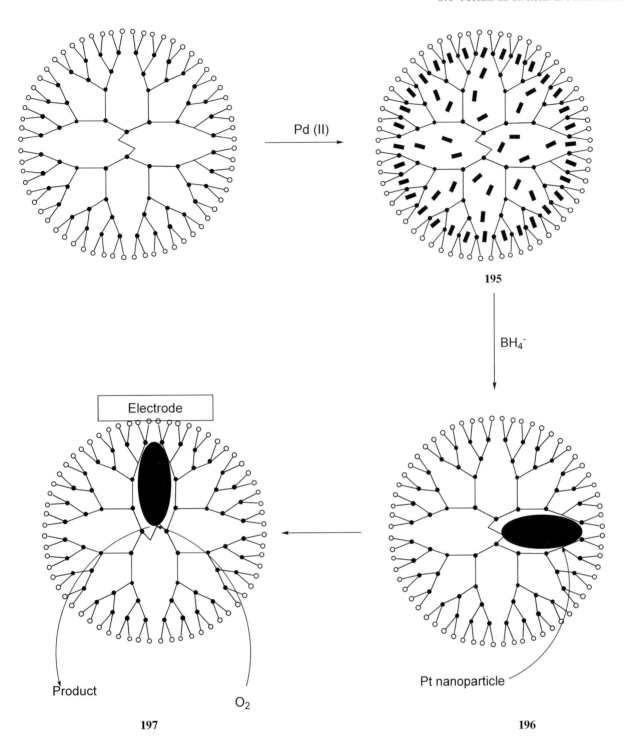

Scheme 8.39 Zhao and Crooks dendrimer-encapsulated Pt nanoparticles.[340]

surfaces were examined as platforms for surface-enhanced Raman scattering measurements. The formation of gold colloids upon reduction of PAMAM gold salt precursors has been proven[348] by means of TEM, SANS, and SAXS experiments, which showed that the gold particles are formed inside the dendrimer and are located offset from the center. Linear polymer analogues have also been used to generate CdS nanocomposites possessing similar optical properties in solution, albeit with reduced emission quantum yield.[349]

Balogh et al.[350] studied the formation of silver and gold PAMAM nanocomposites with respect to the mechanism of formation and structural types. Structural aspects of these materials were found to be a function of the mechanism, chemistry, dendrimer structure, and surface groups. These authors identified three different types of architectures, termed *I*nternal, *E*xternal, and *M*ixed type composites. Examples of nanocomposite formation leading to the different structural types were reported. The binding of

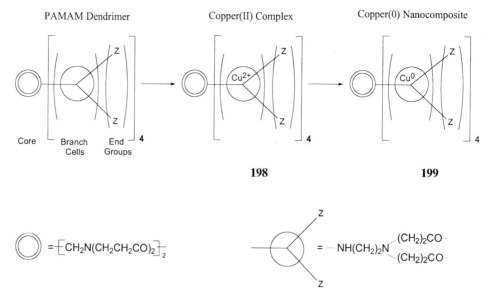

Scheme 8.40 PAMAM copper(II) complexes and copper(0) nanocomposites. Reproduced with permission by the American Chemical Society.[345]

Co(II) ions to PAMAMs in aqueous solution has also been studied;[351] these dendrimers were described as being high capacity chelating agents. Gold–PAMAM nanocomposites, prepared by reduction (H$_2$NNH$_2$) of PAMAM-tetrachloroaurate polysalts, have been electrostatically layered using poly(sodium 4-styrenesulfonate) as a counter polyelectrolyte to afford uniform nanoscale multilayers.[352] Atomic force microscopy supported the high uniformity of the arrays. Nanoparticle formation within dendrimer-polymer hybrids has also been reported[352a] along with photoluminescences characteristics.[352b] Numerous other reports have since appeared.[352c–f]

A Ru(II)-phenanthroline species labeled with a nitroxide radical has further been employed as a probe in an ESR experiment to monitor the binding and dynamics of surface complexes of PAMAMs.[353] The (N–O)· moieties were suggested to reside throughout the various hydration layers at the PAMAM–water interface.

The aggregation kinetics of PAMAM-stabilized CdS nanoclusters have been examined by Ploehn and coworkers.[354] The role of the PAMAMs in the preparation of gold, platinum, and silver nanoparticles has been addressed by Esumi et al.;[355] for the Pt constructs, the particle size (2.4–3.0 nm) was found to be independent of the dendrimer size or concentration. In contrast, the size of the gold nanoparticles (1.5–4.0 nm) decreased with increasing concentration and generation. Very small silver clusters were observed. IR spectroscopy suggested that the metal nanoparticles form at the dendritic surface.

Utilizing a bis(phenol)-modified porphyrin in conjunction with tris(glycidyl)trihydroxymethylethane, Hecht, Emrick, and Fréchet[356] have employed proton-transfer polymerization (see Section 6.2.3) to create unique hyperbranched polyporphyrins.

More recently, PPIs have been modified with donsyl groups and used for internal Co(II) coordination by Vögtle et al.[356a]

8.7 References

[1]   D. Astruc, "Electron and Proton Reservoir Complexes: Thermodynamic Basis for C–H Activation and Applications in Redox and Dendrimer Chemistry", *Acc. Chem. Res.* **2000**, *33*, 287–298.

[2]   D. Astruc, J.-C. Blais, E. Cloutet, L. Djakovitch, S. Rigaut, J. Ruiz, V. Sartor, C. Valério, "The First Organometallic Dendrimers: Design and Redox Functions", *Top. Curr. Chem.* **2000**, *210*, 229–259.

[3]   G. R. Newkome, E. He, C. N. Moorefield, "Suprasupermolecules with Novel Properties: Metallodendrimers", *Chem. Rev.* **1999**, *99*, 1689–1746.

[4]   F. J. Stoddart, T. Welton, "Metal-Containing Dendritic Polymers", *Polyhedron* **1999**, *18*, 3575–3591.

[5] C. M. Casado, I. Cuadrado, M. Morán, B. Alonso, B. Garcia, B. González, J. Losada, "Redox-Active Ferrocenyl Dendrimers and Polymers in Solution and Immobilised on Electrode Surfaces", *Coord. Chem. Rev.* **1999**, *185–186*, 53–79.

[6] I. Cuadrado, M. Morán, C. M. Casado, B. Alonso, J. Losada, "Organometallic Dendrimers with Transition Metals", *Coord. Chem. Rev.* **1999**, *193–195*, 395–445.

[7] M. A. Hearshaw, J. R. Moss, "Organometallic and Related Metal-Containing Dendrimers", *Chem. Commun.* **1999**, 1–8.

[8] J.-P. Majoral, A.-M. Caminade, "Dendrimers Containing Heteroatoms (Si, P, B, Ge, or Bi)", *Chem. Rev.* **1999**, *99*, 845–880.

[9] V. V. Narayanan, G. R. Newkome, "Supramolecular Chemistry within Dendritic Structures", *Top. Curr. Chem.* **1998**, *197*, 19–77.

[10] M. Trollsås, J. L. Hedrick, "Hyperbranched Poly(ε-caprolactone) Derived from Intrinsically Branched AB$_2$ Macromonomers", *Macromolecules* **1998**, *31*, 4390–4395.

[11] P. L. Boulas, M. Gómez-Kaifer, L. Echegoyen, "Electrochemistry of Supramolecular Systems", *Angew. Chem.* **1998**, *110*, 226–258; *Angew. Chem. Int. Ed. Engl.* **1998**, *37*, 216–247.

[12] O. A. Matthews, A. N. Shipway, J. F. Stoddart, "Dendrimers – Branching Out from Curiosities into New Technologies", *Prog. Polym. Sci.* **1998**, *23*, 1–56.

[13] C. B. Gorman, "Encapsulated Electroactive Molecules", *Adv. Mater.* **1997**, *9*, 1117–1119.

[14] C. Gorman, "Metallodendrimers: Structural Diversity and Functional Behavior", *Adv. Mater. (Weinheim, Fed. Repub. Ger.)* **1998**, *10*, 295–309.

[15] M. Venturi, S. Serroni, A. Juris, S. Campagna, V. Balzani, "Electrochemical and Photochemical Properties of Metal-Containing Dendrimers", *Top. Curr. Chem.* **1998**, *197*, 193–228.

[16] F. Zeng, S. C. Zimmerman, "Dendrimers in Supramolecular Chemistry: From Molecular Recognition to Self-Assembly", *Chem. Rev.* **1997**, *97*, 1681–1712.

[17] G. R. Newkome, C. N. Moorefield, "Dendrimers", in *Comprehensive Supramolecular Chemistry* (Ed.: D. N. Reinhoudt), Pergamon, New York, **1996**, pp. 777–832.

[18] N. Ardoin, D. Astruc, "Molecular Trees: From Syntheses Towards Applications", *Bull. Soc. Chim. Fr.* **1995**, *132*, 875–909.

[19] J.-M. Lehn, "Supramolecular Chemistry – Scope and Perspectives. Molecules, Supermolecules, and Molecular Devices (Nobel lecture)", *Angew. Chem.* **1988**, *100*, 91–116; *Angew. Chem. Int. Ed. Engl.* **1988**, *27*, 89–112.

[20] J.-M. Lehn, "Perspectives in Supramolecular Chemistry – From Molecular Recognition Toward Molecular Information Processing and Self-Organization", *Angew. Chem.* **1990**, *102*, 1347–1362; *Angew. Chem. Int. Ed. Engl.* **1990**, *29*, 1304–1319.

[21] J.-M. Lehn, "Supramolecular Chemistry: Concepts and Perspectives", VCH Publishers; Weinheim, **1995**.

[22] J.-M. Lehn, "Perspectives in Supramolecular Chemistry: From Molecular Recognition towards Self-Organization", *Pure. Appl. Chem.* **1994**, *66*, 1961–1966.

[23] G. R. Newkome, "Suprasupermolecular Chemistry: The Chemistry Within the Dendrimer", *Pure Appl. Chem.* **1998**, *70*, 2337–2343.

[24] G. Denti, S. Serroni, S. Campagna, V. Ricevuto, V. Balzani, "Directional Energy Transfer in a Luminescent Tetranuclear Ru(II)-Polypyridine Complex that Contains Two Different Types of Bridging Ligands", *Inorg. Chim. Acta* **1991**, *182*, 127–129.

[25] S. Campagna, G. Denti, S. Serroni, M. Ciano, A. Juris, V. Balzani, "A Tridecanuclear Ruthenium(II)-Polypyridine Supramolecular Species: Synthesis, Absorption and Luminescence Properties, and Electrochemical Oxidation", *Inorg. Chem.* **1992**, *31*, 2982–2984.

[26] G. Denti, S. Campagna, S. Serroni, M. Ciano, V. Balzani, "Decanuclear Homo- and Heterometallic Polypyridine Complexes: Syntheses, Absorption Spectra, Luminescence, Electrochemical Oxidation, and Intercomponent Energy Transfer", *J. Am. Chem. Soc.* **1992**, *114*, 2944–2950.

[27] G. R. Newkome, C. N. Moorefield, G. R. Baker, A. L. Johnson, R. K. Behera, "Alkane Cascade Polymers Possessing Micellar Topology: Micellanoic Acid Derivatives", *Angew. Chem.* **1991**, *103*, 1205–1207; *Angew. Chem. Int. Ed. Engl.* **1991**, *30*, 1176–1178.

[28] S. Serroni, G. Denti, S. Campagna, A. Juris, M. Ciano, V. Balzani, "Arborols Based on Luminescent and Redox-Active Transition Metal Complexes", *Angew. Chem.* **1992**, *104*, 1540–1542; *Angew. Chem. Int. Ed. Engl.* **1992**, *31*, 1493–1495.

[29] V. Balzani, S. Campagna, G. Denti, A. Juris, S. Serroni, M. Venturi, "Bottom-Up Strategy to Obtain Luminescent and Redox-Active Metal Complexes of Nanometric Dimensions", *Coord. Chem. Rev.* **1994**, *132*, 1–13.

[30] S. Campagna, G. Denti, L. Sabatino, S. Serroni, M. Ciano, V. Balzani, "A New Heterotetrametallic Complex of Ruthenium and Osmium: Absorption Spectrum, Luminescence Properties, and Electrochemical Behaviour", *J. Chem. Soc., Chem. Commun.* **1989**, 1500–1501.

[31] S. Campagna, G. Denti, S. Serroni, M. Ciano, V. Balzani, "Hexanuclear Homo- and Hetero-bridged Ruthenium(II) Polypyridine Complexes: Syntheses, Absorption Spectra, Luminescence Properties, and Electrochemical Behavior", *Inorg. Chem.* **1991**, *30*, 3728–3732.

[32] P. Belser, A. von Zelewsky, M. Frank, C. Seel, F. Vögtle, L. De Cola, F. Barigelletti, V. Balzani, "Supramolecular Ru and/or Os Complexes of Tris(bipyridine) Bridging Ligands. Syntheses, Absorption Spectra, Luminescence Properties, Electrochemical Behavior, Intercomponent Energy, and Electron Transfer", *J. Am. Chem. Soc.* **1993**, *115*, 4076–4086.

[33] S. Roffia, M. Marcaccio, C. Paradisi, F. Paolucci, V. Balzani, G. Denti, S. Serroni, S. Campagna, "Electrochemical Reduction of (2,2'-Bipyridine)- and Bis[(2-pyridyl)pyrazine]ruthenium(II) Complexes used as Building Blocks for Supramolecular Species. Redox Series made of 8, 10, and 12 Redox Steps", *Inorg. Chem.* **1993**, *32*, 3003–3009.

[34] A. Juris, V. Balzani, S. Campagna, G. Denti, S. Serroni, G. Frei, H. U. Güdel, "Near-Infrared Luminescence of Supramolecular Species Consisting of Osmium(II)- and/or Ruthenium(II)-Polypyridine Complexes", *Inorg. Chem.* **1994**, *33*, 1491–1496.

[35] P. Ceroni, F. Paolucci, C. Paradisi, A. Juris, S. Roffia, S. Serroni, S. Campagna, A. J. Bard, "Dinuclear and Dendritic Polynuclear Ruthenium(II) and Osmium(II) Polypyridine Complexes: Electrochemistry at Very Positive Potentials in Liquid SO$_2$", *J. Am. Chem. Soc.* **1998**, *120*, 5480–5487.

[36] M. Marcaccio, F. Paolucci, C. Paradisi, S. Roffia, C. Fontanesi, L. J. Yellowlees, S. Serroni, S. Campagna, G. Denti, V. Balzani, "Electrochemistry of Multicomponent Systems. Redox Series Comprising up to 26 Reversible Reduction Processes in Polynuclear Ruthenium(II) Bipyridine-Type Complexes", *J. Am. Chem. Soc.* **1999**, *121*, 10081–10091.

[37] M. A. Fox, W. E. Jones, Jr., D. M. Watkins, "Light Harvesting Polymer Systems", *C. & E. News* **1993**, *71*, 38–48.

[38] V. Balzani, S. Campagna, G. Denti, A. Juris, S. Serroni, M. Venturi, "Designing Dendrimers based on Transition-Metal Complexes. Light-Harvesting Properties and Predetermined Redox Patterns", *Acc. Chem. Res.* **1998**, *31*, 26–34.

[39] S. Campagna, G. Denti, S. Serroni, A. Juris, M. Venturi, V. Ricevuto, V. Balzani, "Dendrimers of Nanometer Size Based on Metal Complexes: Luminescent and Redox-Active Polynuclear Metal Complexes Containing up to Twenty-Two Metal Centers", *Chem. Eur. J.* **1995**, *1*, 211–221.

[40] S. Serroni, A. Juris, M. Venturi, S. Campagna, I. R. Resino, G. Denti, A. Credi, V. Balzani, "Polynuclear Metal Complexes of Nanometre Size. A Versatile Synthetic Strategy Leading to Luminescence and Redox-Active Dendrimers made of an Osmium(II)-based Core and Ruthenium(II)-based Units in the Branches", *J. Mater. Chem.* **1997**, *7*, 1227–1236.

[41] S. Serroni, S. Campagna, G. Denti, T. E. Keyes, J. G. Vos, "A Tetranuclear Ruthenium(II) Complex Containing both Electron-Rich and Electron-Poor Bridging Ligands. Absorption Spectrum, Luminescence, Redox Behavior, and Intercomponent Energy Transfer", *Inorg. Chem.* **1996**, *35*, 4513–4518.

[42] S. Serroni, G. Denti, S. Campagna, M. Ciano, V. Balzani, "A Decanuclear Ruthenium(II)-Polypyridine Complex: Synthesis, Absorption Spectrum, Luminescence and Electrochemical Behaviour", *J. Chem. Soc., Chem. Commun.* **1991**, 944–945.

[43] G. Predieri, C. Vignali, G. Denti, S. Serroni, "Characterization of *mer* and *fac* Isomers of [Ru(2,3-dpp)$_3$][PF$_6$]$_2$ [2,3-dpp = 2,3-bis(2-pyridyl)pyrazine] by ^{1}H and ^{99}Ru NMR Spectroscopy. Proton Assignment by 2D Techniques", *Inorg. Chim. Acta* **1993**, *205*, 145–148.

[44] M. Taddia, C. Lucano, A. Juris, "Analytical Characterization of Supramolecular Species – Determination of Ruthenium and Osmium in Dendrimers by Electrothermal Atomic Absorption Spectrometry", *Anal. Chim. Acta* **1998**, *375*, 285–292.

[45] L. Latterini, G. Pourtois, C. Moucheron, R. Lazzaroni, J.-L. Brédas, A. Kirsch-De Mesmaeker, F. C. De Schryver, "STM Imaging of a Heptanuclear Ruthenium(II) Dendrimer, Mono-Add Layer on Graphite", *Chem. Eur. J.* **2000**, *6*, 1331–1336.

[46] E. Ishow, A. Gourdon, J.-P. Launay, P. Lecante, M. Verelst, C. Chiorboli, F. Scandola, C.-A. Bignozzi, "Tetranuclear Tetrapyrido[3,2-*a*:2',3'-*c*:3",2"-*h*:2"',3"'-*j*]phenazineruthenium Complex: Synthesis, Wide-Angle X-ray Scattering, and Photophysical Studies", *Inorg. Chem.* **1998**, *37*, 3603–3609.

[47] S. Campagna, S. Serroni, S. Bodige, F. M. MacDonnell, "Adsorption Spectra, Photophysical Properties, and Redox Behavior of Stereochemically Pure Dendritic Ruthenium(II) Tetramers and Related Dinuclear and Mononuclear Complexes", *Inorg. Chem.* **1999**, *38*, 692–701.

[48] R. Arakawa, T. Matsuo, K. Nozaki, T. Ohno, M. Haga, "Analysis of Multiply-Charged Ions of Ruthenium(II) Tetranuclear Complexes by Electrospray Ionization Mass Spectrometry", *Inorg. Chem.* **1995**, *34*, 2464–2467.

[49] M.-A. Haga, M. M. Ali, R. Arakawa, "Proton-Induced Switching of Electron Transfer Pathways in Dendrimer-Type Tetranuclear RuOs$_3$ Complexes", *Angew. Chem.* **1996**, *108*, 85–87; *Int. Ed. Engl.* **1996**, *35*, 76–78.

[50] M.-A. Haga, Md. M. Ali, H. Sato, H. Monjushiro, K. Nozaki, K. Kano, "Spectroelectrochemical Analysis of the Intervalence Band in Mixed-Valence Di- and Tetranuclear Ru Complexes by the Flow-Through Method", *Inorg. Chem.* **1998**, *37*, 2320–2324.

[51] C. Moucheron, A. Kirsch-De Mesmaeker, A. Dupont-Gervais, E. Leize, A. Van Dorsselaer, "Synthesis and Characterization by Electrospray Mass Spectrometry of a Novel Dendritic Heptanuclear Complex of Ruthenium(II)", *J. Am. Chem. Soc.* **1996**, *118*, 12834–12835.

[52] H. Suzuki, H. Kurata, Y. Matano, "First Synthesis and Properties of Dendritic Bi$_n$-Bismuthanes", *Chem. Commun.* **1997**, 2295–2296.

[53] Y. Matano, H. Kurata, T. Murafuji, N. Azuma, H. Suzuki, "Synthesis and Properties of a Series of Phenylene-Bridged Bi$_n$-Bismuthanes", *Organometallics* **1998**, *17*, 4049–4059.

[54] M. N. Bochkarev, V. B. Cilkin, L. P. Mayorova, G. A. Razuvaev, U. D. Cemchkov, V. E. Sherstyanux, "Polyphenylenegermane – A New Type of Polymeric Material", *Metalloorg. Khim.* **1988**, *1*, 196–200.

[55] I. B. Myasnikova, V. V. Izvolenskii, A. N. Sundukov, Y. D. Semchikov, M. N. Bochkarev, "The Rheological Properties of Highly Branched Poly(fluorophenylene Germane)", *Vysokomol. Soedin., Ser. A* **1995**, *37*, 1223–1224.

[56] V. Huc, P. Boussaguet, P. Mazerolles, "Organogermanium Dendrimers", *J. Organomet. Chem.* **1996**, *521*, 253–260.

[57] M. Nanjo, A. Sekiguchi, "Group-14-Element-Based Hybrid Dendrimers. Synthesis and Characterization of Dendrimers with Alternating Si and Ge Atoms in the Chains", *Organometallics* **1998**, *17*, 492–494.

[58] J.-F. Nierengarten, L. Oswald, J.-F. Nicoud, "Dynamic *cis/trans* Isomerisation in a Porphyrin-Fullerene Conjugate", *Chem. Commun.* **1998**, 1545–1546.

[59] S. Bodige, A. S. Torres, D. J. Maloney, D. Tate, G. R. Kinsel, A. K. Walker, F. M. MacDonnell, "First-Generation Chiral Metallodendrimers: Stereoselective Synthesis of Rigid D_3-Symmetric Tetranuclear Ruthenium Complexes", *J. Am. Chem. Soc.* **1997**, *119*, 10364–10369.

[60] M.-J. Kim, F. M. MacDonnell, M. E. Gimon-Kinsel, T. DuBois, N. Asgharian, J. C. Griener, "Global Chirality in Rigid Decametallic Ruthenium Dendrimers", *Angew. Chem.* **2000**, *112*, 629–633; *Int. Ed.* **2000**, *39*, 615–619.

[61] G. R. Newkome, F. Cardullo, E. C. Constable, C. N. Moorefield, A. M. W. C. Thompson, "Metallomicellanols: Incorporation of Ruthenium(II)-2,2':6',2"-Terpyridine Triads into Cascade Polymers", *J. Chem. Soc., Chem. Commun.* **1993**, 925–927.

[62] J.-P. Sauvage, J.-P. Collin, C. Chambron, S. Guillerez, C. Coudret, V. Balzani, F. Barigelletti, L. De Cola, L. Flamigni, "Ruthenium(II) and Osmium(II) bis(terpyridine) Complexes in Covalently-Linked Multicomponent Systems: Synthesis, Electrochemical Behavior, Absorption Spectra, and Photochemical and Photophysical Properties", *Chem. Rev.* **1994**, *94*, 993–1019.

[63] K. R. Seddon, "Ruthenium: "A Dance to the Music of Time"", *Platinum Metals Rev.* **1996**, *40*, 128–134.

[64] G. R. Newkome, J. K. Young, G. R. Baker, R. L. Potter, L. Audoly, D. Cooper, C. D. Weis, K. F. Morris, C. S. Johnson, Jr., "Cascade Polymers. pH Dependence of Hydrodynamic Radii of Acid-Terminated Dendrimers", *Macromolecules* **1993**, *26*, 2394–2396.

[65] E. C. Constable, M. D. Ward, "Synthesis and Coordination Behaviour of 6',6"-Bis(2-pyridyl)-2,2':4,4":2",2"'-quaterpyridine; 'Back-to-Back' 2,2':6',2"-Terpyridine", *J. Chem. Soc., Dalton Trans.* **1990**, 1405–1409.

[66] G. R. Newkome, C. N. Moorefield, R. Güther, G. R. Baker, "Novel Building Blocks for Incorporation of Molecular Recognition Sites within Cascade Polymeric Architectures", *Polym. Prepr.* **1995**, *36*, 609–610.

[67] G. R. Newkome, R. Güther, C. N. Moorefield, F. Cardullo, L. Echegoyen, E. Pérez-Cordero, H. Luftmann, "Routes to Dendritic Networks: Bis-Dendrimers by Coupling of Cascade Macromolecules through Metal Centers", *Angew. Chem.* **1995**, *107*, 2159–2162; *Int. Ed. Engl.* **1995**, *34*, 2023–2026.

[68] G. R. Newkome, C. N. Moorefield, G. R. Baker, "Lock and Key Micelles and Monomer Building Blocks Thereof" **1999**, *U. S. Pat.*, 5, 863, 919.

[69] G. R. Newkome, R. K. Behera, C. N. Moorefield, G. R. Baker, "Cascade Polymers: Syntheses and Characterization of One-Directional Arborols Based on Adamantane", *J. Org. Chem.* **1991**, *56*, 7162–7167.

[70] G. R. Newkome, C. D. Weis, "Di-*tert*-butyl 4-[2-(*tert*-butoxycarbonyl)ethyl]-4-aminoheptane-dicarboxylate", *Org. Prep. Proced. Int.* **1996**, *28*, 485–488.

[71] G. R. Newkome, E. He, "Nanometric Dendritic Macromolecules: Stepwise Assembly by Double (2,2':6',2"-Terpyridine)ruthenium(II) Connectivity", *J. Mater. Chem.* **1997**, *7*, 1237–1244.

[72] G. R. Newkome, C. D. Weis, "6,6-Bis(carboxy-2-oxabutyl)-4,8-dioxaundecane-1,11-dicarboxylic Acid", *Org. Prep. Proced. Int.* **1996**, *28*, 242–246.

[73] M. Osawa, M. Hoshino, S. Horiuchi, Y. Wakatsuki, "Palladium-Mediated One-Step Coupling between Polypyridine Metal Complexes: Preparation of Rigid and Dendritic Nano-Sized Ruthenium Complexes", *Organometallics* **1999**, *18*, 112–114.

[74] G. R. Newkome, E. He, L. A. Godínez, "Construction of Dendritic Assemblies: A Tailored Approach to Isomeric Metallomacromolecules by means of Bis(2,2':6',2"-terpyridine)ruthenium(II) Connectivity", *Macromolecules* **1998**, *31*, 4382–4386.

[74a] G. R. Newkome, E. He, L. A. Godínez, G. R. Baker, "Electroactive Metallomacromolecules via Tetrakis(2,2':6',2"-Terpyridine)ruthenium(II) Complexes: Dendritic Networks toward Constitutional Isomers and Neutral Species without External Counterions", *J. Am. Chem. Soc.* **2000**, *122*, 9993–10006.

[75] I. Cuadrado, C. M. Casado, B. Alonso, M. Morán, J. Losada, V. Belsky, "Dendrimers Containing Organometallic Moieties Electronically Communicated", *J. Am. Chem. Soc.* **1997**, *119*, 7613–7614.

[76] U. S. Schubert, C. H. Weidl, C. N. Moorefield, G. R. Baker, G. R. Newkome, "Design, Synthesis, and First Metal Complexes of Dendritic 5,5"-Disubstituted 2,2':6',2"-Terpyridine Ligands", *Polym. Prepr.* **1999**, *40*, 940–941.

[77] U. S. Schubert, C. H. Weidl, A. Cattani, C. Eschbaumer, G. R. Newkome, E. He, E. Harth, K. Müllen, "Metallo-Supramolecular Fullerene Assemblies and Polymers", *Polym. Prepr.* **2000**, *41*, 229–230.

[78] E. C. Constable, A. M. W. C. Thompson, D. A. Tocher, M. A. M. Daniels, "Synthesis, Characterization, and Properties of Ruthenium(II)-2,2':6',2"-Terpyridine Coordination Triads: X-Ray Structures of 4'-(*N,N*-Dimethylamino)- 2,2':6',2"-Terpyridine and *Bis*(4'-(*N,N*-Dimethylamino)-2,2':6',2"-Terpyridine)Ruthenium(II) Hexafluorophosphate Acetonitrile Solvate", *New J. Chem.* **1992**, *16*, 855–867.

[79] E. C. Constable, P. Harverson, "A New Convergent Strategy for High-Nuclearity Metallodendrimers", *Chem. Commun.* **1996**, 33–34.

[80] E. C. Constable, P. Harverson, J. J. Ramsden, "Adsorption of Ruthenadendrimers to Silica–Titania Surfaces Studied by Optical Waveguide Lightmode Spectroscopy (OWLS)", *Chem. Commun.* **1997**, 1683–1684.

[81] E. C. Constable, P. Harverson, "Convergent Synthesis of an Octadecaruthenium Metallodendrimer", *Inorg. Chim. Acta* **1996**, *252*, 9–11.

[82] E. C. Constable, "Metallodendrimers: Metal Ions as Supramolecular Glue", *Chem. Commun.* **1997**, 1073–1080.

[83] E. C. Constable, C. E. Housecroft, M. Cattalini, D. Phillips, "Pentaerythritol-Based Metallodendrimers", *New J. Chem.* **1998**, 193–200.

[83a] S. D. Holmstrom, J. A. Cox, "Electrocatalysis at a Conducting Composite Electrode Doped with a Ruthenium(II) Metallodendrimer", *Anal. Chem.* **2000**, *72*, 3191–3195.

[84] E. C. Constable, P. Harverson, M. Oberholzer, "Convergent and Divergent Approaches to Metallocentric Metallodendrimers", *Chem. Commun.* **1996**, 1821–1822.

[85] B. Boury, R. J. P. Corriu, R. Nuñez, "Hybrid Xerogels from Dendrimers and Arborols", *Chem. Mater.* **1998**, *10*, 1795–1804.

[86] E. C. Constable, A. M. W. C. Thompson, "A New Ligand for the Self Assembly of Starburst Coordination Oligomers and Polymers", *J. Chem. Soc., Chem. Commun.* **1992**, 617–619.

[87] A. J. Amoroso, A. M. W. C. Thompson, J. P. Maher, J. A. McCleverty, M. D. Ward, "Di-, Tri-, and Tetranucleating Pyridyl Ligands which Facilitate Multicenter Magnetic Exchange Between Paramagnetic Molybdenum Centers", *Inorg. Chem.* **1995**, *34*, 4828–4835.

[88] S. Achar, R. J. Puddephatt, "Organoplatinum Dendrimers Formed by Oxidative Addition", *Angew. Chem.* **1994**, *106*, 895–897; *Angew. Chem. Int. Ed. Engl.* **1994**, *33*, 847–849.

[89] S. Achar, J. J. Vittal, R. J. Puddephatt, "Organoplatinum Dendrimers", *Organometallics* **1996**, *15*, 43–50.

[90] S. Achar, R. J. Puddephatt, "Large Dendrimeric Organoplatinum Complexes", *J. Chem. Soc., Chem. Commun.* **1994**, 1895–1896.

[91] S. Achar, R. J. Puddephatt, "Chains and Stars in Organoplatinum Oligomers", *Organometallics* **1995**, *14*, 1681–1687.

[92] G.-X. Liu, R. J. Puddephatt, "Model Reactions for the Synthesis of Organometallic Dendrimers Containing Platinum and Palladium", *Inorg. Chim. Acta* **1996**, *251*, 319–323.

[93] G.-X. Liu, R. J. Puddephatt, "Divergent Route to Organoplatinum or Platinum–Palladium Dendrimers", *Organometallics* **1996**, *15*, 5257–5259.

[94] W. T. S. Huck, F. C. J. M. van Veggel, B. L. Kropman, D. H. A. Blank, E. G. Keim, M. M. A. Smithers, D. N. Reinhoudt, "Large Self-Assembled Organopalladium Spheres", *J. Am. Chem. Soc.* **1995**, *117*, 8293–8294.

[95] W. T. S. Huck, B. H. Snellink-Ruël, J. W. Th. Lichtenbelt, F. C. J. M. van Veggel, D. N. Reinhoudt, "Self-Assembly of Hyperbranched Spheres; Correlation Between Monomeric Synthon and Sphere Size", *Chem. Commun.* **1997**, 9–10.

[96] W. T. S. Huck, F. C. J. M. van Veggel, D. N. Reinhoudt, "Self-Assembly of Hyperbranched Spheres", *J. Mater. Chem.* **1997**, *7*, 1213–1219.

[97] W. T. S. Huck, B. Snellink-Ruel, F. C. J. M. van Veggel, D. N. Reinhoudt, "New Building Blocks for the Noncovalent Assembly of Homo- and Hetero-Multinuclear Metallodendrimers", *Organometallics* **1997**, *16*, 4287–4291.

[98] W. T. S. Huck, F. C. J. M. van Veggel, D. N. Reinhoudt, "Controlled Assembly of Nanosized Metallodendrimers", *Angew. Chem.* **1996**, *108*, 1304–1306; *Int. Ed. Engl.* **1996**, *35*, 1213–1215.

[99] W. T. S. Huck, L. J. Prins, R. H. Fokkens, N. M. M. Nibbering, F. C. J. M. van Veggel, D. N. Reinhoudt, "Convergent and Divergent Noncovalent Synthesis of Metallodendrimers", *J. Am. Chem. Soc.* **1998**, *120*, 6240–6246.

[100] W. T. S. Huck, A. Rohrer, A. T. Anilkumar, R. H. Fokkens, N. M. M. Nibbering, F. C. J. M. van Veggel, D. N. Reinhoudt, "Non-Covalent Synthesis of Multiporphyrin Systems", *New J. Chem.* **1998**, 165–168.

[101] F. C. J. M. van Veggel, W. T. S. Huck, D. N. Reinhoudt, "Nanosize Metallodendrimers", *Macromol. Symp.* **1998**, *131*, 165–173.

[102] B.-H. Huisman, H. Schönherr, W. T. S. Huck, A. Friggeri, H.-J. van Manen, E. Menozzi, G. J. Vancso, F. C. J. M. van Veggel, D. N. Reinhoudt, "Surface-Confined Metallodendrimers: Isolated Nanosize Molecules", *Angew. Chem.* **1999**, *111*, 2385–2389; *Int. Ed.* **1999**, *38*, 2248–2251.

[103] W. T. S. Huck, R. Hulst, P. Timmerman, F. C. J. M. van Veggel, D. N. Reinhoudt, "Noncovalent Synthesis of Nanostructures: Combining Coordination Chemistry and Hydrogen Bonding", *Angew. Chem.* **1997**, *109*, 1046–1049; *Int. Ed. Engl.* **1997**, *36*, 1006–1008.

[104] N. Ohshiro, F. Takei, K. Onitsuka, S. Takahashi, "Synthesis of a Novel Organometallic Dendrimer with a Backbone Composed of Platinum-Acetylide Units", *Chem. Lett.* **1996**, 871–872.

[105] R. R. Tykwinski, P. J. Stang, "Preparation of Rigid-Rod, Di- and Trimetallic, σ-Acetylide Complexes of Iridium(III) and Rhodium(III) via Alkynyl(phenyl)iodonium Chemistry", *Organometallics* **1994**, *13*, 3203–3208.

[106] S. Leininger, P. J. Stang, S. Huang, "Synthesis and Characterization of Organoplatinum Dendrimers with 1,3,5-Triethynylbenzene Building Blocks", *Organometallics* **1998**, *17*, 3981–3987.

[107] N. Ohshiro, F. Takei, K. Onitsuka, S. Takahashi, "Synthesis of Organometallic Dendrimers with a Backbone Composed of Platinum-Acetylide Units", *J. Organomet. Chem.* **1998**, *569*, 195–202.

[108] K. Onitsuka, M. Fujimoto, N. Ohshiro, S. Takahashi, "Convergent Route to Organometallic Dendrimers Composed of Platinum-Acetylide Units", *Angew. Chem.* **1999**, *111*, 751–754; *Angew. Chem. Int. Ed.* **1999**, *38*, 689–692.

[109] F. Vögtle, M. Plevoets, M. Nieger, G. C. Azzellini, A. Credi, L. De Cola, V. De Marchis, M. Venturi, V. Balzani, "Dendrimers with a Photoactive and Redox-Active [Ru(bpy)₃]²⁺-Type Core: Photophysical Properties, Electrochemical Behavior, and Excited-State Electron-Transfer Reactions", *J. Am. Chem. Soc.* **1999**, *121*, 6290–6298.

[110] E. C. Constable, P. Harverson, "A Convergent Approach to Heteroheptanuclear Star Complexes", *Polyhedron* **1999**, *18*, 1891–1901.

[111] H. Takshima, S. Shinkai, I. Hamachi, "Ru(bpy)₃-Based Artificial Receptors toward a Protein Surface: Selective Binding and Efficient Photoreduction of Cytochrome *c*", *Chem. Commun.* **1999**, 2345–2346.

[112] J.-F. Nierengarten, D. Felder, J.-F. Nicoud, "Phenanthroline Ligands Substituted with Fullerene-Functionalized Dendritic Wedges and their Copper(I) Complexes", *Tetrahedron Lett.* **1999**, *40*, 273–276.

[113] V. J. Catalano, N. Parodi, "Reversible C₆₀ Binding to Dendrimer-Containing Ir(CO)Cl(PPh₂R)₂ Complexes", *Inorg. Chem.* **1997**, *36*, 537–541.

[114] C. J. Hawker, J. M. J. Fréchet, "Preparation of Polymers with Controlled Molecular Architecture. A New Convergent Approach to Dendritic Macromolecules", *J. Am. Chem. Soc.* **1990**, *112*, 7638–7647.

[115] R.-H. Jin, T. Aida, S. Inoue, "'Caged' Porphyrin: The First Dendritic Molecule having a Core Photochemical Functionality", *J. Chem. Soc., Chem. Commun.* **1993**, 1260–1262.

[116] J. F. G. A. Jansen, E. W. Meijer, E. M. M. de Brabander-van den Berg, "The Dendritic Box: Shape-Selective Liberation of Encapsulated Guests", *J. Am. Chem. Soc.* **1995**, *117*, 4417–4418.

[117] M. Enomoto, T. Aida, "Self-Assembly of a Copper-Ligating Dendrimer that Provides a New Non-Heme Metalloprotein Mimic: 'Dendrimer Effects' on Stability of the Bis(μ-oxo)dicopper(III) Core", *J. Am. Chem. Soc.* **1999**, *121*, 874–875.

[118] D.-L. Jiang, T. Aida, "A Dendritic Iron Porphyrin as a Novel Haemoprotein Mimic: Effects of the Dendrimer Cage on Dioxygen-Binding Activity", *Chem. Commun.* **1996**, 1523–1524.

[119] D.-L. Jiang, T. Aida, "Dendrimer-Encapsulated Iron Porphyrin as a Novel Hemoprotein Mimic for Dioxygen Binding", *J. Mater. Sci. Pure Appl. Chem.* **1997**, *A34*, 2047–2055.

[120] Y. Tomoyose, D.-L. Jiang, R.-H. Jin, T. Aida, T. Yamashita, K. Horie, E. Yashima, Y. Okamoto, "Aryl Ether Dendrimers with an Interior Metalloporphyrin Functionality as a Spectroscopic Probe: Interpenetrating Interaction with Dendritic Imidazoles", *Macromolecules* **1996**, *29*, 5236–5238.

[121] R. Sadamoto, N. Tomioka, T. Aida, "Photoinduced Electron Transfer Reactions *through* Dendrimer Architecture", *J. Am. Chem. Soc.* **1996**, *118*, 3978–3979.

[122] D. Takasu, N. Tomioka, D.-L. Jiang, T. Aida, T. Kamachi, I. Okura, "Dendrimer Porphyrins for Biomimetic Applications", *J. Inorg. Biochem.* **1997**, *67*, 242.

[123] N. Tomioka, D. Takasu, T. Takahashi, T. Aida, "Electrostatic Assembly of Dendrimer Electrolytes: Negatively and Positively Charged Dendrimer Porphyrins", *Angew. Chem.* **1998**, *110*, 1611–1614; *Int. Ed.* **1998**, *37*, 1531–1534.

[124] D.-L. Jiang, T. Aida, "Morphology-Dependent Photochemical Events in Aryl Ether Dendrimer Porphyrins: Cooperation of Dendron Subunits for Singlet Energy Transduction", *J. Am. Chem. Soc.* **1998**, *120*, 10895–10901.

[125] T. Aida, D.-L. Jiang, "in *Handbook of Porphyrins and Related Macrocycles: Biomaterials for Materials Scientists, Chemists, and Physicists* (Eds.: K. M. Kadish, K. M. Smith, R. Guillard), Academic Press, **1999**.

[125a] M. J. Hannon, P. C. Mayers, P. C. Taylor, "A Dendritic Structure Containing a Designed Cleft which Controls Ligand Coordination Behavior in an Analogous Way to Proteins", *Angew. Chem. Int. Ed.* **2001**, *40*, 1081–1084.

[126] K. W. Pollak, J. W. Leon, J. M. J. Fréchet, "Dendritic Porphyrins by the Convergent Growth Approach: Their Preparation and their Properties", *Polym. Mater. Sci. Eng.* **1995**, *73*, 333–334.

[127] K. W. Pollak, J. W. Leon, J. M. J. Fréchet, M. Maskus, H. D. Abruña, "Effects of Dendrimer Generation on Site Isolation of Core Moieties: Electrochemical and Fluorescence Quenching Studies with Metalloporphyrin Core Dendrimers", *Chem. Mater.* **1998**, *10*, 30–38.

[128] K. W. Pollak, E. M. Sanford, J. M. J. Fréchet, "A Comparison of Two Convergent Routes for the Preparation of Metalloporphyrin-Core Dendrimers: Direct Condensation *vs.* Chemical Modification", *J. Mater. Chem.* **1998**, *8*, 519–527.

[129] P. Bhyrappa, G. Vaijayanthimala, K. S. Suslick, "Shape-Selective Ligation to Dendrimer-Metalloporphyrins", *J. Am. Chem. Soc.* **1999**, *121*, 262–263.

[130] K. S. Suslick, P. Bhyrappa, "Dendrimer Metalloporphyrins as Shape Selective Oxidation Catalysts", *J. Inorg. Biochem.* **1997**, 234.

[131] K. W. Pollak, E. M. Sanford, J. M. J. Fréchet, "A Comparison of Two Convergent Routes for the Preparation of Metalloporphyrin-Core Dendrimers: Direct Condensation *vs.* Chemical Modification", *J. Mater. Chem.* **1998**, *8*, 519–527.

[131a] H. R. Stapart, N. Nishiyama, D.-L. Jiang, T. Aida, K. Kataoka, "Polyion Complex Micelles Encapsulating Light-Harvesting Ionic Dendrimer Zinc Porphyrins", *Langmuir* **2000**, *16*, 8182–8188.

[132] S. Yamago, M. Furukawa, A. Azuma, J. Yoshida, "Synthesis of Optically Active Binaphthols and their Metal Complexes for Asymmetric Catalysis", *Tetrahedron Lett.* **1998**, *39*, 3783–3786.

[133] M. S. Matos, J. Hofkens, W. Verheijen, F. C. De Schryver, S. Hecht, K. W. Pollak, J. M. J. Fréchet, B. Forier, W. Dehaen, "Effect of Core Structure on Photophysical and Hydrodynamic Properties of Porphyrin Dendrimers", *Macromolecules* **2000**, *33*, 2967–2973.

[134] P. Bhyrappa, J. K. Young, J. S. Moore, K. S. Suslick, "Dendrimer-Metalloporphyrins: Synthesis and Catalysis", *J. Am. Chem. Soc.* **1996**, *118*, 5708–5711.

[135] P. Bhyrappa, J. K. Young, J. S. Moore, K. S. Suslick, "Shape-Selective Epoxidation of Alkenes by Metalloporphyrin-Dendrimers", *J. Mol. Cat. A: Chem.* **1996**, *113*, 109–116.

[135a] T. Liwporncharoenvong, R. L. Luck, "Quadruply Bonded Dimolybdenum Atoms Surrounded by Dendrons: Preparation, Characterization, and Electrochemistry", *J. Am. Chem. Soc.* **2001**, *123*, 3615–3616.

[136] P. J. Dandliker, F. Diederich, M. Gross, C. B. Knobler, A. Louati, E. M. Sanford, "Dendritic Porphyrins: Modulating Redox Potentials of Electroactive Chromophores with Pendant Multifunctionality", *Angew. Chem.* **1994**, *106*, 1821–1824; *Int. Ed. Engl.* **1994**, *33*, 1739–1742.

[137] G. R. Newkome, X. Lin, C. D. Weis, "Polytryptophane-Terminated Dendritic Macromolecules", *Tetrahedron: Asymmetry* **1991**, *2*, 957–960.

[138] P. J. Dandliker, F. Diederich, J.-P. Gisselbrecht, A. Louati, M. Gross, "Water-Soluble Dendritic Iron Porphyrins: Synthetic Models of Globular Heme Proteins", *Angew. Chem.* **1995**, *107*, 2906–2909; *Int. Ed. Engl.* **1995**, *34*, 2725–2728.

[139] P. J. Dandliker, F. Diederich, A. Zingg, J.-P. Gisselbrecht, M. Gross, A. Louati, E. Sanford, "Dendrimers with Porphyrin Cores: Synthetic Models for Globular Heme Proteins", *Helv. Chim. Acta* **1997**, *80*, 1773–1801.

[140] J. P. Collman, L. Fu, A. Zingg, F. Diederich, "Dioxygen and Carbon Monoxide Binding in Dendritic Iron(II) Porphyrins", *Chem. Commun.* **1997**, 193–194.

[141] S. A. Vinogradov, L.-W. Lo, D. F. Wilson, "Dendritic Polyglutamic Porphyrins: Probing Porphyrin Protection by Oxygen-Dependent Quenching of Phosphorescence", *Chem. Eur. J.* **1999**, *5*, 1338–1347.

[142] P. Weyermann, J.-P. Gisselbrecht, C. Boudon, F. Diederich, M. Gross, "Dendritic Iron Porphyrins with Tethered Axial Ligands: New Model Compounds for Cytochromes", *Angew. Chem.* **1999**, *111*, 3400–3405; *Int. Ed.* **1999**, *38*, 3215–3219.

[143] M. Kimura, K. Nakada, Y. Yamaguchi, K. Hanabusa, H. Shirai, N. Kobayashi, "Dendritic Metallophthalocyanines: Synthesis and Characterization of a Zinc(II) phthalocyanine[8]³-arborol", *Chem. Commun.* **1997**, 1215–1216.

[144] G. R. Newkome, X. Lin, "Symmetrical, Four-Directional, Poly(ether-amide) Cascade Polymers", *Macromolecules* **1991**, *24*, 1443–1444.

[145] M. Kimura, Y. Sugihara, T. Muto, K. Hanabusa, H. Shirai, N. Kobayashi, "Dendritic Metallophthalocyanines – Synthesis, Electrochemical Properties, and Catalytic Activities", *Chem. Eur. J.* **1999**, *5*, 3495–3500.

[146] A. C. H. Ng, X. Li, D. K. P. Ng, "Synthesis and Photophysical Properties of Non-aggregated Phthalocyanines Bearing Dendritic Substituents", *Macromolecules* **1999**, *32*, 5292–5298.

[147] X. Li, X. He, A. C. H. Ng, C. Wu, D. K. P. Ng, "Influence of Surfactants on the Aggregation Behaviour of Water-Soluble Dendritic Phthalocyanines", *Macromolecules* **2000**, *33*, 2119–2123.

[148] M. Kimura, T. Shiba, T. Muto, K. Hanabusa, H. Shirai, "Intramolecular Energy Transfer in 1,3,5-Phenylene-Based Dendritic Porphyrins", *Macromolecules* **1999**, *32*, 8237–8239.

[149] M. Kimura, T. Shiba, T. Muto, K. Hanabusa, H. Shirai, "Energy Transfer within Ruthenium-Cored Rigid Metallodendrimers", *Tetrahedron Lett.* **2000**, *41*, 6809–6813.

[150] M. Kimura, T. Shiba, T. Muto, K. Hanabusa, H. Shirai, "A Rigid 1,3,5-Phenylene-Based Metallodendrimer containing a Ruthenium(II)-bis(terpyridyl) Complex", *Chem. Commun.* **2000**, 11–12.

[151] M. Kawa, J. M. J. Fréchet, "Self-Assembled Lanthanide-Cored Dendrimer Complexes: Enhancement of the Luminescence Properties of Lanthanide Ions through Site-Isolation and Antenna Effects", *Chem. Mater.* **1998**, *10*, 286–296.

[152] M. Kawa, J. M. J. Fréchet, "Enhanced Luminescence of Lanthanide within Lanthanide-Cored Dendrimer Complexes", *Thin Solid Films* **1998**, *331*, 259–263.

[152a] L. Zhu, X. Tong, M. Li, E. Wang, "Luminescence Enhancement of Tb³⁺ Ion in Assemblies of Amphiphilic Linear – Dendritic Block Copolymer: Antenna and Microenvironment Effects", *J. Phys. Chem. B.* **2001**, *105*, 2461–2464.

[152b] M. Kawa, K. Motoda, "An antenna effect influenced by the focal structure of Tb³⁺-cored dendrimer complexes", *Kobunshi Ronbunshu* **2000**, *57*, 855–858.

[152c] T. Seto, M. Kawa, K. Sugiyama, M. Nomura, "XAFS studies of Tb or Eu cored dendrimer complexes with various properties of luminescence", *J. Synch. Rad.* **2001**, *8*, 710–712.

[153] R. L. C. Lau, T.-W. D. Chan, I. Y. K. Chan, H.-F. Chow, "Fourier Transform Ion Cyclotron Resonance Studies of Terpyridine-Based Polyether Dendrimers and their Iron(II) Metallocomplexes by Liquid Secondary Ion Mass Spectrometry", *Eur. Mass Spectrom.* **1995**, *1*, 371–380.

[154] H.-F. Chow, I. Y. K. Chan, D. T. Chan, R. W. Kwok, "Dendritic Models of Redox Proteins: X-ray Photoelectron Spectroscopy and Cyclic Voltammetry Studies of Dendritic Bis(terpyridine) Iron(II) Complexes", *Chem. Eur. J.* **1996**, *2*, 1085–1091.

[155] H.-F. Chow, I. Y. K. Chan, C. C. Mak, "Facile Construction of Acid-Base and Redox-Stable Polyether-based Dendritic Fragments", *Tetrahedron Lett.* **1995**, *36*, 8633–8636.

[156] H.-F. Chow, I. Y. K. Chan, C. C. Mak, M.-K. Ng, "Synthesis and Properties of a New Class of Polyether Dendritic Fragments: Useful Building Blocks for Functional Dendrimers", *Tetrahedron* **1996**, *52*, 4277–4290.

[157] D. Tzalis, Y. Tor, "Toward Self-Assembling Dendrimers: Metal Complexation Induces the Assembly of Hyperbranched Structures", *Tetrahedron Lett.* **1996**, *37*, 8293–8296.

[158] J. Issberner, F. Vögtle, L. De Cola, V. Balzani, "Dendritic Bipyridine Ligands and their Tris(bipyridine)ruthenium(II) Chelates – Syntheses, Absorption Spectra, and Photophysical Properties", *Chem. Eur. J.* **1997**, *3*, 706–712.

[159] G. R. Newkome, A. Nayak, R. K. Behera, C. N. Moorefield, G. R. Baker, "Cascade Polymers: Synthesis and Characterization of Four-Directional Spherical Dendritic Macromolecules Based on Adamantane", *J. Org. Chem.* **1992**, *57*, 358–362.

[160] G. R. Newkome, C. N. Moorefield, G. R. Baker, R. K. Behera, G. H. Escamilla, M. J. Saunders, "Supramolecular Self-Assemblies of Two-Directional Cascade Molecules: Automorphogenesis", *Angew. Chem.* **1992**, *104*, 901–903; *Int. Ed. Engl.* **1992**, *31*, 917–919.

[161] M. Plevoets, F. Vögtle, L. De Cola, V. Balzani, "Supramolecular Dendrimers with a [Ru(bpy)₃]²⁺ Core and Naphthyl Peripheral Units", *New J. Chem.* **1999**, 63–69.

[162] S. Serroni, S. Campagna, A. Juris, M. Venturi, V. Balzani, G. Denti, "Polyether Arborols Mounted on a Luminescent and Redox-Active Ruthenium(II)-Polypyridine Core", *Gazz. Chim. Ital.* **1994**, *124*, 423–427.

[163] V. V. Narayanan, E. C. Wiener, "Metal-Directed Self-Assembly of Ethylenediamine-Based Dendrons", *Macromolecules* **2000**, *33*, 3944–3946.

[164] C. B. Gorman, B. L. Parkhurst, W. Y. Su, K.-Y. Chen, "Encapsulated Electroactive Molecules Based upon an Inorganic Cluster Surrounded by Dendron Ligands", *J. Am. Chem. Soc.* **1997**, *119*, 1141–1142.

[165] K.-Y. Chen, C. B. Gorman, "Synthesis of a Series of Focally-Substituted Organothiol Dendrons", *J. Org. Chem.* **1996**, *61*, 9229–9235.

[166] C. B. Gorman, M. W. Hager, B. L. Parkhurst, J. C. Smith, "Use of a Paramagnetic Core to Affect Longitudinal Nuclear Relaxation in Dendrimers – A Tool for Probing Dendrimer Conformation", *Macromolecules* **1998**, *31*, 815–822.

[167] C. B. Gorman, W. Y. Su, H. Jiang, C. M. Watson, P. Boyle, "Hybrid Organic-Inorganic, Hexa-Arm Dendrimers based on an Mo₆Cl₈ Core", *Chem. Commun.* **1999**, 877–878.

[168] C. B. Gorman, J. C. Smith, M. W. Hager, B. L. Parkhurst, H. Sierzputowska-Gracz, C. A. Haney, "Molecular Structure–Property Relationships for Electron-Transfer Rate Attenuation in Redox-Active Core Dendrimers", *J. Am. Chem. Soc.* **1999**, *121*, 9958–9966.

[169] C. B. Gorman, J. C. Smith, "Iron–Sulfur Core Dendrimers Display Dramatically Different Electrochemical Behavior in Films Compared to Solution", *J. Am. Chem. Soc.* **2000**, *122*, 9342–9343.

[170] R. Wang, Z. Zheng, "Dendrimers Supported by the [Re₆Se₈]²⁺ Metal Cluster Core", *J. Am. Chem. Soc.* **1999**, *121*, 3549–3550.

[171] H.-F. Chow, C. C. Mak, "Dendritic Bis(oxazoline)copper(II) Catalysts; 2. Synthesis, Reactivity, and Substrate Selectivity", *J. Org. Chem.* **1997**, *62*, 5116–5127.

[172] C. C. Mak, H.-F. Chow, "Dendritic Catalysts: Reactivity and Mechanism of the Dendritic Bis(oxazoline)metal Complex Catalyzed Diels–Alder Reaction", *Macromolecules* **1997**, *30*, 1228–1230.

[173] H.-F. Chow, C. C. Mak, "Facile Preparation of Optically Active Dendritic Fragments Containing Multiple Tartrate-derived Chiral Units", *Tetrahedron Lett.* **1996**, *37*, 5935–5938.

[174] P. B. Rheiner, H. Sellner, D. Seebach, "Dendritic Styryl TADDOLs as Novel Polymer Cross-Linkers: First Application in an Enantioselective Et₂Zn Addition Mediated by a Polymer-Incorporated Titanate", *Helv. Chim. Acta* **1997**, *80*, 2027–2032.

[174a] D. Seebach, A. K. Beck, M. Rueping, J. V. Schreiber, H. Seliner, "Excursions of Synthetic Organic Chemists to the World of Oligomers and Polymers", *Chimia* **2001**, *55*, 98–103.

[175] R. Engel, K. Rengan, C.-S. Chan, "New Cascade Molecules Centered about Phosphorus", *Heteroat. Chem.* **1993**, *4*, 181–184.

[176] N. Armaroli, C. Boudon, D. Felder, J.-P. Gisselbrecht, M. Gross, G. Marconi, J.-F. Nicoud, J.-F. Nierengarten, V. Vicinelli, "A Copper(I) Bis-phenanthroline Complex Buried in Fullerene-Functionalized Dendritic Black Boxes", *Angew. Chem.* **1999**, *111*, 3895–3899; *Int. Ed.* **1999**, *38*, 3730–3733.

[177] G. E. Oosterom, R. J. van Haaren, J. N. H. Reek, P. C. J. Kamer, P. W. N. M. van Leeuwen, "Catalysis in the Core of a Carbosilane Dendrimer", *Chem. Commun.* **1999**, 1119–1120.

[178] C. M. Cardona, A. E. Kaifer, "Asymmetric Redox-Active Dendrimers Containing a Ferrocene Subunit. Preparation, Characterization, and Electrochemistry", *J. Am. Chem. Soc.* **1998**, *120*, 4023–4024.

[179] Y. Wang, C. M. Cardona, A. E. Kaifer, "Molecular Orientation Effects on the Rates of Heterogeneous Electron Transfer of Unsymmetric Dendrimers", *J. Am. Chem. Soc.* **1999**, *121*, 9756–9757.

[180] D. K. Smith, "Branched Ferrocene Derivatives: Using Redox Potential to Probe the Dendritic Interior", *J. Chem. Soc., Perkin Trans. 2* **1999**, 1563–1565.

[181] C. M. Cardona, T. D. McCarley, A. E. Kaifer, "Synthesis, Electrochemistry, and Interactions with β-Cyclodextrin of Dendrimers Containing a Single Ferrocene Subunit Located 'Off-Center'", *J. Org. Chem.* **2000**, *65*, 1857–1864.

[182] J. W. J. Knapen, A. W. van der Made, J. C. de Wilde, P. W. N. M. van Leeuwen, P. Wijkens, D. M. Grove, G. van Koten, "Homogeneous Catalysts based on Silane Dendrimers Functionalized with Arylnickel(II) Complexes", *Nature* **1994**, *372*, 659–663.

[183] R. A. Gossage, L. A. van de Kuil, G. van Koten, "Diaminoarylnickel(II) 'Pincer' Complexes: Mechanistic Considerations in the Kharasch Addition Reaction, Controlled Polymerization, and Dendrimeric Transition Metal Catalysts", *Acc. Chem. Res.* **1998**, *31*, 423–431.

[184] L. A. van de Kuil, D. M. Grove, J. W. Zwikker, L. W. Jenneskens, W. Drenth, G. van Koten, "New Soluble Polysiloxane Polymers Containing a Pendant Terdentate Aryldiamine Ligand

Substituent Holding a Highly Catalytically Active Organometallic Nickel(II) Center", *Chem. Mater.* **1994**, *6*, 1675–1683.

[185] Also know as the "Prins Reaction" and currently described as "Atom Transfer Radical Addition" (ATRA). J. March, *Advanced Organic Chemistry, 4th ed.*, John Wiley and Sons, London, 1992; Sections 5–33.

[186] G. van Koten, D. M. Grove, "Advances in Catalysis with Organonickel(II) Complexes Anchored to Dendrimers and Polymers", *Polym. Mater. Sci. Eng.* **1995**, *73*, 228–229.

[187] R. A. Gossage, J. T. B. H. Jastrzebski, J. van Ameijde, S. J. E. Mulders, A. J. Brouwer, R. M. J. Liskamp, G. van Koten, "Synthesis and Catalytic Application of Amino Acid Based Dendritic Macromolecules", *Tetrahedron Lett.* **1999**, *40*, 1413–1416.

[188] I. P. Beletskaya, A. V. Chuchurjukin, H. P. Dijkstra, G. P. M. van Klink, G. van Koten, "Conjugated G_0 Metallo-Dendrimers, Functionalized with Tridentate 'Pincer'-Type Ligands", *Tetrahedron Lett.* **2000**, *41*, 1081–1085.

[189] J. L. Hoare, K. Lorenz, N. J. Hovestad, W. J. J. Smeets, A. L. Spek, A. J. Canty, H. Frey, G. van Koten, "Organopalladium-Functionalized Dendrimers: Insertion of Palladium(0) into Peripheral Carbon–Iodine Bonds of Carbosilane Dendrimers Derived from Polyols. Crystal Structure of $Si\{(CH_2)_3O_2CC_6H_4I\text{-}4\}_4$", *Organometallics* **1997**, *16*, 4167–4173.

[190] M. Albrecht, R. A. Gossage, A. L. Spek, G. van Koten, "Sulfur Dioxide Gas Detection by Reversible $\eta^1\text{-}SO_2\text{-}Pt$ Bond Formation as a Novel Application for Periphery Functionalised Metallo-Dendrimers", *Chem. Commun.* **1998**, 1003–1004.

[191] M. Albrecht, R. A. Gossage, M. Lutz, A. L. Spek, G. van Koten, "Diagnostic Organometallic and Metallodendritic Materials for SO_2 Gas Detection: Reversible Binding of Sulfur Dioxide to Arylplatinium(II) Complexes", *Chem. Eur. J.* **2000**, *6*, 1431–1445.

[191a] M. Albrecht, N. J. Hovestad, J. Boersma, G. van Koten, "Multiple Use of Soluble Metallodendritic Materials as Catalysts and Dyes", *Chem. Eur. J.* **2001**, *7*, 1289–1294.

[191b] N. J. Hovestad, A. Ford, J. T. B. H. Jastrzebski, G. van Koten, "Functionalized Carbosilane Dendritic Species as Soluble Supports in Organic Synthesis", *J. Org. Chem.* **2000**, *65*, 6338–6344.

[191c] C. Schlenk, A. W. Kleij, H. Frey, G. van Koten, "Macromolecular-Multisite Catalysts Obtained by Grafting Diaminoaryl Palladium(II) Complexes onto a Hyperbranched-Polytriallylsilane Support", *Angew. Chem. Int. Ed.* **2000**, *39*, 3445–3447.

[192] M. Benito, O. Rossell, M. Seco, G. Segalés, "Soluble Iron/Gold Cluster Containing Carbosilane Dendrimers", *Organometallics* **1999**, *18*, 5191–5193.

[193] M. Benito, O. Rossell, M. Seco, G. Segalés, "Carbosilane Dendrimers Functionalized with $AuFe_3$ Clusters", *Inorg. Chim. Acta* **1999**, *291*, 247–251.

[193a] M. Benito, O. Rossell, M. Seco, G. Segalés, "Transition metal clusters containing carbosilane dendrimers", *J. Organomet. Chem.* **2001**, *619*, 245–251.

[193b] M. Zhou, J. Roovers, "Dendritic Supramolecular Assemblies with Multiple Ru(II) Tris(bipyridine) Units at the Periphery: Synthesis, Spectroscopic, and Electrochemical Study", *Macromolecules* **2001**, *34*, 244–252.

[194] N. J. Hovestad, J. L. Hoare, J. T. B. H. Jastrzebski, A. J. Canty, W. J. J. Smeets, A. L. Spek, G. van Koten, "Periphery-Palladated Carbosilane Dendrimers: Synthesis and Reactivity of Organopalladium(II) and -(IV) Dendritic Complexes. Crystal Structure of $[PdMe\{C_6H_4(OCH_2Ph)\text{-}4\}(bpy)]$ (bpy = 2,2'-Bipyridine)", *Organometallics* **1999**, *18*, 2970–2980.

[195] A. W. Kleij, H. Kleijn, J. T. B. H. Jastrzebski, W. J. J. Smeets, A. L. Spek, G. van Koten, "Dendritic Carbosilanes Containing Silicon-Bonded $1\text{-}[C_6H_2(CH_2NMe_2)_2\text{-}3,5\text{-}Li\text{-}4]$ or $1\text{-}[C_6H_3(CH_2NMe_2)\text{-}4\text{-}Li\text{-}3]$ Mono- and Bis(amino)aryllithium End Groups: Structure of $\{[CH_2SiMe_2C_6H_3(CH_2NMe_2)\text{-}4\text{-}Li\text{-}3]_2\}_2$", *Organometallics* **1999**, *18*, 268–276.

[196] G. van Koten, J. T. B. H. Jastrzebski, "Periphery-Functionalized Organometallic Dendrimers for Homogeneous Catalysis", *J. Mol. Catal. A: Chem.* **1999**, *146*, 317–323.

[197] N. J. Hovestad, E. B. Eggeling, H. J. Heidbüchel, J. T. B. H. Jastrzebski, U. Kragl, W. Keim, D. Vogt, G. van Koten, "Selective Hydrovinylation of Styrene in a Membrane Reactor: Use of Carbosilane Dendrimers with Hemilabile P,O Ligands", *Angew. Chem.* **1999**, *111*, 1763–1765; *Int. Ed.* **1999**, *38*, 1655–1658.

[197a] E. B. Eggeling, N. J. Hovestad, J. T. B. H. Jastrzebski, D. Vogt, G. van Koten, "Phosphino Carboxylic Acid Ester Functionalized Carbosilane Dendrimers: Nanoscale Ligands for the Pd-Catalyzed Hydrovinylation Reaction in a Membrane Reactor", *J. Org. Chem.* **2000**, *65*, 8857–8865.

[198] C. Schlenk, A. W. Kleij, H. Frey, G. van Koten, "Macromolecular-Multisite Catalysts Obtained by Grafting Diaminoaryl Palladium(II) Complexes onto a Hyperbranched-Polytriallylsilane Support", *Angew. Chem.* **2000**, *112*, 3587–3589; *Int. Ed.* **2000**, *39*, 3445–3447.

[199] A. W. Kleij, R. A. Gossage, J. T. B. H. Jastrzebski, J. Boersma, G. van Koten, "The 'Dendritic Effect' in Homogenous Catalysis with Carbosilane-Supported Arylnickel(II) Catalysts:

Observation of Active-Site Proximity Effects in Atom-Transfer Radical Addition", *Angew. Chem.* **2000**, *112*, 179–181; *Int. Ed.* **2000**, *39*, 176–178.

[199a] A. W. Kleij, R. A. Gossage, R. J. M. K. Gebbink, N. Brinkmann, E. J. Reijerse, U. Kragl, M. Lutz, A. L. Spek, G. van Koten, "A "Dendritic Effect"in Homogeneous Catalysis with Carbosilane-Supported Arylnickel(II) Catalysts: Observation of Active-Site Proximity Effects in Atom-Transfer Radical Addition", *J. Am. Chem. Soc.* **2000**, *122*, 12112–12124.

[200] S. C. Bourque, H. Alper, L. E. Manzer, P. Arya, "Hydroformylation Reactions Using Recyclable Rhodium-Complexes Dendrimers on Silica", *J. Am. Chem. Soc.* **2000**, *122*, 956–957.

[201] P. Arya, N. V. Rao, J. Singkhonrat, H. Alper, S. C. Bourque, L. E. Manzer, "A Divergent, Solid-Phase Approach to Dendritic Ligands on Beads. Heterogeneous Catalysis for Hydroformylation Reactions", *J. Org. Chem.* **2000**, *65*, 1881–1885.

[201a] P. Arya, G. Panda, N. V. Rao, H. Alper, S. C. Bourque, L. E. Manzer, "Solid-Phase Catalysis: A Biomimetic Approach toward Ligands on Dendritic Arms to Explore Recyclable Hydroformylation Reactions", *J. Am. Chem. Soc.* **2001**, *123*, 2889–2890.

[202] S. B. Garber, J. S. Kingsbury, B. L. Gray, A. H. Hoveyda, "Efficient and Recyclable Monomeric and Dendritic Ru-Based Metathesis Catalysts", *J. Am. Chem. Soc.* **2000**, *122*, 8168–8179.

[203] J. W. Kriesel, S. König, M. A. Freitas, A. G. Marshall, J. A. Leary, T. D. Tilley, "Synthesis of Highly Charged Organometallic Dendrimers and their Characterization by Electrospray Mass Spectrometry and Single-Crystal X-ray Diffraction", *J. Am. Chem. Soc.* **1998**, *120*, 12207–12215.

[204] Y.-H. Liao, J. R. Moss, "Ruthenium-containing Organometallic Dendrimers", *J. Chem. Soc., Chem. Commun.* **1993**, 1774–1777.

[205] Y.-H. Liao, J. R. Moss, "Synthesis of Very Large Organoruthenium Dendrimers", *Organometallics* **1995**, *14*, 2130–2132.

[206] Y.-H. Liao, J. R. Moss, "Organoruthenium Dendrimers", *Organometallics* **1996**, *15*, 4307–4316.

[207] K. J. Naidoo, S. J. Hughes, J. R. Moss, "Computational Investigations into the Potential Use of Poly(benzyl phenyl ether) Dendrimers as Supports for Organometallic Catalysts", *Macromolecules* **1999**, *32*, 331–341.

[208] I. J. Mavunkal, J. R. Moss, J. Bacsa, "Synthesis and Characterization of a First Generation Organorhenium Dendrimer", *J. Organomet. Chem.* **2000**, *593–594*, 361–368.

[208a] H. J. Murfee, T. P. S. Thoms, J. Greaves, B. Hong, "New Metallodendrimers Containing an Octakis(diphenylphosphino)-Functionalized Silsesquioxane Core and Ruthenium(II)-Based Chromophores", *Inorg. Chem.* **2000**, *39*, 5209–5217.

[209] K. Takada, G. D. Stirrier, M. Morán, H. D. Abruña, "Thermodynamics and Kinetics of Adsorption of Poly(amido amine) Dendrimers Surface-Functionalized with Ruthenium(II) Complexes", *Langmuir* **1999**, *15*, 7333–7339.

[210] G. D. Storrier, K. Takada, H. D. Abruña, "Synthesis, Characterization, Electrochemistry, and EQCM Studies of Polyamidoamine Dendrimers Surface-Functionalized with Polypyridyl Metal Complexes", *Langmuir* **1999**, *15*, 872–884.

[211] M. Kimura, K. Mizuno, T. Muto, K. Hanabusa, H. Shirai, "Synthesis and Characterization of a Ligand-Substituted Poly(amidoamine) Dendrimer with External Terpyridine Units and its Iron(II) Complexes", *Macromol. Rapid Commun.* **1999**, *20*, 98–102.

[212] F. Moulines, L. Djakovitch, R. Boese, B. Gloaguen, W. Theil, J.-L. Fillaut, M.-H. Delville, D. Astruc, "Organometallic Molecular Trees as Multielectron and Multiproton Reservoirs: CpFe⁺-Induced Nonallylation of Mesitylene and Phase-Transfer-Catalyzed Synthesis of a Redox-Active Nonairon Complex", *Angew. Chem.* **1993**, *105*, 1132–1134; *Int. Ed. Engl.* **1993**, *32*, 1075–1077.

[213] E. Cloutet, J.-L. Fillaut, Y. Gnanou, D. Astruc, "Hexa-Arm Star-Shaped Polystyrenes by Core-First Method", *J. Chem. Soc., Chem. Commun.* **1994**, 2433–2434.

[214] C. Valério, J.-L. Fillaut, J. Ruiz, J. Guittard, J.-C. Blais, D. Astruc, "The Dendritic Effect in Molecular Recognition: Ferrocene Dendrimers and their Use as Supramolecular Redox Sensors for the Recognition of Small Inorganic Anions", *J. Am. Chem. Soc.* **1997**, *119*, 2588–2589.

[215] F. Moulines, B. Gloaguen, D. Astruc, "One-Pot Multifunctionalization of Polymethyl Hydrocarbon π Ligands. Maximum Space Occupancy by Double Branching and Formation of Arborols", *Angew. Chem.* **1992**, *104*, 452–454; *Int. Ed. Engl.* **1992**, *31*, 458–460.

[216] J.-L. Fillaut, D. Astruc, "Tentacled Aromatics: From Central-Ring to Outer-Ring Iron Sandwich Complexes", *J. Chem. Soc., Chem. Commun.* **1993**, 1320–1322.

[217] J.-L. Fillaut, J. Linares, D. Astruc, "Single-Step Six-Electron Transfer in a Heptanuclear Complex: Isolation of Both Redox Forms", *Angew. Chem.* **1994**, *106*, 2540–2542; *Int. Ed. Engl.* **1994**, *33*, 2460–2462.

[218] S. Rigaut, M.-H. Delville, D. Astruc, "Triple C–H/N–H Activation by O$_2$ for Molecular Engineering: Heterobifunctionalization of the 19-Electron Redox Fe^{I}Cp(arene)", *J. Am. Chem. Soc.* **1997**, *119*, 11132–11133.

[219] E. Cloutet, J.-L. Fillaut, Y. Gnanou, D. Astruc, "Hexa-Arm Polyisobutene Stars by the Core-First Method", *Chem. Commun.* **1996**, 2047–2048.

[220] C. Köllner, B. Pugin, A. Togni, "Dendrimers Containing Chiral Ferrocenyl Diphosphine Ligands for Asymmetric Catalysis", *J. Am. Chem. Soc.* **1998**, *120*, 10274–10275.

[221] B. Pugin, **2000**, *PCT Int. Appl. WO 97 02,232*.

[222] C. Köllner, R. Schneider, A. Togni, "Characterization of Catalytically Active Ferrocenyl Dendrimers", *Chimia* **1999**, *53,* No. 107.

[223] R. Schneider, C. Köllner, I. Weber, A. Togni, "Dendrimers Based on Cyclophosphazene Units and Containing Ferrocenyl Ligands for Asymmetric Catalysis", *Chem. Commun.* **1999**, 2415–2416.

[224] M. Uno, P. H. Dixneuf, "Organometallic Triskelia: Novel Tris[vinylideneruthenium(II)], Tris[alkynylruthenium(II)], and Triruthenium–Triferrocenyl Complexes", *Angew. Chem.* **1998**, *110*, 1822–1824; *Int. Ed.* **1998**, *37*, 1714–1717.

[225] V. Sartor, L. Djakovitch, J.-L. Fillaut, F. Moulines, F. Neveu, V. Marvaud, J. Guittard, J.-C. Blais, D. Astruc, "Organoiron Route to a New Dendron for Fast Dendritic Syntheses Using Divergent and Convergent Methods", *J. Am. Chem. Soc.* **1999**, *121*, 2929–2930.

[226] C. Valério, E. Alonso, J. Ruiz, J.-C. Blais, D. Astruc, "A Polycationic Metallodendrimer with 24 [Fe(η^5-C$_5$Me$_5$)(η^6-*N*-alkylaniline)]$^+$ Termini that Recognizes Chloride and Bromide Anions", *Angew. Chem.* **1999**, *111*, 1855–1859; *Int. Ed.* **1999**, *38*, 1747–1751.

[227] C. Valério, J. Ruiz, J.-L. Fillaut, D. Astruc, "Dendritic Effect in the Recognition of Small Inorganic Anions using a Polycationic Nona-cobalticinium Dendrimer", *C. R. l'Academie Sci., Ser. II Univers.* **1999**, 79–83.

[228] V. Sartor, S. Nlate, J.-L. Fillaut, L. Djakovitch, F. Moulines, V. Marvaud, F. Neveu, J.-C. Blais, J.-F. Létard, D. Astruc, "Activation of Aryl Ether and Aryl Sulfides by the Fe(η^5-C$_5$H$_5$)$^+$ Group for the Synthesis of Phenol Dendrons and Arene-Centered Polyolefin Dendrimers", *New J. Chem.* **2000**, *6*, 351–370.

[229] S. Nlate, J. Ruiz, V. Sartor, R. Navarro, J.-C. Blais, D. Astruc, "Molecular Batteries: Ferrocenylsilylation of Dendrons, Dendritic Cores, and Dendrimers: New Convergent and Divergent Routes to Ferrocenyl Dendrimers with Stable Redox Activity", *Chem. Eur. J.* **2000**, *6*, 2544–2553.

[230] C. Valério, F. Moulines, J. Ruiz, J.-C. Blais, D. Astruc, "Regioselective Chlorocarbonylation of Polybenzyl Cores and Functionalization Using Dendritic and Organometallic Nucleophiles", *J. Org. Chem.* **2000**, *65*, 1996–2002.

[231] E. Alonso, D. Astruc, "Introduction of the Cluster Fragment Ru$_3$(CO)$_{11}$ at the Periphery of Phosphine Dendrimers Catalyzed by the Electron-Reservoir Complex [Fe^{I}Cp(C$_6$Me$_6$)]", *J. Am. Chem. Soc.* **2000**, *122*, 3222–3223.

[232] J. Ipaktschi, R. Hosseinzadeh, P. Schlaf, "Self-Assembly of Quinodimethanes through Covalent Bonds: A Novel Principle for the Synthesis of Functional Macrocycles", *Angew. Chem.* **1999**, *111*, 1765–1768; *Int. Ed.* **1999**, *38*, 1658–1660.

[233] R. Moors, F. Vögtle, "Dendrimere Polyamine", *Chem. Ber.* **1993**, *126*, 2133–2135.

[234] D. Seyferth, T. Kugita, A. L. Rheingold, G. P. A. Yap, "Preparation of Carbosilane Dendrimers with Peripheral Acetylenedicobalt Hexacarbonyl Substituents", *Organometallics* **1995**, *14*, 5362–5366.

[234a] R. Breinbauer, E. N. Jacobsen, "Cooperative Asymmetric Catalysis with Dendrimeric [Co(salen)] Complexes", *Angew. Chem. Int. Ed.* **2000**, *39*, 3604–3607.

[234b] K. Takada, G. D. Storrier, J. I. Goldsmith, H. D. Abruña, "Electrochemical and absorption properties of PAMAM dendrimers surface-functionalized with polypyridyl cobalt complexes", *J. Phys. Chem. B.* **2001**, *105*, 2404–2411.

[235] I. Cuadrado, M. Morán, A. Moya, C. M. Casado, M. Barranco, B. Alonso, "Organometallic Silicon-Based Dendrimers with Peripheral Si–Cyclopentadienyl, Si–Co, and Si–Fe σ-Bonds", *Inorg. Chim. Acta* **1996**, *251*, 5–7.

[236] F. Lobete, I. Cuadrado, C. M. Casado, B. Alonso, M. Morán, J. Losada, "Silicon-Based Organometallic Dendritic Macromolecules Containing {η^6- (Organosilyl)arene}chromium Tricarbonyl Moieties", *J. Organomet. Chem.* **1996**, *509*, 109–113.

[237] I. Cuadrado, M. Morán, C. M. Casado, B. Alonso, J. Losada, "Organometallic Dendrimers with Transition Metals" , *Coord. Chem. Rev.* **1999**, *193–195*, 395–445.

[238] B. Alonso, I. Cuadrado, M. Morán, J. Losada, "Organometallic Silicon Dendrimers", *J. Chem. Soc., Chem. Commun.* **1994**, 2575–2576.

[239] B. Alonso, M. Morán, C. M. Casado, F. Lobete, J. Losada, I. Cuadrado, "Electrodes Modified with Electroactive Films of Organometallic Dendrimers", *Chem. Mater.* **1995**, *7*, 1440–1442.

[240] B. D. Karstedt, "Platinum Complexes of Unsaturated Siloxanes and Platinum-Containing Organopolysiloxanes", **1973**, *U.S. Pat*, 3, 775, 452.

[241] I. Cuadrado, M. Morán, J. Losada, C. M. Casado, C. Pascual, B. Alonso, F. Lobete, "Organometallic Dendritic Macromolecules: Organosilicon and Organometallic Entities as Cores or Building Blocks" in *Advances in Dendritic Macromolecules*, G. R. Newkome, ed., JAI Press, Greenwich, Conn., **1996**, 151–195: see pages 158–161.

[242] M. Morán, C. M. Casado, I. Cuadrado, J. Losada, "Ferrocenyl-Substituted Octakis(dimethylsiloxy)octasilsesquioxanes: A New Class of Supramolecular Organometallic Compunds – Synthesis, Characterization, and Electrochemistry", *Organometallics* **1993**, *12,(11)*, 4327–4333.

[243] J. Losada, I. Cuadrado, M. Morán, C. M. Casado, B. Alonso, M. Barranco, "Ferrocenyl Silicon-Based Dendrimers as Mediators in Amperometric Biosensors", *Anal. Chim. Acta* **1997**, *338*, 191–198.

[244] L.-L. Zhou, J. Roovers, "Synthesis of Novel Carbosilane Dendritic Macromolecules", *Macromolecules* **1993**, *26*, 963–968.

[245] D. Seyferth, D. Y. Son, A. L. Rheingold, R. L. Ostrander, "Synthesis of an Organosilicon Dendrimer Containing 324 Si–H Bonds", *Organometallics* **1994**, *13*, 2682–2690.

[246] R. Rulkens, A. J. Lough, I. Manners, S. R. Lovelace, C. Grant, W. E. Geiger, "Linear Oligo(Ferrocenyl-dimethylsilanes) with Between Two and Nine Ferrocene Units: Electrochemical and Structural Models for Poly(ferrocenylsilane) High Polymers", *J. Am. Chem. Soc.* **1996**, *118*, 12683–12695.

[247] C. M. Casado, I. Cuadrado, B. Alonso, M. Morán, J. Losada, "Silicon-Based Ferrocenyl Dendrimers as Anion Receptors in Solution and Immobilized onto Electrode Surfaces", *J. Organomet. Chem.* **1999**, *463*, 87–92.

[248] I. Cuadrado, M. Morán, C. M. Casado, B. Alonso, F. Lobete, B. Garcia, M. Ibisate, J. Losada, "Ferrocenyl-Functionalized Poly(propylenimine) Dendrimers", *Organometallics* **1996**, *15*, 5278–5280.

[249] K. Takada, D. J. Díaz, H. D. Abruña, I. Cuadrado, C. Casado, B. Alonso, M. Morán, J. Losada, "Redox-Active Ferrocenyl Dendrimers: Thermodynamics and Kinetics of Adsorption, *in situ* Electrochemical Quartz Crystal Microbalance Study of the Redox Process and Tapping Mode AFM Imaging", *J. Am. Chem. Soc.* **1997**, *119*, 10763–16773.

[250] R. Castro, I. Cuadrado, B. Alonso, C. M. Casado, M. Morán, A. E. Kaifer, "Multisite Inclusion Complexation of Redox Active Dendrimer Guests", *J. Am. Chem. Soc.* **1997**, *119*, 5760–5761.

[251] B. González, C. M. Casado, B. Alonso, I. Cuadrado, M. Morán, Y. Wang, A. E. Kaifer, "Synthesis, Electrochemistry and Cyclodextrin Binding of Novel Cobaltocenium-Functionalized Dendrimers", *Chem. Commun.* **1998**, 2569–2570.

[252] A. E. Kaifer, "Interplay between Molecular Recognition and Redox Chemistry", *Acc. Chem. Res.* **1999**, *32*, 62–71.

[253] C. M. Casado, B. González, I. Cuadrado, B. Alonso, M. Morán, J. Losada, "Mixed Ferrocene-Cobaltocenium Dendrimers: The Most Stable Organometallic Redox Systems Combined in a Dendritic Molecule", *Angew. Chem.* **2000**, *112*, 2219–2222; *Int. Ed.* **2000**, *39*, 2135–2138.

[254] C.-F. Shu, H.-M. Shen, "Organometallic Ferrocenyl Dendrimers: Synthesis, Characterization, and Redox Properties", *J. Mater. Chem.* **1997**, *7*, 47–52.

[255] M. Bardaji, M. Kustos, A.-M. Caminade, J.-P. Majoral, B. Chaudret, "Phosphorus-Containing Dendrimers as Multidentate Ligands: Palladium, Platinum, and Rhodium Complexes", *Organometallics* **1997**, *16*, 403–410.

[256] J.-P. Majoral, A.-M. Caminade, "Divergent Approaches to Phosphorus-Containing Dendrimers and their Functionalization", *Top. Curr. Chem.* **1998**, *197*, 79–124.

[257] M. Slany, A.-M. Caminade, J.-P. Majoral, "Specific Functionalization on the Surface of Dendrimers", *Tetrahedron Lett.* **1996**, *37*, 9053–9056.

[258] M. Bardají, A.-M. Caminade, J.-P. Majoral, B. Chaudret, "Ruthenium Hydride and Dihydrogen Complexes with Dendrimeric Multidentate Ligands", *Organometallics* **1997**, *16*, 3489–3497.

[259] M. Slany, M. Bardají, A.-M. Caminade, B. Chaudret, J.-P. Majoral, "Versatile Complexation Ability of Very Large Phosphino-Terminated Dendrimers", *Inorg. Chem.* **1997**, *36*, 1939–1945.

[260] V. Cadierno, A. Igau, B. Donnadieu, A.-M. Caminade, J.-P. Majoral, "Dendrimers Containing Zwitterionic [Phosphonium Anionic Zirconocene(IV)] Complexes", *Organometallics* **1999**, *18*, 1580–1582.

[261] P. Lange, A. Schier, H. Schmidbaur, "Mono-, Di- and Trinuclear Gold(I) Complexes of New Phosphino-Substituted Amides: Initial Steps to Chlorogold(I)diphenylphosphino-Terminated Dendrimers", *Inorg. Chim. Acta* **1995**, *235*, 263–272.

[262] P. Lange, A. Schier, H. Schmidbaur, "Dendrimer-Based Multinuclear Gold(I) Complexes", *Inorg. Chem.* **1996**, *35*, 637–642.

[263] M. T. Reetz, G. Lohmer, R. Schwickardi, "Synthesis and Catalytic Activity of Dendritic Diphosphane Metal Complexes", *Angew. Chem.* **1997**, *109*, 1559–1562; *Int. Ed. Engl.* **1997**, *36*, 1526–1529.

[264] N. Feeder, J. Geng, P. G. Goh, B. F. G. Johnson, C. M. Martin, D. S. Shephard, W. Zhou, "Nanoscale Super Clusters of Clusters Assembled around a Dendritic Core", *Angew. Chem.* **2000**, *112*, 1727–1730; *Int. Ed.* **2000**, *39*, 1661–1664.

[265] H. Alper, P. Arya, S. C. Bourque, G. R. Jefferson, L. E. Manzer, "Heck Reaction using Palladium Complexed to Dendrimers on Silica", *Can. J. Chem.* **2000**, *78*, 920–924.

[266] A. W. Bosman, A. P. H. J. Schenning, R. A. J. Janssen, E. W. Meijer, "Well-Defined Metallodendrimers by Site-Specific Complexation", *Chem. Ber./Recl.* **1997**, *130*, 725–728.

[267] R. J. M. K. Gebbink, A. W. Bosman, M. C. Feiters, E. W. Meijer, R. J. M. Nolte, "A Multi-O_2 Complex Derived from a Copper(I) Dendrimer", *Chem. Eur. J.* **1998**, *5*, 65–69.

[268] T. Mizugaki, M. Ooe, K. Ebitani, K. Kaneda, "Catalysis of Dendrimer-Bound Pd(II) Complex Selective Hydrogenation of Conjugated Dienes to Monoenes", *J. Mol. Catal. A: Chem.* **1999**, *145*, 329–333.

[269] B. A. J. Jansen, J. van der Zwan, J. Reedijk, H. den Dulk, J. Brouwer, "A Tetranuclear Platinum Compound Designed to Overcome Cisplatin Resistance", *Eur. J. Inorg. Chem* **1999**, 1429–1433.

[270] E. C. Wiener, M. W. Brechbiel, H. Brothers, R. L. Magin, O. A. Gansow, D. A. Tomalia, P. C. Lauterbur, "Dendrimer-Based Metal Chelates: A New Class of Magnetic Resonance Imaging Contrast Agents", *Magn. Reson. Med.* **1994**, *31*, 1–8.

[271] D. A. Tomalia, H. Baker, J. Dewald, M. Hall, G. Kallos, S. Martin, J. Roeck, J. Ryder, P. Smith, "A New Class of Polymers: Starburst-Dendritic Macromolecules", *Polym. J. (Tokyo)* **1985**, *17*, 117–132.

[271a] M. Takahashi, Y. Hara, K. Aoshima, H. Kurihara, T. Oshikawa, M. Yamashita, "Utilization of dendritic framework as a multivalent ligand: a functionalized gadolinium(III) carrier with glycoside cluster periphery", *Tetrahedron Lett.* **2000**, *41*, 8485–8488.

[272] C. Bieniarz, "Dendrimers: Applications to Pharmaceutical and Medicinal Chemistry", in *Encyclopedia of Pharmaceutical Technology* (Eds.: J. Swarbrick, J. C. Boylan), Marcel Dekker, Inc., New York, **1999**, pp. 55–89.

[273] E. C. Wiener, F. P. Auteri, J. W. Chen, M. W. Brechbiel, O. A. Gansow, D. S. Schneider, R. L. Belford, R. B. Clarkson, P. C. Lauterbur, "Molecular Dynamics of Ion-Chelate Complexes Attached to Dendrimers", *J. Am. Chem. Soc.* **1996**, *118*, 7774–7782.

[274] B. Jendrusch-Borkowski, J. Awad, F. Wasgestian, "Reactions of Chromium(III)- and Cobalt(III)-amine-complexes with Starburst (PAMAM) Dendrimers", *J. Inclusion Phenom. Macrocycl. Chem.* **1999**, *35*, 355–359.

[275] A. Miedaner, C. J. Curtis, R. M. Barkley, D. L. DuBois, "Electrochemical Reduction of CO_2 Catalyzed by Small Organophosphine Dendrimers Containing Palladium", *Inorg. Chem.* **1994**, *33*, 5482–5490.

[276] P. D. Beer, E. L. Tite, "New Hydrophobic Host Molecules Containing Multiple Redox-Active Centers", *Tetrahedron Lett.* **1988**, *29*, 2349–2352.

[277] R. Deschenaux, E. Serrano, A. M. Levelut, "Ferrocene-Containing Liquid-Crystalline Dendrimers: A Novel Family of Mesomorphic Macromolecules", *Chem. Commun.* **1997**, 1577–1578.

[277a] T. Chuard, R. Deschenaux, "Liquid-Crystalline Dendrimers Based on Ferrocene and Fullerene", *Chimia* **2001**, *55*, 139–142.

[278] H. Zeng, G. R. Newkome, C. L. Hill, "Poly(polyoxometalate) Dendrimers: Molecular Prototypes of New Catalytic Materials", *Angew. Chem.* **2000**, *112*, 1842–1844; *Int. Ed.* **2000**, *39*, 1772–1774.

[279] G. A. Kraus, S. V. Louw, "Synthesis of the First Phthalocyanine-Containing Dendrimer", *J. Org. Chem.* **1998**, *63*, 7520–7521.

[280] N. Maruo, M. Uchiyama, T. Kato, T. Arai, H. Akisada, N. Nishino, "Hemispherical Synthesis of Dendritic Poly(L-lysine) Containing Sixteen Free-Base Porphyrins and Sixteen Zinc Porphyrins", *Chem. Commun.* **1999**, 2057–2058.

[281] T. Kato, M. Uchiyama, N. Maruo, T. Arai, N. Nishino, "Fluorescence Energy Transfer in Dendritic Poly(L-lysine)s Combining Thirty-two Free Base- and Zinc(II)-porphyrins in Scrambling Fashion", *Chem. Lett.* **2000**, 144–145.

[282] C. Kim, I. Jung, "Preparation of Dendritic Carbosilanes with Peripheral Ethynyldicobalt Hexacarbonyl Tetrahedron C_2Co_2", *Inorg. Chem. Commun.* **1999**, *1*, 427–430.

[283] C. Kim, I. Jung, "Preparation of Dendritic Carbosilanes Containing Ethylyl Groups and Dicobalt Hexacarbonyl Clusters on the Periphery", *J. Organomet. Chem.* **1999**, *588*, 9–19.

[284] Q. J. McCubbin, J. F. Stoddart, T. Welton, A. J. P. White, D. J. Williams, "Dithiocarbamate-Functionalized Dendrimers as Ligands for Metal Complexes", *Inorg. Chem.* **1998**, *37*, 3753–3758.

[285] K. Brüning, H. Lang, "Ein einfacher Zugang zu Carbosiloxan-Dendrimeren", *J. Organomet. Chem.* **1999**, *575*, 153–157.

[286] S. J. Hughes, J. R. Moss, K. J. Naidoo, J. F. Kelly, A. S. Batsanov, "Force-Field Parameterisation, Synthesis and Crystal Structure of a Novel Tricarbonylchromium Arene Complex", *J. Organomet. Chem.* **1999**, *588*, 176–185.

[287] N. Brinkmann, D. Giebel, G. Lohmer, M. T. Reetz, U. Kragl, "Allylic Substitution with Dendritic Palladium Catalysts in a Continuously Operating Membrane Reactor", *J. Catal.* **1999**, *183*, 163–168.

[288] D. de Groot, E. B. Eggeling, J. C. de Wilde, H. Kooijman, R. J. van Haaren, A. W. van der Made, A. L. Spek, D. Vogt, J. N. H. Reek, P. C. J. Kamer, P. W. N. M. van Leeuwen, "Palladium Complexes of Phosphine-Functionalised Carbosilane Dendrimers as Catalysts in a Continuous-Flow Membrane Reactor", *Chem. Commun.* **1999**, 1623–1624.

[289] A. W. van der Made, P. W. N. M. van Leeuwen, "Silane Dendrimers", *J. Chem. Soc., Chem. Commun.* **1992**, 1400–1401.

[290] S. C. Bourque, F. Maltais, W.-J. Xiao, O. Tardif, H. Alper, P. Arya, L. E. Manzer, "Hydroformylation Reactions with Rhodium-Complexed Dendrimers on Silica", *J. Am. Chem. Soc.* **1999**, *121*, 3035–3038.

[291] K. Vassilev, W. T. Ford, "Poly(propylene imine) Dendrimer Complexes of Cu(II), Zn(II), and Co(III) as Catalysts of Hydrolysis of *p*-Nitrophenyl Diphenyl Phosphate", *J. Polym. Sci., Part A: Polym. Chem.* **1999**, *37*, 2727–2736.

[292] B. Dardel, R. Deschenaux, M. Even, E. Serrano, "Synthesis, Characterization, and Mesomorphic Properties of a Mixed [60]Fullerene–Ferrocene Liquid-Crystalline Dendrimers", *Macromolecules* **1999**, *32*, 5193–5198.

[293] C. Francavilla, F. V. Bright, M. R. Detty, "Dendrimeric Catalysts for the Activation of Hydrogen Peroxide. Increasing Activity per Catalytic Phenylseleno Group in Successive Generations", *Org. Lett.* **1999**, *1*, 1043–1046.

[294] A. L. Hurley, D. L. Mohler, "Organometallic Photonucleases: Synthesis and DNA-Cleavage Studies of Cyclopentadienyl Metal-Substituted Dendrimers Designed to Increase Double-Stranded Scission", *Org. Lett.* **2000**, *2*, 2745–2748.

[295] H. Beerens, F. Verpoort, L. Verdonck, "Low-Generation Carbosilane Dendrimers as Core for Star Polymers using a Ru-ROMP Catalyst", *J. Mol. Catal. A: Chem.* **2000**, *151*, 279–282.

[296] G. R. Newkome, C. N. Moorefield, "Unimolecular Micelles and Method of Making the Same" **1992**, *U. S. Pat.*, 5, 154, 853.

[297] G. R. Newkome, C. N. Moorefield, "Chemistry Within a Unimolecular Micelle: Metallomicellanoic Acids" *Polym. Prepr.* **1993**, *34*, 75–76.

[298] G. R. Newkome, C. N. Moorefield, "Metallo- and Metalloido-Micellane™ Derivatives: Incorporation of Metals and Nonmetals Within Unimolecular Superstructures", in *International Symposium on New Macromolecular Architectures and Supramolecular Polymers* (Eds.: V. Percec, D. A. Tirrell), Hüthig & Wepf Verlag, Basel, **1994**, pp. 63–71.

[299] H.-W. Marx, F. Moulines, T. Wagner, D. Astruc, "Hexakis(but-3-ynyl)benzene", *Angew. Chem.* **1996**, *108*, 1842–1845; *Int. Ed. Engl.* **1996**, *35*, 1701–1704.

[300] E. C. Constable, C. E. Housecroft, L. A. Johnson, "Dicobalt Cluster-Functionalized 2,2':6',2"-Terpyridine Ligands: Ruthenium(II) Complexes with Covalently Linked $C_2Co_2(CO)_6$ Units", *Inorg. Chem. Commun.* **1998**, *1*, 68–70.

[301] E. C. Constable, O. Eich, C. E. Housecroft, L. A. Johnston, "Towards Organometallic Dendrimers", *Chimia* **1998**, *52*, 452.

[302] E. C. Constable, C. E. Housecroft, "Supramolecular Approaches to Advanced Materials", *Chimia* **1998**, *52*, 533–538.

[303] E. C. Constable, C. E. Housecroft, O. Eich, "Organometallic Metallodendrimers and Metallostars", *Chimia* **1999**, *53*, No. 113.

[304] E. C. Constable, O. Eich, C. E. Housecroft, "High-Nuclearity Cobaltadendrimers", *J. Chem. Soc., Dalton Trans.* **1999**, 1363–1364.

[305] G. R. Newkome, J. Groß, C. N. Moorefield, B. D. Woosley, "Approaches Towards Specifically Functionalized Cascade Macromolecules: Dendrimers with Incorporated Metal Binding Sites and their Palladium(II) and Copper(II) Complexes", *Chem. Commun.* **1997**, 515–516.

[306] G. R. Newkome, C. N. Moorefield, "Metallospheres and Superclusters", **1994**, *U. S. Pat.*, 5, 376, 690.

[307] G. R. Newkome, C. N. Moorefield, "Metallospheres and Superclusters", **1995**, *U. S. Pat.*, 5, 422, 379.

[308] G. R. Newkome, C. N. Moorefield, "Metallospheres and Superclusters", **1996**, *U. S. Pat.*, 5, 516, 810.

[309] G. R. Newkome, C. N. Moorefield, "Metallospheres and Superclusters", **1996**, *U. S. Pat.*, 5, 585, 457.

[310] J. D. Epperson, L.-J. Ming, B. D. Woosley, G. R. Baker, G. R. Newkome, "NMR Study of Dendrimer Structures Using Paramagnetic Cobalt(II) as a Probe", *Inorg. Chem.* **1999**, *38*, 4498–4502.

[311] G. R. Newkome, B. D. Woosley, E. He, C. N. Moorefield, R. Güther, G. R. Baker, G. H. Escamilla, J. Merrill, H. Luftmann, "Supramolecular Chemistry of Flexible, Dendritic-Based Structures Employing Molecular Recognition", *Chem. Commun.* **1996**, 2737–2738.

[312] G. R. Newkome, E. He, L. A. Godínez, G. R. Baker, "Neutral Highly Branched Metallo-macromolecules: Incorporation of (2,2′:6′,2″-Terpyridine)ruthenium(II) Complex Without External Counterions", *Chem. Commun.* **1999**, 27–28.

[313] G. R. Newkome, E. He, L. A. Godínez, G. R. Baker, "Electroactive Metallomacromolecules via Tetrakis(2,2′:6′,2″-terpyridine)ruthenium(II) Complexes: Dendritic Networks towards Constitutional Isomers and Neutral Species without External Counterions", *J. Am. Chem. Soc.* **2000**, *122*, 9993–10006.

[314] G. R. Newkome, A. K. Patri, L. A. Godínez, "Design, Syntheses, Complexation, and Electrochemistry of Polynuclear Metallodendrimers Possessing Internal Metal Binding Loci", *Chem. Eur. J.* **1999**, *5*, 1445–1451.

[315] C. Larré, B. Donnadiu, A.-M. Caminade, J.-P. Majoral, "Regioselective Gold Complexation within the Cascade Structure of Phosphorus-Containing Dendrimers", *Chem. Eur. J.* **1998**, *4*, 2031–2036.

[316] C. C. Mak, N. Bampos, J. K. M. Sanders, "Metalloporphyrin Dendrimers with Folding Arms", *Angew. Chem.* **1998**, *110*, 3169–3172; *Int. Ed.* **1998**, *37*, 3020–3023.

[317] C. C. Mak, D. Pomeranc, M. Montalti, L. Prodi, J. K. M. Sanders, "A Versatile Synthetic Strategy for Construction of Large Oligomers: Binding and Photophysical Properties of a Nine-Porphyrin Array", *Chem. Commun.* **1999**, 1083–1084.

[318] T. Norsten, N. Branda, "A Starburst Porphyrin Polymer: A First Generation Dendrimer", *Chem. Commun.* **1998**, 1257–1258.

[319] A. M. McDonagh, M. G. Humphrey, M. Samoc, B. Luther-Davies, "Organometallic Complexes for Nonlinear Optics; 17. Synthesis, Third-Order Optical Nonlinearities, and Two-Photon Absorption Cross-Section of an Alkynylruthenium Dendrimer", *Organometallics* **1999**, *18*, 5195–5197.

[320] X. Camps, E. Dietel, A. Hirsch, S. Pyo, L. Echegoyen, S. Hackbarth, B. Röder, "Globular Dendrimers Involving a C_{60} Core and a Tetraphenyl Porphyrin Function", *Chem. Eur. J.* **1999**, *5*, 2362–2373.

[321] S. Hecht, H. Ihre, J. M. J. Fréchet, "Porphyrin Core Star Polymers: Synthesis, Modification, and Implication for Site Isolation", *J. Am. Chem. Soc.* **1999**, *121*, 9239–9240.

[322] T. Hayashi, Y. Hitomi, T. Ando, T. Mizutani, Y. Hisaeda, S. Kitagawa, H. Ogoshi, "Peroxidase Activity of Myoglobin is Enhanced by Chemical Mutation of Heme-Propionates", *J. Am. Chem. Soc.* **1999**, *121*, 7747–7750.

[323] C. C. Mak, N. Bampos, J. K. M. Sanders, "Ru(II)-Centred Porphyrin Pentamers as Coordination Building Blocks for Large Porphyrin Arrays", *Chem. Commun.* **1999**, 1085–1086.

[324] S. L. Darling, C. C. Mak, N. Bampos, N. Feeder, S. J. Teat, J. K. M. Sanders, "A Combined Covalent and Coordination Approach to Dendritic Multiporphyrin Arrays Based on Ruthenium(II) Porphyrins", *New J. Chem.* **1999**, *23*, 359–364.

[325] A. Nakano, A. Osuka, I. Yamazaki, T. Yamazaki, Y. Nishimura, "Windmill-Like Porphyrin Arrays as Potent Light-Harvesting Antenna Complexes", *Angew. Chem.* **1998**, *110*, 3172–3176; *Int. Ed.* **1998**, *37*, 3023–3027.

[326] J. Seth, V. Palaniappan, T. E. Johnson, S. Prathapan, J. S. Lindsey, D. F. Bocian, "Investigation of Electronic Communication in Multi-Porphyrin Light-Harvesting Arrays", *J. Am. Chem. Soc.* **1994**, *116*, 10578–10592.

[327] O. Mongin, C. Papamicaël, N. Hoyler, A. Gossauer, "Modular Synthesis of Benzene-Centered Porphyrin Trimers and a Dendritic Porphyrin Hexamer", *J. Org. Chem.* **1998**, *63*, 5568–5580.

[328] M. Ravikanth, J.-P. Strachan, F. Li, J. S. Lindsey, "*trans*-Substituted Porphyrin Building Blocks Bearing Iodo and Ethynyl Groups for Applications in Bioorganic and Materials Chemistry", *Tetrahedron* **1998**, *54*, 7721–7734.

[329] T. Hayashi, Y. Hitomi, H. Ogashi, "Artificial Protein–Protein Complexation between a Reconstituted Myoglobin and Cytochrome *c*", *J. Am. Chem. Soc.* **1998**, *120*, 4910–4915.

[330] U. Michelsen, C. A. Hunter, "Self-Assembled Porphyrin Polymers", *Angew. Chem.* **2000**, *112*, 780–783; *Int. Ed.* **2000**, *39*, 764–767.

[331] M. C. Callama, P. Timmerman, D. N. Reinhoudt, "Guest-Templated Selection and Amplification of a Receptor by Noncovalent Combinatorial Synthesis", *Angew. Chem.* **2000**, *112*, 771–774; *Int. Ed.* **2000**, *39*, 755–758.

[332]   K. Chichak, N. R. Branda, "The Metal-Directed Self-Assembly of Three-Dimensional Porphyrin Arrays", *Chem. Commun.* **2000**, 1211–1212.

[333]   M. Petrucci-Samija, V. Guillemette, M. Dasgupta, A. K. Kakkar, "A New Divergent Route to the Synthesis of Organophosphine and Metallodendrimers via Simple Acid-Base Hydrolytic Chemistry", *J. Am. Chem. Soc.* **1999**, *121*, 1968–1969.

[334]   G. Chessa, A. Scrivanti, L. Canovese, F. Visentin, P. Uguagliati, "Pyridine-Based Dendritic Wedges with a Specific Metal Ion Coordination Site and their Palladium(II) Complexes", *Chem. Commun.* **1999**, 959–960.

[335]   M. F. Ottaviani, S. Bossmann, N. J. Turro, D. A. Tomalia, "Characterization of Starburst Dendrimers by the EPR Technique; 1. Copper Complexes in Water Solution", *J. Am. Chem. Soc.* **1994**, *116*, 661–671.

[336]   M. F. Ottaviani, F. Montalti, M. Romanelli, N. J. Turro, D. A. Tomalia, "Characterization of Starburst Dendrimers by EPR; 4. Mn(II) as a Probe of Interphase Properties", *J. Phys. Chem.* **1996**, *100*, 11033–11042.

[337]   M. F. Ottaviani, F. Montalti, N. J. Turro, D. A. Tomalia, "Characterization of Starburst Dendrimers by the EPR Technique. Copper(II) Ions Binding Full-Generation Dendrimers", *J. Phys. Chem. B.* **1997**, *101*, 158–166.

[338]   M. Zhao, L. Sun, R. M. Crooks, "Preparation of Cu Nanoclusters within Dendrimer Templates", *J. Am. Chem. Soc.* **1998**, *120*, 4877–4878.

[339]   M. E. Garcia, L. A. Baker, R. M. Crooks, "Preparation and Characterization of Dendrimer–Gold Colloid Nanocomposites", *Anal. Chem.* **1999**, *71*, 256–258.

[340]   M. Zhao, R. M. Crooks, "Dendrimer-Encapsulated Pt Nanoparticles: Synthesis, Characterization, and Applications to Catalysis", *Adv. Mater. (Weinheim, Fed. Repub. Ger.)* **1999**, *11*, 217–220.

[341]   M. Zhao, R. M. Crooks, "Intradendrimer Exchange of Metal Nanoparticles", *Chem. Mater.* **1999**, *11*, 3379–3385.

[342]   M. Zhao, R. M. Crooks, "Homogeneous Hydrogenation Catalysis with Monodisperse, Dendrimer-Encapsulated Pd and Pt Nanoparticles", *Angew. Chem.* **1999**, *111*, 375–377; *Int. Ed.* **1999**, *38*, 364–366.

[343]   K. Esumi, A. Suzuki, N. Aihara, K. Usui, K. Torigoe, "Preparation of Gold Colloids with UV Irradiation Using Dendrimers as Stabilizers", *Langmuir* **1998**, *14*, 3157–3159.

[344]   V. Chechik, R. M. Crooks, "Dendrimer-Encapsulated Pd Nanoparticles as Fluorous Phase-Soluble Catalysts", *J. Am. Chem. Soc.* **2000**, *122*, 1243–1244.

[345]   L. Balogh, D. A. Tomalia, "Poly(amidoamine) Dendrimer-Templated Nanocomposites; 1. Synthesis of Zerovalent Copper Nanoclusters", *J. Am. Chem. Soc.* **1998**, *120*, 7355–7356.

[346]   N. C. B. Tan, L. Balogh, S. F. Trevino, D. A. Tomalia, J. S. Lin, "A Small-Angle Scattering Study of Dendrimer-Copper Sulfide Nanocomposites", *Polymer* **1999**, *40*, 2537–2545.

[347]   S. Rubin, G. Bar, R. W. Cutts, T. A. Zawodzinski, Jr., "New Approach for Gold and Silver Colloid Monolayer Preparation on Modified Surfaces", *Proc. Electrochem. Soc.* **1995**, *95–27*, 151–159.

[348]   F. Gröhn, B. J. Bauer, Y. A. Akpalu, C. L. Jackson, E. J. Amis, "Dendrimer Templates for the Formation of Gold Nanoclusters", *Macromolecules* **2000**, *33*, 6042–6050.

[349]   J. Huang, K. Sooklal, C. J. Murphy, H. J. Ploehn, "Polyamine–Quantum Dot Nanocomposites: Linear versus Starburst Stabilizer Architectures", *Chem. Mater.* **1999**, *11*, 3595–3601.

[350]   L. Balogh, R. Valluzzi, K. S. Laverdure, S. P. Gido, G. L. Hagnauer, D. A. Tomalia, "Formation of Silver and Gold Dendrimer Nanocomposites", *J. Nanoparticle Res.* **2000**, *1*, 353–368.

[351]   M. S. Diallo, L. Balogh, A. Shafagati, J. H. Johnson, Jr., W. A. Goddard, III, D. A. Tomalia, "Poly(amidoamine) Dendrimers: A New Class of High Capacity Chelating Agents for Cu(II) Ions", *Environ. Sci. Technol.* **1999**, *33*, 820–824.

[352]   J.-A. He, R. Valluzzi, K. Yang, T. Dolukhanyan, C. Sung, J. Kumar, S. K. Tripathy, L. Samuelson, L. Balogh, D. A. Tomalia, "Electrostatic Multilayer Deposition of a Gold-Dendrimer Nanocomposite", *Chem. Mater.* **1999**, *11*, 3268–3274.

[352a]   F. Gröhn, G. Kim, B. J. Bauer, E. J. Amis, "Nanoparticle Formation within Dendrimer-Containing Polymer Network: Route to new Organic – Inorganic Hybrid Materials", *Macromolecules* **2001**, *34*, 2179–2185.

[352b]   O. Varnavski, R. G. Ispasoiu, L. Balogh, D. Tomalia, T. Goodson, "Ultrafast time-resolved photoluminescence from novel metal-dendrimer nanocomposites", *J. Chem. Phys.* **2001**, *114*, 1962–1965.

[352c]   L. Balogh, D. R. Swanson, D. A. Tomalia, G. L. Hagnauer, A. T. McManus, "Dendrimer – Silver Complexes and Nanocomposites as Antimicrobial Agents", *Nano Letters* **2001**, *1*, 18–21.

[352d]   L. K. Yeung, R. M. Crooks, "Heck Heterocoupling within a Dendritic Nanoreactor", *Nano Letters* **2001**, *1*, 14–17.

[352e]   B. I. Lemon, R. M. Crooks, "Preparation and Characterization of Dendrimer-Encapsulation CdS Semiconductor Quantum Dots", *J. Am. Chem. Soc.* **2000**, *122*, 12886–12887.

[352f] R. G. Ispasoiu, L. Balogh, O. P. Varnavsky, D. A. Tomalia, T. Goodson, III, "Large Optical Limiting from Novel Metal-Dendrimer Nanocomposite Materials", *J. Am. Chem. Soc.* **2000**, *122*, 11005–11006.

[353] M. F. Ottaviani, C. Turro, N. J. Turro, S. H. Bossmann, D. A. Tomalia, "Nitroxide-Labeled Ru(II) Polypyridyl Complexes as EPR Probes of Organized Systems; 3. Characterization of Starburst Dendrimers and Comparison to Photophysical Measurements", *J. Phys. Chem.* **1996**, *100*, 13667–13674.

[354] L. H. Hanus, K. Sooklal, C. J. Murphy, H. J. Ploehn, "Aggregation Kinetics of Dendrimer-Stabilized CdS Nanoclusters", *Langmuir* **2000**, *16*, 2621–2626.

[355] K. Esumi, A. Suzuki, A. Yamahira, K. Torigoe, "Role of Poly(amidoamine) Dendrimers for Preparing Nanoparticles of Gold, Platinum, and Silver", *Langmuir* **2000**, *16*, 2604–2608.

[356] S. Hecht, T. Emrick, J. M. J. Fréchet, "Hyperbranched Porphyrins – A Rapid Synthetic Approach to Multiporphyrin Macromolecules", *Chem. Commun.* **2000**, 313–314.

[356a] F. Vögtle, S. Gestermann, C. Kauffmann, P. Ceroni, V. Vicinelli, V. Balzani, "Coordination of Co^{2+} Ions in the Interior of Poly(propylene amine) Dendrimers Containing Fluorescent Dansyl Units in the Periphery", *J. Am. Chem. Soc.* **2000**, *122*, 10398–10404.

9 Dendritic Networks

9.1 Introduction: Dendritic Assemblies

A full appreciation of dendritic chemistry and hence the iterative method employed for generational construction would be deficient without consideration of 'higher order' macromolecular assemblies; this is especially true in the nano-era. To this end, in this Chapter we start to provide a suitable foundation for a preliminary debate concerning these higher order macromolecular assemblies, or dendritic networks. It should be noted, however, that since a comprehensive review of this subject would be difficult due to "interpretational differences", only representative examples are presented herein. For the purpose of our discussion, a dendritic network shall be defined as the deliberate connection, through covalent or non-covalent means, of multiple (usually preconstructed) dendritic units resulting in architectures with dimensions greater than would be obtained by the preparation of a standard dendrimer. This "deliberate connection" or "positioning" results in at least one fewer degrees of freedom with respect to the relationship of individual dendrimers to other macromolecules in the network. Since there are but a limited number of examples in this area of dendrimerized macromolecular constructs, this Chapter will be unique in its format in that it will include an examination of potential categories of networks as well as modes of formation, which, in part, have been covered in the previous Chapters.

Attention will be focussed on two primary modes of network assembly: (1) *random, uncontrolled connectivity* analogous to 'classical' polymer preparation, where dendrimers act as monomers or building blocks and are orientated in an essentially "unsystematic" manner, corresponding to classical single-pot-type reactions; and (2) *ordered, controlled connectivity* analogous to tessellated dendritic polymer construction, where elements, or building blocks, are "precisely juxtaposed" into a coherent pattern.

Networks composed of cross-linked, linear, classically synthesized polymers have been reviewed[1–5] and will not be dealt with in this Chapter. Further, mathematical treatments of network properties will not be discussed herein, nor do we discuss in detail dendrimerized linear polymers that give rise to cylindrical motifs,[6–10] which have been described earlier in this book (Section 5.3.2.2) and have been reviewed.[11, 12]

As defined above, dendritic networks are considered to result from the one-, two-, or three-dimensional orientation of dendrimers; thus, "ordering" can be geometrically likened to rods; surfaces or sheets; and cubes, tetrahedrons, or spheres, respectively. In view of the broad scope and breadth of potential macromolecular architectures that can be obtained by application of different modes of connectivity, we will concentrate herein on networks that are constructed from the simplest dendritic structures, namely those that are pseudospherical or globular. The principles that are presented here pertaining to network formation should be readily adaptable to non-spheroidal dendritic structures as well as to macromolecular assemblies possessing only limited dendritic character.

Construction of dendritic, nano- or micro-networks is a logical progression of the iterative synthetic method, whereby the desired "positioning" of multiple nuclei components (such as in dendrimers) can be obtained. Realization of networks comprised of dendritic building blocks, as well as individual dendrimers, thus has applications in diverse areas of materials science such as molecular electronics,[13–15] biomolecular engineering,[16] and (liquid) crystal engineering,[17–19] to mention but a few. In short, dendritic networks provide the material scientist with a ready means of constructing molecular devices that are capable of information processing.[18, 20–27] Early on, the intermolecular self-assembly of two-directional arborols[28, 29] (dendrimers) led Lehn[20] to propose the first example of automorphogenesis, and as such demonstrated the supramolecular character of such systems by the molecular encapsulation properties essential to utilitarian purposes such as drug delivery.

Furthermore, and perhaps more importantly, once the molecular weight ceiling had been punctured by the advent of dendrimers, the construction of precise networks and assemblies no longer had a physical limit as molecular chemistry and material sciences approached each other in the nanotechnology regime. Current trends in chemistry and molecular design strongly suggest the potential for the creation of higher-order architectures. This is clearly evident upon examination of current literature; the ubiquitous reports of complex structures contained therein have been compiled in several treatises and reviews (see Appendices 10.1-5).

It should be noted here that characterization, separation, and purification techniques for the larger randomly assembled dendritic networks (discussed in earlier chapters) still continue to test the limits of contemporary instrumentation. Thus, unequivocal characterization of dendritic networks is currently limited and will necessitate the development of new analytical methods and instrumentation. However, dendritic network structural verification and elucidation should be facilitated by the integration of established standard materials science methods, e.g., MS and EM. Examples of precise connectivity of pre-characterized dendrimers or dendrons by means of connector analysis, e.g., a diamagnetic metal center (Figure 8.2),[30] offer a new, accurate insight into the relevant macroassembly. Such classical approaches will eventually be replaced as instrumental techniques amenable to this nanoscopic regime become available.

9.2   Network Formation and Classification

9.2.1   Ordered *versus* Random

Dendritic networks, as defined in Section 9.1, can be considered as falling into two main classes: ordered and random assemblies (Figure 9.1). These two classifications are essentially idealized extremes of a continuum to the dendritic (stepwise) *versus* hyperbranched (one-step) approach to macromolecular synthesis. Hence, many network structures will possess a higher or lower defined degree of "orderliness" (or increased

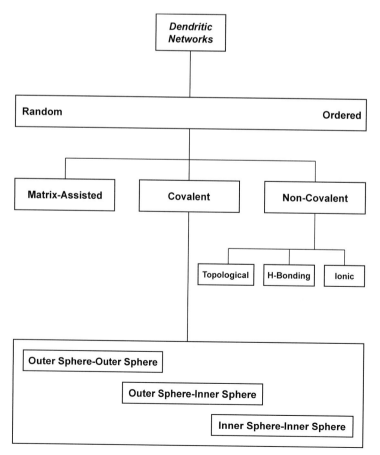

Figure 9.1 Dendritic network classification according to dendrimer connectivity.

"randomness") depending on the degrees of freedom inherent in the method of construction. A randomly prepared network thus lacks patterned connectivity due to the unrestricted manner in which the macromonomers (dendrimers) are positioned. Conversely, an ordered network results from specific restrictions (reduced degrees of freedom) that are fixed by the building block positioning process. It is, of course, realized that the monolayering of a sphere with a smaller spherical object has a precise, mathematically generated, saturation limit. Thus, the random synthetic approach has certain precise constraints that are characteristic of the natural limits of the surface-to-volume relationship and which are illustrated in D'Arcy W. Thompson's classic book[31] entitled "On Growth and Form".

9.2.1.1 Methods of Formation

All ordered and random dendritic networks that are constructed via covalent or non-covalent routes result from the positioning of one dendrimer (sphere) relative to another. Thus, macroassembly positioning can be effected by at least one of three different methods of connectivity. These methods are geometrically rooted in dendritic chemistry.

Since dendrimers are inherently (or can at least be envisaged as) globular or pseudospherical (particularly at higher generations) due to the branching patterns induced by the particular monomer(s) used for their construction, they may be conveniently considered, in essence, as spheres. Spherical geometry dictates that the two major regions where chemical and physical transformations can occur are the outer region (surface

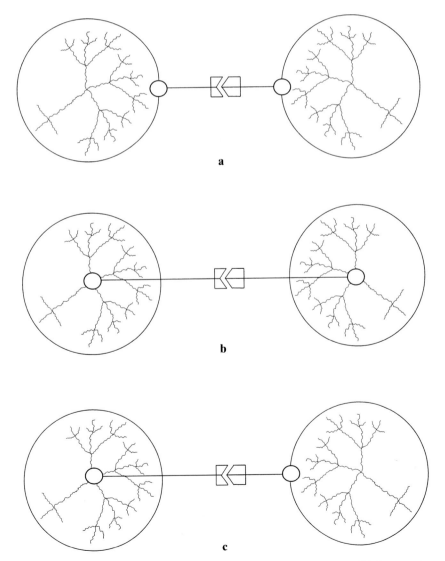

Figure 9.2 Idealized models of dendritic connectivity: (a) outer sphere–outer sphere; (b) inner sphere–inner sphere; and (c) inner sphere–outer sphere.

area) and the inner region (internal superstructure area). Hence, dendritic connection can be brought about through a combination of these regions. "Regional" combinations can thus be classified into three distinct types: (1) outer sphere–outer sphere (tecto-dendrimer); (2) outer sphere–inner sphere (lock and key approach); and (3) inner sphere–inner sphere. Each of these combinations can be employed either separately or in concert (Figure 9.2).

Outer sphere–outer sphere dendritic connectivity can be envisaged as the simple peripheral connection of dendrimers. Surface-to-surface connectivity can be facilitated by direct dendrimer–dendrimer attachment or by dendrimer–bridging unit–dendrimer attachment(s). However, the packing of *solid* (core and attachments) spheres of the same size would organize ideally as six similar spheres around the spherical core, thus only six connections are necessary to create a two-dimensional array, while in a three-dimensional network only twelve connecting spheres afford a structural packing limit. Of course, since these molecular spheres are not solid but merely sphere-like, additional contacts may be possible for identical sized but distorted dendrimers. Further, the number of spheres of uniform size packed around a different-sized core will be greater or smaller depending on the composition of each. There will, however, be a physical packing limit, which can be calculated and, as noted by Thompson[31] in 1917, can be related to Bonanni's approach, published in 1681, to hexagonal cells.

Inner sphere–inner sphere dendritic connection, without the use of a bridging unit, requires branched assemblies capable of interpenetration or molecular distortion, which further dictates the avoidance of critical dense packing limits. The use of an inner sphere-to-inner sphere connector, i.e., an appropriately functionalized bridging unit, obviates the requirement for interdigitation and can be envisaged as two locks and a two-pronged key.

Outer sphere–inner sphere dendritic connection utilizes a combination of dendritic connections, where one unit possesses one or more acceptor(s) and the other an appropriate donor functionality; this can best be envisaged as a lock (inside) and key (outside) approach.

9.2.1.2 Covalent and Non-Covalent Positioning

Covalent dendritic connection can result from any standard synthetic transformation capable of forming a covalent bond. These include nucleophilic, electrophilic, ionic, radical, and carbenoid reactions. Structural elements effecting bond formation include metals, non-metals, and metalloids.

Non-covalent means of dendritic connectivity can be further subdivided into three subcategories, namely topological-, hydrogen-, and ionic-bonding. Topological bonding can be envisaged as any sterically induced association of two units from two different macro-assemblies. Topological, or mechanical bonding[32, 33] can best be envisaged by considering interlocking rings (e.g., catenane- and rotaxane-type), the physical entrapment of units within a designed cavity, or positioning based on other steric factors. Hydrogen- and ionic-bonding methods employed for dendrimer connectivity are unambiguous, as will be demonstrated.

9.3   Random Connectivity

9.3.1   Random, Covalently-Linked Dendrimer Networks

Consider the divergent synthesis of PAMAM dendrimers,[34] where undesired intra- or intermolecular events lead to increased polydispersity and to a loss of ideality. Although these events can be minimized or, in selected cases, be eliminated, it is interesting to note that these processes can give rise to randomly positioned polydendritic systems. Thus, amidation of polyesters (e.g., **1**) with a diamine can lead to *intra*molecular bis-amidation, which can give rise to topological dendrimer connection (Scheme 9.1) as illustrated by

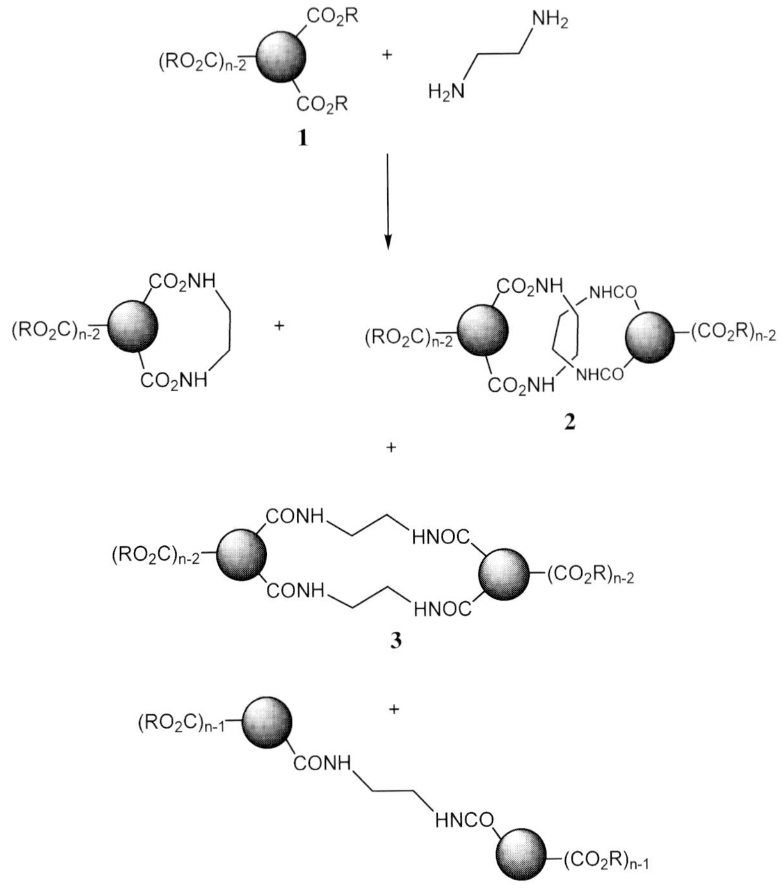

Scheme 9.1 Examples of possible bis(dendrimer) networks formed by the uncontrolled treatment of ester-terminated dendrimers with an alkyane diamine.

the catenated bis(dendrimer) **2**. On the other hand, *inter*molecular bis-amidation can lead to bridged dendrimer connection, as illustrated by bis(dendrimers) **3** and **4**. Statistically based, high-dilution reaction techniques can be employed to enhance the yields of these bridged dendrimers, although a distribution of products will still be generated. Optimal production and isolation of individual components is thus generally more difficult using a statistical preparative method rather than a more directed approach. Random connectivity[34] of amine-terminated PAMAM dendrimers through treatment with di- or trihalides later introduced the general concept of randomly constructed networks. No specific characterization of these bridged dendrimers has been reported,[34] although electron micrographs have been shown[35] to support the linkage of two dendrimers possessing dissimilar surfaces (i.e., surface amines and surface carboxylic acids). To date, evidence for the formation of a bis(dendrimer) by this method has been largely limited to mass spectrometry. The use of PAMAM as a core and as a shell has been reported and structurally evaluated by AFM; from these AFM data, molecular weights of the random assemblies have been estimated.[36]

9.3.2   Coupling through Surface-to-Surface Interaction

Treatment of a dendritic polyester **5** with a dendritic polyamine **6** is perhaps the simplest example of coupling between two dissimilar cascade macromolecules.[37–39] Scheme 9.2 illustrates some potential products (i.e. **7** and **8**) that may be obtained by this general procedure, as well as a continuation of surface amidation to afford polymeric species such as random network **9**. As envisaged, it would be difficult to stop this procedure after the formation of a single amide bond due to the close proximity of adjacent ester amine groups. Potentially, a limited number of juxtaposed bridging amide bonds might be possible at the initial junction locus, depending on the contact surface area defined by the

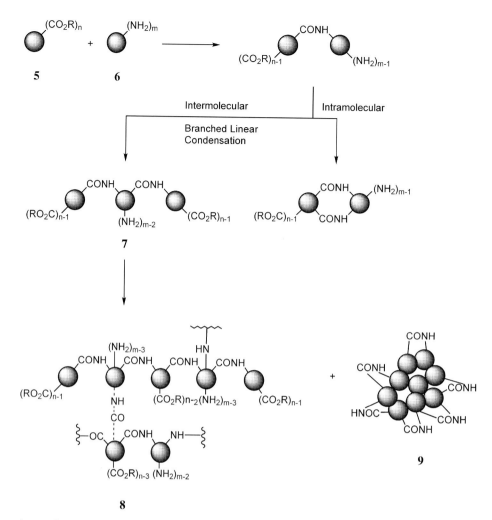

Scheme 9.2 Representations of potential randomly constructed networks obtained by the reaction of an ester-terminated dendrimer with an amine-terminated dendrimer.

interaction of the initial reagents. Other methods of direct surface-to-surface connectivity that have been reported[34, 38] include the reaction of olefin-terminated dendrimers with initiator-terminated dendrimers.

In more recent work, Reetz and Giebel[40] constructed a dendrimer network through the use of scandium as a surface connector. Treatment of amine-terminated dendrimer **10** (Scheme 9.3) with Tf$_2$O/Et$_3$N afforded the sulfonylation product **11**, which was cross-linked by reaction with Sc(OTf)$_3$ to yield the desired material **12**. This new network was shown to be effective in the catalysis of aldol, Diels–Alder, and Friedel–Crafts-type reactions.

9.3.3   Coupling *via* Surface-to-Surface Bridging Units

Connection of dendrimers by treatment with multifunctional cross-linking-type reagents, such as the addition of polyhalides to amine-terminated dendrimers,[37] results in randomly orientated dendritic assemblies. Dendrimers have also been bridged by introducing co-polymerizable units in reactions with dendrimers possessing polymerizable terminal olefins.[37, 38]

Hedrick and coworkers[41] used hyperbranched polyesters, prepared by ROP methods, as templating nanostructures for organosilicates, thus incorporating a degree of controlled porosity. In this way, a route to ultra-low dielectric materials for advanced micro-devices was devised. Similar mesoporous silicas have been obtained by the incorporation of 5th generation PPIs.[42]

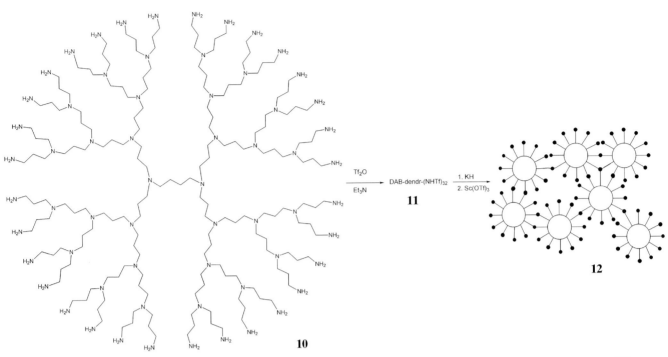

Scheme 9.3 Reetz's scandium cross-linked dendrimers find application as a new class of heterogeneous catalysts.[40]

9.3.4 Random, Non-Covalent

Adronov and Fréchet[43] described the use of mixed, random networks for the construction of a single-layer, light-emitting device (Scheme 9.4). Layer formation in the device was effected by the combined deposition of red and green "emitter" chromophores (i.e., pentathiophene (**13**) and coumarin-343 (**14**) dendrimers,[44, 45] respectively) on an ITO anionic surface. The source of excitation energy could be changed from applied light to applied voltage; critical to the process was terminal dendrimer modification with hole-transporting triarylamines. Light-harvesting self-assembled monolayers (SAMs) comprised of absorbing antennae and emitting components such as coumarin-2-modified dendron **15** and coumarin-343 (**16**) were also discussed.

Similar electroluminescent diodes based on electroactive organic semiconductor cores attached to phenylacetylene dendrons terminated with hole-transporting groups have been reported by Moore and coworkers.[46] The dendritic "layer" was postulated to improve film quality as a result of the globular architecture lowering intermolecular cohesive forces, thereby leading to a more stable amorphous phase.[47]

9.4 Ordered Dendritic Networks

9.4.1 Multilayer Construction

Watanabe and Regen[48] reported the construction of ordered, dendritic multilayers (**17**) by a bridged, outer sphere–outer sphere mode of assembly (Scheme 9.5), where the transition metal Pt was used as a connector moiety. Amine-terminated PAMAM-type dendrimers[34] were employed in this particular case, although the process could easily be extended to other types of macromolecules.

The ordering procedure does not rely on site-specific reactions, but rather on the packing efficiency of the dendritic species, or building blocks. This example possesses characteristics of both random and ordered networks and exemplifies the broad spectrum of the random/ordered network continuum.

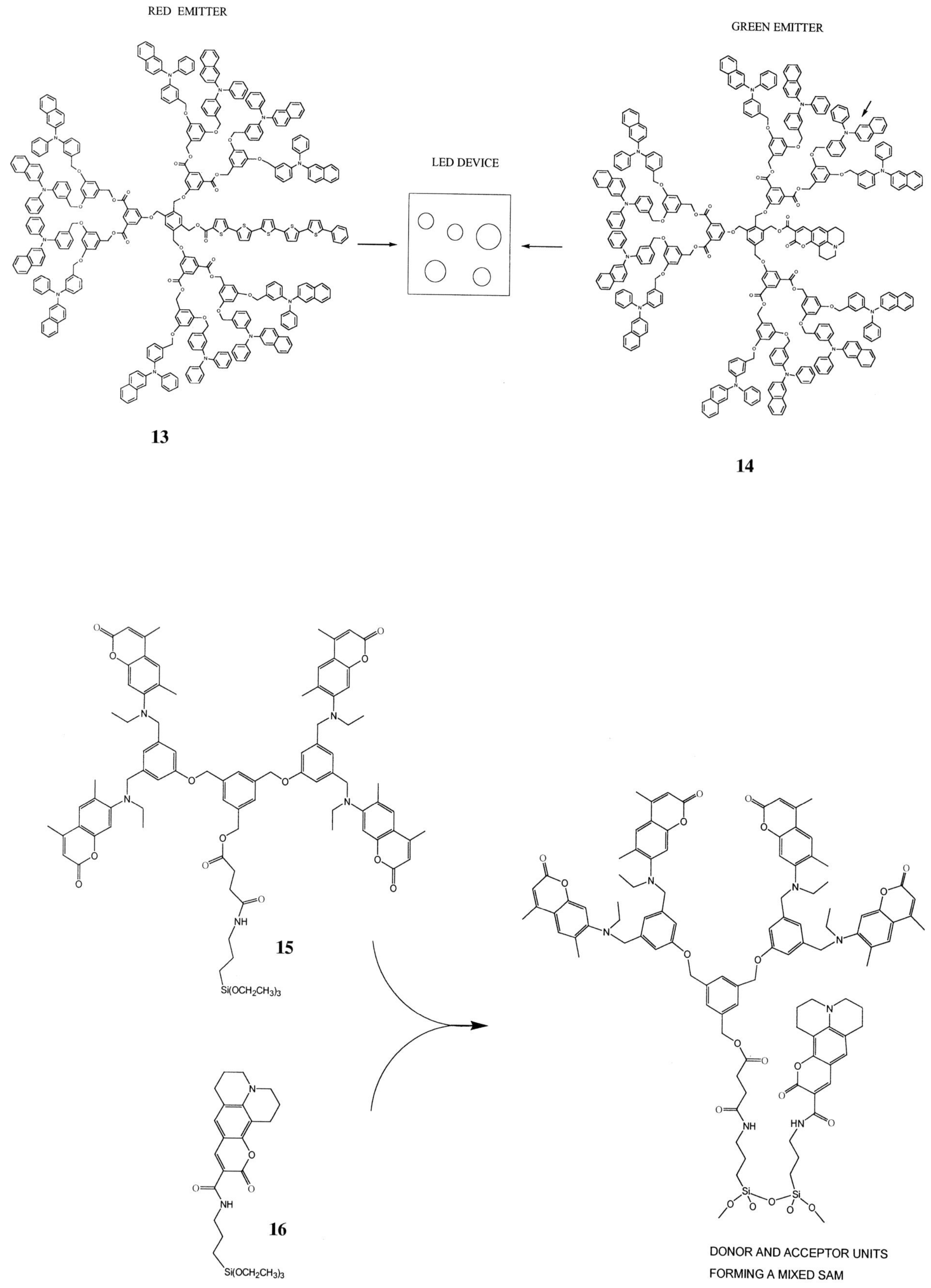

RED EMITTER

GREEN EMITTER

LED DEVICE

13

14

Scheme 9.4 Construction of an LED device and of light-harvesting SAMs through random network formation.[43]

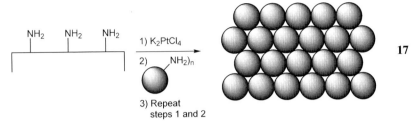

Scheme 9.5 An idealized 2D representation of an ordered dendritic network constructed by the procedure of Watanabe and Regen.[48]

9.4.2 Directed Network Construction

9.4.2.1 Covalent, Metal-Based Assembly

Ideally, from a macromolecular subunit position control perspective, directed approaches towards network construction would be desirable. Indeed, the construction of dendrimers possessing metal-ligating, 4-substituted 2,2':6',2''-terpyridine moieties has been reported.[49–58]

Ligand incorporation was achieved by facile alkyloxylation of 4-chloro-substituted terpyridine by a hydroxy-terminated carboxylic acid followed by divergent dendrimer construction employing amide-based connectivity (see Section 4.1.4) and a 1 → 3 branching multiplicity.[55] Formation of the bis(dendrimer) assembly **18** (Figure 9.3) was achieved by treatment of a larger (3rd tier), cascade-"compartmentalized" terpyridine ligand with a smaller, lower-generation Ru(III) dendritic complex. Although complex **18** is not a multidentate network, it represented the first example of an inner sphere–outer sphere, specifically positioned dendritic assembly.

Metal connectivity is advantageous. Transition metal characterization, e.g. by electrochemical or microscopic analysis, complements standard "organic"-type methods (e.g., ^{13}C and ^{1}H NMR) and thus allows for improved characterization of the large multidendritic structure(s). Cyclic voltammetry data strongly support the connection of branched structures.[55, 58, 59]

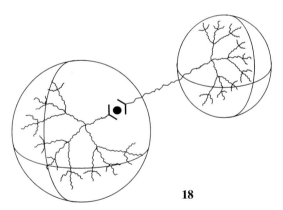

18

Figure 9.3 Bis(dendrimer) assemblies specifically connected through metal ligation.[55]

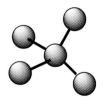

19

Figure 9.4 A representative "dendritic methane" obtained by tetrahedral ordering of four "terminal" dendrimers about a single "core" dendrimer using ligand–metal–ligand connectivity.[58]

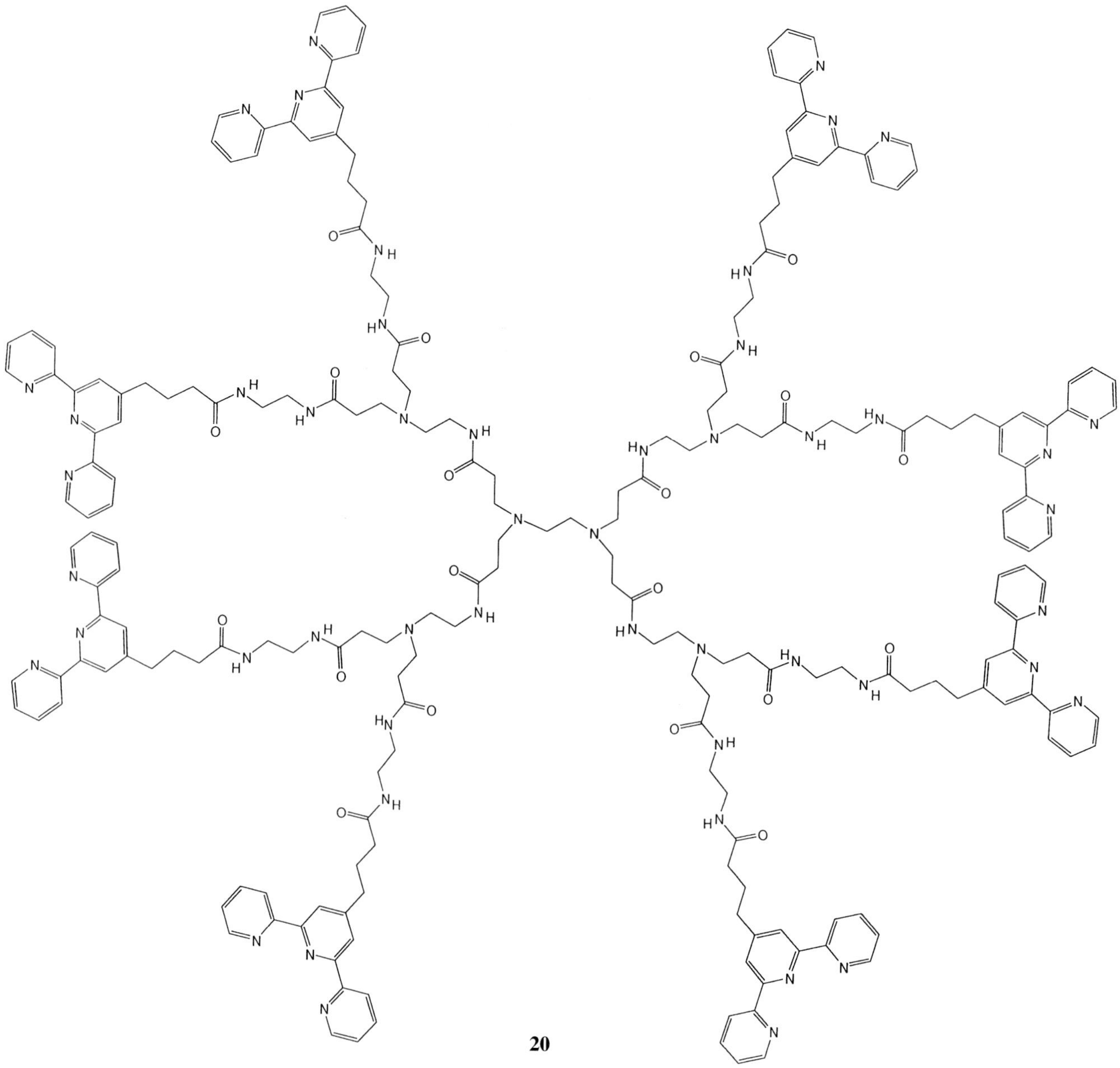

20

Figure 9.5 Abruña's terpyridine-terminated PAMAMs for ordered array formation.[60]

Metal connectivity is adaptable to higher-order and more complex architectures, such as the pseudotetrahedral pentadendrimer **19** depicted in Figure 9.4. Structure **19** is readily accessible through the application of current technology[58] using novel unsymmetrical dendritic building blocks possessing ligating sites and the capacity to continue the branching process.

Abruña and coworkers[60] crafted terpyridine-terminated low generation PAMAMs (Figure 9.5; **20**), which when reacted with Fe^{2+} or Co^{2+} afforded ordered arrays on highly oriented pyrolytic graphite surfaces. Scanning tunneling microscopy revealed ordered 2D hexagonal arrays composed of one-dimensional polymeric strands and repeat units of (terpy–dend–terpy–M)$_x$. These electrochemically active films were found to exhibit reversible waves at formal potentials corresponding to those of the [M(terpy)$_2$]$^{2+}$ complexes.

9.4.2.2 Hyperbranched-Type, Metal-Based Assembly

Extending the concept of metal-based, ordered dendritic network construction, hyperbranched networks can be envisaged. Employing known synthetic methods, complex networks should be accessible by virtue of the propensity of deliberately constructed bridges and dendrimers to self-assemble into intrinsically stable arrays based on isotopi-

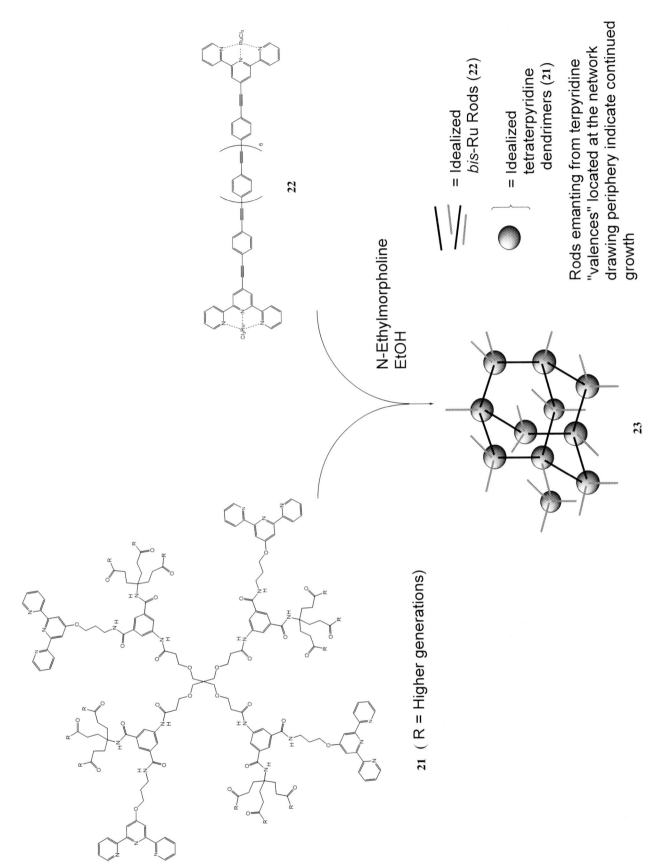

Scheme 9.6 Diamondoid architectures that may potentially be obtained by self-assembly of tetrahedral-based dendrimers.

cally extended connectivity in three dimensions.[61] Thus, treatment of dendrimers[62] possessing four tetrahedrally arranged terpyridine moieties (e.g., **21**) with rigid, bis(ruthenium) connector rods (**22**) of appropriate length should afford "diamondoid" architectures, such as **23** depicted in Scheme 9.6.

Reinhoudt and colleagues[63–69] have reported the preparation of metal-based dendritic assemblies employing a novel self-assembly process. Complementary building block positioning was predicated on labile acetonitrile substitution at tetravalent square-planar Pd(II) complexes with a kinetically inert arylcyanomethyl monomer moiety. Average aggregate diameters, as determined by transmission electron and atomic force microscopies, were found to be 205 nm with a narrow distribution. Energy dispersive X-ray spectroscopy (EDX) confirmed the presence of elemental S and Pd in the aggregates. Triblock and multiblock copolymers possessing polydisperse segments of well-defined architecture capable of forming chain-folded crystallites or metal complex based helices have been reported by Eisenbach et al.[70]

9.4.2.3   Hydrogen-Bonding Assembly

The incorporation of *H*-bonding moieties capable of self-assembly into or onto a dendritic superstructure can lead to ordered networks. One of the first dendrimers to be developed[71] for the purpose of exploring potential network formation[72] is depicted in Scheme 4.16. Essentially an aminopyridinetriester was reacted with a tetraacyl halide core to afford the 1st generation tetrapyridine, which was subsequently deprotected with HCO_2H and treated with an aminotriester to give the 2nd tier, 36-ester. Treatment of the tetrakis(diaminopyridine) polyester with dendrimers connected to either single imide moieties or α,ω-bis(imide) rods can be envisaged as leading to tetrahedral dendritic arrays and "adamantanoid"-type hyperbranched networks (i.e., **24**; Scheme 9.7).

As early as 1986, dumbbell-shaped dendrimers were reported to form linear networks.[28, 29] These surfactant-like dendrimers were found to possess external, branched, ball-shaped architecture connected on either side of a linear alkyl chain interior (Figure 9.6); these dendrimers were termed "arborols" and were discussed in detail in Chapter 4 (Section 4.1.1).

When arborol **25** is added in low concentrations to an aqueous medium, gelation occurs. The tendency to minimize lipophilic/hydrophilic interactions as well as to maximize *H*-bonding and packing effects is proposed as the reason for the formation of the ordered, linear networks (**26**), which can clearly be discerned by electron microscopy of the dried gels.

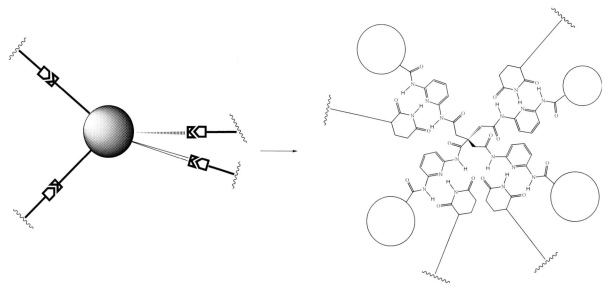

24

Scheme 9.7 Idealized depiction of a network assembled by *H*-bonding interactions.[72]

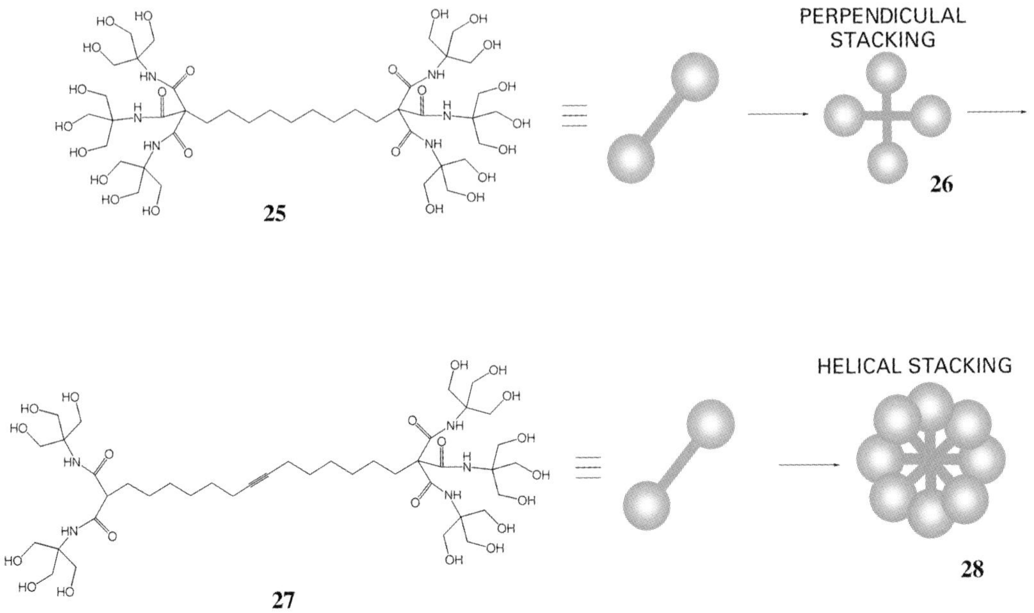

Figure 9.6 Network formation through the stacked aggregation of dumbbell-shaped dendritic molecules. View from the top of the aggregation along the stacking axis.

Incorporation of an alkyne moiety into the lipophilic core (e.g., **27**) imparts helical, scissor-like morphology to the stacked array (e.g., **28**). Disc-like, benzene-based arborols have also been reported[73] to form *H*-bonded aggregates or networks comprised of approximately 50–60 dendrimers. This estimate was based on network diameters determined by electron microscopy.

Zimmerman et al.[74] reported a pioneering effort based on the convergent preparation of dendritic wedges possessing tetraacid moieties (i.e., **29**) that self-assemble into a hexameric, disc-like network **30** (Scheme 9.8). The tetraacid unit **31** is known to form cyclic as well as linear structures in solution as a result of carboxylic acid dimerization. However, when the tetraacid unit is attached to large dendritic wedges, the hexamer form **32** is preferred. It is postulated that the cyclic form is favored due to the less sterically demanding environment than would be the case for linear aggregates. This argument is supported by SEC experiments. Thus, employing the 4th generation wedges, prepared by the method of Fréchet,[75] disc-shaped ordered networks 90 Å in diameter and 20 Å thick (M_w ca. 34,000 amu) are formed as a result of a dual, non-covalent mode of construction comprised of complementary topological and H-bonding aspects.

Zimmerman et al.[76, 77] investigated the self-assembly behavior of their focal tetracarboxylic acid-based dendrons (e.g., **31**) by means of SANS. Data collected for dendron generations 2 and 3 supported the hexameric self-assembly and revealed hexamer aggregation at higher concentrations. Tubular aggregation of the 1st generation dendron was observed. The corresponding tetramethyl ester dendrimer was used as a control, which, as expected, showed no self-assembly. Other notable pioneers in this area of self-assembly include Percec et al.[78–81] with their supramolecular column formation, and Palmans et al.[82] with their use of the "sergeants and soldiers" principle.

Stoddart et al.[83] reported the self-assembly of branched [*n*]rotaxanes in an investigation aimed at the preparation of larger dendritic rotaxanes, as well as the use of small dendrons in conjugation with their bipyridinium rotaxane chemistry for the formation of single-molecule-thick electrochemical junctions at the air–water interface.[84] Fréchet and coworkers also reported the preparation of supramolecular liquid-crystalline networks based on the self-assembly of carboxylic acid-based, trigonally branched, H-bonding donors and bipyridine-type H-bonding acceptors.

Robson and colleagues[85] reported the creation of a hexaimidazole ligand, coordination of which to Cd led to the formation of an infinite α-Po-like network.

Apperloo et al.[86] reported the supramolecular aggregation and concentration-dependent thermochromism of triblock copolymers constructed from oligothiophene

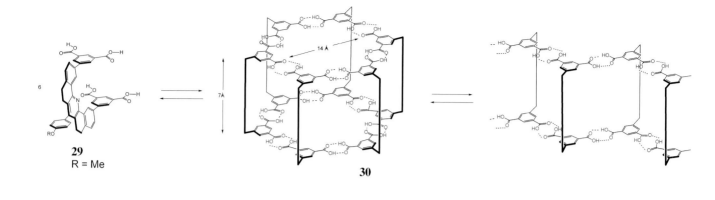

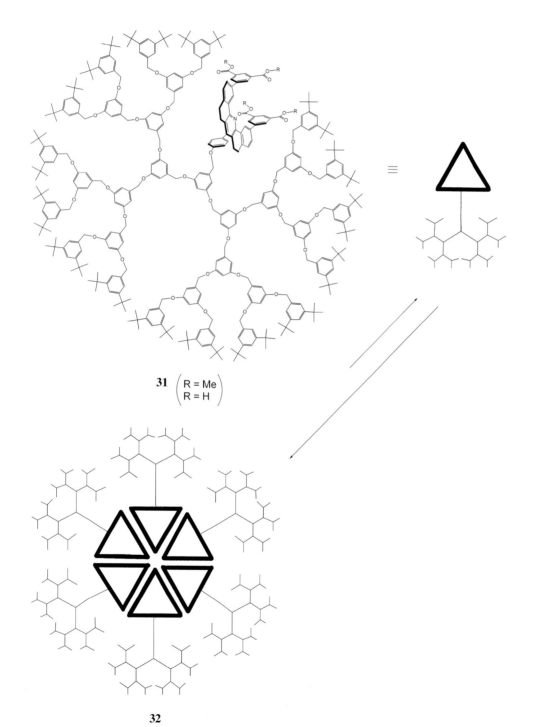

Scheme 9.8 Zimmerman et al.'s dendritic ordering based on well-known carboxylic acid dimer formation.[74]

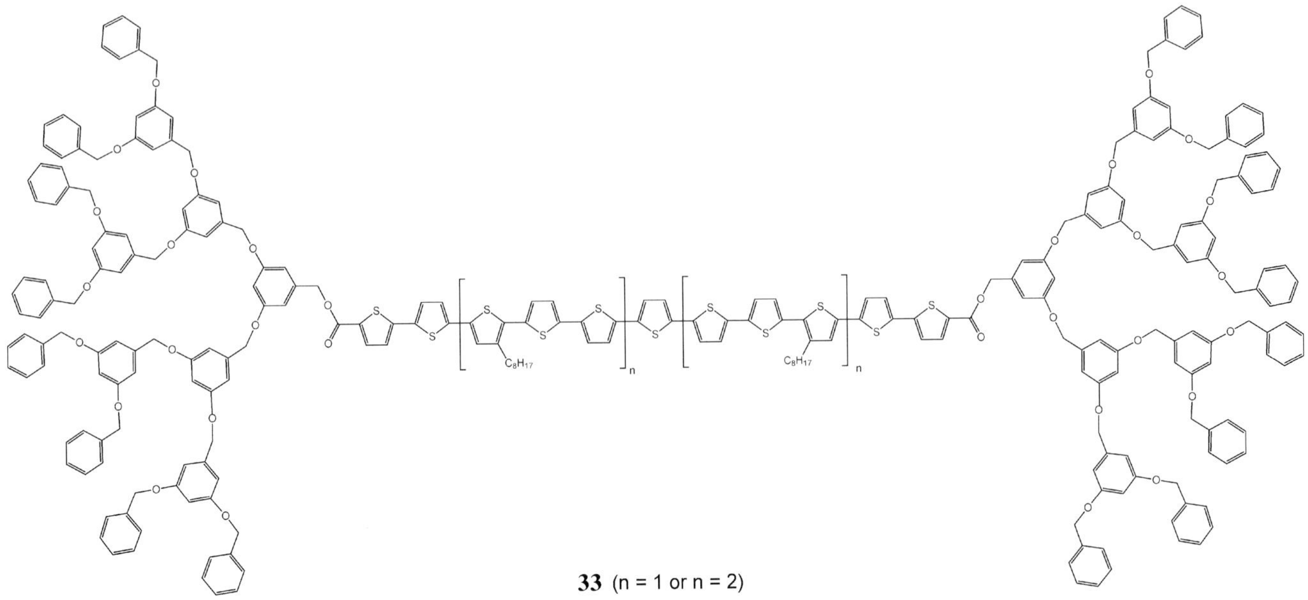

33 (n = 1 or n = 2)

Figure 9.7 Triblock copolymers that exhibit supramolecular aggregation in solution.[86]

cores and poly(benzyl ether) dendrons (Figure 9.7; **33**). Aggregations of four to seven "dumbbell-shaped" dendrimers were observed; the ordering process was attributed to strong π-π interactions. As well, Mong et al.[86a] reported the nanoscale aggregation of β-alanine-based dendrimers and Stupp and coworkers[86b] demonstrated "dendron rod coil molecules" self-assemble into nano ribbons.

Kovvali, Chen, and Sirkar[87] immobilized zeroth generation PAMAMs in the pores of porous hydrophilized polyvinylidene fluoride (PVDF) flat films for the creation of CO_2-selective membranes. The results of CO_2-transport experiments suggested that the higher generation dendrimer membranes were of interest.

Elegant work by Percec et al.[87a, b] utilizing two constitutional libraries of AB_2-type monodendrons and supramolecular dendrimers has delinated predicable self-assembly for poly(aryl ether)constructs.

9.4.2.4 Covalent Assembly

Fréchet and coworkers[88] significantly contributed to the area of dendritic chemistry (see Chapter 5) with the introduction of the convergent method, where dendritic wedges are connected to a core producing a final dendrimer. This method is tantamount to covalent dendritic positioning. Architectures constructed through combining this method with traditional chemistry include the hybrid linear-dendritic block copolymer **34**[89] (Figure 9.8), prepared by free radical copolymerization of styrene-functionalized dendritic wedges with styrene, and the bis(dendrimer) **35**,[90, 91] synthesized by the reaction of brominated dendritic wedges with polyethylene glycol.[92, 93]

Covalent dendritic ordering can thus be realized by using secondary (embedded or latent) protection–deprotection schemes in concert with those already developed for dendrimer construction. For example, consider the preparation of the tetrahedrally-based dendrimer **36** possessing four internal attachment sites, three of which are protected while the fourth is available for connection; additional peripheral and internal functionalities are inert to the chosen attachment and deprotection conditions. Furthermore, consider the connection of two of these dendrimers to afford the bis(dendrimer) **37**. Deprotection of the internal moieties (without nitro reduction) allows further dendrimer attachment, and so on (Scheme 9.9).

Kim, Park, and Jung[94] prepared cylindrical dendrimers using a linear poly(carbosilane) foundation for the divergent growth of regularly spaced silane dendrimers. The 3rd generation dendrimers were propagated from the polymeric backbone using traditional allylation–hydrosilylation repetitive reactions.

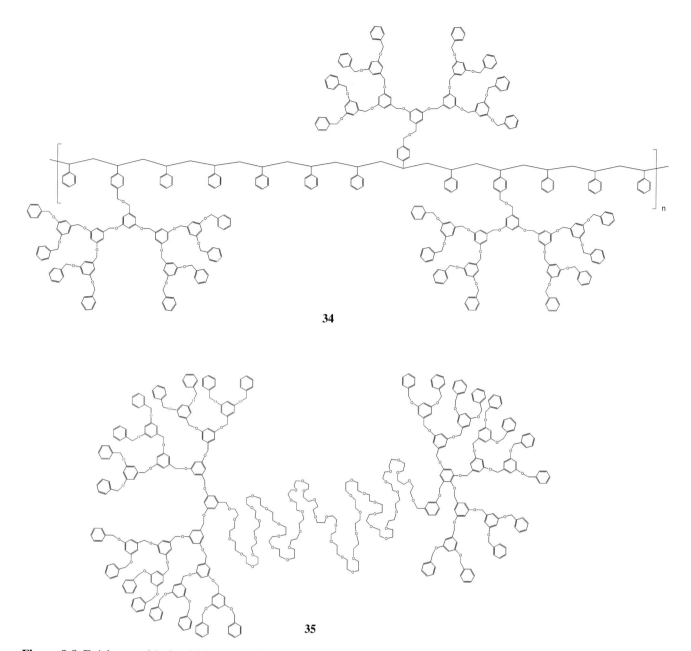

34

35

Figure 9.8 Fréchet et al.'s dendritic assemblies based on a polystyrene backbone and on polyethylene glycol chains.[90, 91]

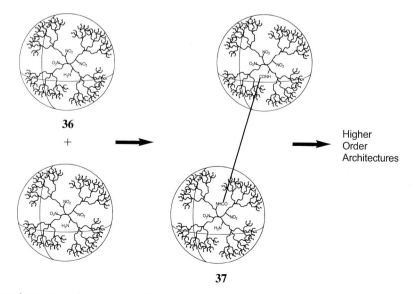

36

+

37

Higher
Order
Architectures

Scheme 9.9 Use of iterative protection–deprotection technology for the construction of network architectures.

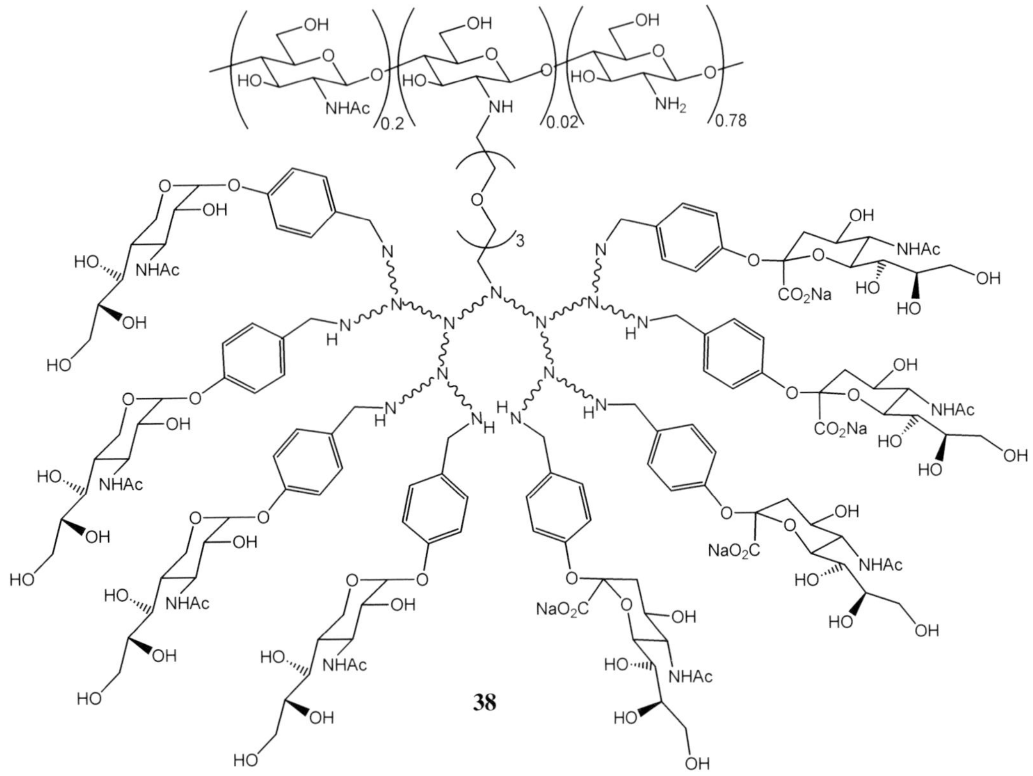

Figure 9.9 A chitosan–sialic acid dendrimer hybrid.[95]

Roy et al.[95] employed a PAMAM scaffolding for the construction of a chitosan–sialic acid dendrimer hybrid (Figure 9.9; **38**). Dendrons were first prepared then attached to the polysaccharide chitosan, which is primarily composed of β-(1→4)-2-amino-2-deoxy-D-glucopyranose repeat moieties. These materials were prepared with a view to examining their potential in inhibiting the hemagglutination of human erythrocytes by influenza virus hemaaglutinin. The grafting of poly(amidoamine)s onto silica[96] and poly(amide)s[97] onto agarose beads has also been reported.

Schlüter and colleagues[8, 10] described the construction of poly(*para*-phenylene)s (PPPs) having 4th generation Fréchet-type dendrons attached to alternate repeat groups (Scheme 9.10). Three different routes to similar dendritic macromonomers, such as **39**, were examined. Dibromide **39** was chosen for polymerization with a bis(aryl borate) due to its greater steric demand and potential to afford more densely packed PPPs. Palladium-mediated Suzuki polycondensation with **40** gave polymer **41** possessing molecular weights of $M_n = 76,000$ and $M_w = 639,000$ amu. A notable feature of these materials is the close proximity of the positioned dendrons on the polymer backbone, which has the effect of stretching the polymer chain to a more linear form, thereby creating a cylindrical motif. An excellent review of this fascinating area of dendrimer chemistry has appeared.[12]

Similar rod-shaped, "worm-like" carbosilanes,[98] "silane arborols",[94, 94a] and siloxane-based[99] polymers have been reported. PAMAM-based rods,[9] dimers, poly(aspartic acid)s,[7] and other "comb-like" architectures[100–103] have also been described.

Li et al.[36, 36a] reported the creation of core–shell tecto(dendrimer)s, where an amine-terminated dendrimer (the core) was reacted with ester-terminated dendrimers (the shell, usually of lower generation than the core) to give the higher-order architecture (Figure 9.10; **42**). Unreacted ester groups were then treated with either ethanolamine or tris(2-hydroxymethyl)aminomethane. Abbreviated designations for these materials were given, for example, as G7G5, where the numbers from left to right indicate the core and shell generations (G), respectively. Notable characteristics include "saturated" or "unsaturated" shells leading to regular as well as irregular motifs. Analysis by means of MALDI-TOF MS and tapping mode ATM was reported. Two general synthetic approaches for accessing these tecto(dendrimer)s were examined: (1) synthesis by direct regioselective bond formation, and (2) synthesis by preorganization followed by *in situ*

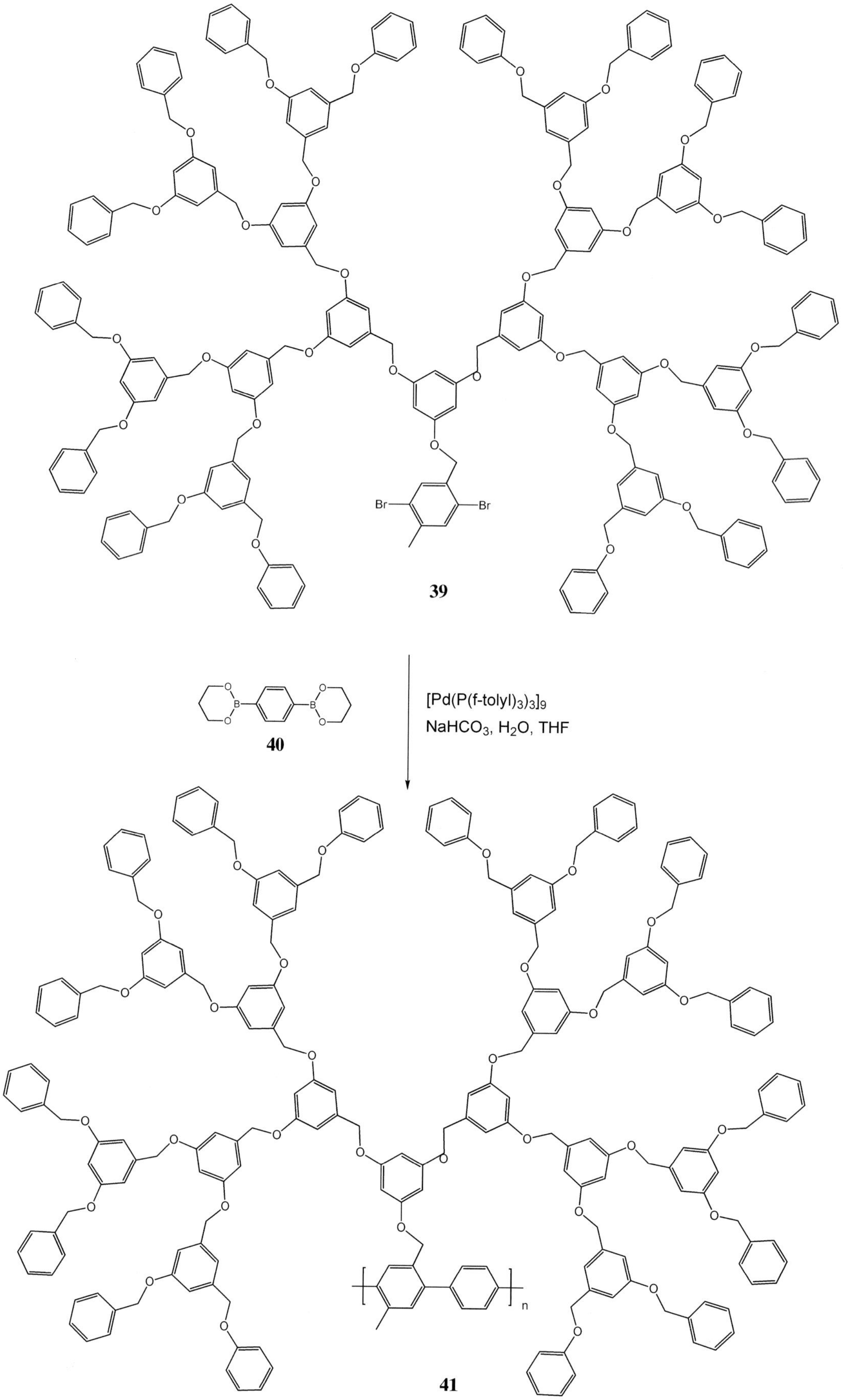

Scheme 9.10 *Suzuki cross-coupling leading to the creation of dendrimerized polymers.*[12]

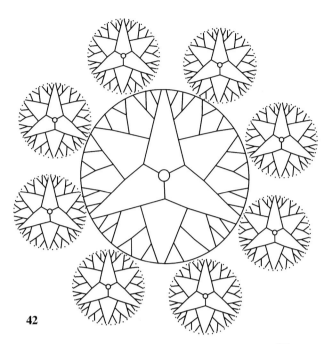

42

Figure 9.10 An idealized representation of a core–shell tecto-dendrimer.[36]

regioselective bond formation. Notably, examples of tris-based architectures had been reported previously;[73] such *H*-bonded assemblies have been characterized and shown to possess unique supramolecular properties, e.g. as unimolecular micelles.[104]

9.4.2.5 Charge-Induced Self-Assembly

Stoddart and coworkers[105] employed the supramolecular self-assembly of bipyridinium dications with crown ethers in order to form 1 → 2 branched dendritic wedges. Thus, reaction of 4,4'-bipyridine with the corresponding bis(chloromethyl) benzocrown ether followed by methylation of the remaining pyridine moieties afforded the desired monomer, which was in equilibrium with its self-associated form. Intermolecular association produced the novel architecture. Notably, the authors also found that "the principle of maximum site occupancy, along with unfavorable entropy factors, conspire in this case against extended supramolecular array formation." Sulfonate-ammonium ionic attraction has been used for dendrimer assembly also[105a] along with dendrimer-surfactant interactions.[105b, c]

The application of this concept in the preparation of networks leads to unlimited architectures. It should facilitate and expand ordered (as well as random) network construction in the future.

9.5 An Observation:

At the end of Chapter 2 of his book entitled *On Growth and Form*,[31] Thompson noted in 1917:[106]

"We draw near the end of this discussion [on magnitude, size, and dimension]. We found, to begin with, that "scale" had a marked effect on physical phenomena, and that increase or diminution of magnitude might mean a complete change of statical or dynamical equilibrium. In the end we begin to see that there are discontinuities in the scale, defining phases in which different forces predominate and different conditions prevail. Life has a range of magnitude narrow indeed compared to that with which physical science deals; but it is wide enough to include three such discrepant conditions as those in which a man, an insect and a bacillus have their being and play their several roles. Man is ruled by gravitation, and rests on mother earth. A water-beetle finds the surface of a pool a matter of life and death, a perilous entanglement or an indispensable support. In the third world, where the bacillus lives, gravitation is forgotten, and the viscosity of the liquid, the resistance defined by Stokes's law, the molecular shocks of the Brownian movement, doubtless also the electric charges of the ionized medium, make up the physical environment and have their potent and immediate influence on the organism. The predominant factors are no longer those of our scale; we have come to the edge of a world of which we have no experience, and where all our preconceptions must be recast."

This conclusion is an interesting overview considering the time at which the original text was written (1st edition, 1917), i.e. prior to modern day instrumentation, prior to penicillin, a polio vaccine, etc. It is interesting to apply a rough scale to Thompson's observations (as he did in the chapter but not in his conclusion). In the three worlds that he envisaged according to his subdivisions, man to trees would fall into the ca. 10^1 to 10^3 cm range, insects to small animals would be in the 10^{-3} to 10^1 range, and the minute bacteria would be in the 10^{-7} to 10^{-4} range. At the end of a chapter on networks, it would seem appropriate to modernize his overview by adding a fourth world, namely the range of the molecular world (10^{-10}–10^{-9} cm) to the limit of microscopic vision $10^{-7(8)}$, in which physical changes will be as great as or even greater than those noted above, and where a vacuum or a space devoid of things plays an ever increasing role. With very limited analytical instrumentation to accurately probe this new nano-regime, assumptions based on larger forms will probably have to be rethought. In view of the very limited number of precisely constructed and accurately analyzed man-made nanomolecules hitherto reported, this new frontier definitely needs to be added to Thompson's list as the fourth world. Interestingly, when one visits the Jefferson Laboratories, where scientists are studying subatomic particles, it becomes obvious that the fifth world is comprised of near nothingness!

9.6 References

[1] M. Adam, J. Bastide, S. Candau, A. Coniglio, M. Delsanti, J. E. Mark, D. Stauffer, A. J. Staverman, "in *Polymer Networks* (Ed.: K. Dusek), Springer, Berlin, **1982**.

[2] S. M. Aharoni, S. F. Edwards, *Rigid Polymer Networks*, Springer, Berlin, **1994**, p. 239.

[3] P. J. Flory, *Principles of Polymer Chemistry*, Cornell University Press, Ithaca, New York, **1953**.

[4] N. K. Müller-Nedebock, S. M. Aharoni, S. F. Edwards, "Networks of Rigid Molecules", *Macromol. Symp.* **1995**, *98*, 701–717.

[5] G. Odian, *Principles of Polymerization*, 3rd ed., Wiley, New York, **1991**.

[6] A.-D. Schlüter, W. Claussen, R. Freudenberger, "Cylindrically-Shaped Dendritic Structures", *Macromol. Symp.* **1995**, *98*, 475–482.

[7] M. Niggemann, H. Ritter, "Polymerizable Dendrimers; Part 2. Mono-methacryl Modified Dendrimers Containing up to 16 Ester Functions via Stepwise Condensations of L-Aspartic Acids", *Acta Polym.* **1996**, *47*, 351–356.

[8] I. Neubert, R. Klopsch, W. Claussen, A.-D. Schlüter, "Polymerization of Styrenes and Acrylates Carrying Dendrons of the First and Second Generation", *Acta Polym.* **1996**, *47*, 455–459.

[9] R. Yin, Y. Zhu, D. A. Tomalia, H. Ibuki, "Architectural Copolymers: Rod-Shaped, Cylindrical Dendrimers", *J. Am. Chem. Soc.* **1998**, *120*, 2678–2679.

[10] Z. Bo, A. D. Schlüter, "Entering a New Level of Use for Suzuki Cross-Coupling: Poly(*para*-phenylene)s with Fourth-Generation Dendrons", *Chem. Eur. J.* **2000**, *6*, 3235–3241.

[11] A.-D. Schlüter, "Dendrimers with Polymeric Core: Towards Nanocylinders", in *Topics in Current Chemistry* (Ed.: F. Vögtle), Springer, Berlin, **1998**, pp. 165–191.

[12] A. D. Schlüter, J. P. Rabe, "Dendronized Polymers: Synthesis, Characterization, Assembly at Interfaces, and Manipulation", *Angew. Chem.* **2000**, *112*, 860–880; *Int. Ed.* **2000**, *39*, 864–883.

[13] *Molecular Electronic Devices* (Ed.: F. L. Carter), Marcel Dekker, New York, **1982**.

[14] *Molecular Electronic Devices II* (Ed.: F. L. Carter), Marcel Dekker, New York, **1987**.

[15] *Molecular Electronics: Biosensors and Biocomputers* (Ed.: F. T. Hong), Plenum, New York, **1989**.

[16] J. G. Tirrell, M. J. Fournier, T. L. Mason, D. A. Tirrell, "Biomolecular Materials", *Chem. Eng. News* **1994**, 40–51.

[17] G. R. Desiraju, "Supramolecular Synthons in Crystal Engineering – A New Organic Synthesis", *Angew. Chem.* **1995**, *107*, 2541–2558; *Int. Ed. Engl.* **1995**, *34*, 2311–2327.

[18] J.-M. Lehn, "Supramolecular Chemistry – Scope and Perspectives. Molecules, Supermolecules, and Molecular Devices (Nobel lecture)", *Angew. Chem.* **1988**, *100*, 91–116; *Int. Ed. Engl.* **1988**, *27*, 89–112.

[19] J. Zhang, J. S. Moore, "Liquid Crystals Based on Shape-Persistent Macrocyclic Mesogens", *J. Am. Chem. Soc.* **1994**, *116*, 2655–2656.

[20] J.-M. Lehn, "Perspectives in Supramolecular Chemistry – From Molecular Recognition Toward Molecular Information Processing and Self-Organization", *Angew. Chem.* **1990**, *102*, 1347–1362; *Int. Ed. Engl.* **1990**, *29*, 1304–1319.

[21] J. S. Lindsey, "Self-Assembly in Synthetic Routes to Molecular Devices. Biological Principles and Chemical Perspectives: A Review", *New J. Chem.* **1991**, *15*, 153–180.

[22] D. Philp, J. F. Stoddart, "Self-Assembly in Organic Synthesis", *Synlett* **1991**, 445–458.

[23] R. J. Pieters, I. Huc, J. Rebek, Jr., "Passive Template Effects and Active Acid-Base Involvement in Catalysis of Organic Reactions", *Chem. Eur. J.* **1995**, *1*, 183–192.

[24] H. Ringsdorf, B. Schlarb, J. Venzmer, "Molecular Architecture and Function of Polymeric Oriented Systems: Models for the Study of Organization, Surface Recognition, and Dynamics of Biomembranes", *Angew. Chem.* **1988**, *100*, 117–162; *Int. Ed. Engl.* **1988**, *27*, 113–158.

[25] H.-J. Schneider, "Mechanisms of Molecular Recognition: Investigations of Organic Host–Guest Complexes", *Angew. Chem.* **1991**, *103*, 1419–1439; *Int. Ed. Engl.* **1991**, *30*, 1417–1436.

[26] L. Yu, M. Chen, L. R. Dalton, "Ladder Polymers: Recent Developments in Syntheses, Characterization, and Potential Applications as Electronic and Optical Materials", *Chem. Mater.* **1990**, *2*, 649–659.

[27] M. D. Ward, "Metal-Metal Interactions in Binuclear Complexes Exhibiting Mixed Valency: Molecular Wires and Switches", *Chem. Soc. Rev.* **1995**, *24*, 121–134.

[28] G. R. Newkome, G. R. Baker, S. Arai, M. J. Saunders, P. S. Russo, K. J. Theriot, C. N. Moorefield, L. E. Rogers, J. E. Miller, T. R. Lieux, M. E. Murray, B. Phillips, L. Pascal, "Synthesis and Characterization of Two-Directional Cascade Molecules and Formation of Aqueous Gels", *J. Am. Chem. Soc.* **1990**, *112*, 8458–8465.

[29] G. R. Newkome, G. R. Baker, M. J. Saunders, P. S. Russo, V. K. Gupta, Z. Yao, J. E. Miller, K. Bouillion, "Two-Directional Cascade Molecules: Synthesis and Characterization of [9]-*n*-[9] Arborols", *J. Chem. Soc., Chem. Commun.* **1986**, 752–753.

[30] G. R. Newkome, F. Cardullo, E. C. Constable, C. N. Moorefield, A. M. W. C. Thompson, "Metallomicellanols: Incorporation of Ruthenium(II)-2,2':6',2''-Terpyridine Triads into Cascade Polymers", *J. Chem. Soc., Chem. Commun.* **1993**, 925–927.

[31] D. W. Thompson, *On Growth and Form*, 1st ed., Cambridge University Press, Cambridge, **1917**.

[32] H. L. Frisch, E. Wasserman, "Organic and Biological Chemistry", *J. Am. Chem. Soc.* **1961**, *83*, 3789–3794.

[33] E. Wasserman, "The Preparation of Interlocking Rings: A Catenane", *J. Am. Chem. Soc.* **1960**, *82*, 4433–4434.

[34] D. A. Tomalia, H. Baker, J. Dewald, M. Hall, G. Kallos, S. Martin, J. Roeck, J. Ryder, P. Smith, "A New Class of Polymers: Starburst-Dendritic Macromolecules", *Polym. J. (Tokyo)* **1985**, *17*, 117–132.

[35] D. A. Tomalia, Presentation at the NATO Workshop, Estes Park, Colorado, USA, 1995.

[36] J. Li, D. R. Swanson, D. Qin, H. M. Brothers, L. T. Piehler, D. Tomalia, D. J. Meier, "Characterizations of Core–Shell Tecto-(Dendrimer) Molecules by Tapping Mode Atomic Force Microscopy", *Langmuir* **1999**, *15*, 7347–7350.

[36a] D. A. Tomalia, S. Uppuluri, D. R. Swanson, J. Li, "Dendrimers as reactive modules for the synthesis of new structure-controlled, higher-complexity megamers", *Pure Appl. Chem.* **2000**, *72*, 2342–2358.

[37] D. A. Tomalia, J. R. Dewald, "Dense Star Polymer and Dendrimers", **1986**, *U. S. Pat.*, 4, 568, 737.

[38] D. A. Tomalia, L. R. Wilson, "Dense Star Polymers for Calibrating/Characterizing Sub-Micron Apertures", *U. S. Pat.* **1987**, *4, 713, 975*.

[39] D. A. Tomalia, "Bridged Dense Star Polymers", **1988**, *U. S. Pat.*, 4, 737, 550.

[40] M. T. Reetz, D. Giebel, "Cross-Linked Scandium-Containing Dendrimers: A New Class of Heterogeneous Catalysts", *Angew. Chem.* **2000**, *112*, 2614–2617; *Int. Ed.* **2000**, *39*, 2498–2501.

[41] C. Nguyen, C. J. Hawker, R. D. Miller, E. Huang, J. L. Hedrick, R. Gauderon, J. G. Hilborn, "Hyperbranched Polyesters as Nanoporosity Templating Agents for Organosilicates", *Macromolecules* **2000**, *33*, 4281–4284.

[42] G. Larsen, E. Lotero, M. Marquez, "Use of Polypropyleneimine Tetrahexacontaamine (DAB-Am-64) Dendrimer as a Single-Molecule Template to Produce Mesoporous Silicas", *Chem. Mater.* **2000**, *12*, 1513–1515.

[43]   A. Adronov, J. M. J. Fréchet, "Light-Harvesting Dendrimers", *Chem. Commun.* **2000**, 1701–1710.

[44]   A. Adronov, S. L. Gilat, J. M. J. Fréchet, K. Ohta, F. V. R. Neuwahl, G. R. Fleming, "Light Harvesting and Energy Transfer in Laser-Dye-Labeled Poly(aryl ether) Dendrimers", *J. Am. Chem. Soc.* **2000**, *122*, 1175–1185.

[45]   A. Adronov, P. R. L. Malenfant, J. M. J. Fréchet, "Synthesis and Steady-State Photophysical Properties of Dye-Labeled Dendrimers Having Novel Oligothiophene Core: A Comparative Study", *Chem. Mater.* **2000**, *12*, 1463–1472.

[46]   P.-W. Wang, Y. J. Liu, C. Devadoss, P. Bharathi, J. S. Moore, "Electroluminescent Diodes from a Single-Component Emitting Layer of Dendritic Macromolecules", *Adv. Mater. (Weinheim, Fed. Repub. Ger.)* **1996**, *8(3)*, 237–241.

[47]   K. Naito, A. Miura, "Molecular Design for Nonpolymeric Organic-Dye Glasses with Thermal Stability: Relations Between Thermodynamic Properties and Amorphous Properties" *J. Phys. Chem.* **1993**, *97*, 6240–6248.

[48]   S. Watanabe, S. L. Regen, "Dendrimers as Building Blocks for Multilayer Construction", *J. Am. Chem. Soc.* **1994**, *116*, 8855–8856.

[49]   G. R. Newkome, C. N. Moorefield, "Metallospheres and Superclusters", **1994**, *U. S. Pat.*, 5, 376, 690.

[50]   G. R. Newkome, C. N. Moorefield, "Metallospheres and Superclusters", **1995**, *U. S. Pat.*, 5, 422, 379.

[51]   G. R. Newkome, C. N. Moorefield, "Metallospheres and Superclusters", **1996**, *U. S. Pat.*, 5, 516, 810.

[52]   G. R. Newkome, C. N. Moorefield, "Metallospheres and Superclusters", **1996**, *U. S. Pat.*, 5, 585, 457.

[53]   G. R. Newkome, C. N. Moorefield, G. R. Baker, "Lock and Key Micelles", **1997**, *U. S. Pat.*, 5, 650, 101.

[54]   G. R. Newkome, C. N. Moorefield, G. R. Baker, "Lock and Key Micelles and Monomer Building Blocks Thereof", **1999**, *U. S. Pat.*, 5, 863, 919.

[55]   G. R. Newkome, R. Güther, C. N. Moorefield, F. Cardullo, L. Echegoyen, E. Pérez-Cordero, H. Luftmann, "Routes to Dendritic Networks: Bis-Dendrimers by Coupling of Cascade Macromolecules through Metal Centers", *Angew. Chem.* **1995**, *107*, 2159–2162; *Int. Ed. Engl.* **1995**, *34*, 2023–2026.

[56]   G. R. Newkome, E. He, "Nanometric Dendritic Macromolecules: Stepwise Assembly by Double (2,2':6',2"-terpyridine)ruthenium(II) Connectivity", *J. Mater. Chem.* **1997**, *7*, 1237–1244.

[57]   G. R. Newkome, E. He, L. A. Godínez, G. R. Baker, "Neutral Highly Branched Metallo-macromolecules: Incorporation of (2,2':6',2"-Terpyridine)ruthenium(II) Complex without External Counterions", *Chem. Commun.* **1999**, 27–28.

[58]   G. R. Newkome, E. He, L. A. Godínez, G. R. Baker, "Electroactive Metallomacromolecules via Tetrakis(2,2':6',2"-terpyridine)ruthenium(II) Complexes: Dendritic Networks toward Constitutional Isomers and Neutral Species without External Counterions", *J. Am. Chem. Soc.* **2000**, *122*, 9993–10006.

[59]   G. R. Newkome, E. He, C. N. Moorefield, "Suprasupermolecules with Novel Properties: Metallodendrimers", *Chem. Rev.* **1999**, *99*, 1689–1746.

[60]   D. J. Díaz, G. D. Storrier, S. Bernhard, K. Takada, H. D. Abruña, "Ordered Arrays Generated via Metal-Initiated Self-Assembly of Terpyridine-Containing Dendrimers and Bridging Ligands", *Langmuir* **1999**, *15*, 7351–7354.

[61]   M. J. Zaworotko, "Crystal Engineering of Diamond Networks", *Chem. Soc. Rev.* **1994**, *23*, 283–288.

[62]   C. N. Moorefield, unpublished results, 1996.

[63]   W. T. S. Huck, F. C. J. M. van Veggel, B. L. Kropman, D. H. A. Blank, E. G. Keim, M. M. A. Smithers, D. N. Reinhoudt, "Large Self-Assembled Organopalladium Spheres", *J. Am. Chem. Soc.* **1995**, *117*, 8293–8294.

[64]   W. T. S. Huck, F. C. J. M. van Veggel, D. N. Reinhoudt, "Controlled Assembly of Nanosized Metallodendrimers", *Angew. Chem.* **1996**, *108*, 1304–1306; *Int. Ed. Engl.* **1996**, *35*, 1213–1215.

[65]   W. T. S. Huck, R. Hulst, P. Timmerman, F. C. J. M. van Veggel, D. N. Reinhoudt, "Noncovalent Synthesis of Nanostructures: Combining Coordination Chemistry and Hydrogen Bonding", *Angew. Chem.* **1997**, *109*, 1046–1049; *Int. Ed. Engl.* **1997**, *36*, 1006–1008.

[66]   W. T. S. Huck, B. Snellink-Ruel, F. C. J. M. van Veggel, D. N. Reinhoudt, "New Building Blocks for the Noncovalent Assembly of Homo- and Hetero-multinuclear Metallodendrimers", *Organometallics* **1997**, *16*, 4287–4291.

[67]   W. T. S. Huck, B. H. Snellink-Ruël, J. W. Th. Lichtenbelt, F. C. J. M. van Veggel, D. N. Reinhoudt, "Self-Assembly of Hyperbranched Spheres; Correlation Between Monomeric Synthon and Sphere Size", *Chem. Commun.* **1997**, 9–10.

[68] W. T. S. Huck, F. C. J. M. van Veggel, D. N. Reinhoudt, "Self-Assembly of Hyperbranched Spheres", *J. Mater. Chem.* **1997**, *7*, 1213–1219.

[69] W. T. S. Huck, L. J. Prins, R. H. Fokkens, N. M. M. Nibbering, F. C. J. M. van Veggel, D. N. Reinhoudt, "Convergent and Divergent Noncovalent Synthesis of Metallodendrimers", *J. Am. Chem. Soc.* **1998**, *120*, 6240–6246.

[70] C. D. Eisenbach, W. Degelmann, A. Göldel, J. Heinlein, M. Terskan-Reinold, U. S. Schubert, "Macromolecules with Specific Constitution and Supramolecular Structures", *Macromol. Symp.* **1995**, *98*, 565–572.

[71] G. R. Newkome, B. D. Woosley, E. He, C. N. Moorefield, R. Güther, G. R. Baker, G. H. Escamilla, J. Merrill, H. Luftmann, "Supramolecular Chemistry of Flexible, Dendritic-Based Structures Employing Molecular Recognition", *Chem. Commun.* **1996**, 2737–2738.

[72] G. R. Newkome, "Molecular Recognition: Smart Macromolecules," Royal Society of Chemistry Perkin Division Symposium on Recognition Processes, Birmingham, England, July 24–29, 1994.

[73] G. R. Newkome, Z. Yao, G. R. Baker, V. K. Gupta, P. S. Russo, M. J. Saunders, "Cascade Molecules: Synthesis and Characterization of a Benzene[9]^3-arborol", *J. Am. Chem. Soc.* **1986**, *108*, 849–850.

[74] S. C. Zimmerman, F. Zeng, D. E. C. Reichert, S. V. Kolotuchin, "Self-Assembling Dendrimers", *Science* **1996**, *271*, 1095–1098.

[75] C. Hawker, J. M. J. Fréchet, "A New Convergent Approach to Monodisperse Dendritic Macromolecules", *J. Chem. Soc., Chem. Commun.* **1990**, 1010–1013.

[76] F. Zeng, S. C. Zimmerman, "Dendrimers in Supramolecular Chemistry: From Molecular Recognition to Self-Assembly", *Chem. Rev.* **1997**, *97*, 1681–1712.

[77] P. Thiyagarajan, F. Zeng, C. Y. Ku, S. C. Zimmerman, "SANS Investigation of Self-Assembling Dendrimers in Organic Solvents", *J. Mater. Chem.* **1997**, *7*, 1221–1226.

[78] V. Percec, J. Heck, D. Tomazos, F. Falkenberg, H. Blackwell, G. Ungar, "Self-Assembly of Taper-Shaped Monoesters of Oligo(ethylene oxide) with 3,4,5-Tris(*p*-dodecyloxybenzyloxy) benzoic Acid and of their Polymethacrylates into Tubular Supramolecular Architectures Displaying a Columnar Mesophase", *J. Chem. Soc., Perkin Trans. 1* **1993**, 2799–2811.

[79] V. Percec, J. Heck, G. Johansson, D. Tomazos, M. Kawasumi, P. Chu, G. Ungar, "Molecular Recognition Directed Self-Assembly of Supramolecular Architectures", *J. Macromol. Sci. – Pure Appld. Chem.* **1994**, *A31*, 1719–1758.

[80] V. Percec, P. Chu, G. Ungar, J. Zhou, "Rational Design of the First Nonspherical Dendrimer which Displays Calamitic Nematic and Smectic Thermotropic Liquid Crystalline Phases", *J. Am. Chem. Soc.* **1995**, *117*, 11441–11454.

[81] V. Percec, C. G. Cho, C. Pugh, D. Tomazos, "Synthesis and Characterization of Branched Liquid-Crystalline Polyethers Containing Cyclotetraveratrylene-Based Disk-Like Mesogens", *Macromolecules* **1992**, *25*, 1164–1176.

[82] A. R. A. Palmans, J. A. J. M. Vekemans, E. E. Havinga, E. W. Meijer, "Sergeants-and-Soldiers Principle in Chiral Columnar Stacks of Disc-Shaped Molecules with C_3 Symmetry", *Angew. Chem.* **1997**, *109*, 2763–2765; *Int. Ed. Engl.* **1997**, *36*, 2648–2651.

[83] D. B. Amabilino, P. R. Ashton, M. Belohradský, F. M. Raymo, J. F. Stoddart, "The Self-Assembly of Branched [*n*]Rotaxanes – The First Step Towards Dendritic Rotaxanes", *J. Chem. Soc., Chem. Commun.* **1995**, 751–753.

[84] E. W. Wong, C. P. Collier, M. Behloradský, F. M. Raymo, J. F. Stoddart, J. R. Heath, "Fabrication and Transport Properties of Single-Molecule-Thick Electrochemical Junctions", *J. Am. Chem. Soc.* **2000**, *122*, 5831–5840.

[85] B. F. Hoskins, R. Robson, D. A. Slizys, "A Hexaimidazole Ligand Binding Six Octahedral Metal Ions to give an Infinite 3D α-Po-like Network through which Two Independent 2D Hydrogen-Bonded Networks Interweave", *Angew. Chem.* **1997**, *109*, 2861–2863; *Int. Ed. Engl.* **1997**, *36*, 2752–2754.

[86] J. J. Apperloo, R. A. J. Janssen, P. R. L. Malenfant, J. M. J. Fréchet, "Concentration-Dependent Thermochromism and Supramolecular Aggregation in Solution of Triblock Copolymers Based on Lengthy Oligothiophene Cores and Poly(benzyl ether) Dendrons", *Macromolecules* **2000**, *33*, 7038–7043.

[86a] T. K. K. Mong, A. Niu, H.-F. Chow, C. Wu, L. Li, R. Chen, "β-Alanine-Based Dendritic β-Peptides: Dendrimers Possessing Unusually Strong Binding Ability Towards Protic Solvents and Their Self-Assembly into Nanoscale Aggregates through Hydrogen-Bond Interactions", *Chem. Eur. J.* **2001**, *7*, 686–699.

[86b] E. R. Zubarev, M. U. Pralle, E. D. Sone, S. I. Stupp, "Self-Assembly of Dendron Rodcoil Molecules into Nanoribbons", *J. Am. Chem. Soc.* **2001**, *123*, 4105–4106.

[87] A. S. Kovvali, H. Chen, K. K. Sirkar, "Dendrimer Membranes: A CO$_2$-Selective Molecular Gate", *J. Am. Chem. Soc.* **2000**, *122*, 7594–7595.

[87a] V. Percec, W.-D. Cho, G. Ungar, "Increasing the Diameter of Cylindrical and Spherical Supramolecular Dendrimers by Decreasing the Solid Angle of Their Monodendrons via Periphery Functionalization", *J. Am. Chem. Soc.* **2000**, *122*, 10273–10281.

[87b] V. Percec, W.-D. Cho, G. Ungar, D. J. P. Yeardley, "Synthesis and Structural Analysis of Two Constitutional Isomeric Libraries of AB₂-Based Monodendrons and Supramolecular Dendrimers", *J. Am. Chem. Soc.* **2001**, *123*, 1302–1315.

[88] J. M. J. Fréchet, C. J. Hawker, K. L. Wooley, "The Convergent Route to Globular Dendritic Macromolecules: A Versatile Approach to Precisely Functionalized Three-Dimensional Polymers and Novel Block Copolymers", *J. Macromol. Sci. – Pure Appld. Chem.* **1994**, *A31*, 1627–1645.

[89] C. J. Hawker, J. M. J. Fréchet, "The Synthesis and Polymerization of a Hyperbranched Polyether Macromonomer", *Polymer* **1992**, *33*, 1507–1511.

[90] I. Gitsov, K. L. Wooley, J. M. J. Fréchet, "Novel Polyether Copolymers Consisting of Linear and Dendritic Blocks", *Angew. Chem.* **1992**, *104*, 1282–1285; *Int. Ed. Engl.* **1992**, *31*, 1200–1202.

[91] I. Gitsov, K. L. Wooley, C. J. Hawker, P. T. Ivanova, J. M. J. Fréchet, "Synthesis and Properties of Novel Linear-Dendritic Block Copolymers. Reactivity of Dendritic Macromolecules toward Linear Polymers", *Macromolecules* **1993**, *26*, 5621–5627.

[92] I. Gitsov, "Hybrid Dendritic Capsules. Properties and Binding Capabilities of Amphiphilic Copolymers with Linear Dendritic Architecture" in "Associative Polymers in Aqueous Solutions", J.E. Glass, Ed., ACS Symp. Series Vol. 765, American Chemical Society, Washington DC, 2000, pp.72–92.

[93] I. Gitsov, "Linear-Dendritic Block Copolymers. Synthesis and Characterization", in *Advances in Dendritic Molecules* (Ed.: G. R. Newkome), JAI Press, **2001**, in press.

[94] C. Kim, E. Park, I. Jung, "Silane Arborols (V). The Formation of Dendrimeric Silane on Poly(carbosilane): Silane Arborols", *J. Korean Chem. Soc.* **1996**, *40*, 347–356.

[94a] Y. Takaguchi, T. Tajima, K. Ohta, J. Motoyoshiya, H. Aoyama, "Photoresponsive Dendrimers: Synthesis and Characterizations of Anthracenes Bearing Dendritic Substituents", *Chem. Lett.* **2000**, 1388–1389.

[95] H. Sashiwa, Y. Shigamasa, R. Roy, "Chemical Modification of Chitosan; 3. Hyperbranched Chitosan–Sialic Acid Dendrimer Hybrid with Tetraethylene Glycol Spacer", *Macromolecules* **2000**, *33*, 6913–6915.

[96] K. Fujiki, M. Sakamoto, T. Sato, N. Tsubokawa, "Postgrafting of Hyperbranched Dendritic Polymer from Terminal Amino Groups of Polymer Chains Grafted onto Silica Surface", *J. Macromol. Sci. – Pure Appld. Chem.* **2000**, *A37*, 357–377.

[97] M. C. Strumia, A. Halabi, P. A. Pucci, G. R. Newkome, C. N. Moorefield, J. D. Epperson, "Surface Modifications of Activated Polymeric Matrices by Dendritic Attachments", *J. Polym. Sci., Part A: Polym. Chem.* **2000**, *38*, 2779–2786.

[98] N. Ouali, S. Méry, A. Skoulios, L. Noirez, "Backbone Stretching of Worm-like Carbosilane Dendrimers", *Macromolecules* **2000**, *33*, 6185–6193.

[99] C. Kim, S. Kang, "Carbosilane Dendrimers Based on Siloxane Polymer", *J. Polym. Sci., Part A: Polym. Chem.* **2000**, *38*, 724–729.

[100] H. Ritter, "Functionalized Comb-like Polymers: Synthesis, Modification and Application", *Angew. Makromol. Chem.* **1994**, *223*, 165–175.

[101] H. Ritter, "Functionalized Polymers with Comb-like, Rotaxanic-, and Dendrimeric Structures", *GIT Fachzeitschrift für das Laboratorium* **1994**, *38*, 615–619.

[102] H. Ritter, "New Comb-like Polymers: Synthesis, Structures and Reactivity", *Makromol. Chem., Macromol. Symp.* **1994**, *77*, 73–78.

[103] G. Draheim, H. Ritter, "Polymerizable Dendrimers. Synthesis of a Symmetrically Branched Methacryl Derivative Bearing Eight Ester Groups", *Macromol. Chem. Phys.* **1995**, *196*, 2211–2222.

[104] G. R. Newkome, Z. Yao, G. R. Baker, V. K. Gupta, "Cascade Molecules: A New Approach to Micelles. A [27]-Arborol", *J. Org. Chem.* **1985**, *50*, 2003–2004.

[105] R. Wolf, M. Asakawa, P. R. Ashton, M. Gómez-López, C. Hamers, S. Menzer, I. W. Parsons, N. Spencer, J. F. Stoddart, M. S. Tolley, D. J. Williams, "A Molecular Chameleon: Chromophoric Sensing by a Self-Complexing Molecular Assembly", *Angew. Chem.* **1998**, *110*, 1018–1022; *Int. Ed. Engl.* **1998**, *37*, 975–979.

[105a] Z. Bo, L. Zhang, Z. Wang, X. Zhang, J. Shen, "Investigation of self-assembled dendrimer complexes", *Mater. Sci. Eng., C* **1999**, *10*, 165–170.

[105b] Y. Li, C. A. McMillan, D. M. Bloor, J. Penfold, J. Warr, J. F. Holzwarth, E. Wyn-Jones, "Small-Angle Neutron Scattering and Fluorescence Quenching Studies of Aggregated Ionic and Nonionic Surfactants in the Presence of Poly(1,4-diaminobutane) Dendrimers", *Langmuir* **2000**, *16*, 7999–8004.

[105c] G. Purohit, T. Skthivel, A. T. Fluorence, "Interaction of cationic partial dendrimers with charged and neutral liposomes", *Int. J. Pharm.* **2001**, *214*, 71–76.

[106] D. W. Thompson, *On Growth and Form*, 2nd ed., The University Press, Cambridge, **1942**.

10 Appendices

10.1 Key Reviews and Highlights Covering Dendrimers and Hyperbranched Materials

10.1.1 2001 Reviews

C. B. Gorman, J. C. Smith, "Structure – Property Relationships in Dendritic Encapsulation", *Acc. Chem. Res.* **2001**, *34*, 60–71.

J. L. Kreider, W. T. Ford, "Quaternary ammonium ion dendrimers from methylation of poly(propylene imine)s", *J. Polym. Sci. , Part A: Polym. Chem.* **2001**, *39*, 821–832.

A. Sekiguchi, V. Y. Lee, M. Nanjo, "Lithiosilanes and their application to the synthesis of polysilane dendrimers", *Coord. Chem. Rev.* **2001**, *210*, 11–45.

H. Y. Wei, W. F. Shi, "Structural characteristics, syntheses and applications of hyperbranched polymers", *Chem. J. Chin. Univ. -Chin.* **2001**, *22*, 338–344.

P. E. Froehling, "Dendrimers and Dyes – A Review", *Dyes and Pigments* **2001**, *48*, 187–195.

A. Juris, M. Venturi, P. Ceroni, V. Balzani, S. Campagna, S. Serroni, "Dendrimers based on electroactive metal complexes. A review of recent advances", *Collect. Czech. Chem. Commun.* **2001**, *66*, 1–32.

R. Haag, "Dendrimers and Hyperbranched Polymers as High-Loading Supports for Organic Synthesis", *Chem. Eur. J.* **2001**, *7*, 327–335.

M. T. Reetz, "Combinatorial and Evolution-Based Methods in the Creation of Enantioselective Catalysts", *Angew. Chem. Int. Ed.* **2001**, *40*, 284–310.

S. Hecht, J. M. J. Fréchet, "Dendritic Encapsulation of Function: Applying Nature's Site Isolation Principle from Biomimetics to Materials Science", *Angew. Chem. Int. Ed.* **2001**, *40*, 75–91.

D. Seebach, A. K. Beck, A. Heckel, "TADDOLs, Their Derivatives, and TADDOL Analogues: Versatile Chiral Auxiliaries", *Angew. Chem. Int. Ed.* **2001**, *40*, 92–138.

N. B. Bowden, M. Weck, I. S. Choi, G. M. Whitesides, "Molecule-Mimetic Chemistry and Mesoscale Self-Assembly", *Acc. Chem. Res.* **2001**, *34*, 231–238.

R. M. Crooks, M. Zhao, L. Sun, V. Chechik, L. K. Yeung, "Dendrimer-Encapsulated Metal Nanoparticles: Synthesis, Characterization, and Applications to Catalysis", *Acc. Chem. Res.* **2001**, *34*, 181–190.

Sekiguchi, V. Y. Lee, M. Nanjo, "Lithiosilanes and their application to the synthesis of polysilane dendrimers", *Coord. Chem. Rev.* **2001**, *210*, 11–45.

U.-M. Wiesler, T. Weil, K. Müllen, "Nanosized Polyphenylene Dendrimers", *Top. Curr. Chem.* **2001**, *212*, 1–40.

D. Muscat, R. A. T. M. van Benthem, "Hyperbranched Polyesteramides – New Dendritic Polymers", *Top. Curr. Chem.* **2001**, *212*, 41–80.

R. M. Crooks, B. I. Lemon, L. Sun, L. K. Yeung, M. Q. Zhao, "Dendrimer-Encapsulated Metals and Semiconductors: Synthesis, Characterization, and Applications", *Top. Curr. Chem.* **2001**, *212*, 81–135.

S. S. Sheiko, M. Möller, "Hyperbranched macromolecules: Soft particles with adjustable shape and persistent motion capability", *Top. Curr. Chem.* **2001**, *212*, 137–175.

M. Ballauff, "Structure of Dendrimers in Dilute Solution", *Top. Curr. Chem.* **2001**, *212*, 176–194.

10.1.2 2000 Reviews

A. Adronov, J. M. J. Fréchet, "Light-harvesting dendrimers", *Chem. Commun.* **2000**, 1701–1710.

A. Archut, F. Vögtle, "Dendritic Molecules – Historic Development and Future Applications" in *Handbook of Nanostructured Materials and Nanotechnology, Vol. 5*, (N. S. Nalwa, ed.), Academic Press, New York **2000**, 333–374.

C. M. Cardona, S. Mendoza, A. E. Kaifer, "Electrochemistry of encapsulated redox centers", *Chem. Soc. Rev.* **2000**, *29*, 37–42.

A. D. Schlüter, J. P. Rabe, "Dendronized Polymers: Synthesis, Characterization, Assembly at Interfaces, and Manipulation", *Angew. Chem. Int. Ed.* **2000**, *39*, 864–883.

D. Seebach, "TADDOLs – from Enantioselective Catalysis to Dendritic Cross Linkers to Cholesteric Liquid Crystals", *Chimia* **2000**, *54*, 60–62.

B. Voit, "New Developments in Hyperbranched Polymers", *J. Polym. Sci., Part A: Polym. Chem.* **2000**, *38*, 2505–2525.

K. L. Wooley, "Shell Crosslinked Polymer Assemblies: Nanoscale Constructs Inspired from Biological Systems", *J. Polym. Sci., Part A: Polym. Chem.* **2000**, *38*, 1397–1407.

S. Nummelin, M. Skrifvars, K. Rissanen, "Polyester and Ester Functionalized Dendrimers" in *Topics in Current Chemistry*, (F. Vögtle, ed.), Springer-Verlag, Berlin **2000**, 1–67.

H. Frey, C. Schlenk, "Silicon-Based Dendrimers" in *Topics in Current Chemistry*, (F. Vögtle, ed.), Springer-Verlag, Berlin **2000**, 69–129.

M. W. P. L. Baars, E. W. Meijer, "Host-Guest Chemistry of Dendritic Molecules", *Top. Curr. Chem.* **2000**, *210*, 131–227.

D. Astruc, J.-C. Blais, E. Cloutet, L. Djakovitch, S. Rigaut, J. Ruiz, V. Sartor, C. Valério, "The First Organometallic Dendrimers: Design and Redox Functions" in *Topics in Current Chemistry*, (F. Vögtle, ed.), Springer-Verlag, Berlin **2000**, 229–259.

W. Krause, N. Hackmann-Schlichter, F. K. Maier, R. Müller, "Dendrimers in Diagnostics" in *Topics in Current Chemistry*, (F. Vögtle, ed.), Springer-Verlag, Berlin **2000**, 261–308.

D. K. Smith, F. Diederich, "Supramolecular Dendrimer Chemistry – A Journey Through the Branched Architecture" in *Topics in Current Chemistry*, (F. Vögtle, ed.), Springer-Verlag, Berlin **2000**, 183–227.

A. Sunder, J. Heinemann, H. Frey, "Controlling the Growth of Polymer Trees: Concepts and Perspectives For Hyperbranched Polymers", *Chem. Eur. J.* **2000**, *6*, 2499–2506.

C. Z. Chen, S. L. Cooper, "Recent Advances in Antimicrobial Dendrimers", *Adv. Mater.* **2000**, *12*, 843–846.

L. Pu, "Novel chiral conjugated macromolecules for potential electrical and optical applications", *Macromol. Rapid Commun.* **2000**, *21*, 795–809.

W. Meier, "Polymer nanocapsules", *Chem. Soc. Rev.* **2000**, *29*, 295–303.

J.-F. Nierengarten, "Fullerodendrimers: A New Class of Compounds for Supramolecular Chemistry and Materials Science Applications", *Chem. Eur. J.* **2000**, *6*, 3667–3670.

S. Mann, "The Chemistry of Form", *Angew. Chem. Int. Ed.* **2000**, *39*, 3392–3406.

V. Balzani, A. Credi, F. M. Raymo, J. F. Stoddart, "Artificial Molecular Machines", *Angew. Chem. Int. Ed.* **2000**, *39*, 3348–3391.

J. M. Tour, "Molecular Electronics. Synthesis and Testing of Components", *Acc. Chem. Res.* **2000**, *33*, 791–804.

M. C. Feiters, A. E. Rowan, R. J. M. Nolte, "From simple to supramolecular cytochrome P450 mimics", *Chem. Soc. Rev.* **2000**, *29*, 375–384.

R. A. T. M. van Benthem, "Novel hyperbranched resins for coating applications", *Prog. Org. Coat.* **2000**, *40*, 203–214.

T. Imae, "Structure and functionality of dendrimers", *Kobunshi Ronbunshu* **2000**, *57*, 810–824.

D. Seebach, "TADDOLs – from Enantioselective Catalysis to Dendritic Cross Linkers to Cholesteric Liquid Crystals", *Chimia* **2000**, *54*, 60–62.

10.1.3 1999 Reviews

Advances in Polymer Science: Branched Polymers I, (J. Roovers) Springer-Verlag, Berlin, Heidelberg, New York **1999**.

Advances in Polymer Science: Branched Polymers II, (J. Roovers) Springer-Verlag, Berlin, Heidelberg, New York, **1999**.

T. Aida, D.-L. Jiang, in *Handbook of Porphyrins and Related Macrocycles: Biomaterials for Materials Scientists, Chemists, and Physicists*, (K. M. Kadish, K. M. Smith, and R. Guillard, ed.), Academic Press, **1999**.

A. J. Berresheim, M. Müller, K. Müllen, "Polyphenylene Nanostructures", *Chem. Rev.* **1999**, *99*, 1747–1785.

C. Bieniarz, "Dendrimers: Applications to Pharmaceutical and Medicinal Chemistry" in *Encyclopedia of Pharmaceutical Technology*, (J. Swarbrick and J. C. Boylan, ed.), Marcel Dekker, Inc., New York **1999**, 55–89.

R. Bischoff, S. E. Cray, "Polysiloxanes in macromolecular architecture", *Prog. Polym. Sci.* **1999**, *24*, 185–219.

A. W. Bosman, H. M. Janssen, E. W. Meijer, "About Dendrimers: Structure, Physical Properties, and Applications", *Chem. Rev.* **1999**, *99*, 1665–1688.

C. M. Casado, I. Cuadrado, M. Morán, B. Alonso, B. Garcia, B. González, J. Losada, "Redox-active ferrocenyl dendrimers and polymers in solution and immobilised on electrode surfaces", *Coord. Chem. Rev.* **1999**, *185–186*, 53–79.

H.-F. Chow, T. K. K. Mong, C.-W. Wan, Z.-Y. Wang, "Chiral Dendrimers" in *Advances in Dendritic Macromolecules*, (G. R. Newkome, ed.), JAI Press Inc., Greenwich, CN **1999**, 107–133.

I. Cuadrado, M. Morán, C. M. Casado, B. Alonso, J. Losada, "Organometallic dendrimers with transition metals", *Coord. Chem. Rev.* **1999**, *193–195*, 395–445.

M. Fischer, F. Vögtle, "Dendrimers: From Design to Application – A Progress Report", *Angew. Chem. Int. Ed.* **1999**, *38*, 885–905.

C. J. Hawker, "Dendritic and Hyperbranched Macromolecules – Precisely Controlled Macromolecular Architectures", *Adv. Polym. Sci.* **1999**, *147*, 114–160.

M. A. Hearshaw, J. R. Moss, "Organometallic and related metal-containing dendrimers", *Chem. Commun.* **1999**, 1–8.

C. E. Housecroft, "Icosahedral Building Blocks: Towards Dendrimers with Twelve Primary Branches?", *Angew. Chem. Int. Ed.* **1999**, *38*, 2717–2719.

A. Hult, M. Johansson, E. Malmström, "Hyperbranched Polymers" in *Advances in Polymer Science: Branched Polymers*, (J. Roovers, ed.), Springer-Verlag, Berlin, Heidelberg, New York **1999**, 2–34.

A. E. Kaifer, "Interplay between Molecular Recognition and Redox Chemistry", *Acc. Chem. Res.* **1999**, *32*, 62–71.

M. Kakimoto, Y. Imai, "Dendritic Molecules by the Divergent Method" in *Star and Hyperbranched Polymers*, (M. K. Mishra and S. Kobayashi, ed.), Marcel Dekker, Inc., New York **1999**, 267–284.

Y. H. Kim, O. W. Webster, "Hyperbranched Polymers" in *Star and Hyperbranched Polymers*, (M. K. Mishra and S. Kobayashi, ed.), Marcel Dekker, Inc., New York **1999**, 201–238.

J.-P. Majoral, C. Larré, R. Laurent, A.-M. Caminade, "Chemistry in the internal voids of dendrimers", *Coord. Chem. Rev.* **1999**, *190–192*, 3–18.

J.-P. Majoral, A.-M. Caminade, "Dendrimers Containing Heteroatoms (Si, P, B, Ge, or Bi)", *Chem. Rev.* **1999**, *99*, 845–880.

L. A. Marcaurelle, C. R. Bertozzi, "New Directions in the Synthesis of Glycopeptides Mimetics", *Chem. Eur. J.* **1999**, *5*, 1384–1390.

G. R. Newkome, E. He, C. N. Moorefield, "Suprasupermolecules with Novel Properties: Metallodendrimers", *Chem. Rev.* **1999**, *99*, 1689–1746.

P. Nguyen, P. Gömez-Elipe, I. Manners, "Organometallic Polymers with Transition Metals in the Main Chain", *Chem. Rev.* **1999**, *99*, 1515–1548.

J. Roovers, B. Comanita, "Dendrimers and Dendrimer-Polymer Hybrids" in *Advances in Polymer Science: Branched Polymers*, (J. Roovers, ed.), Springer-Verlag, Berlin, Heidelberg, New York **1999**, 180–228.

C. Schlenk, H. Frey, "Carbosilane Dendrimers – Synthesis, Functionalization, Application", *Monatsh. Chem.* **1999**, *130*, 3–14.

M. S. Shchepinov, "Oligonucleotide dendrimers: From poly-labelled DNA probes to stable nano-structures", *The Glenn Report* **1999**, *12*, 1–4.

J.-L. Six, Y. Gnanou, "Dendritic Architectures by the Convergent Method" in *Star and Hyperbranched Polymers*, (M. K. Mishra and S. Kobayashi, ed.), Marcel Dekker, Inc., New York **1999**, 239–266.

F. J. Stoddart, T. Welton, "Metal-containing dendritic polymers", *Polyhedron* **1999**, *18*, 3575–3591.

M. Venturi, A. Credi, V. Balzani, "Electrochemistry of coordination compounds: an extended view", *Coord. Chem. Rev.* **1999**, *185–186*, 233–256.

V. W. W. Yam, K. K. W. Lo, "Recent advances in utilization of transition metal complexes and lanthanides as diagnostic tools", *Coord. Chem. Rev.* **1999**, *184*, 157–240.

10.1.4   1998 Reviews

A. Archut, J. Issberner, F. Vögtle, "Dendrimers, Arborols, and Cascade Molecules: Breakthrough into Generations of New Materials" in *Organic Synthesis Highlights III*, (J. Mulzer and H. Waldmann, eds.), Wiley-VCH, **1998**, 391–405.

A. Archut, F. Vögtle, "Functional cascade molecules", *Chem. Soc. Rev.* **1998**, *27*, 233–240.

V. Balzani, S. Campagna, G. Denti, A. Juris, S. Serroni, M. Venturi, "Designing Dendrimers based on Transition-Metal Complexes. Light-Harvesting Properties and Predetermined Redox Patterns", *Acc. Chem. Res.* **1998**, *31*, 26–34.

M. R. Bryce, W. Devonport, L. M. Goldenberg, C. Wang, "Macromolecular tetrathiafulvalene chemistry", *Chem. Commun.* **1998**, 945–951.

A.-M. Caminade, R. Laurent, B. Chaudret, J.-P. Majoral, "Phosphine-terminated dendrimers: Synthesis and complexation properties", *Coord. Chem. Rev.* **1998**, *178–180*, 793–821.

H.-F. Chow, T. K. K. Mong, M. F. Nongrum, C.-W. Wan, "The Synthesis and Properties of Novel Functional Dendritic Molecules", *Tetrahedron* **1998**, *54*, 8543–8660.

J.-P. Collin, V. Heitz, J.-P. Sauvage, "Construction of One-Dimensional Multicomponent Molecular Arrays: Control of Electronic and Molecular Motion", *Eur. J. Inorg. Chem.* **1998**, 1–14.

E. C. Constable, C. E. Housecroft, "Supramolecular Approaches to Advanced Materials", *Chimia* **1998**, *52*, 533–538.

R. M. Crooks, A. J. Ricco, "New Organic Materials Suitable for Use in Chemical Sensor Arrays", *Acc. Chem. Res.* **1998**, *31*, 219–227.

G. H. Escamilla, G. R. Newkome, "Bolaamphiphiles: Golf Balls to Fibers" in *Organic Synthesis Highlights III*, (J. Mulzer and H. Waldmann, eds.), Wiley-VCH, **1998**, 382–390.

N. Feuerbacher, F. Vögtle, "Iterative Synthesis in Organic Chemistry" in *Topics in Current Chemistry*, (F. Vögtle, ed.), Springer-Verlag, Berlin **1998**, 1–18.

H. Frey, "From Random Coil to Extended Nanocylinder: Dendrimer Fragments Shape Polymer Chains", *Angew. Chem. Int. Ed.* **1998**, *37*, 2193–2197.

H. Frey, C. Lach, K. Lorenz, "Heteroatom-Based Dendrimers", *Adv. Mater.* **1998**, *10*, 279–293.

J. W. Goodby, G. H. Mehl, I. M. Saez, R. P. Tuffin, G. Mackenzie, R. Auzély-Velty, T. Benvegnu, D. Plusquellec, "Liquid crystals with restricted molecular topologies: molecules and supramolecular assemblies", *Chem. Commun.* **1998**, 2057–2070.

C. Gorman, "Metallodendrimers: Structural Diversity and Functional Behavior", *Adv. Mater.* **1998**, *10*, 295–309.

R. A. Gossage, L. A. van de Kuil, G. van Koten, "Diaminoarylnickel(II) "Pincer" Complexes: Mechanistic Considerations in the Kharasch Addition Reaction, Controlled Polymerization, and Dendrimeric Transition Metal Catalysts", *Acc. Chem. Res.* **1998**, *31*, 423–431.

N. Hüsing, U. Schubert, "Aerogels – Airy Materials: Chemistry, Structure, and Properties", *Angew. Chem. , Int. Ed. Engl.* **1998**, *37*, 22–37.

N. Jayaraman, S. A. Nepogodiev, J. F. Stoddart, "Synthesis and study of dendritic polysaccharides", *Carbohydrates in Europe* **1998**, 30–33.

Y. H. Kim, "Hyperbranched Polymers 10 Years After", *J. Polym. Sci. , Part A: Polym. Chem.* **1998**, *36*, 1685–1698.

J.-P. Majoral, A.-M. Caminade, "Divergent Approaches to Phosphorus-Containing Dendrimers and their Functionalization", *Top. Curr. Chem.* **1998**, *197*, 79–124.

O. A. Matthews, A. N. Shipway, J. F. Stoddart, "Dendrimers – branching out from curiosities into new technologies", *Prog. Polym. Sci.* **1998**, *23*, 1–56.

F. M. MacDonnell, M.-J. Kim, S. Bodige, "Substitutionally inert complexes as chiral synthons for stereospecific supramolecular syntheses", *Coord. Chem. Rev.* **1998**, *185–186*, 535–549.

M. Müller, C. Kübel, K. Müllen, "Giant Polycyclic Aromatic Hydrocarbon", *Chem. Eur. J.* **1998**, *4*, 2099–2109.

V. V. Narayanan, G. R. Newkome, "Supramolecular Chemistry within Dendritic Structures" in *Topics in Current Chemistry,* (F. Vögtle, ed.), Springer-Verlag, Berlin, Heidelberg, New York **1998**, 19–77.

G. R. Newkome, "Suprasupermolecular chemistry: the chemistry within the dendrimer", *Pure Appl. Chem.* **1998**, *70*, 2337–2343.

R. J. Puddephatt, "Precious metal polymers: platinum or gold atoms in the backbone", *Chem. Commun.* **1998**, 1055–1062.

A.-D. Schlüter, "Dendrimers with Polymeric Core: Towards Nanocylinders" in *Topics in Current Chemistry,* (F. Vögtle, ed.), Springer-Verlag, Berlin, Heidelberg. New York **1998**, 165–191.

D. Seebach, P. B. Rheiner, G. Greiveldinger, T. Butz, H. Sellner, "Chiral Dendrimers", *Top. Curr. Chem.* **1998**, *197,* 125–164.

D. K. Smith, F. Diederich, "Functional Dendrimers: Unique Biological Mimics", *Chem. Eur. J.* **1998**, *4*, 1353–1361.

C. W. Thomas, Y. Tor, "Dendrimers and Chirality", *Chirality* **1998**, *10*, 59.

F. Vögtle, M. Plevoets, G. Nachtsheim, U. Wörsdörfer, "Monofunctionalized Dendrons of Different Generations – as Reagents for the Introduction of Dendritic Substituents", *J. Prakt. Chem.* **1998**, *340*, 112–121.

M. Venturi, S. Serroni, A. Juris, S. Campagna, V. Balzani, "Electrochemical and Photochemical Properties of Metal-Containing Dendrimers" in *Topics in Current Chemistry,* (F. Vögtle, ed.), Springer-Verlag, Berlin, Heidelberg, New York **1998**, 193–228.

10.1.5   1997 Reviews

A. Archut, M. Fischer, J. Issberner, F. Vögtle, "Dendrimere: Hochverzweigte Moleküle mit neuen Eigenschaften", *GIT Labor-Fachzeitschrift* **1997**, *41*, 198–202.

D. Gudat, "Inorganic Cauliflower: Functional Main Group Element Dendrimers Constructed from Phosphorus- and Silicon-Based Building Blocks", *Angew. Chem. Int. Ed.* **1997**, *36*, 1951–1955.

L. J. Hobson, R. M. Harrison, "Dendritic and hyperbranched polymers: advances in synthesis and applications", *Curr. Opin. Solid State Mater. Sci.* **1997**, *2*, 683–692.

N. Jayaraman, S. A. Nepogodiev, J. F. Stoddart, "Synthetic Carbohydrate-Containing Dendrimers", *Chem. Eur. J.* **1997**, *3*, 1193–1199.

Y. H. Kim, "Highly branched polymers: dendrimers and hyperbranched polymers", *Plastics Engineering* **1997**, *40 (Macromolecular Design of Polymeric Materials)*, 365–378.

M. K. Lothian-Tomalia, D. M. Hedstrand, D. A. Tomalia, A. B. Padias, H. K. Hall, Jr., "A Contemporary Survey of Covalent Connectivity and Complexity. The Divergent Synthesis of Poly(thioether) Dendrimers. Amplified, Genealogically Directed Synthesis Leading to the de Gennes Packed State", *Tetrahedron* **1997**, *53*, 15495–15513.

E. Malmström, A. Hult, "Hyperbranched polymers: A review", *J. Macromol. Sci. -Rev. Macromol. Chem. Phys.* **1997**, *C37*, 555–579.

J. S. Moore, "Shape-Persistent Molecular Architectures of Nanoscale Dimension", *Acc. Chem. Res.* **1997**, *30*, 402–413.

C. P. Palmer, "Micelle polymers, polymer surfactants and dendrimers as pseudostationary phases in micellar electrokinetic chromatography", *J. Chromatogr. A* **1997**, *780*, 75–92.

H. W. I. Peerlings, E. W. Meijer, "Chirality in Dendritic Architectures", *Chem. Eur. J.* **1997**, *3*, 1563–1570.

R. Roy, "Recent Developments in the Rational Design of Multivalent Glycoconjugates" , Springer-Verlag, Berlin **1997**, 241–274.

D. A. Tomalia, R. Esfand, "Dendrons, dendrimers and dendrigrafts", *Chem. Ind. (London)* **1997**, 416–420.

K. L. Wooley, "From Dendrimers to Knedel-like Structures", *Chem. Eur. J.* **1997**, *3*, 1397–1399.

F. Zeng, S. C. Zimmerman, "Dendrimers in Supramolecular Chemistry: From Molecular Recognition to Self-Assembly", *Chem. Rev.* **1997**, *97*, 1681–1712.

10.1.6   1996 Reviews

K. Aoi, "Sugar balls", *Kobunshi* **1996**, *45*, 260.

D. Astruc, "Research avenues on dendrimers in molecular biology: from biomimetism to medicinal engineering", *C. R. l'Academie. Sci. , Ser. II Univers* **1996**, *322*, 757–766.

V. Balzani, A. Juris, M. Venturi, S. Campagna, S. Serroni, "Luminescent and Redox-Active Polynuclear Transition Metal Complexes", *Chem. Rev.* **1996**, *96*, 759–833.

M. R. Bryce, W. Devonport, "Redox-active dendrimers, related building blocks, and oligomers" in *Advances in Dendritic Macromolecules,* (G. R. Newkome, ed.), JAI, Greenwich, Conn. **1996**, 115–149.

L. Y. Chiang, L. Y. Wang, "Polyhydroxylated C_{60}: New Synthetic Opportunities towards Dendritic Polymers and Conducting IPN Elastomers", *Trends in Polymer Science* **1996**, *4*, 298–306.

L. Chen, Z. Yao, "Progress on Dendrimers", *Youji Huaxue* **1996**, *16*, 201–208.

I. Cuadrado, M. Morán, J. Losada, C. M. Casado, C. Pascual, B. Alonso, F. Lobete, "Organometallic dendritic macromolecules: organosilicon and organometallic entities as cores or building blocks" in *Advances in Dendritic Macromolecules,* (G. R. Newkome, ed.), JAI, Greenwich, Conn. **1996**, 151–195.

P. R. Dvornic, D. A. Tomalia, "Molecules That Grow Like Trees. Dendrimers: The fourth class of macromolecular architecture", *Science Spectra* **1996**, 36–41.

P. R. Dvornic, D. A. Tomalia, "Recent advances in dendritic polymers", *Current Opinion in Colloid & Interface Science* **1996**, *1*, 221–235.

P. R. Dvornic, D. A. Tomalia, "Dendrimers. The missing link between classical organic chemistry and polymer science?", *J. Serb. Chem. Soc.* **1996**, *61*, 1039–1062.

J. M. J. Fréchet, C. J. Hawker, "Synthesis and Properties of Dendrimers and Hyperbranched Polymers" in *Comprehensive Polymer Chemistry, 2nd Supplement,* (S. L. Aggarwal and S. Russo, ed.), Elsevier, Oxford, UK **1996**, 71–132.

C. J. Hawker, J. M. J. Fréchet, "Comparison of Linear, Hyperbranched, and Dendritic Macromolecules" in *Step-Growth Polymers for High-Performance Materials New Synthetic Methods,* (J. L. Hedrick and J. W. Labadie, ed.), American Chemical Society, Washington, D.C. **1996**, 132–144.

J. M. J. Fréchet, C. J. Hawker, I. Gitsov, J. W. Leon, "Dendrimers and hyperbranched polymers: two families of three-dimensional macromolecules with similar but clearly distinct properties", *J. Macromol. Sci. -Pure Appld. Chem.* **1996**, *A33*, 1399–1425.

Y. Gnanou, "Design and Synthesis of New Model Polymers", *J. Macromol. Sci. Rev. Macromol. Chem. Phys.* **1996**, *C36*, 77–117.

C. J. Hawker, W. Devonport, "Design, Synthesis, and Properties of Dendritic Macromolecules" in *Step-Growth Polymers for High-Performance Materials New Synthetic Methods,* (J. L. Hedrick and J. W. Labadie, ed.), American Chemical Society, Washington, D.C. **1996**, 186–196.

H. Ihre, M. Johansson, E. Malmström, A. Hult, "Dendrimers and hyperbranched aliphatic polyesters based on 2,2-Bis(hydroxymethyl)propionic acid (Bis-MPA)" in *Advances in Dendritic Macromolecules,* (G. R. Newkome, ed.), JAI, Greenwich, Conn. **1996**, 1–25.

M. Johansson, E. Malmström, A. Hult, "The Synthesis and Properties of Hyperbranched Polyesters", *Trends in Polymer Science* **1996**, *4*, 398–403.

J.-F. Labarre, F. Crasnier, M.-C. Labarre, F. Sournies, "The Saga of the Design and Synthesis of Dandelion Dendrimers", *Synlett* **1996**, 799–805.

J.-P. Majoral, A.-M. Caminade, "Arbres moléculaires (dendrimères) phosphorés: une future forêt d'applications", *Actual. Chim.* **1996**, 13–18.

G. R. Newkome, "Heterocyclic Loci within Cascade Dendritic Macromolecules", *J. Heterocycl. Chem.* **1996**, *33*, 1445–1460.

G. R. Newkome, C. N. Moorefield, "Design, Syntheses, and Supramolecular Chemistry of Smart Cascade Polymers" in *From Simplicity to Complexity in Chemistry and Beyond,* (A. Müller, A. Dress, and F. Vögtle, ed.), Vieweg & Sohn, Wiesbaden, Germany. **1996**, 127–136.

G. R. Newkome, C. N. Moorefield, "Dendrimers" in *Comprehensive Supramolecular Chemistry,* (D. N. Reinhoudt, ed.), Pergamon, New York **1996**, 777–832.

S. Serroni, S. Campagna, G. Denti, A. Juris, M. Venturi, V. Balzani, "Dendrimers based on metal complexes" in *Advances in Dendritic Macromolecules,* (G. R. Newkome, ed.), JAI, Greenwich, Conn. **1996**, 61–113.

K. Ute, "High molecular weight monodisperse polymers", *Kobunshi* **1996**, *45*, 104–105.

N. Ventosa, D. Ruiz, C. Rovira, J. Veciana, "Consequences of the fractal character of dendritic high-spin macromolecules on their physicochemical properties" in *Advances in Dendritic Macromolecules,* (G. R. Newkome, ed.), JAI, Greenwich, Conn. **1996**, 27–59.

10.1.7 1995 Reviews

N. Ardoin, D. Astruc, "Molecular trees: from syntheses towards applications", *Bull. Soc. Chim. Fr.* **1995**, *132*, 875–909.

M. N. Bochkarev, M. A. Katkova, "Dendritic polymers obtained by a single-stage synthesis", *Russ. Chem. Rev.* **1995**, *64*, 1035–1048.

H. Brunner, "Dendrizymes: Expanded ligands for enantioselective catalysis", *J. Organomet. Chem.* **1995**, *500*, 39–46.

A.-M. Caminade, J.-P. Majoral, "Main Group Elements-Based Dendrimers", *Main Group Chemistry News* **1995**, *3*, 14–24.

Y. Chen, C. X. Chen, F. Xi, "New Progress of Dendritic Polymers", *Huaxue Tongbao* **1995**, 1–10.

G. Denti, S. Campagna, V. Balzani, "Dendritic Polynuclear Metal Complexes with Made-to-Order Luminescent and Redox Properties" in *Mesomolecules from Molecules to Materials,* (G. D. Mendenhall, A. Greenberg, and J. F. Liebman, ed.), Chapman & Hall, New York **1995**, 69–106.

R. Engel, "Ionic dendrimers and related materials" in *Advances in Dendritic Macromolecules,* (G. R. Newkome, ed.), JAI, Greenwich, Conn. **1995**, 73–99.

G. H. Escamilla, "Dendritic bolaamphiphiles and related molecules" in *Advances in Dendritic Macromolecules,* (G. R. Newkome, ed.), JAI Press, Inc., Greenwich, Conn. **1995**, 157–190.

C. J. Hawker, J. M. J. Fréchet, "Three-dimensional dendritic macromolecules: Design, synthesis, and properties" in *New Methods Polym. Synth.,* (J. R. Ebdon and G. C. Eastmond, ed.), Blackie, Glasgow, UK **1995**, 290–330.

C. J. Hawker, K. L. Wooley, "The convergent-growth approach to dendritic macromolecules" in *Advances in Dendritic Macromolecules,* (G. R. Newkome, ed.), JAI Press, Greenwich, Conn. **1995**, 1–39.

M. Kakimoto, "Chemistry of dendrimer dendron", *Kagaku (Kyoto)* **1995**, *50*, 192–193.

Y. H. Kim, "Highly branched aromatic polymers: Their preparation and applications" in *Advances in Dendritic Macromolecules*, (G. R. Newkome, ed.), JAI Press, Greenwich, Conn. **1995**, 123–156.

J. Michl, "Supramolecular Assemblies from 'Tinkertoy' Rigid-Rod Molecules" in *Mesomolecules: From Molecules to Materials*, (G. D. Mendenhall, A. Greenberg, and J. F. Liebman, ed.), Chapman & Hall, New York **1995**, 132–160.

R. Moors, F. Vögtle, "Cascade molecules: Building blocks, multiple functionalization, complexing units, photoswitches" in *Advances in Dendritic Macromolecules*, (G. R. Newkome, ed.), JAI, Greenwich, Conn. **1995**, 41–71.

G. R. Newkome, C. N. Moorefield, "Cascade Molecules" in *Mesomolecules From Molecules to Materials*, (G. D. Mendenhall, A. Greenberg, and J. F. Liebman, ed.), Chapman & Hall, New York **1995**, 27–68.

J. R. Moss, "New Large Molecules Contain Ruthenium", *Platinum Metals Review* **1995**, *39*, 33–36.

G. R. Newkome, R. Güther, F. Cardullo, "Supramolecular chemistry of cascade polymers: construction, molecular inclusion, and inorganic connectivity", *Macromol. Symp.* **1995**, *98*, 467–474.

G. R. Newkome, G. R. Baker, "„Smart" Cascade Macromolecules. From Arborols to Unimolecular Micelles and Beyond" in *Molecular Engineering for Advanced Materials*, (J. Becher and K. Schaumburg, ed.), Kluwer Academic Press, **1995**, 59–75.

J. P. Tam, J. C. Spetzler, "Chemoselective approaches to the preparation of peptide dendrimers and branched artificial proteins using unprotected peptides as building blocks", *Biomed. Pept. , Proteins Nucleic Acids* **1995**, *1*, 123–132.

D. A. Tomalia, "Dendrimers – nanoscopic supermolecules according to dendritic rules and principles" in *Supramolecular Stereochemistry*, (J. S. Siegel, ed.), Kluwer Academic Publishers, The Netherlands **1995**, 21–26.

D. A. Tomalia, "Dendrimer Molecules", *Sci. Am.* **1995**, *272*, 62–66.

B. I. Voit, "Dendritic polymers: from aesthetic macromolecules to commercially interesting materials", *Acta Polym.* **1995**, *46*, 87–99.

10.1.8   1994 Reviews

K. Akiyoshi, "Cascade polymers as drug carriers", *Kagaku (Kyoto)* **1994**, *49*, 442.

S. C. E. Backson, P. M. Bayliff, W. J. Feast, A. M. Kenwright, D. Parker, R. W. Richards, "Synthesis and properties of aramid dendrimers", *Makromol. Chem. , Macromol. Symp.* **1994**, *77*, 1–10.

V. Balzani, S. Campagna, G. Denti, A. Juris, S. Serroni, M. Venturi, "Bottom-up strategy to obtain luminescent and redox-active metal complexes of nanometric dimensions", *Coord. Chem. Rev.* **1994**, *132*, 1–13.

M. R. Bryce, A. S. Batsanov, W. Devonport, J. N. Heaton, J. A. K. Howard, G. J. Marshallsay, A. J. Moore, P. J. Skabara, S. Wegener, "New materials based on highly-functionalised tetrathiafulvalene derivatives" in *Molecular Engineering for Advanced Materials*, (J. Becher, ed.), Kluwer, Dordrecht, The Netherlands **1994**, 235–250.

E. M. M. de Brabander-van den Berg, A. Nijenhuis, M. Mure, J. Keulen, R. Reintjens, F. Vandenbooren, B. Bosman, R. de Raat, T. Frijns, S. v. d. Wal, M. Castelijns, J. Put, E. W. Meijer, "Large-scale production of polypropylenimine dendrimers", *Makromol. Chem. , Macromol. Symp.* **1994**, *77*, 51–62.

P. R. Dvornic, D. A. Tomalia, "A family tree for polymers", Chem.Br. **1994**, *30*, 641–645.

P. R. Dvornic, D. A. Tomalia, "Starburst® dendrimers: A conceptual approach to nanoscopic chemistry and architecture", *Makromol. Chem. , Macromol. Symp.* **1994**, *88*, 123–148.

G. H. Escamilla, G. R. Newkome, "Bolaamphiphiles: From Golf Balls to Fibers", *Angew. Chem. , Int. Ed. Engl.* **1994**, *33*, 1937–1940.

J. M. J. Fréchet, "Functional Polymers and Dendrimers: Reactivity, Molecular Architecture, and Interfacial Energy", *Science* **1994**, *263*, 1710–1715.

J. M. J. Fréchet, C. J. Hawker, K. L. Wooley, "The convergent route to globular dendritic macromolecules: a versatile approach to precisely functionalized three-dimensional polymers and novel block copolymers", *J. Macromol. Sci. -Pure Appld. Chem.* **1994**, *A31*, 1627–1645.

K. Hatada, K. Ute, N. Miyatake, "Synthetic uniform polymers and their use in polymer science", *Prog. Polym. Sci.* **1994**, *19*, 1067–1082.

C. J. Hawker, K. L. Wooley, J. M. J. Fréchet, "Novel macromolecular architectures: globular block copolymers containing dendritic components", *Makromol. Chem. , Macromol. Symp.* **1994**, *77*, 11–20.

J. Issberner, R. Moors, F. Vögtle, "Dendrimers: From Generations and Functional Groups to Functions", *Angew. Chem. , Int. Ed. Engl.* **1994**, *33*, 2413–2420.

M. Kakimoto, A. Morikawa, "Starburst polymers", *Shinsozai* **1994**, *5*, 75–81.

Y. H. Kim, "Highly branched aromatic polymers prepared by single step syntheses", *Makromol. Chem. , Macromol. Symp.* **1994**, *77*, 21–33.

T. M. Miller, T. X. Neenan, E. W. Kwock, S. M. Stein, "Dendritic analogs of engineering plastics – a general one-step synthesis of dendritic polyaryl ethers", *Macromolecular Symposia* **1994**, *77*, 35–42.

C. N. Moorefield, G. R. Newkome, "A review of dendritic macromolecules" in *Advances in Dendritic Macromolecules*, (G. R. Newkome, ed.), JAI Press, Greenwich, Conn. **1994**, 1–67.

Multiple, *Advances in Dendritic Macromolecules*, (G. R. Newkome) JAI, Greenwich, Conn. **1994**, 1–198.

T. X. Neenan, T. M. Miller, E. W. Kwock, H. E. Bair, "Preparation and properties of monodisperse aromatic dendritic macromolecules" in *Advances in Dendritic Macromolecules*, (G. R. Newkome, ed.), JAI, Greenwich, Conn. **1994**, 105–132.

N. A. Peppas, T. Nagai, M. Miyajima, "Prospects of using star polymers and dendrimers in drug delivery and other pharmaceutical application", *Pharm Tech Japan* **1994**, *10*, 611–617.

A. Rajca, "Organic Diradicals and Polyradicals: From Spin Coupling to Magnetism?", *Chem. Rev.* **1994**, *94*, 871–893.

A. Rajca, "Toward Organic Synthesis of a Nanometer-Size Magnetic Particle", *Adv. Mater. (Weinheim, Fed. Repub. Ger.)* **1994**, *6*, 605–607.

H. Ritter, "Functionalized polymers with comb-like, rotaxanic-, and dendrimeric structures", *GIT Fachzeitschrift für das Laboratorium* **1994**, *38*, 615–619.

T. Seki, ""Chemistry of unimolecular micelle "micellane", *Kagaku (Kyoto)* **1994**, *49*, 586.

I. Stibor, V. Lellek, "Dendrimery", *Chem. Listy* **1994**, *88*, 423–450.

D. A. Tomalia, P. R. Dvornic, "What promise for dendrimers?", *Nature* **1994**, *372*, 617–618.

D. A. Tomalia, "Starburst/Cascade Dendrimers: Fundamental Building-Blocks for a New Nanoscopic Chemistry Set", *Adv. Mater. (Weinheim, Fed. Repub. Ger.)* **1994**, *6*, 529–539.

O. Vogl, J. Bartus, M. F. Qin, P. Zarras, "Molecular architecture of polymers", *J. Macromol. Sci. -Pure Appld. Chem.* **1994**, *A31*, 1329–1353.

10.1.9   1993 Reviews

H. Hart, "Iptycenes, cuppedophanes and cappedophanes", *Pure Appl. Chem.* **1993**, *65*, 27–34.

Y. Li, "A New Category of Dendritic Synthetic Polymers", *Gaofenzi Tongbao (Polymer Bulletin)* **1993**, 155–164.

M. Sprecher, "Unusual Macromolecular Architectures: The Convergent Growth Approach to Dendritic Polymers", *Chemtracts: Org.Chem.* **1993**, <[12] Volume>, 180–185.

D. A. Tomalia, H. D. Durst, "Genealogically Directed Synthesis: Starburst*/Cascade Dendrimers and Hyperbranched Structures", *Top. Curr. Chem.* **1993**, *165*, 193–313.

D. A. Tomalia, "Starburst™/Cascade Dendrimers: Fundamental Building Blocks for a New Nanoscopic Chemistry Set", *Aldrichim. Acta* **1993**, *26*, 91–101.

A. W. van der Made, P. W. N. M. van Leeuwen, J. C. de Wilde, R. A. C. Brandes, "Dendrimeric Silanes", *Adv. Mater. (Weinheim, Fed. Repub. Ger.)* **1993**, *5*, 466–468.

N. Ventosa, D. Ruiz, C. Rovira, J. Veciana, "Dendrimeric hyperbranched alkylaromatic polyradicals with mesoscopic dimensions and high-spin ground states", *Mol. Cryst. Liq. Cryst. Sci. Technol. , Sect. A* **1993**, *232*, 333–342.

Y. Yamamoto, "Silicon and starburst dendrimers", *Organometallic News* **1993**, 40–42.

10.1.10   1992 Reviews

R. Engel, "Cascade Molecules", *Polymer News* **1992**, *17*, 301–305.

C. J. Hawker, K. L. Wooley, J. M. J. Fréchet, "Dendritic macromolecules", *Chem. Austr.* **1992**, *59*, 620–622.

Y. H. Kim, "Highly Branched Polymers", *Adv. Mater. (Weinheim, Fed. Repub. Ger.)* **1992**, *4*, 764–766.

H.-B. Mekelburger, W. Jaworek, F. Vögtle, "Dendrimers, Arborols and Cascade Molecules: Breakthrough to Generations of New Materials", *Angew. Chem. , Int. Ed. Engl.* **1992**, *31*, 1571–1576.

A. Morikawa, M. Kakimoto, Y. Imai, "Starburst polymers (dendrimers)", *Nippon Gomu Kyokaishi* **1992**, *65*, 205–212.

G. R. Newkome, "Unimolecular Micelles" in *Supramolecular Chemistry, Proceedings of the II NATO Forum on Supramolecular Chemistry, Taormina (Sicily) Italy, Dec. 15–18, 1991,* (V. Balzani and L. De Cola, ed.), Kluwer Academic Publishers, Dordrecht, The Netherlands **1992**, 145–155.

G. R. Newkome, C. N. Moorefield, G. R. Baker, "Building blocks for dendritic macromolecules", *Aldrichim. Acta* **1992**, *25*, 31–38.

G. R. Newkome, "The Ins and Outs of Macromolecules" in *Crown Compounds: Towards Future Applications,* (S. R. Cooper, ed.), VCH Publishers, Inc., New York, N. Y. **1992**, 41–49.

D. A. Tomalia, D. M. Hedstrand, "Starburst Dendrimers: Control of Size, Shape, Surface Chemistry, Topology and Flexibility from Atoms to Macroscopic Matter", *Actual. Chim.* **1992**, 347–349.

10.1.11   1991 and Earlier Reviews

G. Denti, S. Campagna, L. Sabatino, S. Serroni, M. Ciano, V. Balzani, "Towards an artifical photosynthesis. di-, tri-, tetra-, and hepta-nuclear luminescent and redox-reactive metal complexes" in *Photochemistry, Conversion and Storage of Solar Energy,* (E. Pelizzetti and M. Schiavello, ed.), Kluwer Academic Publishers, Dordrecht, The Netherlands **1991**, 27–45.

Y. Imai, "Synthesis of new silicon-based condensation polymers", *Kagaku (Kyoto)* **1991**, *46*, 280–281.

Y. Imai, "Synthesis of new functional silicon-based condensation polymers", *J. Macromol. Sci. , Chem.* **1991**, *A28*, 1115–1135.

K. Krohn, "Starburst Dendrimers" and "Arborols", Organic Synthesis Highlights **1991**, 378–383.

A. M. Muzafarov, E. A. Rebrov, V. S. Papkov, "Spatially growing polyorganosiloxanes. Possibilities of molecular construction in highly functional systems", *Usp. Khim.* **1991**, *60*, 1596–1612.

G. Odian, *Principles of Polymerization*, 3rd. ed. Wiley, New York **1991**.

Y. Tezuka, K. Imai, "Synthesis of branched and network state polymers. End-reactive polymers", *Kobunshi* **1991**, *40*, 314–317.

D. Tomalia, "Meet the molecular superstars", *New Scientist* **1991**, *132*, 30–34.

N. J. Turro, J. K. Barton, D. A. Tomalia, "Molecular Recognition and Chemistry in Restricted Reaction Spaces. Photophysics and Photoinduced Electron Transfer on the Surfaces of Micelles, Dendrimers, and DNA", *Acc. Chem. Res.* **1991**, *24*, 332–340.

O. W. Webster, "Living Polymerization Methods", *Science* **1991**, *251*, 887–893.

V. Balzani, F. Barigelletti, L. De Cola, "Metal Complexes as Light Absorption and Light Emission Sensitizers", *Top. Curr. Chem.* **1990**, *158*, 31–71.

Y.-X. Chen, "Cascade – A New Family of Multibranched Macromolecules", *Youji Huaxue* **1990**, *10*, 289–297.

M. Sawamoto, "Recent advances in topologically well-defined polymers", *Kagaku (Kyoto)* **1990**, *45*, 537–539.

D. A. Tomalia, A. M. Naylor, W. A. Goddard, III, "Starburst Dendrimers: Molecular-Level Control of Size, Shape, Surface Chemistry, Topology and Flexibility in the Conversion of Atoms to Macroscopic Materials", *Angew. Chem. , Int. Ed. Engl.* **1990**, *29*, 138–175.

D. A. Tomalia, D. M. Hedstrand, L. R. Wilson, "Dendritic Polymers" in *Encyclopedia of Polymer Science and Engineering*, Wiley & Sons, Inc., New York **1990**, 46.

S. Arai, "Syntheses of Cascade Molecules", *Yuki Gosei Kagaku Kyokaishi* **1989**, *47*, 62–68.

F. H. Kohnke, J. P. Mathias, J. F. Stoddart, "Structure-directed synthesis of new organic materials", *Angew. Chem., Int. Ed. Engl., Adv. Mater.* **1989**, *28*, 1103–1110.

G. Smets, "Synthese van polymeren voor nieuwe materialen", Chemisch Magazine **1989**, 481–483, 485.

A. Juris, V. Balzani, F. Barigelletti, S. Campagna, P. Belser, A. von Zelewsky, "Ru(II) Polypyridine complexes: photophysics, photochemistry, electrochemistry, and chemiluminescence", *Coord. Chem. Rev.* **1988**, *84*, 85–277.

H. Ringsdorf, B. Schlarb, J. Venzmer, "Molecular architecture and function of polymeric oriented systems: models for the study of organization, surface recognition, and dynamics of biomembranes", *Angew. Chem. , Int. Ed. Engl.* **1988**, *27*, 113–158.

K. Krohn, "Starburst Dendrimere und Arborole", *Nachrichten aus Chemie, Technik und Laboratorium* **1987**, *35*, 1252–1255.

S. Shinkai, "Cascade Syntheses", *Kagaku (Kyoto)* **1987**, *42*, 74–75.

F. M. Menger, "Chemistry of Multi-Armed Organic Compounds", *Top. Curr. Chem.* **1986**, *136*, 1–15.

G. R. Newkome, G. R. Baker, "The chemistry of methanetricarboxylic esters. A review", *Org. Prep. Proced. Int.* **1986**, *18*, 117–144.

T. Otsu, T. Matsunaga, "New type of oligomers – starburst and calixarene oligomers", Kagaku (Kyoto) **1986**, *41*, 206–207.

10.2   Advances Series

"Advances in Dendritic Macromolecules" Series (G. R. Newkome, ed.), JAI Press, Greenwich, Connecticut (USA).

Volume 1 (1994):

(1) C. N. Moorefield, G. R. Newkome, "A review of dendritic macromolecules" in *Advances in Dendritic Macromolecules*, (G. R. Newkome, ed.), JAI Press, Greenwich, Conn. **1994**, 1–67.

(2) Z. Xu, B. Kyan, J. S. Moore, "Stiff dendritic macromolecules based on phenylacetylenes" in *Advances in Dendritic Macromolecules*, (G. R. Newkome, ed.), JAI, Greenwich, Conn. **1994**, 69–104.

(3) T. X. Neenan, T. M. Miller, E. W. Kwock, H. E. Bair, "Preparation and properties of monodisperse aromatic dendritic macromolecules" in *Advances in Dendritic Macromolecules*, (G. R. Newkome, ed.), JAI, Greenwich, Conn. **1994**, 105–132.

(4) A. Rajca, "High-spin polyarylmethyl polyradicals" in *Advances in Dendritic Macromolecules*, (G. R. Newkome, ed.), JAI Press, Inc., Greenwich, Conn. **1994**, 133–168.

(5) G. R. Baker, J. K. Young, "A systematic nomenclature for cascade (dendritic) polymers" in *Advances in Dendritic Macromolecules*, (G. R. Newkome, ed.), JAI, Greenwich, CT **1994**, 169–186.

Volume 2 (1995):

(1) C. J. Hawker, K. L. Wooley, "The convergent-growth approach to dendritic macro-molecules" in *Advances in Dendritic Macromolecules*, (G. R. Newkome, ed.), JAI Press, Greenwich, Conn. **1995**, 1–39.

(2) R. Moors, F. Vögtle, "Cascade molecules: Building blocks, multiple functionaliza-tion, complexing units, photoswitches" in *Advances in Dendritic Macromolecules*, (G. R. Newkome, ed.), JAI, Greenwich, Conn. **1995**, 41–71.

(3) R. Engel, "Ionic dendrimers and related materials" in *Advances in Dendritic Macro-molecules*, (G. R. Newkome, ed.), JAI, Greenwich, Conn. **1995**, 73–99.

(4) L. J. Mathias, T. W. Carothers, "Silicon-based stars, dendrimers, and hyperbranched polymers" in *Advances in Dendritic Macromolecules*, (G. R. Newkome, ed.), JAI, Greenwich, Conn. **1995**, 101–121.

(5) Y. H. Kim, "Highly branched aromatic polymers: Their preparation and applica-tions" in *Advances in Dendritic Macromolecules*, (G. R. Newkome, ed.), JAI Press, Greenwich, Conn. **1995**, 123–156.

(6) G. H. Escamilla, "Dendritic bolaamphiphiles and related molecules" in *Advances in Dendritic Macromolecules*, (G. R. Newkome, ed.), JAI Press, Inc., Greenwich, Conn. **1995**, 157–190.

Volume 3 (1996):

(1) H. Ihre, M. Johansson, E. Malmström, A. Hult, "Dendrimers and hyperbranched aliphatic polyesters based on 2,2-Bis(hydroxymethyl)propionic acid (Bis-MPA)" in *Advances in Dendritic Macromolecules*, (G. R. Newkome, ed.), JAI, Greenwich, Conn. **1996**, 1–25.

(2) N. Ventosa, D. Ruiz, C. Rovira, J. Veciana, "Consequences of the fractal character of dendritic high-spin macromolecules on their physicochemical properties" in *Advances in Dendritic Macromolecules*, (G. R. Newkome, ed.), JAI, Greenwich, Conn. **1996**, 27–59.

(3) S. Serroni, S. Campagna, G. Denti, A. Juris, M. Venturi, V. Balzani, "Dendrimers based on metal complexes" in *Advances in Dendritic Macromolecules*, (G. R. New-kome, ed.), JAI, Greenwich, Conn. **1996**, 61–113.

(4) M. R. Bryce, W. Devonport, "Redox-active dendrimers, related building blocks, and oligomers" in *Advances in Dendritic Macromolecules*, (G. R. Newkome, ed.), JAI, Greenwich, Conn. **1996**, 115–149.

(5) I. Cuadrado, M. Morán, J. Losada, C. M. Casado, C. Pascual, B. Alonso, F. Lobete, "Organometallic dendritic macromolecules: organosilicon and organome-tallic entities as cores or building blocks" in *Advances in Dendritic Macromolecules*, (G. R. Newkome, ed.), JAI, Greenwich, Conn. **1996**, 151–195.

Volume 4 (1999):

(1) M. A. Hearshaw, A. T. Hutton, J. R. Moss, K. J. Naidoo, "Organometallic dendri-mers: Synthesis, structural aspects, and applications in catalysis" in *Advances in Dendritic Macromolecules*, (G. R. Newkome, ed.), JAI Press, Inc., Stamford, CN **1999**, 1–60.

(2) M. H. P. van Genderen, E. M. M. de Brabander-van den Berg, E. W. Meijer, "Poly (propylene imine) dendrimers" in *Advances in Dendritic Macromolecules*, (G. R. Newkome, ed.), JAI Press, Inc., Stamford, CN **1999**, 61–105.

(3) H.-F. Chow, T. K. K. Mong, C.-W. Wan, Z.-Y. Wang, "Chiral Dendrimers" in *Advances in Dendritic Macromolecules*, (G. R. Newkome, ed.), JAI Press Inc., Greenwich, CN **1999**, 107–133.

(4) M. V. Diudea, G. Katona, "Molecular topology of dendrimers" in *Advances in Den-dritic Macromolecules*, (G. R. Newkome, ed.), JAI Press, Inc., Stamford, CN **1999**, 135–201.

Volume 5 (2001)

(1) O. Villavicenio, D. V. McGrath, "Azobenzene-containing dendrimers" in *Advances in Dendritic Molecules*, (G. R. Newkome, ed.), JAI Press, Inc., **2001**, in press.
(2) I. Gitsov, "Linear-dendritic block copolymers. Synthesis and characterization" in *Advances in Dendritic Molecules*, (G. R. Newkome, ed.), JAI Press, **2001**, in press.
(3) B. Alonso, E. Alonso, D. Astruc, J.-C. Blais, L. Djakovitch, J.-L. Fillaut, S. Nlate, F. Moulines, S. Rigaut, J. Ruiz, C. Valério, "Dendrimers containing ferrocenyl- or other transition-metal sandwich groups", in *Advances in Dendritic Molecules*, (G. R. Newkome, ed.), JAI Press, Inc., **2001**, in press.
(4) E. Wiener, V. V. Narayanan, "Magnetic resonance imaging contrast agents: Theory and the role of dendrimers" in *Advances in Dendritic Molecules*, (G. R. Newkome, ed.), JAI Press, Inc., **2001**, in press.

10.3 SEARCH Series

Volume 1. "Mesomolecules, From Molecules to Materials", (G. D. Mendenhall, A. Greenberg, J. F. Liebman, eds), Chapman & Hall, New York, **1995**.
(1) G. D. Mendenhall, "Fractal Index and Fractal Notation" (G. D. Mendenhall, A. Greenberg, and J. F. Liebman, ed.), Chapman & Hall, New York **1995**, 181–194.
(2) G. R. Newkome, C. N. Moorefield, "Cascade Molecules" in *Mesomolecules From Molecules to Materials*, (G. D. Mendenhall, A. Greenberg, and J. F. Liebman, ed.), Chapman & Hall, New York **1995**, 27–68.
(3) G. Denti, S. Campagna, V. Balzani, "Dendritic Polynuclear Metal Complexes with Made-to-Order Luminescent and Redox Properties" in *Mesomolecules from Molecules to Materials*, (G. D. Mendenhall, A. Greenberg, and J. F. Liebman, ed.), Chapman & Hall, New York **1995**, 69–106.
(4) J. D. Wuest, "Molecular Tectonics" in *Mesomolecules: From Molecules to Materials*, (G. D. Mendenhall, A. Greenberg, and J. F. Liebman, ed.), Chapman & Hall, New York **1995**, 107–131.
(5) J. Michl, "Supramolecular Assemblies from 'Tinkertoy' Rigid-Rod Molecules" in *Mesomolecules: From Molecules to Materials*, (G. D. Mendenhall, A. Greenberg, and J. F. Liebman, ed.), Chapman & Hall, New York **1995**, 132–160.
(6) J. A. Jaszczak, "Graphite: Flat, Fibrous, and Spherical" in *Mesomolecules: From Molecules to Materials*, (G. D. Mendenhall, A. Greenberg, and J. F. Liebman, ed.), Chapman & Hall, New York **1995**, 161–180.
(7) G. D. Mendenhall, "Fractal Index and Fractal Notation" (G. D. Mendenhall, A. Greenberg, and J. F. Liebman, ed.), Chapman & Hall, New York **1995**, 181–194.

10.4 Topics in Current Chemistry

"Dendrimers I", (F. Vögtle, volume editor), Springer-Verlag, Berlin, **1998**.
(1) N. Feuerbacher, F. Vögtle, "Iterative Synthesis in Organic Chemistry" in *Topics in Current Chemistry*, (F. Vögtle, ed.), Springer-Verlag, Berlin **1998**, 1–18.
(2) V. V. Narayanan, G. R. Newkome, "Supramolecular Chemistry within Dendritic Structures" in *Topics in Current Chemistry*, (F. Vögtle, ed.), Springer Verlag, Berlin/Heidelberg **1998**, 19–77.
(3) J.-P. Majoral, A.-M. Caminade, "Divergent Approaches to Phosphorus-Containing Dendrimers and their Functionalization", *Top. Curr. Chem.* **1998**, *197*, 79–124.
(4) D. Seebach, P. B. Rheiner, G. Greiveldinger, T. Butz, H. Sellner, "Chiral Dendrimers", *Top. Curr. Chem.* **1998**, *197*, 125–164.
(5) A.-D. Schlüter, "Dendrimers with Polymeric Core: Towards Nanocylinders" in *Topics in Current Chemistry*, (F. Vögtle, ed.), Springer-Verlag, Berlin **1998**, 165–191.
(6) M. Venturi, S. Serroni, A. Juris, S. Campagna, V. Balzani, "Electrochemical and Photochemical Properties of Metal-Containing Dendrimers" in *Topics in Current Chemistry*, (F. Vögtle, ed.), Springer-Verlag, Berlin **1998**, 193–228.

"Dendrimers II" (F. Vögtle, volume editor), Springer-Verlag, Berlin, **2000**.
(1) S. Nummelin, M. Skrifvars, K. Rissanen, "Polyester and Ester Functionalized Dendrimers" in *Topics in Current Chemistry,* (F. Vögtle, ed.), Springer-Verlag, Berlin **2000**, 1–67.
(2) H. Frey, C. Schlenk, "Silicon-Based Dendrimers" in *Topics in Current Chemistry,* (F. Vögtle, ed.), Springer-Verlag, Berlin **2000**, 69–129.
(3) M. W. P. L. Baars, E. W. Meijer, "Host-Guest Chemistry of Dendritic Molecules", *Top. Curr. Chem.* **2000**, *210,* 131–227.
(4) D. K. Smith, F. Diederich, "Supramolecular Dendrimer Chemistry – A Journey Through the Branched Architecture" in *Topics in Current Chemistry,* (F. Vögtle, ed.), Springer-Verlag, Berlin **2000**, 183–227.
(5) D. Astruc, J.-C. Blais, E. Cloutet, L. Djakovitch, S. Rigaut, J. Ruiz, V. Sartor, C. Valério, "The First Organometallic Dendrimers: Design and Redox Functions" in *Topics in Current Chemistry,* (F. Vögtle, ed.), Springer-Verlag, Berlin **2000**, 229–259.
(6) W. Krause, N. Hackmann-Schlichter, F. K. Maier, R. Müller, "Dendrimers in Diagnostics" in *Topics in Current Chemistry,* (F. Vögtle, ed.), Springer-Verlag, Berlin **2000**, 261–308.

"Dendrimers III" (F. Vögtle, volume editor), Springer-Verlag, Berlin, **2001**.
(1) U.-M. Wiesler, T. Weil, K. Müllen, "Nanosized Polyphenylene Dendrimers", *Top. Curr. Chem.* **2001**, *212*, 1–40.
(2) D. Muscat, R. A. T. M. van Benthem, "Hyperbranched Polyesteramides – New Dendritic Polymers", *Top. Curr. Chem.* **2001**, *212*, 41–80.
(3) R. M. Crooks, B. I. Lemon, L. Sun, L. K. Yeung, M. Q. Zhao, "Dendrimer-Encapsulated Metals and Semiconductors: Synthesis, Characterization, and Applications", *Top. Curr. Chem.* **2001**, *212*, 81–135.
(4) S. S. Sheiko, M. Möller, "Hyperbranched macromolecules: Soft particles with adjustable shape and persistent motion capability", *Top. Curr. Chem.* **2001**, *212*, 137–175.
(5) M. Ballauff, "Structure of Dendrimers in Dilute Solution", *Top. Curr. Chem.* **2001**, *212*, 176–194.

10.5   Other Reviews that Directly Impact Dendritic and Hyperbranched Chemistry

Supramolecular Chemistry

J.-M. Lehn, "Design of Organic Complexing Agents. Strategies Towards Properties" in *Structure and Bonding,* (J. D. Dunitz, P. Hemmerich, C. K. Jørgensen, J. B. Neilands, D. Reinen, and R. J. P. Williams, ed.), Springer, New York **1973**.

J.-M. Lehn, "Supramolecular chemistry – scope and perspectives. Molecules, supermolecules, and molecular devices. (Nobel lecture)", *Angew. Chem. , Int. Ed. Engl.* **1988**, *27*, 89–112.

C. D. Gutsche, *Monographs in Supramolecular Chemistry,* Royal Society of Chemistry, London **1989**.

J.-M. Lehn, "Perspectives in supramolecular chemistry – from molecular recognition toward molecular information processing and self organization", *Angew. Chem., Int. Ed. Engl.* **1990**, *29*, 1304–1319.

F. Diederich, M. Gómez-López, "Supramolecular fullerene chemistry", *Chem. Soc. Rev.* **1999**, 263–277.

F. M. MacDonnell, M.-J. Kim, S. Bodige, "Substitutionally inert complexes as chiral synthons for stereospecific supramolecular syntheses", *Coord. Chem. Rev.* **1998**, *185–186,* 535–549.

V. V. Narayanan, G. R. Newkome, "Supramolecular Chemistry within Dendritic Structures" in *Topics in Current Chemistry,* (F. Vögtle, ed.), Springer Verlag, Berlin/Heidelberg **1998**, 19–77.

G. R. Newkome, C. N. Moorefield, "Dendrimers" in *Comprehensive Supramolecular Chemistry*, (D. N. Reinhoudt, ed.), Pergamon, New York **1996**, 777–832.

J. W. Goodby, G. H. Mehl, I. M. Saez, R. P. Tuffin, G. Mackenzie, R. Auzély-Velty, T. Benvegnu, D. Plusquellec, "Liquid crystals with restricted molecular topologies: molecules and supramolecular assemblies", *Chem. Commun.* **1998**, 2057–2070.

F. Zeng, S. C. Zimmerman, "Dendrimers in Supramolecular Chemistry: From Molecular Recognition to Self-Assembly", *Chem. Rev.* **1997**, *97*, 1681–1712.

F. Vögtle, *Supramolecular Chemistry*, Wiley, New York, NY **1991**.

E. C. Constable, "Helices, Supramolecular Chemistry, and Metal-Directed Self-Assembly", *Angew. Chem. , Int. Ed. Engl.* **1991**, *30*, 1450–1451.

J.-M. Lehn, "Perspectives in supramolecular chemistry: From molecular recognition towards self-organization", *Pure Appl. Chem.* **1994**, *66*, 1961–1966.

J.-M. Lehn, *Supramolecular Chemistry: Concepts and Perspectives*, VCH, Weinheim **1995**.

P. L. Boulas, M. Gómez-Kaifer, L. Echegoyen, "Electrochemistry of Supramolecular Systems", *Angew. Chem. , Int. Ed. Engl.* **1998**, *37*, 216–247.

Fullerene Chemistry

F. Diederich, R. Kessinger, "Templated Regioselective and Stereoselective Synthesis in Fullerene Chemistry", *Acc. Chem. Res.* **1999**, *32*, 537–545.

L. Dai, "Conjugated and Fullerene-Containing Polymers for Electronic and Photonic Applications: Advanced Syntheses and Microlithographic Fabrications", *J. Macromol. Sci. -Rev. Macromol. Chem. Phys.* **1999**, *C39*, 273–387.

F. Diederich, M. Gómez-López, "Supramolecular fullerene chemistry", *Chem. Soc. Rev.* **1999**, 263–277.

F. Diederich, L. Isaacs, D. Philp, "Syntheses, Structures, and Properties of Methanofullerenes", *Chem. Soc. Rev.* **1994**, *23*, 243–255.

T. D. Ros, M. Prato, "Medicinal chemistry with fullerenes and fullerene derivatives", *Chem. Commun.* **1999**, 663–669.

Topology

S. Busch, H. Dolhaine, A. DuChesne, S. Heinz, O. Hochrein, F. Laeri, O. Podebrad, U. Vietze, T. Weiland, R. Kniep, "Biomimetic Morphogenesis of Fluorapatite-Gelatin Composites: Fractal Growth, the Question of Intrinsic Electric Fields, Core/Shell Assemblies, Hollow Spheres and Reorganization of Denatured Collagen", *Eur. J. Inorg. Chem* **1999**, 1643–1653.

Carbon Networks

U. H. F. Bunz, "Polyynes – Fascinating Monomers for the Construction of Carbon Networks", *Angew. Chem. , Int. Ed. Engl.* **1994**, *33*, 1073–1076.

R. R. Tykwinski, F. Diederich, "Tetraethynylethene Molecular Scaffolding", *Leigigs Ann./Rec.* **1997**, 649–661.

MRI Contrast Agents

P. Caravan, J. J. Ellison, T. J. McMurry, R. B. Lauffer, "Gadolinium(III) Chelates as MRI Contrast Agents: Structure, Dynamics, and Applications", *Chem. Rev.* **1999**, *99*, 2293–2352.

V. W. W. Yam, K. K. W. Lo, "Recent advances in utilization of transition metal complexes and lanthanides as diagnostic tools", *Coord. Chem. Rev.* **1999**, *184*, 157–240.

M. Botta, "Second Coordination Sphere Water Molecules and Relaxivity of Gadolinium(III) Complexes: Implications for MRI Contrast Agents", *Eur. J. Inorg. Chem* **2000**, 399–407.

Radiometal Agents

M. J. Heeg, S. S. Jurisson, "The Role of Inorganic Chemistry in the Development of Radiometal Agents for Cancer Therapy", *Acc. Chem. Res.* **1999**, *32*, 1053–1060.

Boron Neutron Capture Therapy

M. F. Hawthorne, "The Role of Chemistry in the Development of Boron Neutron Capture Therapy of Cancer", *Angew. Chem. , Int. Ed. Engl.* **1993**, *32*, 950–984.

R. F. Barth, A. H. Soloway, "Boron neutron capture therapy of primary and metastatic brain tumors", *Mol. Chem. Neuropath.* **1994**, *21*, 139–154.

Electrospray Mass Spectrometry

W. Henderson, B. K. Nicholson, L. J. McCaffrey, "Applications of electrospray mass spectrometry in organometallic chemistry", *Polyhedron* **1999**, *17*, 4291–4313.

Combinatorial Material Science

B. Jandeleit, D. J. Schaefer, T. S. Powers, H. W. Turner, W. H. Weinberg, "Combinatorial Material Science and Catalysis", *Angew. Chem. Int. Ed.* **1999**, *38*, 2495–2532.

D. Hudson, "Matrix Assisted Synthetic Transformations: A Mosaic of Diverse Contributions. II. The Pattern Is Completed", *J. Combinatorial Chem.* **1999**, *1*, 403–457.

J. K. Borchardt, "Combinatorial chemistry: Not just for pharmaceuticals", *Today's Chemist at Work* **1998**, 35–41.

W. A. Loughlin, "Combinatorial Synthesis: A Heterocyclic Chemist's Perspective", *Aust. J. Chem.* **1998**, *51*, 875–893.

C. Gennari, H. P. Nestler, U. Piarulli, B. Salom, "Combinatorial Libraries: Studies in Molecular Recognition and the Quest for New Catalysts", *Leigigs Ann./Rec.* **1997**, 637–647.

L. A. Thompson, J. A. Ellman, "Synthesis and Applications of Small Molecule Libraries", *Chem. Rev.* **1996**, *96*, 555–600.

Catalysis

K. Matyjaszewski, "Transition Metal Catalysis in Controlled Radical Polymerization: Atom Transfer Radical Polymerization", *Chem. Eur. J.* **1999**, *5*, 3095–3102.

U. Rosenthal, P.-M. Pellny, F. G. Kirchbauer, V. V. Burlakov, "What Do Titano-and Zirconocenes Do with Diynes and Polyynes?", *Acc. Chem. Res.* **2000**, *33*, 119–129.

D. Astruc, "Electron-Transfer Chain Catalysis in Organotransition Metal Chemistry", *Angew. Chem., Int. Ed. Engl.* **1988**, *27*, 643–660.

R. A. Lerner, S. J. Benkovic, P. G. Schultz, "At the Crossroads on Chemistry and Immunology: Catalytic Antibodies", *Science* **1991**, *252*, 659–667.

F. M. Menger, "Enzyme Reactivity from an Organic Perspective", *Acc. Chem. Res.* **1993**, *26*, 206–212.

Molecular Recognition

M. Mammen, S.-K. Choi, G. M. Whitesides, "Polyvalent Interactions in Biological Systems: Implications for Design and Use of Multivalent Ligands and Inhibitors", *Angew. Chem. Int. Ed.* **1998**, *37*, 2754–2794.

J. Rebek, Jr., "Molecular recognition with model systems", *Angew. Chem. , Int. Ed. Engl.* **1990**, *29*, 245–255.

J. Rebek, Jr., "Molecular recognition and biophysical organic chemistry", *Acc. Chem. Res.* **1990**, *23*, 399–404.

N. J. Turro, J. K. Barton, D. A. Tomalia, "Molecular Recognition and Chemistry in Restricted Reaction Spaces. Photophysics and Photoinduced Electron Transfer on the Surfaces of Micelles, Dendrimers, and DNA", *Acc. Chem. Res.* **1991**, *24*, 332–340.

H.-J. Schneider, "Mechanisms of Molecular Recognition: Investigations of Organic Host-Guest Complexes", *Angew. Chem., Int. Ed. Engl.* **1991**, *30*, 1417–1436.

V. Percec, J. Heck, G. Johansson, D. Tomazos, M. Kawasumi, P. Chu, G. Ungar, "Molecular recognition directed self-assembly of supramolecular architectures", *J. Macromol. Sci. Pure Appld. Chem.* **1994**, *A31*, 1719–1758.

Molecular Magnetic Materials

J. S. Miller, A. J. Epstein, "Organic and Organometallic Molecular Magnetic Materials – Designer Magnets", *Angew. Chem. , Int. Ed. Engl.* **1994**, *33*, 385–415.

J. Veciana, C. Rovira, "Stable polyradicals with high spin ground states" in *Magnetic Molecular Materials,* (D. Gatteschi, ed.), Kluwer Academic Publishers, The Netherlands **1991**, 121–132.

Shape Persistent Architectures

J. S. Moore, "Shape-Persistent Molecular Architectures of Nanoscale Dimension", *Acc. Chem. Res.* **1997**, *30*, 402–413.

U. H. F. Bunz, "Poly(aryleneethynylene)s: Synthesis, Properties, Structures, and Applications", *Chem. Rev.* **2000**, *100*, 1605–1644.

R. Ziessel, "Making New Supermolecules for the Next Century: Multipurpose Reagents from Ethynyl-Grafted Oligopyridines", *Synthesis* **1999**, *11*, 1839–1865.

M. Müller, C. Kübel, K. Müllen, "Giant Polycyclic Aromatic Hydrocarbon", *Chem. Eur. J.* **1998**, *4*, 2099–2109.

A. de Meijere, S. I. Kozhushkov, "The Chemistry of Highly Strained Oligospirocyclopropane Systems", *Chem. Rev.* **2000**, *100*, 93–142.

S. Leininger, B. Olenyuk, P. J. Stang, "Self-Assembly of Discrete Cyclic Nanostructures Mediated by Transition Metals", *Chem. Rev.* **2000**, *100*, 853–908.

J. M. Tour, "Conjugated macromolecules of precise length and constitution. Organic synthesis for the construction of nanoarchitectures", *Chem. Rev.* **1996**, *96*, 537–553.

Anion Radicals

L. L. Miller, K. R. Mann, "π-Dimers and π-Stacks in Solution and in Conducting Polymers", *Acc. Chem. Res.* **1996**, *29*, 417–423.

Morphosynthesis of Biomimetic Forms

G. A. Ozin, "Morphogenesis of Biomineral and Morphosynthesis of Biomimetic Forms", *Acc. Chem. Res.* **1997**, *30*, 17–27.

S. Busch, H. Dolhaine, A. DuChesne, S. Heinz, O. Hochrein, F. Laeri, O. Podebrad, U. Vietze, T. Weiland, R. Kniep, "Biomimetic Morphogenesis of Fluorapatite-Gelatin Composites: Fractal Growth, the Question of Intrinsic Electric Fields, Core/Shell Assemblies, Hollow Spheres and Reorganization of Denatured Collagen", *Eur. J. Inorg. Chem* **1999**, 1643–1653.

G. A. Ozin, "Panoscopic materials: synthesis over 'all' length scales", *Chem. Commun.* **2000**, 419–432.

Fractals

W. G. Rothschild, "Fractality and its measurements" in *Fractals in Chemistry,* John Wiley & Sons, Inc., New York **1998**, 170–179.

Fractals, Quasicrystals, Chaos, Knots, and Algebraic Quantum Mechanics, (A. Amann, L. Cederbaum, and W. Gans) Kluwer, Dorrecht, The Netherlands **1988**.

The Fractal Approach to Heterogenous Chemistry: Surface, Colloids, Polymers, (D. Avnir) Wiley, New York **1989**.

B. B. Mandelbrot, *Les Objets Fractals: Forme, Hasard et Dimension*, Flammarion, Paris **1975**.

B. B. Mandelbrot, *Fractals: Form, Chance and Dimension*, Freeman, San Francisco **1977**.

B. B. Mandelbrot, *The Fractal Geometry of Nature*, Freeman, San Francisco **1982**.

H. Takayasu, *Fractals in the Physical Sciences*, Manchester University Press, Manchester **1990**.

A. Harrison, *Fractals in Chemistry*, Oxford University Press, New York **1995**.

D. H. Rouvray, "Similarity in Chemistry: Past, Present, and Future", *Top. Curr. Chem.* **1995**, *Vol. 173*, 1.

T. Wegner, B. Tyler, *Fractal Creations*, 2nd ed. ed. Waite Group Press, Corte Madera, California **1993**, 16.

A. Blumen, H. Schnörer, "Fractals and related hierarchical models in polymer science", *Angew. Chem. , Int. Ed. Engl.* **1990**, *29*, 113–222.

P. Prusinkiewicz, A. Lindenmayer, *The Algorithmic Beauty of Plants*, Springer, New York **1990**.

D. J. Klein, M. J. Cravey, G. E. Hite, "Fractal Benzenoids" in *Polycyclic Aromatic Compounds,* Gordon and Breach, New York **1991**, 163–182.

G. Zumofen, A. Blumen, J. Klafter, "The role of fractals in chemistry", *New J. Chem.* **2000**, *14*, 189–196.

Hydrogen Bonding Aggregates

G. M. Whitesides, E. E. Simanek, J. P. Mathias, C. T. Seto, D. N. Chin, M. Mammen, D. M. Gordon, "Noncovalent Synthesis: Using Physical-Organic Chemistry To Make Aggregates", *Acc. Chem. Res.* **1995**, *28*, 37–44.

G. R. Desiraju, "The C−H−−O Hydrogen Bond: Structural Implications and Supramolecular Design", *Acc. Chem. Res.* **1996**, *29*, 441–449.

D. Philp, J. F. Stoddart, "Self-Assembly in Natural and Unnatural Systems", *Angew. Chem., Int. Ed. Engl.* **1996**, *35*, 1154–1196.

Micellar Electrokinetic Chromatography

K. Otsuka, S. Terabe, "Micellar Electrokinetic Chromatography", *Bull. Chem. Soc. Jpn.* **1998**, *71*, 2465–2481.

C. P. Palmer, N. Tanaka, "Selectivity of polymeric and polymer-supported pseudo-stationary phases in micellar electrokinetic chromatography", *J. Chromatogr. A* **1997**, *792*, 105–124.

Self-assembly

S. Leininger, B. Olenyuk, P. J. Stang, "Self-Assembly of Discrete Cyclic Nanostructures Mediated by Transition Metals", *Chem. Rev.* **2000**, *100*, 853–908.

P. J. Stang, B. Olenyuk, "Self-Assembly, Symmetry, and Molecular Architecture: Coordination as the Motif in the Rational Design of Supramolecular Metallacyclic Polygons and Polyhedra", *Acc. Chem. Res.* **1997**, *30*, 502–518.

G. M. Whitesides, J. P. Mathias, C. T. Seto, "Molecular Self-Assembly and Nanochemistry: A Chemical Strategy for the Synthesis of Nanostructures", *Science* **1991**, *254*, 1312–1319.

D. Philp, J. F. Stoddart, "Self-assembly in organic synthesis", *Synlett* **1991**, 445–458.

J. S. Lindsey, "Self-assembly in synthetic routes to molecular devices. Biological principles and chemical perspectives: a review", *New J. Chem.* **1991**, *15*, 153–180.

V. Percec, J. Heck, G. Johansson, D. Tomazos, M. Kawasumi, P. Chu, G. Ungar, "Molecular recognition directed self-assembly of supramolecular architectures", *J. Macromol. Sci. Pure Appld. Chem.* **1994**, *A31*, 1719–1758.

G. M. Whitesides, "Self-Assembling Materials", *Sci. Am.* **1995**, 146–149.

M. W. Hosseini, A. D. Cain, "Crystal engineering: molecular networks based on inclusion phenomena", *Chem. Commun.* **1998**, 727–733.

P. J. Stang, "Molecular Architecture: Coordination as the Motif in the Rational Design and Assembly of Discrete Supramolecular Species – Self-Assembly of Metallacyclic Polygons and Polyhedra", *Chem. Eur. J.* **1998**, *4*, 10–27.

D. Philp, J. F. Stoddart, "Self-Assembly in Natural and Unnatural Systems", *Angew. Chem. , Int. Ed. Engl.* **1996**, *35*, 1154–1196.

Multiple Antigen Peptides

J. P. Tam, "Recent advances in multiple antigen peptides", *J. Immunol. Methods* **1996**, *196*, 17–32.

Drug Delivery

C. Monfardini, F. M. Veronese, "Stabilization of Substances in Circulation", *Bioconj. Chem.* **1998**, *9*, 418–450.
I. Lebedeva, L. Benimetskaya, C. A. Stein, M. Vilenchik, "Cellular delivery of antisense oligonucleotides", *Eur. J. Pharm. Biopharm.* **2000**, *50*, 101–119.
C. Z. Chen, S. L. Cooper, "Recent Advances in Antimicrobial Dendrimers", *Adv. Mater.* **2000**, *12*, 843–846.

Micelles

H. Hoffmann, G. Ebert, "Surfactants, micelles and fascinating phenomena", *Angew. Chem. , Int. Ed. Engl.* **1988**, *27*, 902–912.
J. F. Rusling, "Controlling Electrochemical Catalysis with Surfactant Microstructure", *Acc. Chem. Res.* **1991**, *24*, 75–81.
F. M. Menger, "Groups of Organic Molecules That Operate Collectively", *Angew. Chem. , Int. Ed. Engl.* **1991**, *30*, 1086–1099.
F. M. Menger, "Quarter Century Progress and New Horizons in Micelles" in *Micelles, Microemulsions, and Monolayers Science and Technology,* (D. O. Shah, ed.), Marcel Dekker, Inc., New York **1998**, 53–71.

Molecular Machines

10.6 Nomenclature

10.6.1 Background on Trivial and Traditional Names

When Vögtle and his coworkers[1] described the first synthetic examples of discrete, branched, polyfunctional molecules prepared via an iterative, step-wise "cascade synthesis," he opened the door to a new vista of discrete meso- and macromolecules, whose structures can be easily envisioned, but are nearly impossible to name based on current nomenclature systems. Thus, researchers in the field resorted to naming their new materials with trivial names, such as: arborols,[2] cascadol,[3] Cauliflower polymers,[4] crowned arborols,[5] dendrimers,[6] molecular fractals,[7–9] polycules,[10] silvanols,[11] and "starburst" dendrimers.[6] Reliance on the *IUPAC* or *Chemical Abstracts* nomenclature resulted in names longer, in most cases, than the associated experimental details and was impossible for researchers to use for retrieval purposes. For example, one on the first macromolecules to be reported was [27]-arborol (Fig. 10.1);[2] the *Chemical Abstracts* name is "1,19-dihydroxy-N,N',N'',N'''-tetra*kis*[2-hydroxy-1,1-*bis*(hydroxymethyl)ethyl]-10-[[4-[[2-hydroxy-1,1-*bis*(hydroxymethyl)ethyl]amino]-3,3-*bis*[[2-hydroxy-1,1-*bis*(hydroxymethyl)ethyl]amino]carbonyl]-4-oxobutoxy]methyl]-2,2,18,18-tetra*kis*(hydroxymethyl)-4,16-dioxo-10-pentyl-8,12-dioxa-3,17-diazanonadecane-5,5,15,15-tetracarboxamide." From the name one cannot readily ascertain the terminal (or surface) groups

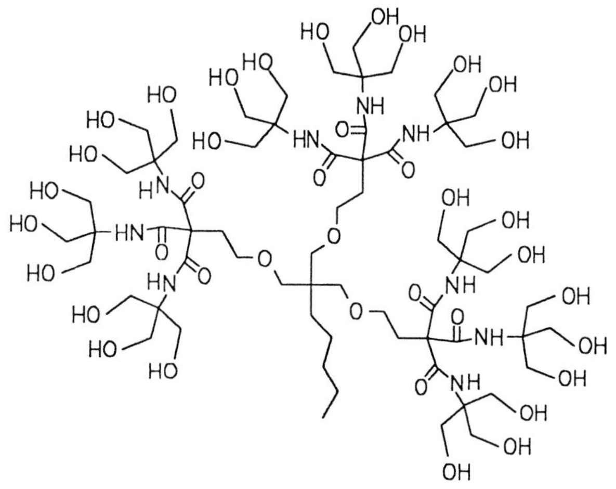

Figure 10.1. 27-Cascade : hexane[3-1,1,1] : (4-oxapentylidene) : (3-oxo-2-azapropylidene) : methanol.

(the 27 alcohol units), the branching multiplicity [three (3)] or the initiator core (a 1,1,1-trisubstituted hexane moiety). Application of the following rules[12,13] for Figure 10.1 leads to 27-Cascade:hexane[3–1,1,1]:(4-oxapentylidyne):(3-oxo-2-azapropylidyne):methanol.

10.6.2   Definition of a Cascade Polymer

A cascade polymer at the *n*th generation has the general formula:

$$C[R_1(R_2(...R_1(...R_n(T)_{N_{b_n}}...)_{N_{b_i}}...)_{N_{b_2}})_{N_{b_1}}]_{N_c} \tag{1}$$

where C is the formula for the core moiety; R_i is the formula for the repeat or branch unit; T is the formula for the terminal moieties; N_{b_i} is the branch multiplicity of the *i*th repeat unit or generations; and N_c is the multiplicity of the branching from the central core. The number of terminal moieties (Z) is calculated :

$$Z = N_c \prod_{i=1}^{n} N_{b_i} \tag{2}$$

If the branching multiplicity remains constant throughout the macromolecule, i.e. $N_{b_i} = N_{b_2} = \cdots = N_{b_n}$, then the relationship simplifies to give: $Z = N_c N_b^G$. For cascade polymers (dendrimers) possessing the same branch unit throughout the structure, the line formula may be more simply represented by:

$$[\text{Core unit}][(\text{Repeat Unit})_{N_b}^G (\text{Terminal Unit})]_{N_c} \tag{3}$$

The proposed nomenclature was derived from these line notations.

10.6.3   Proposed Cascade Nomenclature Rules[12, 13]

1. In the Cascade name, components (names of the units) are separated from each other by colons and are cited in sequence from the core unit out to the terminal unit(s).
2. The name begins with a numeral corresponding to the number of terminal functionalities, followed in succession by "Cascade" (to denote this class of molecules) and names of the core unit, repeat intermediate units, and finally the terminal unit(s).
3. The combination of core and terminal unit names resembles conjunctive nomenclature. The multiplicity of branching (cascading) from the core unit is indicated by a bracketed numeral immediately following the name of the core unit; if locants are

necessary, they are also enclosed within the brackets, following and separated by a hyphen from the multiplicity numeral.

4. A repeat intermediate unit consists of the molecular fragment extending from (but not including) one branch atom (or group) through the next cascade branching site.

5. The parent chain of an intermediate or terminal unit always terminates at a cascade branching site.

6. A superscript on a parenthetical unit denotes the number of successive repetitions of that unit.

7. Numbering of chains of units, including the core and terminal ones, is in descending order in the direction core → terminal unit. This order preserves the well established IUPAC rules that the locant for example -CO_2H is one (1) and that for the point of attachment of the alkyl and related groups is also one (1).

8. Repetition of combinations of repeat units is indicated by enclosure of the component repeat names (inside parentheses and separated by colons) within brackets. A superscript on the bracketed unit denotes the number of repetitions of that sequence.

9. When the repeat unit is composed of nonequivalent branches extending from a cascade branching site, the name of that unit includes a name for each branch. Within the parenthetical name of the repeat unit, the branch names are cited by increasing parent chain length and then alphabetically, as well as are separated by colons. This situation usually results in the attachment of nonequivalent terminal groups at the last cascade branch point; in such cases, the terminal unit names are cited in sequence of increasing parent chain length and separated by colons.

10. For cascade molecules with different arms emanating from the core e. g., a segmental block cascade polymer, the name of each cascade segment, exclusive of the core, is given within square brackets. The segment names are separated by a hyphen, arranged in alphabetical order, and preceded by core locants and group multipliers, when necessary.

10.6.4 General Patterns

10.6.4.1 Similar Repeat Internal Units

Z-Cascade:Core[N_c]:(Internal Units)^n:Terminal Unit

The simplest cascade structure[14] possesses symmetrical branches and identical repeat units. The 2nd generation Micellanoic Acid^{TM} dendrimer has the structure shown in Figure 10.2 and cascade name is given below the structure. The IUPAC name for this structure is 4,4,40,40-tetra*kis*(propylcarboxy)-13,13,31,31-tetra*kis*[12-carboxy-9,9-*bis*(3-propylcarboxy)dodecyl]-22,22-*bis*[21-carboxy-18,18-*bis*(propylcarboxy)-9,9-*bis*[12-carboxy-9,9-*bis*(propylcarboxy)dodecyl]heneicosyl]tritetracontanedioic acid. The cascade name is derived in several steps. First, the number of the terminal groups is determined by calculating the product of bracket subscripts (i.e., 3 ′ 3 ′ 4); this number prefixes the class designation in the name (i.e., 36-Cascade). From the line formula representation, the initiator core is a tetrasubstituted methane (i.e., methane[4]). The next two molecular fragments are an identical nine (9) carbon chain [i.e., (nonylidyne)^2]. Finally, the termini are propanoic acid groups.

$$C[(CH_2)_8C[(CH_2)_8C[(CH_2)_2CO_2H]_3]_3]_4$$

Application of the nomenclature scheme to the 2nd generation cascade, shown in Figure 10.3[15] affords the name 36-Cascade: methane[4]:(3-oxo-6-oxa-2-azaheptylidyne)^2:4-oxapentanoic acid.

The introduction of chirality into dendritic assemblies can be easily addressed as shown for the structure in Figure 10.4[16], which would be named: 6-Cascade:benzene [3–1,3,5]:(5-(3*S*,4*S*-1,6-dioxa-3,4-*O*-isopropylidene-3,4-dihydroxyhexyl)-1,3-phenylene): 4-(3*S*,4*S*-1,6-dioxa-3,4-*O*-isopropylidene-3,4-dihydroxyhexyl)-*tert*-butylbenzene.

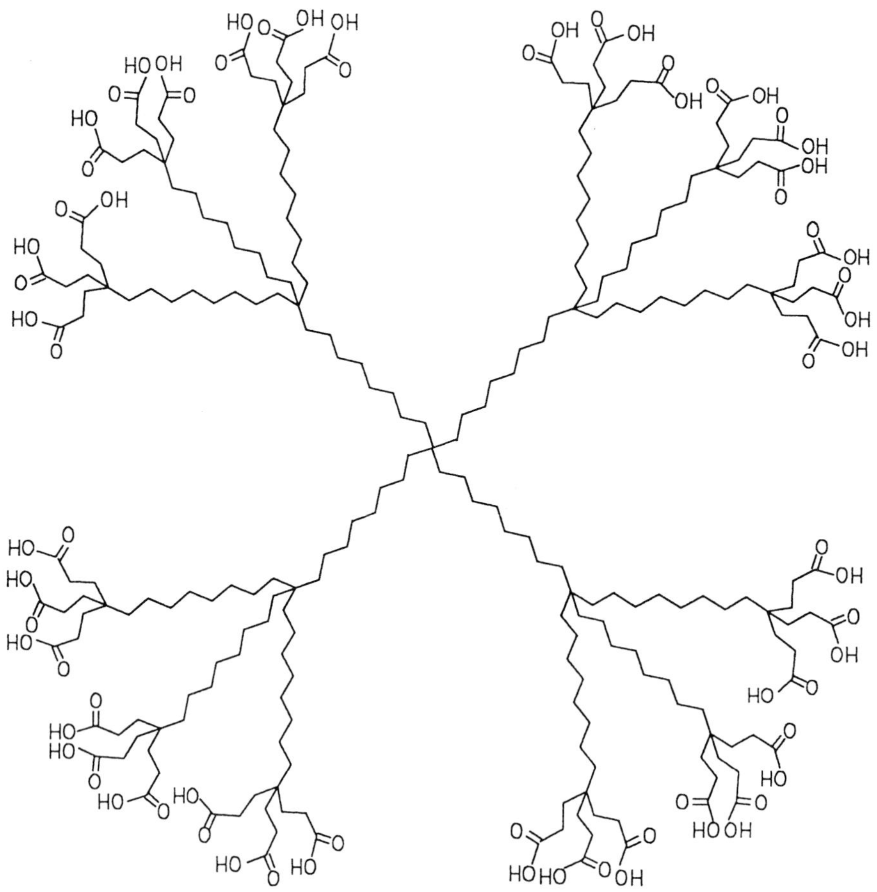

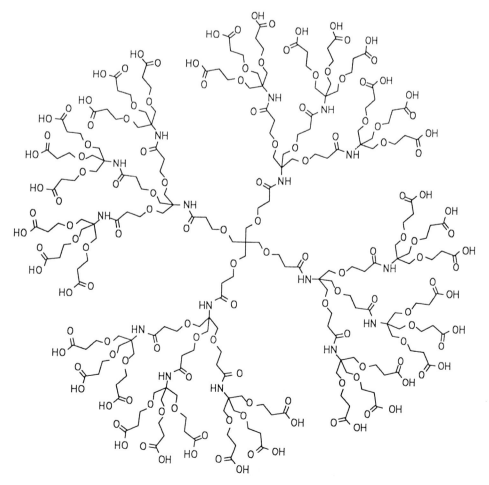

Figure 10.2. 36-Cascade:methane[4]:(nonylidyne)^2:propionic acid.

Figure 10.3. 36-Cascade:methane[4]:(3-oxo-6-oxa-2-azaheptylidyne)^2:4-oxapentanoic acid.

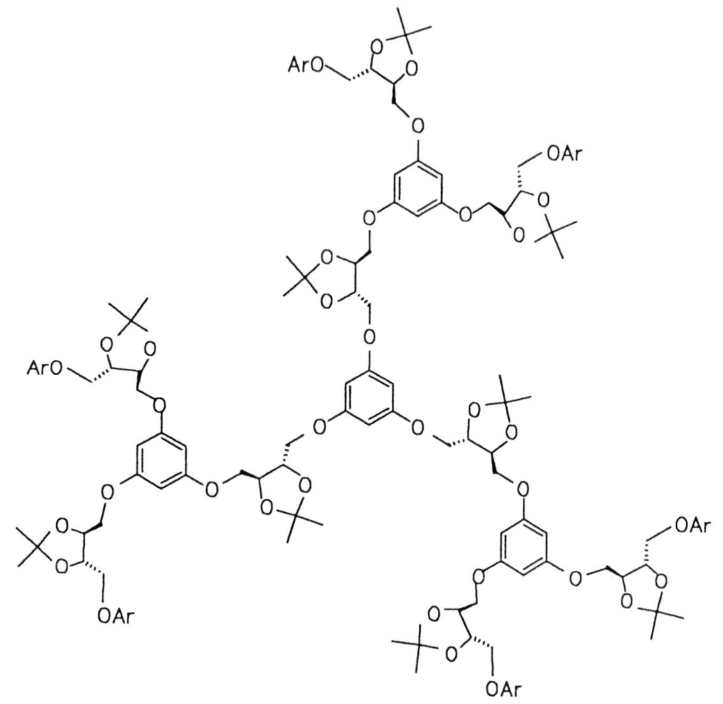

Figure 10.4. 6-Cascade: benzene[3-1,3,5]: (5-(3S,4S-1,6-dioxa-3,4-*O*-isopropylidene-3,4-dihydroxyhexyl)-1,3-phenylene):
4-(3S,4S-1,6-dioxa-3,4-*O*-isopropylidene-3,4-dihydroxyhexyl)-*tert*-butylbenzene.

10.6.4.2 Similar Arms with Dissimilar Internal Units

Z-Cascade:Core[N_c]:(Internal Units A)^m:(Internal Units B)^n: ⋯:Terminal Units
The cascade name for Figure 10.5[17] is derived by the process described above, except
that the attachments to the adamantane core (i.e., 1,3,5,7) must be noted after the core
multiplicity. The repeat units are readily named via *replacement nomenclature*, thus,

← Core (Internal Unit A) Terminus→

$$\text{adamantane} - \overset{3}{C}(=O) - \overset{2}{N}H - \overset{1}{C}R_3$$

(Internal Unit A) (Internal Unit B) Terminus →

$$C -[\overset{5}{C}H_2 \overset{4}{C}H_2 \overset{3}{C}(=O)\overset{2}{N}H\overset{1}{C}R'_3]_3$$

←(Internal Units A & B) Terminus →

$$C - [CH_2CH_2CO_2H]_3$$

Replacement nomenclature emphasizes the length of the repeat unit but masks the functionality (i.e., amide) in these units. Alternatively, a repeat unit name (such as, (propa-
namido)methylidyne), which indicates the functionality but obscures the chain length
may be preferred. The style chosen for the internal (or core or terminal) unit name(s)
does not affect the general form of the proposed cascade nomenclature. Thus, the cascade in Figure 10.5 is 36-Cascade: tricyclo[3.3.1.1^{3,7}]decane[4–1,3,5,7]:(3-oxo-2-azapro-
pylidyne):(3-oxo-2-azapentylidyne): propanoic acid.

10.6.4.3 Dissimilar Arms with Similar Internal Branches

Z-Cascade:Core[N_c]:[(Internal Units)^m:Terminal Unit]:[(Internal Units)^n:Terminal Unit]:[(Internal Units)^p:Terminal Unit]: ⋯
The chiral dendrimer (Figure 10.6)[18] in its racemic form represents an example of the
class, its name is: *rac*-14-Cascade:benzyloxyethane[3–2,2,2,2]:[(5-(2-oxapropyl)-1,3-
phenylene):(2-oxaethyl)benzene]:[(5-(2-oxapropyl)-1,3-phenylene):(2-oxaethyl)-1,3-
phenylene):(2-oxaethyl)benzene]:[(5-(2-oxapropyl)-1,3-phenylene):(2-oxaethyl)-1,3-
phenylene)^2:(2-oxaethyl)benzene].

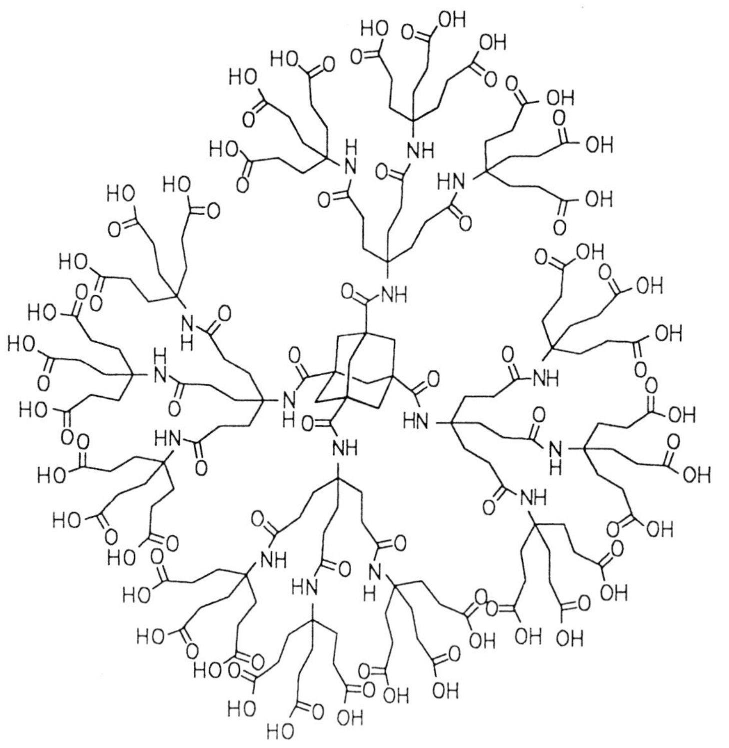

Figure 10.5. 36-Cascade : tricyclo[3.3.1.1^{3,7}]decane[4-1,3,5,7 : (3-oxo-2-azapropylidyne) : (3-oxo-2-azapentylidyne) : propionic acid.

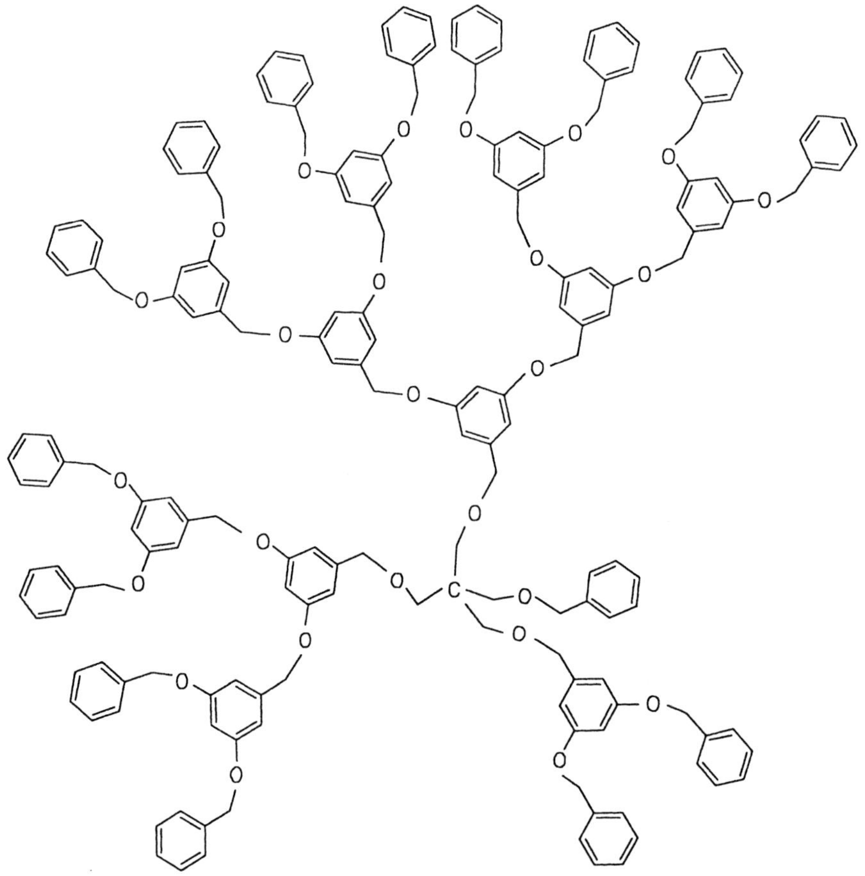

Figure 10.6. *rac*-14-Cascade : benzyloxyethane[3-2,2,2,2] : [(5-(2-oxapropyl)-1,3-phenylene) : (2-oxaethyl)benzene] : [(5-(2-oxapropyl)-1,3-phenylene) : (5-(2-oxaethyl)-1,3-phenylene) : (2-oxaethyl)benzene] : [(5-(2-oxapropyl)-1,3-phenylene) : (5-(2-oxaethyl)-1,3-phenylene)^2 : (2-oxaethyl)benzene].

An alternate name, which is based on methane as the core, would be: *rac*-14-Cascade: methane[4]:[(2-oxapropyl)benzene]:[(5-(2-oxapropyl)-1,3-phenylene):(2-oxaethyl)benzene] : [(-2-oaopropyl)-1,3-phenylene) : (5-(2-oxaethyl)-1,3-phenylene) : (2-oxaethyl)benzene]:[(5-(2-oxapropyl)-1,3-phenylene):(5-(2-oxaethyl)-1,3-phenylene)^2:(2-oxaethyl)benzene].

10.6.4.4 Dissimilar Arms with Dissimilar Internal Branches or Terminal Groups

Z-Cascade:Core[N_c]:[(Internal Units A)^a:(Internal Units B)^b: ⋯ Terminal Units X)]^m-[(Internal Units C)^c:(Internal Units D)^d: ⋯Terminal Units Y)]^n⋯

The cascade depicted in Figure 10.7[19] possesses dissimilar arms with the same internal segments; its name is (where, X = H): 48-Cascade:ethane[3–1,1,1]:*bis*[(5-(2-*p*-phenyl-2-oxaethyl)-1,3-phenylene) : (5-(2-oxaethyl)-1,3-phenylene)^3 : (2-oxaethyl)benzene]-[(5-(2-*p*-phenyl-2-oxaethyl)-1,3-phenylene) : (5-(2-oxaethyl)-1,3-phenylene)^3 : 4-(2-oxaethyl)-1-bromobenzene]; or where (X = Br): 48-Cascade:ethane[3–1,1,1]:[(5-(2-*p*-phenyl-2-oxaethyl)-1,3-phenylene) : (5-(2-oxaethyl)-1,3-phenylene)^3 : (2-oxaethyl)benzene]-*bis*[(5-(2-*p*-phenyl-2-oxaethyl)-1,3-phenylene):(5-(2-oxaethyl)-1,3-phenylene)^3:4-(2-oxaethyl)-1-bromobenzene].

The cascade shown in Figure 10.8 is a hypothetical dendrimer composed of known building blocks (including the core), which possesses different arms comprised of various internal units. The termini also differ. Its name is 36-Cascade:methane[4]:*bis*[(3-oxo-6-oxa-2-azaheptylidyne)^2 : propylamine]-*bis*[(3-oxo-6-oxa-2-azaheptylidyne) : (3-oxo-2-azapentylidyne):propanoic acid].

10.6.4.5 Unsymmetrically Branched Cascades

The cascade in Figure 10.9,[20] based upon lysine, possesses unsymmetrical branches; its name is: 8-Cascade:*N*-(diphenylmethyl)acetamide[2–2,2]:(2-oxo-3-azapropylidene:2-oxo-3-azaheptylidene)^2 : *N*-(*tert*-butoxycarbonyl)amine : *N*-(*tert*-butoxycarbonyl)butyl-amine.

An example of an unsymmetrically branched chiral dendrimer[21] is shown in Figure 10.10, which has the name: 16-Cascade: (*N*-benzyloxycarbonyl)methylamine[2–1,1]:(1*S*-3-oxo-2-azapropylidene:1*S*-3-oxo-2-azapentylidene)^3:ethyl formate:ethyl propanoate.

10.6.5 Fractal Notation

A fractal notation has been proposed by Mendenhall[7–9] to address the complexity of cascade macromolecules, since they possess a high degree of molecular symmetry. This highly condensed description of a symmetrical molecule possessing fractal architecture can be generalized in the following form:

<u>periphery</u> <u>units</u>   <u>subscription</u>   <u>connectors</u>   <u>core</u>

(terminal group)f$_{(outermost hub ... innermost hub)}$(n,n',n'', ... n^z)core

　　　　　<u>fractal</u> <u>notation</u>

Although this notation can be applied to diverse structures, the following rules and examples have been confined to very simple dendritic systems.

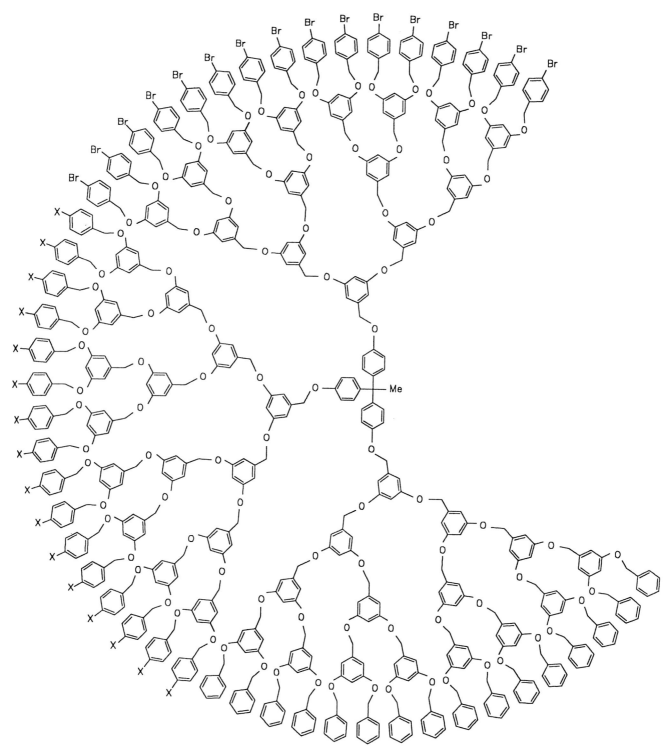

Figure 10.7. 48-Cascade:ethane[3-1,1,1]:*bis*[(5-(2-*p*-phenyl-2-oxaethyl)-1,3-phenylene):(5-(2-oxaethyl)-1,3-phenylene)³:
(2-oxaethyl)benzene]-[(5-(2-*p*-phenyl-2-oxaethyl)-1,3-phenylene):(5-(2-oxaethyl)-1,3-phenylene)³:4-(2-oxaethyl)-1-bromo-
benzene] (where, X = H); or 48-Cascade:ethane[3-1,1,1]:[(5-(2-*p*-phenyl-2-oxaethyl)-1,3-phenylene):(5-(2-oxaethyl)-1,3-
phenylene)3 : (2-oxaethyl)benzene]-*bis*[(5-(2-*p*-phenyl-2-oxaethyl)-1,3-phenylene) : (5-(2-oxaethyl)-1,3-phenylene)³:4-(2-
oxaethyl)-1-bromobenzene] where (X = Br).

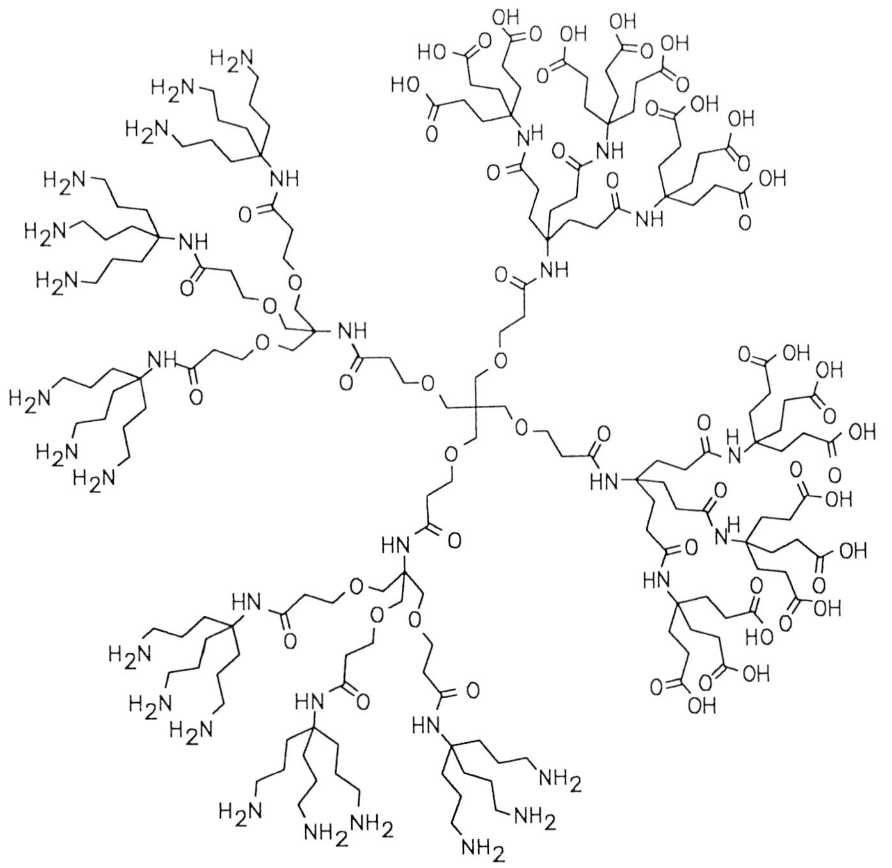

Figure 10.8. 36-Cascade : methane[4] : *bis*[(3-oxo-6-oxa-2-azaheptylidyne)^2 : propylamine]-*bis*[(3-oxo-6-oxa-2-azaheptylidyne) : (3-oxo-2-azapentylidyne) : propanoic acid].

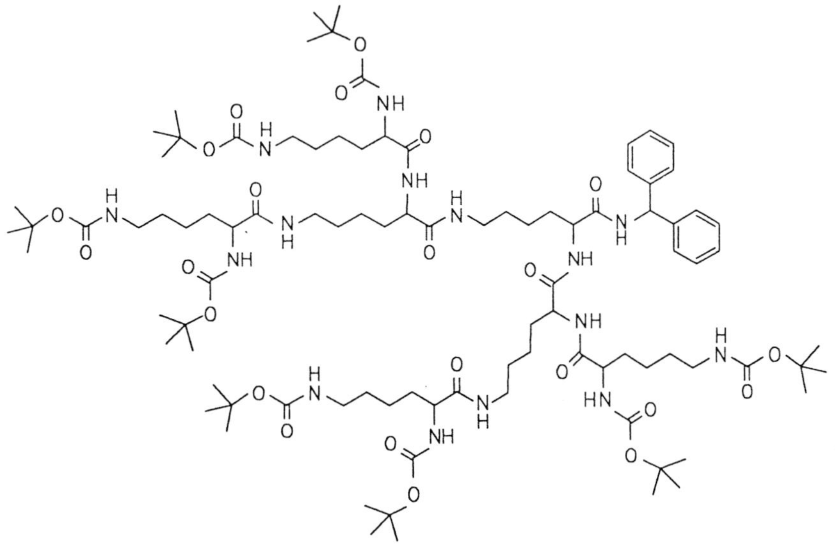

Figure 10.9. 8-Cascade : N-(diphenylmethyl)acetamide[2-2,2] : (2-oxo-3-azapropylidene : 2-oxo-3-azaheptylidene)^2 : N-(*tert*-butoxycarbonyl)amine : N-(*tert*-butoxycarbonyl)butylamine.

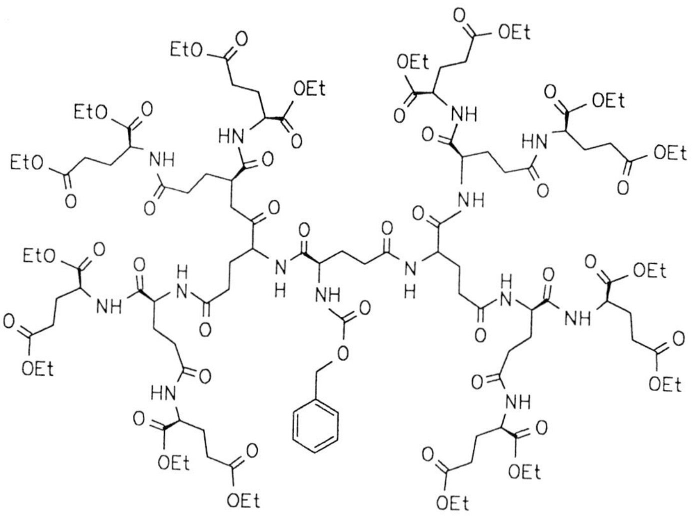

Figure 10.10. 16-Cascade: (*N*-benzyloxycarbonyl)methylamine[2-1,1]:(1*S*-3-oxo-2-azapropylidene:1*S*-3-oxo-2-azapentylidene)^3:ethyl formate:ethyl propanoate.

10.6.5.1   General Rules for the Proposed Fractal Notation

1. The *terminal or peripheral group(s)* is (are) left blank if hydrogen, otherwise it denotes the specific terminal functionality. With diverse functionality, $<X>_n <Y>_m$, where n and m are statistical average values of groups $<X>$ and $<Y>$, respectively. The number ($_\#$) of terminal groups (R) can be denoted by a subscript, e.g. $(R)_\#$. If the molecule has a cyclic structure, the terminal group is absent or incorporated into the core, this is signified by "∅" in the terminal position of the notation.
2. *f* denotes the fractal notation.
3. *Subscripts* refer to the junctures or branching points in the molecule. If the branch point is a tetrahedral carbon, the subscript is omitted. When the branch point is identical throughout the structure, the atom (or group) is only cited once as a subscript. If, however, there are multiple branch points, they are sequentially listed for the outermost → innermost and separated by a period.
4. *Connector groups* are listed again from the outermost → innermost and enclosed in parentheses as well as separated by periods.
5. *Core* is denoted without parentheses at the far right and can be omitted if it is identical to the juncture or branching atom or group.

Thus, a summary of the basic fractal notation is:
(peripheral groups)$_{number}$f$_{junctures\ (coordination\ number)}$(connectors)core.

10.6.5.2   Examples Using the Fractal Rules

The original [27]-Arborol[2] (see Figure 10.1) would be named by this procedure as: the *terminal groups* ≡ $(HO)_{27}$; "f"; *junctures* ≡ tetrahedral carbon, ∴ omitted; *connectors* (from the outside in) ≡ ($CH_2.NHCO.CH_2CH_2OCH_2$); and the *core* ≡ C-C_5H_{11}-n .
Thus, the complete fractal notation for the structure would be:

$HOf(CH_2.NHCO.CH_2CH_2OCH_2)C$-$C_5H_{11}$-n

It was noted[8] that further simplification could be made by a simple number replacement for consecutive methylene groups; therefore, this complete fractal notation can be simplified (or condensed) to:

$HOf(1.NHCO.2O1)C5H$

and for practical utilization the number of terminal groups is appended to afford:

$(HO)_{27}f(1.NHCO.2O1)C5H$

For the original PAMAM amine, depicted in the line notation as:

N[CH$_2$CH$_2$CONHCH$_2$CH$_2$N(CH$_2$CH$_2$CONHCH$_2$CH$_2$N[CH$_2$CH$_2$CONHCH$_2$CH$_2$N
(CH$_2$CH$_2$CONHCH$_2$CH$_2$NH$_2$)$_2$]$_2$)$_2$]$_3$

it would be greatly simplified to

H$_2$Nf$_N$(2NHCO$_2$.)$_4$

where the 4 signifies the 4th generation.

For applications to more complex macromolecules and the use of the additional rules, not cited herein, associated with the fractal notation, one should consult the Mendenhall articles.[7–9]

10.6.6 References

[1] E. Buhleier, W. Wehner, F. Vögtle, ""Cascade" and "Nonskid-Chain-like" Syntheses of Molecular Cavity Topologies", *Synthesis* **1978**, 155–158.

[2] G. R. Newkome, Z. Yao, G. R. Baker, V. K. Gupta, "Cascade Molecules: A New Approach to Micelles. A [27]-Arborol", *J. Org. Chem.* **1985**, *50*, 2003–2004.

[3] A. F. Bochkov, B. E. Kalganov, V. N. Chernetskii, "Synthesis of cascadol – a highly branched, functionalized polyether", *Izvestiya Akademii Nauk SSSR, Seriya Khimicheskaya* **1989**, 2394–2395.

[4] P. G. de Gennes, H. Hervet, "Statistics of <<starburst>> polymers", *J. Phys. Lett.* **1983**, *44*, L351-L360.

[5] T. Nagasaki, M. Ukon, S. Arimori, S. Shinkai, "'Crowned' Arborols", *J. Chem. Soc. , Chem. Commun.* **1992**, 608–610.

[6] D. A. Tomalia, H. Baker, J. Dewald, M. Hall, G. Kallos, S. Martin, J. Roeck, J. Ryder, P. Smith, "A New Class of Polymers: Starburst-Dendritic Macromolecules", *Polym. J. (Tokyo)* **1985**, *17*, 117–132.

[7] G. D. Mendenhall, S. X. Liang, E. H. Chen, "Synthesis and Characterization of the f(1.2) Molecular Fractal, 5,5-Bis(3',3'-dimethylbutyl)-2,2,8,8-tetramethylnonane", *J. Org. Chem.* **1990**, *55*, 3697–3699.

[8] G. D. Mendenhall, "Fractal Index and Fractal Notation" *Search Series*, (Eds.: G. D. Mendenhall, A. Greenberg, J. F. Liebman), Chapman & Hall, New York **1995**, pp. 181–194.

[9] G. D. Mendenhall, "Fractal Notation for Nearly-Symmetrical Dendrimers", *J. Polym. Sci., Part A: Polym. Chem.* **1998**, *36*, 2979–2983.

[10] O. L. Chapman, J. Magner, R. Ortiz, "Polycules" *Polym. Prepr.* **1995**, *36*, 739–740.

[11] G. R. Newkome, Y. Hu, M. J. Saunders, F. R. Fronczek, "Silvanols: water-soluble calixarenes", *Tetrahedron Lett.* **1991**, *32*, 1133–1136.

[12] G. R. Newkome, G. R. Baker, J. K. Young, J. G. Traynham, "A Systematic Nomenclature for Cascade Polymers", *J. Polym. Sci. , Part A: Polym. Chem.* **1993**, *31*, 641–651.

[13] G. R. Newkome, G. R. Baker, "Macromolecular Nomenclature Note No. 7" *Polym. Prepr.* **1994**, *35*, 6–9.

[14] G. R. Newkome, C. N. Moorefield, G. R. Baker, A. L. Johnson, R. K. Behera, "Alkane Cascade Polymers Possessing Micellar Topology: Micellanoic Acid Derivatives", *Angew. Chem. Int. Ed. Engl.* **1991**, *30*, 1176–1178.

[15] G. R. Newkome, X. Lin, "Symmetrical, Four-Directional, Poly(ether-amide) Cascade Polymers", *Macromolecules* **1991**, *24*, 1443–1444.

[16] H.-F. Chow, L. F. Fok, C. C. Mak, "Synthesis and Characterization of Optically Active, Homochiral Dendrimers", *Tetrahedron Lett.* **1994**, *35*, 3547–3550.

[17] G. R. Newkome, A. Nayak, R. K. Behera, C. N. Moorefield, G. R. Baker, "Cascade Polymers: Synthesis and Characterization of Four-Directional Spherical Dendritic Macromolecules Based on Adamantane", *J. Org. Chem.* **1992**, *57*, 358–362.

[18] J. A. Kremers, E. W. Meijer, "Synthesis and Characterization of a Chiral Dendrimer Derived from Pentaerythritol", *J. Org. Chem.* **1994**, *59*, 4262–4266.

[19] K. L. Wooley, C. J. Hawker, J. M. J. Fréchet, "Polymers with Controlled Molecular Architecture: Control of Surface Functionality in the Synthesis of Dendritic Hyperbranched Macromolecules using the Convergent Approach", *J. Chem. Soc. , Perkin Trans. 1* **1991**, 1059–1076.

[20] R. G. Denkewalter, J. F. Kolc, W. J. Lukasavage, "Macromolecular highly branched homogeneous compound based on lysine units" *U. S. Pat.* **1981**, 4,289,872.

[21] L. J. Twyman, A. E. Beezer, J. C. Mitchell, "The Synthesis of Chiral Dendritic Molecules Based on the Repeat Unit L-Glutamic Acid", *Tetrahedron Lett.* **1994**, *35*, 4423–4424.

10.7   Glossary of Terms

acac	acetylacetone
AD	asymmetric dihydroxylation
AFM	atomic force microscopy
AIBN	2,2'-azo*bis*(isobutronitrile)
AMP	adenosine 5'-monophosphate
amu	atomic mass unit
ApcI	atmospheric pressure chemical ionization
arbor [(arbour Brit.)]	Latin "tree" (Arborols)[1]
ATP	adenosine triphosphate
bda	dibenzylideneacetone
BINAP	1,1'-bi-2-naphthol
bis*homo*tris	4-amino-4-[1-(3-hydroxypropyl)]-1,7-heptanediol[2]
BNCT	boron neutron capture therapy
BOC	*tert*-butoxycarbonyl
bpy	2,2'-bipyridine
cascade synthesis	reaction sequences conducted in a repetitive manner[3]
CHP	*N*-cyclohexyl-2-pyrrolidone
convergent synthesis	construction of the (macro)molecule from the outside in[4]
CTAB	cetyltrimethylammonium bromide
CV	cyclic voltammetry
CTAB	trimethylammonium hexadecyl bromide
DABA	3,5-diaminobenzoic acid
DABCO	1,4-diazabicyclo[2.2.2]octane
DB	degree of branching, which is defined[5] as the sum of dendritic units plus terminal units divided by the sum of dendritic units plus terminal and linear units. Frey and Hölter[6] have defined the degree of branching as $DB_{Frey} = \dfrac{2[D_{AB_2}]}{2[D_{AB_2}]+[L_{AB_2}]+[L_{AB}]}$; where the terms $[D_{AB_2}]$, $[L_{AB_2}]$, and $[L_{AB}]$ represent mole fractions of dendritic and linear segments, respectively.[7]
dba	dibenzylideneacetone [e. g., Pd(dba)]
DBOP	diphenyl (2,3-dihydro-2-thioxo-3-benzoxazolyl)phosphonate
DBPO	dibenzoyl peroxide
DBU	1,8-diazabicyclo[5.4.0]undec-7-ene
DCC	dicyclohexylcarbodiimide
DDQ	dichlorodicyanoquinone
DEAD	diethyl azodicarboxylate
DEKC	dendrimer electrokinetic chromatography[8]
dendro-	Greek "tree-like" (dendrimer)[9]
dendron	dendritic building block
DIC	diisopropyl carbodiimide
DIPEA	diisopropylethylamine
Divergent synthesis	construction of the (macro)molecule from the inside out.
DMAc	*N*,*N*-dimethylacetamide
DMAP	4-(dimethylamino)pyridine; *N*,*N*-dimethylaminopyridine
DMF	*N*,*N*-dimethylformamide
DMPU	*N*,*N*'-dimethylpropyleneurea
DMSO	dimethylsulfoxide
DOSY NMR	2-dimenisional diffusion ordered spectroscopy nuclear magnetic resonance (spectroscopy)
DOTA	1,4,7,10-tetraacetic acid-tetraazacyclododecane
DP	degree of polymerization
DPPA	diphenylphosphoryl azide
dppe	(diphenylphosphino)ethane; 1,2-*bis*(diphenylphosphino)ethane
DPPF	diphenylphosphinoferrocene

DPTA	dodecyltrimethylammonium bromide
DSC	digital scanning calorimetry
DTAB	trimethylammonum dodecyl bromide
DV	differential viscometry
EDC	1-ethyl-3-(3-dimethylaminopropyl)carbodiimide hydrochloride
EELS	electron energy loss spectroscopy
EGF	epidermal growth factor
EI-MS	electrospray ionization mass spectroscopy
EMF	electromotive force
EPR	electron paramagnetic resonance
ESI	electron spectroscopic imaging
FAB MS	fast atom bombardment mass spectrometry
focal point	site of connectivity on convergently prepared building block or dendron
GALA peptide	membrane disrupting amphipathic peptides[10]
GPC	gas phase chromatography
1-HOBT	1-hydroxybenzotriazole
HPLC	high performance liquid chromatography
hypercores	large dendrons or wedges
Iteration (or iterative procedure)	repeating a synthetic step with the product of the previous step[11]
ITM	isothermal titration microcalorimetry
IUPAC	International Union of Pure & Applied Chemistry
LAH	lithium aluminum hydride
LALLS	low-angle laser light scattering
LCAA – CPG	long chain alkylamine – controlled pore glass
LED	light-emitting diode
LiTFSI	lithium *bis*(trifluoromethylsulfonyl)imide
MALDI-TOF MS	matrix-assisted laser desorption ionization time-of-flight mass spectrometry
MAP	multiple antigen peptide
MECC	micellar electrokinetic capillary chromatography
MEM	2-methoxyethoxymethyl (group)
Micellane™ series	registered trademark (University of South Florida) descriptor of the all-hydrocarbon based dendrimers
MOTf	(metal) trifluoromethanesulfonates
MRI	magnetic resonance imaging
NAP	*N*-methyl-2-pyrrolidone
NIR	near infrared
NBS	*N*-bromosuccinimide
network	any collection of dendrimers whereby at least one degree of freedom with respect to macromolecular juxtaposition has been removed
NMP	*N*-methylpyrrolidinone
OPV	oligo(*p*-phenylene vinylene)s
o-DCB	*o*-dichlorobenzene
ORD/CD	optical rotatory dispersion / circular dichroism
ordered network	controlled, multiple dendritic positioning
PAA	poly(acrylic acid)
PAHs	polyaromatic hydrocarbons
PAMs	phenylacetylene macrocycles
PAMAM	*polyamidoamine* (dendrimer)[9]
PCS	photo correlation spectroscopy
PEG	polyethylene glycol
PEI	poly(ethylene imine)
PEO	polyethylene oxide
PET	poly(ethylene terephthalate)
PE-TMAI	pentaerythritol trimethylammonium iodide

phen	1,10-phenanthroline
PMMA	poly(methyl methacrylate)
POPAM	polypropylene amine (dendrimer)
PPE	poly(2,6-dimethyl-1,4-phenylene ether)
PPI	polypropylenimine or poly(propylene imine)
RAIR	reflection adsorption infrared
random network	uncontrolled, multiple dendritic positioning
REDOR	rotational-echo double-resonance (NMR)
ROMP	ring opening metathesis polymerization
SANS	small-angle neutron scattering
SAW	surface acoustic wave
SAXS	small-angle X-ray light scattering
SCVP	self-condensing vinyl polymerizations
SDS	sodium dodecylsulfate
SEC	size exclusion chromatography
SEM	scanning electron microscopy
SERS	surface-enhanced Raman scattering
SFM	scanning force microscopy
SPDC	*N*-succinimidyl 3-(2-pyridinyldithio)propionate
SPEI	"*starburst*" *poly*ethylen*i*mine
SPM	scanning probe microscopy
Starburst® dendrimers	registered trademark (DOW) descriptor of PAMAM dendrimers
TADDOLs	■■■■,,',,'-tetraaryl-1,3-dioxolane-4,5-dimethanols
TBDMS	*tert*-butyldimethylsilyl (group)
t-BOC	*tert*-butoxycarbonyl (group)
t-BPB	*tert*-butyl perbenzoate
TCNQ	tetracyanoquinodimethane
TEA	trimethylamine
TEMPO	2,2,6,6-tetramethyl-1-piperidinyloxy
TFA	trifluoroacetic acid
TGA	thermogravimetric analysis
THF	tetrahydrofuran
THT	tetrahydrothiophene
TiPSA	tri*iso*propylsilylacetylene
TMEDA	*N,N,N',N'*-tetramethylethylenediamine
TMS	trimethylsilyl (group)
topological dendritic connectivity	non-covalent, sterically or mechanically induced dendritic attachments such as exhibited by interlocking arms
tpphz	tetrapyridino[3,2-*a*:2',3'-*c*:3", 2"-*h*:2", 3"-*j*]phenazine
tpy	2,2':6',2"-terpyridine
TREN	*tris*(2-aminoethyl)amine
tris	1,1,1-*tris*(hydroxymethyl)aminomethane
TTAB	tetradecyltrimethylammonium bromide
TTF	tetrathiafulvalene
XPS	X-ray photoelectron spectroscopy

References

[1] G. R. Newkome, Z. Yao, G. R. Baker, V. K. Gupta, "Cascade Molecules: A New Approach to Micelles. A [27]-Arborol", *J. Org. Chem.* **1985**, *50*, 2003–2004.

[2] G. R. Newkome, C. N. Moorefield, K. J. Theriot, "A Convenient Synthesis of "Bishomotris": 4-Amino-4-[1-(3-hydroxypropyl)]-1,7-heptanediol and 1-Azoniapropellane", *J. Org. Chem.* **1988**, *53*, 5552–5554.

[3] E. Buhleier, W. Wehner, F. Vögtle, ""Cascade" and "Nonskid-Chain-like" Syntheses of Molecular Cavity Topologies", *Synthesis* **1978**, 155–158.

[4] C. Hawker, J. M. J. Fréchet, "A New Convergent Approach to Monodisperse Dendritic Macromolecules", *J. Chem. Soc. , Chem. Commun.* **1990**, 1010–1013.

[5] C. J. Hawker, R. Lee, J. M. J. Fréchet, "One-Step Synthesis of Hyperbranched Dendritic Polyesters", *J. Am. Chem. Soc.* **1991**, *113*, 4583–4588.

[6] H. Frey, D. Hölter, "Degree of branching in hyperbranched polymers. 3 Copolymerization of AB*m*-monomers with AB and AB*n*-monomers", *Acta Polym.* **1999**, *50*, 67–76.

[7] L. J. Markoski, J. L. Thompson, J. S. Moore, "Synthesis and Characterization of Linear – Dendritic Aromatic Etherimide Copolymers: Tuning Molecular Architecture To Optimize Properties and Processability", *Macromolecules* **2000**, *33*, 5315–5317.

[8] S. A. Kuzdzal, C. A. Monnig, G. R. Newkome, C. N. Moorefield, "Dendrimer Electrokinetic Capillary Chromatography: Unimolecular Micellar Behaviour of Carboxylic Acid Terminated Cascade Macromolecules", *J. Chem. Soc. , Chem. Commun.* **1994**, 2139–2140.

[9] D. A. Tomalia, H. Baker, J. Dewald, M. Hall, G. Kallos, S. Martin, J. Roeck, J. Ryder, P. Smith, "A New Class of Polymers: Starburst-Dendritic Macromolecules", *Polym. J. (Tokyo)* **1985**, *17*, 117–132.

[10] N. K. Subbarao, R. A. Parente, F. C. Szoka, Jr., L. Nadasdi, K. Pongracz, "The pH-dependent bilayer destabilization by an amphiphatic peptide", *Biochem.* **1987**, *26*, 2964.

[11] N. Feuerbacher, F. Vögtle, "Iterative Synthesis in Organic Chemistry" in *Topics in Current Chemistry,* (Ed.: F. Vögtle), Springer-Verlag, Berlin **1998**, pp. 1–18.

Index

DATE DUE